FLORA OF IRAQ

VOLUME SEVEN

T. 78.

FLORA OF IRAQ

VOLUME SEVEN

EDITED BY

SHAHINA A. GHAZANFAR
JOHN R. EDMONDSON
AND
ALI HALOOB

With the collaboration of the staff of
the National Herbarium of Iraq of the
Ministry of Agriculture, Baghdad

Published on behalf of the
MINISTRY OF AGRICULTURE
Republic of Iraq
by
ROYAL BOTANIC GARDENS, KEW
2025

First published in 2025 by
Royal Botanic Gardens, Kew,
Richmond, Surrey, TW9 3AB, UK
www.kew.org

Distributed on behalf of the Royal Botanic Gardens, Kew in North America by the
University of Chicago Press, 1427 East 60th Street, Chicago, IL 60637, USA

ISBN 978-1-84246-840-1
e-ISBN 978-184246-847-0

British Library Cataloguing in Publication Data
A catalogue record for this book is available from the British Library

Design and page layout by Christine Beard
and Fakenham Prepress Solutions

Frontispiece: *Salvia indica* (coloured engraving after F. von Scheidl, 1770)

Printed in the UK by Halston & Co Ltd
Printed in the USA by The University of Chicago Press

EU Authorised Representative: Easy Access System Europe Oü, 16879218. Mustamäe tee 50,
10621, Tallinn, Estonia (email: gpsr.requests@easproject.com).

For information or to purchase all Kew titles please visit shop.kew.org/kewbooksonline or email
publishing@kew.org

Kew's mission is to understand and protect plants and fungi, for the wellbeing of people and the
future of all life on Earth.

Kew receives approximately one third of its funding from Government through the Department
for Environment, Food and Rural Affairs (Defra). All other funding needed to support Kew's vital
work comes from members, foundations, donors and commercial activities, including book sales.

CONTENTS

FOREWORD

Iraq has nurtured a rich botanical heritage that has played a significant role in shaping human civilizations. Many wild plant species that thrived in its fertile lands have been domesticated and have become the origins of some of the most essential crops in the world. Given the importance of this natural resource, conducting a comprehensive scientific study is crucial to provide valuable references for researchers worldwide.

After nearly six decades of collaboration between the National Herbarium of Iraq, Ministry of Agriculture and the Royal Botanic Gardens, Kew, we are proud to present the final volume of this important study. This work, made possible by the unwavering support of the Ministry, offers a detailed scientific exploration of Iraq's plant diversity.

As one of the world's most vulnerable countries to climate change, Iraq is committed to protecting its natural resources. The Flora of Iraq volumes serve as an invaluable resource for understanding and conserving the plant diversity of Iraq. Moreover, this study provides essential knowledge that will contribute significantly in developing sustainable uses of the natural botanical resources in Iraq, and help to develop strategies to mitigate the effects of climate change and ensure a prosperous future for generations to come.

We extend our sincere gratitude to all who contributed to this remarkable achievement, including the Ministry of Agriculture of Iraq, the Royal Botanic Gardens, Kew, and our dedicated Iraqi experts. Their tireless efforts have ensured that Iraq's botanical heritage will continue to inspire and inform scientific research for years to come.

Dr Firas Muzahem Hussein
Director General, Directorate of Seed Testing and Certification,
Ministry of Agriculture, Baghdad, Iraq.

FOREWORD

This final volume of The Flora of Iraq represents a remarkable scientific achievement that began in 1966 as a collaborative venture between the Ministry of Agriculture in Baghdad and the Royal Botanic Gardens, Kew. The first six volumes were published between 1966 and 1985, edited by Cliff Townshend, Evan Guest and Ali al-Rawi. Sadly, by 1985, further progress had to be suspended due to the political situation.

After a hiatus of 25 years, Shahina Ghazanfar and John Edmondson took on the onerous task of completing the Flora, commencing the editing of the final three volumes, with Nicholas Hind's expertise as co-editor for Volume 6, the Asteraceae. They built on the excellent work of Cliff Townsend, Evan Guest and Ali al-Rawi who had persevered with the flora when funding faltered. Dr Ali Haloob, Keeper of the Herbarium in Baghdad, co-edited this final volume. Their collective endeavours ensured these three volumes were published between 2013 and 2025. Great credit is also due to David Mabberley, former Keeper of the Kew Herbarium, who was a potent advocate to re-start this project in 2010. I am deeply grateful to each of them for their long-term vision and commitment to concluding this tremendous work of scholarship.

I wish also to acknowledge the contribution of the Iraqi Government who have been committed to this project from inception. It would not have been possible without their support.

The completion of this Flora addresses a critical gap in the knowledge of vascular plant diversity and distribution in the Middle East. Although focussed upon Iraq, it greatly assists the development of floras for neighbouring countries due to the many species distributed widely throughout the region.

This Flora would not have been possible without access to Kew's extraordinary collections from Iraq. Kew maintains the best-curated collection of Iraqi plants, a testament to the tenacious efforts of both Iraqi and British botanists over more than a century. It demonstrates, again, the incalculable value of herbaria in understanding, documenting and thus helping to conserve our planet's plant diversity.

The Flora of Iraq, to me, exemplifies Kew's invaluable contribution to plant taxonomy. It draws upon our collections, which have been gathered and curated for centuries, our taxonomic expertise and our global network of partnerships. Most saliently, it demonstrates our long-term commitment to working collaboratively to document and conserve global plant diversity. For all these reasons, this remarkable work of scholarship should be commended and celebrated.

Richard Deverell
Director, Royal Botanic Gardens, Kew

LIST OF CONTRIBUTORS
to VOLUME 7

A.D.Q. Agnew, University of Aberystwyth

S.T. Al Kaisi, National Herbarium, Baghdad

A.A. Al Mayah, University of Basra

W.M.T. Al Asadi, University of Basra

M. Bayati, University of Baghdad

F. Celep, Kırıkkale Universitesi
ORCID ID: 0000-0003-3280-8373

T. Dirmenci, Balikesir Üniversitesi
ORCID ID: 000-0003-3038-6904

J.R. Edmondson, Royal Botanic Gardens, Kew
ORCID ID: 0009-0007-0200-4485

S.A. Ghazanfar, Royal Botanic Gardens, Kew
ORCID ID: 0000-0002-0482-2023

A. Haloob, National Herbarium, Baghdad
ORCID ID: 0000-0003-3038-6904

A. Kandemir, Erzincan Binali Yıldırım Üniversitesi, Erzincan
ORCID ID: 0000-0003-1902-9631

L. Musselman, Old Dominion University Norfolk, Virginia

T. Özcan, Balikesir Üniversitesi

A. Patzak, Natural History Museum, Vienna

†K.H. Rechinger, Natural History Museum, Vienna

A.P. Sukhorukov, M.V. Lomonosov State University, Moscow

C.J. Thorogood, University of Oxford Botanic Garden and Arboretum, University of Oxford
  Department of Biology
ORCID ID: 0000-0002-2822-0182

†C.C. Townsend, Royal Botanic Gardens, Kew

PREFACE
to VOLUME 7

In August 1965, Sir George Taylor, the Director of the Royal Botanic Gardens, Kew, wrote in the Preface to vol. 1 of the *Flora of Iraq*, "The compilation of a Flora on the scale of the present work is an onerous undertaking, and it is my earnest hope that its completion will represent a major advance in the progress of botany in the Middle East." So it has proved; but its completion after more than 70 years' work, not including the earlier explorations of a host of plant collectors, undoubtedly marks a point of progress.

We should not overlook the fact that the taxonomic treatments employed in the earlier volumes have to some extent been updated following new research, especially as a result of DNA analysis at the family level. The Hutchinson System (see Bullock in *Flora of Iraq* vol. 2, pp. 1-14, fig. 1) has been superseded by various iterations of the Angiosperm Phylogeny Group classification. One of the consequences of adhering to the scheme adopted at the outset has been a degree of mismatch in the order and circumscription of former and current families. But it is hoped that a synoptic index, currently in preparation for all nine volumes, will enable users to undertake more effective searches.

Since the Flora project was first launched, many investigations have been carried out on the ecology and conservation of the flora of Iraq, and there has been a growing acceptance of the need for more active measures to be taken to preserve the botanical diversity of the country and to sustain the wider ecosystem. Many interventions, some accidental and some deliberate, have adversely affected the flora; in particular, the partial drainage of the southern marshes had a devastating effect, now largely reversed thanks to measures taken to restore the aquatic environment. It remains the case that the aquatic flora of Iraq is still poorly known, compared to its terrestrial counterpart, as is the partly endemic alpine flora occupying the highest levels of Iraq's mountains.

It would be invidious to try to list all the individuals who have contributed, directly or indirectly, to this project; inevitably there would be omissions. For this Volume a list has been prepared of all those who have contributed taxonomic accounts of 517 taxa treated in this Volume (see p. x).

Two institutions have sponsored the project throughout its existence. Making use of the facilities offered by the Royal Botanic Gardens, Kew has been first and foremost, and the Baghdad National Herbarium, Abu Ghraib, Iraq, whose staff have provided the majority of the specimens studied and cited for this Flora, and have thereby added greatly to the coverage now achieved. They have also checked the draft accounts and made many valuable suggestions. Our special thanks go to Shuker Taher Al-Kaisi, who has worked through the final draft of this volume and checked the citations and species distribution in Iraq. A special word of appreciation is due to the former Keeper of the Kew Herbarium, Professor David Mabberley, who oversaw the re-launch of the project following a period of relative quiescence. The Flora has been illustrated by a wide range of drawings, including some commissioned specially from Kew artists and others that have been reproduced, with permission, from other publications. The editors are grateful to all those who supplied these images. Lastly, heartfelt thanks are due to the Ministry of Agriculture, Government of Iraq, who enabled the project to be both initiated and concluded.

Shahina A. Ghazanfar
Royal Botanic Gardens, Kew

John R. Edmondson
Hon. Research Associate, Royal Botanic Gardens, Kew

Ali Haloob
National Herbarium of Iraq, Baghdad

116. SOLANACEAE

Pflanzenfam. 4, 3b: 4–38 (1895)
Barboza *et al.*, Solanaceae in Kubitzki (ser. ed.), J. Kadereit & V. Bittrich, V. (eds),
Fam. Gen. Vasc. Pl. 14: 295 (2016)

C.C. Townsend
Revised by Shahina A. Ghazanfar

Herbs, trees or shrubs, often climbers. Leaves alternate, entire, lobed or more rarely pinnatisect, exstipulate, sometimes becoming opposite in region of inflorescence. Inflorescence typically a cyme or combination of cymes, sometimes racemose, umbellate, fasciculate or flowers solitary. Flowers hermaphrodite, actinomorphic to strongly zygomorphic, hypogynous. Calyx 4–6-lobed, persistent and often considerably enlarged in fruit. Corolla gamopetalous, rotate to long-tubular and funnel-shaped, usually 5-lobed, lobes mostly plicate or convolute, rarely valvate. Stamens inserted on corolla tube and alternate with its lobes, unequal, 2–5; when fewer than 5, a staminode is often present. Anthers 2-locular, loculi parallel, dehiscing longitudinally or by apical pores. Ovary bilocular, sometimes further divided by a false septum; placentation axile, ovules numerous. Style 1, stigma bilobed. Fruit a capsule or berry.

The Solanaceae are a monophyletic group with about 100 genera and 2,500 species found mainly in tropical and temperate areas with a centre of diversity in Central and South America. A family of considerable economic importance, containing such valuable food plants as the potato (*Solanum tuberosum* L.), tomato (*Solanum esculentum* Mill.), strawberry tomatillo or Cape gooseberry (*Physalis* spp.), red peppers (*Capsicum* spp.) etc. It also contains many drug plants such as *Atropa, Datura, Hyoscyamus* and *Nicotiana* (tobacco), and valued garden genera such as *Cestrum, Petunia, Salpiglottis, Schizanthus* etc. 21 native and several cultivated species in Iraq.

Barboza, G.E, Hunziker, A.T., Bernardello, G., Cocucci, A.A., Moscone, A.E., Carrizo García, C., Fuentes, V., Dillon, M.O., Bittrich, V., Cosa, M.T., Subils, R., Romanutti, A., Arroyo, S., Anton, A. (2016). Solanaceae. In: Kubitzki (ser. ed.), Kadereit, J., Bittrich, V. (eds) Flowering Plants. Eudicots. The Families and Genera of Vascular Plants, 14: 295–357. Springer, Cham. https://doi.org/10.1007/978-3-319-28534-4_29.
Chakravarty, H.L. (1964). Solanaceae of Iraq (taxonomy and economics). *Dep. Agr. Iraq Tech. Bull.* 18.
Dupin, J., Matzke, N.J., Särkinen, T., Knapp, S., Olmstead, R.G., Bohs, L. and Smith, S.D. (2017). Bayesian estimation of the global biogeographical history of the Solanaceae. *Journal of Biogeography* 44(4): 887–899.
Olmstead, R.G., Bohs, L., Migid, H.A., Santiago-Valentin, E., Garcia, V.F. and Collier, S.M. (2008). A molecular phylogeny of the Solanaceae. *Taxon* 57(4): 1159–1181.

We have keyed all native and cultivated genera found in Iraq. The cultivated genera are listed alphabetically at the end.

1. Stamens dissimilar, 4 in 2 pairs and one much smaller, sometimes
    abortive (solely cultivated ornamental herbs) 2
   All stamens ± similar, not paired. 3
2. Stamens inserted at middle of corolla tube or below (cultivated) . . . . *Petunia*
   Stamens inserted at top of corolla tube (cultivated) . . . . *Nierembergia*
3. Woody shrubs or small trees with funnel-shaped flowers, corolla with a distinct tube . 4
   Herbs, sometimes woody at base, or if shrubs then corolla rotate or campanulate . . . 6
4. Plant with spinescent branches; flowers < 2 cm, reddish to violet . . . . .4. *Lycium*
   Plant not spinous; flowers > 2 cm, greenish-yellow, large 5
5. Corolla tube hairy, slightly expanded below the limb; lobes of limb
    broad, shallow, scarcely spreading (cultivated) . . . .*Nicotiana*
   Corolla tube glabrous, without any distinct expansion below the limb; lobes
    of limb distinct, narrowly triangular, spreading (cultivated) . . . . *Cestrum*
6. Anthers connivent into a tube around the style, sometimes united. . . . . 1. *Solanum*
   Anthers not connivent around the style. 7
7. Plant shrubby; flowers shortly campanulate, subsessile in axillary
    clusters; calyx inflated and enclosing berry [in Iraq plants]. . . . . 3. *Withania*
   Flowers not in axillary clusters; calyx inflated or not in fruit 8

1. **SOLANUM** L.

Sp. Pl. ed. 1: 14 (1753); Gen. Pl. ed. 5: 85 (1954)

C.C. Townsend
Revised by Shahina A. Ghazanfar

Herbs (annual, biennial or perennial), shrubs or small trees. Stems erect, procumbent or scrambling, terete or angled sometimes climbers, unarmed to densely spinose, glabrous, stellate-hairy or villous, sometimes viscid-glandular. Leaves alternate, undivided, lobed or pinnatifid, imparipinnate or bipinnate, usually with small or larger bract-like leaflets along main axis between larger leaflets. Inflorescence cymose, umbellate, racemose or paniculate, or rarely flowers solitary, with or without bracts. Calyx 5–10-fid, often persistent and ± accrescent in fruit. Corolla yellow, white, purplish or blue, rotate, 5-(rarely 4- or 6-)partite with usually short tube, rarely campanulate, reflexed or not. Stamens isomerous with corolla segments, filaments very short and inserted in corolla throat; anthers connivent or coherent into a tube around gynoecium, dehiscent by pores, with or without a sterile apical beak. Stigma capitate. Fruit a berry, usually bilocular, variable in form.

Solanum with about 1,400 species constitutes approximately half the species in the Solanaceae. It has a worldwide distribution with centres of diversity in tropical America, Africa, and Australia. Molecular studies have shown that several segregate genera, including *Cyphomandra, Lycopersicon, Normania,* and *Triguera,* belong within *Solanum* and that *Lycianthes* does not belong with *Solanum*. Phylogenetic studies indicate that *Solanum* originated in the Americas, most likely in South America, and spread to the Old World and many of the Old World species have spread to the New World (e.g. *S. nigrum, S. villosum*). 12 species (10 of them native) are found in Africa which is also the centre of diversity; fewer species in the Australasia. In Iraq, 3 native and several cultivated species including tomato, potato, egg plant and several ornamentals are present.

Solanum from Lat. *solor,* to comfort, console or relieve, referring to its medicinal uses.

Edmonds, J.M. (2012). *Solanaceae* in H. Beentje (ed.), *Flora of tropical East Africa.* Royal Botanic Gardens. Kew.
Edmonds, J.M. & Chweya, J.A. (1997). Black nightshades: *Solanum nigrum* L. and related species (Vol. 15). Bioversity International.
Knapp, S. (2013). A revision of the Dulcamaroid Clade of Solanum L. (Solanaceae). *PhytoKeys,* (22): 1–432.
Särkinen, T., Poczai, P., Barboza, G.E., van der Weerden, G.M., Baden, M. and Knapp, S. (2018). A revision of the Old World black nightshades (Morelloid clade of Solanum L., Solanaceae). *PhytoKeys* (106): 1–223.
Weese, T.L. & Bohs, L. (2007). A three-gene phylogeny of the genus *Solanum* (Solanaceae). *Systematic Botany,* 32(2): 445–463.

1. Scrambling perennial, woody at base; stems to 2 m; leaves with a large
ovate terminal leaflet and with 1–2 pairs of smaller basal lobes or
pinnae, or undivided with a rounded-truncate to shallowly or deeply
cordate (more rarely cuneate or hastate) base; corolla purple . . . . 3. *S. dulcamara*
     Annual (or short-lived perennial), stems to 60 cm; leaves simple,
undivided; corolla white to greenish-white 2
2. Calyx lobes deltate with acute or rounded tips; free part of the
filaments < 1 mm; mature berries globose, dull black or green; stone
cells 0–4 . . . . .1. *S. nigrum*
     Calyx lobes elliptic to triangular, rounded at tip; free part of the
filaments 1 or >1 mm; mature berries slightly ellipsoid, shiny yellow,
orange or red; stone cells always absent . . . . 2. *S. villosum*

1. **Solanum nigrum** *L.*, Sp. Pl. 1: 186 (1753); Boissier, Fl. Orient. 4: 284 (1879); Zohary, Dep. Agr. Iraq Bull. 31: 131 (1950); Rawi in Dep. Agr. Iraq Tech. Bull. 14: 141 (1964); Chakravarty, Dep. Agr. Iraq. Tech. Bull. 18: 7 (1964); Schönbeck-Temesy in Fl. Iranica [K. H. Rechinger] 100: 7 (1972); Rechinger, Fl. Lowland Iraq: 536 (1964); Baytop in Fl. Turkey [P. H. Davis] 6: 439 (1978); Y. Nasir in Fl. Pakistan [Y. Nasir & Ali] 168: 6 (1985); Collenette, Wild Flow. Saudi Arabia: 705 (1999); Hepper in Fl. Egypt [Boulos] 3: 39 (2002); Abdulridha, Taha & Widad, Ecology & Fl. Basrah: 511 (2016); Taifour & El-Oqlah, Pl. Jordan Annot. Checkl.: 149 (2017).

Solanum nigrum L. var. *atriplicifolium* G.Mey., Chloris Han. 265 (1836).
S. roxburghii Dunal, Prodr. [A. P. de Candolle] 13(1): 57 (1852).
S. nigrum L. var. *macrocarpum* Schur, Enum. Pl. Transsilv. 478 (1866).
S. nigrum L. var. *incisum* Täckh. & Boulos, Publ. Cairo Univ. Cairo Herb. 5: 101 (1974) ["1972"].

Erect or ascending to almost prostrate annual or short lived perennial, (10–)20–40(–60) cm, glabrous to pubescent with glandular or egandular hairs; older stems glabrescent. Leaves ovate or rhomboid, entire to strongly sinuate-dentate, petiolate, lamina to 6–8 × 2–3 cm, rarely more, base long-decurrent along the 10–20 mm petiole. Stem and branches terete or more usually angular with 2 or more smooth or scabrid-tuberculate raised lines. Cymes racemose to umbellate, few- (rarely more than 10-) flowered; pedicels erect in flower, deflexed in fruit. Calyx c. 2 mm, variably 5-fid, lobes elliptic to triangular, rounded at tip, scarcely enlarged in fruit. Corolla white, 6–10 mm in diameter, segments ½–2/3 lobed to base, spreading, finally revolute. Anthers yellow, connivent, free part of anthers < 1 mm. Berry globose, 7–8 mm in diameter, typically dull black or purplish-black or yellowish-green to green, opaque, not shiny; stone cells 0–4. Fig. 1, 1–4.

HAB. A common weedy plant by roadsides and waste places, in shady and moist locations, in clay soil, cultivated land among palm trees, in gardens, rocky mountain slope, on river edge, valley bottom of sandy gravelly soil, mountain slope near water spring, clay soil, by water in orchards, and cultivated land; alt. 5–1500 m; fl. & fr. Mar–Nov.

DISTRIB. Lower hills, alluvial plains and desert regions of Iraq. **MAM**: Sharanish village, 25 km N.E. of Zakho, *Rawi, Tikriti & Nuri* 29033! (BAG); 25 km from Zakho to Kani Masi, *Botany staff* 43803! (BAG); Aradin 15 km W. by N. Amadiya, *Al-Kaisi & K. Hamad* 46068! (BAG). **MRO**: Pushtashan, 15 km N.E. of Rania lower slope of Qandil range, *Rawi & Serhang* 24225 & 26567 (BAG); near Gali Ali Beg, *A. Haloob, Al-Kaisi* & K. Qader 60609 (BAG). **MSU**: Penjwin, *Guest* 13001!; Hauraman mountain, *Rawi, Chakravarty, Alizzi & Nuri* 19809 (BAG); Penjwin, E. *Guest* 13001 (BAG); 15 km to Chuwarta, *Al-Khayat, Al-Kaisi & Thamir* 46318 (BAG). **FNI**: Tel Kaif, *Najib* 5163 (BAG); Eski Kellek, *Chakravarty* 30765 (BAG); mindan bridge on Khazir river bet. Mosul and Aqra, *Alizzi & S. Omar* 35289 (BAG). **FUJ**: Saaira, Alizzi & S. Omar 34653!; bet. Tal Afar and Sinjar, *Alizzi & S. Omar* 35315 (BAG). **FNI**: Nineveh-Mosul, *E. Chapman* 26079! **FPF**: Badra, *Al-Kaisi & Khayat* 50616! **DLJ**: Nr. Rawa, *Alizzi & S. Omar* 35340!; Rawa, *Alizzi & S. Omar* 35355!; 40 km E. of Amarah, *Alizzi & S. Omar* 34729! near Rawa, *Alizzi & S. Omar* 35340 (BAG); **DWD**: Al Qaim, *S. Omar & K. Hamed* 50421!; 3 km W. of Baghdadi, *A. Sharif, K. Hamid & H. Hamed* (BAG); *Al-Qaim, S. Omar & K. Hamad* 50421 (BAG). **LEA**: Basrah, along Shati Al Arab bank, *Alizzi & Sabah* 35063!; Basrah, *Rawi & Gillett* 5951!; Kut, E. Guest 339 (BAG); between Muqdadia and Baquba, in Diyala, *Rechinger & Khudairi* 19544 (BAG); 30 km from Azizya to Kut, *Alizzi & S. Omar* 34633 (BAG); 3 km E. of Kut, *Al-Kaisi & Riyadh* 59778 (BAG). **LCA**: Abu Ghraib, *Alizzi, Thaib & Janan* 32585!; ibid, *Al Baiyar* 21471!; ibid, *Fauzi & Ali* 39509!; around Baghdad, 30/1/1920, *Graham s.n!*; Abu Ghraib, *S. Omar* 37810!, ibid, Hikman *Abbas* 616!; Baghdad, Zaafarina, *Rawi* 19915!; ibid, *Wheeler-Haines* W. 12!; 15 km S. of Diwaniya, in field of tomatoes, *Barkley, Fuad Safat & Agnew* 37/6!; Baghdad, *Paranjpye* 38; *H. Ahmad* 9409 (BAG); Rustam, *Guest* 181 (BAG); Abu Ghreib, *Gillett* 10125 (BAG); Zafaraniya, *Gillett* 5879;

Fig. 1. **Solanum nigrum**. 1, habit, flowering and fruiting branch × ½; 2, flower; 3, calyx (opened) × 5; 4, corolla (opened) × 5; 6, anther (detail); 6, ovary and style (detail).. Reproduced from Fl China Vol. 17, f. 382, 1–4, with permission from Missouri Botanical Garden Press, St. Louis, and Science Press, Beijing.

Diltawa, *Rawi & Gillett* 10215 (BAG); Nasiriyyah, *F. Karim & H. Hamid* 37788 (BAG). **LSM**: Chehala, nr Amara, April, *Henry Field & Yusuf Lazar* s.n.!; 15 km E. of Amarah, *Alizzi & S. Omar* 34718!; 1 km from Chibba, 10 km S.E. of Musharrak, *Rawi & Khatib* 32404!; Hor al Hawixa, 18 km E. of Qalat Salih, *Thamer & Wedad* 46781!; in Hoor of Chibba, 10 km S.E. of Musharrah, *Rawi & Alizzi* 32404 (BAG); 40 km E. of Amarah, *Alizzi & S. Omer* 34729 (BAG); Hor Al-Hawiza (18 km E. of Qalat Salih), *Thamer, Wedad & Hana* 46781 (BAG). **LBA**: Shaibab (nr. Basrah), *Roger* 08 (BAG); Basrah, *Guest* 304; *Rawi & Gillett* 5951; *Chapman* 10034 (BAG); Abu Al-Khasib, *Alizzi & S. Omer* 34664 (BAG); Basrah, *Alizzi & S. Omar* 35063 (BAG); Siba, *K. Haamad, H. Hamid & A. Kadthim* 46355 (BAG).

A widespread weedy species with much morphological variations that have been recognized at various infraspecific levels. Here I (SAG) have followed Särkinen et al. (2018) in their revision of the Old World nightshades, recognising a single species.

Black nightshade. ENAB-ADH-DHIB (Ar.). Of considerable importance in traditional herbal medicine: berries have been used to ease fevers, diarrhea, eye problems. Juice of plant has been used as carthartic, diurectic, for piles, dysentery and for the treatment of an enlarged liver. The berries are reported to be poisonous.

Temperate Eurasia, northern Africa, Australia; introduced in South Africa; naturalized in temperate North America.

2. **Solanum villosum** *Mill*, Gard. Dict. ed. 8, no. 2 (1768); Collenette, Wild Flow. Saudi Arabia: 706 (1999); Hepper in Fl. Egypt [Boulos] 3: 39 (2002); Taifour & El-Oqlah, Pl. Jordan Annot. Checkl.: 149 (2017).

> *Solanum nigrum* L. var. *villosum* L., Sp. Pl. 1: 186 (1753); Y. Nasir in Fl. Pakistan [Y. Nasir & Ali] 168: 7 (1985).
>
> *S. rubrum* Mill., Gard. Dict. ed. 8, no. 4 (1768).
>
> *S. luteum* Mill., Gard. Dict. ed. 8, no. 3 (1768); Blakelock in Kew Bull. 4: 529 (1949); Chakravarty, Dep. Agr. Iraq. Tech. Bull. 18: 11 (1964); Rawi in Dep. Agr. Iraq Tech. Bull. 14: 141 (1964); Schönbeck-Temesy in Fl. Iranica [K. H. Rechinger]100: 12 (1972); Baytop in Fl. Turkey [P. H. Davis] 6: 440 (1978); Feinbrun-Dothan, Fl. Palaest. 3: 165, pl. 273 (1978); Collenette, Wild Flow. Saudi Arabia: 704 (1999).
>
> *S. villosum* Lam., Tabl. Encycl. 2: 18. 1794, nom. illeg., non *S. villosum* Mill. (1768); Boissier, Fl. Orient. 4: 285 (1879).
>
> *S. nigrum* L. var. *rubrum* (Mill.) Aiton, Hort. Kew. [W.Aiton] 1: 234 (1789).
>
> *S. aegyptiacum* J.F.Gmel., Syst. Nat., ed. 13 [bis] 2: 385 (1791).
>
> *S. miniatum* Bernh. Ex Willd., Enum. Pl. [Willdenow] 1: 236 (1809); Rechinger, Fl. Lowland Iraq: 536 (1964).
>
> *S. nigrum* L. var. *miniatum* (Bernh. ex Willd.) Fr., Nov. Flor. Suec.: 111 (1823).
>
> *S. villosum* Lam. var. *miniatum* (Bernh. ex Willd.) Kostel., Clav. Anal. Fl. Bohem. 34 (1824).
>
> *S. nigrum* L. subsp. *miniatum* (Bernh. ex Willd.) Hartm., Sv. Norsk Exc.-Fl.: 35 (1846).
>
> *S. nigrum* L. subsp. *villosum* (Lam.) Hartm., Sv. Norsk Esc.-Fl. 34 (1846).
>
> *S. villosum Mill.* subsp. *miniatum* (Bernh. ex Willd.) Edmonds, Bot. J. Linn. Soc. 89: 166 (1984).

Annual herb; stems branched, erect, decumbent or prostrate, with smooth or dentate ridges, densely pilose to glabrescent with simple glandular- or eglandular or mixed (glandular and eglandular) hairs. Leaves ovate to ovate-lanceolate, occasionally lanceolate, (2–)3–7(–11) × 1.5–4(–7) cm, bases broadly cuneate to truncate and decurrent, occasionally cordate, margins usually sinuate-dentate with obtuse to acute lobes though sometimes entire to sinuate, apices acute; surfaces pubescent as stems; petioles 0.4–3.2(–5) cm. Inflorescences simple, or umbellate or laxly cymose 3–6(–9)-flowered; peduncles erect, 3–14 mm long in flower, 7–12(–26) mm long in fruit; pedicels erect, reflexed in fruit. Calyx campanulate, 1.5–2.5(–3.5) mm long, pilose to villous externally; lobes broadly triangular, persistent and enlarging in fruit. Corolla white to greenish-white, lobes usually triangular, spreading to reflexed after anthesis. Ovary glabrous; style densely pilose for lower three- to one-quarter of length; stigma capitate. Berries ellipsoid, red, orange or yellow, 4.5–11 × 6–9 mm, smooth and shiny, persistent calyx lobes becoming fully reflexed at maturity, eventually falling from calyces; stone cells absent or rarely 1–2.

HAB. Open and disturbed and places, gravely soil, on rocky mountain slope, river bank sandy soil & loam soil with gravels; alt. 410–1100; fl. & fr. Jun.-Oct.
DISTRIB. Mountains and lower hills of N. Iraq. Mountains and lower hills of N. Iraq. **MAM**: nr. Dohuk, *Guest* 1608; Sharanish 25 km N.E. of Zakho, *Rawi* 23694 & *Rawi, Tikriti & Nuri* 29051 (BAG); Zakho, *botany staff* 43788 (BAG); Dairalok village 16 km E. of Amadiya, *Al-Dabbagh* 45862 (BAG); **MSU**: Sulimaniyah, *Agha* 5370!; **MRO/FAR**: between Arbil and Shaqlawa, *Rawi, Chakravarty, Alizzi & Nuri* 19683 (BAG). **FNI**: Mosul,Wadi Aloka, *Barkley* 9006!

NORA ZALA (Kurd.).

Europe to the Middle East and Asia, eastern Africa; introduced and/or cultivated in North America and the Caribbean.

3. **Solanum dulcamara** *L.*, Sp. Pl. 1: 185 (1753); Rawi in Dep. Agr. Iraq Tech. Bull. 14: 141 (1964); Chakravarty, Dep. Agr. Iraq. Tech. Bull. 18: 7 (1964); Baytop in Fl. Turkey [P. H. Davis] 6: 441 (1978); Y. Nasir in Fl. Pakistan [Y. Nasir & Ali] 168: 9 (1985); Zhi-Yun Zhang, Anmin Lu & D'Arcy, Solanancae: Fl. China 17: 318 (1994); Taifour & El-Oqlah, Pl. Jordan Annot. Checkl.: 148 (2017).

Solanum dulcamara var. *indivisum* Boiss., Fl. Orient. [Boissier] 4: 285 (1879).
S. persicum Willd ex Roem. & Schultes, Syst. Veg. ed. 15 bis [Roemer & Schultes] 4: 662 (1819); Fl. SSSR 22: 19 (1955); Schönbeck-Temesy in Fl. Iranica [K. H. Rechinger] 100: 13 (1972).
S. dulcamara var. *persicum* (Willd.) Dinsmore in Post, Fl. Palest. ed. 2, 2: 258 (1933).
S. pseudopersicum Pojark., Not. Syst. (Leningrad) 17: 328 (1955); Fl. SSSR 22: 18 (1955).

Scrambling perennial, woody at base, with a creeping rhizome which is tuberculately thickened at intervals, 30–200 cm, glabrous, pubescent or sometimes tomentose. Leaves 6–8(–12) × 3–4(–8) cm, petioles (1–)1.5–3 cm. Lamina with a large ovate terminal leaflet broadest near base and gradually narrowed to the acute or acuminate apex, and with 1–2 pairs of smaller basal lobes or pinnae, or undivided with a rounded-truncate to shallowly or deeply cordate (more rarely cuneate or hastate) base. Cymes pseudo-terminal, leaf-opposed and finally lateral, pedunculate, divaricately branched, broad, few- to many-flowered. Pedicels erect in flower, recurved in fruit, somewhat thickened below calyx. Calyx 5-dentate with shallow, broadly triangular teeth. Corolla purple, c. 1 mm in diameter, deeply 5-partite, lobes lanceolate, finely pubescent on outer surface above, each with 2 pale-margined green spots at base, at first spreading, finally reflexed. Anthers yellow, adnate into a conical tube from which the style is distinctly exserted. Berry 1 cm, subglobose to ovoid, bright red, shining. Fig. 2, 1–3.

HAB. Lower slopes of mountains, under shade of *Juglans*, by stream and edges of river, in coppiced popular forest, plantation (*Ahrash*); alt. 35–1400 m; fl. & fr. May-Aug.

Fig. 2. **Solanum dulcamara**. 1, habit, shoot × ½; 2, leaf × ½; 3, flower (opened) × 3. Reproduced with permission from Fl. Pakistan 168: f. 1, E–G (1985). Drawn by S. Hameed. © National Herbarium, Pakistan Agriculture Research Council & University of Karachi, Pakistan.

DISTRIB. Mountains and lower alliuvial plains. **MRO**: Pushtashan, 15 km N.E. Rania, lower slope of Qandil Range, *Rawi* 24196!; 23829! **MSU**: Penjiwin, *Rawi* 12159! **FNI**: Nineva-Mosul, *Chapman* 25389! (BAG). **LCA**: Baghdad, *Al-Rathi* 5437!; ibid, *Wheeler-Haines* W1679! (BAG).

ENAB AL-THA'LAB; HAB AL-A'LAM (Ar.). Berries are reported to be poisonous (Chakravarty 1964: p. 8).

Azores, Temperate Eurasia to N. Indo-China, N.W. Africa.

CULTIVATED SPECIES

Solanum aethiopicum *L.*, Cent. Pl. II. 10 (1756).

S. elaeagnifolium Cav., Icon. Pl. 3: 22, t. 243 (1794).

Perennial herb, 10–50 cm, densely covered throughout with closely appressed short-rayed stellate hairs, ± silvery in appearance, with stout or very slender yellowish-brown spines on stems and branches. Flowers solitary or in sessile or pedunculate groups of 2–3. Corolla white to blue. Berry yellow, globose, 7–12 mm in diameter, subtended by persistent and somewhat accrescent calyx.

Native of U.S.A., Mexico and S. America, cultivated in Iraq.

Solanum lycopersicum *L.*, Sp. Pl. 1: 185 (1753).

Lycopersicon esculentum Mill., Gard. Dict. ed. 8 (1754); Chakravarty, Dep. Agr. Iraq Tech. Bull. 18: 24 (1964); Husain & Kasim, Cult. Pl. Iraq 120 (1975).

Annual herb to 2 m high with weak stems, thick below and usually angular, puberulent-hairy above. Corolla lemon-yellow. Fruit 3–10+ cm in diameter, 2–10 or more locular, usually depressed-globose, red or yellow when ripe.

Native of Peru, widely cultivated throughout Iraq. TAMATAM (Ar.); tomato.

Solanum melongena *L.*, Sp. Pl. ed. 1: 186 (1753); Chakravarty, Dep. Agr. Iraq. Tech. Bull. 18: 8 (1964); Husain & Kasim, Cult. Pl. Iraq 122 (1975).

Robust herb 60–200 cm, densely tomentose with greyish stellate hairs. Leaves deeply sinuate or lobed. Inflorescence lateral on stem, solitary or paired fertile flowers or in sessile racemes with lowest flowers fertile and upper functionally male; fertile flowers inclined or nodding. Corolla purple, 3–5 cm in diameter, segments densely stellate-tomentose on outer surface. Berry long, pendulous, ovoid, oblong or obovoid, 5–25 cm.

Native from China to Vietnam, widely cultivated throughout Iraq. BADINJAN (Ar.); egg plant; aubergine; brinjal.

Solanum peruvianum *L.*, Sp. Pl. 1: 186 (1753).

Lycopersicum peruvianum (L.) Miller, Gard. Dict. ed. 8, (1768); Chakravarty, Dep. Agr. Iraq. Tech. Bull. 18: 38 (1964).

Decumbent perennial. Stem slender, weak, terete to somewhat angled. Plant shortly and densely white-canescent throughout, eglandular. Corolla pale- to orange-yellow, divided to about ¾ into 5 lanceolate attenuate lobes, c. 2 cm in diameter. Fruit 1–2 cm in diameter, bilocular uniformly minutely hairy, whitish with bluish-mauve stripes down the midlocular line.

Native from the Galápagos, Ecuador to Chile, cultivated in Iraq at the Horticultural Station at Zafarina.

Solanum pimpinellifolium *Jusl.* in L., Cent. Pl.: 8 (1755); Amoen. Acad. 4: 268 (1759).

Lycopersicon pimpinellifolium (Jusl.) Mill., Gard. Dict. ed. 8 (1768); Chakravarty, Dep. Agr. Iraq. Tech. Bull. 18: 39 (1964).

Annual or perennial herb, stems prostrate, to 2 m in length. Leaves in outline narrowly ovate, to c. 20 cm, pinnate, with 3 pairs of petiolate, lanceolate, entire to crenate pinnae interspersed with shorter, irregularly ovate to rotund segments; Corolla bright lemon-yellow or tinged with orange. Fruit 1(–1.4) cm in diameter, bilocular, glabrous, shining or orange-red when ripe, spherical.

Native to the Galápagos, Ecuador to Chile (Atacama), cultivated in Iraq.

Solanum tuberosum *L.*, Sp. Pl. 1: 185 (1753); Chakravarty, Dep. Agr. Iraq. Tech. Bull. 18: 4 (1964); Husain & Kasim, Cult. Pl. Iraq 123 (1975).

Erect herb grown as an annual from edible underground stem tubers. Leaves petiolate, pinnate with 2–4 pairs of large, ovate, stalked leaflets and many smaller, irregular, sessile intermediate foliage. Corolla white to mauve or purplish with triangular segments. Berry yellowish or green, globose, somewhat flattened.

Native to western South America, widely cultivated throughout Iraq. BATATA (Ar.); potato.

2. **PHYSALIS** L.

Sp. Pl. 1: 182 (1753)

C. C. Townsend
Revised by Ali Haloob and S. T. Al-Kaisi

Annual or perennial herbs. Leaves petiolate, entire or irregularly dentate or sinuate. Inflorescence of solitary flowers, axillary. Calyx campanulate, 5–10-ribbed, 5-teeth convergent at apex, becoming enlarged and inflated in fruit, membranous or leathery and enclosing fruit. Corolla rotate or rotate-campanulate, with short tube and broad flat, sinuate, shallow-lobed limb. Stamens shorter than limb, inserted at base of corolla tube; anthers dehiscing by longitudinal slit. Stigma minutely 2-lobed. Fruit a succulent berry, globose, enclosed in the inflated calyx. Seeds minutely pitted.

A genus of 95 species native to N. and S. America, Afghanistan, India and E. Himalaya; introduced widely into Europe, Africa, S.W. Asia and Australia. Two species have recently been recorded from Iraq; both introduced and possibly naturalized. The two species look very similar in facies and we have not seen any material of either, but *Physalis halicacabum* is photographed in the publication by Abdulridha et al., and is found in Iraq which is within its distribution range.

Physalis, from Gr. φυσαλλις, *fusallis*, bubble, in reference to its inflated calyx.

Leaf margins sinuate; corolla lobes glabrous . . . . .1. *P. halicacabum*
Leaf margins entire or obscurely dentate; corolla lobes pubescent . . . . 2. *P. angulata*

1. **Physalis halicacabum** *Crantz*, Inst. Rei Herb. 2: 370 (1766); Scop. Fl. Carn. ed. 2, 1: 160.

P. divaricata D.Don, Prodr. Fl. Nepal 97 (1825); Schönbeck-Temesy in Fl. Iranica [K. H. Rechinger] 100: 25 (1972); Y. Nasir in Fl. Pakistan [Y. Nasir & Ali] 168: 25 (1985); Abdulridha, Taha & Widad, Ecology & Fl. Basrah: 509 (2016).

Spreading annual herb, up to 60 cm. Leaves petiolate, ovate, lanceolate or rhombic-ovate, apex acute, base cordate to oblique, margins sinuate. Flowers whitish yellow, solitary, axillary; pedicel up to 10 mm. Calyx 2.5–3.5 mm, campanulate, lobed to about a third of its length, becoming inflated, membranous and enlarging to 25 mm in fruit, lobes ciliate. Corolla 5–7 mm. Berry globose, 10 mm in diameter, yellow. Fig. 3, 1–3.

HAB. & DISTRIB. Weedy in sandy places, in sandy soils; fl. Jun./Jul. FUJ: Basra area, cited & photographed in Abdulridha *et al.* (l.c.)

Native from Iran, Afghanistan to Central China.

2. **Physalis angulata** *L.*, Sp. Pl. 1: 183 (1753); Zhi-yun, Lu An-ming & D'Arcy in Fl. China 17: 312 (1994); Al-Allaq, Al-Nahrain J. of Science 15 (4): 31–42 (2012).

Annual herb, 30–77 cm, pubescent to glabrescent. Stems much branched. Leaves petiolate, ovate-rhombic, 30–60 × 20–40 mm, apex obtuse, base cuneate or broadly cuneate,

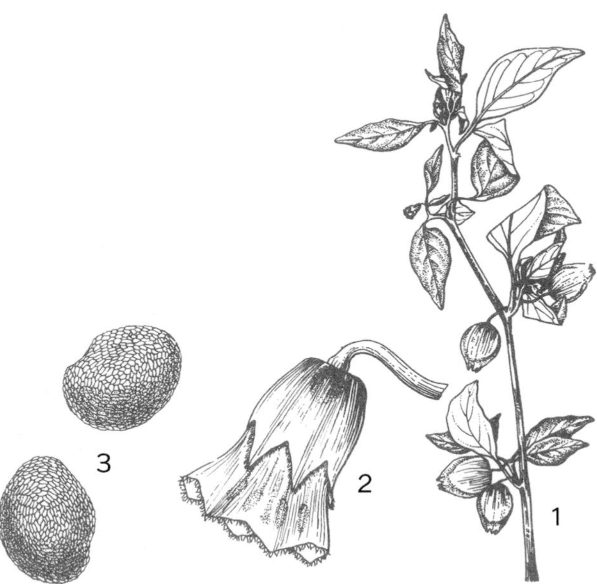

Fig. 3. **Physalis halicacabum**. 1, habit, shoot × ½; 2, flower × 6; 3, seeds × 9. Reproduced with permission from Fl. Pakistan 168: f. 4, C–E (1985). Drawn by S. Hameed. © National Herbarium, Pakistan Agriculture Research Council & University of Karachi, Pakistan.

Fig. 4. **Physalis angulata**. 1, flowering and fruiting branch, habit; 2, flower (opened) × 4. Reproduced from Fl. China Vol. 17, f. 378, 3-4, with permission from Missouri Botanical Garden Press, St. Louis, and Science Press, Beijing.

margins entire or dentate. Flowers white infused with yellow in center, solitary, axillary. Pedicel 4–11 mm. Calyx 3–4.5 mm, campanulate, lobed to about ½ its length, lobes ciliate; fruiting calyx 20–30 mm. Corolla 5–8 mm; lobes acute, pubescent. Anthers bluish. Ovary ovoid. Berry yellowish green, globose, c. 9 mm in diameter. Seeds discoid, yellowish white. Fig. 4, 1–2.

HAB. & DISTRIB. Along roads, in wet areas, orchards and gardens. Recorded from Baghdad (**LCA**), Tuz and Baqouba (**FKI**). No material seen. Description from Al-Allaq (l.c.).

Native to Tropical and Subtropical America, widely introduced into Africa, S.W. & E. Asia.

3. **WITHANIA** Pauquy

Belladone 14 (1825)

C.C. Townsend
Revised by Shahina A. Ghazanfar

Shrubs, ± pubescent or tomentose, more rarely with leaves almost glabrous. Leaves alternate (or upper opposite), petiolate, quite entire. Flowers usually shortly pedicellate or subsessile, in axillary clusters. Calyx campanulate, 5-dentate, considerably accrescent in fruit and embracing the berry. Corolla yellowish or greenish, 3- to 6-partite with valvate lobes, narrowly campanulate. Stamens 5, inserted near base of corolla, filaments flat. Stigma shortly and broadly bilamellate. Fruit a bilocular berry.

Twenty species in tropical & S. Africa, Arabian Peninsula, Turkey to China and Japan; a single species in Iraq.

Withania commemorates the British amateur geologist and landowner Henry Witham FRSE (1779–1844).

1. **Withania somnifera** (*L.*) *Dunal,* Prodr. [A. P. de Candolle] 13 (1): 453 (1852); Boissier, Fl. Orient. 4: 287 (1879); Zohary in Dep. Agr. Iraq Bull. 31: 131 (1950); Chakravarty, Dep. Agr. Iraq. Tech. Bull. 18: 15 (1964); Rechinger, Fl. Lowland Iraq: 535 (1964); Schönbeck-Temesy in Fl. Iranica [K. H. Rechinger] 100: 27 (1972); Baytop in Fl. Turkey [P. H. Davis] 6: 445 (1978); Feinbrun-Dothan, Fl. Palaest. 3: 164, pl. 272 (1978); Y. Nasir in Fl. Pakistan [Y. Nasir & Ali] 168: 30 (1985); Collenette, Wild Flow. Saudi Arabia: 707 (1999); Hepper in Fl. Egypt [Boulos] 3: 41 (2002); Abdulridha, Taha & Widad, Ecology & Fl. Basrah: 511 (2016); Taifour & El-Oqlah, Pl. Jordan Annot. Checkl.: 149 (2017).

Physalis somnifera L., Sp. Pl. 1: 182 (1753).

Shrub, to 1.5 m, with flexuous, canescent-tomentose, terete branches. Leaves long-petiolate, entire, lower alternate, those of inflorescence usually opposite, lamina up to 12 × 9 cm, ± densely shortly stellate-pubescent on both surfaces, but especially on principal veins beneath, upper surface frequently glabrescent at maturity. Flowers shortly pedicellate, in axillary clusters of up to 12. Calyx densely white-lanate in bud and flower, c. 5 mm, teeth linear, glabrescent, green and accrescent in fruit with reticulate venation. Corolla yellowish green, c. 2 × length of calyx, segments narrowly triangular, pubescent on dorsal surface. Berry globose, 5–6 mm in diameter, completely enveloped by loose accrescent calyx. Fig. 5, 1–5.

HAB. Open and disturbed places; alt. 5–100 m; fl. & fr. throughout the year.
DISTRIB. Found on the alluvial plains and estuarine areas of eastern and southern Iraq. **LCA**: Baghdad, *Ghani & Hadi* 5439; Abu Ghraib, *Fawzi & Ali* 39512 (BAG). **LSM**: Amarah, *Omar & Chakravarty,* 37766 (BAG). **LEA**: near Basra, *Rawi & Gillett* 5950!; Abu Al-Khasib, *Omar, Al-Kaisi, Hamad & Hamid* 43985 (BAG).

SAYKARĀN, سيكران, WARAQ AŠ-ŠIFĀ ورق الشفاء. Plant used medicinally for various conditions in its area of distribution. In Iraq, the leaves have been used as a poultice on tumours and ulcers; the roots for reducing inflammation, psoriasis, and bronchitis; the fruit has been used as a diuretic and as a mild sedative.

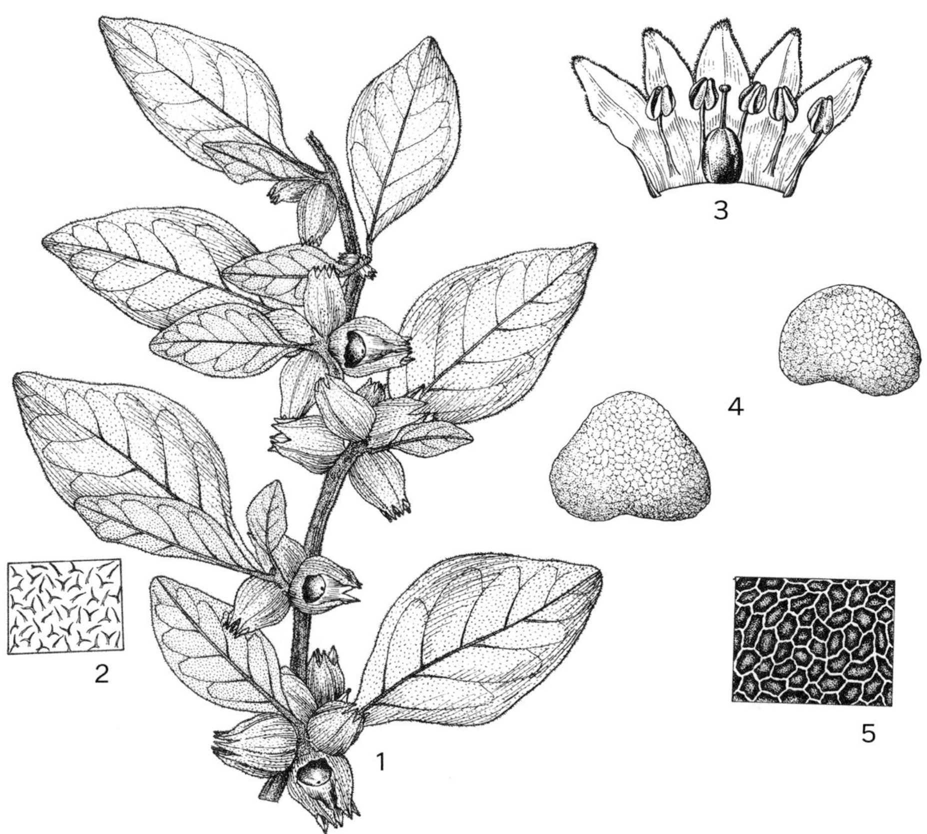

Fig. 5. **Withania somnifera**. 1, habit, shoot with fruit × ½; 2, detail undersurface of leaf showing indumentum; 3, flower (opened) × 5; 4, seeds × 10; 5, seed surface, detail × 25. Reproduced with permission from Fl. Pakistan 168: f. 6, A–E (1985). Drawn by S. Hameed. © National Herbarium, Pakistan Agriculture Research Council & University of Karachi, Pakistan.

4. **LYCIUM** L.

Sp. Pl. ed. 1: 191 (1753); Gen. Pl. ed. 5: 88 (1754)

C.C. Townsend
Revised by Shahina A. Ghazanfar

Deciduous or evergreen shrubs, with an erect or pendulous habit, rigidly spinose or unarmed. Leaves alternate, often fasciculate, shortly petiolate, entire, linear-terete to flat and rhomboid-ovate, mostly greyish green. Flowers axillary, solitary or fasciculate. Calyx (3–)5-dentate, campanulate, not accrescent in fruit. Corolla funnel-shaped, greenish, whitish, purplish or pinkish, usually fading after anthesis; limb 5-partite with spreading lobes, tube expanding upwards, usually longer than limb. Stamens 5, exserted or included, inserted about middle of tube, filaments expanded below or not, with or without a ring of hairs at base. Stigma capitate or bilamellate. Fruit an ovoid or globose bilocular berry.

Lycium (from Gr. λυκιον, *lykion*, the name used by Pliny the elder and Dioscorides for the dyer's buckthorn, so named because it was thought to originate in Lycia, a region of Anatolia.

101 species in the dry tropical and subtropical regions to Mongolia; 2/3 native and 2 cultivated species in Iraq.

Feinbrun, N. (1968). The genus *Lycium* in the Flora Orientalis region". *Collectanea Botanica* 7: 359–379.

1. Filaments bearded at base. 2
    Filaments glabrous at base 3
2. Calyx teeth broadly rounded, finely ciliate with multicellular hairs;
    berries black. . . . . 3. *L. ruthenicum*
    Calyx teeth very unequal, linear or deltoid, calyx usually ± densely
    covered with multicellular hairs; berries red (cultivated). . . . . 5. *L. brevipes*
3. Corolla tube short, 1.5(–2) × as long as lobes of limb. . . . . 1. *L. depressum*
    Corolla tube long, 2.5–3 × as long as lobes of limb. 4
4. Calyx when in flower shortly tubular (4 × 2 mm) . . . . 2. *L. shawii*
    Calyx when in flower poculiform (3 × 3 mm) (cultivated). . . . . 4. L. europaeum

1. **Lycium depressum** *Stocks*, Hooker's J. Bot Gard. Misc. 4: 179 (1852); Schönbeck-Temesy in Fl. Iranica [K. H. Rechinger] 100: 32 (1972); Baytop in Fl. Turkey [P. H. Davis] 6: 447 (1978); Feinbrun-Dothan, Fl. Palaest. 3: 161, pl. 265 (1978); Y. Nasir in Fl. Pakistan [Y. Nasir & Ali] 168: 36 (1985); Taifour & El-Oqlah, Pl. Jordan Annot. Checkl.: 149 (2017).

> *Lycium barbarum* (non L.) auctt.: Dunal, Prodr. [A. P. de Cndolle] 13 (1): 511 (1852); Boissier, Fl. Orient. 4: 289 (1879); Blakelock, Kew Bull. 4: 529 (1949); Zohary in Dep. Agr. Iraq Bull. 31: 131 (1950); Abdulridha, Taha & Widad, Ecology & Fl. Basrah: 507 (2016); Taifour & El-Oqlah, Pl. Jordan Annot. Checkl.: 148 (2017).
> *L. turcomanicum* Turcz. ex Miers, Ann. Mag. Nat. Hist. ser. 2, 14: 183 (1854).

Much-branched shrub 3–4 m; stem and branches armed with many stout lateral spines up to 2 cm, branches and branchlets terminating in spines. Cortex of younger branches yellowish or light brown, scarcely pruinose, in the older parts grey, longitudinally striate, often cracked or peeling. Leaves linear to oblanceolate, entire, long-attenuate below and petiolate, fasciculate, to 50 × 8 mm, glabrous or more usually with scattered clavate hairs, especially in the channeled base. Flowers solitary or fasciculate in groups of up to 10; pedicels 6–10 mm, glabrous or with scattered clavate hairs below. Calyx poculiform, 4 × 4 mm, with broad, shallow, membranous–Margined teeth ciliate at tips when young. Corolla reddish-purple, with a short glabrous tube 5–6 mmm, subequalling to almost twice as long as lobes of limb. Filaments glabrous, anthers long-exserted. Berry shortly ovoid, c. 6 × 5 mm, red. Fig. 6, 1–6.

HAB. Weed in field by roadside, on gravelly soil by roadside, in cultivated land among date-palm trees, on clay and loamy soil, sandstone and conglomerate hills, cultivated fields, rocky clay valley; alt. 2–680 m; fl. & fr. Feb.-Oct.

DISTRIB. Common in the upper plains and foothills, and the desert plateau areas of Iraq. **FUJ**: Ninioeh, *Gillett & Chapman* 11172 (BAG). **FPF**: Diyala, Jabal Hamrin, *Evans* M/153!; Ludlow Hewitt 1069 (BAG); between Shahraban and Sa'adia, *Rechinger* 8079!; Jabal Al-Muaila near Kuwait N. of Amarah, *Guest, Rawi & Rechinger* 17640C (BAG); Injana, *S. Omar & F. Karim* 37833 (BAG); Salmaniya well 10 km W. Rshaida P.O. near Persian border, *H. Alani* 39824 (BAG); Al-Tib, *Hasain Alani* 39693 (BAG); 2 km N. of Sa'adiya, *Al-Kaisi* 42539 & 42538 (BAG); Al-Sudur, *S. Omar & Thamer* 47118 (BAG). **LEA**: between Diyala and Baquba, *coll. ingot.* s.n. **LCA**: Baghdad, nr. Residency, *Rogers* 027!; Babylon, *Guest* 1684!. **LSM/LBA**: road bet. Qalat Salih and Qurna, *Rawi, Chakravarty, Khatib & Tikrity* 29861 (BAG). **LSM:** Qalat Salih, *Botany staff* 42474 (BAG); 20 km N. of Al-Musharrah, *Hasain Al-Ani* 39723 (BAG). **LBA**: Basra, *Rogers* 0466!; 50 km S. Basra, *Khatib & Alizzi* 32652! ; Abu Al-Khasib 20 km E. of Basra, Rawi & Haddad 25962 (BAG); **DLJ**: 48 km N.W. the road between Baiji and Haditha, *Chakravarty, Rawi, Khatib & Alizzi* 31855! **DLJ/DWD**: between Ramadi and Hit, H. Hamid 38978 (BAG);10 km N. of Kubaisa, *H. Hamid* 39034 (BAG); **DGA**: S. of Dor Town, *Al-Khayat, Omar & Widad* 53996 (BAG); **DGJ/DJG**: 20 km from Samarra to Tikrit, *Fawzi & M. Noori* 39668 (BAG). **DWD**: IKa'ra c. 100 km N. of Rutba, *Serkahia* 16481 (BAG); 13 km E. K3, *Rawi, Khatib & Alizzi* 32939 (BAG); Haditha, *S. Omar, H. Abass, Al-Kaisi & K. Hamad* 55814 (BAG); 40 km Rummana to Rawa, *Al-Khayat & K. Hamad* 51682 (BAG); 5 km S.E. of Ain Al-Tamur, *A. Haloob, Ikhla, R. Hamshkan & Riyadh* 59905 (BAG); Al-Majarrah 50 km S.W. of Falluja, *S. Omar, Al-Kaisi, K. Hamad & H. Hamid* 44511 (BAG); Wadi Hauvan near Baghdadi, *S. Omar, Al-Kaisi, K. Hamad & H. Hamid* 44566 (BAG). **DSD**: nr. Zubair, *Rawi & Gillett* 6017.

SARIM (سريم); AWSAJ (عوسج – *Guest* 1684)

Turkey, Syria, Jordan, Palestine, Iran, Transcaucasia, Afghanistan, Pakistan, C. Asia (Kizil Kum to S.W. Pamirs).

2. **Lycium shawii** *Roem. & Schultes*, Syst. Veg. 4: 693 (1819); Feinbrun-Dothan, Fl. Palaest. 3: 160 (1978); Baytop in Fl. Turkey [P. H. Davis] 6: 447 (1978); Y. Nasir in Fl. Pakistan [Y.

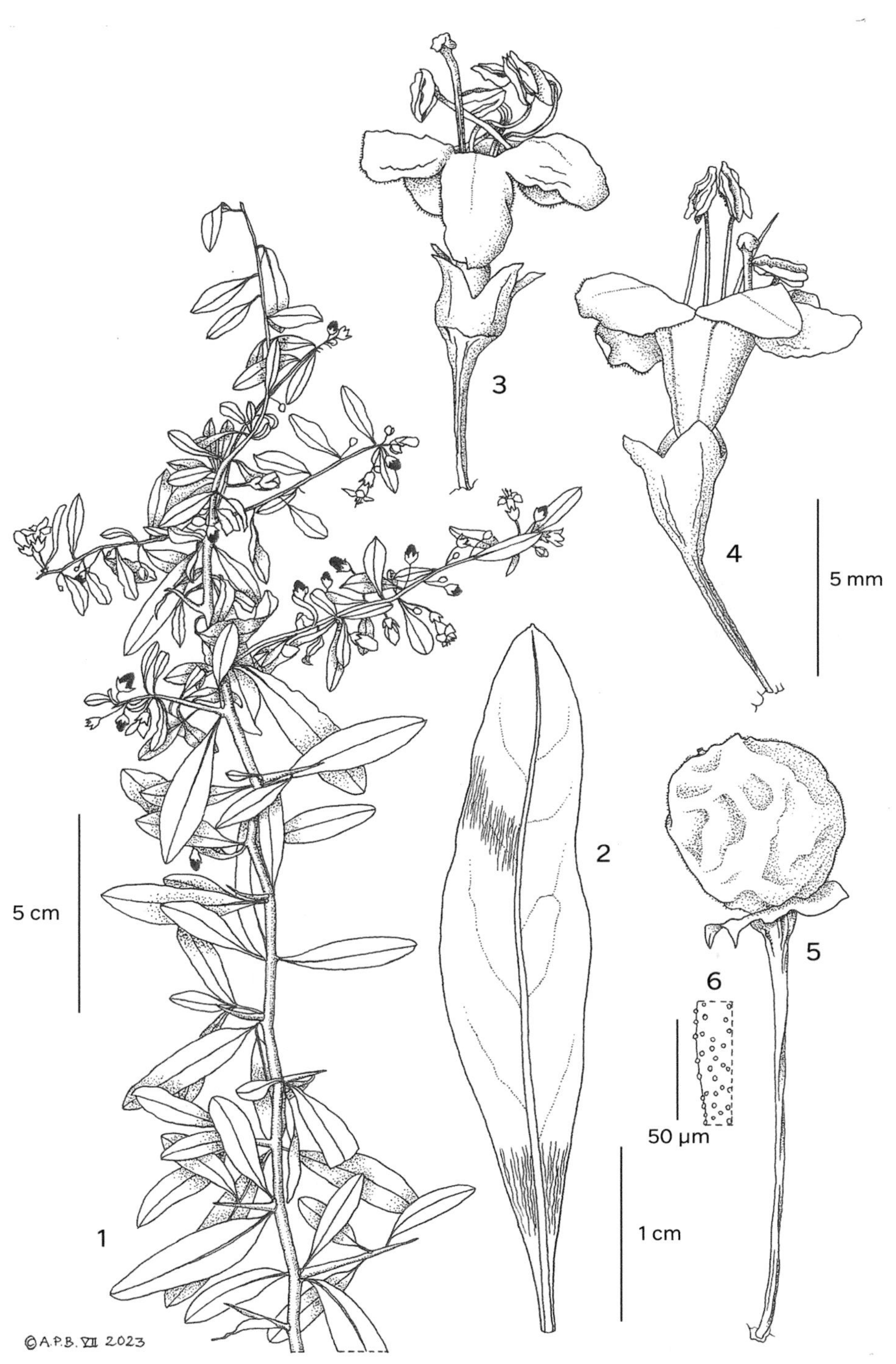

Fig. 6. **Lycium depressum**. 1, habit; 2, leaf; 3, flower; 4, mature flower; 5, fruit; 6, surface of fruit, detail. 1–4 from *Wheeler-Haines* W179; 5, 6 from *Al Ani* et al. 9689. Drawn by © A.P. Brown 2023.

Fig. 7. **Lycium shawii**. 1, habit, portion of shoot × ½; 2, corolla (opened) × 3. Reproduced with permission from Fl. Pakistan 168: f. 7, C–D (1985). Drawn by S. Hameed. © National Herbarium, Pakistan Agriculture Research Council & University of Karachi, Pakistan.

Nasir & Ali] 168: 35 (1985); Collenette, Wild Flow. Saudi Arabia: 701 (1999); Hepper in Fl. Egypt [Boulos] 3: 44 (2002); Taifour & El-Oqlah, Pl. Jordan Annot. Checkl.: 148 (2017).

> *L. barbarum* (non L.) Mill., Gard. Dict. ed. 8 (1768); Chakravarty, Dep. Agr. Iraq. Tech. Bull. 18: 12 (1964); Haloob et al. (eds), Illustr. Fl. Lowlad Iraq 1: 154 (2016).

> *L. arabicum* Schweinf. ex Boiss., Fl. Orient. [Boissier] 4: 289 (1879)

Much-branched shrub, 1–3 m, stem and branches armed with stout, alternate lateral spines to 2.5(–3) cm; branches and branchlets terminating in spines. Cortex brownish, usually with a glaucous-pruinose bloom, in older branches becoming whitish, longitudinally striate and ridged, frequently cracked, glabrescent; younger branches often shortly whitish-tomentellous. Leaves oblong to lanceolate-spatulate, obtuse to subacute, entire, to 20 × 4 mm, narrowed and subpetiolate below, fasciculate, usually shortly cinereous-hairy at least in narrowed basal region. Flowers solitary or in pairs, on tomentellous pedicels 4–10 mm. Calyx 4 × 2 mm, shortly tubular with short, broadly triangular, ± obtuse teeth, shortly puberulent. Corolla purplish with a long glabrous tube of 8–15 mm, much exceeding the short (2–3 mm) lobes of limb. Filaments glabrous throughout, anthers exserted. Berry globose, 5–6 mm in diameter, red. Fig. 7, 1–2.

HAB. On banks of cultivation, on dry mud flats of river, on moderately saline and clay soil, in desert; alt. 105–180 m; fl. & fr. Jan.-Dec.

DISTRIB. Lower hills, alluvial plains and desert regions of Iraq. **FUJ/DLJ**: Jazira, nr. Wadi Tharthir, *Guest* 3556!; S. Jazira *Guets & Rawi & Nuri* 13546! **FPF**: Jalaula [Jalawla formerly Qaraghan], *Guest* 1877!; 2 km N. of Saadiya, *Al-Kaisi* 42538 (BAG). **FNI**: Nineva-Mosul, *Chapman* 25398! **LCA**: Baghdad, nr. Residency, *Rogers* 027!; nr. Falluja, *Guest* 998 & 1126; Babylon, *Guest* 1684! **LEA**: Nr. Kut [Kut as Sayyid], *Guest* 334! 50 km S. Basra, *Khatib & Alizzi* 32652! **LSM**: 75 km N. of Amara, *Hazim & Nuri* 30640! **LBA**: Basra, *Rogers* 0466! **DWD**: Ka'ra c. 100 km N. of Rutba, *Serkahia* 16481 (BAG). **DWD**: N. of Karbala, *Rawi & Gillett* 6330! **DSD**: nr. Zubair, *Rawi & Gillett* 6017; N. of Karbala, *Rawi & Gillett* 6330!; S. desert, 42 km E. by Salman, *Guest, Rawi & Long* 14115!; 150 km S. of As-Salman, *Rawi & Tikriti* 24982!; 17 km W.W.N. of Busaiya, *Chakravarty, Khatib, Rawi & Tikriti*, 29961!; Jabal Sanam, *Rawi & Gillett* 6153 (BAG); *Al-Kaisi, Ghina & Saif* 58428 (BAG);. Al-Takhaded 60 km S. of Salman, *Botany staff* 42194 (BAG); 100 km from Salman to Samawah, *S. Omar, Al-Kaisi, K. Hamad & H. Hamid* 44022 (BAG); Sawa Lake 35 km S.W. of Samawah, *S. Omar, Al-Kaisi, K. Hamad & H. Hamid* 44026 (BAG); 20 km from Salman to Takhadid, *S. Omar, Al-Kaisi, K. Hamad & H. Hamid* 48104 (BAG). **DLJ**: 48 km N.W. the road between Baiji and Haditha, *Chakravarty, Rawi, Khatib & Alizzi* 31855!

AWSAJ; AUSAJ (Ar.).

Palestine, Sinai, Egypt, Arabia, Iran, N. Africa (Libya), tropical E. Africa.

3. **Lycium ruthenicum** *Murr.*, Comm. Goetting. 1779: 9 (1780); Boissier, Fl. Orient. 4: 290 (1879); Feinbrun, Collectanea Botanica 7: 374 (1968); Schönbeck-Temesy in Fl. Iranica [K. H. Rechinger] 100: 31 (1972); Y. Nasir in Fl. Pakistan [Y. Nasir & Ali] 168: 33 (1985).

Much-branched shrub, to c. 2 m, glabrous, stems and branches armed with slender but strong lateral spines, branches and branchlets terminating in spines. Cortex yellowish- to greyish-white, longitudinally striate, often very cracked in older parts. Leaves mostly narrowly oblanceolate, occasionally linear to lanceolate, to 35 × 3 mm, fasciculate, long-attenuate below but scarcely petiolate, obtuse to subacute. Flowers rarely solitary, generally fasciculate in groups of 2–5, on glabrous pedicels 2–5 mm. Calyx short, campanulate-poculiform, c. 4 × 2 mm, soon appearing bi- or trilabiate with obtuse, ciliate teeth. Corolla reddish-purple with a glabrous tube c. 5 mm, equalling or slightly exceeding lobes of limb. Filaments distinctly bearded at base, anthers exserted. Berry globose, 5–8 mm in diameter, black.

DISTRIB. **FAR**: Arbil, Rost, *Haley* 172 (W).

Listed by Gillett (1948, 8) who observes "a specimen from near Baghdad has been identified as belonging to this Central Asian species", but unfortunately does not give the specimen number. Rustam 1126, which is *L. depressum* (q.v.) was originally named as *L. ruthenicum* in the Kew Herbarium, and this may well be the specimen in question. There is only one other record from Arbil, *Haley* 172 (W) cited in Flora Iranica – no other material seen from Iraq.

S. European Russia, Turkey, Transcaucasia, Afghanistan C. Asia (Kazakhstan to Tadjikistan), Mongolia, Tibet, Kashmir.

4. **Lycium europaeum** *L.*, Sp. Pl. ed. 1: 192 (1753); Boissier, Fl. Orient. 4: 288 (1879); Feinbrun, Collectanea Botanica 7: 362 (1968); Baytop in Fl. Turkey [P. H. Davis] 6: 446 (1978); Feinbrun-Dothan, Fl. Palaest. 3: 160, pl. 262 (1978); Hepper in Fl. Egypt [Boulos] 3: 45 (2002);Taifour & El-Oqlah, Pl. Jordan Annot. Checkl.: 148 (2017).

Much-branched shrub, 2–4 m, glabrous save occasionally in young shoots and leaves, armed with stout lateral spines to 2 cm, branches and branchlets greyish to yellowish, in older parts grey, sometimes pruinose, longitudinally striate. Leaves lanceolate to oblanceolate or narrowly elliptic, attenuate and shortly petiolate below, to c. 40 × 8 mm, obtuse to subacute, fasciculate. Flowers solitary or more rarely paired; pedicels 2–6 mm, glabrous. Calyx poculiform, 3 mm, 5-dentate to c. ¼ its length, with broadly triangular, membranous–Margined teeth. Corolla reddish-purple, with a glabrous tube 8–9 mm, 3 × as long as lobes of limb. Filaments glabrous throughout, anthers exserted. Berry globose, 5–6 mm in diameter, red.

Listed by Gillett (1948, 8) as "reported from Baghdad". If it occurred it was certainly as a cultivated shrub or hedge – no specimen has been seen.

A circum-Mediterranean species; S. Europe from Portugal to Greece, Turkey, Syria, Palestine, Egypt, N. Africa from Libya to Morocco.

5. **Lycium brevipes** *Benth.*, Bot. Voy. Sulphur [Bentham] 40 (1844); Chakravarty, Dep. Agr. Iraq. Tech. Bull. 18: 13 (1964).

Divaricately branched shrub 1–3 m, stem and branches armed with spines. Leaves oblanceolate to ± spatulate, 5–25 × 2–6 mm, subglabrous to densely short-pubescent with glandular hairs. Flowers few to several, fasciculate, on 2–10 mm pedicels which are glabrous to ± glandular-pilose. Calyx 3–5 mm, ± densely glandular-pilose, unequal teeth linear to deltoid, equalling or shorter than tube. Corolla lavender or whitish, mostly 9–12 mm, lobes subequalling tube. Filaments pilose at base. Berry ovoid to globose, 5–8 mm, bright red.

LCA: Za'faraniyah, coll. ingot. 20528 (Cult.)

Native to southern U.S.A. (California) and Mexico; cultivated in Iraq.

5. **DATURA** L.

Sp. Pl. 1: 179 (1753); Gen. Pl. ed. 5: 83 (1754)

C.C. Townsend
Revised by Shahina A. Ghazanfar

Annal or more rarely perennial herbs, bushes or small trees with alternate leaves. Flowers solitary or pedicellate, pendulous or erect. Calyx tubular-elongate, regularly or irregularly 5-fid, often pentagonal, splitting longitudinally after flowering, or circumscissile below, upper part deciduous and lower part becoming accrescent under capsule. Corolla tubular to widely funnel-shaped, cream, white or purple, or white and variously purple-diffused or -striped, tube long and slender, limb 5-lobed or forming a 10-toothed border. Stamens 5, included, attached near base of tube. Style filiform, stigma bilobed. Fruit a capsule, four-valved or bursting irregularly, mostly strongly spinous, very rarely smooth.

14 species native to S.W. USA, Mexico and N.W. South America; introduced widely and naturalized in rest of the World; 2 species in Iraq.

Datura is derived from the Sanskrit *dhatūra*, 'thorn-apple', or more specifically *dhattūra*, 'white thorn-apple', *D. metel.*

Avery, A.G., Satina, S. & Rietsema, J. (1959). Blakeslee: The genus *Datura*. Roland Press Company, N.Y.

1. Corolla 6–10(–12) cm, limb 3–4 cm wide; capsule regularly 4-valved at
      maturity, covered with long spines . . . . .1. *D. stramonium*
   Corolla 14–15(–18) cm, limb exceeding 4 cm wide; capsule breaking
      irregularly at maturity, spines long or short to tuberculate. . . . . 2. *D. metel*

1. **Datura stramonium** *L.*, Sp. Pl. ed. 1: 179 (1753); Boissier, Fl. Orient. 4: 292 (1879; Blakelock in Kew Bull. 4: 528 (1949); Avery et al., 18 (1959); Chakravarty, Dep. Agr. Iraq. Tech. Bull. 18: 60 (1964); Schönbeck-Temesy in Fl. Iranica [K. H. Rechinger] 100: 45 (1972); Feinbrun-Dothan, Fl. Palaest. 3: 168, pl. 279 (1978); Baytop in Fl. Turkey [P. H. Davis] 6: 451 (1978); Y. Nasir in Fl. Pakistan [Y. Nasir & Ali] 168: 41 (1985); Collenette, Wild Flow. Saudi Arabia: 698 (1999); Hepper in Fl. Egypt [Boulos] 3: 46 (2002);Abdulridha, Taha & Widad, Ecology & Fl. Basrah: 507 (2016); Taifour & El-Oqlah, Pl. Jordan Annot. Checkl.: 148 (2017).

Stout, usually glabrous, dichotomously branched annual herb up to 2 m tall, but usually less. Stem terete, green, purplish or whitish. Leaves ovate, 5–20 cm, sinuate-dentate or with large, coarse teeth, often unequal at base, long-petiolate, glabrous or sparsely puberulent. Calyx half as long as corolla or slightly more, circumscissile, upper part falling with corolla; teeth rather obtuse to acute, c. $^1/_5$ length of tube. Corolla white, pale purple or purple, rather narrow, 6–10 cm overall, limb not widely expanded, to 4 cm in diameter, with lobes

Fig. 8. **Datura stramonium**., 1, flowering branch with fruit × ½. **Datura metel**. 2, flowering branch × ½; 3, branch showing fruit × ½. Reproduced from Fl China Vol. 17, f. 400 (1, 3, 4), with permission from Missouri Botanical Garden Press, St. Louis, and Science Press, Beijing.

long- or short-acuminate. Capsule ovoid, opening regularly by 4 valves when ripe, 5–7 × 3–5 mm, usually with many long spines (rarely smooth: 'D. inermis' Jacq.). Fig. 8, 1.

HAB. In disturbed places, edges of cultivated fields; alt 220–800 m; fl. Jun.-Sep. & fr. Jul.-Sep.
DISTRIB. Lower shills of N. Iraq. **MUS**: Sulaimaniya, *Al-Radzi* 5279 (BAG). **FNI**: N. of Mosul, *Anders* 36910 (BAG).**FKI**: Kirkuk, Darband-i-Khan *Haines* s.n.

Native from Texas to Central America and the Caribbean; introduced and cultivated or naturalized in S. America and throughout the Old World.

2. **Datura metel** *L.*, Sp. Pl. 1: 179 (1753); Blakelock in Kew Bull. 4: 528 (1949); Avery et al., 32 (1959); Chakravarty, Dep. Agr. Iraq. Tech. Bull. 18: 63 (1964); Hepper in Fl. Egypt [Boulos] 3: 46 (2002); Collenette, Collenette, Wild Flow. Saudi Arabia: 698 (1999); Abdulridha, Taha & Widad, Ecology & Fl. Basrah: 507 (2016).

Datura fastuosa L., Syst. Nat. ed. 10, 2: 932 (1759); Schönbeck-Temesy in Fl. Iranica [K. H. Rechinger] 100: 46 (1972);

D. inoxia Mill., Gard. Dict. ed. 8 (1768); Avery et al., 28 (1959); Feinbrun-Dothan, Fl. Palaest. 3: 168, pl. 280 (1978); Y. Nasir in Fl. Pakistan [Y. Nasir & Ali] 168: 43 (1985).

Stout annual herb, up to 1.5 m tall. Stem erect, terete, yellowish green to purplish, glabrous or soft whitish pubescent. Leaves ovate, entire or sinuate-dentate with a few large teeth, 6–15 cm, often unequal at base, long-petiolate. Calyx not attaining half the length of corolla, upper part deciduous with corolla; teeth long, acute. Corolla white, yellow, purple or purple outside and white within, 14–15(–18) cm overall, with a widely expanded limb 8–12+ cm across, lobes long- or short-acuminate. Capsule globose, inclined, 4–6 cm in diameter, covered with short, broad-based spines or tubercles, breaking irregularly at maturity. Fig. 8, 2–3.

HAB. Roadsides, disturbed and waste places, weedy in orchards and gardens, on clay soil; alt. 35–100 m; fl. & fr. Apr.-Aug.

DISTRIB. In the eastern and central alluvial plains and southern marsh district. **LEA**: Kut, *Guest* 3479!; E. of Kut, *Al-Kaisi & Riyadh* 59776 (BAG). **LCA**: near Baghdad, *Lazar* 90! 298!; Abu Ghraib, *Sahira & Salim* 34557 (BAG); Baghdad, Karada, *Guest & Haidar* 19485 (BAG). **LSM**: Amarah exper. Station, *Chakravarty & Omar* 37768 (BAG). **LBA**: Basrah, *Guest* 1610 (BAG).

Native from Texas to Colombia, introduced and naturalized in Africa, Turkey, Arabian Peninsula, Afghanistan to China and Australia.

6. **HYOSCYAMUS** L.

Sp. Pl. 1: 179 (1753); Gen. Pl. ed. 5: 84 (1754)

C. C. Townsend

Revised by Shahina A Ghazanfar, Ali Haloob & Shuker T. Al-Kaisi

Annual or biennial, more rarely perennial herbs, hairy to villous, rarely glabrous, frequently viscid. Leaves petiolate or sessile, sinuate-dentate or pinnatifid, rarely entire. Flowers bracteate, pedicellate or sessile, lower axillary, upper in frequently secund, scorpioid racemes or spikes. Calyx 5-toothed, tubular-campanulate or urceolate, rigid and accrescent in fruit. Corolla weakly or more strongly zygomorphic, campanulate or funnel-shaped, limb oblique, 5-fid with broad obtuse lobes, purplish or yellow, usually darker at throat, often with reticulate dark venation. Stamens inserted around middle of corolla tube, included or exserted. Stigma broadly capitate. Fruit a bilocular circumscissile capsule, concealed within the accrescent calyx. Seeds very rough, tuberculate or scrobiculate.

32 species distributed from Macaronesia to N.E. Tropical Africa, the Arabian Peninsula, Europe and temperate Asia; 9 species in Iraq.

Hyoscyamus from Gr. υς, *hus*, pig and κυαμος, *kuamos*, bean.

The genus *Hyoscyamus* includes some of the most important medicinal plants belonging to the Solanaceae as a natural source of tropane alkaloids such as hyoscyamine, scopolamine, and tropine.

Hajrasouliha, S., Massoumi, A.A., TaherNejadSattar, S.M., Hamdi, M. and Mehregan, I. (2014). A phylogenetic analysis of *Hyoscyamus* L.(Solanaceae) species from Iran based on ITS and trnL-F sequence data. J. Biodiv. Env. Sci. 5(1): 647–654.

Sanchez-Puerta, M.V. & Abbona, C.C. (2014). The chloroplast genome of *Hyoscyamus niger* and a phylogenetic study of the tribe Hyoscyameae (Solanaceae). PLoS One 9(5), p.e98353.

Sheidai, M., Mosallanejad, M. & Khatamsaz, M. (1999). Karyological studies in *Hyoscyamus* species of Iran. Nordic Journal of Botany 19(3): 369–374.

1. Plant covered with compound and simple hairs, leaves always succulent;
   fruiting calyx cylindrical to campanulate, leathery in texture . . . . .9. *H. insanus*
   Plant covered with simple hairs only, leaves not or slightly succulent;
   fruiting calyx variable in shape and texture. 2
2. Annual; corolla small, 3.5–10 mm across; petals yellow, without
   reticulate venation . . . . .6. *H. pusillus*

1. **Hyoscyamus niger** *L.*, Sp. Pl. 1: 179 (1753); Boissier, Fl. Orient. 4: 294 (1879); Chakravarty, Dep. Agr. Iraq. Tech. Bull. 14: 141 (1964); Chakravarty, Dep. Agr. Iraq. Tech. Bull. 18: 45 (1964); Schönbeck-Temesy in Fl. Iranica [K. H. Rechinger] 100:66 (1972); Baytop in Fl. Turkey [P. H. Davis] 6: 454 (1978); Y. Nasir in Fl. Pakistan [Y. Nasir & Ali] 168: 50 (1985).

H. persicus Boiss. & Bushe, Nouv. Mém. Soc. Imp. Naturalistes Moscou 12: 158 (1860).

Erect biennial (occasionally annual) with a nauseating smell, to 80 cm, ± densely covered throughout with viscid glandular hairs considerably variable in length. Stem stout, woody below, terete, sometimes almost glabrous near base. Lower leaves petiolate, large, oblong-ovate, 15–20 × 8–12 cm; cauline leaves sessile, amplexicaul, somewhat diminishing above; all leaves with a few large acuminate teeth, the sinuses between teeth wide and rounded, or more rarely subentire or subpinnatifid. Flowers subsessile, in 2 rows, forming a ± secund raceme which elongates greatly as flowering advances. Bracts leaflike, diminishing above. Calyx at flowering time campanulate, c. 12 mm, 5-dentate with broadly triangular, acuminate teeth c. ⅓ length of tube; in fruit becoming strongly accrescent, tube to 20 × 10 mm, firm and with ribs thickening, with prominent reticulate intermediate venation, distinctly urceolate, teeth rigidly spinose. Corolla c. 2 cm across, creamy yellow, with dense reticulate purple venation, throat purple. Anthers purple. Capsule entirely concealed by accrescent calyx, c. 15 × 12 mm, circumscissile in upper ¼ with an apiculate lid. Seeds compressed, strongly reticulate-rugose, ± 1 mm in diameter. Fig. 9, 1–4.

HAB. Mountain slopes, on serpentine rocks, on slopes, on clay soils; alt. 900–1640 m; fl. Apr.-Jun. & fr. Apr.-Aug.

DISTRIB. Mountains and lower hills of N. Iraq. **MAM**: Dori village 5 km E. Kani Masi, *S. Omar & Al-Kaisi* 45453 (BAG !). **MRO**: Arbil, N. Rost, *Thesiger* 950!; Helgord, *Al-Kaisi, Ikhlas, Noor, Saif* 58341 (BAG). **MSU**: Sulimaniyah, nr. Penjwin, *Rechinger* 10552!; Kajan mt near Penjwin, *Rawi* 12168! Penjwin, *Rawi* 12168!; Khormal, *Rawi* 8888 (BAG). Cultivated at Abu Ghraib farm, *Rawi* 29691 (BAG) and at Amarah, exp. Station (**LSM**), *S. Omar & Chakravarty* 37767 (BAG).

CHABAND (Kurd.)

Europe, eastwards to Siberia.

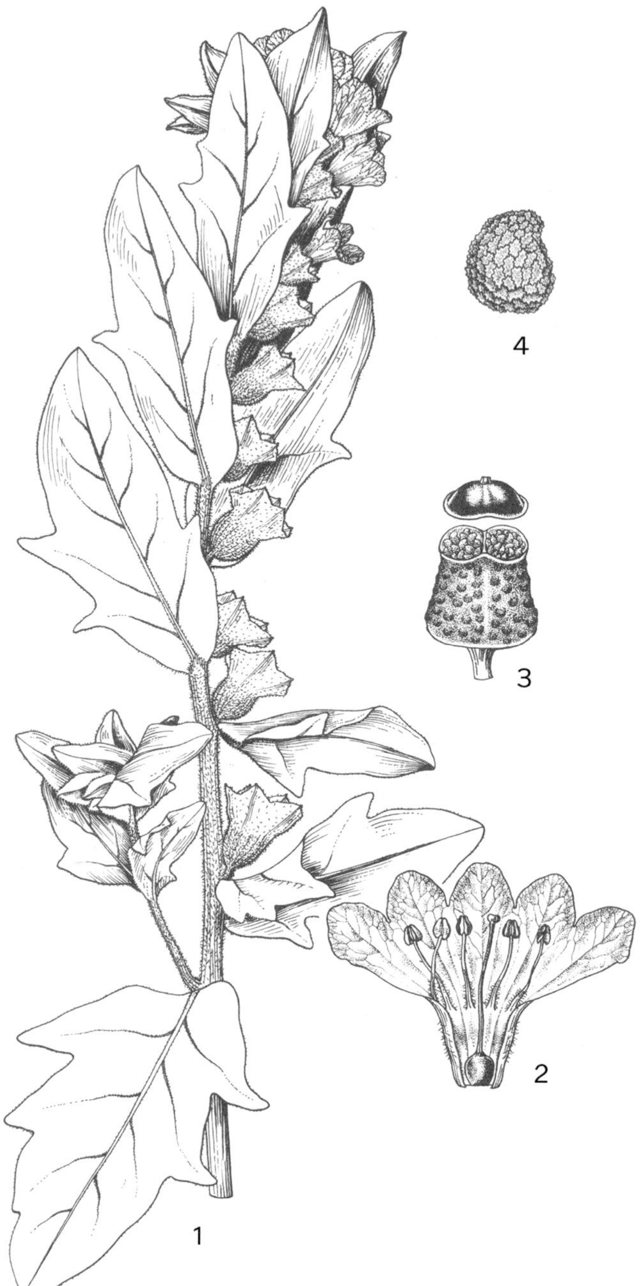

Fig. 9. **Hyoscyamus niger**. 1, habit, portion of shoot × ½; 2, corolla (opened) × 1; 3, fruit (showing dehiscence) × 1 ½; 4, seed × 10. Reproduced with permission from Fl. Pakistan 168: f. 11, A–D (1985). Drawn by S. Hameed. © National Herbarium, Pakistan Agriculture Research Council & University of Karachi, Pakistan.

2. **Hyoscyamus kurdicus** *Bornm.*, Verh. K.K. Zool.-Bot. Ges. Wien 48: 611 (1898). Type: Iraq, Arbil, Rawandiz, Algurd Dagh, 3100 m, *Bornmüller* 1568 (W! holo.); Schönbeck-Temesy in Fl. Iranica [K. H. Rechinger] 100: 57 (1972).

Erect perennial (or biennial?) with a thick rootstock, stem 25–50 cm, densely covered throughout with viscid eglandular hairs, very short to very long hairs intermingled. Stem stout, woody and shortly hairy below, terete. All but lowest leaves sessile, lowest attenuate into a winged petiole, upper amplexicaul, only slightly diminishing above, passing into leaflike bracts. Leaves all similar, lanceolate to ovate-lanceolate in outline, lower leaves conspicuous sharply or obtusely dentate-pinnatifid, upper less divided. Lowest flowers shortly pedicellate, upper subsessile, in 2 rows in a ± secund raceme which elongates during flowering. Calyx with long, multicellular, eglandular hairs, in flower narrowly campanulate, c. 15 mm, 5-dentate to about halfway with narrowly lanceolate, acute teeth; in fruit becoming strongly accrescent, ribs thickening and with intermediate reticulate venation, teeth strongly spinose and somewhat spreading, tube as long as to somewhat longer than broad, c. 12 × 10 mm, not strongly urceolate above. Flowers ± 2 cm across, petals creamy yellow with purple reticulate venation, not purple at throat. Anthers yellow. Capsule entirely concealed by accrescent calyx, 10 × 6 mm, circumscissile in upper ⅓ with an apiculate lid. Seeds strongly reticulate-rugose, compressed, c. 1.2 mm in diameter.

HAB. Along mountain roads, between rocks, on mountains slopes in valley, in Astragalus zone; alt. 1800–3100 m; fl. & fr. Jun.-Sep.

DISTRIB. Mountains of N. Iraq. **MRO**: Haji Omran, *Chapman* 12288!; above Damar, *Gillett* 11866!; Algurd Dagh, *Guest & Wheeler-Haines* 2953!; Algird range, Goum Tawerah, *Rawi & Serhang* 24721!; Serin mt., on road to Qandil, *Rawi* 24016!; S. Haji Omran, lower slope of Kudi mt., *Al Shahbaz & Al Mayah* 76299!; Seri mountain, on road to Qandil, *Rawi* 24016 (BAG); Arbil, Rawandiz, Algurd Dagh, *Bornmüller* 1568 (type).

Iran.

3. **Hyoscyamus reticulatus** *L.*, Sp. Pl. 2: 257 (1762); Boissier, Fl. Orient. 4: 295 (1879); Blakelock in Kew Bull. 4: 529 (1949); Zohary in Dep. Agr. Iraq Bull. 31: 131 (1950); Rawi, Dep. Agr. Iraq. Tech. Bull. 14: 141 (1964); Chakravarty, Dep. Agr. Iraq. Tech. Bull. 18: 141 (1964); Rechinger, Fl. Lowland Iraq: 534 (1964); Schönbeck-Temesy in Fl. Iranica [K. H. Rechinger] 100: 54 (1972); Feinbrun-Dothan, Fl. Palaest. 3: 163, pl. 268 (1978); Baytop in Fl. Turkey [P. H. Davis] 6: 454 (1978); Hepper in Fl. Egypt [Boulos] 3: 48 (2002); Taifour & El-Oqlah, Pl. Jordan Annot. Checkl.: 148 (2017).

> *H. pojarkovae* Schoenback-Temsey in Fl. Iranica [K. H. Rechinger] 100: 56 (1972). Type: Iraq, between Kirkuk and Arbil *Thesiger* 450 (BM!, holo.).

Erect annual (more rarely biennial), 15–60 cm, covered throughout with short glandular-puberulent hairs and with scattered longer hairs interspersed. Stem stout or rather slender, terete, simple or somewhat branched. Lower leaves petiolate, lamina oblong or lanceolate in outline, to 13 × 5 cm, usually ± deeply pinnatifid, more rarely sinuate-dentate or subentire (var. integrifolius Boiss.); petiole to 4 cm, stem leaves rapidly becoming sessile and less divided, uppermost frequently lanceolate or ovate-lanceolate, subentire, amplexicaul, passing into the leaflike bracts. Flowers shortly pedunculate, ± secund in two rows in an elongating raceme. Calyx when in flower narrowly campanulate, c. 14 mm, 5-dentate to about halfway, teeth narrowly lanceolate-acuminate; in fruit becoming rigid and accrescent, but the tube remaining short and broad, ± parallel-sided, c. 12 × 10 mm, not ventricose, ribs thickening, with intermediate reticulate venation, teeth often patent or recurved. Corolla ± 2 cm across, whitish yellow (often suffused with purple) and with dense dark purple reticulate venation, throat not conspicuously darker. Capsule entirely concealed by accrescent calyx, 9 × 6 mm, circumscissile in upper ⅓ with an apiculate lid. Seeds strongly reticulate-rugose, compressed, c. 1.5 mm in diameter.

HAB. Mountains, lower hills, cultivated fields, on red stony soils, sandy clay, in depressions, near roadside in gravelly soil; alt. 100–2200 m; fl. & fr. Mar.-Jul.

DISTRIB. Common in the mountains, foothills and western desert in Iraq. **MAM**: nr. Dohuk, *Guest* 2234!; Shaikh Adi [Shaikhad] 10 km N.W. Ayn Sifini *Thesiger* 686 (BM!); Zewita N.E. of Zakho, *Rawi* 23587 (BAG!); 7 km N. of Zakho, Botany staff 43343 (BAG)! **MRO**: Haji Omran, *Chapman* 12288!; Rawandiz, *Thesiger* 8441 (BM!); Arl Gird Dagh, *Ludow-Hewit* 2953 (BAG); Karoukh mountain, *Al-kas, Serhang & Nuri* 27418 (BAG); **MJS**: Jabal sinjar, *Al-Kaisi, Al-Khayat & F. Karim* 50849 (BAG); Bet. Karsi & Shimai, *Al-Kaisi & K. Hamad* 54287 (BAG); Asi, *Rawi* 8534! **MSU**: Hawraman, *Gillett* 11866!; Sur Dash Village, *S. Omar, F. Karim, R. Hamza & H. Hamid* 37378 (BAG); **FUJ**: Mosul, *Guest* 1338!; *Lazar* 3364!; Mosul Dist., *Bradburne* 8995!; Jazira (N. of Sinjar), *Guest* 13301 (BAG); 14 km S. of Sinjar, *Chakravarty, Rawi, Khatib & Alizzi* 33112 (BAG); Beginning of Hatra road, *Rawi* 41360 (BAG). **FAR**: Arbil, [Erbil] *Ludow-Hewit*, 620 A & 2102!; nr. Erbil, *Guest* 620B! **FKI**: Kirkuk, *Rogers* 0494!; between Tuz Khormato and Tauq, *Gillett & Rawi* 7458!; **FPF**: 7 km from Jalaula to Qara Tepe, *Al-Kaisi* 48653 (BAG); **DGA**: 20 km between

Samarra and Tikrit, *Botany staff* 49894 (BAG). **DWD**: 50 km N. of Rutba, *Chakravarty, Rawi, Khatib &* *Alizzi* 32901 (BAG); 13 km E. of K3, *Chakravarty, Rawi, Khatib & Alizzi* 32970 (BAG); Manayf, *S. Omar & H. Hamid* 36570 (BAG).

BINJ (Arabic), CHAUHAR, GIAMARANA (Kurd.). A poisonous plant especially so during later fruiting period containing the narcotic principle hyoscyamine.

Egypt (Sinai), Jordan, Palestine, Syria, Iran, Transcaucasus, Uzbekistan.

4. **Hyoscyamus albus** *L.*, Sp. Pl. ed. 1: 180 (1753); Boissier, Fl. Orient. 4: 295 (1879); Blakelock in Kew Bull. 4: 529 (1949); Rawi, Dep. Agr. Iraq. Tech. Bull. 14: 141 (1964); Chakravarty, Dep. Agr. Iraq. Tech. Bull. 18: 43 (1964); Schönbeck-Temesy in Fl. Iranica [K. H. Rechinger] 100: 69 (1972); Feinbrun-Dothan, Fl. Palaest. 3: 163, pl. 269 (1978); Baytop in Fl. Turkey [P. H. Davis] 6: 455 (1978); Collenette, Wild Flow. Saudi Arabia: 698 (1999); Hepper in Fl. Egypt [Boulos] 3: 50 (2002); Taifour & El-Oqlah, Pl. Jordan Annot. Checkl.: 148 (2017).

Erect annual or biennial, 15–60 cm, covered throughout with long, white, multicellular hairs and a very mixed but shorter indumentum of viscid-glandular hairs. Stem stout to rather slender, terete, simple or somewhat branched, indumentum dense. Leaves all petiolate, diminishing upwards, lamina broadly ovate to suborbicular, shallowly cordate or shortly cuneate at base, coarsely and bluntly sinuate-dentate, to 17 × 12 cm; petioles of lower leaves equaling or exceeding lamina; those of cauline leaves becoming shorter above; leaf surfaces sparsely to moderately hairy, petioles densely so. Bracts leaflike, diminishing above, lower petiolate, upper sessile. Inflorescence generally lax, ± secund with flowers in 2 rows; fruiting calyces rarely overlapping the next above in the same row. Lowest flowers distinctly pedunculate, upper sessile or subsessile. Calyx when in flower broadly campanulate, 12 × 6 mm, 5-dentate with short, broad teeth; in fruit becoming strongly accrescent, tube broadly campanulate, almost parallel-sided below, to 15 × 10 mm, but variable according to size of plant, sometimes somewhat urceolate below the usually ± expanded limb; indumentum very dense, usually with plentiful long, white hairs; nerves thickened, with intermediate reticulate venation. Flowers variable with size of plant, corolla 5–20 mm across, conspicuously zygomorphic, pale whitish yellow with no reticulate venation, throat greenish or purplish. Capsule concealed in accrescent calyx, 12 × 8 mm, circumscissile in upper ⅓ with an apiculate lid. Seeds strongly reticulate-rugose, c. 1.2 mm in diameter, compressed.

HAB. Mountain and mountain slopes, crevices and cracks in walls; alt. 650–1800 m; fl. May-Jun & fr. May-Jul.
DISTRIB. Mountains of N. Iraq. **MAM**: Amadiyah, *Haines s.n.*; ibid., *Rawi* 8569 (BAG); Shaikh Adli, 900 m, *Guest* 3673!; Dinarta village, *Chapman* 26151 (BAG); Aqra, *Rawi* 11302 (BAG). **MRO**: Algurd Dagh, *Guest & Ludlow Hewitt* 2953! **MSU**: Diyala, Mandali, *Haines* 544!

SHIRRABAND (Kurd.).

N. Africa, Arabian Peninsula, Jordan, Palestine, Syria, Turkey, S. Europe.

5. **Hyoscyamus desertorum** (*Aschers.* ex *Boiss.*) *V.Täckh.*, Svensk. Bot. Tidskr. 36 (2–3): 252 (1942); Feinbrun-Dothan, Fl. Palaest. 3: 163, pl. 270 (1978); Collenette, Wild Flow. Saudi Arabia: 699 (1999); Hepper in Fl. Egypt [Boulos] 3: 50 (2002); Taifour & El-Oqlah, Pl. Jordan Annot. Checkl.: 148 (2017).

H. albus L. var. *desertorum* Ascher. ex Boiss., Fl. Orient. [Boissier] 4: 296 (1879).
H. cylindrocalyx Rech.f., Ark. Bot. ser. 2, 2: 420 (1952); Rawi, Dep. Agr. Iraq. Tech. Bull. 14: 141 (1964); Chakravarty, Dep. Agr. Iraq. Tech. Bull. 18: 42 (1964); Rechinger, Fl. Lowland Iraq: 534 (1964);

Erect annual or biennial (?perennial), 15–45 cm, covered throughout with long and short glandular-viscid hairs, long whitish eglandular hairs absent or very rare. Stem moderately stout to slender, terete, simple or somewhat branched, densely glandular. Lower leaves petiolate, lamina broadly ovate to suborbicular, shallowly cordate to shortly cuneate at base, decurrent on petiole, broadly and shallowly sinuate-dentate, to 6 × 4 cm; surfaces ± densely glandular, margins closely glandular-ciliate. Upper leaves diminishing, more shortly petiolate to subsessile, passing into leaflike bracts. Inflorescence dense, ± secund with flowers in 2 rows, fruiting calyces (excepting the lowest) attaining or overlapping the next above in the same row. Lowest flowers distinctly pedunculate, upper sessile or almost so. Calyx in flower broadly campanulate, c. 7 × 4 mm, 5-dentate to ¼ with broadly triangular teeth; in fruit becoming strongly accrescent, tube elongate-conical, gradually expanding from base to apex, (12–)15(–20) × 6–10 mm below limb, densely glandular with long glandular hairs most frequent at base, limb scarcely or not expanded, nerves thickening but slender, with finely reticulate intermediate venation. Corolla c. 10 mm across, strongly zygomorphic, yellow with a greenish or purplish throat, not reticulate-veined. Capsule concealed by accrescent calyx, c. 8 × 5 mm, circumscissile in upper ⅓ with an apiculate lid. Seeds strongly reticulate-rugose, compressed, c. 1 mm in diameter.

HAB. In hard gravel desert (*haswa*) in depressions, beside road, and silty depressions, on clay and gravelly soils; alt. 40–750 m; fl. & fr. Nov.-Apr.

DISTRIB. Western and southern desert regions of Iraq. **DLJ**: Jazira, c. 17 km N.W. of Falluja, *Guest & Rawi* 15977!; nr. Falluja, *Guest & Rawi* 13504!; Albaiyder, 78 km S. of Sinjar, *Chakravarty, Rawi, Khatib & Alizzi* 32102!; near Rawa, *Alizzi & Hussin 33988* (BAG); 10 km S. of Manayf, *S. Omar & H. Hamid 36599* (BAG). **DWD**: nr. Karbala, *Rawi* 15761!; Albaiyder, 78 km S. of Sinjar, *Chakravarty, Rawi, Khatib & Alizzi* 32102!; 35 km W. of Nukhaib, *Rawi* 14754 (BAG); 75 km W. of Rutba (Wadi Al-Walaj), *Rawi* 21187(BAG); 100 km S.W.W. of Rutba, *Rawi* 21199 (BAG); 145 km N.E. of K3, *Chakravarty, Rawi, Khatib & Alizzi* 33018 (BAG)!; Al-Misih 15 km E. of Awaj, *Al-Dabagh & Al-Kaisi* 41951 (BAG); 10 km W. of Kubaisa, *S. Omar, K. Hamad & H. Hamid* 44531 (BAG).

Egypt, Palestine, Jordan, Syria, N.E. Saudi Arabia.

6. **Hyoscyamus pusillus** *L.*, Sp. Pl. 1: 180 (1753); Boissier, Fl. Orient. 4: 294 (1879); Chakravarty, Dep. Agr. Iraq. Tech. Bull. 18: 44 (1964); Schönbeck-Temesy in Fl. Iranica [K. H. Rechinger] 100: 67 (1972); Feinbrun-Dothan, Fl. Palaest. 3: 162, pl. 267 (1978); Baytop in Fl. Turkey [P. H. Davis] 6: 453 (1978); Collenette, Wild Flow. Saudi Arabia: 700 (1999); Y. Nasir in Fl. Pakistan [Y. Nasir & Ali] 168: 48 (1985); Hepper in Fl. Egypt [Boulos] 3: 48 (2002); Taifour & El-Oqlah, Pl. Jordan Annot. Checkl.: 148 (2017).

Erect annual, 5–30 cm, sparsely to more densely covered with very short glandular hairs, occasionally subglabrous. Stem slender, wiry, terete, simple to much-branched from base. Lower leaves petiolate, lamina ovate to oblong-lanceolate, subentire to repand-sinuate or pinnatifid with long-acute segments, to 5 × 5 cm, long-attenuate below, decurrent on petiole, which in radical leaves sub equaling the lamina. Upper cauline leaves diminishing, less divided, more shortly petiolate, passing into the leaflike bracts. Inflorescence becoming lax, ± secund, flowers in two rows. Flowers all sessile or almost so. Calyx in flower broadly campanulate, c. 12 × 5 mm, divided to ⅓ with triangular-subulate teeth, in fruit becoming much accrescent, elongate-conical, to 12 × 7 mm, excluding the narrowly triangular, spinous teeth which are spreading to subsquarrose; nerves slender, with fine intermediate reticulate venation. Corolla small, 3–10 mm across, yellow with a purple throat, lacking reticulate venation. Capsule concealed by accrescent calyx, c. 8 × 4 mm, circumscissile in upper ⅓ with an apiculate lid. Seeds finely rugose, compressed, c. 1 mm in diameter. Fig. 10, 1–6.

HAB. On sandy and gravelly soils; alt. ± 120 m; fl. Feb.-Mar.

DISTRIB. Rare and found only once in the western desert in Iraq. **DSD**: Al-Samah, *Alizzi & S. Omar* 35731! (BAG).

S. European Russia to Mongolia, Pakistan, Afghanistan, Iran, Turkey, Jordan Palestine, Kuwait and Saudi Arabia.

7. **Hyoscyamus longipedunculatus** *C.C.Towns.*, Kew Bull. 19: 417 (1965). Type: Iraq, nr. Amadia [Ama'diya], 1372 m, *O.V. Polunin* 5151 (K!, holo.); Schönbeck-Temesy in Fl. Iranica [K. H. Rechinger] 100: 73 (1972).

Dwarf perennial, c. 8 cm, covered throughout with whitish viscid-glandular hairs. Stem slender, terete, covered at base with the scaly remains of previous years' growth. Leaves all petiolate, lower with petioles sub equaling lamina; lamina orbicular to deltoid-orbicular with large, irregular teeth, at base cuneate, decurrent along petiole, c. 2 × 2 cm. Flowers axillary, long-pedunculate on peduncles up to 4 cm. Bracts leaflike. Calyx in flower tubular-campanulate, c. 7 mm, 5-dentate to about ¼, with broad obtuse teeth. Corolla pale yellow, without reticulate venation, c. 1.5 cm in diameter. Anthers pale yellow. Fruit unknown.

Fig. 10. **Hyoscyamus pusillus**. 1, habit; 2, young flower; 3, flower, face view; 4, calyx at fruiting stage; 5, calyx with front sepal removed; 6, seed. 1–6 from *Alizzi et al.* 35731. Drawn by © A.P. Brown 2023.

HAB. On perpendicular cliffs and ledges; alt. ±1370 m; fl. Apr.
DISTRIB. N. Iraq. **MAM**: Ama'diya, N.E. of Mosul, *Polunin* 5151! (type).

Endemic.

8. **Hyoscyamus aureus** *L.*, Sp. Pl. 1: 180 (1753); Boissier, Fl. Orient. 4: 296 (1879); Zohary in Dep. Agr. Iraq Bull. 31: 131 (1950); Rawi, Dep. Agr. Iraq. Tech. Bull. 14: 141 (1964); Chakravarty, Dep. Agr. Iraq. Tech. Bull. 18: 44 (1964); Schönbeck-Temesy in Fl. Iranica [K. H. Rechinger] 100: 70 (1972); Feinbrun-Dothan, Fl. Palaest. 3: 163, pl. 271 (1978) Baytop in Fl. Turkey [P. H. Davis] 6: 456 (1978); Hepper in Fl. Egypt [Boulos] 3: 52 (2002); Taifour & El-Oqlah, Pl. Jordan Annot. Checkl.: 148 (2017).

Decumbent perennial, 15–50 cm, densely covered throughout with viscid glandular hairs considerably variable in length. Stem stout to rather slender, simple or somewhat branched from near base, terete. Leaves all similar, petiolate, lamina suborbicular to ovate in outline, sharply, jaggedly and irregularly lobed and dentate, with a ± deeply cordate base, lowest to 8 × 6 cm, with a petiole of equal length or sometimes more; upper leaves gradually diminishing and more shortly petiolate, passing into the leaflike petiolate bracts. Flowers all pedunculate, peduncles in lower part of inflorescence to c. ⅓ length of fruiting calyx. Calyx in flower broadly campanulate, c. 12 mm, 5-dentate to about ¼ with broadly triangular, finely acuminate teeth; in fruit strongly accrescent, tube elongating to 15–25 mm, with prominent ribs and feebler intermediate reticulate venation, gradually expanded above, not urceolate. Corolla c. 2 cm across, strongly zygomorphic, upper segments much shorter than lower, golden-yellow with a lilac throat, not reticulate-veined. Anthers yellow, long-exserted. Style long-exserted. Capsule entirely concealed by accrescent calyx, 6 × 4 mm, circumscissile in upper ⅓ with an apiculate lid. Seeds strongly reticulate-rugose, compressed, c. 1 mm in diameter.

HAB. Rocky hills in shady location, limestones mountain slopes; a typical plant of ancient walls and ruins; alt. 450–1200 m; fl. & fr. Apr.-Jul.
DISTRIB. Mountains and lower hills of N. Iraq. **MAM**: nr. Sharanish, *Rechinger* 11470; nr. Zakho, *Rechinger* 10726! **MJS**: Jabal Sinjar, *Gillett* 11143!; Kursi (Karsi), *Gillett* 10889!; Jabal Sinjar, S. Omar, *Al-Khayat* & *Al-Kaisi* 52577 (BAG). **FNI**: Jabal Maklub, *Polunin* 861 & 19545 (BAG).

Greece, Cyprus, East Aegean Is., Turkey, Syria, Jordan, Palestine, Sinai, Egypt.

9. **Hyoscyamus insanus** *Stocks*, Hooker's J. Bot. Kew Gard. Misc. 4: 178 (1852); Schönbeck-Temesy in Fl. Iranica [K. H. Rechinger] 100: 78 (1972); Y. Nasir in Fl. Pakistan [Y. Nasir & Ali] 168: 47 (1985).

Plant perennial, pilose-tomentose or lanate. Stem ascending, pendent or decumbent, thick, much branched at base and above. Leaves 3–15 cm, succulent, elliptical, elliptic-ovate to obovate or suborbicular, dentate or sinuate, rarely entire, acute or acuminate, cuneate or attenuate, rarely sub-cordate at base; petiole shorter than to as long as lamina, narrowly winged. Bracts shorter than calyx, elliptical, ovate to linear-lanceolate or oblanceolate, entire, sessile or very shortly petiolate. Flowers pedicellate in lax helicoid cymes. Calyx 1.5–3 cm, somewhat thick and succulent, campanulate to tubular; lobes variable. Corolla 2.5–5 cm, obliquely-infundibuliform, variegated white and purple with indistinct reticulation; lobes variable in shape and size. Stamens much exserted, inserted at or below the middle of corolla tube; filaments hairy; anthers oblong or obovate to elliptic-oblong. Ovary cylindrical to subcylindrical or obovate; style hairy; stigma capitate, indistinctly bilobed. Fruiting calyx 15–35 mm, leathery, campanulate to tubular or infundibuliform, dilated or narrow at limb; lobes variable in shape and size, acute, the stalk often long and slender sometimes deflexed. Capsule 6–17 mm, variable in shape. Seed c. 1 mm, rectanguler to subreniform, light to dark brown.

HAB. On mountains, in sand and gravel soils, gravelly valley floor; alt. 90–140 m; fl. & fr. Mar.-May.
DISTRIB. Rare on mountains (found once in the mountain region), occasional in S.E. of upper plains and foothills in Iraq; **MSU**: Sartek gorge, 12 km S. of Darbendkhan, *Al-Khayat, Widad & Ikhlas* 58025 (BAG). **FPF**: Jabal Giska near Al-Hashimah, *Al-Kaisi* 56736 (BAG); Al-Hashimah between Badra & Mandli, *Al-Kaisi & Khayat* 50713 (BAG); 25 km N. of Tip, *Al-Kaisi & H. Hamid* 46560 (BAG); 10 km E. of Zurbatiya, *Al-Kaisi & Khayat* 46549 (BAG); 10 km E. of Tip, *Al-Kaisi & Khayat* 46554 (BAG); Mandali, *Wheeler-Haines* W. 544!

Oman, Iran, Afghanistan, Pakistan, N. India, N.W. Nepal.

Hyoscyamus senecionis Willd. was recorded for Iraq by Blakelock (Kew Bull. 4: 529 (1949)). The specimen in question, however, was stunted *H. albus*, as Blakelock himself later realised. *H. muticus* L. was mentioned by Chakravarty, Dep. Agr. Iraq. Tech. Bull. 18: 42, but with the observation that "no record of its occurrence in Iraq has yet been made". This is still true.

CULTIVATED SPECIES

ALKEKENGI Mill. Gard. Dict. Abr., ed. 4. [unpaged] (1754)

Annual or perennial herbs, glabrous or simple or stellate-hairy. Leaves alternate, simple, sinuate or more rarely pinnatifid. Flowers usually small, rarely more showy, solitary in leaf axils, pedicellate, white to purple, more rarely yellowish. Calyx red, campanulate, 5-fid, in fruit greatly inflated, membranous, angled, enclosing and often much exceeding the berry, teeth connivent. Corolla white, rotate or very broadly campanulate, limb plicate, broadly 5-lobed. Stamens 5, filaments filiform, inserted at base of corolla, anthers with parallel loculi, longitudinally dehiscent. Stigma bilamellate. Fruit a bilocular, globular, yellowish or greenish berry.

A single species distributed from C. & S. Europe to Japan.

Alkekengi is separated from *Physalis* based on molecular studies and in its distinctive red inflated calyx, lobed white flowers and its Eurasian geographic distribution.

Alkekengi originally from Persian *kakanj*, كاكانج, adopted by Andalusian Arabic and into Spanish *akquequeje* before being adopted as the Latin name of the genus by Philip Miller citing Tournefort's *Alkekengi officinarum* as the source.

Alkekengi officinarum *Moench*, Suppl. Meth. (Moench) 177 (1802).

Physalis alkekengi L., Sp. Pl. 1: 183 (1753); Husain & Kasim, Cult. Pl. Iraq 122 (1975); Abdulridha, Taha & Widad, Ecology & Fl. Basrah: 509 (2016).
P. franchetii Mast., Gard. Chron. 2: 434, 441, f. 57 (1894); Chakravarty, Dep. Agr. Iraq. Tech. Bull. 18: 14 (1964).

Perennial with long-creeping rhizomes, 20–75 cm. Stems slender, erect, ridged, zigzag, sparsely hairy to glabrous, usually simple, more rarely branched from near base. Leaves petiolate, lamina deltoid-ovate, to 10 × 8 cm, weakly cordate to truncate or shortly attenuate at base, ± decurrent on 2–4 cm petiole, subacute at apex, margins weakly sinuate, superficially glabrous or very sparingly pilose. Calyx shortly campanulate, densely clothed with multicellular hairs especially round margins of teeth, 4–5 mm, 5-dentate to c. ⅓, greatly accrescent and inflated in fruit to 4 cm, glabrescent, membranous and reticulate-veined, ± truncate below, contracting above to the connivent teeth, becoming bright crimson at maturity. Corolla 2–3 cm in diameter, white, lobed to less than 1/2. Anthers yellow. Berry globose, c. 1.5 cm in diameter, red, completely concealed within base of accrescent calyx.

A showy plant, frequently cultivated in Iraq.

S. & C. Europe from France and Spain to Balkans and S. Russia (incl. Crimea), Turkey, Caucasus, Iran, Pakistan, C. Asia (Turkmenia to Pamir mts.), Mongolia, China, Korea and Japan; introduced into N. America and elsewhere.

ATROPA L., Sp. Pl. 1: 181 (1753)

Perennial herbs. Leaves entire. Flowers solitary, axillary. Calyx deeply 5-lobed, slightly accrescent in fruit. Corolla tubular to campanulate, 5-lobed; lobes imbricate. Stamens 5, included; filaments curved, attached near the base of the corolla, densely villous at base; anthers dehiscing longitudinally. Ovary bilocular; stigma ± bilobed. Fruit a globose berry. Seed many, compressed.

Five species distributed in Europe, Turkey, Iran and Pakistan as well as some countries of N. Africa; single species cultivated in Iraq for medicinal use in the 1970's.

Atropa, from the name of a Gr. goddess Ατροπος, *Atropos,* one of the three Fates.

Atropa bella-donna *L.,* Sp. Pl. 1: 181 (1753).

Perennial herb with thick rootstock. Stems 50–150(–200) cm, branched, green or somewhat dull purple, glandular-pubescent. Lower leaves alternate, with short petiole, upper sessile and opposite, ovate, oblong-ovate, or elliptic, 50–150(–220) × 30–80 mm, acuminate or acute, with densely minute sessile glands and sparsely puberulent on veins. Flowers solitary on glandular-pubescent pedicels. Calyx campanulate, 10–17 mm, 5-lobed; lobes 4–5 × 6–8 mm, ovate or oblong-ovate, acuminate, glandular-puberulent. Corolla purple-reddish to dark purple, 20–35 mm, lobes broad, triangular-ovate, obtuse or subacute, slightly recurved. Filaments hairy in lower part. Berry 10–20 mm in diameter, initially green, later becoming black, glossy. Seeds numerous, reniform.

HAB. & DISTRIB. Only a single specimen at BAG in cultivated plants (LCA: Abu Ghraib, *M. Shahid* C. 707!). According to Chakravarty (1976) the plant was cultivated in the Abu Ghraib Experimental Field and showed good promise. Fl. & Fr. Jun.-Sep.

CAPSICUM L., Sp. Pl. 1: 188 (1753); Gen. Pl. ed. 5: 86 (1754)

Perennial herbs, woody below, commonly cultivated as annuals, almost or quite glabrous. Leaves simple, entire or sinuate-repand. Flowers pedicellate, solitary or 2–3 in leaf axils, erect or deflexed. Corolla rotate, 5-lobed, white or greenish-white, sometimes suffused with violet. Calyx campanulate, subentire to shallowly 5-dentate, somewhat accrescent in fruit. Stamens 5, filaments filiform, inserted at base of corolla; anthers with parallel loculi, longitudinally dehiscent. Stigma clavate or slightly expanded. Fruit a pod-like indehiscent berry wtih a firm integument, 2–3-locular (often multilocular in cultivated forms), from sublinear to globose or large and ovoid in shape.

About 25 species, almost all in tropical S. & C. America; one in eastern Asia.

Capsicum from Gr. καπτο, *kapto,* to gulp or swallow; or Lat. *capsa,* a box, with reference to its pods.

H.C. Irish, "A revision of the genus *Capsicum* with especial reference to garden varieties". Rep. Missouri Bot. Gard. 9: 53–110 (1898).

Capsicum annuum *L.,* Sp. Pl. 1: 188 (1753); Chakravarty, Dep. Agr. Iraq. Tech. Bull. 18: 18 (1964); Husain & Kasim, Cult. Pl. Iraq: 119 (1975);

Herb, 30–90 cm, cultivated as an annual in temperate regions, maybe biennial or perennial in tropics. Stems woody below, zigzag, erect. much-branched, terete or with narrow raised lines, glabrous below, sometimes sparsely hairy above especially at nodes. Leaves petiolate, oblong to elliptic or ovate, entire, attenuate at both ends and acuminate above, variable in size, from 2–18 cm, decurrent on the 0.6–2.0 cm petiole. Peduncles axillary, solitary or paired, lengthening in fruit to c. 1.5 cm and somewhat thickened below calyx. Calyx poculiform, c. 2 mm, somewhat enlarged in fruit, obscurely 5-dentate to almost truncate. Corolla white or dirty-white, rarely purplish-suffused, 1.0–1.5 cm across, 5-lobed to about ½, lobes triangular. Fruit very variable in form and size, from narrow and tapering to large, tumid and widely umbilicate at each end, green to straw-coloured or bright red.

The two specimens seen from Iraq seem best referred to the cultivars 'conoides' (Cone Pepper, with shortly tapering fruits) and possibly cv. 'grossum' (Sweet Pepper strains, with large tumid fruits widely impressed at each end).

CESTRUM L., Sp. Pl. 1: 191 (1753)

Shrubs or small trees, glabrous to simply- or stellate-hairy. Leaves simple, entire, alternate. Flowers numerous, in axillary racemes, cymes or fascicles, or forming terminal panicles. Corolla 5-lobed, greenish, white, yellow or red, funnel-shaped or salver-shaped with an elongate tube which is ± adnate to ovary at base and expanded or contracted at throat. Calyx poculiform to shortly tubular, 5-fid. Stamens 5, included; filaments inserted about middle of corolla tube or above, often pilose at base, or incrassate, or with small tooth-like appendages; anthers short, longitudinally dehiscent. Stigma expanded, subpeltate, entire or emarginate. Fruit a small bilocular berry, globose to oblong, often scarcely succulent.

About 250 species in the West Indies and the warmer regions of N. & S. America. 2 species cultivated ion Iraq.

Cestrum from Gr. Κεστρος, *kestros*, sharpness.

Francey, P. (1935). Monographie du genre *Cestrum*, Candollea 6: 46–398 (1935) & 7: 1–132 (1936).

Flowers ± scattered along inflorescence axis, c. 2.5 cm, lowest in each
       raceme distinctly pedicellate. . . . . 1. *C. nocturnum*
Flowers mostly ± clustered at ends of peduncles, only 1–3 more remote,
       never exceeding 2 cm, all sessile or almost so . . . . 2. *C. diurnum*

1. **Cestrum nocturnum** *L.*, Sp. Pl. 1: 191 (1753); Francey, Candollea 7: 67 (1936); Chakravarty, Dep. Agr. Iraq. Tech. Bull. 18: 41 (1964).

Shrub 1.5–4 m, glabrous throughout its vegetative parts when mature, puberulent when young. Branches flexuose when young, cortex brownish to green, striate with raised, often interrupted ridges. Leaves of thin texture, alternate, petiolate, lamina oblong-ovate to elliptic, rather shortly acuminate above, shortly attenuate below, 5–12 × 2–4 cm, ± shining on both surfaces, veins prominent on lower surface; petiole 1–2 cm. Inflorescences axillary and subracemose, often equalling leaves, and terminal panicles to 30-flowered but often less shortly pedunculate, peduncles up to c. 12 mm; pedicels of lower flowers 4–7(–9) mm. Calyx shortly campanulate, c. 3 mm, 5-dentate to c. ⅓ with deltoid teeth, slightly puberulent. Corolla funnel-shaped, c. 2.5 cm, greenish-yellow, expanded at throat, segments of limb lanceolate, erect or spreading, puberulent at least on margins, tube glabrous. Berry shortly ellipsoid, 8–10 × 5 mm.

DISTRIB. Cultivated in Iraq. **FUJ**: Mosul, *Alizzi & S. Omar* 35296 (BAG); LCA: Baghdad, *Sahira* 306 (cult.) (BAG).

Native of the West Indies and Central America (Mexico, El Salvador, Guatemala). Frequently cultivated in warmer regions of the world.

2. **Cestrum diurnum** *L.*, Sp. Pl. 1: 191 (1753); Francey, Cndollea 6: 284 (1934); Chakravarty, Dep. Agr. Iraq. Tech. Bull. 18: 40 (1964).

Shrub 1–3 m, young shoots and leaves somewhat puberulent, but glabrescent with age. Branches erect, whitish- or yellowish-green, or sometimes brownish-tinged, striate with obscure to distinct ridges. Leaves alternate, petiolate, lamina oblong or elliptic-oblong, shortly narrowed at each end with apex acute or obtuse, 1–12 × 2–4 cm, shining above but duller on lower surface, with nerves more prominent; petiole 4–12 mm. Inflorescences axillary racemes, condensed and cluster-like with only 1–3 lower flowers more remote, and terminal panicles, 7–10(–20)-flowered on each branch, long lower peduncles 2–5 cm or more; flowers sessile or almost so. Calyx poculiform, c. 3 mm, very shallowly 5-dentate, with throat ciliate. Corolla funnel-shaped, pure white or creamy white, (12–)14–18(–20) mm, tube clavate and gradually expanding upwards, glabrous, teeth broadly deltoid-ovate, obtuse, tomentose along margins, finally reflexed. Berry shortly ellipsoid, c. 7 × 5 mm.

Native of the West Indies and Mexico (Yucatan). Frequently cultivated in the tropics and subtropics.

NICANDRA Adans., Fam. Pl. (Adanson) 2: 219, 528 (1763)

Perennial herbs. Stems and branches with simple and glandular hairs. Leaves simple. Flowers solitary, axillary. Calyx longer than corolla tube, 5-lobed, lobes cordate-sagittate at base, acute; fruiting calyx much enlarged, chartaceous. Corolla campanulate. Stamens inserted near base of corolla-tube, included. Ovary 3–5-celled, with numerous ovules; style slender, stigma capitate. Fruit a berry, globose, surrounded by the enlarged calyx.

Three species, native from Peru to S. America; single species in Iraq, cultivated.

Nicandra commemorates the botanist and poet Nicander of Colophon, who wrote a book about poisons and their antidotes.

Nicandra physalodes (*L.*) *Gaertn.*, Fruct. Sem. Pl. ii. 237. t. 131. f. 2. (1791).

Atropa physalodes L., Sp. Pl. 1: 181 (1753);
Physalis peruviana L., Sp. Pl. ed. 2: 16700 (1763); Chakravarty, Dep. Agr. Iraq. Tech. Bull. 18: 14 (1964); Schönbeck-Temesy in Fl. Iranica [K. H. Rechinger] 100: 2 (1972); Hepper in Fl. Egypt [Boulos] 3: 43 (2002).

Perennial herb, 10–90 cm (to 2 m in the wild). Stems erect and simple to diffuse and bushy with many branches, ridged, densely clothed with multicellular simple and glandular hairs. Leaves similarly densely hairy, lamina deltoid-ovate, remotely sinuate-dentate to almost entire, to 11 × 9 mm, truncate to cordate at base, shortly acuminate above. Calyx at anthesis 4 mm, 5-dentate to about ⅓, indumentum similar to that of leaves, greatly accrescent and inflated in fruit, 3–4 cm, membranous and reticulate-veined, remaining considerably hairy, contracting above to the connivent teeth, greenish-yellow at maturity. Corolla 1.5–2 cm in diameter, yellow with large purplish to black blotch at base of each lobe, 5-dentate to c. ⅓. Anthers bluish. Ovary 1.25–1.5 cm across, yellow, completely enclosed within the accrescent calyx.

A native of tropical America (Colombia, Peru and Bolivia). Elsewhere naturalized – Macaronesia, China, India, Australia, tropical Africa, Arabian Peninsula etc. and becoming invasive in some countries. Frequently cultivated for its fruit and as an ornamental.

NIEREMBERGIA Ruiz & Pav., Fl. Peruv. Prodr. 23 (1794)

Herbs or sometimes small shrubs with prostrate to erect stems. Flowers often showy. Calyx tubular to campanulate, 10-ribbed, deeply 5-lobed with linear or lanceolate lobes, lobes in fruit somewhat enlarged. Corolla white, pink to purple, with a yellow or purple centre, infundibuliform to salviform, unequally 5-lobed. Stamens 5, often the uppermost shorter than the other 4, inserted at top of corolla tube, exserted. Ovary 2-locular, ovules many. Style slender; stigma reniform. Capsule bilocular, septicidal.

Twenty three species, native to Mexico, W. & S. South America to S.E. & S. Brazil; a single species in Iraq, recorded to be occasionally cultivated there.
Nierembergia commemorates Juan Eusebio Nieremberg y Ottín (1595–1658), a German-born Jesuit classicist who taught natural history in the Imperial College in Madrid.

Nierembergia linariifolia *Graham*, Edinburgh New Philos. J. 378 (1831); Husain & Kasim, Cult. Pl. Iraq: 121 (1975).

Perennial to c. 20 cm tall with short narrow to linear leaves. Flowers purple with a yellow center, cup-shaped.

A native of Bolivia, Brazil and Argentina, recorded to be cultivated in Iraq. No material at BAG or K.

NICOTIANA L., Sp. Pl. 1: 180 (1753); Gen. Pl. ed. 5: 84 (1754)

Tall shrubs to diminutive annuals, often with a viscid-glandular indumentum, rarely glabrous. Leaves entire, alternate, petiolate or sessile. Inflorescence various (in Iraqi species a cylindrical or ovoid panicle). Calyx 5-partite, usually much shorter than corolla, often accrescent in fruit. Corolla regular or somewhat zygomorphic, tubular, funnel-shaped or salver-shaped, limb 5-partite, shallowly 5-lobed or almost entire, spreading, expanding at dusk. Stamens 5, usually included, equal or unequal, insertion variable; anthers with or without connective, dehiscent along a longitudinal suture. Ovary bilocular. Capsule firm, dry, membranous to slightly woody, dehiscent above.

About 60 species in Australia, Polynesia and the Americas.

Nicotiana commemorates Jean Nicot de Villemain (1530–1604), a French diplomat who first brought tobacco from Portugal to France in 1560, possibly in the form of snuff.

1. Leaves distinctly petiolate, petioles unwinged; corolla limb only about
    twice as wide as tube immediately above calyx. . . . . .1. *N. glauca*
   Leaves sessile, or lamina decurrent along petiole to form a wing; corolla
    limb widely expanded, much more than twice as wide as slender tube
    immediately above calyx . . . . 2. *N. tabacum*

1. **Nicotiana glauca** *Graham*, Edinburgh New Philos. J. (Apr.–Jun.) 175 (1828); Bot. Mag. 55: t. 2837 (1828); Blakelock in Kew Bull. 4: 529 (1949); Chakravarty, Dep. Agr. Iraq. Tech. Bull. 18: 57 (1964).

Loosely branched shrub or small tree, 3–6–10) m. Stem glabrous, glaucous or green to purplish, becoming reddish-brown with age. Leaves thick, fleshy, ovate to elliptic, with a petiole from ½ length of lamina to subequalling it in upper leaves; lamina cordate, truncate or shortly attenuate below, ovate to lanceolate, up to 25 cm, glabrous. Panicles short, lower branches becoming much elongate. Pedicels 3–10 mm, slender, glabrous. Calyx 8–12 mm, divided to c. ¼, glabrous or minutely pubescent with swollen-based hairs, teeth broadly triangular. Corolla 25–35 mm, greenish to yellow, narrowly cylindrical, limb not much expanded, pubescent on outer surface; tube not dilated into a broad gaping throat above, slightly widened below limb. Stamens included, reaching almost to top of tube, subequal. Capsule 7–15 mm, broadly elliptic, included.

Culivated. **LCA**: Abu Ghraib, *S. Omar & Al-Kaisi*, 904 (cult.) (BAG). **LEA**: Basra, Kut as Sayyid, *Dowson* 1558; Hilla, in garden, *Guest* 3168.

Native from Bolivia to Brazil, widely naturalized in many of the tropical and subtropical regions as well as the warmer temperate areas.

2. **Nicotiana tabacum** *L.*, Sp. Pl. 1: 180 (1753); Blakelock in Kew Bull. 4: 529 (1949); (1954); Chakravarty, Dep. Agr. Iraq. Tech. Bull. 18: 41 (1964).

Stout annual or short-lived perennial, clothed ± abundantly throughout with short to very short stipitate viscid glands. Stems robust, somewhat angular, branches few. Leaves large, attaining 60 cm or more below, radical leaves ovate to oblong or lanceolate, upper decurrent on stem, sessile or winged-petiolate, broadly ovate, uppermost diminishing, narrow. Calyx cylindrical to cylindrical-campanulate, 12–20 mm, divided to halfway or somewhat less; teeth triangular, long-acuminate. Corolla c. 45 mm, limb whitish to pink or red, tube paler, greenish cream below, expanded at halfway or somewhat above into a wide, gaping, cup-shaped throat; lobes of limb triangular-acuminate. Four stamens equal, slightly exserted, the fifth shorter, included. Capsule 15–20 mm, narrowly elliptic, ovoid or orbicular, exserted or included.

DISTRIB. Cultivated in Iraq. **MJS**: Karsi, *Al-Kaisi & Hamad* 1106 (cult.) (BAG). **MRO**: Rawanduz, *S. Omar, Sahira, Karim & Hamid* 582 (cult.) (BAG). **MUS**: Halabja, *Noori & Hamid* 1831 (cult.) (BAG).

Native to Bolivia. Its various cultigens are cultivated in many tropical, subtropical and temperate regions and become naturalised here and there.

Nicotiana rustica L. is mentioned by Chakravarty, op. cit. 57, but without clear indication of its being grown in Iraq. No specimens of this species or of *N. alata* have been seen from Iraq, but both may be cultivated and are included in the key. *N. alata* is the common white-flowered sweet-scented species of gardens in Europe and elsewhere.

PETUNIA Juss., Ann. Mus. Paris 2: 215 (1803)

Annual or perennial viscid-pubescent herbs or subshrubs with entire, alternate (or upper opposite) leaves. Flowers solitary, pedunculate, arising from leaf axils. Calyx deeply 5-partite with oblong or linear lobes. Corolla white or in shades of pink or purple, funnel-shaped or salver-shaped, with a long slender tube, limb broad, 5-lobed, regular or oblique. Stamens

5, 4 in pairs and the fifth much smaller, inserted at or below middle of tube. Style slender, curved above, stigma capitate. Ovary bilocular, multiovulate. Capsule bilocular, septicidal.

About 40 species, native to Bolivia to southern America and S. E. & S. Brazil; cultivated in Iraq.

Petunia is derived from the Tupi-Guarani language of Brazil, from *petun* meaning tobacco.

1. **Petunia × hybrida** *Vilm.*, Fl. pleine terre ed. 1: 615 (1863); Chakravarty, Dep. Agr. Iraq. Tech. Bull. 18: 59 (1964); Husain & Kasim, Cult. Pl. Iraq: 121 (1975).

Decorative annual, mostly 10–45 cm, glandular-viscid with ovate, shortly petiolate leaves. Flowers white to deep purple, or with stripes from the throat to margins of limb, or otherwise marked, mostly 5–7.5 cm, with a wide limb and a long tube, funnel-shaped, lobes sometimes irregularly dentate-lacerate.

Baghdad, *Shakir Sabir* 16685; Za'faraniyah Hort. Station, *Shakir Sabir* 20194; *Alkas* 20497.

The common Garden Petunia is believed to be a series of hybrid forms derived from *P. axillaris* (Lam.) Britton, Sterns & Poggenb. *P. violacea* Lindl.

117. CONVOLVULACEAE

Benth. & Hook.f., Convolvulaceae in Gen. Pl. 2: 865–881 (1873)
Peter, Convolvulaceae in Pflanzenfam. 4: 1–40 (1891)

K. H. Rechinger & A. Patzak
Revised by A. Kandemir & Shahina A. Ghazanfar

Herbs or shrubs, usually climbing or trailing, sometimes parasitic. Leaves alternate, entire, divided, or compound, petiolate or sessile, reduced to scales in parasitic species. Flowers solitary, axillary or in cymes, racemes, panicles, umbels, or capitula, hermaphrodite, regular or slightly zygomorphic, bracteate or not. Sepals 4–5, free or joined at base, imbricate. Corolla sympetalous, usually 5-pleated or -lobed, funnel-form, salver-form, campanulate, tubular or urceolate. Stamens 5, epipetalous, included or exserted. Ovary superior, (1–)2(–5)-locular with 1–2 ovules in each loculus; styles 1, bifid or 2; stigmas entire or 2-lobed. Fruit a capsule, variously dehiscent.

Convolvulaceae is represented by 57 genera and 1660 species in tropical, subtropical and temperate regions of both hemispheres; 34 native and 1 cultivated species in Iraq.

Cuscuta has been placed in the Convolvulacae in the APG IV (2016) system, but in Hutchinson's classification (followed for this Flora) it is treated in the monogeneric family Cuscutaceae Dum.

Several members of the family are of economic importance. Some species supply food and medicine while several others are valuable ornamentals. The sweet potato, *Ipomoea batatas* (L.) Lam. is one of the world's most important root crop. Some powerful drugs are obtained from species of *Convolvulus* and *Ipomoea*. A number of species, particularly those of *Convolvulus*, *Ipomoea* and *Merremia* are used as ornamental climbers. Some species belonging to *Convolvulus*, *Calystegia* and *Cuscuta* tend to become weedy, with *Cuscuta* sometimes a serious parasite of cultivated fields.

Manos, P.S., Miller, R.E. & Wilkin, P. (2001). Phylogenetic analysis of *Ipomoea, Argyreia, Stictocardia,* and *Turbina* suggests a generalized model of morphological evolution in morning glories. Systematic Botany 26(3): 585–602.

Muñoz-Rodríguez, P., Carruthers, T., Wood, J.R., Williams, B.R., Weitemier, K., Kronmiller, B., Goodwin, Z., Sumadijaya, A., Anglin, N.L., Filer, D. & Harris, D. (2019). A taxonomic monograph of *Ipomoea* integrated across phylogenetic scales. Nature plants 5(11): 1136–1144.

Wood, J.R., Williams, B.R., Mitchell, T.C., Carine, M.A., Harris, D.J. & Scotland, R.W. (2015). A foundation monograph of *Convolvulus* L. (Convolvulaceae). PhytoKeys, (51) 1.

Wood, J.R.I., Carine, M.A., Harris, D., Wilkin, P., Williams, B. & Scotland, R.W. (2015). *Ipomoea* (Convolvulaceae) in Bolivia. Kew Bull. 70: 30–122. DOI https://doi.org/10.1007/s12225-015-9592-7

1. Corolla lobed to ½ its length; stamens distinctly exserted; styles 2 . . . . .5. **Cressa**
   Corolla scarcely lobed; stamens included or just exserted; style 1 2
2. Bracteoles large, partly or wholly concealing calyx .3. **Calystegia**
   Bracteoles small, never concealing calyx 3
3. Stigma lobes linear, cylindric or clavate . . . . .4. **Convolvulus**
   Stigma lobes globose 4
4. Anthers straight; pollen spiny . . . . .1. **Ipomoea**
   Anthers spirally twisted; pollen smooth, not spiny .2. **Distimake**

1. **IPOMOEA** L.

Sp. Pl.: 159 (1753); Gen. Pl., ed. 5, 76 (1754)

Annual or perennial herbs with woody or herbaceous twining or prostrate stems. Leaves petiolate, entire or lobed. Flowers mostly axillary, solitary or in cymes. Bracts various; bracteoles absent or inconspicuous. Sepals 5, equal to unequal, ± enlarged in fruit. Corolla funnel-shaped, or narrowly campanulate, inconspicuously lobed. Stamens included or exserted; anthers straight; pollen spiny. Ovary 2–4 (–5)-locular; stigma of 1–3 globose lobes. Capsule globose to ovoid, 3–10-valved.

Ipomea L. is represented by about 500 species distributed in the tropical and warm temperate regions of the world; 4 species in Iraq, cultivated and/or naturalized in parts of Iraq.

Ipomoea, from Gr. ιψ, ips, woodworm, found in vines (which are twirling) and ομοιος, *homoios*, resembling.

1. Shrubs to 3 m tall . . . . .4. *I. carnea* subsp. *fistulosa*
   Herbaceous, usually climbers 2
2. Leaves palmately divided to base . . . . 3. *I. cairica*
   Leaves entire or lobed, not divided to base 3
3. Pedicels and sepals spreading hirsute . . . . 1. *I. purpurea*
   Pedicels and sepals glabrous . . . . 2. *I. tricolor*

1. **Ipomoea purpurea** (*L.*) *Roth*, Bot. Abh. 27 (1787); Cat. Bot. 1: 36 (1797); Hall.f. in Engl., Bot. Jharb. 18: 137 (1893); Verdcourt in Taxon 6: 231 (1957); Meeuse in Bothalia 6: 734 (1957); Austin in Ann. Missouri Bot. Gard. 62: 189 (1975); Parris in Fl. Turkey [P. H. Davis] 6: 222 (1978); Austin in Fl. Pakistan [Nasir & Ali] 126: 47 (1979); Meikle, Fl. Cyprus 2: 1165 (1985); Gonçalves in Fl. Zamb. 8(1): 87 (1987); Rhui-cheng & Staples in Fl. China 16: 305 (1995); Boulos in Fl. Egypt 2: 261 (2000); Wood et al. in Kew Bull. 70: 99, f. 31B (2015).

Convolvulus purpureus L., Sp. Pl. ed. 2: 219 (1762).

Annual herbs. Stems 2–3 m, twining; axial parts short pubescent and long retrorse hirsute. Leaves ovate or broadly ovate in outline, entire or 3 lobed, 4–11 × 3.5–10 cm, base cordate, apex acuminate, ± strigose; petiole 2–19 cm. Flowers solitary or in few flowered cymes; peduncle 4–18 cm. Bracts linear to 4 mm, much shorter than pedicels, spreading hirsute. Pedicels 5–18 mm, hirsute. Sepals subequal, ± 14 mm long, spreading hirsute; outer sepals oblong, acuminate; inner linear-lanceolate. Corolla funnel-shaped, 4–6 cm long, red to reddish-purple, or blue-purple, glabrous. Stamens unequal; filaments glabrous above, hairy towards base. Ovary glabrous; stigma glabrous. Capsule globose, enclosed within the persistent calyx.

HAB. Usually cultivated in gardens and sometimes found as an escape.
DISTRIB. **MRO**: Shaqlawa, *Dabbagh & Hamad* 858! & 859!; *Sahin* 317!; Kani Masi, *Botany Staff* 43895! **DGA**: between Khalis and Kirkuk, *Rawi, Alizzi & Nuri* 19692!

Native to Bolivia, naturalized throughout the warm temperate and subtropical regions of the world; it is cultivated in many warm countries for its beautiful flowers. *I. purpurea* is sometimes confused with *I. nil*, but *I nil* can be distinguished by its unlobed leaves, shorter oblong-lanceolate sepals and pink flowers.

2. Ipomoea tricolor *Cav.*, Icon. Descr. Pl. 3: 5, t. 208 (1795); Choisy in A.DC., Prodr. [de Candolle] 9: 359 (1845); Verdcourt in Fl. Trop. E. Afr. Convolv. 82 (1963); Austin in Fl. Pakistan [Nasir & Ali] 126: 49 (1979); Wood et al. in Kew Bull. 70: 104, f. 30E (2015).

I. rubro-caerulea Hook., Bot. Mag. 61: t. 3397 (1834).

Herbaceous glabrous twiner; stems to 3 m long. Leaves ovate, up to 7 cm long and 6 cm broad, base cordate, apex acuminate, margins entire. Inflorescence few-flowered; peduncle ± 7 cm long; bracteoles minute. Pedicels 10–30 mm. Sepals subequal, 4–5 mm long, triangular to ovate-lanceolate, glabrous, green with white margins, carinate along midrib. Corolla funnel-shaped, ± 7 cm long, purple or violet-blue with white tube, glabrous. Capsule ovoid, glabrous.

HAB. Cultivated.
DISTRIB. **MRO**: Shaqlawa, *Dabbagh & Hamad* 857!

Native to Mexican/Caribbean region, widely cultivated in temperate and subtropical regions including Iraq.

3. Ipomoea cairica (*L.*) *Sweet*, Hort. Brit. 287 (1826); Hall.f. in Engl., Bot. Jharb. 18: (1893); Verdcourt in Kew Bull. 15: 13 (1961); Meeuse in Bothalia 6: 761 (1957); Austin in Fl. Pakistan [Nasir & Ali] 126: 40 (1979); Gonçalves in Fl. Zamb. 8, 1: 105 (1987); Rhui-cheng & Staples in Fl. China 16: 309 (1995); Boulos in Fl. Egypt 2: 259 (2000); Wood et al. in Kew Bull. 70: 96, f. 30D (2015).

Convolvulus cairicus L., Syst. Nat. 10: 922 (1759).
C. tuberculata Desr. in Lamarck, Encyl. 3: 545 (1791).
Ipomoea palmata Forssk., Fl. Aegyp.-Arab. 43 (1775); Baker & Rendle in Fl. Trop. Africa 4, 2: 178 (1905).

Perennial, glabrous twiner or sometimes prostrate with a tuberous root. Stems to 5 m, tuberculate or smooth. Petiole 0.5–7 cm, base with leafy pseudostipules; leaves ovate to orbicular in outline, 2–8 × 2–7 cm, palmately divided into 4–5 lanceolate to oblong-

Fig. 11. **Ipomoea cairica.** 1, habit × ½ ; 2, flower dissected × 2; 3, gynoecium ×3. Reproduced with permission from Fl. Pakistan 126: f. 5, G–I (1979). Drawn by S.Y. Salim. © National Herbarium, Pakistan Agriculture Research Council & University of Karachi, Pakistan.

lanceolate, ovate-elliptic or ovate-lanceolate segments; segments entire or minutely undulate, apex acute or obtuse, mucronulate; basal pairs parted or not; middle segment larger. Inflorescences axillary, 1- or several flowered; peduncle 2–8 cm. Bracts and bracteoles early deciduous, squamiform, small. Pedicels 0.5–4 cm. Sepals unequal; outer 2 slightly smaller, 4–6.5 mm; inner sepals 5–9 mm, glabrous. Corolla broadly funnel-shaped, 3.5–7 cm, pink, purple, or reddish-purple, with a dark centre, rarely white. Stamens unequal, included. Ovary glabrous; stigma 2 lobed. Capsule ± globose, glabrous. Fig. 11, 1–3.

HAB. Cultivated places, loam soils; cultivated in gardens and sometimes found as an escape; alt. ± 50 m; fl. Jun.-Dec.
DISTRIB. **LCA**: Baghdad, *Sahina* 202! **LBA**: *Omar* & *Kaisi* 43503!; Basra, *Guest* 286! **LEA**: Siba, 50 km E.S.E. of Basra, *Kazim, Nuri, Hamid* & *Kadhim* 806!

Ipomoea cairica has been separated into few varieties by some authors based on leaf and flower size; these characters are variable and do not justify subspecific rank.

Probably native to the Old World, widespread throughout tropical and subtropical regions of the world.

4. **Ipomoea carnea** *Jacq.*, Enum. Syst. Pl. 13 (1760) subsp. **fistulosa** (*Mart.* ex *Choisy*) *D.F. Austin* in Taxon 26: 237 (1977); Wood et al. in Kew Bull. 70: 64, f. 31B (2015).

Ipomoea fistulosa Martius ex Choisy in A.DC., Prodr. [de Candolle] 9: 349 (1845); Bahatta, Journ. Bombay Nat. Hist. Soc. 73: 318–319 (1973).
I. crassicaulis (Benth.) Rob., in Proc. Amer. Acad. Arts 51: 530 (1916).

Shrubs. Stems woody at base, herbaceous towards tips, 1–3 m, glabrous or minutely pubescent. Leaves ovate or ovate-oblong, 7–10 × 4–9 cm, ± glabrous, base cordate or truncate, apex long acuminate, margins entire, mucronulate; petiole 2.5–12 cm. Inflorescences axillary; flowers few to 14 in cymose-paniculate clusters; peduncle stout, 4–8 cm, glabrescent to puberulent; bracts early deciduous, elliptic-oblong. Pedicel 0.8–1.5 cm, puberulent. Sepals subequal; outer 2 suborbicular, ± 5.5 mm long, puberulent; inner 3 slightly longer than outer sepals. Corolla lilac or pink, dark pink inside, funnel-shaped, 4.5–7 cm, puberulent. Stamens included. Stigma 2-lobed. Ovary puberulent above with silvery hairs. Capsule not seen.

HAB. Cultivated; fl. Jul.-Nov.
DISTRIB. **LBA**: Fao, *Alizzi* & *Omar* 35868!

Native of the American tropics now cultivated and naturalized in most subtropical and tropical countries of the world, sometimes becoming weedy.

2. **DISTIMAKE** Raf.

Fl. Tellur. 4: 82 (1838); Simões & Staples in Bot. J. Linn. Soc. 183(4): 571 (2017).
Astromerremia Pilg., Notizbl. Bot. Gart. Berlin- Dahlem 13: 107 (1936);
Davenportia R.W.Johnson, Austrobaileya 8(2): 171–176 (2010)

Revised by Shahina A. Ghazanfar

Herbaceous climbers (in Iraq). Leaves usually 5–7-palmately lobed or compound (rarely simple or reduced to scales). Flowers axillary, solitary or in few- to many-flowered. Bracts small, linear or lanceolate. Sepals mostly flat (not convex), accrescent in fruit later the sepals reflexing. Corolla white or pale yellowish, with or without a dark red centre, glabrous, drying with dark lines in mid-petaline bands. Anthers spirally dehiscing. Capsule 4-valved. Seeds glabrous (less commonly shortly velvety puberulent with dehiscent hairs).

About 35 + species, widespread in tropical America and tropical Africa with disjunct species in Asia and northern Australia.
Distimake, like some of Constantine Rafinesque's other generic names, could have been invented.

1. **Distimake dissectus** (*Jacq.*) *A.R.Simões* & *Staples*, Bot. J. Linn. Soc. 183(4): 574 (2017).

Convolvulus dissectus Jacq., Observ. Bot. [Jacquin] 2: 4, tab. 28 (1767).

Merremia dissecta (Jacq.) Hallier.f., 18 (1894); Austin in Fl. Pakistan [Nasir & Ali] 126: 55 (1979); Rhui-cheng & Staples in Fl. China 16: 294 (1995).
Operculina dissecta (Jacq.) House, Bull. Torrey Bot. Club 33: 500 (1906).

Herbs. Stems twining with woody base, herbaceous towards tip, tuberculate, hirsute to glabrescent. Leaves 4.5–6 × 5–8 cm, palmately lobed; lobes elliptic-lanceolate, margin coarsely dentate to irregular pinnately lobed, glabrous or puberulent, middle one larger; petiole 5 cm with yellowish spreading hairs. Flowers axillary, 1-several; penduncle to 8 cm, glabrous. Pedicels glabrous, thicker above, tuberculate, ± 2.2 cm. Sepals, subequal, 13–18 mm, ovate-lanceolate, apex acute or obtuse, mucronulate, enlarged in fruit, glabrous. Corolla funnel-form, ± 3 cm long, white with purple throat and midpetaline bands. Anthers spirally twisted; filaments white hairy, dilated at base. Ovary glabrous; style inserted, glabrous; stigma papillate. Capsule depressed globose, glabrous. Seeds usually 4. Fig. 12, 1.

Hab. Cultivated places; fl. May to December.
Distrib. **LCA**: Abu Ghuraib, *Alizzi* 34817!

A native of the Americas cultivated and naturalized in Africa, Asia, Malaysia and Australia.

3. **CALYSTEGIA** R.Br.

Prodr. Fl. Nov. Holl., 483 (1810) nom. cons.

Rhizomatous herbs. Stems trailing or twinning, glabrous. Leaves petiolate, sagittate or cordate. Flowers solitary, axillary; bracteoles 2, enclosing the calyx. Sepals subequal. Corolla large, funnel-form. Ovary glabrous. Style 1; stigmas 2. Capsule glabrous.

Fig. 12. **Distimake dissectus.** 1, habit × ½. Reproduced with permission from Fl. Pakistan 126: f. 7, D (1979). Drawn by S. Hameed. © National Herbarium, Pakistan Agriculture Research Council & University of Karachi, Pakistan.

Fig. 13. **Calystegia sepium.** Habit × ½. Reproduced from Flora of China 16: f. 296 (1995) with permission from Missouri Botanical Garden Press, St. Louis, and Science Press, Beijing.

About 25 species (and about 70 subspecies) mostly distributed in temperate regions; a single species in Iraq.

Calystegia, from Gr. καλνχ, *kalux,* cup and στεγε, *stege,* covering, describing the way the bracts that enclose the sepals.

1. **Calystegia sepium** (*L.*) *R.Br.,* Prodr. Fl. Nov. Holl. 483 (1810); Boissier, Fl. Orient., 4: 111 (1875); Post in Dinsmore, Fl., ed. 2, 2: 211 (1933); Rechinger, Fl. Iranica 22 (1963); Brummit in Fl. Turkey [P. H. Davis] 6: 220 (1978); Meikle, Fl. Cyprus 2: 1165 (1985).

Convolvulus sepium L., Sp. Pl., 153 (1753).
Calystegia sepium subsp. *spectabilis* Brummitt, J. Linn. Soc., Bot. 64: 73 (1971); Fang Rhui-cheng, & Brummitt in Fl. China 16: 286 (1995).

Perennial herb. Stems twining, ± 5 m, glabrous or pubescent. Leaves petiolate, 3.5–8 × 1–6 cm, sagittate or cordate, apex acute or acuminate, glabrous or sparsely pubescent. Flowers axillary, solitary; peduncle ± 7 cm; bracteoles ovate-cordate, apex acute, longer than calyx, ± 19 mm long, glabrous. Sepals ovate, apex acute, 11–16 mm, glabrous. Corolla 4–5.5 cm, white. Filaments glabrous above, glandular hairy near base. Ovary conical, glabrous; style longer than ovary, glabrous; stigma with lobes ± 2 mm. Capsule subglobose, glabrous. Fig. 13, 1.

HAB. Rocky clay soils, near streams, roadsides; alt. 0–500 m; fl. Jun.-Aug.
DISTRIB. **MRO**: 35 km to Haji-Umran, *Kaisi* & *Hamad* 43536!; Galâla, *Omar* 37649! **LSM**: *Alkahla* & *Nuri* 41308!

A polymorphic species where several subspecies are recognized; widespread in temperate regions.

4. **CONVOLVULUS** L.

Sp. Pl.: 153 (1753); Gen. Pl. ed. 5, 76 (1754)

Ali Kandemir & Shahina A. Ghazanfar

Annual or perennial herbs or subshrubs. Stems stiff or spinescent or herbaceous, prostrate, trailing or twining, or ascending to erect. Leaves simple, exstipulate, petiolate or sessile, margin entire or ± lobed. Flowers solitary or in few-flowered cymes or in dense involucrate heads. Pedicels often bracteolate. Sepals 5, free, obtuse or acute. Corolla funnel-shaped, with spreading limb, 5-angled or -lobed or repand–5-plaited. Ovary glabrous or pubescent, bilocular, ovules 2 per locule; style 1, hairy or glabrous, filiform, with 2 linear-cylindrical often revolute stigmas. Capsule valved or dehiscing irregularly.

About 250 species mostly distributed in temperate and subtropical regions; 18 species in Iraq.

In *Convolvulus* cauline leaves and bracts have been used to mean the same by some authorities. In this treatment cauline leaves refer to the upper leaves that do not have axillary inflorescence and bracts refer to small leaf-like or scale-like structures with an inflorescence branch in their axil.

Convolvulus, from Lat. *convolvere*, to coil around, and the diminutive suffix *-olus.*

Sa'ad, Fātima az-Zahrā' Maḥmūd 'Abdallāh (1967). The *Convolvulus* species of the Canary Isles, the Mediterranean region and the Near and Middle East. Bonder-Offset.
Wood, J.R., Williams, B.R., Mitchell, T.C., Carine, M.A., Harris, D.J. and Scotland, R.W. (2015). A foundation monograph of *Convolvulus* L. (Convolvulaceae). PhytoKeys, 51: 1–282.

Species found in Iraq belong to three Sections:

1. Young and older branches either ending in a spine or bearing spines
   which represent the upper sterile peduncles or peduncles of fallen
   flowers. . . . . .Sect. Acanthocladi
   Young and older branches not spinescent. 2
2. Stems and branches erect, ridgid or prostrate, not twining . . . . Sect. Inermes
   Stems and branches (or at least the branches) twining . . . . Sect. Convolvulus

1. Annual herbs; flowers with blue limb and yellow tube . . . . 17. *C. pentapetaloides*
   Perennials; flowers variously coloured 2
2. Cauline leaves and at least lower bracts distinctly petiolate, base hastate,
   sagittate or cordate; filaments with glands on expanded part 3
   Cauline leaves and bracts sessile, base cuneate to attenuate, sometimes
   rounded or cordate; filaments not glandular 6
3. Ovary and capsule hairy . . . . 14. *C. betonicifolius*
   Ovary and capsule glabrous 4
4. Leaf margins crenate-dentate to irregularly lobed . . . . .16. *C. stachydifolius*
   Leaf margins entire 5
5. Outer sepals 4–4.5 mm, subequal to slightly shorter than inner sepals. . .13. *C. arvensis*
   Outer sepals 7–9 mm, distinctly shorter than inner sepals. . . . . 15. *C. scammonia*
6. Small shrubs with rigid ± spiny branches. 7
   Herbaceous, sometimes suffrutescent perennials, never spiny 8
7. Dwarf shrubs 10–20 cm; leaves elliptic or obovate; style glabrous . . . . .1. *C. hamrinensis*
   Shrubs 20–45 cm; leaves lanceolate to oblong-elliptic; style hairy . . . . . . 2. *C. oxyphyllus*
8. Flowers congested, 3-many together in dense clusters, sometimes 1–2, then sessile . . 9
   Flowers either in lax inflorescences or solitary, pedicels distinct. 14
9. Inflorescence terminal . . . . .6. *C. commutatus*
   Inflorescence axillary 10
10. Ovary glabrous, sometimes slightly hairy at top; style glabrous 11
    Ovary and style distinctly hairy. 13
11. Leaves reticulate beneath; bracts cordate or subcordate at base
    . . . . .11. *C. reticulatus* subsp. *reticulatus*
    Leaves not reticulate beneath; bracts cuneate to attenuate at base 12
12. Peduncles to 2 mm long or wanting; upper bracts lanceolate . . . . 10. *C. kotschyanus*
    Peduncles distinct, usually much longer than 5 mm long; upper bracts
    ovate . . . . .9. *C. euphraticus*
13. Stem silky tomentose with hairs longer than diameter of stem; at least
    some peduncles longer than bracts . . . . .8. *C. cephalopodus*
    Stem villous-tomentose with shorter than diam. of stem; peduncles
    much shorter than bracts or wanting. . . . . 2. *C. oxyphyllus*

14. Outer sepals longer than 10 mm, pouched at base
      . . . . 7. *C. holosericeus* subsp. *macrocalycinus*
      Outer sepals shorter than 8 mm, not pouched 15
15. Sepals glabrous 16
      Sepals hairy 17
16. Ovary hairy at top; style hairy . . . . 5. *C. sarothrocladus*
      Ovary and style glabrous . . . . 4. *C. leptocladus* subsp. *glabrosepalus*
17. Stem nearly leafless, glabrous except near base; sepals to 2.5 mm
      long . . . . 3. *C. chondrilloides* var. *chondrilloides*
      Stem leafy, hairy throughout; sepals longer than 5 mm . . . . 12. *C. pilosellifolius*

SECT. ACANTHOCLADI Boiss.

1. **Convolvulus hamrinensis** *Rech.f.*, Anz. Österr. Akad. Wiss. Math.-Naturwiss. Kl. 98: 11 (1961). Type: Iraq, Jabal Hamrin, Muqdadiya to Sa'dia, *Rechinger* 8083 (W!, holo., E, iso.); Rechinger, Fl. Lowland Iraq 482 (1964); Sa'ad, 74 (1967) as "*hamarinensis*"; Wood et al.: 235 (2015).

  Convolvulus oxyphyllus subsp. *sheilae* R.R.Mill, Edinburgh J. Bot. 70: 376 (2013).
  C. infantispinosus R.R.Mill, Edinburgh J. Bot. 70: 374 (2013).

Suffrutescent, 10–20 cm high. Stems rigid divaricately branched, spinescent, tomentose-canescent with densely adpressed hairs. Leaves coriaceous, petiolate; leaves of flowering branches obovate-spatulate, 4–6 × 2–3 mm, rounded or indistinctly apiculate at tip, entirely or nearly nerveless; leaves of sterile branches, 7–10 × 5–7 mm, acute, cuneate at base, nerves prominent below; petiole nearly as long as lamina. Flowers axillary, solitary, sessile. Bracts small, suborbicular to elliptic; bracteoles oblanceolate, obtuse, shorter than sepals. Calyx 5–7 mm, hirsute-tomentose; outer sepals oblong, acute, with green upper parts; inner sepals lanceolate, acuminate. Corolla 8–10 mm, white or pale pink, with hairy bands on the outside. Ovary hirsute; style and stigmas glabrous.

HAB. In sandy-gypsum soils; alt. ± 100 m; fl. Mar.-Oct.
DISTRIB. FPF: Jabal Hamrin, Muqdadiya to Sa'dia, *Rechinger* 8083 (type). DSD: Tuqaiyid, *Mahalhal* 19504!; *Guest, Rawi & Rechinger* 16151B! DLJ: Jazira, 16 km N.W. of Falluja, *Guest & Rawi* 15961!

Saudia Arabia.

2. **Convolvulus oxyphyllus** *Boiss.*, Diagn. Pl. Orient. ser. 1, 7: 26 (1846); Boissier, Fl. Orient. 4: 88 (1875); Guest in Dep. Agr. Iraq Bull. 27: 25 (1933); Blakelock in Kew Bull. 4: 527 (1949); Zohary in Dep. Agr. Iraq Bull. 31: 117 (1950); Rechinger, Fl. Iranica 10 (1963); Rechinger, Fl. Lowland Iraq 483 (1964); Rawi in Dep. Agr. Iraq Tech. Bull. 14: 133 (1964); Sa'ad, Convol. Sp. Canar. Is. Medit. 76 (1967); Wood & al. 234 (2015).

  C. oxyphyllus Boiss. subsp. *cateniflorus* Rech.f., Anz. Österr. Akad. Wiss. Math.-Naturwiss. Kl. 1961: 23 (1961). Type: Iraq, Diyala River, near Mandali, *Rechinger* 9639 (W!, holo.); Rechinger, Fl. Iranica 10 (1963) & Fl. Lowland Iraq 484 (1964).
  C. cateniflorus (Rech.f.) Sa'd, Meded. Bot. Mus. Herb. Rijks Univ. Utrecht 281: 154(1967). Type as for *C. oxyphyllus* subsp. *cateniflorus* Rech.f.
  C. oxyphyllus Boiss. subsp. *oxycladus* Rech.f., Anz. Österr. Akad. Wiss. Math.-Naturwiss. Kl. 1961: 24 (1961). Type: Iraq, between Ramadi and Rutbah, *Rechinger* 9886 (W!, holo., E!, iso.). Wood & al. 235 (2015).

Suffrutescent, 20–45 cm high; indumentum pubescent or tomentose with admixture of spreading hairs. Stems erect or arcuate-ascendent, branching from base; branches stiff or ± flexuous, dichotomously and divaricately branched, ending in a spine or not. Lower leaves sessile, oblong or elliptic, 10–25 × 4–6 mm, gradually attenuate towards base, apex acute or mucronate, margins entire, adpressed or tomentellous hairy; upper much smaller, ovate, lanceolate or linear-lanceolate, apex acute with a spinescent tip, sometimes reduced to scales. Bracts similar to cauline leaves but smaller, 3–7 mm. Flowers sessile, 1–3 in axillary inflorescences, sometimes 3–5 together at base of twigs and in axils; peduncles to 11 mm long. Sepals unequal, silky or tomentose with hairs ± 1 mm long, lanceolate, 7–8 mm; outer sepals oblong, acute, or acuminate; middle one with right and left half unequal; inner sepals narrowly ovate, shorter than outer sepals. Corolla ± 2 × as long as calyx with

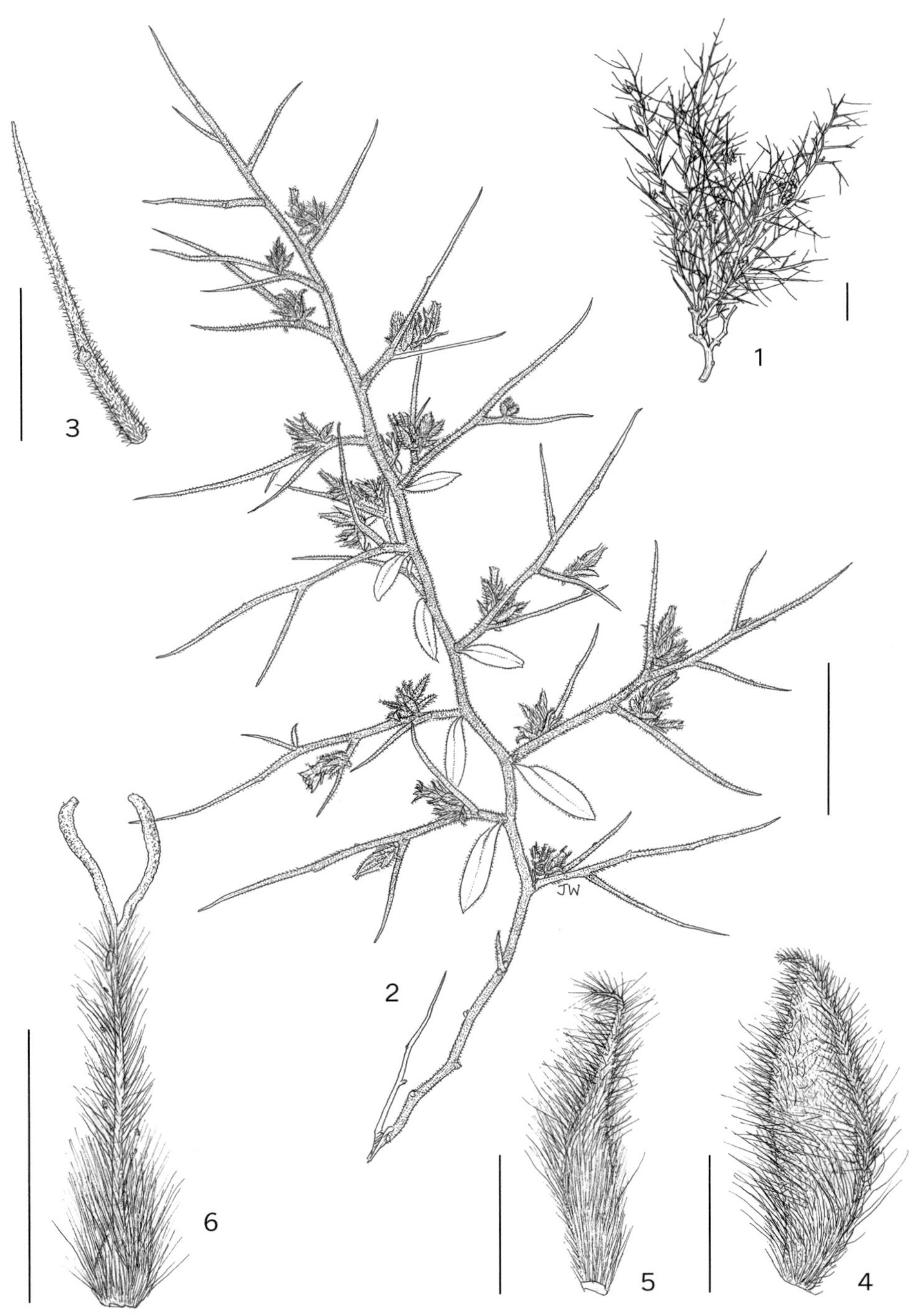

Fig. 14. **Convolvulus oxyphyllus.** 1, sketch of habit; 2, habit; 3, detail of spinescent lateral branch; 4, outer sepal; 5, inner sepal (abaxial view); 6, pistil. 1,4–6 from *Barkley et al.* 3606; 2,3 from *Rawi & Alizzi* 34452. Scale bars: 1, 2 = 3 cm; 3 = 1 cm; 4, 5, 6 = 3 mm. © Drawn by Juliet Beentje.

adpressed-silky bands on the outside. Filaments glabrous. Ovary and style hairy. Fig. 14, 1–6.

HAB. Sandy or stony ground, dunes, calcareous mountains; alt. 20–1000 m; fl.: Jan.-Dec.
DISTRIB. Common in the foothills and a characteristic shrub of desert regions. **FKI**: Baba Gurgur, nr. Kirkuk, *Guest* 4016 & 4017!; 2 km N. of Kirkuk, *Rawi* 21548!; Kirkuk, *Rechinger* 9978!; Altun Kupri, on Lesser Zab river, *Guest* 4020! **FPF**: 75 km N. of Amara, *Hazim & Nuri* 30635!; Eastern end of Jabal Hamrin, *Barkeley* 1800!; Buskaya, *Rawi & Haddad* 25561!; 10 km E. of Mandali, *Rechinger* 9639!; Mandali, *Rawi* 20642!; Mandali, *Guest* 868!; Injana, Jabal Hamrin, *Gillett* 12485!; N. of Diltawa, *Robertson* 13499! **DSD**: 2 km S. of Busaiya, *Guest* 14220! & 14234!; S. of Zubair, *Rawi & Gillett* 6085!; 15 km S.W. of Zubair, *Guest, Eig & Zohary* 14347!; Zubair, *Guest & Rawi* 14341!; 5 km E.S.E. of Zubair, *Guest, Rawi & Rechinger* 16819!; Jabal Sanam, *Guest, Rawi & Schwan* 14387!; 40 km W. of Basra on road to Khadhar-al-Mai, Barucha, *Rawi & Tikriti* 29335!; 50 km E. of Samarra, *Rawi* 20462!; Makatu nr. Mandali, *Guest* 8680!; Mandali, *Rawi* 20642! & 20680!; Inaiza nr Tursak, *Haines* W.1127!; 35 km S.W. of Ur, nr Salaibiyat al-Hamir, *Guest* 16058! **DWD**: 5 km S.W. of Rutba, *Rechinger* 9886!; 10 km S.W. of Rutba, *Rawi* 21054!; 25 km W. of Ana, *Omar & Hamad* 50412!; 40 km E. of Rutba, *Omar, Kaisi & Hamad* 43961!; Ramadi, 40 km N. of Falluja, *Barkeley & Ani* 3606!; Jabal Anaiza, *Barkley* 2430! **LCA:** 70 km N.N.W. of Falluja, *Rawi* 20247!

ADHGHRIS Ar. Subsp. *oxycladus* had been separated on the basis of its rigid and spinescent shoots, and subsp. *cateniflorus* on the basis of flexuous stems and flower clusters on the stems. These characters are not well defined and there is considerable overlap in the rigidity and spiny character of the stems and the inflorescence to justify independent subspecific ranks. Plants at elevation tend to be with rigid spinescent stems and those at lower altitudes, especially in desert areas tend to have soft ± spiny stems. Also some of the material at K is of young plants that often have flexuous stems.

Saudi Arabia, Iran.

SECT. INERMES Boiss.

3. **Convolvulus chondrilloides** *Boiss.*, Diagn. Pl. Orient. ser. 1, 11: 83 (1849) var. **chondrilloides** *Rech.*f., Fl. Iranica 13 (1963); Rawi in Dep. Agr. Iraq Tech. Bull. 14: 133 (1964); Sa'ad, Meded. Bot. Mus. Herb. Rijks Univ. Utrecht 87 (1967); Parris in Fl. Turkey [P. H. Davis] 6: 204 (1978); Wood & al.: 178, f. 15, 35–43 (2015).

 Evolvulus virgatus Choisy, Prodr. [A.P. de Candolle] 9: 446 (1845), nom illeg., non *Evolvulus virgatus*   Willd. ex Spreng. 1824. (Choisy 1845: 446). Type: Baghdad, *Aucher-Eloy* 1410 (K!, G, P, syn.).
 Convolvulus chondrilloides subsp. *eriocalycinus* Bornm. & Gauba, Repert. Spec. Nov. Regni Veg. 51: 215   (1942).

Suffrutescent, 40–80 cm high. Stems much-branched, virgate, erect, nearly leafless, glabrous except near puberulous base. Basal leaves spatulate to oblanceolate, 50–80 × 10–20 mm, adpressed hairy or puberulous; upper leaves linear-lanceolate, sessile. Flowers axillary and terminal in up to 8-flowered cymes or solitary; peduncles much longer than the small filiform bracts; bracteoles minute. Pedicels 2–6 mm, hairy or glabrescent, longer in fruit. Sepals unequal, ovate or elliptic, to 2.5 mm long, obtuse, sometimes mucronulate, adpressed hairy; outer sepals ovate mucronulate; middle sepal ovate with right and left half unequal; inner sepal broadly ovate with right and left half subequal. Corolla white, 3 × longer than calyx, hairy bands on outside. Filaments glabrous. Ovary and style hairy; stigmas filiform, as long as style or slightly shorter, glabrous. Capsule ovoid, hairy above.

HAB. Dry fields; alt. 550–1300 m; fl. May-Jul.
DISTRIB. Occasional in lower forest zone of N. Iraq: **MAM**: Gorge of Begma, *Gillett* 8223! **MRO**: Mt. Handarin, *Bornmüller* 1534!; Soran (Diana), nr Rowanduz, *Guest & Husham* 15877!; Rowanduz, *Haussknecht* s.n.! **MSU**: Jarmo, nr Chamchamal, *Wheeler-Haines* W. 338! **LCA**: Baghdad, *Aucher* 1410! (type).
 Iran, Turkey.

4. **Convolvulus leptocladus** *Boiss.*, Diagn. Pl. Orient. ser. 1, 7: 25 (1846); Boissier, Fl. Orient. 4: 91 (1875); Rechinger, Fl. Iranica 11 (1963); Rawi in Dep. Agr. Iraq Tech. Bull. 14: 133 (1964); Sa'ad, Meded. Bot. Mus. Herb. Rijks Univ. Utrecht. 99 (1967); Wood & al. 182, f. 16, 31–36 (2015).

 subsp. **glabrosepalus** *Kandemir* **subsp. nov.** Type: Qarati, Khanaqin, *Rawi* 5720! (K, holo.)

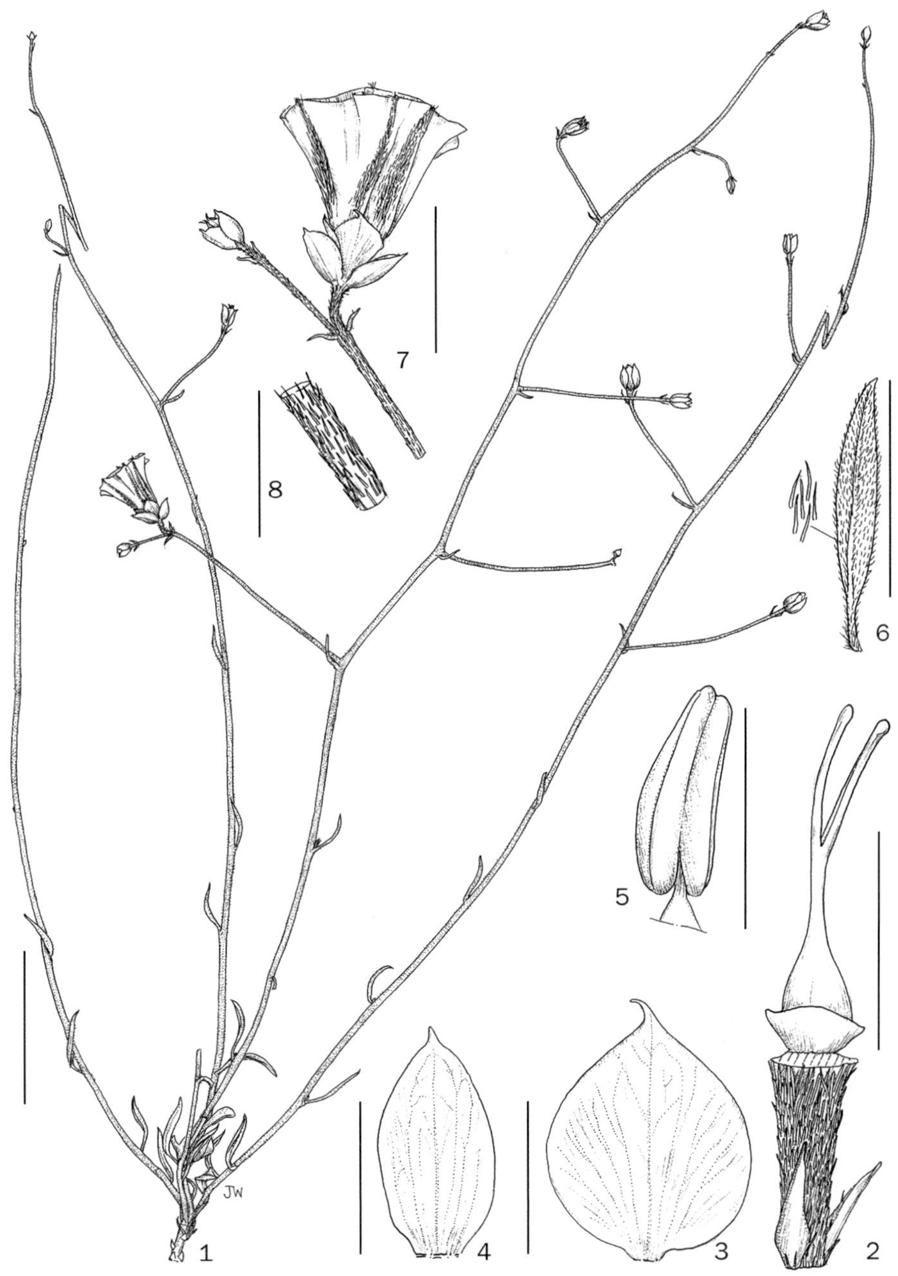

Fig. 15. **Convolvulus leptocladus** subsp. **glabrosepalus.** 1, habit; 2, detail indumentum of stem; 3, leaf (underside); 4, flower (side view); 5, outer sepal; 6, inner sepal; 7, stamen; 8, pistil on stalk with bracts. 1–8 from *Rawi* 5720. Scale bars: 1 = 3 cm; 3, 4 = 1 cm; 2, 5, 6, 7, 8 = 3 mm. © Drawn by Juliet Beentje.

Suffrutescent. Stems 26–30, densely adpressed-pilose. Leaves sessile; lower leaves linear to linear-spatulate, 10–18 × ± 2 mm, acute, margins entire, adpressed-pilose; stem leaves decreasing in size, linear, ± revolute; upper leaves reduced to small narrow linear scales to 8 mm long. Flowers usually solitary on slender patent peduncles, rarely 2-flowered. Bracts much shorter than peduncles, ± 3 mm, with glabrous tip; bracteoles minute, ± 1 mm. Pedicels 2 mm, shorter than calyx, adpressed hairy. Sepals convex, unequal, glabrous (shortly pubescent type subsp.) ± 4 mm long, mucronate to short acuminate; outer sepals ovate; inner sepals broadly ovate. Corolla ± 12 mm long, with hairy bands on outside. Filaments glabrous, shorter than anthers; anthers 2.8 mm long. Ovary glabrous, conical; style glabrous; stigma filiform, as long as style, glabrous. Capsule not seen. Fig. 15, 1–8.

HAB. In deserts; alt. ± 50 m; fl. May.
DISTRIB. Rare, found only once in Iraq close to the Iranian frontier. FPF: Qarati, Khanaqin, *Rawi* 5720!

The new subspecies is different from subsp. *leptocladus* in its glabrous sepals and indumentum with one arm. In subsp. *leptocladus* the sepals are densely adpressed hairy and the indumentum throughout the plant is with two arms and branches and branchlets are more flexuous.

The original description of *C. leptocladus* Boiss. is based on Auchers's specimens (*Aucher* 4942!). According to Boissier's description *C. leptocladus* has glabrous sepals. However, in the type, and other examined from Iran, the sepals are distinctly hairy (*Aucher* 4942; *Grey-Wilson* & *Hewer* 105; *Rechinger* 3389; *Leonard* 5921).

The new subspecies is close to *C. sarothrocladus* Boiss. in its facies and the indumentum is similar in both taxa. However *C. sarothrocladus* differs from the new taxon in its hairy ovary and style. Besides, its sepals have a broad transparent border which is lacking in *C. leptocladus* subsp. *glabrosepalus*.

Endemic.

5. **Convolvulus sarothrocladus** *Boiss.* & *Hausskn.* ex *Boiss.*, Fl. Orient. [Boissier] 4: 92 (1875). Type: N. Iraq, Derbent-i-Basian, from Kirkuk to Sulaimaniya, *Haussknecht* 653 (G!, lecto.). Zohary in Dep. Agr. Iraq Bull. 31: 117 (1950); Rechinger, Fl. Iranica 13 (1963); Rawi in Dep. Agr. Iraq Tech. Bull. 14: 133 (1964); Sa'ad, Meded. Bot. Mus. Herb. Rijks Univ. Utrecht. 105 (1967); Wood & al. 191, f. 17, 15–20 (2015).

Virgate shrub 30–40 cm high. Stems and branches pale green, erect, rigid, dichotomously branched at base, adpressed-hairy. Leaves chartaceous, sessile, adpressed to sub-adpressed hairy; lower leaves shortly petiolate, obovate, spatulate-oblong to elliptic-lanceolate, 18–60 × 4–10 mm, base attenuate, obtuse or acute at apex; upper leaves narrowly linear, acute. Flowers axillary, solitary or rarely in pairs. Bracts linear, adpressed hairy, shorter than peduncle; bracteoles shorter than pedicels. Pedicels adpressed hairy, sometimes glabrescent above, shorter than calyx, with bracts at base. Sepals 6 mm long, with broad scarious margins, glabrous; outer sepals obovate to elliptic, mucronate; middle one mucronate; inner sepals oblanceolate, mucronate. Corolla ± 14 mm long, 3 × shorter than calyx, with hairy bands on outside. Filaments glabrous. Ovary hairy at top; style hairy. Capsule glabrous.

HAB. In sand; alt. 700–1000 m; fl. May-Jun.
DISTRIB. Rare, occurring only in a small area of N. Iraq. MSU: Darband-e-Bazian, from Kirkuk to Sulaimaniya, *Haussknecht* 653! (type). MSU: Jarmo, nr. Chamchamal, *Helbaek* 1726!

Iran.

6. **Convolvulus commutatus** *Boiss.*, Diagn. Pl. Orient. ser. 1, 11: 81 (1849). Type: Iraq, near Mosul, *Aucher-Éloy* 1411 (G, holo., iso. BM!, OXF, P); Boissier, Fl. Orient. 4: 94 (1875); Zohary in Dep. Agr. Iraq Bull. 31: 117 (1950); Rechinger, Fl. Iranica 14 (1963); Rawi in Dep. Agr. Iraq Tech. Bull. 14: 133 (1964); Rechinger, Fl. Lowland Iraq 484 (1964); Sa'ad, Meded. Bot. Mus. Herb. Rijks Univ. Utrecht. 116 (1967); Wood & al.: 209 (2015).

C. modestus Boiss., Diagn. Pl. Orient. ser. 1, 11: 82 (1849).

Suffrutescent, to 25 cm high, branched from base. Stems several, ascending, thick with short strict branches, densely grey adpressed hairy. Basal leaves spatulate or oblong-lanceolate, 25–50 × 3–5 mm, attenuate, acute; cauline leaves sessile, elliptic-linear, attenuate, acute. Flowers 3 to several in compact terminal cymes, usually sessile. Bracts similar to cauline leaves; bracteoles oblong, long acuminate, with long spreading hairs. Sepals 13–15 mm long, with long spreading hairs: outer sepals lanceolate, long acuminate; middle one with right and the left unequal, one half membranous; inner sepals ovate, long acuminate, with sides glabrous and membranous. Corolla twice as long as calyx, with silky bands on outside. Filaments glabrous. Ovary ovoid, with long hairs; style hairy. Capsule hairy.

HAB. Rocky mountain slopes; alt. 100–1000 m; fl. Jun.-Jul.
DISTRIB. **MAM**: Zawita, *Omar, Kaisi & Wedad* 49666!; *Omar, Dabbagh & Kaisi* 45344! **FUJ**: nr. Mosul, *Aucher* 1411! (type); in siccis Mesopotamiae, *Olivier & Bruguière* s.n.!

N. & S. Iran, Armenia, Turkey?

7. **Convolvulus holosericeus** *M.Bieb.*, Fl. Taur.-Cauc. 1: 147 (1808) subsp. **macrocalycinus** *Hausskn. & Bornm.*, Mitth. Thüring. Bot. Vereins 6: 66 (1894); Sa'ad, Meded. Bot. Mus. Herb. Rijks Univ. Utrecht. 133 (1967) as "*macrosepalus*"; Parris in Fl. Turkey [P. H. Davis] 6: 208 (1978); Wood & al.: 213 (2015).

Suffrutescent, to 30 cm. Stems branched or not, ascending to procumbent, adpressed sericeous hairy. Leaves sessile, entire, sericeous; basal leaves linear-lanceolate or spatulate, 18–40 × 3–9 mm, acute or obtuse; cauline leaves ± linear, base attenuate, apex acute. Bracts linear-lanceolate, 6–9 × 1–3.5 mm. Flowers 1–9, in axillary and terminal cymes. Peduncles shorter than bracts; bracteoles linear, longer or shorter than pedicels. Sepals short adpressed hairy; outer sepals ovate-orbicular, 9–12 × 8–12 mm, gibbous; middle one with right and left halves unequal; inner sepals suborbicular, mucronate. Corolla 20–26 mm long, segments with hairy bands on outside. Filaments glabrous. Ovary ovoid, hairy; style hairy. Capsule 6–8 × 5–7 mm, hairy.

HAB. On mountains, calcareous hillsides, sandy hills; alt. 840–1500 m; fl. May-Jun.
DISTRIB. Occurs nr. Turkish frontier: **MAM**: between Sandur and Zawita, *Al Kas* 18640!; Zawita, *Omar, Dabbagh & Kaisi* 45344!; *Omar, Kaisi & Wedad* 49666!

Turkey.

8. **Convolvulus cephalopodus** *Boiss.*, Diagn. Pl. Orient. ser. 1, 7: 24 (1846) subsp. **buschiricus** *(Bornm.) J.R.I.Wood & R.W.Scotland*, Phytokeys 51: 245 (2015).

C. buschiricus Bornm., Dan. Sci. Invest. Iran 4: 35 (1945); Rechinger, Fl. Iranica 15 (1963); Rechinger, Fl. Lowland Iraq 486 (1964); Sa'ad, Meded. Bot. Mus. Herb. Rijks Univ. Utrecht. 151 (1967).

Perennial, 20–30 cm high. Stems prostrate to ascending or erect, branching at base, silky tomentose with short and long hairs. Basal leaves oblong-lanceolate, 40–60(–70) × 8–10 mm, base attenuating into the petiole, apex obtuse or subacute, margins somewhat undulate; petiole nearly ¼ as long as lamina; cauline leaves sessile, smaller, veins immersed above, prominent below. Flowers axillary, 3–5 together in compact cymes; peduncles as long as or shorter than the subtending bracts. Bracts ovate-lanceolate or broadly lanceolate, 3 × longer than broad; bracteoles lanceolate, acuminate, not exceeding calyx. Sepals unequal, covered with long sericeous hairs; outer 2 sepals lanceolate, 8–10 mm, acuminate, upper part green; inner 3 smaller, oblong, acuminate. Corolla pink, 15–19 mm, with hairy bands on outside, 1.5 × longer than bracts. Filaments glabrous. Ovary ovoid, hairy; style hairy; stigmas hairy, longer than style.

HAB. Sandy gravely soil; alt. 120–180 m; fl. Mar.-Apr.
DISTRIB. Local in the Southern Desert. **DSD**: Chilawa, 110 km S.S.W. of Basra, *Rechinger* 8803!; E. of Ghazlani, 105 km S.W. of Basra, *Rechinger* 8812!; *Rechinger, Guest & Rawi* 17242!; Busaiya, *Fawzi, Hazim & Hamid* 38949!

Differs from the type subsp. *cephalopodus* in its relatively larger leaves, longer velvety pubescence and a hairy style (in subsp. *cephalopodus* leaves < 50 mm, indumentum asperous, style glabrous or with a few hairs, Fig 16, 1).

Subsp. *cephalopodus* is found in Pakistan, Iran, Bahrain, Qatar, UAE, Oman, Saudi Arabia.

Iran, Kuwait, Saudi Arabia.

9. **Convolvulus euphraticus** *Bornm.*, Beih. Bot. Centralbl. 20 (2): 181 (1906). Type: Iraq, inter Arrah et Deïr, *Strauss* s.n. (B!); Zohary in Dep. Agr. Iraq Bull. 31: 117 (1950); Rechinger Anz. Österr. Akad. Wiss. Math.-Naturwiss. Kl. 1961: 24 (1961); Rechinger, Fl. Lowland Iraq

Fig. 16. **Convolvulus cephalopodus**: 1, habit × ½. **Convolvulus pilosellifolius**: 2, habit × ½. **Convolvulus arvensis**. 3, habit × ½. Reproduced with permission from Fl. Pakistan 126: f. 3, B, C; f. 4, H (1979). Drawn by S.Y. Salim. © National Herbarium, Pakistan Agriculture Research Council & University of Karachi, Pakistan.

485 (1964); Rawi in Dep. Agr. Iraq Tech. Bull. 14: 133 (1964); Sa'ad, Meded. Bot. Mus. Herb. Rijks Univ. Utrecht. 157 (1967); Wood & al.: 246 (2015).

Suffrutescent, with stems ascending, branching from base, to 45 cm long densely tomentose-villous, with up to 2(–5) mm long ± spreading hairs.. Basal leaves oblong-spatulate, 30–70 × 5–10(–20) mm, base narrowing into a 20–40 mm petiole, apex obtuse; cauline leaves sessile, oblong-lanceolate to obovate-oblong, 20–30(–40) × 10–15(–25) mm, obtuse to subacute, tomentose with long spreading hairs. Flowers in many-flowered clusters on 3–55 mm peduncles. Bracts as long as or longer than peduncles; lower bracts elliptical to lanceolate, upper bracts ovate. Sepals 7–10 mm with densely long hairy; outer sepals lanceolate, with a green upper part; inner sepals lanceolate, acuminate. Corolla pink, ± 20 mm, with hairy bands on outside. Filaments glabrous. Ovary and style glabrous. Capsule ovoid, glabrous.

HAB. Rocky mountain slopes, rocky hills, gravelly sandy soils; alt. 120–1000 m; fl. April-May.
DISTRIB. Localised, but frequent in the Western Desert. **MAM**: Zawita, *Omar, Dabbagh & Kaisi* 45319!
DWD: foot of Wadi Hauran, 12 km S.E. of H.1, *Rawi* 31643!; 2 km N.W. of Rutba, *Rawi* 21221!; 20 km E.S.E. of Rutba, *Rawi* 31299!; 75 km E. of Rutba, *Rawi & Khatib* 32292!; road side nr Rutba, *Alizzi & Husain* 34096!; 40 km E. of Rutba, *Omar, Kaisi & Hamad* 43956!; 3 km S.E. of Rutba towards Ramadi, *Rechinger* 9859!; 45 km W. of Rutba, *Rechinger* 9959!; 160 km W. of Ramadi towards Rutba, *Rechinger*

9797!; 260 km N.W. of Ramadi, *Rawi* 20975!; 160 km N.W. of Ramadi, *Rawi* 20899!; 140 km W. of Ramadi, *Rawi* & *Nuri* 27028!

Iran.

10. **Convolvulus kotschyanus** *Boiss.*, Diagn. Pl. Orient. ser. 1, 7: 23 (1846); Boissier, Fl. Orient. 4: 101 (1879); Rechinger, Fl. Iranica 17 (1963); Sa'ad, Meded. Bot. Mus. Herb. Rijks Univ. Utrecht. 162 (1967); Ghazanfar in Fl. Pakistan [Nasir & Ali] 126: 19 (1979); Wood & al.: 236, f. 24, 20–27 (2015).

 C. gonocladus Boiss., Diagn. Pl. Orient. 7 : 22. 1846.
 C. onvolvulus Boiss., Fl. Orient. [Boissier] 4: 102. 1875, p.p. illegit. Superfl. Name for both *C. gonocladus* Boiss. and *C. pyrrotrichus* Boiss. (1846) cited in synonymy.

Perennial, 17–28 (–43) cm high, with a woody base, tomentellous with hairs longer than diameter of stems. Stems usually simple, sometimes branched below. Leaves crowded near base, sessile, oblong, lanceolate or oblanceolate, 12–40 (–80) × 3–8 mm, base gradually attenuate, apex obtuse or acute, margins entire, sometimes slightly rugose, tomentellous with spreading hairs. Flowers axillary, up to 3 in a compact scorpioid cymes. Bracts lanceolate to elliptical, 7–20 (–80) mm long. Peduncles wanting or to 2 mm long; bracteoles to 11 mm long, with long hairs. Sepals unequal, hirsute; outer sepals lanceolate, ± 11 mm long; middle one lanceolate, acuminate, ± 9 mm long; inner sepals as the middle one. Corolla pink, 11–20 mm long, with hairy bands on outside. Filaments glabrous. Ovary glabrous, rarely with weak hairs at top; stigma and style glabrous. Capsule ovoid, glabrous. Fig. 17, 1–7.

HAB. Clay gypsum hillsides, sandy soils; alt. 75–130 (–700) m; fl. Mar.-Jun.
DISTRIB. Not common. **FPF**: 10 km to Tib from Ali al-Gharib, *Khayat & Kaisi* 50556!; Zurbatiya, *Haddad & Mohammad* 9979!; 5 km N. of Zurbatiya, *Kaisi & Khyat* 50623! lea: 5 km W. of Tib, *Thamer* 47695!

S.W. Pakistan (Baluchistan), Iran.

11. **Convolvulus reticulatus** *Choisy*, Prodr. [A.P. de Candolle] 9 : 399 (1845) ; Boissier, Fl. Orient. 4: 100 (1875); Handel-Mazzetti in Ann. Naturh. Mus. Wien 27: 343 (1913); Blakelock in Kew Bull. 4: 527 (1949); Zohary in Dep. Agr. Iraq Bull. 31: 117 (1950); Rechinger, Fl. Iranica 16 (1963); Rawi in Dep. Agr. Iraq Tech. Bull. 14: 133 (1964); Sa'ad, Meded. Bot. Mus. Herb. Rijks Univ. Utrecht. 165 (1967); Parris in Fl. Turkey [P. H. Davis] 6: 203 (1978); Wood & al.: 248, f. 24, 28–37 (2015).

subsp. **reticulatus**

Type: Iraq (Mesopotamia) *Aucher-Eloy* 1408! (G-DC, lecto.).

Perennial, 30–60 cm high. Stems very thick, prostrate, ascending or rarely erect densely tomentose-villous or woolly. Leaves thick, sessile; lower leaves oblanceolate 3.5–5.5 × 1–1.8 cm, attenuate at base, apex acute to obtuse; cauline leaves broadly ovate or ovate-rounded, 2–5 × 3–4 cm, cuneate or nearly truncate at base, rounded or nearly so at tip; upper leaves sessile or subsessile, ovate, 2–4 × 1.5–3.5 cm, cuneate, truncate or sometimes cordate at base, obtuse or rather rounded at tip, tomentellous, rugose above, reticulate beneath. Flowers fasciculate, 4–6(–8) together in dense clusters on thick peduncles, axillary. Bracts oblong or ovate-orbicular, 10–40 × 7–28 mm, truncate or cordate at base; bracteoles oblong, to 11 mm, acuminate. Sepals with long sericeous hairs; outer sepals ovate-lanceolate, 8–12 mm long; middle one linear, acuminate; inner sepals as the middle one. Corolla white, ± 13 mm long, with hairy bands on outside. Filaments glabrous. Ovary glabrous; style and stigmas glabrous. Capsule ovoid, glabrous.

HAB. Fallow fields, steppes, roadsides, sandy and calcareous soils; alt. 130–1300 m; fl. May-Jul.
DISTR. Widespread in foothills and upper plains in N. Iraq. **MAM**: 8 km S. of Zakho, *Rechinger* 10681!; Amadiya, *Kotschy* 348!; 6 km E. of Aqra, *Chapman* 26081!; Atrush, *Chapman* 1352!; Sawara Tuka, *Robertson R.C.* 65!; Sirsang, *Omar* 37699!; Khaira Mountain, nr Zakho, *Rawi* 23040! **MRO**: Rowanduz, *Omar, Sahina, Kazım & Hamed* 38377!; 50 km S.S.E. of Rowanduz, *Rawi & Serhang* 18234!; Kew-a Rash, nr Rania, *Rawi* 23750!; Khanzad pass, *Guest* 3007! **MSU**: between Kirkuk and Chamchamal, *Katib & Tikriti* 29730!; Sulaimaniya-Dukan, *Rechinger* 10087!; 20 km N.W. of Sulaimaniya, *Rawi* 21765!; Sulaimaniya to Dukan, *Haussknecht* s.n.!; 16 km N.W. of Darbandi Khan, *Rawi, Nuri & Alizzi* 29437! **FNI**: Ain Sifni, *Guest* 4044! **FAR**: Jabal Makhmur, *Gillett* 11243!; Ankawa, *Bornmüller* 1525! **FKI**:

Fig. 17. **Convolvulus kotschyanus**. 1, habit; 2, leaf (underside); 3, flower (side view); 4, outer sepal; 5, inner sepal; 6, pistil; 7, capsule with sepals. 1–6 from *Khayat & Al-Kaisi* 50556; 7 from *Sharif, Haddad & Nori Mohammed* 9979. Scale bars: 1 = 3 cm; 2, 3, 7 = 5 mm; 4, 5, 6 = 3 mm. © Drawn by Juliet Beentje.

8 km S. of Kirkuk, *Rawi & Gillett* 11590!; Kirkuk, *Ali Haidari* 3947!; Jarmo, nr Chamchamal, Haines W.294!; Kani Domlan, *Guest* 4289! & 4370!; 36 km from Tuz Khurmatu to Injana, *Rechinger* 10614!; "Mesopotamia, Kurdistan and Mosul", *Haussknecht* s.n.!; Kirkuk, *Kazim* 39386!; 3 km from Kirkuk to Sulaimaniya, *Kazim, Hamid & Jasim* 40492!; **FPF**: nr Lajama, *Rechinger* 9663! **DLJ**: Tikrit, *Lester-Garland* s.n.!

C. reticulatus subsp. *waltherioides* (Boiss. & Hausskn.) Sa'ad, Meded. Bot. Mus. Herb. Rijks Univ. Utrecht 281: 166 (1967), [=*Convolvulus waltherioides* Boiss. & Hausskn., Pl. Orient. Nov. (dec. prim.) 1: 6 (1875) was described from Iran on the basis of its thick and densely woolly stems, elliptic to obovate-elliptic bracteoles and outer sepals. Specimens: MSU: *Quest & Kas* 18609; *R.M. Gorrie* 13498; FKI: *Rawi* 22867; FPF: *Guest* 880; *Kaisi & Yahya* 45202; *Rechinger* 20701; LCA: *Thamer* 50461 have densely woolly stems, but as there are no flowers on the specimens it not possible to fully identify them. *Rawi* 20701, 22867, *Rechinger* 10614, *Kaisi & Yahya* s.n. [19/10/1977]) have been identified by Wood & al. as subsp. *waltherioides* (2015: p. 248). More flowering material is needed to establish the status of subsp. *waltherioides* in Iraq.

Iran, Syria, Turkey.

12. **Convolvulus piloselllifolius** *Desr.*, Encycl. [Lamarck & al.] 3: 551 (1792); Boissier, Fl. Orient. 4: 103 (1875); Handel-Mazzetti in Ann. Naturh. Mus. Wien 27: 393 (1913); Nábělek in Publ. Fac. Sci. Univ. Masaryk 72: 14 (1926); Guest in Dep. Agr. Iraq Bull. 27: 24 (1933); Zohary in Dep. Agr. Iraq Bull. 31: 117 (1950); Rechinger, Fl. Iranica 17 (1963); Rechinger, Fl. Lowland Iraq 486 (1964); Rawi in Dep. Agr. Iraq Tech. Bull. 14: 133 (1964); Sa'ad, Meded. Bot. Mus. Herb. Rijks Univ. Utrecht. 190 (1967); Migahid & Hammouda, Fl. Saudi Arabia 236 (1974); Parris in Fl. Turkey [P. H. Davis] 6: 209 (1978); Ghazanfar in Fl. Pakistan [Nasir & Ali] 126: 22 (1979); Boulos in Fl. Egypt 2: 248 (2000); Karim & Fawzi, Fl. United Arab Emir. 2: 125 (2007); Wood & al.: 176, f. 15, 17–25 (2015).

var. **piloselllifolius**

Perennial low shrub, 50–80 cm high. Stems simple or branched, ascending or prostrate, ± sparingly adpressed-hirsute, bearing flowering branches from middle upwards. Leaves pale green, hirsute, margin often repand-wavy; lower leaves oblong or oblong-lanceolate, 30–70 × 15–20 mm, attenuate; upper ones sessile, lanceolate, smaller, acute, sometimes subcordate at base. Flowers axillary and terminal, solitary or in 2–5 (–8) cymes. Bracts like cauline leaves. Bracteoles linear, as long as, longer or shorter than pedicels. Pedicels shorter than calyx, pilose. Sepals unequal, 6–7 mm, hirsute with short adpressed and longer spreading hairs, with herbaceous tips; outer sepals obovate to oblong acute; middle one with right and left half unequal, one half glabrous; inner sepals ovate, acuminate. Corolla pink, with hairy bands on outside, 10–13 mm, 2–3 × as long as calyx. Filaments glabrous. Ovary and style glabrous. Capsule ovate, glabrous. Fig.16, 2.

HAB. Sandy gravelly soil, cultivated areas, on mountains, in steppes; alt. 6–2000 m; fl. Apr.-Jun.
DISTRIB. Widespread and common in desert, steppe and foothills. **MRO**: Chinaruk road to Qandil, *Rawi* 26474!; Qurnaqo valley N. of Pushtashan, *Rawi* 26612!; Bradost Mountain, *Thesiger* 1215!; Rawanduz, *Omar, Sahina, Kazim & Hamid* 38383! **MSU**: 20 km N.W. of Sulaimaniya, *Rawi* 21763A!; 20 km W. of Sulaimaniya, *Rechinger* 12493!; Sulaimaniya to Dukan, *Rechinger* 12478!; Jarmo, nr Chamchamal, *Helbaek* 1683!; *Helbaek* 1816! & 1888! **FNI**: Mosul, *Aucher* 1412!; Mosul to Sinjar, *Haussknecht* 654! **FAR**: S. of Ser Kurawa, *Gillett* 9650!; Ankawa, *Bornmüller* 1530! **FKI**: Kirkuk, *Rechinger* 12788! **FPF**: Badra, *Rechinger* 9186!; Al-Hashimah, between Badra and Mandali, *Kaisi & Haidari* 3928!; 10 km E. of Mandali & *Khayat* 50714!; Saadiya, *Kaisi* 47210! **LEA**: Ba'quba, *Graham* s.n.!; *Rechinger* s.n.!; nr. Kut, *Alizzi & Omar* 34902! **LCA**: Diltawa, *Guest* 2344! & 2432! & 2475!; Baghdad, *Sahina* 38156!; between Baghdad & Falluja, *Omar & Wedad* 47510!; 50 km N. of Diwaniya, *Thamer* 47654!; Falluja desert N. of Baghdad, *Haines* W.153!; Rustam nr Baghdad, *Gillett* 5874!; *Rawi* 19939!; Qasr Naqib (Governor's House) nr. Baghdad, *Handel-Mazzetti* 929!; Hilla, *Graham* 220!; 50 km from Basra to Nasiriya, *Husham & Muhammad* 29622!; **LBA**: Basra, *Haussknecht* s.n.! **DLJ**: Sumaicha, *Rechinger* 8099!; Rawa, *Alizzi & Omar* 35349!; 4 km E. of Samarra, *Rechinger* 13499! **DWD**: 20 km W. of Rutba, *Rechinger* 9898!; 2 km E. of Rutba, *Rechinger* 12804!; 2 km W. of Rutba, *Rawi* 21136!; Rutba, *Alizzi & Omar* 36189!; 5 km W. of Ana, *Omar & Hamad* 50404!; 40 W. of Ramadi, *Barkley, Safwat & Salih* 5055! **DSD**: 55 km N.W. of Shabicha, *Guest, Rawi & Long* 14048! & 14117!; Shabicha, *Guest & Rawi* 14060E!; 134 km S.S.W. of Salman, *Guest, Rawi & Rechinger* 19069!; 13 km from Salman to Shamiya, *Kasim, Nuri, Hamid & Kadhim* 40145!; 95 km S.W. of Salman, *Guest* 19163!; 30 km N.W. of Busaiya, *Guest, Rawi & Long* 14157!; 70 km S.W. of Samawa, *Guest* 18728!; S. of Khidar Al-Mai, *Weinert, Fawzi, Adel & Hussain* 47594!

E. Mediterranean, Turkey, Caucasus, Iran, Turkmenia, Afghanistan, Pakistan.

SECT. **CONVOLVULUS**

13. **Convolvulus arvensis** *L.*, Sp. Pl.: 153 (1753); Boissier, Fl. Orient. 4: 108 (1875); Handel-Mazzetti in Ann. Naturh. Mus. Wien 27: 3 (1913); Nábělek in Publ. Fac. Sci. Univ. Masaryk 70: 14 (1926); Guest in Dep. Agr. Iraq Bull. 27: 24 (1933); Zohary in Dep. Agr. Iraq Bull. 31 : 117 (1950); Rechinger, Fl. Iranica 20 (1963); Rechinger, Fl. Lowland Iraq 487 (1964); Rawi in Dep. Agr. Iraq Tech. Bull. 14: 133 (1964); Sa'ad, Meded. Bot. Mus. Herb. Rijks Univ. Utrecht. 214 (1967); Migahid & Hammouda, Fl. Saudi Arabia 236 (1974); Parris in Fl. Turkey [P. H. Davis] 6: 213 (1978); Ghazanfar in Fl. Pakistan [Nasir & Ali] 126: 28 (1979); Meikle, Fl. Cyprus 2: 1172–1173 (1985); Rhui-cheng & Staples in Fl. China 16: 291 (1995); Boulos in Fl. Egypt 2: 249 (2000); Karim & Fawzi, Fl. United Arab Emir. 2: 122–123 (2007); Wood & al.: 60, f. 3, 19–27 (2015).

    C. hastatus Forssk., Fl. Aegypt.-Arab.: 203 (1775).
    C. auriculatus Desr., Encycl. [Lamarck et al.] 3: 540 (1792).
    C. prostratus F.W.Schmidt, Fl. Boëm. Cent. 2: 93. 1793 [pub. 1794], nom. illeg., non *Convolvulus prostratus* Forssk. (1775).
    C. sagittifolius Salisb., Prodr. Stirp. Chap. Allerton 123. 1796, nom. superfl. for *Convolvulus arvensis* L.
    C. hastifolius Poir., Encycl. [Lamarck et al.] Suppl. 3: 467 (1814), lapsus [spelling mistake] for *C. hastatus* Forssk.
    C. longipedicellatus Sa'ad, Meded. Bot. Mus. Herb. Rijks Univ. Utrecht.: 233 (1967).
    (for full synonymy see Wood & al. 2015: 60).

Perennial with prostrate or climbing stems arising from a slender rootstock, 30–100 cm, gabrous or pubescent. Leaves glabrous to subglabrous; lower leaves with petioles shorter than lamina; lamina sagittate or hastate, triangular-ovate or oblong, 15–30(–50) × 10–15(–30) mm, entire or repand-dentate, upper ones usually smaller. Flowers solitary, axillary, sometimes in pairs; peduncles shorter or as long as or longer than subtending bracts. Bracts like leaves, but smaller; bracteoles filiform or linear, shorter than pedicels. Pedicels as long as calyx or longer. Sepals subequal; outer sepals obovate to broadly oblong, scarious–margined, rounded or obtuse, truncate or mucronulate, ± 4–4.5 mm long, glabrous or scattered hairy with ciliate margin; middle one longer and wider; inner sepals ovate, retuse, mucronulate. Corolla white or pink, 20–30 mm, 6–7 × longer than calyx, slightly hairy on outside. Filaments with glands on expanded part. Ovary and style glabrous. Capsule glabrous. Fig. 16, 3.

HAB. Cultivated places, roadsides, wastelands; alt. 250–1000(–2610) m; fl. Apr.-Sep.
DISTRIB. Common throughout Iraq, except for desert regions. **MAM**: 15 km E. of Amedie, *Dabbagh & Hamid* 45867!; 8 km S. of Dohuk, *Botany Staff* 43361!; 25 km N.E. of Zako, *Rawi, Nuri & Tikriti* 29064! **MRO**: Ser-i Hasan Beg, nr Rowanduz, *Guest* 3037!; nr Karokh Mountain, *Kass, Nuri & Serhang* 27548!; 12 km N. of Haibat Sultan Dagh, *Rawi, Nuri & Kass* 28386!; Germasur Lake on Qandil range, *Rawi & Serhang* 24127!; Galala, *Haley* 3!; Baba Chichak, *Nábělek* 692!; Pishtashan, *Rechinger* 11757!; between Sirwan and Halabja, *Botany Staff* 43097! **MSU**: Sulaimaniya, *Haussknecht* s.n.!; *Salim & Gook* 5371!; Bakrajo, *Feddou* 5351!; Tawila nr Sosokan in Avroman mts., *Rechinger* 10195!; Azmir Mountain, *Kazim* 39367!; 10 km N. of Sacid Sadip, *Kaisi & Hamad* 43516! **FUJ**: Tal Afar, *Guest* 13363!; *Guest* 13366! **FNI**: Nimrud nr Mosul, *Helbaek* 871!; Mosul, *Thesiger* 786! **FAR**: 7 km from Altun-Kopri to Erbil, *Botany Staff* 43261B! **FKI**: Qaranjir, *Rawi & Gillett* 7494! **FPF**: Mandali, *Guest* 887!; Badra, *Safar* 5529! **LEA**: Baquba, *Rechinger* 9749!; **LCA**: Abu Ghuraib, *Omar* 34499!; *Rawi* 10753!; *Alizzi* 32596!; *Alizzi & Jenan* 32815!; Baghdad, *Guest* 176!; Rustamiya, *Lazar* 1170!; Karada Mariam, *Haines* W.191!; Babylon, *Reuter* 699!; Zafariniya, *Sahina & Sakim* 34775!; 5 km from Kut to Amara, *Alizzi & Omar* 34904!; 3 km E. of Khan Bani Sa'd, *Barkley & Askari* 1243! **LSM**: Musaida nr Amara, *Field & Lazar* 71!; Amara, *Botany Staff* 42489! **LBA**: Abu-Al-Khasib, 20 km E. of Basra, *Rawi* 25938!; Basra, *Mameriyan & Jharah* 5149!; *Alizzi* 35076!; 17 km N.W. of Fao, *Khatib & Alizzi* 33413! **DLJ**: 5 km above Rawa, *Rawi & Gillett* 7015! **DWD**: 5 km W. of Ana, *Omar & Hamad* 50406!;

Throughout the temperate regions of both hemispheres.

14. **Convolvulus betonicifolius** *Mill.*, Gard. Dict. ed. 8. no. 20 (1768); Meded. Bot. Mus. Herb. Rijks Univ. Utrecht.. 219 (1967); Parris in Fl. Turkey [P. H. Davis] 6: 216 (1978); Meikle, Fl. Cyprus 2: 1171 (1985). Wood & al.: 67, f. 4, 4–9 (2015).

    C. pubescens Sol., in Russell, Aleppo, ed. 2, 2: 246 (1794), nom. illeg. superfl.
    C. hirsutus M.Bieb., Fl. Taur.-Caucas 1: 422 (1808).

C. atriplicifolius Poir., Encycl. (Lamarck), Suppl. 3 (2): 467 (1814).
Convolvulus amoenus K.Koch. Linnaea 19: 19 (1847), nom. illeg., non *Convolvulus amoenus* Dietrich (1816).
C. peduncularis Boiss., Diagn. Pl. Orient. 11: 84 (1849).
C. hirsutus M.Bieb. var. *tomentosus* Boiss., Fl. Orient. [Boissier] 4: 105 (1875).
C. betonicifolius subsp. *peduncularis* (Boiss.) Parris in Fl. Turkey [P. H. Davis] 6: 217 (1978).
C. hirsutus var. *virescens* Boiss., Fl. Orient. [Boissier] 4: 105 (1875).
C. armenus Boiss. & Kotschy ex Boiss., Fl. Orient. [Boissier] 4: 105 (1875).
C. betonicifolius var. *armenus* (Boiss. & Kotschy ex Boiss) Sa'ad, Meded. Bot. Mus. Herb. Rijks Univ. Utrecht 281: 221 (1967).
C. aleppensis Sa'ad, Meded. Bot. Mus.Herb. Rijks Univ. Utrecht 281: 209 (1967).

Perennial, prostrate or climbing with a slender creeping rootstock. Stems simple or slightly branched, ± 1 m long, ± densely and softly pilose with spreading hairs. Leaves sagittate to hastate or cordate, margin entire, auricles entire or dentate, puberulous-pubescent to velvety; basal and lower cauline leaves petiolate, 30–70 × 15–30(–60) mm, lamina longer than petiole or rarely equal, acute or sometimes obtuse, obscurely undulate or nearly entire; upper leaves smaller. Inflorescence axillary; peduncles longer than bracts, 1–3-flowered. Bracts similar to leaves. Pedicels as long as, longer or shorter than calyx; bracteoles filiform. Sepals pubescent in lower and upper green part; outer sepals ovate-oblong, acuminate, 8–18 mm long; middle one with unequal halves, one half hairy; inner sepals ovate, acuminate, with both halves hairy. Corolla whitish-pink or yellowish, 30–38 mm long, with pilose bands on outside. Filaments with glands on expanded part. Ovary densely covered with long hairs; style glabrous above, hairy towards base. Capsule about 10 mm long, very hairy.

HAB. Cultivated fields, clay and stony soils, roadsides; alt. 350–1200 m; fl. Apr.-Jul.
DISTRIB. Occasional in northern forest zone of Iraq. **MAM**: Mesopotamia, Kurdistan, Mosul *Aucher*1392!; *Kotschy* 76!; 7 km N.E. of Sarsang to Amadia, *Kaisi, Khayat & Karim* 51085! **MSU**: Diyala, *M.E.D.P.* 349!; Halabja, *Rawi* 8825!; Tainal, *Rawi & Gillett* 11638!; 20–25 km E. of Qaranjir, *Rawi* 21663!; Sulaimaniya to Dukan, *Rechinger* 10085 p.p.!; Dohuk to Zakho, *Guest* 2236! **FNI**: Eski Kellek, *Gillett* 8210!; Olaka, Mosul, *Rawi* 8441! **FUJ**: Gali Mazurka, *Dabbagh & Jasim* 46898!

Eastern Mediterranean region eastwards to the Caucasus and Iran.

15. **Convolvulus scammonia** *L.*, Sp. Pl: 153 (1753); Rechinger, Fl. Iranica 20 (1963); Rawi in Dep. Agr. Iraq Tech. Bull. 14: 133 (1964); Sa'ad, Meded. Bot. Mus. Herb. Rijks Univ. Utrecht. 241 (1967); Parris in Fl. Turkey [P. H. Davis] 6: 217 (1978); Boulos in Fl. Egypt 2: 249 (2000); Wood & al.: 57, f. 3, 1–9 (2015).

Trailing or twining perennial herbs. Stem simple or branched, glabrous, ± 3 m long. Lower cauline leaves ovate, 20–80 × 10–50 mm, with 10–50 mm long petiole, base cordate-auriculate, sagittate or hastate, acute to acuminate, glabrous; upper cauline leaves similar to lower ones. Flowers axillary, solitary or 2–3 (–6) in cymes. Peduncle usually longer than subtending bracts, 1.5–11 cm long. Bracts like cauline leaves; bracteoles ± 6 mm long. Pedicels 5–9 mm, longer in fruit. Sepals unequal, glabrous, scarious; outer sepals clearly shorter than inner sepals, 7–10 mm long, apex retuse, mucronulate; middle one, 9.5–13 mm long, right and left half unequal, apex retuse, mucronulate; inner sepals to 13 mm, apex retuse, mucronulate. Corolla cream to pale yellow, 28–32 mm long, glabrous, sometimes sparsely hairy near apex on band. Filaments with glands on expanded part. Ovary and style glabrous. Capsule ovoid, glabrous.

HAB. Cultivated places, roadsides, mountain slopes, open woods, rocky clay soils, fallow fields; alt. 450–1200 m; fl. Apr.-Jul.
DISTRIB. Occasional in lower forest zone of N. Iraq. **MAM**: Kahantur Mountain, nr. Zakho, *Rawi* 23411!; 35 km N.E. of Zakho, *Omar & Dabbagh*!; 5 km S. of Zakho, *Rechinger* 10708!; Zawita, *Omar* 37674!; *Robertson* 1031!; *Omar, Kaisi & Wedad* 49648!; 20 km N.W. of Sarsang, *Sharief & Hamad* 50304!; *Kaisi & Hamad* 45946!; 10 km E. of Kani Mazi, *Kaisi, Hamid & Hamad* 45924!; between Dihok and Amadiya, *Rechinger* 11670!; Bekma, *Gillett* 8236! **MRO**: Haibat Sultan Dagh, *Karim, Hamid & Jasim* 40750!; Kuh-i Safin, *Bornmüller* 1563! **MSU**: Surdash, Sulaimaniya to Dukan, *Rechinger* 10107! **FUJ**: Jabal Khanuqa, *Rechinger* 12038!; Gara Mountain, *Dabbagh & Jasim* 46947!

Aegean Islands, Greece, Turkey, Syria, Lebanon, Palestine.

16. Convolvulus stachydifolius *Choisy*, Prodr. [A.P. de Candolle] 9 : 408 (1845); Boissier, Fl. Orient. 4: 107 (1875); Handel-Mazzetti in Ann. Naturh. Mus. Wien 27: 393 (1913); Nábělek in Publ. Fac. Sci. Univ. Masaryk 70: 14 (1926); Guest in Dep. Agr. Iraq Bull. 27: 221 (1932); Rechinger, Fl. Iranica 19 (1963); Rechinger, Fl. Lowland Iraq 487 (1964); Rawi in Dep. Agr. Iraq Tech. Bull. 14: 133 (1964); Sa'ad, Meded. Bot. Mus. Herb. Rijks Univ. Utrecht. 243 (1967); Parris in Fl. Turkey [P. H. Davis] 6: 215 (1978); Boulos in Fl. Egypt 2: 251 (2000); Wood & al.: 71, f. 24, 28–37 (2015).

Incl. *C. stachydifolius* Choisy var. *villosus* Hallier f., Bot. Jahrb. Syst. 18: 107 (1894) [publ. 1893]; Sa'ad, Meded. Bot. Mus. Herb. Rijks Univ. Utrecht. 246 (1967).

C. quadriflorus Hochst., in J.A.Lorent, Wanderungen 335 (1845).

Prostrate perennial herbs. Stems flexuous, up to 100 cm, pubescent with spreading or retrorse hairs (villous to tomentose in var. *villosus*). Basal and lower cauline leaves long petiolate, ovate-cordate, 30–60 × 30–50 mm, base cordate, apex obtuse, margins undulate, crenate-dentate to sinuate-dentate, adpressed-pubescent, sometimes glabrescent above; upper leaves smaller, acute. Flowers solitary or up to 9 in axillary monochasium or dichasium; peduncles 22–80 mm long, as long as or longer than subtending bracts. Bracts like cauline leaves; bracteoles filiform, 2–8 mm long, shorter than pedicel. Pedicels 2–3 × as long as calyx, at length refracted. Sepals elliptic, with scarious margins, obtuse or retuse, slightly hairy; outer sepals 6–7 mm long, obovate, retuse and mucronulate, with scattered retrorse hairs; middle one obovate, scattered hairy on midrib above; inner sepals like outer sepals, scattered hairy on midrib above. Filaments with glands on expanded part. Corolla purple, 20–25 mm, (*c.* 15 mm in var. *villosus*), 4 × as long as calyx, glabrous or with hairy bands on outside. Ovary and style glabrous. Capsule glabrous. Fig. 18, 1–8.

HAB. Roadsides, in fallow fields, edge of cultivated fields, in calcareous vineyards; alt. 140–810 (–1750) m; fl. Apr.-June.
DISTRIB. Widespread in the foothills and lower forest zone of northern Iraq. **MAM**: Mosul to Aqra, *Rawi* 11350!; 8 km S. of Dohuk, *Botany Staff* 43366! **MRO**: Haibat Sultan Dag, *Omar, Kazim, Hamza & Hamid* 37196!; 27 km from Koi Sanjaq to Erbil, *Omar, Kaisi & Wedad* 49524!; Koi Sanjaq, *Rawi, Nuri & Kass* 28127!; 2 km N. of Darband, *Botany Staff* 43205! **MSU**: Jarmo nr Chamchamal, *Haines* W.258!; W. 362!; 60 km from Chamchamal to Sulaimaniya, *Alkas* 18548!; 20 km N.W. of Sulaimaniya, *Rawi* 21763!; Sulaimaniya to Dukan, *Rechinger* 10084!; 12 km S.E. of Arbat, *Poore* 350! **MJS**: 40 km E. of Karsi, *A. & E. Barkley* 7999!; Jabal Sinjar, *Kaisi & Hamad* 49171! **FUJ**: 30 km W. of Balad Sinjar, *Field & Lazar* 692!; Tal Afar to Sinjar, *id.* 509!; Hammam Ali, *Omar & Hamid* 36392!; 12 km S. of Ba'aj, *Kaisi & Hamad* 49038!; Tal-Afar, *Rawi & Hamad* 34166!; Mosul, *Lazar* 3357!; 28 km from Mosul to Kasik, *Dabbagh* 46736!; Ain Ghazal nr Mosul, *Guest* 4059!; Humaidat, on bank of Tigris, *Handel-Mazzetti* 1333!;15 km from Erbil to Kerkuk, *Kaisi & Hamad* 49305! **FNI**: 4 km to Tal Kaif, *Botany Staff* 43393! **FAR**: Arbil, *Guest* 1467!; *Shahwani* 25110!; Ankawa, *Bornmüller* 1527!; 7 km from Altun-Kopri to Erbil, *Botany Staff* 43261A! **FKI**: 10 km from Tuz to Kirkuk, *Botany Staff* 42920; 9 km from Kirkuk to Altin Kupri, *Rechinger* 15503!; 10–15 km N. of Kirkuk, *Rawi* 21519! **FPF**: 10 km N. of Qaraghan (Jalaula), *Rawi & Gillett* 7369!; 3 km N. of Sa'adiya, *Kaisi* 46352!; Jabal Hamrin, *Bornmüller* 1528!; Mandali, *Guest* 1809!; Mandali to Naft Khana, *Rawi* 12680!; Badra, *Rechinger* 9187!

Egypt, Jordan, Lebanon, Syria, Turkey.

17. Convolvulus pentapetaloides *L.*, Syst. Nat. ed. 12, 3: 229 (1768); Boissier, Fl. Orient. 4: 110 (1875); Rechinger, Fl. Iranica 21 (1963); Rawi in Dep. Agr. Iraq Tech. Bull. 14: 133 (1964); Sa'ad, Meded. Bot. Mus. Herb. Rijks Univ. Utrecht. 188 (1967); Parris in Fl. Turkey [P. H. Davis] 6: 212 (1978); Meikle, Fl. Cyprus 2: 1174 (1985); Wood & al.: 150, f. 12, 31–39 (2015).

C. tricolor subsp. *pentapetaloides* (L.) O. Bolòs & Vigo in Collect. Bot. (Barcelona) 14: 90 (1983).

Annual herbs. Stems ascending or prostrate, 5–30 cm high, sparingly pubescent, simple or branched. Leaves thinly pubescent, rarely glabrescent; lower leaves petiolate, oblong-spatulate, 15–30 × 2–4 mm, base attenuate, apex obtuse; upper leaves sessile, oblong-spatulate or broadly linear-lanceolate, 15–30 × 4–8 mm, sessile. Flowers axillary, solitary; peduncles ± 2.5 cm long, longer than flower and shorter than the subtending bracts, hairy, reflexed in fruit. Upper bracts attenuate or rounded at base or sometimes amplexicaul. Pedicels as long as sepals or longer. Bracts minute. Sepals scarious, glabrous, ovate-oblong,

Fig. 18. **Convolvulus stachydifolius.** 1, habit, lower part; 2, habit, upper part; 3, flower closed (side view); 4, outer sepal; 5, middle sepal; 6, inner sepal; 7, pistil; 8, capsule with sepals. 1, 3–7 from *Al-Kaisi & K. Hamid* 49171; 2 from *Botany staff* 43393; 8 from *Guest* 13364. Scale bars: 1, 2 = 3 cm; 3, 4, 5, 6, 7, 8 = 5 mm. © Drawn by Juliet Beentje.

mucronulate. Corolla blue, 8 mm, 2–3 × as long as calyx, with hairy bands on outside. Ovary and style glabrous. Filaments with glands on expanded part. Capsule glabrous, ovoid-globose.

HAB. Cultivated fields, mountains, fallow slopes, dry eroded sand, calcareous cliffs; alt. 700–800 m; fl. Mar.-May.
DISTRIB. Local in the foothills of E. Iraq. **MSU**: Khormal nr Sulaimani, *Rawi* 8853!; Halabja, *Rawi* 8875!; Sulaimani plain, *Thesiger* 370!; above Dukan, *Barkley* & *Haddad* 7592B!

Widespread in the Mediterranean region, extending to Iran.

CONVOLVULUS CANTABRICA *L.* (1753) has been recorded by Rechinger in Fl. Lowland Iraq (1964: 484), but we have not seen any specimens from Iraq. This species can look very similar to forms of *C. pilosellifolius* but differs in the larger corolla and pilose ovary and capsule; it is probably erroneously recorded from Iraq.

5. **CRESSA** L.

Sp. Pl. ed. 1, 223 (1753); Gen. Pl. ed. 5, 104 (1754)

Much-branched annual or perennial herbs, sometimes woody at base, apparently short-lived. Stems at first erect than becoming decumbent, grey adpressed pilose to sericeous. Leaves sessile, simple, entire. Flowers terminal in congested racemes or solitary in axils upper leaves. Sepals free, imbricate. Corolla deeply 5 lobed. Stamens 5, exserted. Ovary 2 locular; styles 2, exserted; stigmas capitate. Capsule ovoid.

Cressa is a cosmopolitan genus with four species; a single species in Iraq.

Cressa, Gr. χρυσος, *chrusos*, gold.

Austin, D.F. (2000). A revision of *Cressa* L.(Convolvulaceae). Bot. J. Linn. Soc. 133(1): 27–39 (2000).

1. **Cressa cretica** *L.*, Sp. Pl.: 223 (1753); Rechinger, Fl. Iranica 1 (1963); Rawi in Dep. Agr. Iraq Tech. Bull. 14: 133 (1964); Rechinger, Fl. Lowland Iraq: 481 (1964); Migahid & Hammouda, Fl. Saudi Arabia 232 (1974); Parris in Fl. Turkey [P. H. Davis] 6: 198 (1978); Meikle, Fl. Cyprus 2: 1176 (1985); Gonçalves in Fl. Zamb. 8. 1: 21 (1987); Austin in Bot. J. Linn. Soc. 133: 33 (2000); Boulos in Fl. Egypt 2: 263 (2000); Karim & Fawzi, Fl. United Arab Emir. 2: 131 (2007).

C. *arabica* Forssk., Fl. Aegypt.-Arabica 54 (1775).
C. *humifusa* Lam., Fl. Francoise 2: 268 (1788), nom. illeg.
C. *villosa* Hoffmanns. & Link, Fl. Portugaise 1: 372 (1820).
C. *microphylla* St-Lag., Annales de la Soc. Bot. De Lyon 7: 63, 123 (1880), nom. illeg.

Perennial with a woody base. Stems ascending to erect or then becoming decumbent, 7– 23 cm high, divaricately branched, pubescent. Leaves sessile, rarely short pedicellate, ovate or lanceolate, 1.5 – 8(–12) × 1–4 (–6) mm, acute to acuminate, rounded or cuneate at base, adpressed pubescent. Flowers terminal in congested racemes or solitary in axils of upper leaves, subsessile. Calyx ovoid, 2–3.5 mm long; sepals as long as corolla tube, concave, obovate, obtuse, pubescent. Corolla white, 5 mm long with spreading or recurved lobes; lobes hairy at apex. Filaments glabrous. Ovary ovoid, long hairy above; styles 2, glabrous; stigmas capitate. Capsule ovoid. Fig. 19, 1–2.

HAB. Gravelly sandy and saline soils; alt. ± 150 m; fl. Jun.-Oct.
DISTRIB. Common in the plains and desert areas of Iraq. **FUJ**: Hatra *Alizzi* & *Omar* 35245! **FKI**: Tuz, *Roger* 8395! **DLJ**: Beiji, *Haines* W1596! **DWD**: Habbaniya, *Guest* 3541!; *Guest* 3542!; 20 km N.W. of Kerbela, *Rawi, Guest* & *Serkahiya* 16282!; Shihatha, *Rawi* 26912! **DSD**: Kuraiz al-Mali, ± 15 km E.N. of Jaliba *Rawi* & *Rechinger* 16143! **LEA**: Fakka, 70 km E. of Amara, *Rawi* 25812!; Kut, *Guest* 193! **LCA**: 5 km W. of Shamiya, *Alizzi* & *Muhammad* 32147!; Sudur, *Botany Staff* 47163!; Mahmudiya, *Guest* 13110!; Zafariniya, *Haines* W48!; on Hille road, nr. Dora, *Rechinger* & *Naib* 80! **LSM**: Amara, *Rawi* 14935!; Mesaida, nr. Amara, *Field* & *Lazar* 65!; 30 km E. of Amara, *Bharucha, Rawi* & *Tikriti* 29307! **LBA**: Basra, *Muemariyan* 5162!

Mediterranean region and throughout the tropics of the Old and New Worlds.

Fig. 19. **Cressa cretica**: 1, habit × ½; 2, flower × 4. Reproduced with permission from Fl. Pakistan 126: f. 4, J–K (1979). Drawn by S. Hameed. © National Herbarium, Pakistan Agriculture Research Council & University of Karachi, Pakistan.

118. **SCROPHULARIACEAE** Juss.

Gen. Pl. [Jussieu] 117 (1789), nom. cons.
Fischer in Kubitzki (ser. ed.), J.W. Kadereit (ed.), Fam. & Gen. Vasc. Pl. 7: 333 (2004)

Shahina A. Ghazanfar

Herbs, undershrubs, sometimes climbers or shrubs, rarely trees. Leaves exstipulate, simple, alternate or opposite (especially lower ones), rarely all basal or the upper-most whorled, often variously lobed or dissected, toothed or entire. Flowers hermaphrodite, zygomorphic (sometimes very slightly) usually in bracteate spikes or racemes, less often solitary in leaf-axils, or in cymes. Calyx persistent; calyx-tube campanulate, tubular or almost obscure; calyx-teeth usually 4 or 5, rarely 3, sometimes calyx 2-lipped, valvate, imbricate or open in bud. Corolla united; tube campanulate, cylindric or ventricose or enlarged above, straight or curved or bent, sometimes very short, sometimes with 1 or 2 spurs or sacs at the base, usually 4- or 5-lobed with the lobes more or less equal and all spreading, or distinctly 2-lipped, with the upper lip entire or 2-lobed, lower lip usually 3-lobed. Stamens 4, didynamous or 2, rarely 3, 5 or 6–8; staminodes sometimes present. Ovary superior, 2-locular, ovules numerous. Fruit usually capsular, septicidal or loculicidal, rarely a berry and indehiscent, or fruit separating into 2 cocci.

About 220 genera with ± 3000 species worldwide, mostly herbaceous, especially in temperate regions; 77 species in Iraq.

Recent molecular study of a limited number of species of the Scophulariacae s.l. by Olmstead *et al.* (Amer. J. Bot. 88(2): 348–361 (2001); see also Oxelman *et al.* in Taxon 54(2): 411–425 (2005)) have shown that the family consists of 5 monophyletic groups with *Mimulus* not belonging to any of the groups. The family-level classification accepted by them include (in our Flora area): Orobanchaceae, Scrophulariaceae s.str. Stilbaceae and

Veronicaceae. Fischer in Kubitzki (ed. J.W. Kadereit), Fam. & Gen. of Vasc. Pl. Vol. 7 (2004), accepts all genera traditionally included in the Scrophulariaceae except *Lindenbergia*.

For the treatment of the Scrophulariaceae in this Volume, I have transferred some genera previously included in the Scrophulariaceae s.l. to the Orobanchaceae (Family 120 in this Volume) These are: *Bartsia, Bungea, Euphrasia, Odontites, Parentucellia, Pedicularis* and *Rhynchocorys*.

Since Plantaginaceae for the Flora of Iraq has already been described in Volume 5 part 2 (2013) (including only *Plantago*), genera that have been transferred to the Plantaginaceae & Phyrmaceae (according to the APG IV, 2016 classification), are here described in the Scrophulariaceae s.l. These are: *Albraunia, Bacopa, Kickxia, Linaria, Chaenorhinum, Veronica* (all now in the Plantaginaceae), and *Peplidium* (now in the Phyrmaceae).

Nanorrhinum ramosissimum (Wall.) Betsche, Courier Forschungsinst. Senckenberg 71: 132 (1984): (1984).

(Syn. *Linaria ramosissima* Wall., Pl. Asiat. Rar. (Wallich). 2: 43, t. 153 (1831) has been listed in Rawi (1964) but I have not seen any material from Iraq. Its occurrence in Iraq is not certain.

Albach, D. C., Meudt, H. M. & Oxelman, B. (2005). Piecing Together the "New" Plantaginaceae. *American Journal of Botany* 92(2), 297–315. http://www.jstor.org/stable/4123876

Ghebrehiwet, M., Bremer, B., & Thulin, M. (2000). Phylogeny of the tribe Antirrhineae (Scrophulariaceae) based on morphological and ndhF sequence data. *Plant Systematics and Evolution* 220(3/4), 223–239. http://www.jstor.org/stable/23643825

Olmstead, R.G., DePamphilis, W., Wolfe, A.D., Ellisons, N.D., Reeves, P.A. (2001.). Disintegration of the Scrophulariaceae. *Amer. J. Bot.* 88(2): 348–361.

Oxelman, B., Kornhall, P., Olmstead, R.G., Bremer, B. (2005). Further disintegration of Scrophulariaceae. *Taxon* 54(2): 411–425.

Speta, F. (1982). Drei neue Antirrhineen-Gattungen aus dem Orient: *Holzneria, Hueblia* und *Albraunia* (Scrophulariaceae).

1. Stamens 2 2
   Stamens 4 or 5 (sometimes 4 fertile and 1 staminode). 3
2. Annual or perennial, not aquatic; calyx usually 4-lobed; corolla rotate;
   usually with 4 unequal lobes, usually glandular inside; style short . . . . .9. *Veronica*
   Aquatic annual; calyx 5-lobed; corolla not rotate, lobes ± equal, not
   glandular inside; style expanded above and curved over the stamens;
   seeds rugose with 4 longitudinal ridges . . . . 10. *Peplidium*
3. Corolla actinomorphic; stamens 4 or 5; filaments villous with yellowish
   or purplish hairs (rarely glabrous) . . . . 1. *Verbascum*
   Corolla zygomorphic; stamens 4; filaments not villous. 4
4. Corolla with a distinct abaxial spur 5
   Corolla without spur 8
5. Anthers ciliate; capsule dehiscence operculate (circumsessile lateral
   lid), or capsule indehiscent. . . . . 4. *Kickxia*
   Anthers not ciliate; capsule not operculate 6
6. Calyx lobes oblong to linear, ciliate, as long as or slightly longer than
   corolla tube, persistent and covering capsule . . . . 3. *Albraunia*
   Calyx lobes not persistent and covering capsule. 7
7. Upper lip of corolla almost erect; palate almost closing corolla throat;
   capsule dehiscing apically . . . . 5. *Linaria*
   Upper lip of corolla not erect; palate not closing throat of corolla;
   dehiscing irregularly at base or each locule opening above by a
   pore. . . . . .6. *Chaenorhinum*
8. Aquatic herbs; leaves submerged and aerial; submerged leaves
   verticillate, finely divided; aerial leaves opposite or verticillate,
   undivided or variously laciniate . . . . 8. *Limnophila*
   Herbs terrestrial, or aquatic or in moist locations; leaves not submerged and aerial. . 9
9. Flowers usually small, dark red to purple to brownish or greenish-
   purple; corolla with 5 short orbicular lobes; corolla tube ventricose,
   inflated; capsule acute or beaked at apex . . . . 2. *Scrophularia*

Flowers not small, whitish, blue to pinkish-purple; corolla lobes equal
or the upper 2 connate; corolla tube not ventricose nor inflated;
capsule not beaked at apex . . . . .7. *Bacopa*

1. **VERBASCUM** L.

Sp. Pl.: 177 (1753); Gen. Pl. ed. 5, 83 (1754)
Murbeck, Monogr. der Gatt. *Verbascum*. Lunds Univ. Arsskr. nov. ser. 29(2), (1933)
Fischer in Kubitzki (ser. ed.), J.W. Kadereit (ed.), Fam. & Gen. Vasc. Pl. 7: 361 (2004)
incl. *Rhabdotosperma* Hartl, Beitr. Biol. Pflanzen. 53: 57 (1977)
Celsia L., Sp. Pl. 621 (1753); Gen. Pl. ed. 5, 272 (1754)

Shahina A. Ghazanfar

Annual, biennial or perennial herbs, glabrous or hairy; indumentum of glandular or
eglandular, simple or branched hairs. Leaves alternate (rarely opposite), entire or divided,
the basal ones forming a rosette. Inflorescence of terminal racemes, spikes or panicles.
Flowers bracteate, usually yellow, one or more in the axils of bracts. Calyx 5-lobed almost
to base. Corolla actinomorphic or ± zygomorphic, 5-lobed. Stamens 4 or 5 (sometimes 4
fertile and 1 staminode); filaments villous with yellowish or purplish hairs (rarely glabrous),
equal or 2 lower ones longer and thicker; anthers various, reniform, medifixed, obliquely
or longitudinally inserted. Style slender to filiform; stigma rounded, obovate or spatulate.
Capsule globose to oblong-ovoid or cylindrical; dehiscence septicidal. Seeds ovoid or
obovoid, longitudinally grooved or pitted.

A large genus of ± 360 species distributed mainly in Asia, Europe and N. America; 19
species in Iraq, found mostly in the mountains and upper plains of Northern Iraq, with a
few species are found in the desert areas.

Recent molecular analyses confirmed the monophyly of *Verbascum* including
Rhabdotosperma, and strongly supported the reinstatement of the genus *Rhabdotosperma* into
Verbascum (Alzahrani et al. 2024).

Rhabdotosperma was separated as a new genus from the closely related *Verbascum* L. by Hartl
in 1977 on the absence of accessory flowers, longitudinally grooved seeds and a dilated to
disciform stigma. *Rhabdotosperma* has been recognized by E. Fischer with about 7 species in
Africa and Arabia in Kubitzki (ser. ed.), The Families and Genera of Vascular Plants 7: 361
(2004).

In *Verbascum*, hybrids are often found when two or more species grow together, they are
intermediate in their morphological characters between the parents and are always sterile;
several hybrid taxa have been recorded in Iraq (Bermani & Musawi 1987).

Verbascum, from Lat. *verbascum*, mullein.

Alzahrani, A.M., Brehm, J.M., Ghazanfar, S.A. & Maxted, N. (2024). DNA barcoding of the genus
    Verbascum (Scrophulariaceae) in the Arabian Peninsula. Taxon. https://doi.org/10.1002/
    tax.13156
Bermani, A.K. 1981. Systematic study of the genus *Verbascum* (Scrophulariaceae) as it occurs in Iraq.
    M.Sc. Thesis in Biology/ Botany, University of Baghdad, Baghdad, Iraq. Unpublished.
Bermani, A. K. & Musawi, A. H. E. (1987). Zanoc 5(1): 39–50.
Musawi, A. H. E. & Bermani, A. K. (1986). A new species of *Verbascum* (Scrophulariaceae) from Iraq.
    Feddes Repertorium 97, 7–8: 393–396.
Musawi, A. H. E. & Bermani, A. K. (1986). Five *Verbascum* for Iraq. Journal of Biological Sciences
    Research (17)3: 1–10.
Hadeethy, M., Mashhadani, A., Khesraji, T., Barusrux, S., Jewari, H., Theerakulpisut, P. and
    Pornpongrungrueng, P. (2014). Pollen morphology of *Verbascum* L. (Scrophulariaceae) in
    Northern and Central Iraq. Bangladesh Journal of Plant Taxonomy 21(2): 159–165.
Hadeethy, Muazaz, Jawad Mohammed, M., Nantawan Kanawapee, Jewari Hazim, Mshhdani
    Athiya, Khesraji Talib, Barusrux Sahapat & Piyada Theerakulpis (2014). Genetic Diversity and
    Relationships among *Verbascum* species in Iraq by RAPD-PCR Technique. Global Journal of
    Molecular Evolution and Genomics 2(2): 63–74.
Sh. Saeidi-Mehrvarz, Atter, F., Hamdi, S.M.M., Sharifnia, F., Assadi, M., Naanaie, S. Y. and Mehregan,
    I. (2011). Scrophulraiaceae in Flora of Iran (in Persian), eds. M. Assadi, A.A. Maassoumi, P.
    Babakhanlou & V. Mozaffarin. Vol. 68.

1. Flowers single in the axils of bracts 2
   Flowers more than 1 in the axils of bracts 14
2. Pubescence stellate somewhere on plant, but usually on stems, leaves and pedicels . . 3
   Pubescence simple, glandular or mixed 7
3. Only corolla lobes stellate-pubescent on the outside 4
   Corolla lobes not stellate-pubescent on the outside 5
4. Bracts lanceolate; pedicel 8–17 mm . . . . .8. *V. alceoides*
   Bracts subulate to linear; pedicel 1–2 mm . . . . 9. *V. pseudo-digitalis*
5. Flowers in spikes; pedicel 3–4 mm, sparsely stipitate-glandular to
     glabrous; calyx stipitate-glandular; corolla yellow with purple centre;
     capsule oblong, 6–8 × 3 mm, glabrous . . . . 2. *V. laetum*
   Flowers in lax paniculate inflorescence; pedicel more than 4 mm; calyx
     tomentose; corolla yellow; capsule ovoid to globose, smaller than 6 mm long 6
6. Pedicel stout, patent, to 11 mm; capsule globose, 3 × 4 mm, white-
     tomentose . . . . 10. *V. lanceolatum*
   Pedicel 8–17 mm, not patent; capsule ovoid to globose, 4–6 × 5 mm,
     densely stellate-tomentose . . . . .4. *V. oreophilum*
7. Basal leaves pinnatisect to deeply lobed 8
   Basal leaves not pinnatisect or deeply lobed 9
8. Bracts linear to oblong, the lower with 1–2 lobes, the upper usually
     entire; pedicel 3–4 mm, glandular; calyx glandular . . . . .1. *V. orientale*
   Bracts ovate, cordate, acuminate or caudate, the lower dentate or
     entire; pedicel 10–12 mm, glabrous; calyx glabrous . . . . 6. *V. agrimonifolia*
9. Stems densely white-pubescent, glandular or egandular 10
   Stems not densely white-pubescent; hairs long and short, or long
     egalndular and short glandular 12
10. Bracts, calyx and capsule glandular; pedicel erecto-patent . . . . 3. *V. alepense*
    Bracts, calyx and capsule tomentose, but not with glandular hairs;
      pedicel not erecto-patent 11
11. Bracts lanceolate; pedicel 8–17 mm . . . . .8. *V. alceoides*
    Bracts subulate to linear; pedicel 1–2 mm . . . . 9. *V. pseudo-digitalis*
12. Plant biennial; stems eglandular below, glandular and glabrous above;
      leaves oblong to oblong-lanceolate, gland-dotted . . . . .7. *V. froedinii*
    Robust biennial or perennial plants with stout woody stems;
      pubescence of dense long and short ± glandular hairs; leaves ovate to
      ovate-lanceolate, not gland-dotted 13
13. Whole plant covered with long articulated eglandular hairs and short
      glandular hairs; stamens 4, the two anterior anthers decurrent;
      filaments with purplish hairs right up to the anthers . . . . 11. *V. bornmuellerianum*
    Plant with only glandular hairs; stamens 5, anthers reniform; filaments
      with purplish woolly hairs, the 2 anterior glabrous near apex . . . . . 5. *V. macrocarpum*
14. Basal leaves with margins crenate to dentate 15
    Basal leaves with margins entire, sinuate, sometimes obscurely crenate 17
15. Stems white pubescent, eglandular; bracts linear, ± glabrous; calyx
      lobed almost to base; lobes oblong . . . . 18. *V. hausknechtianum*
    Stems stellate-tomentose; bracts ovate to ovate-lanceolate; calyx 2/3
      lobed or lobed almost to base, lobes ovate to oblong 16
16. Flowers 1–8 in clusters in axils of bracts; calyx 6–8 mm, lobed almost
      to base, lobes lanceolate, acute to acuminate, stellate-tomentose;
      filaments with yellow woolly hairs reaching to apex . . . . 17. *V. aqranse*
    Flowers 1–4 in clusters in axils of bracts (lower clusters forming a
      3-flowered cyme with or without accessory flowers); calyx 4–6 mm,
      lobed to about 2/3, lobes lanceolate, acute; filaments with whitish
      woolly hairs, the 2 anterior filaments glabrous near apex . . . . 19. *V. geminiflorum*
17. Basal leaves linear-lanceolate to oblong (7–30 × 1.5–8 cm); cauline
      leaves lanceolate to ovate to suborbicular . . . . .13. *V. cheiranthifolium*
    Basal leaves ovate-lanceolate to elliptic to lanceolate 18
18. Basal leaves with margins sinuate or sinuate and obscurely crenate 19
    Basal leaves with margins entire, not sinuate 20

19. Basal leaves grey-green, broadly lanceolate to oblanceolate (15–36 ×
      5–10 cm), stellate-pubescent on both surfaces; flowers 2–3 in axils of
      bracts; pedicel 4 mm . . . . .12. *V. sinuatum*
    Basal leaves yellowish-green, ovate to ovate-lanceolate (7–16 × 6.5–10
      cm), densely tomentose; flowers 2–6 in axillary clusters, sessile or very
      shortly pedicellate . . . . .14. *V. carduchorum*
20. Robust perennial to 2 m tall with many stems; bracts ovate-lanceolate
      with an acuminate apex; calyx 3–5 mm with linear-lanceolate lobes . .15. *V. speciosum*
    Perennial with simple, unbranched stem to 50+ cm; bracts ovate with a
      long acuminate apex; calyx 6–10 mm, about ½ lobed, lobes ovate . 16. *V. songaricum*

1. **Verbascum orientale** (*L.*) *All.*, Fl. Pedem. 1: 106 (1785); Boissier, Fl. Orient. 4:
360 (1879); Murbeck, Monogr. 22: 100 (1925); Dinsmore in Post, Fl. Syria, Palest. &
Sinai 2: 280 (1933); Huber-Morath in Fl. Turkey [P. H. Davis] 6: 480 (1978); Feinbrun-
Dothan, Fl. Palaest. 3: 181 (1978); Huber-Morath in Fl. Iranica [K. H. Rechinger] 147:
10 (1981); Boulos, Fl. Egypt 3: 72 (2002); Gabrielian & Fragman-Sapir, Fl. Transcaucasus
& Adjacent Areas: 384 (2008); Saeidi-Mehrvarz et al. in Fl. Iran [Assadi et al.] 68: 11
(2011).

Celsia orientalis L., Sp. Pl. 621 (1753).

Annual, to 25 cm (in Iraq material seen, up to 80 cm elsewhere). Stems erect, branched
from base, pubescent below, glandular-pubescent above, hairs simple. Basal leaves petiolate,
pinnatisect to deeply lobed to coarsely crenate; petiole 5–8 mm; lamina 10–11 cm in outline,
glandular; cauline leaves oblong, pinnatisect to lobed with linear segments. Flowers in lax
spikes, single in the axil of bracts. Bracts oblong, linear to oblong, lower with 1–2 lobes,
the upper usually entire. Pedicel 3–4 mm, ebracteolate, glandular. Calyx lobed to base,
lobes oblong, 6–7 mm, acute, glandular. Corolla pale yellow, lobes broadly ovate, 6–8 mm,
rounded at apex. Stamens 4; anthers reniform; filaments with yellow woolly hairs right up
to anthers. Capsule ellipsoid, 6–7 mm, compressed, keeled above, with faint reticulate veins,
glabrous. Fig. 20, 1–3.

HAB. Limestone hillslopes, amongst *Quercus*, fallow fields; locally frequent; alt. 900–1200 m; fl. May-
Jun.
DISTRIB. **MAM**: Sharanshah, N. of Zakho, 1200 m, *Rechinger* 12088! (W). **MSU**: Kurdistan, *Coll. ignot.*!;
MJS: J. Sinjar, *Gillett* 11094!

Armenia, C.W. & S. Transcaucasia, E. Mediterranean, Turkey, Lebanon, Palestine, Jordan,
Syria, Iran.

2. **Verbascum laetum** *Boiss. & Hausskn.*, Fl. Orient. [Boissier] 4: 338 (1879); Rawi in
Dep. Agr. Iraq Tech. Bull. 14: 146 (1964); Huber-Morath in Fl. Turkey [P. H. Davis]. 6: 508
(1978); Huber-Morath in Fl. Iranica [K. H. Rechinger] 147: 32 (1981).

Verbascum microcarpum Benth., Prodr. [A. P. de Candolle] 10: 230 (1846).
Verbascum arbelense Bornm., Allg. Bot. Z. Syst. xiii. 95 (1907). Type: Iraq, Ankawa, *Bornmüller* 1617
    (B); Rawi in Dep. Agr. Iraq Tech. Bull. 14: 146 (1964); Huber-Morath in Fl. Iranica [K. H.
    Rechinger] 147: 33 (1981).
Celsia leatherdalei Rech.f., Ost. Akad. Wiss. Math.-Nat.Kl. Anz. 9: 191 (1950). – Type: Iraq, Sawara
    Tuka, *Leatherdale* 56 (W!, holo.).

Biennial, up to 45 cm. Stems several arising from a woody stock, erect, purplish, stellate-
pubescent, becoming glabrescent and stipitate glandular above. Leaves basal, grey-green,
petiolate, densely stellate-pubescent on both surfaces; petiole 1.5–3(–14) cm; lamina 6–18 ×
3–8 cm, ovate to ovate-lanceolate to oblanceolate, base cuneate and tapering into the petiole,
apex acute, margins entire to obscurely dentate, nerves pale green, prominent on the lower
surface. Inflorescence a branched panicle; flowers single in the axil of bracts. Bracts 2–3
mm, lanceolate, sparsely stipitate glandular, sometimes stellate-pubescent at base. Flowers
pedicellate, in spikes; pedicel ebracteolate, 3–4 mm, sparsely stipitate-glandular to glabrous.
Calyx 3 mm, oblong, obtuse, stipitate-glandular. Corolla yellow with purple centre, 7–8 mm,
lobes rounded. Stamens 5; filaments with whitish yellow and purple woolly hairs; anthers
reniform, medifixed. Capsule oblong, 6–8 × 3 mm, glabrous. Fig. 21, 1–12.

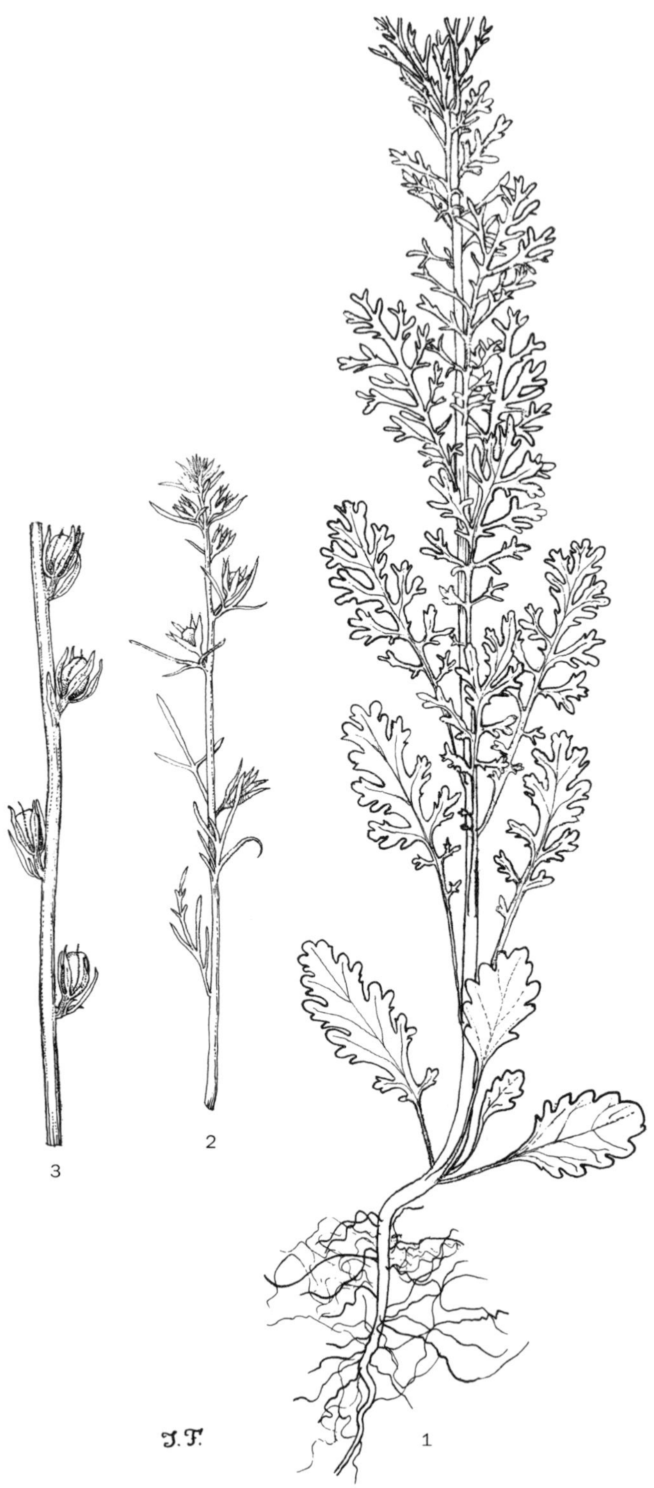

Fig. 20. **Verbascum orientale**. 1, habit × ⅓; 2, portion of shoot with flowers × 1; 3, portion of shoot with fruit × 1. Reproduced with permission from Feinbrun-Dothan, Fl. Palaestina 3: Plates, f. 300 (1977). Drawn by Illana Ferber. © The Israel Academy of Sciences and Humanities.

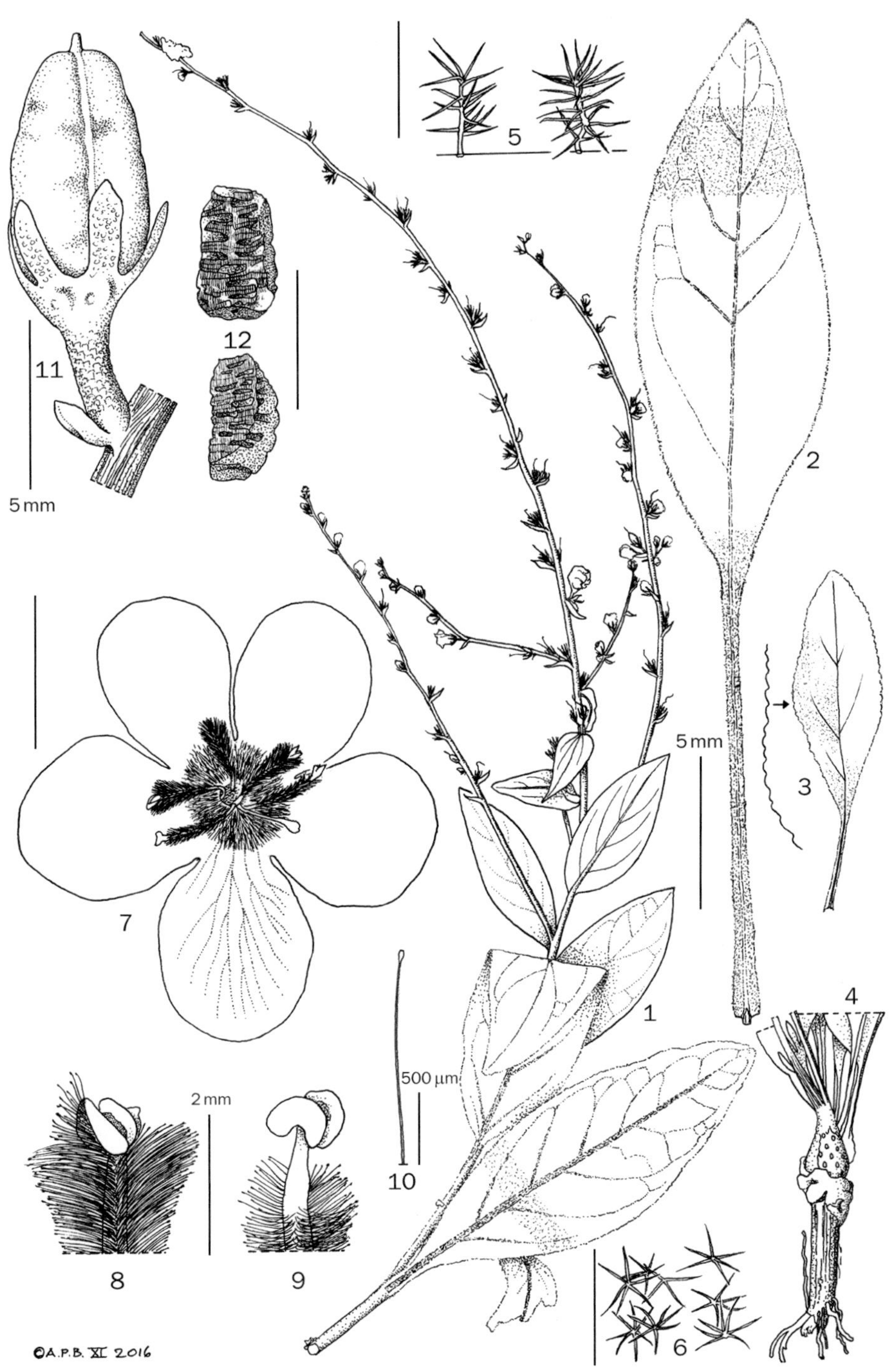

Fig. 21. **Verbascum laetum**. 1. habit; 2, basal leaf detail; 3, leaf margin detail (cauline leaf); 4, rootstock; 5, stellate hair detail from basal leaf; 6, simple stellate hair detail from upper cauline leaf; 7, flower; 8, posterior stamen, detail; 9, anterior stamen, detail; 10, staminal hair; 11, capsule; 12, seed (two views).1–2, 5–6 from *Gillett* 11013; 3 from *Al Kaisi et al.* 51984; 4 from *Rawi* 8517; 7–10 from *Bradburne* 8999; 11–12 from *Omar et al.* 49495. Drawn by © A.P. Brown, Nov. 2016.

HAB. Rocky mountains and hillsides; fallow and wheat fields; alt. 200–1100 m; fl. Apr.-Jun.
DISTRIB. Mountains and upper plains in N. Iraq. **MAM**: Bakirma, 800 m, *Rawi* 8517!; Mosul distr., *Mrs Bradburne* 9000!; Jebel Baykhair nr. Zakho, *Field & Lazar* 789!; Aqra, 1100 m, *Rawi* 11311!; nr. Zakho, *Rechinger* 10679! **MRO**: Haibat Sultan, 920 m, *Omar, Al Kaisi & Wedad* 49495!; Hasar, Omar et al. 37069!; Khahzad, 780 m, *Omar, Al Kaisi & Wedad* 49618!; Salah-ad-Din, *Karim* 37538! **MSU**: Ankawa, *Bornmüller* 1617 (B, type of *V. arbelense*); between Rawandaz and Erbil, *Nábělek* 1935 (SAV). **MJS**: Jebel Sinjar, 800–900 m, *Gilllett* 11013!; Jebel Sinjar, N. slope, 1100 m, *Al Kaisi, Al Khayat & Karim* 50958! **FUJ**: 10 km to Mosul, 200 m, *Al Kaisi & Wedad* 51984!; Mosul District, in desert, *Mrs Bradburne* 8999!

Turkey.

3. **Verbascum alepense** *Benth.*, Prodr. [A.P. de Candolle] 10: 241 (1845); Rawi in Dep. Agr. Iraq Tech. Bull. 14: 145 (1964); Huber-Morath in Fl. Turkey [P. H. Davis] 6: 507 (1978); Huber-Morath in Fl. Iranica [K. H. Rechinger] 147: 31 (1981); Saeidi-Mehrvarz et al.in Fl. Iran [Assadi et al.] 68: 45 (2011).

Robust perennial to 45 cm. Stem erect, simple or branched, densely white-pubescent, ± glandular. Basal and lower leaves with long petioles 2–7 cm, upper leaves sessile with cordate base. Basal leaves lanceolate, 3–10 × 1.5–4 cm, base tapering into the petiole, apex obtuse, margins irregularly dentate, undulate. Inflorescence a lax panicle; flowers single in the axil of bracts. Bracts ovate-lanceolate, 2–6 mm, glandular. Pedicel 3–5 mm, in fruit erecto-patent, ebracteolate. Calyx lobed to about 1/3, lobes lanceolate, 6–8 mm, acute, glandular-pubescent. Corolla yellow, purple at base, c. 20 mm across, without pellucid glands. Stamens 5; anthers reniform medifixed, filaments with yellow hairs right up to anthers or glabrous above. Capsule subglobose, 5–6 mm, densely glandular.

HAB. Rocky hillsides, wadi beds, in cornfield, on alluvial, sandy and clayey soil; common; alt. 120–400 m; fl. Mar.-Aug.
DISTRIB. Upper plains and foothills of N. Iraq. **FAR**: Arbil, *Haussknecht* 1867!; *Bornmüller* 1641! **FKI**: Tuz, *Rogers* 0172!; Baba Gurgur, *Guest* 4015!; Jabal Hamrin, *Guest* 689A!; Hamrin, *Al-Kaisi* 50512!; Kirkuk, *Rechinger* 9974! **FPF**: 10 km N. of Khanaqin, *Hamid* 4660!; 10 km N. of Jalula (formerly Qaraghan) *Rawi & Gillett* 7366!; 20 km from Saadiya to Khanaqin, *H. Hamid* 46617!; Diyala, c. 10 km E. of Mandali, *Rechinger* 9641!; 28 km N. of Mandali, *Al-Kaisi & Yahya* 45233! **FPF/LEA**: Daimah, *Guest* 1713!

Turkey, Syria, Iran.

4. **Verbascum oreophilum** *K.Koch*, Linnaea 22(6): 726 (1849) var. **joannis**; (Bordz.) Hub.-Mor., Denkschr. Schweiz. Naturforsch. Ges. 87: 136 (1971); Huber-Morath in Fl. Turkey [P. H. Davis]. 6: 489 (1978); Huber-Morath in Fl. Iranica [K. H. Rechinger] 147: 24 (1981); Saeidi-Mehrvarz et al.in Fl. Iran [Assadi et al.] 68: 31(2011).

Celsia aureiformis Murb., Monogr. Celsia 83, f. 11 (1925)
C. joannis Bordz., Monit. Jard. Bot. Tiflis, N.S. 5: 49 (1931)
Verbascum aureiforme (Murb.) Murb., Monogr. Verbascum 541 (1933)
V. joannis Bordz. ex Murb., Monogr. Verbascum 542 (1933)
V. aureum (C.Koch) O.Kuntze var. *joannis* (Bordz.) Murb., Bot. Not. 1940: 6 (1940); Rawi in Dep. Agr. Iraq Tech. Bull. 14: 145 (1964)

Perennial, to 50 cm. Stem erect, branched, adpressed-pubescent, pubescence stellate or simple below, stellate above. Basal leaves petiolate with petiole 2–5 cm; upper leaves shortly petiolate or sessile; lamina lanceolate-oblong 10–17 × 0.5–5 cm, base cuneate, apex obtuse, margins crenate-dentate; upper leaves smaller in size. Inflorescence a lax panicle to 15 cm; flowers single in the axil of bracts. Bracts oblong, shorter than pedicels. Pedicel 8–17 mm, ebracteolate. Calyx lobed to base, lobes oblong to oblong lanceolate, 3–4 mm, obtuse to acute. Corolla yellow, to 30 mm across, lobes broadly ovate, 6–8 mm, rounded at apex, stellate tomentose outside. Stamens 4; hairs on filaments purplish (var. *oreophilum*) or whitish yellow (var. *joannis*); anthers reniform, medifixed. Capsule ovoid to globose, 4–6 × 5 mm, densely stellate-tomentose.

Two varieties are recognized: var. *oreophilum* and var. *joannis* (Bordz.) Hub.-Mor. separated on colour of filament hairs; the former endemic to Turkey and the latter found in Armenia, Iraq, Syria and Iran.

Hab. Stony sides of mountains, in forest; alt. 1950–2800 m; fl. Jul.
Distrib. Forest zone of N. Iraq. **MRO**: Algird Dag, *Guest* 2919!, 2931!; Jabal E.N.E. Seri Hassan Beg, *Guest & Ludlow-Hewitt* 3220! Chia Mandali, *Guest* 2720!; Mt. Qandil, *Rechinger* 11145! (W); Mt. Sakri Sakran, *Bornmüller* 1644! (W).

Armenia, Syria, Iran.

5. **Verbascum macrocarpum** *Boiss.*, Diagn. Fl. Orient. Ser. 1, 12: 6 (1853); Rawi in Dep. Agr. Iraq Tech. Bull. 14: 146 (1964); Huber-Morath in Fl. Turkey [P. H. Davis] 6: 492 (1978); Huber-Morath in Fl. Iranica [K. H. Rechinger] 147: 23 (1981); Saeidi-Mehrvarz et al.in Fl. Iran [Assadi et al.] 68: 29 & fig. p. 30 (2011); Mill in Fl. Pakistan [Ali & Qaiser] 220: 267 (2015).

Robust biennial or perennial, up to 150 cm (in Iraq material seen). Stems purplish at base, erect, simple, woody, ribbed, with dense long and short, glandular hairs. Leaves large, basal leaves petiolate, the upper sessile to shortly petiolate; petiole up to 10 cm; lamina 8–20(–45) × 4–12 cm, ovate to ovate-lanceolate, base cuneate to truncate to ± cordate, apex acute, margins dentate, glandular-pubescent on both surfaces. Inflorescence a simple spike, densely glandular-pubescent, viscid; flowers single in the axil of bracts. Bracts 3–6 cm, reducing in size upward, ovate with a long drawn apex, margins serrate, entire towards the apex. Pedicel 4 mm, ebracteolate, glandular. Calyx lobed to base, lobes ovate-oblong, 5–7 mm, acute, glandular-pubescent. Corolla yellow, lobes broadly ovate, ± 6 mm, rounded at apex. Stamens 5; anthers reniform; filaments with purplish woolly hairs, the 2 anterior glabrous near apex. Capsule globose, 4.5–5 × 5 mm, beaked, glandular-pubescent.

Hab. In fields; alt. ± 1400–1800 m; fl. Jun-Aug.
Distrib. **MAM**: Jabal Khantur, *Rechinger* 12099 (W!). **MRO**: Seri Hasan Beg, 1800 m, *Guest* 3028! **MSU**: Avroman mt., 1400 m, *Rechinger* 10260 (W!); Kurdistan, nr. Avroman, 1868, *Haussknecht*!

Turkey, Armenia, Iran, Pakistan.

6. **Verbascum agrimoniifolium** (*C.Koch*) *Hub.-Mor.*, Fl. Turkey [P. H. Davis] 6: 487 (1978); Huber-Morath in Fl. Iranica [K. H. Rechinger] 147: 23 (1981); Saeidi-Mehrvarz et al. in Fl. Iran [Assadi et al.] 68: 24 (2011); Breckle et al., Vasc. Pl. Afghanistan Checklist: 456 (2013).

Celsia agrimoniaefolia C.Koch in Linnaea 22: 732 (1849); *C. heterophylla* Desf. In Pers., Syn. Pl. 2: 161 (1807); *Verbascum heterophyllum* (Desf.) O.Kuntze, Rev. Gen. Pl. 2: 469 (1891) non Miller (1860) nec Moretti (1822) nec Griseb. (1841) nec Velen. (1888). *V. agrimoniifolium* subsp. *agrimoniifolium* Hub.-Mor., Bauhinia 5(3): 151 (1975) invalid publ. without basionym reference.

Biennial, 20–60 cm. Stems erect, branched, pubescent and glandular below, glabrous above, hairs simple. Basal leaves to 45 mm, pinnatisect to deeply lobed with 4–10 pairs of serrate to dentate lobes, the terminal lobe largest, pellucid punctate, glabrous; cauline leaves oblong, lobed or unlobed. Inflorescence spicate; flowers single in the axil of bracts. Bracts ovate, 5–7 mm, cordate, acuminate or caudate, the lower dentate or entire, glabrous. Pedicel 10–12 mm, ebracteolate, glabrous. Calyx lobed to base, lobes elliptic, 3–3.5 mm, acute, glabrous, persistent. Corolla yellow, lobes broadly ovate, 4–5 mm, rounded at apex. Anthers reniform, filaments with yellow whitish woolly hairs right up to anthers, the 2 anterior filaments glabrous near the apex. Capsule ellipsoid, 3–4 × 4 mm, compressed, glabrous. Fig. 22, 1–12.

Hab. In flood plain, stony ground near water, along irrigation channels and in forest near streams; alt. 750–1500 m; fl. Apr.-Jun.
Distrib. **MAM**: Zawita, *Omar* 37673!; 5 km nr. Zakho, *Rechinger* 10706!; 7 km N.E. of Sarsang to Amadia, 950 m, *Kaisi, Khayat & Karim* 51038!; Jebel Baykhair nr. Zakho, *Field & Lazar* 837! **MRO**: Rowandaz Gorge, 550 m, *Guest* 2991!; Bekhal, *Omar* 37637!; Kuh-Sefin region, 900 m, *Bornmüller* 1639!; Gali Ali Beg, *Anders* 1432 (W!). **MSU**: Hawraman mt., 22–26/1961, *Rawi et al.* 19821!; Pir Omar Gudrun, 1500 m, *Rawi* 12128!; Qaradagh village, 850 m, *Gillett* 7960!; Sulimaniyah, *Field & Lazar* 961 & 963!; Tasluja, 1000 m, *Omar & Karim* 47066!; Sosakan, nr. Tawila, *Rechinger* 10170 (W!). **FKI**: Kirkuk, Chamchamal, *Rechinger* 10062 (W!). **FPF**: 5 km N. of Saadiya, 119 m, *Kaisi* 44343!; Saadiya, 110 m, *Kaisi* 46707! **LCA**: Lesser Zab flood plain, 1955, *Robertson*, RB54!

Fig. 22. **Verbascum agrimoniifolium**. 1. habit; 2, leaf detail, adaxial surface; 3, leaf detail; 4, part of lower stem showing indumentum detail; 5, flower viewed from top; 6, posterior anther and filament detail; 7, anterior anther and filament detail; 8, hair of filament detail; 9, upper part of inflorescence showing capsules; 10, mature capsule; 11, capsule, dehisced; 12, seed. 1–3 from *Al Kaisi* 46707; 4–8 from *Furse* 12128; 10 from *Al-Kaisi et al.* 51038; 9,11,12 from *Rawi et al.* 19821. Drawn by © A.P. Brown, Sept. 2016.

GIA RASHA (*Rawi* 12128).

Transcaucasia, Turkey, Iran, Afghanistan.

7. **Verbascum froedinii** *Murb.* in Lunds Univ. Arsskrift, n. f. xxxii. No. 1 (Nachtr. Verbascum) 17, f. 19 (1936); Huber-Morath in Fl. Turkey [P. H. Davis] 6: 533 (1978); Huber-Morath in Fl. Iranica [K. H. Rechinger] 147: 49 (1981).

Biennial to 100 cm. Stems branched, eglandular below, glandular and glabrous above. Lower cauline leaves oblong to oblong-lanceolate, (22 x 6 cm), coarsely crenate, gland-dotted. Inflorescence a large panicle, without accessory flowers. Bracts lanceolate, acuminate. Bracteoles linear-lanceolate. Calyx densely glandular, lobes mucronate. Corolla yellow 20–25 mm across, pellucid-punctate, glandular outside. Stamens 5; anthers reniform, filaments with whitish yellow hairs up to the anthers. Capsule stellate-tomentose. (No material seen for Iraq; description from Fl. Turkey l.c.).

HAB. Not recorded; alt. ± 1000 m; fl. not recorded.
DISTRIB. Known only from two gatherings in N. Iraq; **MRO**: Mt Qandil, *Rechinger* 11006 (W); Kani Mazu Shirin, *Haines* 2140 (cited in Fl. Iranica).

S.E. Turkey.

8. **Verbascum alceoides** *Boiss. & Hausskn.*, Fl. Orient. [Boissier] 4: 306 (1879); Murbeck, Murb., Lunds. Univ. Arsskr. N.F. Avd. 2, 29, 2: 522 (1933); Rawi in Dep. Agr. Iraq Tech. Bull. 14: 145 (1964); Huber-Morath in Fl. Iranica [K. H. Rechinger] 147: 29 (1981); Saeidi-Mehrvarz et al. in Fl. Iran [Assadi et al.] 68: 40, f. 10 (2011).

Perennial to 60 cm with a woody stock. Stem erect, simple, several arising from stock, white-tomentose, eglandular. Leaves grey-purplish (in herbarium material seen). Basal leaves petiolate, lanceolate to lanceolate-oblong, 5–10 × 2–5 cm, base tapering into the petiole, apex obtuse, margins shallow crenate-dentate; petiole 4–7 cm; upper leaves smaller in size, shortly petiolate. Flowers single in axils, in lax to congested spikes. Bracts triangular lanceolate. Pedicel 8–17 mm, ebracteolate. Calyx lobed to about half, lobes ovate-triangular, 6–8 mm, acute, densely tomentose. Corolla yellow, c. 30 mm across, lobes not pellucid punctate, stellate tomentose outside. Stamens 5; anthers reniform, medifixed, filaments with purplish woolly hairs right up to anthers. Capsule ovoid to subglobose, 5–6 mm, densely tomentose.

HAB. Mountains of N. Iraq; alt. 220–1400 m; fl. & fr. Apr.
DISTRIB. **MRO**: Kani Domlan hills, *Guest* 4284! The following specimens are cited in Fl. Iranica: **MSU**: Pira Magrun, *Haussknecht* 713; Sulimaniyah, *Thesiger* 416!; 18 km W. Sulimaniyah *Eig & Zohary*; Penjwin, *Rechinger* 12249; Dukan, *Haines*. **FKI**: 37 km E. Kirkuk, *Rechinger* 10035! nr. Tauq, *Eig & Zohary*.

Iran.

9. **Verbascum pseudo-digitalis** *Murbeck*, Murb., Lunds. Univ. Arsskr. N.F. Avd. 2, 29, 2: 522 (1933); Huber-Morath in Fl. Iranica [K. H. Rechinger] 147: 30 (1981); Saeidi-Mehrvarz et al. in Fl. Iran [Assadi et al.] 68: 42 (2011).

> *V. pseudo-digitalis* var. *chrysandrum* Murb., Lunds. Univ. Arsskr. N.F. Avd. 2, 32, 1: 7 (1936); Huber-Morath in Fl. Iranica [K. H. Rechinger] 147: 30 (1981).
> *V. pseudo-digitalis* var. *phoenicandrum* Murb., Lunds. Univ. Arsskr. N.F. Avd. 2, 32, 1: 6 (1936); Huber-Morath in Fl. Iranica [K. H. Rechinger] 147: 30 (1981).

Perennial, 40–90 cm with a woody stock. Stem several arising from stock, white-tomentose, eglandular. Leaves greyish green. Basal leaves petiolate, elliptic to lanceolate, 5–13 × 2–7 cm, base tapering into the petiole, apex obtuse, margins entire to obscurely irregularly dentate; petiole 6–8 cm; upper leaves smaller in size, shortly petiolate. Flowers single in axils, in lax spikes. Bracts subulate to linear, longer than the flowers. Pedicel 1–2 mm, ebracteolate. Calyx lobed to about half, lobes ovate-triangular, 6–8 mm, acute, densely tomentose. Corolla yellow, 25–35 mm across, lobes not pellucid punctate, stellate tomentose outside. Stamens 5; anthers reniform medifixed, filaments with purplish or whitish yellow woolly hairs. Capsule globose, 5–7 mm, densely tomentose.

The two subspecies described by Murbeck are based on the colour of hairs on the filaments (var. *chrysandrum* with whitish yellow hairs & var. *phoenicandrum* with purplish hairs)

HAB. Mountains of N. Iraq; alt. 1400–1600 m; fl. & fr. Apr.
DISTRIB. **MRO**: Avroman nr. Tawilla *Rechinger* 10241 (var. *chrysandrum*); Kirkuk, 110 km N. Jaula, *Barkley* 4444; Darband-i Khan, Haines; *Barkley & al.* 7730; 7 km S. Darband-I Khan *Barkley & Haddad* 5102-B (var. *phoenicandrum*).

Iran.

10. **Verbascum lanceolatum** (*Vent.*) *O.Kuntze*, Revis. Gen. Pl. 2: 469 (1891) non Kit.: Linnaea 32(4–5): 442 1863; Murbeck, Monogr. 31: (1933); Dinsmore in Post, Fl. Syria, Palest. & Sinai 2: 280 (1933).

Celsia lanceolata Vent. Descri. Pl. Nouv. t. 27 (1800)
C. layardii Ball ex Turrill, Bull. Misc. Inform. Kew 1921(5): 220 (1921). Type: in montibus Kurdistan, *Layard* 327! (K, holo. [K000975846]).
V. assurense Bornm. & Hand.-Mazz., Ann. Nat. Hofmus. Wien. xxvii. 402 (1913) var. *assurens* Hub. -Mor. in Fl. Iranica [K. H. Rechinger] 147: 25 (1981). Type: Iraq, Western Desert District, inter Abukemal et Ramadi, *Handel-Mazzetti* 772 (WU, holo.).
V. assurense Bornm. & Hand.-Mazz. var. *singaricum* Hub. -Mor. in Fl. Iranica [K. H. Rechinger] 147: 25 (1981).
Celsia lanceolata Vent. var. *singarica* Murb., Lunds Univ. Arsskr. N.F. Avd. 2, 35, 1:58 (1939). Type: Iraq, Foothills of Jebel Sinjar, *Uvarov* (HUJ, holo.).

Perennial, 20–50(–70) cm with a woody branched stock. Stems arising from stock, branched, erect, stout, greyish, white-tomentose to adpressed stellate white-pubescent. Leaves greyish, somewhat forming a basal rosette, oblong to oblong-lanceolate, 20–70 × 5–11 mm (15 × 3 cm in *E. & F. Barkley* 5458), base cuneate, tapering into a long petiole, apex obtuse, margins irregularly dentate-serrate to dentate-crenate; upper leaves oblong to oblong-lanceolate, shortly petiolate or sessile, margins entire or crenate, white-pubescent on both surfaces. Flowers in a spreading paniculate inflorescence, single in the axils of bracts. Bracts 5–10 mm, linear, reducing in size upwards. Pedicel stout, patent, to 11 mm, bracteolate. Calyx lobed to base, lobes ovate-oblong, 4–5 mm, acute, tomentose. Corolla yellow, lobes broadly ovate, ± 6 mm, rounded at apex, pubescent outside. Stamens 4; anthers reniform, filaments with yellow or purple woolly hairs (filament hairs white below purple above *Rawi & Gillett* 7125; filament hairs purple *Gillett & Rawi* 6820; 7396; *Field & Lazar* 476 & 386, *Bayliss* 84; filament hairs yellow *Gillett* 11159). Capsule globose, 3 × 4 mm, white-tomentose.

HAB. On mountain and hill slopes, in subdesert, open *Achillea* and *Poa* shrubland, on sandy and clay soils, and gravelly and sandy soils; alt 70–400 m; fl. & fr. Mar.-May.
DISTRIB. Upper and Lower Jazira Districts, lower plains, central alluvial plains and western desert of Iraq. **FUJ**: 26 km E. of Belad Sinjar, *Gillett* 11159!; 20 km S.W. Humid, *Alizzi & Hussain* 33955!; Hadar district, *Rawi & Hamid* 34154!; Ain al Hasan (nr. Sinjar) *Guest* 4194!; nr. Sharqal, J. Makhul, *Guest* 13453!; Mountains of Kurdistan, *Layard* 327 K000975846! (type of *V. layardii*). **FNI**: Telegraph Pole, M90, between Baiji & Mosul, *Field & Lazar* 376 & 386!; Qaiydrah, Mosul, *Bayliss* 84! **FKI**: Kirkuk, Kifri, *Rawi & Gillett* 7396!; Tuz, *Rogers* 0157! **FPF**: J. Hamrin nr. Samara, *Omar & Karim* 37841! J. Hamrin, *Guest* 689!; J. Hamrin nr. Samara, *Rechinger* 9573! Injana, J. Hamrin, *Omar & Karim* 37841!; Qaraghan, *Rogers* 0172!; 17 km E. of Zurbatita, *Al-Kaisi & Al-Khayat* 50650! **DLJ**: Southern border of Lake Tharkar, *Rawi & Chakravarty* 30444!; nr. Lake Tharkar, *F. A. Barkley & J. Brahim* 1690!; Dulein, 4 km W. of Tharkar, *Rawi & Gillett* 7125!; nr. Tikrit, *Omar & Hamid* 36299!; 20 km W. of Tikrit, *Alizzi & Hussain* 33770!; Schbaichan, 80 km S. of J. Sinjar, *Rawi* 5732!; Jazira, between Ba'ag & Talail al Nil, *Guest* 13420!; Jazira desert, *Maj. Edmonds* 3797!; Iraq-Syria border, date1867, *Hausbknecht* s.n.! **DLJ/ FUJ**: 5 km N. of Police station of Baka, *Alizzi & Hussain* 33846!; nr. Baka, *Alizzi & Hussain* 33805! **DWD**: 3 km E. of Hit, *Gillett & Rawi* 6820!; Ramadi district, 5.9.1935, *G. A. Watson* s.n.!; Ghubrah plain, *Guest, Eig & Zohary* 5079! **LCA**: nr. Daltawa, *Rogers* 0326!; 155 km N.E. K3 (Skaabani), *Chakravarty, Rawi, Khalid & Alizzi* 32986;

Syria, N.W. Iran.

11. **Verbascum bornmuelleranum** *Hub.-Mor.*, Bauhinia 5(1): 10 (1973); Huber-Morath in Fl. Turkey [P. H. Davis] 6: 489 (1978); Huber-Morath in Fl. Iranica [K. H. Rechinger] 147: 21 (1981); Saeidi-Mehrvarz et al.in Fl. Iran [Assadi et al.] 68: 28 (2011).

Celsia bornmuelleri Murb., Lunds Univ. Arsskrift, n. f. xxii. No. 1, 182 (1925).

Biennial, to 70 cm. Stems stout, simple (in Iraq material), the whole plants covered with long articulated hairs and short glandular hairs. Basal leaves large, petiolate, ovate, 9–13 × 5–6 cm, base cuneate to truncate, apex acute, margins crenate-dentate; petiole 8–11 cm; upper leaves sessile, ovate-lanceolate, 4–7 × 1–3 cm, base cordate, acute. Flowers single in axils of bracts. Bracts 10–20 mm, ovate with a long acuminate apex. Calyx lobed to base, lobes ovate-oblong, 5–6 mm. Corolla yellow, to c. 40 mm across [in herbarium material], lobes rounded, glandular outside. Stamens 4, the two anterior anthers decurrent, filaments with purplish hairs. Capsule immature c. 6 mm across.

HAB. Rocky and clayey soil, clearings in oak woodland; alt. 1300–1700 m; fl. Apr.-May.
DISTRIB. Mountains and upper plains of N. Iraq. **MSU**: Qara Dagh, *M.E.D. Poore* 427!; Erbil, Kuh Safin, *Bornmüller* 1367 (cited in Fl. Iranica); **FUJ**: Gara mt., *Al-Dabbagh & Jasim* 47003!

Turkey, Iran.

12. **Verbascum sinuatum** *L.*, Sp. Pl. 178 (1753) var. **adenosepalum** *Murb.*, Lunds. Univ. Arsskr. N.F. Avd. 2, 29, 2: 371 (1933); Rawi in Dep. Agr. Iraq Tech. Bull. 14: 146 (1964); Huber-Morath in Fl. Turkey [P. H. Davis] 6: 538 (1978); Huber-Morath in Fl. Iranica [K. H. Rechinger] 147: 45 (1981); Saeidi-Mehrvarz et al.in Fl. Iran [Assadi et al.] 68: 65 (2011).

Robust biennial, 50–80 cm (in Iraq material seen). Stems greenish purple at base, erect, simple, woody, white stellate-pubescent to glabrescent. Leaves grey-green, large, basal leaves petiolate, the upper sessile or shortly petiolate; basal leaves: petiole 2–8 cm; lamina broadly lanceolate to oblanceolate, 15–36 × 5–10 cm, base cuneate and tapering into the petiole, apex obtuse, margins sinuate, stellate-pubescent on both surfaces; nerves prominent on the lower surface; cauline leaves decreasing in size upwards, ovate to obovate-lanceolate, 2–7 × 1–2.5 cm, ± cordate, margins sinuate to entire, stellate-pubescent. Inflorescence a large panicle; flowers 2–3 in axils of bracts. Bracts triangular, cordate, acute to mucronate. Pedicel 4 mm. Bracteoles ovate. Calyx lobed to base, lobes ovate-lanceolate, 5–7 mm, acute, stellate-pubescent with ± glandular hairs. Corolla yellow, 2 cm across, lobes 10–15 mm, rounded, with brown pellucid glands, stellate-pubescent on the outside. Stamens 5; filaments with purple woolly hairs right up to anthers; anthers reniform. Capsule globose, 3–4 mm, densely stellate-pubescent. Fig. 23, 1–12.

HAB. Rocky mountain slope, in sandy and gravely soils, nr water and in fields; alt. 580–1700 m; fl. & fr. Apr.-Aug.
DISTRIB. Mountains, upper and lower plains of Iraq. **MAM**: Village 16 km S. Amadiya, *Al Dabbagh & H. Hamid* 45868!; Kani Masi, *Al Kaisi & H. Hamid & Hamad* 45736!; Zawita, *Guest* 1973!; 3747!; Kuh Sefin, Bornmüller 1645! **MRO**: Sefin Dagh, *F. Karim* 39410! **MSU**: Arbat [Arbet], *F. Karim* 39305!; Hawraman mt., *Rawi & Chakravarty* 12822!; Halabja, *Guest* 12931! **MJS**:, Qilat (J. Golat), between Ain Tellawi & Balad Sinjar, *Field & Lazar* 477! **FUJ**: Ser Amadiya, Gali Muazurka, *Al Dabbagh, Al Kaisi & H. Hamid* 46009!; Gali Mazurka, *Al Dabbagh & M. Jasim* 46913! **FNI** : Hinnis, nr. Ain Sifna, *Guest* 3615!; Mahad, *Selim Effendi* 2615! **FKI/MSU**: Qara Hasan, *F. Karim, H. Hamid & Jasim* 40462! **FPF**: Kani Masi, Botany Staff 43894!; 3 km N. of Saadiya, Al Kaisi 43488!; Makatu nr. Mandali, *Guest* 881! **LEA**: Kut, *Guest* 204! **LCA**: Za'farinyah, *S. Omer & Sahira* 37814!; *Sahira* 39478!; Mahmudiya, *Guest* 2384!; Badah, N. of Baghdad, Guest 2504! **LBA**: 5m of Al-Asisiya, *Al Khayat & Th. Redha* 52278!

SIMRAH (*Guest* 2384); MASI JARRAK (Kurd. *Selim Effendi* 2615); KAVALPIR (Kurd., *Guest* 881).

Guest 3747 record that meat is wrapped and cooked in the basal leaves to make a "dolma".

Egypt, Sinai, Syria, Turkey, Iran, Turkmenistan and Afghanistan.

13. **Verbascum cheiranthifolium** *Boiss.*, Diagn. Ser. 1(4): 56 (1844); Hub.-Mor., Candollea 12: 206, f.11 (1949); Huber-Morath in Fl. Turkey [P. H. Davis] 6: 587 (1978); Huber-Morath in Fl. Iranica [K. H. Rechinger] 147: 40 (1981).

Biennial. Stems slender, branched, striate, tomentose stellate, hairs yellowish or greyish. Basal leaves petiolate, linear-lanceolate to oblong, 7–30 × 1.5–8 cm, apex obtuse to acute, margins entire, rarely crenate; petiole 2–6 cm; Cauline leaves lanceolate to ovate to

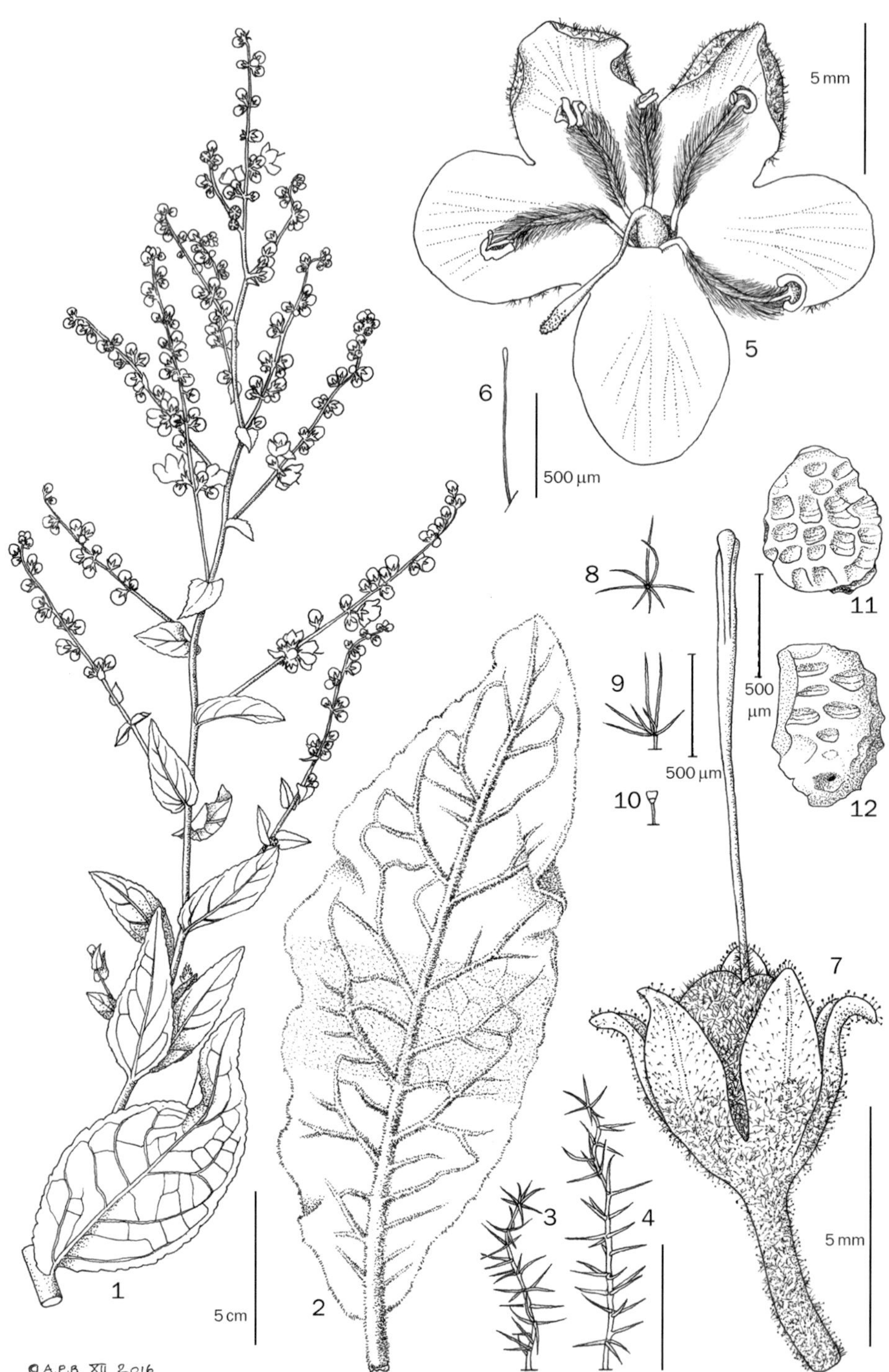

Fig. 23. **Verbascum sinuatum** var. **adenosepalum. 1,** habit, lateral branch of inflorescence; 2, basal leaf detail, abaxial surface; 3, hair from abaxial surface (basal leaf); 4, hair from midrib (abaxial surface basal leaf); 5, flower; 6, hair from staminal filament; 7, mature capsule; 8 & 9, stellate hairs from capsule; 10, glandular hair deom sepal; 11 & 12, seed, convex surface & lateral surface).1 from *Bornmüller* 1643; 2–4 from *Guest* 3747; 5–6 from *Al-Khayat et al.* 52278; 7–12 from *Guest* 3615. Drawn by © A.P. Brown, Jan. 2017.

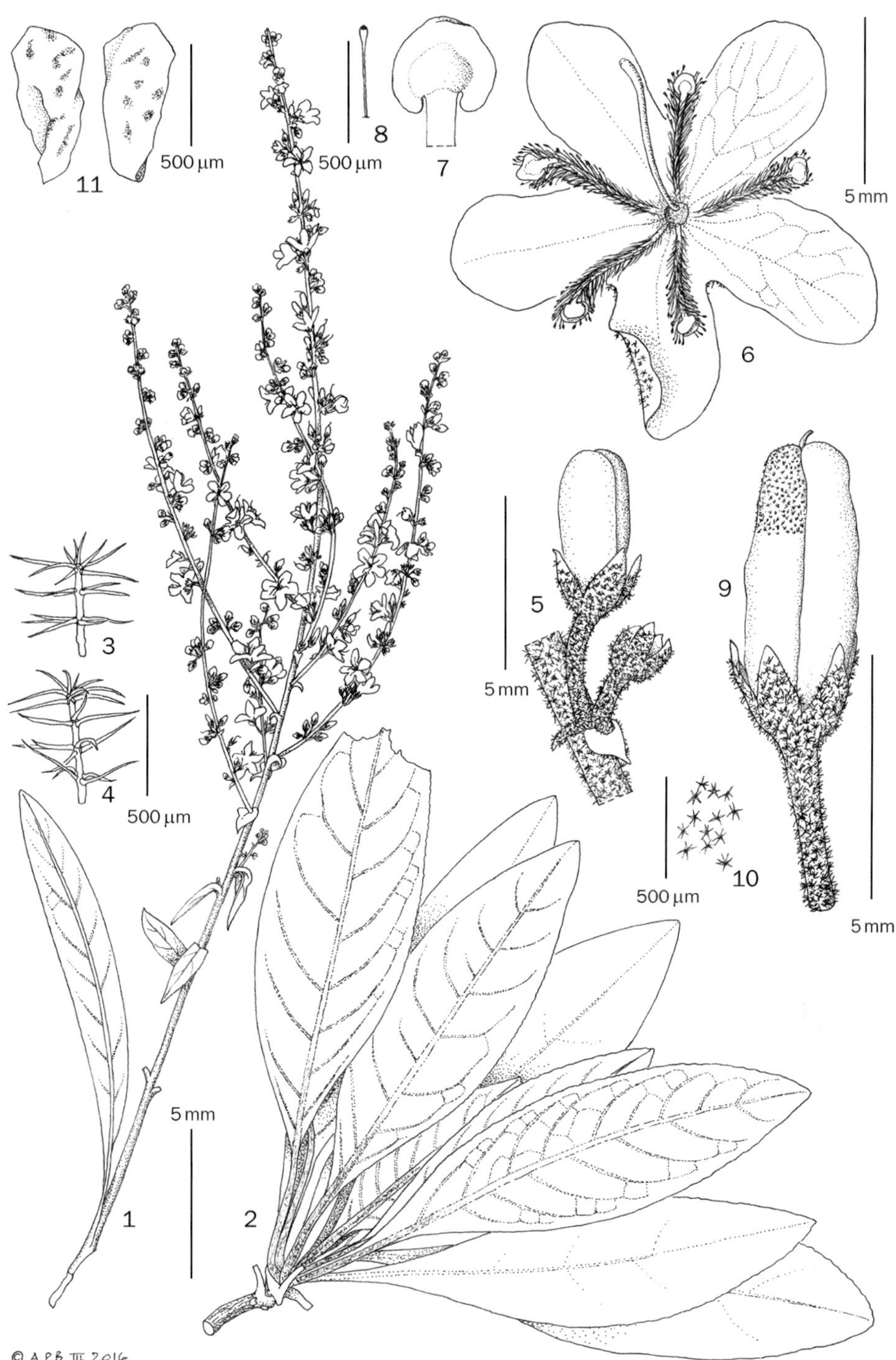

Fig. 24. **Verbascum cheiranthifolium** var. **asperulum.** 1, habit, showing branched inflorescence; 2, habit, basal leaf rosette; 3, hair from abaxial surface (basal leaf); 4, hair from adaxial surface (basal leaf); 5, glomerule with 2 subtending bracts; 6, flower (showing stellate hairs on abaxial surface of petal); 7, anther; 8, hair from staminal filament; 9, mature capsule; 10, stellate hairs from capsule; 11, seed.1, 5–11 from *Bornmüller* 14472; 2 from *Davis* 13641. Drawn by © A.P. Brown, Mar. 2017.

suborbicular, base rounder to subcordate, acute to acuminate. Inflorescence a spreading panicle; flower clusters loose, of 2–7 flowers. Bracts triangular to linear-lanceolate. Pedicels to 12 mm. Bracteoles lanceolate. Calyx1.5–4.5 mm, lobes linear to triangular lanceolate, acute to acuminate. Corolla yellow, 20–25 mm across, without pellucid glands, stellate-tomentose outside. Stamens 5, anthers reniform, filaments with whitish-yellow hairs up to anthers. Capsule cylindrical to oblong-elliptic, densely stellate tomentose. (No Iraq material seen; description from Fl. Turkey).

A widespread species in Transcaucasia, Turkey and Iran.

Two varieties are recorded from Iraq, based on single collections, distinguished on the length and shape of calyx and capsule; both varieties also occur in Turkey and Iran.

    Calyx 2 mm; capsule cylindrical . . . . var. *asperulum*
    Calyx 3–4.5 mm; capsule oblong-cylindrical, 2–2.5 x calyx. . . . . var. *cataonicum*

var. **asperulum** (*Boiss.*) *Murb.*, Lunds. Univ. Arsskr. N.F. Avd. 2, 29, 2: 277 (1933); Huber-Morath in Fl. Iranica [K. H. Rechinger] 147: 40 (1981). Fig. 24, 1–11.

    V. asperulum Boiss., Diagn. Pl. Orient. Ser. 1, 4: 57 (1844).

MRO: Haji Omran, 1700 m, *Rechinger* 11326-b (cited in Fl. Iranica).

Iran.

var. **cataonicum** (*Hand.Mazz.*) *Murb.*, Lunds. Univ. Arsskr. N.F. Avd. 2, 29, 2: 277 (1933); Huber-Morath in Fl. Turkey [P. H. Davis] 6: 589 (1978).

    V. shepardii Post, Fl. Syr. 20 (1896); *V. cataonicum* Hand.-Mazz., Ann Nat. Hofmus. Wien. 27: 400 (1913).

MAM: between Sharanish and Marsis, 1300 m, *Rechinger* 10871 (cited in Fl. Iranica).

Turkey.

14. **Verbascum carduchorum** *Bornm.*, Allg. Bot. Z. Syst. xiii. 94 (1907); Murb., Lunds. Univ. Arsskr. N.F. Avd. 2, 29, 2: 226 (1933); Rawi in Dep. Agr. Iraq Tech. Bull. 14: 145 (1964); Huber-Morath in Fl. Iranica [K. H. Rechinger] 147: 35 (1981).

Biennial. Stem simple, woody, purplish below, stellate-tomentose. Basal leaves yellowish-green, petiolate, ovate to ovate-lanceolate, 7–16 × 10–6.5 cm, apex obtuse to ± slightly notched at apex, base unequal, margins entire, sinuate, densely tomentose; nerves prominent on lower surface; petiole 2–6 cm; Cauline leaves sessile, ovate to ovate-lanceolate, base subcordate. Inflorescence a large spreading panicle; flowers 2–6 in axillary clusters, sessile or very shortly pedicellate. Bracts as cauline leaves, reducing in size upwards. Bracteoles lanceolate. Calyx 3–4 mm, lobes to about ½, lobes ovate, acute. Corolla yellow, 20–25 mm across, without pellucid glands, sparsely stellate-tomentose outside. Stamens 5; anthers reniform, medifixed; filaments with whitish-yellow hairs up to anthers. Capsule ovoid, 3–4 mm, densely white tomentose.

HAB. Rocky mountain slope, rocky ledges, open mountain valleys and rocky ravines; alt. 450–1400 m; fl. & fr. Jul.-Sep.
DISTRIB. Mountains and upper valleys of northern Iraq. **MAM**: Sandur, *S. Omar, Al-Kaisi & Wedad* 49693!; Sarsang, *Al-Dabbagh* 46176!; *Wheeler-Haines* W548!; Atrush, *Guest* 3649!; nr. Zakho, Jabal Khantur, *Rechinger* 10741!; Zawita, *Guest* 3739! **MRO**: Rowanduz gorge, *Guest & Husham* 15919!; Kuf-Sefin nr. Schaqlawa, *Bornmüller* 1615 (identified as *Carducorum × sinuatum*). **MJS**: Meer Khasim, balad Sinjar & Tell Afar, *Henry Field & Yusaf Lazar* 543A! **FNI**: Hinnis, *Guest* 3616A.

W. Iran; near endemic.

15. **Verbascum speciosum** *Schrad.*, Index Seminum [Gottingen] 2: 22 (t. 16) (1811); Huber-Morath in Fl. Turkey [P. H. Davis] 6: 586 (1978); Huber-Morath in Fl. Iranica [K. H. Rechinger] 147: 39 (1981 Saeidi-Mehrvarz et al.in Fl. Iran [Assadi et al.] 68: 56 (2011).

    V. longifolium DC., Fl. Franc. [de Candolle & Lamarck], ed. 3. 6: 414 (1815).

Robust biennial, to 2 m tall. Stems many, branched, angular, whitish to yellowish stellate-tomentose. Basal leaves lanceolate to lanceolate-oblong, 10–45 × 3–10 cm, base tapering into the petiole, margins entire to crenulate, apex acuminate to caudate; petiole 2–7 cm, ± winged; cauline leaves similar but smaller. Inflorescence a wide robust panicle with erect branches; flowers congested, 2-several in axils of bracts. Bracts ovate-lanceolate, acuminate. Pedicels c. 12 mm. Bracteoles lanceolate. Calyx 3–5 mm, lobes linear-lanceolate. Corolla yellow, 20–30 cm in diameter, pellucid glands absent, stellate tomentose outside. Stamens 5, reniform; anthers medifixed; filaments with whitish-yellow hairs up to anthers. Capsule oblong-ovate, densely stellate-tomentose.

HAB. Rocky mountain; alt. ± 1550 m; fl. & fr. Jul.-Aug.
DISTRIB. Mountains of northern Iraq. **MAM**: Zakho near Sharanshah, *Al Bermani* 37868!

Central and S.E. Europe, Turkey, Iran.

16. **Verbascum songaricum** Schrenk, Enum. Pl. Nov. i. 26 (1841) var. **songaricum**; Rawi in Dep. Agr. Iraq Tech. Bull. 14: 146 (1964); Huber-Morath in Fl. Iranica [K. H. Rechinger] 147: 37 (1981); Deyuan et al. in Fl. China, Scrophulariaceae [Z. Y. Wu & I. Al-Shehbaz] 18: (1998).

V. polystachyum Kar. & Kir., Bull. Soc. Imp. Naturalistes Moscou 14: 716 (1841).
V. daënense Boiss., Diagn. Pl. Orient. ser. 1, 7: 38 (1846).

Perennial. Stem simple, unbranched to 50+ cm. Basal leaves yellowish green, sessile to shortly petiolate, ovate-lanceolate to lanceolate, base tapering into the petiole, apex obtuse, margins entire, densely stellate tomentose; nerves prominent on the lower surface. Inflorescence a congested panicle; flowers 2–3 in axils of bracts; pedicel 5–12 mm. Bracts sessile, 6–7 mm, ovate, apex long acuminate. Calyx 6–10 mm, about ½ lobed, lobes 3–4 mm, triangular-lanceolate, acute. Corolla yellow, c. 30 mm across, lobes not pellucid punctate, stellate tomentose outside. Stamens 5; filaments with yellow woolly hairs; anthers reniform medifixed. Capsule ovoid, 5–7 mm, densely stellate tomentose. Fig. 25, 1–5.

HAB. Rocky mountain slope, by stream; common; alt. 1000–2100 m; fl. & fr. Jul.-Aug.
DISTRIB. Mountains of northern Iraq. **MAM**: Bikhair mt. nr. Zakho, *Rawi* 22975! **MRO**: Algurd Dagh, *Gillett* 9500!, *Guest* 2825! **MJS**: Jabal Sinjar, *Al-Kaisi, Al Khayat, & F. Karim* 50973!

Turkey, Iran, Afghanistan, C. Asia to Xinjiang.

17. **Verbascum aqranse** *Al-Bermani & Al-Musawi*, Feddes Repert. 97(7–8): 393 (1986). Type: Iraq, Nainawa, [Nineveh], 4 km S. Aqra, *Al Bermani* 38119 (BHU, holo.)

Biennial herb. Stems cylindrical, branched above, erect, 40–100 cm, stellate-tomentose. Basal leaves petiolate, lanceolate to elliptic-lanceolate, 8–48 cm long, base attenuate, apex acute to acuminate, margins crenate to crenate-dentate; lower cauline leaves petiolate, the upper sessile, lanceolate to oblong-lanceolate, 6–42 cm long, base attenuate, apex acute to caudate, margins dentate-crenate. Inflorescence a lax panicle; flowers 1–8 in clusters in axils of bracts; pedicel 2–5 mm. Bracts sessile, ovate to lanceolate, base cordate. Bracteoles 2, lanceolate. Calyx 6–8 mm, lobed almost to base, lobes lanceolate, acute to acuminate, stellate-tomentose. Corolla yellow, c. 3.5 cm across, lobes densely stellate-hairy on the outside. Stamens 5; filaments with whitish yellow woolly hairs reaching to apex; anthers reniform, medifixed. Ovary densely white hairy; style stellate hairy at base. Capsule subglobose, 4–6 mm, densely stellate hairy or glabrous. Seed with grove and transversely foveolate. (Description from Al-Bermani & Al-Musawi, l.c.).

HAB. Rocky mountain; alt. 700–820 m; fl. & fr. Jun., Aug.
DISTRIB. Mountains of northern Iraq. **MAM**: Nineveh, c. 50 km S.W. Aqra, *Al Bermani* 37829 (BHU); Nainawa, [Nineveh], 4 km S. Aqra, *Al Bermani* 38119 (Type:); c. 4 km S. Aqra, *Al Musawi & Al Bermani* 38094 (BHU); 25 km S.W. Aqra, *Baghdad Salahuddin taxonomists* 39703 (BHU).

Endemic.

Fig. 25. **Verbascum songaricum** var. **songaricum**. 1, lower portion of plant; 2, middle part of plant with leaves; 3, fruiting upper part; 4, flower (opened); 5, fruit with calyx. 1-5 × 1 Reproduced from Fl. China 18: f. 4 (1988), with permission from Missouri Botanical Garden Press, St. Louis, and Science Press, Beijing. Drawn by Ji Chaozhen.

18. **Verbascum haussknechtianum** *Hub.-Mor.*, Fl. Iranica [K. H. Rechinger] 147: 44 (1981) var. **haussknechtianum**; Saeidi-Mehrvarz et al.in Fl. Iran [Assadi et al.] 68: 64 (2011).

V. persicum Hausskn. ex Bornm., Beih. Bot. Centralbl., Abt. 2. 22(2): 103 (1907) non Kuntze (1891).

Perennial. Stems simple to branched, erect, 40–70 cm, pubescent, eglandular. Basal leaves petiolate, oblong to oblong-lanceolate, base cuneate, apex obtuse, margins crenate to crenate-dentate, whitish-pubescent; cauline leaves similar to basal leaves but smaller; upper leaves sessile, acuminate, margins dentate. Inflorescence a panicle; flowers (1–)2–4 in clusters in axils of bracts; pedicel 5–10 mm. Bracts sessile, linear, ± glabrous. Calyx 2–3 mm, lobed almost to base, lobes oblong, acute to obtuse. Corolla yellow, 12–18 mm across, lobes densely pellucid punctate, glabrous on the outside. Stamens 5; filaments with whitish woolly hairs, not reaching to apex; anthers reniform, medifixed. Capsule ellipsoid, 3–4 mm, ± pubescent. (No material seen. Description from Flora Iranica.)

HAB. Rocky mountain slope; alt. not known; fl. & fr. not known.
DISTRIB. Mountains of northern Iraq. **MSU**: Qara Dagh, *Hadač* & al. 6287 (cited in Fl Iranica); Qara Dagh. *Haines* 1174; Penjwin, *Hadač* & al. 4894 (cited in Fl. Iranica).

Near endemic; W. Iran.

19. **Verbascum geminiflorum** *Hochst.*, in Lorent, Wanderungen 336 (1845); Huber-Morath in Fl. Turkey [P. H. Davis] 6: 553 (1978); Huber-Morath in Fl. Iranica [K. H. Rechinger] 147: 44 (1981).

> *V. cestroides* Boiss. & Hausskn., Fl. Orient. [Boissier] 4: 331 (1879); Rawi in Dep. Agr. Iraq Tech. Bull. 14: 145 (1964).

Biennial. Stems branched, erect, 30–100 cm, adpressed yellow stellate tomentose. Basal leaves oblong-lanceolate, 5–15 × 2–7 cm, base cuneate, apex obtuse, margins crenate to crenate-dentate, stellate tomentose; petiole 3–8 cm; cauline leaves similar to basal leaves but smaller; upper leaves sessile, ovate-lanceolate, acuminate, margins obscurely crenulate. Inflorescence a branched spreading panicle; flowers 1–4 in clusters in axils of bracts (lower clusters forming a 3-flowered cyme with or without accessory flowers). Bracts sessile, ovate, acuminate. Calyx 4–6 mm, lobed to about 2/3, lobes lanceolate, acute. Corolla yellow, 20–30 mm across, lobes not pellucid punctate, stellate tomentose on the outside. Stamens 5; filaments with whitish woolly hairs, the 2 anterior filaments glabrous near apex; anthers reniform, medifixed. Capsule ellipsoid to subglobose, 5–6 mm, stellate-tomentose. (No material seen. Description from Flora Turkey).

HAB. Not recorded; alt. ± 250 m; fl. & fr. not recorded.
DISTRIB. N. Iraq. **MAM**: between Mosul and Zakho, 86 km N. Mosul, *Rechinger* 10630; Mosul, *Thesiger* s.n.; S. Mosul, 250 m, *Anders* 1235 (cited in Fl Iranica).

S.E. Turkey, N. Iran

Species recorded from Iraq by Huber-Morath in Flora Iranica 147: 1981.

These species are based on single collections. I have not seen any material of these specimens and no material is present at BAG or K.

Verbascum blattaria L. is reported from Kurdistan in Flora Iranica (p. 22), from Zaribar (Zrewar); this locality is in the Iranian Zagros mountains. No material present from Iraq.

Verbascum phyllostachyum Boiss. & Hausskn. ex Boiss., Fl. Orient. [Boissier] 4(2): 331 (1879).

Specimen cited in Flora Iranica (1981: 37, 147), from Mt. Avroman et Schahu, 1800 m, *Haussknecht* 712 (JE), type of *Verbascum phyllostachyum* Boiss. & Hausskn. ex Boiss. is most likely from Iran. There is no other material cited from Iraq, and the only one other specimen cited is from Iran. It is listed by Rawi (Rawi in Dep. Agr. Iraq Tech. Bull. 14: 146 (1964), probably based on Boissier's Fl. Orientalis where it was first published. I have not included this in the key as characters are not fully visible on the JSTOR image https://plants.jstor.org/stable/viewer/10.5555/al.ap.specimen.je00007342).

Description from Fl. Iranica: Biennial, 40–50 cm tall, stems yellowish-tomemtose, egalndular. Basal leaves with petiole 3–5 cm, lamina elliptic oblong, c. 16 x 6 cm, margins crenate; cauline leaves sessile to shortly petiolate, with an amplexicaule to cordate base and an acuminate-cuspidate apex. Inflorescence with floral clusters of 4; pedicel 6–13 mm. Calyx 7–10 mm, with lanceolate lobes almost to base, tomentose. Corolla yellow, 18–20 mm across, sparsely pellucid punctate, stellate-tomentose on the outside. Stamens 5, filaments villous. Ovary densely tomentose. Capsule not seen.

2. **SCROPHULARIA** L.

Sp. Pl. 2: 619 (1753); Gen. Pl. ed. 5, 271 (1754); Stiefelhagen in Bot. Jahrb. 44: 406 (1910)
Fischer in Kubitzki (ser. ed.), J.W. Kadereit (ed.), Fam. & Gen. Vasc. Pl. 7: 362 (2004)

Shahina A. Ghazanfar

Annual, biennial or perennial herbs or small shrubs; glabrous or hairy. Leaves alternate or opposite (especially lower ones opposite), entire to pinnatisect, translucent gland-dotted. Inflorescence of lax thyrse, terminal or axillary cymes or spikes. Flowers usually small, dark red to purple to brownish or greenish-purple. Calyx 5-lobed almost to base; calyx-tube campanulate to tubular. Corolla 2-lipped, with 5 short rounded lobes, the upper erect, the lower spreading; tube ventricose, inflated. Stamens 4, didynamous, included or excluded; scale-like staminode present. Disc oblique. Style slender, stigma retuse or sometimes capitate. Capsule ovoid to subglobose, acute or beaked at apex, dehiscence septicidal, valves entire or bifid above. Seeds ovoid or oblong, wrinkled or pitted.

A large genus of ± 200 species distributed mainly in Asia, Europe and N. America; 18 species in Iraq, found mostly in the mountains and upper plains with a few species in desert areas.

Scrophularia, from Lat. *scrofula*, with suffix *-aria* to indicate that it possesses medicinal properties to treat the disease known as scrofula, the tuberculosis of lymph nodes.

1. Basal leaves deeply lobed, incised or pinnatilobed 2
   Basal leaves not lobed or incised or pinnatilobed 6
2. Plant glaucous, to 100 cm with ± rigid, erect stems; calyx lobes oblong
   to oblong-obovate, 2–3 mm; corolla dark purplish red, to 5 mm, the
   upper lobes narrower at base; staminode white, as large as the upper
   corolla lobe, margin crenulate . . . . .3. *S. xanthoglossa*
   Plant green, not glaucous 3
3. Stems sparsely to densely glandular 4
   Stems glabrous to puberulent, not glandular 5
4. Cymes 5–7-flowered, sparsely glandular; calyx lobes glabrous; staminal
   filaments glabrous . . . . .4. *S. sulaimanica*
   Cymes 1–2-flowered, glandular; calyx lobes with sessile glands at base
   (sometimes glabrous especially in fruit); staminal filaments glandular
   12. *S. atroglandulosa*
5. Young and flowering branches glabrous, not appearing zigzag; pedicels
   1–2 mm, glabrous; staminode smaller than the posterior corolla lobe . . .1. *S. desertii*
   Young and flowering branches glabrous to puberulent appearing
   somewhat zig-zag; pedicels 2–3 mm minutely puberulent; staminode
   about as long as the posterior corolla lobe . . . . 2. *S. striata*
6. Stems and branches white to greyish white, smooth, stout, branching
   divaricately, the lower spreading, the upper erect to erect-horizontal;
   lateral branches becoming spinescent . . . . 18. *S. hypericifolia*
   Stems and branches dark brown to greenish brown, branched but not
   divaricate, laterals not becoming spinescent 7
7. Plant villous; leaves villous on both surfaces or sometimes the lower
   surface sparsely villous to glabrous
   14. *S. pegaea*
   Plant pubescent, glandular or ± glabrous, not villous. 8
8. Plant densely glandular throughout, especially the leaves and lower parts of stem . . . 9
   Plant in part pubescent, glandular or ± glabrous, never densely glandular 13
9. Basal and lower leaves subreniform to ovate-orbicular. 10
   Basal and lower leaves ovate 11
10. Leaves 10–25 × 8–20 mm; petiole 5–10 mm; cymes 3–7-flowered;
    capsule glabrous. . . . . .16. *S. gracilis*
    Leaves 8–10 × 6–12 mm, petiole 3–5 mm; cymes 1–2-flowered; capsule
    glandular-pubescent. . . . . .17. *S. amadiyana*
11. Glands disctinctly white (in Iraq plants); fruiting pedicel of central
    flower in cyme stout, 3–5 mm, glandular . . . . .11. *S. catariifolia*

Glands brown; fruiting pedicel of central flower in cyme not stout, ... 12
12. Stems white pruinose and glandular; flowers 5–11 in lax cymes
      (sometimes with lateral branches developing in one plane,
      alternately in opposite directions, the main axis appearing zig-zag);
      pedicel 1–3 mm . . . . 15. *S. pruinosa*
   Stems glandular-pubescent; lower cymes 8–25-flowered, upper
      2–5-flowered, never appearing zig zag; pedicel 15 mm . . . . 13. *S. crenophila*
13. Stems and leaves glabrous; stems distinctly quadrangular, narrowly
      winged. . . . . 6. *S. umbrosa*
   Plants glabrous to obscurely puberulous or glandular; stems not
      distinctly quadrangular, nor winged. 14
14. Stems and leaves dark green; leaves sessile to very shortly petiolate . . . . . . 9. *S. kurdica*
   Stems and leaves green; leaves petiolate 15
15. Leaves lanceolate 35–65 × 5–12 mm; peduncles and pedicels pruinose
      to glandular . . . . 10. *S. nervosa*
   Leaves ovate to ovate-oblong (or sometimes pinnatipartite), wider than 5 mm 16
16. Large perennial herb to 1 m; leaves ovate to ovate-oblong, 20–90 ×
      13–60 mm, ± glabrous . . . . 7. *S. macrophylla*
   Perennial herbs to 75 cm; leaves 10–45 × 8–25 mm, sparsely to densely
      glandular or glabrous 17
17. Leaves glabrous; calyx margins scarious, not lacerate; corolla greenish
      with lateral lobes pinkish; staminode broadly ovate to elliptical . . . . . . 5. *S. libanotica*
   Leaves sparsely to densely glandular; calyx margins scarious, lacerate;
      corolla dark pink with lateral lobes pinkish; staminode obcordate . . . . . . 8. *S. kollakii*

1. **Scrophularia derserti** *Delile*, Descr. Egypte Hist. Nat. 240, t. 33, f.1 (1813); Boissier, Fl. Orient 4: 414 (1879); Stiefelhagen, Bot. Jahrb. 44: 473 (1910); Post, Fl. Syria, Palest. & Sinai 2: 295 (1933); Eig in Palest. J. Bot. Jerusalem 3: 79 (1944); Blakelock in Kew Bull. 4(4): 531 (1949); Rechinger, Fl. Lowland Iraq 548 (1964); Rawi in Dep. Agr. Iraq Tech. Bull. 14: 144 (1964); Feinbrun-Dothan, Fl. Palaest. 3: 199 (1978); Wood, Handb. Fl. Yemen: 262 (1997); Grau in Fl. Iranica [K. H. Rechinger] 147: 264 (1981); Boulos, Fl. Egypt 3: 72 (2002); Saeidi-Mehrvarz et al. in Fl. Iran [Assadi et al.] 68: 268 (2011); Ghazanfar, Fl. Oman 3: 144 (2015); Qaiser, Khatoon & Hamidullah in Fl. Pakistan [Ali & Qaiser] 220: 256 (2015).

S. *marginata* Boiss., Diagn. Pl. Orient. Nov. ser. 1, 4: 72 (1844).
S. *sinaica* Benth., in DC., Prodr. 10: 314 (1846).
S. *deserti* Del. var. *foliata* Eig, Palest. J. Bot. Jerusalem 3: 79 (1944).

Perennial herb, up to 60 cm. Stems ascending to erect, branched from base and expanding into a branched thyrsoid inflorescence, quadrangular, glabrous. Cauline leaves opposite, the lower alternate and opposite, petiolate; petiole 10–30 mm; lamina of upper leaves ovate to obovate, 8–20 × 5–7 mm, base cuneate, apex rounded, margins irregularly dentate; lamina of lower leaves 20–55 × 7–25 mm, spatulate to oblanceolate, margins dentate or shallow or deeply lobed to pinnatilobed; all leaves glabrous. Bracts linear, 1–2 mm. Flowers 5–7 in lax axillay cymes; pedicel 1–2 mm, glabrous. Calyx lobes oblong, ± 2 mm, obtuse, margins white scarious, glabrous. Corolla dark red to purplish red, 2–5 mm, the upper lobes small. Staminode white, broadly ovate, smaller than the corolla lobe. Stamens included. Capsule globose, 2–3 mm long, beaked at apex, glabrous. Fig. 26, 1–4.

HAB. Sand and gravelly soils, in desert and desert gravel plains; common; alt. 40–1600 m; fl. Feb.-Jun. (-Sep.)
DISTRIB. **MAM**: Atrush [Hatrush], 795 m, *Guest* 3628!; **MRO**: Khuhzad, 780 m, *Omar, Kaisi & Wedad* 49593!; Pusht Ashwar, 150 m, *Rawi* 23921! **MSU**: Kajan mt., 1590 m, *Rawi* 22747!; Penjwin, 1280 m, Rawi 22525! **FKI**: Tuz, 6 Apr. 1931, *Guest* 1402!; nr. Kirkuk, 300 m, *Guest* 1351!; N. of Fetha, 23 Mar. 1963, *Barkley & Brahim* 4667! **FPF**: 8 km from Tib to Sharhany, 60 m, *Kaisi, Thamer & Salah* 51223!; 20 km to Khanaqin, *Fawzi & Noori* 39585!; Khanaqin, *Fawzi & Noori* 39583!; **DLJ**: N. of Baji, Hikmat Abbas 9704! **DGA**: Samara, 110 m, *Botany Staff* 49833! **DWD**: 35 km W. of Nukhaib, 250 m, *Rawi* 14758; 40 km W. Ramadi to Tutba, 100 m, *Kaisi & Hamad* 48039; 70 km S.W. of Khader Alma, 180 m, *Khatib & Alizzi* 32746! **DSD**: 41 km S. & W. of Busaiya, 215 m, *Guest & Ibrahim* 15271; 42 km E. of S. Salman, 180 m, *Guest & Rawi* 14113; 14 km ESE Khadr at Mai, 185 m, 28/03/1956, *Guest & Ibrahim* 15245; 75 lm from Shibcha to Salman, *Saadun* 151134; 44 km S. & W. of Busaiya, 215 m, *Guest & Rawi & Rechinger* 16075; Jabal Sanam, *Guest, Rawi & Schwaun* 14370A; 30 km W. of J Sanam, *Khatib &*

Fig. 26. **Scrophularia deserti**. 1, habit with portion of stem; 2 & 3, upper portion of stems showing flowers and fruits; 4, flower × 4. Reproduced with permission from Feinbrun-Dothan, Fl. Palaestina 3: Plates, f. 335 (1977). Drawn by Esther Huber. © The Israel Academy of Sciences and Humanities.

Alizzi 32683! **LEA**: Fakka, 40 m, *Rawi* 25819!; Amarah, Fakka-Laqlaq, *Rawi* 25834!; Al-Tib, 6/iii/1973, *Husain Al Ani* 39710! **LCA**: nr. Faluja, 40 m, *Guest & Rawi* 13635!; 15 km w of Falluja, *Barkley & Abbas* 147!; Um Qasr, *Khatib & Alizzi* 32674!; nr. Dargala village, 950 m, *Rawi, Nuri & Kass* 28893!; Sudur, 68 m, *Omar et al.* 44325!

ALGA (*Saadun* 15134); SULAJA (*Guest & Rawi & Rechinger* 16075).

Egypt, Palestine, Jordan, Syria, the Arabian Peninsula, Iran, Afghanistan and Pakistan.

2. **Scrophularia striata** *Boiss.*, Fl. Orient. [Boissier] 4: 413 (1879); Stiefelhagen, Bot. Jahrb. 44: 473 (1910); Lall & Mill in Fl. Turkey [P. H. Davis] 6: 642 (1978); Grau in Fl. Iranica [K. H. Rechinger] 147: 279 (1981); Saeidi-Mehrvarz et al.in Fl. Iran [Assadi et al.] 68: 296 (2011); Qaiser, Khatoon & Hamidullah in Fl. Pakistan [Ali & Qaiser] 220: 255 (2015).

S. juncea C.Richter ex Stapf, Denkschr.Akad. Wiss. Wien, Math.-Nat. Kl. 50: 24 (1885).
S. persica Benth., in DC., Prodr. 10: 314 (1846) p.p.
S. sinjariflora Sawah ined.

Perennial herb, up to 70 cm, very similar in facies to *S. deserti*. Stems ascending to erect, angular, branched from base and expanding into a much branched thyrsoid inflorescence, young and flowering branches appearing somewhat zig-zag, glabrous to puberulent. Cauline leaves alternate, the lower alternate and opposite, all leaves petiolate; petiole 10–30 mm; lamina of upper leaves linear to narrow oblong, 6–35 × 1–3 mm, base cuneate, apex rounded, margins entire to 2–5 lobed; lamina of lower leaves 3–6 × 4–12 mm, pinnatisect; all leaves glabrous. Bracts linear, 1–2 mm. Flowers 3–7 in lax axillary cymes; pedicel 2–3 mm, minutely puberulent. Calyx lobes oblong, ± 2.5 mm, obtuse, margins white scarious, ± glabrous. Corolla dark purplish red, 2–5 mm, the upper lobes small. Staminode white, broadly ovate, about as long as the posterior corolla lobe. Stamens exserted. Capsule globose, 2–3 mm long and wide, shortly mucronate at apex, glabrous. Fig. 27, 1–4.

HAB. In the mountains, mountain slopes amongst stones; common; alt. 450–1800 m; fl.& fr. Apr.-Jun. (-Oct.).

DISTRIB. **MAM**: Bikhair mt., nr. Zakho, *Rawi, Tikriti & Nuri* 28956!; 5 km S. of Zakho, *Rawi* 23109!; Atrush, 800 m, *Guest* 3628! **MRO**: Shaqlawa, *Weeler-Haines* W781!; Rowanduz, *Gillett* 8300!; Pushtashan, *Rawi & Serhang* 24194!; Gali Ali Beg, *Rawi & Serhang* 24262!; Jindian, *Guest & Husham* 15878!; Chinaruk-Qandil Range, *Rawi* 25355!; Kaiwa rush, *Rawi & Serhang* 23762!; Sulimaniyah, *Barkley*, 8491!; S. of Jabtak bridge, *Rawi, Nuri & Kass* 28082!; Gali Warta, *Rawi, Nuri & Kass* 28860! **MSU**: Tawila, *Rawi* 21883!; Hauraman mt., *Rawi* 22114!; Jarmo, *Weeler-Haines* W273!; 20 km N.W. of Sulimaniyaa, *Rawi* 22767! Darbandi Basian, *Rawi* 22796! **MJS**: Jabal Sinjar, nr. TV Tower, *Sawah* 81.64 (BUH, type of *S. sinjariflora*). **FUJ**: Nr. Baaj, *Omar & Hamid* 36540!; c. 15 km N. E. of Mosul, *Hossain* 514!; Tawila nr. Sosakan, *Rechinger* 10191! **FNI**: Nimrud, *Helbaek* 985! **FKI**: 3 km from Kirkuk to Sulimaniya, *Karim, Hamid & Jassim* 40488!; Qaran jir, 14/iv/1947, *Rawi & Gillett* 7525!; Kirkuk-Chamchmal, *Guest* 13603. **FPF**: 55 km S.E. Mandali, *Rawi* 20748!

Widespread; *Rechinger* 10191 from Sosakan near Tawila placed under *S. azerbaijanica* Grau is referable to *S. striata*.

Russia, Turkey, Iran, Afghanistan, Pakistan.

3. **Scrophularia xanthoglossa** *Boiss.*, Diagn. Pl. Orient. ser. 1, 12: 38 (1853); Biossier, Fl. Orient. 4: 413 (1879); Post, Fl. Syria, Palest. & Sinai 2: 295 (1933); Eig in Palest. J. Bot. Jerusalem 3: 86 (1944); Blakelock in Kew Bull. 4(4): 533 (1949); Rechinger, Fl. Lowland Iraq 547 (1964); Rawi in Dep. Agr. Iraq Tech. Bull. 14: 145 (1964); Feinbrun-Dothan, Fl. Palaest. 3: 197 (1978); Boulos, Fl. Egypt 3: 72 (2002).

S. decipiens Boiss. & Kotschy, Diagn. Pl. Orient. II, 3: 156 1856.
S. hispidula Boiss. & Balansa, Diagn. Pl. Orient. II, 6: 157 1859.
S. orientalis Ehrenb. ex Boiss., Fl. Orient. [Boissier] 4: 413 (1879).
S. gileadense Post, J. Linn. Soc., Bot. 24: 438 (1888).
S. xanthoglossa Boiss. var. *typica* Eig., Palest. J. Bot. 3: 86 (1944).

Perennial herb, 35–100 cm, glabrous, dark bluish-green. Stems erect, ± rigid, angular, branched from base and expanding about below the middle into a branched thyrsoid or paniculate inflorescence. Lower leaves alternate or opposite; cauline leaves alternate; all leaves petiolate; leaves deeply lobed to pinnatisect, lobes linear, with margins entire to 2–5 lobed or denticulate; all leaves glabrous. Bracts linear, 1–2 mm. Cymes with peduncles 5–10 mm, glabrous. Flowers 3–5 in lax axillary cymes; pedicel ± 1mm, glabrous (in Iraq material). Calyx lobes oblong to oblong-obovate, 2–3 mm, obtuse, margins white scarious, glabrous. Corolla dark purplish red, to 5 mm, the upper lobes narrower at base. Staminode white, as

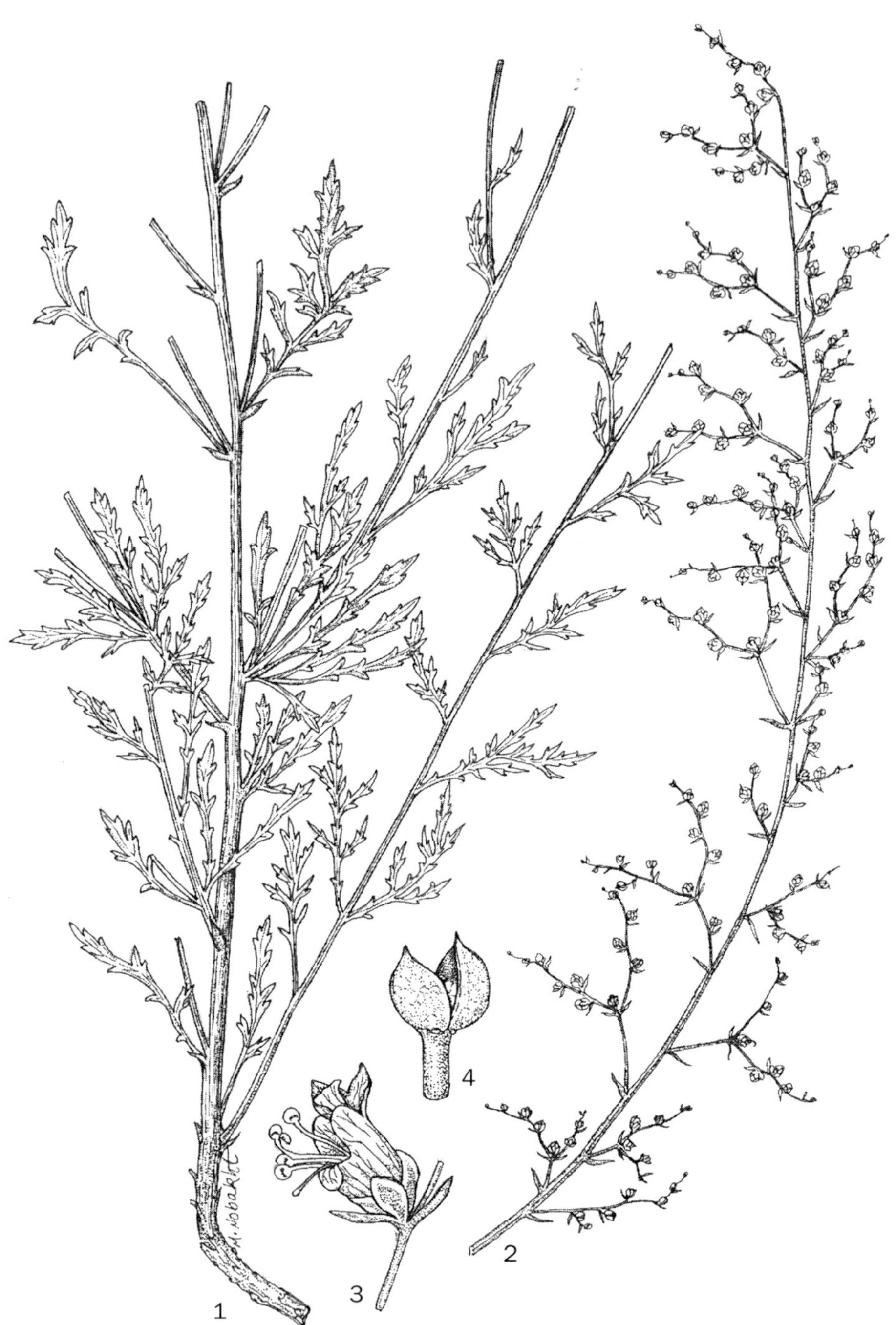

Fig. 27. **Scrophularia striata**. 1. Habit, lower part with leaves × ½; 2, habit inflorescence × ½; 3, flower × 5; 4, capsule × 5. Reproduced with permission from Flora of Iran 68: f. 80 (2011). Drawn by M. Nobakht. © Ministry of Jihad-e-Agriculture.

large as the upper corolla lobe, margin crenulate. Stamens exserted. Capsule globose, ± 3 mm long and wide, narrowed at apex, mucronate, glabrous.

HAB. In the mountains, limestone slopes, amongst rocks and stones, in wheat and corn fields and in remnants of *Quercus* forests; not uncommon; alt. 200–1400 m; fl.& fr. Mar.-July.

DISTRIB. **MAM**: Aqra, *Rawi* 11335 & 11508!; Atrush, *Mrs Bradshaw* 9042!; Dohuk, *Guest* 2287! 25 km from Zawita to Atrush, *Kaisi, Hamad & Hamid* 46089!; Sandur, *Kaisi & Khalid*! **MRO**: Rowandaz, *Guest* 586! **MSU**: 5 km N. of Darband, *Botany Staff* 43194! Darband-i-Khan, *Poore* 559! **FUJ**: Jebel Sinjar, Kursi, *Gillett* 10959!; Karya Sheikh Khamis, nr. Bilad Sinjar, Jun., *Field & Lazar* 599!; Mosul, *Lazar* 3393!; Sinjar, *Kaisi & Hamad* 49062!; Ain Sifni, *Salim Effendi* 2558!; Atrush, *Guest* 3628!; **FPF**: Khanaqin, *Guest* 1841! **FAR**: Erbil, *Guest* 2128!; 8 km E. Zurbatiyn, *Kaisi & Hamad* 46536!; 25 km N. of Erbil, *Barkley* 5634! **FKI**: Chamchamal, *Rawi & Gillett* 7585! Nr. Kirkuk, on rd. to Baba Gurgur oil wells, *Guest* 1351! Taq, *Omar et al.* 37137!; 6 km E. of Kirkuk, *Botany Staff* 42995! **DSD**: 44 km S. of Zubair, *Thamer* 47419! **LEA**: Amarah, Fakka-Laqlaq, *Rawi* 25834!

KASHISHAT (Kur.); a form of fodder plant (fide *Effendi* 2558).

Turkey, Armenia, Syria, Palestine, Jordan, Egypt (Sinai).

Scrophularia peyroni Post, Bull. Herb. Boissier 1: 28 (1893) has been distinguished from *S. xanthoglossa* on its smaller flowers and pruinose inflorescence (see Fl. Palaest. 3: 198, 1978). I have seen a single collection from Kurdistan & Mosul, 1841, *Kotschy* 341, identified as *peyroni*, but which falls within the range of flower size in *S. xanthoglossa* and lacks the pruinose peduncles; this collection is referable to *S. xanthoglossa*.

4. **Scrophularia sulaimanica** *S.A.Ahmad* in Kew Bull. 69(2): 9509 (2014). Type: Iraq, Kurdistan, Sulaimani Province, Hawraman Mts, between Darashesh and Hawar, 1069 m, 16 June 2012, *Saman A. Ahmad* 12-320 (SUFA, holo.; K!, MO, SUFA, iso.).

Perennial herb, sparsely to densely glandular, not glaucous. Stems (50–)75–145 cm, several to many arising from a woody caudex, erect to ascending, branching below the inflorescence. Leaves petiolate; petiole 35–70 mm. Lamina of basal leaves broadly lanceolate in outline, 20–50 × 10–25 mm, margins deeply incised to pinnatifid, sparsely to densely glandular; middle and upper cauline leaves alternate, petiolate, lanceolate, pinnatifid to deeply incised or coarsely dentate, gradually reduced in size upwards and becoming dentate. Inflorescence a branched thyrse; cymes 5–7-flowered, sparsely glandular. Bracts short petiolate, elliptic, 5–15 × 2–5 mm, becoming subulate-linear upwards; bracteoles similar to distal bracts but smaller. Peduncle 7–20 mm; pedicel slender, 2–4 mm, longer than subtending bracteoles; fruiting pedicel of central flower somewhat thick, 4–6 mm, sparsely glandular or glabrous. Calyx lobes orbicular to broadly ovate, in fruit, 3–4 mm, glabrous, margins scarious, undulate, lacerate. Corolla 6–7 mm, maroon; lateral lobes with white margins; upper lip not margined, emarginate, lobes suborbicular; lower lip white; corolla tube white, ± 3 mm. Staminode subsessile, broadly ovoid, not reaching notch of upper lip. Staminal filaments glabrous. Capsule globose, 3.5–5 mm in diameter, apiculate, glabrous.

HAB. Hawraman Mts, between rocks, on cliffs, and on eroded places; alt.1050–1990 m; fl. Jun.
DISTRIB. **MSU**: Hawraman Mts, between Darashesh and Hawar, *Saman A. Ahmad* 12-320 (type); Hawraman Mts, Lase Marf, 1895 m, *Saman A. Ahmad* 12–1077 (K, MO, SUFA).

Endemic. Recorded as extremely rare and known from the above specimens; its conservation status remains uncertain, currently assessed as Data Deficient (DD).

Scrophularia sulaimanica, is closely related to the Iranian endemic *S. valida* Grau and the Anatolian-Caucasian *S. thesioides* Boiss., but differs from the former in its solid, glandular stems, smaller maroon flowers and glabrous filaments, and from the latter in its glandular, terete stems, broadly ovoid staminodes and globose capsules.

5. **Scrophularia libanotica** *Boiss.*, Diagn. Pl. Orient. ser. 1, 12: 36 (1853) subsp. **libanotica** Grau in Fl. Iranica [K. H. Rechinger] 147: 251 (1981); Saeidi-Mehrvarz et al.in Fl. Iran [Assadi et al.] 68: 264 (2011).

S. urvilleana sensu DC., Ann. Sc, Nat. Ser. 2, 2: 252 (1834), non Wydl. (1928).
S. variegata M.B. var. *libanotica* (Boiss.) Boiss., Fl. Orient. [Boissier] 4: 418 (1879).

Fig. 28. **Scrophularia libanotica**. 1, habit; 2, leaf, underside; 3, inflorescense; 4, inflorescense detail; 5, flower; 6, capsule. 1 from *Balansa* 679; 2, 4 from *Samuelsson* 4664; 3 from *Kas, Nuri & Sarhang* 27381; 5, 6 from *Bornmüller* 1620. Scale bars: 1, 2, 3=3 cm; 4, 5, 6=5 mm. Drawn by © Juilet Beentje, Jan. 2016.

S. rupestris M.B. var. *libanotica* (Boiss.) Lall, Notes RBG Edinburgh 30: 136 (1970).
S. libanotica Boissier subsp. *libanotica* var. *libanotica* Lall & Mill in Fl. Turkey 6: 629 (1978).

Perennial herb; stems 15–60 cm, few arising from a woody caudex, erect, sparingly branching or unbranched below the inflorescence, glabrous to sparsely glandular-puberulent above. Leaves petiolate below to ± sessile above; petiole 0–17 mm; lamina ovate, rarely pinnatipartite, 10–45 × 8–25 mm, gradually reduced in size upwards, glabrous on both surfaces, base tapering, margins coarsely dentate to serrate to irregularly lobed. Inflorescence a branched thryse, not leafy; cymes 5–9-flowered, peduncle glandular; bracts and bracteoles linear to subulate. Pedicels stout, 2–7 mm, longer than subtending bracteoles, glandular; fruiting pedicel of central flower, stout, ± 2 mm, glandular. Calyx lobes broadly ovate, 2–2.5 mm, glabrous, rarely sparsely glandular at base, margins scarious. Corolla ± 5 mm, greenish; lateral lobes pinkish. Staminode broadly ovate to elliptical. Stamens ± exserted. Capsule globose, 3–4 mm in diameter, apiculate, glabrous. Fig. 28, 1–6.

HAB. Mountain slopes, shaded places, amongst *Quercus*; alt. 1000–2000 m; fl. May-Jun.
DISTRIB. **MAM**: Gara Dagh, *Rawi* 9262!, *Rawi* 9272!; Sarsang, slopes of Gara Dagh, *Chapman* 26427!; Bamerny, 20 km N.W. Sersang, *Kaisi & Hamad*!; Khantur Mt., N.E. of Zalko, *Rawi, Nuri & Tikriti* 29017!; ibid., 1490 m, *Rawi* 23429!, 23468!; ibid., *Rawi* 8638!; Ser Amadia, *Guest* 4991; ibid., *Guest* 3777! **MRO**: Karoukh Mt. (Doli Sarkai), *Kas, Nuri & Serhang* 27381!; southern slope of Karoukh Mt., *Kas, Nuri & Serhang* 27519!; Sefindagh, *Gillett* 8086!; Kuh Sefin, Schaklawa, *Bornmüller* 1620!; Saran village, *Kas, Nuri & Serhang* 27296!; Shaqlawa, *Wheeler-Haines* W782! **MSU**: Mela Kowa, *Rawi, Hosham & Nuri* 29573!; Zangla Bichik, Pir Omar Gudrun, *Rawi* 11530!; Pir Omar Gudrun, *Gillett* 7756!; Kamarspa, *Rawi* 22212!; Hawraman Mts., *Rawi, Hosham & Nuri* 29347!; ibid., *Gillett* 11852!; **MSU**: Jebel Khatchra nr. Bilad Sinjar, *Field & Lazar* 648!

A widely distributed and variable taxon that requires revision throughout its range. *S. variegata* M.Bieb. and its subspecific taxa found in Eastern Turkey, Talish, the Caucasus and Iran, is probably conspecfic with *S. libanotica* with which it differs mainly in its fewer flowered cymes, as is probably *S. sulaimanica* S.A.Ahmad which differs mainly in its larger maroon flowers. Iraq material identified as *S. variegata* subsp. *libanotica* and var. *libanotica* are referable to *S. libanotica*.

Turkey, Lebanon, Palestine, Iran.

6. **Scrophularia umbrosa** (*L.*) *Dumort.*, Fl. Belg. 37 (1827); Richardson in Fl Europ. 3: 219 (1972); Lall & Mill in Fl.Turkey [P. H. Davis] 6: 622 (1978); Grau in Fl. Iranica [K. H. Rechinger] 147: 235 (1981); Saeidi-Mehrvarz et al.in Fl. Iran [Assadi et al.] 68: 235 (2011); Qaiser, Khatoon & Hamidullah in Fl. Pakistan [Ali & Qaiser] 220: 243 (2015).

S. alata Gilib., Fl. Lit. Inch. 1: 127 (1782), invalid name; Boiss., Fl. Orient. 4: 399 (1879).
S. alata var. *cordata* Gilib.; Rawi in Dep. Agr. Iraq Tech. Bull. 14: 144 (1964).

subsp. **umbrosa**

Perennial herb, up to 80 cm (as measured on the only sheet from Iraq). Stems erect, usually unbranched, quadrangular, narrowly winged, glabrous. Leaves petiolate; petiole 5–30 mm; lamina ovate to elliptic, 25–35 × 8–15 mm, base cuneate, apex obtuse, margins crenate to crenulate, sometimes irregularly dentate, glabrous. Bracts linear, 1–2 mm. Inflorescence thyrsoid, forming almost half the length of stem, glabrous. Flowers 5–7 in lax axillary cymes; pedicel 6–7 mm, glabrous, winged, ± glandular. Calyx lobes suborbicular, 2–2.5 mm, obtuse, margins white scarious, glabrous. Corolla purplish red to brownish, 3–6 mm. Staminode transversely oblong, entire, smaller than the posterior corolla lobe. Stamens included. Capsule globose, 3–6 mm long, glabrous.

HAB. In the mountains with *Quercus* & *Crataegus*; not common; alt. 600–1800 m; fl. May.
DISTRIB. **MAM**: Aqrah, *Chapman* 26160!; Sersang, *Haines* 436! **MSU**: Azmir, *Omer & Karim* 38039!; 50 km from Sulimaniya, *Katih & Tikriti* 29743!

Europe, Turkey, Armenia, Iran, Afghanistan, Pakistan.

7. **Scrophularia macrophylla** *Boiss.*, Diagn. Pl.Orient. ser. 1, 12: 32 (1853); Boissier, Fl. Orient. 4: 413 (1879); Post, Fl. Syria, Palest. & Sinai 2: 293 (1933); Eig, Palest. J. Bot. Jerusalem 3, 2: 82 (1944); Rawi in Dep. Agr. Iraq Tech. Bull. 14: 145 (1964); Feinbrun-Dothan, Fl. Palaest. 3: 196 (1978).

Scrophularia guestii Eig, Palest. J. Bot. Jerusalem 3: 80 (1944). Type: Kurdistan, Arl Gird Dagh, 24.vii.1932, *Guest* 2943 (K!, holo. & iso.); Blakelock in Kew Bull. 4(4): 532 (1949).

S. rafidainsis Sawah ined. Type: Iraq, Dohuk Province, Raqaba above Sharanish, 1440 m, 24/6/1981, *Sawah & Musawi* 44256 (BUH!, holo.).

Perennial herb, up to 1 m. Stems erect, quadrangular, leafy, usually simple, expanding about the middle into a branched thyrsoid inflorescence, glabrous or obscurely puberulous except the inflorescence. Leaves opposite, in fascicles at nodes, petiolate; petiole 10–40 mm; lamina ovate to ovate-oblong to oblong-ovate, 20–90 × 13–60 mm, base cuneate to cordate, apex acute, margins coarsely dentate, glabrous or obscurely puberulous. Bracts and bracteoles, 1–2 mm. Inflorescence terminal, of axillary cymes, pedunculate, glandular. Flowers 5–7 in lax cymes; pedicel 4–13 mm, (longest in fruit), glandular. Calyx lobes broadly ovate, ± 3 mm, obtuse, with white scarious margins, glabrous. Corolla 6–7 mm. Stamens included. Capsule ovoid, 4–5 mm, acute to shortly beaked at apex, glabrous.

HAB. Rocky mountain side, under trees, in dry water steams and near water; common; alt. 700–1250; fl. May, Nov., fr. Jun.-Aug.

DISTRIB. **MAM**: Surlaf, *Al-Kaisi, Al-Khayat & Karim* 51061!; Aradin, 15 km W. of Byn, *Al-Kaisi & Hamad* 46049!; Sorroka vilage, 12 km N.E. Sarsang, *Al-Dabbagh & Hamad* 46156; N.E. of Rania, *Rawi & Serhang* 18304; Raqaba above Sharanish, *Sawah & Musawi* 44256 (type of *S. rafidainsis*) & ibid. 44255 (BUH). **MRO**: Haj Umran, *Al-Kaisi, Karim & Hamad* 49774 & 49776! **MSU**: Ahmad Aawa, *Khayat, Kaisi & Thamir* 46307A!; ibid., *Omar, Kaisi & Wedad* 49398!; Qara Dagh, around village, *Eig & Feinbrun* (cited in Eig, l.c.).

Grau in Fl. Iranica (p. 234) has placed *S. guestii* as synonym to *S. chlorantha* Kotschy & Boiss. *S. chlorantha* is found in Turkey and the Caucasus, and distinct in being a large stout plant with leaves up to 200 cm, often glandular beneath, and ascending cymes with up to 30 flowers. Attar in Fl. Iran 68: 232 (2011) cites two specimens from Kurdistan, which I have not seen, may be referable to *S. macrophylla*. *S. guestii* is a smaller plant with lax cymes that are not ascending, and with glandular peduncles and pedicels. It is similar to *S. macrophylla* and is treated under this taxon in this Flora. *S. guestii* has been treated as synonym under *S. scopolii* var. *scopolii* Hoppe ex Pers. by Lall (Fl. Turkey p. 616) a widespread species with several infraspecific taxa in Turkey. Material of *S. scopolii* var. *scopolii* that I have seen at K is similar to *S. macrophylla* in facies, but has serrate leaves (on the whole larger than *macrophylla*) and the whole plant is pubescent to puberulous. In my opinion, the *S. macrophylla* complex (including *S. chlorantha, S. macrophylla, S. scopolii* and *S. guestii*) form an aggregate of species distributed from Europe to Iran (and including Syria, Lebanon and Palestine) that requires a thorough study which is beyond the scope of this Flora.

Turkey, Syria, Palestine, Lebanon, Jordan.

Wheeler-Haines W1360 from Sarsang, (coll. 26/viii/1957), is a robust plant with glandular inflorescence is probably referable to *S. macrophylla*.

8. **Scrophularia kollakii** *S.A.Ahmad*, Harvard Pap. Botany 21(1): 93 (2016). Type: Iraq, Kurdistan, Sulaimani Province, Azmar Mt., on the road to Khamza village, dry sandy place, grassland, 25 April 2015, *S. A. Ahmad A. Hama, S. Babarasul, & S. R. Fayaq* 15–1227 (KBFH, holo. & iso.).

Perennial herb. Stems erect to ascending, several to many from woody caudex, 30–75 cm, terete below quadrangular above, glabrous, glandular. Basal leaves forming a rosette, petiole 1–4 cm; lamina ovate to ovate-oblong, 2–4 × 1–2 cm, crenate, sparsely to densely glandular; lower and middle cauline leaves ovate to oblong-lanceolate in outline, usually opposite, deeply incised to pinnatifid, 1.5–3 × 0.7–1.5 cm, sparsely to densely glandular; upper cauline leaves alternate, lanceolate, slightly pinnatifid, double serrate, 1–5 × 1–2 cm. Inflorescence (thyrsoid) branched, 20–45 cm; cymes 4–12-flowered, glabrous; peduncle 0.5–1.5 cm; lowermost bracts leafy, short petiolate, ovate, 10–15 × 2–4 mm, becoming elliptic upwards; bracteoles lanceolate, 1–2 × 0.4–0.7 mm; flowering pedicels slender, 1–5 mm, substantially longer than subtending bracteoles; fruiting pedicels of central flower somewhat thick, angled, 3–7 mm, sparsely glandular. Calyx lobes suborbicular to broadly reniform, 3–5 × 3.5–5 mm, glabrous; scarious margin undulate, lacerate, 1–1.6 mm wide, usually purplish or dark yellow. Corolla urceolate, 7–9 mm, dark pink; lateral lobes pinkish, truncate; upper lip emarginate, 3.5–5 mm, lobes suborbicular; lower lip pinkish, obtuse;

corolla tube white, c. 3 mm; staminode subsessile, obcordate, slightly exerted from corolla throat; filaments of fertile stamens, glandular; style glabrous. Fruit globose, 2-lobed, 3.5–5 mm in diameter, glabrous; seeds oblong, black, slightly curved, c. 2 × 1 mm, transversely sulcate.

HAB. On mountain, dry sandy place, in grassland, near road; alt. ± 1570 m; m; fl. Apr.
DISTRIB. Rare, known only from Azmar Mt., where it is restricted to a small area near the road to Khamza Village. **MRO.** Azmar Mt., on the road to Khamza village, *S. A. Ahmad A. Hama, S. Babarasul, & S. R. Fayaq* 15–1227 (KBFH, type); Azmar Mt., *S. A. Ahmad & S. R. Fayaq* 15–1448 (KBFH).

Endemic.

9. **Scrophularia kurdica** *Eig,* Palestine J. Bot., Jerusalem Ser. 3: 81 (1944). Type: Iraq, Kurdistan, Sulimaniya District, Pira Magrun [Pir-i-Mukurun Dagh], 19 ix 1933, *Eig & Duvdevani* (HUJ, syn.); ibid, Mergapan, 19 ix 1933, *Zohary & Amdursky* (HUJ, syn.). Rawi in Dep. Agr. Iraq Tech. Bull. 14: 145 (1964); Lall & Mill in Fl. Turkey [P. H. Davis] 6: 614 (1978); Grau in Fl. Iranica [K. H. Rechinger] 147: 223 (1981); Saeidi-Mehrvarz et al.in Fl. Iran [Assadi et al.] 68: 217 (2011).

(incl. subsp. **glabra** Grau in Fl. Iranica 147: 223 (1981).

Perennial herb, to 100 cm, glandular pubescent or pubescent or glabrous (subsp. **glabra)**, dark green. Stems erect, angular, leafy, usually simple or branching above. Leaves opposite, sessile to very short petiolate, ovate to narrowly ovate, 70–30 × 35–10, base subcordate to truncate, apex acute, margins dentate or serrate. Inflorescence leafy with axillary cymes; peduncles 15–20 mm, slender. Flowers 3–20; pedicel ± 1mm; Bracts linear to narrowly lanceolate, Calyx lobes broadly ovate to oblong-ovate, 2–3 mm, obtuse, margins narrowly white scarious, glabrous. Corolla dark purplish red or green, to 6 mm. Staminode wide at top. Stamens included. Capsule subglobose, 3–5 mm long, narrowed at apex, glabrous.

HAB. Rocky mountain side, in crevices, in shaded and moist locations; alt. 1200–2250; fl.& fr. Aug., Sep.
DISTRIB. **MRO**: Mergapan, *Zohary & Amdursky* (type HUJ); Ser Kurawa, *Gillett* 9718!; **MSU**: Pira Magrun [Pir-i-Mukurun Dagh], *Eig & Duvdevani* (type HUJ); Pira Magrun, *Thesiger* 1118 (BM); Hawraman *Rawi et al.* 19738!, 19754!, 19766!

Turkey, Iran, Iraq.

10. **Scrophularia nervosa** *Benth.,* Prodr. [A. P. de Candolle] 10: 303 (1846); Boissier, Fl. Orient. 4: 392 (1879); Rawi in Dep. Agr. Iraq Tech. Bull. 14: 145 (1964); Grau in Fl. Iranica [K. H. Rechinger] 147: 227 (1981); Saeidi-Mehrvarz et al.in Fl. Iran [Assadi et al.] 68: 210 (2011).

subsp. **boissierana** (Jaub & Spach) Grau in Fl. Iranica 147: 228 (1981).

S. boissierana Jaub. & Spach, Illustr. Fl. Orient. 3, 29 tab. 223.
S. orientalis Boiss. in Kotschy exs. non L.

Perennial herb, to ?60 cm, glabrous to pruinose and minutely glandular, dark reddish-brown. Stems erect, angular, leafy, simple or with sparse branching above. Leaves opposite, shortly petiolate; petiolate 8–15 mm; lamina lanceolate 35–65 × 5–12 mm, base cuneate, apex acute, margins serrate to crenate to crenulate to irregularly serrate or sometimes ± entire. Inflorescence aphyllus, in 3–7-flowered cymes. Bracts linear. Peduncles to 20 mm, pruinose to glandular; pedicels ± 5 mm, pruinose to glandular. Calyx lobes oblong to oblong-ovate, 2–3 mm, obtuse, margins narrowly white scarious, glabrous. Corolla dark purplish red or green, to 6 mm. Staminode absent. Stamens exserted. Capsule subglobose, 2–3 mm long, narrowed at apex, glabrous.

HAB. Rocky mountain side, in crevices, in shaded and moist locations; alt. 1200–2250; fl.& fr. Aug., Sep.
DISTRIB. **MSU**: Kamarspa, on rd. between Halabja & Tawela, *Rawi* 22184!; Tawela, *Rechinger* 10330 (BUH).

Grau cites a specimen collected by Haussknecht from Avroman [Hawraman] which he places under subsp. **nervosa**, differening in its broader leaves (up to 4 cm) with ± regularly serrate margins. I have not seen this specimen, but have seen the type of *nervosa* from Iran. Whereas the leaves are somewhat larger than in *S. boissierana*, there is little difference in leaf margins. With more material from Iraq, *S. boissierana* is likely to be conspecific with *S. nervosa.*

A near endemic found in Iran and Iraq.

11. **Scrophularia catariifolia** *Boiss. & Heldr.*, Diagn. Pl. Orient. ser. 1, 12: 36 (1853); Boissier, Fl. Orient 4: 407 (1879); Lall & Mill in Fl. Turkey [P. H. Davis] 6: 627 (1978); Grau in Fl. Iranica [K. H. Rechinger] 147: 238 (1981); Saeidi-Mehrvarz et al.in Fl. Iran [Assadi et al.] 68: 229 (2011).

> *S. rimarum* Bornm., Repert. Spec. Nov. Regni Veg. 7: 202 (1909). Type: Iraq, Kurdistan, Algurd Dagh, [Helgurd], 3000 m, 26/vi/1893, *Bornmüller* 1624 (?B, ?JE) p.p.; Lall & Mill in Fl. Turkey [P. H. Davis] 6: 627 (1978).
> *S. rimarum* Bornm. var. *farinea,* var. *glabrescens* & var. *pubescens* Bornm., Repert. Spec. Nov. Regni Veg. 7: 202 (1909).

Perennial herb, densely glandular-puberulent throughout, glands distinctly white (in Iraq plants). Stems 8–45(–60) cm, several arising from a woody caudex, ascending, sparingly branching below the inflorescence. Leaves petiolate to sessile; petiole 0–5 mm; lamina ovate, 8–30 × 5–25 mm, gradually reducing in size upwards, densely glandular, base cordate to tapering, margins coarsely dentate to serrate. Inflorescence thyrsoid, not leafy; cymes 5–7-flowered, glandular; bracts and bracteoles linear to subulate. Pedicel stout, 3–7 mm, longer than subtending bracteoles; fruiting pedicel of central flower stout, 3–5 mm, glandular. Calyx lobes broadly ovate, 3.5–4 mm, glandular and puberulent or sometimes glabrous (especially in fruit), margins scarious, undulate, ± lacerate. Corolla 6.5–7 mm, greenish brown to brownish red, tube somewhat ventricose, lateral lobes with white margins. Staminode reniform. Stamens ± exserted. Capsule globose, 5–6 mm in diameter, apiculate, glabrous. Fig. 29, 1–3.

HAB. Mountain side, rocky slopes, on rocks, and nr. streams; alt.2200–3100 m; fl. Jun.-Sep.
DISTRIB. **MRO**: Algurd Dagh, *Rechinger* 11433!; ibid, *Guest* 3051!; ibid., *Rawi & Serhang* 24830!; Algurd Dagh, Goum Tawerah, *Rawi & Serhang* 24719!; Qandil Range, Kermasur lake, *Rawi* 24131!; ibid., Baisar village, *Rawi & Serhang* 24169!; N.E. of Rania, *Rawi & Serhang* 18256!; top of Qandil Range, between Perrish & Bardanas, *Rawi & Serhang* 24579 & 24554!; southern part of Karoukh mt., *Kass, Nuri & Serhang* 27477!; N.E. of Qandil, *Rawi & Serhang* 24414!; Qandil Range, *Rawi & Serhang* 24073!

A species variable in its size of leaves and indumentum in its entire distribution range, but more so in Turkey than in N.W. Iraq. All Iraq plants are quite similar in habit and vegetative features, usually with densely glandular stems and leaves, and distinct white glands. *S. catariifolia* is a Turkish species found rarely in N.W. Iran, and is localised in N. Iraq as well.

Turkey, Iran.

12. **Scrophularia atroglandulosa** Grau in Fl. Iranica [K. H. Rechinger] 147: 244, tab. 173 (1981). Type: Iraq, Kurdistan, Arbil, N.E. Qandil, 2850–3100 m, 26.viii.1957, *Rawi & Serhang* 24438 (K!, holo).

Robust perennial herb, glandular-puberulent throughout. Stems 20–45 cm, several arising from a woody caudex, ascending, sparingly branching below the inflorescence. Leaves opposite or the upper alternate, sessile or petiolate; petiole 0.5–12 mm; lamina lanceolate-ovate in outline, 8–50 × 5–15 mm, reducing in size upwards, glandular, base tapering, margins coarsely dentate to shallow to deeply pinnatilobed; basal leaves shallow to deeply pinnatilobed. Inflorescence thyrsoid, not leafy; cymes 1–2-flowered, glandular; bracts and bracteoles linear to subulate. Pedicel 3–5 mm, longer than subtending bracteoles, glandular. Calyx lobes broadly ovate, 3–4 mm, with sessile glands at base, or sometimes glabrous (especially in fruit), margins white scarious, not entire, undulate, ± lacerate. Corolla 6–7 mm, greenish red, lateral lobes with white margins. Staminode oblong, adnate to the corolla. Stamens ± exserted, filaments glandular. Capsule globose to ovoid, 5–6 mm in diameter, apiculate, glabrous.

HAB. Mountain side, rocky slopes, on rocks, and nr. streams; alt.2200–3100 m; fl. Jun.-Sep.
DISTRIB. **MRO**: Arbil, N.E. Qandil, *Rawi & Serhang* 24438 (type); and ibid. 24409!; Algurd Dagh, *Gillett* 9564!; Kani Khanjer Khan, N.E. Helgord range, *Rawi* 24705!; top of Qandil range between Persian

Fig. 29. **Scrophularia catariifolia**. 1, habit × 1; 2, flower with calyx × 3; 3, capsule × 3. Reproduced with permission from Flora of Iran 68: f. 62 (2011). Drawn by M. Nobakht. © Ministry of Jihad-e-Agriculture.

border & Bardanas, *Rawi & Serhang* 24570!; Warshanka-Magar range, *Rawi & Serhang* 24324!; nr. Bermasand, *Rawi & Serhang* 24794!

A species quite similar to *S. catariifolia* but distinct in its habit and glandular pubescence.

Endemic to Iraq.

13. **Scrophularia crenophila** *Boiss.*, Diagn. Pl. Orient Nov. ser. 1, 7: 41 (1846); Grau in Fl. Iranica [K. H. Rechinger] 147: 232 (1981); Saeidi-Mehrvarz et al.in Fl. Iran [Assadi et al.] 68: 234 (2011).

Perennial herb, glandular-pubescent. Stems 40–75 cm, several to many arising from a woody caudex, erect to ascending, quadrangular, branching below the inflorescence. Leaves petiolate; petiole to 40 mm; lamina ovate, 15–130 × 10–60 mm, base cordate, apex acute, margins crenate, glandular-pubescent. Inflorescence thyrsoid; lower cymes 8–25-flowered, upper 2–5-flowered. Peduncle to 5–20 mm; pedicels slender, 3–15 mm, glandular to glabrescent. Calyx lobes elliptic, 3–4 mm, glabrous, margins white scarious, undulate, ± lacerate. Corolla 6–7 mm, tube greenish; Staminode ovate, retuse, greenish, ± equalling the lateral lobes. Stamens included; filaments glandular; Capsule ovoid, 5–6 mm in diameter, apiculate, glabrous.

HAB. By water, in shade, mountain slopes, in *Quercus* forest; comon; alt. 950–3250 m; fl. Jun.-Aug.
DISTRIB. **MAM**: Sersang, *Wheeler-Haines* W1229!; **MRO**: Pushtashan, N.E. Rania, lower slopes of Qandil range, *Rawi & Serhang* 26545!; Qandil Range, *Rawi & Serhang* 24468!; on rd. to Qandil, *Rawi* 24018!; Haji Omran, *Gillett* 12447!; **MSU**: Mt. Avroman, Biyara, *Gillett* 11747!; Biyara, *Rawi, Hosham & Nuri* 29483!; nr. Khormal, *Rawi, Hosham & Nuri* 29415!; Kopi Qaradagh, *Wheeler-Haines* W1519; Khalana, *Rawi* 13822!

Iran.

14. **Scrophuaria pegaea** *Hand.-Mazz.*, Ann. Nat. Hofmus. Wien 27: 403, t. 18 (1913); Lall & Mill in Fl. Turkey [P. H. Davis] 6: 615 (1978); Grau in Fl. Iranica [K. H. Rechinger] 147: 233 (1981).

Perennial herb, villous. Stems 40–75 cm, several arising from a woody caudex, erect to ascending, branched. Leaves ovate, 20–45 × 10–25 mm, base cordate, margins crenate, villous on both surfaces or sometimes the lower surface sparsely villous to glabrous; petiole to 40 mm. Inflorescence leafy; bracts ovate, gradually reduced in size upwards; the uppermost lanceolate to linear. Inflorescence a branched thyrse with cymes 5–many-flowered, glandular; peduncle 15–20 mm; bracteoles linear-lanceolate, 7–10 mm, glandular. Pedicel slender, 4–5 mm, longer than the subtending bracteoles, glandular. Calyx lobes ovate-oblong (broadly ovate in fruit), 3–5 mm, glandular, margins scarious, narrow, white. Corolla brownish purple to maroon, urceolate, 6–7 mm; lower lip green. Staminode transversely oblong, retuse. Capsule broadly ovoid, ± 4 mm in diameter, apiculate, glabrous.

HAB. By springs, on metamorphic rocks, in *Quercus* forest; alt. 1150–1440 m; fl. Aug., Sep.
DISTRIB. **MRO**: Erbil, Sairo, *Gillett* 9665! **FPF**: Kani Baz, *Rawi & Serhang* 24861!

E. Turkey.

15. **Scrophularia pruinosa** *Boiss.*, Diagn. Pl. Orient. Nov. Ser. 1, 12: 38 (1853); Boissier, Fl. Orient. 4: 416 (1879); Grau in Fl. Iranica [K. H. Rechinger] 147: 272 (1981).

S. *haematantha* Boissier & Heldr., Fl. Orient. [Boissier] 4: 415 (1875). Type: Iraq, Mt Avroman [Hawraman Mts] Haussknecht 734 (JE, W, syn.).
S. *pruinosa* subsp. *iraquensis* Eig, Palest. J. Bot. Jerusalem 3: 90 (1944). Type: Iraq, E.N.E. of Sari Hasan Beg, 24.vii.1932, *Guest* (BAG-2920! lectotype chosen here).
S. *pruinosa* subsp. *avromanica Sawah* ined.
S. *pruinosa* subsp. *aqransis Sawah* ined.

Perennial or biennial herb, up to 1 m. Stem arising from a rootstock, erect, quadrangular, leafy at base, simple or branching at base, expanding below the middle into a branched thyrsoid inflorescence, white pruinose and glandular. Leaves opposite, simple or sometimes in fascicles at nodes, petiolate, petiole 5–40 mm; lamina ovate to ovate-lanceolate, 15–40 × 10–30 mm, base cuneate tapering into the petiole, apex acute, margins coarsely dentate to deeply incised, pruinose and glandular-punctate. Bracts lanceolate to linear-ovate, ± 10 mm; bracteoles to 2 mm. Inflorescence terminal, composed of axillary cymes, pedunculate, glandular with yellow-brown sessile glands. Flowers 5–11 in lax cymes (sometimes with lateral branches developing in one plane, alternately in opposite directions, the main axis appearing zig-zag); peduncle ± 10 mm; pedicel 1–3 mm, glandular. Calyx lobes broadly ovate, ± 3 mm, obtuse, with wide white scarious margins, ± lacerate, glabrous. Corolla dark purplish red, 7–8 mm. Staminode obtuse. Stamens exserted. Capsule globose, 4–5 mm in diameter, shortly beaked at apex, glabrous.

HAB. On mountains, limestone scree slopes and hillsides, destroyed *Quercus* forest; alt. 1000–2500 m; fl. May-Aug.

DISTRIB. **MAM**: Aqra, 1100 m, *Rawi* 11415!; Zawita, *Rechinger* 10961 (BUH, type of *S. pruinosa* subsp. *aqransis*). **MAM/MRO**: Mergan, nr. Bardanas, *Rawi & Serhang* 24348! **MRO**: E.N.E. OF SARI HASAN BEG, *Guest* (BAG-2920! type of *S. pruinosa* subsp. *iraquensis*); Mt Marmarut, Rowanduz, 1200 m, *Guest* 2060!; Qandil Range, N.E. of Rania, 2300 m, *Rawi* 26825!; Kanirush [Kani Rash], *Serhang & Rawi* 26647!; Haj Omran, *Rawi* 42482!; Sefin, *Rawi* 9111! **MSU**: Penjwin, *Rawi* 8832!; Taweela, *Rawi* 22303!; Gwaija Dagh, *Rawi* 8879!; ibid., *Rawi & Gillett* 11696!; Avroman [Hawraman] Mt, *Rechinger* 10243!; Hawraman Mt, *Rawi et al. 19722*!; Mt Avroman, N. of Halabja, *Rawi* 22079 (BAG, type of *S. pruinosa* subsp. *avromanica* Sawah).

N.W. Iran.

16. **Scrophularia gracilis** Blakelock, Kew Bull. 4(4): 531 (1949). Type: Jindian nr. Rowanduz, on rocky cliff wall of Sayyid Taha's cave, 18.4.1932, *Guest* 2039 (K, holo.; G, S, iso.).

> *S. rimarum* Bornm. var. *pubescens* Bornm., Repert. Spec. Nov. Regni Veg. 7: 203 (1909). Type: Al Gurd Dagh, *Bornmüller* 1624 p.p.
> *S. ericiflora* Rech.f., Anz. Math.-Nat. Kl. Oestrr. Akad. Wiss. 87: 92 (1950). Type: Zakho, 3600 m, 16.iii.1947, *Leatherdale* 6! (BM, holo.).

Perennial herb; stems 10–35 cm, sparingly branched, several arising from a woody stock, whole plant glandular pubescent. Leaves alternate and opposite, petiolate; petiole 5–10 mm; lamina subreniform to ovate-orbicular, 10–25 × 8–20 mm, base tapering to truncate, apex acute, margins coarsely dentate to dentate-crenate, densely pubescent to glandular on both surfaces. Inflorescence leafy, in 3–7-flowered cymes. Bracts linear. Peduncles to 10 mm; pedicels ± 5 mm, glandular. Calyx lobes oblong-ovate, 2.5–3 mm, obtuse, margins white scarious, glandular. Corolla dark purplish to reddish-brown, 7–9 mm. Staminode absent. Stamens ± exserted. Capsule subglobose, 5–6 mm long, narrowed at apex, apiculate, glabrous.

HAB. Rocky mountain side, in crevices, on stone walls, in shaded and moist locations; alt. 1000–2100; fl. & fr. Jul.- Sep.

DISTRIB. **MAM**: 25 km N.E. of Zakho Masir village, *Rawi* 23446!; Sharanrush, N.E. Zakho, *Rawi* 23693; ibid, *Rawi et al.* 29050 & 29040; Lower Sawera Gauric valley, *Rawi* 23676; ibid, *Rawi & Rechinger* 16687; Khantur mt., *Rawi et al.* 23311 & 23367; Khantur mt., *Rawi, Tikriti & Nuri* 28988 & 28985! **MRO**: Jindian nr. Rowanduz, on cliff face by a spring at Sayyid Taha's cave, *Guest* 730! **MSU**: Hauraman Mts, *Rawi et al.* 19756!

Endemic to N. E. Turkey and N. Iraq.

17. **Scrophularia amadiyana** Ghaz. & Haloob, Kew Bull. 72: 17 (2017). Type: Upper Jazira District, Gali Mazurka, 2 km N.W. Amadiyah, 1700 m, 1/6/1977, *Al-Dabbagh & M. Jasim* 46915 (K, holo.; BAG, iso.).

> *S. amatiana* Sawah ined.

Perennial herb; stems 10–150 mm, simple, erect, almost 4-angled, with glandular-pubescent hairs. Leaves opposite, petiolate; petiole 3–5 mm; lamina reniform or suborbicular-ovate, 8–10 × 6–12 mm, base truncate or rounded, margins dentate, densely glandular-puberulent. Inflorescence leafy, in 1–2-flowered cymes. Lower bracts reniform or ovate-suborbicular with irregularly dentate margins; middle snd upper bracts ovate-lanceolate, with entire margins; petioles of bracts 0.5–3 mm. Bracteoles linear, 2–3 mm long. Peduncles 6–6.5 mm; pedicels 2–3 mm, almost equal to bracteoles, glandular-hairy. Calyx lobes oblong-lanceolate, 3–3.5 mm long, acute-obtuse, without scarious margins, glandular on both surfaces. Corolla purple or red, 7.5–9.5 mm, lobes subequal, upper and lateral lobes orbicular, median broadly ovate, smaller than the upper lobe. Staminode absent. Stamens ± exserted. Capsule globose, 3.5–4 × 3.5–4 mm, apiculate, glandular-pubescent. Seeds oblong, 0.9–1.3 × 0.4–0.6 mm, reticulate. Fig. 30, 1–9.

HAB. Rocky mountain slope; alt. ± 1700 m.

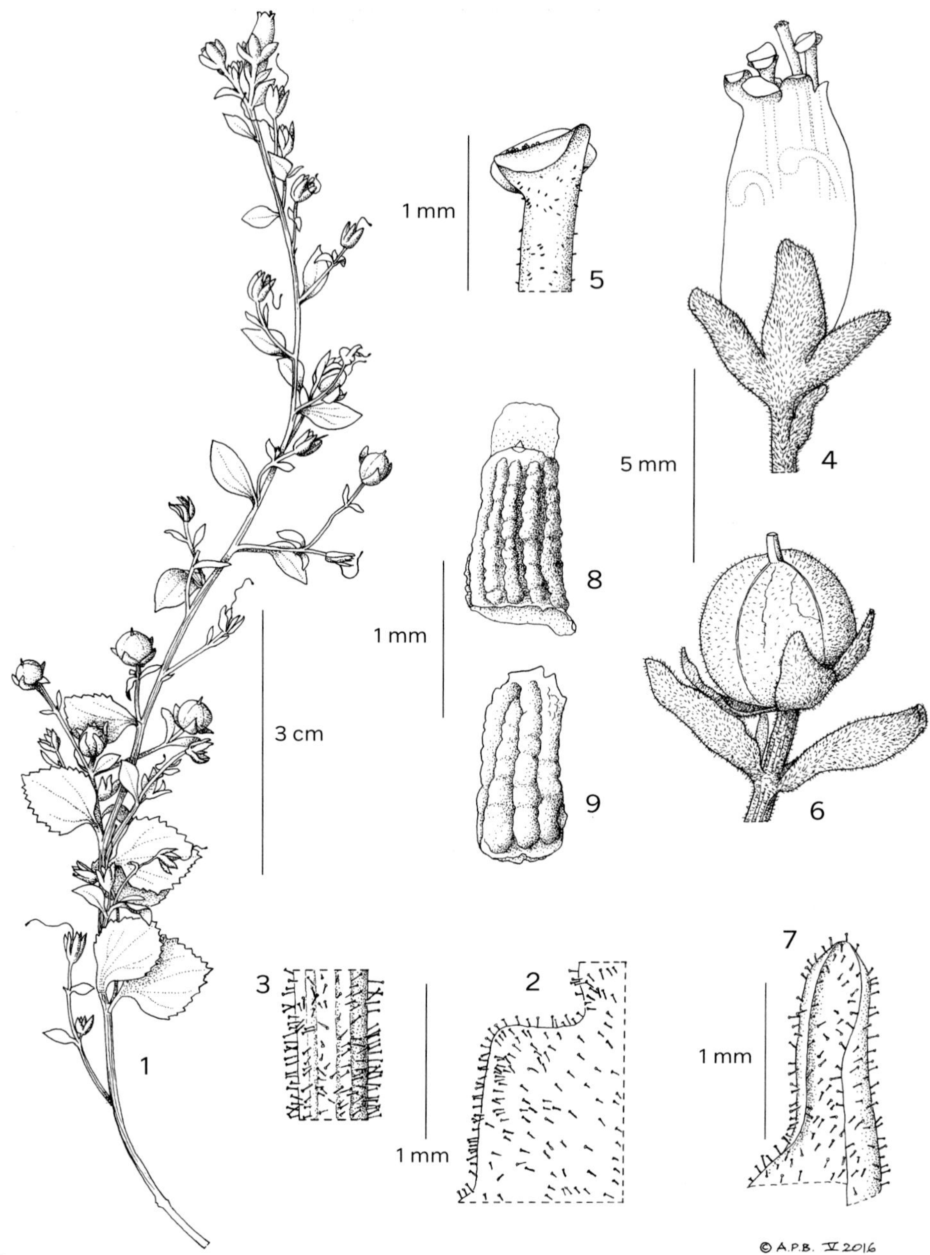

Fig. 30. **Scrophularia amadiyana**. 1, habit; 2, indumentum on leaf (abaxial surface and margin); 3, indumentum on stem; 4, flower (showing two anthers inside tube); 5, anther and distal portion of filament; 6, capsule with calyx and bracteoles; 7, tip of calyx lobe showing indumentum on outer and inner surfaces; 8 & 9, seeds. 1–9 from *Al-Dabagh & Jassim* 46915. Drawn by © A.P. Brown, 2017.

DISTRIB. **FUJ**. Upper Jazira District, Gali Mazurka, 2 km N.W. Amadiyah, *Al-Dabbagh & M. Jasim* 46915 (BAG; K; type of *S. amatiana*).

Scrophularia amadiyana, is closely related to *S. gracilis* Blakelock and *S. kotschyana* Benth. but differs from *S. gracilis* which is an endemic Iraqi species, in its oblong calyx lobes, without scarious margins, glandular on both surfaces, and smaller glandular capsules. It differs from *S. kotschyana* which is distributed in Anatolia, Caucasus and western Transcaucasia in its erect stems, smaller, densely glandular reniform to suborbiculate-ovate leaves and 1–2-flowered cymes.

18. **Scrophularia hypericifolia** *Wydler*, Mém. Soc. Phys. Genève 4: 166, t. 5 (1828); Boissier, Fl. Orient. 4: 421 (1879); Post, Fl. Syria, Palest. & Sinai 2: 297 (1933); Eig, Palest. J. Bot. Jerusalem 3, 2: 91 (1944); Rech., Fl. Lowland Iraq 547 (1964); Rawi in Dep. Agr. Iraq Tech. Bull. 14: 145 (1964); Feinbrun-Dothan, Fl. Palaest. 3: 200 (1978).

Scrophularia syriaca Benth., Prodr. [A. P. de Candolle] 10: 316 (1846).

Small shrub, 25–40 cm. Stems and branches white to greyish white, glabrous, smooth, stout, branching divaricately, the lower spreading, the upper erect to erect-horizontal; branches persistent, becoming spinescent. Upper stem ending in simple raceme or few-flowered cymes. Leaves alternate, petiolate; petiole 1–4 mm; lamina fleshy, oblong to ovate-oblong to obovate, 6–20 × 3–8 mm, base tapering into the petiole, apex obtuse, margins entire to 1–2-denticulate, glabrous. Cymes to 3-flowered; pedicels ± 1 mm. Bracts linear; bracteoles minute. Calyx rounded, ± 1.5 mm, margins white scarious, entire, glabrous. Corolla purplish, ± 3 mm, lobes small. Staminode linear, acute. Stamens exserted. Capsule depressed-globose, ± 4 mm in diameter, shortly mucronate, glabrous. Fig. 31, 1–4.

HAB. In desert, on "haswa" plain, on sand and gravel, along banks of stony bed; common in desert; alt. 40–150 m; fl. Mar.; fr. remaining on stems until Sep.

DISTRIB. **DLJ**: 6 km E. of Thirthar, *Rawi & Gillett* 7156!; nr LakeThirthar, *Barkley & Abbas* 3622!; 70 km N.W. of Faluja, *Chakravarty & Rawi* 30343!; 70 km N.W.N. of Faluja, *Rawi* 20256! **DWD**: E. of Shithatha, nr. Bahr al Milh, *Rawi* 26922!; c. 70 km S.W. by W. of Karbala, *Guest, Rawi & Rechinger* 19476!; c. 55 km WSW Karbala, 70 m, *Guest, Rawi & Long* 14015!; ibid., *Rawi* 30783!; **DSD**: 12 km W. of Karbala, 50 m, *Chakrawarty & Rawi* 29773!; 15 km W. of Karbala, *Rawi, Alizzi & Omar* 36235!; 40 km E. of Busaiya, 90 m, *Rawi* 26020!; Bisaiya Abu Ghor, *Fawzi, Hazim & Hamid* 38961!; 35 km from Najaf, *Fauzi, Hazim & Hamid* 38598!; 30 km N. of Najaf, 60 m, *Rawi & Serkakia* 16271!; nr. Salman, 210 m, *Guest, Rawi & Rechinger* 18817!; Salman, *Rawi & Gillett* 6193!; Abu Ghrar, *Guest & Ibrahim* 15300!; 5 km S. of Busaiya, *Weinert et al.* 47593!; 7 km W. by N. of Shabicha, *Guest, Rawi & Rechinger* 19249!; 2 km S. of Busaiya, *Omar et al.* 43989!; ouside city on Baghdad rd., *Wheeler-Haines* W703!;

ARGED (*Rawi & Gillett* 7156); ARAIFIYĀN (*Guest & Ibrahim* 15300). Recorded to be eaten by sheep (*Rawi & Gillett* 7156).

Sinai, Saudi Arabia, Syrian desert, Palestine, Jordan.

SPECIES DOUBTFULLY RECORDED

Scrophularia amplexicaulis *Benth.*, Prodr. [A. P. de Candolle] 10: 310 (1846).

Scrophularia hajariana Parsa, Kew Bull. 217 (1948).

Rawi in Dep. Agr. Iraq Tech. Bull. 14: 144 (1964) & Lall & Mill in Fl. Turkey 6: 611 (1978) have recorded this species from N. Iraq, but without locality. It is found in Turkey, the Caucasus and Iran. I have not seen any specimens from Iraq.

3. **ALBRAUNIA** Speta

Bot. Jahrb. Syst. 103(1): 32 (1982); Fischer in Kubitzki (ser. ed.), J.W. Kadereit (ed.), Fam. & Gen. Vasc. Pl. 7: 378 (2004)

Shahina A. Ghazanfar

Annual herbs. Leaves oblong, obtuse, base attenuate, lower leaves petiolate, the upper shortly petiolate. Flowers in racemes with short pedicels. Calyx lobed with oblong to linear lobes, ciliate, as long as or slightly longer than corolla tube, persistent and covering capsule. Corolla 2-lipped; upper lip bifid, the lower shortly trifid; tube shortly spurred. Capsule cylindrical, many-seeded. Seeds ovoid to ellipsoid, irregularly cristate or reticulate, tuberculate.

Three species found in Turkey, Iran and Iraq; 2 species in Iraq.

Fig. 31. **Scrophularia hypericifolia**. 1, habit; 2, habit fruiting; 3, flower side view; 4, capsule. 1, 3 from *Rawi & Gillett* 6193; 2 from *Guest, Ibrahim & Rawi* 15300; 4 from *Guest, Rawi & Rechinger* 11817. Scale bars: 1, 2=3 cm; 4, 5=5 mm. Drawn by © Juilet Beentje, Mar. 2016.

Corolla bright-blue, lower lip orange; tube white with 2 orange stripes;
　　seeds tuberculate . . . . .1. *A. fugax*
Corolla reddish-purple, tube yellow, upper lip with purple stripes, lower lip
　　paler; seeds not tuberculate . . . . 2. *A. psilosperma*

Speta, F. (1982). Drei neue Antirrhineen-Gattungen aus de Orient: *Holzneria, Hueblia* und *Albraunia* (Scrophulariaceae). Bot. Jahrb. Syst. Pflanzengeschichte und Pflanzengeographie.

1. **Albraunia fugax** (*Boiss. & Noë*) *Speta,* Bot. Jahrb. Syst. 103(1): 35 (1982); Podlech & Iranshahr in Fl. Iranica [K. H. Rechinger] 180: 15 (2015); Haloob & al., Illustr. Fl. Lowland Iraq 1: 130 (2016).

Antirrhinum fugax Boiss., Diagn. Pl. Orient. ser. 2, 3: 160 (1856).
A. ceratotheca Nábělek, Spisy Přír. Fak. Masarykovy Univ. 70: 31 (1926); Rawi in Dep. Agr. Iraq Tech. Bull. 14: 142 (1964); Rechinger, Fl. Lowland Iraq: 546 (1964).

Dwarf annual herb to 3 cm high. Stems simple or branched from base, stout, densely hairy. Leaves sessile, greenish above, purplish below; lower leaves oblong, 25–35 × 5–7 mm, upper leaves smaller than lower ones, leaves with entire margins, glandular to ± glabrous. Racemes 1–12-flowered, axis glandular-pilose. Flowers sessile, solitary, axillary. Calyx lobed, 5–6 mm, glandular-hairy to glabrescent, lobes elongating to 25 mm and becoming accrescent in fruit. Corolla 2-lipped, 11–12 mm, blue-purple, palate orange-yellow, tube white with 2 orange lines, glandular-pilose. Capsule erect, cylindrical, mucronate, many-seeded, scabrid to pilose. Seeds black-brown, cylindrical-obovate, tuberculate. Fig. 32, 1–3. (See also Haloob & al. for illustr.).

HAB. Dry foothills; alt. ± 70 m; fl. Mar.-Apr.
DISTRIB. Foothills, alluvial plains and southern marsh districts of S.E. Iraq. **FPF**: Kirkuk, Khanaqin, *Rechinger* 9043 (W); Diyala, Jabal Hamrin, *Haines* 2157! **LSM**. Shatt at Tibb, 70 km N. Amarah, *Rechinger* 15849 (W, cited in Rechinger l.c.)

Turkey, Iran.

2. **Albraunia psilosperma** *Speta,* Bot. Jahrb. Syst. 103(1): 39 (1982). Type: Iraq, Mosul Liwa, Jabal Makhul nr. Ain Dibis, 480 m, *Rawi & Gillett* 7204 (K, holo. [K001070125]); Podlech & Iranshahr in Fl. Iranica [K. H. Rechinger 180: 15 (2015).

Annual herb. Stems simple or branched, 1–5 cm, scabrid, eglandular. Leaves sessile, oblong lanceolate to spatulate, 13–28–35 × 3–7 mm, apex obtuse, margins entire, sparsely glandular-pilose. Racemes 3–9-flowered. Calyx lobed, 4–8 mm, lobes linear, elongating to 20 mm in fruit, ciliate. Corolla 2-lipped, 7–10 mm, reddish-purple, tube yellow, upper lip with purple stripes, lower lip paler. Capsule cylindrical, 4–6 mm. Seeds not tuberculate, glabrous.

HAB. Occasional in dry mixed shrubland on lower hills; alt. ± 480 m; fl. Apr.
DISTRIB. Plains and foothills of N. Iraq. **FKI**: Kirkuk, *Rawi & Gillett* 7392; Jabal Makhul nr. Ain Dibis, *Rawi & Gillett* 7204 (type).

Endemic.

4. **KICKXIA** Dumort.

Fl. Belg. (Dumortier) 35 (1827) (1827); Fischer in Kubitzki (ser. ed.), J.W. Kadereit (ed.), Fam. & Gen. Vasc. Pl. 7: 379 (2004)

Shahina A. Ghazanfar

Annuals (in Iraq) and perennials. Leaves alternate, petiolate. Flowers pedicellate, solitary, axillary or in bracteate racemes. Calyx 5-lobed to base, lobes ± equal. Corolla 2-lipped, spurred; upper lip erect, 2-lobed, lower lip 3-lobed. Stamens 4, didynamous, included. Capsule globose to ovoid, 2-locular, dehiscence by a circumsessile lateral lid or capsule indehiscent. Seeds many, rugose or tuberculate.

24 species distributed from Europe to Central Asia, Iran to Pakistan, N. Africa, Palestine, Lebanon, Syria, Arabian Peninsula; 3 species in Iraq.

Fig. 32. **Albraunia fugax**. 1, habit × ½; 2, flower × 4½; 3, capsule × 4½. **Kickxia aegyptiaca**. 4, habit; 5, portion of stem showing indumentum; 6, flower × 2. 1–3, reproduced with permission from Flora of Iran 68: f. 92 (2011), © Ministry of Jihad-e-Agriculture. 4–6, reproduced with permission from Feinbrun-Dothan, Fl. Palaestina 3: Plates, f. 305 (1977). Drawn by Illana Ferber. © The Israel Academy of Sciences and Humanities.

Kickxia commemorates the Belgian botanist Jean Kickx (1775–1831), professor of botany and pharmacy and author of *Flora Bruxellensis* (1812).

Medhanie Ghebrehiwet (2000). Taxonomy, phylogeny and biogeography of *Kickxia* and *Nanorrhinum* (Scrophulariaceae). Nord. J. Bot. 20(6): 655–689.
Sutton, D.A. (1988). A revision of the tribe Antirrhineae. Oxford University Press.

1. Annual herb, glandular-pubescent to white-villous; pedicels not rigid;
     capsule puberulent to velutinous, glandular or not 2
   Perennial subshrub, densely glandular-pubescent; pedicels rigid;
     capsule villous. . . . . 2. *K. aegyptiaca*
2. Pedicels filiform, curving upwards below the flowers; calyx lobes not
     accrescent, glabrous; capsule puberulent to villous . . . . *K. elatine*
   Pedicels not filiform, erect to patent; calyx lobes accrescent, glandular-
     ciliate to villous; capsule glandular-pubescent at apex . . . . *K. spuria*

1. **Kickxia elatine** (*L.*) *Dumort.*, Fl. Belg. (Dumortier) 35 (1827); Boulos and Snogerup in Fl. Egypt 3: 66 (2002); Podlech & Iranshahr in Fl. Iranica [K. H. Rechinger] 180: 21 (2015); Taifour & El-Oqlah, Pl. Jordan Annot. Checklist 118 (2017).

Antirrhinum elatine L., Sp. Pl. 2: 612 (1753).
Linaria sieberi Rchb., Fl. Germ. Excurs. 374 (1830–1832); Rawi in Dep. Agr. Iraq Tech. Bull. 14: 142 (1964).
Kickxia sieberi (Rchb.) Dörfl. & Allan, Bull. Dept. Sc. & Indust. Research, N. Z. No. 83 (Handb. Nat. Fl. N. Z.:Bot. Div. Publ. No. 4) 300 (1940); Rechinger, Fl. Lowland Iraq: 544 (1964).

Annual herb. Stems slender, much-branched, prostrate to decumbent to ascending, up to 60 cm long, white-villous, hairs spreading. Basal leaves ovate to oblong-ovate to suborbicular, 15–25 × 10–20 mm, base hastate, apex acute to obtuse, margins usually entire; upper leaves ovate-lanceolate to ovate-hastate 3–4 × 1–2 mm; petiole 3–10 mm, patent. Flowers axillary; pedicels filiform, longer than most of the leaves, curving upwards below the flowers. Calyx deeply 5-lobed, lobes 4–6 mm, lanceolate with narrow scarious margins, glabrous. Corolla 2-lipped extending at the base into a long spur, 6–10 mm long including the spur, yellow, purple on the inside of the upper lip and base of the lower lip, hairy on the outside, palate hairy inside; spur slender, 3–4 mm. Anthers ciliate. Capsule globose to subglobose, 3.5–5 mm, puberulent to velutinous.

HAB. Low hills and mountains; alt. ±1160 m; fl. Mar.-Apr.
DISTRIB. **MSU**: Sersang, *Haines* 409!; Mosul, *Kotschy* 448; Diyalah district, *Rechinger* 9731 (W, cited in Fl. Lowland Iraq & Fl. Iranica).

A widespread species of which very little material is present for Iraq. Podlech & Iranshahr place Iraq species under subsp. *crinita* (Mabille) W.Greuter, Boissiera 13: 108 (1967) [syn. *Linaria crinata* Mabille, Rech. Pl. Corse 1 : 30 (1867).

N. and N.E. Africa, Egypt, Arabian Peninsula, Europe, Turkey to C. Asia and Pakistan.

2. **Kickxia aegyptiaca** (*L.*) *Nábělek*, Spisy Přír. Fak. Masarykovy Univ. 70: 31 (1926); Feinbrun-Dothan, Fl. Palaest. 3: 184 (1978); Collenette, Wild Flow. Saudi Arabia: 675 (1999); Taifour & El-Oqlah, Pl. Jordan Annot. Checklist 117 (2017).

Antirrhinum aegyptiacum L., Sp. Pl. 2: 613 (1753).
Linaria aegyptiaca Dum.Cours., Bot. Cult. 2: 92 (1802).
Elatinoides aegyptiaca (L.) Wettst., Nat. Pflanzenfam. [Engler & Prantl] 4(3b): 58 (1891).

Subshrub, 20–50 cm, densely glandular-pubescent. Stems branched from base, decumbent to ascending, herbaceous at first, becoming hard, woody and spinescent later. Lower leaves ovate to oblong-ovate, 10–12 × 7–10 mm, base rounded, apex acute to obtuse, margins entire; upper stem leaves smaller than lower leaves, hastate, 3–6 × 2–3 mm; petiole 5–8 mm. Flowers axillary; pedicels stout, curving upwards, below flowers, 15–20 mm. Calyx deeply 5-lobed, lobes 3–5 mm, linear-lanceolate. Corolla 2-lipped, spurred, yellow, 15–20 mm long including the spur; spur slim, slightly curved inwards. Anthers ciliate. Capsule globose to subglobose, 5 mm across, villous. Fig. 32, 1–3.

HAB. In dry and desertic areas; alt. ± 700 m; fl. Feb.-Mar.

DISTRIB. **DWD**: Wadi Al Walaj, 75 km S.W. Rutbah, *Rechinger* 9935 (cited in Fl. Lowland Iraq, p. 543).

N. & N.W. Africa, Sinai, Palestine, Jordan, Syria, Saudi Arabia.

3. **Kickxia spuria** (*L.*) *Dumort.*, Fl. Belg. (Dumortier) 35 (1827); Feinbrun-Dothan, Fl. Palaest. 3: 187 (1978). Podlech & Iranshahr in Fl. Iranica [K. H. Rechinger 180: 23 (2015); Taifour & El-Oqlah, Pl. Jordan Annot. Checklist 118 (2017).

Antirrhinum spurium L. Sp. Pl.: 612 (1753).

Annual herb. Stems slender, branched, prostrate to decumbent to ascending, up to 50 cm long, glandular-pubescent. Basal leaves ovate to suborbicular, 10–35 × 8–30 mm, base cordate to rounded, apex acute to obtuse, margins usually entire; upper leaves ovate-lanceolate; petiole 2–10 mm. Flowers axillary; pedicels 3–24 mm, erect to patent, villous. Calyx deeply 5-lobed, lobes accrescent, margins narrow scarious, glandular-ciliate to villous, base cordate. Corolla 2-lipped, extending at the base into a long spur, 6–8 mm long, yellow, purple on the inside of the upper lip and base of the lower lip, glandular villous or eglandular villous on the outside. Capsule globose to subglobose, 3.5–4 mm, glandular-pubescent at apex. (No material at K, description from Iran material). Fig. 33, 1–3.

According to Podlech & Iranshahr in Fl. Iranica (p. 24) Iraq material falls under subspecies *integrifolia* (Brot.) R.Fernandes in Heywood, Bot. J. Linn. Soc. 64: 74 (1971).

HAB. Mountain slopes, gravel and sandy clay depression, in between small rocks; alt. ± 1000 m; fl. Apr.
DISTRIB. Mountains of N. Iraq. **MSU**: Mosul, *Rechinger* 12179! (W); above Sersang *Rechinger* 11910! (W); Diyalah, Jabal Hamrin, *Sutherland* 20, 560 (cited in Fl. Iranica).

Iran.

5. **LINARIA** Mill.

Gard. Dict. Abr. ed. 4: (1754); Sutton, Rev. Tribe Antirrhineae 260 (1988); Fischer in
Kubitzki (ser. ed.), J.W. Kadereit (ed.), Fam. & Gen. Vasc. Pl. 7: 380 (2004)

Ali Haloob & S.T. Al-Kaisi

Annual, biennial or perennials herbs, glabrous or glandular-pubescent to villous. Leaves sessile to petiolate, simple, entire, opposite, alternate (above) or whorled (below). Inflorescence spicate, racemose, or capitate. Flowers sessile or pedicellate. Calyx 5-lobed, adaxial lobe usually longest. Corolla white, yellow, light purple or pinkish-purple, palate usually yellow, 2-lipped; tube base spurred, spur conical to linear; upper lip 2-lobed, almost erect, lower 3-lobed, with a palate, almost closed at throat. Stamens 4, didynamous, anterior pair longer. Ovary globose to ovoid. Stigma 2-lobed. Capsule glabrous, oblong or globose, dehiscing apically. Seeds numerous, winged or not, flat, discoid, or reniform or angled, reticulate, tuberculate or smooth.

The genus includes 187 species, native to temperate Eurasia, Canary Islands to N. Africa; 12 species in Iraq.

Linaria, from Lat. *linum*, flax and -*aria* (resembling).

1. Annual; leaves at least in lower part opposite or whorled. 2
   Perennial; leaves alternate 6
2. Corolla 16–20 mm long (incl. spur); spur 9–12 mm, curved . . . . .3. *L. chalepensis*
   Corolla less than 11 mm long; spur up to 5.5 mm, straight 3
3. Inflorescence a raceme; corolla yellow; spur longer than 2 mm 4
   Inflorescence capitate; corolla white or blue; spur less than 2 mm 5
4. Calyx lobes 4–5 mm; corolla 8–11 mm; seed angled, without wing, cristate . 1. *L. tenuis*
   Calyx lobes 2–4.5 mm; corolla 4.5–9 mm; seeds discoid, winged,
       tuberculate, papillose in centre . . . . 2. *L. simplex*
5. Calyx glabrous; corolla white with yellow palate and purplish veins; seed
       angled. . . . . 5. *L. albifrons*

Fig. 33. **Kickxia spuria**. 1, habit × 1; 2, flower × 5; 3, seed × 15. Reproduced with permission from Feinbrun-Dothan, Fl. Palaestina 3: Plates, f. 310 (1977). Drawn by Esther Huber. © The Israel Academy of Sciences and Humanities.

Calyx glandular-pubescent; corolla blue or blue with yellow palate; seed
    discoid, winged. . . . . 4. *L. micrantha*
6. Corolla > 25 mm 7
   Corolla < less than 25 mm. 9
7. Bract deflexed; calyx lobes 10–15 × 4–6 mm, glabrous. . . . . 12. *L. grandiflora*
   Bract straight; calyx lobes equal to or less than 9 × 4 mm, hairy 8
8. Leaves longer than 40 mm; bract 10–20 mm long; calyx lobes 3–4 mm
    broad; corolla throat ± 10 mm wide . . . . 10. *L. pyramidalis*
   Leaves shorter than 35 mm; bract up to 5.5 mm; calyx lobes less than 3
    mm broad; corolla throat 6–8 mm wide. . . . . 11. *L. confertiflora*
9. Leaves less than 3 mm wide; corolla 10–12 mm, spur 3–4 mm . . . . 9. *L. iraqensis*
   Leaves more than 3 mm wide, corolla more than 17 mm, spur longer than 6 mm . . 10
10. Leaves with acuminate apex; calyx lobes without membranous
    margins. . . . . 8. *L. genistifolia* subsp. *qandilensis*
   Leaves with acute apex; calyx lobes with membranous margins 11
11. Leaves with 3–7 veins; bracts 2–3 mm, with membranous margins,
    shorter than or as long as pedicels; seeds with testa smooth . . . . 6. *L. fastigiata*
11– Leaves with 3(–5) veins; bracts 3–7 mm, without or with narrow
    membranous margins, longer than pedicels; seeds papillate in middle . 7. *L. kurdica*

1. **Linaria tenuis** *Spreng.*, Syst. Veg. ed. 16, 2: 795 (1825); Post, Fl. Syria, Palest. & Sinai
ed. 1: 586 (1883–1896); Feinbrun-Dothan, Fl. Palaest. 3: 191, pl. 320 (1978); Boulos in Fl.
Egypt 3: 63 (2002); Collenette, Wild Flow. Saudi Arabia: 680 (1999); Taifour & El-Oqlah, Pl.
Jordan Annot. Checklist 119 (2017).

    Antirrhinum tenue Viv. Fl. Libyc. Spec.: 33 (1824).
    Linaria ascalonica Boiss. & Kotschy, Diagn. Pl. Orient. ser. 2, 3: 165 (1856); Rawi in Dep. Agr. Iraq
        Tech. Bull. 14: 143 (1964); Rechinger, Fl. Lowland Iraq: 545 (1964).
    L. hellenica Turrill, Kew Bull. 10(3): 356 (1955).
    L. tenuis var. *ascalonica* (Boiss. & Kotschy) Feinbrun, Israel J. Bot. 25(1–2): 81 (1976).
    L. tenuis var. *brachyloba* (Post) Täckh. & Boulos, Publ. Cairo Univ. Herb. 5: 97 (1974).

Annual herb; stems erect, 10–35 cm, glabrous. Leaves glabrous, alternate except in lower
part, oblong-linear, 3–20 × 1–3 mm, apex obtuse-subacute; upper leaves linear-filiform,
20–50 × 0.5–1.5 mm, apex obtuse-acute. Inflorescence lax racemes, sometimes become
dense at the apex, 3–12-flowered. Bracts linear-subulate, 2–4 × 0.5 mm, as long as or longer
than pedicels, glabrous. Pedicel 2–4 mm long, with glandular hairs. Calyx lobes oblong-
lanceolate, 4–5 × ± 1 mm, shorter than corolla, apex acute-obtuse, margins membranous,
glabrous, ± glandular. Corolla 8–11 mm, yellow, upper lip 3–4 mm, lower lip 3–4 × 3 mm,
spur conical, straight, 3–5.5 mm, corolla glabrous with tomentose hairs on palate. Capsule
glabrous, globose, 3.5–5 × 2.5–4 mm long, slightly longer than calyx. Seed black, angled
subglobose, 0.7–0.8 × ± 0.5, tuberculate. Fig. 34, 1–4.

HAB. Sandy desert, sandy clay in depression, sandy soil & sandy gravelly soil; alt. 400–500 m.; fl. & fr.:
Feb.-Apr.
DISTRIB. Occasional in the desert region of Iraq. **DLJ**: Rawah-Sinjar 115 S.W. of Sinjar, *Chakravarty,
Rawi, Khatib & Alizzi* 31980! (BAG); 23 km E. of Haditha, *K. Hamad* 49970! (BAG). **DWD**: 12 km W. of
Ukhaider, *Rawi* 30879! (BAG); 60 km N. of Rutba, *Rawi & Khatib* 32309! (BAG). **DSD**: Um Qasr (c. 40
km S.E. by S. of Zubair), *Guest, Rawi & Rechinger* 16892! (BAG); near Um Qasir, *Khatib & Alizzi s.n.!*
(BAG); Samah, *S. Omar, Al-Kaisi, K. Hamad & H. Hamid* 44005! (BAG); Al-Samah, *Al-Kaisi, K. Hamad &
H. Hamid* 48465! (BAG).

Tunisia, Egypt, Sinai, Palestine, Jordan, Syria, Saudi Arabia, Greece.

2. **Linaria simplex** *Desf.*, Tabl. École Bot. 65 (1804); Dinsmore in Post, Fl. Syria, Palest.
& Sinai ed. 2, 2: 286 (1933, as *L. parviflora*); Feinbrun-Dothan, Fl. Palaest. 3: 189, pl. 314
(1978); Davis in Fl. Turkey 6: 670 (1978); Collenette, Wild Flow. Saudi Arabia: 680 (1999);
Boulos, Fl. Egypt 3: 63 (2002); Sarwar & Ghazanfar in Fl. Pakistan [Ali & Qaiser] 220: 91,
f. 12 A-E (2015); Podlech & Iranshahr in Fl. Iranica [K. H. Rechinger] 180: 64 (2015);
Abdulridha, Taha & Widad, Ecology & Fl. Basrah: 503 (2016); Taifour & El-Oqlah, Pl.
Jordan Annot. Checklist: 119 (2017).

    Antirrhinum parviflorum Jacq., Collectanea 4: 204 (1791).

Fig. 34. **Linaria tenuis**. 1, habit × 1; 2, flower × 3; 3, fruit with calyx × 4; 4, seed × 20. Reproduced with permission from Feinbrun-Dothan, Fl. Palaestina 3: Plates, f. 320 (1977). Drawn by Illana Ferber. © The Israel Academy of Sciences and Humanities.

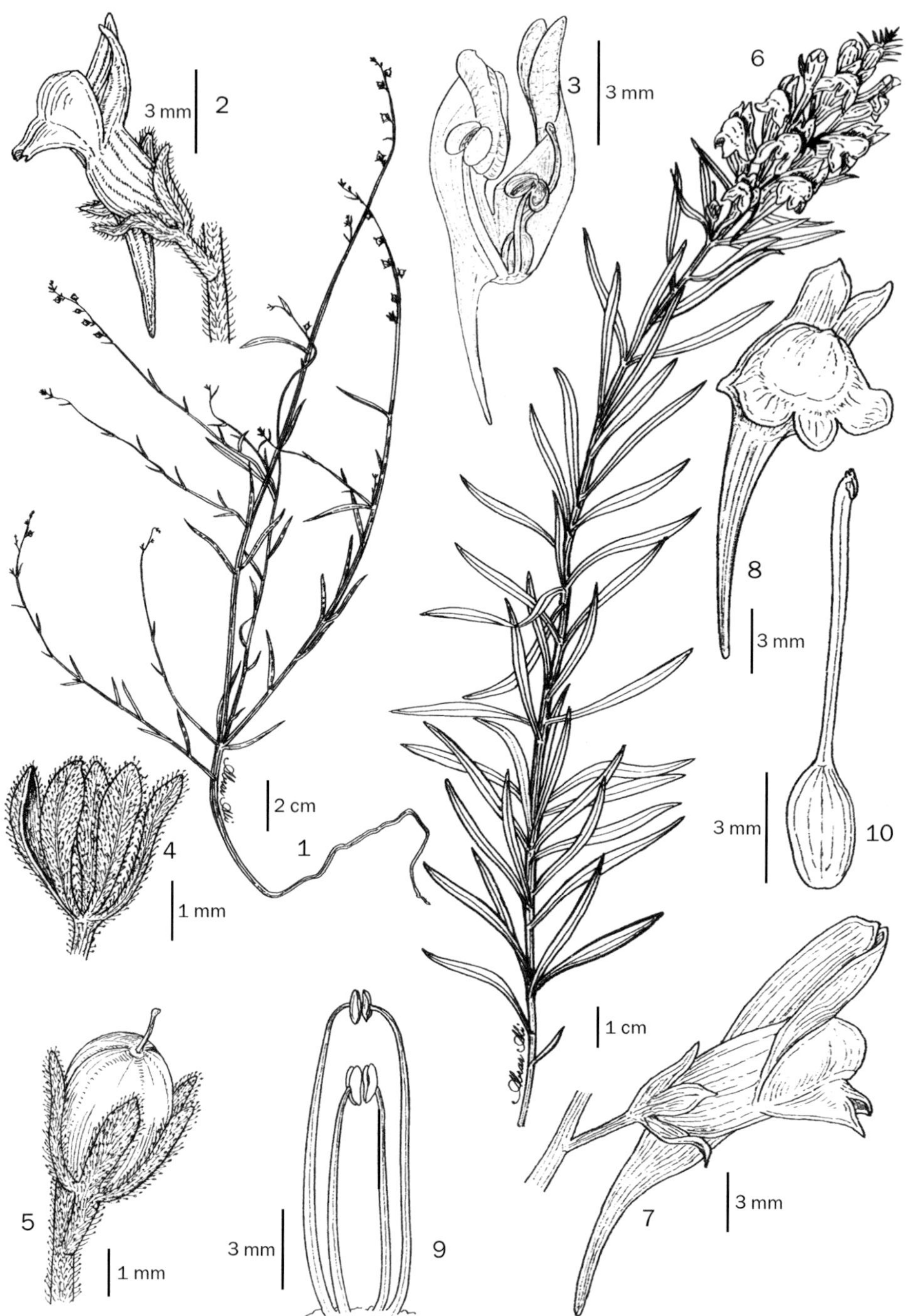

Fig. 35. **Linaria simplex**. 1, habit; 2, flower; 3, dissected corolla showing stamens and carpel; 4, calyx; 5, fruit. Reproduced with permission from Fl. Pakistan 220: f. 12, A-E (1985). Drawn by Abrar Ali. © University of Karachi, Pakistan.

Antirrhinum simplex Willd., Sp. Pl., ed. 4, 3: 243 (1800), non Link (1799);

Linaria simplex (Willd.) DC. In Lam. & DC., Fl. Fr. Ed. 3, 3: 588 (1805); Davis, Fl. Turkey 6: 670 (1978); Sutton, Rev. Tribe Antirrh. : 405 (1988), nom. illeg.

L. flaviflora (Boiss.) Cufod., Bull. Jard. Bot. État Bruxelles 33(Suppl.): 890 (1963).

L. parviflora (Jacq.) Halácsy, Oesterr. Bot. Z. 56: 278 (1906).

L. quadrifolia Hance, J. Bot. 9: 133 (1871).

L. turcomanica Kuprian., Trudy Bot. Inst. Akad. Nauk S.S.S.R., Ser. 1, Fl. Sist. Vyssh. Rast. 9: 69 (1950).

L. teheranica (Rech.f. & Aellen) Wendelbo, in Arbok Univ. Berg., Mat.-Nat. 1962, No. 1, 45 (1962).

Annual herbs; fertile stems erect, (6–)12–40 cm, simple, glabrous except the inflorescence Leaves linear, 3–30 × 0.5–2.5 mm, flat, whorled in lower part, alternate in upper part, acute or obtuse; sterile stems with similar leaves, whorled. Inflorescence a raceme, in flowering dense, lax in fruit, glandular. Bracts linear-narrowly oblong, 1.5–3.5 mm, longer than pedicel. Pedicels 0.5–2.5 mm, erect. Calyx lobes subequal, linear-spatulate, 2–4.5 × 0.4–1 mm, obtuse-subacute, glandular-pubescent. Corolla bright yellow, often with purplish or blue-veins, 4.5–9 mm; tube narrow, spur straight, 2–4.5 mm. Capsule globose, 4–6 mm in diameter, slightly longer than calyx. Seeds discoid, 1.5–2.3 mm, blackish brown, wing entire 0.3–0.5 mm, tuberculate and papillose in centre. Fig. 35, 1–5.

HAB. Mountains and the upper plains and foothills; fl. Apr.-Jun.(-Jul.).
DISTRIB. Rare, found only twice. **MAM**: Sersang, *Haines* 1117. **FKI**: 9 km N. Kirkuk, Altin Kopri, *Erdtman & Goedemans* 15532. (No specimens at BAG, description and records from Fl. Iranica (2015).

Mediterranean region to Central Asia, Arabian Peninsula.

3. **Linaria chalepensis** (*L.*) *Mill.*, Gard. Dict., ed. 8. n. 12 (1768); Dinsmore in Post, Fl. Syria, Palest. & Sinai ed. 2, 2: 287 (1933); Rawi in Dep. Agr. Iraq Tech. Bull. 14: 143 (1964); Rechinger, Fl. Lowland Iraq: 545 (1964); Feinbrun-Dothan, Fl. Palaest. 3: 190, pl. 317 (1978); Davis in Fl. Turkey 6: 664 (1978); Collenette, Wild Flow. Saudi Arabia: 678 (1999); Boulos in Fl. Egypt 3: 61 (2002); Podlech & Iranshahr in Fl. Iranica [K. H. Rechinger] 180: 69 (2015); Taifour & El-Oqlah, Pl. Jordan Annot. Checklist 118 (2017).

Antirrhinum chalepense L., Sp. Pl. 2: 617 (1753).

Annual herbs; stem erect, 18–45 cm, simple or branched from the base, glabrous. Leaves opposite or whorled, narrowly lanceolate-elliptic or lanceolate; leaves of flowering stem opposite or whorled in lower part and alternate in upper part, linear, acute, glabrous; middle leaves larger than other leaves, 25–60 × 1–3 mm. Inflorescence a lax raceme, 4–30 cm. Bracts linear, 5–30 × 1–2 mm, longer than flowers, glabrous. Pedicels 1–2 mm, elongating in fruit to 3 mm, glabrous. Calyx lobes linear, 6–10 × 0.5–1 mm, apex acute-acuminate, without membranous margins, glabrous. Corolla white to yellowish, 16–20 mm long, glabrous, palate white, tomentose; lower lip 3–5 mm; upper lip 4–6 mm; throat 1.5–2 mm broad; spur curved, 9–12 mm and 0.6–0.8 mm broad at base. Capsule globose-subglobose, 4–5 × 3–4.5 mm, shorter than calyx lobes, glabrous. Seed brownish, angled, 0.8–1.5 × ± 0.5, cristate. Fig. 36, 1–3.

HAB. Cultivated fields, shaded hills, clay-soil in cultivated land, wet soil, clay with organic matter on rocky mountain, gravely loamy soil, sandy soil; alt. 110–1580 m, fl. & fr. Mar.-Apr.
DISTRIB. Frequently found in almost all regions of Iraq except in Lower Mesopotamian region.
MAM: Amadiya, *Botany staff* 43411! (BAG). **MRO**: Rawanduz, *E. Guest* 2043! (BAG); Salahaddin, *Chapman* 11995! (BAG). **MSU**: on way to Darband, *Al-Shahwani* 25315! (BAG). **MJS**: Sinjar, *S. Omar & H. Hamid* 36500! (BAG); Jabal Sinjar (around the area), *Al-Kaisi & K. Hamad* 49217! (BAG). **FNI**: Mosul, *Y. Lazar* 3325! (BAG); 30 km N. Mosul, *Al-Dabbagh* 46761! (BAG); 2 km below Eski killek, *Rawi & Gillett* 10511! (BAG). **FAR**: Qish Tafa (near Erbil), *Guest* 1471! (BAG); Arbil, *Al-Shahwani* 25117! (BAG). **FKI**: Kirkuk, *Rawi & Gillett* 7463! (BAG); 10 km N. of Alton Kupri, *Rawi & Gillett* 10543! (BAG). **FPF**: Qizil Robat, *Guest* 1769! (BAG); Saadiya water project, *Al-Kaisi* 46343! (BAG). **DLJ**: Rawa, *Rawi & Gillett* 7027! (BAG). **DGA**: Mussaige Al-Chizlan in Dour, *Janan, Sahira, Wedad & Kalaf* 49919! (BAG). **DWD**: 17 km S.E. of K3 (Albu Hayat), *Al-kaisi & K. Hamad* 53117! (BAG).

HANDAKUKI حندكوكي (Mosul, *Lazar* 3325).

N. and E. Mediterranean region, Arabian Peninsula, Iran, E. Caucasus.

Fig. 36. **Linaria chalepensis**. 1, habit × 1; 2, flower × 3; 3, fruit with calyx × 3. Reproduced with permission from Feinbrun-Dothan, Fl. Palaestina 3: Plates, f. 317 (1977). Drawn by Esther Huber. © The Israel Academy of Sciences and Humanities.

4. **Linaria micrantha** (*Cav.*) *Hoffmanns. & Link,* Fl. Portug. 1: 258 (1811); Dinsmore in Post, Fl. Syria, Palest. & Sinai ed. 2, 2: 286 (1933); Rawi in Dep. Agr. Iraq Tech. Bull. 14: 143 (1964); Rechinger, Fl. Lowland Iraq: 545 (1964); Feinbrun-Dothan, Fl. Palaest. 3: 189, pl. 315 (1978); Collenette, Wild Flow. Saudi Arabia: 679 (1999); Boulos in Fl. Egypt 3: 60 (2002); Podlech & Iranshahr in Fl. Iranica [K. H. Rechinger] 180: 64 (2015).

Taifour & El-Oqlah, Pl. Jordan Annot. Checklist 118 (2017).

Antirrhinum micranthum Cav., Icon. 1: 51 (1791).
Antirrhinum parviflorum (Desf.) Willd., Sp. Pl., ed. 4, 3: 245 (1800).
Linaria parviflora Desf., Fl. Atlant. 2: 44 (1798).

Annual herbs. Stem erect, 5–35 cm long, branched near the base, glabrous. Lower leaves opposite or whorled (of 3 leaves), upper alternate, narrowly lanceolate, 5–25 × 1–5 mm, apex acute, glabrous. Inflorescences capitate, elongate when mature to lax racemes, glabrous or with glandular hairs. Bracts lanceolate-elliptic, 2–8 × 0.8–2 mm, longer than pedicels, glabrous, margin ciliate with glandular hairs. Pedicels less than 1 mm long, elongate in fruit to 2 mm, glandular hairy. Calyx lobes lanceolate, 2–3 × 1 mm, apex obtuse, margins somewhat membranous, ciliate, glandular-hairy. Corolla blue or blue with yellow palate, 3–5 mm; lower lip 1.5–2 mm; upper lip 1–1.5 mm; spur straight, ± 1.5 mm long; palate tomentose. Capsule globose, 4–5 × 3–3.5 mm, glabrous. Seeds discoid, 1–1.2 mm in diameter, winged, smooth, dark brown. Fig. 37, 1–4.

HAB. Gravely soil with sand, hill slopes, dry gravel land by road side, saline soil in low depression, silty plain, rocky hill side, sandy and rocky soil; alt. 60–800 m; fl. & fr. Feb.-Apr.
DISTRIB. Common in almost all regions in Iraq, but not found in the Lower Mesopotamian region.
MSU/FAR: 10–15 km from Arbil to Darband, *A. Shahuani* 25275! (BAG). **MJS**: Kursi, *Gillett* 10987 (BAG). **FUJ**: Albaiyder 78 km S. of Sinjar, *Chakravarty, Rawi, Khatib & Alizzi* 32087! (BAG). **FNI**: Ninevah, *Roger* 552! (BAG). **FAR**: Bastura Chai, *Rawi & Gillett* 10490 (BAG). **FKI**: 25 km S. of Kirkuk, *Rawi & Gillett* 10339! (BAG); **FPF**: 16 km S. of Badra, *Rawi* 18179! (BAG). **FPF/DLJ/DWD**: 6 km above Ana, *Rawi & Gillett* 6963! (BAG). **DLJ**: Rawa, *Rawi & Gillett* 6993 (BAG). **DWD**: Nakhaib, *Rawi* 14772! (BAG); Wadi Al-Ghadaf, *Rawi* 15727! (BAG); 50 km W. of H3, *Rawi* 31163! (BAG); Wadi Al-Ajrumiya, *Rawi* 31253! (BAG); Bottom of Hawran valley (12 km S.E. of H1), *Chakravarty, Rawi, Khatib & Alizzi* 31642! (BAG); 48 km S.E. of T1, *Chakravarty, Rawi, Khatib & Alizzi* 31825! (BAG); 10 km W. of K3, *S. Omar, Al-Kaisi, K. Hamad & H. Hamid* 44750! (BAG); 17 km S.E. of K3 (Albu Hayat), *Al-Kaisi & K. Hamad* 53119! (BAG). **DSD**: 10 km S. of Zubair, *Rawi & Gillett* 6107! (BAG); Shahicha, *Rawi & Gillett* 6265! (BAG); 5 km E. of

Fig. 37. **Linaria micrantha**. 1, habit × 1; 2, flower × 8; 3, fruit with calyx × 8; 4, seed × 10. Reproduced with permission from Feinbrun-Dothan, Fl. Palaestina 3: Plates, f. 315 (1977). Drawn by Ruth Kopel. © The Israel Academy of Sciences and Humanities.

Jumaima, *Al-Kaisi, K. Hamad & H. Hamid* 48209! (BAG). **LCA**: 15 km W. of Falluja, *Chakravarty & Rawi* 30321! (BAG).

Canary Islands, N. Africa, southern Europe, Jordan, Palestine, Syria, Arabian Peninsula, Iran to central Asia.

5. **Linaria albifrons** *Spreng.*, Syst. Veg., ed. 2: 793 (1825); Dinsmore in Post, Fl. Syria, Palest. & Sinai ed. 2, 2: 289 (1933); Rawi in Dep. Agr. Iraq Tech. Bull. 14: 143 (1964); Rechinger, Fl. Lowland Iraq: 546 (1964); Feinbrun-Dothan, Fl. Palaest. 3: 192, pl. 321 (1978); Collenette, Wild Flow. Saudi Arabia: 678 (1999); Boulos in Fl. Egypt 3: 60 (2002); Podlech & Iranshahr in Fl. Iranica [K. H. Rechinger] 180: 68 (2015).

Antirrhinum albifrons Sm., Fl. Graec. Prodr. 1(2): 432 (1809).
Linaria glauca Ehrenb. ex Boiss., Fl. Orient. [Boissier] 4(2): 382 (1879).
Linaria minutiflora C.A.Mey., Verz. Pfl. Casp. Meer. 109 (1831).
Linaria modesta Boiss., Diagn. Pl. ser. 1, 12: 43 (1853).

Annual herb; stems erect or ascending, 12–14 cm, glabrous. Leaves opposite or whorled, ovate-lanceolate or oblong-lanceolate, 5–15 × 3–5 mm, apex acute to obtuse, glabrous, somewhat glaucous. Inflorescence capitate, elongated in fruit. Bracts lanceolate to narrow lanceolate or elliptic, 4–10 × 1–4 mm, longer than pedicels, glabrous. Pedicel 1–2 mm. Calyx 4–5 mm, glabrous, slightly shorter than corolla; lobes lanceolate, 1–1.5 mm wide, apex acute. Corolla 5.5–7 mm long, white with yellow palate and purplish veins; upper lip deeply bifid, 3–4 mm; spur conical, straight, 1.5–2 mm. Capsule glabrous, oblong, 5 mm. Seed oblong, angled, punctate-pitted.

HAB. Sandy gravelly soil; alt. 240–460 m; fl.: Feb.-Mar.
DISTRIB. Rare, found twice in the W. sector in the desert region of Iraq. **DWD**: Uglat hauran, *Al-Kaisi & K. Hamad* 53016! (BAG); Ga'ra (N. of Rutba), *S.Omar, H. Abass & Al-Kaisi* 56046! (BAG).

Cyprus, Egypt, Libya, Saudi Arabia, Palestine, Jordan, Syria, Iran, Azerbaijan.

6. **Linaria fastigiata** *Chav.*, Monogr. Antirrh. 125, t. 7 (1833). Type: Inter Baghdad & Kermanschah, *Olivier* (P, holo.). Rawi in Dep. Agr. Iraq Tech. Bull. 14: 143 (1964); Sutton, Rev. Tribe Antirrh.: 302 (1988); Sarwar & Ghazanfar in Fl. Pakistan [Ali & Qaiser] 220: 98 (2015); Podlech & Iranshahr in Fl. Iranica [K. H. Rechinger] 180: 42 (2015).

Perennial herbs; stem ascending or erect, 25–80 cm, branched in middle and upper parts, glabrous. Leaves lanceolate or broadly elliptic, 15–75 × 5–30 mm, 3–7-viened, apex acute, glabrous; middle leaves larger than other leaves, 30–70 × 10–30 mm. Flowers crowded in paniculate inflorescence. Bracts lanceolate-narrowly, 2–3 mm, shorter than or as long as pedicels, with membranous margins, glabrous. Pedicels 1–2 mm, elongate in fruit to 4 mm, glabrous. Calyx lobes lanceolate, 3–4 × 1.2–2 mm, apex obtuse to obtuse, margins membranous, sometimes glandular ciliate. Corolla yellow with purple veins, 17–20 mm; lower lip 6–7 × 5–7 mm; upper lip 6–8 mm long; throat 5–6.5 mm wide; spur 6–8 mm long and 1–1.5 mm wide at base, palate tomentose. Capsule globose, 5.5–7 mm in diameter with minute glandular hairs. Seeds oblong to oblong-reniform, 3–3.5 × 2.5–3 mm, winged, smooth.

HAB. Clay soil, in field, hill side, loamy gravely soil, mountain side; alt. 800–1800 m; fl. May-Jul. fr. Jun.-Aug.
DISTRIB. Occasional in the in the lower forest zone of the mountain region of Iraq.
MRO: Hero, *Rawi* 9386! & 9390! (BAG); Gwaija Dagh, *Rawi & Gillett* 11704! (BAG); Haj Omran, *Rawi* 24280! (BAG). **MSU**: Bakrajo, *S. Omar & F. Karim* 38012! (BAG); Sulaimaniya Hawara-Barzay, *Al-Khayat* 54498! (BAG); Zewiya, Pira Magrun, *Botany staff* 58936! (BAG); 9 km from Choarta to Serser, *M. Noori & K. Hamad* 41262! (BAG); Sulaimaniya, Hawara-Barzay, *Al-Khayat* 54498! (BAG); Zewi, Pira Magrun, *Botany staff* 58936! (BAG).

Iraq, Iran, Afghanistan, Pakistan.

7. **Linaria kurdica** *Boiss. & Hohen.*, Diagn. Pl. Orient. ser. 1, 4: 73 (1844) subsp. **kurdica**. Type: N. Iraq, Monte Gara, Kurdistan, *Kotschy* 391 (G-Boiss., holo.; BM, E, H, JE, LE, M, W, iso.); Rawi in Dep. Agr. Iraq Tech. Bull. 14: 143 (1964); Davis in Fl. Turkey 6: 666 (1978); Podlech & Iranshahr in Fl. Iranica [K. H. Rechinger] 180: 46 (2015).

Linaria araratica Tzvelev, Bot. Mater. Gerb. Bot. Inst. Komarova Akad. Nauk S.S.S.R. 19: 17 (1959).
L. kurdica subsp. *araratica* (Tzvelev) P.H.Davis, Notes Roy. Bot. Gard. Edinburgh 36(1): 8 (1978).
L. megrica Tzvelev, Bot. Mater. Gerb. Bot. Inst. Komarova Akad. Nauk S.S.S.R. 19: 15 (1959).
L. zargariana Parsa, Kew Bull. 3(2): 216 (1948).

Perennial herbs; stem erect, 35–80 cm, simple or branched in upper parts, glabrous. Leaves lanceolate or elliptic, apex acute, glabrous; middle leaves larger than other leaves, 40–80 × 5–25 mm, 3(–5)-veined; upper leaves smaller, narrowly elliptic to narrowly lanceolate, 6–30 × 1–7 mm. Inflorescence paniculate, dense. Bracts lanceolate to narrowly lanceolate, 3–7 mm, longer than pedicels, with or without narrow membranous margin, glabrous. Pedicels 1.5–2.5 mm, elongating to 4 mm in fruit, glabrous. Calyx lobes lanceolate, 3–4 × 1–2 mm, apex acute-obtuse, margins membranous, adaxial surface glabrous or with glandular hairs near the base, abaxial surface with short glandular hairs, margins ciliate with glandular hairs or ciliate near base. Corolla yellow with purple veins, 17–21 mm; lower lip 5–6 × ± 5 mm; upper lip 6–7 mm; throat 4–5 mm wide; spur 6–8 mm, 1–1.5 mm broad at base, palate tomentose. Capsule globose-subglobose, 5–5.5 × 4–5 mm, with minute glandular hairs. Seeds discoid-oblong, papillate in the middle, 2.5–3 × 2–2.5 mm, winged.

HAB. Rocky mountain, among rocks, *Astragalus* zone on mountain, on black-brownish rock, mountain slope near stream, cultivated mountain slope, *Quercus* forest on mountain slope, in valley; alt. 1000–2900 m; fl. & fr. Jun.-Sep.
DISTRIB. Occasional in the lower forest zone of the mountainous regions of Iraq.
MRO: Arl Gird Dagh, *Guest & Ludlow-Hewitt* 2959! (BAG); Arl Gird Dagh, *Gillett* 9579! (BAG); Rayat, *Guest* 13072! (BAG); Sula Khel, *Rawi & Serhang* 24676! (BAG); Magar range, Haj Omaran, *Rawi & I. Serhang* 24301! (BAG); Surade village nr. Pushtashan, *Rawi & Serhang* 26519! (BAG); Qerna Qaw Valley, n of Pushtashan, *Rawi & Serhang* 26643! (BAG); lower slope of Qandil range, *Rawi & Serhang* 26707! (BAG); Sakran mountain 15 km S.E. Chuman, *Al-Dabbagh & K. Hamad* 46212! (BAG). **MSU**: Hauraman mountain, *Rawi, Chakravarty*, Nuri & *Alizzi* 19734! & 19843! (BAG).

Syria, Iran, Turkey, Caucasus.

8. **Linaria genistifolia** (*L.*) *Mill* subsp. **qandilensis** *Haloob & Al-Kaisi* **subsp. nov.** Type: Iraq, N.E. of Qandil, 25/8/1957, *Al-Rawi* and *Serhang* 24367 (holo., BAG).

L. genistifolia subsp. *qandilensis* is closely related to *L. genistifolia* subsp. *sofiana* (Velen.) Chater & D.A.Webb in having narrow leaves, the middle leaves > 10 × longer than broad, but differs in its calyx indumentum. It is also closely related to *L. genistifolia* subsp. *confertiflora* (Boiss.) P.H.Davis in having a dense inflorescence but differs in its narrow leaves and hairy calyx. *Linaria genistifolia* (*L.*) *Mill* subsp. *qandilensis* is different from the other subsp. in its pilose calyx while all other subspecies of *L. genistifolia* subspecies have an almost glabrous calyx.

Perennial herbs; stem erect, up to 50 cm long, branched above, glabrous. Leaves alternate, linear to narrowly lanceolate, 50–70 × 3–5 mm, upper leaves much shorter, somewhat curved, apex acuminate, glabrous. Flowers in dense spike-like racemes, 50–90 mm long. Bracts narrow lanceolate, 3–8 mm, as long as or longer than pedicels, without membranous margins, glabrous. Pedicels 3–4 mm, glabrous or with glandular hairs. Calyx lobes lanceolate, 4–5 × 1 mm, apex acute, margins ciliate, not membranous, outer surface pilose, sparsely puberulent within. Corolla yellow, 17–20 mm; lower lip 5–6 × 4.5–5.5 mm; upper lip 7–9 mm; throat 4–5 mm broad; spur 6–7 mm long and 1–1.5 mm broad at base, straight, palate tomentose.

HAB. In *Astragalus* zone on mountain sides; alt. 2650–2850 m; fl. Aug.
DISTRIB. Very rare, collected only once in area between the alpine and thorn-cushion zone in the N. Iraq. **MRO**: N.E. of Qandil, *Al-Rawi* and *Serhang* 24367! (type BAG).

Endemic

9. **Linaria iraqensis** *Al-Kaisi & Haloob* **sp. nov.**

Affinis *L. kurdica* et *L. guilanensis* sed folia calinum (1–)1.5–2.5 mm lata, acuminata. Inflorescentia 20–25 mm longa, bracts 2–4 mm longa. Pediceli 2–3 mm longi. Calycis lobi ovati-lanceolati, 2.5–3 × 1.5–2 mm, Corolla tubo ore 4–4.5 mm lato, calcarae 3–4 mm longo, 1.5–2 mm lato ad basem. Typus: Iraq, Serva, 9/8/1947, *Gillett* 9697 (BAG!, holo.).

L. iraqensis is similar to *L. kurdica* in having calyx with membranous margins, glandular hairs and in its narrow leaves to 1 mm wide but it differs in having a shorter inflorescence, corolla and spur. The new species is also similar to *L. guilanensis* in its narrow leaves, calyx lobes with membranous margins, glandular hairs and in corolla length but differs in having acuminate leaves, longer pedicel, and shorter calyx lobes, and a wider spur and corolla throat.

Perennial herbs; stem erect, 25–30 cm, branched about the middle, glabrous. Leaves dense on the stems, linear, 35–50 × 1.5–2.5 mm, upper leaves smaller, 15–30 × 1 mm, apex acuminate, glabrous. Flowers in a dense raceme, 20–25 mm long. Bracts linear, 2–4 mm, longer than or as long as pedicels, glabrous. Pedicels 2–3 mm, glabrous. Calyx lobes ovate-lanceolate, 2.5–3 × 1.5–2 mm, apex acute-obtuse, margins entire, glabrous, membranous, glandular hairy, sparsely hairy within. Corolla yellow with reddish-purple veins, 10–12 mm; lower lip 4 × 4 mm; upper lip 4.5–5 mm long; throat 4–4.5 mm broad; spur 3–4 mm and 1.5–2 mm broad at base; palate tomentose. (Ovary not mature in the only specimen collected).

HAB. Mountain region of N. Iraq; alt. ± 2000 m; fl. Aug.
DISTRIB. Very rare, collected only once. **MRO**: Serva, *Gillett* 9697! (type BAG)

Endemic.

10. **Linaria pyramidalis** *F.Dietr.*, Nachtr. Vollst. Lex. Gärtn. 4: 417 (1818) subsp. **pyramidalis**; Podlech & Iranshahr in Fl. Iranica [K. H. Rechinger] 180: 56 (2015).

> *Antirrhinum pyramidale* Vent., Encycl. [J. Lamarck & al.] 4(1): 360 (1797).
> *Antirrhinum pyramidatum* (Tourn. ex Spreng.) Benth., Prodr [A. P. de Candolle]10: 272 (1846).
> *Linaria pyramidata* Tourn. ex Spreng., Syst. Veg. ed. 16, 2: 796 (1825); Davis in Fl. Turkey 6: 668 (1978);

Perennial herbs; stem erect, up to 70 cm, simple, glabrous. Leaves lanceolate-elliptic, 40–95 × 5–10 mm, apex acuminate, glabrous. Flowers in dense spike-like racemes, ± 13 cm long. Bracts lanceolate, (8–)10–20 × 4–5 mm, apex acuminate, much longer than pedicels, entire, ciliate, glabrous. Pedicels 1–2(–3) mm. Calyx lobes unequal, ovate to broadly lanceolate, 6–8 × 3–4 mm, apex subrounded-obtuse, without membranous margins, outer surface with dense glandular hairs. Corolla yellow, 30–32(–34) mm long; lower lip 10–12 × ± 15 mm; upper lip 10–13 mm long; throat ± 10 mm wide; spur 8–10 mm and 3 mm broad at base, palate tomentose.

HAB. Slope of mountain in *Astragalus* zone; alt 1820–2800 m; fl. Jul.-Aug.
DISTRIB. Rare, collected from thorn-cushion zone in N.E. sector of mountain region in Iraq. **MRO**: Malakh (or Gomasure) N.E. of Rania (lower slopes of Qandil range), *Rawi & Serhang* 18243! (BAG); Serin Mt. on road to Qandil, *Rawi & Serhang* 24021 A! (BAG); Arbil, mt. Qandil 2800 and between Pushtashan and lake Goam-e-Kirmosaran, *Rechinger* 11111 (W, cited in Fl. Iranica)

Turkey, Caucasus, Iran.

11. **Linaria confertiflora** *Benth.*, Prodr. [A.P. de Candolle] 10: 272 (1846). Type: In glareosis montis Gara, Kurdestan, *Kotschy* 170! (K, lecto.; W, iso.). Davis in Fl. Turkey 6: 669 (1978); Podlech & Iranshahr in Fl. Iranica [K. H. Rechinger] 180: 41 (2015).

Perennial herbs; stem erect or ascending, 20–32 cm, glabrous. Leaves on fertile stems narrowly ovate, 14–35 × 3–8(–13) mm, apex acute, glabrous. Inflorescence of 10–20 flowers, dense, puberulent. Bracts elliptic, 3.5–5.5, apex acute, longer than pedicels. Pedicels 1–2 mm. Calyx lobes unequal, the longer 8–9 × 1.3–2 mm, shorter 4–6 × 1.4–2.5 mm, (without membranous margins), glandular-villous. Corolla yellow, 25–32 mm; throat 6–8 mm wide; spur straight, 8–11 × 2–2.3 mm at base, glabrous or glandular-puberulent, palate tomentose, Capsule 4.6–8 × 5–7 mm, glabrous. Seed 3–3.3 mm, oblong to reniform, brown, wing 0.6-0.9 mm broad, densely tuberculate in centre .

HAB. Mountain regions; alt 2000–3300 m; fl. (Apr.-) May-Jun.
DISTRIB. Rare, collected only three times from the mountain region of N. Iraq. **MAM**: mt. Gara, *Kotschy* 170 (K, type). **MRO**: mt. Qandil, *Thesiger* 1160! (K). **MSU**: E. Qala Diza, *Thesiger* 1163 (K).

Syria and Turkey.

12. **Linaria grandiflora** *Desf.*, Ann. Mus. Hist. Nat. 11: 51, t. 2 (1908); Davis, Fl. Turkey 6: 661 (1978); Sutton, Rev. Tribe Antirrh.: 321 (1988); Podlech & Iranshahr in Fl. Iranica [K. H. Rechinger] 180: 66 (2015); Sarwar & Ghazanfar in Fl. Pakistan [Ali & Qaiser] 220: 89 (2015).

Linaria calycina Boiss. & Balansa, Diagn. Pl. Orient. ser. 2, 6: 130 (1859).

Perennial herbs; stem erect-ascending, up to 75 cm, simple, glabrous. Leaves alternate, dense, lanceolate-ovate, apex acute; middle and upper leaves somewhat amplexicaul, 35–45 × 15–20 mm; lower leaves 15–25 × 10 mm. Flowers crowded in elongate spike-like raceme, up to 35 cm. Bracts deflexed, lanceolate-ovate, 10–20 × 4–10 mm, much longer than pedicels, glabrous. Pedicels 1–3 mm, glabrous. Calyx lobes lanceolate, 10–15 × 4–6 mm, apex acute, without membranous margins, glabrous. Corolla bright yellow, 30–35 mm; lower lip 12–15 mm long; upper lip 18–20 mm; throat 8–10 mm wide; spur 10–15 mm long and 2–3 mm wide at base; palate tomentose.

HAB. River banks; fl. May.
DISTRIB. Very rare, recorded only once from the upper plains and foothills. FNI: Eski Kelek, *Anderson* 41403! (BAG)

Turkey, Azerbaijan, Bulgaria, Iran.

Species doubtfully recorded

Linaria pelisseriana (*L.*) *Mill.* Gard. Dict. ed. 8, no. 11, (1768) is listed in Rawi, Dep. Agr. Iraq Tech. Bull. 14: 143 (1964) from Jabal Sanam citing Anthony (J. Ind. Bot. Soc. Vol. 17, 1932). We have not seen any material of this species and it is likely to be a misidentification.

6. **CHAENORHINUM** (DC.) Rchb.

Consp. Regn. Veg. [H.G.L.Reichenbach] 123 (1828); Sutton, Revis. tribe Antirrhineae: 97 (1988); Fischer in Kubitzki (ser. ed.), J.W. Kadereit (ed.), Fam. & Gen. Vasc. Pl. 7: 378 (2004)

Shahina A. Ghazanfar

Annual or perennial herbs. Stems erect to ascending, glabrous or glandular-pubescent. Leaves entire, shortly petiolate, the lower leaves opposite. Flowers axillary, solitary. Calyx deeply 5-lobed. Corolla bilabiate, spurred, throat open; upper lip 2-lobed, lower lip 2-lobed. Stamens 4, didynamous, included. Ovary ovoid to globose. Capsule dehiscing irregularly at base or each locule opening above by a pore. Seeds many, oblong-ovoid, smooth, ribbed longitudinally, tuberculate-papillate.

Twenty seven species in north western Africa, Somalia; Europe, Turkey, Arabian Peninsula to W. Himalaya; 2 species in Iraq.

Formerly included in the Scrophulariaceae, *Chaenorhinum* is now included in the Plataginaceae.

Chaenorhinum, from Gr. χημη, *cheme*, gaping and ρυγχος, *rhinchos*, snout, alluding to open throat of corolla (closed in *Linaria*).

Plant glandular-pubescent; corolla 7–9 mm, white suffused purplish
    pink; seeds tuberculate . . . . 1. *C. rubrifolium*
Plant tomentellous, not glandular; corolla 10–12 mm, blue; seeds
    ribbed and foveolate between ribs . . . . 2. *C. calycinum*

1. **Chaenorhinum rubrifolium** (*Robill. & Castagne ex DC.*) *Fourr.*, Ann. Soc. Linn. Lyon sér. 2, 17: 127 (1869).

subsp. **gerense** (*Stapf*) *D.A.Sutton*, Revis. tribe Antirrhineae: 114 (1988); Podlech & Iranshahr in Fl. Iranica [K. H. Rechinger] 180: 6 (2015).

Linaria gerensis Stapf, Bull. Misc. Inform. Kew (1906)3: 75 (1906).

Annual herb; stems slender, 5–20 cm, glandular-pubescent. Lower leaves petiolate, 3.5–20 × 2–9 mm, ovate, obtuse; upper leaves shortly petiolate to sessile, narrow elliptic to lanceolate,, 4–12 × 1.5–4 mm, obtuse to acute. Racemes lax. Pedicel 6–20 mm, erect to patent. Calyx lobes unequal, linear, 3–9 × 0.5–1 mm, recurved in fruit. Corolla 7–9 mm, white suffused purplish pink; tube c. 2 mm in dimeter. Spur conical, c. 1.5 mm, acute, smaller than corolla. Capsule 3 mm. Seeds tuberculate.

HAB. Mountains; alt. 600–1200m; fl. Apr.-Jun.
DISTRIB. Mountains of N. Iraq. **MRO**: Arbil, Safin Dagh, *Gillett*; Rawandiz, *Emberger & al.* 15525; Kani Mazu, *Agnew, Hadac & Haines* W2136; ibid, *Agnew et al.* 6016; prope Shaklawa, *Bornmüller* 1651.

N.W. Africa, Cyprus, Iran.

2. **Chaenorhinum calycinum** (*Banks & Sol.*) *P.H.Davis*, Notes Roy. Bot. Gard. Edinburgh 36(1): 4 (1978); Davis in Fl. Turkey 6: 653 (1978); Feinbrun-Dothan, Fl. Palaest. 3: 194, pl. 325 (1978); Podlech & Iranshahr in Fl. Iranica [K. H. Rechinger] 180: 11 (2015).

Antirrhinum calycinum [Soland.], Nat. Hist. Aleppo, ed. 2 [A. Russell] ii. 256. (1794).
Linaria persica Chav., Monogr. Antirrh.: 173 (1833); Rawi, Dep. Agr. Iraq Tech. Bull. 14: 143 (1964).
Antirrhinum rytidospermum Fisch. & C.A.Mey., Index Seminum (LE, Petropolitanus) 2 : 27 (1836).
A. rytidocarpum G.Don, Gen. Hist. 4: 516 (1838).
Linaria rytidosperma (Fisch. & C.A.Mey.) Boiss., Diagn. Pl. Orient. 4: 73 (1844).
Chaenorhinum persicum O.Fedtsch. & B.Fedtsch., Consp. Fl. Turkestanicae [O.A. Fedchenko & B.A. Fedchenko] 5: 84 (1913).
C. rytidospermum (Fisch. & C.A.Mey.) Kuprian. in Komarov (ed.) Fl. URSS 22: 228 (1955); Feinbrun-Dothan, Fl. Palaest. 3: 194, pl. 325 (1978).

Annual herb; stems erect simple or branched at base, 3–20 cm, tomentellous. Leaves somewhat fleshy, oblong to lanceolate, 10–20 × 4–12 mm, base tapering into a short petiole. Racemes lax. Pedicel 2–4 mm, erect. Calyx 8–10 mm, lobes, linear-lanceolate, 8–13 ×1–2 mm, obtuse. Corolla blue, 10–12 mm, not including spur, Spur whitish, 2–5 mm, slender, slightly

Fig. 38. **Chaenorhinum calycinum**. 1, habit × 1. Reproduced with permission from Feinbrun-Dothan, Fl. Palaestina 3: Plates, f. 325 (1977). Drawn by Esther Huber. © The Israel Academy of Sciences and Humanities.

curved downwards. Capsule globose, 4–5 mm in diameter, beaked at apex, velutinous, breaking irregularly near base. Seeds ribbed and foveolate between ribs. Fig. 38, 1.

HAB. Stony and rocky mountain sides; alt. ±600 m; fl. Apr.
DISTRIB. Mountains of N. Iraq. **MRO**: Arbil, 2 km infra Eski Kellek, *Chapman* 10512 (cited in Fl Iranica).

Iran, Azerbaijan, Turkey.

7. **BACOPA** Aubl. nom. cons.

Hist. Pl. Guiane 128 (1775); Fischer in Kubitzki (ser. ed.), J.W. Kadereit (ed.), Fam. & Gen. Vasc. Pl. 7: 384 (2004)
Monniera P.Br., Hist. Jamaica: 269 (1756)

Shahina A. Ghazanfar

Annual or perennial herbs, submerged aquatics or in moist locations, usually glabrous. Leaves opposite, entire or dentate or variously dissected, sometimes punctate. Flowers axillary, solitary or in racemes. Bracteoles present or absent. Calyx unequally 5-lobed, outer lobes the broadest, inner ones narrow. Corolla with a short tube, lobed above; lobes equal or the upper 2 connate. Stamens 4, didynamous, included. Stigma usually shortly 2-lobed. Capsule globose or ovoid, 2 or 4-celled, loculicidal or septicidal. Seeds numerous.

About 57 species distributed in the tropics and subtropics, usually aquatic or in moist locations; a single species in Iraq.

Bacopa, originally in the Scrophulariaceae, now included in the Plantaginaceae.
Bacopa: Aublet offers no explanation for the origin of the name, which may be Guianan Creole.

1. **Bacopa monnieri** (*L.*) *Wettst.*, Nat. Pflanzenfam. [Engler & Prantl] 4(3b, lief. 67): 77 (1891); Rawi in Dep. Agr. Iraq Tech. Bull. 14: 59 (1964); Rechinger, Fl. Lowland Iraq: 549 (1964); Boulos and Snogerup in Fl. Egypt [Boulos] 3: 76 (2002); Abdulridha, Taha & Widad, Ecology & Fl. Basrah: 503 (2016); Taifour & El-Oqlah, Pl. Jordan Annot. Checklist 117 (2017).

Perennial herb. Stems 10–30 cm, branched, ascending to prostrate and creeping, somewhat fleshy, rooting at nodes, 4-angled, glabrous. Leaves sessile, 8–15 × 4–12 mm, the upper ones smaller, oblong to obovate to spatulate, apex obtuse, margins entire to crenate-dentate. Flowers solitary in the axils of leaves. Pedicels 5–10 mm. Bracteoles 2, ± 5 mm, ovate, present below the calyx. Calyx 5-lobed to base, 3–4 mm, glabrous, the outer 3 lobes broadly ovate, with the upper lobe largest; 2 lateral lobes smallest, lanceolate. Corolla bluish to white, tube ± 5 mm, upper lip emarginate, 3–5 mm, lower lip equally 3-lobed. Anthers sagittate. Capsule 2–3 mm, ovoid, glabrous. Fig. 39, 1.

HAB. In moist locations, at edges of creeks, river banks, in marshes; alt. 50–450 m; fl. Mar., Apr.
DISTRIB. Western desert, Basrah esturine, and southern marsh districts of S.E. Iraq. **DWD**: Shithatha, *Rawi* 26911!; ibid. *Chakrawarty & Rawi* 29787! **LEA**: Basrah district, *Rawi* 16550; ibid. *Alizzi & S. Omar* 34738! **LSM**: Qurnah, *Hadac & Haines* W1874!; Chahala nr. Amara, *Field & Lazar* 43!; Al Musharrah, Shat al Malh, *Alizzi & Saleh* 35799!; 40 km S. of Al Musharrah, *Alizzi & Saleh* 34714!; Al Agair, 50 km N. of Qurnah, *Rawi* 16544!; Al Chabaish, *S. Omar & F. Karim* 36771!; Abu Al Khasb, *Alizzi & S. Omar* 34696!; 20 km E. of Al Kahla, *Rawi & Alizzi* 32422!

SHEHAIMA, GARNA, AHWAR, TANOMA (Basra region; Albulridha, l.c.)

S. America, Mexico, western and eastern Africa, Arabian Peninsula to Vietnam.

8. **LIMNOPHILA** R.Br. nom. cons.

Prodr.: 442 (1810); Philcox in Kew Bull. 24: 101 (1970)
Fischer in Kubitzki (ser. ed.), J.W. Kadereit (ed.), Fam. & Gen. Vasc. Pl. 7: 386 (2004)

C.C. Townsend
Revised by Shahina A. Ghazanfar

Aquatic herbs with simple or branched stems, erect or prostrate, ± submerged, rooting at nodes. Leaves submerged and aerial; submerged leaves verticillate, finely divided; aerial

Fig. 39. **Bacopa monnieri**. 1, habit × 1. Reproduced with permission from Feinbrun-Dothan, Fl. Palaestina 3: Plates, f. 339 (1977). Drawn by Esther Huber. © The Israel Academy of Sciences and Humanities.

leaves opposite or verticillate, undivided or variously laciniate; ± punctate. Flowers solitary in leaf axils, or in terminal or axillary spikes or racemes. Bracteoles 2 or absent. Calyx tubular, 5-lobed, lobes unequal, tube with prominent nerves. Corolla tubular or funnel shaped, bilabiate with upper lip entire or 2-lobed, lower lip 3-lobed. Stamens 4, didynamous, included. Ovary glabrous; stigma bilamellate; style filiform. Capsule ellipsoid to globose, 4-valved, septicidal. Seeds numerous.

35 species in the tropics and subtropics of the Old World; a single species in Iraq.

Limnophila, from λιμνη, *limne,* pool or lake, and fem. of φιλυς, filus, loving.

1. **Limnophila indica** (*L.*) *Druce* in Rep. Bot. Excl. Club Brit. Isles 1913, 3: 420 (1914); Rawi in Dep. Agr. Iraq Tech. Bull. 14: 142 (1964); Fischer in Fl. Ethiop. & Eritr. 5: 260 (2006); Philcox in Flora Zambesiaca 8(2): 44 (1990); Abdulridha, Taha & Widad, Ecology & Fl. Basrah: 503 (2016).

Hottonia indica L., Syst. Nat., ed. 10. 2: 919 (1759).

Perennial aquatic herb; aerial stem simple to branched, with sessile or stipitate glands above becoming subglabrous; submerged stem branched, glabrous. Leaves on aerial stem verticillate, lanceolate, margins crenateserrate, 1–3-veined leaves towards the apex, sessile glandular to subglabrous; submerged leaves verticillate, pinnatisect, up to 3 cm long. Flowers solitary, axillary, shortly pedicellate to 15 mm, sessile glandular to stipitate glandular. Bracteoles 2, linear to linear-oblong, acute, entire to irregularly serrate-dentate to deeply incised, glandular to subglabrous. Calyx 3.5–6 mm, lobes 2–3 mm, broadly ovate to lanceolate, shortly acuminate, sessile glandular, not striate at maturity. Corolla white to pale

Fig. 40. **Limnophila indica**. 1, habit × 2/3; 2, stem with capsule × 1; 3, flower × 6; 4, capsule × 6; 5, seed × 32. Reproduced with permission from FTEA Scrophulariaceae: f. 16, 2008. Drawn by © Juilet Beentje.

yellow or yellow at base of tube, pink-purple above, about twice as long as calyx, externally glabrous; lobes entire. Stamens with anthers contiguous, glabrous. Capsule compressed ellipsoid to subglobose, ± 3.4 mm long, dark brown. Fig. 40, 1–5.

HAB. & DISTRIB. Submerged in shallow water on mudflat; rare; alt. 0–20 m; fl. Mar.-Apr. Abdulridha (l.c.) reported it to be common on the banks of the Euphrates before the draining of marshes in Southern Iraq, and has not been collected since.

QURNA (Basra region), Abdulridha (l.c.)

Widespread throughout the Old World tropics.

9. **VERONICA** L.

Sp. Pl. 1: 9 (1753); Gen. Pl. ed. 5, 83 (1754)
Fischer in Kubitzki (ser. ed.), J.W. Kadereit (ed.), Fam. & Gen. Vasc. Pl. 7: 398 (2004)

J. R. Edmondson

Annual to perennial, prostrate to erect, glabrous, glandular or densely villous herbs. Leaves sessile to petiolate, lanceolate to orbicular, entire or variously lobed, opposite, upper cauline leaves usually alternate, basal leaves in rosettes. Flowers in terminal or axillary racemes, pedicellate, zygomorphic. Calyx usually 4-lobed, lobes basally connate, two larger and two smaller, sometimes 5-lobed with uppermost lobe much smaller than the others, lobes entire or dentate. Corolla rotate, usually with 4 unequal lobes, usually glandular inside, bluish, purple, reddish or white, tube broader than long. Stamens 2, inserted on tube, usually exserted. Ovary 2-locular, each cell with numerous or 1–2 ovules; style 1, short; stigma capitate. Capsule usually smooth, laterally flattened or turgid, 2-celled, unlobed or bilobed. Seeds numerous or 1–2, weakly to strongly rugose, flat or excavate on both sides.

A large genus with 460 species; cosmopolitan; 18 species in Iraq.

Veronica, originally a Gr. personal name Φερενικη, *ferenike*, "bringer of victory".

Ellmouni, F. Y., Karam, M. A., Ali, R. M., & Albach, D. C. (2017). Molecular and morphometric analysis of *Veronica* L. section Beccabunga (Hill) Dumort. Aquatic Botany 136: 95–111.
Musselman, L.J. (2011); Checklist of Plants of Lebanon and Syria http://ww2.odu.edu/~lmusselm/ plant/lebsyria/Checklist of Lebanon Plants.pdf.

1. Corolla white . . . . .9. *V. cymbalaria*
    Corolla blue or lilac, sometimes with a white or yellow throat. 2
2. Annuals. 3
    Perennials, often with a woody rootstock or rhizomatous .12
3. Flowers borne singly in leaf axils 4
    Flowers forming terminal or lateral racemes or spikes, bracts not resembling leaves . 7
4. Stem with 1–6 pairs of leaves 5
    Stem with numerous leaves, > 6 pairs. 6
5. Stem branched from base; sinus of capsule acute. . . . . .5. *V. arvensis*
    Stem branched in upper half; capsule with broad deep sinus . . . . .2. *V. syriaca*
6. Plant decumbent, leaves broader than long, crenate . . . . .8. *V. polita*
    Plant ascendent; leaves as broad as long, deeply serrate. . . . . 10. *V. hederifolia*
7. Plant densely glandular-hairy; fruiting pedicels 2x length of bracts . . . . 4. *V. viscosa*
    Plant eglandular and/or sparsely glandular-hairy; fruiting pedicels
      equal to or less than twice length of bracts 8
8. Lower bracts deeply 5-palmatifid, upper trifid . . . . 6. *V. triphyllos*
    Lower bracts ovate to lanceolate to linear, not 5-palmatifid. 9
9. Leaves with entire margins, glabrous. . . . . .3. *V. hispidula*
    Leaves with crenate to weakly serrate-dentate to subentire margins 10
10. Corolla pale-blue, 2–4 mm across. . . . . .7. *V. intercedens*
    Corolla bright blue with white centre 11
11. Annual 4–10 cm with unbranched stems; pedicels enclosed by bracts 11. *V. avromanica*
    Annual 10–15 cm with branched stems; pedicels exceeding bracts. . . . . .1. *V. bozakmanii*
12. Inflorescence a single raceme, terminal (alpine plants) 13
    Inflorescence much-branched, with numerous flowers 14

13. Plant 5–10 cm, stems decumbent, creeping at base; leaves not forming
a basal rosette; corolla deep blue with yellow eye, 7–9 mm across . . . . . . 12. *V. davisii*
Creeping mat-forming plant; leaves forming a basal rosette; cauline
leaves 1–5 pairs; corolla pale to deep blue, 8–15 mm across . . . . . . 13. *V. gentianoides*
14. Corolla blue to pinkish-blue with a white throat 15
Corolla deep blue to bluish to pinkish-blue without a white throat 16
15. Stems short curved appressed pubescent; peduncle 10–30 mm; corolla
deep blue, 6–10 cm across . . . . 14. *V. microcarpa*
Stems puberulent or pubescent, or glabrous; peduncle 2–6 cm; ; corolla
blue to pinkish-blue 8–14 mm across . . . . 15. *V. orientalis*
16. Stems creeping in lower part; plant to 20 cm tall; leaves shortly but
distinctly petiolate; racemes usually with fewer than 15 flowers; plant
glabrous, eglandular . . . . 18. *V. beccabunga* subsp. *abscondita*
Stems erect, or creeping only at base; plant 20–100 cm tall; upper stem
leaves sessile, or lamina tapering into a very short petiole; racemes
usually with more than 15 flowers; plant glabrous or pubescent or
racemes glandular hairy 17
17. Leaves 10–35 mm, ± sessile, obovate to oblanceolate or elliptic, margins
irregularly crenate or serrate, glandular and eglandular.
. . . . 16. *V. macrostachya* subsp. *schizostegia*
Leaves 8–100 mm, lower leaves ovate to ovate-lanceolate, petiolate to
subsessile; margins weakly crenate, denticulate or serrulate, glabrous 18
18. Corolla blue with violet veins, 5–10 mm across; capsule, orbicular to
elliptical, usually shorter than calyx . . . . 17. *V. anagallis-aquatica*
Corolla blue to pale blue to pinkish to white; capsule suborbicular to
ovate, about as long as or longer than calyx . . . . .18. *V. anagalloides* subsp. *heureka*

1. **Veronica bozakmanii** *M.A.Fisch.*, Oesterr. Bot. Z. 120(4–5): 418 (1972); Fischer in Fl. Turkey [P. H. Davis] 6: 707 (1978); Fischer in Fl. Iranica [K. H. Rechinger] 147: 66 (1981).

Annual, erect, 10–15 cm, branched, blackening when dry. Stems sparsely hairy with short eglandular hairs and longer glandular hairs. Leaves subsessile below, sessile above, ovate, base truncate to cuneate, margins weakly toothed. Raceme 7–28-flowered, glandular-pubescent. Bracts oblanceolate. Pedicels erecto-patent, exceeding the bracts, pubescent and glandular-hairy. Calyx 4-lobed. Corolla bright blue with white eye, 4–6 mm across. Capsule heart-shaped, 1.5x calyx, with acute sinus and broadly truncate base. Style equal in length to depth of sinus. Stamens blue. Seeds 10–20, elliptic, flat, smooth, sticky when wet.

HAB. Pastures, stream sides, shallow stony soil, damp places, hillsides; alt. 760–1520 m; fl. & fr. Mar.-May.
DISTRIB. Occasional; scattered across the forested zone of northern Iraq. **MAM**: S.W. of Aqra, *N. Polunin et al.* 88!; Sarsang, *Haines* W. 1061! 1062!; near Rabatki, *anon.* 4306! **MRO**: Shaqlawa, *Guest* 2015!; Rowanduz, *Guest* 2050! **MSU**: Sulaymaniya, *Graham* 752! Kuh-e Safin, Bornm. 1623. nr Penjwin, *Rechinger* 1047, *Hadač* 4962.

Related to *V. acinifolia*, but differing in having stem usually unbranched at base and fruiting pedicels curved upwards.

Greece, Turkey, Armenia, Iran, Afghanistan.

2. **Veronica syriaca** *Roem. & Schult.*, Syst. Veg., ed. 15 bis [Roemer & Schultes] 1: 116 (1817); Dinsmore in Post, Fl. Syria, Palest. & Sinai ed. 2, 2: 304 (1933); Rawi in Dep. Agr. Iraq Tech. Bull. 14: 147 (1964); Fischer in Fl. Turkey [P. H. Davis] 6: 711 (1978); Feinbrun-Dothan, Fl. Palaest. 3: 204, pl. 345 (1978); Boulos & Snogerup in Fl. Egypt (Boulos) 3: 79 (2002); Taifour & El-Oqlah, Pl. Jordan Annot. Checklist 120 (2017).

V. debilis Freyn in Bull. Herb. Boiss. 4: 56 (1896).

Annual, erect, 8–12 cm, scarcely blackening when dry. Stems puberulent, eglandular, branching in upper half or unbranched. Leaves subsessile, linear-oblong to suborbicular, 5–15 × 3–10 mm, weakly crenate to entire, glabrous. Terminal racemes with pedicels patent to erect, tending to curve upwards in fruit, clothed with fine glandular hairs. Bracts

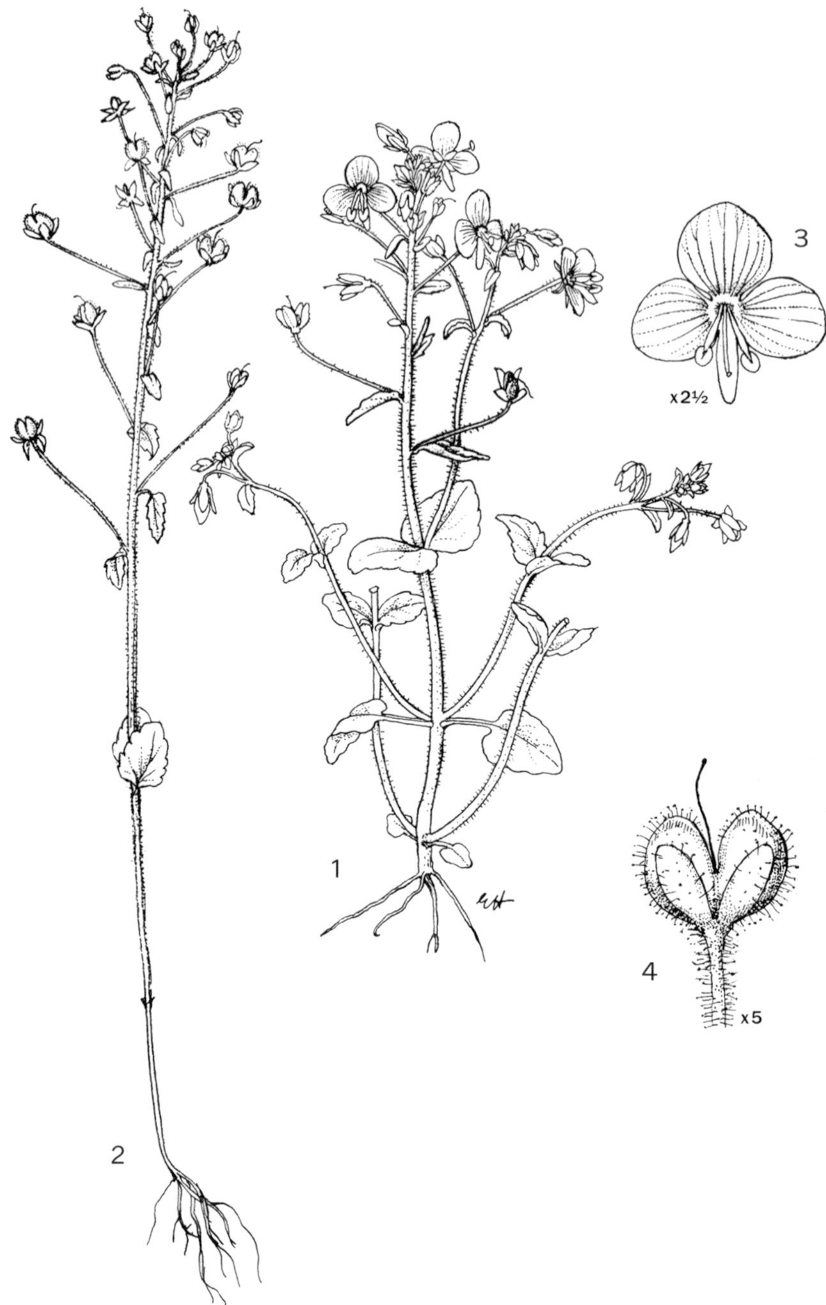

Fig. 41. **Veronica syriaca**. 1, habit with flowers × 1; 2, habit with fruit; 3, flower × 2 ½ ; 4, fruit × 5. Reproduced with permission from Feinbrun-Dothan, Fl. Palaestina 3: Plates, f. 345 (1977). Drawn by Esther Huber. © The Israel Academy of Sciences and Humanities.

becoming narrower above. Corolla blue with white eye, 7–9 mm across. Capsule 2–3.5 × 3–5.5 mm, glandular-pubescent to glabrous, ± truncate at base, with a deep sinus. Seeds (in ours) with chalazal podium on ventral size. Fig. 41, 1–4.

HAB. Open vegetation, fallow land, hillsides; alt. 760–1520 m; fl. & fr. Mar.-Apr.

DISTRIB. Scattered in the lower mountain zone. **MAM**: 5 km E. of Sarsang, *F. & E. Barkley*!; Amadiya,

Guest 1263!; near Rabatki, *anon.* 4306a! **MRO**: Salahaddin, Pirnum Dagh, *Gillett* 10483!; Jindian nr. Rowanduz, *Guest* 751! **FUJ**: Sharqat (Ashur), iv. 1920, *R. Thomas* s.n.! **FPF**: Khanaqin, *Haines* W. 919!

Turkey (S. Anatolia), Syria, Palestine, Jordan, Egypt.

3. **Veronica hispidula** *Boiss. & Huet* in Boiss., Diagn. Pl. Orient. ser. 3, 3: 173 (1856); Fischer in Fl. Turkey [P. H. Davis] 6: 709 (1978); Fischer in Fl. Iranica [K. H. Rechinger] 147: 68 (1981).

V. nudicaulis Kar. & Kir., Bull. Soc. Nat. Mosc. 15: 416 (1842) non Lam. (1805).

Annual, erect, 3–15 cm, unbranched or nearly so. Stems puberulent with upward-curving hairs, eglandular. Petiole very short, lamina linear to oblanceolate, 4–12 × 1.5–3.5 mm, entire, glabrous, margins entire. Raceme 5–20-flowered, axis straight, with appressed eglandular and glandular-pubescent hairs. Bracts 2–5 mm, lanceolate to linear. Fruiting pedicels 2–4 mm, erecto-patent with the greatest deflexion below fruit. Corolla blue, 2–3 mm across. Capsule enclosed by or slightly exceeding calyx, 1.5–2.5 × 3.5–4.5 mm, sparsely glandular-ciliate to glabrous; sinus subacute or to 90 deg. wide. Seeds elliptical to oblong, ± convex, smooth, pale brownish, sticky when damp. (Description adapted from Flora Iranica & Flora of Turkey).

HAB. Damp sandy places, by streams, below melting snow patches; alt. 2000–2250 m; fl. ?May. DISTRIB. Rare in the alpine zone. **MRO**: Mt. Qandil nr Pushtashan, *Rechinger* 11829 p.p.; Halgord Dagh near Tenji Saipiyan (Sarchal), *Hadač* 2092, 2166, 2199.

Lebanon, Turkey, Iran, Pakistan, C. Asia.

4. **Veronica viscosa** *Boiss.*, Diagn. Pl. Orient. ser. 1, 12: 47 (1853); Rawi in Dep. Agr. Iraq Tech. Bull. 14: 147 (1964); Fischer in Fl. Turkey [P. H. Davis] 6: 712 (1978); Fischer in Fl. Iranica [K. H. Rechinger] 147: 69 (1981); Taifour & El-Oqlah, Pl. Jordan Annot. Checklist 120 (2017).

Annual, erect, 8–20 cm, little-branched, blackening when dry. Stems crispate-hairy, eglandular. Petiole very short, lamina broadly ovate, 10–18 × 6–15 mm, subglabrous, margins coarsely crenate-serrate. Raceme 12–16-flowered, densely glandular- and glandular-pubescent. Lower bracts leaf-like, serrate, upper lanceolate, entire. Fruiting pedicels up to 9 mm, to twice as long as bracts. Corolla bright blue, 3–4 mm across, enclosed by calyx. Capsule enclosed in calyx, 4–5 × 6–8 mm, glabrous with prominent veins, sinus 90 deg. or wider. Seeds suborbicular, yellowish, smooth, becoming sticky when wet.

HAB. Hillsides, limestone scree; alt. 900–2000 m; fl. & fr. May. DISTRIB. Very occasional in the montane zone. **MRO**: Plinyan 17 km N.W. of Rania, *Rawi, Nuri & Kass* 28658!; Haji Umran, *Chapman* 11182! **MJS**: Jabal Sinjar, Kursi, *Gillett* 11032!

Turkey, Iran, Lebanon, Jordan.

5. **Veronica arvensis** *L.*, Sp. Pl. 1: 13 (1753); Fischer in Fl. Turkey [P. H. Davis] 6: 713 (1978); Feinbrun-Dothan, Fl. Palaest. 3: 204, pl. 344 (1978); Fischer in Fl. Iranica [K. H. Rechinger] 147: 75 (1981); Abedin, Qaiser & Sarwar in Fl. Pakistan [Ali & Qaiser] 220: 280 (2015); Taifour & El-Oqlah, Pl. Jordan Annot. Checklist: 120 (2017).

Erect annual, to 25 cm, ± strongly branched, stem with 3–6 pairs of leaves, ± villous with mainly eglandular hairs which can be in two rows. Lower leaves shortly petiolate, upper sessile; lamina broadly ovate, sparsely pubescent, margins crenate-serrate. Inflorescence 15–30-flowered, densely pubescent with both glandular and eglandular hairs. Bracts coarsely serrate at base. Flowers subsessile; corolla blue, 2–3 mm across. Capsule obcordate, 2.5–3.5 × 3–4 mm, sinus acute, glandular-ciliate on keel, elsewhere glabrous. Seeds 10–25, ovate, slightly rugulose, yellowish. Fig. 42, 1–2.

HAB. Clearings in *Quercus* forest, grassy places, fields, riverbanks; alt. 250–930 m; fl. Mar.-May. DISTRIB. **MRO**: Shaqlawa, *Haines* 774! **MSU**: Darband-i Khan, *Haines* s.n.!; Takiya, *Hadač* 1136; Benawa Suta nr Penjwin, *Hadač* 4975. **FUJ**: S. of Mosul, *Anders* 2609!

Europe, Anatolia, Caucasus, C. Asia, N. Africa; widely introduced and naturalized.

Fig. 42 **Veronica arvensis**. 1, habit × 1; 2, fruit with calyx × 5. Reproduced with permission from Feinbrun-Dothan, Fl. Palaestina 3: Plates, f. 344 (1977). Drawn by Levy Benjamini. © The Israel Academy of Sciences and Humanities.

6. **Veronica triphyllos** *L.*, Sp. Pl. 14 (1753). Fischer in Fl. Turkey [P. H. Davis] 6: 716, figs. 22 & 23 (1978); Fischer in Fl. Iranica [K. H. Rechinger] 147: 77 (1981).

Annual, erect, 5–15 cm, with several spreading branches. Stems densely puberulent with patent or retrorse straight hairs 0.05– 0.2 mm and scattered longer glandular hairs. Lower leaves ovate to triangular, 5–7 × 3–6 mm, coarsely crenate, the upper 5–12 × 4–12 mm, deeply palmate-pinnatifid with 5–7 lobes. Racemes lax, 5–15-flowered, clothed with very short eglandular hairs and much longer glandular hairs, tips often dark purple. Lower bracts deeply 5-palmatifid, upper trifid, lobes linear to subspatulate, glandular-pubescent. Pedicels 4–12 mm, patent or ± arcuate-ascending. Calyx 4–6(–8) mm long in fruit, lobes linear to oblanceolate, obtuse. Corolla deep blue, 5–9 mm across. Capsule 4–5 × 5–6 mm, suberect, surface weakly nerved, glandular-pubescent towards keel, sinus acute. Style 0.7–1.3 mm, ± equal to sinus. Seeds 10–18, 1.3–2 × 0.9–2 mm, suborbicular, lightly rugulose on the dorsal side, dark brown to blackish. (No material at K; description adapted from Fl. Iranica & Fl. Turkey).

HAB. In meadows and on stony slopes, sandy fields, edges of paths; alt. not stated on labels; fl. & fr. Feb.-May.
DISTRIB. Very local, recorded only twice from Iraq. **MAM**: Zawita, *Hadač et al.*, 3741; Zakho, by ancient bridge, *Hadač et al.* 3720.

Europe, S.W. & C. Asia; introduced in N. America. Doubtfully native to Iraq.

7. **Veronica intercedens** *Bornm.*, Beih. Bot. Centrabl., Abt. 22(2): 112 (1907). Type: in den Gebirgen östlich von Erbil (Assyrien), 12.5.1893, *Bornmüller* 1638 (JE, holo., WU, iso.). Rawi in Dep. Agr. Iraq Tech. Bull. 14: 147 (1964); Fischer in Fl. Turkey [P. H. Davis] 6: 720 (1978); Fischer in Fl. Iranica [K. H. Rechinger] 147: 91 (1981); Abedin, Qaiser & Sarwar in Fl. Pakistan [Ali & Qaiser] 220: 286 (2015).

Annual, with 2–4 subsessile leaves on a 5–15 cm puberulent unbranched stem. Leaves linear to linear-oblong, 25–30 × 2–3mm, sparsely pubescent to glabrous, margins entire to subserrate. Racemes 2–4, subterminal, 5–20-flowered, puberulent with short eglandular hairs. Bracts lanceolate. Pedicels 7–12 mm in fruit, recurved and deflexed. Calyx lobes partly connate, ovate to lanceolate, 6–9 mm, margins ciliate, otherwise glabrous. Corolla pale blue, 2–4 mm across. Capsule 3–4 × 4.5–5 mm, glandular-hairy, 2-lobed, with an acute sinus with style c. 1 mm slightly exceeding it. Seeds 6–10, subglobose, cymbiform, slightly rugulose.

HAB. *Quercus* and *Juniperus* woodland, rocky slopes, roadsides, fields and waste places; alt. 1500–2400 m; fl. & fr. Apr.-Jun.
DISTRIB. Lower montane zone of north-eastern Iraq. **MAM**: Sarsang, *Haines* 1060! **MRO**: Hisar-e Sakran, *Hadač & Kader* 5639 p.p!; Sakri Sakran, *Hadač & Kader* 5549, 5575; Ser-i Rost, *Thesiger* 953, 931 p.p; Zaita, *Hadač et al.* 5874. **MSU**: Benawa, *Hadač* 4931 p.p.; Mt. Hawraman near Tawila, *Rechinger* 12439 p.p. ;above Penjwin, *Hadac et al.* 4906. **FAR**: Kuh Sefin E. of Arbil, *Bornmüller* 1638 (type).

A single Iraqi specimen is cited in Flora Iranica 147: 84 (1981) as *V. argute-serrata* × *campylopoda*: **MSU**: M. Avroman (Hawraman) nr Tawilla (Tawila), *Rechinger* 12439 p.p. It is likely to be a morphological variant; the entire complex is of hybrid origin according to Fischer (1975), *Untersuchungen über den Polyploidekomplex Veronica cymbalaria agg. (Scrophulariaceae)*. Pl. Syst. Evol. 123: 97.

Turkey, Iran, Afghanistan, Pakistan, C. Asia.

8. **Veronica polita** *Fr.*, Novit. Fl. Suec. Alt. 1 (1828); Dinsmore in Post, Fl. Syria, Palest. & Sinai ed. 2, 2: 305 (1933); Fischer in Fl. Turkey [P. H. Davis] 6: 720 (1978); Feinbrun-Dothan, Fl. Palaest. 3: 206, pl. 348 (1978); Fischer in Fl. Iranica [K. H. Rechinger] 147: 99 (1981); Hong-Deyuan & Fischer in Fl. China [Ahsan Al-Shahbaz et al.] 18: 70 (1998); Collenette, Wild Flow. Saudi Arabia: 696 (1999); Boulos, Fl. Egypt 3: 79 (2002); Abedin, Qaiser & Sarwar in Fl. Pakistan [Ali & Qaiser] 220: 288 (2015); Ghazanfar, Fl. Oman 3: 153 (2015); Taifour & El-Oqlah, Pl. Jordan Annot. Checklist: 120 (2017).

V. didyma Prod. Fl. Napol. 6 (1811); Rawi in Dep. Agr. Iraq Tech. Bull. 14: 146 (1964).
V. agrestis auct: Rawi in Dep. Agr. Iraq Tech. Bull. 14: 146 (1964).

Decumbent annual, much-branched at base, Stems 5–30 cm, clothed with long eglandular hairs, with 1–3 pairs of leaves the lowest of which are wider than long and regularly crenate-serrate, equal to bracts; leaves sub orbicular to ovate, 6–12 × 10 mm, more hairy on lower surface. Flowers single in leaf axils, pedicels 6–15 mm, lower pedicels recurved in fruit. Calyx segments broadly ovate, overlapping, 4–7 mm in fruit. Corolla azure blue to pinkish

Fig. 43 **Veronica polita**. 1, habit × 1; 2, fruit with calyx × 5. Reproduced from Flora of China 18: f. 106. 1988, with permission from Missouri Botanical Garden Press, St. Louis, and Science Press, Beijing. Drawn by Ji Chaozhen.

purple, 3–6 mm across. Capsule 3–4 × 4–6 mm, with long glandular hairs and shorter crispate eglandular hairs, lobes parallel with deep narrow sinus, style ± exerted from sinus. Seeds 16–24, broadly elliptical, cymbiform, rugose-costate, pale brown in colour. Fig. 43, 1–3.

HAB. Woodland, dry rocky slopes, irrigated fields, meadows, by irrigation ditches, vineyards, rice fields, olive groves, date groves, gardens, waste places; alt. ± 250 m and above; fl. Sep.- Jun.

DISTRIB. Mountain slopes, lower hills and alluvial plains of northern and central Iraq. **MRO**:

Rowanduz, *Guest* 602! **MJS**: Jabal Sinjar, *Widad & Khayat* 53560! **FAR**: 42 km S. of Hajji Omran to Arbil, *Shahwani* 25263!; Altun Kupri, *Hadač* 1330, *Guest* 381!; bank of river Khazir between Kirkuk and Tuz Khurmatu, *Hadač et al.* 3849; Tuz Khurmatu, *Rawi & Gillett* 10285! **FUJ**: between Hadhar and Sharqat, *Hadač & Haines* 3151!; Qara Sharqat, Maresch 93; S. of Mosul, *Anders* 2609, 1724!; Balad Sinjar, *Omar & Hamid* 36504! **FNI**: Bashiqah, *Hadač* 1567; Nineveh, *Barkley & Brahim* 4613!, *Anders* 1026!; 30 km W. of Arbil, near Eski Kellek, *Rawi* 25166! **FPF**: Khanaqin, *Haines* s.n.!; Mansur (Table Mountain), *Graham* 710!; 10 km from of Jalaula, *Hadač* 2983; Mirjana, *Sutherland* 25! **DSD**: Karbala, *Rawi & Gillett* 6333! **LCA**: 25 km N. of Baghdad, *Omar* 47526!; Baghdad, *Rogers* 049! *Guest* 1112!; Khadimiya botanic garden, 16.3.70, *Sheikh s.n.!*; Falluja, *Guest* 980! **LEA**: Kut al-Imara, *Gillett* 6574!; Abul-Khasib, *Thamar* 47410!

HAUWA ZATTA (Ir.); a weed of date gardens.

Veronica persica Poiret is cited from two localities in Iraq by Fischer in Flora Iranica 147: 102 (1981), but is not specifically mentioned for Iraq in his earlier account in Flora of Turkey 6: 721 (1978). Like *V. polita*, it is a weed suspected of having resulted from the hybridisaion of *V. polita* and *V. ceratocarpa*: see Fischer (1975), *Untersuchungen über den Polyploidekomplex Veronica cymbalaria agg.(Scrophulariaceae).* Pl. Syst. Evol. 123: 97.

Europe, N. Africa, Asia; widespread and naturalized in irrigated and waste fields.

9. **Veronica cymbalaria** *Bodard,* Mem. Veronique Cymbalaire 3 (1798); Dinsmore in Post, Fl. Syria, Palest. & Sinai ed. 2, 2: 306 (1933); Rawi in Dep. Agr. Iraq Tech. Bull. 14: 146 (1964); Fischer in Fl. Turkey [P. H. Davis] 6: 723 (1978); Feinbrun-Dothan, Fl. Palaest. 3: 206, pl. 349 (1978); Qaiser in Jafri & El Gadi, Fl. Libya 88: 13 (1982); Husseini & Shams, Bull. Fac. Sci. Cairo Univ. 75: 157 (2007); Albach, New Phytologist 176: 481 (2007); Taifour & El-Oqlah, Pl. Jordan Annot. Checklist: 120 (2017). Incl. *V. cymbalarioides* C.I.Blanche ex Boiss., Fl. Orient. [Boissier] 4(2): 468 (1879).

> *V. panormitana* Tineo ex Guss., Fl. Sicul. Prodr. Suppl.(1): 4 (1832).
> *V. cymbalaria* var. *baradostensis* M.A.Fisch., Pl. Syst. Evol. 123(2): 99 (1975).
> *V. panormitana* subsp. *baradostensis* (M.A.Fisch.) M.A.Fisch., Nouv. Fl. Liban & Syrie [P. Mouterde] 3(3): 265 (1980); Fischer in Fl. Iranica [Rechinger] 147: 108 (1981).
> *V. panormitana* subsp. *panormitana* M.A.Fisch., in Fl. Iranica [K. H. Rechinger] 147: 108 (1981).
> *V. cymbalaria* subsp. *panormitana* (Tineo ex Guss.) O.Bolòs & Vigo, Collect. Bot. (Barcelona) 14: 98 (1983).

Decumbent to ascending annual, open much-branched, to 30 cm; stems sparsely hairy. Lower leaves long-petiolate, margins deeply crenate, broadly ovate to ovate-triangular, 8–12 × 7–15 mm. Flowers single in leaf axils; pedicels 10–30 mm, patent or recurved, pilose to glabrous. Calyx segments ovate to orbicular, eglandular or glandular-pubescent or glabrous. Corolla white, 4–12 mm across. Capsule transversely ellipsoid, 2.5–3.5 × 3.5–4.5 mm, apex emarginate but lacking a deep sinus, glabrous or pubescent. Seeds 3–4, subglobose, rugulose, brownish. Fig. 44, 1–3.

HAB: Weed of fields, shaded gardens, under shade of limestone cliffs, loam on conglomerate, stony hillsides, by streams and irrigation ditches, grassy slopes; alt. 130–1400 m; fl. Mar.-May, fr. Apr.-Jun. DISTRIB. Mainly in the humid mixed shrubland of the foothills in N. Iraq. **MAM**: Aqra, *Rawi* 11394!; *Polunin & Naib* 89!; 11 km S. of Dohuk, *F. Barkley & Brahim* 4626! 4628!; Bekma, *Helbaek* 770! **MRO**: Sefin Dagh, *Gillett* 8159!; Shaqlawa, *Wheeler-Haines* W. 777!; *Rawi* s.n.!; Rowanduz gorge, *Guest* 603! 2056! 2116! *Guest et al.* 15468! 15522! **MJS**: Jabal Sinjar, 6.5.1867, *Haussknecht*!; Kursi, Jabal Sinjar, *Gillett* 10901!; 70 km N.E. of Shaqlawa, *Shahwani* 25224!; Shaqlawa, *Rawi* 10442! 10443! *Wheeler-Haines* W. 776! *Barkley & Brahim* 7092! **MSU**: 7 km S. of Darband-i Khan, *F & E. Barkley & Shalli* 5154!; E. of Darband-i Khan, *F. Barkley & Haddad* 7752!; 110 km N. of Jalaul, *F. & E. Barkley* 4442! Tawila, *Rawi* 21877! **LCA**: Baghdad, Karrada Mariam, *Wheeler-Haines* W. 589!

A complex polyploid taxon showing morphological variations in pubescence and other characters throughout its range of distribution. I have included *V. panormitana* subsp. *panormitana* and *V. panormitana* subsp. *baradostensis* under *V. cymbalaria* which have been treated under V. *panormitana* by Fischer in Fl. Iranica (l.c.).

S. Europe, Crimea, Turkey, Syria, N.W. Africa.

10. **Veronica hederifolia** *L.*, Sp. Pl. 1: 13 (1753); Dinsmore in Post, Fl. Syria, Palest. & Sinai ed. 2, 306 (1933); Rawi in Dep. Agr. Iraq Tech. Bull. 14: 147 (1964); Feinbrun-Dothan,

Fig. 44. **Veronica cymbalaria**. 1, habit × 1; 2, flower × 10; 3, fruit × 10. Reproduced with permission from Feinbrun-Dothan, Fl. Palaestina 3: Plates, f. 349 (1977). Drawn by Ruth Koppel. © The Israel Academy of Sciences and Humanities.

Fl. Palaest. 3: 206, pl. 350 (1978); Fischer in Fl. Turkey [P. H. Davis] 6: 725 (1978); Fischer in Fl. Iranica [K. H. Rechinger] 147: 109 (1981); Taifour & El-Oqlah, Pl. Jordan Annôt. Checklist: 120 (2017).

Diminutive annual with ascending habit, single-stemmed or little-branched from base; stems 10–20 cm, pubescent. Lower leaves and bracts orbicular, 3- or 5-lobed, with a large terminal lobe. Flowers solitary in leaf-axils; pedicels short, exceeded by leaves. Calyx segments cordate-ovate, 4–5 mm, acute, ciliate, ± erect in fruit. Corolla blue or pinkish-purple, shorter than calyx, 4–8 mm across. Capsule 4 × 4 mm, subglobose, shallowly 4-lobed, glabrous. Seeds suborbicular, cymbiform.

HAB: Damp soil; alt. 1050–1220 m; fl. & fr. Apr.-May.
DISTRIB: Rare in forest zone of N. Iraq. **MAM**: Amadia (Amedi), *Guest* 1220! **MRO**: Shaqlawa, *Wheeler-Haines* W. 778!; Jindian near Rowanduz, *Guest* 743!

Native to Europe, widely introduced to eastern and western N. America, E. Asia, Australia and Africa.

11. **Veronica avromanica** *M.A.Fisch.*, Fl. Iranica [K. H. Rechinger] 147: 73 (1981). Type: Mt. Avroman ad confines Persiae in ditone pagi Tawilla (Tawila), in glareosis calc., 1800–2000 m, 15–18.6.1957, *Rechinger* 10334 (W, holo.). Fischer in Fl. Iranica [K. H. Rechinger] 147: 74 (1981).

Annual, 4–10 cm, stem unbranched, shortly pubescent. Leaves broadly to narrowly lanceolate, cuneate at base, subsessile, 10–20 × 2–8 mm, margins weakly serrate-dentate or subentire, sparsely glandular and eglandular-hairy or almost glabrous, darkening when dried. Racemes 1–15, dense, 15–40-flowered, on 3–15 mm glandular peduncles; pedicels in fruit 4–8 mm, enclosed by linear to oblanceolate bracts, patent. Corolla bright blue with white centre. Capsule in fruit 3.5–5.5 × 3.8–4.7 mm, shorter than calyx, clothed in short glandular hairs; style greatly exceeding the sinus. Seeds 4–6, suborbicular, deeply cymbiform, smooth, yellowish.

(See Flora Iranica, op. cit. p. 74, for a discussion of its similarities to other species).

HAB: Limestone and serpentine screes, clearings in *Quercus* woodland, mountain slopes; alt. (900–) 1400–3000 m; fl. Jun.
DISTRIB: Alpine zone of N.E. Iraq. **MRO**: mt. Qandil above Pushtashan, *Rechinger* 11829 p.p.; Mt. Helgord N. of Rost, *Thesiger* 936; Shaqlawa, *Wheeler-Haines* W. 772. **MSU**: Mt. Hawraman near Tawila, *Rechinger* 10334 (type); near Penjwin, *Rechinger* 10487 (W!); Malakawa range, *Rechinger* 12319 (W).

Near endemic; found in W. Iran close to the frontier with Iraq.

12. **Veronica davisii** *M.A.Fisch.*, Pl. Syst. Evol. 128: 237, f. 1 (1977); Fischer in Fl. Turkey [P. H. Davis] 6: 705 (1978).

Perennial. Stem 5–10 cm, ± decumbent, creeping at base, branched, glabrous. Leaves oblong-spatulate to obovate-spatulate, 8–15 × 3–7 mm, glabrous, apex truncate, base cuneate, tapering into a short petiole, margins crenulate above, elsewhere entire. Raceme single, erect, terminal, (3–)5–8-flowered, densely patent-glandular hairy, ± bracteate at base, bracts oblanceolate. Pedicels slightly exceeding the bracts. Calyx 1.5–2.5 mm, 4-lobed, oblong, obtuse. Corolla deep blue with yellow eye, 7–9 mm in diameter. Ovary hairy. Capsule splitting lengthways (from apex).

HAB. Alpine flushes below melting snow patches, grassy slopes on calcareous, igneous and metamorphic rocks, among small stones where snow has melted; alt. 2750–3500 m.; fl. Jul.- Sep.
DISTRIB. Confined to the highest mountains of N.E. Iraq. **MRO**: Ser Kurawa, *Gillett* 9722!; mt. Helgord, *Gillett* 9609!; *Rawi & Serhang* 24742! *Guest* 2832! 3071! *Guest & Alizzi* 15199! *Rechinger* 11438a, 11877; mt. Qandil, *Rechinger* 11795! *Rawi & Serhang* 26783!; Hisar-i Sakran, *Hadač* 5617.

Near endemic, found outside Iraq only on Sat Dağ and Cilo Dağ, in Turkey's Hakkâri province, but likely to occur in Iran close to the Iraqi frontier.

Turkey (S.E. Anatolia).

13. **Veronica gentianoides** *Vahl*, Symb. 1: 1 (1790); Fischer in Fl. Turkey [P. H. Davis] 6: 705 (1978); Fischer in Fl. Iranica [K. H. Rechinger] 147: 140 (1981).

Creeping, often mat-forming, perennial. Flowering stems unbranched, ascending to erect, glabrous or sparsely hairy above. Basal leaves in a rosette, short-petiolate, lamina 15–40 × 2–12 mm, obovate to narrowly lanceolate, tapering at base, margins remotely serrate to subentire, clothed with short sparse glandular hairs or subglabrous. Cauline leaves sessile, 1–5 pairs, ± pubescent with glandular hairs. Raceme single, terminal, 10–40-flowered, densely glandular-pubescent, with lanceolate bracts. Pedicels 4–15 mm, ± exceeding bracts. Calyx with 4 or occasionally 5 lanceolate lobes, 2–4 mm. Corolla pale to deep blue, 8–15 mm across. Style 4–8 mm. Capsule orbicular, 4–7 × 5–7 mm, twice as long as calyx. Seeds 10–20 per capsule, 1.8–2.4 × 1.5–2.0 mm, flattened, smooth, brownish.

HAB. Alpine marshes and damp places; alt. 3000–3800 m; fl. May-Aug.
DISTRIB. Confined to the highest mountains of N.E. Iraq. **MRO**: Ser Kurawa, *Gillett* 9751; western slopes of mt. Halgurd, *Rechinger* 11438-b.

Turkey, Crimea, Caucasia, N. & W. Iran.

14. **Veronica microcarpa** *Boiss.*, Diagn. ser. 1, 4: 76 (1844); Fischer in Fl. Turkey [P. H. Davis] 6: 736 (1978); Fischer in Fl. Iranica [K. H. Rechinger] 147: 122 (1981).

Woody perennial, branches from the base, stems ascending or decumbent, 10–20 cm, clothed with short curved appressed pubescence. Leaves 4–15 × 4–15 mm, orbicular to rhomboid or oblanceolate, rounded at apex; petioles 1–4 mm. Racemes 1–4(–6), 12–40-flowered, densely pubescent with (usually) glandular and eglandular hairs; peduncles 10–30 mm; pedicels 4–5 mm, equaling to twice length of bracts. Corolla deep blue, with a white throat, 6–10 mm across. Capsule orbicular, 2–3 × 2–3 mm, emarginate, pubescent. Seeds 2–4 per capsule, c. 1.5 × 0.8 mm, smooth, brownish.

HAB: Stony mountain slopes; alt. 2100+ m; fl. May-Jul.
DISTRIB: Known in Iraq only from two gatherings from **MRO**: Mr. Botin nr Rawandiz, *Agnew et al.* 6120, 6250!.

Turkey (E. Anatolia), Armenia, N.W. Iran.

15. **Veronica orientalis** *Miller*, Gard. Dict. ed. 8, no. 10 (1768). Fischer in Fl. Turkey [P. H. Davis] 6: 748 (1978); Fischer in Fl. Iranica [K. H. Rechinger] 147: 130 (1981).

Strongly branched from the woody base, stems suberect to decumbent or ascending, puberulent or pubescent with eglandular hairs, or glabrous. Leaves subsessile, petiole less than 1 mm; lamina suborbicular to broadly or narrowly elliptic, 4–35 × 2–15 mm, margins serrate, crenate or subentire. Racemes 2–6, 10–50-flowered, axis and pedicels thinly or thickly clothed with eglandular hairs; peduncle 2–6 cm. Bracts entire, linear-lanceolate to oblanceolate. Pedicels 2–5 mm in flower, 3–6 mm in fruit, erect to erecto-patent. Calyx 2–4.5 mm in flower, expanding up to 6 mm in fruit, ± retrorsely puberulous or glabrous, sepals 4, equal or two of them up to 5 × longer, linear to narrowly lanceolate. Corolla blue to pinkish-blue with a white throat, 8–14 mm across. Capsule obcordate to broadly transverse elliptic, flattened, 3–4 × 4–5 mm, somewhat exceeding the calyx lobes, with a distinct apical sinus, surface smooth, densely eglandular-pubescent or glabrous. Seeds 4–6 per capsule, 1.6–2.6 × 1.1–1.8 mm, broadly ovate, deeply rugose, brownish.

HAB: Rocky and stony slopes, grassland, meadows, vineyards, *Quercus* woodland and scrub, below melting snows; alt. 1050–3800 m; fl. Apr.-Aug.
DISTRIB: montane and alpine zones of N.E. Iraq, widespread. **MAM**: Chalki, 30.4.1951, *Mooney* s.n.!; Galli Zawita near Turkish border, *Rawi* 23637! Suwara Tuka between Dihok and Amadiya, *Rechinger* 11584; mt. Botin nr Shirwan Mazu, *Haines* W. 2096!; Zawita N. of Zakho, *Rechinger* 10945; between Sarsang and Zawita, *Anders* 2392; S.W. of Sarsang, *Anders* 1358; Sarsang, *O. Polunin* 5082!; Jabal Khantur nr Sharnish, *Rechinger* 10824, *Rawi* 8555!, 8586!; Mt. Gara, *Kotschy* 426,; Qara Dagh, *Alkas* 19520!, *Buthaina Makki* 434, *Hadač* et al. 5225, *Barkley* et al. 8498, *Gillett* 7917! 7919!, *Chapman* 26404!, *Poore* 353!; Pir Omar Gudrun, *Haussknecht* s.n., *Rawi* 12092!, *Gillett* 7086!; **MRO**: Siwuka, S. foot of Karoukh, *Alkas, Nuri & Serhang* 27361!, 27415!, 27455!, 27580!, 27587!; Marmarut, *Guest* 2150!; Ser Kurawa, *Gillett* 9721!; Mergan nr Bardanas, *Rawi* 24347a!; nr Rania, *Rawi* 18246!; mt. Kharokh between Saman and Kilkil, *Kass, Nuri & Serhang* 273401; mt. Baradost, *Thesiger* 886; N. of Rost, *Thesiger* 952; Halgurd Dagh, *Guest & Husham* (Alizzi) 15821! *Rawi* 13743! 13753!, *Rechinger* 11437, *Guest* 2830!, 2945!, *Hadač* 2077,

2079, 2200, 2201, 2249, 2389, 2646; Qandil range, *Rawi & Serhang* 24037! 24106! 24386! 24435! 24491! 24496! 24591!; above Pushtashan, *Rechinger* 11068; by lake Goam-e Kirmosoran, *Rechinger* 11119; mt. Potin (Botin), *Agnew* et al. 2096, *Hadač* et al. 6162, *Haines* W. 2096!; above Zeita, *Agnew* et al. 5950; nr Shaqlawa, *Haines* W. 780! *Bornmüller* 1629, 1630!; Chiya-i Marmarut, *Guest* 2150; Hisar-e Sakran, *Hadač* et al. 5619; mt. Sakri-Sakran, *Bornmüller* 1631; **MSU:** Pira Magrun, *Haussknecht* 6.1867, s.n.!, *O. Polunin* 5157!, *Gillett* 7786! mt. Hawraman & Shahu, *Haussknecht* s.n. **MJS:** Jabal Sinjar, *Gillett* 10950!, 11049!; **FAR:** Kuh Sefin, *Gillett* 8098!

Turkey (E. Anatolia), Syria, Lebanon, W. Iran, Transcaucasia.

16. **Veronica macrostachya** *Vahl*, Enum. Pl. 1: 71 (1804); Fischer in Fl. Turkey [P. H. Davis] 6: 742 (1978). Fischer in Fl. Iranica [K. H. Rechinger] 147: 135 (1981).

subsp. **schizostegia** (*Bornm.*) *M.A.Fisch.* in Fl. Turkey [P. H. Davis] 6: 744 (1978) is the local subspecies.

Stems 15–30 cm, woody at base, branched, creeping to ascending, ± densely pubescent. Leaves ± sessile, obovate to oblanceolate or elliptic, 10–35 × 6–9 mm, cuneate at base, margins irregularly crenate or serrate, with glandular and eglandular hairs. Racemes 1–5, 12–50-flowered, dense, clothed with glandular hairs. Bracts 4–10 mm, linear, pinnatifid below, entire above. Pedicels 3–7 mm in fruit, erect, overtopped by bracts. Calyx 3–5 mm in flower, 4–6 mm in fruit, with 4 or more rarely 5 linear lobes. Corolla dark blue, violet, pink or whitish, 5–12 mm across. Capsule obcordate, flattened, 3.5–4.0 × 4.0–4.5 mm, shorter than calyx, glandular-pubescent. Style 2.5–3.0 mm. Seeds few per capsule, 1.7–1.9 × 1.0–1.3 mm, ovoid, rugose, shining, brownish. (Description adapted from Fl. Iranica & Fl. Turkey).

HAB. Rocky slopes and screes; alt. 800–2000 m; fl. May-Jun.
DISTRIB. **MAM:** Jabal Khantur, nr Sharanish, *Rechinger* 10847. **MRO:** mt. Baradost, nr Diana, *Field & Lazar* 921; Kurrak Dagh near Rawandiz, *Nábělek* 2030; mt. Handren nr Rawandiz, *Nábělek* 2031; foot of mt. Botin at Kani Mazu Shirin, Agnew et al. 6007; Kuh Safin nr Shaklawa, Bornm. 1628, *Haines* 779; **MSU:** Pira Magrun, *Haussknecht* 747; 12 km from Dihok towards Amadiya, *Rechinger* 11504; Ain Nuni between Aradin and Amadiya, *Nábělek* 2017; above Aradin, *Nábělek* 2000.

A near endemic subspecies, occurring outside Iraq only in Kermanshah province, W. Iran. Subsp. *macrostachya* occurs in Syria and Lebanon, and subsp. *sorgerae* M.A.Fisch. is found mainly in Turkey (S. Anatolia); subsp. *mardinensis* (Bornm.) M.A.Fisch. is endemic to S.E. Turkey. *V. aleppica* Boiss., Diagn. Pl. Orient. ser. 2, 3: 169 (1856) listed in Rawi in Dep. Agr. Iraq Tech. Bull. 14: 147 (1964) is a synonym of *V. macrostachya* subsp. *macrostachya*.

W. Iran.

17. **Veronica anagallis-aquatica** *L.*, Sp. Pl. 1: 12 (1753); Rawi in Dep. Agr. Iraq Tech. Bull. 14: 147 (1964); Fischer in Fl. Turkey [P. H. Davis] 6: 726 (1978); Fischer in Fl. Iranica [K. H. Rechinger] 147: 150 (1981); Collenette, Wild Flow. Saudi Arabia: 693 (1999); Hong-Deyuan & Fischer in Fl. China [Ahsan Al-Shahbaz et al.] 18: 79 (1998); Boulos, Fl. Egypt 3: 82 (2002); Ghazanfar, Hepper & Philcox, Scrophulariaceae, in Fl. Trop. East Africa [Beentje & Ghazanfar] 101 (2008); Abedin, Qaiser & Sarwar in Fl. Pakistan [Ali & Qaiser] 220: 300 (2015); Taifour & El-Oqlah, Pl. Jordan Annot. Checklist: 120 (2017).

Veronica oxycarpa Boiss. in Kotschy, Pl. Pers. no. 639 (1845); Rawi in Dep. Agr. Iraq Tech. Bull. 14: 147 (1964)
V. anagallis-aquatica subsp. *oxycarpa* (Boiss. in Kotschy), A.Jelen., Bull. Soc. Nat. Mosc. Biol. 74, 6: 76 (1969).
V. beccabungoides Bornm., Beih. Bot. Centralbl. 22(2): 111 (1907).

Perennial; Stems 30–50 cm, erect, simple or branched, glabrous or glandular-pubescent. Leaves 20–100 mm, glabrous, lower leaves ovate, petiole, upper ovate-lanceolate, subsessile, weakly crenate towards apex. Racemes 3x subtending leaves; pedicels 4–7 mm, erect or arcuate-ascending, exceeding the linear bracts. Calyx segments spreading, erect in fruit. Corolla blue with violet veins, 5–10 mm across. Capsule 2.5–4 × 2.5–3.5 mm, orbicular to elliptical. Style short. Seeds 40–70, elliptic to suborbicular, plano-convex.

HAB. Growing near water; alt. not stated; fl. & fr. Jul.
DISTRIB. Occasional. **FAR:** Taqtaq, *Omar, Karim & Hamza* 37149! Unlocalised: "Jundesan" (?Jindian), *Omar, Sahir, Karim & Hamid* 38420! "Grganek village", *Omar, Karim, Maza & Hamid* 37296!

V. anagallis-aquatica subsp. *lysimachioides* (Boiss.) M.A.Fisch., Fl. Iranica [K. H. Rechinger] 147: 154 (1981) and V. *anagallis-aquatica* subsp. *michauxii* (Lam.) A.Jelen., Bull. Soc. Nat. Mosc. Biol. 74, 6: 77 (1969) are reported from Iraq by Fischer in Fl. Iranica (l.c.), but I have not seen any specimens from Iraq. *V. anagallis-aquatica* is a widespread and variable taxon – where more material from Iraq is necessary to verify infraspecific taxa as cited by Fischer in Flora Iranica.

N., E. and C. Africa, Jordan, Palestine, the Arabian Peninsula, Temperate Europe and Asia; introduced into N. & S. America.

18. **Veronica anagalloides** *Guss.*, Pl. Rar. 5, t. 3, (1826) subsp. **heureka** *M.A.Fisch.*, Fl. Iranica [Rechinger] 147: 158 (1981). Type: Iraq, Mt. Avroman prope Tawilla, 1400 m, 15.– 18, VI, 1957, *Rechinger* 10290 (W, holo.).

V. heureka (M.A.Fisch.) Tzvelev, Bot. Zhurn. (Moscow & Leningrad) 69 (9): 1257 (1984).

Perennial or annual. Stems 5–20(–60) cm, erect, branched, glabrous. Leaves ovate to ovate-lanceolate, (8–)15–35(–60) mm, upper leaves sessile, lower subsessile, serrulate to denticulate, glabrous. Racemes axillary, branched, 15–30-flowered; pedicel 3–7 mm, erecto-patent, in fruit 5–8 mm. Peduncle 10–20 mm. Corolla blue to pale blue to pinkish to white. Capsule 2.5–4 × 2.5–3.5 mm, suborbicular to ovate, about as long as or longer than the calyx lobes. Style short. Seeds elliptic to suborbicular, brownish. Fig. 45, 1–2.

HAB. Dry river beds, ditches, near springs; alt. 450–1400 m; fl. & fr. Jul.
DISTRIB. **MAM**: Suwara Tuka, *Polunin* 5148. MRO: Arbil, Gali Ali Beg, *Agnew & al.* 6243! **MSU**: Mt Avorman [Hawraman] near Tawila *Rechinger* 10290 (type, W). **MSU**: Danbandikhan, *Haines* 1951. **LSM**; Dibis-Altin Kopru, *Agnew & al.* 5736.

Quite difficult to differentiate from *V. anagalloides.*

Turkey, Iran, Palestine, Syria, Lebanon, Turcomania, Afghanistan, Pakistan, W. Himalayas.

19. **Veronica beccabunga** *L.*, Sp. Pl. 12 (1753); Fischer in Fl. Turkey [P. H. Davis] 6: 731 (1978).

subsp. **abscondita** *M.A.Fischer* in Flora Iranica [K. H. Rechinger] 147: 146 (1981).

Perennial with creeping rhizome; Stems 5–20 cm, decumbent to ascending, thick, often reddish, pruinose but otherwise glabrous. Leaves shortly but distinctly petiolate, petiole 2–6 mm; lamina broadly oblong to elliptic, rounded or truncate at base, margins minutely crenulate-serrulate or subentire. Racemes 2–6, 5–15-flowered, glabrous.. Fruiting peduncles 20–40 mm, twice exceeding length of subtending bracts. Pedicels 4–8 mm in fruit, ± erect. Corolla twice as long as calyx, deep purplish-blue, 6–9 mm across. Capsule 3–4.5 × 3–4 mm, orbicular or broadly elliptic, base truncate, rounded-obtuse at apex, subglabrous. Style 2–3.5 mm.

HAB. Wet places, marshy areas by streams, pools, springs; alt. 350–3000 m; fl. Apr.-Aug.
DISTRIB. Widespread in montane areas of N.E. Iraq. **MRO**: Qandil above Pushtashan, *Rechinger* 11753; Halgurd Dagh, *Bornmüller* 1627, *Hadač* 2129, 2283; Hisar-i Sakran valley nr Rawandiz, *Hadač* et al. 5692; Tanji Sai[aiyan, *Hadač* 2202; Nowanda valley, *Hadač* 2383. **MSU**: nr Sulaimaniya, *Barkley* 8494! E. of Qala Diza, *Thesiger* 1162. **FKI**: by river Zab nr Kirkuk, *Grigg* 63! 10 km N. of Altin Kupri, *Rawi & Gillett* s.n.

Subsp. *muscosa* (Korsh.) A. Jelen. (1971) differs from the Iraqi subspecies in having a somewhat shorter style (1.2–1.9 mm), shorter fruiting peduncles (3–15 mm) encased by the subtending bracts, and pedicels patent in fruit. It occurs mainly in Afghanistan and further east . The type subspecies, subsp. *beccabunga*, occurs in Europe and N. Africa.

Turkey, Lebanon, Transcaucasia, Iran.

VERONICA PERSICA Poir., Encycl. [J. Lamarck & al.] 8: 542 (1808), native to Iran, North and Transcaucasus is introduced into Iraq found as a plant in gardens and in cultivated ground.

V. persica is an annual, branching from base with procumbent stems, pubescent crispate hairy. Leaves all petiolate, ovate to elliptic, subcordate at base, margins crenate-serrate. Flowers solitary in leaf axils. Pedicels longer than bracts, nodding in fruit. Calyx lobes in

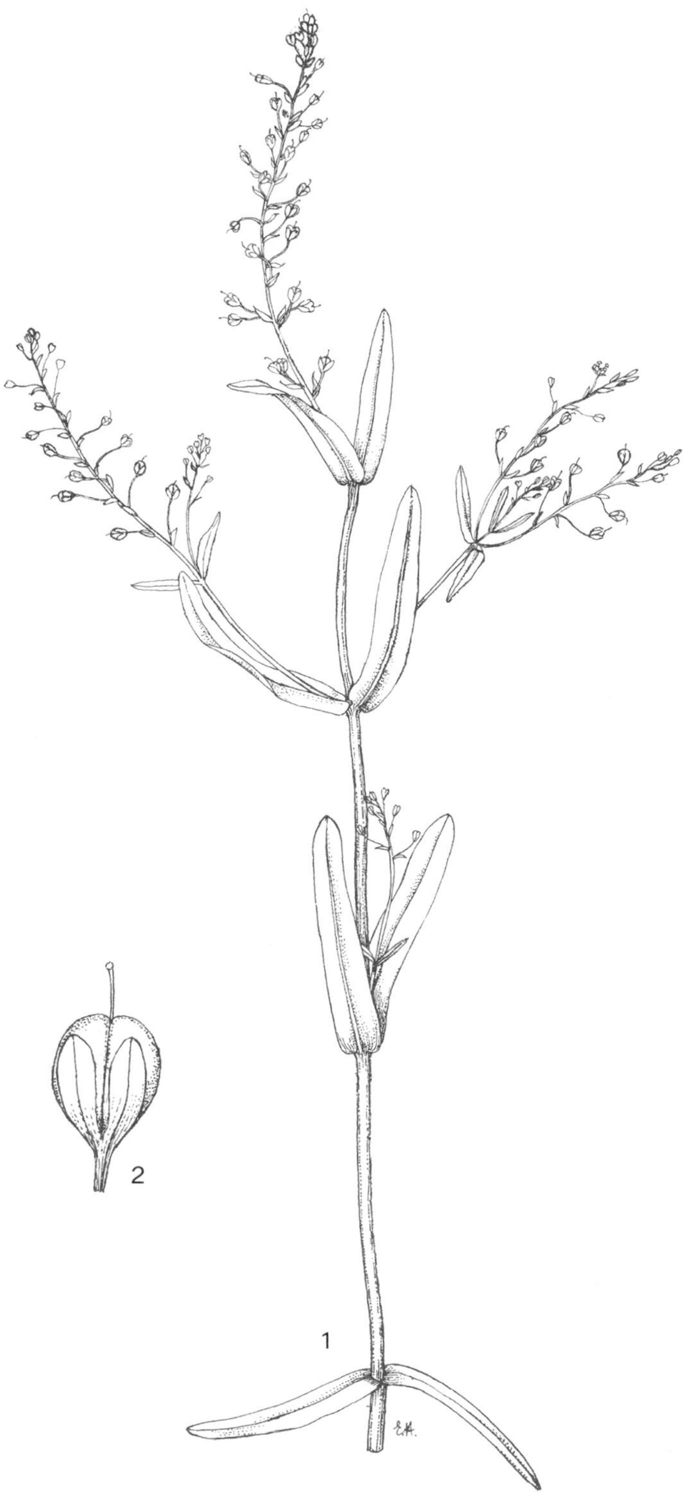

Fig. 45. **Veronica anagalloides**. 1, habit × 1; 2, fruit × 7. Reproduced with permission from Feinbrun-Dothan, Fl. Palaestina 3: Plates, f. 341 (1977). Drawn by Esther Huber. © The Israel Academy of Sciences and Humanities.

pairs, ciliate. Corolla blue, c. 10 mm across, longer than calyx. Capsule 2-lobed, the lobes wide, glandular-tomentellous. Seeds rugose.

VERONICA BILOBA is treated in Flora Iranica, p. 79. with one Iraqi record cited, but in the words of Fischer in Flora of Turkey 6: 718, "no literature records have been accepted for *V. biloba*, *V. bornmuelleri* and *V. campylopoda* because of frequent misidentifications".

10. **PEPLIDIUM** Delile

Descr. Egypte, Hist. Nat. 148, t. 4 (1813)
Fischer in Kubitzki (ser. ed.), J.W. Kadereit (ed.), Fam. & Gen. Vasc. Pl. 7: 402 (2004)

Shahina A. Ghazanfar

Aquatic annual with prostrate to creeping stems, rooting at nodes. Leaves opposite. Flowers axillary, ebracteate. Calyx 5-angled, 5-lobed. Corolla with tube shorter than calyx, 2-lipped, lobes ± equal. Stamens 2, filaments expanded at base; anthers 1-celled. Style expanded above and curved over the stamens. Capsule globose to ovoid, dehiscing irregularly or indehiscent. Seeds rugose with 4 longitudinal ridges.

Four species distributed from Egypt to Sinai, the Indian Subcontinent and Australia; a single species in Iraq.

Formerly included in the Scrophulariaceae, *Peplidium* is now included in the Phyrmaceae.

Peplidium: its author, Alire Raffeneau Delile (1778–1850), author of *Description de l'Égypte* (1812) was a member of Napoleon's brief campaign in Egypt and worked at Cairo's botanic garden. The name is possibly related to Gr. πεπλις, *peplis;* the suffix -ιδιον, *idion,* is the diminutive form of the word.

1. **Peplidium maritimum** (*L.f.*) Asch., Beitr. Fl. Aethiop. [Schweinfurth] 275 (1867); Rawi in Dep. Agr. Iraq Tech. Bull. 14: 144 (1964); Boulos & Snogerup in Fl. Egypt [Boulos] 3: 77 (2002); Abdulridha, Taha & Widad, Ecology & Fl. Basrah: 503 (2016).

Hedyotis maritima L.f., Suppl. Pl. 119 (1782).

Aquatic annual, 5–10 cm. Stems prostrate, rooting at nodes, glabrous. Leaves glandular-punctate, ovate to obovate, elliptic or ± orbicular, apex obtuse, margins entire, base attenuating into a short petiole, petioles of opposite leaves connected by membranous ridge across the node. Flowers purplish. Calyx tubular, 2–3 mm, 5-ribbed, shallowly 5-lobed. Corolla about as long or slightly longer than calyx, shallowly 2-lipped, upper lip 2-lobed, lower lip 3-lobed, lobes rounded. Capsule ovoid, rupturing at base. Fig. 46, 1–2.

HAB. Edges of marshes, in rice fields; rare; alt. ± 50 m (?); fl. Mar., Apr.
DISTRIB. Central alluvial plains and southern Iraq. **LCA**: Shatra, in rice field, *S. Omar & F. Karim* 36760!
LSM: Hor al Hammar, *Rawi* 16591!

Egypt, Sinai, Iran to India and Sri Lanka, S.E. Asia.

119. **ACANTHACEAE** Juss.

Gen. Pl. [Jussieu] 102 (1789) nom. cons.

Shahina A. Ghazanfar

Herbs, shrubs, twiners or (rarely) trees; stems and branches with or without cystoliths. Leaves mostly opposite, stipules absent. Flowers solitary or in dichasial cymes or cymules often aggregated into axillary and/or terminal paniculate to racemoid cymes. Bracts foliaceous; bracteoles usually present. Flowers bisexual. Calyx 4–5-lobed, fused at least at base, lobes equal to unequal lobes or 2-lipped. Corolla with a narrow basal tube, limb 5-lobed to distinctly 2-lipped, more rarely 1-lipped. Stamens 4, usually didynamous. Ovary 2-locular, surrounded by an annular disc at base; style simple, filiform, with 2 (one often reduced) linear to oblong or ovoid or capitate stigmatic lobes. Fruit a 2-locular loculicidal capsule (rarely a drupe), usually explosively dehiscent.

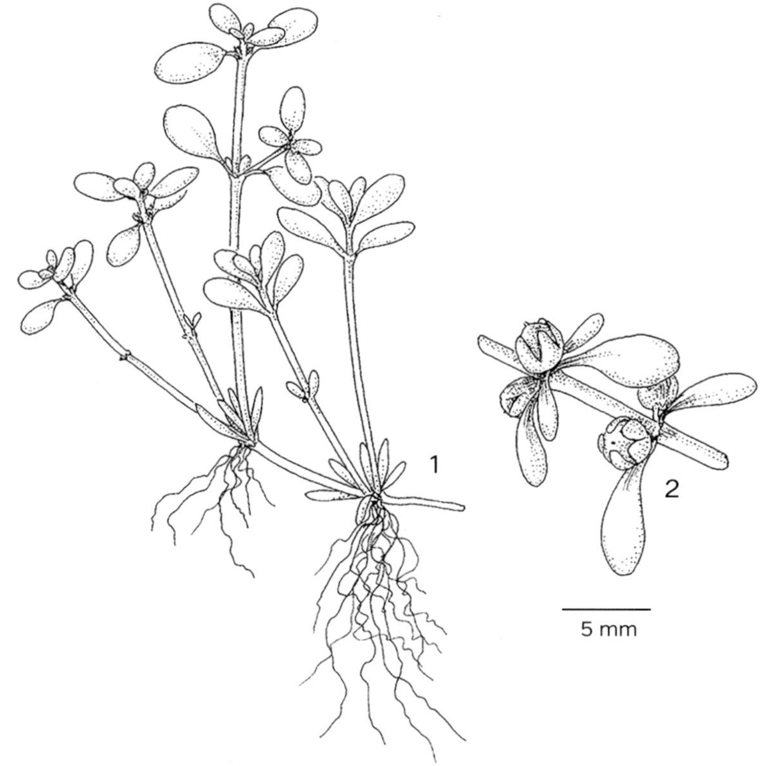

Fig. 46. **Peplidium maritimum**. 1, habit ; 2, fruiting branch. Reproduced with permission from Flora of Egypt 3: pl. 19, f. 7 (2002). Drawn by © M Tebbs.

About 250 genera and 2500–3000 species, widely distributed in all tropical and subtropical regions.

1. **ACANTHUS** L.

Sp. Pl.: 639 (1753) & Gen. Pl.: 286 (1754); Nees in Prodr. [A. P. de Candolle] 11: 270 (1847)

Perennial herbs, shrubs or small trees; cystoliths absent. Leaves opposite or in whorls of 4. Flowers in terminal or axillary racemoid cymes. Bracts large, with spinose margins or entire; bracteoles 2. Calyx divided to base, lobes thickened and horny at the base, 1–3-veined, entire or with 2–3 small teeth. Corolla tube short, white, retrorse hairy above; limb split dorsally into a 3–5-lobed lip; lobes broadly oblong, rounded or central emarginate. Stamens 4, exserted, inserted just inside throat; anthers 1-thecous, oblong, stiff hairy along the ventral side. Ovary 2-locular; style linear; stigma 2-lobed, lobes lanceolate, acute. Capsule 2–4-seeded, woody, ellipsoid, sessile, glossy. Seeds discoid, covered with hygroscopic hairs or glabrous to puberulous.

A genus of 25 species distributed from southern Europe to the Pacific and south through Africa to Angola, Zambia and Malawi; a single species in Iraq.

Acanthus from Ancient Greek ἄκανθος (ákanthos), from ἀκή (akḗ, "thorn") + ἄνθος (ánthos, "flower").

1. **Acanthus dioscoridis** *L.*, Cent. Pl. ll. 23 (1756); Rawi in Dep. Agr. Iraq Tech. Bull. 14: 148 (1964); Rechinger, Fl. Iranica 24: 3 (1966); Eneyat Hussain in Fl. Turkey [P. H.

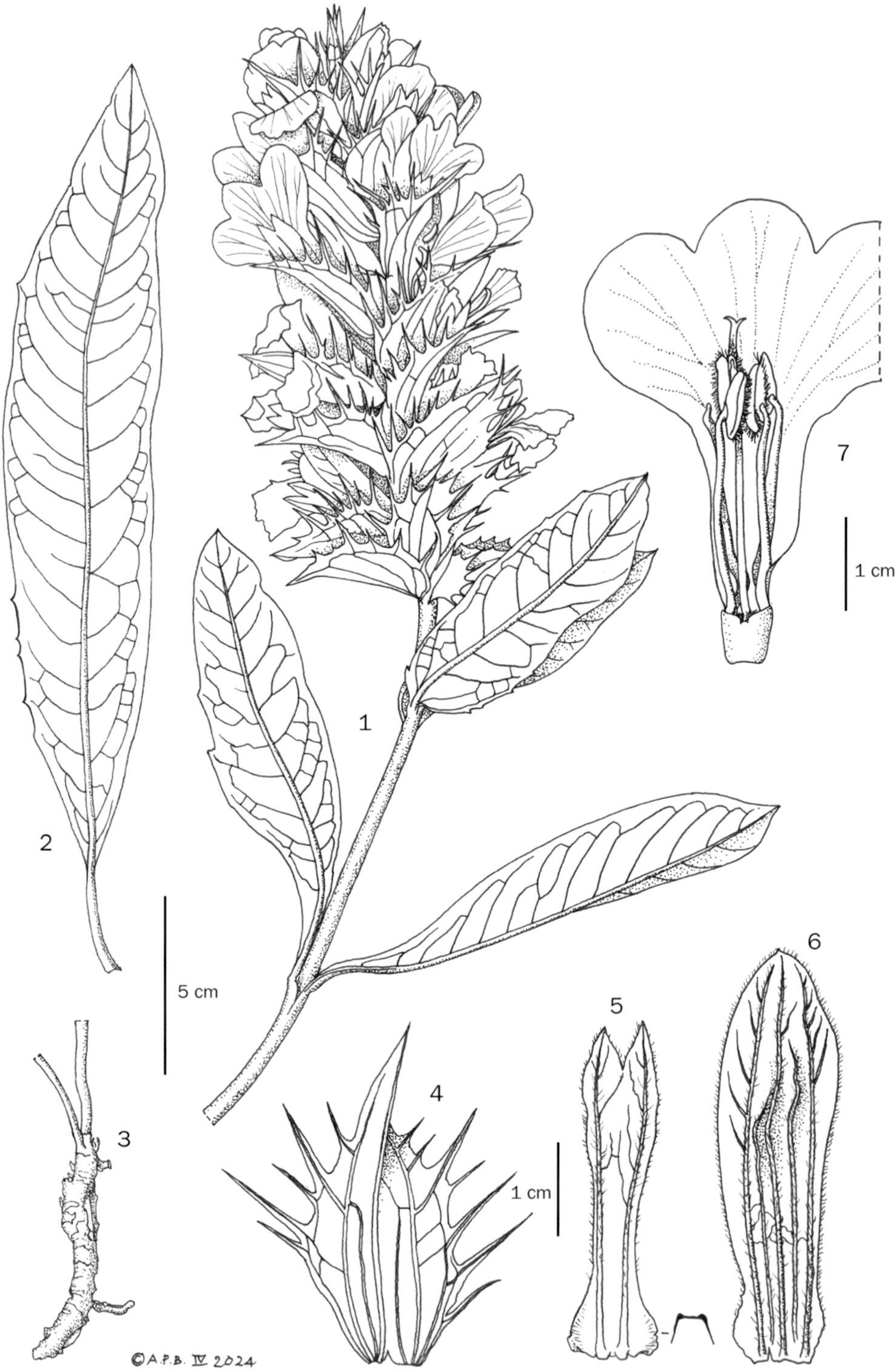

Fig. 47. **Acanthus dioscorides**. 1, habit; 2, basal leaf, abaxial view; 3, rootstock; 4, floral bract, abaxial surface; 5, outer face of lower part of calyx, with TS at base; 6, calyx, dorsal view of upper part; 7, corolla with stamens and style, dorsal view. 1, 4–7 from Nuri et al. 27775 (K004696926); 2, 3 from Rechinger 109826 (K00469694). Drawn by © A.P. Brown, Apr. 2024.

Davis] 7: 23 (1982); Musselman, Checklist of Plants of Lebanon: 93 (2011) [http://ww2.
odu.edu/~lmusselm/plant/lebsyria/Checklist of Lebanon Plants.pdf].

> *A. dioscoridis* L. var. *grandiflorus* Bornm. in Mitt. Thur. Bot. Ver. N.F. 6: 67 (1894).
> *A. dioscoridis* L. var. *boissieri* Freyn & subsp. *longistylis* Freyn in Bull. Herb. Boiss. 4:140 (1896).
> *A. dioscoridis* L. var. *boissieri* Bornm. & var. *straussii* Hausskn. ex Bornm. in Beih. Bot. Centr. 22 (2): 117 (1876).
> *Acanthus grandiflorus* Bornm., Beih. Bot. Centralbl., Abt. 2. 28(2): 483 (1911).

Erect herb, 15–40 cm. Basal leaves lanceolate to ovate-lanceolate, 7–25 × 1–7 cm, shallow dentate to ± entire, rarely 2–3-lobed. Bracts dentate, margins ± spinescent; bracteoles linear to linear lanceolate. Corolla purple or reddish-purple lobes pubescent or glandular-pubescent outside, glandular-pubescent inside; tube glabrous or pubescent outside, hirsute within. Ovary hairy at apex. Fig. 47, 1–7.

HAB. Rocky mountain slopes, in *Quercus* forest between trees, near springs, in clayey and muddy soils; alt. 920–2500 m; fl. & fr. Jul.

DISTRIB. **MAM**: Mosul, Jabal Khantur, *Rechinger* 10826!; Maya Mt. 25 km to Kani Masi, *Hamid & Fadhil* 45508!; Kani Masi, *Hamid & Fadhil* 45620!; Dainka village, 53 km N.E. Zakho, *Omar & Al Dabbagh* 45378!; Khantur Mt. N.E. of Zakho, *Rawi* 23392!; Sharanish, *Rechinger* 10826!; Matina, *Rawi* 8739A!; 8679A! **MRO**: Helgurd Dagh, on way to Rawandaz, *Kass, Nuri & Serhang* 27219!; Haibat Sultan, *Omar, Al-Kaisi & Wedad* 49466!; Haji Omran, *Nuri & Kass* 27775!; Haji Omran, *Sahira, Karim & Hamid* 38505!; Kawriesh side of Karoukh Mt., Kass & Nuri 27676!; Helgurd, *Rawi* 13742!; Rawandaz gorge, *Gillett* 8332!; Serin rd. on way to Qandil, *Rawi* 24009!; Arbil, Mt. Qandil above Pushtashan, *Rechinger* 11816; inter Shahidan et Pushtashan, *Rechinger* 11016; Chiya-i Mandau, *Guest* 2765!; Ser-i Hasan Beg, *Guest & Ludlow-Hewitt* 2908; Rawandiz, *Low* 369; Mt. Sakri Sakran, *Bornmuller* 1561; Mt. Baradost, *Thesiger* 857; Kuh Safarin, *Bornnüller* 1559; Mt. Handren, *Regel* 30; *Bornnüller* 1560, 1562; Chiya-i-Mandau, Guest 2765!; **MSU**: Sulimaniyah, *Agha* 5670!; Chuwarta to Sersea, *Noori & Hamod* 41269!; Qala Diza, *Thesiger* 1155 (BM!); Sulimaniyah, Penjwin, *Rechinger* 10481 (W); Qara Dagh, *Makki* 435 & *Najwan* 57863 (BAG); Diyala, Mandali *Nöe* 1157!; Kopi Qaradagh, *Wheeler-Haines* W1170!: Mesopotamia, Kurdistan and Mosul, *Kotschy* 171!

Turkey, Lebanon, Iran.

DOUBTFUL RECORD

A single specimen from Kurdistan, without locality, collected by Zohary & Amdursky (19 Sept. 1933) is identified as *Rungia* sp. [https://rbge.org.herb/E00271853]. There is no other material from Iraq at BAG, K or E; more material is required to fully identify this specimen.

CULTIVATED TAXA

Eranthemum pulchellum *Andrews* Bot. Repos. 2, t. 88 (1800).

Perennial, spreading shrub to c. 1 m, with large, ovate-elliptic (to 20 cm) hairy dark green leaves. Flowers blue to purplish-blue, salverform, in simple or branched spikes; tube slender, cylindrical, twice as long as or equal to the limb; limb-petals imbricate. Common name: Blue sage.

LCA: Adhamiya, Baghdad, *Sahira* C 942! (BAG); Zaffaraniya Botanical Garden, *H. Hamid* C 474! (BAG).

Native from N. Pakistan to Indo-China; introduced elsewhere for its use as an ornamental medicinal plant.

Pseuderanthemum laxiflorum (*A. Gray*) *F.T.Hubb.*, Rhodora 18: 159 (1916).

Perennial, spreading bushy shrub to c. 1.5 m, with ovate-elliptic, leaves. Flowers pinkish blue to purplish pink, salverform, in simple or branched spikes; tube slender, cylindrical, twice as long as or equal to the imbricate limb-petals. Common name: Shooting star.

LCA: Baghdad Zaffaraniya, *Shakir Sabir* C 16673! (BAG); Baghdad, Karada, *Al-Mokhtar* C 33156! (BAG).

Native to Fiji, introduced into several tropical and subtropical countries as an ornamental plant.

120. **OROBANCHACEAE** Vent.

Tabl. Regn. Vég. 2: 292 (1799), nom. cons.

Chris J. Thorogood

Annual, biennial, or perennial herbs, often monocarpic. Root parasites of other plants, either without chlorophyll (holoparasites) or photosynthetic (hemiparasites). Specimens often turning blackish on drying. Stems unbranched, or branched sparingly. Leaves simple, in holoparasites reduced to scale-like bracts, arranged spirally or subimbricately. Inflorescences racemose, spicate, or subcapitate, rarely 1-flowered; bract 1, usually similar to those of the stems; bractlets (bracteoles) 2, adnate to base of calyx or pedicel. Flowers cosexual, sometimes showy, subsessile or weakly pedicelled. Calyx actinomorphic to zygomorphic, tubular or campanulate, (3–)4–6-lobed, or spathe-like. Corolla bilabiate, usually curved, tubular-campanulate or infundibulate with 5 unequal or subequal lobes; upper lip entire, emarginate, or 2-lobed; lower lip 3-lobed. Stamens 4, didynamous, inserted at or near the base of the corolla tube, variably pubescent; filaments slender; anthers 2-celled, dehiscing longitudinally, sometimes 1 cell fertile and another sterile or reduced to spur. Pistil 2- or 3-carpellate; ovary superior; placentas 2–4 or 6(–10), parietal or axile at the ovary base; ovules usually numerous, anatropous. Style long; stigma inflated, orbiculate, discoid, or 2–4-lobed. Fruit a septicidal (sometimes also loculicidal) capsule. Seeds small to minute (dust-like), often wind-dispersed (possibly water-dispersed in some *Cistanche*); testa ornamented, pitted or reticulate.

The Orobanchaceae has a cosmopolitan distribution and includes about 100 genera and 2000 species; 32 species in Iraq.

Several genera included in the Scrophulariaceae are now placed under the Orobanchaceae. In its current circumscription (used in this Flora), Orobanchaceae includes all genera with a parasitic habit, though that may not always be obvious. Orobanchaceae can sometimes be confused with Verbenaceae, Acanthaceae and Lamiaceae, but can be differentiated by the number of ovules (many in these three families and one in Orobanchaceae) in addition to its parasitic habit.

Al-Asadi, W.M. & Al–Mayah, A.A. (2016). Three new records of *Orobanche* (Orobanchaceae) to the flora of Iraq. Am. Sci. Res. J. Eng. Technol. Sci. 22 (1): 63–70.

Al–Mayah, A.A. & Al-Asadi, W.M. (2017). An Account of *Phelipanche* Pomel (Orobanchaceae) in Iraq. Am. Sci. Res. J. Eng. Technol. Sci. (ASRJETS) 34(1): 237–251.

Aldughayman, M., Thorogood, Al–Mayah, A.A. C.J. & Hawkins, J.A. (2024a). An account of the genus Cistanche (Orobanchaceae) in Iraq and taxonomic considerations in the Middle East. *Phytokeys* 238: 281–294.

Aldughayman, M., Thorogood, C.J. & Hawkins, J.A. (2024b). Neotypification of *Cistanche tubulosa* (Schenk) Wight ex Hook.f.: a name applied to a widely distributed, polyphyletic group of plants. *Phytotaxa* 633(1): 009-016.

Ataei, N. (2017). Molecular systematics and evolution of the non-photosynthetic parasitic *Cistanche* (Orobanchaceae). bonndoc.ulb.uni-bonn.de.

Beck-Mannagetta, G. (1890). *Monographie der Gattung Orobanche* (Vol. 19). Fischer.

Li, X., Feng, T., Randle, C. & Schneeweiss, G.M. (2019). Phylogenetic relationships in Orobanchaceae inferred from low-copy nuclear genes: Consolidation of major clades and identification of a novel position of the non-photosynthetic Orobanche clade sister to all other parasitic Orobanchaceae. *Frontiers in Plant Science* 10: 902.

Moreno Moral, G., Sánchez Pedraja, O., Piwowarczyk, R. (2017). Contributions to the knowledge of *Cistanche* (Orobanchaceae) in the Western Palearctic. Phyton-Annales Rei Botanicae 57: 19–36. DOI:10.12905/0380.phyton57-2018-0019.

Sánchez Pedraja, Ó., Moreno Moral, G., Carlón, L., Piwowarczyk, R., Laínz, M., & Schneeweiss, G. M. (2016). [continuously updated]: Index of Orobanchaceae. Published at http://www.farmalierganes. com/0trospdf/publica/0orobanchaceae% 201ndex. htm.

Thorogood, C.J. & Rumsey, F. (2021). 1003. OROBANCHE RAPUM-GENISTAE: Orobanchaceae. *Curtis's Botanical Magazine* 38(4): 487–499.

Thorogood, C. & Rumsey, F. (2021). *Broomrapes of Britain and Ireland.* Botanical Society of Britain and Ireland.

Thorogood, C.J., Rumsey, F.J. & Hiscock, S.J. (2009). Host-specific races in the holoparasitic angiosperm *Orobanche minor*: implications for speciation in parasitic plants. *Annals of Botany*, 103(7): 1005–1014.

Thorogood, C. J., Rumsey, F. J., Harris, S. A., & Hiscock, S. J. (2009). Gene flow between alien and native races of the holoparasitic angiosperm *Orobanche minor* (Orobanchaceae). *Plant Systematics and Evolution*, 282(1): 31–42.

1. Flowers usually solitary (–2), terminal, bright red; stout single stems, often in a clump, lacking functional leaves but with conspicuous sheathing bracts . . . . .1. **Phelypaea**
   Flowers in spikes or racemes. 2
2. Calyx lobes rounded; corolla lobes subequal. .2. **Cistanche**
   Calyx truncate or dentate; corolla distinctly bilabiate. 3
3. Flowers subtended by a bract and 2 bracteoles; calyx 5-dentate; mature capsule valves free . . . . .3. **Phelipanche**
   Flowers subtended by a bract only, bracteoles absent; calyx divided into 2 lateral segments; capsule valves remaining joined distally . . . . .4. **Orobanche**

1. **PHELYPAEA** Tourn. ex L.

Opera Var. 237 (1758); Fischer in Kubitzki (ser. ed.), ed. J.W. Kadereit, Fam. & Gen. Vasc. Pl. 7: 409 (2004)

Chris J. Thorogood

Simple-stemmed parasitic herbs with large, sheathing bracts at the base of the stem, and solitary (rarely 2), terminal flowers. Calyx connate, 5-toothed with posterior teeth more or less connate. Corolla broad, tubular and flattened, bright red, with broad, patent, rounded lobes with black villous folds on the lower lip. Stigma capitate-bilobed to discoid. Fruit a bivalved capsule.

Four species from the eastern Mediterranean to Iran and Armenia; a single species in Iraq.

1. **Phelypaea coccinea** (*M.Bieb.*) *Poir.*, Encycl. [J. Lamarck & al.] 5 : 268 (1804); Salih, J. Garmian Uni. 5(2): 433 (2018) [https://jgu.garmian.edu.krd/article_67190.html].

Orobanche coccinea M.Bieb., Tabl. Prov. Casp.: 58 (1798).

Conspicuous, parasitic herb with stout, reddish-brown stems to 30(–50) cm; stems smooth, tubular and slender, grooved, with sparse bracts. Stems often clumped; stem bracts semi-amplexicaul, to 20 mm long. Calyx irregular, more or less 5-toothed, glandular-pubescent; teeth tapered, fused. Corolla zygomorphic, 25–40 mm long, with a short, curved tube, bright red, often paler externally; lobes broad, oval, sub-equal, with two black-villous folds on the inner surface of the lower lip. Stamens inserted below the middle of the corolla tube; filaments short, villous below; anther thecae parallel, acuminate, sharply apiculate-mucronate, with a lanate suture, glabrescent above. Style glabrous, stigma orbiculate. Capsule ovoid. Seeds densely porose-reticulate.

HAB. Parasitic on the Asteraceae, especially *Centaurea*; mountains; alt. 1500–2000 m; fl. not recorded.
DISTRIB. Rare and local, restricted to a small area in the Pira Magrun Mountain range.

Central Greece and Turkey east to N. and N.W. Iran.

2. **CISTANCHE** Hoffmannsegg & Link

Fl. Port. 1: 319 (1809); Fischer in Kubitzki (ser. ed.), ed. J.W. Kadereit, Fam. & Gen. Vasc. Pl. 7: 409 (2004)

Chris J. Thorogood, Julie A. Hawkins & Majed Zaal M. Aldughayman

Parasitic herbs lacking functional leaves and chlorophyll. Stems fleshy, unbranched. Inflorescence long and spicate; bract 1; bracteoles 2, rarely absent. Calyx tubular or campanulate, apex (4 or) 5-lobed, rarely 5-parted, lobes usually equal. Corolla tubular-campanulate, apex 5-lobed; lobes subequal. Stamens 4, inserted in corolla tube; anthers 2-celled, all fertile, often pubescent. Ovary 1-locular, parietal placentas 4, rarely 2 or 6. Style

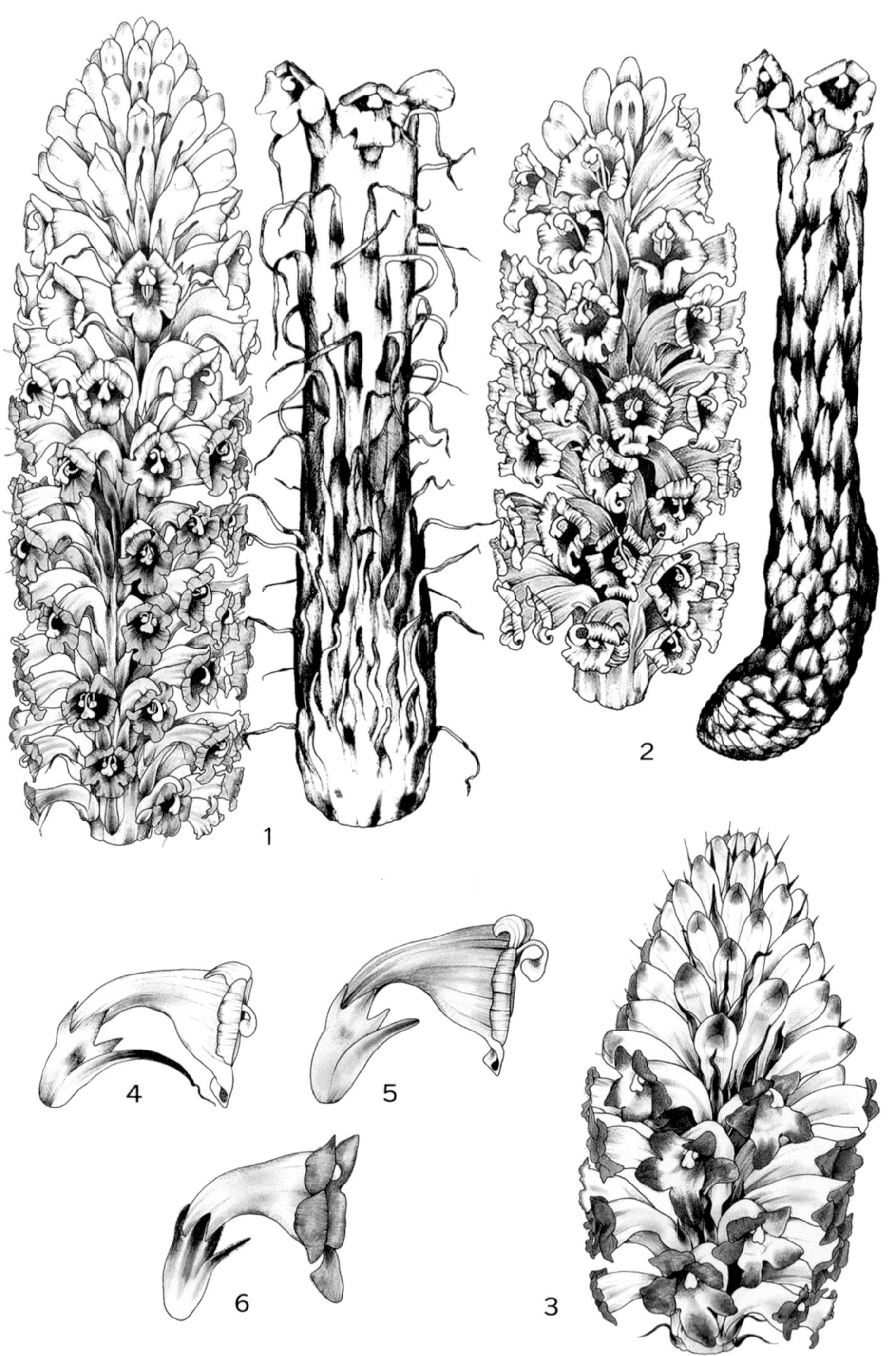

Fig. 48. Illustrations of **Cistanche** species putatively found in Iraq and adjacent territories (inflorescences). 1, **C. flava**; 2, **C. tubulosa**; 3, **C. salsa**. Corollas in profile, showing calyx and bract characteristics. 4, **C. flava**; 5, **C. tubulosa**; 6, **C. salsa**. Drawn by © C. Thorogood.

slender, persistent; stigma 2-lobed. Capsule ovoid-globose or globose, dehiscing by 2(or 3) valves. Seeds subglobose; testa reticulate.

Most accounts report only *C. tubulosa* from Iraq (Aldughayman et al., 2024a). However, species limits remain confused in the genus and several species are reported in adjacent regions. Therefore protologues and types, where available, were examined for the following: *C. eremodoxa, C. salsa, C. ridgewayana, C. fissa, C. laxiflora, C. flava, C. ambigua, C. lutea* and *C. phelypaea*. All were examined alongside material from Iraq during the preparation of this account. A specimen representing an unpublished species from Iraq (Ataei, 2017) was also examined, but proved doubtful. The only species identified with certainty to occur in Iraq is *C. tubulosa* s.l. (Aldughayman et al., 2024a).

About twenty species in the Mediterranean region, N. and W. Africa, Ethiopia, to western India and north-western China; a single species in Iraq.

1. **Cistanche tubulosa** (*Schenk*) *Wight ex Hook*.f., Fl. Brit. India [J. D. Hooker] 4(pt. 11): 324 (1884); Dinsmore in Post, Fl. Syria, Palest. & Sinai ed. 2, 2: 312 (1933); Rawi in Dep. Agr. Iraq Tech. Bull. 14: 148 (1964); Rechinger, Fl. Lowland Iraq: 554 (1964); Schiman-Czeika in Fl. Iranica [K. H. Rechinger 5: (1964); Jafri in Fl. Pakistan [Nasir & Ali] 98: 5 (1976); Feinbrun-Dothan, Fl. Palaest. 3: 209, pl. 354 (1978); Collenette, Wild Flow. Saudi Arabia: 586 (1999); Haloob et al., Illustr. Fl. Lowland Iraq, 1: 124 (2016); Abdulridha, Taha & Widad, Ecology & Fl. Basrah: 447 (2016).

Orobanche tinctoria Forssk., Fl. Aegypt.-Arab.: 112 (1775).
Phelypaea tubulosa Schenk, Pl. Spec. Schubert: 23 (1840).
Cistanche tubulosa Wight, Icon. Pl. Ind. [Wight] t. 1420 bis. (1849).
Cistanche tinctoria (Forssk.) Beck, Bull. Herb. Boissier sér. 2, 4: 685 (1904).

The name *Cistanche tubulosa* s.l. is used from Africa and the Middle East to South and Central Asia and China, however as applied, the name refers to a widely distributed, polyphyletic group of plants. Because the type specimen of *C. tubulosa* was lost, a specimen from South Sinai near the type locality was recently designated as the neotype (Aldughayman et al. 2024b). This neotypification, alongside further phylogenetic work, is necessary to re-evaluate whether the name *Cistanche tubulosa* is a synonym for the name *C. tinctoria*, as has been proposed by Moreno Moral et al. (2017), and also to confirm whether the name *C. tubulosa* is the correct name for any Iraqi entity (Aldughayman et al. 2024a, 2004b).

Robust, glabrous herb, 20–50 cm. Lower stem bracts sinuate, imbricate, broadly lanceolate, up to 3 cm long. Upper stem bracts sinuate, dense, ovate-lanceolate, grey, 4–10 m long; bracts deeply sinuate, ovate-lanceolate, equal or slightly longer than calyx, grey, 14–22 mm long. Bracteoles sinuate, oblong-lanceolate, equal or shorter than calyx, grey, 2–3 mm long. Calyx tubular, pentamerous, usually ½ the corolla length, lobes subequal or one slightly shorter, oblong, obtuse; corolla tube tubular-campanulate, pentamerous, pale lemon to deep yellow-orange, often with a pink or violet-tinted limb (especially in bud), 34–52 mm long, lobes equal, rounded. Stamens didynamous, epipetalous, densely lanate at the base. Anthers cordate, rounded at the base and acute at the apex, densely lanate. Ovary ovate. Style cylindrical, oblique. Stigma bilobate. Fruit a splitting capsule. Seeds small, black and numerous. Fig. 48, 2, 5; Fig. 49, 4, 10.

HAB. Parasitic on shrubby Amaranthaceae and occasionally Capparaceae and Zygophyllaceae; hosts reported reliably include: *Haloxylon salicornicum* (Moq.) Bung, *Capparis spinosa* L., *Zygophyllum coccineum* L., *Anabasis* spp.; potential hosts reported (with doubt) include: *Ephedra* spp. (Ephedraceae) and *Limonium* spp. (Plumbaginaceae); sand dunes, sandy gravel substrates, clayey and stony soils, and saline areas at low altitude; widespread; alt. 10–300 m; fl. Jan.-May.
DISTRIB. In the foothills, central and southern desert areas of Iraq. MSU Diyala: Hamrin, near Shahraban, 8 May 1958, *Coll. Ingot. s,n.* (E). MJS: Nineveh: Faidah Al-Rbaswi, *F. Karim, M. Noori, H. Hamid & H. Kadhim* 40279 (K). FUJ?: 6 km from Rabia, *F. Karim, H. Hamid & H. Kadhim* 39944 (K). FPF: Mandali, *Guest* 1742 (K, BAG); 30 km N.E. of Mandali, *Al-Kaisi & Khayat* 50782 (K); Near Jalibah (Jalaula?), 8 April 1933, *anon.* 5065 (K); 5 km from Badra to Kut, *Al-Kaisi & H. Hamid* 46525 (K, BAG); 51 km N.E. of Kut between Jassan and Badrah, *Hikmat Abbas & F. R. Bharucha* 2613 (K,W). DLJ: Ramadi east of Lake Tharthar, *Fred A. Barkley & Ramdan Eljumaili* 7263 (K); Shibaichan road 10 km N. of Rawah, *Khatib & Alizzi* 31967 (K);
DWD: 10 km N. of Rutba, *Rawi & Gillett* 6326 (K). DWD: 10 km from Hit to Kubaysah, *S. Omar, Alkaisi, K. Hamad & H. Hamid* 44354 (K, BAG); 8 km W. of Karbala, *Rawi & Gillett* 6415 (K); Razazza, *A. Haloob,*

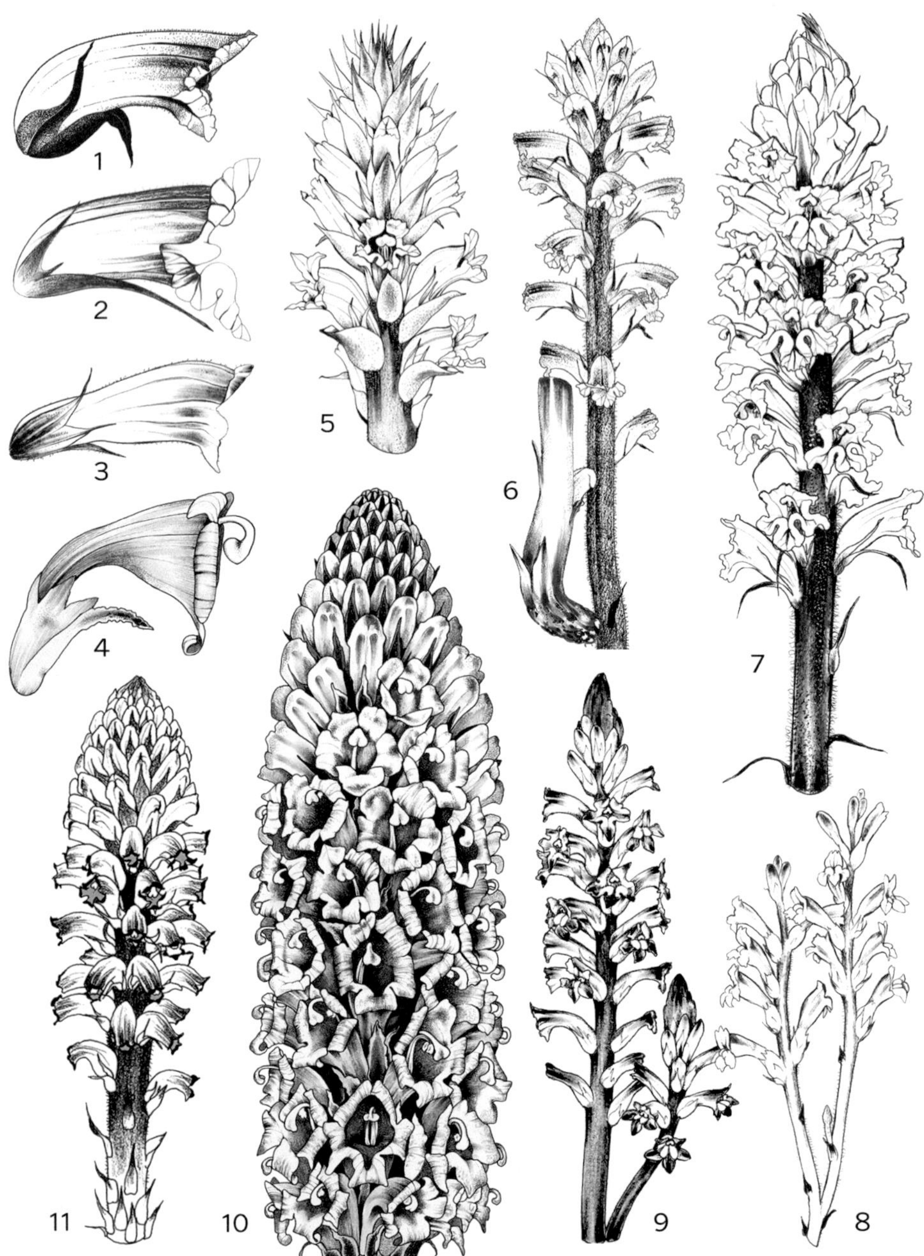

Fig. 49. Corollas in profile (1-4) and habit (5-11 of the holoparasitic Orobanchaceae of Iraq (*Cistanche, Orobanche, Phelipanche*). 1, **Orobanche minor**; 2, **O. crenata**; 3, **Phelipanche ramosa**; 4, **Cistanche tubulosa**; 5, **Orobanche camptolepis**; 6, **O. minor**; 7, **O. crenata**; 8, **Phelipanche ramosa**; 9. **P. aegyptiaca**; 10, **Cistanche tubulosa**; 11, **Orobanche cernua**. Drawn by © C. Thorogood.

Ikhlas, R. Hamshkan & Riyadh 59879 (BAG); 2 km W. of Ukhaidhir, 12 March 1980, *Coll. Ingot* 51219 (BAG). **DSD**: 18 km W. of Karbala, *Martin L. Grant* 18228 (W); Eridu, *Seton Lloyd* 6328 (K, BAG); **DSD**: 77 km N.W. of Zubair, *F. Barkley & Hikmat Abbas Al-Ani* 6499 (K,W); 28 km SE.E. by S. of Zubair, *Guest, Rawi & Rechinger* 16875 (K, BAG); 70 km E. of Zubair, *Turner* 47457 (K); Zubair, *Rechinger* 5247 (W); Between Zubair and Safwan, *Alizzi* 34341 (K); Rumaila, Toba railway station 20 km W. of Ghubaishiyia, *Sharif Y. Haddad* 9535 (K); 30 km W. of Jabal Sanam, *Alizzi* 32684 (K, BAG); Umm Qasr Port, *Husain*

Al-Ali 39929 (K); Jabal Sanam, *anon.* 29889 (BAG); Shaib Al-batin (Batin), Jarishan, *Rechinger* (W); 6 km S.E. of Safwan, *Rechinger* 5245 (W); Najaf: Al-Hira, *Riyadh, Yasin, Dhya'a, Adel & Sinan* 59291 (BAG); Muthanna, 10 km S. of Samawah, *Rawi & Gillett* 6125 (K); 15 km W. of Samawa, *Rawi* 14880 (K); 25 km to Busaiya from Khadhar al-Mai, *Alkaisi, K. Hamad & H. Hamid* 48514 (K); Khadhar al-Mai, enclosure, *F. Karim, A. Sharief, K. Hamad & H. Hamid* 48066 (K); 50 km E. of Busaiya to Khadhar al-Mai, *F. Karim, A. Sharief, K. Hamad & H. Hamid* 48034 (K); 13 km west Samawa, *Ibrahim Al-Mahallal* 15204 (K,BAG); Wasit: Kut, *Alizzi & S. Omar* 34893 (K). **LEA**: Muqdadiya (Shahraban), *Haines* W. 847 (E, K); 55 km E. of Kut al-Imara, *F.A. Barkley* 4055 (K). **LCA**: Between Fallujah and Wadi Tharthar, *Rechinger* 11247 (W); 80 km W. of Shaikh Sa'd, *Rawi & S. Haddad* 25520 (K); 10 km E. of Zurbatiyah, *Al-Kaisi & H. Hamid* 46551 (BAG); **LCA**: 20 km N.W. of Falluja, *Omar & Alkhayat* 31967 (BAG);

Northern and Eastern Africa to the Arabian Peninsula, Iran to India, Turkmenistan.

3. **PHELIPANCHE** Pomel

Nouv. Mat. Fl. Atl. 1: 102 (1874).
Orobanche sect. *Trionychon* Wallr., Sched. Crit. 314 (1822).

Chris J. Thorogood, Abdulridha A. A. Al–Mayah, Widad M. Al-Asadi and Shahina A. Ghazanfar

Stem simple or branched, ± slender. Bracts and bracteoles subtending the flowers in a spike or raceme; calyx 4(–5)-dentate, gamosepalous, campanulate. Corolla waisted at level of insertion of stamens, inflated below, gradually dilated above into a limb, usually blue-violet, rarely whitish or yellowish. Anthers glabrous or lanate.

A recent report by Al–Mayah & Al-Asadi (2017) increased the species count for the *Phelipanche* species to 9 taxa in Iraq. Further work is required across the genus in Iraq and elsewhere in the Middle East.

1. Flowers subtended by a bract and 2 bracteoles; calyx 5-dentate; mature
    capsule valves free (*Phelipanche* group) 2
    Flowers subtended by bract only; bracteoles absent; calyx dived into 2
    lateral segments; capsule valves remaining joined distally 10
2. Corolla ≤17 mm long. 3
    Corolla ≥17 mm long. 5
3. Calyx teeth ± shorter than their tube . . . . 1. *P. ramosa*
    Calyx teeth equalling or longer than their tube 4
4. Corolla narrowly campanulate in profile, often purple-tinted . . . . 2. *P. nana*
    Corolla narrowly tubular in profile, whitish. . . . . 3. *P. umqasrensis*
5. Corolla not more than 20 mm long 6
    Corolla large, 22–44 mm long 9
6. Corolla lower lip lobes acuminate 7
    Corolla lower lip lobes blunt. 8
7. Calyx teeth equalling or longer than their tube . . . . 4. *P. oxyloba*
    Calyx teeth conspicuously longer than their tube. 5. *P. orientalis*
8. Stem indumentum glandular, corolla arch smooth in profile . . . . 6. *P. mutelii*
    Stem indumentum arachnoid-tomentose; corolla with distal hump in
    profile . . . . 7. *P. hypertomentosa*
9. Stem often branched, growing in cultivated habitats . . . . 8. *P. aegyptiaca*
    Stem usually simple, growing in natural habitats . . . . 9. *P. coelestis*

1. **Phelipanche ramosa** (*L.*) *Pomel*, Nouv. Mat. Fl. Atl. 1: 103 (1874). Schiman-Czeika in Fl. Iranica [K. H. Rechinger] 5: 6 (1964); Chaudhary & Musselman in Fl. Kingdom Saudi Arabia 2(2): 516 (2001); Musselman, Checkl. Pl. Lebanon & Syria (2011) http://ww2.odu. edu/~lmusselm/plant/lebsyria/Checklist of Lebanon Plants.pdf; Ghazanfar, Fl. Oman 3: 159 (2015); Al–Mayah & Al-Asadi, Am. Sci. Res. J. Eng. Technol. Sci. 34(1): 244 (2017); Taifour & El-Oqlah, Pl. Jordan Annot. Checkl.: 115 (2017).

> *Orobanche ramosa* L., Sp. Pl. 2: 633 (1753); Dinsmore in Post, Fl. Syria, Palest & Sinai 2: 314 (1933); Schiman-Czeika in Fl. Iranica [K. H. Rechinger] 5: 4 (1964); Gilli in Fl. Turkey [P. H. Davis] 6 (1982).
> *Phelypaea ramosa* (L.) C.A.Mey., Verz. Pfl. Casp. Meer. (C.A. von Meyer). 104 (1831).

Small annual herb 6–15 cm tall; stems slender (c. 5 mm across), yellowish or purplish, simple or branching from (or just above) the base, glandular-pubescent. Stem bracts 5–15 mm, ovate to lanceolate, pubescent to glabrescent. Inflorescence of ± cylindrical, few-flowered spikes. Bracts 6–7 mm; bracteoles 2, linear-lanceolate, about as long as calyx. Calyx 6–8 mm, campanulate, 4-toothed, glandular-pubescent. Corolla white to pale yellow with a bluish to purplish limb, 10–15 mm long, tubular-infundibuliform; tube conspicuously constricted above the ovary, curving outwards, broader above, glandular-pubescent outside; lobes shallow, those of the upper lip rounded, recurved, minutely denticulate or entire, finely pubescent. Stigma lobes white or bluish. Stamens inserted 3–6 mm above the base of the corolla tube (just below constriction); filaments virtually hairless to slightly pubescent at base near base; anthers hairy or glabrous. Capsule 8 mm. Parasitic on cultivated plants such as crops and bedding plants, especially in the Solanaceae. Also reported (with doubt) to grow on apricots in the region. Fig. 49, 3, 8.

HAB. Recorded to be a parasite on tomato; in irrigated fields, wheat fields, and gardens on clay and alluvial soils; alt. 30–1350 m; fl. & fr. Apr., May, Jul., Oct., Nov.

DISTRIB. In the mountain region and the central alluvial plains. **MAM**: collector not known, Zawita, 1671!; Khantur N.E. Zakho, *Rawi* 23302!; Aqra, *Rawi* 11434! Galy Zakho, near Zakho, *S.H.S. Miran, Al-Khayat & J. Saeed*, 1273; 1274. **MRO**: Chewa Rash N.E. of Rania, *Al-Rawi, Nuri & Kassi* 28448; Kuh Sefin prope Shaqlawa, *Bornmüller* 16427 (W). **MSU**: Jarmo, *Wheeer-Haines* W295!; Qaraghan, *Rogers* 0174! (on grass); Darbandikan, *Al-Shehbaz & Husian* 1624; Tawela, *Al-Rawi*, 21868!; Hasa Galla on the road, towards Qala Chawlan, Shahrabashir, Sul., *Al–Mayah & Al-Asadi* 1444 (BSRA). **MJS**: Jabal Sinjar, *Gillett* 111123!; *Al-Shahbaz & W. Al-Hashimi* 0030516; Kursi, *Gillett* 10885!; nr. Derbendikan Dam-Sul., *Hikmat Abbas Al-Ali, Elizabeth D. Barkley & Fred A. Barkley*, 0019599. **FPF**: near Talwasat (Nasr) police post, about 21 km N.N.W. of Mandali, *Al-Shahbaz & Al–Mayah* 0029749. **LCA**: Baghdad, *Lazar*, 3593; 3741!; Baghdad *Wheeler-Haines* W. 211 in garden! **LEA**: 60 km S. of Baghdad to Kut, *Rawi* 18384! Aski-Kalak, 2 May 1987, Y.I. Elia s.n.

HALUK-RIHI (Ar.).

S. Europe to C. Asia, North and tropical east Africa, Sinai, Palestine, Arabian Peninsula, Turkey, Iran to Pakistan and China.

2. **Phelipanche nana** (Noë & Rchb.f.) Soják, Čas. Nár. Mus., Odd. Přír. 140(3–4): 130 (1972); Al-Mayah & Al-Asadi, Am. Sci. Res. J. Eng. Technol. Sci. 34(1): 243 (2017).

Phelypaea nana (Noë ex Rchb.) Rchb.f., Icon. Fl. Germ. Helv. [H.G.L. Reichenbach] 20: 88 (1862).
Orobanche nana Noë ex Rchb., Exsicc. no. 1352 (1842); DC., Prodr. [A.P. de Candolle] 11: 9 (1847); Boissier, Fl Orient. 4: 499 (1879); Schiman-Czeika in Fl. Iranica [K. H. Rechinger] 5: 7 (1964); Gilli in Fl. Turkey [P. H. Davis] 7: 7 (1982).

Small annual herb 7–14 cm tall, glandular-pubescent, similar in appearance to *P. ramosa*. Stem 5–10 cm, simple or branched, slender. Stem bracts 6.5–9 mm, ovate-lanceolate, glandular-pubescent. Inflorescence short-cylindrical, 4–5 cm, few flowered (4–7 flowers). Bract 6.5–9 mm, lanceolate, glandular-pubescent. Bracteoles linear, 4.5–6 mm, glandular-pubescent. Calyx 5–7 long, glandular-pilose, calyx teeth 2.5–4 mm, lanceolate-sublate, equalling or exceeding the calyx tube. Corolla 10–15 mm, blue, glandular-pubescent outside, lobes of lower lip rounded or broadly oval, folds white-hairy. Stamens inserted 4.5–6 mm above base of corolla; filaments 6–9.5 mm, glabrous at base; anther 1.3–2 mm, glabrous or sparsely hairy at the base, whitish-violet. Style 6–8 mm, glandular under stigma.

HAB. Parasitic on the Brassicaceae; mountain slopes, rocky lime stone on dry soil; alt. 1600 m; fl. & fr. Mar.-Jun.

DISTRIB. In the Rowanduz district of N. Iraq. **MSU**: Jabal Samaralwa, Dara Tri above Twiala 3–5 km, *Al–Mayah & Al-Asadi* 15067 (BSRA).

Mediterranean Basin and northern Africa east to Turkey and Russia; introduced into the U.S.A.

3. **Phelipanche umqasrensis** Al-Mayah & Al-Asadi, Global J. Biol. Agric. & Health Sci. 5(3): 107 (2016). Type: Iraq, 5 km N.W. Um-Qasr, near the Iraq-Kuwait borders, *Al–Mayah & Al-Asadi*, 1410 (BSRA, holo).

Al-Mayah & Al-Asadi, Am. Sci. Res. J. Eng. Technol. Sci. 34(1): 245 (2017).

Annual herb, 10–30 cm tall. Stem thick, erect, simple or occasionally branched underground, beige-coloured. Inflorescence many flowered, (34–89 flowers) in a dense, thick spike, 4–11 cm long. Bracts 5–11 mm, ovate-lanceolate, glandular. Bracteoles 2, 7–10 mm, lanceolate-subulate, glandular. Flowers small, 14–16 mm, white. Calyx 7–9.5 mm, glandular-puberulent; calyx teeth, 4–5.5 mm, acute to acuminate nearly equalling the calyx tube. Corolla infundibuliform, white, 12–15 (–16) mm, 2-lipped, the upper lip 2-lobed, the lower lip 3-lobed, variable in shape. Stamens 4, inserted at 3–4 mm above the base; filaments glabrous-pubescent at base; anther glabrous to sub-glabrous. Ovary oblong. Style glabrous; stigma lobed, white to yellowish.

HAB. Parasitic on *Rhanterium epapposum*; on sandy, sandy gravel and desert soil; alt. fl. & fr. Mar.-Apr. DISTRIB. Common in the *arfaj* (*Rhanterium epapposum*) community near the Iraq-Kuwait frontier in S.E. Iraq. **DSD**: on road to port of Khor al-Zubair, 5–10 km N.E. Um-Qasr, *Al-Mayah & Al-Asadi* 1423 (BSRA); on road of Safwan-Um-Qasr 15 km N.W. Um-Qasr near the Kuwait border, *Al-Mayah & Al-Asadi* 1414 (BSRA); Jarishan-Khadhar al-Mai road, about 15 km before Khidhr al-Mai *Al-Mayah & Al-Asadi* 1446, 1652 (BSRA); 8 km N.W. Um-Qasr, near the Iraq-Kuwait Border, *Al-Mayah & Al-Asadi*, 1410 (type); ibid., 1648 (BSRA).

Endemic.

4. **Phelipanche oxyloba** (*Reut.*) Soják, Čas. Nár. Mus., Odd. Přír. 140 (3–4): 130 (1972); Al-Mayah & Al-Asadi, Am. Sci. Res. J. Eng. Technol. Sci. 34(1): 243 (2017).

Orobanche oxyloba (Reut.) Beck, Monogr. Orobanche Beck, 114, t. 2, f. 21 (1890); Schiman-Czeika in Fl. Iranica [K. H. Rechinger] 5: 9 (1964); Jafri in Fl. Pakistan [Nasir & Ali] 98: 11 (1976); Gilli in Fl. Turkey [P. H. Davis] 7: 9: (1982).
Phelipanche mutelii var. *oxyloba* (Reut.) Rätzel & Uhlich, Carinthia II 207/127(2): 649 (2017).
Phelypaea oxyloba Reut. Prodr. [A. P. de Candolle] 11: 9 (Orobanches sp.) (1847).

Annual herb 18–30 cm with stems 5–11 mm across, erect, and thickened at the base, brownish, glandular-pubescent. Stem bracts lanceolate, 12–16 × 3–5 mm, Inflorescence spike 5–10 cm, shorter or equalling the lower part of the stem, dense above, laxer below. Bracts 6–10 ×1.5–3.5 mm, light brown, lanceolate, glandular-hairy. Calyx (3–)7(–9) mm, campanulate with 4 sub-equal segments, basally ovate with acuminate teeth. Corolla 10–18(–20) mm, campanulate, bluish-purple, yellowish at base, glabrous to sparsely hairy, dorsal line curved, upper lip curved, often slightly bilobed, lobes broadly rounded; lower lip large, patent or deflexed, lobes rounded, with broad folds between the lobes; all lobes irregularly denticulate at the margins. Stamens inserted 2–5 mm above the base of corolla tube; filaments 6–10 mm, glandular hairy; anthers mucronate, brown, sometimes pubescent along the suture. Stigma, yellow or brownish. Capsule ellipsoid, 5–8 mm.

HAB. Parasitic on Asteraceae, especially *Centaurea* spp.; alluvial and sandy soils; alt. 250–270 m; fl. Apr. DISTRIB. Mountains of N. Iraq. **MSU:** 10 km N. of Dorbendi-khan on road to Sulimaniyah, *Agnew* 5926 (K, BAG).

The species has been considered to be conspecific with *P. nana* (Reut.) Soják (1972) and *P. dalmatica* auct. non Tzvelev (1958) (Sánchez Pedraja et al., 2016).

Balkan Peninsula, Turkey, Armenia, Kurdistan, Arabian Peninsula, Caucasus, Pakistan.

5. **Phelipanche orientalis** (Beck) Soják, Čas. Nár. Mus., Odd. Přír. 140 (3–4): 130 (1972); Al-Mayah & Al-Asadi, Am. Sci. Res. J. Eng. Technol. Sci. 34(1): 244 (2017).

Orobanche orientalis Beck, Monogr. Orobanche 1: 110 (1890); Jafri in Fl. Pakistan [Nasir & Ali] 98: 11 (1976);

Annual herb up to 7–31(–40) cm tall. Stem simple or branched, 2–10 mm thick, purplish or yellowish. Stem bracts 5.5–10 mm, lanceolate, glandular-pilose. Inflorescence cylindrical, 3–24 cm long, typically dense and many-flowered (16–38 flowers). Bract 8.5–12 mm, as long as calyx or shorter, lanceolate, glandular-pubescent. Bracteoles lanceolate to linear-subulate, 8–10 mm, glandular-pubescent. Calyx 9–13 mm, glandular-pilose, calyx teeth 7–9 mm, distinctly longer than calyx tube, calyx tube 4–5 mm, lanceolate or filiform at the apex. Corolla 15–17(–19) mm, limb pale purple, corolla tube whitish purple, sparsely glandular-pubescent outside, lobes of lower lip acuminate, folds white-hairy. Stamens inserted 4–6 mm above base of corolla tube; filaments 9–13 mm, glabrous, anther 1–2 mm, hairy at base, whitish-purple. Style 8–13 mm, glandular beneath the stigma. Reported to be parasitic on *Galium* spp. (Rubiaceae) and *Vicia* spp. (Fabaceae).

HAB: Reported (doubtfully) on *Prunus* (Rosaceae); mountains in N. Iraq; alt. ; fl. Apr.-Jun.
DISTRIB. **MAM**: Amadiya graveyard, about 5–10 km N.W. of Amadiya, *Al–Mayah & Al-Asadi* 1473 (BSRA); **MSU:** Pira-Magrun, *Al–Mayah & Al-Asadi* 1417; 15017 (BSRA); J Samaralwa, Dara Tri above Tawila, 3–5 km, *Al–Mayah & Al-Asadi* 15044 (BSRA).

Syria, Jordan, Oman, Iran, Turkey, India, Afghanistan, Pakistan.

6. **Phelipanche mutelii** (F.W.Schultz) Pomel, Nouv. Mat. Fl. Atl. 1: 106 (1874); Al–Mayah & Al-Asadi, Am. Sci. Res. J. Eng. Technol. Sci. 34(1): 242 (2017).

Orobanche mutelii F.W.Schultz, Fl. Franç. (Mutel) 2: 353, t. 43 (1835); Schiman-Czeika in Rechinger in Fl. Lowland Iraq: 553 (1964); Rawi in Dep. Agr. Iraq Tech. Bull. 14: 148 (1964); Schiman-Czeika in Fl. Iranica [K. H. Rechinger] 5: 7 (1964); Feinbrun-Dothan, Fl. Palaest. 3: 211, pl. 356 (1978); Gilli in Fl. Turkey [P. H. Davis] 7: 7 (1982).
Phelypaea mutelii (F.W.Schultz) Reut., Prodr. [A. P. de Candolle] 11: 8 (1847).

An annual herb 9–22 cm tall, glandular-pubescent. Stem 3.5–8 cm, simple or branched, slender, 3.5–6 mm across, purplish. Stem bracts few, ovate-lanceolate, up to 6.5–11 mm, glandular-pilose. Inflorescence cylindrical, lax and typically few-flowered (4–25 flowers). Bract ovate-lanceolate, 5.5–8 mm, glandular-pubescent. Bracteoles linear, 4.5–7 mm, shorter than calyx, glandular-pubescent. Calyx 6–8 mm, glandular-pilose, calyx teeth lanceolate or filiform at apex, 3–4 mm, equalling calyx tube, 3–4 mm. Corolla 15–20 mm, purple-bluish, sparsely glandular-pubescent externally; lobes of lower lip rounded, folds white-hairy. Stamens inserted 3.25–4 mm above base of corolla tube; filaments 8.5–11 mm, glandular at base with few shorter hairs; anther 1.2–1.7mm long, glabrous or hairy. Style 8.5–10 mm, glandular.

HAB. Parasitic mainly on the Asteraceae; mountain slope, on clay and gravelly soil, stony clay hillside, rocky limestone dry soil; alt. 100–1200 m; f. & fr. Apr.-Aug.
DISTRIB. **MAM:** Khatur Mt., N.E. of Zakho, *Al-Rawi* 23302!. **MRO:** Haibat Sultan Dagh Mt. N. of Korsanjaq, *Al-Rawi, Nuri & Kass* 28175A. **MSU:** Jajla village, 3–5k m sfter Mawit towards Basi, *Al–Mayah & Al-Asadi*, 15072 (BSRA). **MJS:** Jabal Sinjar, S.E. slope curve No.126, *Widad & Al-Khayat* 53383; Jabal Sinjar, *Gillett* 11113. **FPF:** near general road c. 15 km, *S.H.S. Miran & Kh. Musa* 1272 (BSRA). **DWD:** 40 km W. Ramadi to Rutba, *Al-Kaisi & K.Hamad* 48934. **DSD:** S.E. of Ashuriya (50 km W.N.W. of Shabicha), *Guest Al-Rawi & Rechinger*, 19359. **LCA:** 60 km S. of Baghdad to Kut, *coll ingot.* 18384.

Mediterranean Basin, North Africa, Turkey to Afghanistan, Armenia; introduced into Australia.

7. **Phelipanche hypertomentosa** (*M.J.Y.Foley*) *M.J.Y.Foley*, Edinburgh. J. Bot. 64(2): 209 (2007); Al-Mayah & Al-Asadi, Am. Sci. Res. J. Eng. Technol. Sci. 34(1): 241 (2017).

Orobanche hypertomentosa M.J.Y.Foley, Edinburgh J. Bot. 55(2): 232 (1998).

Annual herb 12–25 cm tall, densely arachnoid-tomentose with glandular and eglandular hairs. Stem 10–14.5 cm, simple or sparingly branched, slender, 2.5–4 mm across, purplish. Stem bracts few, lanceolate, up to 10–15 mm, arachnoid-tomentose. Inflorescence cylindrical, 4–12 cm, laxly few-flowered (9–29 flowers). Bracts lanceolate, 6–11 mm, densely arachnoid-tomentose. Bracteoles linear, 4–6 mm, shorter than calyx, arachnoid-tomentose glandular or eglandular hairs. Calyx 7–11 mm, densely arachnoid-tomentose; calyx teeth lanceolate at apex, 2–3.5 mm, equalling calyx tube, 2–3.5 mm. Corolla 15–22 mm, pale blue, sparsely glandular-pubescent outside, lobes of lower lip rounded, folds white-hairy. Stamens inserted 4–6 mm above base of the corolla tube; filaments 9–10 mm, glandular at base with few shorter hairs; anther 2–2.5 mm long, glabrous or hairy. Style 6.5–9 mm long, glandular.

HAB. Parasitic on *Reseda* spp. (Resedaceae) and *Eremobium aegyptiacum* (Brassicaceae); on sandy or compact sandy substrates; alt.: ± **60** m; fl & fr.: Mar.-Apr.
DISTRIB. **DSD:** Southern desert of Iraq on Jrishan Khadhr al-Mai road 15 km before Khadhr al-Mai, *Al–Mayah & Al-Asadi* 1690 (BSRA).

Arabian Peninsula (Bahrain, Saudi Arabia).

8. **Phelipanche aegyptiaca** (*Pers.*) *Pomel,* Nouv. Mat. Fl. Atl. 1: 107 (1874); Al-Mayah & Al-Asadi, Am. Sci. Res. J. Eng. Technol. Sci. 34(1): 239 (2017).

> *Orobanche aegyptiaca* Pers., Syn. Pl. [Persoon] 2(1): 181 (1806); Dinsmore in Post, Fl. Syria, Palest & Sinai 2: 315 (1933); Blakelock in Kew Bull. 4(4) 536 (1949); Al-Rawi, Dep. Agr. Iraq Tech. Bull. 12:148 (1964); Schiman-Czeika in Fl. Iranica [K. H. Rechinger] 5: 7 (1964); Jafri in Fl. Pakistan [Nasir & Ali] 98: 8 (1976); Feinbrun-Dothan, Fl. Palaest. 3: 211, pl. 357 (1978); Gilli in Fl. Turkey [P. H. Davis] 7: 8 (1982).
> *Phelypaea aegyptiaca* Walp., Repert. Bot. Syst. (Walpers) iii. 463 (1844).

Annual herb 10–45 cm tall, stems simple or branching from, or just above the base, ± slender (to 6 mm across), glandular-pubescent, similar in form to *P. ramosa* but with larger flowers and usually darker in colour. Stem bracts 5–7 mm, ovate-lanceolate, acute, glandular-pubescent to glabrescent. Inflorescence paniculate, flowers sessile or the lowers shortly pedicellate, flowers in lax spikes. Bracts lanceolate, 6–8 mm, shorter than calyx; bracteoles 2, linear-lanceolate, shorter than calyx. Calyx 9–11 mm, (including lobes) campanulate, 4-lobed, lobes subulate to filiform, tips acuminate, glandular-pubescent. Corolla bluish, 20–28(–35) mm long, tubular-infundibuliform, somewhat inflated, lilac, whitish at base; tube constricted above the ovary, curved outwards, broader above, pubescent within; lobes ovate, obtuse, the lower lobes larger than upper. Stigma lobes white or bluish. Stamens weakly pubescent below, inserted 3–6 mm above the base of the corolla tube; anthers villous. Capsule ovoid, ± 8 mm. Fig. 49, 9.

HAB. Parasitic on various cultivated crops, especially in the Solanaceae, including tomatoes, potatoes, tobacco; also *Visnaga daucoides* (Apiaceae)*; Vicia* spp. (Fabaceae); *Plantago boissieri* (Plantaginaceae); cultivated fields, clearings, nr. villages in degraded *Quercus* forest and under *Populus*, on clayey and loamy soils; alt. 100–1800 m; fl. & fr. May, Sep.

DISTRIB. Mountain, upper plains and foothills region of N. Iraq; common in the region.. **MAM**: Zakho to Sharanish, *Karim, Hamod & Jassom* 41128!; Sharanish, *Rawi* 8656!; 25 km from Zakho to Kani Masi, *Botany Satff* 43802!; Zawitah, *Guest* 1671; Zawitah, *Guest* 3741! Aqra, *Rawi* 11433!; Bekme gorge, *Helbaek* 772!; nr. Greater Rab, *F. & E. Barkley* 5533! **MRO**: *Bornmüller* 16427 (W); Gali Ali Beg, *Gillett* 9435! **MSU**: Halabja, *Noori & Hamad* 41201; Penjwin, *Rawi* 12263; Qara Dagh, *Gillett* 7902!; Sarchinar, nr. Sulimaniyah, *Raddi* 5267!; Siwan river, 7 km S. of Derbenikhan, *F. Barkley* 5112! **MJS**: Jebal Golat, between Tellawi & Belad Sinjar, *Field & Lazar* 415; Kursi, W. of J. Sinjar, *Gillett* 11024. **FUJ**: Ad Raba city, *Noori* 40736!; nr. Ain Diba, *Rawi & Gillett* 7234! **FNI**: Halook, *Tikrutt* 16373! **FPF**: Sa'adiya, *Al-Kaisi* 47222! **DLJ**: Rawa, *Alizzi & Omar* 35351! **LCA**: nr. Baghdad, *Lazar* 88; 496; 533; Bekhair, *Rawi* 8631.

Two collections **MRO**: Kermasur, Lake Qandil, *Rawi & Serhang* 24131! and **MSU**: Jarmo, in wheat field, *Helbaek* 1869! fall between *P. aegyptiaca* and *P. ramosa* in having a more lax spike as in the former and smaller flowers as in the latter. Both are treated here as *P. aegyptiaca.*

LISĀN AT THŌR, QURQAH (*Lazar* 3593).

SE Europe to Central Asia, Arabian Peninsula, Iran to Pakistan and India and Bangladesh; N. Africa to Sahara.

9. **Phelipanche coelestis** (*Reut.*) *Beck,* Monogr. Orobanche Beck, 114, t. 2, f. 21 (1890); Al-Rawi, Dep. Agr. Iraq Tech. Bull. 12:148 (1964); Schiman-Czeika in Fl. Iranica [K. H. Rechinger] 5: 9 (1964); Gilli in Fl. Turkey [P. H. Davis] 7: 10 (1982); Al–Mayah & Al-Asadi, Am. Sci. Res. J. Eng. Technol. Sci. 34(1): 241 (2017).

> *Phelipaea coelestis* Reut., Prodr. [A. P. de Candolle 11: 5 (1847).
> *Orobanche coelestis* Boiss. & Reut. ex Reut., Prodr. [A. P. de Candolle] 11: 5, 37 (1847); Rawi in Dep. Agr. Iraq Tech. Bull. 14: 148 (1964); Jafri in Fl. Pakistan [Nasir & Ali] 98: 12 (1976).

Annual herb 15–40 cm tall. Stems unbranched (or rarely branched), glandular pubescent, especially below. Bracts ovate-lanceolate or lanceolate, 1–1.5 cm, numerous below. Flowers rather large relative to the stem; floral bracts ovate-lanceolate, 8–18 mm; bracts, calyx, bracteoles, and corolla all glandular-pubescent abaxially; bracteoles narrowly lanceolate to linear, shorter than calyx, subsessile or shortly-pedicellate. Calyx shortly campanulate, 10–15 mm, 4-lobed; lobes narrowly lanceolate, about 2/3 as long as calyx. Corolla inclined, blue, tubular; whitish below, pubescent adaxially, 18–26(–28) mm; upper lip 2-lobed; lower lip longer than upper with acute to rounded lobes. Stamens pubescent basally, sparsely glandular pubescent upward (rarely glabrous); anthers sparsely villous, base mucronate;

filaments inserted 4–6 mm above corolla base. Ovary ellipsoid-globose. Style short, glandular pubescent; stigma 2-lobed. Capsule ellipsoid-globose, 9–11 cm.

HAB. Recorded doubtfully to be parasitic *Quercaus aegilops* (Fagaceae); host range unknown; rocky mountain, slope, cultivated fields, near streams, on clayey and limestone soils; alt. 800–2500 m.; fl. May-Jun.

DISTRIB. **MAM**: Zawita, *Rechinger* 11997 (W). **MRO**: Qandil, Pustanshan, *Rechinger* 11759; Mt. Qandil, Goam-e-Kirmosoran, *Rechinger* 11124 (W); Handaris, Rawandaz, *Thesiger* 1088! **MSU**: Pira Magrun [Pir Omar Gudrun], *Haussknecht* s.n.

S.W. & C. Asia, east to China (distribution improperly known due to widespread confusion with similar species)

4. **OROBANCHE** L. nom. cons.

Tabl. Regn. Vég. 2: 292 (1799); Fischer in Kubitzki (ser. ed.), ed. J.W. Kadereit, Fam. & Gen. Vasc. Pl. 7: 410 (2004)

Chris J. Thorogood, Abdulridha A. A. Al-Mayah, Widad M. Al-Asadi and Shahina A. Ghazanfar

Holophrastic herbs lacking chlorophyll. Typically annuals or monocarpic biennials (rarely perennial). Stems succulent, simple or branched, typically pubescent. Inflorescence spicate, bracteate. Bracts scale-like; bracteoles, if present 2, free, about as long as calyx. Calyx 4(–5)-toothed, or split into 2 lateral segments, each segment entire or bidentate. Corolla bilabiate, tubular-campanulate, typically, whitish, reddish, purplish or yellowish; corolla tube curved and spreading to sub-erect; upper lip 2-lobed, the lower lip 3-lobed with the middle lobe often larger than the lateral ones; all lips often notched. Stamens 4, included, variably hairy (an important feature). Ovary 1-locular; stigma variously coloured (an important feature). Capsules ovoid, beaked by the persistent style base. Seeds numerous, dust-like.

A large genus with 197 accepted species found in temperate and tropical regions of both hemispheres; 10 species have been reported from Iraq (one of them with doubt); a further 3 species are doubtfully recorded and need confirmation.

Like other holoparasites, *Orobanche* preserve poorly in herbaria, leading to taxonomic confusion. Size, colour and pubescence are very variable and difficult to use as reliable key characters in isolation. Other characters such as bracts and calyx morphology, corolla shape, and position of insertion and pubescence of the stamens are more dependable. Host species can also be an important taxonomic character however often unrecorded or recorded in error (the parasite may appear several metres away from the host) (Thorogood & Rumsey, 2021). The diversity of *Orobanche* in Iraq may be under-estimated, especially in the complex *minores* group, in which cryptic taxa are likely to exist (Thorogood et al., 2009).

HALUK (Ar.) هالوك

1. Corolla strongly curved, almost downward-pointing; filaments inserted
      in the middle of corolla tube . . . . .1. *O. cernua*
   Flowers ± evenly curved (never downward-pointing); filament inserted
      below the middle of corolla tube 2
2. Upper lip of corolla deeply bilobed 3
   Upper lip of corolla emarginate to entire, but not deeply bilobed 4
3. Corolla yellowish (flushed red); stigma yellow . . . . 2. *O. kurdica*
   Corolla brownish-purple; stigma reddish. . . . . 3. *O. singarensis*
4. Flowers narrowly tubular and small, rarely exceeding 15 mm . . . . . 4. *O. minor* (group)
   Flowers tubular to campanulate, exceeding 15 mm 5
5. Bracts ovate, much shorter than corolla tube . . . . 5. *O. camptolepis*
   Bracts equalling or longer than corolla tube. 6
6. Corolla margins crenate; flowers strongly fragrant, white; stamens
      inserted 2–5 mm above corolla base; calyx teeth filiform . . . . 6. *O. crenata*
   Corolla margins not crenate; flowers strongly fragrant and white; calyx
      teeth attenuated but not filiform 7

1. **Orobanche cernua** *Loefl.*, Iter Hispan. 152 (1758); Dinsmore in Post, Fl. Syria, Palest & Sinai 2: 316 (1933); Schiman-Czeika in Fl. Iranica [K. H. Rechinger] 5: 13 (1964); Al-Rawi, Dep. Agr. Iraq Tech. Bull. 12:148 (1964); Rechinger, Fl. Lowland Iraq: 553 (1964); Jafri in Fl. Pakistan [Nasir & Ali] 98: 15 (1976); Feinbrun-Dothan, Fl. Palaest. 3: 213, pl. 360 (1978); Gilli in Fl. Turkey [P. H. Davis] 7;13 (1982); Zhi-Yun Zhang & Nikolai N. Tzvelev in Fl. China 18: 2 (1998); Chaudhary & Musselman in Fl. Kingdom Saudi Arabia 2(2): 518 (2001); Ghazanfar, Fl. Oman 3: 161 (2015); Taifour & El-Oqlah, Pl. Jordan Annot. Checkl.: 115 (2017); Al-Maya & Al-Asadi, Biol. Appl. Environm. Res. 2(1): 27 (2018).

Herb 12–23(65) cm tall. Stem yellowish, 8–39 cm, the base slightly thickened, 8–11 mm across. Stem bracts 7–10 mm, broadly lanceolate-ovate, glandular-pilose. Inflorescence 4–25 cm, usually cylindrical with numerous flowers (8–83) crowded in a dense spike, becoming somewhat laxer in fruit. Bract 6–9 mm, brownish-yellow, glandular-pubescent, lanceolate with ovate base. Calyx segments free, 7–9 mm, bifid or bidentate with two unequal teeth, ovate-lanceolate or subulate, glandular-pilose. Corolla 10–15(–18) mm, somewhat downward-pointing, tubular, inflated above the insertion of the stamens, constricted in the middle, inflated again near the upper lip, sparsely glandular-pubescence, whitish or pale yellow, with dark blackish or bluish-purple at the margins. Stamens inserted 6–7 mm above the base of corolla tube; filaments 7–10 mm, glabrous and sparsely glandular-pubescent below anther; anther 1.75–2 mm, glabrous. Style 7–9 mm, subglabrous; stigma 0.6–1.5 mm long, whitish-yellowish. Fig. 49, 11.

HAB. Parasitic on *Artemisia scoporia* (possibly other Asteraceae); in cultivated fields, in depressions in desert in sands, and sandy and gravely soil, on summit plateaus on clay rocky soil, by road side on loam; alt. 150–550m; fl. & fr. Mar.-Jun.

DISTRIB. Throughout Iraq. **MSU**: Spring, Pira-Magrun, *Hashimi & Waleed* 637; Siwia-Sulamaniyhm, *Hashimi & Waleed* 0019609!; 25 km N. Penjawin towards Iranian border, *Karzan, Omer & Kader* 17088. **FUJ**: 4 km W. of K2, Al-Jezira Desert, *Burkley, Palmutier & Jumaa Brahim* 001960; 0019608; Jezira, *Gusr*, 13411; 2 km from Shargat to Mosul, *Widad & Al-Khagat* 53692. **FKI**: Jadida, W. of Nukhaib, *Al-Rawi* 31052. **FPF**: Chlat police station, on Persian border, *Al-Rawi & Sh. Haddadd*, 25707; nr. Chlat, about 60 km N.W. Wadi Tip, *Al–Mayah & Al-Asadi* 1689; 60 km from Tip police station, *Al–Mayah & Al-Asadi* 1697. **DLJ**: Baiji, *Al-Hilli* 0025068. **DGA**: Ghurfa near Injana, *Al-Bermani* s.n.; Adaim, the third wall at the range management, *Widad & Salah* 54399; Adim, Ghurfa, 5 km from Baghdad, *Al-Shahbaz & Al-Mousawi* 0025444. **DWD**: 4 km S. Rutba, *Khider Wanni* 7451; 20 km E.S.E. of Rutba, *Al-Rawi*, 31297; 40 km S. of Rutba, *Al-Rawi* 21245; Al-Karra–Rutba, *Al-Rawi* 23746; bottom of Hawran wadi, *Chakravarty, Al-Rawi & Al-Izzi* 31600! **DSD**: Ruhaba 35 km N. of Najef, *Waines, Hader, Waleed & Agnew* 5510; S. Desert, Salman, *Guer, Al-Rawi & Sillus* 14078!; S. Desert, 42 km E. by S. Salman, *Guer, Al-Rawi & Sillus* 14110! **LCA**: 20 km N. of Taji comp. Baghdad, *Agnew, Hader & R.N. Halnss* 0019611; Baghdad-Hit high way roadside, *Wanni, Sheikh & Batanouny* 7523.

Europe, N. Africa and Canary Islands, Arabian Peninsula, Jordan, Lebanon, Syria, Palestine, Iran, Turkey, Pakistan to India, China and Australia (introduced).

2. **Orobanche kurdica** *Boiss. & Hausskn.* ex Boiss., Fl. Orient. [Boissier] 4(2): 505 (1879); Schiman-Czeika in Fl. Iranica [K. H. Rechinger] 5: 17 (1964); Al-Rawi, Dep. Agr. Iraq Tech. Bull. 12:148 (1964); Gilli in Fl. Turkey [P. H. Davis] 7: 20 (1982); Al-Maya & Al-Asadi, Biol. Appl. Environm. Res. 2(1): 29 (2018).

Monocarpic biennial herb, 28–32cm. Stem 11–18 cm, up to 5–12 mm across, glandular-pubescent. Stem bracts 25–28 mm, lanceolate, sparsely glandular-pilose. Inflorescence 10–16 cm, cylindrical, lax, 12–34-flowered. Bract 20–27mm long, lanceolate, glandular-pubescent, as long as corolla or shorter. Calyx segments 20–23 mm, bidentate or entire, glandular-pilose, teeth lanceolate. Corolla 23–26 mm, tubular-campanulate, wide towards apex, inflated above insertion of the stamens, upper lip deeply 2-lobed, pale yellowish, flushed pinkish-red externally (light brown in dried material), glandular-pilose outside. Stamens inserted 4–6 mm above base of corolla tube; filaments 17–20 mm long, pilose from base to middle, glabrous above, but glandular below anther; anther 2–3 mm, glabrous. Style 10–13 mm, sparsely glandular-pilose; stigma 1.25–1.5 mm, yellow.

HAB. Host not confirmed (usually reported on the Apiaceae elsewhere); reported from Iraq to grow on *Tanacetum* and *Phlomis* but further work required to determine the host range; summit plateau, rocky mountain, limestone hill, on rich clay soil; alt. ± 900 m; fl. & fr. Jun.-Jul.

DISTRIB. **MRO**: Potine Mt. N. of Shirwan Mazin, Erbil Liwa, *Agnew, Haines, Hader & Kadir* 0019620. **MSU**: Avraman Mt., *Al-Rawi & Hisham* 29355; Avraman, Tawela, *Weinert & Mousawi*, 0027443.

Turkey, Transcaucasus, Iran.

3. **Orobanche singarensis** *G.Beck ex Hand.-Mazz.*, Ann. Nat. Hofmus. Wien xxvii. 407 (1913); Rawi in Dep. Agr. Iraq Tech. Bull. 14: 148 (1964); Salih, J. Garmian Uni. 5(2): 433 (2018). [https://jgu.garmian.edu.krd/article_67190.html].

Herb to 22 cm. Stem with numerous oblong bracts; bracts glandular-pubescent along the margins, denticulate. Spike oblong-cylindrical, dense, 7 cm. Flowers suberect; floral bracts 17–20 mm. Calyx 2-parted, very unequal, with narrow teeth, apically filiform, more or less equalling the corolla, glandular-hairy. Corolla tubular, gradually expanded at the throat (not at the point of filament insertion), reddish- to brownish-purple, sparsely glandular-pubescent; upper lip deeply bilobed with broad, rounded lobes; lower lip 3-lobed with rounded lobes, all lobes deeply and unevenly crenulate and hairless. Stamens inserted 3–4 mm above the base of the corolla tube; filaments sparsely glandular-pubescent below. Stigma bilobed, reddish; style sparsely glandular-pubescent. A poorly-known species.

HAB. Host unknown; rocky mountains and associated plains; alt. ± 1400 m; fl. not recorded.
DISTRIB. Recorded from Chuwarta and Piramagrun mountain (**MSU**), Kalar (**FPF**) and Sinjar mountain (**MJS**) (Salih, l.c.).

Endemic to N.W. Iraq.

4. **Orobanche minor** *Sm.*, Engl. Bot. 6: 422 (1797); Dinsmore in Post, Fl. Syria, Palest & Sinai 2: 316 (1933); Chaudhary & Musselman in Fl. Kingdom Saudi Arabia 2(2): 519 (2001); Frost & Musselman, Weed Science 28(1): 119 (1980); Taifour & El-Oqlah, Pl. Jordan Annot. Checkl.: 115 (2017); Al-Maya & Al-Asadi, Biol. Appl. Environm. Res. 2(1): 29 (2018).

Variable, slender annual herb, 6.5–52 cm. Stem simple, 12–35 cm, slender, red-purple, glandular-pubescent, middle part 1.5–4 mm across. Stem bracts 19–21 mm, lanceolate, glandular-pilose. Inflorescence cylindrical, 5–17 cm, with 5–32 small flowers, lax at base of spike. Bract 10–16 mm, lanceolate, glandular-pilose. Calyx segments 5–12 mm, entire or bidentate, glandular-pilose, broad at base and filiform at tip, about half as long as corolla or shorter. Corolla small, 10–15(20) mm, narrowly tubular, inflated near throat, yellowish at base with purplish or purple reddish near upper lip. Lower lip of corolla of 3 rounded equal, crisped lobes, glandular-pilose outside; corolla curved evenly in profile. Stamens inserted 2–3.5 mm above the base of corolla tube; filaments 7–9 mm, sparsely pilose at base and glandular up to anther; anther 1–1.5mm, pubescent. Style 5.5–8.5 mm long, sparsely glandular-pilose; stigma 1–2 mm, purplish, brown or reddish purple. Fig. 49, 6.

HAB. Parasitic on numerous herbs (and occasionally shrubs), especially in the Asteraceae, Fabaceae and Apiaceae; on mountains; alt. ± 1700 m; fl. & fr. May.-Jun.
DISTRIB. **MAM**: Aqra, *Al-Rawi*, 11433.

Small Broomrape, Lesser Broomrape. A highly variable species with infraspecific taxa. Photographs from the region depict species from the *minores* group that do not correspond well to *Orobanche minor* s.s. These may represent novel taxa, and require further work.

Europe and the Mediterranean Basin; introduced to temperate regions worldwide.

5. **Orobanche camptolepis** *Boiss. & Reut.* ex *Boiss.*, Fl. Orient. [Boissier] 4(2): 515 (1879); Feinbrun-Dothan, Fl. Palaest. 3: 213, pl. 361 (1978); Al-Maya & Al-Asadi, Biol. Appl. Environm. Res. 2(1): 25 (2018).

Orobanche latisquama sensu Reut. ex Boiss.: Fl. Orient. [Boissier] 4(2): 516 (1879).

Orobanche platylepis Reut. ex Boiss. Fl. Orient. [Boissier] 4(2): 516 (1879).

O. cernua Loefl. var. *latebracteata* G.Beck forma *camptolepis* (Boiss. & Reut.) G.Beck in Engler, Pflanzenr. 96: 126 (1930).

O. ovata Blakelock in Kew Bull. 4: 537, f. 5 (1949). Type: Iraq, Jabel E.N.E. of Seri Hassan Beg (Rowanduz Area) *Guest* 2912!; Al-Rawi, Dep. Agr. Iraq Tech. Bull. 12: 148 (1964).

Herb, 7–10 cm tall, glandular-pubescent. Stem simple, 3.5–5 mm across. Stem bracts 8–10 mm, broadly ovate, tip attenuated with short, lanceolate cups, hooked when dry. Inflorescence a dense terminal spike, with flowers produced from the base. Bract 6–8 mm, broadly-ovate, glandular-pilose, shorter than the corolla tube. Calyx segments free, 9.5–11 mm, bidentate, segments oblong at base, teeth 2.5–3 mm. Corolla upward-infected, 10–18 mm, tubular, tinted dull blue, whitish at the base, glandular-pubescent outside. Stamens inserted 3.5–4 mm above corolla tube base; filaments 10–12 mm, glabrous; anther 2.5–3 mm, subglabrous. Style 4–6 mm, sparsely glandular; stigmas orange to purple, 2.5–3 mm. Fig. 48, 5.

HAB. Parasitic on the Polygonaceae; rocky mountains; alt. 1500–2000 m; fl. & fr. Jun-Aug.

DISTRIB. **MRO**: Jabal Seri Hassan Beg, 1800 m, *Guest* 2912 (type of *O. ovata*); **MSU**: Pira-Magrun, east-facing, *Miran & Faris* 1271.

W. & C. Asia, Turkey, Saudi Arabia, Syria, Lebanon, Iran, Afghanistan.

6. **Orobanche crenata** *Forssk.*, Fl. Aegypt.-Arab.: 113 (1775); Dinsmore in Post, Fl. Syria, Palest & Sinai 2: 315 (1933).Blakelock in Kew Bull. 4(4) 537 (1949); Schiman-Czeika in Fl. Iranica [K. H. Rechinger], 5: 19 (1964); Al-Rawi, Dep. Agr. Iraq Tech. Bull. 12:148 (1964); Feinbrun-Dothan, Fl. Palaest. 3: 213, pl. 362 (1978); Gilli in Fl. Turkey [P. H. Davis] 7:14 (1982); Chaudhary & Musselman in Fl. Kingdom Saudi Arabia 2(2): 518 (2001); Taifour & El-Oqlah, Pl. Jordan Annot. Checkl.: 115 (2017); Al-Maya & Al-Asadi, Biol. Appl. Environm. Res. 2(1): 28 (2018).

Robust herb 14–45 cm tall. Stem simple, stout, reddish, purplish or brownish, glandular-pubescent. Stem bracts 10–17 mm, lanceolate, glandular-pilose. Inflorescence long, cylindrical, and lax below, dense above, with 11–numerous flowers, fragrant (smelling strongly of carnations). Bracts 15–20 mm lanceolate. Calyx segments free, 10–16m long, bifid or unequally bidentate, lanceolate-filiform or subulate, almost glabrous. Corolla 19–23 mm, campanulate, inflated above the insertion of stamens, white with purplish veins, especially distally, dorsal line of corolla profile curved forward at the base, upper lip deeply crenate. Stigma lobes whitish, yellow-orange or pink (rarely purplish). Stamens inserted 2–4(5) mm above the base of corolla tube; filaments 8–10 mm, pubescent or pilose at base and glandular up to the anther; anther 1.5–2 mm, often pubescent. Style 6–7mm long, glandular-pubescent. Fig. 49, 7.

HAB. Parasitic on a wide variety of plants, especially cultivated Fabaceae; sometimes appearing in large numbers for example on bean crops; stony, clay hillside, summit plateau, rocky clay soil, rocky valley; alt. 850–1830 m; fl. & fr. May-Jul.

DISTRIB. **MRO**: Potine Mt. N. of Shirwan Mazin, Arbil Liwa, *Agnew, Haines, Hader & Kadir* 6230a; Haibat Sultan Dagh Mt., N. of Kis Sanjaq, *Al-Rawi & Kass*, 28175!; Rowanduz, *Blakelock* 2718!. **MSU**: Avraman Mt. of Halabcha on Persian border, *Al-Rawi* 22086; beginning of Khomal-Halabcha rd. towards Halabcha, *Al-Mayah & Al-Asadi* 1504.

Europe, N. Africa, Turkey, Armenia Georgia, Syria, Palestine, Lebanon, Iran.

7. **Orobanche anatolica** *Boiss. & Reut. ex Reut.*, Prodr. [A. P. de Candolle] 11: 17 (1847); Boissier, Fl. Orient. 4: 504 (1879); Dinsmore in Post, Fl. Syria, Palest & Sinai 2: 315 (1933); Blakelock in Kew Bull. 4(4) 536 (1949); Schiman-Czeika in Fl. Iranica [K. H. Rechinger] 5: 18 (1964); Al-Rawi, Dep. Agr. Iraq Tech. Bull. 12:148 (1964); Gilli in Fl. Turkey [P. H. Davis] 7: 21 (1982); Al-Maya & Al-Asadi, Biol. Appl. Environm. Res. 2(1): 25 (2018).

Herb 14–39 cm tall, glandular-white hairy. Stem 8.5–26 cm, dark red, thick and succulent with the middle part 7–15mm in diameter, glabrous. Stem bracts 9.5–15 x 0.5–7.5 mm, brown-red (or reddish-brown), broadly ovate to lanceolate, pilose. Inflorescence terminal, cylindrical, with 10–46 flowers. Bract equalling or longer than the corolla tube, broadly ovate to lanceolate, vary sparsely white pilose. Calyx 10–7 mm, campanulate, glandular-pilose at margins. Corolla 18–27 mm, tubular, brownish- red or pink or reddish-brown, densely glandular-pilose outside and inside, lanate on upper lip, dorsal line curved, upper lip curved, undivided, limb long pilose or whitish lanate. Stamens inserted 0.3–1.25mm above the base of corolla; filaments 14–20 mm, pilose or pubescent near base and sparsely glandular below anther; anther 2–3.25 mm, glabrous. Style 8–17 mm long, glabrous; stigma lobes yellow, 3.5–5 mm.

HAB. Apparently parasitic on *Astragalus* (Fabaceae) (*Guest & Ludlow- Hewitt* 2892; 2939) and on *Salvia* spp. (Lamiaceae); elsewhere reported on *Salvia* (apparently the main host); mountain, slopes, in *Quercus* woodland, on most soil; alt. 800–3000 m; fl. & fr. Apr.-Aug.

DISTRIB. **MAM**: Aqra, *Rawi,* 11434 (BAG!); Dohuk, *Al-Rwai,* 8761!; Bakirma, *Al-Rwai,* 8521!; Mt. Khantur N.E. of Zakho, *Rawi* 23325; **MRO**: Susa 15 km E. Arbil!; Batas, *Rawi,* 8574!; Shaqlawa, Erbil Liwa, *Hadae, Lukman & Waleed* 1370!; Shaqlawa, *Gillett* 8060!; Qandil Range, *Serhang & Rawi* 26762!; Ebsnman sherin valley, W.N. of Potine Mt. N. of Mt. Shirran *Agnew, Hader & Kader,* 0019612!; Algurd Dag, *Guest & Ludlow- Hewitt* 2892!; 2939! **MSU**: Qara Mt., *Al-Kas* 19515!; *Al-Mayah & Al-Asadi,* 1457!; Bayana, before Zalan on Shahrabashir-Chawarta road, *Al-Mayah, Al-Asadi & Al-Mousawi* 1458!; Delaisher, Sul. 1 km before Pira-Magron road, *Al-Mayah & Al-Asadi* 1456!; Pira-Magron, *Al-Mayah & Al-Asadi,* 1462!; Pira-Magron, Summer resort *Al-Mayah & Al-Asadi* 15048!

Syria, Lebanon, Iran, Azerbaijan, Turkey, Caucasia.

8. **Orobanche armena** *Tzvelev* in Fl. URSS [Schischkin] 23: 686 (1958); Gilli in Fl. Turkey [P. H. Davis] 7: 21 (1982).

Herb 28–31 cm. Stem unbranched, brown, 7.5–8.5 mm across. Bracts alternate-spiral, with hairs on the lower surface and margins, apically acuminate, basally truncate, brown; lower stem bracts 14–16 mm, upper bracts 14–17 mm. Inflorescence a dense spike, 16–19 cm; floral bracts narrowly oblong, the margins minutely dentate, apically acuminate, basally truncate, glandular, brown, 15–20 mm. Flowers numerous. Calyx connate at the base, 2-toothed, the teeth narrowly ovate, glandular, brown, 21–25 mm. Corolla tubular-campanulate, glandular-pilose, brown, 10–13 mm, the upper lip emarginate, the lower lip 3-lobed with margins entire. Stamens 4, inserted above the corolla tube base; filaments densely pilose below, sparingly glandular-pilose above, brown, 6.5–9.0 mm, anthers narrowly oblong, brown. Stigma rough. Capsule, ovoid- narrowly ovoid, 8–11 mm.

HAB. Parasitic individuals on *Astragalus* spp. (Fabaceae), on rocky-clay substrates; alt. ± 2350 m; fl. Aug.
DISTRIB. **MRO**: Qandil mountain (North-East of Erbil) within Rowanduz district.

Similar to *O. kurdica* but with a smaller corolla and calyx connate at the base.

Turkey.

9. **Orobanche kotschyi** *Reut.,* Prodr. [A. P. de Candolle] 11: 33 (1847); Schiman-Czeika in Fl. Iranica [K. H. Rechinger] 5: 15 (1964); Jafri in Fl. Pakistan [Nasir & Ali] 98: 18 (1976).

Biennial or perennial herb, up to 60 cm; stems stout, < 10 mm across, densely pilose. Bracts oblong-lanceolate, 10(–15) mm, distantly placed. Inflorescence a dense spike; flora bracts c. 15 mm, as long or slightly shorter than corolla tube. Bracteoles absent. Calyx as long as bracts, divided into 2 segments, each segment deeply 2-fid to below middle; teeth linear. Corolla blue, (20–)25–30(–35) mm, tube constricted about the middle, inflated below; lips obtuse. Filaments and anthers minutely pubescent. Stigma 2-lobed. Capsule c. 10 mm, ovoid.

(Only seen the type material from Iran. Description from Fl. Pakistan l.c.)

HAB. Parasitic on the Apiaceae; alt. ± 1800 m; fl. Aug.
DISTRIB. **MSU** Hawraman mts. [Avroman mts], prope Tawilla, *Rechinger* 10325 (W).

Turkey, Transcaucasus, Iran, Afghanistan to Pakistan; C. Asia, Xinjiang.

10. **Orobanche reticulata** *Wallr.,* Orobanches Gen. Diask. 42. (1825).

Herb 15–70(–81) cm tall. Stem glandular-hairy, ± yellowish or purplish, slightly swollen below. Inflorescence rather dense above, laxer below. Bracts 12–25 mm, narrowly triangular equal to, to shortly exceeding corolla, rather dark (contrasting the flowers). Calyx 7–12 mm, bifid, upper segment largest. Corolla 12–22 mm, pale creamy-yellowish, purple to reddish-brown distally (especially the margins), with sparse, dark glands (especially distally), broadly cylindrical-campanulate; upper lip with 2 spreading lobes, the lower lip with 3 equal lobes, the middle rather square; all lobes more or less toothed. Corolla tube somewhat broadened above the insertion of the stamens, the back curved strongly above the base and behind the upper lip. Stamens inserted 2–4 mm above the base of the corolla; filaments hairless to sparsely hairy below, somewhat glandular to hairless above. Stigma lobes dark purple, just touching at base.

HAB. Parasitic on the Asteraceae, mainly subfamily Carduoideae.
DISTRIB. Reported from the region (Al-Asadi & Al-Mayah 2016), growing on *Vicia monantha* based on purple pigmentation and hairless filaments however photographs depict a species in the *Minores* group. It is very possible that it occurs in the region based on global distribution.

Widespread from Europe to Central Siberia to Turkey.

TAXA DOUBTFULLY RECORDED

Orobanche alba *Stephan ex Willd.*, Sp. Pl., ed. 4 [Willdenow] 3(1): 350 (1800).

Herb to 25 cm. Stems reddish, short-glandular-hairy, slightly swollen with numerous reddish, scale-like stem bracts at base. Inflorescence lax, slightly fragrant. Bracts 12–25 mm, lanceolate, acuminate, glandular, slightly shorter or equalling the length of the corolla. Calyx 8–16 mm usually entire, occasionally bidentate, about equalling the corolla tube. Corolla 15–25 mm, cream, veined or suffused reddish-purple, glandular-pubescent especially distally, erecto-patent, slightly curved, more strongly so at the basal and distal ends; tube campanulate-cylindrical; upper lip entire or 2-lobed with somewhat spreading margins; lower lip glandular-ciliate, the 3 lobes about equal, the middle just the largest. Stamens inserted 1–3 mm above the base of the corolla tube; filaments slightly hairy below, glandular above. Stigma lobes reddish or purplish, the lobes touching.

HAB. Parasitic on the Lamiaceae.

Reported from the region (Al-Asadi & Al-Mayah 2016), based on coloration and hairy filaments, though photographs do not correspond with this species. However, it is quite likely that the species occurs in the region based on global distribution.

Orobanche caryophyllacea *Sm.*, Trans. Linn. Soc. London 4: 169 (1798); Zhi-Yun Zhang & Nikolai N. Tzvelev in Fl. China 18: 237 (1998); Rumsey in Curtis's Bot. Mag. 26 (4): 369 (2010); Musselman, Checklist of Plants of Lebanon and Syria (2011) http://ww2.odu. edu/~lmusselm/plant/lebsyria/Checklist of Lebanon Plants.pdf.; Al–Maya & Al-Asadi, Biol. Appl. Environm. Res. 2(1): 26 (2018).

Herb 24–47 cm tall, sometimes clumped. Stem 14–25.5 cm, middle part up to 4–11 mm thick, glandular-pubescent, slender to stout, yellowish-purple to reddish-brown. Stem bracts 10–25 mm long, lanceolate, glandular-pilose. Inflorescence 8–23 cm, with 12–35 fragrant (clove-scented) flowers in a lax spike. Bracts 15–22 mm long, about as long as corolla, glandular-hairs deflexed in the middle. Calyx segments free, 9.5–16 mm, bidentate, glandular-pilose, teeth lanceolate. Corolla 20–24 mm, pale pinkish-brown to purplish, often tinged red, densely glandular-pilose outside, tube broadly campanulate, upper lip very broad, lower lip of corolla deflexed with three crenate lobes of equal size. Stamens inserted 1.5–2.5(–3) mm from base of corolla tube; filaments 12–13 mm, hairy for 1/3 length from base, glandular below; anther 2–2.3 mm long, glabrous or subglabrous. Style 6.5–8.5 mm long, sparsely glandular-hairy; stigma dark purple to brownish-purple, 3–3.5 mm long; lobes distant. Parasitic on *Galium* spp. (Rubiaeae).

Photographs of living material identified as this species (Al-Asadi & Al-Mayah 2016) based on presumed host (*Galium*) and bifid calyx depict a species in the *Minores* group. Material unequivocally belonging to *C. caryophyllacea* from the region has not been observed and further work is required to confirm its presence in Iraq.

N.W. Africa, Europe, Russia, Armenia, Turkey, Iran, Pakistan, China,

Orobanche arenaria *Borkh.*, in Roem. Neues Mag. i. 6 (1794) & ex Gaertn. Mey. & Scherb. Fl. Wett. ii. 405.

Has been reported from Mt. Piramagrun mt. (alt. 1250–1550 m) by Salih, J. Garmian Uni. 5(2): 433 (2018) [https://jgu.garmian.edu.krd/article_67190.html], but without details of collector or collector's number. The record needs confirmation.

5. **EUPHRASIA** L.

Sp. Pl. 2: 604 (1753); Yeo, Bot. J. Linn. Soc. 77(4): 223 (1978); Fischer in Kubitzki (ser. ed.), ed. J.W. Kadereit, Fam. & Gen. Vasc. Pl. 7: 423 (2004)

Shahina A. Ghazanfar

Annual semi-parasitic herbs. Stems branched, glabrous to pubescent, often glandular. Leaves opposite, upper irregularly alternate, margins crenate or dentate. Flowers solitary, in axils of leaf-like bracts, shortly pedicellate. Calyx 4-lobed to about 2/3 its length. Corolla 2-lipped, hairy outside, lower lip 3-lobed, lobes emarginate, upper lip hooded, shortly 2-lobed. Stamens 4, didynamous, included; filaments inserted near corolla-throat, curved with anthers held in the upper lip; anthers hairy, spurred, posterior anther with a longer spur. Ovary bilocular, laterally compressed, oblong. Stigma capitate. Capsule dehiscent in the upper half. Seeds fusiform, reticulate.

224 species distributed in N.W. Africa, Eurasia, Australia, New Zealand, N. America, Peru; a single species in Iraq.

Yeo, P.E. (1978). A taxonomic revision of *Euphrasia* in Europe, Bot. J. Linn. Soc. 77 (4): 223–334.

1. **Euphrasia pectinata** *Ten.*, Prodr. Fl. Nap. 36 (1811); Yeo in Fl. Iranica [K. H. Rechinger] 147: 175 (1981); Yeo in Fl. Turkey [P. H. Davis] 6: 759 (1978); Siddiqui & Qaiser in Fl. Pakistan (Ali & Qaiser) 220: 42 (2015).

Euphrasia regelii Wettst., Monogr. Euphrasia [Wettstein] 81 (1896) p.p.; Rawi in Dep. Agr. Iraq Tech. Bull. 14: 142 (1964).
E. kurdica Rech.f., Sym Bot. Upsal. 11(5): 43 (1952).

Erect annual. Stems simple, slender, 2–2.5 cm (in a few plants seen from the same location), retrorse hispid-hairy. Leaves 2–3 pairs, elliptic to obovate, 5–7 × 2–3 mm, setulose, apex obtuse, margins serrate to dentate, hispid. Lower bracts ovate, often adpressed to calyx. Calyx c. 6 mm (incl. lobes), lobes c. 2–3 mm, lanceolate, acuminate to aristate, purple veined. Corolla slightly exceeding calyx, white suffused purple, with blackish purple streaks, lower lip 3-lobed, margins emarginate, hairy on the outside. Capsule c. 5 mm, oblong, emarginate. Fig. 50, 1–6.

HAB. In damp ground, subalpine; alt. ±2900 m; fl. Aug.
DISTRIB. A few plants at a single location in N. Iraq; rare. **MRO**. Arbil, Algurd Dagh, *Gillett* 12329!

S. Europe, Turkey to Pakistan, Central Asia to S.W. Siberia and China.

6. **PARENTUCELLIA** Viv.

Fl. Lib. 31 (1824); Fischer in Kubitzki (ser. ed.), ed. J.W. Kadereit, Fam. & Gen. Vasc. Pl. 7: 424 (2004)

Shahina A. Ghazanfar

Annual herbs. Stems erect, glandular-viscid. Leaves opposite, alternate above, sessile, entire or pinnatifid with linear segments, margins entire or dentate to serrate. Flowers pedicellate. Calyx tubular, 4(–5)-lobed to almost ½ its length; lobes ± equal. Corolla strongly 2-lipped; upper lip hooded; lower lip spreading, 2-lobed. Stamens 4, included; anthers mucronate. Ovary narrow oblong. Capsule oblong. Seeds numerous, smooth or reticulate.

Two species in Western Europe, Mediterranean region to Central Asia and Iran; a single species in Iraq.

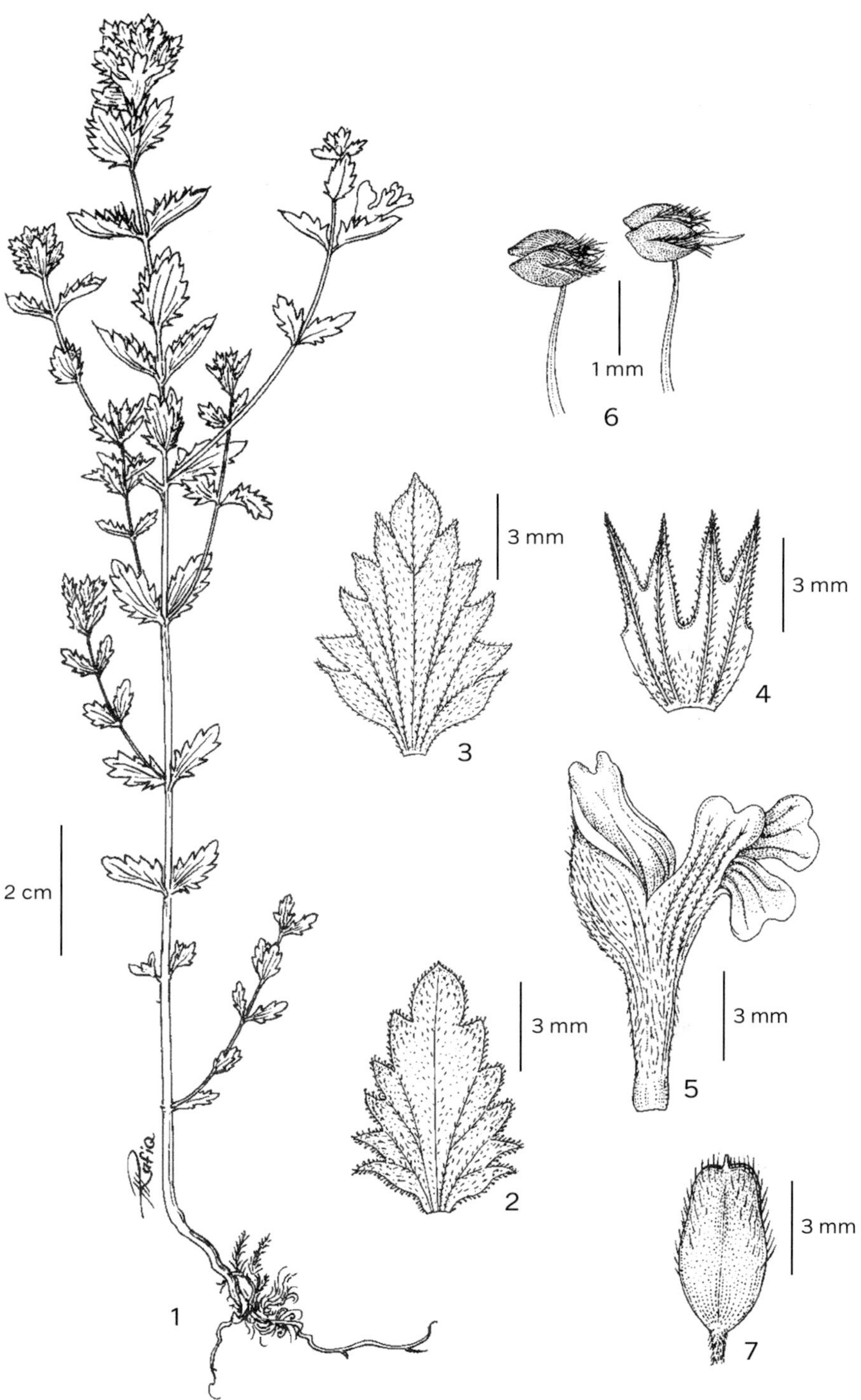

Fig. 50. **Euphrasia pectinata**. 1, habit; 2, cauline leaf; 3, floral leaf; 4, calyx; 5, corolla; 6, anthers. Reproduced with permission from Fl. Pakistan 220: f. 4, I-O. Drawn M. Rafiq. © University of Karachi, Pakistan.

1. **Parentucellia latifolia** (*L.*) *Caruel*, Fl. Ital. (Parlatore) 6: 480 (1885); Dinsmore in Post, Fl. Syria, Palest. & Sinai ed. 2, 2: 307 (1933); Rawi in Dep. Agr. Iraq Tech. Bull. 14: 144 (1964); Hedge in Fl. Turkey [P. H. Davis] (1978).

Euphrasia latifolia L., Sp. Pl. 2: 604 (1753).

Annual herb. Stems slender, simple, 15–30 cm, glandular-viscid. Leaves sessile, ovate, 6–10 × 4–5 mm, apex obruse, margins serrate to dentate, often revolute, hispid. Calyx tubular, 13–14 mm (incl. lobes), glandular; lobes lanceolate, to 5 mm, 1-veined, glandular. Corolla purplish, exceeding calyx; corolla tube c. 5 mm; upper lip c. 4 mm; lower lip c. 4 mm, shortly 3-lobed. Anthers glabrous. Fig. 51, 1–6.

HAB. Gravelly foothill of mountain, in loamy soil; alt. ± 200 m; fl. May.
DISTRIB. N. Iraq; known from a single collection from Iraq. **MRO**. 19 km S.W. Salahuddin, *F.A. Barkley* 5629!

Turkey, Syria.

7. **ODONTITES** Ludw.

Inst. Regn. Veg., ed. 2. 120 (1757); Fischer in Kubitzki (ser. ed.), ed. J.W. Kadereit, Fam. & Gen. Vasc. Pl. 7: 422 (2004)

Shahina A. Ghazanfar

Annual herbs. Stems erect, hispid or glandular-viscid. Leaves opposite, sessile, margins entire to dentate or crenate. Flowers shortly pedicellate. Inflorescence racemose. Calyx campanulate, unequally 4-lobed to ½ its length, weakly 2-lipped. Corolla 2-lipped, tube shorter or slightly longer than calyx; lower lip 3-lobed, upper 2-lobed. Stamens 4, didynamous, included. Ovary narrowly elliptic to oblong, emarginate. Seeds 2–20 per locule, reticulate, with longitudinal ridges.

32 species in Madeira, West and South Europe, North Africa to West Asia (Himalayas); a single species in Iraq.

1. **Odontites aucheri** *Boiss.*, Diagn. Pl. Orient. ser. 1, 4: 74 (1844); Dinsmore in Post, Fl. Syria, Palest. & Sinai ed. 2, 2: 309 (1933); Rawi in Dep. Agr. Iraq Tech. Bull. 14: 143 (1964);

Annual herb 10–40 cm. Stems simple or branched, slender, white-hispid. Leaves sessile, opposite to subopposte, linear, 8–30 × 1 mm, the upper leaves smaller, acute, margins sometimes revolute. Calyx c. 8 mm, persistent; lobes unequal, lanceolate, c. 4 mm, hispid. Corolla yellow, exceeding calyx, hispid; upper lip hooded (but not strongly); lower lip 3-lobed. Anthers glabrous. Capsule emarginate, c. 7 mm, hispid. Fig. 51, 1–7.

HAB. Mountains, on metamorphic rocks, stony and rocky open hillsides, in *Astragalus* zone; alt. ± 2200 m; fl. Jun.-Jul.
DISTRIB. Mountains of N. Iraq. **MAM**. Zaita [Zawita], nr. Shirwan, *Angew, Hatim & Haines* W. 2138! **MRO**. S.E. of Sairo, *Gillett* 9698!; Chai Mandali [Mandau], *Guest* 2701! **MSU** Hawraman mts. [Avroman mts], *Rawi, Charavarty, Nuri & Alizzi* 19711!

Turkey, Syria, Lebanon, Iran, Transcaucasus, Turkmenistan.

8. **BARTSIA** L. nom. cons.

Sp. Pl. 2: 602 (1753); Fischer in Kubitzki (ser. ed.), ed. J.W. Kadereit, Fam. & Gen. Vasc. Pl. 7: 422 (2004)

Shahina A. Ghazanfar

Annual or perennial herbs, densely pubescent, often glandular. Stems erect to prostrate to ascending. Leaves opposite, sessile, with revolute, crenate, rarely entire margins. Inflorescence terminal, spicate. Flowers solitary, axillary, shortly pedicellate. Calyx campanulate, unequally 4-lobed, divided to about ½ its length. Corolla tubular, 2-lipped, tube longer than the calyx. Stamens 4, inserted at the top of the corolla tube, included;

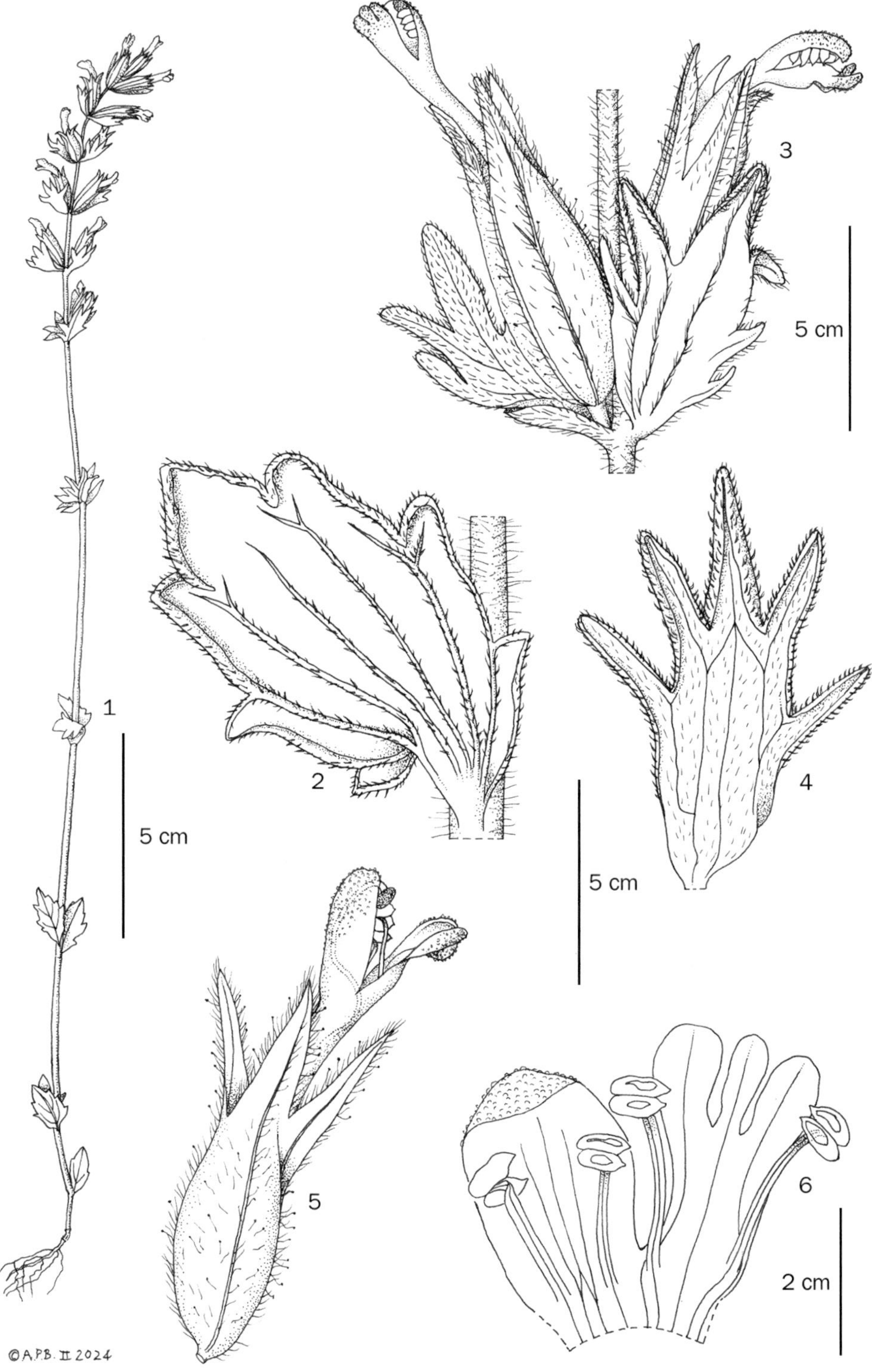

Fig. 51. **Parentucellia latifolia**. 1, habit; 2, leaf, abaxial surface; 3, pair of flowers; 4, floral bract, abaxial surface; 5, flower, side view; 6, corolla lobes opened (dorsal lobe on left). 1–6 from *Barkley* 5629. Drawn by © A.P. Brown, Feb. 2024.

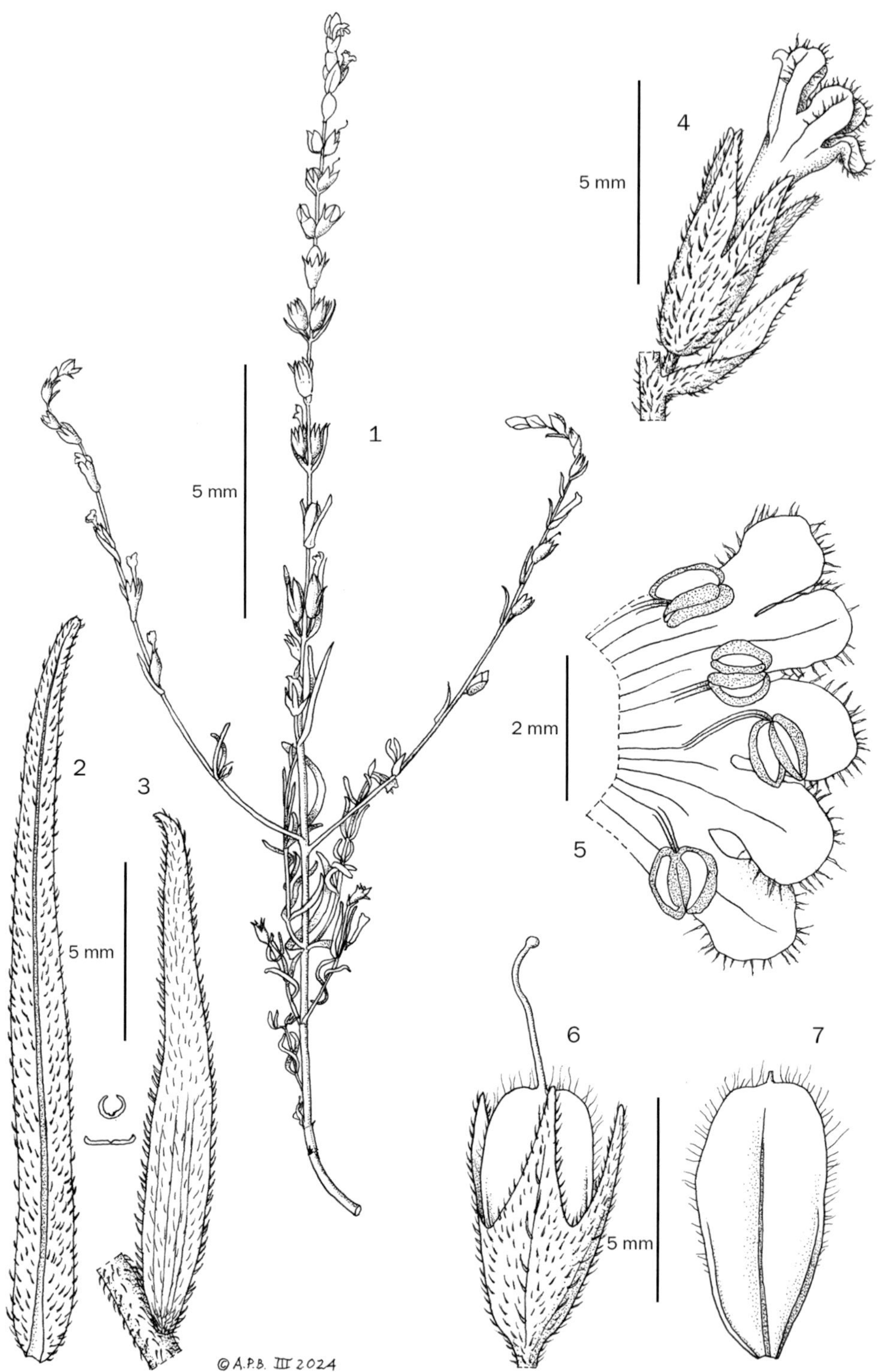

Fig. 52. **Odontites aucheri**. 1, habit; 2, leaf, adaxial view; 3, leaf, abaxial view with leaf TS when rolled up and when flat; 4, flower, side view; 5, dorsal and ventral lobes of corolla (flattened); 6, fruit with remains of style, in situ in calyx; 7, mature fruit after abscission of style. 1–7 from *Guest* 2701. Drawn by © A.P. Brown, Feb. 2024.

anthers bearded. Ovary oblong to subglobose, not emarginate. Capsule oblong. Seeds numerous, testa reticulate, with longitudinal wings, wings striate.

Fifty species mainly S. and C. America; a few in tropical Africa, Europe and N. America; a single species in Iraq.

1. **Bartsia trixago** *L.*, Sp. Pl. 602 (1753); Rawi in Dep. Agr. Iraq Tech. Bull. 14: 142 (1964); Fischer in Fl. Ethiop. & Eritr. 5: 307 (2006).

Bellardia trixago (L.) All., Fl. Pedem. 1: 61 (1785); Dinsmore in Post, Fl. Syria, Palest. & Sinai ed. 2, 2: 308 (1933); Feinbrun-Dothan, Fl. Palaest. 3: 208, pl. 353 (1978).

Annual herb. Stems erect, simple (in our material), leafy throughout, to 40 cm, pubescent to hispid. Leaves sessile, linear-lanceolate, 10–35 × 1–5, apex obtuse, margins coarsely serrate, hispid and glandular-pilose. Inflorescence terminal. Flowers sessile to very shortly pedicellate. Bracts similar to leaves but smaller in size, glandular-pubescent. Calyx campanulate, 5–6 mm, unequally 5-lobed with 2 main lobes; lobes broadly triangular, glandular-pilose. Corolla white to pinkish, tubular, 2-lipped; corolla-tube, 6–7 mm, glandular-pubescent outside; upper lip c. 7 mm; lower lip 3-lobed, lobes, c. 7 mm, rounded, deflexed, dorsally glabrous. Anther transverse, mucronate at apex, bearded along the anther suture. Ovary oblong, ± 4 mm, pilose; Stigma clavate. Capsule ovoid, 8–9 mm, pilose. Seeds longitudinally and transversely striate, ± narrowly winged. Fig. 53, 1–6.

HAB. On hillside in cultivated field, in rich loamy soil; alt. ± **1000 m**; fl. May-Jun.
DISTRIB. **MAM.** 19 km E. of Amadia, *Sharief & Hamad* 50319! **MSU.** 38 km S. of Sulimaniyah, *F. & E. Barkley* & *Haddad* 5856!

Europe, N.W. Africa (mountains), Yemen, Palestine, Syria, Turkey to Iran, Afghanistan and Pakistan; Greenland, N.E. Canada.

9. **PEDICULARIS** L.

Sp. Pl. 2: 607 (1753); Fischer in Kubitzki (ser. ed.), ed. J.W. Kadereit, Fam. & Gen. Vasc. Pl. 7: 424 (2004)

Shahina A. Ghazanfar

Annual, biennial or perennial (Iraq species), often rosulate, herbs. Stems erect to ascending, sometimes winged, glabrous to densely villous. Cauline leaves or leaves of basal rosette opposite, alternate or verticillate, sessile or petiolate, usually 3–4-pinnatifid or pinnatisect, rarely entire with dentate to serrate margins. Inflorescence terminal. Flowers sessile to pedicellate. Calyx not split anteriorly, campanulate or tubular, 5-lobed to 2–5-dentate, lobes entire, dentate, serrate or pinnatifid, accrescent after anthesis. Corolla 2-lipped; tube cylindrical to obconical; upper lip laterally compressed forming a hood (galea); lower lip spreading. Stamens 4, included. Ovary ovoid, compressed. Capsule with acute or acuminate apex. Seeds numerous, globose to oblong, straight or curved, reticulate.

About 670 species in the subarctic and temperate N. America, Europe, Asia, high tropical mountains in N. W. Africa and N. W. South America; 3 species in Iraq, rare and found only at high elevation.

1. Calyx densely glandular-pilose; galea rostrate; capsule with acuminate
    beak. . . . . 2. *P. caucasica*
  Calyx pubescent, not glandular; galea not rostrate; capsule without
    acuminate beak 2
2. Calyx lobes mucronate, glabrous; galea curved at the back; capsule
    oblique . . . . 1. *P. sibthorpii*
  Calyx lobes acute, sparsely villous; galea not cured at back; capsule
    symmetrical, covered by the inflated calyx. . . . . 3. *P. pycnantha*

1. **Pedicularis sibthorpii** *Boiss.*, Diagn. Pl. Orient. Ser. 1, 4: 83 (1844); Wendelbo in Fl. Iranica [K. H. Rechinger] 147: 193 (1981).

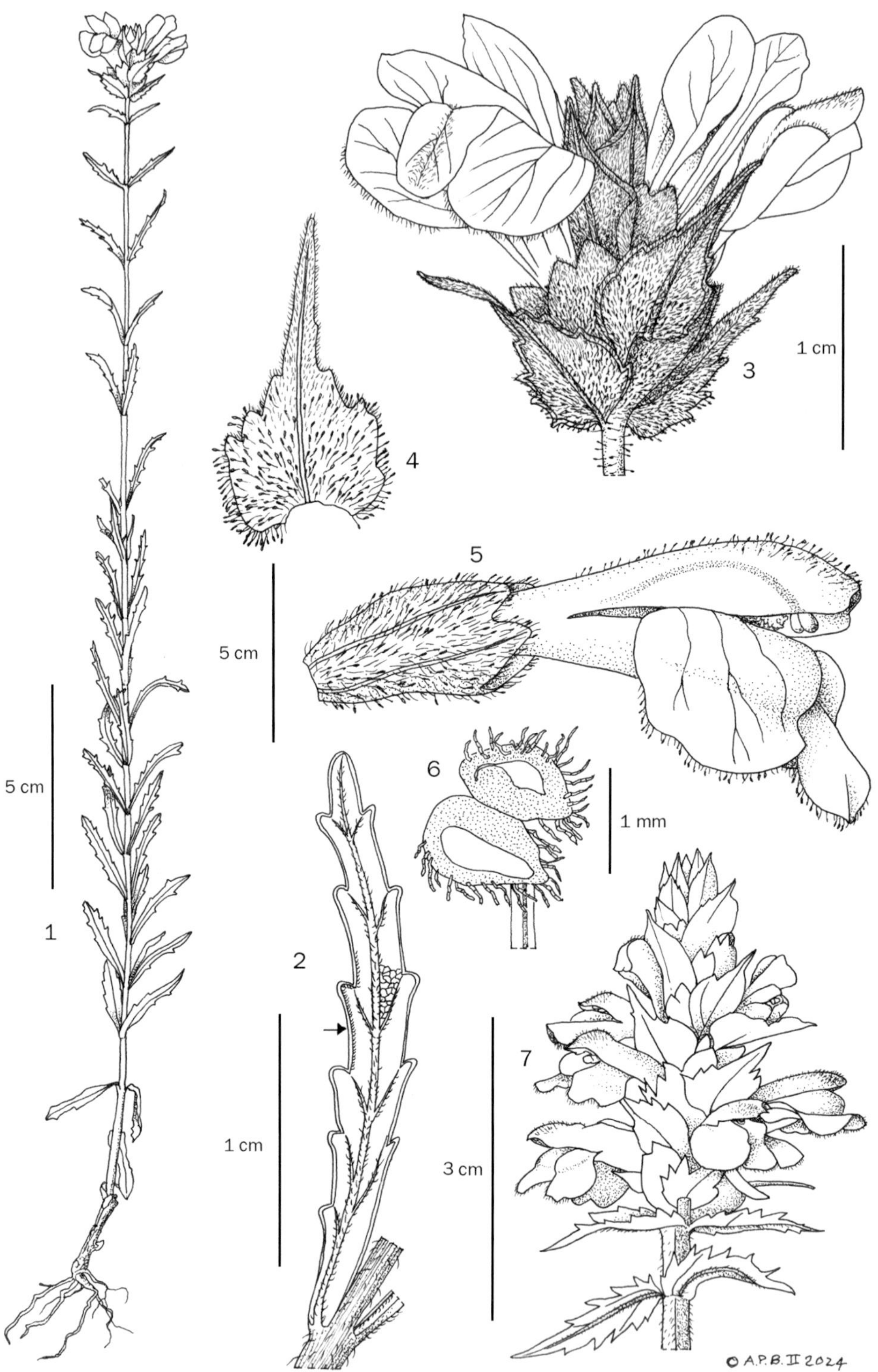

Fig. 53. **Bartsia trixago**. 1, habit; 2, mid-stem leaf abaxial surface; 3, inflorescence; 4, floral bract, adaxial surface (midrib sunken); 5, flower, hydrated and reconstructed; 6, anther showing multicellular hairs; 6, inflorescence (drawn from live material, Myrthios, Crete, Greece). 1–6 from *Barkley & Barkley* 5856; Drawn by © A.P. Brown, Feb. 2024.

P. comosa L. var. *sibthorpii* (Boiss.) Boiss., Fl. Orient. [Boissier] 4: 492 (1879).
P. comosa L. auct.: Rawi in Dep. Agr. Iraq Tech. Bull. 14: 144 (1964)

Perennial herb, stems, erect to ascending arising from a woody stock, glabrous to villous in the inflorescence. Basal leaves with petiole 3–10 cm; lamina pinnatisect, segment margins irregularly dentate; upper leaves shortly petiolate. Flowers whitish-yellow, in compact racemes. Lower bracts pinnatisect; upper bracts lanceolate. Calyx campanulate, somewhat inflated, c. 8 mm; lobes ovate, acute to mucronate. Corolla tubular c. 20 mm; galea curved at the back, inclined at apex ; lower lip shorter than galea, obscurely 3-lobed. Capsule ovoid, oblique.

HAB. On mountains, alpine growing on rocks; alt. 3100–3300 m; fl. Jul.-Aug
DISTRIB. Rare in mountains of N. Iraq. MAM. Ser Kurawa, *Gillett* 9748!; Sai Koh, *Ludlow-Hewitt* 1531A! Between Darband and Hajji Omran, *Hadač* 2579 (cited in Fl. Iranica).

Turkey, Iran, North Caucasus.

2. **Pedicularis caucasica** *M.Bieb.*, Fl. Taur.-Caucas. 2: 72 (1808); Rawi in Dep. Agr. Iraq Tech. Bull. 14: 144 (1964); Wendelbo in Fl. Iranica [K. H. Rechinger] 147: 199 (1981).

Perennial herb, caespitose. Stems arising from the woody stock, erect to ascending, 7–15 cm. Leaves of the basal rosette 2–4 cm long, pinnatifid, margins irregularly dentate, ± villous. Bracts 5–7 mm, margins irregularly dentate. Flowers yellow-purplish, pedicellate with pedicels patent, racemes compact. Calyx persistent, membranous, 5–6 mm, tubular below, lobes 2mm, lanceolate, acute, densely glandular-pilose. Pedicels to 2 mm. Corolla tube c. 10 mm; galea c. 5 mm, rostrate; lower lobe smaller than galea, sparsely villous to glabrous. Capsule 16–17 mm, with an acuminate beak.

HAB. On mountains, in crevices of rocks; alt. ± 3000 m; fl. Apr., Jul.
DISTRIB. Rare in mountains of N. Iraq. MAM. Sai Koh, *Ludlow-Hewitt* 1531! Algurd Dagh, *Guest* 3063!; Mt. Qandil, *Rechinger* 11168 (W, cited in Fl. Iranica).

Turkey to Caucasus, Iran.

3. **Pedicularis pycnantha** *Boiss.*, Diagn. Pl. Orient. ser. 1, 12: 45 (1853); Rawi in Dep. Agr. Iraq Tech. Bull. 14: 144 (1964); Wendelbo in Fl. Iranica [K. H. Rechinger] 147: 200 (1981); Mill in Fl. Pakistan [Ali & Qaiser] 220: 154, f. 15 A–E (2015).

Perennial herb with conical roots and a short stem, villous. Basal leaves clustered, petiolate with petiole 3–20 mm, lamina bipinnatipartite, segments decurrent along the rachis, margins of segments irregularly dentate, upper surface villous, lower villous to glabrous. Lower bracts lanceolate, ± toothed, upper bracts linear. Flowers whitish yellow, in compact racemes. Calyx campanulate, inflated, 7–12 mm, tubular below, lobes c. 5 mm, ovate, acute, sparsely villous. Corolla 18–25 mm; galea 3–7 mm; lower lobe shorter than galea. Capsule ovoid, symmetrical, covered by the inflated calyx. Fig. 54, 1–5.

HAB. On mountains, alpine growing on rocks; alt. ±3350 m; fl. Jul.-Aug
DISTRIB. Rare in mountains of N. Iraq. MAM. Mt Qandil, *Thesiger* 1159 (BM); Mt Sakri Sakran, *Bornmüller* 1652 (W, cited in Fl. Iranica).

Iran, Afghanistan, Pakistan, Turkmenistan, West Himalaya.

10. **RHYNCHOCORYS** Griseb. nom. cons.

Spic. Fl. Rumel. 2(4): 12 (1844); Fischer in Kubitzki (ser. ed.), ed. J.W. Kadereit, Fam. & Gen. Vasc. Pl. 7: 425 (2004)

Shahina A. Ghazanfar

Annual or perennial herbs. Stems erect, glabrous or glandular-pubescent. Leaves opposite, rarely alternate below, sessile to rarely shortly petiolate, with dentate to crenate or serrate margins. Inflorescence paniculate. Flowers ± shortly pedicellate. Calyx 2-lipped, laterally compressed, upper lip emarginate, lower lip 2-lobed, labrous or villous, accrescent after anthesis. Corolla tubular, 2-lipped; upper lip compressed laterally (galeate) and

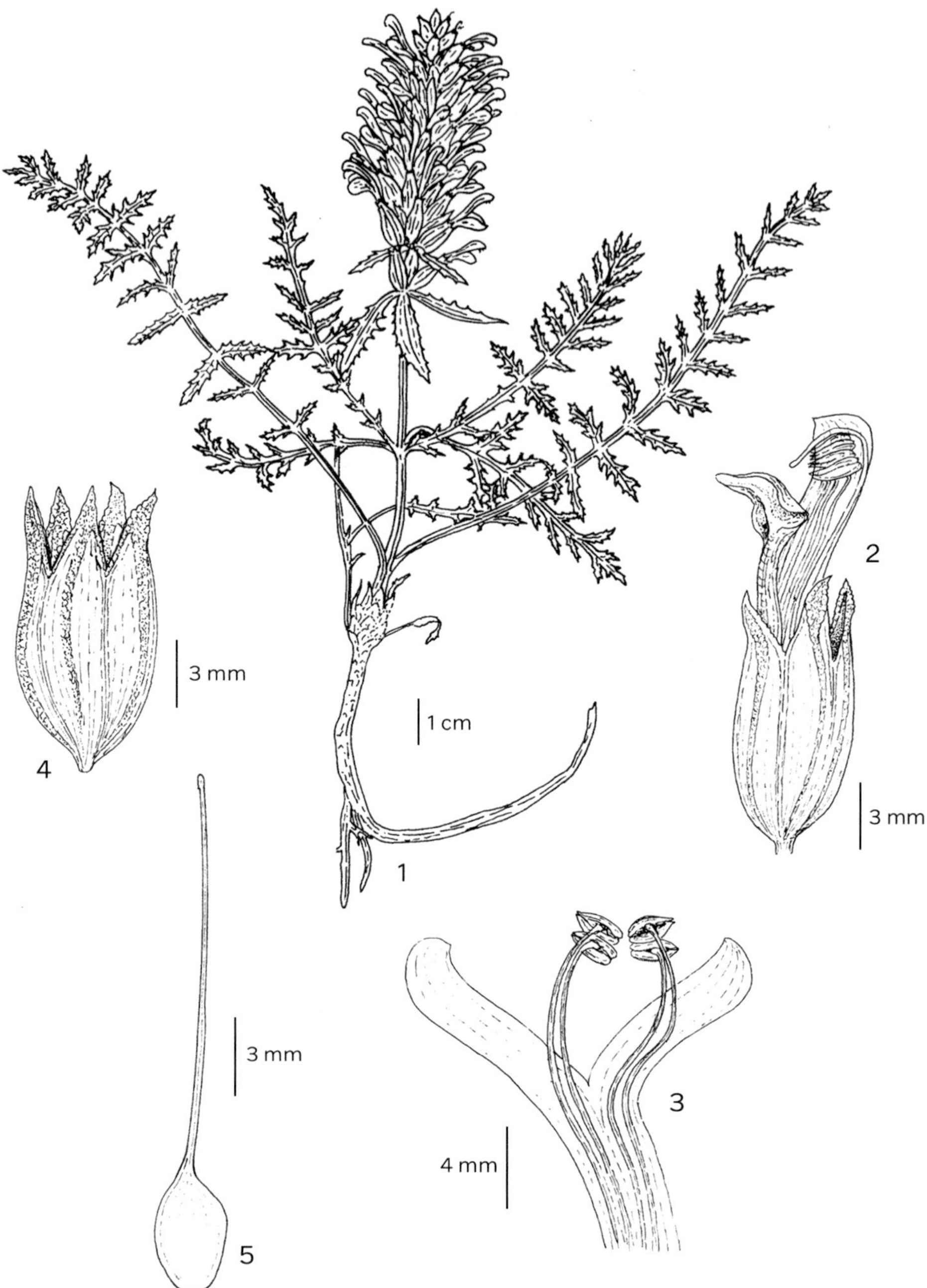

Fig. 54. **Pedicularis pycnantha**. 1, habit; 2, flower; 3, corolla with stamens (dissected); 4, calyx; 5, gynoecium. Reproduced with permission from Fl. Pakistan 220: f. 15, A-E. Drawn Abrar Ali. © University of Karachi, Pakistan.

prominently rostrate, lower lip 3-lobed. Stamens 4, included; anthers glabrous. Ovary compressed laterally. Seeds discoid, flattened, reticulate-striate, not winged.

Six species in Morocco, Algeria, southeast Europe, Turkey, Syria to Iran; 3 species in Iraq.

Abdullah Sh. Sardar. (2017). A New Record of *Rhynchocorys odontophylla* Burbidge & Richardson (Scrophulariaceae) for Flora of Iraq. 13(2): 274. P-ISSN: 2222-8373 E-ISSN: 2518-9255.

Burbidge, R.B. & Richardson, I.B.K. (1970). Revision of the genus *Rhynchocorys*. Edinb. Roy. Bot. Gard Notes.

1. Leaves opposite, broadly ovate to almost orbicular. 3. *R. elephas*
   Leaves alternate below, opposite above, lanceolate to narrow lanceolate 2
2. Leaf margins crenate to serrate . . . . 1. *R. odontophylla*
   Leaf margins entire . . . . 2. *R. kurdica*

1. **Rhynchocorys odontophylla** *R.B.Burb & I.Richardson*, Notes Roy. Bot. Gard. Edinburgh 30:102 (1970); Hedge in Fl. Turkey [P. H. Davis] 6: 780 (1978).

Perennial herb, 45–60 cm; stems erect to ascending, villous and glandular. Leaves sessile, alternate distichous below, opposite-decussate above, narrowly ovate-oblong, 30–50 × 10–30 mm, base truncate, apex acute, margins crenate to serrate. Inflorescence paniculate. Bracts narrowly ovate, lanceolate or narrow lanceolate, crenate to serrate, upper bracts entire. Flowers greenish-yellow; pedicel to 6 mm, glandular. Calyx campanulate, tube 3.5–4.5 mm, 3-lobed, a folded upper lobe and 2 triangular lower lobes, apex truncate in upper lobe, acute in lower lobes. Corolla golden yellow, 2-lipped; tube 5.5–6.5 mm; galea 6–7 mm, base swollen (containing anthers); lower lip 3-lobed, 5–6 mm, margins undulate, glabrous. Stamens inserted about the middle of corolla tube. Ovary villous; style filiform, pilose; stigma capitate, exserted, yellow. Capsule green-brown or black, broadly oblong, 3–5 mm. Seeds numerous, yellowish brown discoid, reticulate-striate.

HAB. On mountains, in wet locations, on rocky soil; alt. 1700–1800 m; fl. Jul.-Aug.
DISTRIB. Qandil range in Rowanduz district; present singly, not common. **MRO**: Mt. Qandil mountain, N.E. of Erbil, *Sardar, Al-Dabagh & Rasul* 7453.

This is a recent new record for Iraq, extending the distribution of this species into Iraq. The difference between the *R. odontophylla* and *R. kurdica* seems to be in the leaf margins, crenate to serrate in the former and entire in the latter. However in the material that I have examined of *R. kurdica* leaf margins are entire to sometimes irregularly serrate, showing that the character of leaf margins is not a good character. I have not seen any material of *R odontophylla* from Iraq, perhaps with more material the two could be better separated.

Turkey.

2. **Rhynchocorys kurdica** *Nábělek*, Spisy Přír. Fak. Masarykovy Univ. 70: 36 (1926); Rawi in Dep. Agr. Iraq Tech. Bull. 14: 144 (1964); Hedge in Fl. Turkey [P. H. Davis] 6: 779 (1978); Rechinger, Fl. Iranica 147: 208 (1981).

Perennial herb; stems stout, simple, branched above, to 50 cm, glandular-viscid. Leaves alternate below, opposite above, sessile, lanceolate, 30–50 × 15–23 mm, margins entire to sometimes irregularly serrate, glabrous to sparsely hispid. Inflorescence of a narrow panicle. Flowers shortly pedicellate; pedicel to 6 mm, erect to patent. Calyx tubular to campanulate, 6–16 mm, upper lobes c. 7 mm, emarginate. Corolla yellow, sometimes with a red spot on each side of beak, upper lip c. 5 mm, with 2 lateral projections; lower lip ± 3-lobed, shorter than the upper lip. Style pilose. Capsule broadly oblong, pilose. Fig. 55, 1–7.

HAB. On mountains, on rocky and stony places, in *Astragalus* zone; alt. 1800–2900 m; fl. Jul.-Aug.
DISTRIB. Qandil range; not common. **MRO**: Mt. Qandil mountain, N.E. Ranya, *Rawi & Serhang* 18293! Above Pustanshan, *Rechinger* 11069!; Malakh, N.E. Rania, *Rawi & Serhang* 18244!; Chia Mandali, *Guest* 2853!; Ser Kurawa, *Gillett* 9785!

S.E. Turkey, W. Iran.

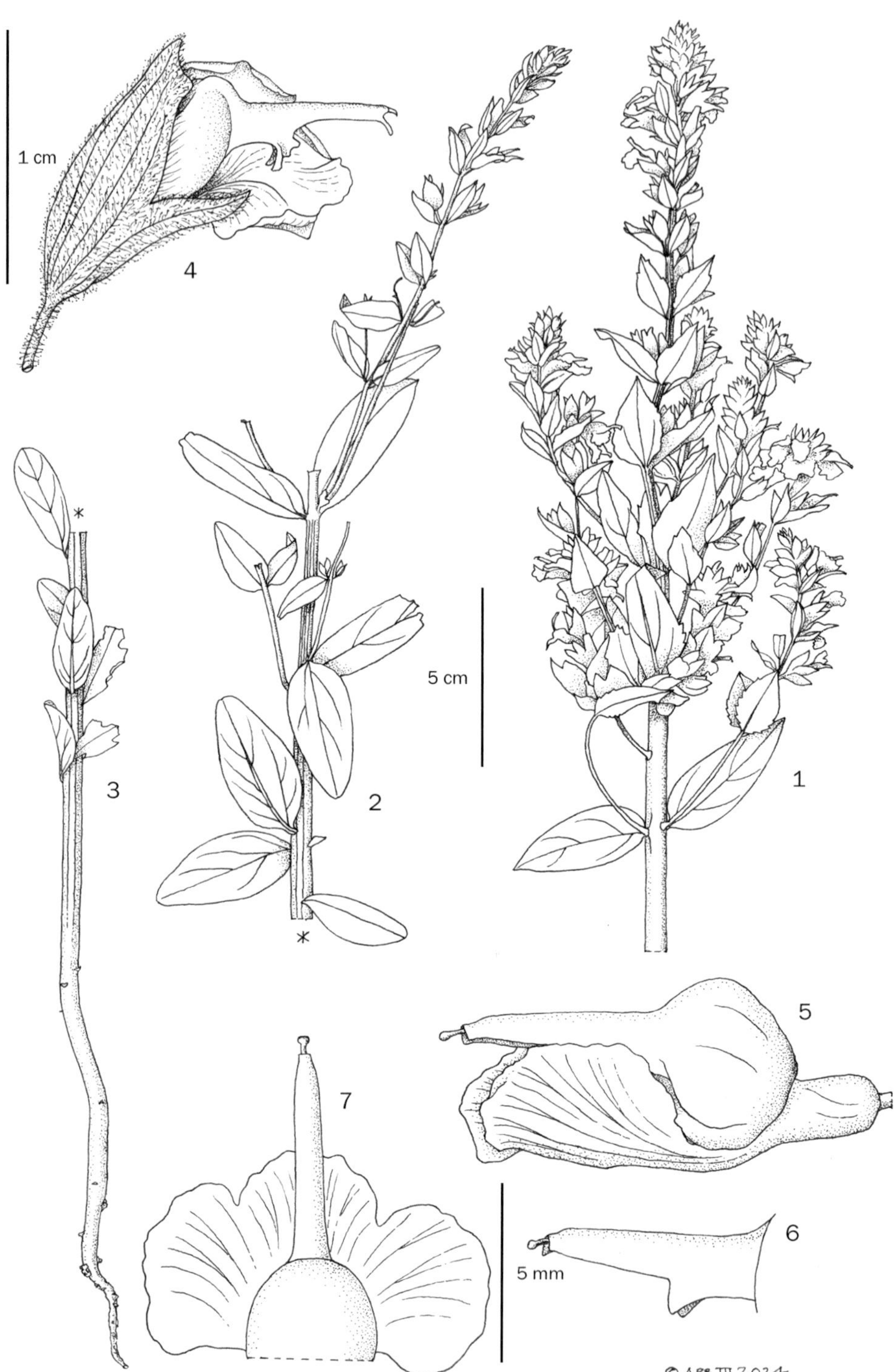

Fig. 55. **Rhynchocorys kurdica**. 1, upper part of flowering stem; 2, mid section of stem with lowest flowering branch; 3, base of stem; 4, flower, side view; 5, side view of corolla after removal of calyx; 6, detail of corolla hood; 7, dorsal view of corolla with lobed flattened. 1, 4, from *Rawi et al.* 18293; 2, 3, 5, 6, 7 from *Rawi et al.* 18244. Drawn by © A.P. Brown, Feb. 2024.

3. **Rhynchocorys elephas** (*L.*) *Griseb.*, Spic. Fl. Rumel. 2(4): 12 (1844); Rawi in Dep. Agr. Iraq Tech. Bull. 14: 144 (1964); Hedge in Fl. Turkey [Davis] 6: 780 (1978); Rechiger in Fl. Iranica [K. H. Rechinger] 147: 210 (1981);

Rhinanthus elephas L., Sp. Pl. 2: 603 (1753).

subsp. **carduchorum** *R.B.Burb & I.Richardson*, Notes Roy. Bot. Gard. Edinburgh 30:106 (1970). Type: Iraq, Algurd Dagh, 2300–2700 m, July, *Guest & Ludlow-Hewitt* 2941 (K, holo.)

Perennial herb, 10–35 cm; stems simple, glandular-pubescent. Leaves opposite, sessile to shortly petiolate, broadly ovate to almost orbicular, 15–25 × 10–16 mm, reducing in size upwards, base truncate, apex obtuse, margins serrate to crenate, sparsely hispid on surfaces and margins. Flowers shortly pedicellate; pedicel to 8 mm, erect to patent. Calyx campanulate, 6–11 mm (larger in fruit), unequally divided to about 1/2, glabrous, margins of lobes ciliate. Corolla yellow, 2-lipped; tube 4–6 mm; upper lobe c. 7 mm; lower lip longer than the upper. Ovary globose; style glabrous. Capsule suborbicular, pubescent or glabrous, not emarginate.

HAB. On mountains, amongst rocks, by stream; alt. 2100–3000 m; fl. Aug.
DISTRIB. Endemic; rare in mountains of N. Iraq. **MRO**: Algurd Dagh, *Guest & Ludlow-Hewitt* 2941!; Algurd Dagh, *Gillett* 9549!; E. Rost, *Thesiger* (BM!).

Distribution of subsp. *elephas* : N.W. Africa, S.E. Europe, Turkey, Transcaucasus, Iran.

Endemic.

11. **BUNGEA** C.A.Mey.

Verzeichniss Pfl. Caucasus: 108 (1831); Fischer in Kubitzki (ser. ed.), ed. J.W. Kadereit, Fam. & Gen. Vasc. Pl. 7: 419 (2004)

Shahina A. Ghazanfar

Perennial herbs. Stems woody, ascending to erect, tufted, densely pilose. Leaves opposite, sessile or petiolate, lower leaves entire, linear-lanceolate, the upper trifid, with linear segments. Inflorescence spicate. Flowers sessile. Bracteoles 2. Calyx with a short tube, deeply 4-lobed, persistent. Corolla strongly 2-lipped, upper lip hooded, acuminate, with 2 short teeth at base, lower lip 3-lobed. Stamens 4, transverse, included; anthers mucronate, glabrous. Stigma capitate. Ovary ovoid, acuminate. Capsule ovoid, enclosed in the membranous calyx, dehiscence loculicidal. Seeds few.

Two species: one distributed in Southwest, and the other in Central Asia and China; a single species in Iraq.

1. **Bungea trifida** (*Spreng.*) *C.A.Mey.*, Verz. Pfl. Casp. Meer. (C.A. von Meyer). 108 (1831); Dinsmore in Post, Fl. Syria, Palest. & Sinai ed. 2, 2: 307 (1933); Rawi in Dep. Agr. Iraq Tech. Bull. 14: 142 (1964); Hedge in Fl. Turkey [P. H. Davis] 6: 782 (1978).

Rhinanthus trifidus Vahl, Symb. Bot. (Vahl) 1: 44 (1790).

Caespitose perennial with a woody rootstock. Stems leafy, erect to ascending 10–20 cm, hirsute. Leaves sessile, lower leaves linear, 10–20 × 1–2 mm, upper leaves with trifid linear segments, pilose. Inflorescence spicate. Flowers solitary, axillary. Bracteoles 2, linear, as long as calyx. Pedicels 1–3 mm. Calyx 28–40 mm (incl. lobes), membranous between nerves, persistent; calyx teeth linear, acuminate, 25–30 mm, exceeding corolla. Corolla pale yellow, 25–35 mm, pilose; upper lip hooded, acute; the lower lip longer than the upper, 3-lobed. Capsule ovoid, 12–20 × 4–9 mm, with a long beak. Fig. 56, 1–6.

HAB. Mountains of N. Iraq; known from two collections; alt. 1200–1600 m; fl. Jul (?)
DISTRIB. N. Iraq. **MAM**. Jabel Khantur, Sharanish, *Rechinger* 10771. **MSU**. Nr. Penjwin, *Rechinger* 10468!; *Rawi* 23278!.

Rare in Iraq; known from a few collections.

Turkey, Armenia, Syria, Iran.

Fig. 56. **Bungea trifida.** 1, habit, young plant; 2, mid-stem trifid leaf from young plant; 3, flowering shoot of large plant; 4, flower (hydrated) side view; 5, corolla opened to show style and stamens; 6, fruit, dehisced. 1, 2, 6 from *Rechinger* 10468; 3, 4, 5, from *Rawi* 23278. Drawn by © A.P. Brown, Feb. 2024.

121. **LENTIBULARIACEAE** Rich.

Fl. Paris. [Poiteau & Turpin] 1: 23 (ed. fol.) (1808) (as "Lentibulariae"); Fischer, E., Barthlott, W., Seine, R. & Theisen, I. in Kubitzki (ser. ed.), J.W. Kadereit (ed.), Fam. & Gen. Vasc. Pl. 7: 276 (2004)

J.R. Edmondson

Herbaceous, terrestrial, aquatic, or epiphytic carnivorous plants (aquatic in ours). Inflorescences terminal or lateral, racemose or simple, often reduced to a single flower. Leaves often much divided. Flowers zygomorphic, bisexual; calyx 2-, 4- or 5-lobed (2 in ours). Corolla sympetalous, usually spurred; upper lip entire or 2-lobed, lower often gibbous, entire or 2–5-lobed. Stamens 2. Ovary superior, 1-locular, formed of 2 fused carpels. Fruit usually a capsule.

The family contains three carnivorous genera, *Pinguicula* and *Genlisea* (not recorded from Iraq) and *Utricularia*.

Recent research has confirmed the relationship between this family and members of the Lamiales, particularly *Byblis* which has now been assigned its own family, but there is doubt as to whether Martyniaceae is a sister family.

1. **UTRICULARIA** L.

Sp. Pl. 1: 18 (1753).

Lentibularia Séguier, Pl. Veron. 3: 128 (1754).

Aquatic floating perennials, with greatly divided leaves; roots absent (in ours). Leaves verticillate at nodes of stolon; stalked traps present on branches of pseudostem. Inflorescence racemose, bracteate, emergent. Calyx 2-lobed, gamosepalous. Corolla bilabiate, with upper lip usually smaller than lower; lower lip spurred. Stamens 2, inserted on corolla at base of saccate part of upper lip; anthers dorsifixed. Ovary globose or ovoid; style short. Fruit a capsule, dehiscing by longitudinal slit(s).

A large genus of 284 species, cosmopolitan; a single species in Iraq.

Taylor, P. (1989). The genus *Utricularia* – a taxonomic monograph. Kew Bull. Add. Series no. XIV.

1. **Utricularia australis** *R.Br.*, Prodr. Fl. Nov. Holland. 430 (1810).

U. major Schmidel, Ic. Pl. ed. Bischof 80: t. 21 a-l (1771), nom illegit.
U. vulgaris (non L.) Pollich, Hist. Pl. Palat. Elect. 1: 21 (1776). Buch.-Ham. Ex D.Don, Prodr. Fl. Nepal. 84 (1825).
U. neglecta Lehm., Linnaea 5(3): 386 (1830).

Floating aquatic perennial, rootless but with a few filiform rhizoids. Stolons terete, filiform, branched, to 50 cm. Leaves 1.5–4 cm, divided at base into two primary segments that are ± pinnately subdivided, ultimate segments capillary. Traps dimorphic, arising laterally from penultimate segments and at base of primary segments; internal glands of traps 2- and 4-armed, arms subulate. Inflorescence emergent, 10–30 cm; peduncle filiform, with scales in upper half. Bracts basifixed; bracteoles absent. Flowers 4–10, borne on filiform pedicels. Calyx lobes slightly unequal. Corolla yellow, lower lip with basal swollen part much darker and with reddish lines and spots. Ovary clothed with sessile glands. Fig. 57, 1–10.

H𝙰𝙱. In shallow stagnant pools with *Alisma*; alt. ± 50 m; fl. Apr.
D𝙸𝚂𝚃𝚁𝙸𝙱. Scattered in lowland Iraq. 𝙻𝙲𝙰: Daltawa (Khalis), *Haines* W. 1666!

Throughout Europe and temperate Asia, tropical and South Africa, tropical Asia, Australia, New Zealand.

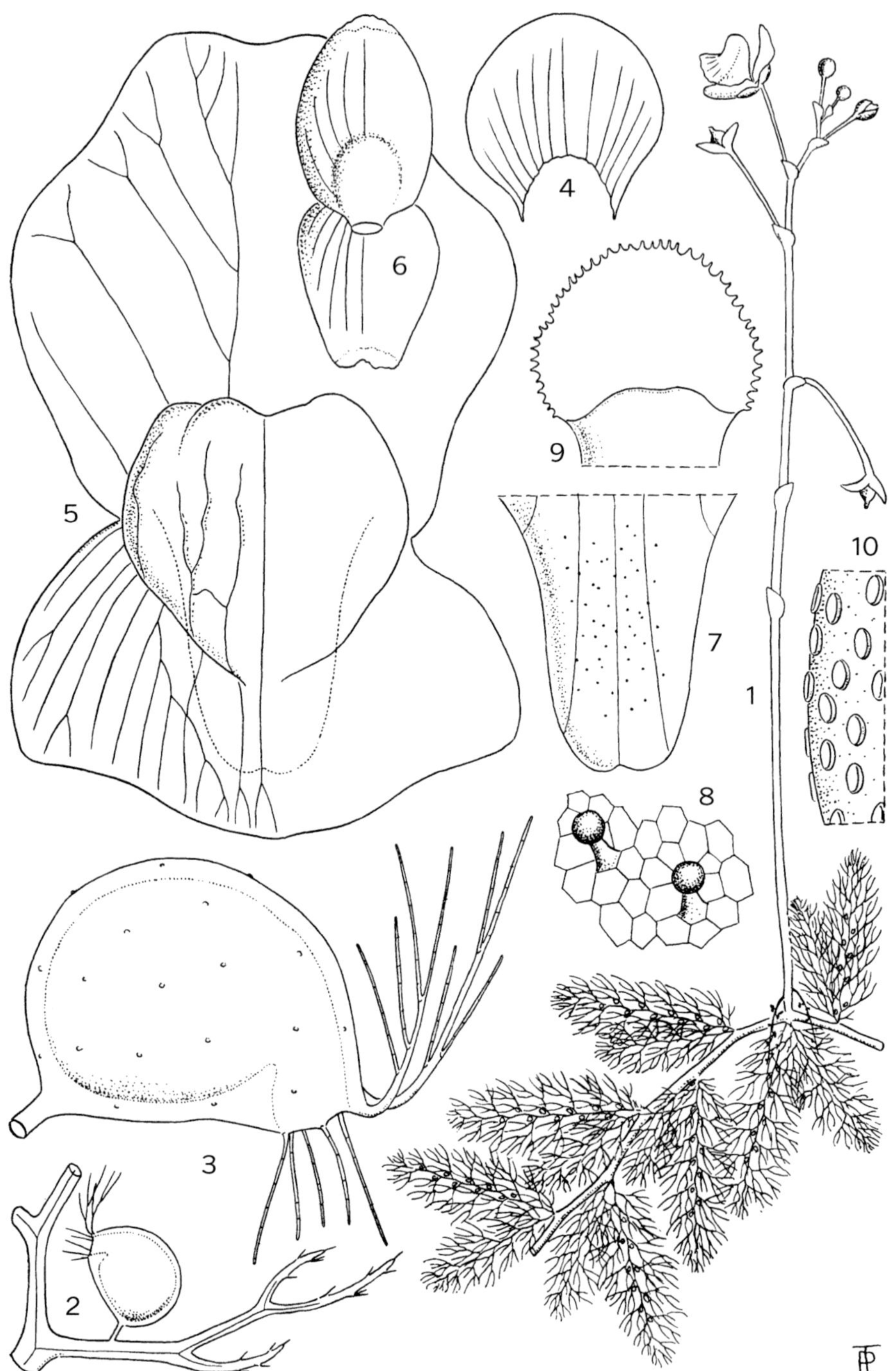

Fig. 57. **Utricularia australis**. 1, part of flowering plant, × 1; 2, part of leaf × 8; 3, trap, lateral view × 30; 4, bract × 8; 5, flower, abaxial view × 8; 6, calyx × 8; 7, spur adaxial view, showing disposition of glands on inner surface × 8; 8, glands on inner surface of spur × 150; 9, stigma × 30; 10, surface of ovary × 150. Illustration from Kew Bull. 18: f. 71. Reproduced with permission.

122. **GERANIACEAE** Adans. nom. cons.

Fl. Pl. 2: 384 (1763); Kunth in Engler, Pflanzenr. IV, 129 (1912); Albers & Van der Walt in Kubitzki (series ed.) Fam. Gen. Vasc. Pl. 9: (2009)

C.C. Townsend & A. Agnew
Revised by Shahina A. Ghazanfar

Annual or perennial herbs or shrublets with opposite or alternate, lobed or compound, stipulate leaves. Inflorescence cymose forming pseudoumbels or flowers solitary. Flowers pentamerous, hermaphrodite, actinomorphic with a tendency towards zygomorphy in some genera. Sepals free or united at base, persistent. Petals free, the latter often fugacious. Stamens sometime shortly connate at base, in 2 or more whorls of 5, sometimes some sterile. Carpels 5(4); ovary 5-lobed at base often produced above into a long beak; style with 5 free capitate stigmas. Ovules 2 per locule, of which only one matures. Fruit schizocarpic, dehiscing into awned mericarps that separate from a central beak (rostrum).

Eleven genera found throughout the World, but mainly temperate and subtropical; 4 genera and 23 species in Iraq.

Biebersteina has been placed in the Geraniaceae by several authors (see Bakker et al. 1998 for summary), but recognised as a separate order and family by Takhtajan (1997) based on pollen characters, one ovule per locule and in having a gynophore. Based on molecular evidence it is placed as a distinct family Biebersteiniaceae in the Order Sapindales by APG IV 2016 classification, and is treated as such in this Flora.

Bakker, F.T., Vassiliades, D.D., Morton, C. & Savolainen, V. (1998). Phylogenetic relationships of *Biebersteinia* Stephan (Geraniaceae) inferred from rbc L and atp B sequence comparisons. *Bot. J. Linn. Soc.* 127(2): 149–158.

Jeiter, J., Cole, T. C., & Hilger, H. H. (2017). Geraniales Phylogeny Poster (GPP). PeerJ Preprints, 5, 3127v1. https://doi.org/10.7287/peerj.preprints.3127v1

Jeiter, J., Weigend, M., & Hilger, H. H. (2017). Geraniales flowers revisited: evolutionary trends in floral nectaries. *Ann. Bot.* 119(3), 395–408.

Muellner, A. N., Vassiliades, D. D., & Renner, S. S. (2007). Placing Biebersteiniaceae, a herbaceous clade of Sapindales, in a temporal and geographic context. *Pl. Syst. Evol.* 266 (3–4), 233–252.

Price, R.A. & Palmer, J.D. (1993). Phylogenetic relationships of the Geraniaceae and Geraniales from rbcL sequence comparisons. *Ann. Missouri Bot. Gard.* 80: 661–671.

1. Flowers in groups of 10 or more, forming an umbel, zygomorphic; pedicel containing a nectariferous tube leading to the posterior sepal; only two conspicuous petals . . . . .4. *Pelargonium*
   Flowers mostly regular; pedicels lacking nectariferous tube; all petals evident 2
2. Stamens 15, all fertile . . . . .3. *Monsonia*
   Stamens 10, all fertile or 5 fertile and 5 sterile 3
3. Leaves palmately nerved or divided; carpel beaks curved upwards on dehiscence . . . . 1. *Geranium*
   Leaves pinnately nerved or divided; carpel beaks coiling spirally on dehiscence . . . . .2. *Erodium*

1. **GERANIUM** L.

Sp. Pl.: 676 (1753), pro parte; Gen. Pl. ed. 5, 832 (1737), pro parte

C.C. Townsend
Revised by Shahina A. Ghazanfar

Herbs with palmate or palmately lobed leaves, alternate below, opposite above. Flowers solitary or (in Iraq) in pairs, actinomorphic, without a spur. Stamens 10, all fertile, or rarely 5 sterile and 5 fertile. Seeds one in each carpel. Dehiscence by the carpel rolling upwards along the beak, throwing out the loosely held seed. Beaks of carpels never coiling spirally and never hairy internally.

A large genus of about 353 species, mostly in temperate regions of the world and mountains in the tropics; 9 species in Iraq, confined to wet places and to the northern part of the country; none in desert areas.

Geranium is the Latin diminutive of Gr. γερανος, *geranos*, crane, referring to its long-beaked fruit.

Aedo, C. (1996). Revisión of *Geranium* subgenus Erodioidea (Geraniaceae). *Syst. Bot. Monogr.* 49.
Aedo, C., Garmendia, F.M. & Pando, F. (1998). World checklist of *Geranium* L. (Geraniaceae). *Anales del Jardín Botánico de Madrid* 56(2): 211–252.

1. Perennials; flowers >1.5 cm diameter. 2
   Annuals; flowers 1 cm or less in diameter 3
2. Plant rhizomatous; flowers borne in widely separated pairs. . . . . .9. *G. kurdicum*
   Plant with deep spherical root tubers; flowers in pairs at each node of a
     ± dense cymose inflorescence . . . . 8. *G. tuberosum*
3. Stems and leaves polished, often reddish; sepals with small transverse
     ridges, glabrous . . . . .6. *G. lucidum*
   Stems and leaves hairy, not polished; sepals without ridges, hairy. 4
4. Leaves polygonal in outline 5
   Leaves circular, semicircular or reniform in outline. 6
5. Leaves divided to base; carpel beak thickened in fruit . . . . .7. *G. robertianum*
   Leaves lobed only; carpel beak unthickened and filiform in fruit. . . . .4. *G. divaricatum*
6. Mature fruit 9–11 mm long (including beak) 7
   Mature fruit 15–20 mm long. 8
7. Carpel valves hairy with glabrous centre, smooth, without transverse
     ridges . . . . 1. *G. pusillum*
   Carpel valves glabrous, with transverse ridges. . . . . .5. *G. molle*
8. Leaf lamina divided nearly to petiole. . . . . .2. *G. dissectum*
   Leaf lamina divided to half-way or less. . . . . 3. *G. rotundifolium*

1. **Geranium pusillum** *L.*, Syst. Nat. ed. 10: 1144 (1759); Burm. f., Spec. Geran.: 27 (1759); Boissier, Fl. Orient. 1: 880 (1867); Blakelock in Kew Bull. 3: 409 (1948); Bobrov, Fl. SSSR 14: 52 (1949); Rawi, Dep. Agr. Iraq Tech. Bull. 14: 59 (1964); Schönbeck-Temesey in Fl. Iranica [K. H. Rechinger] 69: 27 (1970); Mouterde, Nouv. Fl. Lib. et Syrie 2: 438 (1970); Y. Nasir in Fl. Pakistan [Nasir & Ali]149: 9 (1983); Taifour & El-Oqlah, Pl. Jordan Annot. Checkl.: 97 (2017).

Erect annual herb, branching at base and above, sparsely hairy at nodes and on leaves. Leaves reniform, bifid or trifid divided to about ½ into 5–9 obtuse, mucronate lobes. Peduncles 2-flowered, shorter than pedicels, glandular-hairy. Sepals ovate-oblong, ± 3 mm, acuminate, long-hairy at margins, finely glandular. Petals pink, oblong, retuse at apex, scarcely longer than calyx, ciliate at base. Stamens 5 fertile + 5 sterile; filaments lanceolate,

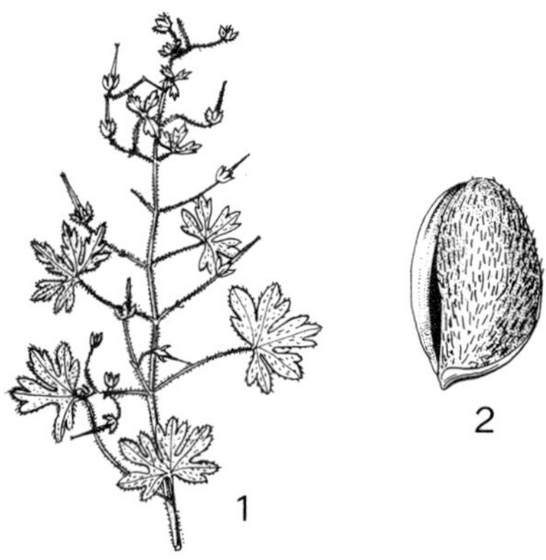

Fig. 58. **Geranium pusillum**. 1, habit × ½; 2, mericarp × 10. Reproduced with permission from Fl. Pakistan 149: f. 2, A–B (1985). Drawn by M.Y. Saleem. © National Herbarium, Pakistan Agriculture Research Council & University of Karachi, Pakistan.

ciliate at margins. Fruit 9–11(–13) mm, carpel valves smooth, short-hairy except for a glabrous dorsal ridge, beak narrowing at apex. Seeds without ornament, smooth. Fig. 58, 1–2.

HAB. A ruderal of irrigated fields; alt. 50–300 m; fl. & fr. Mar.
DISTRIB. Rare, but widely distributed throughout Iraq. **MRO**: Rowanduz, *Guest* 2048!; **FUJ**: Tel Afar, *Mudir of Zummar* 3142!; **LCA**: 6 km S.E. of Kut al Amara, *Gillett* 6744!; **LSM**: between Amara and Qurna, *Rechinger* 8957!

Often mistaken for *G. dissectum* in field, from which it differs in its shorter fruit and smooth seeds.

Throughout Europe and the Mediterranean area to Turkey, Syria, Caucasia, N.W. Iran, Iraq, Afghanistan, Pakistan, Turkmenistan and Kashmir.

2. **Geranium dissectum** *L.*, Cent. 1: 21 (1755); Boissier, Fl. Orient. 1: 881 (1867); Anthony in Notes Roy. Bot. Gard. Edinburgh 27: 37 (1933); Blakelock in Kew Bull. 3: 409 (1948); Bobrov in Fl. SSSR 14: 14 (1949); Zohary in Dep. Agr. Iraq Bull. 31: 96 (1950); Rechinger, Fl. Lowland Iraq: 390 (1964); Rawi in Dep. Agr. Iraq Tech. Bull. 14: 58 (1964); Mouterde, Nouv. Fl. Lib. et Syrie 2: 439 (1970); Schönbeck-Temesey in Fl. Iranica [K. H. Rechinger] 69: 30 (1970); Zohary, Fl. Palaest. 2: 229, pl. 326 (1987); Boulos in Fl. Egypt 2: 2 (2000); Abdulridha, Taha & Widad, Ecology & Fl. Basrah: 405 (2016); Taifour & El-Oqlah, Pl. Jordan Annot. Checkl.: 97 (2017).

Annual herb, 15–50 cm tall, often much-branched. Branches ascending, with reflexed hairs. Lower leaves long-petiolate, 2–7 cm in diameter, very deeply divided (to ⅘ of lamina or more) into 5–7 lobes; lobes 3–7 mm across at base, often pinnately lobed or trifid or dichotomous, ultimate lobes oblong. Peduncles 1.5–3 cm. Flowers in pairs on 5–10 mm pedicels which lengthen to 10–15 mm in fruit. Peduncles, pedicels and often sepals with glandular hairs. Sepals ovate-lanceolate, 5–7 mm, tapering gradually into a 1–2 mm awn, pilose. Petals pink, obovate, emarginate, as long as or slightly shorter than sepals. Carpels pubescent, 15–20 mm in fruit including beak (8–12 mm) and persistent style (3–4 mm). Seeds reticulate pitted.

HAB. A weed of irrigated and cultivated land, on clay soil, near river and humid rocky soils, tolerating subsaline conditions in C. Iraq; alt. up to 600 m; fl. & fr. Mar.-May (rarely later).
DISTRIB. Occasional in forest zone of mountains and on Mesopotamian plains.
MAM: S.W. of Aqrah, *Anders* 1218; **MRO**: Sisawa, *Helbaek* 756; 11 km N.W. of Rania, *Rawi, Nuri & Kass* 28563! **MSU**: Qara Dagh, *Gillett* 7869!; Halabja, *Barkley* 7538; Tinar 15 km before Dukan, *S. Omar* 42775 (BAG); spring betweeb Erbit & Said Sadiq, *Al-Kaisi et al* 43157 (BAG). **FKI**: Kirkuk, *Guest* 1990! **FPF**: 2 km N.W. of Saadiya, *Al-Kaisi* 42565 (BAG). **LEA**: Shahraban, *Gillett & Rawi* 7310! 10 km S. of Baquba, *S. Omar, Fawzi, Al-Kaisi & M. Nori* 44261 (BAG). **LCA**: Baghdad, *Guest* 1115!; Hilla, *Civil Veterinary Surgeon of Hilla* 2524! **LSM**: 80 km S.E. of Amara, *Guest, Rawi & Rechinger* 17666! Amara to Kurna, *Rechinger* 8957; **LBA**: 20 km S.E. of Basra, *Rawi & Haddad* 25954! below Basra, *Rechinger* 8604.

Europe and Asia eastwards to Iran and Afghanistan, western N. Africa; introduced in N. & S. America.

3. **Geranium rotundifolium** *L.*, Sp. Pl. 653 (1753); Boissier, Fl. Orient. 1: 880 (1867); Nábělek, Publ. Fac. Sci. Univ. Masaryk 35: 58 (1923); Blakelock in Kew Bull. 4: 49 (1949); Bobrov in Fl. SSSR 14: 53 (1949); Zohary in Dep. Agr. Iraq Bull. 31: 96 (1950); Rechinger, Fl. Lowland Iraq 390 (1964); Rawi, Dep. Agr. Iraq Tech. Bull. 14: 58 (1964); Davis in Fl. Turkey 2: 459 (1967); Mouterde, Nouv. Fl. Lib. et Syrie 2: 438 (1970); Schönbeck-Temesey in Fl. Iranica [K. H. Rechinger] 69: 28 (1970); Y. Nasir in Fl. Pakistan [Nasir & Ali] 149: 8 (1983); Zohary, Fl. Palaest. 2: 229, pl. 327 (1987); Collenette, Wild Flow. Saudi Arabia: 338 (photo) (1999); Breckle et al.: Vasc. Pl. Afghan. Checkl. 488 (2013); Taifour & El-Oqlah, Pl. Jordan Annot. Checkl.: 97 (2017).

Annual herb resembling *G. dissectum* in habit but often more erect. Leaves reniform in outline, lower divided to ½ or less; lobes cuneate at base, often 3-lobed at apex, sometimes obovate. Branches and pedicels often glandular-pubescent. Peduncles 2–3 cm. Flowers in pairs on 15–25 mm long pedicels. Flowers and fruit similar to *G. dissectum* but sepals abruptly ending in a small (to 1 mm) awn, or awn absent; petals slightly longer than sepals, rounded or more often slightly retuse at apex; seeds more coarsely reticulate-pitted. Fig. 59, 1–3.

Fig. 59. **Geranium rotundifolium.** 1, habit × ½; 2, flower × 5; 3, seed × 8. Reproduced with permission from Fl. Pakistan 149: f. 1, M–O (1985). Drawn by M.Y. Saleem. © National Herbarium, Pakistan Agriculture Research Council & University of Karachi, Pakistan.

HAB. A weed of shaded non-saline, irrigated land and gardens in central Iraq; in the foothills growing in open moist situations on limestone hillsides; alt. 100–1200 m; fl. Mar.-May, rarely in autumn.

DISTRIB. Scattered throughout the foothills and Mesopotamian plain: **MAM**: Sharanish, 25 km N.E. of Zakho, *Rawi* 23696!; Shaikh Adi, 10 km N.W. of Ain Sifni, *Thesiger* 663; 8 km S. of Zakho, *Rechinger* 1216. **MRO**: Rowanduz gorge, *Guest* 388! *ibid.*, 2546a; 30 km W.N.W. of Rania, *Rawi, Nuri & Kass* 28854!; foot of Baradost nr. Shanidar, *Erdtman & Goedemans* (*Rechinger* 15695)!; Kuh Sefin, *Bornmüller* 994. **MSU**: Dukan, *Meade* 131; Mt. Avroman [Hawraman], Susakan near Tawila, *Rechinger* 10183; Qara Dagh, *Gillett* 7887! **MJS**: Sinjar, *S. Omar & H. Hamid* 36495 (BAG). **FPF**: 2 km N. of Saadiya, *Al-Kaisi* 47243 (BAG). **FNI**: Jabal Maqlub, *N. Polunin* 2434! **DWD**: Ana, *Al-Khayat, S. Omar & Adel* 55469 (BAG). **LCA**: Baghdad, Karrada, *Guest & Haidar* 19490!

Europe and Asia eastwards to the Himalayas, western N. Africa; introduced in N. America.

4. **Geranium divaricatum** *Ehrh.*, Beitr. Naturk. 7: 164 (1792); Boissier, Fl. Orient. 1: 881 (1867); Bobrov in Fl. SSSR 14: 57 (1949); Rawi, Dep. Agr. Iraq Tech. Bull. 14: 59 (1964); Davis in Fl. Turkey 2: 460 (1967); Schönbeck-Temesey in Fl. Iranica [K. H. Rechinger] 69: 32 (1970); Zohary, Fl. Palaest. 2: 227 (1987); Collenette, Wild Flow. Saudi Arabia: 338 (photo) (1999).

A prostrate or erect annual branching from base, to 60 cm tall, glandular-pubescent and sparsely long-hairy throughout. Leaves ¾ divided into 5–7 rhomboidal lobes with large blunt minutely mucronate teeth, hairs on upper surface almost hispid. Peduncles 2-flowered, ± as long as pedicels. Sepals ± 5 mm, ovate, mucronate, sparsely but coarsely hairy, shortly glandular. Petals obcordate, equaling calyx. Stamens 10, all fertile; filaments filiform, glabrous. Fruit 9–11 mm, carpel valves with a few anastomosing ridges, sparsely stiff-hairy, especially on ridges, beak remaining filiform in fruit and not dehiscing. Seeds smooth.

HAB. Waste places; alt. 800–1200 m; fl. May-Jun.

DISTRIB. Rare, found only in northernmost foothills: **MAM**: Matina, *Rawi* 8711! **MSU**: slopes above Penjwin, *Hadač* 4910! **MJS**: Jabal Sinjar, above Sinjar, *Gillett* 11134!

C. & S. Europe, N. Africa, Turkey, Caucasus, N.W. & N. Iran, Iraq, Afghanistan, Turkmenia, W. Siberia (Altai mts.).

5. **Geranium molle** *L.*, Sp. Pl. 682 (1753); Boissier, Fl. Orient. 1: 882 (1867); Anthony in Notes Roy. Bot. Gard. Edinburgh 90: 284 (1935); Blakelock in Kew Bull. 4: 409 (1949); Bobrov in Fl. SSSR 14: 55 (1949); Zohary in Dep. Agr. Iraq Bull. 31: 96 (1950); Rechinger, Fl. Lowland Iraq: 391 (1964); Rawi in Dep. Agr. Iraq Tech. Bull. 14: 59 (1964); Davis in Fl. Turkey 2: 459 (1967); Mouterde, Nouv. Fl. Lib. et Syrie 2: 438 (1970); Schönbeck-Temesey in Fl. Iranica [K. H. Rechinger] 69: 31 (1970); Zohary, Fl. Palaest. 2: 230, pl. 328 (1987); Collenette, Wild Flow. Saudi Arabia: 337 (photo) (1999); Boulos in Fl.Egypt 2: 2 (2000); Taifour & El-Oqlah, Pl. Jordan Annot. Checkl.: 97 (2017).

Small (to 25 cm) erect or procumbent annual, with a rosette of long-petiolate leaves, branching from base, covered with soft spreading hairs and short glandular hairs. Leaves reniform (usually up to 2–5 cm in diameter) divided to about ½ into 5–9 trifid obtuse mucronate lobes. Peduncles 2-flowered, longer than pedicels. Sepals 3–5 mm, ovate, shortly mucronate (mucro less than 0.5 mm). Petals pinkish-purple, 1.5 × as long as sepals, bifid at apex. Stamens 10, all fertile; filaments narrowly lanceolate. Fruit 9–11 mm, carpel valves glabrous with many anastomosing transverse ridges especially towards apex; beak constricted about 1.5 mm below stigmas. Seeds smooth.

HAB. Dry limestone hillsides; alt. 600–1000 m; fl. Mar.-May.
DISTRIB. Locally common in the northern foothills of Iraq: **MAM**: 18 km S. of Zakho, *Agnew* 5707!; **MRO**: Rowanduz gorge, *Guest* 2027! **MJS**: Kizil Khan, above village, *Agnew, Hadač & Al-Hashimi* 5362!; Sinjar, *Polunin, Naib et al.* (BUH Exc. no. 65!); Jabal Sinjar, W. slope near Karsi, *Al-Kaisi & Wedad* 52231 (BAG). **FUJ**: Mosul, *Shahwani* 25178!; S.W. of Aqrah, *Anders* 1313! **FNI**: Eski Kellek, *Gillett* 10403!

Easily confused with *G. rotundifolium* with which it often grows, the length and hairiness of the fruit being the best way to separate them in the field.

Europe and temperate Asia eastwards to Iran, N. Africa; introduced to N. & S. America and New Zealand.

6. **Geranium lucidum** *L.*, Sp. Pl. 682 (1753); Boissier, Fl. Orient. 1: 884 (1867); Blakelock in Kew Bull. 4: 409 (1949); Bobrov in Fl. SSSR 14: 36 (1949); Zohary in Dep. Agr. Iraq Bull. 31: 96 (1950); Rawi in Dep. Agr. Iraq Tech. Bull. 14: 59 (1964); Davis in Fl. Turkey 2: 456 (1967); Mouterde, Nouv. Fl. Lib. et Syrie 2: 440 (1970); Schönbeck-Temesey in Fl. Iranica [K. H. Rechinger] 69: 36 (1970); Y. Nasir in Fl. Pakistan [Nasir & Ali] 149: 4 (1983); Zohary, Fl. Palaest. 2: 231, pl. 331(1987); Collenette, Wild Flow. Saudi Arabia: 336 (photo) (1999); Taifour & El-Oqlah, Pl. Jordan Annot. Checkl.: 97 (2017).

An erect reddish-stemmed annual, branching from base and above, to 30 cm tall, glabrous except for sparse hairs on leaves. Leaves reniform, divided to ½ into 5–7, 3–6-fid obtuse mucronate lobes. Peduncles 2-flowered, longer than glabrous pedicels. Sepals ovate-lanceolate, 5–7 mm, sparsely minutely gland-dotted, shortly awned, with 3 winged nerves with prominent transverse ridges between them. Petals pink, clawed, claw as long as sepals, spreading above into an oblong limb, 3–4 mm. Stamens 10, all fertile; filaments subulate, glabrous, a little longer than sepals. Fruit 13–16 mm; carpel valve ± 2 mm, with many minutely pubescent lateral wrinkles above, beak constricted and filiform 6–7 mm above valves. Seeds smooth. Fig. 60, 1–4.

HAB. Mesic habitats, always in deep shade, often in rocky soil or clefts of rocks by sides of stream; alt. ± 1000 m; fl. May-Jun.
DISTRIB. Occasional in mountainous areas of N.E. Iraq: **MAM**: Sharifa, N.E. of Amadiya, *O. Polunin* 5128!; **MRO**: Karoukh Mt., *Nuri & Kass* 27761!; Shaqlawa, *Gillett* 8095!; Rowanduz, *Guest* 589, 2057, 2110; Kuh-i Sefin, *Bornmüller* 1000; **MSU**: Qarachitan, nr. Pira Magrun, *Gillett* 7711!; Khormal, *Rawi* 8934!; Darband-i Khan, *Barkley* 144a!; 18 km N.E. of Chemchemal, *Barkley* 5827!

Throughout Europe, Madeira, Morocco, Saudi Arabia, Syria, Turkey, Caucasus, Iran, Afghanistan, Pakistan, Russia and temperate Himalayas.

7. **Geranium robertianum** *L.*, Sp. Pl. 681 (1753); Boissier, Fl. Orient. 1: 883 (1867); Blakelock in Kew Bull. 1948: 409 (1949); Bobrov in Fl. SSSR 14: 35 (1949); Rawi in Dep.

Fig. 60. **Geranium lucidum.** 1, leaf × 1; 2, petal × 3; 3, fruiting calyx × 2; 4, seed × 5. **Geranium robertianum. 5**, habit × ½; 6, fruiting calyx × 2. Reproduced with permission from Fl. Pakistan 149: f. 1, A–F (1985). Drawn by M.Y. Saleem. © National Herbarium, Pakistan Agriculture Research Council & University of Karachi, Pakistan.

Agr. Iraq Bull. 14: 59 (1964); Davis in Fl. Turkey 2: 458 (1967); Mouterde, Nouv. Fl. Lib. et Syrie 2: 439 (1970); Schönbeck-Temesey in Fl. Iranica [K. H. Rechinger] 69: 37 (1970); Y. Nasir in Fl. Pakistan [Nasir & Ali] 149: 5 (1983); Collenette, Wild Flow. Saudi Arabia: 337 (photo) (1999); Taifour & El-Oqlah, Pl. Jordan Annot. Checkl.: 97 (2017).

Erect, pale green or often reddish annual to 50 cm tall; lower leaves and stems often ± glabrous, upper parts, pedicels and calyx glandular-hirsute. Leaves opposite, peduncles ± equaling lamina; lamina 3–6 cm, palmate or very deeply pinnatisect into 5 lobes, distal one the longest so that leaf is polygonal in outline and may appear pinnate at first glance; lobes pinnatifid, each pair of lateral lobes with a common stipe for a short distance. Peduncles longer than leaves. Pedicels shorter than flowers. Sepals oblong-ovate, obtuse, mucronate (mucro 1–1.5 mm), pilose. Petals pink, obovate, 9–12 mm (twice length of calyx). Fruit 20–23 mm; carpels reticulately ridged, separating from beak and hanging from fruit by a long tuft of white hairs near apex. Seed retained in carpel, smooth. Fig. 60, 5–6.

HAB. Wet rock faces and damp ravine sides; alt. 750–1050 m; fl. May-Jun.

DISTRIB. Rare, found only in central mountain area of Iraq. **MAM**: Aqra, *Rawi* 11390B!; **MRO**: Shaqlawa, *Haines* 397!

Canary Islands, C. & S. Europe, Turkey, Caucasus; Saudi Arabia, Iraq, Iran, Pakistan and W. Himalayas.

8. Geranium tuberosum *L.*, Sp. Pl.: 680 (1753); Boissier, Fl. Orient. 1: 872 (1867); Handel-Mazzetti in Ann. Naturh. Mus. Wien 26: 62 (1913); Anthony in Notes Roy. Bot. Gard. Edinburgh 90: 284 (1935); Blakelock in Kew Bull. 4: 409 (1949); Bobrov in Fl. SSSR 14: 60 (1949); Zohary in Dep. Agr. Iraq Bull. 31: 96 (1950); Davis in Notes Roy. Bot. Gard. Edinburgh 22: 26 (1955); Rechinger, Fl. Lowland Iraq: 291 (1964); Rawi in Dep. Agr. Iraq Tech. Bull. 14: 59 (1964); Davis in Fl. Turkey 2: 462 (1967); Zohary, Fl. Palaest. 2: 228, pl. 323 (1987); Gabrielian & Fragman-Sapir, Fl. Transcaucasus & Adj. Areas: 202 (2008); Taifour & El-Oqlah, Pl. Jordan Annot. Checkl.: 97 (2017).

> *G. radicatum* M.Bieb., Fl. Taur.-Cauc. 2: 134 (1808).
> *G. tuberosum* L. var. *micranthum* Schönbeck-Temesy in Fl. Iranica 69: 7 (1970).

Perennial herb with solitary, unbranched stems and 1–2 long-petiolate leaves arising from ovoid root tubers up to 1.5 cm diameter and 5–15 cm long. Stem 20–30 cm, pubescent, sometimes sparsely hirsute, with or without a pair of leaves on its lower ⅓, terminating in a dichotomously branched inflorescence with 2 flowers at each dichotomy. Basal leaves with lamina palmate into 5–9 elliptical segments, pinnatifid, ultimate lobes obtuse; cauline leaves with much narrower segments, incised-serrate, ultimate lobes acute. Petioles of leaves below dichotomy much longer than lamina; those at the dichotomy subsessile. Inflorescence of 3–4 dichotomies, lower flower pairs pedunculate, upper pedicels sessile. Sepals 5–8 mm, hairy, elongating in fruit. Petals pinkish to pinkish purple, 1–2 × length of sepals, obovate, notched apically, with a densely ciliate claw. Filaments alternately glabrous and hirsute. Fruit 17–19 mm. Seeds finely pitted.

A variable and widespread species; Iraq specimens agree with the typical subspecies.

subsp. **tuberosum**

Stem usually naked below the dichotomy; leaf segments deeply pinnatifid; peduncles and pedicels without, or with few glands; beak of carpels not sharply narrowed below stigmas, hairy to apex.

HAB. Fertile sites of the lower forest zone with *Quercus* spp., not at higher altitude in *Q. libani* forest, on rocky and clayey soils, in between rocks and in rock crevices; alt. 500–1800 m; fl. Mar.-May.
DISTRIB. Occasional in mountains and higher foothills of N.E. Iraq. **MAM**: Zawita, *Emberger, Guest et al.* 15395!; **MRO**: Diana, *Guest* 705!; Haji Umran, *Chapman* 11180!; Penjwin, *Rawi* 8837!; Jabal Sacran, *Al-Kaisi, Ghina, Shaima'a & Saif* 58572 (BAG). **MSU**: Mela Kowa, *Rawi* 22508!; Jarmo, *Helbaek* 387; Sulaimaniya, *Thesiger* 353; Azmar, *Al-Kaisi & K. Hamad* 49312 (BAG). **MJS**: above Kursi, *Gillett* 11041! Jabal Sinjar, *Al-Kaisi & K. Hamad* 49205 (BAG). **FAR**: Khanzad valley, *Guest* 2334!; Erbil, *Shahwani* 11180! **FNI**: Dohuk, *Guest* 1294!

Throughout Mediterranean Europe, Crimea, Caucasus, Turkey, Palestine, Lebanon, Syria, Iraq, Iran, Afghanistan, N. Africa (Morocco, Algeria, Tunisia).

Where this species occurs outside the forest, it may be an indicator of former forest conditions.

9. Geranium kurdicum *Bornm.* in Feddes Repert. 8: 82 (1910). Type: Iraq, in partis alpinis montis Helgurd, 1893, *Bornmüller* 998 (K! syn.). Blakelock in Kew Bull. 4: 409 (1949); Zohary in Dep. Agr. Iraq Bull. 31: 96 (1950); Davis in Notes Roy. Bot. Gard. Edinburgh 22: 25 (1955); Rawi in Dep. Agr. Iraq Tech Bull. 14: 59 (1964); Davis in Fl. Turkey 2: 470 (1967).

Perennial herb, rhizomes clothed with brown persistent stipules. Flowering stems to 25 cm, with 2–4 pairs of opposite leaves surrounded by a tuft of radical leaves, all parts glabrous to sparsely strigose-hairy. Radical leaves with petiole 2–5 × longer than lamina, reniform, palmatisect into 5–7 pinnatifid mucronate lobes, ultimate segments oblong or linear. Flowers borne in pairs on 2–5 cm pedicels; peduncles variable, longer or shorter than pedicels. Sepals 7–9 mm excluding awn, ovate, obtuse, aristate, the awn 1.5–2 mm. Petals blue-violet, nearly twice as long as calyx, broadly ovate and showy. Fruit 22–26 mm; carpel valve 6–7 mm. Seed coat with minute reticulations.

HAB. Streamsides and flushes, damp subalpine scree, igneous or metamorphic rocks; alt. 2200–3600 m; fl. Jun.-Jul.
DISTRIB. Locally abundant in a restricted area of the alpine zone of the Helgurd massif. **MRO**: Helgord, *Bornmüller* 998 (type); ibid., *Guest & Ludlow-Hewitt* 2805!; ibid., *Haley* 136; Kodo, nr. Haji Umran, *Rawi*

9191!; Ari Gird Dagh, *Gillett* 12359 (BAG); N. of Helgurd nr. Bermasand lack, *Al-Rawi & Serhang* 24776 (BAG); Helgord range, *Al-Rawi & Serhang* 24823 (BAG).

Turkey (S.E. Anatolia); possibly also in W. Iran but no good material has been seen from that area.

2. **ERODIUM** L'Her.

Geraniologia t. 1–6 (1787–88)
Geranium L., Sp. Pl.: (1753), pro parte.

C.C. Townsend
Revised by Shahina A. Ghazanfar

Annual or perennial herbs, leaves usually opposite, sometimes alternate in lower part. Flowers hermaphrodite or sometimes unisexual, actinomorphic or slightly zygomorphic, in umbels of 2–6 flowers subtended by small scarious bracts. Pedicels (in Iraq) strongly geniculate in fruit. Stamens 10, those opposite the petals sterile. Carpels with a long beak in fruit, becoming first detached either apically or basally and then falling away attached to the solitary seed on dehiscence; beak acting as an awn, twisting and untwisting.

About 80 species from Europe to C. Asia, mainly Mediterranean in distribution; a few species in N. America, S. Africa and Australia; 12 species in Iraq belonging to two sections.

Erodium is derived from Gr. ερoδιoς, *erodios*, heron, on account of its long-beaked fruit.

Fiz, O., Vargas, P., Alarcón, M. L., & Aldasoro, J. J. (2006). Phylogenetic relationships and evolution in *Erodium* (Geraniaceae) based on trnL-trnF sequences. *Systematic Botany* 31(4): 739–763.

Lobe of carpel dehiscent from beak downwards; beak not usually
    coiling spirally, or it may do so for a short distance only, with long
    silky hairs right to the tip of the beak; lobe of carpel without pits,
    usually with semicircular grooves just below insertion of beak . . . . .Sect. **Plumosa**
Lobe of carpel not dehiscent from beak downwards; beak coiling
    spirally after dehiscence, with long reddish hairs only on the spiral,
    not on the long (>1 cm) straight tip of the beak; carpel lobes with 2
    pits just below the beak . . . . Sect. **Barbata**

1. Beak separating tip first and uniformly plumose within; desert plants (sect. *Plumosa*) 2
   Beak separating from base first, with tuft of long bristles within, as well
     as short hairs; habitats various (sect. *Barbata*) 4
2. Outer sepals with 3 broad veins; carpel valves without or with only a
     single ridge above . . . . 3. *E. oxyrhinchum*
   Outer sepals with 4–5 narrow veins; carpel valves with 2 or more
     concentric ridges above. 3
3. Leaves deeply pinnatisect, spreading-hairy . . . . .2. *E. hirtum*
   Leaves pinnatifid or lobed only, hairs (if any) adpressed . . . . 1. *E. glaucophyllum*
4. Leaves completely pinnate, without lobes or wings on rachis between pinnae. 5
   Leaves pinnatifid or pinnatisect, with a winged or lobed rachis between pinnae. 6
5. Sepals reticulately veined, with many minor veins joining major ones .12. *E. moschatum*
   Sepals simply veined . . . . .11. *E. cicutarium*
6. Perennial, dioecious; high-alpine plant . . . . .10. *E. trichomanefolium*
   Annuals with hermaphrodite flowers; mostly lowland or desert plants 7
7. Mature fruits over 6 cm 8
   Mature fruits under 5 cm 9
8. Leaves pinnatisect with lobes on ± winged rachis; carpel valve with no
     groove below terminal pit . . . . .9. *E. cicutarium*
   Leaves pinnatifid only, rachis never free; carpel valves with grooves
     below terminal pit . . . . 8. *E. gruinum*
9. Fruit 2–3 cm . . . . .6. *E. malacoides*
   Fruit more than 3 cm 10
10. Carpel valves with groove below terminal pit. . . . . 7. *E. neuradifolium*
    Carpel valves with no grooves 11

11. Pedicels ± glabrous; umbel bracts 2, entire, ovate, brownish, glabrous . .4. *E. laciniatum*
    Pedicels at least pubescent; umbel bracts of many joined lobes, hyaline,
        or at most pale brown only, ciliate, often pubescent . . . . .5. *E. pulverulentum*

SECT. PLUMOSA Boiss.

1. **Erodium glaucophyllum** (*L.*) *L'Her.*, Geraniol. n. 25 (1787–88); Boissier, Fl. Orient.
1: 895 (1867); Blakelock in Kew Bull. 4: 408 (1948); Zohary in Dep. Agr. Iraq Bull. 31: 97
(1950); Burtt & Lewis in Kew Bull. 403 (1954); Rawi in Dep. Agr. Iraq Tech. Bull. 14: 58
(1964); Rechinger, Fl. Lowland Iraq: 393 (1964); Mouterde, Nouv. Fl. Lib. et Syrie 2: 446
(1970); Schönbeck-Temesey in Fl. Iranica [K. H. Rechinger] 69: 45 (1970); Y. Nasir in Fl.
Pakistan [Nasir & Ali] 32 (1983); Zohary, Fl. Palaest. 2: 234, pl. 333 (1987); Collenette,
Wild Flow. Saudi Arabia: 332 (1999); Boulos in Fl.Egypt 2: 6 (2000); Abdulridha, Taha &
Widad, Ecology & Fl. Basrah: 403 (2016); Taifour & El-Oqlah, Pl. Jordan Annot. Checkl.:
96 (2017).

   Geranium glaucophyllum L., Sp. Pl.: 679 (1753).

Perennial or annual herb, with ascending branches to 40 cm or more and with a swollen
storage taproot, at least in young specimens. Indumentum various but usually glabrous
on stems, sometimes shortly hairy on leaves. Leaves 2–5 cm, usually almost simple but
sometimes 3-lobed, triangular in outline, often glaucous. Flowers 2–4 in each umbel.
Bracts broadly triangular, ciliate at edges, central vein not well defined. Sepals oblong, 6–9
mm, acute, mucronate, with spreading hairs especially at base, sometimes glabrous distally;
outer sepals at least with 4–5 narrow veins. Petals deep pink to purplish, oblong or obovate,
10–12 mm. Carpel lobes 6–7 mm, densely hairy at least at base, bases of hairs not very
prominent, with 2–3 pairs of semicircular concentric ridges at apex; carpel beak 6–9 cm.

HAB. A common plant of the open desert, especially in gypsiferous soils, depressions, in clayey and
rocky soils, throughout lower Iraq; alt. 10–300 m; fl. (Feb.-)Mar.-Jun.
DISTRIB. Almost throughout the desert and steppe zones of Iraq except on fine deep soils of the
N. Jezira: **FPF**: Koma Sank, nr Mandali, *Rawi* 20592!; 12 km E. of Samarra, *Rechinger* 9561; 10 km
E. of Mandali, *Rechinger* 9635, 12762, 12797; 16 km S. of Zurbatiya, *Al-Kaisi & Yahya* 45275 (BAG);
Jabal Qiska near Al-Hashimah, *Al-Kaisi* 56735 (BAG). **FKI**: Kirkuk, *Bornmüller* 997; Tuz, *Rogers* 0153;
Inganah, *S. Omar, F. Karim, R. Hamza & H. Hamid* 37033 (BAG). **DLJ**: *Rawa, Gillett & Rawi* 6990!. **DGA**:
11 km E. of Samarra, *Rechinger* 9561! **DWD**: 15 km W. of Ramadi, *Guest & Rawi* 13925!; Al-Qaim to Hit,
Handel-Mazzetti 663!; Ana – Wadi Al-Qaser, *S. Omar, Al-Kaisi, K. Hamad & H. Hamid* 44361 (BAG). **DSD**:
Salman, *Rechinger* 9297, 9308; Chilawa, 112 km S.S.W. of Basra, *Rechinger* 8788; Nukhaila, 40 km S.W. of
Basra, *Rechinger* 8841; 30 km E. of Salman, *Guest, Rawi & Rechinger* 18742!; Jabal Sanam, *Gillett & Rawi*
6145! *Rechinger* 8598; 10 km SSE of Jiraibiyat, 125 km S.W. of Basra, *Rechinger* 14414. **LEA**: Shahraban
(Muqdadiya), *Strauss s.n.*; 15 km S.E. of Badra, *Gillett* 6728!; 10 km E. of Falluja, *Gillett & Rawi* 6751!

N. Africa, Sudan, Sinai to Iran and the Arabian Peninsula.

2. **Erodium hirtum** *Willd.*, Sp. Pl., ed. 4 [Willdenow] 3: 632 (1800); Boissier, Fl. Orient.
1: 894 (1867); Rawi, Dep. Agr. Iraq Tech Bull. 14: 58 (1964); Rechinger, Fl. Lowland Iraq:
393 (1964); Zohary, Fl. Palaest. 2: 233, pl. 332 (1987); Collenette, Wild Flow. Saudi Arabia:
333 (1999).

   Geranium hirtum Forssk., Fl. Aegypt.-Arab.: 123 (1775) nom. illeg.

Perennial herb, with brown ovoid tubers up to the size of a walnut on the much branched
root system. Stems branched, springing from rosette of leaves on rootstock, to 20 cm,
hirsute throughout with stiff spreading hairs. Radical leaves with petiole 1–2 × length of
leaves; lamina 2–5 cm, cauline leaves shorter, petiolate, all with laminas ± triangular in
outline, bipinnatifid, segments oblong, acute or obtuse. Umbels with 4–7 scarious bracts;
pedicels shorter than peduncles. Sepals 5–8 mm, oblong, obtuse, mucronate, 3–5-nerved,
covered with short spreading hairs especially at base. Petals bright purple-pink, with a black
spot within at base, obovate, twice as long as sepals. Stamens with basally expanded, ciliate
filaments, nearly as long as calyx. Fruit 5–7 cm; carpel valves 5–6 mm, stiffly long-hairy, but
not verrucose, with 2 pairs of semicircular concentric glabrous grooves above.

HAB. On hammada desert soils; alt. 350–750 m; fl. Mar.-Apr.
DISTRIB. Rare; found only on the stony plateau extending from Jordan into Iraq in the Western Desert

region: **DWD**: 10 km S. of Rutba, *Rawi* 21055!; 5 km S.E. of Rutba, Rechinger 9896; 73 km N. of Rutba, *Rawi & Nuri* 27173!

N. Africa from Morocco to Egypt, Sinai, Syria, Saudi Arabia.

3. **Erodium oxyrhinchum** *M.Bieb.*, Fl. Taur.-Cauc. 2: 133 (1808); Rawi in Dep. Agr. Iraq Tech. Bull. 14: 58 (1964); Mouterde, Nouv. Fl. Lib. et Syrie 2: 447 (1970); Schönbeck-Temesey in Fl. Iranica [K. H. Rechinger] 69: 42 (1970); Y. Nasir in Fl. Pakistan [Nasir & Ali] 31 (1983);

Fig. 61. **Erodium oxyrhinchum.** 1, habit × ½; 2, 3, filaments showing variation × 5; 4, mericarp × 4; 5, plumose beak × ½. Reproduced with permission from Fl. Pakistan 149: f. 7, J–N (1985). Drawn by M.Y. Saleem. © National Herbarium, Pakistan Agriculture Research Council & University of Karachi, Pakistan.

Zohary, Fl. Palaest. 2: 234 (1987); Collenette, Wild Flow. Saudi Arabia: 335 (2000); Boulos in Fl.Egypt 2: 4 (1999); Taifour & El-Oqlah, Pl. Jordan Annot. Checkl.: 97 (2017).

> *Geranium oxyrhinchum* Poir., Encycl. [J. Lamarck & al.] Suppl. 2. 743 (1812).
> *Erodium hohenackeri* Ledeb., Fl. Ross. (Ledeb.) 1(2): 478 (1842).
> *E. bryoniifolium* Boiss., Diagn. Pl. Orient. ser. 1, 1: 61 (1843); Boissier, Fl. Orient. 1: 896 (1867); Blakelock in Kew Bull. 4: 408 (1948); Vvedenskiy in Fl. SSSR 14: 64 (1949); Zohary in Dep. Agr. Iraq Bull. 31: 97 (1950); Burtt & Lewis in Kew Bull. 1954: 400 (1954) (as "E. byroniifolium", sphalm.); Rechinger, Fl. Lowland Iraq: 393 (1964); Abdulridha, Taha & Widad, Ecology & Fl. Basrah: 401 (2016).

Annual herb with flowering branches prostrate from a central rosette, or sometimes erect. Leaves triangular in outline, 1–3.5 cm, variable, 3-lobed, similar to *E. glaucophyllum* but often more divided, usually densely pubescent on both surfaces. Flowers 2–4 in each umbel on the pubescent flowering stems. Involucral bracts lanceolate, with a well marked central nerve. Sepals 5–7 mm, usually adpressed hairy, all with 3 broad green veins, obtuse, mucronate, inner with broad scarious margins. Petals bluish pale purple, oblong, 8–10 mm. Carpel lobes 6–7 mm, sparsely hairy, bases of hairs enlarged and often pigmented, appearing verrucose; apex of lobe with one semi-circular ridge, or none; carpel beak (6–)8–11 cm. Fig. 61, 1–5.

HAB. Open stony ground in desert and steppe areas, usually associated with gypsum, sometimes on sandy soil; alt. 150–600 m; fl. Mar.-May.

DISTRIB. Occasional from the Southern and Western Deserts to Jabal Hamrin. **FKI**: Jabal Hamrin on Adhaim river, *Agnew* 5473! **FUJ**: Qal'a Sharqat, *Handel-Mazzetti* 1122; **FPF**: Pilkana, nr Khanaqin, *Rawi* 12767!; Table Mountain (nr Mansur), *Guest* 1822; Wadi Tib police post 70 km N. of Amara, *Guest, Rawi & Rechinger* 17520A! 22 km S.S.E. of Badra, *Rechinger* 13923 p.p. **DWD**: 5 km W. of Karbala, *Gillett & Rawi* 6368!; 35 km W. of Rutba, *Rawi* 21163!; 15 km W. of Ramadi on Hit road, *Gillett & Rawi* 6786! 20 km S. of H3, *Rechinger* 12568; **DSD**: 5 km N.W. of Sharaf, *Rechinger* 13687; 95 km S.S.E. of Salman, *Guest, Rawi & Rechinger* 18902! Chilawa, 110 km S.S.W. of Basra, *Rechinger* 14391; **LCA**: Babylon, *Rawi & Walter* 18362! *Reuter* 4444; Iskandariya, *Rawi & Walter* 18369!

N. Africa from Libya to Egypt, Sinai, Palestine, Jordan, Caucasus, Iraq, Iran, Afghanistan, Pakistan, Turkmenia, Kazakhstan, Tadjikistan, W. Siberia.

SECT. **BARBATA** Boiss.

4. **Erodium laciniatum** (*Cav.*) *Willd.*, ed. 4 [Willdenow] 3(1): 633 (1800); Boissier, Fl. Orient. 1: 893 (1867) pro parte; Zohary in Dep. Agr. Iraq Bull. 31: 97 (1950); Burtt & Lewis in Kew Bull. 1954: 404 (1954) pro parte; Rechinger, Fl. Lowland Iraq: 394 (1964); Rawi in Dep. Agr. Iraq Tech. Bull. 14: 58 (1964); Mouterde, Nouv. Fl. Lib. et Syrie 2: 445 (1970); Schönbeck-Temesey in Fl. Iranica [K. H. Rechinger] 69: 54 (1970); Y. Nasir in Fl. Pakistan [Nasir & Ali] 36 (1983); Zohary, Fl. Palaest. 2: 240, pl. 345 (1987); Collenette, Wild Flow. Saudi Arabia: 333 (photo) (1999); Boulos in Fl.Egypt 2: 7 (2000); Taifour & El-Oqlah, Pl. Jordan Annot. Checkl.: 97 (2017).

> *Geranium laciniatum* Cav., Dis. 228.
> *G. triangulare* Forssk., Fl. Aegypt.-Arab. 123 (1775).
> *G. maritimum* auct. non Burm.f. (1759); Forssk., Fl. Aegypt.-Arab. 123 (1775),
> *Erodium strigosum* Karel. in Ledeb., Fl. Ross. 1: 475 (1842).
> *E. triangulare* (Forssk.) Muschl., Man. Egypt 1: 558 (1912); Vvedenskiy in Fl. S.S.S.R. 14: 70 (1949).
> *Erodium triangulare* subsp. *laciniatum* (Cav.) Maire, Cat. Pl. Maroc 2: 446 (1932).

Annual herb with trailing, often reddish, stems and often a central rosette of leaves. Pubescence of coarse short spreading hairs, sometimes glabrescent, without glands. Leaves ± triangular in outline, deeply or shallowly pinnatisect or pinnatifid, 1.5–4 cm, the apex of the lobes and teeth ± acute. Umbel of 2–5 flowers subtended by 2(–3) broad ovate, entire, glabrous brownish scarious bracts, sometimes undulate parallel with margin. Sepals oblong, 5–6 mm, obtuse, with a short (0.5–1 mm) mucro, pubescent with longer spreading hairs at base. Petals light pink or mauve, 1.5–2 × calyx length. Carpel lobes pubescent, ± 4 mm, with small triangular pits and no furrow below them; beak 35–45 mm, often reddish at base, with internal bristles at and above carpel lobe. Fig. 62, 1.

HAB. Only on sandy soil, in places where the vegetation is good, sandy gypseous subdesert, on sloping ground, sandy stony hillside, sandy island in river; alt. 10–600 m; fl. Mar.-Apr.

Fig. 62. **Erodium laciniatum.** 1, habit × ½. **Erodium cicutarium.** 2, habit × ½; 3, sepal × 2; 4, petal × 3; 5, stamen × 8; 6, staminode × 8; 7, pistil × 4; 8, fruit × 2; 9, seed × 5. Reproduced with permission from Fl. Pakistan 149: f. 8, J, K–R (1985). Drawn by M.Y. Saleem. © National Herbarium, Pakistan Agriculture Research Council & University of Karachi, Pakistan.

DISTRIB. Occasional in foothills and steppe zone of Iraq. **FPF**: nr Wadi Tib police post, 70 km N. of Amara, *Guest, Rawi & Rechinger* 17521!; ibid., *Rechinger* 8884!; 22 km S.E. of Badra, *Rechinger* 13923 p.p. Chlat police station near Persian boarder, *Al-Rawi* 25675A (BAG). **FKI**: nr. Injana, Jabal Hamrin, *Guest* 678 (BAG). **DLJ**: 5 km S. of Beiji, *Gillett* 10714! **LEA**: 3 km E. of Shahraban (Muqdadiya), *Agnew* 1675!; Shahraban (Muqdadiya), *Rechinger* 13375!. **DWD**: 260 km W. of Ramadi, *Al-Rawi & Nuri* 27059 (BAG); 17 km S. of k3 (Albu Hayat), *Al-Kaisi & K. Hamad* 53127 (BAG). **DSD**: Shabicha, *Gillett & Rawi* 6300 (BAG).

Mediterranean area extending east to Syria and Iran.

5. **Erodium pulverulentum** (*Cav.*) *Willd.*, Sp. Pl. ed. 4 [Willdenow] 3(1): 632 (1800); Zohary in Dep. Agr. Iraq Bull. 31: 97 (1950); Abdulridha, Taha & Widad, Ecology & Fl. Basrah: 403 (2016).

Geranium pulverulentum Cav., Diss. 5: 272, t. 125 f. 1 (1788).
Erodium bovei Delile, Ind. Sem. Horti Monsp.: 6 (1838).
E. laciniatum var. *pulverulentum* (Cav.) Boiss., Fl. Orient. 1: 893 (1867); Zohary, Fl. Palaest. 2: 241, pl. 345a (1987); Blakelock in Kew Bull. 1948: 408 (1949); Rawi in Dep. Agr. Iraq Tech. Bull. 14: 58 (1964); Boulos, Fl. Egypt 2: 7 (2000).
E. laciniatum subsp. *pulverulentum* (Cav.) Burtt & Lewis in Kew Bull. 1954: 405 (1954); Mouterde, Nouv. Fl. Lib. et Syrie 2: 445 (1970); Collenette, Wild Flow. Saudi Arabia: 334 (1999).
E. pulverulentum subsp. *bovei* (Delile) Schönbeck-Temesey in Fl. Iranica 69: 53 (1970).

Annual, similar to *E. laciniatum* but uniformly dense pubescence covers whole plant and is all crisped and of fine hairs; leaves less divided, tips of lobes and segments obtuse; pedicels crisped pubescent, often glandular, subtended by 4–6 acute or obtuse bracts, each with a central nerve, usually fused at base into 2 opposite rows, tips free, ciliate on margins and ± pubescent on outer side. Sepals 4.5–5 mm, mucronulate or acute. Fruit often a little shorter than *E. laciniatum*, not reddish, with inner face of beak glabrous immediately below carpel valve, bristles commencing 6–10 mm above insertion of beak into valve.

HAB. Gypsum and stony deserts, seldom on sand; alt. 50–200 m; fl. Feb.-Apr.
DISTRIB. Common in all desert regions of Iraq as far as S. Jazira and Jabal Hamrin. **FPF**: Wadi Tib, 70 km N. of Amara, *Rechinger* 8927, *Guest, Rawi & Rechinger* 17520!; 3o km S.E. of Badra, *Rechinger* 9138, 9139. **DLJ**: Hatra, *Agnew, Hadač & Al-Hashimi* 5205! **DGA**: 50 km E. of Samarra, *Rawi* 20455! **DWD**: 100 km W. of Karbala, *Rawi* 20090! **DSD**: Salman, *Guest, Rawi & Rechinger* 18823!; 75 km W. of Khidhr el Mai, *Rawi & Khatib* 2913! **LEA**: between Shahraban (Muqdadiya) and Jalaula, *Rechinger* 8998; Falluja, *Guest* 960A!; **LCA**: Sudur, Mansuriyat al-Jabal, *Barkley* 6793.

Mediterranean area, S.W. Asia.

6. **Erodium malacoides** (*L.*) *L'Hér.*, Ait. Hort. Kew. ii. 415 (1789); Boissier, Fl. Orient. 1: 893 (1867); Zohary, Fl. Pal. ed. 2, 1: 261 (1932); Vedenskiy in Fl. S.S.S.R. 14: 69 (1949); Zohary in Dep. Agr. Iraq Bull. 31: 97 (1950); Rechinger, Fl. Lowland Iraq: 395 (1964); Rawi in Dep. Agr. Iraq Tech. Bull. 14: 58 (1964); Schönbeck-Temesey in Fl. Iranica [K. H. Rechinger] 69: 50 (1970); Y. Nasir in Fl. Pakistan [Nasir & Ali] 33 (1983); Zohary, Fl. Palaest. 2: 242, pl. 348 (1987); Collenette, Wild Flow. Saudi Arabia: 332 (1999); Boulos in Fl.Egypt 2: 6 (2000) Abdulridha, Taha & Widad, Ecology & Fl. Basrah: 403 (2016); Taifour & El-Oqlah, Pl. Jordan Annot. Checkl.: 97 (2017).

Geranium malacoides L., Sp. Pl.: 680 (1753).

Annual herb with erect or ascending stems to 40 cm, glandular-pubescent in all vegetative parts. Leaves mostly borne above ground level, adpressed-hairy, triangular in outline, cordate at base, shallowly lobed into 3 segments with lobed margins. Umbel of 4–6 flowers subtended by 4–5 whitish scarious bracts with ciliate margins. Sepals 4–5 mm, ovate, acute, mucronate, mucro up to 0.5 mm, tipped by 1–3 bristles, sparsely hairy along the 4–5 veins and with spreading hairs at base. Glandular hairs usually present on sepals and pedicels. Petals ± as long as sepals, oblong, pink. Carpels with beak 2–3 mm and a large, usually glandular pore and furrow on the lobe. Beak twisting spirally, pilose from just above the carpel lobe. Fig. 63, 1–3.

HAB. Wet irrigated places, often growing as a weed amongst crops, under palms and in waste ground; alt. 150–500 m; fl. Mar.-Apr.
DISTRIB. Occasional, mainly confined to the Mesopotamian plain and foothills, rocky hillside, clayey gypsum soil, and gravelly soil. **MSU**: Diyala, *Rechinger* 9025!; Khormal, *Hadač* s.n.! **FNI**: Jabal Maqlub, *Haines* 1498! **FPF**: Mandali, *Guest* 897A!; Badra, *Rechinger* 13996! 30 km S.E. of Badra, *Rechinger* 9158! **DWD**: Haditha, below Ana, *Handel-Mazzetti* 788; Ana, *S. Omar, H. Abass & Al-Kaisi* 56380 (BAG); 80 km N. of Rutba, *Al-Rawi & Khatib* 32372 (BAG). **DSD**: Shabicha, *Haines* s.n.! Al-Takhaded, 60 km S. of Salman, *Al-Kaisi et al.* 42198 A (BAG); Al-Mitautia, 50 km S. of Ansab, *Al-Kaisi et al.* 42253 (BAG). **LEA**: Baghdad, *Guest* 1085A!; along Shatt al-Arab, S.E. of Basra, *Dept. of Botany* 8605! Amara to Qurna, *Rechinger* 8956. **LSM**: Azair, 80 km S.E. of Amara, *Guest, Rawi & Rechinger* 17667! **LBA**: below Basra, *Rechinger* 8605!.

Macaronesia to Pakistan and Somalia; introduced to S. America and southern Africa.

7. **Erodium neuradifolium** *Delile* ex *Godr.* in Mem. Acad. Montpellier (sect. Medec.) 1: 425 (1853); Boulos in Fl. Egypt 2: 6 (2000); Taifour & El-Oqlah, Pl. Jordan Annot. Checkl.: 97 (2017).

E. subtrilobum Jord., Pugill. Pl. Nov. 42 (1852); Rechinger, Fl. Lowland Iraq 395 (1964);
E. aegyptiacum Boiss., Diagn. Pl. Orient. ser. 2, 1: 111 (1854); Boissier, Fl. Orient. 1: 894 (1867); Zohary in Dep. Agr. Iraq Bull. 31: 97 (1950); Rawi in Dep. Agr. Iraq Tech. Bull. 14: 58 (1964); Schönbeck-Temesey in Fl. Iranica [K. H. Rechinger] 69: 51 (1970); Collenette, Wild Flow. Saudi Arabia: 332 (1999);
E. malacoides subsp. *aragonense* (Loscos) O.Bolòs & Vigo, Fl. Països Catalans 2: 303 (1990): (1990).
E. neuradifolium var. *aegyptiacum* (Boiss.) Täckh. & Boulos, Publ. Cairo Univ. Herb. 5: 47 (1974).

Annual, resembling *E. malacoides*, but with a denser pubescence, often almost crispate. Sepals a little larger than that of *E. malacoides* with a longer (to 1.5 mm) mucro. Fruit 4–5 cm, the pore and furrow at tip of carpel valve being (in ours) eglandular.

HAB. Silty soils in grassland, usually in wetter situations; alt. 200–800 m; fl. Mar.-Apr.
DISTRIB. Northern Jezira and Jabal Hamrin; rare. **FUJ**: Hatra (Hadhr), *Agnew, Hadač & Al-Hashimi*

Fig. 63. **Erodium malacoides.** 1, habit × ½; 2, mericarp with beak × 2; 3, mericarp with basal furrows and pits × 10. Reproduced with permission from Fl. Pakistan 149: f. 8, A–B (1985). Drawn by M.Y. Saleem. © National Herbarium, Pakistan Agriculture Research Council & University of Karachi, Pakistan.

5219! **FKI**: Injana, *Polunin* 2052! **FPF**: Pilkana nr Khanaqin, *Rawi* 12768!; **FPF/LEA**: Shahraban to Jalaula, *Rechinger* 8998; Table Mountain on Jabal Hamrin, *Guest* 1815! **DLJ**: Rawa, *Rawi & Gillett* 7002! **DWD**: 50 km N. of Rutba, *Rawi & Nuri* 27133!

Mediterranean area from Spain to Egypt, Somalia, through Arabia, Syria, Palestine, Iran, Afghanistan to Pakistan.

The preceding four species are easily confused in the field, since the chief diagnostic characters of the fruit need careful examination of mature material with a lens or microscope for accurate identification, and the form of their leaves may be almost indistinguishable. The following points may assist:

E. laciniatum usually has acute leaf lobes and red stems and fruits, with two brown, entire, glabrous umbel bracts; *E. pulverulentum* has sepals with the mucro short or absent, or if present without terminal bristles; the umbel bracts are numerous, pubescent; *E. malacoides* has an extremely short fruit, and is a weed of cultivation;

E. neuradifolium has the sepal mucro with its terminal bristles, as well as the grooved carpel valve.

8. **Erodium gruinum** (*L.*) *L'Her.* ex *Aiton*, Hort. Kew. ed. 1, 2: 415 (1789); Boissier, Fl. Orient. 1: 892 (1867); Fl. Pal. ed. 2, 1: 260 (1932); Blakelock in Kew Bull. 1948: 408 (1949); Zohary in Dep. Agr. Iraq Bull. 31: 97 (1950); Rawi in Dep. Agr. Iraq Tech. Bull. 14: 58 (1964); Schönbeck-Temesey in Fl. Iranica [K. H. Rechinger] 69: 47 (1970); Boulos in Fl. Egypt 2: 8 (2000); Taifour & El-Oqlah, Pl. Jordan Annot. Checkl.: 97 (2017).

Geranium gruinum L., Sp. Pl.: 680 (1753).

Annual herb with procumbent or shortly erect, sparingly branched stems to 30 cm arising from a basal rosette of leaves, sparsely pilose to glabrous, stems occasionally with strigose hairs, eglandular. Radical leaves with petiole 1–2 × longer than lamina; lamina 2.5–5 cm, ovate-lanceolate in outline, pinnatifid, usually with 2 lobes at base and a much larger terminal portion, but occasionally more uniformly divided, lobes coarsely dentate. Cauline leaves with shorter petioles and more deeply divided. Umbels of 1–3 flowers, peduncles equal to or longer than pedicels, subtended by 4–6 free, scarious, glabrous bracts. Sepals oblong-ovate, scarious at edges and between the pubescent nerves, aristate (awn 2–3 mm), 9–12 mm in flower, up to 15 mm in fruit. Petals obovate, 1.5 × length of sepals, purplish-blue. Stamens with filaments markedly expanded and ovate at base, ½ as long as calyx. Fruit 7–10 m, carpel valve 10–14 mm, densely hirsute, with apical pits and a (sometimes obscure) semicircular groove below each pit, both groove and pit being themselves coarsely pitted. Fig. 64, 1–3.

HAB. Open annual vegetation in over-grazed areas, on stony or sandy soil, rocky clay mountain slope, irrigated field, often associated with *Trifolium cherleri*. alt. 200–1000 m; fl. & fr. Feb.-Apr.
DISTRIB. Mixed shrubland zone in foothills of N.E. Iraq. **MJS**: Jabal Sinjar, *Al-Kaisi & K. Hamad* 54222 (BAG); **MAM**: Dohuk, *Makki Beg* 3245! **MSU**: Chemchamal, *Rogers* 0192!; Kirkuk, *Rogers* 1918! *Ludlow-Hewitt* 1918!; Jarmo, *Helbaek* 329. **FUJ**: Qala Shirgat, *Gillett & Rawi* 7196A (BAG). **FPF**: Pilkana nr Khanaqin, *Rawi* 12843!; Darband-i Khan, *Haines* s.n.; Khanaqin, *Rechinger* 14103. **DGA**: Adhaim, *A. H. Al-Mosawi* 43712 (BAG)

E. Mediterranean, Egypt, Palestine, Syria, Greece, Turkey, Iran.

9. **Erodium ciconium** (*L.*) *Willd.*, Sp. Pl., ed. 4 [Willdenow] 3(1): 629 (1800); Boissier, Fl. Orient. 1: 891 (1867); Vvedenskiy in Fl. S.S.S.R. 14: 71 (1949); Blakelock in Kew. Bull. 1948: 408 (1949); Zohary in Dep. Agr. Iraq Bull. 31: 96 (1950); Burtt & Lewis in Kew Bull. 1954: 401 (1954); Rawi in Dep. Agr. Iraq Tech. Bull. 14: 58 (1964); Schönbeck-Temesey in Fl. Iranica [K. H. Rechinger] 69: 48 (1970); Y. Nasir in Fl. Pakistan [Nasir & Ali] 149: 35 (1983); Zohary, Fl. Palaest. 2: 238, pl. 340(1987); Boulos in Fl. Egypt 2: 8 (2000) Taifour & El-Oqlah, Pl. Jordan Annot. Checkl.: 96 (2017).

Geranium ciconium L., Cent. Pl. I. 21 (1755).
Erodium ciconium var. *edomeum* Eig in Beih. Bot. Centralbl. 50, 2: 232 (1932).

Glandular-hairy annual herb of variable habit, sometimes 5–10 cm tall, occasionally to 80 cm (smaller plants not exceeding 15 cm have been separated as var. *edomeum*), indumentum short-glandular and inconspicuous except on peduncles, pedicels and calyx (indumentum hirsute-glandular throughout in var. *edomeum*). Stems ascending to erect, branching above. Radical leaves with petiole equal to or longer than lamina, cauline often nearly sessile; lamina 2–10 cm (lamina 3–5 cm, bipinnatisect in var. *edomeum*), ovate in outline, pinnatisect, with deeply pinnatifid lobes, ultimate segments oblong, or even somewhat linear, obtuse; rachis winged and with small lobules between pinnae. Umbels of (2–)3–6 flowers, subtended by 4–6 pubescent bracts, peduncles longer than pedicels and both glandular-hairy. Sepals 6–9mm (without awn) in flower, enlarging in fruit to up to 14 mm, oblong-ovate, aristate (awn 2–4 mm), scarious at edges and between the glandular-pubescent nerves. Petals obovate, a little longer than calyx, pale blue. Stamens ⅔ as long as sepals (excluding awns),

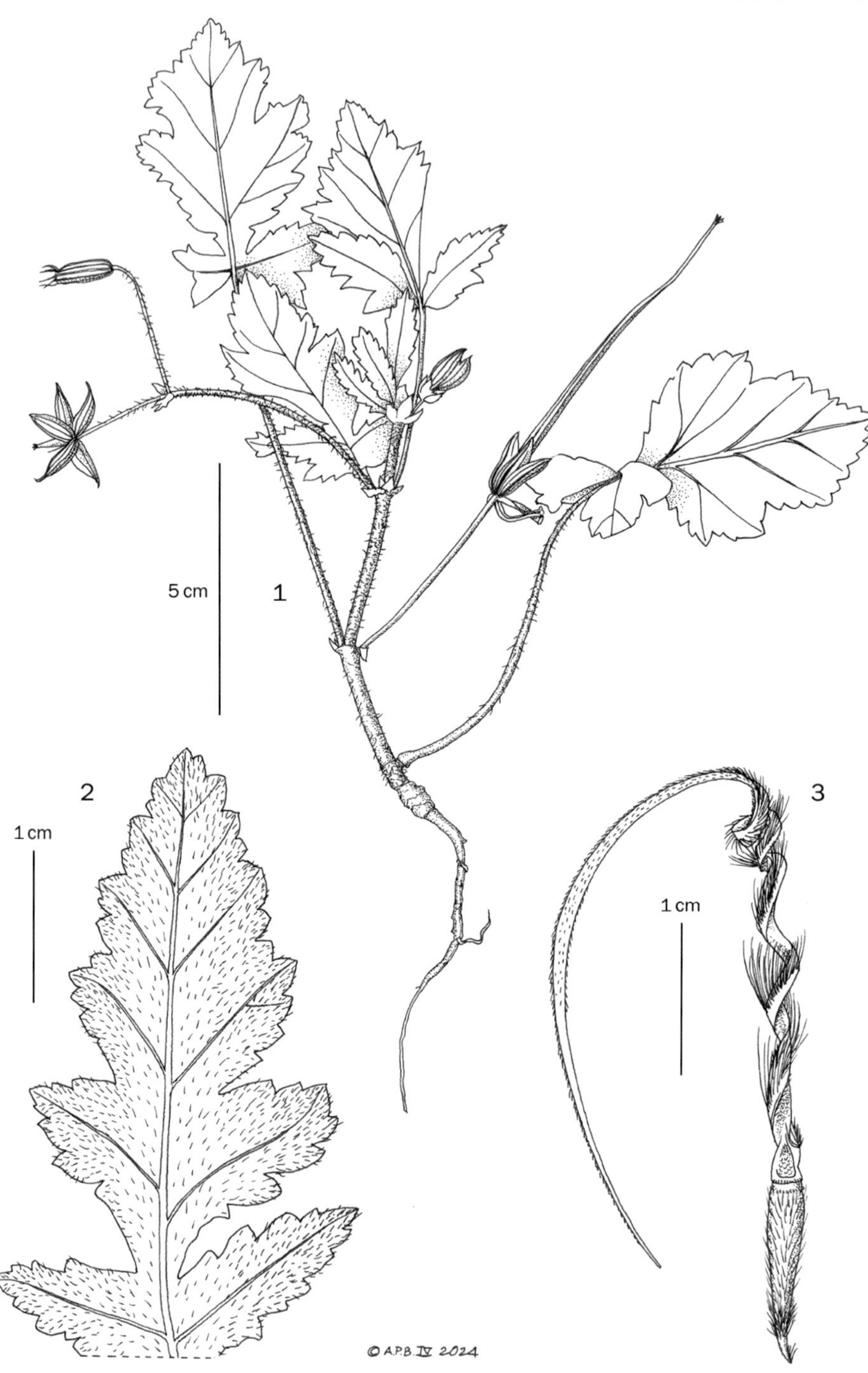

Fig. 64. **Erodium gruinum**. 1, habit of fruiting plant; 2, adaxial surface of distal portion of leaf; 3, detached mature fruit. 1 from *Omar et al.* 36405; b from Ludlow-Hewit 1918; 3 from *Gillett et al.* 7531. Drawn by © A.P. Brown, April 2024.

Fig. 65. **Erodium ciconium.** 1, habit × ½; 2, sepal × 2; 3, mericarp with beak × 1; 4, mericarp × 2; 5, mericarp with pits × 5 Reproduced with permission from Fl. Pakistan 149: f. 8, C–F (1985). Drawn by M.Y. Saleem. © National Herbarium, Pakistan Agriculture Research Council & University of Karachi, Pakistan.

filaments lanceolate at base. Fruit 7–9mm, carpel valves 8–11 mm, hirsute, with 2 pubescent pits at apex of each, without concentric grooves below. Fig. 65, 1–5.

HAB. Sand or silty soils in desert depressions and stony open habitats in the foothills, in orchards, on sandy clay soil; alt. 120–640 m; fl. Feb.-Apr.

DISTRIB. Widespread in the Central and Western Deserts of Iraq, extending into the foothills. **FUJ**: Tal Kochak, *Gillett* 10840! **FAR**: Sharqat, on Ziggurat mound at Ashur, *Guest* 404! **FKI**: Kirkuk, *Guest* 1201! Jabal Hamrin on Diltawa-Tuz Khurmatu road, *Gillett & Rawi* 10263! **FPF**: Pilkhana, 10 km E. of Khanaqin, *Rawi* 12766!; Jabal al Muwaila, 70 km N. of Amara, *Guest, Rawi & Rechinger* 17591!; c. 30 km S.E. of Badra, *Rechinger* 9139! **DLJ**: 6 km above Ana, *Gillett & Rawi* 6949! **DWD**: 18 km S. of Rutba, *Rawi* 14645!; 15 km W. of Rutba, *Rawi & Uthman* 27098!; 17 km S.E. of Haditha, *Al-Kaisi & K. Hamad* 48953 (BAG); 17 km S.E. of K3 (Albu Hayat), *Al-Kaisi & K. Hamad* 53132 (BAG). **DSD**: nr Al Salman, *Springfield & Al-Jaf* 21377! Wadi Al-Ajramiya, 73 km N. of Rutba, *Rawi & Nuri* 27174!; 76 km N.W. of Shubaicha, *Haines, Hadač, Al-Hashimi & Agnew* 5574!; 15 km N. of Aujab to Khider *Al-Mai, Al-Kaisi, K. Hamad & H. Hamid* 48429 (BAG).

Distributed across the Mediterranean area, Palestine, Jordan, Balkans, Hungary, Crimea, eastwards to Afghanistan; introduced in N. America.

10. Erodium trichomanefolium *L'Her.* ex *DC.*, Prodr. [A. P. de Candolle] 1: 645 (1824); Boissier, Fl. Orient. 1: 887 (1867);

Stemless dioecious perennial, often forming cushions, glandular-hairy throughout, with a thick branching rootstock covered with persistent stipules of previous years' leaves. Leaves grey-green, spreading and glandular-hairy, with petiole c. 1.5 × length of lamina; lamina 20–40 × 8–20 mm, oblong-lanceolate in outline, 2- to 3-pinnatisect into linear-oblong lobules, rachis with lobules between the pinnae, and whole surface densely hirsute with viscid glands under the hairs. Umbels solitary, or 2–3 together on a short stem with reduced leaves, 3–5-flowered, peduncle equal to or shorter than pedicels. Sepals 3.5–4 mm, oblong-ovate, obtuse, ± apiculate, 3–5- nerved. Petals obovate, c. 1.5 × length of sepals, pale purple (in sicco) to white. Fruit 26–29 mm, carpel valves hirsute, ovate, with obscure pit and no groove below, beak 21–23 mm.

HAB. Rocky mountain slopes; alt. ± 2500 m; fl. ?Jun.
DISTRIB. Known only from the Helgord massif. **MRO**: Helgord, *Rawi & Serhang* 24747!

Syria, Lebanon, N. Palestine.

11. Erodium cicutarium (*L.*) *L'Her.*, Hort. Kew. [W.Aiton] 2: 414 (1789); Boissier, Fl. Orient. 2: 890 (1867); Vvedenskiy in Fl. S.S.S.R. 14: 71 (1949); Blakelock in Kew Bull. 1948: 408 (1949); Zohary in Dep. Agr. Iraq Bull. 31: 96 (1950); Rawi in Dep. Agr. Iraq Tech. Bull. 14: 58 (1964); Schönbeck-Temesey in Fl. Iranica [K. H. Rechinger] 69: 57 (1970);Y. Nasir in Fl. Pakistan [Nasir & Ali] 149: 36 (1983); Zohary, Fl. Palaest. 2: 236, pl. 337 (1987); Boulos in Fl. Egypt 2: 8 (2000); Abdulridha, Taha & Widad, Ecology & Fl. Basrah: 401 (2016); Taifour & El-Oqlah, Pl. Jordan Annot. Checkl.: 96 (2017).

Geranium cicutarium L., Sp. Pl.: 680 (1753).

Annual, with ascending branches arising from the base, sparsely clothed with spreading hairs, sometimes glandular-pubescent. Lower leaves with petiole as long as or longer than lamina, somewhat shorter in upper leaves; lamina oblong in outline, pinnate with 5–7 pinnae, terminal lobe larger than laterals, pinnae simply or bipinnatisect, with acute lanceolate lobes. Umbels of 4–6 flowers subtended by 5–7 ciliate bracts fused in lower ⅓. Pedicels 3–4 × shorter than peduncles. Sepals 4–5 mm, oblong-ovate, 4–5-nerved, obtuse, mucronate, pubescent with 2–3 erect conical hairs on mucro. Petals slightly zygomorphic, obovate, twice as long as sepals, purple-pink. Stamens equaling sepals; filaments ovate-acuminate, glabrous. Fruit usually (2–)3(–4) cm; carpel valves 3–4 mm, pubescent, with a pair of apical eglandular pits and one concentric groove below each; carpel beak bristly within to base. Fig. 61, 2–9.

HAB. Fine soil, silt or clay in desert depressions, cultivated areas in Mesopotamian plain, and open vegetation of the foothills and mountains, tolerating subsaline conditions, sandy soil, on hillsides between rocks, under *Quercus*; alt. 10–1500 m; fl. Dec.-May.
DISTRIB. Widespread throughout Iraq except for sandy deserts, fully developed forest, and the alpine zone. **MJS**: Shimal, *Mudir of Shimal* 3128!; Jabal Sinjar, *Al-Kaisi, Al-Khayat & F. Karim* 50858 (BAG). **MAM**: Amadiya, *Guest* 1275! **MRO**: Mirowa pass nr Shaqlawa, *Guest* 570!; Salah ad-Din, *Erdtman & Goedemans* (*Rechinger* 15553). **MSU**: Jarmo, *Helbaek* 336!; Dolasur, *Gowan* 2400! Dukan junction, *S. Omar* 42618 (BAG); Sai kanian mountain, *S. Omar* 42634 (BAG). **MSU/FKI**: Kani Hinjer, *S. Omar* 42637 (BAG); 17 km N. of Chuwarta, *Al-Kaisi & A. Haloob* 60594 (BAG). **FUJ**: Jezira S. of Sinjar, *Guest* 13372! **FNI**: Jabal Makhul, *Gillett & Rawi* 7197! **FAR**: Arbil, *Guest* 630! **FKI**: Tuz to Kifri, *Guest* 411!; Kirkuk, *Bornmüller* 995, *Guest* 4337!, *Hadari* 3927! **FPF**: Pilkhana, 10 km N.E. of Khanaqin, *Rawi* 12805! 30 km S.E. of Badra, *Rechinger* 14028; Saadyiya, *Al-Kaisi* 42521 (BAG). **DLJ**: Rawa, *Gillett & Rawi* 6995! **DGA**: 11 km E. of Samarra, *Rechinger* 9562! **DWD**: 20 km W. of Rutba, *Rawi* 14676! **DSD**: nr Zubair, *Gillett & Rawi* 6037! 35 km W. of Rutba, *Rechinger* 9914. **LCA**: Rustam, *Rogers* 066!; Falluja, *Rechinger* 8373; Ali Gharbi, Hor Saniyah area, *Khudairy* 1138! Hafriyah, 60 km S.E. of Baghdad, *Rechinger* 9094! **LSM**: between Amara and Qurna, *Rechinger* 8959! **LBA**: Basra, *Rogers* 0480A!

Europe, temperate Asia to N. W. Himalaya and N. China, N. Africa; introduced in N. & S. America.

12. Erodium moschatum (*L.*) *L'Her.* ex *Aiton*, Hort. Kew. [W. Aiton] 2: 414 (1789); Boissier, Fl. Orient. 1: 891 (1867); Blakelock in Kew Bull. 1948: 409 (1949); Zohary in Dep. Agr. Iraq Bull. 31: 96 (1950); Rawi in Dep. Agr. Iraq Tech. Bull. 14: 58 (1964); Schönbeck-Temesey in Fl. Iranica [K. H. Rechinger] 69: 55 (1970); Zohary, Fl. Palaest. 2: 236, pl. 338

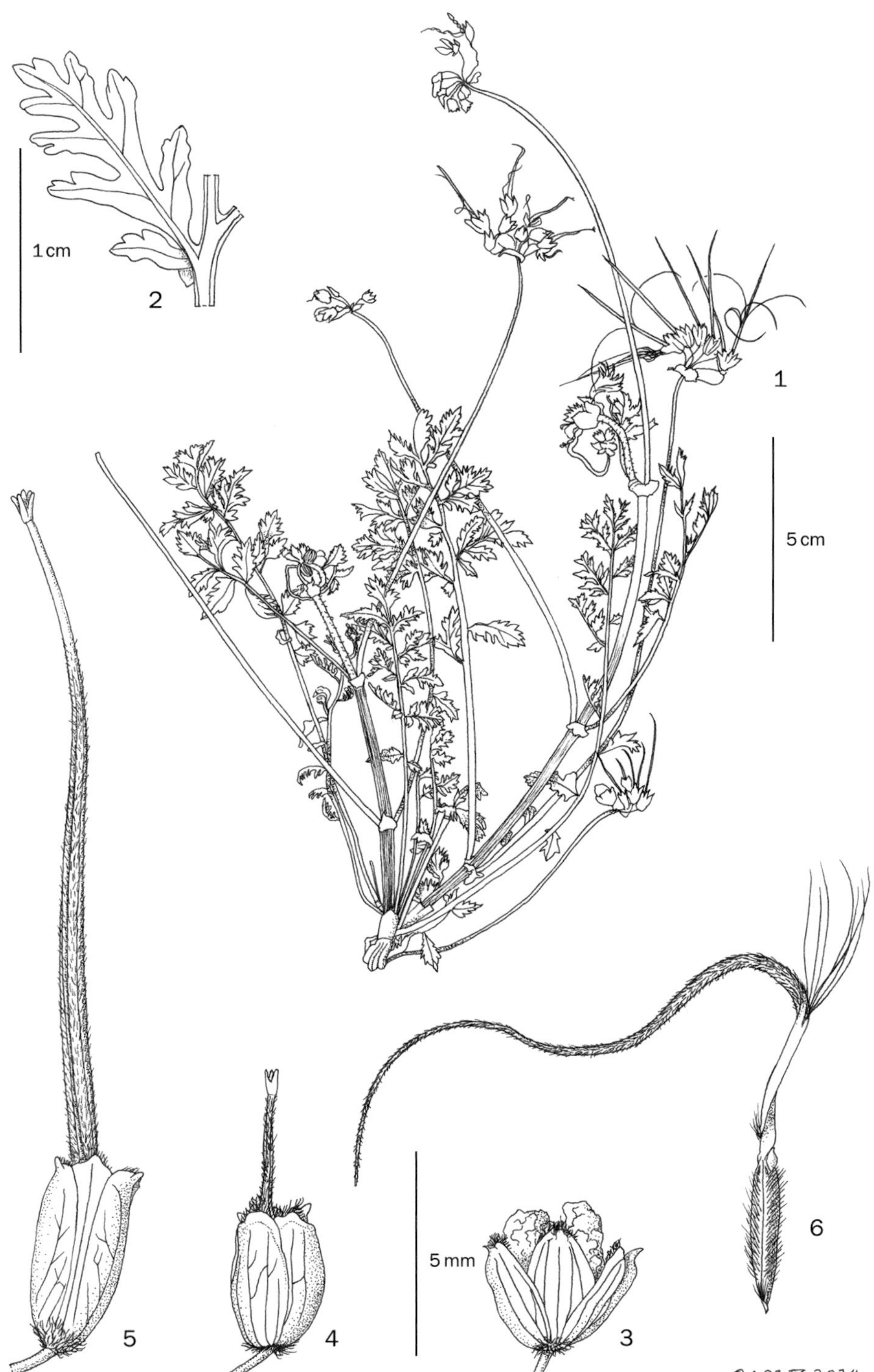

Fig. 66. **Erodium moschatum** subsp. **moschatum**. 1, habit; 2, abaxial surface of typical leaf lobe;, 3, flower, side view; 4, flower after fertilisation showing extending style; 5, flower with mature ovary; 6, fruit. 1–5 from *Al-Kaisi et al.* 48339; 6 from 36244. Drawn by © A.P. Brown, April 2024.

(1987); Collenette, Wild Flow. Saudi Arabia: 334 (photo) (1999); Boulos in Fl. Egypt 2: 9 (2000); Taifour & El-Oqlah, Pl. Jordan Annot. Checkl.: 97 (2017).

> *Geranium moschatum* L., Sp. Pl.: 680 (1753).
> *G. cicutarium* var. *moschatum* L., Sp. Pl. 2: 680 (1753).
> *Erodium robertianum* C.A.Mey. ex Nyman, Consp. Fl. Eur. 1: 140 (1878).

Annual herb with ascending stems, sparsely crisped-pubescent to almost glabrous. Leaves with lamina longer than petiole, oblong in outline, pinnate, pinnae ovate, dentate to bipinnatisect. Umbel of 5–6 flowers, pedicels much shorter than peduncles, the 4–6 bracts obtuse, pubescent, scarious, with or without a median nerve. Sepals 5–5.7 mm, ovate, obtuse, mucronate, reticulately veined, pubescent or glabrous above, with a conspicuous tuft of hairs at base. Petals 1.5 × as long as sepals, obovate,. purple. Stamens as long as sepals, staminodes ovate, half as long. Fruit 3–4.5 cm, valves 4.5–5.5 mm, hirsute, with 2 glandular pits and a glandular furrow below each pit.

subsp. **moschatum**.

Plant ± glabrous except on sepals; pinnae of leaves dentate, or at most lobed, or pinnatisect. Fig. 66, 1–6.

HAB. Wadi bottoms, irrigated fields on sandy or silty soils, exposed places in areas of high rainfall, usually in closed, mesic vegetation; alt. 120–850 m; fl. Mar.-May.
DISTRIB. Foothills and mountain valleys of N.E. Iraq. MSU: Qarachitan, nr Pir Omar Gudrun, *Gillett* 7717! FAR: Koi Sanjaq, *Rawi, Nuri & Kass* 28121!; Arbil, *Guest* 1448! FKI: Chemchemal, *Gillett & Rawi* 7598!; Kirkuk, *Rogers* 0220! FUJ: Balad Sinjar, *Agnew, Hadač & Al-Hashimi* 5292! FPF: Badra, *Gillett & Rawi* 6628! LCA: between Baquba and Shahraban (Muqdadiya), *Rechinger* 8965!

Mediterranean area eastwards to Iran, Canary Isles, Ethiopia, S. Africa, N. & S. America, New Zealand.

subsp. **touchyanum** (Delile ex Godr.) **Ghaz.**, comb. nov.

> *Erodium touchyanum* Delile ex Godr., Mem. Acad. Montp. (Sect. Medic.) i. 423 (1853); Taifour & El-Oqlah, Pl. Jordan Annot. Checkl.: 97 (2017).
> *E. moschatum* subsp. *deserti* (*Eig*) *Eig* in Beih. Bot. Centralbl. 50(2): 236 (1932).
> *E. deserti* Eig in Palestine J. Bot., Jerusalem Ser. 1: 311 (1939); Zohary in Dep. Agr. Iraq Bull. 31: 96 (1950); Burtt & Lewis in Kew Bull. 1954: 401 (1955); Rechinger, Fl. Lowland Iraq 397 (1964).

Plant pubescent with crisped hairs, especially on leaves; leaflets 1–2-pinnatisect.

HAB. Any open vegetation with sufficient rainfall; desert depressions, stony hillsides; alt. 120–380 m; fl. Mar.-Apr.
DISTRIB. Occasional and scattered in lowland Iraq. FKI: Adhaim river gorge through Jabal Hamrin, *Prentice* 5223! FPF: Pilkhana nr Khanaqin, *Rawi* 12806! 60 km N. of Amara, *Rechinger* 14254; DLJ: 4 km E. of Samarra, *Rawi* 20338! DGA: 4 km E. of Samarra, *Rechinger* 13495; DWD: 18 km S. of Rutba, *Rawi* 14622!; K3 in Wadi Hauran, *Gillett & Rawi* 6878! DSD: nr Zubair, *id.* 6066!; 110 km S.E. of Salman, *Guest, Rawi & Rechinger* 18904! Wadi al-Khirr, 15 km W. of Lussuf, *Rechinger* 13566; 40 km W.S.W. of Shabicha, *Rechinger* 9448; 30 km E. of Salman, *Rechinger* 9341!

Canary Islands (Fuertaventura), N. Africa, Jordan, Syria, Arabian Peninsula, Iran.

E. touchyanum has been considered as a distinct species, but the absence of any character other than leaf shape to distinguish it, together with the occurrence of much intermediate material (e.g. *Agnew, Hadač & Al-Hashimi* 5292, above) persuaded me to treat it as a subspecies dependent on ecological conditions. Experimental evidence is required before it can be accepted as a distinct species in Iraq.

3. **MONSONIA** L.

Mant. Pl. 14, 105: (1767)

A. Agnew
Revised by Shahina A. Ghazanfar

Annual or biennial herbs or low shrubs. Leaves opposite or alternate, petiolate, pinnately-lobed or nerved, stipulate. Flowers in umbels or by abortion solitary or in pairs, regular.

Sepals and petals 5. Nectar glands 5, alternating with petals. Stamens 15, all fertile, shortly fused at base into a ring and into 5 bundles of 3 stamens each opposite the petals. Ovary of 5 carpels, beaked; styles and stigmas 5. Fruit schizocarpic, mericarps 1-seeded, separating from axis, beak plumose or with bristles.

A small genus of 40 species centered in S. Africa; a single species in Iraq.

Monsonia commemorates Lady Anne Monson (1726–1776), an English botanist and plant collector who helped James Lee to promote the use of Linnaeus' sexual system of classification. The surname is that of her second husband, Colonel George Monson.

Touloumenidou, T., Bakker, F. T., & Albers, F. (2007). The phylogeny of *Monsonia* L. (Geraniaceae). Pl. Syst. Evol. 264(1–2), 1.

1. **Monsonia nivea** (*Decne.*) *Webb,* Fragm. Fl. Aethiop.-Aegypt. 59 (1854); Boissier, Fl. Orient. 1: 897 (1867); Rechinger, Fl. Lowland Iraq: 397 (1964); Schönbeck-Temesey in Fl. Iranica [K. H. Rechinger] 69: 60 (1970); Zohary, Fl. Palaest. 2: 243 (1971); Ghafoor in Fl. Libya 63: 2 (1978); Y. Nasir in Fl. Pakistan [Nasir & Ali] 149: 40 (1983); Zohary, Fl. Palaest. 2: 244, pl. 350 (1987); Collenette, Wild Flow. Saudi Arabia: 339 (photo) (1999); Boulos, Fl. Egypt 2: 9 (2000); Taifour & El-Oqlah, Pl. Jordan Annot. Checkl.: 98 (2017).

Erodium niveum Decne., Fl. Sinaica: 61 (1834).

Annual or perennial herb or low shrub, all parts canescent. Leaves ovoid-oblong in outline, base rounded or almost cordate, margin obtusely crenate. Umbels of 2–6 flowers. Sepals 5–6 mm, spatulate, mucronulate. Petals scarcely longer than sepals, obovate, entire, red. Mature fruit 4 cm, beak plumose within.

Reported from Iraq from the southern desert without locality.

N. Africa, Arabian Peninsula, Palestine, Jordan, Iran to Pakistan.

4. **PELARGONIUM** L'Her. ex Aiton

Hort. Kew. 2: 417 (1789)

A. Agnew
Revised by Shahina A. Ghazanfar

Perennial (rarely annual) herbs or shrubs, with opposite, rarely alternate, stipulate leaves. Inflorescence umbellate, of f2-many flowers, rarely flowers solitary. Flowers zygomorphic, with a tube leading from the posterior sepal within the pedicel ending in a nectariferous, often swollen, chamber, so that the pedicels often appear jointed. Sepals and petals 5. Stamens 10, filaments obliquely connate at base, not all fertile. Ovary 5-lobed, stigmas 5, free. Fruit beaked, its 1-seeded mericarps separating as in *Erodium* and *Geranium.*

A genus of over 280 species centered in S. Africa with a very wide range of growth forms including succulents and shrubs. The cultivated "geraniums" of Baghdad gardens are species of *Pelargonium.*

Pelargonium is derived from Gr. πελαργος, *pelargos;* its fruits resemble the beak of the stork.

1. **Pelargonium quercetorum** *Agnew* in Kew Bull. 21(2): 227 (1967). Type: Iraq, Kani Mazu Shirin (Kani-mam Sheereen), *Al-Sabry* 859! (K, holo.). Schönbeck-Temesey in Fl. Iranica [K. H. Rechinger] 69: 62 (1970).

Herb with thick perennial rootstock, stems 50–100 cm, with many leaves and old stipules at base, sparsely leafy and 1–2-branched above, sparingly tomentose at least below. Cauline leaves long-petiolate, lamina ± circular in outline, 10–20 cm in diameter, palmatifid divided to halfway or less into 5–7 ovate, very coarsely toothed, obtuse lobes. Stem leaves shorter, upper subsessile, lamina reniform, almost palmatisect. Peduncles 13–17 cm, glabrous. Umbels of 15–30 flowers, subtended by numerous pubescent bracts, on 20–28 mm pedicels. Sepals 13–17 mm, oblong-lanceolate, acuminate, ciliate and pubescent above, usually reddish-tinged; 2 upper petals 25–30 mm, spatulate, undulate at tip, pink or red above,

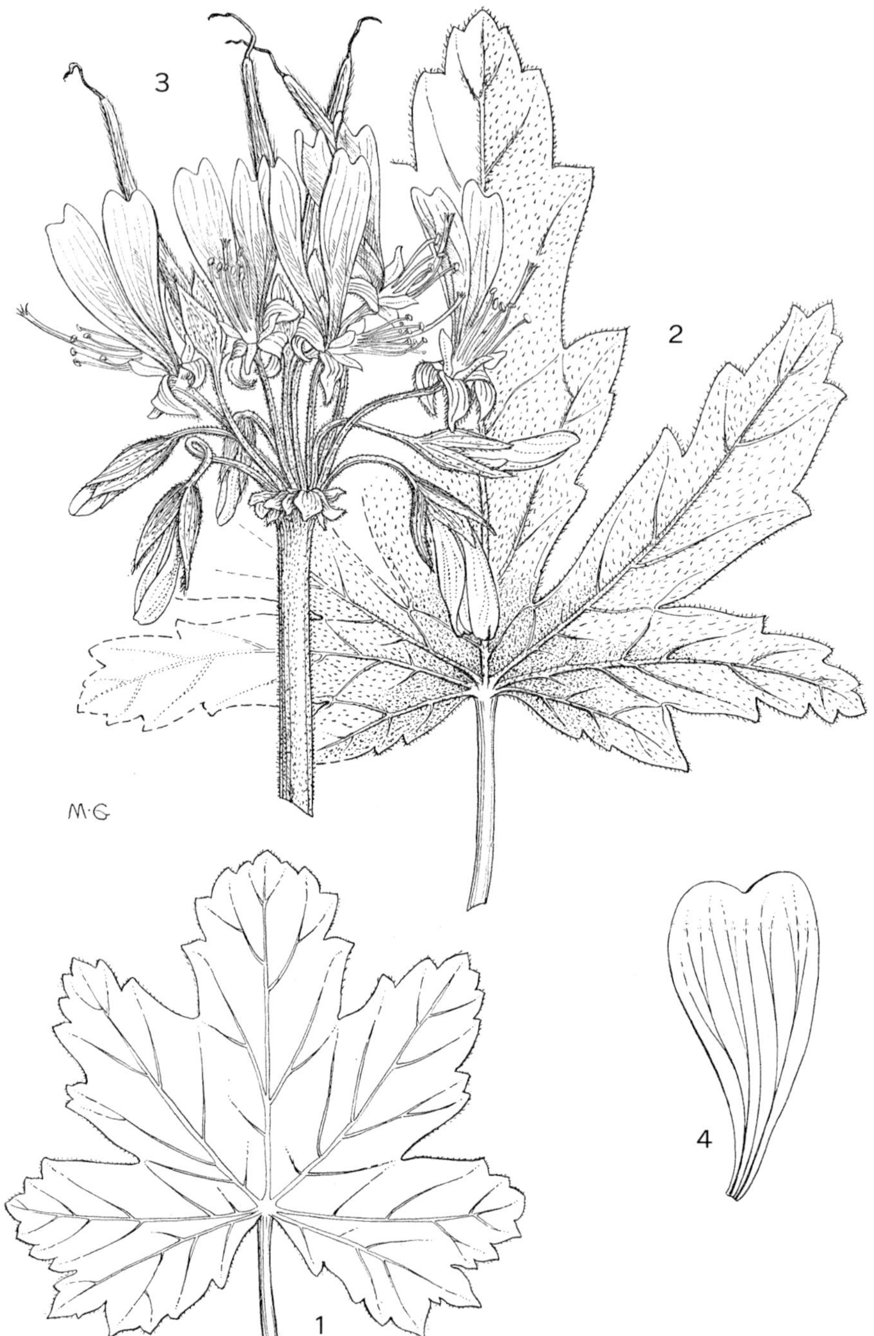

Fig. 67. **Pelargonium quercetorum**. 1, lower leaf, adaxial surface × 1; 2, upper stem leaf, abaxial surface ×1; 3, inflorescence ×1; 4, upper petal × 1.3. Reproduced from Kew Bull. 21: f. 1, p. 226 (1967).

paler with purple nerves and spots within below; lower 3 petals reduced or absent. Fertile stamens 7, with 3 short sterile filaments, all fused at expanded bases into a short tube, filiform, glabrous. Fruit 4–5.7 cm, carpel valves 8–12 mm, oblong, hirsute. Beak of fruit sparsely hairy and glandular-pubescent outside, long-bristly within except for the upper ⅓ which is sericeous. Seeds 6–7 mm. Fig. 67, 1–4.

HAB. Rich soil of *Quercus* forests; alt. ±1300 m; fl. May-Jun.
DISTRIB. A rare species of the upper forest zone of N.E. Iraq. **MAM**: Sarsang, N. slopes of Qara Dagh, *Chapman* 26374! (E); *Rechinger* 11904! **MRO**: mt. Potine (Botin) N. of Shirwan Mazin, *Agnew, Hadač, Haines & Al-Sabry* 6215A! (E); Kani Mazu Shirin [Kani-mam Sheereen], *Al-Sabry* 859! (type).

Turkey (S.E. Anatolia). Reported by *Davis* 45381(from Şine Dere and Çukurca, in the Hakkari province of Turkey), to be "very fragrant" and "with a strong distinctive scent". Related to *P. endlicherianum* (native to C. & E. Anatolia) with which it has been hybridized in cultivation, the hybrid being named as *Pelargonium* 'Kavushan'; it was produced by Richard Reidy in Los Lunas, New Mexico.

123. **OXALIDACEAE** R.Br.

Narr. Exped. Zaire 433 (1818), nom. cons.
Cocucci in Kubitzki (ed.), Fam. Gen. Vasc. Pl. 6: 285 (2004)

J. R. Edmondson

Perennial or annual herbs. Leaves alternate or whorled, pinnate or palmate (Iraq), petiolate, exstipulate; leaflets sessile, often folded together at night, margins entire. Inflorescence cymose or umbellate, or flowers solitary; bracteate. Flowers bisexual, regular, 5-merous. Sepals 5, free or fused below. Petals 5, free or fused above base, convolute. Stamens 10, in 2 whorls of 5, the outer whorl usually with shorter filaments. Ovary superior, 5-locular; styles 5; stigmas capitate or obscurely 2-fid. Fruit a loculicidal capsule (in Iraq plants). Seeds provided with a basal aril, seeds ejected explosively from capsule.

Six to eight genera with about 780 species mostly in the tropics and subtropics of the world.

1. **OXALIS** *L.*

Sp. Pl. 1: 433 (1753); Gen. Pl. ed. 5, 198 (1754).

Annual or perennial herbs; roots usually with tubers, bulbs or rhizomes. Stems creeping or absent, apart from stalk bearing inflorescence. Leaves 3-foliolate, ± obcordate, notched at apex, sessile. Inflorescence with a long peduncle; bracts 2, small. Flowers pink or yellow, with clawed petals. Styles erect or curved. Fruit explosively dehiscent. Seeds flattened; aril translucent.

About 700 species in the tropics and subtropics of both hemispheres, extending into temperate regions; 3 species in Iraq of which 2 are escapes from cultivation and spread as weeds of cultivated areas.

Oxalis, from Gr. οξινης, *oxines,* sharp (a reference to its acidic juice).

1. Flowers violet or pink . . . . 2. *O. articulata*
   Flowers yellow 2
2. Plants lacking bulbils. . . . . 1. *O. corniculata*
   Plants arising from bulbils. . . . . .3. *O pes-caprae*

1. **Oxalis corniculata** *L.,* Sp. Pl.1: 435 (1753); Dinsmore in Post, Fl. Syria, Palest. & Sinai ed. 2, 2: 253 (1933); Rawi in Dep. Agr. Iraq Tech. Bull. 14: 59 (1964); Rechinger, Fl. Lowland Iraq: 398 (1964); Cullen in Fl. Turkey [P. H. Davis] 2: 488 (1967); Rechinger in Fl. Iranica 40: 3 (1967); Y. Nasir in Fl. Pakistan (Nasir & Ali) 4: (1971); Zohary, Fl. Palaest. 2: 225, pl. 321 (1978); Collenette, Wild Flow. Saudi Arabia: 591 (1999); Boulos in Fl. Egypt 2: 375 (2000); Abdulridha, Taha & Widad, Ecology & Fl. Basrah: 449 (2016); Taifour & El-Oqlah, Pl. Jordan Annot. Checklist 115 (2017).

O. corniculata L. var. *villosa* (Bieb.) Hohen., Bull. Soc. Imp. Naturalistes Moscou 11(4): 395 (1838).

Annual, 5–40 cm, lacking bulbils. Stems procumbent or ascending, usually rooting below, clad with long spreading hairs. Petioles 1–8 cm; leaflets 5–15 × 6–18 mm, lateral leaflets smaller than terminal, obcordate, base cuneate, apex obtuse with emarginate apex, pilose

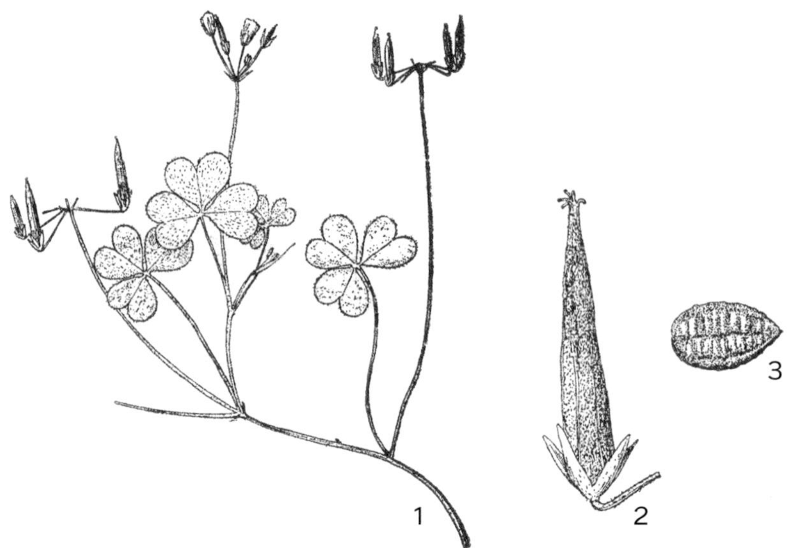

Fig. 68. **Oxalis corniculata**. 1, habit × 1; 2, fruit × 2 ; 3, seed × 8. Reproduced with permission from Fl. Pakistan 4: f. 2, A-C (1971). Drawn by S.Y. Salim. © National Herbarium, Pakistan Agriculture Research Council & University of Karachi, Pakistan.

on margins and undersurface. Flowers yellow, 1–6 in pseudo-umbels; peduncles axillary; bracteoles linear-lanceolate. Pedicels c. 1 cm, appressed hairy or ± glabrous, deflexed. Sepals lanceolate, acute. Petals yellow, clawed, 8–10 mm. Capsule oblong-cylindrical, 12–20 mm, 5-angled, pubescent. Fig. 68, 1–3.

HAB. Cultivated areas, date orchards, small gardens, clay soil, by irrigation canals; alt. 10–770 m; fl. Nov.-May.
DISTRIB. Mainly found around Baghdad and in irrigated areas of the Mesopotamian plain. **MSU**: Zalum nr Khormal, *Rahim Husham & Nuri* 29494! **LCA**: Baghdad, *Omar & Sahira* 34846! *Barkley & Brahim* 1324! *Lazar* 480! *Graham* s.n. (24.3.1920)! *Schlafli* s.n.!; *S.Omar* 47527; Abu Ghraib, *Noori* 39910!; *Sahira* 35833; Diyala district, *Rechinger* 9746; Zafaraniya, *Guest & Rawi* 15330!; Karrada, *Guest* 1114! *Rogers* 0114!; Diltawa, *Gillett & Rawi* 10916. **LEA**: Aziziya, *Rawi & Alizzi* 32390!; Ba'quba, Raaya, *Alizzi & Hazim* 33694! **LSM**: Amara, *Nábelek* 4454! **FPF**: Badra, *Kaisi & Yahya* 45285!; *Al-.Kasisi* 53203! **DWD**: Ana, *Al-Khayat & K. Hamad* 51772; *S. Omar,Al-Khayat & Adel* 55447; *Rawi & Gillett* 7031.

KARAM MARTAB, HAMEDH. The leaves and stems are sour to taste. The plant has been used medicinally.

Native to from Mexico to Venezuela and Peru and the Caribbean; introduced and naturalised almost throughout the tropics, subtropics of the world.

2. **Oxalis articulata** *Savigny*, Encycl. [J. Lamarck & al.] 4(2): 686 (1798); Ghahremaninejad & Gholamian, Iran. Journ. Bot. 12(1): 55 (2006).

Perennial herb, bulbiferous at base. Leaves basal, with petiole to 30 cm; leaflets 12–18 mm. Inflorescence a pseudumbel, with up to 10 flowers; peduncle longer than petioles. Sepals to 5 mm, broadly lanceolate, orange-tipped. Petals purple-pink to violet. Fruit to 8 mm, ovoid.

HAB. Cultivated areas; alt. 30–50 m; fl. Nov.
DISTRIB. Recorded only twice from cultivation. **LCA**: Baghdad, *Sahira* C633! *Jenan* C409!; Zaffaraniya, *Al-Khayat, Jenan, Sahira & Wedad* C1074.

Native to Argentina, Brazil and Uruguay, introduced and naturalised in N. Africa, Europe, Turkey, Iran; USA, Japan, Korea; sometimes grown as an ornamental plant, but now a widespread agricultural weed.

3. **Oxalis pes-caprae** *L.*, Sp. Pl. 1: 434 (1753); Zohary, Fl. Palaest. 2: 225, pl. 320 (1978); Boulos in Fl. Egypt 2: 375 (2000); Taifour & El-Oqlah, Pl. Jordan Annot. Checkl.: 115 (2017).

Plant reproducing mainly by bulbils. Flowers yellow, heterostylous, self-incompatible (only one form is normally found in cultivation, so plants do not set seed). Petals slightly recurved at anthesis.

HAB. Riverside; alt. not recorded; fl. Mar.

DISTRIB. Recorded only once from the Mesopotamian alluvial plain. **LCA**: Hinaidi, nr Baghdad, *Lazar* s.n. (6.3.1936)!

Native of Cape Provinces and Namibia, widely introduced in the Mediterranean region, N. Africa, S.E. China and N. & S. America.

124. **CUSCUTACEAE** Dumort

Anal. Fam. Pl. 20 (& 25) (1829), nom. cons.

C. C. Townsend
Revised by Lytton J. Musselman & A. Kandemir

Parasitic annuals, very rarely perennials. Roots withering and absent in mature plants. Stem yellow or reddish, rarely greenish, glabrous, leafless, filiform, climbing or trailing. [Species in subgenus Grammica with searching and coiling stems, those in subgenera Monogyna and Cuscuta with only coiling stems]. Leaves reduced to minute scales. Flowers hermaphrodite, actinomorphic, usually clustered in axillary capitate heads. Calyx and corolla each united into a tube in lower part, usually 4 or 5 lobed, white or reddish. Stamens as many as corolla lobes, inserted above scales. Ovary superior, 2-locular; ovules 2 per locule; styles entire or partially to completely divided; stigmas elongate, capitate or ovoid. Fruit a capsule, dehiscing irregularly or circumscissile near base.

Monotypic; about 170 species, cosmopolitan, and in a wide range of habitats; 14 species in Iraq.

CUSCUTA L.

[Tournefort, Inst. Rei Herb. 1: 652. pl. 422 (1700)]; Sp. Pl. 1: 124 (1753); Gen. Plant. 5: 60 (1754); Choisy, Mém. Soc. Phys. & Hist. Nat. Genève 9: 268 (1841).

Description same as that of the family.

Cuscuta is subdived into 3 subgenera, Grammica, Monogyna and Cuscuta based on the number of styles and morphology of stigma.

Cuscuta, from Lat. *cuscuta*, a plant that hugs.

Dawson, J. H., Musselman, L. J., Wolswinkel, P. & Dörr, I. (1994). Biology and control of *Cuscuta*. Reviews of Weed Science 6: 265–317.

Stefanović, S., Kuzmina, M. & Costea, M. (2007). Delimitation of major lineages within *Cuscuta* subgenus *Grammica* (Convolvulaceae) using plastid and nuclear DNA sequences. American Journal of Botany 94: 568–589.

Yuncker, T.G. (1932). The genus *Cuscuta*. Memoirs Torrey Bot. Club 18: 113–331.

1. Style 1 (Subgenus *Monogyna*) 2
   Styles 2 3
2. Calyx about enclosing corolla; scales scarcely reaching anthers . . . . . 13. *C. monoygyna*
   Calyx mostly reaching up to middle of corolla tube; scales covering
     anther base . . . . 14. *C. lehmanniana*
3. Stigma capitate (Subgenus *Grammica*) 4
   Stigma elongate (Subgenus *Cuscuta*) 5
4. Calyx lobes ovate to orbicular; scales exserted between corolla lobes;
     capsule irregularly opening . . . . 1. *C. campestris*

Calyx lobes triangular-ovate; scales not exserted; capsule circumscissile. . 2. *C. chinensis*
5. Flowers distinctly pedicellate 6
     Flowers sessile or subsessile. 8
6. Flowers mostly 4-merous . . . . 5. *C. pedicellata*
     Flowers 5-merous 7
7. Calyx truncated with very short and broad obtuse lobes . . . . 6. *C. babylonica*
     Calyx not truncate; lobes distinct, triangular-ovate . . . . 4. *C. pulchella*
8. Flowers mostly 5-merous 9
     Flowers mostly 4-merous 12
9. Style + stigma shorter than ovary 10
     Style + stigma as long as or longer than ovary 11
10. Corolla lobes shorter than tube; capsule regularly circumscissile . . . . . . 12. *C. brevistyla*
     Corolla lobes ± equal or longer than tube; capsule irregularly
        opening . . . . 3. *C. kotschyana* var. *caudata*
11. Flowers 2–3 mm long; calyx deeply divided. . . . . 11. *C. planiflora*
     Flowers 3–4 mm; calyx turbinate, not deeply divided . . . . 10. *C. approximata*
12. Calyx lobes obtuse . . . . 7. *C. europaea*
     Calyx lobes acute to acuminate. 13
13. Calyx exceeding corolla tube, lobes membranous; capsule with 4
        longitudinal furrows . . . . 8. *C. kurdica*
     Calyx about enclosing corolla tube, lobes thickened or fleshy; capsule
        not with 4 longitudinal furrows. . . . . 9. *C. palaestina*

1. **Cuscuta campestris** *Yuncker* in Mem. Torrey Bot. Club. 18: 138, f. 14 (1932); Meeuse in Bothalia 6: 648 (1957); Rechinger, Fl. Lowland Iraq 489 (1964); Feinbrun in Israel Journ. Bot. 19: 20 (1970); Plitmann in Fl. Turkey [P. H. Davis] 6: 225 (1978); Meikle, Fl. Cyprus 2: 1178 (1985); Gonçalves in Fl. Zamb. 8. 1: 131 (1987); Rajput & Tahir in Fl. Pakistan [Nasir & Ali] 189: 21 (1988); Rhui-cheng & Staples in Fl. China 16: 323 (1995); Boulos in Fl. Egypt 2: 265 (2000).

  C. arvensis auct.: Fiori & Paol, Fl. Anal. Ital. 2: 929 (1928), non Beyrich, nom. illegit.
  C. pentagona Rugel is sometimes considered synonymous with this.
  Grammica campestris (Yuncker) Hadać & Chrtek in Folia Geobot. Phytotax. Bohem. 5: 445 (1970).

Stems slender, yellowish. Flowers 2–3 mm, smooth, 5-merous in compact, globose clusters. Pedicels mostly shorter than flowers. Calyx broadly campanulate, enclosing corolla-tube; lobes ovate to orbiculate, obtuse to acute, ± overlapping when young. Corolla lobes triangular, acute, tips patent to usually reflexed, shorter than or as long as campanulate tube. Stamens slightly shorter than corolla lobes; filaments longer than or equalling anthers. Scales ovate, elliptic or obovate, fimbriate, slightly exserted between corolla lobes. Ovary globose; styles 2, suberect to divaricate; stigmas capitate. Capsule depressed-globose, ± 2.5 mm long, dehiscing irregularly.

HAB. Parasitic on many herbs and woody species in most places. alt. s.l.– 1500 m; fl. Jun.-Oct.
DISTRIB. Lower hills of N. Iraq and the central alluvial plains. **FUJ**: Mosul, *Omar & Kazim* 36969! & 36970! **FNI**: Eski Kellek, *Nuri* 30762!; *Mamluk* OM9! & OM10! **LEA**: Aziziya, *Rawi & Alizzi* 32389!; Hor Suwaicha, *Mamluk* OM 33! & OM 26! **LCA**: Gharraf, Kut al-Imara, *Kazim & Hamid* 37808!; Abu Ghuraib, *Janan* 38159! & 38160!; Hurriya, *Haines* W802!; Kuwarish, *Mamluk* OM24!

Native to North America, widely introduced and naturalized in many regions of the Old World.

A serious threat in cultivated fields especially of Medicago sativa (*berseem*).

2. **Cuscuta chinensis** *Lam.*, Encycl. Meth. Bot. 2: 229 (1786); Rawi in Dep. Agr. Iraq Tech. Bull. 14: 134 (1964); Rechinger, Fl. Lowland Iraq: 489 (1964); Yuncker in Mem. Torrey Bot. Club 18: 209, f. 80 A-G (1932); Rajput & Tahir in Fl. Pakistan [Nasir & Ali] 189: 20 (1988); Rhui-cheng & Staples in Fl. China 16: 323 (1995); Boulos in Fl. Egypt 2: 265 (2000); Karim & Fawzi, Fl. U.A.E. 2: 137 (2007).

  C. ciliaris auct.: Hohen. in Boiss., Diagn. Pl. Orient. ser. 3, 3: 129 (1856), non Kotschy (1845).
  Grammica chinensis (Lam.) Hadać & Chrtek in Folia Geobot. Phytotax. Bohem. 5: 445 (1970).

Stems slender, yellow. Flowers 2–3.5 mm, 5-merous, sessile to shortly pedicellate in lateral dense clusters. Calyx cupular, about reaching corolla lobes; lobes triangular-ovate, apex acute to obtuse, partly thickened, slightly overlapping. Corolla subglobular; lobes triangular-ovate, or oblong-ovate, apex acute or obtuse. The acute corolla lobe is usually a defining character, often carinate. Stamens inserted at throat, shorter than lobes; filaments longer than anthers. Scale oblong, reaching stamens, long fimbriate. Ovary subglobose; styles 2; stigmas globose. Capsule depressed-globose, circumscissile. Fig. 69, 4–7.

HAB. Parasitic usually on members of the Fabaceae and Asteraceae; alt. 100–2500 m; fl. Jun.-Oct. DISTRIB. Lower hills of N. Iraq. **FUJ**: Mosul, *Kotschy* 431!

Afghanistan, Indonesia, China, Japan, Kazakhstan, Korea, Russia, Sri Lanka, Africa, S.W. Asia & Australia.

3. **Cuscuta kotschyana** *Boiss.*, Diagn. Pl. Orient. ser. 1, 7: 29 (1846); Zohary, Fl. Iraq & its Phytogeo. Subdivis. 118 (1946); Rawi in Dep. Agr. Iraq Tech. Bull. 14: 134 (1964); Yunker & Rechinger in Fl. Iranica 8: 8 (1964); Karim, Flow. Parasitic Pl. of Iraq 22 (1978).

C. persica Decne. in Engelm. Trans. Acad. Sci. St. Louis 1: 470 (1859).
C. stapfiana Pabil. in Bot. Zhurn. 3: 25 (1995).

var. **caudata** *Bornm. & Schwarz* in Feddes Rep. Beih. 26: 57 (1924); Yuncker & Rechinger in Fl. Iranica 8: 7 (1964); Karim, Flow. Parasitic Pl. of Iraq 22 (1978); Plitmann in Fl. Turkey [P. H. Davis] 6: 226 (1978) as "subsp. *caudata* Bornm. & Schwarz".

Stems thick, not slender. Flowers 3.5–4 mm, sessile to shortly pedicellate in compact, glomerulate clusters, somewhat fleshly and often papillate. Calyx more or less loose about the corolla which it encloses. Perianth lobes long, attenuated-caudate, equal to or longer than the tube. Ovary and fruit conical-globose. (Description from Fl. Iranica.)

HAB. Parasitic mainly on *Peganum harmala*, *Phlomis* and *Marrubium*; alt. ±1000m; fl. not given on label DISTRIB. In the Arbil District of Iraq. **FAR**: Arbil, Mt Quandil supra lactum Goam-e Kirmosoran, *Rechinger* 11122 (W).

Turkey, Iraq, Iran and Afghanistan.

The species was recorded from Erbil in Flora Iranica. It has also been recorded from Sulaimaniya Region by Zohary (op. cit.), but no specimens are present at K.

4. **Cuscuta pulchella** *Engelm.*, Trans. Acad. Sci. St. Louis 1: 472 (1859); Yunker in Mem. Torrey Bot. Club 18: 271, f. 140 (1932); Yunker & Rechinger in Fl. Iranica 8: 8 (1964); Rajput & Tahir in Fl. Pakistan [Nasir & Ali] 189: 19 (1988).

C. pulchella Engelm. var. *afghana* Engelm. in Trans. Acad. Sci. St. Louis. 1: 472 (1859).

Stems very slender, violet to dark pinkish. Flowers reddish to dark pinkish, sometimes yellowish, 2.5–3 mm long, 5-merous, with 1.5–3 mm long pedicels in umbellate-cymose clusters, sometimes solitary. Calyx about reaching corolla lobes; lobes triangular-ovate. Corolla lobes ovate, acute, tip mostly glandular, shorter than tube. Stamens shorter than tubes. Scales reaching stamens, oblong, slightly fringed. Ovary globose; styles very short, 2; stigmas elongated, cylindrical, shorter than ovary. Capsule globular, opening intrastylar. Seeds mostly 2.

HAB. Parasitic on *Artemisia*, *Acantholimon* and *Alhagi*; alt. ± 2000 m; fl. Aug. DISTRIB. Kurd., S. Maydan, *Gysel* s.n.

Afghanistan, Pakistan.

This species is recorded from Iraq in Flora Iranica. The description was written by using the type specimen in K (*Griffith* 690! described from Afghanistan). No specimens at K.

5. **Cuscuta pedicellata** *Ledeb.*, Fl. Altaic. [Ledebour] 1: 293, t. 234 (1829); Yuncker in Mem. Torrey Bot. Club 18: 271, f. 141 (1932); Rechinger, Fl. Lowland Iraq: 490 (1964); Feinbrun in Israel Journ. Bot. 19: 21 (1970); Migahid & Hammouda, Fl. Saudi Arabia

232 (1974); Plitmann in Fl. Turkey [P. H. Davis] 6: 226 (1978); Boulos, Fl. Egypt 2: 267 (2000).

> *C. arabica* Fresen., Mus. Senckenberg. 1: 165 (1834); Wight, Icon. Pl. Ind. Orient. [Wight] t. 1371 (1840).
> *C. laxiflora* auct.: Aznav. in Magyar Bot. Lapok 4: 135 (1905), non Benth., Bot. Voy. Sulphur.: 138 (1845).

Stem slender, yellowish. Flowers 2–3 mm long, pedicellate, rarely some subsessile, mostly 4-merous, 4–13 flowers in umbellate clusters 4–11 mm in diameter. Calyx rotate or campanulate, about reaching corolla tube; lobes as long as tube, triangular-ovate, acute, not overlapping. Corolla globular-campanulate, clearly exceeding calyx; lobes as long as tube, ovate or triangular, obtuse or acute. Stamens shorter than or as long as corolla lobes. Scales oblong, reaching filaments, dentate to fimbriate. Styles 2, obsolete; stigmas, shorter than ovary. Capsule globose-depressed, dehiscing irregularly.

HAB. Parasitic on mainly ephemeral plants, sometimes on herbs or small shrubs; alt. ± 2285 m; fl. Apr.-Aug.
DISTRIB. Mountains of N. Iraq; also found in the southern district. **MAM**: Bikhair Mountain nr. Zakho, *Rawi* 23024 A! **DSD**: 42 km S. E. of Salman, *Guest, Rawi & Gilbert* 14116!

Arabia, N. Africa, S.W. & C. Asia to West Siberia.

6. **Cuscuta babylonica** *Aucher* ex *Choisy* in Mèm. Soc. Phys. Hist. Nat. Genève 9: 270, t. 1, f. 1 (1841). –Type: Baghdad, *Aucher* 1420 (MO, lecto.; K!, P, isolecto.); Yuncker in Mem. Torrey Bot. Club 18: 273, f. 142 A-E (1932); Rawi in Dep. Agr. Iraq Tech. Bull. 14: 134 (1964); Yunker & Rechinger in Fl. Iranica 8: 8 (1964); Rechinger, Fl. Lowland Iraq 490 (1964); Plitmann in Fl. Turkey [P. H. Davis] 6: 227 (1978).

Stem slender, yellowish to reddish. Flowers 2–3 mm on pedicels as long as or slightly longer than flowers, 5-merous, smooth or papillate in few- to many flowered, densely or loosely cymose-umbellate clusters. Calyx hemispherical or cup-shaped, shorter than corolla tube, truncated with very short and broad obtuse lobes. Corolla campanulate; lobes shorter than or as long as tube, ovate-triangular or ovate-oblong, obtuse to acute, overlapping at base. Filaments shorter than or as long as anthers. Scales reaching filaments, obovate or oblong lanceolate, entire or shallowly dentate at top. Styles 2; stigmas as long as styles. Capsule globose, irregularly dehiscing below. Seeds 1–4 per capsule.

> Calyx shorter than corolla tube; flowers smooth or sparsely papillate. . . . var. *babylonica*
> Calyx equaling corolla tube; flowers densely papillate . . . . var. *elegans*

var. **babylonica**

> *C. viticis* Hand.-Mazz. in Ann. Naturh. Mus. Wien 27: 394 (1913).

HAB. Parasitic on perennial herbs and low shrubs; alt. 700–110 m; fl. June-Dec.
DISTRIB. In the mountains, lower hills and the central alluvial plains of Iraq. **MAM**: Khair Mountain nr. Zakho, *Rawi* 23036!; Zakho, nr. Turkish frontier, *Rechinger* 10727! **MRO**: between Erbil and Rowanduz, *Bornmüller* 1536!; Shaikan, *Rawi & Serhang* 24887! **FNI**: Eski Kellek, *Gillet* 8197! **LCA**: nr. Baghdad, *Haines* s.n.! Mesopotamia, *Graham* 113!

C. Asia, S. Iran, Turkey, Palestine, Syria.

var. **elegans** (*Boiss. & Balansa*) *Engelm.* in Trans. Acad. Sci. St. Louis 1: 461 (1859); Yuncker in Mem. Torrey Bot. Club 18: 274, f. 142, F (1932); Plitmann in Fl. Turkey [P. H. Davis] 6: 227 (1978).

> *C. elegans* Boiss. & Bal. in Boissier, Diagn. Pl. Or. ser. 2, 3: 129 (1856).

HAB. Parasitic on perennial herbs and low shrubs; alt. 700–1700 m; fl. Jun.-Dec.
DISTRIB. In the mountains and lower hills of N. Iraq. **MAM**: 10 km N. of Zakho, *Rawi* 23161! **MRO**: Handarin Mt. Rowanduz, *Bornmüller* 1538!; Gali Ali Beg, *Gillet* 9429!; Ser-i Hassan Beg, *Guest* 3043! Kawriesh E. side of Karoukh mount., *Kass, Nuri & Serheng* 27680 (BAG). **FUJ**: Mosul, *Kotschy* 388! **FNI**: 85 km N.W. of Mosul, *Rawi* 23137!

E. Turkey (Anatolia), W. Iran

7. **Cuscuta europaea** *L.*, Sp. Pl.: 124 (1753); Yuncker in Mem. Torrey Bot. Club 18: 274, f. 143 (1932); Yuncker & Rechinger in Fl. Iranica, 8: 10 (1964); Feinbrun in Israel Journ. Bot. 19: 21 (1970); Plitmann in Fl. Turkey [P. H. Davis] 6: 228 (1978); Rajput & Tahir in Fl. Pakistan [Nasir & Ali] 189: 15 (1988); Rhui-cheng & Staples in Fl. China 16: 325 (1995).

C. tetrandra Moench, Meth. 461 (1794).
C. vulgaris Persoon, Syn. Pl. 1: 289 (1805).
C. viciae Koch & al. in Allg. Thüring. Gartenzeitung 4: 118 (1845).
C. brachystyla C.Koch in Linnaea 22: 747 (1849).

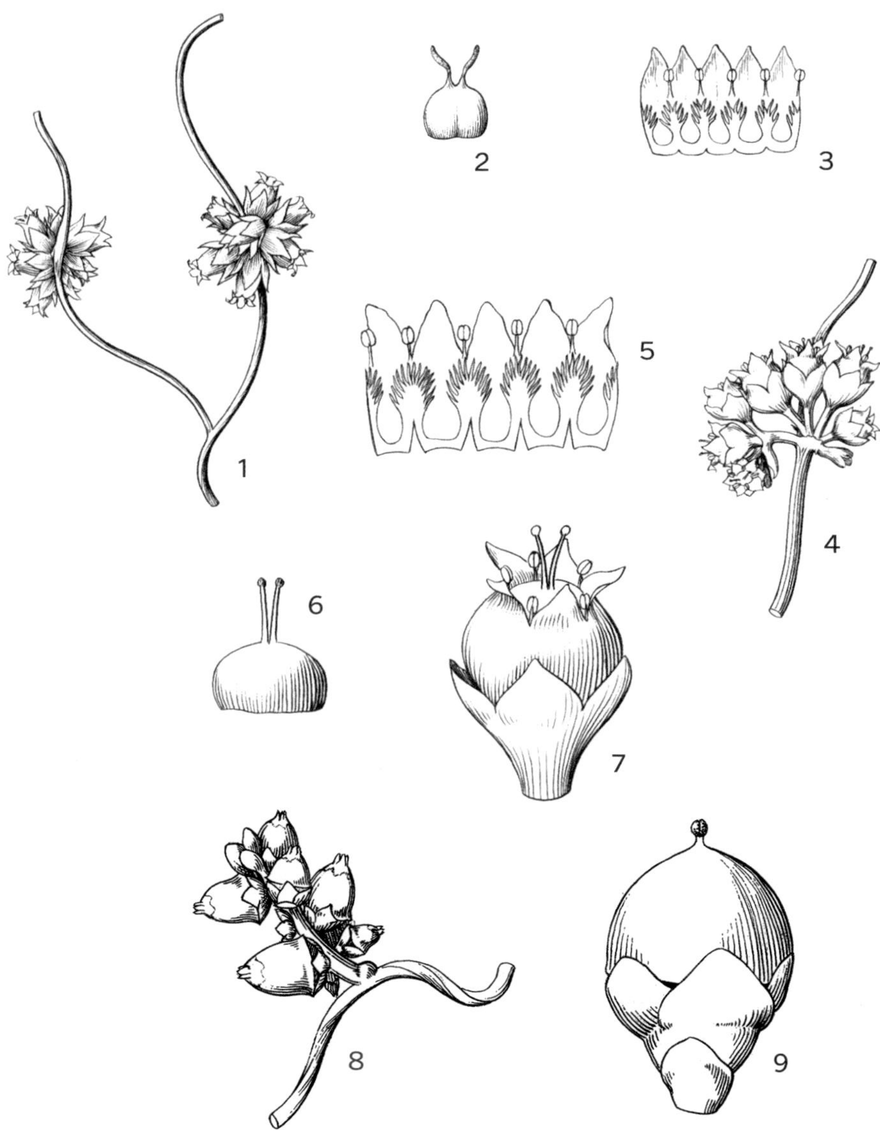

Fig. 69. **Cuscuta europaea**. 1, inflorescence; 2, pistil; 3, opened corolla showing stamens 1and scales. **Cuscuta chinensis**. 4, inflorescence; 5, opened corolla showing stamens and scales; 6, pistil; 7, fruit with calyx, corolla and stamens. **Cuscuta monogyna**. 8. inflorescence; 9, fruit with calyx. (1-9 enlarged, showing detail). Reproduced from Flora of China 16: f. 317 & 318 (1995), with permission from Missouri Botanical Garden Press, St. Louis, and Science Press, Beijing. Drawn by Li Xichou.

Stem slender, reddish or yellowish. Flowers mostly 4-merous, 2.5–3 mm, sessile to shortly pedicellate, in globose clusters 6–13 mm in diameter. Calyx as long as corolla tube; triangular-ovate, obtuse at apex. Corolla campanulate, becoming urceolate in fruit; lobes about as long as tube, ovate, triangular, obtuse or subacute. Stamens included; anthers exceeding corolla tubes. Scales shorter or reaching filaments, commonly bifid, sometimes entire, mostly sparingly fringed towards top. Styles 2, shorter than ovary. Capsule 2.5–3 mm long, ovoid. Seeds 3–4 mm. Fig. 69, 1–3.

HAB. Parasitic on members of the Fabaceae, *Pedicularis* and *Phlomis* spp; alt. ± 3000 m; fl. Jul.-Aug.
DISTRIB. Mountains and lower hills of N. Iraq. **MRO**: Algurd Dag, *Haley* 146!

Europe, N. Africa, Turkey to C. Asia, China; occasionally in N. and S. America.

8. **Cuscuta kurdica** *Engelm.* in Trans. Acad. Sci. St. Louis 1: 470 (1859). Type: Iraq, Kurdistan, Gara mt., *Kotschy* 388b (BM, K!, G, GOET, NY, syn.). Yuncker in Mem. Torrey Bot. Club 18: 278, f. 146 (1932); Rawi in Dep. Agr. Iraq Tech. Bull. 14: 134 (1964); Plitmann in Fl. Turkey [P. H. Davis] 6: 229 (1978).

Stems slender, reddish. Flowers 2–3 mm, mostly 4-merous, some flowers 5-merous, sessile, in dense clusters 3–8 mm in diameter. Calyx longer than corolla tubes; lobes, ovate-lanceolate, acute, longer than tube. Corolla lobes as long as tube, ovate-triangular, acute. Stamens shorter than corolla lobes; anthers as long as filaments. Scales hardly reaching filaments or mostly shorter, oblong, bifid, sparingly fringed above. Styles 2, shorter than ovary, equaling to a little shorter than stigmas. Capsule depressed-globose, with 4 longitudinal furrows. Seeds mostly 4.

HAB. Parasitic on perennial herbs; alt. 1000–3000 m; fl. Jun.-Dec.
DISTRIB. **MAM/MRO**: Kurdistan, *Brant* (K!, syn.). **MRO**: Chiya-I Mandau, *Guest* 2688!; *Guest* 2697!; Halgurd, *Rawi & Serhang* 24293! **MJS**: Jabal Sinjar S.E. slope, *Al-Khayat & Wedad* 53362 (BAG!).

Turkey & W. Iran.

9. **Cuscuta palaestina** *Boiss.*, Diagn. Pl. Orient. ser. 1, 11: 86 (1849); Yuncker in Mem. Torrey Bot. Club 18: 279, f. 147 (1932); Rawi in Dep. Agr. Iraq Tech. Bull. 14: 134 (1964); Rechinger, Fl. Lowland Iraq: 491 (1964); Feinbrun in Israel Journ. Bot. 19: 22 (1970); Plitmann in Fl. Turkey [P. H. Davis] 6: 229 (1978); Meikle, Fl. Cyprus 2: 1178 (1985); Boulos in Fl. Egypt 2: 267 (2000).

C. capillaries Rchb., Icon. Bot. 5: 64 (1827), nom. nud.
C. globularis Bertoloni, Fl. Ital. 7: 625 (1851).

Stems very slender, reddish or yellow. Flowers 1.5–2.5 mm, more or less fleshy and glandular, sessile to subsessile, mostly 4, rarely 3 or 5-merous in few-flowered compact clusters 3–7 mm in diameter. Calyx about enclosing corolla tube; lobes triangular-ovate, with thickened acute apex. Corolla lobes as long as cylindrical tube, ovate-triangular, acute, mostly spreading with cucullate tip. Stamens shorter than lobes; anthers slightly longer than filaments. Scales reaching filaments, oblong, obtuse to truncate, fimbriate above. Styles 2; stigmas as long as styles. Capsule depressed-globose, circumscissile at base. Seeds mostly 4.

HAB. Parasitic on a wide variety of herbs and shrubs; alt. s.l.–3000 m; fl. Apr.-Oct.
DISTRIB. Mountains and lower hills of N. Iraq. **MRO**: Halgurd Dagh, *Gillet* 12336! **MSU**: Khurmal, *Rawi* 8842!; Jarmo, *Helbaek* 1898! **FNI**: Eski Kellek, *Rawi & Gillet* 10398!

N. Africa, Arabia, S. & W. Iran, Syria, Turkey, Cyprus, Greece.

10. **Cuscuta approximata** *Bab.*, Ann. Mag. Nat. Hist. 13: 253 (1844); Yuncker in Mem. Torrey Bot. Club 18: 295, f. 158 (1932); Rawi in Dep. Agr. Iraq Tech. Bull. 14: 133 (1964); Feinbrun in Israel Journ. Bot. 19: 27 (1970); Plitmann in Fl. Turkey [P. H. Davis] 6: 232 (1978); Rhui-cheng & Staples in Fl. China 16: 325 (1995); Boulos in Fl. Egypt 2: 268 (2000).

C. urceolata Kuntze in Flora (Resensburg) 4: 651 (1846).
C. planiflora Tenore var. *approximata* (Bab.) Engelm. in Trans. Acad. Sci. St. Louis 1: 465 (1859); Gonçalves in Fl. Zamb. 8. 1: 135 (1987).
C. approximata var. *urceolata* (Kuntze) Yuncker in Mem. Torrey Bot. Club 18: 297 (1932).

Stem medium, yellowish to reddish. Flowers 3.5(–4) mm long, 5-merous, sessile in dense, globose, few to many-flowered clusters 5–11 mm in diameter. Calyx as long as to longer than corolla tube; lobes usually as long as tube, sometimes shorter or longer, triangular to ovate or obovate, acute, fleshy or turgid towards apex. Corolla campanulate; lobes triangular-ovate, shorter than tube, obtuse. Anthers as long as or shorter than filaments. Scales reaching stamens or shorter, oblong, ovate or obovate, simple or bifid, fimbriate at apex. Styles 2; stigmas as long as or slightly longer than style. Capsule subglobose, circumscissile. Seeds 4.

HAB. Mainly parasitic on members of the Asteraceae and Fabaceae; alt. 40–2200 m; fl. Jun.-Sep.
DISTRIB. Mountains, lower hills and the central alluvial plains of N. Iraq. **MRO**: Halgurd Dagh, *Guest* 2865!; Rowanduz, *Bornmüller* 1535! **LCA**: Abu Ghuraib, *Al-Baiyar* 21472 (BAG!) & 21481 (BAG!).

Macaronesia, Mediterranean to S.W. Siberia and Indo-China; Arabian Peninsula, Egypt, Eritrea to Zambia; introduced to N. & S. America, Caribbean, Australia, and parts of northern Europe.

11. **Cuscuta planiflora** *Ten.* Fl. Nap. 3: 250 (1824–1829); Rechinger, Fl. Lowland Iraq: 491 (1964); Yuncker in Mem. Torrey Bot. Club 18: 292, f. 157 (1932); Meeuse in Bothalia 6: 655 (1958); Rawi in Dep. Agr. Iraq Tech. Bull. 14: 134 (1964); Feinbrun in Israel Journ. Bot. 19: 27 (1970); Migahid & Hammouda, Fl. Saudi Arabia 232 (1974); Plitmann in Fl. Turkey [P. H. Davis] 6: 234 (1978); Meikle, Fl. Cyprus 2: 1179 (1985); Gonçalves in Fl. Zamb. 8. 1: 134 (1987); Boulos in Fl. Egypt 2: 267 (2000); Karim & Fawzi, Fl. U.A.E. 2: 140 (2007).

> *C. planiflora* var. *tenorii* Engelm. in Trans Acad. Sci. St. Louis 1: 466 (1859).
> *C. epithymum* subsp. *planiflora* (Ten.) Rouy, Fl. France 10: 359 (1908).
> *C. planiflora* subsp. *godronii* (Desmoulins) Jovet & R. de Vilmorin in Coste & al., Fl. France, Suppl.: 366 (1977) [comb. inval.].

Stems slender, reddish or yellowish. Flowers 1.5–2.5 mm, 5-merous, sessile, in dense, globose, few to many-flowered clusters, up to 8 mm in diameter. Calyx slightly longer than corolla tube, deeply divided; lobes longer than tube, ovate-deltoid, obtuse to subacute, fleshy, turgid at apex. Corolla campanulate-globose; lobes as long as or slightly shorter than tubes, lanceolate to ovate-deltoid, acute or subacute, fleshy and turgid towards apex. Anthers exserted, as long as or shorter than filaments. Scales reaching stamens, or shorter, simple or infrequently bifid, fimbriate at apex. Styles 2; stigmas a little longer than styles. Capsule depressed-globose, circumscissile at base. Seeds 2–4.

HAB. Parasitic on annual or perennial herbs and low shrubs; alt. 100–2500 m; fl. Mar.-Aug.
DISTRIB. Mountains and lower hills of N. Iraq and the western and southern desert districts. **MAM**: Acra, *Rawi* 11329!; Zawita, *Guest* 4868A! **MSU**: Balkha village, 7 km W. of Tawila, *Rawi* 22357!; Jarmo, *Helbaek* 1801!; Avroman, above Biyara, *Gillet* 11748! **MRO**: Halgurd Dagh, *Gillet* 9509!; Warshanka- Magar Range *Rawi* & *Serhang* 24330! **FKI**: Kani Domlan, *Guest* 4357!, 4373! **FPF**: Jabal Muwaila, nr. Kuwait, *Guest, Rawi* & *Rechinger* 17629A! & 17629B! **DGA**: Samara, *Bornm.* s.n.! **DWD/DSD**: Faidhat al-Gaiyara, *Karim, Nuri, Hamid* & *Kadhim* 40338! **DSD**: 30 km N.W. of Shabicha, *Guest, Rawi* & *Gillet* 14054!; Jabal Sanam, *Haines* W1032!; Umm Qasr, *Guest, Rawi* & *Rechinger* 16939!; 10 km E. of Salman, *Guest, Rawi* & *Rechinger* 18794! & 18795!; 50 km N. of Salman, *Rawi* 14868!; 40 km W. of Basra, *Bharucha, Rawi* & *Tikriti* 29335A! . Mesopotamia, *Kotschy* 104!

Mediterranean region, Arabia, Egypt, Iran, Afghanistan.

12. **Cuscuta brevistyla** *A.Braun* ex *A.Rich.*, Tent. Fl. Abyss. 2: 79 (1850); Rawi in Dep. Agr. Iraq Tech. Bull. 14: 134 (1964); Rechinger, Fl. Lowland Iraq: 491 (1964); Yuncker in Mem. Torrey Bot. Club 18: 289, f. 153 (1932); Feinbrun in Israel Journ. Bot. 19: 27 (1970); Plitmann in Fl. Turkey [P. H. Davis] 6: 234 (1978).

Stems slender, reddish or whitish. Flowers 1.5–3 mm, mostly 5-merous, sessile, in compact, globose clusters 3–7 mm in diameter. Calyx as long as corolla tube; lobes shorter or longer than tube, ovate or ovate-triangular, obtuse, thickened at apex. Corolla campanulate; lobes triangular-ovate, shorter than tube, obtuse or acute. Anthers as long as filaments or slightly longer. Scales reaching to filaments or shorter, obovate-oblong, truncate or obtuse, fimbriate above. Styles 2, clearly shorter than stigmas. Capsule depressed-globose. Seeds 4.

HAB. Parasitic on annual or perennial herbs and low shrubs; alt. 600–1500 m; fl. Apr.-Aug.
DISTRIB. Mountains and lower hills of N. Iraq and the southern marsh desert districts. **MAM**: 10 km S. of Sarsang, *Al- Kaisi, Al-Khayat* & *F. Karim* 51025!; Dori village 5km E. Kani Masi, *S. Omar* & *Al-Kaisi* 45405!

MRO: Haji Umran, *Al-Dabbagh & K. Hamad* 46286! **FPF**: 17 km E.S. of Zurbatiya, *Al-Kaisi & Khayat* 50661!; 30 km from Mandali to Badra, *Al-Kaisi & Yahya* 45246! **LSM**: 5 km N. of Qurna, *Bharucha, Rawi & Tikriti* 29313!; nr. Majar, *Alizzi* 34433! Mesopotamia, *Graham* 113!

N. Africa, S.W. & C. Asia to Tibet, W. Europe.

13. **Cuscuta monogyna** *Vahl*, Symb. Bot. (Vahl) 2: 32 (1791); Boissier, Fl. Orient. 4: 121 (1875); Rawi in Dep. Agr. Iraq Tech. Bull. 14: 134 (1964); Rechinger, Fl. Lowland Iraq: 489 (1964); Yuncker in Mem. Torrey Bot. Club 18: 256, f. 128 (1932); Feinbrun in Israel Journ. Bot. 19: 28 (1970); Plitmann in Fl. Turkey [P. H. Davis] 6: 236 (1978); Meikle, Fl. Cyprus 2: 1177 (1985); Rhui-cheng & Staples in Fl. China 16: 324 (1995); Rajput & Tahir in Fl. Pakistan [Nasir & Ali] 189: 10 (1988); Boulos in Fl. Egypt 2: 265 (2000).

Monogynella monogyna (Vahl) Hadač, Iraq Nat. Hist. Mus. Publ. 18 (Fam. Cuscutac. Iraq): 21 (1960). *M. vahliana* Des Moul., Étud. Cuscut. 65 (1853), nom. illeg.

Stems medium to thick, usually reddish, sometimes whitish. Flowers 3–5 mm long, 5-merous in loosely or densely spike-like, 1–6-flowered cymules, sessile or short pedicellate, sometimes fruiting pedicels as long as capsule, more or less fleshy. Calyx cupular, about enclosing corolla; lobes as long as or longer than tube, orbicular-ovate, obtuse or subacute, overlapping. Corolla tube ± exceeding calyx; lobes shorter than cylindrical tubes, ovate, obtuse or subacute, sometimes crenulate. Anthers sessile. Scales scarcely reaching anthers, simple, mostly truncate, shortly dentate. Ovary subglobose-conic; style 1, as long as or shorter than capitate stigma. Capsule ovate-conic or elongated-globose, circumscissile. Seeds 1 or 2. Fig. 69, 8–9.

HAB. Parasitic on trees, shrubs or perennial herbs; alt. ± 2000 m; fl. Jun.-Oct.
DISTRIB. Mountains of N. Iraq. **MAM**: Zawita, *Guest* 4869!; Sarsang, *Haines* W1583!; Bekhair Mountain nr Zakho, *Rawi, Tikriti & Nuri* 28962! **MSU**: Caradagh, *Rawi* 37771! **MRO**: Bole Village, 16 km N.E. of Rowanduz, *Rawi & Serhang* 20217!; Pishtashan, 15 km N.E. of Rania, *Rawi & Serhang* 23863! & 23873!; nr Haruna, *Gillet* 9643!; Shirwan Mazin, *Gillet* 9628!

C. & S. Europe, N.E. Africa, W. & C. Asia, S. Russia.

It is difficult to distinguish some dried specimens of *C. monygyna* from *C. lehmaniana*, however the seed morphology is quite different than those in other subgenera. [See Knepper, Creager & Musselman (1990), Seed Science and Technology 18: 731–741].

14. **Cuscuta lehmanniana** *Bunge*, Beitr. Fl. Russl. 220 (1852); Yuncker in Mem. Torrey Bot. Club 18: 257, f. 128 (1932); Rawi in Dep. Agr. Iraq Tech. Bull. 14: 134 (1964); Rajput & Tahir Fl. Pakistan [Nasir & Ali] 189: 11 (1988).

Monogynella lehmanniana (Bunge) Hadač, in Publ. Iraq Nat. Hist. Mus. No. 18 (Fam. Cuscutac. Iraq) 20 (1960).
M. lehmaniana (Bunge) Hadać & Chrtek in Folia Geobot. Phytotax. 5: 444 (1970).

Stem medium to thick, whitish to reddish. Flowers 4–5 mm long, 5-merous, sessile or subsessile in thyrsoid clusters. Calyx deeply divided, as long as or shorter than half of corolla tube; lobes longer than tube, ovate, obtuse, carinate, broadly overlapping. Corolla tube cylindrical or funnel-form, longer than calyx; lobes much shorter than tube, ovate-orbicular, obtuse, crenulate. Anthers oblong, sessile. Scales covering anther base, sometimes scarcely reaching anthers, rotund, fimbriate. Ovary globose-conic; style 1, usually as long as capitate stigma. Capsule globose-ovoid, circumscissile.

HAB. Usually parasitic on woody species; alt. ± 1000 m; fl. Aug.-Oct.
DISTRIB. Mountains and lower hills of N. Iraq, and the central alluvial plains. **MAM**: Batanura village 7 km E. Kani Masi, *Al-Kaisi et al.* 43859 (BAG!). **MSU**: nr Chuwarta, *Khayat, Kaisi & Thamir* 46320 (BAG!). **MRO**: Pushtashan, N. E. of Rania, lower slope of Qandil Range, *Ali Rawi & Serhang* 26546 (BAG!). **LCA**: Rustamiya, *Guest* 3498!; *Lazar* 300!

Iran, Afghanistan, Pakistan, Turkestan, C. Asia, China, Russia.

125. **BORAGINACEAE** Juss.

Gen. Pl.: 128 (1789) ('Borragineae') nom. cons.
Gürke in Engler & Prantl., Pflanzenfam. ed, 1, IV (3a): 71 (1897); Weigend, Selvi, Thomas & Hilger in Kubitzki (ser. ed.), Kadereit & Bittrich (vol. eds.), Fam. Gen. Vasc. Pl. 14: 41 (2016)

C.C. Townsend
Revised by Shahina A. Ghazanfar, J. R. Edmondson and Ali Haloob

Annual, biennial and perennial herbs, rarely shrubs or trees. Indumentum often hispid, usually of tubercle-based trichomes, often glochidiate, stellate or uncinate. Leaves mostly alternate, exstipulate, entire. Flowers generally 5-merous, entomophilous, hermaphrodite (rarely unisexual and then plant polygamous), actinomorphic (rarely zygomorphic), usually in scorpioid or ± circulate cymes, or inflorescence rarely thyrsoid; bracts present or absent. Calyx gamosepalous, but usually ± deeply 5-lobed (occasionally irregular or 9-lobed), lobes usually imbricate in bud, often accrescent after anthesis. Corolla 5-lobed, cylindrical, salver-shaped, rotate or funnel-shaped, lobes imbricate or contorted in bud, usually clearly differentiated into a tube and a distinctly lobed limb; throat often with 5 appendages, invaginations, or tufts of hairs or ring of hairs. Stamens epipetalous, alternating with corolla lobes, equal or unequal; filaments sometimes appendiculate. Ovary mounted on a nectariferous disk, superior, bicarpellate, becoming 2- or 4-locular with a false septum at maturity, each locule with one ovary; style gynobasic or rarely apical, usually undivided, stigmas capitate or conical or 2(–4)-lobed. Fruit borne on a flat or conical gynobase, the area of attachment forming an areole; nutlets usually 4, rarely 2 corky mericarps or a drupe, sessile or with a short stalk, beaked or not, keeled or rounded, often with a marginal spreading or incurved wing which may be glochidiate or spiny-glochidiate; endosperm absent or scanty.

The Boraginaceae has been variously classified at the familial level from being recognised as a single, widely variable family to recognition of several distinct families. In a recent study, Leubert et al. (2016) recognised eleven families that are monophyletic and morphologically distinct within the traditionally recognised Boraginaceae. These are: Boraginaceae s.str., Codonaceae, Coldeniaceae, Cordiaceae, Ehretiaceae, Heliotropiaceae, Hoplestigmataceae, Hydrophyllaceae, Lennoaceae, Namaceae, and Wellstediaceae. Of these, genera belonging to Boraginaceae s.str., Cordiaceae and Heliotropiaceae are found in Iraq. Since the Flora of Iraq follows the classification of Hutchinson, Boraginaceae is treated here in the broad sense to include *Cordia, Heliotropium* and *Euploca*.

A large family of about 95 genera and 2100–2200 species (including Ehretiaceae with about 10 genera and 500 species); tropical and subtropical, largely absent from the wet tropics; represented by 31 genera and 98 species in Iraq.

Chacón, J., Luebert, F., Hilger, H.H., Ovchinnikova, S., Selvi, F., Cecchi, L., Guilliams, C.M., Hasenstab-Lehman, K., Sutorý, K., Simpson, M.G. and Weigend, M. (2016). The borage family (Boraginaceae s. str.): A revised infrafamilial classification based on new phylogenetic evidence, with emphasis on the placement of some enigmatic genera. Taxon 65(3): 523–546.
Johnson, I.M. (1924). The old world genera of the *Boraginoideae*. Contr. Gray Herb. 73: 42–73.
Luebert, F., Cecchi, L., Frohlich, M.W., Gottschling, M., Guilliams, C.M., Hasenstab-Lehman, K.E., Hilger, H.H., Miller, J.S., Mittelbach, M., Nazaire, M. & Nepi, M. (2016). Familial classification of the Boraginales. Taxon 65(3): 502–522.
Weigend, M., Selvi, F., Thomas, D.C. & Hilger, H.H. (2016). Boraginaceae. In: The families and genera of vascular plants (K. Kubitzki, ser. ed.). Flowering plants: Eudicots – Boraginaceae (J.W. Kadereit & V. Bittrich, vol. eds). Vol. 14. pp. 41–102.
Zhu, Gelin, Riedl, H. & Kamelin, R.V. (1995). Boraginaceae. In: Wu, Z. Y. & P. H. Raven, eds., Flora of China. 16: 329.
Nasir, Y. (1989). Boraginaceae. In: S.I. Ali & Y.J. Nasir, eds, Flora of Pakistan. Fasc. 191: 1–200.

1. Fruit drupaceous, enclosed by the persistent calyx . . . . 1. **Cordia**
   Fruit of usually 4 (rarely 2) nutlets 2
2. Style 1, terminal (without a central column) 3
   Style gynobasic (nutlets separating or inseparable from a persistent central column) 4
3. Inflorescence bracteate; anthers protracted, apex pubescent, coherent
   apically, closing the corolla tube; nutlets 4 . . . . 15. **Euploca**

Inflorescence ebracteate; anthers usually included, not protracted nor coherent apically; nutlets usually 4, rarely 1–2 . . . . .16. **Heliotropium**
4. Calyx longer than corolla; faucal scales absent but corolla throat with ciliate appendages; stigma terminal . . . . 2. **Ogastemma**
Calyx shorter than corolla; faucal scales present or throat with cilia 5
5. Flowers nodding at anthesis; corolla lobes small and reduced; anther connective forming an acute apical appendage, exserted from corolla 6
Flowers erect, not nodding at anthesis; corolla lobes distinct, not reduced; anther connective not forming an apical appendage 7
6. Corolla blue with yellowish apex (in Iraq plants); plants of rock crevices, crevices in walls, in moist locations . . . . 4. **Podonosma**
Corolla yellow, whitish or cream-coloured; plants of various habitats, not necessarily of rock crevices . . . . 7. **Onosma**
7. Corolla distinctly zygomorphic, infundibuliform; faucal (throat) scales absent; staminal filaments long, unequal, at least two exserted. . . . . .8. **Echium**
Corolla actinomorphic; staminal filaments equal 8
8. Calyx circumscissile above the base at maturity; faucal scales absent but with 5 patches of trichomes in throat; stamens included, attached at different heights, with 3 filaments longer; nutlets 2–3 . . . . 10. **Moltkiopsis**
Calyx regular, not circumscissile above the base at maturity; stamens attached at same height in corolla throat. 9
9. Nutlets small (1–3(–5) mm, obliquely tetrahedral, lentil-shaped or with flat abaxial side and ventrally keeled, dark brown to black, usually smooth and shiny. . . . . .21. **Myosotis**
Nutlets mostly larger (>5 mm), variously shaped and coloured, but never tetrahedral or lentil-shaped 10
10. Nutlets gibbous at the sides; rhizomatous plants. . . . . 11. **Buglossoides**
Nutlets various shapes, but not gibbous at sides 11
11. Nutlets without well-developed, plug-shaped elaiosome at base; attachmemt scar not thickened. 12
Nutlets with well-developed, plug-shaped elaiosome and ± thickened basal ring around attachment scar 15
12. Nutlets 2; anthers with appendages at base and apex; roots without dye . . .6. **Cerinthe**
Nutlets 4 or fewer; anthers without appendages at base and apex; roots containing a purplish dye 13
13. Nutlets 4, glabrous, depressed-ovoid, globose or trigonal, with acute to acuminate beak and ventral keel; annuals herbs . . . . .5. **Arnebia**
Nutlets 1–2 (by reduction), or 4, tuberculate, reticulate, rugose or finely papillose; perennial herbs, woody at base 14
14. Corolla yellow (in ours), throat with a ring of hairs and sometimes small invaginations. Stamens included (filaments short, inserted in a whorl or spiral within tube); nutlets 1–2 (by reduction), reniform to obliquely ovoid, tuberculate, reticulate or rugose, beaked. . . . . 3. **Alkanna**
Corolla blue, with a ± cylindrical tube; faucal scales absent; stamens distinctly exserted; nutlets 4 or fewer, ovoid to ovoid-trigonal, weakly tuberculate and finely papillose; beak horizontally incurved, subacute . . .9. **Moltkia**
15. Nutlets smooth or rugose-tuberculate, but without irregular crests, distinct tubercles or glochidia. 16
Nutlets mostly with irregular crests, wings, distinct tubercles or glochidia 20
16. Plants usually with glandular hairs; faucal scales hairy or replaced by tufts of hairs; fruiting calyx strongly accrescent and deflexed in fruit . . . . .17. **Nonea**
Plants lacking glandular hairs; faucal scales triangular-oblong to oblong, not replaced by hairy tufts; fruiting calyx ± accrescent 17
17. Basal ring around attachment scar on nutlets toothed; corolla white to yellowish; faucal scales equalling stamens (in our species); robust perennial, basal leaves long petiole (up to 8 cm) . . . . 18. **Symphytum**
Basal ring around attachment scar on nutlets thick, collar-like or inconspicuous, not toothed. 18
18. Perennial herbs, rhizomatous; inflorescence a lax panicle or corymb; cymes ebracteate; faucal scales trapezoid, ciliate; stamens exserted . . . 14. **Brunnera**

Annual or perennials, not rhizomatous; inflorescence of ± lax scorpioid
cymes or dichotomously branched cymes, bracteate; faucal scales
triangular-oblong, not ciliate; stamens included 19
19. Annual herbs, setose-hispid; calyx cup-shaped, divided to almost base
into linear-lanceolate lobes; corolla pink or blue, never yellow; faucal
scales emarginate, papillose . . . . .13. **Gastrocotyle**
Annual, biennial or perennial herbs; indumentum of dense short hairs
and sparse tubercle-based trichomes or hispid-setose; calyx divided to
1/3 or to base; corolla yellow, blue to purple; faucal scales triangular-
oblong. . . . . 12. **Anchusa**
20. Calyx deeply divided into 5 lobes narrowed towards tip, ribbed or
cordate-winged near base, anthers with twisted appendages at tips;
nutlets often with a narrow dentate wing . . . . 19. **Trichodesma**
Calyx 2–5 lobed, not ribbed or cordate winged near base; anthers
regular, without twisted appendages 21
21. Calyx 2-lobed almost to base, strongly accrescent, 2-lipped in fruit, the
lips 7- and 8-dentate respectively, enclosing nutlets. . . . . 23. **Asperugo**
Calyx 5-lobed 22
22. Calyx accrescent in fruit to a stellate form; corolla throat broadening
abruptly; faucal scales arcuate, closing throat; staminal filaments
unequal with lateral and anterior anthers smaller and posterior
anther long-exserted; style long-exserted, forming a cup in fruit
which encloses the nutlets. . . . . 20. **Caccinia**
Calyx accrescent but not a stellate form in fruit; corolla throat straight;
faucal scales different shapes; staminal filaments and anthers equal;
style exserted or included not forming a cup which encloses the nutlets 23
23. Nutlets at apex free from gynobase, not attached along their entire
length to the gynobase 24
Nutlets parallel, attached along their entire adaxial surface to the gynobase. 25
24. Calyx lobes minute in flower, accrescent to 3.5 mm in fruit, falcately
incurved from middle, overtopping calyx in fruit; nutlets 1–2 by
reduction, attached obliquely near base, sometimes nutlets united
with gynophore and upper nutlet wing cup-shaped, others triangular
and wingless . . . . .25. **Rochelia**
Calyx lobes not minute in flower, oblong to linear, not incurved; nutlets 2–4 27
25. Nutlets narrowly ovate to oblong, glochidiate, differentiated into two
opposite pairs, nutlets of one pair distinctly wider than those of the
other . . . . 24. **Pseudoheterocaryum**
Nutlets ovate, triangular or conical, marginally rimmed (not winged)
or margin indistinct; rim with stalked glochidia or smooth or
tuberculate with few glochidia 26
26. Leaves up to 5(–7) mm wide; pedicel at fruiting up to 4 mm long,
straight or recurved; nutlets 2–6 mm long, smooth or distinctly
glochidiate; their attachment scar obvious, lanceolate to ovoid. 25.**Lappula**
At least basal leaves 5–11(–15) mm wide; fruiting pedicel longer (up to
8 mm long); nutlets c. 2 mm long, inconspicuously glochidiate, their
attachment scar linear (strip-like) . . . . .26. **Pseudolappula**
27. Nutlets ovoid to orbicular, dorsal surface (disc) convex or
depressed, glochidiate. 28
Nutlets flat, with a broad membranous outer wing and rarely also a
narrower ± incurving inner wing 30
28. Calyx divided ± to base; corolla cylindrical or funnel-shaped; faucal
scales subquadrate to triangular, inserted near middle of tube, not closing
throat; stamens exserted; anthers orbicular-elliptic. Style greatly
exceeding calyx, usually exserted; stigma capitate; nutlets clothed
with spinulose glochids . . . . .31. **Solenenthus**
Calyx about 2/3-lobed; corolla rotate, tube shorter than calyx; faucal
scales oblong or subquadrate, ± closing throat; stamens included;
nutlets glochidiate . . . . 28. **Cynoglossum**
30. Nutlets conspicuously winged, the central portion smooth, aculeolate

or glochidiate; wing inflexed, membranaceous, margin entire or
glochidiate . . . . 29. **Paracaryum**
Nutlets with a broad membranous outer wing and rarely also a
narrower ± incurving inner wing . . . . 30. **Rindera**

1. **CORDIA** L.

Sp. Pl. 1: 190 (1753)

J.R. Edmondson

Shrubs and trees. Leaves alternate, entire, petiolate. Inflorescence mostly terminal, ebracteate. Flowers pentamerous, rarely tetramerous, bisexual and often distylous, occasional unisexual and dioecious. Calyx cylindrical to campanulate, subentire to shallowly lobed, becoming crateriform-indurated in fruit. Corolla tubular to funnel-shaped, lobes oblong to spatulate. Stamens exserted; filaments generally adnate to the corolla tube, at least at the base; anthers hastate. Ovary entire, 4-locular; style terminal; stigma 4-fid; nectar disc usually at base of ovary. Fruit drupaceous, enclosed by the persistent calyx.

The woody, drupaceous genus *Cordia* (formerly recognized as a family or as part of the Heliotropiaceae or Boraginaceae) is now placed in a distinct family Cordiaceae (2/350) with the main distinction of having the gynoecium with 4 stigmatic branches (if 2, then fruit completely enclosed in accrescent calyx), orthotropous ovules and plicate cotyledons.

Cordia, commemorates Euricius Cordus (1486–1535), a Lutheran physician and one of Germany's earliest botanists, and his son Valerius Cordus (1515–1544), an accomplished pharmacist and botanist.

1. **Cordia myxa** *L.*, Sp. Pl. 1: 273 (1753); Meikle, Fl. Cyprus 2: 1120 (1985).

Tree to 6 m tall; leaf blade to 8 × 7 cm, broadly ovate, obovate or suborbicular, upper surface glabrous and glossy, margins entire or weakly lobed, apex apiculate, petiole 1–3 cm. Inflorescence cymose, first branching at a wide angle, terminating in a corymb or thyrse. Calyx campanulate, ± 4 mm wide in flower, accrescent and becoming discoid in fruit. Corolla white, tubular with 5 recurved lobes; stamens exserted, anthers yellow. Ovary narrowly ovoid, acuminate, with short style 1.5–2 mm divided into 4 linear stigmas. Fruit ± globose, 1.5 cm in diameter, pinkish when first ripe, becoming brownish with age; mesocarp fleshy, sticky; endocarp ± 1.3 cm. Fig. 70, 1–4.

HAB. On clay soil; alt. 2 m; fl. Dec.
DISTRIB. Basra region, said to be rare. **LBA**: 10 km S. of Fao, *Hamad, Hamid & Kadhim* 46353!

Native of Tropical Asia, widely planted in the sub-tropics but possibly native in S. Iran. The sticky fruits are used to make a preparation for trapping birds ("bird-lime"); the bark has medicinal uses, and the flesh of the fruit is edible.

2. **OGASTEMMA** Brummitt

Kew Bull. 36: 679 (1982); Weigend & al. in Kubitzki, Fam. Gen. Vasc. Pl. 14: 62 (2016)
Megastoma (Benth. & Hook. f.) Coss. & Durieu ex Bonnet & Barratte (1895), nom. illeg.

Ali Haloob and S.T. Al-Kaisi

Annual herb, stems erect, branched from base, appressed hairy, hairs arising from tubercles. Leaves sessile, the lowest pairs opposite, the upper alternate, densely appressed with tubercle-based hairs. Inflorescences terminal with flowers produced along the whole length of branches (the first flowers of the young shoots developing in the axil of the first bifurcating stems). Bracts foliose. Flowers erect, shortly pedicellate. Calyx 5-lobed to base, longer than corolla, calyx-lobes unequal, lanceolate, margins membeanous, elongating in fruit, base of calyx becoming subglobose in fruit and lobes adhering over fruit. Corolla white, subcylindrical, ± equalling the longest calyx-lobe, indistinctly lobed above; faucal scales absent; throat densely pubescent. Stamens included, attached above the base of

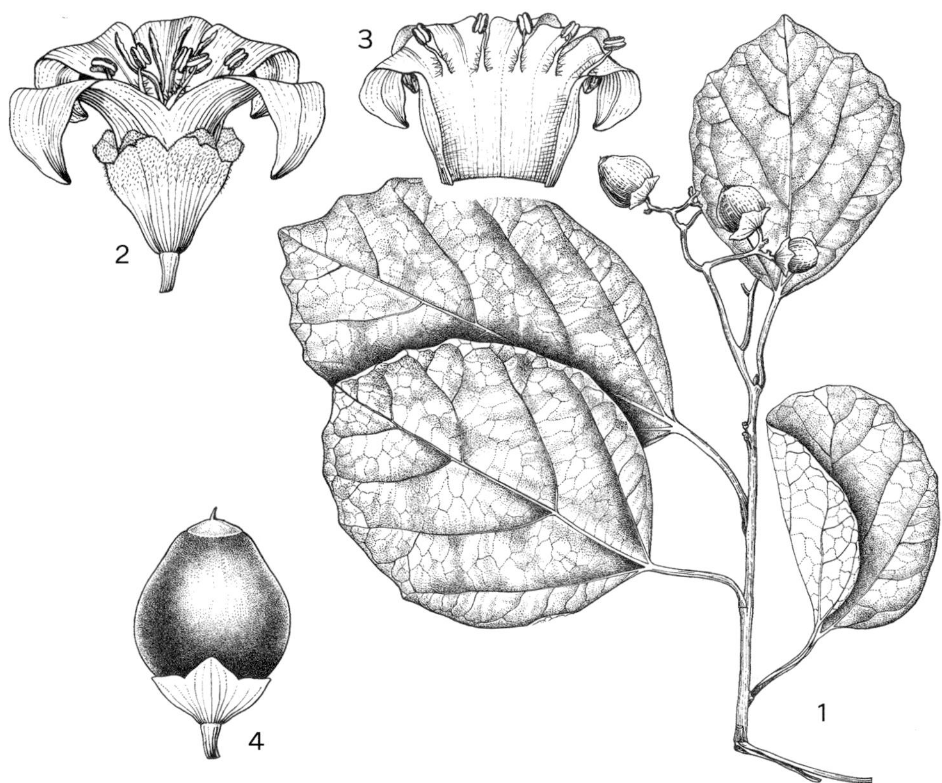

Fig. 70. **Cordia myxa**. 1, habit portion of shoot × ½; 2, flower × 4; 3, flower dissected × 3; 4, drupe × 1.5. Reproduced with permission from Fl. Pakistan 191: f. 1, F-I (1989). Drawn by S. Hameed. © National Herbarium, Pakistan Agriculture Research Council & University of Karachi, Pakistan.

corolla tube. Ovary deeply 4-lobed. Style gynobasic, filiform; stigmas 2, globose. Nutlets erect, ovoid, apex conical to rostrate, verrucose, ventral side with narrow areole.

Monotypic; distributed in the Canary Islands, northern Africa to the Arabian Peninsula.

Lönn, E., 1999. Revision of the three Boraginaceae genera *Echiochilon, Ogastemma* and *Sericostoma*. Bot. Journ. Linn. Soc. 130(3): 185–259.

Mousa, M.O. & Shahatha, S.S. (2021). Taxonomic study for the new record, *Ogastemma pusillum* (Boraginaceae) in Iraq. The Iraqi Journal of Agricultural Science, 52(3): 724–735.

1. **Ogastemma pusillum** (*Coss. & Durieu ex Bonnet & Barratte*) Brummitt, Kew Bull. 36: 679 (1982); Boulos, Fl. Egypt ; Ghazanfar, Fl. Oman 3: 82 (2015); Collenette, Wild Flow. Saudi Arabia: 94 (1999); Taifour & El-Oqlah, Pl. Jordan Annot. Checklist: 53 (2017); Mousa & Shahatha, Iraq J. Agric. Sc. 52(3): 724 (2021).

Megastoma pusillum Coss. & Durieu ex Balansa, PI. d'ALgkrie 1853: 1035 (1853), nom. nud.; ex Bonnet & Barratte, Explol: Sci. Tunisie, Ill. Bat: t. 11, figs 4–11 (1892–1895); Expl. Sci. Tunisie, Cat. Pl.: 301 (1896); Feinbrun-Dothan, Fl. Palaest. 3: 59 (1978).

Annual herb. Stems erect or ascending, much branched, the lowest branches opposite, densely covered with appressed tubercle-based hairs and minute glandular hairs. Leaves sessile, linear to narrowly lanceolate, 8–17 × 1–2 mm, base attenuate, apex obtuse, margins entire, densely hairy with appressed, tubercle-based egaldular hairs on both surfaces. Pedicel c. 1 mm at anthesis, to 2 mm in fruit. Bracts as leaves becoming smaller towards the top of branches. Calyx lobes narrow-lanceolate, unequal, ± 3 mm, with white dense strigose eglandular and minute glandular hairs, margins membranous, ciliate with spreading hairs.

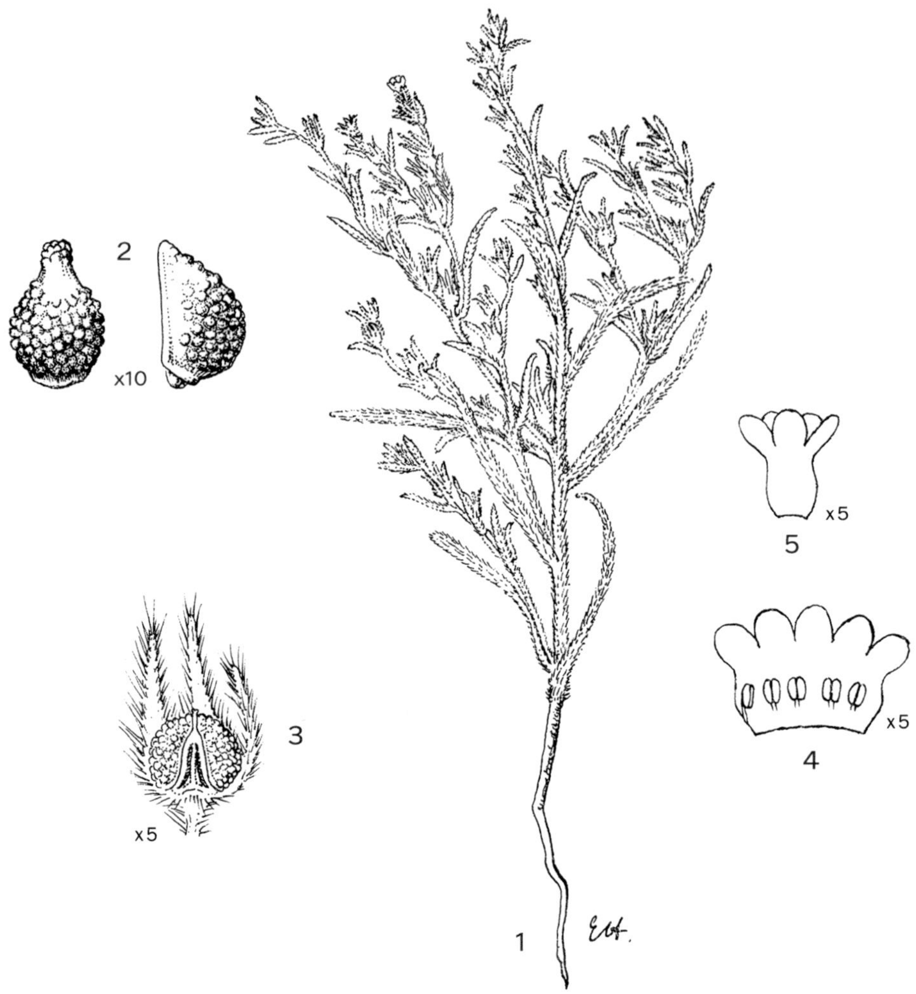

Fig. 71. **Ogastemma pusillum**. 1, habit; 2, corolla × 5; 3, coralla opened × 5; 4, calyx × 5; 5, mature nutlets x10. Reproduced with permission from Feinbrun-Dothan, Flora Palaestina, 3: Plates, f. 96 (1977). Drawn by Esther Huber. © The Israel Academy of Sciences and Humanities.

Corolla white, 2–3 mm, tube about 1.9 mm, glabrous, shorter than calyx. Stamens short, inserted in the lower half of corolla tube. Style gynobasic, glabrous, shorter than stamens, stigma bilobed, capitate. Nutlets 1.5–1.7 mm, dull yellow, densely verrucose. Fig. 71, 1–5

HAB. Gravelly and sandy soils; alt. 275–280 m 2 m; fl. Apr.-Jun.
DISTRIB. Rare in the western desert plateau region of Iraq. **DWD**: 155–160 km W. of Ramadi, *Mousa* 60265!; 60264! (BAG); 130 km W. of Ramadi, *Mousa* 60266! (BAG).

Canary Islands, northern Africa, Jordan, Palestine, Arabian Peninsula.

3. **ALKANNA**

Tausch, Flora 7(1): 234 (1824) nom. cons., non Adans. (1763); Kew Bull. 36: 679 (1982); Weigend & al. in Kubitzki, Fam. Gen. Vasc. Pl. 14: 70 (2016)

Shahina A. Ghazanfar

Perennial herbs with sweet-scented flowers, roots often with purple dye. Indumentum variable, but often composed of setae, eglandular and glandular hairs. Inflorescence terminal, with 1-many cymes, bracteate. Calyx 5-fid ± to base; lobes accrescent in fruit. Corolla yellow (in ours), funnel- shaped or salver–shaped, with nectariferous ring at base of tube. Throat with a ring of hairs and sometimes small invaginations. Annulus small, glabrous or ciliate. Stamens included, filaments short, inserted in a whorl or spiral within tube. Style included, stigma entire or bilobed. Nutlets 1–2 (by reduction), reniform to obliquely ovoid, tuberculate, reticulate or rugose, beaked.

About 40 species, Mediterranean and S.W. Asia; 6 species in Iraq.

Alkanna is derived from Ar. الحناء *al-hanna*, with reference to the red dye in the roots.

1. Corolla pubescent outside, glandular within. . . . . .5. *A. trichophylla*
   Corolla glabrous 2
2. Plant densely patent-setose hairs, with long and short, eglandular hairs . . .6. *A. frigida*
   Plant with glandular and eglandular hairs. 3
3. Leaves with margins wavy; cymes few-flowered .1. *A. orientalis*
   Leaves margins not wavy; cymes usually dense, elongating in fruit 4
4. Plant canescent, eglandular hispid-tomentose, and short glandular
   hairy; nutlets with beak strongly deflexed . . . . .3. *A. kotschyana*
   Nutlets with beak curved, but not strongly deflexed. 5
5. Plant patent-setose and glandular-pubescent throughout; cymes few-
   flowered; flowers fragrant . . . . .2. *A. hirsutissima*
   Plant sparsely hispid with short tubercle-based setae and glandular-
   pubescent; cymes dense, elongating in fruit; flowers not distinctly
   fragrant . . . . .4. *A. bracteosa*

1. **Alkanna orientalis** (*L.*) *Boiss.*, Diagn. Pl. Orient., ser. 1, 4: 46 (1844); Dinsmore in Post, Fl. Syria, Palest. & Sinai ed. 2, 2: 243 (1933); Riedl in Fl. Iranica [K. H. Rechinger] 48: 216 (1967); Feinbrun-Dothan, Fl. Palaest. 3: 78 (1978); Huber-Morath in Fl. Turkey [P. H. Davis] 6: 420 (1978); El-Hadidi & Boulos, Fl. Egypt 2: 295 (2002); Khatamsaz, Flora Iran 39: 172 (2008); Ahmed, Feddes Rep. 124: 66 (2013); Taifour & El-Oqlah, Pl. Jordan Annot. Checkl.: 49 (2017).

Perennial, with a branched stock with short hairs and longer tubercle-based hairs at base. Stems 20–40 cm, branched. Leaves lanceolate, 3–6 × 0.6–1.5 cm, base tapering into a long petiole, apex obtuse margins wavy. Cymes dense, elongating in fruit. Bracts shorter than cauline leaves, sessile. Pedicels obsolete. Calyx 6–7 mm in flower, accrescent in fruit, lobes lanceolate, divided almost to base. Corolla yellow or whitish yellow, ± 12 mm, glabrous. Nutlet 1, 3–4 mm wide, horizontal, densely reticulate-rugose; beaked. Fig. 72, 1–8.

HAB. Subalpine areas, between rocks; alt. 470–2000 m; fl. May.
DISTRIB. Sulimaniyah district, where it is not uncommon. MSU. Tawila, nr weather station, *Ahmad* 11-578 (SUFA); Zallm river, *Ahmad* 11-681 (SUFA); Ahmad Awa, *Ahmad* 11-727 (SUFA).

Egypt, Syria, Palestine, Lebanon, Turkey, Caucasus, Iran; Greece.

2. **Alkanna hirsutissima** (*Bertol.*) *A.DC.*, Prodr. [A. P. de Candolle] 10: 101 (1846); Dinsmore in Post, Fl. Syria, Palest. & Sinai ed. 2, 2: 243 (1933); Rawi, in Dep. Agr. Tech. Bull. 14: 134 (1964); Rechinger, Fl. Lowland Iraq: 506 (1964); Riedl in Fl. Iranica [K. H. Rechinger] 48: 218 (1967).

Lithospermum hirsutissimum Bertol., Misc. Bot. 1: 13 (1842). Type: Mesopotamia, near the Castle of Sedjim Kala, 4th April 1836, *Chesney Expedition* 75 (BM-013839848), E, GH).

Perennial, 15–20 cm, patent-setose and glandular-pubescent throughout. Stems erect or ascending, branched. Leaves 1–5 × 2–5 cm, linear-lanceolate, subacute, the lowest shortly petiolate, the upper sessile. Cymes few-flowered, slightly accrescent in fruit; bracts similar to leaves. Calyx 6–7 mm in flower, accrescent in fruit, deeply divided, lobes lanceolate, acute. Corolla yellow, 8–10 mm, glabrous, fragrant. Nutlet 1 (by reduction), transversely ovoid, reticulate-rugose, beaked.

HAB.: Rocky places, limestone gravel in wadi, sandy clay soil; alt. 400–750 m; fl. Mar.-May.
DISTR.: Mainly in the moist zone of the Upper Jazira of Iraq.
MJS. Jabal Sinjar, *Rawi* 41379! *Gillett* 11100! Sinjar, *Haussknecht* (May 1867)! FUJ. 40 km E. of Sinjar,

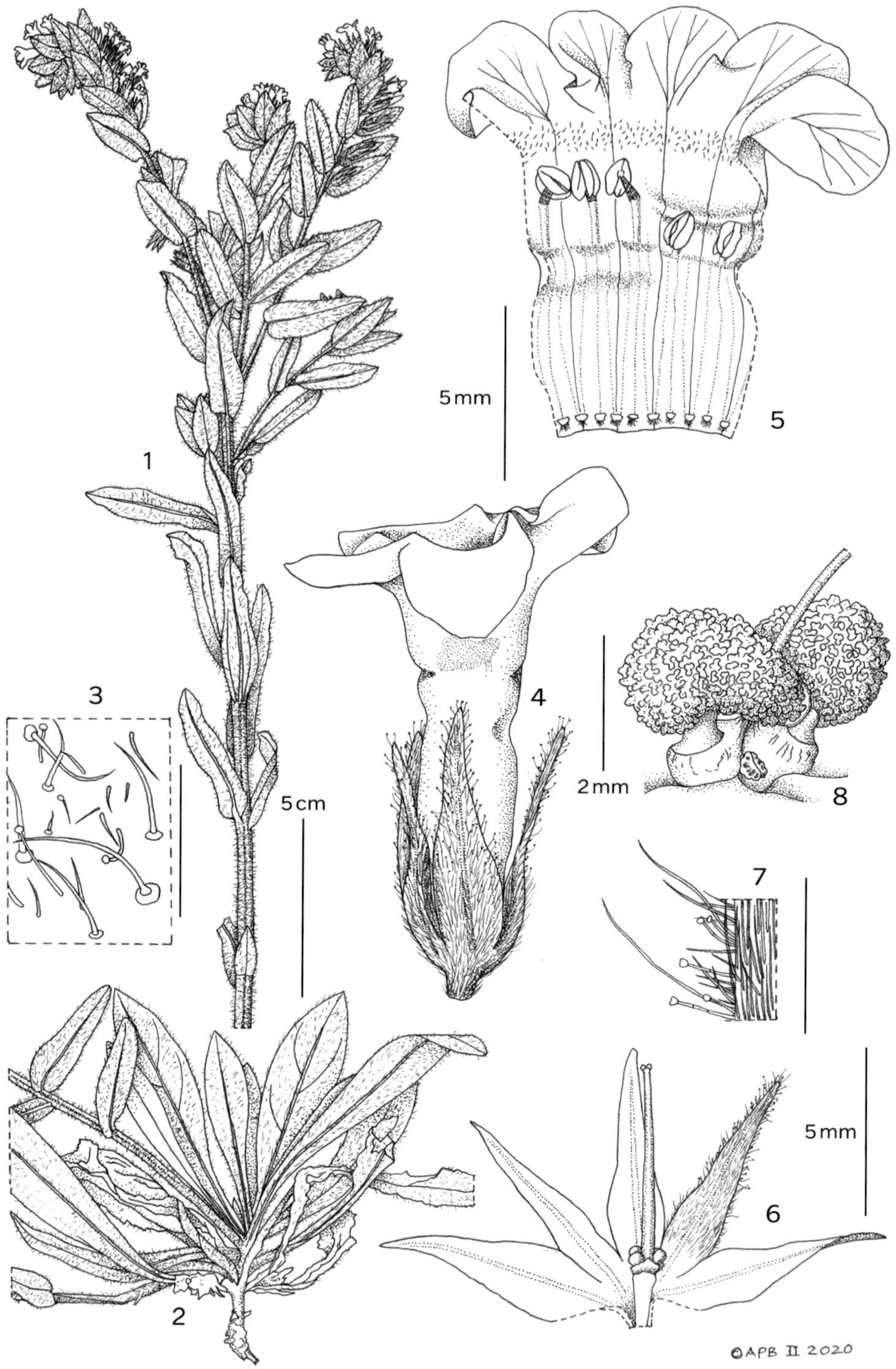

Fig. 72. **Alkanna orientalis.** 1, habit, erect flowering shoot; 2, basal rosette of leaves; 3, indumentum (adaxial surface of cauline leaf); 4, flower; 5, corolla, opened; 6, calyx; 7, detail indumentum calyx; 8, mature nutlets. 1, 3 from *Samuelson* 2314; 2, 4–8 from *Davis* 41482 (Turkish material). Drawn by © A.P. Brown 2020.

Chakravarty, Rawi, Khatib & Alizzi 33142! 1 km N. of Sinjar, *Barkley* 7965!; *Barkley & Haddad* 6717!; 5 km W. of Aqra, *Barkley* 9148!; Ain Tallawi, nr Sinjar, *Rawi* 5726!; Mosul, *Lazar* 3362! *Leatherdale* s.n.; Ain al–Husan (nr Sinjar), *Guest* 4196!; Mosul, *Graham* 776!; Tal Afar to Sinjar, *Guest* 13625!; 12 km N. of Tel Afar, *Ani* 9369!; 1 km N. of Sinjar, *Barkley & Haddad* 33225! **DSD**. Shuaiba, *Haddad* 9500!; 'Mesopotamia', *Aucher* 9978! unlocalised ('Mesopot., Kurdistan & Mosul'), *Kotschy* 20!

The type specimen from Sedjim Kale, *Chesney* (Apr. 1836) is probably from Syria; there is no reference to Sedjim in Chesney's narrative but it may be a mistranscription of Kalat-en Nejm (castle of the stars) which was reached before the descent to Kara Bamburge on the 30th (of March). The party was at Beles in April. Ainsworth, the expedition's geologist, is described as collecting botanical specimens on p. 289 of Chesney's account, but the main natural history observations were carried out by Dr J.W. & Mrs Helfer (see Chesney p. 353).

KHUZAIMAH (Ir., Mosul), *Lazar* 3362.

Syria.

3. **Alkanna kotschyana** *A.DC.*, Prodr. [A. P. de Candolle] 10: 98 (1846); Dinsmore in Post, Fl. Syria, Palest. & Sinai ed. 2, 2: 244 (1933); Rawi, in Dep. Agr. Tech. Bull. 14: 134 (1964); Huber-Morath in Fl. Turkey [P. H. Davis] 6: 422 (1978).

   A. amplexicaulis A.DC., op. cit. 101 (1846); *A. brachysolen* Boiss., Fl. Orient. [Boissier] 4: 230 (1875).

Perennial, 15–34 cm, canescent, eglandular hispid-tomentose, and with short glandular hairs. Basal leaves oblong-cuneate, 4–8 × 0.7–1.2 cm, cauline linear-lanceolate, much shorter and narrower. Cymes elongating to 20 cm in fruit. Bracts linear-oblong, longer than calyx. Calyx 5–8 mm in flower, 9–13 mm in fruit, lobes linear-lanceolate, villous. Corolla yellow, 10–15 mm, tube glabrous outside, limb 4–6 mm in diameter. Nutlet 3–4 mm, scrobiculate-reticulate; beak strongly deflexed.

HAB. Limestone cliffs, limestone with coppiced *Quercus aegilops*, on clay soil near stream, rocky mountainside, *Quercus* forest, rocky clay soil; alt. 930–1580 m; fl. Apr.-Jul.
DISTRIB. Locally common in the northern sector of the forest zone of Iraq, rare elsewhere.
MAM. 13 km from Amadiya to Sarsang, *Sharif & Hamad* 50327!; Bamerny 20 km N.W. of Sarsang, *Kaisi & Hamad* 45948!; Maya, 20 km W. of Kani Masi, *Hamid & Fadhil* 45474!; Amadiya, *Botany Staff* 43413! SER Amadiya (Gali Mazurka), *Dabbagh, Kaisi & Hamid* 46018!; between Zakho and Turkish frontier, *Karim, Dabbagh & Hamad* 44773!; 10 km S. of Sarsang, *Kaisi & Hamad* 45817!; Tirwanish 10 km E. of Kani Masi, *Kaisi, Hamid & Khalid* 45748!; Sulaf to Amadiya, *Kaisi, Khayat & Karim* 51072!; mt. Bikhair nr Zakho, *Rawi* 22995!; Atrush, *Guest* 3625! **MSU**: N. of Biyara, *Gillett* 11819! FUJ: mt. Gara, *Dabbagh & Jasim* 47002!

Turkey (S. Anatolia); erroneously recorded as endemic to Turkey by Huber-Morath in Davis, Fl. Turkey (1978).

4. **Alkanna bracteosa** *Boiss.*, Diagn. Pl. Orient. ser. 1, 11: 118 (1849); Riedl in Fl. Iranica [K. H. Rechinger] 48: 218 (1967); Khatamsaz, Flora Iran 39: 178 (2008).

   A. heterophylla Vatke in Zeitschr. Naturwiss. 45: 126 (1875).

Perennial, sparsely hispid with short tubercle-based setae and glandular-pubescent. Stems erect, little-branched. Basal leaves long-petiolate, to 20 cm, blade lanceolate, acute, tapering to base; cauline oblong to ovate, 3–6 mm, sessile. Cymes elongating in fruit, densely bracteate. Calyx 7–9 mm in flower, accrescent in fruit, divided almost to base, lobes lanceolate, acute. Corolla yellow, 10–11 mm, glabrous, funnel-shaped or salver-shaped, limb spreading. Nutlet usually 1 (by reduction), reniform, reticulate-rugose, beaked, not strongly deflexed. Fig. 73, 1–3.

HAB.: *Juglans* forest; alt. 1550–1700 m; fl. Jun.
DISTRIB.: Occasional in eastern sector of forest zone of Iraq.
MRO. mt. Mollakhort, *Rawi, Alizzi & Nuri* 29571!; Goum Tawera, *Rawi, Alizzi & Nuri* 29560! **MSU**. mt. Hawraman, *Gillett* 11819! *Haussknecht* s.n.; mt. Hawraman, at Tawila, *Rechinger* 10326!; Tawila, *Rawi* 29560!

W. Iran.

Fig. 73. **Alkanna bracteosa**. 1, habit × 1; 2, corolla, opened × 4; 3, nutlet, side view × 10. Reproduced with permission from Jamzad, Flora of Iran 76: f. 40 (2012). © Ministry of Jihad-e-Agriculture.

5. **Alkanna trichophila** *Hub.-Mor.* in Bauhinia 2: 201 (1963); Riedl in Fl. Iranica [K. H. Rechinger] 48: 217 (1967); Khatamsaz, Flora Iran 39: 174 (2008).

subsp. **trichophila**

Perennial, clothed throughout with short to long patent setae, and with minute glandular and eglandular hairs. Stems 10–25 cm, prostrate to ascending, branched from base. Basal leaves linear-lanceolate to spatulate-lanceolate, 2–12 cm × 0.3–1.5 cm; cauline 2.5–5 cm, oblong, sessile. Cymes elongating to 10–15 cm in fruit, bracteate. Calyx 7–10 mm in flower, to 15 mm in fruit. Corolla yellow, 11–14 mm, tubular to funnel-shaped, hairy externally, glandular within, limb small. Nutlet 1 (by reduction), 3.5–5 mm, scrobiculate-reticulate; beak deflexed.

HAB. Disturbed ground under *Quercus* scrub, *Pinus brutia* forest, earthy places in fields, rocky ledges, limestone cliffs, rocky hillside; alt. 800–1220 m; fl. Apr.-Jul.
DISTRIB. Locally common in the northern sector of the forest zone of Iraq. **MAM**. Sarsang, *O. Polunin* 5090! *Haines* W. 527! *Alkas* 18262!; Aqra, *Rawi* 11532!; jebel Khair, 25 miles W. of Aqra, *Thesiger* (BM 013839845); Zawita gorge, *Guest* 2204! Zawita, *Gillett* 8358! Zakho valley, *Guest* 2262!; Jabal Bekhair, *Rawi* 9419!; Jabal Khantur N.E. of Zakho, *Rawi, Tikriti & Nuri* 29020!; nr Dohuk, *Chapman* 2260!, *Emberger, Guest & Long* 15349!; Suwara Tuka, *Hunting Aerosurveys* 59.RC.61.2!; Atrush, *Guest* 3625!; Kaira (Qaiyara?) near Shar, *Rawi* 5661!

Regionally endemic to N. Iraq and N.W. Iran, rare in S.E. Anatolia. Subsp. MARDINENSIS Hub.-Mor. is more widely distributed in S.E. Anatolia.

6. **Alkanna frigida** *Boiss.*, Diagn. Pl. Orient. ser. 1(7): 32 (1846); Rawi in Dep. Agr. Tech. Bull. 14: 134 (1964); Riedl in Fl. Iranica [K. H. Rechinger] 48: 217 (1967); Khatamsaz, Flora Iran 39: 176 (2008).

Perennial, with a woody stock covered with remains of old bases of leaves, clothed throughout with short to long patent hairs, eglandular. Stems few, 8–15 cm, ascending. Basal leaves oblong to linear-oblong, 3–6(–10) × 0.5–0.7 (–1.2) cm, base tapering into a long petiole, apex obtuse; cauline leaves sessile. Cymes terminal, Bracts similar to leaves, sessile. Calyx ± 5 mm, lobed almost to base, lobes linear. Corolla yellow, ± 10 mm, glabrous, tubular to funnel-shaped. Nutlet 1 (by reduction), tuberculate, beak deflexed. (Description from specimen from Iran, 1868 Exped., *Haussknecht* s.n., K).

Cited by Riedl (l.c.) from Mt Hawraman, *Haussknecht* (no specimens seen).

Iran.

4. **PODONOSMA** Boiss.

Diagn. Pl. Orient. ser. 2, 11: 113 (1849); Johnston, J. Arnold Arbor. 35: 1 (1954);
Weigend & al. in Kubitzki, Fam. Gen. Vasc. Pl. 14: 70 (2016)

Shahina A. Ghazanfar

Perennial herbs, hispid, often glandular. Leaves sessile, elliptic to ovate-lanceolate. Inflorescence bracteose. Flowers pedicellate, pendulous. Calyx 5, lobed to base, accrescent in fruit. Corolla blue with yellowish apex (in Iraq plants), cylindrical, lobed at apex, lobes triangular; faucal scales absent. Annulus 10-lobed, pubescent. Stamens ± exserted; connective forming an acute apical appendage. Style exserted; stigma obscurely bilobed. Nutlets 4 or 1–2 by abortion, incurved, beaked apically, keeled ventrally, base shortly stipitate, minutely tuberculate or smooth.

Three species in N.E. Africa, E. Mediterranean and S.W. Asia; 2 species in Iraq.

Chacón, J., Luebert, F., Hilger, H.H., Ovchinnikova, S., Selvi, F., Cecchi, L., Guilliams, C.M., Hasenstab-Lehman, K., Sutorý, K., Simpson, M.G. & Weigend, M. (2016). The borage family (Boraginaceae s. str.): A revised infrafamilial classification based on new phylogenetic evidence, with emphasis on the placement of some enigmatic genera. Taxon 65(3): 523–546. http://dx.doi.org/10.12705/653.6
Cecchi, L., & Hilger, H.H. (2022). A prickly matter: nomenclatural synopsis of *Onosma* L. and its segregates (Boraginaceae). Plant Biosystems – An International Journal Dealing with all Aspects of Plant Biology 156 (5): 1076–1084, DOI: 10.1080/11263504.2021.1998241.

1. Leaves basal and cauline, elliptic to ovate-lanceolate, surfaces hispid,
   margins and median vein not glandular; nutlets tuberculate. . . . . 1. *P. orientalis*
   Leaves mostly cauline, broadly lanceolate, pubescent with long setae
   from glabrous tubercles, lower surface with short, dense hairs,
   margins and median vein with glandular hairs; nutlets smooth,
   sometimes with 3 inconspicuous longitudinal wrinkles on the back . 2. *P. sindjarensis*

1. **Podonosma orientalis** (*L.*) *Feinbrun* in Israel J. Bot. 26: 80 (1976); Riedl in Fl. Iranica [K. H. Rechinger] 48: 175 (1967); Feinbrun-Dothan, Fl. Palaest. 3: 73 (1978); Plant Biosystems, Appendix S1: 101 (2022).

Cerinthe orientalis L., Cent. Pl. I: 7 (1755).
Onosma orientalis (L.) L., Sp. Pl. ed. 2: 196 (1762); Riedl in Fl. Turkey 6: 337 (1978); Khatamsaz, Flora Iran 126: (2002); Taifour & El-Oqlah, Pl. Jordan Annot. Checkl.: 38 (2017).
Podonosma syriaca (Labill.) Boiss., Diagn. Pl. Orient. ser. 1, 11: 114 (1849); Rechinger, Fl. Lowland Iraq: 626 (1964).
Onosma syriaca Labill., Icon. Pl. Syr. 3: 8, t. 5 (1809); Rawi in Dep. Agr. Tech. Bull. 14: 140 (1964); Dinsmore in Post, Fl. Syria, Palest. & Sinai ed. 2, 2: 230 (1933).
Onosma rostellata Lehm., Pl. Asperif. Nucif. 2: 374 (1818); Rawi in Dep. Agr. Tech. Bull. 14: 139 (1964); Riedl in Fl. Iranica [K. H. Rechinger] 48: 174 (1967); Riedl in Fl. Turkey [P. H. Davis] [P. H. Davis] 6: 336 (1978); Khatamsaz, Flora Iran 39: 125 (2002). (See nomenclature remarks in Plant Biosystems, Appendix S1: 101 (2022)).
Onosma orientalis var. *palmyrena* Mouterde ex Charpin & Greuter in Mouterde, Nouv. Fl. Liban et Syrie 3(1): cover p. 3 (1978).

Perennial herb of rocks crevices and walls, 20–40 cm. Stems woody at base, many, simple and branched, ascending to prostrate or pendulous, often brittle, leafy covered with hispid long patent and shorter adpressed setae. Leaves green to greyish-green, elliptic to ovate-lanceolate, 15–35 × 5–10 mm, base sessile or tapering to a very short petiole, margins entire, apex acute, surfaces hispid. Flowers in terminal cymes, becoming pendulous. Bracts linear, 5–6 mm. Pedicel 2–5 mm. Calyx lobed to base, lobes 7–8 mm, densely hispid. Corolla blue with yellow apex, longer by ½ of calyx, apical lobes triangular, erect or reflexed. Anther connective well exserted from corolla, forming an acute apical appendage. Nutlets 2 (in most specimens seen), c. 3 mm, whitish grey, apically beaked, tuberculate. Fig. 74, 1–4.

HAB. Mountains and lower hills of northern Iraq, amongst and between rocks, limestone cliffs and in crevices in gorge; alt. 650–1220 m; fl. Apr.-Jul.
DISTRIB. **MAM**: Zawita, Omar, *Dabbagh & Qaisi* 45348!, 45337!; Jabal Bekher, *Omar* 37640!, *Rawi* 8443!; Zawita gorge, *Guest* 2182, 2224, *Emberger, Guest, Long, Schwan & Jusuf* 15379!; Aqra, *Rawi* 11463!; Dohuk gorge, 2 km from the town, *E. Chapman* 26282!; Ser Amadiya, Kani Sinji, *Dabbagh, Kaisi & H. Hamid* 46028!; Gali Mazurka (gorge), 2 km N.W. of Amadiya, *Dabbagh & Jasim* 46910!; Bamerne 20 km N.W. of Sarsang, *Kaisi & H. Hamid* 45942!; Zawita gorge, *Guest* 3714!; Jabal Khantur, *Rechinger* 10759!; Zawita, *Guest* 4940!; 21 km from Dohuk, *Rechinger* 11539 (W!); Kirkuk, *Rechinger* 12502! (W). **MRO**: Gali Ali Beg, *Karim, Hamid & Jasim* 40914!; Haji Umran, *Chapman* 11928!; Mt Avroman, nr Tawila, *Rechinger* 10132! (W); Darimar, Hauraman mt., *Gillett* 11830! Shakh-i-Harir nr Harir, Emberger, *Guest et al.* 15437A!; Handren nr Rowanduz, *Guest* 13008!; Shaqlawa, Haines W. 768!; Sefin Dagh above Shaqlawa, *Gillett* 8082!; Sefin Dagh, *Rawi* 9083!; Kuh-e Sefin, above Shaqlawa, *Bornmüller* 1597!**MSU**: Azmir, *Karim* 39325!; E. of Darband-i Khan, *Hussein & Barkley* 7725!; Pir Omar Gudrun (Pira Magrun), *Gillett* 7758!; Qopi Qara Dagh, *Poore* 370!; Betas, *Rawi* 8446!; Khurmal, *Rawi* 8940!; ?Fakirma, (Bakurman?) *Rawi* 8510!; Amoret, nr Qaradagh, *Haines* W.1116!; Qara Dagh, *Poore* 437!; Darband-i Bazian nr Chemchemal, *Rechinger* 10604! (W); **MJS**: Jabal Sinjar, *Qaisi & Wedad* 52064!, *Omar, Khayat & Qaisi* 52617!, *Gillett & Rawi* 5930!, *Gillett* 11144!; Jabal Sinjar, *Kaisi, Khayat & Karim* 50868!; Jabal Sinjar, *Gillett* 1113225 km N.W. of Rania, Zewa village, *Rawi, Nuri & Kass* 28934!; Warta village c. 30 km N.W. by N. of Rania, *Rawi, Nuri & Kass* 28864!;

Iran, Lebanon, Palestine, Jordan, Syria, Turkey.

2. **Podonosma sindjarensis** (*Riedl*) *L.Cecchi & Hilger*, Plant Biosystems, Appendix S1 (2022).

Onosma sindjarensis Riedl, Anz. Österr. Akad. Wiss., Math.-Naturwiss. Kl. 101: 361 (1964). Type: Mesopotamia: Jebel Sinjar, above the town of Sindjar, 8. VI. 1910, *Handel-Mazzetti* No. 1378 (WU, holo. & iso.).

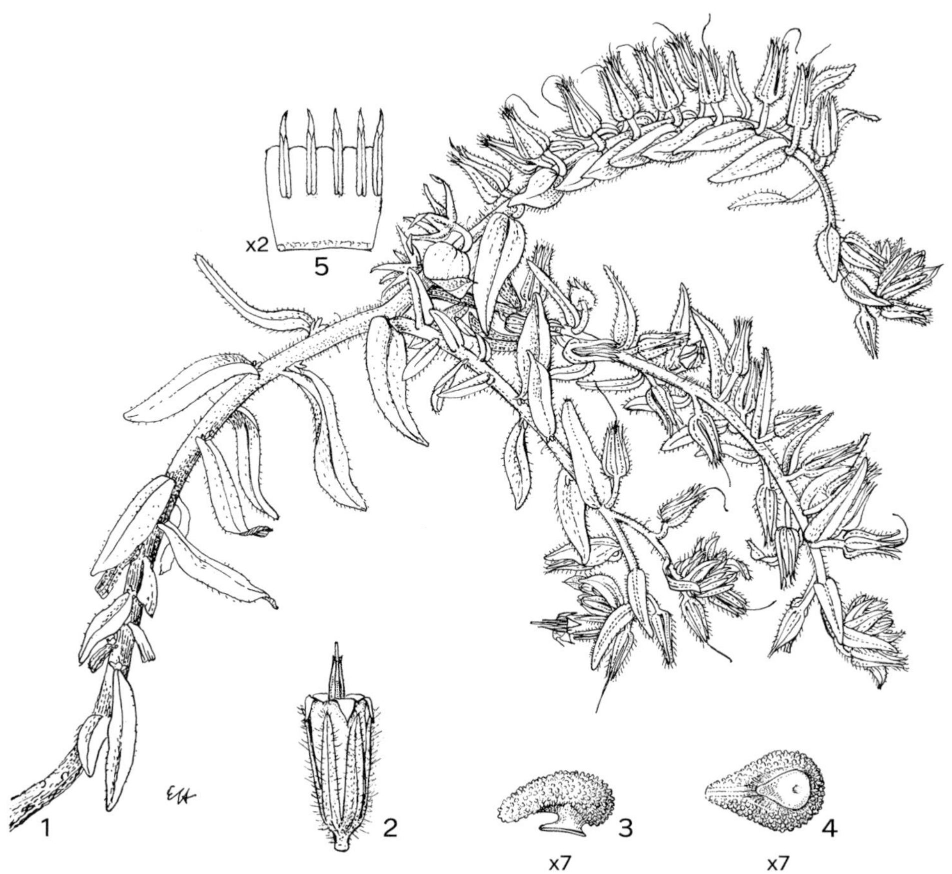

Fig. 74. **Podonosma orientalis.** 1, habit × ½, flowering branch × 1; 2, flower × 2; 3, corolla opened × 2; 4, nutlet side view and from below × 7. Reproduced with permission from Feinbrun-Dothan, Flora Palaestina, 3: Plates, f. 119 (1977). Drawn by Esther Huber. © The Israel Academy of Sciences and Humanities.

Perennial, with a branched woody base. Stems numerous, up to 35 cm, erect to ascending with dense indumentum of long setae arising from glabrous tubercles, eglandular, and shorter hairs, usually glandular. Leaves mostly cauline, broadly lanceolate, 15–32 × 5–15 mm, base attenuate, apex acute to acuminate, pubescent with long setae from glabrous tubercles and shorter hairs, lower surface with short, dense hairs, margins and median vein with glandular hairs. Inflorescence scorpioid, becoming lax and elongating to 10 cm. Bracts similar to leaves, base ± amplexicaul. Flowers pendulous. Pedicel 2–3(–4) mm, in fruit elongating to 10 mm. Calyx lobed to base, lobes ± 6 mm, triangular-lanceolate, elongating to 8 mm, densely hispid with long and short hairs, glandular. Corolla 7–8 mm, longer than calyx, cylindrical, glabrous, lobes c. 1 mm, triangular, reflexed. Annulus sparsely pubescent. Stamens inserted about the middle of corolla; anthers exserted to about 3 mm above corolla, connective tip sharp, 2 mm. Nutlets c. 2.5 mm, smooth, sometimes with 3 inconspicuous longitudinal wrinkles on the back, white, shiny, glabrous, beak short, curved downwards, very short-stalked, surrounded by a very distinct flattened disciform neck, reaching 2/3 of the length of the nutlet. [No material seen; description from the protologue].

HAB. Known from a single collection on Jebel Sinjar, on mountain, amongst stones; alt. ± 500 m (?); fl. Jun.

DISTRIB. **MJS**: Jabal Sinjar, above the town of Sindjar, *Handel-Mazzetti* 1378 (type).

Endemic and known only from the type locality.

5. **ARNEBIA** Forssk.

Fl. Aegypt.-Arab. 62 (1775)
Weigend et al. in Kubitzki, Fam. Gen. Vasc. Pl. 14: 71 (2016)

Shahina A. Ghazanfar

Annual herbs, hispid; root containing a purplish dye. Leaves usually oblong, entire, blade tapering into petiole. Inflorescence much-branched from upper part of stem; cymes bracteate. Flowers heterostylus or not. Calyx deeply divided, lobes linear to lanceolate, ± accrescent in fruit. Corolla funnel- or salver-shaped, yellow; throat lacking scales and folds; tube with a nectariferous ring at base; annulus present or absent Stamens included. Style filiform, divided; stigmas 2 or 4. Nutlets 4, glabrous, depressed-ovoid, globose or trigonal, with acute to acuminate beak and ventral keel, elaiosome present; attachment scar flat.

About 30 species, mainly S.W. and C. Asia, Himalayas; also N.E. Africa and S.E. Europe; 4 species in Iraq.

Arnebia is recorded by Forsskål (op. cit. p. 38) as coming from the Arb. *sajaret el arneb*, meaning the arneb tree.

1. Corolla pale blue to pinkish blue, almost glabrous; nutlets almost
   smooth, not keeled on ventral side. . . . . .4. *A. tinctoria*
   Corolla yellow, hairy; nutlets sparsely verrucose to tuberculate-rugose,
   keeled on ventral side 2
2. Base of calyx base prominently angled, nerves projecting as 5 sparsely
   setose wings; stigmas 4. . . . . .3. *A. decumbens*
   Base of calyx not angled; stigmas 2. 3
3. Calyx 5–8 mm, hardly elongating in fruit; flowers heterostylous . . . . . .2. *A. hispidissima*
   Calyx 8–10 mm, elongating in fruit to c. 30 mm; flowers not
   heterostylous . . . . .1. *A. linearifolia*

1. **Arnebia linearifolia** *A.DC.*, Prodr. 10: 95 (1846) incl. subsp. *desertorum* H.Riedl.; Oesterr. Bot. Z. 109: 74 (1962). Dinsmore in Post, Fl. Syria, Palest. & Sinai ed. 2, 2: 239 (1933); Rawi in Dep. Agr. Tech. Bull. 14: 135 (1964); Rechinger, Fl. Lowland Iraq: 509 (1964); Feinbrun-Dothan, Fl. Palaest. 3: 70 (1978); Y. Nasir in Fl. Pakistan [Ali & Y. Nasir] 191: 60 (1989); El-Hadidi & Boulos in Fl. Egypt 2: 294 (2002); Abdulridha, Taha & Wedad, Ecology and Fl. Basra: 235 (2016); Taifour & El-Oqlah, Pl. Jordan Annot. Checkl.: 50 (2017).

A. flavescens Boiss., Diagn. ser. 1, 11: 117 (1849).
Echioides linearifolium Rothm. in Feddes Rep. 49: 56 (1940).
Lithospermum aucheri Johnston in J. Arn. Arb. 33: 328 (1952).

Annual herb. Stems to 20 cm, densely covered with long spreading tubercle-based and short appressed hairs. Leaves oblanceolate to oblong, obtuse, base tapering into the petiole. Inflorescence remaining dense in fruit; bracts as long as the calyx. Calyx 8–10 mm; lobes lanceolate, hispid; fruiting calyx elongating to 30 mm, midvein not thickened, base inflated. Corolla yellow; tube 10–12 mm; limb ± 2.5 mm in diameter, lobes obtuse. Annulus present. Style bifid; stigmas 2. Nutlets flattened-trigonal, glossy, sparsely verrucose. Fig. 75, 1–4.

HAB. Sandy and gravelly desert, sandy soil on rocky limestone plateaux; alt. 50–620 m; fl. Mar.-Apr.
DISTR. Widespread, mainly in the desert and subdesert zones of Iraq. **FUJ.** ruins of Al-Hadhr, *Ani* 9394! **FPF.** 25 km E. of Badra, *Rawi* 20756! **LEA.** Garmashaya, 70 km N.E. of Kut, *Rawi* 17992! **DLJ.** 20 km from Falluja, *Rawi & Gillett* 6704! **DLJ.** 65 km N.N.W. of Falluja, *Rawi* 20240!Tayrat, 60 km N.W. of Rawa, *Chakrawarty, Rawi, Khatib & Alizzi* 31250!; 94 km NNW of Falluja, *Rawi* 20316! Falluja desert, *Haines* W. 158! **DWD.** 5 km S.E. of Ramadi, *Mokhtar* 33332!; nr Ana, *Hamid* 39082!; 18 km S. of Rutba, *Rawi* 14612!; 3 km E. of Rutba, *Rechinger* 12624 (type of subsp. *desertorum* H. Riedl)!; 5 km W. of Rutba, *Rawi & Nuri* 27064!; 15 km S. of Rutba, *Khatib & Hazim* 32439!; 30 km N. of Rutba, *Chakrawarty, Rawi, Khatib & Alizzi* 31512!; 43 km N. of Rutba, *Rawi & Khatib* 32230!; 60 km N. of Rutba, *Rawi & Khatib* 32304!; 95 km N.E. of Rutba, *Rawi & Khatib* 32321!; 60 km N. of Rutba, *Rawi & Khatib* 32307!; 130 km W. of Ramadi, R*awi & Khatib* 32160!; 12 km N. of T1, on road to Husabah, *Chakr., Rawi, Khatib & Alizzi* 31727! 10 km N. of Kubaisa, *Hamid* 39024! **DSD.** near Chilawa, Guest, *Rawi & Rechinger* 17201!

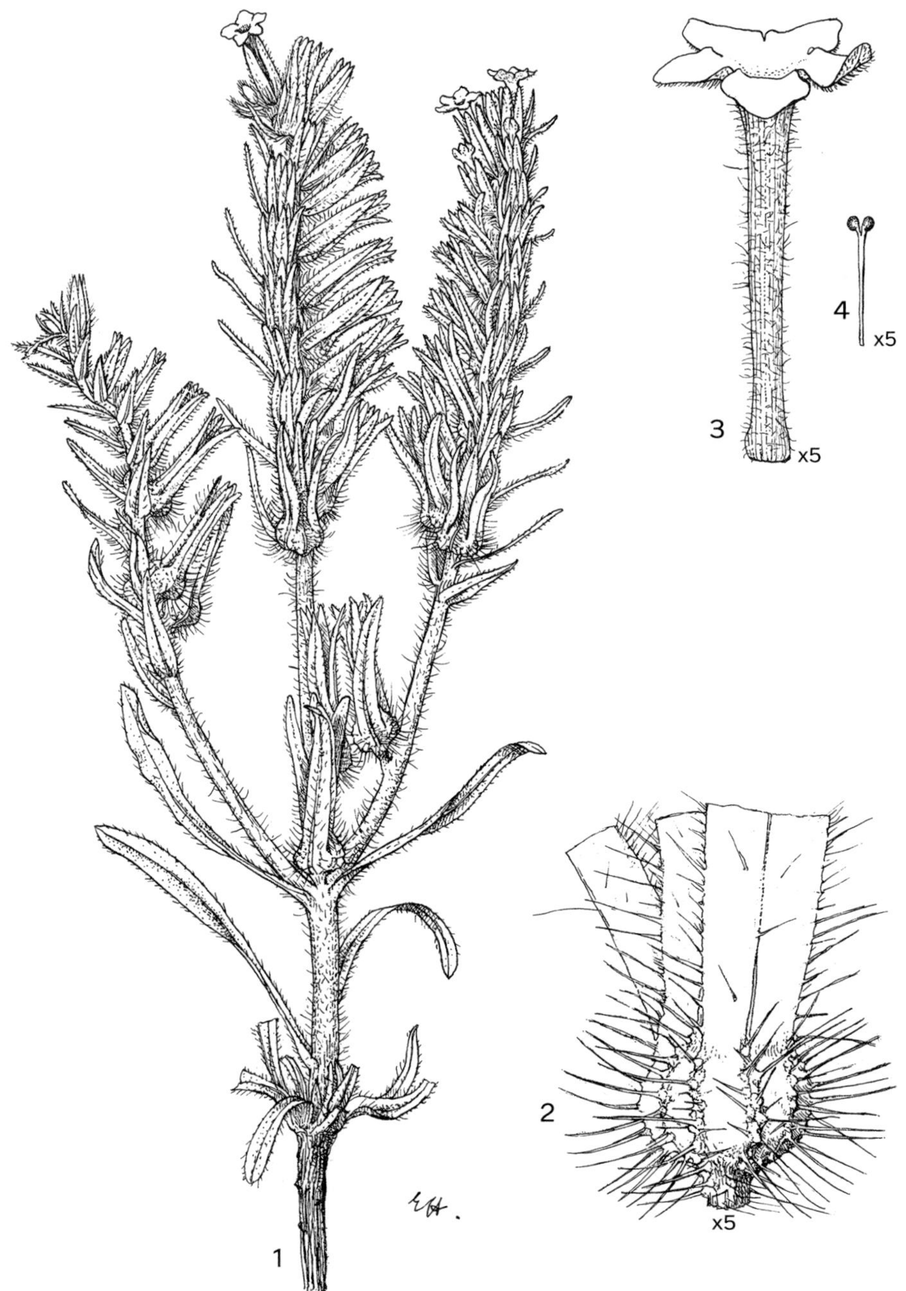

Fig. 75. **Arnebia linearifolia**. 1, habit × 1; 2, flower × 5; 3, stigma × 5; 4, base of fruiting calyx × 5. Reproduced with permission from Feinbrun-Dothan, Flora Palaestina, 3: Plates, f. 114 (1977). Drawn by Esther Huber. © The Israel Academy of Sciences and Humanities.

17217A!; Zarqa, 12 km S.W. of Samawa, *Alkas* 17676!; Jabal Sanam, *Haines* W.1115! *Guest, Rawi & Rechinger* 16981!; 50 km N. by W. of Aidaha, *Guest, Rawi & Rechinger* 19204!; 59 km from Ahidhan to Ansab, *Fawzi, Hazim & Hamid* 38837!; midway between Shabicha and Nukhail, *Alizzi & Omar* 35495!; 23 km S.E. by S. of Zubair, *Guest, Rawi & Rechinger* 16843!; 35 km from Najaf, *Fawzi, Hazim & Hamid* 38600!; 33 km WNW of Ansab, on Saudi border, *Guest, Rawi & Rechinger* 19037!; Wadi al-Khirr, 90 km N.W. by W. of Shabicha, *Guest, Rawi & Rechinger* 19426!; Umm Qasr, *Alizzi & Omar* 35006! (subsp.

desertorum); *Ani* 39922!; nr Ansab, *Guest, Rawi & Rechinger* 18958!; Safai al-Moghif, 100 km WSW of Basra, *Guest, Rawi & Rechinger* 17253!: Nukhaila station, 40 km W. by N. of Basra, *Guest, Rawi & Rechinger* 17357!; nr Safwan, *Alizzi & Omar* 34949! *Guest, Rawi & Rechinger* 17022!; 59 km from Akidha to Ansab, *Fawzi, Hazim & Hamid* 38835!; 10 km from Busaiya, *Karim, Nuri, Hamid & Kadhim* 40378!; 12 km ESE of Salman, *Guest, Rawi & Rechinger* 18842!; 15 km from Ansab to Samah, *Fawzi, Hazim & Hamid* 38855!; Umm Qasr, *Ani* 39922!

CHAHAL (Ir.-Safwan & Salman), *Guest, Rawi & Rechinger* 17022 & 18842.

N. Africa, Egypt (Sinai), Arabia (Saudi Arabia, Kuwait), Iran, Afghanistan, Pakistan.

2. **Arnebia hispidissima** (*Sieber ex Lehm.*) *A.DC.*, Prodr. [A.P. de Candolle] 10 : 94 (1846); Riedle in Fl. Iranica [K. H. Rechinger] 48: 155 (1967); Y. Nasir in Fl. Pakistan [Ali & Y. Nasir] 191: 65 (1989).

Lithospermum hispidissimum Sieber ex Lehm., Icon. Descrip. Nov. Stirp.: 1: 23, t. 39 (1821).
Anchusa hispidissima Sieber ex A.DC., Prodr. [A.P. de Candolle] 10 : 94 (1846), nom. nud., not validly published.

Annual, (rarely perennial) herb 5–25 cm, sometimes with a woody base. Stems erect to ascending, branched, hispid with short ± appressed and long tubercle-based hairs. Leaves lanceolate or linear-lanceolate, 10–50 × 2–7 mm, sessile, base tapering or shortly petiolate, apex acute to obtuse, margins entire, white-hispid. Bracts linear-lanceolate, equalling the calyx. Inflorescence of short cymes, elongating in fruit to 10 cm; flowers heterostylus. Calyx 5–8 mm, scarcely increasing in fruit; lobes linear, divided almost to base, white-hispid. Corolla bright yellow, 10–15 mm; tube cylindrical; limb ± third of the tube, with broadly rounded 5 lobes, hairy outside, glabrous or hairy inside. Annulus present. Tip of style bifid; stigmas 2. Nutlets 4, with small tubercles. Roots giving a blue dye.

HAB. Stony and sandy desert; alt. 50–150 m; fl. Feb.-Apr.
DISTR. Western desert. **DWD**: Wadi al Mira, *Mohammed U. Musa* 57784!, 57785!, ibid., *Mohammed U. Musa* 22 Mar 2000, s.n.; al Salihi, *Mohammed U. Musa* 57786!; Wadi Ubaidh, *Mohammed U. Musa* 57787!

Africa (Nigeria to Somalia, Egypt, the Arabian Peninsula, and from Iraq, Iran, Pakistan, India to Tibet.

3. **Arnebia decumbens** (*Vent.*) *Cosson & Kralik* in Bull. Soc. Bot. Fr. 4: 402 (1857); Dinsmore in Post, Fl. Syria, Palest. & Sinai ed. 2, 2: 239 (1933); Rawi, in Dep. Agr. Tech. Bull. 14: 135 (1964); Rechinger, Fl. Lowland Iraq: 509 (1964); Feinbrun-Dothan, Fl. Palaest. 3: 69 (1978); El-Hadidi & Boulos in Fl. Egypt 2: 292 (2002); Abdulridha, Taha & Wedad, Ecology & Fl. Basra: 235 (2016); Haloob & Al-Kaisi, Illustr. Fl. Lowland Iraq 1(1): 64 (2016); Taifour & El-Oqlah, Pl. Jordan Annot. Checkl.: 50 (2017).

Lithospermum decumbens Vent., Descr. Pl. Hort. Cels. t. 37 (1801). Type 'inter Baghdad et Hit', *Olivier & Bruguière.*
L. micranthum Viv., Fl. Lib. Spec. 10: t. 1 f. 4 (1824).
L. cornutum Ledeb., Fl. Petrop. 1: 22 (1835).
Arnebia cornuta (Ledeb.) Fisch. & Mey. in Ind. Sem. Hort. Petrop. 1: 22 (1835).
Lithospermum tubatum Bertol., Misc. Bot. 1: 15 (1842).
Arnebia viviani Cosson & Dur. in Ann. Sci. Nat. ser. 4, 1: 240 (1854).
Arnebia decumbens var. *macrocalyx* Coss. & Kralik in Bull. Soc. Bot. France 4: 403 (1857).
Echioides decumbens (Vent.) Rothm. in Feddes Repert. 49: 56 (1940).

Annual much-branched herb, 15–25 cm, densely patent-hispid. Leaves lanceolate, linear to oblanceolate, to 30 × 8 mm, sessile, adpressed to patent-hispid. Inflorescence compact before and during anthesis, greatly extending in fruit. Bracts equalling or longer than calyx. Calyx 5–9 mm at anthesis, 6–10 mm in fruit, divided to more than 3/4, lobes linear, acute, converging in fruit to enclose nutlets; lobes oblong-lanceolate, base more prominently angled, nerves projecting as 5 sparsely setose wings; accrescent in fruit. Corolla pale yellow; tube 10–12 mm; limb ± 2.5 mm in diameter; annulus present. Tip of style twice bifid; stigmas 4. Nutlets about 2.5 mm, ovoid-trigonal, with acuminate beak, tuberculate-rugose. Fig. 76, 1–9.

HAB. Stony hillsides, stony desert with sandy and silty soil; alt. 400–550 m; fl. Feb.-Apr.

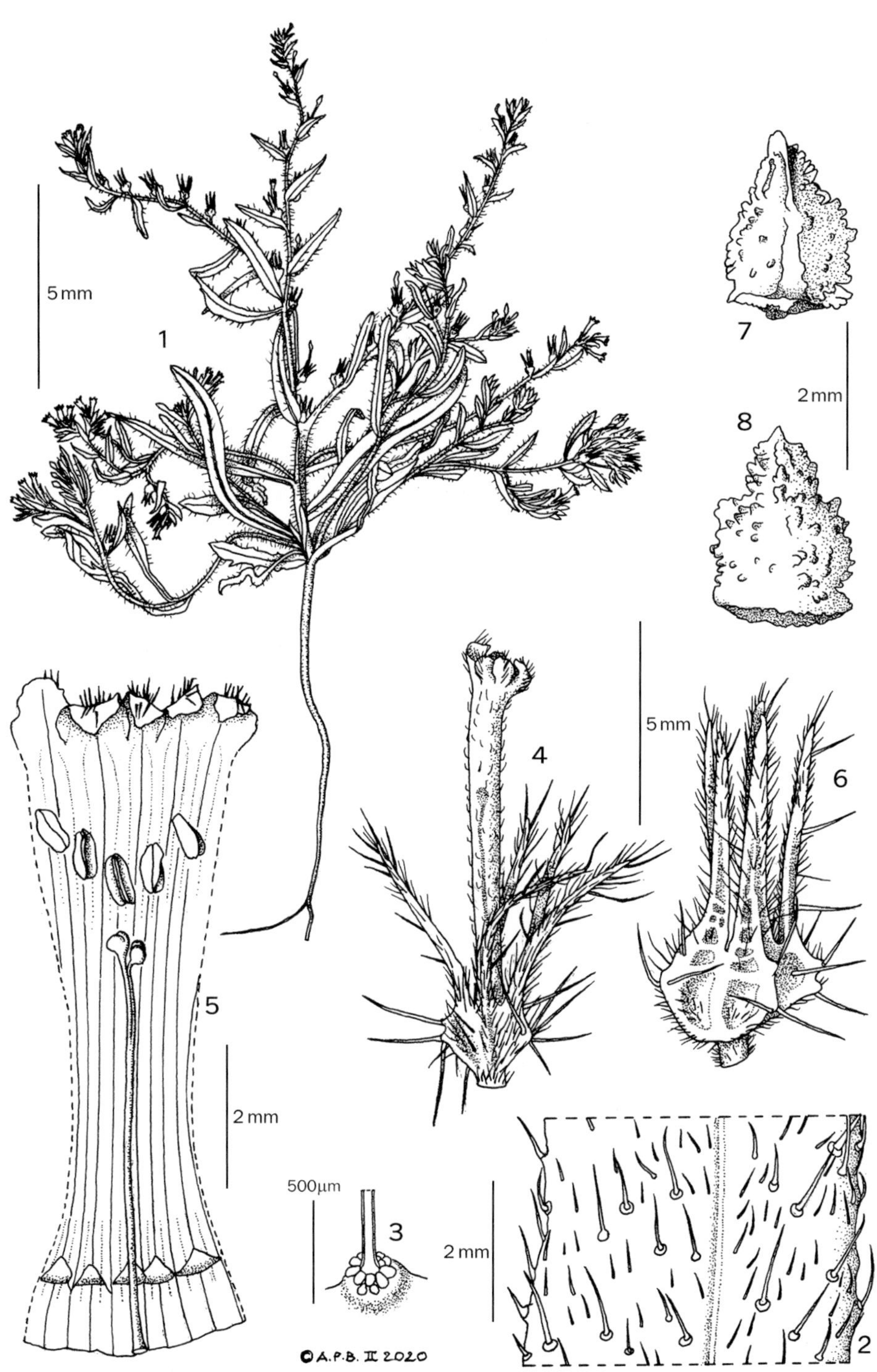

Fig. 76. **Arnebia decumbens**. 1, habit; 2, indementum on adaxial surface of leaf; 3, base of large leaf hair; 4, flower; 5, corolla, opened; 6, fruiting calyx; 7, nutlet, inner face; 8, nutlet, outer face. 1–3 from *Guest* 1394; 4–5 from *Hunting Aerosurveys* 61 RA 39; 6–8 from *Rogers* 293. Drawn by © A.P. Brown 2020.

DISTR. Widespread, mainly in the desert and subdesert zones of Iraq; alt. 5-450 m; fl. Feb.-Apr.
MSU. Darbandikhan, *Haines W.*1920! **FUJ**. Qaiyara, *Bayliss* 22! **FUJ**: ruins of Al-Hadhr, *Ani* 9395! **FNI**.
Mosul, hill of Nineveh, *Lazar* 3361!; **FKI**. Tuz, *Guest* 1394!; N.E. of Sar Qa'lah, *Poore* 516!; Kirkuk, road
to Bab Gurgur oil wells, *Guest* 1363!; Jabal Hamrin on Baghdad-Kirkuk road, *Rawi & Gillett* 10248!;
5 km S. of Shwana, *Rawi* 14816!; banks of river Zab nr Kirkuk, *Grigg* 41! **KI/FPF**. Jabal Hamrin, *Guest*
682! **FPF**. 25 km S.E. of Mandali, *Rawi* 20717!; Laqlaq, *Kaisi, Thamer & Salah* 51285!; Jabal Darawishka,
nr Khanaqin, *Guest* 1788!; Pilkana, 10 km E. of Khanaqin, *Rawi* 12733! **DLJ**. 55 km N.W. of Falluja,
Rawi 20225!; Takrit, *Omar & Hamid* 36280!; Rawa, *Hamid* 39111!; 75 km N.W. of Baiji, *Barkley &*
Haddad 5418!; Manaiyif, between Hadhr and Rawa, *Hashimi* 43630!; 25 km N. of Baiji to Mosul, *Ani*
9708! **DGA**. 50 km E. of Samara, at Asila, *Rawi* 20449!; 4 km E. of Samarra, *Rawi* 20324!; Mussaige
al-Ghislan in Dour, *Jenan, Shahira, Wedad & Kalaf* 49903! **DWD**. 43 km N. of Rutba, *Rawi & Khatib*
32229!; 5 km W. of Rutba, *Rawi & Nuri* 27065!; 18 km S. of Rutba, *Rawi* 14626!; 13 km E. of K3, *Rawi,*
Chakrawarty, Khatib & Alizzi 32949!; 20 km W. of Nukhaib, *Rawi* 30989!; 50 km S.E. of Rutba, *Jenan*
al-Mokhtar 33365!; 138 km W. of Ramadi, *Gounot & Barkley* 5265!; between Ana and Qa'im, *Hamid*
39181! 39202! 39229! 39207!; Wadi al-Ghadaf, *Rawi* 14840!; Jazira S. of Sinjar, *Guest* 13374! **DSD**. 70
km S.W. of Samawah, *Guest, Rawi & Rechinger* 18731!; 20 W. by S. of Safwan, *Guest, Rawi & Rechinger*
17023!; W.of Zubair, *Guest, Eig & Zohary* 5017!; Golaib, 30 km E. of Salman, *Guest, Rawi & Rechinger*
18756!; Jumaima, *Guest, Rawi & Rechinger* 19104!; Ansab, *Guest, Rawi & Rechinger* 18957!; Shabicha,
Rawi & Gillett 62284!; Zarqa, *Alkas* 17676!; Salman, al 'Ain al Hilwa, *Karim, Nuri, Hamid & Kadhim*
40113!; between Najaf and Shahbicha, *Omar, Karim & Hamid* 36991!; 85 km S.W. of Basra, *Guest, Rawi*
& Rechinger 17096!; W. of Jarishan, 90 km S.W. of Basra, *Guest, Rawi & Rechinger* 17139!; 15 km E. by
S. of Busaiya, *Guest & Rawi* 14211!; 29 km S. of Busaiya, *Botany Staff* 42372! 65 km W. of Basra, *Botany*
Staff 42413!; Samawa, *B. & Y. Matti* 346!; 50 km from Ansab to Samah, *Fawzi, Hazim & Hamid* 38866!;
20 km from Rumana to Rawa, *Khayat & Hamad* 51644!; Nukhaila station, 40 km W. by N. of Basra,
Guest, Rawi & Rechinger 17358!; Chilawa, 110 km S.W. by S. of Basra, *Guest, Rawi & Rechinger* 17201!;
Um Qasr, *Guest, Rawi & Rechinger* 16885!; 85 km S.W. of Basra, *Guest, Rawi & Rechinger* 17096!; W.
of Jarishan, 90 km S.W. of Basra, *Guest, Rawi & Rechinger* 17139!; 15 km E. by S. of Busaiya, *Guest &*
Rawi 14211! **LEA**. Abulkhasib, *Khatib & Alizzi* 33487!; Ba'quba, *H.H.*, 21/4/1920!; Nahrwan, *Hunting*
Aero Surveys 61 (R.A. 39)! **LCA**. Babylon, *Bornmüller* 530! *Rogers* O293!; Daltawa (Khalis), *Rogers* O293!;
Babylon, *Guest* 1683!; Baghdad, *Aucher* 2389!; Iskandariya, *Rawi & Walter* 18367!; 70 km N.W. of
Fallujah, *Chakrawarty & Ali* 30379!

KHUZAIMAH (Ir.-Mosul), *Lazar* 3361; CHAHALL or HELIMA (Ir.-Shahbicha), *Rawi*
& Gillett 6284; CHAHAL (Ir.-Basra), *Guest et al.* 17096, (Ir.-Busaiya), *Guest & Rawi* 14211;
HASSAR (Ir.-Sinjar), *Guest* 13374.

Var. **macrocalyx** Coss. & Kral., Bull. Soc. Bot. Fr. 4: 403 (1857) is recognised on its larger
fruiting calyx (15 mm) as apposed to var. **decumbens** where the fruiting calyx is 8–10
mm (Feinbrun-Dothan, l.c.) This distinction is not clear in our material. I have therefore
included var. *macrocalyx* in the type variety.

Africa through the Arabian Peninsula, Palestine, Syria, Turkey, Caucasus to Iran,
Afghanistan, Pakistan and China.

4. **Arnebia tinctoria** *Forssk.*, Fl. Aegypt.-Arab: 63 (1775); Dinsmore in Post, Fl. Syria, Palest.
& Sinai ed. 2, 2: 239 (1933); Rawi in Dep. Agr. Tech. Bull. 14: 135 (1964); Rechinger, Fl.
Lowland Iraq: 509 (1964); Feinbrun-Dothan, Fl. Palaest. 3: 69 (1978); El-Hadidi & Boulos
in Fl. Egypt 2: 291 (2002); Abdulridha, Taha & Wedad, Ecology & Fl. Basra: 235 (2016);
Taifour & El-Oqlah, Plants Jordan Annot. Checkl.: 50 (2017).

A. *tetrastigma* Forssk., Fl. Aegypt.-Arab. 62 (1775) nom. nud.

Annual herb, 4–15 cm, branching from base, with adpressed, tuberculed based hairs. Leaves
oblong-linear-lanceolate, 10–40 × 2–8 mm, base tapering, apex obtuse. Inflorescence compact
in flower and fruit. Bracts linear. longer than calyx. Calyx ± 1 cm, hirsute, lobes linear, acute,
accrescent in fruit. Corolla pale blue; tube glabrous, about as long as the calyx; annulus
absent. Tip of style twice bifid; stigmas 4. Nutlets about 2 mm, smooth, glossy, not keeled
ventrally.

HAB. Compact sandy soil; sandy hills; alt. 70–140 m; fl. Mar.
DISTR. Rare, only recorded from the southern desert zone. **DSD**. 2 km S.S.W. of Tal Laku (near Ur),
Guest 15211D!; Al-Batin 2 (Kuwait border 105 km S.E. of Busaiya), *Guest & Mahallal* 15256A!; Al-Zarqa,
12 km S. of Samarra, *Alkas* 17676!

CHAHĀL (Ir.-Al Batin), *Guest & Mahallal* 15256A.

Arnebia deserti-syriaci Zohary listed in Rawi (1964, p. 135) is not recorded in IPNI; it is likely a misidentification and has never been published.

Libya, Egypt, Kuwait, Palestine, Arabia.

6. **CERINTHE** L.

Sp. Pl. 1: 136 (1753)
Weigend et al. in Kubitzki, Fam. Gen. Vasc. Pl. 14: 71 (2016)

C.C. Townsend
Revised by Shahina A. Ghazanfar

Biennial or perennial herbs, subglabrous, ± glaucous. Leaves often with calcareous tubercles, subglabrous. Inflorescence terminal, branched, cymes somewhat nodding at anthesis, elongating, becoming erect, and straightening in fruit; bracts conspicuous. Calyx divided to at least 1/2, with unequal lobes; accrescent in fruit. Corolla yellow, maroon or violet-purple, tube lacking scales and folds. Stamens included or reaching mouth of corolla tube, anthers sagittate to hastate, with appendages at base and apex; connate at base; style long, exserted; stigmas 1 or 2. Nutlets 2, united into 2-seeded carpids, ± ovoid-globose, 3–5 × 3–4 mm, shortly beaked, chestnut brown to black, smooth.

A genus with six species in Europe, Caucasus, N. Africa to Iran; a single species in Iraq.

Cerinthe owes its name from the belief in ancient Greece that bees obtain wax from its blossoms to make honeycomb: Gr. κηρος, *kiros*, wax and ανθος, *anthos*, flower.

I.M. Johnston, Studies in the Boraginaceae. J. Arnold Arbor. 35: 1–4, 65–69 (1954).

1. **Cerinthe minor** *L.*, Sp. Pl.: 137 (1753); Dinsmore in Post, Fl. Syria, Palest. & Sinai ed. 2, 2: 223 (1933); Rawi in Dep. Agr. Tech. Bull. 14: 135 (1964); Riedl in Fl. Iranica [K. H. Rechinger] 48: 213 (1967); Khatamsaz, Fl. Iran 39: 181 (2002); Gabrielian & Fragman-Sapir, Flow. Transcaucasus & Adj. Areas: 100 (2008).

> *C. auriculata* Ten., Ind. Sem. Horti Bot. Neap. 10 (1830).
> *C. maculata* All., Fl. Pedem. 1: 178 (1785).
> *C. minor* subsp. *auriculata* (Ten.) Domac in Bot. J. Linn. Soc. 65: 260 (1972); Edmondson in Fl. Turkey
> [P. H. Davis] 6: 376 (1978).

Stems erect, 25–40 cm, simple or branched, with many leaves. Lower leaves spatulate to oblanceolate, shrivelled at anthesis; cauline leaves lanceolate-elliptic to obovate, to 8 × 3 cm, obtuse, sessile and amplexicaul with a somewhat cordate base. Pedicels pilose. Calyx divided almost to base, lobes lanceolate, acute, margins ciliate. Corolla broadening towards apex, colour not recorded on Iraqi specimens but elsewhere variously yellow, maroon, purple or violet; tube divided to 1/2, lobes lanceolate, acute, erect. Nutlets 2, 3.5 × 3 mm, ovoid-globose. Fig. 77, 1–9.

HAB. Shady rocks on north-facing slope in *Quercus* forest; alt. ±1500 m; fl. Jul.-Aug.
DISTRIB. Occasional in the upper forest zone of Iraq. **MAM.** Sarsang, *Haines* W.1325! **MRO.** mt Halgurd, *Rawi* 13741!; Kani Rash in Qurnago valley, *Rawi* 26671.

C. and S. Europe, Balkans, Caucasia, N. Turkey N. & N.W. Iran; W. Syria.

7. **ONOSMA** L.

Sp. Pl. ed. 2: 196 (1762)
Johnston in J. Arnold Arbor. 32: 201–225 (1951); Weigend & al. in Kubitzki, Fam. Gen. Vasc. Pl. 14: 72 (2016)
Choriantha Riedl in Österr. Bot. Z. 108: 399 (1961)

Shahina A. Ghazanfar

Biennial or perennial herbs or small shrubs covered with dense hispid setae; setae (rarely hairs) arising form rounded, slightly raised, multicellular tubercles with small spinules or setules arranged stellately around the base of setae or tubercles glabrous. Leaves linear

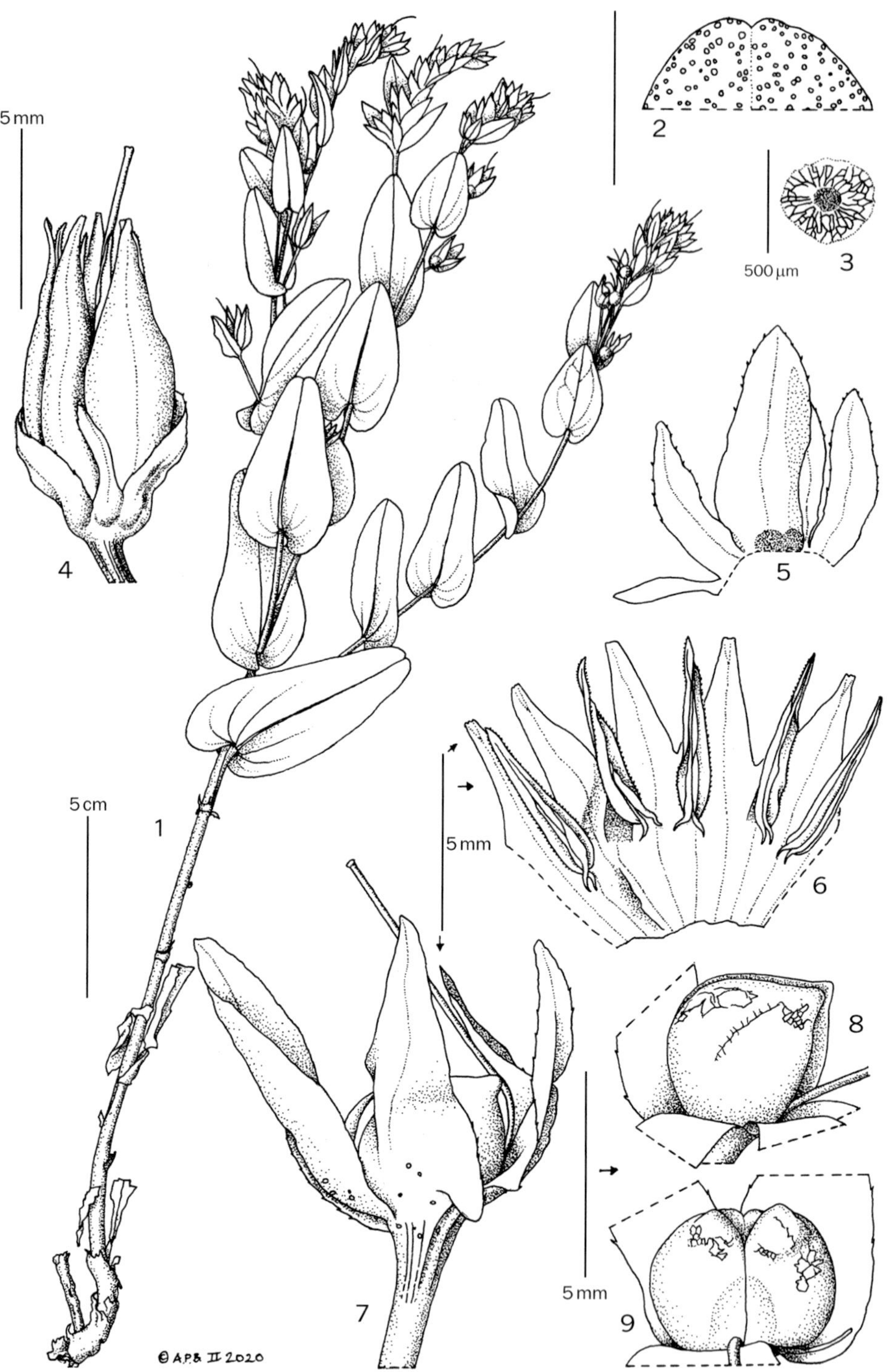

Fig. 77. **Cerinthe minor**. 1, habit, fruiting plant; 2, detail leaf tip, adaxial surface showing calcareous tubercles; 3, tubercle detail; 4, flower; 5, calyx (opened); 6, corolla, opened; 7, fruit with calyx; 8, nutlet, side view; 9, nutlet face view. 1–9 from *Rawi* 13741; ! from *Rawi* 26671 (rootstock). Drawn by © A.P. Brown 2020.

to narrow obovate. Inflorescence, cymose, scorpioid, bracteate; bracts leaf-like. Flowers pedicellate with pedicel elongating in fruit, pendulous. Calyx 5, usually lobed to base, often accrescent and enveloping the fruit. Corolla tubular, lobes small, yellow, whitish or yellowish-white. Annulus present, irregularly lobed, hairy or glabrous. Staminal filaments adnate to corolla tube in the lower half; anthers connivent and forming tube, connective forming a distinct appendage at apex. Style usually exserted; stigmas 1–2. Nutlets 4, or 1–2 by abortion, ovoid, smooth or tuberculate, apically beaked, dorsally convex, ventrally keeled.

About 150–180 species found in Europe, N.W. Africa and tropical and temperate Asia; 21 species in Iraq.

Nasrollahi et al. (Nordic J. Bot. 37: 2019) showed that *Onosma* as currently circumscribed do not form a monophyletic group. Majority of the species they analyzed belonged to a single clade, *Onosma* s.s. This included all species of the subsections *Onosma*, *Haplotricha* and *Heterotricha*, however none of the subsections were monophylectic. *O. rostellata* Lehm. appeared outside the *Onosma* clade (in a subclade of Sino-Indian species and *Maharanga emodii*). As also shown earlier (Thomas et al. 2008, Cecchi & Selvi 2009, Chacón et al. 2016, 2019) *Podonosma orientalis* (= *Onosma orientalis*) was more closely related to *Alkanna* and belonged to the 'Alkanna clade' only distantly to *Onosma*. Their results suggested that *O. rostellata* and the Sino-Indian species should be treated under a new genus (*Protonosma* or *Maharanga*).

Onosma is derived from ονος, *onos*, ass or donkey, and οσμη, *osme*, smell (meaning a foul smell).

The indumentum of leaves and stems are important in the identification of subsections in *Onosma*. Species with setae (rarely hairs) arising form rounded, slightly raised, multicellular tubercles with small spinules or setules arranged stellately around the base of setae are traditionally classified in subsection Asterotricha. Small hairs are present in between the setae. Species with glabrous tubercles are placed in subsection Haplotricha. Both subsections are artificial, but help in the identification of groups of species. Some species fall intermediate between these two subsections. The setae give the plant the typical hispid indumentum in *Onosma*.

Chacón, J., Luebert, F., Hilger, H.H., Ovchinnikova, S., Selvi, F., Cecchi, L., Guilliams, C.M., Hasenstab-Lehman, K., Sutorý, K., Simpson, M.G. & Weigend, M. (2016). The borage family (Boraginaceae s. str.): A revised infrafamilial classification based on new phylogenetic evidence, with emphasis on the placement of some enigmatic genera. Taxon 65(3): 523–546. http://dx.doi.org/10.12705/653.6

Cecchi, L., & Hilger, H. H. (2022). A prickly matter: nomenclatural synopsis of *Onosma* L. and its segregates (Boraginaceae). Plant Biosystems 156(5): 1076–1084.

Cecchi, L. & Selvi, F. (2009). Phylogenetic relationships of the monotypic genera *Halacsya* and *Paramoltkia* and the origins of serpentine adaptation in circum-mediterranean Lithospermeae (Boraginaceae): insights from ITS and matK DNA sequences. Taxon, 58(3): 700–714.

Lacaita, C.C. (1924). The Onosmas of Linnaeus and Sibthorp, with a note on those of Tournefort's Herbarium. Bot. J. Linn. Soc. 46: 387–400.

Nasrollahi, F., Kazempour Osaloo, S., Saadati, N., Mozaffarian, V. and Zare Maivan, H. (2019). Molecular phylogeny and divergence times of *Onosma* (Boraginaceae ss) based on nrDNA ITS and plastid rpl32 trnL (UAG) and trnH–psbA sequences. Nordic J. Bot. 37(1):

Peruzzi. L., Passalacqua, N.G. (2008). Taxonomy of the *Onosma echioides* (L.) L. complex (Boraginaceae) based on morphometric analysis. Bot. J. Linn. Soc. 157:763–774.

Riedl, H. (1962). Beiträge zur Kenntnis der Gattung *Onosma* in Asien (Vorarbeiten zu K. H. Rechinger, Flora Iranica lll). Österreichische botanische Zeitschrift 109: 213–249.

– (1965). [Das korr.Mitglied K.H.Rechinger übersendet eine kurze Mitteilung, und zwar] Notizen zur Orient–Flora 74. Über *Onosma* L. sect. *Onosma* ser. *Aleppica* H.Riedl. Anzeiger der Österreichischen Akademie der Wissenschaften. Mathematisch–naturwissenschatliche Klasse 102(3): 49–55.

– (1978, publ. 1979). *Onosma* L. In: Davis, P.H. (ed.) Flora of Turkey and the Aegean Islands, vol 6. Edinburgh University Press, Edinburgh, pp. 326–376.

Stroh, G. (1939). Die Gattung *Onosma* L. Bieh. Bot. Centrabl. 59B: 430, 454.

Thomas, D.C., Weigend, M. & Hilger, H.H. (2008) Phylogeny and systematics of *Lithodora* (Boraginaceae–Lithospermeae) and its affinities to the monotypic genera *Mairetis*, *Halacsya* and *Paramoltkia* based on ITS1 and trnLUAA–sequence data and morphology. Taxon 57(1): 79–97.

Weigend. M., Gottsschling, M., Selvi, F., Hilger, H.H. (2009). Marbleseeds are gromwells–Systematics and evolution of *Lithospermum* and allies (Boraginaceae tribe Lithospermeae) based on molecular and morphological data. Mol. Phylogenet. Evol. 52: 755–768.

1. Setae arising from stellately setulose tubercles 2
   Setae arising from glabrous tubercles 7
2. Corolla pink to purple and bluish, longer than calyx, 25–28 (–35)
   mm (35 mm var. *macrocalycina*), minutely lobed at apex glabrous,
   pubescent near the apex . . . . 17. *O. alborosea*
   Corolla reddish-yellow, yellow, or whitish. 3
3. Base of upper leaves auriculate; calyx lobes oblanceolate to spatulate, c.
   10 mm, in fruit to 20 mm . . . . .19. *O. auriculata*
   Base of upper leaves cuneate or tapering into the petiole; calyx lobes
   linear-lanceolate, acute, > 10 mm 4
4. Inflorescence of compact scorpioid cymes; corolla reddish-yellow,
   or white flushed with pink, the lower part often yellowish, upper
   reddish, turning orange when dying, campanulate, 18–20 mm,
   longer than calyx, apical lobes reflexed, glabrous; nutlets glabrous 18. *O. rascheyana*
   Inflorescence of lax few cymes; corolla golden-yellow or whitish, narrow
   campanulate to tubular; nutlets smooth or hairy 5
5. Corolla golden-yellow 6
   Corolla whitish; calyx c. 15 mm . . . . .21. *O. ovalifolia*
6. Leaves ovate to obovate, 30–90 × 18–50 mm, somewhat reducing in size
   upwards; calyx 25–26 mm . . . . 16. *O. dasytricha*
   Leaves oblanceolate to lanceolate–oblong, 100–220 × 10–15 mm,
   margins often revolute; calyx 11–12 mm . . . . 20. *O. gigantea*
7. Corolla greenish; annulus absent; nutlets densely pubescent . . . . .11. *O. popoviana*
   Corolla yellow, whitish or reddish-purple; annulus always present;
   nutlets mostly glabrous 8
8. Corolla reddish to purplish. (Densely setaceous plant; calyx linear-
   lanceolate, 5–8 mm, densely setose, with yellowish setae; anthers
   included; nutlets glabrous) . . . . .3. *O. angustiloba*
   Corolla yellow to orangish or whitish with purplish lobes 9
9. Nutlets tuberculate or densely pubescent 10
   Nutlets smooth. 13
10. Calyx 15–17 mm in flower; corolla 17–20 mm, deep yellow. . . . . .7. *O. haussknechtii*
    Calyx < 15 mm in flower; corolla; < 17 mm 11
11. Corolla white with purplish lobes; nutlets densely pubescent . . . . 10. *O. striata*
    Corolla yellow; nutlets tuberculate to rugose, not densely pubescent. 12
12. Corolla campanulate, longer than calyx, 18–20 mm, pubescent;
    annulus glabrous . . . . .12. *O. olivieri*
    Corolla cylindrical, 11–13 mm, glabrous . . . . 15. *O. tinctoria*
13. Corolla glabrous on the outside. 14
    Corolla pubescent on the outside. 16
14. Bracts lanceolate to ovate-lanceolate, as long or longer than calyx;
    annulus glabrous . . . . .14. *O. cardiostegia*
    Bracts oblong to oblong-lanceolate or lanceolate, shorter than calyx;
    annulus hairy 15
15. Leaves linear to linear-spatulate, 40–120 × 8–18 mm; corolla
    18–20 mm. . . . . 6. *O. bulbotricha*
    Leaves oblong to oblanceolate, 30–80 × 5–10 mm; corolla 13–18 mm . . . 2. *O. echinata*
16. Annulus hairy. 17
    Annulus glabrous. 18
17. Calyx 15–18 mm; corolla 18–20 mm . . . . 6. *O. bulbotricha*
    Calyx 5–6 mm; corolla 10–12 mm . . . . 1. *O. rechingeri*
18. Corolla 20–24 mm . . . . .5. *O. bodeanum*
    Corolla < 20 mm 19
19. Cymes 1–2, lax . . . . 9. *O. microcarpa*
    Cymes compact 20

20. Cauline leaves broadly elliptic, 30–40 × 15–20 mm..............13. *O. hawramanensis*
    Cauline leaves lanceolate, obovate to oblanceolate to linear-lanceolate...........21
21. Calyx lobes distinctly blackish at flowering, linear-lanceolate, 9–15 mm 4. O. *nemoricola*
    Calyx lobes not blackish at flowering.....................................22
22. Bracts ovate-lanceolate, 12 × 2 mm; calyx lobes, linear-lanceolate;
    corolla 20–24 mm ....5. *O. bodeana*
    Bracts linear to linear-lanceolate, 6–7 mm; calyx lobes lanceolate,
    becoming somewhat membranous when dry; corolla 14–16 mm........8. *O. sericea*

1. **Onosma rechingeri** *H.Riedl*, Österr. Bot. Z. 108(4/5): 404 (1961); Rawi in Dep. Agr. Tech. Bull. 14: 139 (1964). Type: Distr. Erbil, montes Qandil ad confines Persiae, supra Pushtanshan, 1500–2000 m, 28.vii. 1956, *Rechinger* 11182! (W, holo.). Riedl in Fl. Iranica [K. H. Rechinger] 48: 176 (1967); Riedl in Fl. Turkey [P. H. Davis] 6: 338 (1978); Khatamsaz, Flora Iran 39: 134 (2002); Cecchi & Hilger, Plant Biosystems, Appendix S1, 101 (2022).

Perennial. Stems several arising from a woody stock, erect to ascending, 10–25 cm, simple or ± branched, brown, densely strigose-pubescent with long patent tubercule-based hairs to 2 mm; tubercle glabrous. Leaves mostly basal, oblanceolate to ovate-lanceolate, 30–60 × 10–20 mm, apex obtuse, base tapering into a short petiole or leaves sessile, densely hairy on both surfaces with long ± adpressed hairs. Cymes lax, up to 5 cm, terminal on main branches, paniculate. Bracts linear-lanceolate, 5–10 × 4–6 mm. Pedicels slender, 5–10 mm. Calyx lobed to base, lobes linear-lanceolate, 5–6 mm, densely strigose-hairy with long ± adpressed hairs; hairs white to yellowish. Corolla yellow, urceolate-tubular, 10–12 mm, lobed at apex, minutely pubescent. Annulus hairy. Filaments adnate to about half the length of the corolla tube; anthers included. Style slender, exserted, minutely 2-lobed above. Nutlets grey-brown, ± 2 mm, beaked, surface smooth, glossy.

Hab. On limestone, on dry mountain slope, in shade of *Juglans*, and near streams; alt. 1000–1500 m; fl. Jun.-Aug.
Distrib. In the Rawanduz district of N. Iraq. **MRO**: Mt. Berrog on road to Qandil, *Rawi & Serhang* 23944!; ushtashan village N.E. of Rania, *Rawi & Serhang* 23888!; Kani Rash, Qurnago valley N. of Pushtashan, *Serhang & Rawi* 26663!; Saran village (Kani Kawan spring), Karoukh region, *Nuri & Kass* 27280!

Turkey, N.W. Iran.

2. **Onosma echinata** *Desf.*, Fl. Atlant. 1: 161, t. 43 (1798); Cecchi & Hilger, Plant Biosystems, Appendix S1, 47 (2022). including *O. xanthotricha* Boiss., Diagn. Pl. Orient. ser. 1, 11: 109 (1849).

> *O. echinata* Aucher ex A.DC., Prodr. [A.P. de Candolle] 10: 58. (1846); Rawi in Dep. Agr. Tech. Bull. 14: 139 (1964); Taifour & El-Oqlah, Pl. Jordan Annot. Checkl.: 53 (2017).
> *O. aleppica* Boiss., Diagn. Pl. Orient. ser. 1, 11: 107 (1849); Dinsmore in Post, Fl. Syria, Palest. & Sinai ed. 2, 2: 231 (1933); Mouterde, Nouv. Fl. Liban et Syrie 3: 69 (1978); Feinbrun-Dothan, Fl. Palaest. 3: 71 (1978); Riedl in Fl. Turkey [P. H. Davis] 6: 354 (1978); Taifour & El-Oqlah, Pl. Jordan Annot. Checkl.: 53 (2017).
> *O. aleppica* Boiss var. *xanthotricha* (Boiss.) Boiss., Fl. Orient. [Boissier] 4: 184 (1875).
> *O. setosa* var. *dichroantha* (Boiss.) Boiss., Fl. Orient. [Boissier] 4: 181 (1875).
> *O. dichroantha* auct. non. Boiss. (1849): Rawi in Dep. Agr. Tech. Bull. 14: 139 (1964).

Perennial or biennial small shrub to 60 cm, branched. Stems erect to ascending, whitish-yellow, densely hispid, hairs 1–4 mm long, patent with a glabrous tuberculate base. Leaves sessile, oblong to oblanceolate, 30–80 × 5–10 mm, leaves reducing in size upwards; apex acute to obtuse, base tapering, margins entire, sometimes revolute; densely hispid with white or yellowish patent hairs on both surfaces. Cymes up to 7 cm, terminal on main and lateral branches. Bracts similar to leaves but smaller. Pedicels 4–5 mm, densely sericeous. Calyx lobed to base, lobes 13–15 mm, elongating to 18 mm in fruit, linear-lanceolate, densely setose-sericeous, setae white or yellowish (in *O. xanthocarpa*). Corolla pale yellow, narrow campanulate, 13–18 mm, minutely lobed at apex, glabrous; lobes c. 2 mm; reflexed, glabrous, sometimes with few hairs near apex. Annulus ciliate. Filaments adnate to about half the length of the corolla tube; anthers coherent at base; sterile tip exserted. Style slender, exserted. Nutlets 3.5–4 mm, brownish-yellow, glossy, ± smooth. Fig. 78, 1–9.

Fig. 78. **Onosma echinata**. 1, habit; 2, leaf indumentum, adaxial side; 3, indumentum on stem; 4, flower; 5, corolla, opened; 6, outer face of corolla; 7, stigma; 8, details of inner face of stamen; 9, mature nutlet. 1–9 from Guest 1390 A. Drawn by © A.P. Brown 2022.

HAB. On hillside, in degraded *Quercus* forest, corn fields, by road verges, on limestone, loamy, clayey and gravelly soil; alt. 80–1400 m; fl. & fr. Mar.-Jun.

DISTRIB. On lower altitudes of mountains, in the Upper Jazira District and parts of the Western desert areas. **MAM**: nr Dohuk, *Chapman* 26261! **MAM/FNI**: nr Chobur (Khabur), Mesopotamia, *Haussknecht* s.n. (1867). **MAM / FUJ**: Between Mosul and Aqra, *Rawi* 11366! **MRO**: Rowanduz, *Sahira, Karim & Hamid* 38382!; HAB-al Sultan Dagh nr Qusan, *Omar, F. Karim, Hamza & Hamid* 37187!; between Sei Waka and Dargala, *Kass & Nuri* 27619! **MSU**: 16 km N.W. of Darband-I Khan, *Rawi, Alizzi & Nuri* 29442!; Silwan river 7 km S. of Darband-I Khan, *F. & E. Barkley & Haddad* 5100!; 43 km S. of Sulaimaniya, *F. Barkley & Haddad* 5066!; Gweija (Goizha) Dagh, *Rawi & Gillett* 11668; Qara Anjir, *Rawi & Gillett* 7511!; return from Groomah through Kurdistan mts., 1859, *Capt. Garden* s.n.! **MJS**: Jabal Sinjar, *Sharif & Hamad* 50229! Kursi, Jabal Sinjar, *Gillett* 10869! **FUJ**: 43 km S. of Sinjar, *Chakravarty, Rawi, Khatib & Alizzi* 33101! Nr Ba'aj, *Omar & Hamid* 36545! Between Mosul and Tal Afar, *F. Karim* 37468! Nr Tal Afar, *Guest* 13371!; Mosul, *Guest* 1390A!; Jarmo, *Wheeler Haines* W225!; Kurdistan, 1840, *Brant & Strangeways* s.n.!; 50 km N. of Mosul towards Tal Afar, *Hikmat Abbas Al-Ani* 9780!; 35 km S. of Mosul towards Qaiyara, *Hikmat Abbas Al-Ani* 9853!; Haimed, S. of Sinjar, *Alizzi & Husain* 33941! **FAR**: Qosh Tepe, *Guest* 1479!; 7 km from Altun-Kupri to Arbil, *Botany Staff* 43245; Ankawa nr Arbil, *Bornmüller* 1606!; Arbil, *Gillett* 8012! **FKI**: 10 km N. of Tuz to Kirkuk, *Botany Staff* 42924!; Tuz Khurmatu, *Guest* 1405. [aberrant specimen]!; nr Bawanur, *Poore* 395! **FPF**: 2 km N.E. of Mandali, *Kaisi, Thamer & Salah* 51492! **DWD**: 160–190 km W. by N. of Ramadi on road to Rutba, *Rawi* 31321!; 6 km above Ana, *Rawi & Gillet* 6976!; Wadi Hawran nr Baghdad, *Omar, Kaisi, Hamad & Hamid* 44579!

A widespread species and to some extent variable in habit and leaf characters. *O. dichroantha* and *O. auriculata* are similar in facies, however, *O. dichroantha* is a more robust plant, with a longer, glabrous corolla (25–35 mm) and glabrous annulus. The filaments are attached more than half way up the corolla tube and anthers are included or only just exserted from the corolla. *O. dichroantha* has a more eastern distribution, found in Kazakhstan, Kirgizstan, Tadzhikistan, Transcaucasus, Turkmenistan, Uzbekistan, Pakistan, Afghanistan, Iran and Turkey, overlapping with *O. auriculata* in Turkey and N.W. Iran.

KHASIAN AL-BAGHL (Ar.) (*Guest* 13371, used to heal wound when castrating mules as recorded on label); BERDUSHKA (Kurd.) (*Gillett* 8012).

Turkey, N.W. Iran, Palestine, Jordan, Lebanon, Syrian Desert.

3. **Onosma angustiloba** *Rech.f. & Riedl* in Österr. Bot. Z. 109: 223 (1962). Type: Iraq, Kurdistan, Mt. Hendren (E. of Rowandaz), 1951-05-27, *Thesiger* 1087! (BM, holo. [BM000810231]); Riedl in Fl. Iranica [K. H. Rechinger] 48: 177 (1967); Cecchi & Hilger, Plant Biosystems, Appendix S1: 23 (2022).

Perennial or ?biennial to 40 cm. Stems simple, erect, densely setose with ± patent setae arising from a glabrous tuberculate bases and shorter pubescence. Leaves reducing in size upwards, oblanceolate to lanceolate, 10–55 × (2–)5–10 mm, apex acute to obtuse, base of lower leaves tapering into the petiole, upper leaves sessile, densely setaceous with white or yellowish patent hairs on both surfaces. Inflorescence of several compact scorpioid cymes on terminal and lateral branches. Bracts similar to leaves but smaller, c. 7 mm. Pedicels 1–2 mm, densely setaceous. Calyx lobed to base, lobes linear-lanceolate, 5–8 mm, densely setose, setae white and yellowish. Corolla reddish to purplish, campanulate, 8–10 mm, minutely lobed at apex, puberulous. Annulus glabrous. Stamens about equalling the corolla; filaments adnate to about half the length of the corolla tube; anthers included. Style slender, exserted. Nutlets brownish, glabrous.

HAB. On stony slope of mountain, rare; alt. 1300–2000 m; fl. May.

DISTRIB. Rare and only known from 2 gatherings. **MRO**: Between Sewok and Dargala villages, Karoukh mt., *Kass & Nuri* 27612!; Mt. Hendren (E. of Rowandaz), 1951-05-27, *Thesiger* 1087! (BM-type).

Endemic.

4. **Onosma nemoricola** *Hausskn. & Bornm.* in Repert. Spec. Nov. Regni Veg. 8: 541 (1910) (as *nemoricolum*). Type: Iraq, Erbil, "Kuh Safin prope Shaqlawa", 14 May 1893, *Bornmüller* It. Pers-Turc. 1892–1893 no. 1609, (B, Lecto. designated by Riedl 1967: 197; BM, JE, K!, PH, isolecto.). Rawi in Dep. Agr. Tech. Bull. 14: 139 (1964); Riedl in Fl. Turkey [P. H. Davis] 6: 341 (1978); Cecchi & Hilger, Plant Biosystems, Appendix S1: 90 (2022).

    O. froedinii Rech.f in Ann. Naturhist. Mus. Wien 49: 276 (1939). Type: Iraq, Kurdistan, Kuh-i-Sefin supra pag. Schaklawa, 1500–1600 m, 14. V. 1893. *Bornmüller* 1609 (B), nom. illegit.

Fig. 79. **Onosma nemoricola**. 1, habit; 2, leaf indumentum, adaxial side; 3, indumentum on stem; 4, hair from leaf, adaxial surface; 5, flower; 6, corolla, opened; 7, outer face of corolla; 8, stigma; 9, details of inner face of stamen. 1–4 from *Guest* 2037; 5–9 from *Ludlow-Hewitt* 2037A. Drawn by © A.P. Brown 2020.

O. *cornuta* Riedl in Oesterr. Bot. Z. 109: 228 (1962). Type: Iraq, Kurdistan, Ari Gird Dagh prope Rust, on rocky mountain side, 6500 ft., 24. VII. 1932, *Guest & Ludlow-Hewitt* 2918 (K!, holo.); Riedl in Fl. Iranica 48: 197 (1967).

O. *nervosa* Riedl. in Österr. Bot. Z. 109: 234 (1962). Type: Iraq, Kajan mt., nr Penjwin, 1590 m, 21.vi.1957, *Rawi* 22705 (K!, holo.).

Perennial. Stems 16–50 cm, simple or sparsely branched, erect to ascending, whitish yellow, densely pubescent with long tubercule-based hairs, ± patent intermixed with simple hairs; tubercles glabrous. Leaves sessile, lanceolate to oblanceolate, 10–70 × 5–20 mm, apex acute, base tapering, margins entire, densely hispid on both surfaces with long ± adpressed hairs intermixed with short hairs. Cymes up to 7 cm, terminal on main branches. Bracts 15–20 × 7–10 mm, ovate to ovate-lanceolate, hispid. Pedicels 2–3 mm. Calyx lobed to base, lobes distinctly blackish at flowering, linear-lanceolate, 9–15 mm, densely appressed setose-sericeous especially at base, setae white. Corolla yellow to orange-yellow, narrow campanulate, 14–16 mm, minutely lobed at apex, pubescent; lobes c. 2 mm; reflexed, pubescent. Annulus glabrous. Filaments adnate to about half the length of the corolla tube; anthers coherent at base; sterile tip shortly exserted. Style slender, exserted. Nutlets brownish ± 4 mm, smooth. Fig. 79, 1–9.

HAB. On mountains, scree slopes, in open *Quercus* forest, on limestone, loamy and silty soil; alt. 600–2200 m; fl. & fr. Apr.-May.

DISTRIB. Northern mountains and gorges. **MAM**: Ser Amadiya (Gali Mazurka), *Dabbagh, Kaisi & Hamid* 45995!; mt. Maya 25 km W. of Kani Masi, *H. Hamid & Fadhil* 45500! Aqra, *Rawi* 11441!; Sawara Tuka, *Chapman* 26322!; Ser Kuwara *Gillett* 9790! **MRO**: nr Dergala village, ± 35 km N.W. by N. of Rania, *Rawi, Nuri & Kass* 28877, 28884!; Shaqlawa, *Haines* W.767!; Algurd Dag, nr. Rost *Guest* 2918!; Sefin Dagh, *Gillett* 8154!; Mt. Hauraman, N. of Biyara, *Gillett* 11750!; Mt Hauraman, above Tawila, *Rechinger* 15895 (W!); Baradost, *Rechinger* 15617 (W!); between Sandul and Zawitha, *Joseph Alkas* 18635! Kuh-e Sefin, *Bornmüller* 1609 (type). **MSU**: Rowanduz gorge, *Helbaek* 804, *Guest* 2037!, 2037A! Pira Magrun, *Poore* 626!; Mt. Kajan, nr Penjwin, *Rawi* 22704!, 22705!

Guest & Ludlow-Hewitt 2918 (type of *O. cornuta*) is treated by Riedl as a separate taxon, *O. cornuta* Riedl, endemic to Iraq with *Gillett* 9790, and *Chapman* 26322 cited under that taxon. I have looked at the type and other material and cannot justify this as a new species. *O. nemoricola* shows a fair amount of variation in its vegetative and floral characters in its range of distribution and *O. cornuta* falls within that variation

Turkey.

5. **Onosma bodeana** *Boiss.*, Fl. Orient. [Boissier] 4: 187 (1875); Riedl in Fl. Iranica [K. H. Rechinger] 48: 195 (1967); Khatamsaz, Flora Iran 39: 149 (2002); Cecchi & Hilger, Plant Biosystems, Appendix S1: 34 (2022).

O. *elwendica* Wettst. Ex Stapf in Denkschr. Kaiserl. Akad. Wiss., Wien. Math.-Naturwiss. Kl. 50: 28 (1885). Riedl in Fl. Iranica 48: 196 (1967); Khatamsaz, Flora Iran 39: 153 (2002).

O. *gaubae* Bornm., Repert. Spec. Nov. Regni Veg. 41: 325 (1937).

O. *froedinii* Rech.f. in Ann. Nat. Mus. Wien 49: 276 (1939), p.p.

O. *polioxantha* Rech.f. in Ann. Naturhist. Mus. Wien 55: 7 (1947).

O. *iranica* Parsa, Kew Bull. 1948: 212 (1948).

O. *pabotii* Riedl, Österr. Bot. Z. 109: 237 (1962).

Perennial. Stems 25–40 cm, simple or sparsely branched, erect, 1–3 arising from a woody stock, leafy throughout, appressed white pubescent, tubercles with minute stellate setae or glabrous. Leaves lanceolate to oblanceolate, 20–80 × 8–18 mm, reducing in size upwards, apex acute, base tapering into a short petiole or leaves sessile, margins entire, densely pubescent on both surfaces with adpressed hairs; midvein prominent on the lower surface. Cymes compact, terminal on main branches; flowering axis appearing whitish, white pubescent. Bracts ovate-lanceolate, 12 × 2 mm, sericeous, especially at base. Pedicels 2–3 mm. Calyx lobed to base, linear-lanceolate, lobes 12–20 mm, densely appressed sericeous. Corolla yellow to white, narrow campanulate, 20–24 mm, minutely lobed at apex, pubescent; lobes c. 2 mm; ± reflexed. Annulus glabrous. Filaments adnate to about half the length of the corolla tube; anthers coherent at base; sterile tip included. Style slender, exserted. Nutlets brownish ± 4 mm, glossy, ± smooth.

HAB. On limestone mountain slopes, in degraded *Quercus* forest; alt. 700–2000 m; fl. & fr. May-Jul.

DISTRIB. Endemic to the Northern mountains of Iraq. **MAM**: Jabal Bekher, *Rawi* 8450!; Atrush, N. of

Mosul, *Guest* 3622!; Jabal Khantur, *Rechinger* 10832! **MRO**: Shaqlawa, *Gillett* 8053!; Molla Khort mountain, *Rawi, Alizzi & Nuri* 29468!; Sai Watka village (nr S. foot of Karokh), *Kass & Nuri* 27545!; Chiya-i Mandau (nr. Walza), *Guest* 2678; mt Helgord, *Rawi* 13739!; Pushtashan, *Rechinger* 11197 (W). **MSU**: Khurmal, *Rawi* 8915!; Gweija, above Sulaimaniya, *Rawi* 9419-H!; Sulimaniyah, *Rechinger* 10104 (W).

MISMISOH (*Gillett* 8053).

N.W. Iran.

6. **Onosma bulbotricha** *A.DC.*, Prodr. [A.P. de Candolle] 10: 64 (1846); Rawi in Dep. Agr. Tech. Bull. 14: 139 (1964); Riedl in Fl. Iranica [K. H. Rechinger] 48: 191 (1967); Riedl in Fl. Turkey [P. H. Davis] 6: 354 (1978); Khatamsaz, Flora Iran 39: 137 (2002); Breckle, Hedge, Podlech & Rafiqpoor, Vasc. Pl. Afghan. Checkl.: 211 (2013); Cecchi & Hilger, Plant Biosystems, Appendix S1: 37 (2022).

> *Onosma wheeler-hainesii* Riedl, Fl. Iranica [K. H. Rechinger], 48: 189 [189–190, tab. 31 fig. 2]. 1967. Type: Iraq, Khanaqin, 3rd range of Jebel Hamrin,4.5.1957, *Wheeler-Haines s.n.* (E, holo.).
> *O. sulaimaniaca* Riedl, Fl. Iranica [K. H. Rechinger], 48: 190 [190, tab. 32 fig. 1]. 1967 ('sulaimaniacum'). Type: Iraq, 48 km S. Sulimaniyah, *Barkley* 5866 (W, holo.).

Annual or biennial or perennial. Stems to 15–40 cm, branched above, erect to ascending, 1-several arising from a woody stock, densely setaceous. Leaves sessile, linear to linear-spatulate, 40–120 × 8–18 mm, reducing in size upwards, apex acute, densely pubescent on both surfaces with tubercle-based setae or lower surface with setae on midrib only; midrib prominent on lower surface; tubercles glabrous. Cymes compact, terminal on main branches. Bracts lanceolate. Pedicels 1–3 mm, elongating after flowering, densely setose; setae white or yellowish. Calyx lobed to base, lobes linear-lanceolate, 15–18 mm, densely sericeous. Corolla pale yellow to white, narrow campanulate, 18–20 mm, minutely lobed at apex, glabrous to pubescent; lobes reflexed at apex. Annulus hairy. Filaments adnate to about half the length of the corolla tube; anthers coherent at base, included; sterile tips exserted. Style slender, exserted. Nutlets greyish, smooth, ± 6 mm, ventral keel prominent, beak strongly incurved, obtuse (seen in specimens from Iran).

HAB. In shade of rock, on limestone, with coppiced *Quercus*; alt. 120–1200 m; fl. Apr.-Jun.
DISTRIB. Sulaymaniyah District. **MSU**: Qarachitan nr. Pir Omar Gudrun, *Gillett* 7724!; Mt Avroman, nr Tawila, *Rechinger* 10133! (W); between Mosul and Dohuk, *Rechinger* 10617! (W); 48 km S. Sulimaniyah, *Barkley* 5866! (W, type of *O. sulaimaniaca*). **FKI**: Kirkuk, 37 km E. Kirkuk, *Rechinger* 12502 (W); Kirkuk, *Rawi & Gillett* 7470!; Qara Anjir, Kirkuk district, *Rechinger* 12517! (W); **FPF**: Diyala, Jabal Hamrin, *Bornmüller* 1606! (W); Khanakin, *Nábělek* 633 (W); Khanaqin, 3rd range of Jebel Hamrin, *Wheeler-Haines s.n.*! (E, type of *Onosma wheeler-hainesii*). **DWD**: ad Euphratem medium inter Abukemal et Ramadi, *Handel-Mazzetti* 738! (W).

Onosma wheeler-hainesii Riedl, and *Onosma sulaimaniaca* Riedl, are described from single collections each. Both are very close to *O. bulbotricha*, and difficult to tell apart from *O. bulbotricha*. At present I am treating these two species under *O. bulbotricha*, but more material may confirm their status as good species.

Turkey, N.W. Iran, Syria(?).

7. **Onosma haussknechtii** *Bornm.*, Repert. Spec. Nov. Regni Veg 8: 543 (1910). Type: N. Iraq, Kordestan "(Assyria orientalis): ad fines Persiae in montis Helgurd regione alpina, 2700 m s. m", 26 June 1893, *J. F. N. Bornmüller* It. Pers.-Turc. 1892-93 no. 1608 ["1609"], (B: B 10 0365359, ex Herb. Bornmüller; Riedl, Fl. Iranica [K. H. Rechinger] 48: 203 (1967); Riedl, Fl. Turkey [P. H. Davis] 6: 342 (1978); Cecchi & Hilger, Plant Biosystems, Appendix S1: 68 (2022).

Perennial. Stems to 40 cm, unbranched with 1–2 branches, erect, arising from a short woody stock, appressed white pubescent. Leaves lanceolate to linear-lanceolate, 50–100 × 10–20 mm, apex acute, base tapering into a long petiole in lower leaves, the upper sessile, lamina appressed pubescent on both surfaces with tubercle-based setae; tubercles glabrous. Cymes compact, becoming lax, terminal on 6–10 cm flowering stem. Pedicels 2–4 mm. Calyx lobed to base, lobes linear, 15–17 mm, appressed white pubescent. Corolla deep yellow, longer than calyx, narrow campanulate, 17–20 mm, minutely lobed at apex, pubescent; lobes reflexed at apex. Annulus glabrous. Filaments adnate to just more than half the length of the corolla tube; anthers included; sterile tips just exserted or not. Style slender, exserted. Nutlets brown, ± 2 mm (immature), tuberculate.

HAB. North facing stony hillside, amongst rocks. ± 1800–2200 m; fl. Apr., Sep.
DISTRIB. Known from two collections in the Rowandaz and Sulaymaniyah District in N. Iraq. **MRO**: Sua Khal,

Rawi & Serhang 24678! **MSU**: Pira Magrun, *O. Polunin* 5156!

Turkey.

8. **Onosma sericea** *Willd.*, Sp. Pl., ed. 4 [Willdenow] 1(2): 774 (1798); Dinsmore in Post, Fl. Syria, Palest. & Sinai ed. 2, 2: 231 (1933); Rawi in Dep. Agr. Tech. Bull. 14: 139 (1964); Riedl in Fl. Iranica [K. H. Rechinger] 48: 193 (1967); Riedl in Fl. Turkey [P. H. Davis] 6: 339 (1978); Mouterde, Nouv. Fl. Liban et Syrie 3: 70 (1983); Khatamsaz, Flora Iran 39: 147 (2002); Gabrielian & Fragman-Sapir, Flow. Transcaucasus & Adj. Areas: 104 (2008); Cecchi & Hilger, Plant Biosystems, Appendix S1: 104 (2022).

Including *Onosma trachytricha* Boiss., Diagn. Pl. Orient. ser. 1, 11: 103 (1849).

Colsmannia flava Lehm., Mag. Ges. Nat. Berlin 8: 92 (1818).
Onosma sericea Boiss., Diagn. Pl. Orient. Ser. 1, 11: 101 (1849), pl. Heldr. Anat. Exs. non Willd.
O. elegans C.Koch in Linnaea 17: 306 (1843), incl. var. *gundelsheimeri* C.Koch in Linnaea 17: 306 (1843).
O. brachysolen Boiss., Diagn. Pl. Orient. Ser. 1, 11: 104 (1849).
O. flava (Lehm.) Boiss., [Boissier] Fl. Or. 4: 186 (1875).
O. subsericea Freyn, Bull. Herb. Boiss., ser. 2, 1: 274 (1901).

Perennial. Stems 1–3, 15–40 cm, erect, arising from a short woody stock, appressed white pubescent. Leaves obovate to oblanceolate to linear-lanceolate, 20–60 × 3–10 mm, apex acute, base of lower leaves tapering into a long petiole to 30 mm, upper leaves sessile, lamina appressed pubescent on both surfaces with long setae; tubercles obscure, glabrous. Cymes compact, terminal. Bracts linear to linear-lanceolate, 6–7 mm. Pedicels 2–8 mm. Calyx lobed to base, lobes lanceolate, 13–15 mm, appressed white sericeous, becoming somewhat membranous when dry. Corolla yellow, narrow campanulate, slightly longer than calyx, 14–16 mm, minutely lobed at apex, pubescent; lobes reflexed at apex. Annulus glabrous. Filaments adnate to below the middle of corolla; anthers included, sterile tips exserted. Style slender, exserted. Nutlets black, ± 4 mm, dorsally and ventrally keeled, surfaces ± smooth.

HAB. Rocky mountain slopes and hillsides, near water, under *Quercus*; common; alt. 260–1180 m; fl. Apr.-Jul.
DISTRIB. Common on the hills and foothills of N. Iraq. **MAM**: Zawita, *Guest* 4881, *Omar* 37681, *Hunting Aero Surveys RC.* 22.2.57, *Robertson* 368 (2 sheets)!; Sarsang, *Omar* 37745!; Sarsang to Amadiya, *Karim, Hamid & Jasim* 40939!; below Amadiya, *Rechinger* 11639!; 60 km from Mosul to Dohuk, *Karim* 37502!; Atrush to Rabetki, *Chapman* 9353!; Atrush, *Guest* 4388!; Aqra, *Rawi* 11381!; Atrush N. of Mosul, *Guest* 3622, 4288!; 10 km E. of Aqra, *Chapman* 26089! **MRO**: Sefin Dagh, *Rawi* 9103!; Salah ad-Din, *Rawi* 36222!; Pir Dagh, *Gillett* 8016!; Shaqlawa, *Gillett* 11292! 31 km N. of Kirkuk towards Koi Sanjaq, *Rawi, Nuri & Kass* 27997! **MSU**: 4 km N.W. of Sulaimaniya on road to Dokan, *Rechinger* 10081!; Jarmo, *Helbaek* 1744!; Pira Magrun, *Haussknecht* s.n., Jun. 1867. ("Onosma flavida Boiss. var."); *Rawi* 12133!; Pira Magrun, anon. 5806!; Qaradagh, Timara village, *Karim, Hamid & Jasim* 40618!; Mt. Hawraman, Hawara Birza, *Rawi, Alizzi & Nuri* 29529!; Nuragora 2 km N.E. of Balkha, *Rawi, Alizzi & Nuri* 29543!; Penjwin, *Khalil Feddo* 3446!; *Guest* 10973 (2 sheets)!; Mesopotamia, Kurdistan & Mosul (1841), *Kotschy* s.n.!; Qara Anjir, *Rawi & Gillett* 7511!; Jarmo, *Haines* W.222!; Tisluja, *Omar & Karim* 47043! Pira Magrun, *Haussknecht*, s.n., Jun. 1867 ("sp. Nov.!"). **MJS**: Karsi, Jabal Sinjar, *Gillett* 10952!; Jabal Sinjar, Karsi, *Gillett* 10952! **FNI**: 40 km N. of Mosul towards Ain Sifni, *Hossain* 1650! **FAR**: Taktak, *Omar, Karim, Hamzi & H. Hamid* 37140! **FKI**: Near Bawanur, *Poore* 578! **FKI** /**FPF**: Between Sarkal and Qal'a Shirwana, on right bank of the Diyala near Jalaula, *Poore* 371! **FPF**: Sar Qal'a near Jalaula, *Poore* 371!; Qara Tu (?), nr Iranian border, *Rawi* 5731!

Onosma trachytricha Boiss. (1849) has been treated as a separate species by Riedl in his treatment in Flora Iranica (p. 194) and Flora Turkey (p. 340) noting that it was closely related to *O. sericea* and possibly of hybrid origin. I have examined material of *O. trachytricha* from Iraq (*Kotschy* 399, *Rawi, Nuri & Kass* 27997, *Haussknecht* Jun. 1867, s.n. & *Guest* 3622, 4288) and seen that apart from setae that arise from more prominent tubercles, especially on lower surface of lower leaves, other vegetative and floral characters are similar to those of *O. sericea*. A more detailed study may show that these two are distinct taxa, but for the present Flora account I have included Iraq material under *O. sericea*.

Turkey, Transcaucasus (Georgia, N. & C. Armenia, Nachitchevan), Iran, Syria, Lebanon.

9. **Onosma microcarpa** *Steven* ex *A.DC.*, Prodr. [A.P. de Candolle] 10: 62 (1846); Rawi, in Dep. Agr. Tech. Bull. 14: 139 (1964); Riedl in Fl. Iranica 48: 181 (1967); Riedl in Fl. Turkey [P. H. Davis] 6: 350 (1978); Khatamsaz, Flora Iran 39: 130 (2002); Cecchi & Hilger, Plant Biosystems, Appendix S1: 86 (2022).

> *O. microsperma* Steven, Bull. Soc. Imp. Nat. Mosc. 11: 305 (1838), nomen, Boissier, Fl. Orient. 4: 191 (1879), descr. nom. illeg.; Rawi in Dep. Agr. Tech. Bull. 14: 139 (1964).
> *O. gmelini* var. *microparpa* Ledeb., Fl. Ross. 3: 126 (1849).
> *O. spathulata* Wettst. ex Stapf, Denkschr. Akad. Wiss. Wien. Math.-Naturwiss. Kl. 50: 29 (1885).
> *O. stapfii* Wettst. ex Stapf, Denkschr. Akad. Wiss. Wien. Math.-Naturwiss. Kl. 50: 29 (1885).
> *O. microcarpa* var. *spathulata* (Wettst.) Bornm., Bull. Herb. Boiss. Ser. 2, 7: 783 (1907).
> *O. microcarpa* var. *stapfii* (Wettst.) Bornm, Beih. Bot. Centrbl. 28/B: 468 (1911).

Perennial. Stems 10–25 cm, erect to ascending, branched, several arising from a woody stock, densely white patent setose mixed with shorter hairs, setae with glabrous tubercles. Leaves narrow elliptic to oblanceolate to oblong-lanceolate, 10–50 × 5–15 mm, apex acute, base of lower leaves tapering into a long petiole to 40 mm, upper leaves sessile, lamina patent setose intermixed with shorter hairs on both surfaces. Cymes 1–2, terminal, lax. Bracts lanceolate. Pedicels 1–3 mm. Calyx lobed to base, lobes linear, 9–10 mm, acute, densely white patent setose intermixed with shorter hairs. Corolla yellow, campanulate, longer than calyx, 14–15 mm, minutely lobed at apex, sparsely pubescent; lobes reflexed at apex. Annulus glabrous. Filaments adnate to below the middle of corolla; anthers and sterile tips included. Style slender, exserted, obscurely bilobed at apex. Nutlets brown, ± 4 mm, surfaces smooth.

HAB. Rocky mountain slopes and hillsides, on clayey soil; alt. 750–2000 m; fl. Apr.-Aug.
DISTRIB. Mountains of N. Iraq. **MRO**: 31 km N. of Kirkuk to Koi Sanjaq, *Rawi, Kass & Nuri* 27959!; Kaiwa Rush, nr Rania, *Rawi & Serhang* 23781!; Qandil nr Pushtanshan, *Rechinger* 11200 (W!); Mt Avroman, above Tawila, *Rechinger* 10333 (W!); Malakawa, *Rechinger* (W!). **MSU**: Penjwin, *Rawi* 8835!; Qopi Qaradagh, *Poore* 442!; Pl. Mesopot., Kurdistan & Mosul. 1841, *Kotschy* s.n.!

Turkey, Iran.

10. **Onosma striata** *Riedl*, Oesterr. Bot. Z. 109: 243 (1962). Type: Iraq, Mosul Liwa, Dohuk, 1300 m, 13.V.1947, *Rawi* 8701 (K!, holo.). Riedl in Fl. Iranica [K. H. Rechinger] 48: 192 (1967); Cecchi & Hilger, Plant Biosystems, Appendix S1: 115 (2022).

Annual or (?)perennial. Stems to 30 cm, erect to ascending, simple to very sparsely branched from base, densely white patent setose mixed with shorter hairs, setae with glabrous tubercles. Leaves oblanceolate to lanceolate, 15–35 × 3–7 mm, apex acute, base of lower leaves tapering into a short petiole or sessile, upper leaves sessile, lamina patent setose intermixed with shorter hairs on both surfaces. Cymes 1–2, to 9 cm, terminal, lax. Bracts lanceolate 8–10 mm. Pedicels 3–10 mm. Calyx lobed to base, lobes linear-lanceolate, 7–10 mm, acute, densely white patent setose intermixed with shorter hairs. Corolla white with purplish lobes, cylindrical-campanulate, 10–11 mm, minutely lobed at apex, pubescent on the outside; lobes reflexed at apex. Annulus cilate. Filaments adnate to about the middle of corolla; anthers and sterile tips included. Style slender, included, obscurely bilobed at apex. Nutlets mature not seen (immature with densely pubescent surfaces).

HAB. Not given on the single specimen seen; alt. ± 1300 m; fl. May.
DISTRIB. N. Iraq, known only from the type specimen. **MAM**: Mosul Liwa, Dohuk, *Rawi* 8701 (type).

Endemic.

11. **Onosma popoviana** (*Riedl*) *L.Cecchi & Hilger*, Plant Biosystems, Appendix S1: 96 (2022).

> *Choriantha popoviana* Riedl, Österr. Bot. Z. 108: 400 (1961). Type: Iraq, Mt. Qandil, Qura Shina nr. N.E. Rania, 1060 m, *Rawi & Serhang* 23818! (K, holo.); Riedl in Fl. Iranica [K. H. Rechinger] 48: 168 (1967).

Perennial herbs; stem erect, 20–30 cm, covered with dense white patent setae, simple, with a bulbous base. Leaves not seen. Inflorescence cymose, scorpioid, bracteate. Pedicel

2–3 mm, patent setose, elongating in fruit, pendulous. Calyx 5, lobed to base, lobes c. 4 mm, lanceolate, densely white setose. Corolla greenish-brown (in herbarium specimen), tubular, 3–4 mm, folded in between lobes; lobes acute. Annulus absent. Staminal filaments adnate to corolla tube in the lower half with ± small scales at base; anthers connivent and forming tube, connective forming a distinct sterile appendage at apex. Style pubescent, longer than corolla. Nutlets densely pubescent.

HAB. *Quercus* forest on mountain slope, by stream; alt. ±1060 m; fl. Jul.
DISTRIB. Mountain of N. Iraq; known from a single collection. MRO: Mt. Qandil, Qura Shina nr. N.E. Rania, *Rawi & Serhang* 23818 (type:).

Endemic.

12. **Onosma olivieri** *Boiss.*, Pl. Orient. [Boissier] Nov. Ser. 2: 2 (1875). Type: "Inter Bagdad at Kermanschah", *Olivier*, Lectotype designated by Riedl 1962: 236. (G), erroneously cited as "Holotypus". Boissier, Fl. Orient. 4: 195 (1879); Rawi in Dep. Agr. Tech. Bull. 14: 139 (1964); Riedl in Fl. Iranica [K. H. Rechinger] 48: 181 (1967); Khatamsaz, Flora Iran 39: 159 (2002); Cecchi & Hilger, Plant Biosystems, Appendix S1: 86 (2022).

Including *O. lanceolata* Boiss. & Hausskn. Type: Iraq, Pira Magrun, *Haussknecht* 677 (JE, holo., W!, iso.)
O. macrophylla Bornm., Repert. Spec. Nov. Regni Veg. 8: 539 (1910); Khatamsaz, Flora Iran 39: 151 (2002).
O. macrophylla var. *angustifolia* Bornm. In Beih. Bot. Centrabl. 28B: 470 (1911); Riedl in Fl. Iranica 48: 201 (1967); Riedl in Fl. Turkey [P. H. Davis] 6: 343 (1978).
O. farsica Ponert in Feddes Repert. 68: 503 (1975).

Robust perennial. Stems 30–50 (or more), 1–2, branching above in inflorescence, erect, arising from a woody stock, densely white patent setose mixed with shorter hairs, setae with tubercles with short hairs. Leaves forming a basal rosette (seen in only 1 specimen with basal parts), greyish green, lanceolate to linear-lanceolate, 16–18 × 2–2.5 cm, reducing in size upwards, apex acute, base of lower leaves tapering into a long petiole to 60 mm, upper leaves sessile, lamina ± patent setose intermixed with shorter hairs on both surfaces. Inflorescence paniculate, flowering pedicels flexuous, with conspicuous yellow to whitish yellow patent indumentum; cymes terminal, ± compact. Bracts lanceolate, 15–40 mm. Pedicels slender, flexuous, elongating to 20 mm, densely yellow-white patent hairy. Calyx lobed to base, lobes linear, 9–10 mm, acute, densely white and yellowish patent setose intermixed with shorter hairs. Corolla yellow, campanulate, longer than calyx, 18–20 mm, minutely lobed at apex, pubescent; lobes reflexed at apex. Annulus glabrous. Stamens included, filaments attached to above of middle of corolla, included. Style slender, exserted, obscurely bilobed. Nutlets greyish black, ± 4 mm, surfaces ± tuberculate. Fig. 80, 1–9.

HAB. Fields and clearings in *Quercus* woodland, under *Quercus*, and on heavily grazed hillsides; common; alt. 700–1200 m; fl. May-Jun.
DISTRIB. Common in the lower hills and clearings in N. Iraq. MAM: Zawita, *Robertson* 60!; Ispindari saddle, Swaratuka, *E. Chapman* 26334!; Jabal Bekher, *Rawi* 8456A! MRO: N. slopes of Handren Dagh, *Gillett* 8302!; Salah ad-Din, *Gillett* 11268!; Saran, Karoukh region, *Rawi, Nuri & Kass* 27269; Jabal Karoukh between Saran and Kilkil, *Kass & Nuri* 27347! MSU: Qaradagh, *Haines* W. 1303!; Qaradagh, Timara village, *Karim, Hamid & Jasim* 40602!

O. olivieri is described from N.W. Iran (Type: Montes Avroman et Schahu, *Haussknecht* 679, G) which is close to the border N.W. Iraq. The distribution maps for *O. macrophylla* and *O. olivieri* in Khatamsaz (l.c.) (maps 72 & 77, pp. 454 & 455 respectively) show both species overlap in N.W. Iran, however, *O. macrophylla* is more widely distributed, extending in distribution to south western Iran. Khatamsaz includes var. *angustifolia* in the type variety. Reidl (l.c.) recognizes var. *macrophylla* (leaves elliptic to lanceolate, to 50 mm wide) endemic to N.W. Iran. I have not been able to find any distinction between *O. olivieri* and *O. macrophylla* and believe that *O. olivieri* is a widespread taxon found in S.E. Turkey, N.W. Iran and N. Iraq. Riedl (l.c., page 203) has included *O. lanceolata* Boiss. & Hausskn. for Iraq, citing a single specimen (type) from Pira Magrun, *Haussknecht* s.n. (JE, holo., W, iso.) differing from *O. olivieri* Boiss. in its longer bracts (20 mm) and hirsute annulus. With the lack of material and the fact that *O. olivieri* is a widespread species, I am including it under *O. olivieri*.

Turkey, Iran.

Fig. 80. **Onosma olivieri**. 1, habit; 2, leaf indumentum, adaxial side; 3, indumentum on stem; 4, flower; 5, corolla, opened; 6, outer face of corolla; 7, stigma; 8, details of inner face of stamen; 9, nutlet, two views. 1, from *Wheeler-Haines* W1303 (rootstock); 1, 2, 3, from *Gillett* 11268; 4–8 from *Karim et al.* 40602. Drawn by © A.P. Brown 20.

13. **Onosma hawramanensis** *S.A.Ahmad* in Harvard Pap. Bot. 19(2): 201, f. 1 (2014). Type: Iraq, Kurdistan, Sulaimani Province, Rangin Mt., subalpine rocky grassland, 2004 m, 8 June 2012, *Saman A. Ahmad* 12-997 (SUFA, holo. & iso.); Cecchi & Hilger, Plant Biosystems, Appendix S1: 86 (2022), [type erroneously cited as Iran in Cecchi & Hilger].

Perennial herb, woody at base, canescent. Stems 20–30 cm, several branches at base, hispid, with spreading white setae 1–2 mm. Basal leaves and lowermost cauline ones soon withered. Leaves (middle) broadly elliptic, 3–4 × 1.5–2 cm, base cuneate, apex subacute, densely pubescent with ascending trichomes to 2 mm and with sparsely pubescent tuberculate base with simple trichomes; uppermost leaves elliptic-oblanceolate, smaller. Inflorescence compact, cymes 5–10-flowered. Bracts lanceolate, 10–20 × 2–5 mm. Flowering pedicels 1–2 mm, slightly elongated and 3–4 mm in fruit. Calyx lobed to base, lobes linear, c. 10 mm in flower, 12–15 mm in fruit, densely white pubescent. Corolla yellow, tubular, 13–15 mm, slightly expanded at apex, pubescent outside, glabrous inside; teeth broadly triangular. Annulus sparsely pubescent. Filaments c. 4 mm, inserted at middle of corolla tube; anthers with sterile apex bidentate, to 1.5 mm, included. Nutlets 5–5.5 × 3.5–4 mm, broadly ovoid, glossy, slightly reticulate, straight, glabrous, keeled adaxially, slightly so abaxially, apex straight, flattened.

HAB. Subalpine rocky grassland; alt. ± 2000 m; fl. Jun.
DISTRIB. Known only from the type gathering. **MSU**: Sulaimani Province, Rangin Mt., *Saman A. Ahmad* 12-997 (SUFA, type:).

Onosma hawramanensis is known only from the type gathering collected in the Rangin Mountain, a range that was explored botanically for the first time in 2012. It is easily distinguished by the broadly elliptic cauline leaves, basal tubercles of leaves sparsely pubescent with small simple hairs, yellow corolla and glossy nutlets.

Endemic.

14. **Onosma cardiostegia** *Bornm.*, Beih. Bot. Centralbl., Abt. 2. 33(2): 171 (1915); Riedl in Fl. Iranica [K. H. Rechinger] 48: 207 (1967); Khatamsaz, Flora Iran 39: 161 (2002); Cecchi & Hilger, Plant Biosystems, Appendix S1: 39 (2022).

Perennial. Upper stems erect, simple to very sparsely branched (in the specimen seen) covered with dense white patent long setae mixed with shorter setae, arising from glabrous tubercles. Leaves oblong to oblanceolate, the upper sessile, 30–40 × 5–7 mm, apex acute margins entire, revolute, both surfaces with long and short setae. Inflorescence lax to 10 cm, terminal. Bracts lanceolate to ovate-lanceolate, about as long or longer than calyx. Calyx lobed to base, lobes linear-lanceolate, 18–20 mm, acute, margins with longer setae, surfaces with shorter setae. Flowers shortly pedicellate. Corolla yellow, narrow campanulate, longer than calyx, ± 24 mm, minutely lobed at apex, glabrous on the outside; lobes reflexed at apex. Annulus glabrous. Stamens included; filaments attached to about middle of corolla, included. Style slender, exserted, obscurely bilobed. Nutlets greyish with brown blotches, ± 5 mm, surfaces smooth, glossy, beaked.

HAB. Limestone mountain in *Quercus* forest; alt. 1500–2000 m; fl. May.
DISTRIB. N. Iraq. **MRO**: Sefin Dagh, *Gillett* 8104; Mt Avroman nr. Taweela, *Rechinger* 10370! (W).

N.E. Iran.

15. **Onosma tinctoria** var. **flavicoma** *A.DC.*, Prodr. [A.P. de Candolle] 10: 64 (1846). Type: Iraq, Mossul, *Aucher-Eloy* 3170! (G: G00201971. Herb. De Candolle); Cecchi & Hilger, Plant Biosystems, Appendix S1: 122 (2022).

Biennial. Stems erect, branched to 40 cm, densely patent-setose and short hairy. Leaves basal narrow-linear, 15–17 mm, sessile, obtuse; cauline leaves oblong with short petiole. Inflorescence scorpioid, compact, of short, few-flowered cymes, becoming straight at maturity. Bracts ovate to ovate-lanceolate, subglabrous beneath. Pedicels 1–2 mm; Calyx 8–10 mm, elongating to 20 mm in fruit, 5-lobed, lobes linear, densely hispid near base, (hairs on calyx and bracts yellowish, var. *flavicoma*). Corolla yellow, 11–13 mm, glabrous. Anthers included or shortly excluded with elongate bidentate sterile tips. Nutlets with an obtuse beak and acute dorsal and ventral keels, slightly rugose.

Based on a single collection from Mossul, collected by *Aucher-Eloy* in 1836. I have not seen any material of *O. tinctoria* from Iraq, but the var. *falvicoma* is very similar to the type variety (from the single collection seen). In the holotype, the inflorescence is very compact, which is suggestive that the specimen is young. With more material the presence of *O. tinctoria* could be established in Iraq. (Description of *O. tinctoria* from Fl. Turkey, p. 345).

HAB. N. Iraq; alt. not given on label; fl. not given on label.
DISTRIB. Kown from a single collection from N. Iraq. **MSU**: Mossul, *Aucher- Eloy* 3170 (type).

E. Europe to S.W. Siberia, S.E. Turkey (distribution of *O. tinctoria*).

16. **Onosma dasytricha** *Boiss.*, Diagn. Pl. Orient. Ser. 1(7): 33 (1846); Riedl in Fl. Iranica [K. H. Rechinger] 48: 210 (1967); Khatamsaz, Flora Iran 39: 161 (2002); Cecchi & Hilger, Plant Biosystems, Appendix S1: 45 (2022).

> *O. latifolia* Boiss. & Hausskn. in Boissier, Diagn. Pl. Orient. Ser. 2: 3 (1875). Type: Inter Kurdistan, Avroman et Schahu, 1867, *Haussknecht* s.n. (K!, B, W, syn.); Rawi in Dep. Agr. Tech. Bull. 14: 139 (1964); Riedl in Fl. Iranica [K. H. Rechinger] 48: 207 (1967).
> *O. xanthocalyx* (Bornm.) Vatke in Z. Gesammten Naturwiss. (Halle) 45: 124 (1875).
> *O. qandilica* Rech.f. & Riedl. in Österr. Bot. Z. 109: 239 (1962). Type: Erbil (Kurdistan), Mt. Qandil ad confines Persiae, in faucibus supra Pushtashan, 28. VII. 1957. *K. H. Rechinger* 11205 (W!, holo., K!, iso.). Riedl in Fl. Iranica [K. H. Rechinger] 48: 209 (1967).

Perennial. The whole plant greyish-green, covered with dense white adpressed setae arising from stellately setulose tubercles. Stems 1–2, ± erect, arising from a woody stock. Leaves ovate to obovate, 30–90 × 18–50 mm, somewhat reducing in size upwards, apex acute, base tapering into the petiole; petiole 10–20 mm. Inflorescence of few flowers in a lax scorpioid cyme, terminal; pedicels 5–7 mm. Bracts lanceolate, about as long as the calyx. Calyx lobed to base, lobes linear-lanceolate, 25–26 mm, acute. Corolla golden-yellow, narrow campanulate, longer than calyx, ± 20 mm, minutely lobed at apex; lobes reflexed at apex (?). Annulus glabrous. Stamens included, filaments attached to about middle of corolla, included. Style slender, exserted, obscurely bilobed. Nutlets ± 4 mm, surfaces hairy, keel acute.

HAB. Mountain and lower slopes, in valleys, on limestone, with coppiced *Quercus*; and river edges; alt. 1100–2100 m; fl. Jun.-Aug.
DISTRIB. N. Iraq. **MRO**: Kanirush, Qerna Qaw valley, N. of Pushtashan, *Serhang & Rawi* 26667A!, 26685! Pushtashan, 15 km N.E. of Rania, *Rawi & Serhang* 24211!; nr Pushtashan, *Rechinger* 11205! (type of *O. qandilica*). **MSU**: Avroman (Hauraman) N. of Biyara, *Gillett* 11762!; mt. Avroman (Hauraman), nr Tawila, Sosakan, *Rechinger* 10181!; Molla Khort, *Rawi* 29572!; Pira Magrun, *Haussknecht* s.n. (W!).

Onosma dasytricha is described from Iran (*Kotschy* 146!, W, holo; lectotype designated by Riedl 1967: 207. IRAN, Fars, "Kazerun", February 1845, Kotschy pl. Pers. austr. no. 146, W.) and according to Riedl (Fl. Iranica, p. 207), endemic to N.W. Iran. However this and two other taxa that I have placed in synonymy extend the distribution of this species to N. Iraq.

Iran.

17. **Onosma alborosea** *Fisch. & C.A.Mey.* in Index Seminum (LE) 5: 38 (1839); Rawi in Dep. Agr. Tech. Bull. 14: 139 (1964); Riedl in Fl. Iranica [K. H. Rechinger] 48: 205 (1967); Riedl in Fl. Turkey [P. H. Davis] 6: 364 (1978); Khatamsaz, Flora Iran 39: 165 (2002); Cecchi & Hilger, Plant Biosystems, Appendix S1: 21 (2022).

Incl. *O. alborosea* var. *macrocalycina* Hausskn. & Bornm., Mitt. Thur. Bot. Ver. 20: 39 (1904). Type: Iraq, Kuh-Sefin, Schaklava, Erbil, *Bornmüller* 1603 (B, [B100365366]).

Perennial to 30 cm. The whole plant greyish-green, covered with dense white adpressed setae arising from stellately setulose tubercles. Stems 1–several, erect to ascending, arising from a woody stock. Leaves lanceolate to oblanceolate, 20–50 × 5–15 mm, somewhat reducing in size upwards, apex acute, base tapering into the petiole; petiole ± 5 mm. Lower leaves brownish when dry and attached to lower part of stem. Inflorescence of few flowers in a lax scorpioid cyme, terminal; pedicels 5–7 mm. Bracts lanceolate, c. 15 mm. Calyx lobed to base, lobes linear-lanceolate, 18–24 (–27) mm (27 mm in var. *macrocalycina*), acute. Corolla pink to purple and bluish, campanulate, longer than calyx, 25–28 (–35) mm (35 mm var. *macrocalycina*), minutely lobed at apex, glabrous, pubescent near the apex; lobes ± reflexed

Fig. 81. **Onosma alborosea**. 1, habit; 2, leaf indumentum, adaxial side; 3, indumentum on stem; 4, flower; 5, corolla, opened; 6, outer face of corolla; 7, stigma; 8, details of inner face of stamen; 9, mature nutlet. 1 from *Bot. staff* 43204; 2–8 from *Tawfiq et al.* 44123; 9 from *Rawi et al.* 7626. Drawn by © A.P. Brown 2019

at apex. Annulus glabrous. Stamens included, filaments attached to about middle of corolla, included or anthers just exserted. Style slender, exserted, obscurely bilobed. Nutlets black [white], ± 4 [8mm] mm, surfaces glabrous, smooth [minutely tuberculate]. Fig. 81, 1–9.

HAB. On limestone and clay soils on mountain slopes, valley bottoms, crevices and clefts in rocks, in fields, amongst rocks and stones and relictual *Quercus* forest; alt. 200–2300 m; fl. Apr.-Aug.

DISTRIB. Mountains and foothills of N. Iraq. **MAM**: Amadiya, *Guest* 1241!, *Ludlow-Hewitt* 1502!, *Botany staff* 43401!; Zawita Gorge, *Guest* 4468!; Zawita Valley, *Guest* 4444!; Zawita, *Robertson* 369!; *Guest* 4925!, *Guest* 4830!, 4904!; Sarsang, *Botany staff* 43477!; 10 km E. of Aqra, *E. Chapman* 26088!; Zawita to Sawara Tuka, Guest 13256!; Zawita, *Hunting Aero Surveys* 58!; Zawita gorge, *Guest* 2193! **MRO**: 2 km N. of Darband, *Botany staff* 43204! 5 km N. of Darband, *Botany staff* 43192!; Salah ad-Din mt., *Karim, Hamid & Jasim* 40870!; Jabal Karokh, *Anon.* 27480!; Salah ad-Din, *Rawi & Gillett* 10427!; 1 km N. of Haji Umran, *Alkas* 18608!; Salah ad-Din, *Chapman* 11963!; Shaqlawa, *Haines* W. 769! (var. *macrocalycina*); Kuh Sefin (Sefin Dagh) above Shaqlawa, *Bornmüller* 1603 (type of var. *macrocalycina*); **MRO / MSU**: Bestash junction, *Omar, Karim, Hamza & Hamid* 37292! **MSU**: Sehkanian, mt. left to Dokan, *Botany staff* 43121!; Qopi Qaradagh, *Barkley* 8501!; 49 km W. of Sulaimaniya, *F. & E. Barkley* 4477!; Darband-I Basian, *Rawi & Gillett* 7626!; Azmir, *Qaisi & Hamad* 49310!; Sehkanian left to Dokan, *Botany staff* 43073!; Jasana, *Omar* 42761!; Palegawra, W. of Sagirma Dagh, 25 km S.W. of Pir I Mukhrun (Pira Mugrun), *Helbaek* 670!; Darband-I Basian, *Ludlow-Hewitt* 1195!; 4 km N. of Darband-I Khan dam, *Barkley & Haddad* 7465!; Qara Dagh, *Gillett* 7939! **MJS**: Jabal Sinjar, *Qaisi & Hamad* 49133!; Balad Sinjar, *Qaisi & Hamad* 49348!; Chanarock, *Omar, Karim, Hamza & Hamid* 37225!; Asi, *Rawi* 8491! **FUJ**: Mosul, Balkhshakur, *Graham* 671!; 5 km before Baidiha, *Karim, Nuri & Hamad* 44808!; 4 km W. of Sa'adiya, *Qaisi* 42569! **FKI**: 6 km E. of Kirkuk to Sulaimaniya, *Botany staff* 42988! **FUJ**: S. of Jabal Sinjar, *Rawi* 41376!; Mosul, *Shahwani* 25184! **FNI**: 3 km before Mindan bridge, *Taufiq, Dabbagh & Lafta* 44123! **FPF**: Kuh-I Bamu, *Poore* 313!

I have included *O. alborosea* var. *macrocalycina* in the type variety, as it was distinguished only in its larger calyx and corolla from just 2 specimens and bordering within the upper range of calyx and corolla var. *alborosea*.

MĪSHMAIZHŪK (*Guest* 1241).

Turkey, Iran, Lebanon, Syria.

18. **Onosma rascheyana** *Boiss.*, Diagn. Pl. Orient. Ser. 1, 11: 110 (1849); Dinsmore in Post, Fl. Syria, Palest. & Sinai ed. 2, 2: 233 (1933); Rawi in Dep. Agr. Tech. Bull. 14: 139 (1964); Riedl in Fl. Iranica [K. H. Rechinger] 48: 207 (1967); Riedl in Fl. Turkey [P. H. Davis] 6: 366 (1978); Mouterde, Nouv. Fl. Liban et Syrie 3: 72, pl. 35, f. 4, (1983); Khatamsaz, Flora Iran 39: 163 (2002); Cecchi & Hilger, Plant Biosystems, Appendix S1: 100 (2022).

Perennial to 25 cm. The whole plant greyish green covered with dense white ± adpressed setae arising from stellately setulose tubercles. Stems yellowish, 1–several, erect to ascending, arising from a woody stock. Lower part of stems covered with old leaves. Leaves oblanceolate to oblong-lanceolate, 20–40 × 3–5 mm, apex obtuse, margins often revolute, base of lower leaves tapering into the petiole; petiole ± 8 mm, upper leaves sessile. Inflorescence of compact scorpioid cymes, terminal; pedicels 3–5 mm. Bracts lanceolate to somewhat cordate, 15–18 mm, often covering the inflorescence head. Calyx lobed to base, lobes linear-lanceolate, 10–16 mm, acute, the setae sometimes yellowish. Corolla reddish-yellow, or white flushed pink, the lower part often yellowish, upper reddish, turning orange when dying, campanulate, longer than calyx, 18–20, apical lobes reflexed, glabrous. Annulus glabrous. Stamens included, filaments attached to above the middle of corolla, anthers included. Style slender, exserted, obscurely bilobed. Nutlets brown, ± 4 mm, with a ventral keel lateral ridges, beak acute, surfaces glabrous, smooth.

HAB. On rocky mountain sides, limestone screes and ledges, and in denuded *Quercus* forest; alt. 1500–2700 m; fl. Apr.-Aug.

DISTRIB. Mountains and limestone screes of N. Iraq. **MRO**: Chiya-i Mandau, *Guest* 2728!; Kudu nr Haji Umran, *Rawi* 9202!; Sefin Dagh, Shaqlawa, *O. Polunin* 5015!, 5029!; Haji Umran, *Mooney* 4263!, *Chapman* 11184!; mt. Helgord, *Bornmüller* 1602!, *Rechinger* 11412!; HAB-as Sultan Dagh N. of Koi Sanjaq, *Rawi, Nuri & Kass* 28189! **MSU**: Azmir pass, *Poore* 345!; Penjwin, *Rawi* 9419!; Qara Dagh, *Gillett* 7938!

Turkey, Iran, Syria.

19. **Onosma auriculata** *Auch.* ex *A.DC.*, Prodr. [A.P. de Candolle] 10: 61 (1846). Type: Mosul, Nineveh, *Aucher-Eloy* 2313, G-DC, lectotype selected here ; K!, BM!, FI, LE, MPU, isolecto). Dinsmore in Post, Fl. Syria, Palest. & Sinai ed. 2, 2: 234 (1933); Rawi in Dep. Agr.

Tech. Bull. 14: 139 (1964); Riedl in Fl. Turkey [P. H. Davis] 6: 374 (1978); Mouterde, Nouv. Fl. Liban et Syrie 3: 74, pl. 36, f. 2, (1983); Cecchi & Hilger, Plant Biosystems, Appendix S1: 31 (2022).

Biennial or perennial, robust shrub to 1.5 m. The whole plant greyish green, covered with dense white ± adpressed setae arising from stellate setulose tubercles. Stems woody, several arising from base, erect. Leaves linear-lanceolate, 40–180 × 5–20 mm, apex ± obtuse, margins often revolute, base of lower leaves tapering into the petiole to 60 mm, upper leaves sessile, base auriculate. Inflorescence of many compact scorpioid cymes in large panicles to 30 cm, cymes becoming lax in fruit. Pedicels 4–5 mm, elongating to 12–13 mm in fruit, with dense patent setae. Bracts as cauline leaves but smaller. Calyx lobed to base, lobes oblanceolate to spatulate, c. 10 mm, elongation to c. 20 mm in fruit. Corolla yellow, campanulate, longer than calyx, 18–20, apical lobes obtuse, reflexed, glabrous. Annulus glabrous. Stamens included; anthers as long as the filaments, included. Style slender, exserted, obscurely bilobed. Nutlets brown, ± 4 mm, ovoid, dorsally keeled, surfaces glabrous, smooth.

HAB. Fallow field in valleys, open grassland, on calcareous clay on limestone; alt. 750–1200 m; fl. Apr.-Jun.

DISTRIB. In the Sulimaniyah and Nineveh districts of north eastern Iraq. MSU: Baranand range, Qara Dagh road, nr Sulaimaniya, *O. Polunin & Mooney* 4340!; Tanjero valley, 15 km S. of Sulaimaniya, *Poore* 363! FNI: Nineveh, *Aucher* 2313! (type).

Turkey, Syria.

20. **Onosma gigantea** *Lam.*, Tabl. Encycl. 1: 407 (1792), Boissier, Fl. Orient. 4: 202 (1879); Post, Fl. Syria, Palest. & Sinai, 554 (1896); Riedl in Fl. Turkey [P. H. Davis] 6: 375 (1978).

Biennial. Stem branched below, erect, 35–60(–95) cm, densely setae arising from stellately setulose tubercles. Leaves oblanceolate to lanceolate-oblong, 100–220 × 10–15 mm, apex acute, margins often revolute, lower leaves tapering into the petiole; petiole up to 80 mm, upper leaves sessile, somewhat crisped at base, densely setae arising from stellately setulose tubercles. Cymes numerous, lax, up to 5 cm, with 14–30 flowers, forming a paniculate inflorescence. Peduncle 15–40 mm. Bracts sessile, lanceolate or ovate, attenuate with acute apex, broad subcordate base, 60–20 × 20–11 mm. Pedicels up to 15 mm. Calyx lobed to base, lobes linear-oblanceolate, 11–12 mm, apex obtuse, dense white patent setose adaxially, very dense long white simple hairs abaxially. Corolla golden yellow, narrow campanulate, 17–21 mm, glabrous, lobed, lobes acute-obtuse at apex, recurved. Annulus glabrous. Filaments adnate to about half the length of the corolla tube; anthers included, ± 7 mm long. Style slender, exserted, minutely 2-lobed above.

HAB. Between rocks; alt. ± 1100 m; fl. Apr.-May.

DISTRIB. Very rare found once in Jabal Sinjar N.W. Iraq. MJS: Jabal Sinjar, S.E. slope, *Widad & Al-Khayat* 53543! (BAG).

Cyprus, Lebanon, Syria, Palestine, Turkey.

21. **Onosma ovalifolia** *Kotschy & Boiss.*, Fl. Orient [Boissier] 4(1): 196 (1879); Riedl in Fl. Turkey [P. H. Davis] 6: 357 (1978); Cecchi & Hilger, Plant Biosystems, Appendix S1: 92 (2022).

Perennial. Stems simple or ± branched, erect, to 25 cm, pubescent setose. Leaves lanceolate, 30–60 × 6–18 mm, base cuneate, apex acute, margins entire, upper leaves sessile, basal and cauline leaves with petiolate, setose. Inflorescence of 1–2 cymes. Bracts linear to linear-filiform. Pedicels c. 1 mm. Calyx lobed to base, lobes linear-lanceolate, c. 15 mm, yellowish hairy. Corolla whitish, tubular, 16–20 mm, puberulous. Stamens included; anthers longer than filaments. Nutlets not seen. (Description from Fl. Turkey; no material seen from Iraq).

HAB. Mountains; alt. 2000 m; fl. not given on label.

DISTRIB. Kown from a single collection from N. Iraq. MSU: Sulaymaniyah, Pira Magrun, *Poore* 581 (W).

Turkey.

8. **ECHIUM** L.

Sp. Pl. 1: 139 (1753); Gen. Pl. ed. 5, 68 (1754)
Bramwell in Lagascalia 2: 37 (1972); Weigend & al. in Kubitzki, Fam. Gen. Vasc. Pl. 14: 73 (2016)
Magacaryon Boiss., Pl. Orient. Nov. 1, 7 (1875)

Shahina A. Ghazanfar

Annual or perennial herbs (in Iraq) covered with dense hispid tubercle-based hairs, sometimes sericeous. Leaves narrowly obovate to braoadly ovate. Inflorescence bracteate, with densely branched thyrsoids; bracts leaf-like. Flowers shortly pedicellate, erect. Calyx 5, lobed to base, often accrescent in fruit. Corolla blue (in Iraq plants), rarely pink, red, yellow or white, zygomorphic, infundibuliform; faucal (throat) scales absent. Annulus of 5–10 lobes or swellings present, glabrous. Staminal filaments long, unequal, at least two exserted. Style filiform, exserted; stigmas 2. Nutlets 4 or 1–2 by abortion, ovoid, rugose to tuberculate.

A genus with 68 species found in Macaronesia, N. Africa, Europe to Central Asia; a single species in Iraq.

Echium, from Gr. εχιον, *echion*, Pliny's name for the plant, derived from εχιδνα, *echidna*, a snake.

1. **Echium italicum** *L.*, Sp. Pl.: 139 (1753); Dinsmore in Post, Fl. Syria, Palest. & Sinai ed. 2, 2: 235 (1933); Rawi, in Dep. Agr. Tech. Bull. 14: 136 (1964); Riedl in Fl. Iranica [K. H. Rechinger] 48: 214 (1967); Edmondson in Fl. Turkey [P. H. Davis] 6: 321 (1978); Khatamsaz, Flora Iran 39: 186 (2008); Breckle, Hedge, Podlech & Rafiqpoor, Vasc. Pl. Afghan. Checkl.: 205 (2013); Taifour & El-Oqlah, Pl. Jordan Annot. Checkl.: 51 (2017).

Including *Echium italicum* subsp. *biebersteinii* (Lacaita) Greuter & Burdet in Willdenowia 11: 37 (1981).

E. pyramidatum A.DC., Prodr. 10: 23 (1846).
E. glomeratum Ledeb. Fl. Ross. (Ledeb.) 3(1,8): 106 (1847).
Echium italicum var. *biebersteinii* Lacaita in J. Linn. Soc. Bot. 44: 408 (1919).

Hispid-setose biennial, with basal rosette and much-branched stems, 30–70 cm. Basal leaves narrowly oblong to lanceolate, to 25 × 2 cm, tapering at base into short petiole; cauline leaves smaller, subsessile. Inflorescence ± pyramidal, lowest branches arising from lower cauline leaves' axils, or narrower and spike-like. Calyx deeply divided, to 8 mm in flower, accrescent to 12 mm in fruit. Corolla pink (turning blue), lilac or whitish, 7–9.5 mm, narrowly funnel-shaped, sparsely hairy or glabrescent outside. Stamens long-exserted. Style bifid. Nutlets 4 × 2 mm, ± oblong, with incurved acute beak and raised ribbed shoulders, rugose-tuberculate and reticulate. Fig. 82, 1–7.

HAB. Abandoned vineyard, in orchard, under *Populus*, destroyed *Quercus* forest, rocky clay, near stream, stony hillside; alt. 600–2300 m; fl. May-Aug.

DISTR. Northern and eastern sectors of the lower forest zone of Iraq. **MAM**. Maya, 20 km from Kani Masi, *Hamid & Fadhil* 45471!; Zawitah, *Guest* 3895! *Chapman* 26308! Sarsang, *Haines* W. 571!; Atrush, *Guest* 3642!; Aradin, 15 km W. by N. of Amadiya, *Kaisi & Hamad* 46045! Daimka 35 km N.E. of Zakho, *Kaisi, Hamid & Hamad* 45363!; Olaka, *Rawi* 8434! **MRO**. Haji Omran, *Dabbagh & Hamad* 46260!; 2 km below Darband, *Gillett* 9449!; Rayat, *Guest* 13085! **MSU**. Dukan road, *Omar & Karim* 38015!; Tasluja, *Omar & Karim* 47023!; Khormal, *Rawi* 8941!; Qaradagh, *Gillett* 7953! **MJS**. Kursi, Jabal Sinjar, *Gillett* 10853!; Kurdistan & Mosul, *Kotschy* 437!

According to *Gillett* 7953 (Qaradagh) the young stems are peeled and eaten.

Echium italicum subsp. *biebersteinii* has been separated from the type subsp. based on its wider inflorescence and longer stems and treated as such by Riedl in Flora Iranica (p. 214). *Echium italicum* is widespread in Iraq in a range of habitats and shows variation in its vegetative characters especially habit and size of plant. The differences separating sub. *italicum* and subsp. *biebersteinii* do not hold good for Iraq material.

S. & C. Europe through Turkey, Iraq and Iran to Afghanistan.

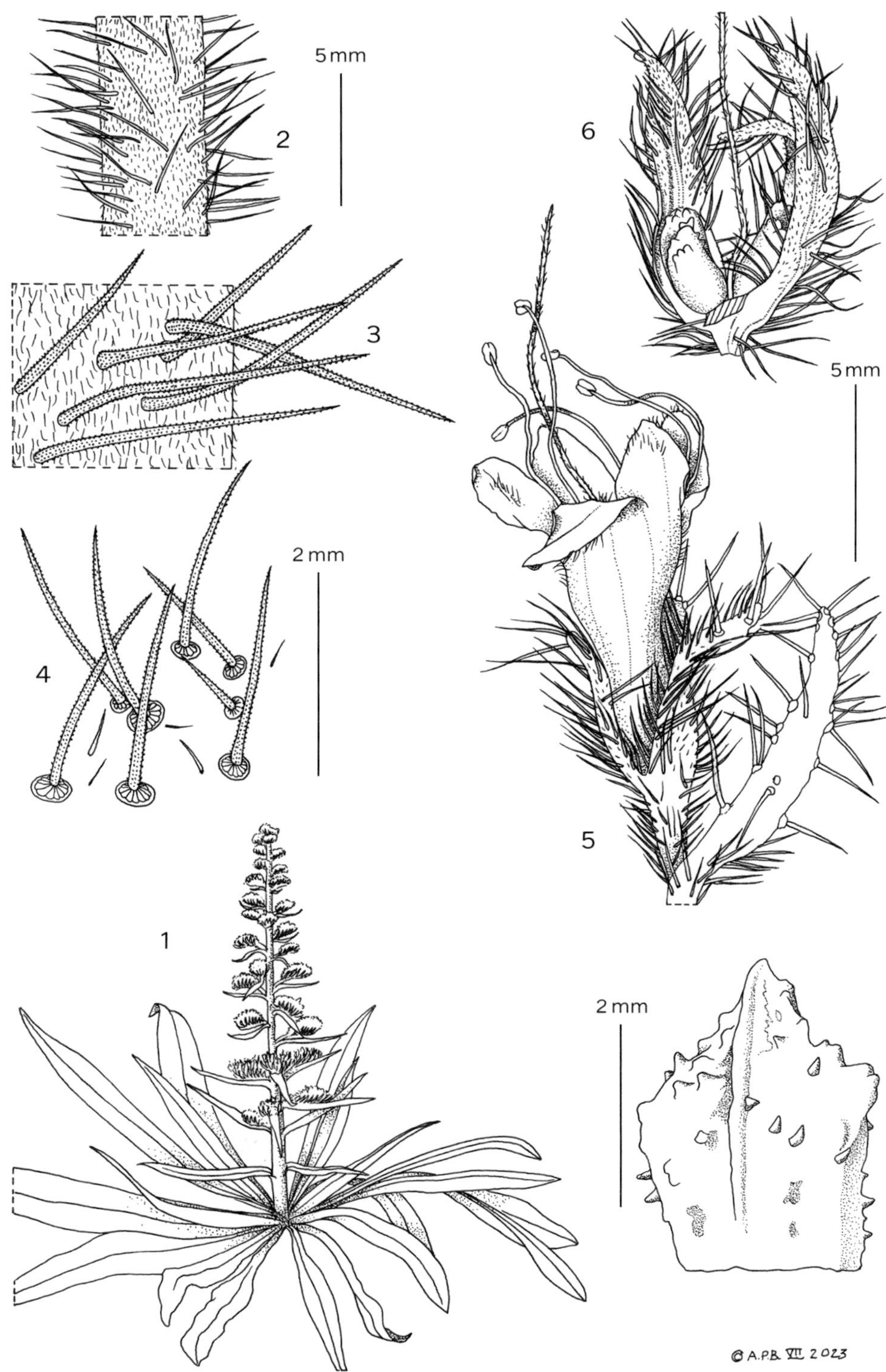

Fig. 82. **Echium italicum**. 1, habit; 2, indumentum of stem; 3, detail indumentum; 4, detail leaf indumentum; 5, flower; 6, young fruits (sepals removed); 7, fruit outer face. 1, from photo of plant from Greece, Mt. Hymettos by S. Apostolon; 2–4, 6, 7 from *Dabbagh et al.* 46260; 5 from *Hamid et al.* 45471. Drawn by © A.P. Brown 2023.

9. **MOLTKIA** Lehm.

Neue Schr. Naturf. Ges. Halle 3 (2): 3 (1817)

C.C. Townsend

Perennial herbs and shrublets with woody rootstock containing purple dye. Stems densely leafy. Cymes terminally crowded, bracteate. Calyx divided to base. Corolla funnel-shaped, with a ± cylindrical tube; faucal scales absent. Stamens distinctly exserted. Nutlets 4 or fewer, ovoid to ovoid-trigonal, weakly tuberculate and finely papillose; beak horizontally incurved, subacute.

A genus with six or seven species; a highly derived member of the Lithospermeae found mainly in the E. Mediterranean and Irano-Turanian phytochoria, though our species tends towards the Saharo-Arabian in its distribution; a single species in Iraq.

Moltkia, is supposed to commemorate Joachim Godske Moltke (1746–1818), latterly Prime Minister of Denmark, who died the year after the name was first published.

1. **Moltkia angustifolia** *A.DC.*, Prodr. 10: 72 (1846); Rawi, in Dep. Agr. Tech. Bull. 14: 138 (1964); Riedl in Fl. Iranica [K. H. Rechinger] 48: 147 (1967).

Echium longiflorum non Dum.-Cours. (1814): Bertol., Misc. Bot. 1: 15 (1842).
Moltkia longiflora (Bertol.) Wettst., Oesterr. Bot. Z. 67: 368 (1918). Type: Syria/Iraq: 'ex oris Euphraticis', *Chesney Expedition* 188 (leg. *Helfer?*); Mesopotamia, Persia & Syria, *Aucher*s.n.! (BM, syn.); Rechinger, Fl. Lowland Iraq: 506 (1964); Mouterde, Nouv. Fl. Liban et Syrie 3: 61, pl. 33, f. 2 (1983); Khatamsaz, Flora Iran 39: 168 (2002) as syn. of *M. coerulea*.

Branched perennial herb, woody at base, clothed with adpressed setae. Stem 10–30 cm. Leaves narrowly obovate to linear, 25–30 × 2.5–4 mm, often arcuate, apex obtuse, margins involute. Bracts linear, scarcely overtopping calyx. Calyx deeply divided; lobes linear, obtuse. Corolla blue, salver-shaped, glabrous, long-exserted from calyx. Anthers yellow, exserted. Nutlets weakly rugose. Fig. 83, 1–9.

HAB. Dry steppe on limestone, saline and sandy clay soil, sandy gravelly soil underlain with gypsum, stony banks, silt covered by rocky layer, alt. 40–680 m; fl. Mar.-May.
DISTRIB. Western part of the steppe zones of Iraq. **FUJ**: Hadhr district, *Rawi & Hamada* 34155!; Qaiyara, nr Mosul, *anon.* 101!; Serida, 7 km S.W. of Ba'aj, *Widad & Khayat* 53534!; Qara Sharqat, *Rao* 120!; 26 km E. of Balad, *Gillett & Chapman* 11156!; Umm Shabaish area E. of Balad Sinjar, *Gillett* 11156! **DLJ**: Wadi Thirthar canal, Jazira, *Guest & Haidar* 18315a!; 6 km N. of Tal Baka police station, *Alizzi & Husain* 33847!; 48 km N.W. of Baiji towards Haditha, *Chakravarty, Rawi, Khatib & Alizzi* 31862!; Raisha hill nr Shibaichan, 78 km S. of Sinjar, *Chakravarty, Rawi, Khatib & Alizzi* 32046!; S. edge of Lake Thirthar, *Rawi & Chakravarty* 30441!; 10 km S.W. of Manayif, *Alizzi & Husain* 33978!; 10 km S. of Ana, *Alizzi & Husain* 34001!; Wadi Thirthar canal, Jazira, *Guest & Haidar* 18315b!; opposite Rawa, *Rawi & Gillett* 7051! **DWD**: 13 km E. of K3, *Chakravarty, Rawi, Khatib & Alizzi* 32940! **DSD**: Malha 160 km N. of Haditha, *Rawi* 5725! **LCA**: 30 km N.W. of Falluja, *Rawi & Chakravarty* 30345! unlocalised: 'Mesopotamia', *Aucher* (syntype).

According to Mouterde (l.c.) the species also occurs in Turkey, but it was treated as 'doubtfully recorded' by Riedl in Davis, Fl. Turkey 7: 326 (1978), as "M. longifolia" (sphalm.).

Syria, Iran.

10. **MOLTKIOPSIS** I.M. Johnston

J. Arnold Arbor. 38: 2 (1953); Weigend et al. in Kubitzki, Fam. Gen. Vasc. Pl. 14: 75 (2016)

Shahina A. Ghazanfar

Setose perennial shrub with long rootstock; indumentum hispid. Leaves sessile. Inflorescence of 1–2 scorpioid cymes, lengthening in fruit. Calyx lobed almost to base; fruiting calyx circumscissile. Corolla salver-shaped, bluish purple to red, rarely white; annular nectary at base of corolla tube. The plant exhibits floral dimorphism. Corollas either 4–7 mm or 10–15 mm; short corollas have anthers and stigma at same level within throat, while in short flowers stigma is exserted. Nutlets 2–3, ventrally keeled, smooth or with few tubercles, with wide attachment scar.

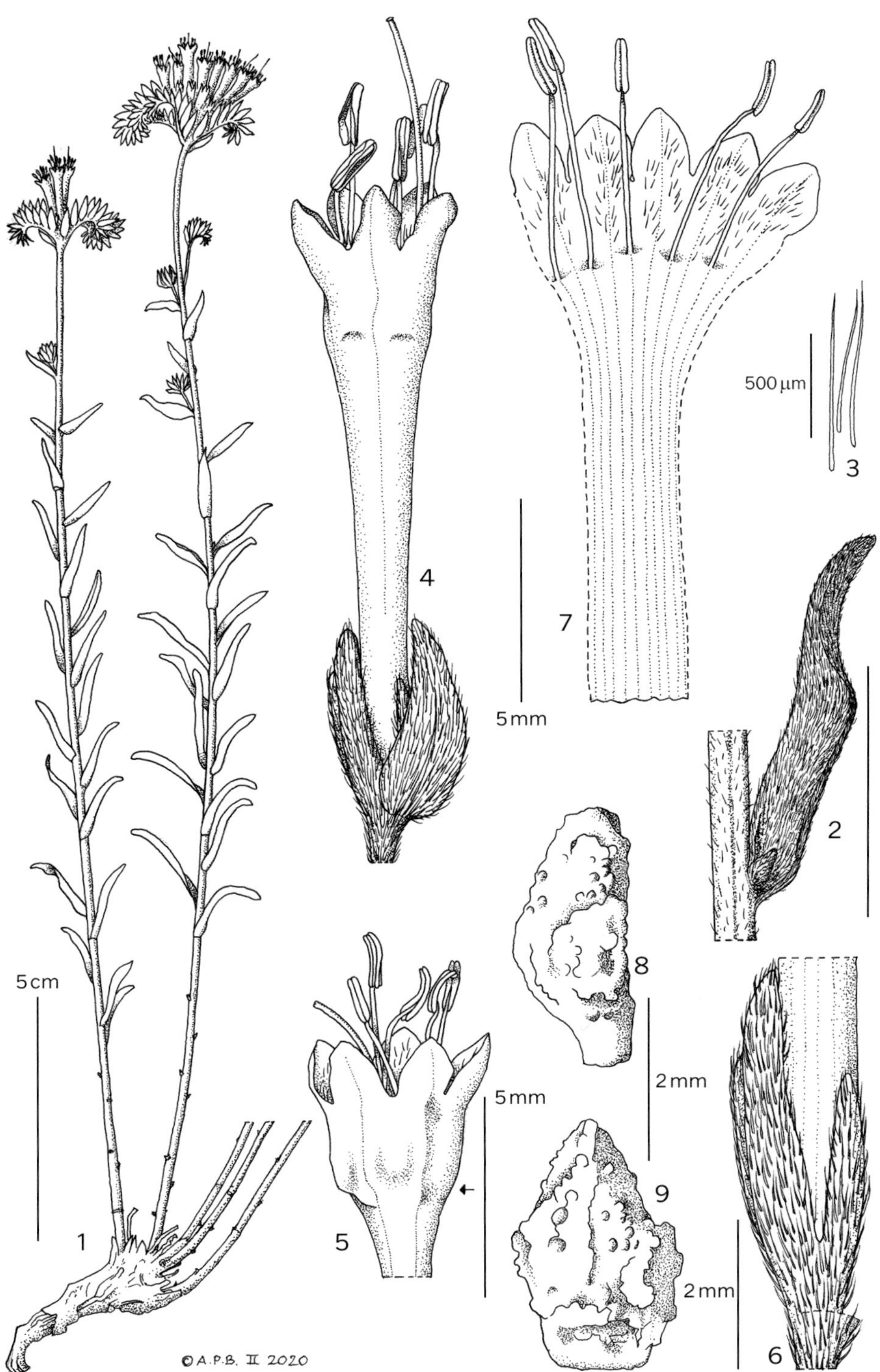

Fig. 83. **Moltkia angustifolia**. 1, habit; 2, detail leaf and stem indumentum; 3, setae on leaf; 4, flower; 5, distal portion of older flower; 6, calyx (bract removed); 7, corolla, opened; 8, nutlet, side view; 9, nutlet outer face. 1–3 from *Rawi et al.* 3044; 4–9 from *Guest* 18315. Drawn by © A.P. Brown 2020.

One species; N.E. Africa, S.W. Asia.

Moltkiopsis means 'resembling *Moltkia*', the suffix *-opsis* meaning 'looking like'.

1. **Moltkiopsis ciliata** (*Forssk.*) *I.M.Johnston* in J. Arnold. Arbor. 34: 3 (1953); Rechinger, Fl. Lowland Iraq: 508 (1964); Riedl in Fl. Iranica [K. H. Rechinger] 48: 54 (1967); Feinbrun-Dothan, Fl. Palaest. 3: 59 (1978); Khatamsaz, Flora Iran 39: 111 (2002); Collenette, Wild Flow. Saudi Arabia: 93 (1999); El-Hadidi & Boulos in Fl. Egypt 2: 296 (2002); Al-Turki & Thomas in Turkey J. Bot. 34: 367–377 (2010); Ghazanfar, Fl. Oman 3: 82 (2015); Abdulridha, Taha & Wedad, Ecology and Fl. Basra: 239 (2016); Taifour & El-Oqlah, Plants Jordan Annot. Checkl.: 52 (2017).

Lithospermum ciliatum Forssk., Fl. Aegypt.-Arab: 39 (1775).
L. angustifolium Forssk., Fl. Aegypt.-Arab: 39 (1775).
L. callosum Vahl, Symb. Bot. 1: 14 (1790); Boiss., Fl. Orient. 4: 219 (1879); Dinsmore in Post, Fl. Syria, Palest. & Sinai ed. 2, 2: 241 (1933).
L. callosum Vahl var. *asperrimum* Bornm., Mitt. Thur. Bot. Ver. ser., 6: 58 (1894).
Moltkia callosa (Vahl) Wettst., Osterr. Bot. Z. 67: 368 (1918), nom. illeg.
M. ciliata (Forssk.) Maire, Cat. Pl. Maroc. 4: 1102 (1941).

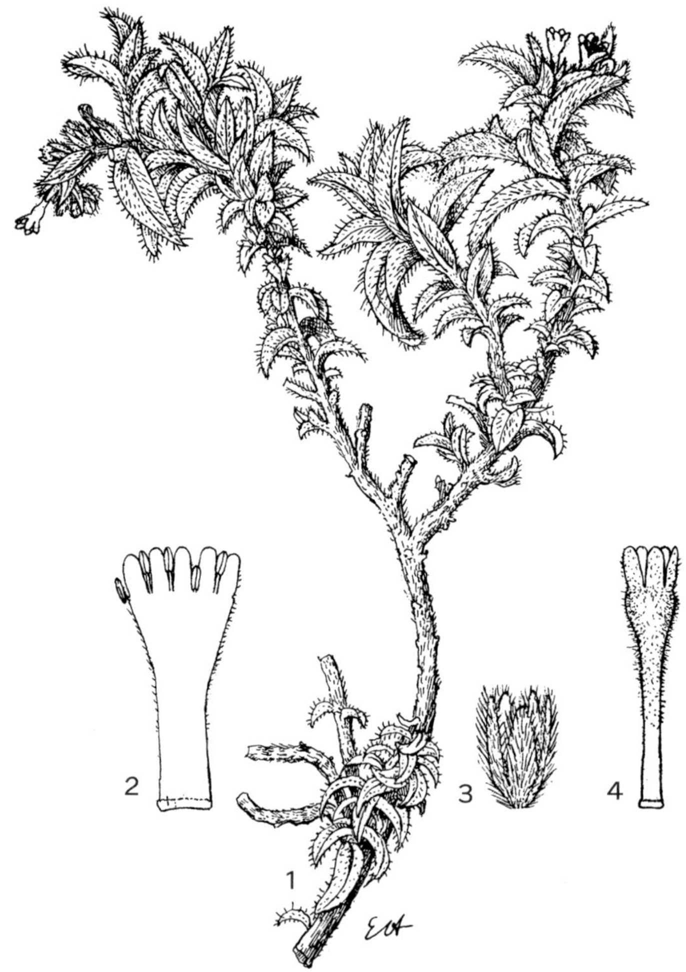

Fig. 84. **Moltkiopsis ciliata**. 1, habit x1; 2, calyx × 3; 3, flower × 3; 4, corolla opened × 3. Reproduced with permission from Feinbrun-Dothan, Flora Palaestina, 3: Plates, f. 95 (1977). Drawn by Esther Huber. © The Israel Academy of Sciences and Humanities.

Dwarf, perennial shrub, woody at base, hirsute throughout with white appressed hairs and long, ± patent, tubercule-based setae. Stems much-branched from base, 8–15 cm, erect or ascending, spreading. Leaves oblong- to ovate-lanceolate, 5–15 × 3–6 mm, sessile, apex acute, margins ciliate. Inflorescence dense, of terminal scorpioid cymes, elongating to 25 mm in fruit. Flowers bracteate, sessile or pedicel short. Calyx 4–5 mm, lobes linear-lanceolate, hispid with patent long and short adpressed hairs. Corolla blue when young, turning pale blue after anthesis, 9–10 mm, lobes c. 2 mm, triangular, tube and lobes hairy outside, glabrous inside. Nutlets ovoid-triquetrous with an acute beak, dorsally rounded and ventrally keeled. Fig. 84, 1–4.

HAB. Sandy rocky hillsides, sandy *haswa* desert, sandy gravel and gypsum soils; alt. 25–230 m; fl. Mar.-Apr.

DISTR. Predominantly a desert plant, but also found in dry sandy and gravelly areas of Iraq.
FKI/FPF. Jabal Hamrin, *Barkley* 7688! **FPF.** Chilat police station near Iranian frontier, *Rawi & Haddad* 25656! **LEA.** 5 km S.E. of Al-Shihabi police station, *Rawi & Haddad* 25624!; 78 km N.E. of Fakka, *Hazim* 30654!; Fakka-Laqlaq valley on Iranian frontier, *Rawi* 25832!; **LCA.** Baghdad to Falluja, *Wedad* 47549!; Sudur, *Isaac* 323! **DLJ.** W. of Fallujah, near S. shore of lake Habaniya, *Haidar* 15189!; 5 km S.E. of Falluja, *Omar, Kaisi, Hamad & Hamid* 44498!; 65 km N.W. of Falluja, *Rawi* 30241!; E. of lake Thirthar, *Barkley & Jumaili* 72524! **DWD.** Karbala, *Haines* W. 372! **DWD/DSD.** Faidhat al Rhazmi, *Karim, Nuri, Hamid & Kadhim* 40289!; Busaiya, *Fawzi, Hazim & Hamid* 38947! *Guest & Rawi* 14238!; Zarqa, 12 km S.E. of Samawa, *Alkas* 17675!; 30 km N. of Khidr al-Mai, *Weinert, Fawzi, Adel & Hussain* 47617!; Al-Rikhaimiya, 41 km N. of Samah, *Botany Staff* 42327!; Ichrisi, 35 km E. by N. of Busaiya, *Guest & Rawi* 14190!;15 km S. by W. of Zubair, *Guest, Rawi & Schwann* 14348!; 25 km from Khidr al-Mai to Busaiya, *Kaisi, Hamad & Hamid* 48515!; 40 km from Khidr al-Mai to Busaiya, *Kaisi, Hamad & Hamid* 48499!; Umm Qasr, *Ani* 39919!; 23 km N.N.W. of Najaf, *Guest, Rawi & Long* 14002!; Al-Layah 34 km W. of Busaiya, *Botany Staff* 42363!; 69 km from Khadhar al-Mai, *Kazim, Nuri, Hamid & Kadhim* 40390!; Jabal Sanam, *Kaisi, Hamad & Hamid* 48582! *Omar, Kaisi, Hamad & Hamid* 43979!; 55 km from Karlila to Nukhail, *Rawi, Alizzi & Omar* 36241! Sand dune desert pavement E. of Lake Tharthar, *Rechinger* 13524 (W!).

HAMĀT (Ir.-Ichrisi), *Guest & Rawi* 14190; HABĀT (Ir.-Najaf), *Guest et al.* 14002. The name, *hamāt* is also used in northern Saudi Arabia for this species.

N. Africa through the Arabian Peninsula, Syria, Jordan, Palestine to Iran.

11. **BUGLOSSOIDES** Moench

Meth. 418 (1794)
I.M. Johnston in J. Arn. Arb 38: 38 (1954); Weigend et al. in Kubitzki (ser. ed.), Fam. Gen. Vasc. Pl. 14: 76 (2016)

Shahina A. Ghazanfar

Annuals (in Iraq) or perennials, ± branched, rhizomatous. Stem sparsely pilose or densely villous. Leaves linear to oblong-lanceolate, basal shrivelled at flowering time, cauline rounded at apex, base of lamina sessile. Inflorescence of many-flowered cymes. Calyx 3–4 mm at flowering time, lobed almost to base, somewhat accrescent in fruit, indumentum yellowish. Corolla minute, 5–6 mm, 5-lobed, white, pink or blue; throat of corolla tube with 5 longitudinal folds covered with velvety hairs. Stamens inserted below or at the middle of the corolla tube. Stigmas 2. Nutlets 1(–2), smooth (perennials) or 4, verrucose (annuals), gibbous at sides.

About 10 species found in the Mediterranean and S.W. Asia: 3 species in Iraq.

Buglossoides is derived from *Buglossum*, a Medieval Fr. name from Lat. *buglossa* and originally from Gr. βους, *bous*, cow and γλωσσα, glossa, tongue; the suffix *-oides* denotes resemblance.

1. Calyx hairs yellowish; corolla blue; nutlets strongly bigibbous. . . . . .1. *B. tenuiflora*
   Calyx hairs white; corolla whitish, pale yellow or pinkish; nutlets not
      strongly bigibbous 2
2. Fruiting pedicels not swollen, cylindrical; receptacle not oblique in fruit .2. *B. arvensis*
   Fruiting pedicels swollen, obconical to clavate; receptacle oblique in
      fruit. . . . . 3. *B. incrassata*

1. **Buglossoides tenuiflora** (*L.f.*) *I.M.Johnst.* 35: 42 (1954); Y. Nasir in Fl. Pakistan [Ali & Nasir] 191: 69 (1989); Feinbrun-Dothan, Fl. Palaest. 3: 68 (1978); Edmondson in Fl. Turkey [P. H. Davis] 6: 316 (1978); Khatamsaz, Flora Iran 39: 105 (2002); El-Hadidi & Boulos in Fl. Egypt 2: 290 (2002); Ghazanfar, Fl. Oman 3: 95 (2015); Taifour & El-Oqlah, Pl. Jordan Annot. Checkl. 50 (2017).

> *Lithospermum tenuiflorum* L.f., Suppl. 130 (1781); Boiss., Fl. Orient 4: 217; Dinsmore in Post, Fl. Syria, Palest. & Sinai ed. 2, 2: 240 (1933); Rechinger, Fl. Lowland Iraq: 507 (1964); Rawi, in Dep. Agr. Tech. Bull. 14: 138 (1964); Riedl in Fl. Iranica [K. H. Rechinger] 48: 150 (1967).
> *Myosotis tenuiflora* (L.f.) Viv., Fl. Lib. 9 (1824).

Annual herb with main stem erect and branches prostrate to ascending, adpressed to patent-villous, 10–25 cm tall. Cauline leaves oblong-lanceolate, 15–40 × 2–8 mm, sessile, pilose. Inflorescence bracteate, bracts much smaller than leaves but similar in shape and indumentum. Pedicels very short, thickening after anthesis. Calyx 3–4 mm in flower, clothed with yellowish hairs, divided to base into linear segments, becoming swollen at base in fruit and enclosing persistent nutlets. Corolla blue. Style filiform. Nutlets ovoid, ± 2 mm, strongly bigibbous, minutely verrucose.

HAB. Dry slopes in denuded *Quercus* forest, sandy gravelly depressions, wheat fields, disturbed grasslands; alt. 150–1000 m; fl. Mar.-Apr.

DISTR.: Not uncommon in the mountains and the Upper and Lower Jazira districts of Iraq. **MAM**. Dohuk, *Guest* 1308!; Sarsang, *Haines* W. 1113! **MRO**. Razan nr Rowanduz, *Mooney* 4285!; Shaqlawa, *Haines* W.1112!; Haji Umran, *Mooney* 4278! **MSU**. Darband-i Khan, *Haines* W.1687!; Gweija (Goizha) Dagh, *Rawi & Gillett* 10620!; Sulaimaniya, *Graham* 666!; Jarmo, *Helbaek* 328! 441!; mt. Sekanian, *OImar* 42631!; Penjwin, *Rawi* 9419J! **MJS**. Jabal Sinjar, Kursi (Karsi), *Gillett* 10928! **FUJ**. Hatra (Hadhr), *E. & F. Barkley* 6324! 4164! *Ani* 9399!; Balad Sinjar, *Guest* 4126! **FAR**. Bastura Chai, *Rawi & Gillett* 10491! **FKI**. Kirkuk, *Bornm.* 1621!; Kani Domlan, *Guest* 4323!; 25 km S. of Kirkuk, *Rawi & Gillett* 10345! **FPF**. Pilkana nr Khanaqin, *Rawi* 12735! 12796!; Third range of Jabal Hamrin, *Haines* W.1318! **DLJ**. 12 km W. of Rawa, *Omar, Kaisi, Hamad & Hamid* 44459!; Rawa, *Rawi & Gillett* 6985! 7036! 4 km W. of Haditha, *Rawi & Gillett* 6921! **DWD**. 8 km E. of Ana, *Omar, Kaisi, Hamad & Hamid* 45104!; Wadi Hawran nr Baghdad, *Omar, Kaisi, Hamad & Hamid* 44570!; Ana, Wadi al Qaiser, *Omar, Kaisi, Hamad & Hamid* 44377!; 10 km W. of K3, *Omar, Kaisi, Hamad & Hamid* 44734!; Tall al-Nisr, 48 km N.E. of Rutba, *Rawi* 31196!; Wadi al Ajrumiya, *Rawi* 31250!; 8 km E. of Ana, *Omar, Kaisi, Hamad & Hamid* 44347!; 10 km from Rutba to H3, *Botany Staff* 41766!; 40 km N. of Rutba, *Rawi & Khatib* 32276!; 65 km N. of Rutba, *Rawi & Khatib* 32248!; 25 km N.E. of Rutba, *Rawi & Khatib* 32359!; 12 km S.E. of K3, *F. & E. Barkley* 4282! **LCA**. Daltawa (Khalis), *Rogers* 0291!

CHAMAL (*Gillett* 10928).

Oman, Saudi Arabia.

2. **Buglossoides arvensis** (*L.*) *I.M.Johnston* in J. Arn. Arb. 35: 42 (1954); Y. Nasir in Fl. Pakistan [Ali & Nasir] 191: 691 (1989); Feinbrun-Dothan, Fl. Palaest. 3: 67 (1978); Edmondson in Fl. Turkey [P. H. Davis] 6: 316 (1978); Collenette, Wild Flow. Saudi Arabia: 84 (1999); Khatamsaz, Flora Iran 39: 107 (2002); Taifour & El-Oqlah, Pl. Jordan Annot. Checkl: 50 (2017).

> *Lithospermum arvense* L., Sp. Pl.: 132 (1753); Boissier, Fl. Orient 4: 216; Dinsmore in Post, Fl. Syria, Palest. & Sinai ed. 2, 2: 240 (1933); Rawi, in Dep. Agr. Tech. Bull. 14: 137 (1964); Riedl in Fl. Iranica 48: 151 (1967).

Stem normally undivided, erect, branched from base, 25–40 cm tall, adpressed strigose. Cauline leaves linear-oblong, 10–30 × 4–6 mm, sessile, rounded at apex; uppermost leaves enclosing lax inflorescence with single flowers in lower axils. Pedicel pubescent, not thickening in fruit. Calyx deeply divided with lanceolate-linear lobes, 5–6 mm, acute. Corolla minute, tube ± equalling limb, whitish, pale yellow or pinkish, antrorsely hairy outside. Nutlets 2.5 mm, ovoid-lanceolate, somewhat gibbous at the sides, rugose-scrobiculate, ridges forming a network, sometimes tuberculate. Fig. 85, 1–4.

HAB. Hillsides, dry silty slopes towards wadi, on limestone with coppiced *Quercus* forest; alt. 1000–2000 m; fl. Mar.-Jun.

DISTR. Occasional in the northern forest zone and adjacent foothills. **MAM**: Aqra, *Rawi* 11396!; Bakirma (Bakurman), *Rawi* 9419!; Matina, *Rawi* 8714! **MRO**. Haji Omran, *O. Polunin* 5058! *Rawi* 12082!, *Chapman* 11188!; Shaqlawa, *Rawi* 10454!; Dera, 16 km N.W. of Rania, *Rawi, Nuri & Kass* 28601!; Salah ad-Din, *Chapman* 11970! **MSU**. Halabja, *Rawi* 8827!; Jarmo, *Helbaek* 439!, 570! **DLJ**. Rawa, *Rawi & Gillett* 7035! **FNI**. Jabal Maqlub nr Mosul, *Shahwani* 25208! **FAR**. Bastura Chai, *Rawi & Gillett* 10498!

Fig. 85. **Buglossoides arvensis**. 1, habit × ½ ; 2, flower × 5 ; 3, corolla, opened × 5; 4, fruit × 5. Reproduced with permission from Fl. Pakistan 191: f. 18, E–H (1989). Drawn by S. Hameed. © National Herbarium, Pakistan Agriculture Research Council & University of Karachi, Pakistan.

A variable species of disturbed habitats, weedy in habit.

N. Africa; Saudi Arabia; S. Europe, Turkey, Syria, Jordan, Lebanon, Palestine, Iran, Afghanistan, Pakistan, eastward to Kashmir; Japan; introduced to E. Asia and N. & S. America.

3. **Buglossoides incrassata** (*Guss.*) *I.M.Johnston* in J. Arn. Arb. 35: 43 (1954); Feinbrun-Dothan, Fl. Palaest. 3: 67 (1978); Edmondson in Fl. Turkey [P. H. Davis] 6: 317 (1978); Khatamsaz, Flora Iran 39: 109 (2002); Taifour & El-Oqlah, Pl. Jordan Annot. Checkl: 50 (2017).

Lithospermum incrassatum Guss., Index Seminum (Boccadifalco) 1826: 6 (1826); Boiss., Fl. Orient 4: 217; Dinsmore in Post, Fl. Syria, Palest. & Sinai ed. 2, 2: 240 (1933); Rawi, in Dep. Agr. Tech. Bull. 14: 137 (1964); Riedl in Fl. Iranica 48: 152 (1967).
Lithospermum gasparrinii Heldr. ex Guss. Fl. Sicul. Syn. 1: 217 (1843).
Buglossoides arvensis subsp. *gasparrinii* (Heldr. ex Guss.) R.Fern. Bot. J. Linn. Soc. 64: 379 (1971).
Buglossoides arvensis subsp. *incrassata* (Guss.) Kerguélen, Lejeunia, n.s., 120: 59 (1987).

Very similar to *B. arvensis* but differs in its stems that greatly elongate in fruit, and in its pedicels which become thickened and obconical in fruit. Corolla bluish (rarely white). Receptacle oblique in fruit.

HAB. In grassy area on top of hill and slope of mountain, under oak trees; alt. 1100–1800 m; fl. May-Jun. DISTR. Occasional in the north forest zone. **MAM.** Shaklawa, *Haines* W757; Kani Mazi, *Agnew* s.n. **MRO.** Haji Omran, *Chapman* 11956!; Kani Rush (Kanirush) *Kass & Nuri* 27674!; Arbil, Kuh-e-Safin, *Bornmüller* 1619, 1620; **MSU.** Penjwin, *Rechinger* 12321 (W!); Kara Dagh, *Haines* W1112!

Lithospermum babylonicum Zohary listed in Rawi (l.c., p. 137) is not recorded in IPNI and possibly not validly published. All our plants are referable to the above 3 species.

Iran, Turkey, Syria, Jordan.

12. **ANCHUSA** L.

Sp. Pl.: 133 (1753)
Weigend & al. in Kubitzki, Fam. Gen. Vasc. Pl. 14: 78 (2016)

C.C. Townsend
Revised by Shahina A. Ghazanfar & J. R. Edmondson

Annual, biennial or perennial herbs. Indumentum with dense short hairs and sparse tubercle-based trichomes or hispid-setose, lacking glandular hairs. Leaves basal and cauline. Inflorescence of ± lax scorpioid cymes, elongating in fruit, bracteate. Calyx divided to 1/3 or to base, ± accrescent in fruit. Corolla yellow, blue to purple, hypocrateriform; tube long; limb spreading. Faucal scales triangular-oblong, at throat of tube. Stamens included, inserted near or above middle of tube. Style with bilobed stigma. Nutlets erect with apex rounded or transversely ovoid with lateral beak, chestnut brown to black, rugose-tuberculate.

About 35 species (when delimited in a narrow sense) distributed mainly in the Mediterranean basin and the Middle East with a single species in the Ethiopian-West Arabian highlands and two in South Africa; 5 species in Iraq.

Anchusa, from Gr. αγχουσα, *anchousa*, of unknown meaning.

Gusuleac, M. (1929). Species Anchusae generis Linn. hucusque cognitae. Repert. Spec. Nov. Regni Veg. 26: 286–322.
Selvi, F. & Bigazzi, M. (2003). Revision of genus *Anchusa* (Boraginaceae-Boragineae) in Greece. Bot. J. Linn. Soc. 142(4): 431–45

1. Corolla-tube bent just below the middle . . . . .5. *A. arvensis*
   Corolla tube straight 2
2. Corolla-tube exceeding the calyx 3
   Corolla as long as or shorter than calyx. 4
3. Calyx lobes 4–5 mm; corolla yellowish white, corolla tube 7 mm, well
     exserted from calyx . . . . 3. *A. aegyptiaca*
   Calyx lobes 6–8 mm; corolla bright blue, rarely pink; corolla tube 9–12
     mm, exceeding the calyx. . . . . 2. *A. strigosa*
4. Annual, 5–20 cm, sparsely tuberculate-strigose; leaves oblong to linear-
     oblong; calyx c. 4 mm in flower, enlarging to c. 10 mm in fruit;
     corolla yellowish, whitish, pink or purple. . . . . 4. *A. aucheri*
   Perennial, 20–95 cm, coarsely hispid; leaves linear-elliptic to lanceolate
     or oblanceolate; calyx 7–8 mm in flower, 12–15 mm in fruit; corolla
     blue (rarely whitish). . . . . .1. *A. azurea*

1. **Anchusa azurea** *Mill.*, Gard. Dict. ed. 8: no. 9 (1768); Dinsmore in Post, Fl. Syria, Palest. & Sinai ed. 2, 2: 224 (1933); Chamberlain in Fl. Turkey [P. H. Davis] 6: 393 (1978); Y. Nasir in Fl. Pakistan [Ali & Nasir] 191: 80 (1989); Selvi & Bigazzi, Bot. J. Linn. Soc. 142: 445 (2003); Abdulridha, Taha & Wedad, Ecology & Fl. Basra: 235 (2016); Haloob & Al-Kaisi, Illustr. Fl. Lowland Iraq 1(1): 62 (2016); Taifour & El-Oqlah, Plants Jordan Annot. Checkl.: 50 (2017).

A. macrocarpa Boiss. & Hohen. in Boissier, Diagn. Pl. ser. 1, 4: 42 (1844).

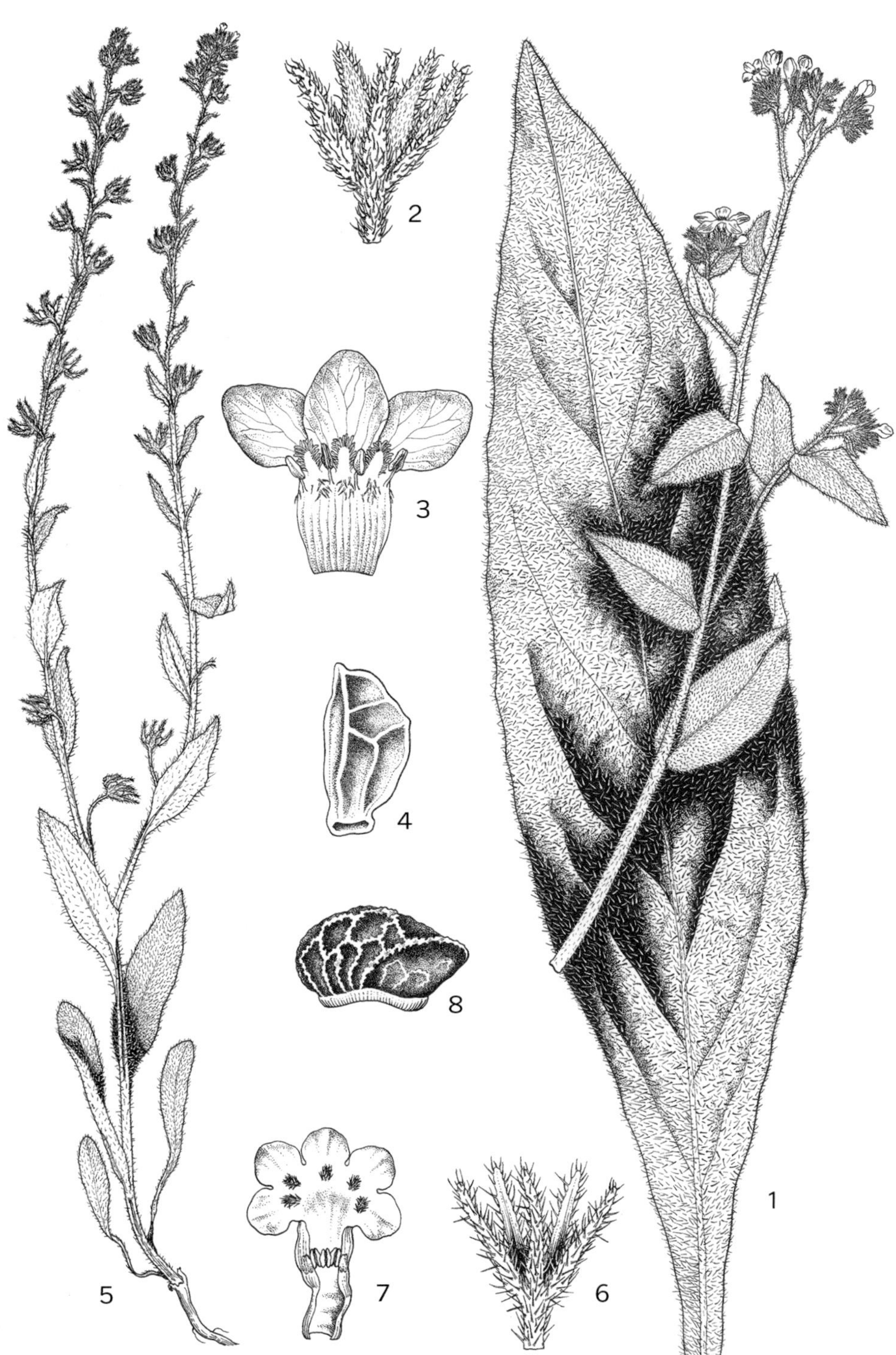

Fig. 86. **Anchusa azurea**. 1, habit, portion of shoot and leaf × ½; 2, flower × 2; 3, corolla, opened × 4; 4, nutlet × 5. **Anchusa arvensis** subsp. **orientalis**. 5, habit, portion of shoot and leaf × ½; 6, flower × 2; 7, corolla, opened × 4; 8, nutlet × 5. Reproduced with permission from Fl. Pakistan 191: f. 20, A–H (1989). Drawn by S. Hameed. © National Herbarium, Pakistan Agriculture Research Council & University of Karachi, Pakistan.

A. italica Retz., Obs. Bot. 1: 12 (1779); Boissier, Fl. Orient. 2: 154 (1875); Rawi, in Dep. Agr. Tech. Bull. 14: 135 (1964); Rechinger, Fl. Lowland Iraq: 503 (1964); Riedl in Fl. Iranica [K. H. Rechinger] 48: 233 (1967); Feinbrun-Dothan, Fl. Palaest. 3: 83 (1978); Khatamsaz, Flora Iran 39: (2008); Breckle, Hedge, Podlech & Rafiqpoor, Vasc. Pl. Afghan. checklist: 203 (2013).

A. paniculata Aiton, Hort. Kew. 1: 172 (1789).

A. italica Retz var. *kurdica* Gusuleac in Bul. Fac. Stiinte Cern. 1: 274 (1927); Rawi, in Dep. Agr. Tech. Bull. 14: 135 (1964).

A. italica Retz var. *macrocarpa* (Boiss. & Hohen.) Gusuleac in Bul. Fac. Stiinte Cern. 1: 273 (1927).

A. italica subsp. *macrocarpa* (Boiss. & Hohen.) Chamberlain in Notes Roy. Bot. Gard. Edinburgh. 35: 298 (1977).

A. azurea var. *kurdica* (Gusul.) Chamberlain in Notes Roy. Bot. Gard. Edinburgh 35: 338 (1977).

Coarsely hispid perennial herb; indumentum monomorphic or dimorphic. Stems erect, branched, 20–95 cm. Basal and lower leaves linear-elliptic to lanceolate or oblanceolate, 50–90 × 6–30 mm, base tapering into an indistinct petiole, apex acute to obtuse, margins ± entire to weakly crenate, ± densely setose; upper leaves ovate to ovate-lanceolate to linear. Inflorescence branched, paniculate, ultimate cymes monochasial, elongating in fruit. Bracts minute, narrow-lanceolate. Pedicel 2–3 mm. Calyx 7–8 mm in flower, 12–15 mm in fruit, divided almost to base into linear, lobes acute. Corolla blue (rarely whitish); tube 6–10 mm, limb 5–8 mm, obtuse. Stamens inserted near top of tube and overtopping scales. Stigma with two spreading lobes. Nutlets greyish-brown, oblong-ovoid, erect, 6–10 mm, reticulate-rugose, with a thick basal ring. Fig. 86, 1–4.

HAB. By bubbling spring, in irrigated cereal fields, roadside, waste land by garden, stony ground; alt. 100–1700 m; fl. Mar.-Jun.

DISTR. Mainly in the lower forest zone of N. E. Iraq. **MAM**: Asi, *Rawi* 480!; Zakho to Turkish frontier, *Dabbagh & Hamad* 44777!; Benata 16 km N.W. of Sarsang, *Dabbagh & Hamid* 45798!; Maya 20 km W. of Kani Masi, *Hamid & Fadhil* 45485!; mt. Bekhair, nr Zakho, *Rawi* 23010!; Gali Sergali 9 km E. of Amadiya, *Dabbagh & Hamid* 45893!; 10 km S. of Sarsang, *Kaisi, Khayat & Karim* 51014!; Ser Amadiya, Gali Mazurka, *Dabbagh, Kaisi & Hamid* 46008!; 20 km to Dohuk from Mosul, *Kaisi & Wedad* 51995!; mt. Gara, *Kotschy* 364 (type of *A. macrocarpa*)!; Zawita, *Guest* 3743!; Suwaratuka, before Ispindari Saddle, *Chapman* 26367! **MRO**: Khanzad, *Omar, Kaisi & Wedad* 49612!; Haji Omran, *Dabbagh & Hamad* 46285!; Pishtashan to Surade, *Serhang & Rawi* 26523!; on way to Darband, *Shahwani* 25312!; Gali Ali Beg, *Omar, Kaisi & Wedad* 49559!; 10 km from Koi Sanjaq to Arbil, *Omar, Kaisi & Wedad* 49511! **MSU**: Sulaimaniya to Arbat, *Poore* 341!; Sirwan to Halabja, *Botany Staff* 43099!; Dukan, *Omar & Karim* 38032!; Jarmo, *Helbaek* 1218! 1283! 10 km E. of Chemchemal, *Alkas* 18614!; N. of Chemchemal, *Robertson* R.B. 66!; Rayat, *Guest* 13028! **FUJ**: 10 km from Mosul, *Kaisi & Wedad* 51971!; between Mosul and Hammam Ali, *Chakravarty, Rawi, Khatib & Alizzi* 32132!; Mosul, *Guest* 1337!; Midway between Tal Afar and Sinjar, *Omar & Hamid* 36479!; nr Ba'aj, *Omar & Hamid* 36532! **FUJ/FKI**: Al-Fatha, Jabal Hamrin, *Khatib & Tikriti* 29715! **FNI**: 30 km from Mosul to Aqra, *Kaisi* 49721!; Mindan, *Rawi* 11361! **FAR**: 10–15 km from Arbil to Darband, *Shahwani* 25294! **FKI**: 19 km E. of Kirkuk, *Rawi & Gillett* 10586!; Kirkuk, *Guest* 536A!; Kifri, *Rawi & Gillett* 7441!; Tuz Khurmatu, *N. Polunin & Khudairi* 81! **FPF**: 25 km S.W. of Badra, *Kaisi, Thamer & Saleh* 51356!; Buskaya, *Rawi & Haddad* 25557! **DLJ**: nr Takrit, *Omar & Hamid* 36281!; 25 km N.W. of Baiji, *Barkley & Haddad* 5407! **DWD**: Hit to Haditha, *Hamid* 39050!; 10 km W. of Haditha, *Omar & Hamad* 50372!; Wadi Hauran to K3, *Rawi & Gillett* 6879!; 15 km E. of Rutba, *Khalil & Hazim* 32445!; Samadan 95 km N.E. of Haditha, *Khayat & Hamad* 51850! **DSD**: Wadi Al-Ajrumiya, *Rawi* 31252! **LEA**: Ba'quba, *Graham* s.n.! **LCA**: Baghdad, *Colvill* s.n.! ?? Kalala junction, *Alkas* 18582! ??Qaser Halgom, *Rawi, Khatib, Alizzi & Hamid* 32802! **DLJ**: 12 km N.W. of Haditha, *Chakravarty, Rawi, Khatib & Alizzi* 31829! 15 km E. of Rutba, *Khalil & Hazim* 32445!; Samadan 95 km N.E. of Haditha, *Khayat & Hamad* 51850! **DSD**: Wadi Al-Ajrumiya, *Rawi* 31252! **LEA**: Ba'quba, *Graham* s.n.! **LCA**: Za'faraniya, *Jenan & Sahira* 47567!; Kifri, *Rawi & Gillett* 441!

(There are about 60 other specimens at K and BAG that are not cited)

QOLA ZWAN (Kurd.-Kifri *Rawi & Gillett* 7441); ZWAN (Ir.-Wadi Hauran, *Rawi & Gillett* 6879); KUZRIK (Kurd.-Zawita), *Guest* 3743.

Turkey (SE Anatolia), W. Iran.

2. **Anchusa strigosa** *Banks & Sol.* [Soland.] in Russell, Nat. Hist. Aleppo, ed. 2, 2: 246. (1794); Dinsmore in Post, Fl. Syria, Palest. & Sinai ed. 2, 2: 225 (1933); Rawi, in Dep. Agr. Tech. Bull. 14: 135 (1964); Chamberlain in Fl. Turkey [P. H. Davis] 6: 396 (1978); Rechinger, Fl. Lowland Iraq: 503 (1964); Riedl in Fl. Iranica [K. H. Rechinger] 48: 235 (1967); Feinbrun-Dothan, Fl. Palaest. 3: 84 (1978); Selvi & Bigazzi, Bot. J. Linn. Soc. 142: 445 (2003); Khatamsaz, Flora Iran 39: 229 (2008); Haloob & Al-Kaisi, Illustr. Fl. Lowland Iraq 1(1): 63 (2016); Taifour & El-Oqlah, Plants Jordan Annot. Checkl.: 50 (2017).

A. strigosa Labill., Icon. Pl. Syr. 3: 7, t. 4 (1809).
A. strigosa Labill. var. *decalvata* Bég. & A.Vacc., Atti R. Ist. Veneto Sci. Lett. Arti 72: 325 (1912).

Stout tuberculate-strigose perennial. Stems erect, branched from base, 30–75 cm. Basal leaves forming a rosette, narrowly elliptic to oblanceolate, 7–22 × 3–10 mm, obtuse, sparsely tuberculate-strigose, margins crenate; lower leaves tapering to a short petiole; cauline leaves few, sessile. Inflorescence lax, paniculate, branches subtended by minute ovate to linear bracts, elongating in fruit. Bracts narrow-lanceolate, shorter than calyx. Pedicels 2–3 mm. Calyx deeply divided almost to base into 5 linear-oblong lobes, lobes 6–8 mm, accrescent in fruit. Corolla bright blue, rarely pink, salver-shaped; tube 9–12 mm, greatly exceeding the calyx, limb 3–4 mm in diameter. Faucal scales white hairy. Stamens inserted between and a little below faucal scales at top of tube; anthers shortly exserted. Style glabrous; stigma capitate or club-shaped. Nutlets often reduced to 2 by abortion, 5–8 × 2.5–3 mm, oblong-ovoid, erect, densely and minutely papillose, verruculose with sharp longitudinal ridges, basal ring inconspicuous.

HAB. Sandy, clay, and gravelly soil, wheat fields, edges of barley fields, stony hillside, dry hillside, sandstone and conglomerate hillside, alt. 90–950 m; fl. Feb.-May.
DISTRIB. Mainly in the lower mountain grassy zone of Iraq, extending into the lower forest zone and western desert. **MAM**: Dohuk, *Guest* 1588!; Zawita gorge, *Emberger, Guest, Long, Schwann & Jusef* 15382! *Guest* 2189! (flowers pink); 2190! (flowers blue); Sarsang to Amadiya, *Karim, Hamid & Jasim* 41000! **MSU**: Jarmo, *Helbaek* 457! **FUJ**: Qaiyara, *Guest* 13141!; 30 km N. of Mosul, *Dabbagh* 46744!; Qal'a Sharqat, *Rao s.n.*!; 60 km S. of Sinjar, *Alizzi & Husain* 33927! **FNI**: Ba'ashiqa, Tikriti 16389! *Alizzi & Omar* 35273!; N. of Mosul, *Hossain* 168! (flowers pink); 28 km from Mosul to Kasik, *Dabbagh* 46734!; 14 km S.E. of Mosul, Tawfiq, *Dabbagh & Lafta* 44093!; N.E. of Mindan bridge, *Chapman* 26198! **FKI**: Kirkuk, Graham s.n.! *Bornmüller* 1593!; 3 km from Kirkuk to Sulaimaniya, *Karim, Hamid & Jasim* 40493!; Injana (Jabal Hamrin), *Omar & Karim* 37839!; Tuz, *Rogers* O217!; Kirkuk, *Rogers* 1585!; 10 km N. of Tuz to Kirkuk, *Botany Staff* 42937!; Al Adhaim, *Karim, H. Hamid & Jasim* 40441! **FPF**: 20 km from Sa'adiya to Khanaqin, *Hamid* 46618! 10 km to Tib from Ali al-Gharbi, *Kaisi & Khayat* 50570! Sa'diya, *Kaisi* 42518!; Naft Khina, *Guest* 1849!; Koma Sank police station near Iranian frontier, *Rawi* 200629!; Jabal Hamrin, on Khanaqin road, *Rawi & Gillett* 7344!; Jabal Hamrin, *Haines* W. 587!; Jabal al-Muwaila, nr Kuwait frontier, *Guest, Rawi & Rechinger* 17616!; Badra, *Gillett* 6634! 6668!; Jabal Hamrin, at Table Mountain (Mansur), *Guest* 2621!; Sa'adiya water project, *Kaisi* 46340!; 8 km from Zurbatiya, *Kaisi & Hamid* 46537!; few km N.E. Mandali bet. Smood & Numan, *Al-Kaisi & K. Hamad* 43576! **DLJ**: nr Baiji, *Rawi & Hamada* 34151!; 25 km N. of Baiji towards Mosul, *Ani* 9706!; 5 km before Baiji, Tawfiq, *Dabbagh & Lafta* 44040! **DWD**: 10 km W. of K3, *Omar, Kaisi, Hamad & Hamid* 44724!; 160–190 km from Ramadi to Rutba, *Rawi* 31361!; 12 km W. of K3, *Barkley & Palmatier* 2143! unloc., "Mespot., Kurdistan & Mosul", *Kotschy* 381!

LISĀN AL-THŪR, HAMHAN (Ir.-Qaiyara), *Guest* 13141; WARD LISSAN THOR (Ir.), KULA ZWANA (Kurd.), Jabal Hamrin, *Rawi & Gillett* 7344.

Turkey, Cyprus, East Aegean Is., Iran, Syria, Jordan, Palestine.

3. **Anchusa aegyptiaca** (*L.*) *A.DC.*, Prodr. [A. P. de Candolle] 10: 48 (1846); Dinsmore in Post, Fl. Syria, Palest. & Sinai ed. 2, 2: 227 (1933); Rawi in Dep. Agr. Tech. Bull. 14: 134 (1964); Chamberlain in Fl. Turkey [P. H. Davis] 6: 399 (1978); Riedl in Fl. Iranica [K. H. Rechinger] 48: 236 (1967); Feinbrun-Dothan, Fl. Palaest. 3: 84 (1978); Collenette, Wild Flow. Saudi Arabia: 81 (1999); El-Hadidi & Boulos in Fl. Egypt 2: 298 (2000); Khatamsaz, Flora Iran 39: 230 (2008); Taifour & El-Oqlah, Plants Jordan Annot. Checkl.: 50 (2017).

Lycopsis aegyptiaca L., Sp. Pl.: 138 (1753).
Asperugo aegytiaca L., Sp. Pl., ed. 2. 1: 198 (1762).
Anchusa flava Forssk., Fl. Aegypt.-Arab. 40 (1775).

Annual, 10–35 cm; stems branched at or above base, ± erect to ascending, white tuberculate-strigose. Leaves 20–55 × 4–8 mm, oblong-lanceolate, base tapering, apex acute, margins sinuate-denticulate, tuberculate-strigose on both surfaces. Inflorescence lax, elongating in fruit, flowers arising from axils of bracts. Bracts lanceolate. Pedicels 2 mm, elongating to 12 mm in fruit, nodding. Calyx 4–5 mm, enlarging in fruit to 12 mm, divided almost to base into linear lobes, acute, white-strigose. Corolla yellowish-white, tube 7 mm, well exserted from calyx, lobes c. 1 mm, rounded. (Faucal scales exserted, oblong, tips reflexed – not seen – dscript. from Feinbrun-Dothan). Nutlets ± 4 mm, straight, outer face reticulate-ribbed and granulate.

HAB. Rocky mountain, between crevices of rocks; recorded to be common; alt. 100–1000 m; fl. Mar., May.

DISTR. Localised in N.E. of Iraq. **MRO**: mt. Hab as Sultan [Haibat Sultan], *Karim, Hamid & Tasim* 40754! **MJS**: Jabal Sinjar, *Al Kaisi & Wedad* 52136!; Sinjar, *Haussknecht* 682! **DWD**: Jabal Ana, *Al Khayat & K. Hamad* 51701!

Although recorded to be common, it is localised and found at only a few of locations in Iraq.

Greece, Crete, Cyprus, Turkey, Syria, Iran; Jordan, Palestine, Arabia (Saudi Arabia), Egypt (Sinai), Libya.

4. **Anchusa aucheri** *A.DC.*, Prodr. [A. P. de Candolle] 10: 49 (1846); Dinsmore in Post, Fl. Syria, Palest. & Sinai ed. 2, 2: 226 (1933); Rawi, in Dep. Agr. Tech. Bull. 14: 134 (1964); Chamberlain in Fl. Turkey [P. H. Davis] 6: 402 (1978).

Phyllocara aucheri (DC.) Gusuleac in Bull Fac. Stiinte Cern. 1: 119 (1927).

Annual, 5–20 cm; stems branched at or above base, sparsely tuberculate-strigose. Leaves 20–50 × 8–20 mm, oblong to linear-oblong, base tapering, apex obtuse, margins weakly dentate, tuberculate-strigose. Inflorescence compact in flower, elongating in fruit, flowers arising from axils of leafy bracts. Calyx c. 4 mm in flower, enlarging to c. 10 mm in fruit, divided to more than ½ into acute lobes, densely strigose. Corolla yellowish, whitish, pink or purple; tube 4–7 mm, slightly exserted from calyx, lobes c. 2 mm. Stamens inserted near top of tube, overlapping the triangular papillose ciliate scales. Nutlets c. 5 × 2.5 mm, transversely reniform to oblong, obliquely beaked, reticulate-rugose.

HAB. Grassy slopes in cleared *Quercus* forest, clay soil, serpentine rocks, dry slopes, earth banks; alt. 1000–1590 m; fl. Apr.-Jun.

DISTR. Northern and eastern sectors of the forest zone of Iraq. **MAM**: Jabal Khantur, Sharanish, *Rechinger* 12102 (W); pass between Dagh al Rajiem and Sharanish, *Rechinger* 12122 (W!); Sarsang, *Haines* W. 1110!; Amadiya, *Guest* 1443!; Alho, *Rawi* 8667A!; Sharifa nr Amadiya, *O. Polunin* 5144! **MRO**: 8 km W. of Diana (Soran), *Helbaek* 793!; Kuh Sefin, *Bornmüller* 1630!; mt. Hawraman, Susakan nr Tawila, *Rechinger* 10193!; Serderrian pass, *Nabélek* 653. **MSU**: Halgord Dagh above Nowanda, *Rechinger* 11854!; Malakawa pass near Penjwin, *Rechinger* 12335, 10543!; Mt. Kajan nr Penjwin, *Rawi* 22685!; Mt. Pira Magrun, *Hausskenecht s.n.* **FPF**: Khanaqin, *Haines* 1111.

Sinai, Syria, Turkey, Iran.

5. **Anchusa arvensis** (*L.*) *M.Bieb.*, Fl. Taur.-Caucas. 1: 123 (1808) subsp. **orientalis** (*L.*) *Nordh.* in Norsk. Fl. 526 (1940).

Lycopsis orientalis L., Sp. Pl. 1: 139 (1753).
Anchusa ovata Lehm., Pl. Asperif. Nucif. 1: 222 (1818); Riedl in Fl. Iranica [K. H. Rechinger] 48: 237 (1967); Feinbrun-Dothan, Fl. Palaest. 3: 85 (1978).
A. orientalis (L.) Reichb., Icon. Fl. Germ. Helv. 18: 63 (1858), nom. illeg., non L., Sp. Pl. 133 (1753); Dinsmore in Post, Fl. Syria, Palest. & Sinai ed. 2, 2: 228 (1933); Rawi, in Dep. Agr. Tech. Bull. 14: 135 (1964).
Lycopis arvensis L. subsp. *orientalis* (L.) Kuntze in Trudy Imp. S.-Peterburgsk. Bot. Sada 10: 216. 1887.

Annual, 20–30 cm sparsely tuberculate-strigose. Stem erect, simple. Leaves oblong to ovate-oblong, base tapering, apex obtuse, margins weakly repand-dentate, tuberculate-strigose. Inflorescence lax, flowers arising from axils of leafy bracts. Calyx c. 5 mm in flower, divided to more than ½ into acute lobes, strigose. Corolla blue; corolla tube bent just below the middle, lobes c. 2 mm, rounded. Stamens inserted middle of tube; faucal scales papillose. Nutlets transversely ovoid, reticulate-rugose, with a broad basal ring. Fig. 86, 5–8.

HAB. Western Desert; alt. ± 200 m; fl. Apr.

DISTR. Rare; only known from a single collection in the western desert of Iraq. **DWD**: S.E. of Ana, *Rawi & Gillett* 6932!

Eastern Mediterranean, Eastern Europe to W. Asia.

13. **GASTROCOTYLE** Bunge

Ann. Sci. Nat. Bot. 3, 12: 363 (1949); Weigend & al. in Kubitzki, Fam. Gen. Vasc. Pl. 14: 80 (2016)

C.C. Townsend
Revised by Shahina A. Ghazanfar

Annual herbs, indumentum setose-hispid. Leaves narrow obovate. Inflorescence of dichotomously branched cymes, bracteate. Calyx cup-shaped, divided to almost base into linear-lanceolate lobes, subacute. Corolla pink or blue, campanulate to subrotate. Faucal scales oblong, emarginate, papillose. Stamens 5, included. Style included. Nutlets transversely ovoid, minutely tuberculate, with a thickened collar-like ring at base.

Two species, distributed in northern Africa, Arabia to W. China and N.W. Himalaya; a single species in Iraq.

Gastrocotyle, from Gr. γαστ῾ρ, *gaster,* stomach and κοτυλ῾α, *kotulea,* hollow or cup-shaped.

Bigazzi, M., Selvi, F., & Fiorini, G. (1999). A reappraisal of the generic status of *Gastrocotyle, Hormuzakia* and *Phyllocara* (Boraginaceae) in the light of micromorphological and karyological evidence. Edinb. J. Bot. 56(2): 229–251.

1. **Gastrocotyle hispida** (*Forssk.*) *Bunge,* Ann. Sci. Nat., Bot. sér. 3, 12: 363 (1849); Rawi, in Dep. Agr. Tech. Bull. 14: 136 (1964); Rechinger, Fl. Lowland Iraq: 502 (1964); Riedl in Fl. Iranica [K. H. Rechinger] 48: 254 (1967); Feinbrun-Dothan, Fl. Palaest. 3: 89 (1978); Y. Nasir in Fl. Pakistan [Ali & Nasir] 191: 82 (1989); Gelin Zhu, Riedl & Kamelin, Fl. China 16: 359 (1995); Collenette, Wild Flow. Saudi Arabia: 88 (1999); Khatamsaz, Flora Iran 39: 214 (2002); Ghazanfar, Fl. Oman 3: 94 (2015); Abdulridha, Taha & Wedad, Ecology & Fl. Basra: 237 (2016); Taifour & El-Oqlah, Pl. Jordan Annot. Checkl.: 51 (2017).

Anchusa hispida Forssk., Fl. Aegypt.-Arab.: 40 (1775); Dinsmore in Post, Fl. Syria, Palest. & Sinai ed. 2, 2: 226 (1933).

Annual herbs, 15–30 cm, clothed with tubercle-based setae. Stems greyish, branched from base, decumbent to erect or spreading. Lower leaves with long petioles; leaves oblong to oblong-lanceolate, 4–9(–12) × 2–6 cm, apex obtuse, margins wavy or weakly denticulate, setulose. Flowers solitary in upper leaf axils, minute, later nodding. Calyx with 5 lobes, 1.5 – 2.5 mm; lobes lanceolate, densely strigose inside and outside, accrescent. Corolla blue or purple, tubular, 2–2.5 mm, equalling calyx; limb 5-lobed with rounded lobes. Faucal scales papillose, emarginate. Stamens inserted below faucal scales. Nutlet transversely ovoid, minutely tuberculate, beak acute-angled, with a thickened-like collar at base. Fig. 87, 1–4.

HAB. Sandy hillside, rocky gravelly desert, clay soil; alt. 10–500 m; fl. Feb.-Apr.
DISTR. Mainly in the southern and (to a lesser degree) western desert zones of Iraq, extending to the eastern foothills.

FPF: Koma Sank police station, nr Mandali, *Rawi* 20630!; 38 km S.E. of Mandali, *Rawi* 20702! **DGA**: 4 km E. of Samarra, *Rawi* 20353! **DSD**: 30 km E. of Salman, nr Golaib, *Guest, Rawi & Rechinger* 18746!; Safai al-Maghif, N.E. of Ghazlani, *Guest, Rawi & Rechinger* 17258!; Al-Baniya, 65 km W.S.W. of Basra, *Guest, Rawi & Rechinger* 17321!; 6 km S.E. of Shabicha, *Guest, Rawi & Rechinger* 19243!; 8 km N. by W. of Aidaha, *Guest, Rawi & Rechinger* 19152!; 10 km S.E. by S. of Jiraibiyat, *Guest, Rawi & Rechinger* 17174!; Nukhaila station, *Guest, Rawi & Rechinger* 17345!; Jabal Sanam, 20 km W. by S. of Safwan, *Guest, Rawi & Rechinger* 16970! 17019!; W. of Zubair, *Guest, Eig & Zohary* 5030!; 23 km S.E. by S. of Zubair, *Guest, Rawi & Rechinger* 16848! **DSD**: W.N.W. of Jarishan, *Guest, Rawi & Rechinger* 17108!; 65 km S.W. of Basra, *Guest, Rawi & Rechinger* 17078!; Jumaima, *Guest, Rawi & Rechinger* 19118!; 12 km SSW of Busaiya, *Guest & Mahallal* 15227B!; 80 km ESE of Busaiya, *Guest & Rawi* 14296!; Salman, *Guest, Rawi & Long* 14079A!; 25 km S.E. of Salman, *Rawi* 14909!; nr Samah, *Alizzi* 34398!; 5 km S.E. by E. of Zubair, *Guest, Rawi & Rechinger* 16784!; Umm Qasr, *Alizzi & Omar* 35017! **DWD**: Amrah, 120 km N.W. of Rutba, *Omar* 34237!; 90 km W. of Ramadi, *Rawi* 20888!; 60 km N.W. of Ramadi, *Alizzi* 35148!; 30 km W. of Ramadi, *Rawi* 20839!; 20 km N.W. of Rutba, *Rawi* 21129!; 10 km S.W. of Rutba, *Rawi* 21074!; Nukhaib, *Rawi* 14797!

LBA: Shahraban, *Haines* W. 305!

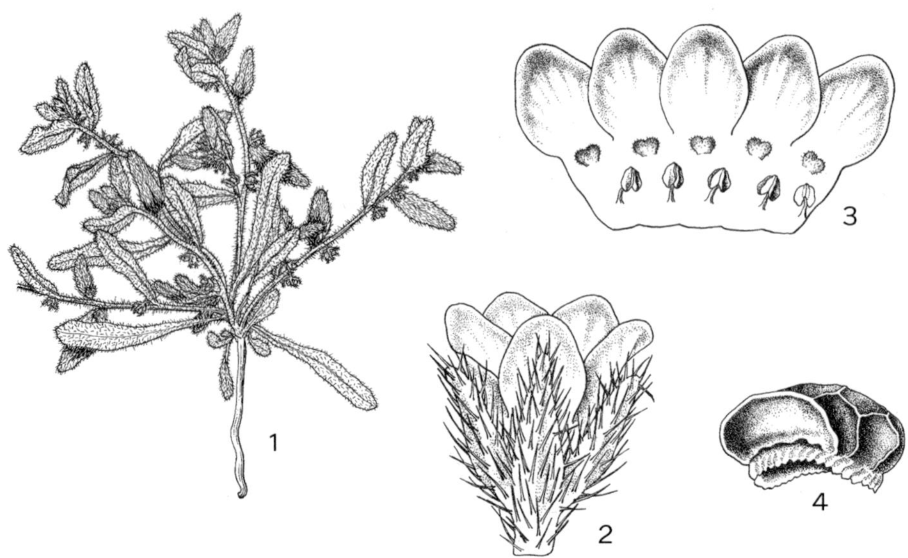

Fig. 87. **Gastrocotyle hispida**. 1, habit × 1.5; 2, flower × 8; 3, corolla opened × 8; 4, nutlet × 6. Reproduced with permission from Fl. Pakistan 191: f. 21, A–D (1989). Drawn by S. Hameed. © National Herbarium, Pakistan Agriculture Research Council & University of Karachi, Pakistan.

CHAHAL (Ir.-Aidaha), *Guest et al.* 19152; ZURRAIJA (Ir.-Jarishan), *Guest et al.* 17108.

N. Africa, Sinai, Arabian Peninsula, Palestine, Jordan, Iran, Afghanistan, Pakistan, India, C. Asia, Xinjiang.

14. **BRUNNERA** Steven

Bull. Soc. Imp. Naturalistes Moscou 24 (1): 582 (1851); Weigend & al. in Kubitzki, Fam. Gen. Vasc. Pl. 14: 80 (2016)

C.C. Townsend
Revised by Shahina A. Ghazanfar

Robust sparsely hispid-setose perennial herbs, rhizomatous. Basal leaves long-petiolate. Inflorescence a lax panicle or corymb; cymes ebracteate. Calyx cup-shaped, divided to almost base into linear-lanceolate lobes, subacute. Corolla blue, salver-shaped, limb rotate; faucal scales trapezoid, ciliate, white. Stamens 5; filaments inserted near top of corolla tube, exserted. Style included. Nutlets oblong-obovoid, reticulate-rugose, ribbed and verruculose, with a thickened collar-like ring at base.

Three species, distributed from E. Medit. to Siberia; a single species in Iraq.

Brunnera commemorates a Swiss botanist by the name of S. Brunner of Berne, 'egregii botanici' (an outstanding botanist), who collected plants in Crimea.

1. **Brunnera orientalis** (*Schenk.*) *I.M.Johnst.* in Contr. Gray Herb. 73: 55 (1924).
Riedl in Fl. Iranica [K. H. Rechinger] 48: 239 (1967); Feinbrun-Dothan, Fl. Palaest. 3: 86, pl. 140 (1978); Edmondson in Fl. Turkey [P. H. Davis] 6: 388 (1978); Khatamsaz, Flora Iran 39: 221 (2002); Taifour & El-Oqlah, Pl. Jordan Annot. Checkl.: 50 (2017).

Myosotis orientalis Schenk, Pl. Sp. It. Aegypto-Pers. 26 (1840).
Anchusa neglecta A.DC. in DC., Prodr. 10: 49 (1846); Dinsmore in Post, Fl. Syria, Palest. & Sinai ed. 2, 2: 225 (1933); Rawi, in Dep. Agr. Tech. Bull. 14: 135 (1964).

Perennial, rhizomatous, glandular-pubescent. Stems perennial; stems 15–35 cm, sparsely

Fig. 88. **Brunnera orientalis.** 1, habit × 1; 2, flower, opened × 7; 3, nutlet × 7. Reproduced with permission from Feinbrun-Dothan, Flora Palaestina, 3: Plates, f. 140 (1977). Drawn by Esther Huber. © The Israel Academy of Sciences and Humanities.

setose with tubercled-based setae. Basal leaves ovate-lanceolate to elliptic, 8–19 × 1.5–8.5 cm, base attenuating into a long petiole, 2–12 cm, apex acute, margins irregularly minutely crenate; cauline leaves ± elliptic, with short petioles. Inflorescence sparsely branched, ultimate branches clustered. Pedicels 4–6 mm, reflexed. Calyx 2–3 mm at anthesis, strongly accrescent to 8 mm in fruit, enclosing nutlets, setose. Corolla bright blue, tube 2.5–4 mm, limb 3.5–7 mm diameter, rotate. Stamens exserted. Nutlets 2–3 mm, obliquely oblong-obovoid, prominently rugose. Fig. 88, 1–3.

HAB.: *Quercus* thickets on limestone, waste ground, shady rocks, rocky slope above trees; alt. 600–1700 m; fl. Apr.-Jul.

DISTR.: Northern and Eastern sectors of the upper forest zone of Iraq. **MAM**: Sarsang, *O. Polunin* 5071!; between Dohuk and Amadiya, *Rechinger* 11913 (W!); Jabal Rusi, *Mudur of Gilli* 3151!; Basingera (Bazingrah), *Rechinger* 11508 (W!); Gali Zawita, N.E. of Zakho, *Rawi* 23555! *Rechinger* 11916!; Jabal Khantur, N. of Zakho, *Rawi* 8561! *Rechinger* 10857! **MRO**: Mt. Botin, *Agnew, Hadač & Haines*, W. 2162!; Pira Magrun, *Haussknecht* s.n. **MJS**: Jabal Sinjar, *Haussknecht*, May 1867!; N.slope, *Kaisi, Khayat & Karim* 50937!; Rashed, N. side of Jabel Sinjar, *Anders* 2712, 2147 (W!).

GŪRISK (Kurd.-Jabal Rusi), *Mudur of Gilli* 3151.

Turkey, N. & W. Iran, Syria (Latakia), Lebanon, Palestine, Jordan.

15. **EUPLOCA** Nutt.

Trans. Amer. Philos. Soc. ser. 2, 5: 189 (1836)
Diane, Hilger, Förther, Weigend & Luebert in Kubitzki, Fam. Gen. Vasc. Pl. 14: 208 (2016)

Shahina A. Ghazanfar

Annual or perennial herbs or small shrubs. Indumentum strigose (in Iraq species). Leaves alternate, linear to narrow oblong. Inflorescences unbranched or sparsely branched, bracteate. Calyx lobed 1/3 to 1/2 of length, unequal. Corolla tubular, 5-lobed above, lobes rounded to ovate. Stamens 5, anthers protracted, apex pubescent, coherent apically, closing the corolla tube. Fruit dry, separating into 4 nutlets, each with two pits on the abaxial side.

The genus *Euploca*, established by Nuttall in 1836, was reduced to a synonym of *Heliotropium* L. by Gray in 1874, but re-established as a genus by Hilger & Diane in 2003 based on molecular and morphological evidence. *Euploca* and *Heliotropium* are separated based on the presence (*Euploca*) or absence (*Heliotropium*) of bracteate inflorescences, coherent anthers (*Euploca*) or not (*Heliotropium*), the number of nutlets (1–4 in *Helitropium* & 4 in *Euploca*), curved embryo (*Euploca*) and kranz leaf anatomy.

About 120 species distributed in Africa, Australia, and tropical America; a single species in Iraq.

Euploca, from the Gr. πλεχω, *plecho*, plait with reference to the unusual shape of the corolla.

Hilger, H.H. & Diane, N. (2003). A systematic analysis of Heliotropiaceae (Boraginales) based on trnL and ITS1 sequence data. *Bot. Jahrb. Syst.* 125(1): 19–52.

1. **Euploca strigosa** (*Willd.*) *Diane & Hilger*, Bot. Jahrb. Syst. 125(1): 49 (2003).

Heliotropium myosotoides Lehm. in Russell, Aleppo, ed. II. ii. 245 (1794); Rawi in Dep. Agr. Iraq Tech. Bull. 14: 136 (1964).
H. strigosum Willd., Sp. Pl., ed. 4 [Willldenow] 1(2): 743 (1798); Y. Nasir in Fl. Pakistan [Ali & Y. Nasir] 191: 25 (1989); Ghazanfar in Fl. Oman 3: 79 (2015).

Annual or perennial herb; stems erect or prostrate, spreading, up to 40 cm, pubescent with appressed to ± patent hairs. Leaves linear to narrow oblong or narrow elliptic, 3–40 × 1–5 mm, base tapering into short petiole to 2 mm, apex acute, margins entire to ± revolute. Inflorescence simple or branched, cymes short or long to 10 cm. Bracts 1–5 mm. Calyx lobes unequal, linear-lanceolate, 1.5–3 mm. Corolla white with sometimes a yellowish centre, tubular, pubescent in the upper part on the outside; lobes to 2 mm. Stigmatic head shortly conical. Nutlets 4, globose to ovoid, c. 1.5 mm, pubescent on the back, with a cavity on each inner surface. Fig. 89, 1–6.

Fig. 89. **Euploca strigosa**. 1, habit portion of shoot × ½; 2, flower × 5; 3, corolla opened × 5; 4, carpel × 20; 5, nutlets with calyx × 8; 6, nutlet × 10. Reproduced with permission from Fl. Pakistan 191: f. 7, G-L (1989). Drawn by S. Hameed. © National Herbarium, Pakistan Agriculture Research Council & University of Karachi, Pakistan.

HAB & DISTRIB. Known from a single collection from the Western desert of Iraq; alt. ± 620 m; fl. & fr. Nov. **DWD**.

E. & N.E. Tropical Africa, Arabian Peninsula to tropical Asia and Australia.

16. **HELIOTROPIUM** L.

Sp. Pl.: 130 (1753)
Diane, Hilger, Förther, Weigend & Luebert in Kubitzki, Fam. Gen. Vasc. Pl. 14: 209 (2016)

Shahina A. Ghazanfar

Annual or perennial herbs or subshrubs. Indumentum strigose, villous, tomentose, tuberculate or glandular, rarely stems glabrous. Leaves alternate to subopposite, linear, broadly ovate to orbicular or elliptic, margins entire to sinuate or crenate, often revolute, petiolate or sessile. Inflorescence of terminal or axillary scorpioid cymes, ebracteate. Flowers usually sessile. Calyx divided nearly to base, lobes linear to ovate. Corolla tubular, 5-lobed above, lobes obtuse. Stamens 5, filaments adnate to the corolla tube; anthers usually included. Ovary usually 4-locular. Style 1, terminal; stigmatic head conical, sometimes 2-lobed apex. Nectar disc present at base of ovary. Fruits dry or fleshy (dry in Iraq species), usually 4-seeded, rarely 1–2-seeded, separating into 1–4 nutlets with 1–2 seeds each.

About 300 species, cosmopolitan, but most abundant in Iran, Turkey and the Caucasus; 11 species in Iraq.

The genus *Heliotropium* (with genera *Euploca* Nutt., *Ixorhea* Fenzl and *Myriopus* Small.) are treated in a separate family Heliotropiaceae in Kubitzki (ed), Fam. Gen. Vasc. Pl. 14: 209 (2016). These genera have previously been treated as a subfamily of Boraginaceae included in Ehretieae, or split into Heliotropioideae and Ehretioideae s.l.

Heliotropium refers to the phenomenon by which plants move to face the sum; Gr. ηλιος, helios, sun and τρεπειν, *trepein*, to turn.

Akhani, H. (2007). Diversity, biogeography, and photosynthetic pathways of *Argusia* and *Heliotropium* (Boraginaceae) in South-West Asia with an analysis of phytogeographical units. Bot. J. Linn. Soc. 155(3): 401–425.
Akhani, H., Förther, H. (1994). The genus *Heliotropium* L. (Boraginaceae) in Flora Iranica area. Sendtnera 2: 187–276.
Brummitt, R.K. (1971). The diagnostic characters of the European *Heliotropium* species. Bot. J. Linn. Soc. 64: 60–67.

von Bunge, A.A. (1869). Über die Heliotropien der mittelländischen-orientalischen Flor. Bull. Soc. Imp. Nat. Moscou 42: 279–332.

Chaudhary, S.A. (1985). Studies on *Heliotropium* in Saudi Arabia. Arab J. Sci. Res. 3: 33–53.

Diane, N., Förther, H. & Hilger, H. (2002). A systematic analysis of *Heliotropium*, *Tournefortia*, and allied taxa of the Heliotropiaceae (Boraginales) based on ITS1 sequences and morphological data. American J. Bot. 89: 287–295.

Förther, H. (1998). Die infragenerische Gliederung der Gattung *Heliotropium* L. und ihre Stellung innerhalb der subfam. Heliotropioideae (Schrad.) Arn. (Boraginaceae). Sendtnera 5: 35–241.

Ge-Ling Z, Riedl H, Kamelin R. (1995). *Heliotropium* and *Tournefortia*. In: Zhengyi W, Raven PH, eds. Flora of China, Vol. 16. St. Louis and Beijing: Missouri Botanical Garden Press and Science Press, 338–342.

Luebert, F., Cecchi, L., Frohlich, M.W., Gottschling, M., Guilliams, C.M., Hasenstab-Lehman, K.E., Hilger, H.H., Miller, J.S., Mittelbach, M., Nazaire, M. & Nepi, M. (2016). Familial classification of the Boraginales. Taxon 65(3): 502–522.

Nasir Y. 1989. Heliotropium. In: Nasir E, Ali S. I., eds. Flora of Pakistan, Vol. 191. Karachi: University of Karachi, 18–50.

Riedl, H. (1966). Die Gattung *Heliotropium* in Europa. Ann. Naturh. Mus. Wien 69: 81–93.

Thulin M. (2005). Three new species of *Heliotropium* (Boraginaceae) from the horn of Africa region. Nordic Journal of Botany 23: 527–532.

1. Calyx shallow lobed to 1/4, enlarging at maturity and enveloping fruit, persistent and dropping off with nutlets; one nutlet developing . . . . . . 1. *H. supinum*
   Calyx 5-lobed almost to base; fruit not completely enveloped by calyx; mostly 4 nutlets developing 2
2. Plants glabrous to sparsely pubescent or with tuberculate-based hairs or glandular 3
   Plants with white- villous to tomentose indumentum 7
3. Indumentum of strigose tuberculate and sometimes glandular hairs 4
   Indumentum appressed pubescent or plants glabrous. 5
4. Lamina ovate to elliptic, 10–25 × 6–15 mm, margins entire or slightly sinuate; inflorescence terminal and lateral, simple to branched, biseriate . . . . .5. *H. suaveolens*
   Leaves linear to linear-lanceolate, 5–60 × 2–16 mm, margins entire to crenulate-undulate and revolute; inflorescence usually terminal, simple, uniseriate, with few sessile flowers .9. *H. bacciferrum*
5. Plants glabrous. . . . . 10. *H. curassavicum*
   Plants pubescent, indumentum appressed or mixed with patent hairs. 6
6. Grey-green annual; indument appressed or patent; corolla without intercalary lobes; nutlets pilose or glabrous. . . . . 6. *H. europeanum*
   Perennial herb with a woody stock; stems densely appressed pubescent mixed with sparse long, patent hairs; corolla between lobes folded inwards; nutlets covered by long, white, silky hairs . . . . 7. *H. digynum*
7. Corolla hyporcrateriform, plicate or not; corolla lobes with margins entire or crenulate. 8
   Corolla tubular to infundibuliform 9
8. Corolla 3–4 mm, hyporcrateriform, plicate, appressed pilose outside; corolla lobes with margins entire or obscurely crenulate . . . . .2. *H. noeanum*
   Corolla 6–8 mm, hypocrateriform, pubescent outside below the lobes, inside pilose above anther base, in throat with five longitudinal appendage-like ridges; corolla lobes margins slightly crenulate, ± papillate . . . . .8. *H. crassifolium*
9. Perennial with a woody base, with several herbaceous stems ascending from base; inflorescence scorpoid at first, becoming lax, 2(–2.5) cm; corolla yellow; stigma, ovary and nutlets glabrous. . . . . 11. *H. lasianthum*
   Annuals; inflorescence > 2 cm; corolla white or yellowish; stigma, ovary and nutlets pilose 10
10. Corolla white, 6–12 mm (including lobes); corolla lobes triangular-subulate, coiled at tips; intercalary lobes smaller than lobes; stigma subulate-conical, beaked near the middle, pilose; nutlets, distinctly rugose, glabrous. . . . . 4. *H. circinatum*
    Corolla yellowish white, 4–6 mm; lobes c. 1 mm, ovate-triangular, not

coiled at tips, margins ± crenulate, densely pilose in the lower part on
the outside; stigma pilose above the middle; nutlets rugose, sparsely
pilose to glabrous. . . . . .3. *H. bovei*

1. **Heliotropium supinum** *L.*, Sp. Pl. 1: 130 (1753); Rawi in Dep. Agr. Iraq Tech. Bull. 14: 136 (1964); Dinsmore in Post, Fl. Syria, Palest. & Sinai ed. 2, 2: 220 (1933); Riedl in Fl. Iranica [K. H. Rechinger] 48: 52 (1967); Riedl in Fl.Turkey [P. H. Davis] 6: 255 (1978); Feinbrun-Dothan, Fl. Palaest. 3: 54, (1978), pl. 84 (1977); Y. Nasir in Fl. Pakistan [Ali & Y. Nasir] 191: 21 (1989); Rechinger, Fl. Lowland Iraq: 495 (1964); Akhani & Förther, Sendtnera 265 (1994); El-Hadidy & Boulos in Fl. Egypt 2: 281 (2000); Khatamsaz, Fl. Iran 39: 72 (2002); Taifour & El-Oqlah, Pl. Jordan Annot. Checkl.: 52 (2017).

Annual. Stems branching from base, branched, ascending to prostrate, 10–50 cm long, grey-green, appressed tomentose intermixed with long tomentose hairs. Leaves petiolate, grey-green, ovate to ovate-orbicular, 1–2 × 1–2 cm, base cuneate, apex acute to obtuse, margins entire to slightly undulate; nerves prominent on the lower surface, impressed on the upper; petiole 1–2 cm. Inflorescence lateral and terminal, elongating at maturity to 6 cm. Calyx 2 mm, shallowly lobed to ¼, enlarging to 3–4 mm at maturity, enveloping fruit, persistent and dropping off with nutlets. Corolla white, 2–3 mm, tubular, expanded at the base, inside glabrous, outside appressed pilose; corolla lobes minute, ovate-oblong, intercalary lobes missing or if present extremely minute. Anthers inserted just above the base of corolla base. Stigma pilose at apex. Fruit with only one nutlet developing, glabrous, ridged at edges. Fig. 90, 1–6.

HAB. Valleys, river banks, dry fields and wasteland, on sandy and gravelly soils; alt. 50–100 m; fl. Oct.-Apr.
DISTRIB. Common in the lower hills and semi-desert of the central and eastern alluvial plains.
MSU/FKI: Banks of R. Diyala, *Guest* 3512! **FNI**: 1 km N.E. Mindan Bridge, *Chapman* 26191!
FKI: Hills and lake, 10 km W. of Kirkuk, *Jumaa Rahim* 6150!; Hill near old stone bridge, 45 km S.E. of Kirkuk, *Jumaa Rahim* 6124! **DWD**: Al Qaim, *Al Sharif & Hemida & Hamad* 47313!; Rutba, bottom of wadi Hamran, *Alizzi & S. Omar* 35364! **LEA**: Azizyah, *Guest* 3571!; 30 km from Azizya to Kut, *Alizzi & Omar* 34625!; 50 km S. Baghdad of Kut, *Rawi* 20792!; Musharrah, side of river bank, *Janan* 47113!; Kut, *Lazar*

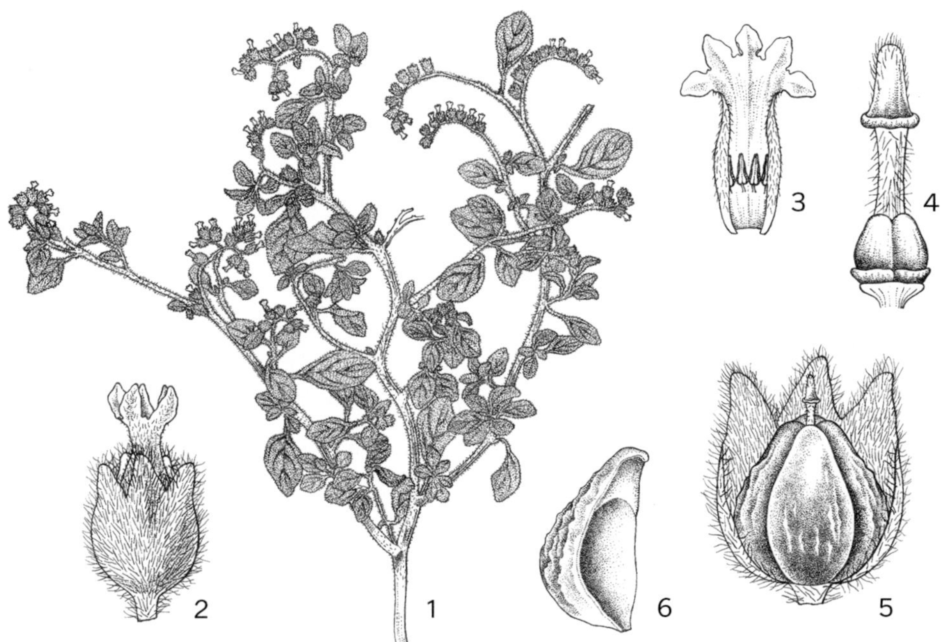

Fig. 90. **Heliotropium supinum**. 1, habit portion of shoot × ½; 2, flower × 5; 3, corolla opened × 5; 4, carpel × 20; 5, nutlets with calyx × 8; 6, nutlet × 10. Reproduced with permission from Fl. Pakistan 191: f. 6, A–F (1989). Drawn by S. Hameed. © National Herbarium, Pakistan Agriculture Research Council & University of Karachi, Pakistan.

3480! **lca**: Rustam, *Lazar* 3587!; Abu Ghraib, *Gillett* 12507! Baghdad, *Wheeler-Haines* W4! Zafraniyah, *Gillett* 5861; *Rawi* 19901!; Nasiriyah, *Abu Rizaq Barbuti* 2540! Baghdad, flood banks of Tigris River, *Agnew & Showqi* 174!; Daurah, Baghdad-Hillah Road, *Rechinger* 8107 (W). **lsm**: Garnat Bani Saad, *Rawi* 16609! **lsm**: Al Arathmiat, 2 km from Al-Fahood (Fuhud), *Noori* 41337!

SATIH (*Lazar* 3587); INI'MAH (*Abu Rizaq Barbuti* 2540). According to the local herdsmen, grazed by all livestock which get fat on it.

Southern Europe, Africa, Saudi Arabia, Syria, Iran, Russia, Central Asia, Pakistan, India.

2. **Heliotropium noeanum** *Boiss.*, Diagn. Pl. Orient., ser. 2, 3: 132 (1856). Type: In subalpinis Haneky [Khanaqin, 9.1851], *Noë* 981 (G-BOIS! lecto. designated by Akhani & Förther (1994); F, GOET!, LE!, P, isolecto.); Rawi in Dep. Agr. Iraq Tech. Bull. 14: 136 (1964); Riedl in Fl. Iranica [K. H. Rechinger] 48: 42 (1967); Akhani & Förther, Sendtnera 257 (1994); Khatamsaz, Fl. Iran 39: 58 (2002).

> *H. haussknechtii* Vatke, Z. Gesammte Naturwiss. (Halle) 45: 128 (1875), non Bunge 1869, nom. illegit.. Type: In monte Asmirdagh pr. Sulimanieh Kurdistaniae, 4000', 7.1867, *Haussknecht* 667 (lecto. JE! designated by Akhani & Förther,1994; LE!, M!, P!, W!, isolecto.).
>
> *H. noeanum* var. *edentulum* Boiss., Fl. Orient [Boissier] 4: 128 (1879). Syntypes. in cultis vallis Chyrsan Susianae, pr. Bebehan, VIII. 1868, *Haussknecht* (BM!, G-BOIS!, JE!, P!); in ruderatis montes Avroman et Schahu et circa Sulimanieh, 3000–6000', 7.1867, *Haussknecht* (G-BOIS!, JE!, K!); Darrian ad ped. Schahu, 5000', 7.1867, *Haussknecht* (JE, M!, P!); ? Masibin, 7000', 8.1867, *Haussknecht* (JE!).
>
> *H. schahpurense* Bornm., Beih. Bot. Centralbl. 61 (2): 88 (1941).
>
> *H. borasdjunense* Rech.f, Ann. Naturhist. Mus. Wien 55: 6 (1947).

Annual or perennial, grey-green, with many erect or ascending branches arising from a woody base to 50 cm; indumentum whitish villous to subtomentose. Leaves, grey-green, ovate to broadly ovate, 10–30 × 7–25 mm, base rounded, apex obtuse to slightly acute at apex, margins entire; nerves impressed on upper surface, prominent on lower surface; petiole ± 8 mm. Inflorescence single or branched of two cymes, usually uniseriate, terminal and lateral, each cyme elongating to 8 cm at maturity. Calyx lobes almost free to base, lanceolate, 2–3 mm, villous inside and out, deciduous. Corolla white, 3–4 mm, hyporcrateriform, plicate, appressed pilose outside; lobes obtuse, margins entire or obscurely crenulate; intercalary lobes usually present. Anthers inserted above corolla base. Stigma long conical with a pilose apex; style pilose. Nutlets ovate, almost glabrous, surface rugose.

hab. Dry hills, mountain slopes, in degraded Quercus forest, sandy wadis, beside roads, on sandy, stony and gravelly soils; alt. 500–1400 m; fl. Jun.-Aug.

distrib. Mountains and lower hills of north eastern Iraq. **mam**: Bikhair (Bekher) Mt. near Zakho, *Rawi* 22982!; Kakari, Dohuk to Amadiya, 15 km W. of Amadiya, *Rechinger* 11633 (M, W, Herb. PODL.). **msu**: Halabja, *Rechinger* 10128 (M, W). msu: jugi inter Dukan et Mirza Rustam, *Rechinger* 10992 (M, W); Mt. Avroman (Hawraman), Tawilla, *Rechinger* 12396 (W). **mro**. Balow Kaiter, rd. to Kandul, *Rawi & Serhang* 26478!; Rist, hillside back of village, *Guest & Husham* 15845!; Qerna Waw valley, N. or Pushtanshan, *Rawi & Serhang* 26605!; Khemzad Pass, *Guest* 3005!; Rawanduz, rocky mountain slope, *Dabbagh & Hamad* 46185!; Haji Umran, *Al-Kaisi & K. Hamad* 43532!; Kaiwa rush near Raina, *Rawi & Serhang* 23768!; Rawanduz gorge, *Guest* 15922!; Rowanduz, 1893-21-Vl, *Bornmüller* s.n.; Ropwanduz, *Bornmüller* 1583!; Sefin Dagh, *Rawi* 9096!; Rowanduz gorge, in open degraded *Quercus* forest, *Gillett* 9443! **msu**. Ahmad Kulwan, dry hills, *Guest* 15795!; mro: Mt. Qandil, below Pushtashan towards Shahidan, *Rechinger* 11938 (W); infra Rowanduz, *Rechinger* 11236 (M, W); inter Rowanduz et Bersorin, *Rechinger* 11271 (M, W); valley between Rayat and Haji Umran, *Rechinger* 11277 (W). **mro/dsd**: Between the slope of Karukh Mt. & Dargala village, *Kass & Nuri* 27709!; 25 km N. of Bisbas, *Rawi, Husham, Nuri & Alizzi* 29429! **msu**. Ahmad Kulwan, dry hills, *Guest* 15795!; 22 km E. of Sulimaniyah, *Al Dabagh* 47006!; Mt. Hawraman, *Rawi, Charavarty & Alizzi* 19828! **fuj**: Between Mosul & Tel Kotchek (Tal Kochak), *Mrs Dickson* 367! **fni**: Dry hills above R. Great Zab nr. Aski Kelak (Eski Kellek), *F.A. Barkley & Jumaa Rahim* 3371A! **fpf**: Jalaula, on roadside, *F. Karim* 39239!; 2.5 km E. of Badra, *Rawi* 20766!; 2.5 km E. of Badra, *Rawi* 20766!; near Khanaqain (Khanaqin), *Rechinger* 2146 (W); Jabal Hamrin between Sharaban and Jalaula, *Rechinger* 9963 (M, W). **fki**: 45 km southeast of Kirkuk, Kirkuk Liwa, *Brahim* 6124 (W); 10 km north of Kirkuk, Kirkuk Liwa, Brahim 6150 (W). **lca/fki**: Between Khalis & Kirkuk, *Rawi, Charavarty & Alizzi* 19690!

KHATONAN (Kurd. *Al Dabagh* 47006)

W. & S.W. Iran.

3. **Heliotropium bovei** *Boiss.*, Diagn. Pl. Orient. 11: 87 (1849); Rawi in Dep. Agr. Iraq Tech. Bull. 14: 136 (1964); Dinsmore in Post, Fl. Syria, Pales. & Sinai 2: 221 (1933); Riedl in Fl. Iranica [K. H. Rechinger] 48: 42 (1967); Riedl in Fl.Turkey [P. H. Davis] 251 (1967); Feinbrun-Dothan, Fl. Palaest. 3: 56, pl. 88 (1978); Akhani & Förther, Sendtnera 212 (1994); Khatamsaz, Fl. Iran 39: 55 (2002); Haloob et al., Illustr. Fl. Lowland Iraq, 1: 66 (2016).

> *Heliotropium mamanense* Bunge, Bull. Soc. Imp. Naturalistes Moscou 42(1): 308 (1869); Riedl in Fl. Iranica [K. H. Rechinger] 48: 38 (1967).
> *H. kotschyanum* Bunge, Bull. Soc. Imp. Naturalistes Moscou 42: 312 (1869).
> *H. albo-villosum* Riedl., Fl. Iranica [K. H. Rechinger] 48: 44 (1967), nom inval. (no holotype designated).

Annual. Stems greyish-green, 6–30 cm, branching from base, erect to ascending; indument of whitish villous patent and appressed hairs. Leaves with petiole to 20 mm; lamina 10–25 × 8–20 mm, broadly ovate, elliptic to suborbicular, base attenuate to rounded, apex obtuse, margins somewhat thickened; upper side of leaves appressed-villous, nerves slightly impressed; lower side patent-villous, nerves prominent. Inflorescence terminal or lateral, unbranched; cymes unilateral and uniseriate, up to 8 cm long. Flowers white, sessile to the lower ones shortly pedicelled up to 1 mm. Calyx lobed nearly to base, lobes 2–3 mm, oblong, acute; calyx dropping with nutlets. Corolla yellowish-white, 4–6 mm, infundibular; lobes c. 1 mm, ovate-triangular, margins ± crenulate, densely pilose in the lower part on the outside. Anthers with a beaked apex, placed slightly above base of corolla. Style glabrous; stigma pilose above the middle. Nutlets ovoid to ellipsoid, 1–2 mm, rugose, sparsely pilose to glabrous. Fig. 91, 1–5.

HAB. On mountains, foothills and central alluvial plains; alt. 50–700 m; fl. May-Aug.
DISTRIB. Mountains and lower hills of northern Iraq. **MAM**. Dohuk, 52 km N. Mosul, *Rechinger* 10626 (W, cited in Fl. Iranica under *H. mamaense*). **MRO**: nr. Shaklawa, *Salim Effendi* 2600!; Mt. Qandil, Chinarak, *Rawi* 25344; Kirkuk Prov., sine loc., *Ali Effendi Hadari* 3936! **MSU**: Diala, *Rechinger* 12726 (W); **FPF**: Khanaqin, *Rechinger* 15491 (M, W). **LCA**: Baghdad, Jazira, *Rechinger* 10608 (M, W).

Recorded to be a grazing plant; KIA BAR RUZH HOR (*Salim Effendi* 2600), NAIMAH (*Ali Effendi Hadari* 3936)

Palestine, Jordan, Lebanon, Syria, Turkey, Iran.

4. **Heliotropium circinatum** *Griseb.*, Spic. Fl. Rumel.: 78 (1844); Rawi in Dep. Agr. Iraq Tech. Bull. 14: 136 (1964); Dinsmore in Post, Fl. Syria, Palest. & Sinai ed. 2, 2: 221 (1933); Riedl in Fl. Iranica [K. H. Rechinger] 48: 38 (1967); Riedl in Fl.Turkey [P. H. Davis] 6: 250 (1967); Feinbrun-Dothan, Fl. Palaest. 3: 78, pl. 129 (1978); Akhani & Förther, Sendtnera 222 (1994); Khatamsaz, Fl. Iran 39: 53 (2002)

Annual. Stems branching sparingly from base or stems simple, erect to ascending, 15–40 cm; indumentum whitish villous, ± appressed, mixed with long patent hairs. Leaves grey-green, sometimes purplish-green on the underside; upper leaves shortly petiolate, lower leaves with petiole to 10 mm; lamina villous to tomentose on both surfaces, broadly-ovate to suborbicular, 20–45 ×15–25 mm, base truncate, apex obtuse, margins entire, ± thickened. Inflorescence lateral and terminal, the terminal branched; cymes 2–20 cm, flowers uiniseriate, unilateral. Calyx lobed to base; lobes linear-oblong, 3–4 mm, pilose inside, villous outside, dropping easily with flowers and nutlets. Corolla white, infundibular, 6–12 mm (including lobes), tube pilose in the lower part; corolla lobes triangular-subulate, coiled at tips; intercalary lobes smaller than lobes. Anthers inserted slightly above the corolla base. Stigma subulate-conical, beaked near the middle, pilose. Nutlets ovoid to subglobose, distinctly rugose, glabrous. Fig. 92, 1.

HAB. On rocky mountain, wasteland, cultivated fields, gravelly wadis, sandy and gravelly soils near river and stream banks; alt. 280–950 m; fl. Oct.
DISTRIB. Mountains and lower hills of northern Iraq. **MAM**: Kani Masi Bridge, *Botany Staff* 43850! 25 km from Zakho to Kani Masi, *Botany Staff* 43813!; Sharanish village, 25 km N.E. of Zakho, *Rawi, Nuri & Tikriti* 29042!; Zakho, *Botany Staff* 43796!; Dohuk, *Wheeler-Haines* W. 456!, W. 1219!; Dohuk, Bikhair, *Rawi, Nuri & Tikriti* 28954! **MRO**: Saruchawa (small village 110 km N. of Raina), *Omar, Sahira, F. Karim & H. Hamid* 38243!; Rowanduz, *Omar, Sahira, F. Karim & H. Hamid* 38350! **MJS**: Niniveh, *Bornmüller* 1582! Kurdistan, *Kotschy* 1841!

Turkey, Iran.

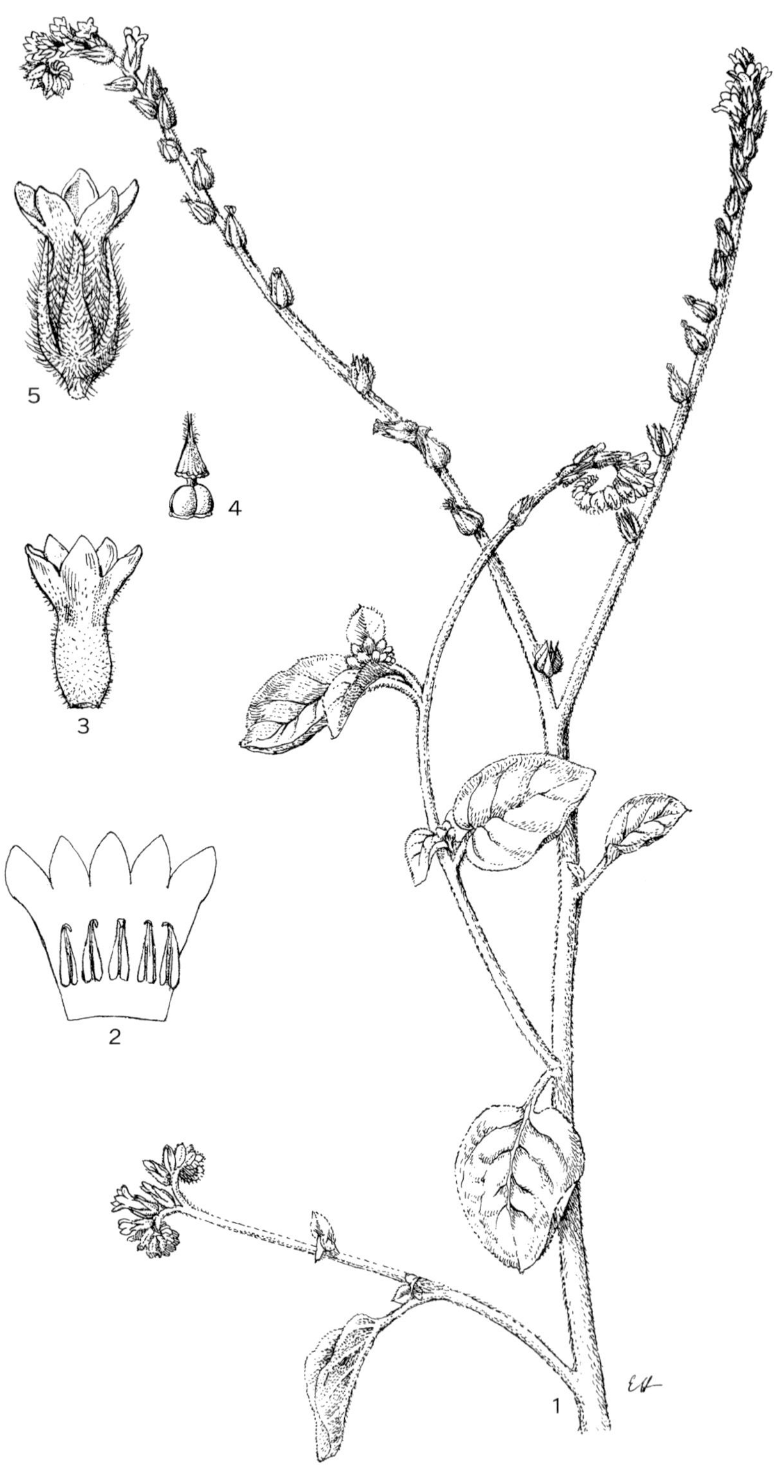

Fig. 91. **Heliotropium bovei**. 1, habit x1; 2, flower × 5; 3, corolla, 5; 4, corolla, opened × 5; 5, carpel × 5. Reproduced with permission from Feinbrun-Dothan, Flora Palaestina, 3: Plates, f. 88 (1977). Drawn by Esther Huber. © The Israel Academy of Sciences and Humanities.

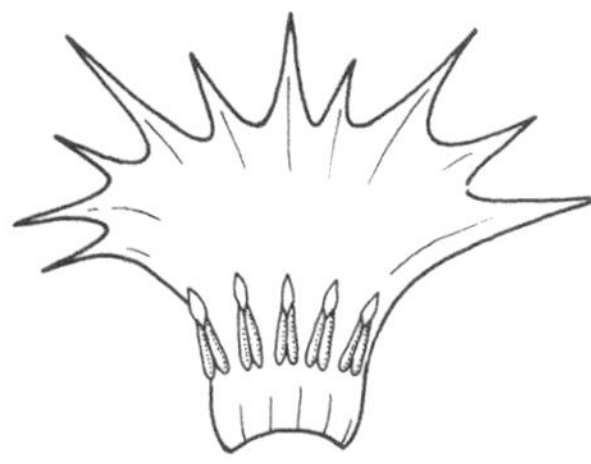

Fig. 92. **Heliotropium circinatum**. 1, corolla. Reproduced with permission from Fl. Turkey, 6: 249, fig 11, 1 Corolla.

5. **Heliotropiuin suaveolens** *M.Bieb.*, Fl. Taur.-Cauc. 3: 116 (1819); Rawi in Dep. Agr. Iraq Tech. Bull. 14: 137 (1964); Dinsmore in Post, Fl. Syria, Palest. & Sinai ed. 2, 2: 220 (1933); Riedl in Fl. Iranica [K. H. Rechinger] 48: 51 (1967); Feinbrun-Dothan, Fl. Palaest. 3: 56, pl. 89 (1978); Akhani & Förther, Sendtnera 264 (1994); Khatamsaz, Fl. Iran 39: 70 (2002); Taifour & El-Oqlah, Pl. Jordan Annot. Checkl.: 52 (2017).

> *H. confertiflorum* Boiss. & Noë, Diagn. Pl. Orient. Ser.2, 3: 132 (1856); Riedl in Fl. Iranica [K. H. Rechinger] 48: 551 (1967).
> *H. boissieri* Gürke, Nat. Pflanzenfam. [Engler & Prantl] 4(3a): 96 (1894) nom. illegit. – Type: In subalpinis pr. Haneky [Khanaqin, 9.1851], *Noë* (G-BOIS!, holo., GOET. iso) LE, P).

Annual. Stems thick, ± woody, erect, usually branched, 10–60 cm, erect; indument of tuberculate and sometimes glandular hairs. Leaves with petiole up to 10 mm; lamina ovate to elliptic, 10–25 × 6–15 mm, base ± truncate, apex acute or obtuse, margins entire or slightly sinuate; upper side appressed-villous, nerves slightly impressed, lower side densely covered with tuberculate patent hairs, nerves prominent. Inflorescence lateral and terminal, simple or branched. Cymes biseriate 6–10(–20) cm. Calyx lobed to base, linear-lanceolate, 2–3 mm, villous, persistent. Corolla white, 4–7 mm, tube 2–3 mm, tuberculate and sometimes glandular hairy outside, glabrous within; corolla lobes obtuse, rarely obscurely crenulate, without distinct intercalary teeth. Anther inserted just above the corolla base. Style glabrous, hidden by stigma; stigma pilose, apex bilobed. Nutlets 1.2 mm, ovate, rugose, glabrous.

HAB. In bottom of wadi, depressions, in desert on sandy and gravelly soils; alt. 80–110 m; fl. Aug.-Oct. DISTRIB. Foothills and central alluvial plains. **FUJ/FKI**: 17 km N.E. of Fatha, on road to Kirkuk, *J. Brahim* 6101!; Hatra, *Guest* 3559! **DLJ**: c. 20 km N.E. Ramadi, *Guest & Rawi & Nuri* 13543!; c. 17 km W. of Ramadi, *Guest & Rechinger* 15955! **DWD**: Rutbah, botton of Wadi Hamran, *Alizzi & Omar* 35368; Rutbah, Al Masad depression, *Alizzi & Omar* 35378! **LCA**: Above Baghdad, *Lester-Garland* 1918!; Baghdad, *V.C. Robertson* 13491!; Baghdad, *Wheeler Haines* W35!

ZURRAIJ (*Guest* 3559).

Balkans, Armenia, Azerbaijan, Turkey, Palestine, Iran.

6. **Heliotropium europaeum** *L.*, Sp. Pl. 1: 130 (1753) sensu lato.– Lectotype (Förther in Jarvis 1993: 53): Herb. Clifford: 45, Heliotropium nr. 1 (BM!); Rawi in Dep. Agr. Iraq Tech. Bull. 14: 136 (1964); Dinsmore in Post, Fl. Syria, Palest. & Sinai ed. 2, 2: 220 (1933); Feinbrun-Dothan, Fl. Palaest. 3: 55, pl. 87 (1978); Akhani & Förther, Sendtnera 242 (1994); Khatamsaz, Fl. Iran 39: 64 (2002); Taifour & El-Oqlah, Pl. Jordan Annot. Checkl.: 51 (2017).

> *Heliotropium ellipticum* Ledeb. in Eichw., PI. Nov. 1: 10, tab. 4 (1831); Riedl in Fl. Iranica [K. H. Rechinger] 48: 48 (1967).
> *H. europaeum* L. var. *ellipticum* (Ledeb.) Regel, Bull. Soc. Imp. Naturalistes Moscou 41: 75. (1868);
> *H. eichwaldi* Steud., Nomencl. Bot. ed. 2, 1: 744. 1840, nom. illegit
> *H. lasiocarpum* Fisch. & C.A.Mey., Index sem. hort. Petrop. 4: 38 (1837); Riedl in Fl. Iranica [K. H. Rechinger] 48: 49 (1967)
> *H. eichwaldi* Steud. var. *lasiocarpum* (Fisch. & C.A.Mey.) C.B.Clarke in Hook.f, Fl. Brit. India 4(10): 150 (1883)
> *H. ellipticum* Ledeb. var. *lasiocarpum* (Fisch. & C.A.Mey.) Popov, Trudy Glavn. Bot. Sada 42: 221 (1931)

H. europaeum L. var. *lasiocarpum* (Fisch. & C.A.Mey.) Kazmi, J. Arnold Arb. 51: 176 (1970); Y. Nasir in Fl. Pakistan [Ali & Y. Nasir] 191: 39 (1989).

H. strictum Ledeb., Fl. Ross. 3: 100 (1847), non Humb., Bonpl. & Kunth 1818, nom. illegit.

H. tenuiflorum Bunge, Bull. Soc. Imp. Naturalistes Moscou 42: 293 (1869) non Colla 1835, nom. illegit.

H. dolosum De Not., Repert. Legur.: 284 (1844); Riedl in Fl. Iranica [K. H. Rechinger] 48: 47 (1967).

Grey-green annual. Stems erect to ascending, 7–60 cm; indument appressed or patent. Leaves 15–50(–80) × 5–40 mm, ovate to elliptic, base cuneate to rounded, apex rounded or acute at top; petiole up to 40 mm. Inflorescence often branched; cymes dense, in fruit biseriate and bilateral. Calyx lobed to base, lobes lanceolate to oblong-linear, 2–3 mm; calyx persistent after nutlets falling. Corolla 2–5 mm; lobes rounded, without intercalary lobes, sometimes shortly denticulate. Anthers inserted just above corolla base. Stigma conical to linear subulate, pilose or glabrous. Nutlets pilose or glabrous.

HAB. Rocky mountains slope, hillsides, in shade of *Quercus*, wadi bottoms, depressions, cultivated land, irrigation ditches, beside roads, on sandy, clayey and gravelly soils; alt. 35–1300 m; fl. Apr.-Dec.

DISTRIB. Mountains, upper plains, foothills and the alluvial plains of Iraq; common. **MAM**: 20 km N.W. of Sarsang, *Al-Kaisi & K. Hamad* 45947!; Sersand, *Wheeler-Haines* W578!; Zawitah, *Guest* 3755! **MSU**: Hawraman, *Rawi, Alizzi, Chakravarty & Nuri* 19825!; Avroman et Schahu, June/July 1867, *Haussknecht s.n.*!; Sulimaniyah, *Abd al Majid & Salim Agha* 5365! **MRO**: Pushtanshah, N.E. of Raina, *Serhang & Rawi* 26480!; Jabel Rayat, *Abd al Jabbar* 5219! **FUJ**: Mosul Harim al Khamesa, *Tikrity* 16381! **FKI**: 25 km from Kirkuk, *Khatib & Tikrity* 29721!; Kirkuk, *Ali Borhanaddin* 5523! **FPF**: Al Sudur, *Omar & Thamer* 47117! **DLJ**: Ishaqi, *Robertson* RB 46!; On road to Amara, *Hazim & Nuri* 30612! **DSD**: Zubair, *Rawi* 25971!; 10 km S.W. of Manayit, *Alizzi & Omar* 33982!; **DGA**: Hamri Saqia, Guest 5609! **DWD**: Rutbah, Al Masad depression, *Alizzi & Omar* 34624!; Rutba, bottom of Wadi Hamran, *Alizzi & Omar* 35370!; 10 km S.W. of Rutbah, *Rawi* 21057!; 55 km W. of Ramadi, *Rawi* 20861! Hit, *Guest* 3529! **LEA**: Kut, *Guest* 199! 4 km N. of Sa'adiya, *Al Kaisi* 42893! **LCA**: Rustam expt. Farm, *Guest* 2391!; *Lazar* 1190!; *Guest* 155!; N. of Falluja, *Agnew & Barkley* 6023!; 50 km N. of Diwaniyah, Thamer 47658!; 60 kn S.E. Al Musaiyib, *Thamer & Wedad* 47268!; 15 km N.E. Baghdad, *Andrews* 2010!; Baghdad, *Mukhtar* 35848!; Abu Ghraib, *Alizzi* 32795!; Suaira, *Alizzi & Omar* 34659!; Badah, *Guest* 2498!, 2498!; Baghdad, *Sahira* 38157!; Zaffraniya, *coll. Ingot.* 26061! **LSM**: E. of Amara, *Rawi* 16640!

ZURRAIJ (*Guest* 2391, 3529, *Tikrity* 16381); BUHNSHINK, Kurd. (*Guest* 3755).

Europe, southwest and central Asia to India.

7. **Heliotropium digynum** (*Forssk.*) *C.Christ.*, Dansk Bot. Ark. 4 (3): 14 (1922); Dinsmore in Post, Fl. Syria, Palest. & Sinai ed. 2, 2: 243 (1933); Riedl in Fl. Iranica [K. H. Rechinger] 48: 26 (1967); Feinbrun-Dothan, Fl. Palaest. 3: 55 (1978), pl. 85 (1977); Akhani & Förther, Sendtnera 234 (1994); Abdulridha, Taha & Widad, Ecology & Fl. Basrah: 237 (2016); Taifour & El-Oqlah, Pl. Jordan Annot. Checkl.: 51 (2017).

Lithospermum digynum Forssk., Fl. Aegypt.-Arab.: 40 (1775).

Heliotropium luteum Poir, Encycl. [J. Lamarck & al.] Suppl. 3. 22 (1813); Rawi in Dep. Agr. Iraq Tech. Bull. 14: 136 (1964);

Perennial herb with a woody stock. Stems yellowish to whitish, woody at base, erect to ascending, up to 50 cm, densely appressed pubescent mixed with sparse long, patent hairs. Leaves ovate to oblong-ovate, 10–30 × 7–20 mm, base rounded, apex obtuse, margins revolute; petiole up to 10 mm, sometimes ± sessile; upper side of leaves pubescent to tomentose, nerves distinctly impressed; lower side densely patent hairy, nerves distinctly prominent. Inflorescence ± divided into two branches; cymes congested, up to 3 cm. Calyx divided to base, calyx-lobes oblong-linear, acute, 2– 3.5 mm, densely hairy. Corolla dull yellow, tubular, 3–4 mm; corolla lobes ± 9 mm, acute, between lobes folded inwards, densely pilose in the lower part on the outside. Anthers inserted just above corolla base. Style pilose. Stigma shortly conical, apex pilose. Nutlets 2 mm, ± ellipsoid, covered by long, white, silky hairs.

HAB. In subdesert, in sandy hollow, on sandy, stony, gravelly and gypsum soils; alt. 20–150; Mar.-Apr.; Jun., Oct.

DISTRIB. Foothills, the western desert and the central alluvial plains of Iraq. **FPF**: Chlat Police stn., border with Iran, *Rawi & Haddad* 25696!; nr. Wadi Tib police post nr. Kuwait, *Guest, Rawi & Rechinger* 17530! **DLJ**: 3 km E. of Hit, *Rawi & Gillett* 6826!, 6827! **DWD**: Habaniya, *Rawi & Alizzi* 34451! **DSD**: nr.

Salaibiqat at Hamer, 3 km S.W. of Ur, *Guest, Rawi & Rechinger*16054!; inter Ur et AI Busaiya, *Rechinger* 8191 (W). **lea**: 18 km N.E. Fakka, *Hazim* 30660!

North Africa, Arabian Peninsula, Iraq and Southwestern Iran.

8. **Heliotropium crassifolium** *Boiss. & Noë*, Diagn. Pl. Orient. ser. 2, 3: 131 (1856). Type: Lectotype (hoc loco designatus): In subalpinis Persiae occidentalis prope Haneky [Khana-qin], IX. 1 85 1, Noë 1039 (G-BOISS, lectotype designated by Akhani & Förther in Sendtnera 223 (1994); iso. P).

Erect annual of 60–100 cm, ± branched from base, divaricately branched above. Indumentum greyish tomentose mixed sparsely with long hairs. Leaves ovate, elliptic, 20–40 ×15–30 mm, relatively thick, obtuse to subacute at top, attenuate or subtruncate at base, with short petiole up to 8 mm, upper ones sessile, margins somewhat undulate; both sides tomentose, nerves in upper surface impressed, in lower surface prominent. Inflorescence elongated, ± simple; cymes lax, unilateral, up to 50 cm. Calyx easily dropping; calyx lobes 4–5 mm, linear, obtuse, tomentose. Corolla 6–8 mm, hypocrateriform, pubescent outside below the lobes, inside pilose above anther base, in throat with five longitudinal appendage-like ridges; corolla lobes subrounded with slightly crenulate margins, ± papillate; intercalary lobes absent. Anthers inserted just above corolla base. Stigma elongate-conical, apex densely pilose, disc subglabrous, nearly sessile on ovary. Nutlets ovoid, 1.3–1.4, glabrous, rugose. (Desription from Akhani & Förther in Sendtnera 223 (1994)).

HAB. Lower hills; alt. not given on label; fl. Sep.
DISTRIB. Known from the type gathering. **FPF**: Kirkuk: in subalpinis Persiae occidentalis prope Haneky [Khanaqin] 9.1851, *Noë* 1039 (type).

Near Endemic. S.W. Iran.

9. **Heliotropium bacciferum** *Forssk.*, Fl. Aegypt.-Arab.: 38 (1775); Riedl in Fl. Iranica [K. H. Rechinger] 48: 19 (1967); Feinbrun-Dothan, Fl. Palaest. 3: 58, pl. 94 (1978); Y. Nasir in Fl. Pakistan [Ali & Y. Nasir] 191: 43 (1989); Akhani & Förther, Sendtnera 205 (1994); Khatamsaz, Fl. Iran 39: 37 (2002); Abdulridha, Taha & Widad, Ecology & Fl. Basrah: 237 (2016); Taifour & El-Oqlah, Pl. Jordan Annot. Checkl.: 51 (2017).

> *H. tuberculosum* (Boiss.) Boiss., Fl. Orient. [Boissier] 4: 147 (1879).
> *H. undulatum* Vahl var. *tuberculosum* Boiss., Diagn. Pl. Orient. ser. 1, 11: 89 (1849).
> *H. ramosissimum* (Lehm.) DC. Prodr. [A. P. de Candolle] 9: 536 (1845); Rawi in Dep. Agr. Iraq Tech. Bull. 14: 136 (1964) ; Riedl in Fl. Iranica [K. H. Rechinger] 48: 18 (1967)
> *H. kotschyi* Bunge, Bull. Soc. Imp. Naturalistes Moscou 42: 331 (1869), nomen nudum.
> *H. persicum* auct.: Boiss., Fl. Orient. [Boissier] 4: 147. 1879, non Lam. (1789).
> *H. lignosum* Bomm., Repert. Spec. Nov. Regni Veg. 41: 323 (1937) nomen nudum.

Perennial subshrub with erect, ascending, or prostrate branches arising from a woody base, sometimes spreading to form low large bushes, often on low sandy dunes. Stems and leaves covered by strigose tuberculate hairs. Leaves linear to linear-lanceolate, 5–60 × 2–16 mm, margins entire to crenulate-undulate and revolute, mature leaves sometimes somewhat fleshy. Inflorescence cymose, usually terminal, at anthesis elongating up to 7 cm, uniseriate, with few, sessile flowers. Calyx lobed to base, persistent, lobes linear, 2–3 mm, acute, hairy outside. Corolla white, 2.5–4 mm, tubular, slightly constricted at throat, pilose outside; lobes ovate, oblong or rounded with entire or sinuate margins, ± imbricate. Anthers inserted slightly above corolla base. Stigma conical, acute, pilose or glabrous. Fruit subglobose, ± 2 mm in diameter; nutlets 4, sometimes slightly winged at edges, rugulose, glabrous or pilose. Fig. 93, 1–6.

A common and widespread species in the desert and "haswa" plains of Iraq, extremely variable in the form of leaves, nutlets and indument. Several species and infraspecific taxa have been described, but from the many intermediate forms seen it is difficult to recognise distinct taxa in Iraq. (For full synonymy see Akhani & Förther (1994)).

HAB. Sandy desert, low rocky hills, sandy plateau, depressions, "haswa" plains, on sandy, gravelly, clayey and gypsum soils; alt. 10–400 m; fl.; Feb.-Oct.
DISTRIB. Throughout the southern and western deserts, lower hills, and the central alluvial plains; common. **FKI**: J. Hamrin, *Gillett* 12487! **FPF**: Badra, *Gillett* 6641!; 25 km S.W. of Badra, *Al Kaisi, Thamer & Salah* 51354!; Koma Sank Police stn. nr. Mandali on Persian border, Rawi 20610!; Fakki-Laglag valley

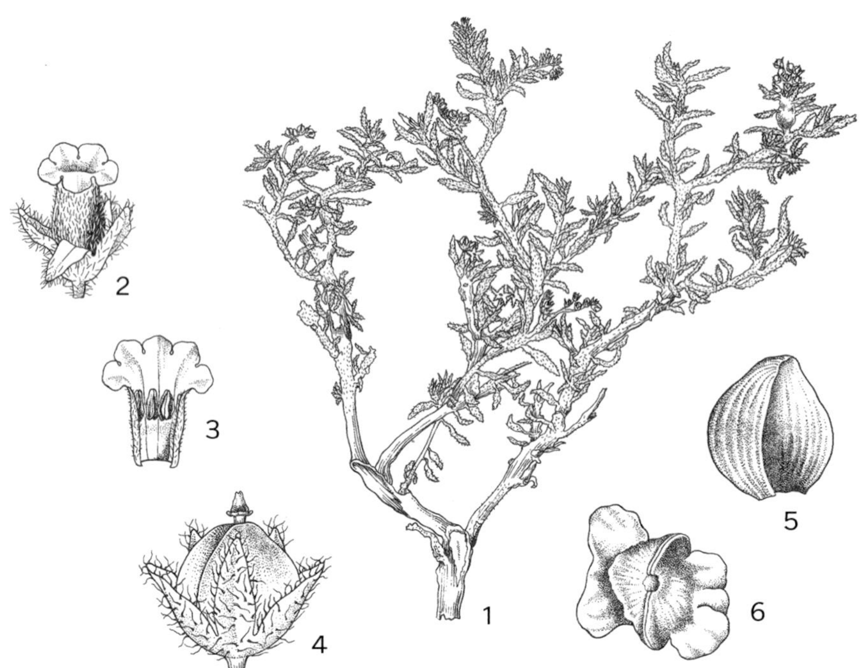

Fig. 93. **Heliotropium bacciferum**. 1, habit portion of shoot × ½; 2, flower × 5; 3, corolla opened × 5; 4, carpel × 20; 5, nutlets with calyx × 8; 6, nutlet × 10. Reproduced with permission from Fl. Pakistan 191: f. 11, I–M (1989). Drawn by S. Hameed. © National Herbarium, Pakistan Agriculture Research Council & University of Karachi, Pakistan.

on Persian border, *Rawi* 25839!; Buskaya, *Rawi & Haddad* 25565!; Chlat police stn., *Rawi & Haddad* 25679!; The Sudur, *Botany Staff* 47131!; Jabal al Muwaila, *Guest, Rawi & Rechinger* 17578!; 25 km S.E. Mandali, *Rawi* 20707! **DLJ**: 70 km N.N.W. of Falluja, *Rawi* 20270!; 65 km NNW of Falluja, *Rawi* 20237!; 2 km W. of Falluja, *Charavarty & Rawi* 30245!; nr. Falluja, `13503!; 70 km N.W. of Falluja, *Charavarty & Rawi* 30350!; Ramadi, *E. & F. Barkley & Fuad Safwat* 5049!; 24 km N.E. of Haditha, *Charavarty, Rawi, Khatib & Alizzi* 31831!; 35 km from Ramadi to Rutba, *Jenan Al Mokhtar* 33384!; 22 km W. of Ramadi, *E. & F. Barkley & Fuad Safwat* 5049! **DGA**: 4 km E. of Samara, *Rawi* 20333! **DWD**: 5 km W. of Ana, *S. Omar & Hamad* 50407!; 260 km N.W. of Ramadi, *Rawi* 20988!; Wadi al Khir, 15 km N. of Lussaf, *Guest, Rawi & Rechinger* 19458!; 31 km W. of Ramadi, *Rawi* 20829!; c. 17 km W. by S. of Ramadi, *Guest, Rawi & Rechinger*15952!; Ramadi–Rutbah rd., *Omar* 32499!; 35 km W. of Nubhaib, *Rawi* 14761!; 75 km W. of Rutba, *Rawi* 14711!; E. of Shithatha nr. Bahr, *Rawi* 26923! **DSD**: 87 km from Bussaiya to Salman, *Khatib, Rawi & Tikrity* 29994!; nr. Bussaiya, *Guest & Rawi* 14235!; al Bussaiya, *Haddad* 9565!; 10 km S.E. of Bussaiya, *Rawi* 26010!; 20 km to Busaiya, *Fauzi, Hamid, & Khadim* 40362!; 25 km to Busaiya from Khader al mai, *Al-Kaisi, K. Hamad & H. Hamid* 48507!; Khader al mai, *Alizzi* 34377!; Khadar al mai enclosure, *Khatib, Rawi & Tikrity* 29121!; 14 km ESE Khader al mai, *Guest & Ibrahim Mahalla* 19505!; 50 km W. of Khadar al mai enclosure, *Khatib, Rawi & Tikrity* 29137!; 2 km N. of Khader al mai, *Weinert, Fauzi, Adel & Hussain* 47626!; Khader al mai, *Al-Kaisi, K. Hamad & H. Hamid* 48490!; between Nassriya & Dewaniyah, *Hosham & Alizzi* 29629!; from Basra to Nassriya, *Hosham & Alizzi* 29626!; 40 km W. of Basra, *Rawi & Tikrity* 29330!; 99 km E. of Shabicha, *Khatib, Rawi & Tikrity* 29282!; Shabicha, *Rawi & Gillett 6288*!; 51 km N. of Neutral zone, *Khatib, Rawi & Tikrity* 29188!; 12 km WNW of Ansab, *Guest, Rawi & Rechinger* 18990!; 27 km WNW of Ansab on Saudi frontier, *Guest, Rawi & Rechinger* 19011!; 40 k NWN of Sabicha, *Guest, Rawi & Rechinger* 19294!; Southern desert nr. Juhara, *Guest, Rawi & Rechinger* 16090!; Southern desert, N. aspect of neutral zone, *Guest, Rawi & Rechinger* 16081!; 5 km SSW of Bussaiya, *Guest, Rawi & Rechinger* 16098!; Southern desert *Guest & Rawi* 14299!; Zubair, *Mustapha* 3158!; between Zubair and Safwan, *Alizzi* 34343!; S. of Zubair, *Rawi* 25867!; Jumaria police stn., *Al-Kaisi, K. Hamad & H. Hamid* 48251!; Al Baniya, c. 65 km WSW Basrah, *Guest, Rawi & Rechinger* 17315!; 45 km S.W. of Busaiya, *Guest & Ibrahim* 15277!; 8 km N. by W. of Aidaba (Al-Aida) *Guest, Rawi & Rechinger* 19149!; 59 km between Aida & Ansali, *Fauzi, Hazim & H. Hamid, 38822*!; 17 km from Shabicha to Salman, *Karim, Nuri, Hamid & Khadum* 40047!; 30 km E. of Salman, (nr. Golaib), *Guest, Rawi & Rechinger* 18740!; 12 Km ESE of Salman, *Guest, Rawi & Rechinger* 18851!; 20 km N.W. of Salman, *Fauzi, Hazim & H. Hamid* 38700!; 25 km between Aida and Ansab, *Fauzi, Hazim & H. Hamid* 38801!; c. 50 km WSW of Karbala, *Rawi* 30784!;

Ukaidre, Karbala, *Underwood & F. Barkley* 4335!; Chuwaibah walls nr. Zubair, *Guest, Eig & Zohary* 5051!; 5 km to Ma'aniya, *Hazim* 32499!, 32516!, 32533!; Um Qasr, *Khatib & Alizzi* 33438!; ibid., *Alizzi & Omar* 35859!; between Najaf & Shabicha, *Omar, Karim & Hamid* 36989!; 46 km from Najaf nr. Rahba, *Fauzi, Hazim, & H. Hamid* 38619!; nr. Samah, *Alizzi* 34388!; Wadi al Tabat, 100 km N. of Nukhaib, Rawi 31094!; Salman, *Alizzi & Omar* 34950!; Butain, 60 km S.W. of Zubair, *Alizzi & Omar* 35095! **LEA**: 77 km N.E. of Kut, *Rawi* 18067! **LCA**: 10 km E. of Falluja, *Rawi & Gillett* 6762!; Haswa between Abu Ghraib Farm & Falluja, *Rawi* 15106!; Falluja, *Wheeler-Haines* W. 150!; 25 km E. of Falluja, *H. Abbas Al Ani* 9937!; Beni Saad Railway Stn., 3 km NNEof Baghdad, *Anderson* 1990!

RIM RĀM (*Guest & Rawi* 14235; *Guest, Rawi & Rechinger* 18851). An infusion made of this plant is taken after snake or scorpion bite. It is hung in dwellings to keep off insects and "put in milk to stop snakes from drinking it" (*Rawi & Gillett* 6288).

North Africa, Arabian Peninsula, Iran, Afghanistan, Pakistan, Turkmenistan.

10. **Heliotropium curassavicum** *L.*, Sp. Pl. 1: 130 (1753); Y. Nasir in Fl. Pakistan [Ali & Y. Nasir] 191: 23 (1989); Riedl in Fl. Turkey [P. H. Davis] 6: 255 (1967); al-Rudainy & al–Maya in Bull. Iraq Nat. Hist. Mus. (2023).

Perennial herb, much branched with prostrate to decumbent stems, 20–46 cm, glabrous. Leaves sessile, succulent, glaucous, lamina lanceolate to oblanceolate-ovate, 13–40 × 3–12 mm, apex obtuse, margins entire. Inflorescence terminal and lateral, one-sided, forked, dense, ebracteate, 2.5–5 cm long. Flowers sessile. Calyx 2–3 mm, sepals lobes lanceolate, 1.3–2 mm. Corolla white with yellow patches at the center on the inner side, funnel form, longer than calyx, corolla throat without scales; corolla tube 1.5–2.3 mm, corolla lobes ovate to oblong, margins sinuate. Stamens with short filaments, inserted in middle of corolla-tube. Stigma short, conical. Style absent. Ovary glabrous. Fruits globose, breaking up into four wedged-shaped nutlets, 1.5–2.5 mm. Fig. 94, 1–6.

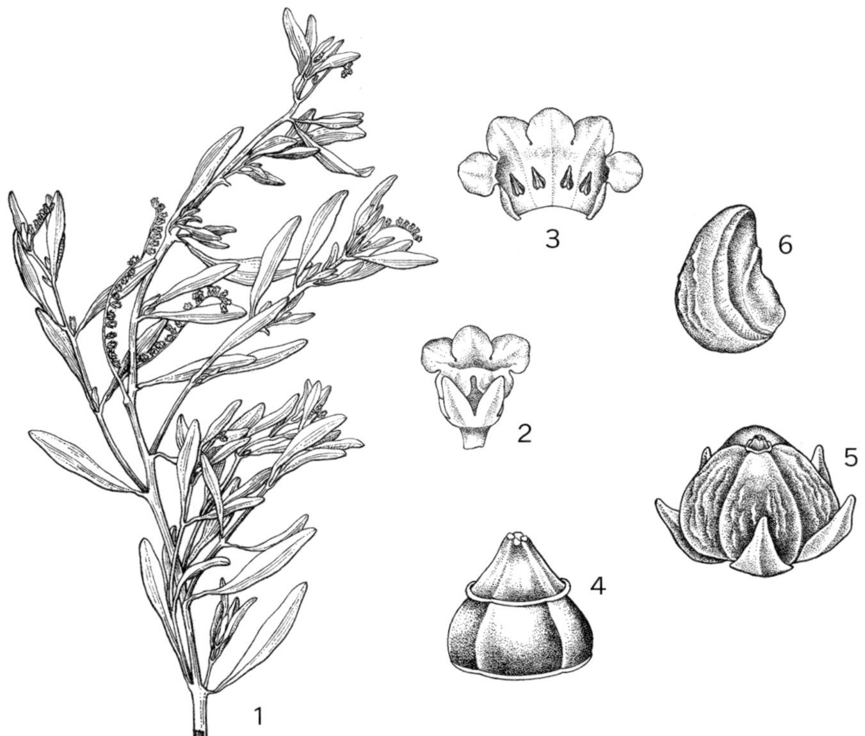

Fig. 94. **Heliotropium curassavicum**. 1, habit portion of shoot × ½; 2, flower × 5; 3, corolla opened × 5; 4, carpel × 20; 5, nutlets with calyx × 8; 6, nutlet × 10. Reproduced with permission from Fl. Pakistan 191: f. 6, G–L (1989). Drawn by S. Hameed. © National Herbarium, Pakistan Agriculture Research Council & University of Karachi, Pakistan.

HAB & DISTRIB. A new record for Iraq; known from the south of Iraq in several locations along the coast of Shatt al Arab (al-Faw, al-Mekhrag, al-Dotrah, al-Duweeb, al-Sheeba), in moist and moderately saline soils; alt. ± 2 m; fl. & fr. May-Nov.

Widespread; native to Australia, the Hawaiian Islands, Tropical & Subtropical America; introduced into northern and southern Africa, southern Europe, Turkey, Arabia, Pakistan and India; a new record for Iraq.

11. **Heliotropium lasianthum** *Riedl,* Kew Bull. 33(3): 517 (1979). Type: Iraq, Ramadi liwa, Jebel Thebaa, 30 km from Rutba 18 Nov. 1964, *F.A. Barkley* & *Hikmat Abbas al Ani* 9217 (W, holo.; MSB, NY, iso.).

Perennial with a woody base with several herbaceous stems ascending from base, up to 30 (–35) cm, covered with dense, white patent hairs. Leaves dense on the upper part of stems, c. 10 × 9–12 mm, broadly ovate, base obliquely truncate or ovate-cordate, veins distinct on the underside; petiole 2–3 mm. Inflorescence scorpoid at first, becoming lax, 2(–2.5) cm. Calyx almost lobed to base, lobes linear-lanceolate 3–3.5 mm, 4.5 mm at anthesis, densely hairy, persistent. Corolla yellow, twice as long as the calyx, tube cylindrical, 4.5–5 mm, limb funnel-shaped reaching to 3 mm in diameter, margins undulate. Anthers subsessile, rounded at the apex, included. Ovary, subconical, glabrous. Nutlets 4, free, ovoid, brown, wrinkled, glabrous. (Description from Kew Bull 33(3): 517 (1979)).

HAB & DISTRIB.. Known from a single collection from the Western desert of Iraq; alt. ± 620 m; fl. & fr. Nov.

DWD. Jebel Thebaa, 30 km from Rutba, *F.A. Barkley* & *Hikmat Abbas al Ani* 9217 (type).

Endemic.

SPECIES DOUBTFULLY RECORDED FROM IRAQ

Heliotropium myosotoides *Banks* & *Sol.* in A. Russell, Nat. Hist. Aleppo, ed. 2, 2: 245 (1794); Riedl in Fl. Turkey [P. H. Davis] 6: 252 (1967; Riedl in Fl. Iranica 48: 45 (1967).

Heliotropium aleppicum Boiss., Diagn. Pl. Or. ser. 1(11): 88 (1849).

Annual herb. Stems erect, branched from base, up to 20 cm, densely strigose-tomentose, hairs stiff, patent, arising on small tubercles. Leaves ovate-lanceolate, 18–27 × 7–9 mm, base tapering into short petiole or leaves sessile, apex acute, margins entire, strigose-tomentose. Inlorescence lax, unilateral; bracteate. Calyx divided to base, calyx-lobes oblong-linear, acute, c. 2 mm, densely hairy. Corolla dull yellow, tubular, c. 2–3 mm, hirsute on the outside, glabrous inside; corolla lobes small, erect; intercalary lobes small. Anthers inserted just above corolla base. Stigma subulate from a disc-like base, penicillate above. Nutlets with small warts and pits, hispid or glabrous.

S.E. Turkey to Iraq and Palestine.

Mentioned by Zohary (1946) in "The Flora of Iraq and its Phytogeographical Subdivision" from Amadiyah in N. Iraq, but neither Riedl (in Fl. Iranica, p. 45) nor I have seen any material from Iraq.

Heliotropium aucheri *A.DC.,* Prodr. [A. P. de Candolle] 9: 533 (1845), with Type: ? in Turcorum Asia prope Mossul, [1836], *Aucher* (G-BOIS!). Rawi (Dep. Agr. Iraq Tech. Bull. 14: 136. 1964). *H. halame* Boiss. & Buhse, Nouv. Mem. Soc. Imp. Naturalistes Moscou 12: 150. 1860. *H. aucheri* is not found in Iraq. According to Akhani and Förther (1994), the type locality of *H. aucheri* DC. is doubtful; De Candolle mentioned the locality with question mark in his original diagnosis "? in Turcorum Asia pr. Mossul". Boissier (1879) correctly stated that it must have been collected from Iran. Material studied by Akhani & Förther found it neither in Turkey nor Iraq; it is limited in its distribution to central Iran and adjacent parts of Pakistan.

Heliotropium ferugineo-griseum Näbelek, Spisy Prir. Fak. Masarykovy Univ., No. 70: 16, t. 3. 1926.

Syntypes: Mesopotamia superiore (Gebel et Tur, Tur Abdin) ad monasterium Der el-

Ahmar dit. Midiat, ca. 1000 m, 7.7.1910, Nábĕlek 613 (SAV); in Kurdistania Turcica ad pagum Mar Jakub dit. S'ert, ca. 920 m, \%.1.\9\0, Näbelek 66 (SAV).

This species was described from Turkish Kurdistan. It is listed in Rawi (Dep. Agr. Iraq Tech. Bull. 14: 136. 1964) but is not found in Iraq.

17. **NONEA**

Medik., Philos. Bot. 1: 31 (1789)
Weigend et al. in Kubitzki, Fam. Gen. Vasc. Pl. 14: 81 (2016)

C.C. Townsend
Revised by Shahina A. Ghazanfar

Annual or perennial herbs, hispid and usually glandular. Stems leafy up to terminal racemes. Calyx 5-fid, accrescent and deflexed in fruit. Corolla salver-shaped, funnel-shaped, bell-shaped or nearly tubular, with 5 short ± hairy scales or tufts of hairs or a hairy ring in throat. Stamens included. Style included; stigma entire or 2-lobed. Nutlets 4, free, reticulate-rugose to almost smooth, with or without prominent costae, usually puberulent, transversely ovate-oblong (in ours), with entire or denticulate basal ring.

About 35 species in Europe, N. Africa and W. Asia; 5 species in Iraq.

Cecchi, L., Coppi, A. & Selvi, F. (2009). *Nonea palmyrensis* (Boraginaceae): morphology and phylogenetic affinities of a rare endemic of the Syro–Iraqi desert. Nord. J. Bot. 27(5): 381–387.
Selvi, F. & Bigazzi, M. (2001). The *Nonea pulla* group (Boraginaceae) in Turkey. Plant Syst. & Evol. 227(1–2): 1–26.
Selvi, F. & Bigazzi, M. (2002). Chromosome studies in Turkish species of *Nonea* (Boraginaceae): the role of polyploidy and descending dysploidy in the evolution of the genus. Edinburgh J. Bot. 59: 405.
Selvi, F., Papini, A. & Bigazzi, M. (2002). Systematics of *Nonea* (Boraginaceae-Boragineae): new insights from phenetic and cladistic analyses. Taxon 51: 719–730.
Selvi, F., Bigazzi, M., Hilger, H.H. & Papini, A. (2006). Molecular phylogeny, morphology and taxonomic re-circumscription of the generic complex *Nonea/Elizaldia/Pulmonaria/Paraskevia* (Boraginaceae-Boragineae). Taxon 55(4): 907–918.

1. Perennials. Nutlets with prominent costae ending in teeth 2
   Annuals. Nutlets without prominent costae. 3
2. Inflorescence glandular-viscid; faucal scales scabrid; style short, stout,
      not exserted from corolla-throat; stigma divided into 2 rounded
      lobes; nutlets reticulate-rugose, with 14–16 vertical costae terminating
      into obtuse teeth . . . . .1. *N. persica*
   Inflorescence pubescent with sparse longer setae, not glandular; faucal
      scales densely long-hairy; style slightly exserted from corolla-throat;
      stigma bilobed; nutlets, papillose, reticulate, with 12–18 vertical
      costae terminating into short teeth . . . . .2. *N. macrantha*
3. Corolla white; nutlets reniform . . . . 5. *N. echioides*
   Corolla orange, pink, yellow to pale yellow; nutlets transversely ovoid 4
4. Stems hispid-setose with long and short hairs; leaves oblong- to linear-
      lanceolate, 25–80 × 5–20 mm, margins entire to denticulate; calyx
      5–6 mm; anthers overlapping faucal scales; nutlets ± reticulate
      with whitish dots, glabrous or puberulent at base, basal ring not
      thickened, teeth 11–14 . . . . 3. *N. capsica*
   Stems hispid and glandular-viscid; leaves linear-lanceolate, 2–5 × 3–6
      mm, margins repand-dentate; calyx 10.5 mm at anthesis; anthers
      below faucal scales; nutlets smooth, slightly rugose towards base,
      basal ring thick with 10–12 prominent teeth . . . . 4. *N. palmyrensis*

1. **Nonea persica** *Boiss.*, Diagn. Pl. Orient. Nov. 1, 7: 32 (1846); Boissier, Fl. Orient. 4: 167 (1879) [pro. syn. *N. pulla* (L.) DC.)] non Steven: Bull. Soc. Nat. Mosc. 24(1): 574 (1851). Selvi & Bigazzi in Pl. Syst. Evol. 227(1–2): 2, fig. 1a,b; 2a; 3a-d (2001); Khatamsaz, Flora Iran 39: 205 (2002).

Fig. 95. **Nonea persica**. 1, habit × 1; 2, flower with bract × 5; 3, open corolla with pistil and stamens × 5; 4, corolla in lateral view × 5; 5, mericarp in lateral view × 10. Reproduced from Selvi & Bigazzi, Plant Syst. & Evol. 227(1–2): f. 1b (2001).

N. pulla (L.) DC. var. *persica* Boiss., Fl. Orient. 4: 167 (1879).
N. pulla (L.) DC. subsp. *monticola* Rech.f., Ann. Nat. Mus. Wien 55: 15 (1947), p.p. excl. typ.
N. pulla (L.) DC. var. *persica* (Boiss.) M.Pop., Flora SSSR 19: 254 (1953).
N. pulla (L.) DC. subsp. *scabrisquamata* A.Baytop, Notes Roy. Bot. Gard. Edinburgh 35: 299 (1977).

Hispid and glandular-viscid perennial. Stems several arising from base, 15–35 cm, erect to ascending, branched mainly above middle. Basal leaves forming a rosette, narrowly ovate to ovate-oblong, to 3 × 8 cm, ± covered with whitish tubercles; cauline leaves lanceolate, sessile, ± amplexicaul, apex acute, margins somewhat undulate, entire, both surfaces densely hispid. Inflorescence densely glandular, bracteate. Bracts lanceolate, acute, subcordate at the base, equalling or longer than calyx. Calyx 5–8 mm, divided to less than ½; lobes triangular, acute; fruiting calyx 8–12 mm, accrescent, reflexed at maturity. Corolla purple, 6–12 mm, funnel-shaped, limb exserted from calyx, shorter than tube. Faucal scales scabrid, not conspicuous. Annulus glabrous. Anthers 1.2–2 mm, inserted at different heights in the upper half of corolla-tube. Style short, stout; stigma deeply divided into 2 lobes. Nutlets 4, c. 2 × 4 mm sometimes reduced to 1, obliquely ovoid, reticulate-rugose, with 14–16 vertical costae terminating into obtuse teeth. Fig. 95, 1–5.

HAB. Stony mountainsides, in *Astragalus* zone, among rocks; alt. (200–)1060–3100 m; fl. Apr.-Aug.
DISTR. In the upper forest and thorn-cushion zones of N.E. Iraq. **MRO**: between Perrish & Bardanas, Qandil range, *Rawi & Serhang* 24586!; N. of Rania, *Rawi & Serhang* 18260! 18253!; Malakh (Gomasur), N.E. of Rania, *Rawi & Serhang* 18241!; Pishtashan, *Rawi & Serhang* 24234!; Haji Omran, *Shahwani* 25256! *Emberger, Guest, Long, Schwann & Jusuf* 15513! *Mooney* 4259!; Jabal Kodo (Kudu), *Rawi* 9186!; Jabal Karoukh, *Kass & Nuri* 27339! 27413! 27437!; E. side, Kawriesh, *Kass & Nuri* 27666!; S. part, *Kass & Nuri* 27452! **MSU**: Ab-i Tanjaro valley, *Poore* 363! **LCA**: Daltawa (Khalis), *Rogers* O825! Mt Potina nr. Shistwan Moza, *Agnew, Hadac & Haines* W.2163

N. persica is a widespread, polymorphic species with variations in plant size, shape of leaf and corolla characters and is genetically heterogenous. All Iraqi material determined as *N. pulla* subsp. *monticola* Rech.f., *N. pulla* subsp. *rudbarensis* Rech.f. and *N. pulla* subsp. *scabrisquamata* A.Baytop are referable to *N. persica*. Plants from dry areas tend to be smaller and densely glandular-pubescent. *N. pulla* is a C. & E. European species and *N. monticola* is endemic of the mountains of Paphlagonia, C. & N. Anatolia.

C. and E. Anatolia, eastwards through C. and N. Iran to Pakistan and Afghanistan.

2. **Nonea macrantha** (*H.Riedl*) *A.Baytop*, Notes Roy. Bot. Gard. Edinburgh 35(3): 299 (1977). Type [Iraq] *Nonnea armena* Boiss. & Huet, Iraq, Kurdistan, in montis Helgurd, (ditionis Riwanduz), *Bornmüller* 1629! (W lecto; G, K, iso.). Baytop in Fl. Turkey [P. H. Davis] 6: 411 (1978); Selvi & Bigazzi in Pl. Syst. Evol. 227(1–2): 15, f. 7f–h (2001).

N. pulla (L.) DC. subsp. *macrantha* H.Riedl, Oesterr. Bot. Z. 110: 531 (1963).

Perennial, ± caespitose. Stems several arising from woody base, erect to ascending, up to 45 cm, yellowish, angular, densely pubescent with sparse longer setae; root woody. Basal leaves mostly brown at anthesis; cauline leaves lanceolate to ovate-lanceolate, 30–100 × 6–25 mm, amplexicaul, apex acute, margins entire. Inflorescence paniculate, shortly branched with densely bracteate cymes. Bracts as leaves, longer than flowers. Calyx tubular, 8–12 mm, divided to about 1/3, teeth triangular, acute; fruiting calyx up to 18 mm. Corolla campanulate; tube purplish up to 8 mm, minutely pubescent inside; limb dark-purple, 5–6 mm, with ± erect, rounded lobes. Faucal scales densely long-hairy. Annulus glabrous. Anthers over-lapping the scales. Style slightly exserted from throat; stigma bilobed. Nutlets dark brown, 4 mm, obliquely ovoid, papillose, reticulate, with 12–18 vertical costae terminating into short teeth. Fig. 96, 1–4.

HAB. Mountain sides in *Astragalus* zone in N. Iraq, amongst damp rocks by lake; alt. 3000–3880 m; fl. Jul.-Aug.
DISTR. High mountain range in northern Iraq. **MRO**: mt. Halgord, *Bornmüller* 1629! (type); mt. Halgord *Guest & Ludlow-Hewitt* 2861! 2863A! Qandil range, *Rawi & Serhang* 24063!

S.E. Turkey, N.W. Iran.

3. **Nonea caspica** (*Willd.*) *G.Don*, Gen. Hist. 4: 336 (1838); Riedl in Fl. Iranica [K. H. Rechinger] 48: 250 (1967); Baytop in Fl. Turkey [P. H. Davis] 6: 409 (1978); Y. Nasir in

Fig. 96. **Nonea macrantha**. 1, habit × 0.7; 2, flower with bract × 3; 3, open corolla with pistil and stamens × 3.5; 4, mericarp in lateral view x7. Reproduced from Selvi & Bigazzi, Plant Syst. & Evol. 227(1–2): f. 8 (2001).

Fig. 97. **Nonea caspica**. 1, habit × ½; 2, corolla, opened × 4; 3, nutlet, × 6. Reproduced with permission from Fl. Pakistan 191: f. 19, D–F (1989). Drawn by S. Hameed. © National Herbarium, Pakistan Agriculture Research Council & University of Karachi, Pakistan.

Fl. Pakistan [Ali & Nasir] 191: 74 (1989); Breckle, Hedge & Rafiqpoor, Vasc. Pl. Afghan. checklist: 146 (2013) as subsp. *caspica*; Khatamsaz, Flora Iran 39: 208 (2002).

Onosma caspica Willd., Sp. Pl., ed. 4 [Willdenow] 1(2): 775 (1798).
Anchusa picta M.Bieb., Fl. Taur.-Caucas. 1: 127 (1808).
Nonea picta (M.Bieb) Sweet, Hort. Brit. [Sweet] 292 (1826).
N. picta (M.Bieb.) Fisch. & C.A.Mey. in Ind. Sem. Horti Petrop. 2: 43 (1835) nom. superfl.
N. melanocarpa Boiss., Diagn. Pl. Orient. Nov. 1, 11: 96 (1849); Dinsmore in Post, Fl. Syria, Palest. & Sinai ed. 2, 2: 229 (1933); Rawi in Dep. Agr. Tech. Bull. 14: 138 (1964); Baytop in Fl. Turkey [P. H. Davis] 6: 408 (1978); Feinbrun-Dothan, Fl. Palaest. 3: 341 (1978); Taifour & El-Oqlah, Pl. Jordan Annot. Checkl.: 52 (2017).
N. diffusa Boiss. & Buhse in Mém. Soc. Nat. Mosc. 12: 152 (1860).
N. melanocarpa Boiss. f. *macra* Boiss. ex Lipsky in Acta Hort. Petrop 26: 471 (1910).
N. caspica (Willd.) G.Don subsp. *melanocarpa* (Boiss.) H.Riedl in Oesterr. Bot. Z. 110: 251 (1963); Rechinger, Fl. Lowland Iraq: 505 (1964); Mouterde, Nouv. Fl. Liban et Syrie 3: 84 (1983); Y. Nasir in Fl. Pakistan 191: 77 (1989); Breckle, Hedge & Rafiqpoor, Vasc. Pl. Afghan. checklist: 146 (2013).

Annual, hispid-setose with long and short hairs, longer setae arising from a tuberculate base. Stems 6–30 cm, branching from base. Leaves oblong- to linear-lanceolate, 25–80 × 5–20 mm, acute to obtuse, base tapering, margins entire to denticulate; cauline leaves sessile, decurrent. Calyx 5–6 mm, divided to ½, lobes narrowly triangular, acuminate; fruiting calyx 7–9 mm, accrescent. Corolla deep orange at first (rarely white), becoming violet, brownish or purple, narrowly funnel-shaped, 9–12 mm; throat yellowish, limb slightly zygomorphic. Faucal scales ± inflated, hairy on upper margins. Annulus densely hairy. Anthers 1.5 mm, below faucal scales. Stigma obscurely bilobed. Nutlets reniform, blackish, ± reticulate with whitish dots, glabrous or puberulent at base, beak lateral, basal ring not thickened, costae ending in 11–14 teeth. Fig. 97, 1–3.

HAB. Dry steppe, cultivated fields, irrigated cornfields, dry foothills, eroded silt & pebbles, clay and gypseous soil, disturbed gravel; common; alt. 100–305(–3100) m; fl. Mar.-Jun.
DISTR. Moist and dry foothill zone of N. and E. Iraq. **MAM**: Sarsang to Amadiya, *Karim, Hamid & Jasim* 41004!; 8 km S. of Dohuk, *Botany Staff* 43368! **MRO**: Ser Kurawa, *Gillett* 9750! **FUJ**: 6 km S. of Qal'a Sharqat, *Rawi & Hamada* 33608!; Hadhr, *Botany Staff* 41602!; Qaiyara, *Bayliss* M339!; 36! **FNI**: Tal Kaif, *Kaisi & Hamad* 49273! **FAR**: bank of r. Zab 6 km below Gau Gossik, *Rawi & Gillett* 10518! **FKI**: Kirkuk

on road to Baba Gurgur oil wells, *Guest* 1365! 25 km S. of Kirkuk, *Gillett & Rawi* 10338! **FKI/FPF**: Jabal Hamrin, *Haines* W. 873! **FPF**: Jabal al-Muwaila, nr Kuwait, 70 km N. of Amara, *Guest, Rawi & Rechinger* 17638!; middle range of Jabal Hamrin nr Qasrabat (Kizil Robat), *Haines* 1318A!; Pilkana, 10 km E. of Khanaqin, *Rawi* 12736!; Mandali, *Guest* 883! *Rogers* 883! **DLJ**: 7 km S. of Baiji, *Barkley & Palmatier* 466B!; Opposite Rawa, *Rawi & Gillett* 7065! **DWD**. 62 km Husaiba to Akashat, *Al-Khayat & K. Hamad* 51587!

A polymorphic species showing variations in size of plant and size of leaves in Iraq. Subsp. *melanocarpa* (Boiss.) Riedl and subsp. *zygomorpha* H.Riedl have been separated mainly on the colour of the corolla. This character cannot be relied upon as the corolla is deep orange at first and becomes, violet, brownish or purple after anthesis. This and several intermediate vegetative characters make it difficult to clearly separate these taxa.

Palestine, Turkey, Caucasia, Iran, Afghanistan, Pakistan.

4. **Nonea palmyrensis** (*Post*) *Sam.*, Svensk Bot. Tidskr. 29: 379 (1935); Rechinger, Fl. Lowland Iraq: 505 (1964); Mouterde, Nouv. Fl. Liban et Syrie 3: 83 (1983).

Anchusa palmyrensis Post, Pl. Postiana 4: 9 (1892); Dinsmore in Post, Fl. Syria, Palest. & Sinai ed. 2, 2: 228 (1933).

Annual, 5–20 cm, hispid and glandular-viscid with sparse setae along leaf margins. Stems branched from base, prostrate to erect. Basal leaves mostly brown at or before anthesis; lower and cauline leaves linear-lanceolate, 20–50 × 3–6 mm, lowest petiolate, cauline sessile, apex acute, margins repand-dentate. Inflorescence of leafy cymes. Flowers sessile to shortly pedicellate, pedicels elongating to 4–5 mm in fruit. Bracts lanceolate, acute, longer than calyx. Calyx 10.5 mm at anthesis, 1/3 to ½ divided, lobes triangular, glandular-pubescent, pubescent inside; fruiting calyx inflated to 9 mm wide, patent to deflexed. Corolla pink or pale yellow, becoming deeper-coloured and with violet stripes towards limb, 14–16 mm; tube exceeding calyx; limb spreading with age, lobes 1.5–3 mm. Faucal scales hairy. Annulus glabrous. Stamens inserted below scales; anthers overlapping faucal scales. Stigma globose. Nutlets transversely ovoid, blackish to dark brown, smooth, slightly rugose towards base, with dorsal keel, basal ring thick with 10–12 prominent teeth. Fig. 98, 1–5.

HAB. Sandy hilltop, sandy clay soil, silty soil in depressions; alt. 280–560 m; fl. Mar.-Apr.
DISTR. Scattered in the dry foothills and desert zones of Iraq. **FUJ**: Between Qal'a Sharqat and Hadhr, *Omar & Hamid* 36316! **DLJ**: Buga (Tal Baqqa) police station, Jazeira desert, *Barkley* 7884! Ain Dibs to Baiji, *Rawi & Gillett* 7166! Shibaichan, 115 km S.W. of Sinjar, *Chakravarty, Rawi, Khatib & Alizzi* 32012!; Shibaichan rd., 90 km N. of Rawah, *Chakravarty, Rawi, Khatib & Alizzi* 31971! 50 km from Haditha to Baiji, *Omar, Noori, Hamid, Muhamad & Jasim* 41485! **DWD**: Rutba wells, *Dickson* s.n. (Apr. 1933)!; 19 km N.E. of Rutba, *Rawi* 31182!; 30 km N. of Rutba, Rai & Khatib 32216! 45 km E. of Rutba towards Ramadi, *Rechinger* 12575!; Ubaila, 12 km N. of Rutba, *Chakravarty, Rawi, Khatib & Alizzi* 31445! **LEA**: Shahraban (Muqdadiya), *Rechinger* 13373!; Tib, 70 km N. of Amara, *Rechinger* 14184!

Cecchi et al. (2009) who have collected material from type locality in Syria, describe the flower colour as tinged dark purple in upper half of throat; corolla-lobes white at the apex with ten dark-purple and ten bright green oblong spots, regularly alternating at base near throat.

Syria.

5. **Nonea echioides** (*L.*) *Roem. & Schult.*, Syst. Veg., ed. 15 bis [Roemer & Schultes] 4: 71 (1819).

Lycopis echioides L., Sp. Pl. ed. 2. 1: 199 (1762).
Anchusa ventricosa Sm., Fl. Graec. Prodr. 1(1): 117 (1806).
Anchusa echioides (L.) M.Bieb., Fl. Taur.-Caucas. 1: 123 (1808).
Nonea ventricosa (Sm.) Griseb., Spic. Fl. Rumel. 2(4): 98 (1844); Boissier, Fl. Orient. 6: 169 (1875); Dinsmore in Post, Fl. Syria, Palest. & Sinai ed. 2, 2: 229 (1933); Rawi in Dep. Agr. Tech. Bull. 14: 139 (1964); Rechinger, Fl. Lowland Iraq: 505 (1964); Baytop in Fl. Turkey [P. H. Davis] 6: 413 (1978); Feinbrun-Dothan, Fl. Palaest. 3: 341 (1978); Mouterde, Nouv. Fl. Liban et Syrie 3: 84 (1983); El-Hadidi & Boulos in Fl. Egypt 2: 301 (2000); Taifour & El-Oqlah, Pl. Jordan Annot. Checkl.: 52 (2017).

Fig. 98. **Nonea palmyrensis. 1,** habit; 2, flower; 3, open corolla; 4, limb of corolla seen from above; 5, mericarpid in lateral view. Scale bars: (A)=30 mm, (B)–(D)=5 mm, (E)=2 mm. Original drawing by A. Maury based on collection HB 07.14 (FI). Reproduced here with permission.

Annual, 10–35 cm, appressed hairy with sparse tubercle-based setae Stem branching from base, prostrate to ascending. Leaves oblong to oblanceolate, 18–50 × 5–10 mm, lower sessile with base attenuate, upper ± amplexicaul, apex obtuse, margins entire. Inflorescence of leafy cymoid racemes. Flowers shortly pedicellate to sessile. Bracts lanceolate, acute, as long as calyx. Calyx 4–5 mm at anthesis, ¼ divided, lobes triangular; fruiting calyx inflated and ventricose. Corolla white; tube exceeding calyx; limb-lobes ± 1 mm. Faucal scales reduced to tufts of hairs. Annulus hairy. Stamens inserted below throat. Stigma bilobed. Nutlets reniform, blackish to dark brown, slightly rugose, keel at back not distinct, beak absent.

HAB. In silty and sandy loamy soils, farms; alt. 280–560 m; fl. Apr.-May.
DISTR. Foothill zone of northern Iraq. **MSU**: Darbandikhan, *Wheeler-Haines* W. 1939! **FAR**: 25 km N.W. of Arbil (Erbil), *Barkley & Brahim* 89! **FUJ**: Mosul Liwa, Nenawa farm, *Hikmat Abbas Al-Ani* 9765! Chesney's Expedition to the Euphrates, 1837, *Chesney* 209!

Similar and closely related to *N. palmyrensis*, but distinguishable by its nutlets that are reniform and lack the prominent teeth of nutlets of *N. palmyrensis*.

S. Europe, Cyprus, East Aegean Is., Turkey, Syria, Jordan, Palestine, Sinai.

18. **SYMPHYTUM** L.

Sp. Pl.: 136 (1753); Gen. Pl. ed. 5, 66 (1754)
Weigend & al. in Kubitzki, Fam. Gen. Vasc. Pl. 14: 82 (2016)

Shahina A. Ghazanfar

Perennial herbs, hispid. Leaves basal and cauline; basal leaves long petiolate, cauline sessile or decurrent. Inflorescence of bracteate thyrsoids; flowers with long pedicels. Calyx cylindric to campanulate, nearly to base, accrescent in fruit. Corolla cylindrical to campanulate, with a short tube and spreading lobes; faucal scales present. Stamens included. Style filiform, exserted; stigma entire. Nutlets ovoid, obliquely keeled, attachment scar surrounded by a toothed ring.

A genus of about 35 species in Europe and W. Asia; 2 species in Iraq.

Symphytum, from Gr. συμφυτομ, *sumfutom*, symbiont (growing together).

Al-Zubaidy, A.M.A. & Tobakari, S.R.A. (2018). A comparative systematic study of the genus *Symphytum* L. (Boraginaceae) with new first record of the species *Symphytum tuberosum* L. from Iraq. Plant Archives 18(2): 2068–2076 e-ISSN:2581-6063 (online), ISSN:0972-5210.
Kurtto, A. (1982). Taxonomy of the *Symphytum asperum* aggregate (Boraginaceae), especially in Turkey. *Ann. Bot. Fennici,* 177–192.
Hacıoğlu, B.T. & Erik, S. (2011). Phylogeny of *Symphytum* L. (Boraginaceae) with special emphasis on Turkish species. *African J. Biotech.* 10(69), 15483–15493.

Non-rhizomatous perennial; calyx 6–11 mm, accrescent to c. 13 mm in
    fruit, divided to ¼–1/3 of its length . . . . 1. *S. kurdicum*
Rhizomatous perennial with a creeping rhizome and tuberous sections;
    calyx c. 5 mm, accrescent to c. 7.5 mm in fruit, divided almost
    to base . . . . 2. *S. tuberosum*

1. **Symphytum kurdicum** *Boiss. & Hausskn.* ex Boiss., Pl. Orient. Nov. Dec. ii. 5 (1875); Boissier, Fl. Orient. 4: 174 (1879); Rawi, in Dep. Agr. Tech. Bull. 14: 140 (1964); Riedl in Fl. Iranica [K. H. Rechinger] 48: 241 (1967); Wickens in Fl. Turkey [P. H. Davis] 6: 378 (1978); Khatamsaz, Fl. Iran 39: 219 (2002).

Robust perennial, with membranous hispid leaves, 60–100 cm tall. Basal leaves ovate, 6–20 × 2–12 cm, long-petiolate to 8 cm, base cordate; cauline leaves smaller towards apex, repand-denticulate, with fine hairs and tubercle-based setae. Calyx 6–11 mm, lengthing to c. 13 mm in fruit, divided to ¼–1/3 of its length; teeth triangular acute. Corolla white to yellowish, campanulate, 16–20 mm; scales c. 6 mm, broadly subulate, obtuse, equalling stamens. Nutlets narrowed above base, tuberculate. Fig. 99, 1–6.

HAB.: *Quercus* woodland, thickets in leaf litter, under *Quercus* scrub, hillside on clayey soil; alt. 850–1800 m; fl. & fr. Jun.-Aug.
DISTR.: Mainly in the lower forest zone of northern Iraq. **MAM**: Baidh to Kani Masi, *Karim, Dabbagh & Hamad* 44853!; Bakarma, *Rawi* 9419G!; Jabal Khantur, *Rawi* 8547!; Jabal Bekher, *Rawi* 9419P!; Zawita gorge, *Emberger, Guest, Long, Schwann & Serkahia* 15356! **MRO**: Sarsang, *Alkas* 18692!; Sarsang, N. side of

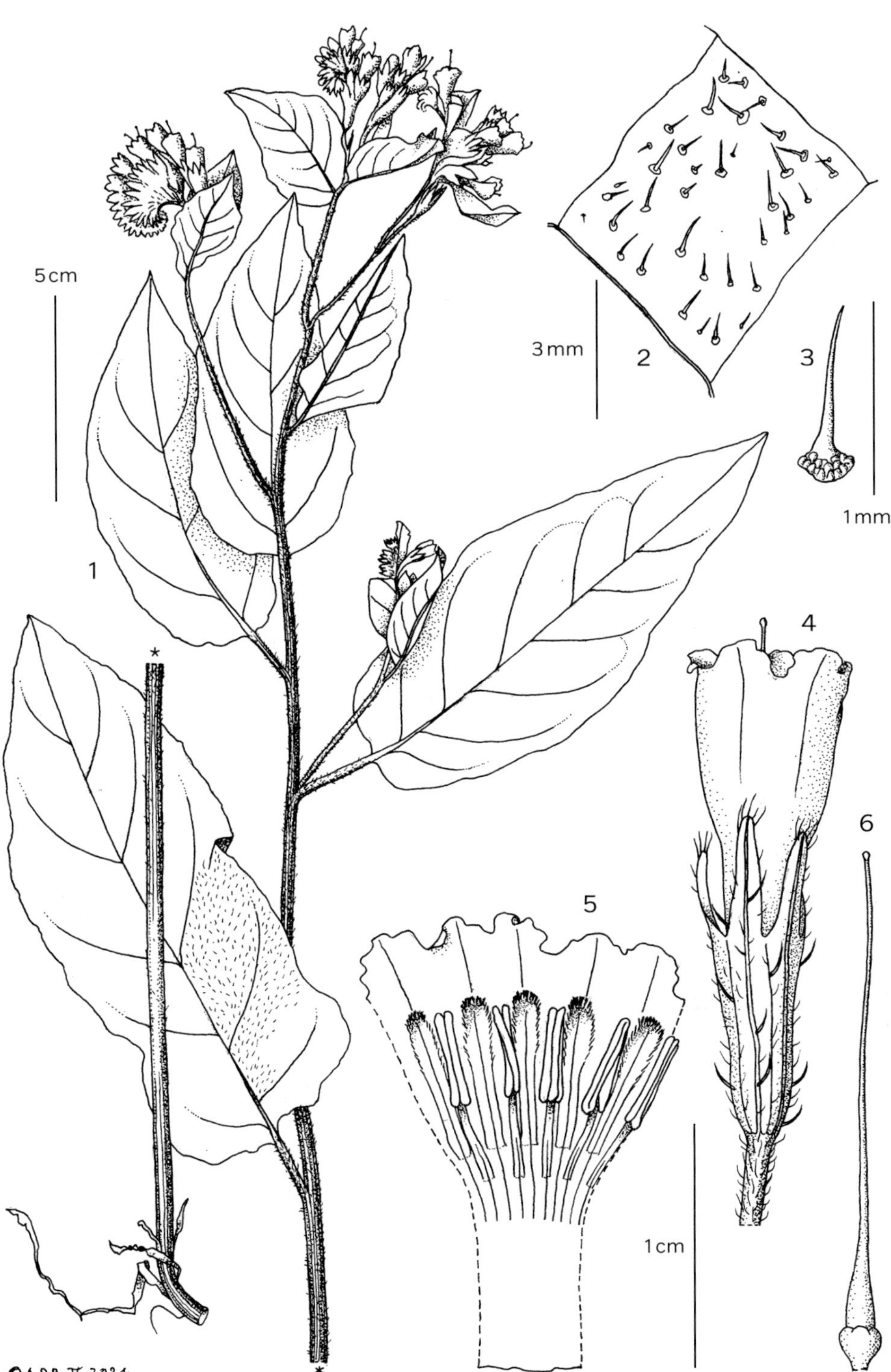

Fig. 99. **Symphytum kurdicum**. 1, habit; 2, adaxial leaf indumentum; 3, indumentum detail; 4, flower, side view; 5, corolla dissected; 6, ovary style and stigma. Drawn by ©A. P. Brown 2024.

Gara mountains, *Anders* 2363 (W!); Gali Warta, 30 km N.W. by N. of Rania, *Rawi, Nuri & Kass* 28734!; Shaqlawa, *Haines* W. 379!; Shaqlawa, Sefin Dagh, *O. Polunin* 5028!; Kuh-e Sefin, above Shaqlawa, *Bornmüller* 1633!; Barsarini gorge, *Guest* 2038!; Pushtashan, on lower slopes of Mt. Qandil, *Rawi & Serhang* 26570!; Rowanduz, *Nábělek* 581, 582 (SAV!). **MSU**: Between Qarachitan and Zewiya, *Gillett* 7766!; Jasana, *Botany Staff* 43138!; Penjwin, *Rawi* 8812! 9419K!; Kopi Qara Dagh, *Poore* 426! 645!; Azmir, *Kaisi & Hamad* 49319!; Kani Watman 20 km S.W. of Dukan, *Kaisi & Hamad* 49337!; above Suwara Tuka, *Rechinger* 11962 (W!); Jabal Khantur, *Rechinger* 12051 (W!); above Sirsank [Sarsang], between Dohuk and Amadiya, *Rechinger* 11652 (W!).

Turkey, Iran.

2. **Symphytum tuberosum** *L.*, Sp. Pl. 1: 136 (1753); Wickens in Fl. Turkey [P. H. Davis] 6: 385 (1978).

Rhizomatous perennial herb; rhizome creeping, with small tuberous sections; stems to 32 cm, pilose to hispid. Basal leaves elliptic to elliptic-ovate, 6–8 × 4 cm, long-petiolate to 4.5 cm, base cuneate; cauline leaves smaller than basal leaves, elliptic to lanceolate, margins entire, puberulent; upper leaves sessile. Calyx 5.5–5 mm, lengthing to c. 7.5 mm in fruit, divided almost to base; lobes linear-lanceolate, obtuse. Corolla yellowish, campanulate, c. 14 mm; scales triangular-subulate, obtuse, exceeding stamens. Nutlets tuberculate. (Description from Fl. Turkey).

HAB. N. Iraq, in woodland, on clayey and rocky soil; alt. 695–1950 m; fl. not recorded.
DISTR. In forest zone of northern Iraq. Recorded from (**MAM**) Piramagroon and (**MRO**) Rawanduz (no material seen).

W. & Central Europe, Turkey.

19. **TRICHODESMA** R.Br., nom. cons.

Prodr. Fl. Nov. Holl. 496 (1810)
Weigend & al. in Kubitzki, Fam. Gen. Vasc. Pl. 14: 83 (2016)

J. R. Edmondson & Shahina A. Ghazanfar

Perennial herbs and shrubs, hispid to sericeous adpressed indumentum. Leaves all cauline, sessile. Inflorescence in cymes; bracteate. Pedicels deflexed. Calyx deeply divided into 5 lobes narrowed towards tip, ribbed or cordate-winged near base, accrescent in fruit. Corolla funnel-shaped or salver-shaped, tube shorted than limb. Faucal scales absent (rarely present). Stamens 5–6; filaments short; anthers with often twisted appendages at tips. Style filiform, stigma capitate. Nutlets 1–4, often with a narrow dentate wing, smooth or rugose, often hairy.

40 to 50 species from Africa to S. and S.E. Asia and Australia; a single species in Iraq.

Trichodesma, from Gr. τριχ-, hairy, and δεσμη, *desme*, bundle.

1. **Trichodesma incanum** (*Bunge*), *A.DC.*, Prodr. [A.P. de Candolle] 10: 174 (1846); Rawi, in Dep. Agr. Tech. Bull. 14: 140 (1964); Riedl in Fl. Iranica [K. H. Rechinger] 48: 244 (1967); Edmondson in Fl. Turkey [P. H. Davis] 6: 435 (1978); Y. Nasir in Fl. Pakistan [Ali & Y. Nasir] 191: 90 (1989);

Friedrichsthalia incana Bunge, Del. Sem. Hort. Dorp. 7 (1843).
Trichodesma molle A.DC., Prodr. [A.P. de Candolle] 10: 174 (1846).
T. strictum Aitch. & Hems., J. Linn. Soc., Bot. 19: 178 (1882).
T. molle var. *virescens* Bornm., Beih. Bot. Centrbl. 20B: 195 (1906).

Perennial herb, clothed with short greyish hairs; stems at length glabrescent, erect or ascendent to procumbent, much-branched, to 90 cm. Leaves sessile, ovate to elliptic or oblong, apex mucronate, tapering or rounded at base, 3–8 × 1.5–3 cm, both sides silky-grey pubescent, ± velutinous on veins of lower surface. Inflorescence lax, many-flowered, with terminal scorpioid cymes. Calyx ovate-campanulate, divided to 2/3, 14–20 mm diam., slightly accrescent in fruit, lobes ovate, tapering to a mucro that is recurved in fruit. Corolla blue, with white throat, tube 7–8mm, limb 20–25 mm diam., lobes triangular-ovate, narrowed at apex. Anthers 5, yellow, connivent to form an exserted cone, spirally-twisted appendages

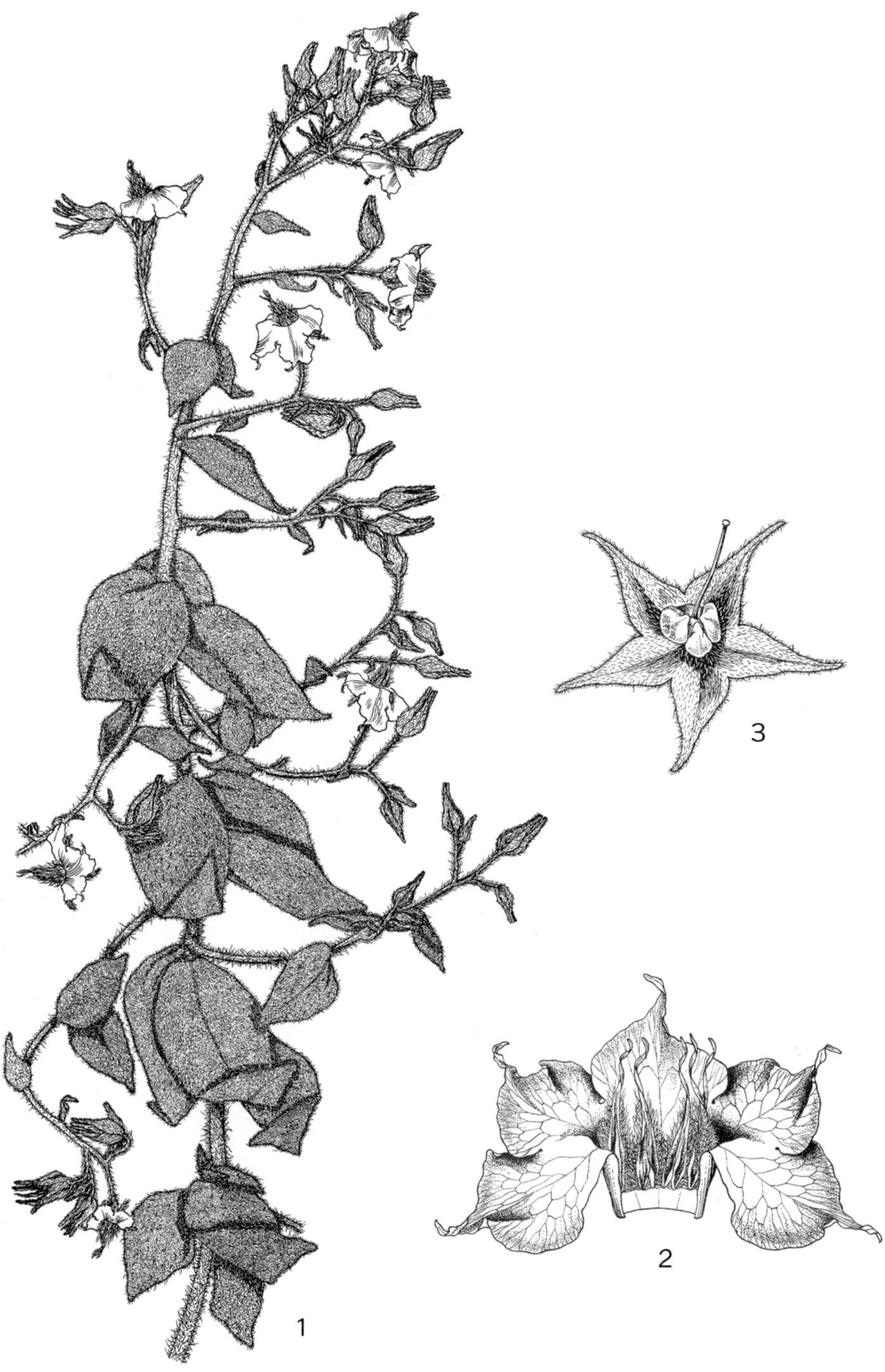

Fig. 100. **Trichodesma incanum.** 1, habit portion of shoot × ½; 2, corolla, opened × 1½; 3, fruiting calyx × 1. Reproduced with permission from Fl. Pakistan 191: f. 23, F–G & f. 24, A (1989). Drawn by S. Hameed. © National Herbarium, Pakistan Agriculture Research Council & University of Karachi, Pakistan.

equalling length of thecae. Nutlets ovoid, 5–6 mm, greyish-brown, emarginate or with a narrow margin, rugose externally, ± minutely pubescent. Fig. 100, 1–3.

HAB. On screes, *Quercus* forest on mountain slope by stream, metamorphic rocks, dry bare slopes, by stream; alt. 1100–2030 m; fl. Jun.

DISTR. Mainly in the thorn-cushion zone of the mountains of N.E. Iraq. **MRO**: N. of Pishtashan, *Rawi & Serhang* 18302!; above Pishtashan, *Rechinger* 11191; Kani Rash, *Rawi* 26667! 26695!; Halgord Dagh, *Gillett* 9624! **MSU**: nr Sulaimaniya, *Haines* W. 1068!; Qara Dagh, *Makki* 490!; Azmir, *Karim* 39341! Pira Magrun, *Thesiger* 1112, *Hausskn.* s.n.!

Iran, Afghanistan, Pakistan, Turkmenistan.

20. **CACCINIA** Savi

Cose Bot. 1. t.1 (1832)
Weigend & al. in Kubitzki, Fam. Gen. Vasc. Pl. 14: 83 (2016)

C.C. Townsend
Revised by Shahina A. Ghazanfar

Strigose biennial and perennial herbs. Leaves all cauline, obovate-lanceolate to oblong-lanceolate leaves, often glaucous. Inflorescence thyrsoid; bracteate. Flowers pedicellate. Calyx divided to more than 1/2, accrescent in fruit to a stellate form. Corolla salver-shaped, tube longer than calyx, throat broadening abruptly, lobes spreading, blue or purple, somewhat unequal, linear to oblong. Faucal scales arcuate, closing throat. Stamens inserted near top of tube; filaments unequal; anthers exserted, lateral and anterior smaller, subsessile, posterior long-exserted. Style filiform, long-exserted, forming a cup in fruit which encloses the nutlets. Nutlets 2(–1) by reduction, flattened, ovate or orbicular.

About six species in S.W. and C. Asia; a single species in Iraq.

Caccinia possibly commemorates a musical family, Caccini, from Florence (Firenze).

1. **Caccinia macranthera** (*Banks & Sol.*) *Brand* in Engler, Pflanzenr. [Engler] Borag. Cynogloss. 78, 252: 90 (1921).

var. **crassifolia** (*Vent.*) *Brand* in Engler, Pflanzenr. [Engler] 78, 252: 92 (1921); Rawi in Dep. Agr. Iraq Tech. Bull. 14: 135 (1964); Y. Nasir in Fl. Pakistan 191: 86 (1989); Riedl in Fl. Iranica [K. H. Rechinger] 48: 227 (1967); Khatamaz, Fl. Iran 39: 417 (2002).

Borago macranthera Banks & Sol. in Russell, Aleppo ed. 2, 2: 246 (1794).
Borago crassifolia Vent., Descr. Pl. Nouv. t. 100 (1803).
Caccinia rauwolfii C.Koch in Linnaea 17: 303 (1843); A.DC., Prodr. [A.P. de Candolle] 10: 167 (1846).
C. crassifolia (Vent.) C.Koch in Linnaea 22: 647 (1849).

Perennial herb (in ours), bushy, 20–50 cm. Stem branched from base, glabrous. Basal leaves obovate to elliptic, 4.5–10.5 × 2–4 mm, petiolate, obtuse, leathery in texture, margins smooth; cauline leaves lanceolate to oblong, similar in size, sessile, obtuse to subacute, margins ± aculeolate. Inflorescence branched in upper half of stem. Calyx 8–12 mm in flower, to 20 mm diameter in fruit. Corolla blue (occasionally pinkish); tube 8–15 mm; lobes 6–9mm. Nutlets 5–11 mm, depressed-ovoid, puberulous, margins dentate. Fig. 101, 1–5.

HAB. Soft soil in floor of deep gorge, gravelly and sandy clay soil, clay soil of wheat field, forest plantation, sandy 'haswa' desert; alt. 50–400 m.; fl. Mar.-Apr.
DISTR. Widespread in foothills and desert areas of Iraq. **FNI**: Eski Kallek, *Botany Staff* 43288! **FPF**: 20 km S. of Khanaqin, *Haines* W. 1906!; 10 km from Sa'adiya to Khanaqin, *Kaisi* 48752!; Sa'adiya water project, *Kaisi* 44200! 44203!; Jabal Hamrin, nr Table Mountain (Mansur), *Guest* 1896!; Laqlaq, *Kaisi, Thamer & Salah* 51283! **LEA**: Fakki-Laqlaq valley, by Iranian frontier, *Rawi* 25825! **DLJ**: Tayarat, 60 km N.W. of Rawa, *Chakravarty, Rawi, Khatib & Alizzi* 31954! **DWD**: 35 km S. of Rutba, *Chakravarty, Rawi, Khatib & Alizzi* 31432!; Wadi Tibul, 100 km N. of Nukhaib, *Rawi* 30497!; 50 km S. of Rutba, *Chakravarty, Rawi, Khatib & Alizzi* 32833!; 8 km N. of Rutba, *Rawi & Khatib* 32377!; W. of Falluja, *Guest* 15190!

HEMHEM (Ir.-Wadi al-Tibal), *Rawi* 30497.

Turkey, Iran, Afghanistan, Pakistan, Tajikistan. The distribution of the type species is similar, with the addition of Syria.

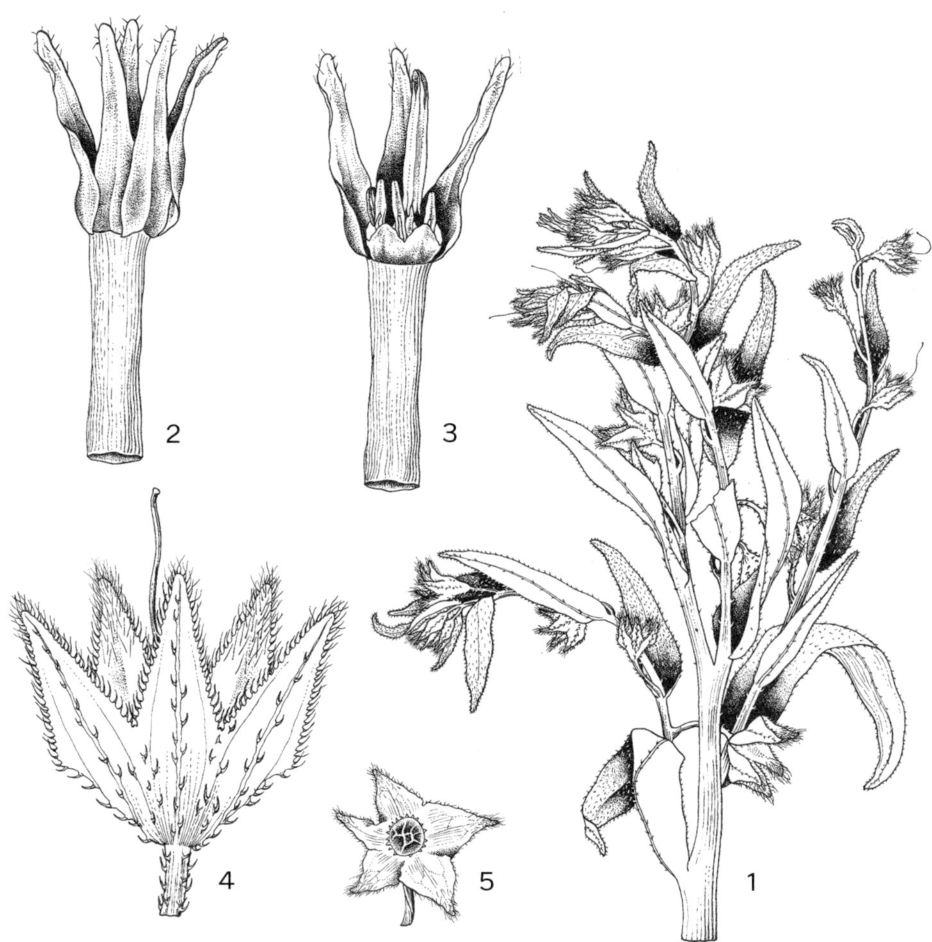

Fig. 101. **Caccinia macranthera**. 1, habit portion of shoot × ½; 2, flower × 3; 3, corolla, opened × 3; 4, fruiting calyx × 3; 5, nutlet × ½. Reproduced with permission from Fl. Pakistan 191: f. 21, E-I (1989). Drawn by S. Hameed. © National Herbarium, Pakistan Agriculture Research Council & University of Karachi, Pakistan.

21. **MYOSOTIS** L.

Sp. Pl.: 131 (1753); Gen Pl. ed. 5, 63 (1754)
Weigend & al. in Kubitzki, Fam. Gen. Vasc. Pl. 14: 86 (2016)

C.C. Townsend
Revised by Shahina A. Ghazanfar

Annual to perennial herbs; pubescent. Leaves basal and cauline, entire, elliptic, oblong or lanceolate. Inflorescence racemose or scorpioid cymes, ebracteate or bracteate only at base. Flowers pedicellate; pedicel often recurved in fruit. Calyx campanulate, 1/3-½-lobed, accrescent in fruit and enclosing nutlets. Corolla blue, more rarely white or pale pink, with short tube and rotate 5-lobed limb. Faucal scales bright yellow, largely reduced. Stamens inserted in corolla tube, included; anthers with small apical appendages. Style included. Nutlets (1–)4,, small, erect, tetrahedral, smooth, with weak ventral keel and basal or sub-basal areole.

About 80–100 species in Africa, Asia, Australia, Europe and North America; 6 species in Iraq.

Myosotis, from Gr. μυς, *mus*, mouse and ωτις, otis, genetive of ear.

Stroh, G. (1941). Die Gattung *Myosotis* L. Versuch einer system- atischen Ubersicht uber die Arten. Beih. Bot. Centralbl. 61: 317–345.

Winkworth, R.C., Grau, J., Robertson, A.W. & Lockhart, P.J. (2002). The origins and evolution of the genus *Myosotis* L. (Boraginaceae). Mol. Phylogen. Evol. 24(2): 180–193.

1. Calyx with appressed hairs, sometimes hooked hairs present 2
   Calyx with hooked hairs always present at least in lower half. 3
2. Dwarf perennial; corolla limb to 8 mm in diameter; nutlets broadly
   ovoid to ellipsoid, with large elliptic attachment scar and lateral
   grooves, without appendages . . . . 1. *M. alpestris*
   Dainty annual or biennial, hairs adpressed; corolla limb to 5 mm in
   diameter; nutlets ovoid, obtuse, glossy, truncate at base, attachment
   scar with appendage. . . . . 6. *M. caespitosa*
3. Pedicels erecto-patent in fruit. 4
   Pedicels usually deflexed, sometimes patent 5
4. Calyx lobes with patent and deflexed hooked hairs; calyx to 5 mm in
   fruit. . . . . 2. *M. lithospermifolia*
   Calyx lobes with hooked hairs only; calyx to 2.5 mm in fruit. . . . . 3. *M. kurdica*
5. Pedicel deflexed at maturity; calyx lobes divided to c. 1/3, elongating
   to 3.5 mm in fruit, connivent and accrescent, with strongly deflexed
   hooked hairs on segments and straight hairs at base . . . . 4. *M. refracta*
   Pedicels patent or recurved; calyx lobes divided to 1/2, elongating to 5
   mm in fruit, with many hooked deflexed hairs at base . . . . 5. *M. ramosissima*

1. **Myosotis alpestris** *F.W.Schmidt*, Fl. Boëm. Cent. 3: 26, t. 281 (1794); Rawi in Dep. Agr. Iraq Tech. Bull. 14: 138 (1964); Khatamaz, Fl. Iran 39: 255 (2002).

M. suaveolens Waldst. & Kit. ex Willd., Enum. Pl. [Willdenow] 176 (1809).
M. sylvatica Hoffm. var. *alpestris* (F.W.Schmidt) A.DC., Prodr. [A.P. de Candolle] 10: 108 (1846).
M. olympica Boiss. var. *laxa* Boiss., Fl. Orient. 4: 237 (1875).
M. sylvatica Hoffm. subsp. *alpestris* (F.W.Schmidt) Gams in Hegi, Ill. Fl. Mittel-Eur. 5(3): 2168 (1927).
M. sylvatica Hoffm. var. *suaveolens* (Waldst. & Kit. ex Willd.) Cincović & Kojić, Fl. SR Srbije 6: 48 (1974).
M. sylvatica Hoffm. var. *lithospermifolia* (Hornem.) Cincovic & Kojić, Fl. Srbije 6: 48 (1974).
M. alpestris F.W.Schmidt subsp. *suaveolens* (Waldst. & Kit. ex Willd.) Strid, Mountain Fl. Greece [Strid] 2: 55 (1991).

Dwarf perennial herbs with several rosettes. Stems 5–24 cm, hairy or glabrescent below, patent-hairy or adpressed-hairy in upper part. Basal leaves ± petiolate, ovate to elliptic, 3–5 × 0.7–1.6 cm, apex acute, base attenuate, margins entire, hairy on both surfaces or glabrescent beneath, hairs discoid at base; cauline leaves smaller, ovate to linear. Inflorescence clustered in flower, ebracteate; pedicels erecto-patent in fruit, densely short-strigose. Calyx to 7 mm in fruit, lobed nearly to base; lobes lanceolate, acuminate, densely hairy, with or without hooked hairs; not deciduous. Corolla blue, rotate, limb to 8 mm in diameter. Nutlets to 1.8 mm, broadly ovoid to ellipsoid, dark brown to black, smooth, ± obtuse, smooth, shiny, with large elliptic attachment scar and lateral grooves (base without appendages).

The more elongate forms of this variable species have been named *M. suaveolens* Willd., but little separates them from the typical dwarf alpine form. It may be an ecotype adapted to wetter places.

HAB. In gravel of late snow patch, rocky ledges, on schist and serpentine; alt. 2100–3660 m; fl. Jul.-Sep. DISTR. In the alpine zone of N.E. Iraq. **MRO**: Halgord Dagh, *Bornmüller* 1622! *Guest* 2833! 2954! 3072!, *Gillett* 9556! 9598! 12343!; Zawita, Sharanish, *Rechinger* 10946; W. side, *Rechinger* 11435!; near Bermasand Lake, *Rawi & Serhang* 24797!; Siah Kuh, *A.V.R. Ludlow-Hewitt* 1513!; Ser Kurawa, *Gillett* 9770!; Sula Khal, *Rawi & Serhang* 24926!; N.E. of Qandil, *Rawi & Serhang* 24416!; E. side of Qandil range, *Rawi & Serhang* 24091! 24508!; mt Qandil, *Thesiger* 11561; Kani Khanjer Khan, N.E. of mt Halgord, *Rawi & Serhang* 24699!; Kodo (Kudu), *Rawi* 9183; mt Qandil, Chiya-i Mandau, *Guest* 2731!;

C. & S. Europe, Caucasus, Turkey, Iran, W. & C. Asia, N. America.

2. **Myosotis lithospermifolia** *Hornem.*, Hort. Bot. Hafn. 1: 173 (1813); Riedl in Fl. Iranica [K. H. Rechinger] 48: 257 (1967); Grau in Fl. Turkey [P. H. Davis] 6: 276 (1978); Musselman, Checkl. Pl. Lebanon & Syria (2011) http://ww2.odu.edu/~lmusselm/plant/lebsyria/Checklist of Lebanon Plants.pdf.

Perennial herb, branched from base, grey green, densely hairy. Basal leaves in a rosette, narrow obovate to elliptic, 6 × 1 cm, apex obtuse, base attenuate, margins entire; cauline leaves narrow obovate to linear, decreasing in size upwards. Inflorescence branched, ebracteate; pedicels to 6 mm, erecto-patent. Calyx to 5 mm in fruit; lobes lanceolate, acuminate, densely hairy, with patent to deflexed hooked setae; persistent. Corolla bright blue, rotate, limb to 6 mm in diameter. Nutlets to 1.8 mm, narrow ovoid, greyish black, with large attachment scar and short lateral grooves.

HAB. Open ground; alt. 650–3000 m; fl. Jun.-Jul.
DISTRIB. Forest zone of N. Iraq. **MRO**: Zawita, Sharanish, *Rechinger* 10946; Mt Qandil, *Thesiger* 11561; Kalak, *Low* 87.

Iran, Lebanon, Syria, North Caucasus, Transcaucasus, Turkey.

3. **Myosotis kurdica** *Riedl*, Oesterr. Bot. Z. 110: 522 (1963). Type: Iraq, Kurdistan, Arbil, Mt Qandil, above Pushtanshan, *Rechinger* 11750 (W, holo.); Riedl in Fl. Iranica [K. H. Rechinger] 48: 263 (1967).

Perennial herb, 10–15 cm, adpressed-hairy. Basal leaves oblong-lanceolate, 35–75 × 8–16 mm, apex obtuse, base attenuate, margins entire; cauline leaves smaller, oblong to lanceolate. Inflorescence racemose scorpioid; pedicels c. 1 mm, in fruit to 4 mm, erecto-patent. Calyx 1.–1.5 mm, elongating to 2–2.5 mm in fruit, with hooked hairs. Corolla bright blue, rotate, 2–2.5 mm. (Description from Fl. Iranica).

HAB. Known only from the type collection; alt. 2000–2200 m; fl. May?
DISTRIB. Alpine, N.E. Iraq. **MRO**: Arbil, Mt Qandil, above Pushtanshan, *Rechinger* 11750 (W, type:);

Endemic.

4. **Myosotis refracta** *Boiss.*, Voy. Bot. Espagne 2: 433 (1841); Rawi in Dep. Agr. Iraq Tech. Bull. 14: 138 (1964); Riedl in Fl. Iranica [K. H. Rechinger] 48: 264 (1967); Grau in Fl. Turkey [P. H. Davis] 6: 272 (1979); Khatamaz, Fl. Iran 39: 264 (2002).

Annual, diminutive plant with simple stem. Leaves mostly basal, narrowly lanceolate to obovate, apex obtuse, base attenuate, lamina bearing hooked hairs on lower surface. Flowers distantly placed below, crowded above. Pedicels usually deflexed at maturity. Calyx lobes divided to c. 1/3, to 3.5 mm in fruit, connivent and accrescent, with strongly deflexed hooked hairs on segments and straight hairs at base. Corolla bright blue, rotate, limb to 1.5 mm in diameter. Nutlets c. 2 × 1 mm, ellipsoid or narrowly ovoid, with small lateral attachment scar and distinct rim, ventrally furrowed along entire length.

HAB. Open slopes, between rocks, shady ground in forest clearing; alt. 1065–1600 m.; fl. May.
DISTRIB. Occasional in upper forest zone of Iraq. **MAM**: Sersang, *Haines* W. 1119! **MRO**: Shaqlawa, *Haines* W. 766! *Gillett* 8085! **MSU**: Malakawa pass, *Rechinger* 4544.

S. Europe, Turkey, Iran, Afghanistan, Pakistan, Turkmenia, C. Asia.

5. **Myosotis ramosissima** *Rochel* ex *Schultes*, Oestr. Fl., ed. 2, 1: 366 (1814). Riedl in Fl. Iranica [K. H. Rechinger] 48: 257 (1967); Grau in Fl. Turkey [P. H. Davis] 6: 270 (1978); Khatamaz, Fl. Iran 39: 261 (2002).

> *M. hispida* D.F.K.Schltdl., Mag. Neuesten Entdeck. Gesammten NaTurkey Ges. Naturf. Freunde Berlin 8: 230 (1818).

Annual, 10–40 cm, stem erect, softly patent-hairy, hairs adpressed above. Leaves c. 4 × 1 cm, oblanceolate to spatulate, apex obtuse, base attenuating into a long petiole, lamina with patent straight hairs. Inflorescence lax, ebracteate; pedicels 3–5 mm in fruit, patent or recurved. Calyx to 5 mm in fruit, divided to 1/2, with many hooked deflexed hairs at base.

Corolla minute, bright blue, limb of corolla to 4 mm in diameter. Nutlets ovoid, c. 1.5 × 1 mm, brown; attachment scar with spongy tissue.

HAB. Sandbank in river, damp situations, shady gardens, under *Populus* trees; alt. 250–950 m; fl. Mar.-May.
DISTRIB. Moist steppe and montane parts of N.E. Iraq. **MAM**: nr Khabur bridge, *Karim, Nuri & Hamad* 44861! **MRO**: Jindian nr Rowanduz, *Guest* 2040!; Kuh-e Sefin at Shaqlawa, Bornmuller 1632! *Rawi & Gillett* 10462! **MSU**: Sulaimaniya, *Graham* s.n. **FAR**: r. Zab, about 5 km S. of Gau Gossik, *Rawi & Gillett* 10522!

Europe, Turkey, N. Africa, C. & S. Russia, Caucasia, N.W. Iran.

6. **Myosotis caespitosa** *C.F.Schultz*, Prodr. Fl. Stargard. Suppl. 1: 11 (1819); Boissier, Fl. Orient. 4: 235 (1875); Rawi in Dep. Agr. Iraq Tech. Bull. 14: 138 (1964); Riedl in Fl. Iranica [K. H. Rechinger] 48: 257 (1967); Y. Nasir in Fl. Pakistan [Ali & Y. Nasir] 191: 108 (1989); Khatamaz, Fl. Iran 39: 248 (2002).

Myosotis sparsifolia Kirschl. ex Mutel, Fl. French. (Mutel) 2: 428, 446 (1835).
M. laxa Lehm. subsp. *caespitosa* (C.F.Schultz) Hyl. ex Nordh., Norsk Fl.: 529 (1940).
M. scorpioides L. subsp. *caespitosa* (C.F.Schultz) Herman in Ill. Fl. Mitteleur. 5, 3: 2164 (1927).

A rather dainty annual or biennial, to 40 cm. Stem branched with branches ascending, bearing antrorse adpressed hairs. Leaves to 8 × 1 cm, with short adpressed setae, glabrescent on lower surface, extending up to lowest flowers; uppermost leaves gradually reducing in size. Lowermost pedicels to 1 cm, patent to slightly deflexed. Calyx to 5 mm in fruit, divided to ½; lobes obtuse, appressed hairy. Corolla bright blue, tube c. 4 mm, limb to 5 mm in diameter. Style shorter than calyx. Nutlets to 1.7 mm, blackish brown, ovoid, obtuse, glossy, truncate at base, attachment scar with appendage. Fig. 102, 1–4.

HAB. Wet open ground by spring, in ditch by cornfield; alt. 850–1830 m; fl. Jun.-Jul.
DISTRIB. Occasional in the forest zone of Iraq.
MAM: Zawita, N. of Zakho, *Rawi* 23611! **MRO**: Saire S. of Ser Kurawa, *Gillett* 9666!; Seri Hasan Beg, nr Rowanduz, *Guest* 3022! **MSU**: Tainal Chai, between Darband-i Basian and Sulaimaniya, *Rawi & Gillett* 11649!

Europe, N. Africa, Asia, N. America.

SPECIES DOUBTFULLY RECORDED

Myosotis sparsiflora *J.C.Mikan ex Pohl*, Bot. Taschenb. Anfanger Wiss. Apothekerkunst 74 (1807).

M. pseudopropinqua M.Pop., Not Syst. Leningrad 14: 307 (1951).

Annual, 10–40 cm, stem erect, branched at base, sparsely hirsute. Basal leaves few; cauline leaves 15–70 × 4–20 mm, oblong-ovate, apex acute to acuminate, the lower with short petiole. Inflorescence lax, ebracteate; pedicels 5–7 mm in fruit, the lower pedicels longer, patent or slightly recurved. Calyx to 2 mm, in fruit to 5 mm, divided to 1/2, sparsely appressed pilose to hooked hairy. Corolla minute; tube included in calyx, bright blue, limb of corolla to 3 mm in diameter. Nutlets ovoid, c. 2 mm, attachment scar with white appendage.

M. sparsifolia is not recorded from Iraq. The two collections: **MAM**: Shaqlawa, 900 m, *Haines* 765 and **MRO**: Sirsang, *Haines* 1114 (cited as *M. sparsifolia*) belong to *M. caespitosa.*

22. **OMPHALODES** Miller

Gard. Dict. Abr. ed. 4., [968] (1754)

Shahina A. Ghazanfar

Perennial herbs; leaves ovate to oblong, petiolate. Flowers in terminal bracteate or ebracteate cymes. Calyx divided to more than 1/2, accrescent in fruit. Corolla blue or white (rarely pink), ± rotate; throat with 5 whitish saccate invaginations. Anthers small, included; filaments inserted at middle of tube. Style included; stigma capitate. Nutlets depressed-globose to discoid, with incurved wing.

A genus with 11 species found in S. & E. Europe to the Caucasus, Iran Afghanistan and Pakistan; a single species in Iraq

Fig. 102. **Myosotis caespitosa**. 1, habit × ½; 2, calyx (opened) × 6; 3, corolla, opened × 10; 4, seed × 20. Reproduced with permission from Fl. Pakistan 191: f. 30, A–D (1989). Drawn by S. Hameed. © National Herbarium, Pakistan Agriculture Research Council & University of Karachi, Pakistan.

Omphalodes, from ομφαλος, *omfalos*, navel or belly-button, with reference to the incurved wings of the nutlet margin, and -*odes*, resembling.

1. **Omphalodes luciliae** *Boiss.*, Diagn. Pl. Orient. ser. 1, 4: 41 (1844) subsp. **kurdica** *Rech.f. & Riedl*, Fl. Iranica [K. H. Rechinger] 48: 97 (1967). Type: Mt Qandil E. of Qala Diza, *Thesiger* 1158! (BM); Rawi, in Dep. Agr. Tech. Bull. 14: 139 (1964); Riedl in Fl. Iranica [K. H. Rechinger] 48: 97 (1967); Edmondson in Fl. Turkey [P. H. Davis] 6: 281 (1978).

Perennial herb. Stem 5–20 cm, ascending. Leaves ovate to oblong, to 4.5 × 3.5 mm, rounded or subacute at apex, narrowed or truncate at base, petiolate. Inflorescence

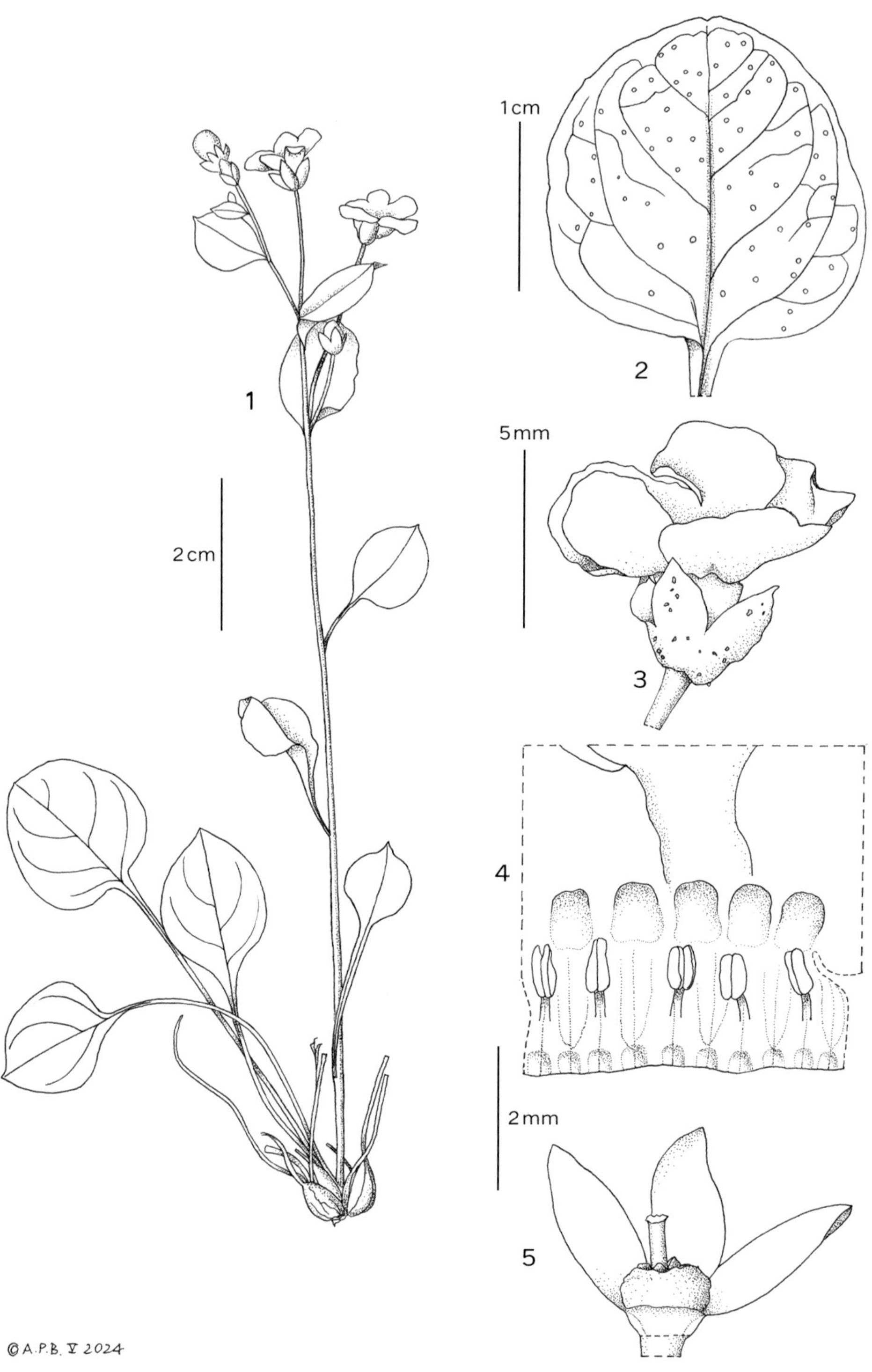

Fig. 103. **Omphalodes luciliae**. 1, habit; 2, large leaf distal part, adaxial surface; 3, flower, side view; 4, inner face of corolla, lower part, showing stamens and saccate invaginations; 5, calyx after removal of two sepals showing ovary, style and stigma in situ. 1-5 from *Hewer* H2093. Drawn by © A.P. Brown 2024.

bracteate, bracts ovate. Calyx lobes narrowly elliptic, subacute. Corolla sky-blue, tube 3–4 mm, limb 4.5–8 mm in diameter. Nutlets to 2.5 × 1.7 mm, triangular-ovate in outline, apex subacute, wing narrow, indistinct, c. 0.4 mm broad. Fig. 103, 1–6.

HAB. Serpentine rocks on mountain slope, rocks and cliffs in thorn-cushion zone; alt. 2900–3390 m; fl. Jul.-Aug.
DISTR. Local in the alpine zone of the Qandil range of N.E. Iraq. **MRO**: Mt Qandil E. of Qala Diza, *Thesiger* 1158! (type of subsp. *kurdica*); Qandil range, N.E. of Rania, *Rawi & Serhang* 18268!; Mt. Perrish, *Rawi & Serhang* 24543!

Subsp. *luciliae* is restricted to Greece and Turkey. One of the most beautiful alpine plants of Iraq; its specific name commemorates Lucile Boissier, wife of Edmond Boissier (author of *Flora Orientalis*).

Turkey (S.E. Anatolia), W. Iran.

23. **ASPERUGO** L.

Sp. Pl. 1: 138 (1753); Gen. Pl. ed. 5: 67 (1754)
Weigend & al. in Kubitzki, Fam. Gen. Vasc. Pl. 14: 88 (2016)

C.C. Townsend
Revised by Shahina A. Ghazanfar

Straggling hispid annual. Stems branched, indumentum retrorsely aculeolate. Leaves basal and cauline, alternate below, paired on branches above. Inflorescence cymose, very lax. Flowers with pedicel elongating and recurved in fruit. Calyx bilobed almost to base, strongly accrescent, 2-lipped in fruit, the lips 7- and 8-dentate respectively, enclosing the nutlets. Corolla funnel-shaped, ± equalling calyx. Faucal scales small, white. Stamens included; anthers subsessile. Style included; stigma capitate. Nutlets 4, dorsally keeled, attached asymmetrically to carpophore along lower part of ventral surface.

Monotypic genus distributed in Europe and Asia; a single species in Iraq.

Asperugo, from Lat. *asper,* rough, and suffix *-ugo* which denotes a thin coating.

1. **Asperugo procumbens** *L.,* Sp. Pl. 138 (1753); Rawi in Dep. Agr. Iraq Tech. Bull. 14: 135 (1964); Riedl in Fl. Iranica [K. H. Rechinger] 48: 96 (1967); Y. Nasir in Fl. Pakistan [Ali & Y. Nasir] 191: 105 (1989).

Annual herb, trailing or decumbent. Stems 10–50 cm, much branched. Leaves oblong to elliptic or oblanceolate, 2–7 x 0.8–2 cm, obtuse to subacute, margins hispid; basal leaves petiolate, cauline leaves subsessile. Calyx c. 3 mm at anthesis, lobes triangular; expanding to 12 mm in fruit. Corolla blue, turning purple or violet. Nutlets c. 3 mm, pear-shaped, laterally compressed, minutely tuberculate. Fig. 104, 1–5.

HAB. Weed in orchard, hedges, damp place under cliff, irrigated fields, sandy loam, rocky mountain slope; alt. 170–1800 m; fl. Mar.-May.
DISTR.: Scattered across a wide range of zones, but rare in the alluvial zone of southern Iraq. **MAM**: Rubal waterfall N.E. of Amadiya, *Barkley* 8207!; Sharifa nr Amadiya, *O. Polunin* 5146!; Amadiya, *Guest* 1217! 13279!; Aqra, *Rawi* 11391! **MRO**: Kamarspa between Halabja and Tawila, *Rawi* 22268!; Haji Umran, *Chapman* 12296!; Jabal Karoukh, between Soran and Kilkil, *Kass & Nuri* 27370! 27436! **MSU**: Qara Dagh, *Gillett* 7851!; Penjwin, *Rawi* 12164!; Sulaimaniya, *Graham* 743! **MJS**: Sinjar, *Khudairi, Kaisi, Shahwani et al.* 83! *Omar & Hamid* 36502! **FUJ**: Hadhr, *Barkley & Brahim* 4165! *Chakravarty, Khatib, Rawi & Alizzi* 33050! **FNI**: Nenawa (Nineveh?) farm, Ani 9774! **FKI**: Tuz Khurmatu, Gillett 10291! **LEA**: nr Ba'quba, *Rogers* s.n.! **LCA**: Baghdad, *Hausskn.* s.n.! *Guest* 1117! **LCA**: Karrada, *Graham* s.n.! **DLJ**: Between Rawa and Haditha, *Omar, Kaisi, Hamad & Hamid* 44626! **DWD**: 17 km S.E. of K3, *Kaisi & Hamad* 53125!; Ana, *Khayat & Hamad* 51778!; 8 km E. of Ana, *Omar, Kaisi, Hamad & Hamid* 44346!; Al Kalah island, Ani, *Palmatier & Barkley* 642!; Warta, 30 km N.W. by W. of Rania, *Rawi, Nuri & Kass* 28775! 28783! **DSD**: 17 km W.N.W. of Busaiya, *Chakravarty, Khatib, Rawi & Tikriti* 29972!

DUTALAN (Kurd.-Qara Dagh), *Gillett* 7851.

N. Africa, Egypt, Arabian Peninsula to temperate Eurasia; widespread.

Fig. 104. **Asperugo procumbens**. 1, habit × ½; 2, flower × 8; 3, corolla, opened × 8; 4, fruiting calyx × 3; 5, seed × 5. Reproduced with permission from Fl. Pakistan 191: f. 29, A–E (1989). Drawn by S. Hameed. © National Herbarium, Pakistan Agriculture Research Council & University of Karachi, Pakistan.

24. **PSEUDOHETEROCARYUM** Kaz.Osaloo & Saadati
in Austr. Syst. Bot. 30(1): 109 (2017)

J. R. Edmondson & Shahina A. Ghazanfar

Annual decumbent or erect herbs, 5–40 cm high, covered by subpatent to appressed hairs. Stems, branches and leaves covered loosely or densely with trichomes usually arising from tuberculate bases. Leaves linear to linear-lanceolate. Pedicel long (absent in *P. subsessile*), thickened. Calyx divided at the base into lanceolate-linear lobes, elongating in fruit. Corolla small, 5-lobed. Nutlets four, unequal, zygomorphic, tapering, attached to the columnar gynobase throughout their length and not detachable, with a single series of appendages at margins, dorsal middle area not pubescent. (Description from Saadati et al. 2017).

Genus with four species found from E. Mediterranean to N.W. China and the Arabian Peninsula; 3 species in Iraq.

Pseudoheterocaryum is recently split from *Heterocaryum* on the basis of molecular analysis; it differs from *Heterocaryum* in being pubescent (plants glabrous in *Heterocaryum*), fruiting pedicels erect to slightly deflexed and nutlets (fruiting pedicels often strongly reflexed in *Heterocaryum*) with a single series of appendages at the margins with glabrous dorsal middle area (nutlets with two series of appendages at the margins with pubescent middle area in *Heterocaryum*). All Iraq material of *Heterocaryum* is now identified as *Pseudoheterocaryum*.

Pseudoheterocaryum, a compound word formed from ψευδο, *pseudo*, false, ετερος, *eteros*, different and καρυου, *karuon*, seed.

Saadati, N., Mozaffar, M.K., Sherafati, M. & Osaloo, S.K. (2017). *Pseudoheterocaryum*, a new genus segregated from *Heterocaryum* (Boraginaceae) on the basis of molecular data. Austr. Syst. Bot. 30(1): 105–111.

1. Lateral pair of nutlets with margin prolonged into cup-shaped wing bearing 4–6 glochidiate spines on each side; second pair narrower, lacking marginal wing . . . . 1. *P. szovitsianum*
 Two larger nutlets with margin usually prolonged into a spiny-glochidiate wing, 2 smaller nutlets lacking wing, or with narrow usually entire margin 2
2. Pedicel developing to 1.5 mm in fruit; corolla blue, campanulate . . . . . . 2. *P. subsessile*
 Pedicel elongating, becoming stout and recurved in fruit; corolla pale to deep blue, funnel-shaped; throat with distinct narrowly triangular appendages. . . . . 3. *P. rigidum*

1. **Pseudoheterocaryum szovitsianum** (*Fisch. & C.A.Mey.*) *Kaz.Osaloo & Saadati*, Austral. Syst. Bot. 30(1): 109 (2017).

Echinospermum szovitsianum Fisch. & C.A.Mey. in Index Seminum [St.Petersburg (Petropolitanus)] 2: 36 (1835).
Heterocaryum szovitsianum (Fisch. & C.A.Mey.) A.DC., Prodr. [A.P. de Candolle] 10: 145 (1846); Riedl in Fl. Iranica [K. H. Rechinger] 48: 86 (1967); Edmondson in Fl. Turkey [P. H. Davis] 6: 261 (1978); Y. Nasir in Fl. Pakistan [Ali & Y. Nasir] 191: 183 (1989); Khatamaz, Fl. Iran 39: 279 (2002).
Heterocaryum pachypodum A.DC., Prodr. [A.P. de Candolle] 10: 144 (1846).
Lappula echinophora var. *szovitsiana* (Fisch. & C.A.Mey.) Kuntze, Trudy Imp. S.-Peterburgsk. Bot. Sada 10: 214 (1887).
Lappula echinophora var. *pachypoda* (A.DC.) Kuntze, Trudy Imp. S.-Peterburgsk. Bot. Sada 10: 214 (1887).
Lappula szovitsiana (Fisch. & C.A.Mey.) Druce, List Brit. pl. 50 (1908); Rawi in Dep. Agr. Iraq Tech. Bull. 14: 137 (1964).

Annual herb. Stems procumbent to erect, 12–15 cm, much-branched from base, densely patent-hispid. Leaves adpressed-hispid; basal to 9 × 4 cm, linear-spatulate to linear; cauline 20–45 × 2–5.5 mm, linear, ± densely crowded. Flowers becoming widely spaced on bracteate inflorescence, subsessile at anthesis; pedicels becoming strongly thickened and recurved in fruit. Calyx lobes linear, acute. Corolla deep blue; tube c. 2 mm; limb saucer-shaped, c. 2.2 mm in diameter. Nutlets distinctly unequal, lateral pair to 8 × 5 mm, disc tuberculate, margin prolonged into cup-shaped wing bearing 4–6 glochidiate spines on each side; second pair narrower, to 6 × 2 mm, lacking marginal wing. Fig. 105, 1–4.

HAB. Dry cliffs (limestone), sandy soil over gypsum, fields on dry gravel soil, margins of rain-fed barley patches, clay soil, wheat field, alt. 150–640 m; fl. Feb.-Mar.
DISTRIB. Widespread in desert and subdesert. **MAM**: Dohuk, *Guest* 1305! 2295! **MSU**: Jarmo, *Helbaek* 446! **FUJ**: Mosul, *Aucher* 2337, 2735. **FKI**: Kirkuk, on road to Baba Gurgur oil wells, *Guest* 1360!;14 km E. of Kirkuk, *Rawi & Gillett* 10652!; 26 km S. of Kirkuk, *Barkley & Brahim* 4492!; 13 km N. of Kirkuk, *Rawi & Gillett* 10577!; 10 km N. of Tuz Khormatu, *Ani & Barkley* 7015!; Tuz Khurmatu, *Rawi & Gillett* 10310! **FPF**: Jabal Hamrin, Table Mountain (Mansur), *Guest* 1900!; 25 km S.W. of Badra, *Kaisi, Thamer & Salah* 51355!; Qaraulos nr Mandali, *Rawi* 12646!; Qasr rabat, 17.3.1961, 2nd range Jabal Hamrin, *Haines* s.n.! (E). **LCA**: c. 20 km E. of Falluja, *Guest & Rawi* 13636!; Haswa, on road to Falluja, *Rawi* 15100!; Tigris banks after Adhaim flood, Baghdad, Tadriyah, *Haines* 3.4.1961 s.n. (E!). **DLJ**: Rawa, *Rawi & Gillett* 6984! 7042!; 12 km W. of Rawa, *Omar, Kaisi, Hamad & Hamid* 44462! **DWD**: 18 km W. of Rutba, *Rawi* 14633!; 40 km from Rumana to Rawa, *Khayat & Hamad* 51676!; 10 km W. of Ana, *Omar, Kaisi, Hamad & Hamid* 45016!; midway between T1 and Ana, *Botany Staff* 41674!; 8 km E. of Ana, *Omar, Kaisi, Hamad & Hamid* 45112!; 18 km N.W. of T1 to al-Qaim, *Omar, Kaisi, Hamad & Hamid* 45079! **DSD**: Shabicha, *Rawi & Gillett* 6278!; 25 km N. of Busaiya to Salman, *Kaisi, Hamad & Hamid* 48534!

JEDHA or JADHA (Ir.-Shabicha), *Rawi & Gillett* 6278!

Oman, Syria, Turkey, Caucasia, Iran, Afghanistan, Pakistan, C. Asia, Tien Shan.

2. **Pseudoheterocaryum subsessile** (*Vatke*) *Kaz.Osaloo & Saadati*, Austral. Syst. Bot. 30(1): 109 (2017).

Heterocaryum subsessile Vatke, Z. Gessammten Naturwiss. (Halle) 45: 129 (1875); Riedl in Fl. Iranica [K. H. Rechinger] 48: 85 (1967); Y. Nasir in Fl. Pakistan [Ali & Y. Nasir] 191: 182 (1989); Khatamaz, Fl. Iran 39: 275 (2002);

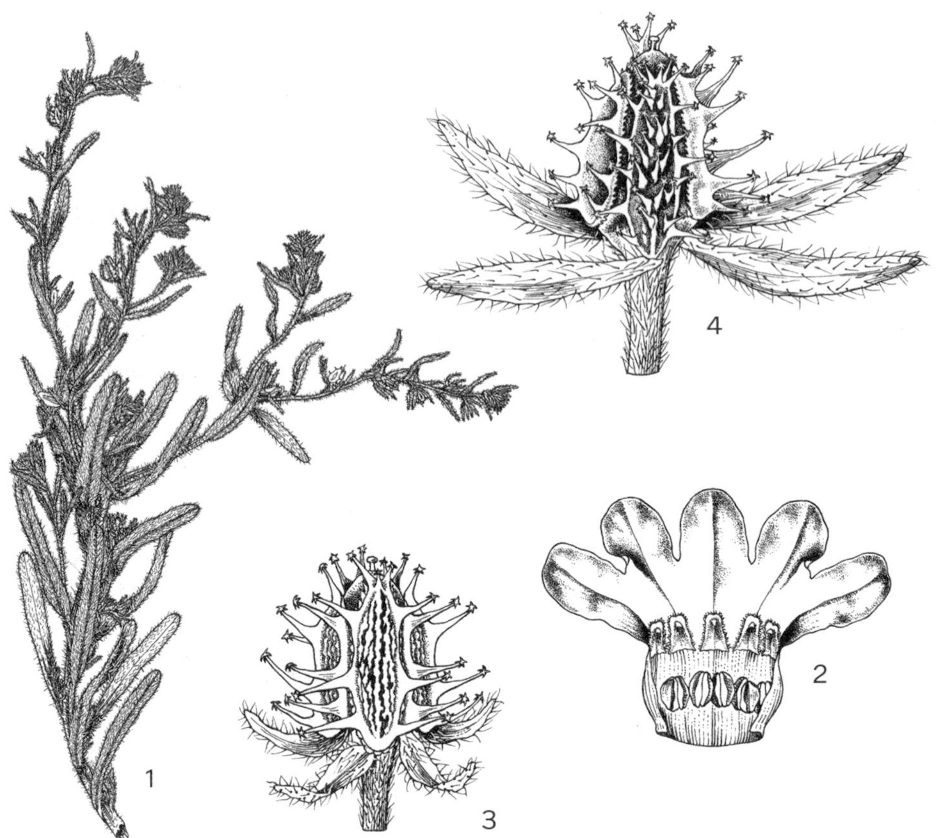

Fig. 105. **Pseudoheterocaryum szovitsianum**. 1, habit × ½; 2, corolla, opened × 10; 3, fruit. **Pseudoheterocaryum rigidum**. 4, fruit × 5. Reproduced with permission from Fl. Pakistan 191: f. 37, F–I (1989). Drawn by S. Hameed. © National Herbarium, Pakistan Agriculture Research Council & University of Karachi, Pakistan.

Echinospermum oligacanthum Boiss., Fl. Orient. [Boissier] 4: 248. (1875), nom. illeg., non Ledeb. (1847).
Lappula echinophora var. *sessilis* Kuntze, Trudy Imp. S.-Peterburgsk. Bot. Sada 10: 215 (1887).
Cynoglossospermum oligacanthum Kuntze, Rev. Gen. Pl. 2: 436 (1891).
Echinospermum echinophorum var. *sessile* (Kuntze) Lipsky, Trudy Imp. S.-Peterburgsk. Bot. Sada 26: 539 (1910).
Heterocaryum echinophorum var. *oligocanthum* (Kuntze) Brand, Pflanzenr. 4. 252: 97 (1931).
H. oligocanthum (Kuntze) Bornm., Beih. Bot. Centralbl., 59-B: 306 (1939).
Lappula subsessilis (Vatke) Greuter & Burdet, Willdenowia 11: 37 (1981).

Annual. Stem simple or branched from base, hispid with patent setae. Basal and lower cauline leaves linear, narrowed into petiole, 15–40 × 1–3 mm, obtuse or subacute; upper cauline similar but sessile. Inflorescence bracteate. Flowers sessile at anthesis; pedicel developing to 1.5 mm in fruit. Calyx lobes linear, accrescent to 4 mm. Corolla blue, campanulate. Nutlets 4, unequal; one pair with margin winged with glochids 1–2 mm long; the smaller lateral pair with an entire wing or wing margins with very short glochids.

HAB. Slope of sandy hill, sandy soil, dry steppe, rocky areas; alt. 170–260 m; fl. Feb.-Mar.
DISTRIB. Subdesert of S. Iraq. **DLJ**: 36 km N.E. of Haditha, *Chakravarty, Rawi, Khatib & Alizzi* 31834!; Jabal Makhul, *Rawi & Gillett* 7107!; Raisha hill nr Shibaichan, 78 km S. of Sinjar, 260 m, *Chakravarty, Rawi, Khatib & Alizzi* 32025!; Jabal Hamrin, *W. Edgar Evans* no. M/110 (grown from seed on 29.11.1918) (E!).

Syria, Iran, Afghanistan, Turkestan, Pakistan.

3. **Pseudoheterocaryum rigidum** (*A.DC.*) *Kaz.Osaloo & Saadati,* Austral. Syst. Bot. 30(1): 109 (2017).

Heterocaryum rigidum A.DC., Prodr. [A.P. de Candolle] 10: 145 (1846).
Echinospermum heterocaryum Bunge, Mem. Sav. Etr. Acad. St. Petersbourg 7: 411 (1852).
E. minimum Lehm., Pl. Asperif. Nucif. 126 (1818); Rawi in Dep. Agr. Iraq Tech. Bull. 14: 137 (1964).
E. szovitsianum auct. non Fisch. & C.A.Mey.: Boissier, Fl. Orient. 4: 248 (1875), p.p.].
E. minimum auct. non Lehm.: C.B.Clarke, Fl. Brit. India 4: 162 (1883).
Heterocaryum echinophorum var. *minimum* (Lehm.) Brand, Pflanzenr. 4, 252: 95 (1931), p.p.
Heterocaryum szovitsianum (Fisch. & C.A.Mey.) A.DC. subsp. *rigidum* Y.Nasir in Fl. Pakistan [Ali & Y. Nasir] 191: 184 (1989).

Annual herbs; stems coarsely pilose to hispid. Basal leaves shortly petiolate; cauline sessile. Flowers sessile at anthesis; pedicel elongating, becoming stout and recurved in fruit. Calyx deeply divided, accrescent in fruit. Corolla pale to deep blue, funnel-shaped; throat with distinct narrowly triangular appendages. Anthers subsessile, inserted below middle of tube, included. Style included; stigma capitate. Nutlets 4, unequal, ± oblong to linear, fused to narrowly pyramidal gynobase and to each other: 2 larger with margin usually prolonged into spiny-glochidiate wing, 2 smaller lacking wing, with narrow usually entire margin. Fig. 105, 5.

HAB. Slope of sandy hill, sandy soil, dry steppe, rocky areas; alt. 150–300 m; fl. Feb.-Mar.
DISTR. Subdesert of S. Iraq. **DLJ**: Baghdad, Jadriyah, Tigris bank following Adhaim flood, 31.3.1961, *Haines* s.n.! (E); Jabel Hamrin, 1932, *Anthony*.

Iran, Afghanistan, C. Asia to Pakistan and Xinjiang, Saudi Arabia.

25. **LAPPULA** Moench,

Methodus: 416 (1794)
Cynoglossospermum Kuntze, Revis. Gen. Pl. 2: 436 (1891), *Echinospermum* Sw. ex Lehm., Pl. Asperif. Nucif.: 113 (1818), *Omphalolappula* Brand, Pflanzenr. (Engler), Borrag.-Borraginoid.-Cryptanth.: 135 (1931), *Sclerocaryopsis* Brand, Pflanzenr. (Engler), Borrag.-Borraginoid.-Cryptanth.: 98 (1931), *Lepechiniella* Popov, Fl. URSS 19: 713 (1953)

A.P. Sukhorukov

Annual, biennial or perennial herbs with a taproot, usually branched from the base, ± densely covered with simple persistent hairs having a prominent conical podium. Rosulate leaves present and usually withered at fruiting. Stem leaves numerous, similar to basal leaves, green or greyish due to present of {semi)appressed hairs, decreasing in size towards the inflorescence. Inflorescence spike-like (scorpioid), loose. Flowers bracteate, sessile or short-stalked, hermaphrodite. Calyx of 5 free segments, slightly accrescent in fruit, spreading or clasping. Corolla campanulate with a tube and 5 spreading lobes forming a limb; limb 2–8 mm across, blue, rarely white or cream, with small appendages near the throat. Stamens 5, attached in tube, included. Style 1, gynobasic, terminating with a capitate stigma, exserted or included. Fruit breaking into 4 stout nutlets (mericarps) with obvious, lanceolate to ovoid scar; nutlets marginally rimmed (not winged), ovate, triangular or conical, adaxial keel expressed not exceeding their top; rim with the stalked glochidiate outgrowths located in 1–3 rows and terminating with 3–6 recurved hooks, rarely mericarps smooth or tuberculate without glochidiate projections; central part of the nutlets (between the rims) with small subulate prickles with or without a podium (seen at higher magnifications). Seed with a basal embryo; endosperm scanty.

About 60 species in temperate, mountain and arid regions of Eurasia, North America, subtropical parts of Africa and Australia. The composition of the genus has been changed after the separation of *Heterocaryum* A.DC. with heteromorphic nutlets, e.g. Nazaire & Hufford (2012), Chacón & al. (2016). At the same time, the merger of *Lepechiniella* Popov with *Lappula* has been proposed. However, only three out of ~20 species of *Lepechiniella* were involved in the phylogenetic analysis (Khoshsokhan-Mozaffar & al. 2018), and its members are not uniform in the nutlet structure. *Lepechiniella* has a distribution in Central Asia and Iran, and no species is reported from Iraq. *Lappula sinaica* is currently considered within a new genus *Pseudolappula sinaica* (DC.) Khoshsokhan-Mozaffar & al. (2018). Here I follow

Khoshsokhan & al. (2018) separating *Lappula sinaica* from other species of the genus. See also other comments under *Pseudolappula*; 3 species in Iraq.

Lappula, diminutive of Lat. *lappa,* a bur, referring to the hooked hairs on its nutlets.

Popov M.G. (1953). Boraginaceae, Fl. USSR 19: 97–691 (1953).
Nazaire M. & Hufford L. (2012). A broad phylogenetic analysis of Boraginaceae: Implications for the relationships of *Mertensia*. Syst. Bot. 37: 758–783.
Chacón J., Luebert F., Hilger H.H., Ovchinnikova S., Selvi F., Cecchi L., Guilliams C.M., Hasenstab-Lehman K., Sutorý K., Simpson M.G., Weigend M. (2016). The borage family (Boraginaceae s.str.): A revised intrafamilial classification based on new phylogenetic evidence, with emphasis on the placement of some enigmatic genera Taxon 65: 523–546.
Khoshsokhan-Mozaffar M., Sherafati M. & Kazempour-Osaloo, S. (2018). Molecular phylogeny of the tribe Rochelieae (Boraginaceae, Cynoglossoideae) with special reference to *Lappula*. Ann. Bot. Fenn. 55: 293–308.

1. Annuals; nutlets smooth or tuberculate, without glochidia . . . . 3. *L. spinocarpos*
    Biennials or short-leaved perennials forming leaf rosettes in the first
    year; nutlets glochidiate. 2
2. Plants at fruiting often forming hemispherical or tumble-weed habit;
    corolla 4–5 mm long, limb 3.5–8 mm across; nutlets 3.5–4 mm long,
    marginally with 2 (rarely 3) rows of glochidia on each side of nutlet . . . 1. *L. barbata*
    Plants not forming hemispherical or tumble-weed habit; corolla 2.5–3
    mm long; limb 3–4 mm across; nutlets 2.5–3 mm long, marginally
    with one row of glochidia on each side of nutlet . . . . .2. *L. microcarpa*

1. **Lappula barbata** (*Bieb.*) *Gürke*, Nat. Pflanzenfam. [Engler & Prantl] 4, 3a: 107 (1897); Popov, Fl. USRR 19: 461 (1953); Riedl in Fl. Iranica [K. H. Rechinger] 48: 72 (1967).

Myosotis barbata Bieb., Fl. Taur.-Cauc. 1: 121 (1808).
Echinospermum barbatum (Bieb.) Lehm., Pl. Asperif. Nucif. 1: 128 (1818).
Echinospermum filiforme Godet in DC., Prodr. 10: 140 (1846).
Lappula saxatilis Pall. ex Kusn., Fl. Cauc. Crit. 4, 2: 183 (1913).

Biennial or short-leaved perennial, at fruiting often forming hemispherical or tumble-weed habit with obliquely oriented and horizontally spreading branches. Leaves rosulate and cauline, up to 25 mm long and up to 7 mm wide, villous; rosulate leaves oblanceolate or lanceolate, shortly petiolate, often withered in anthesis; cauline leaves lanceolate, sessile, often folded at ventral side. Inflorescence elongate, lax. Pedicels absent or very short (up to 3 mm long). Calyx 2.5–3.0 mm long. Corolla blue or white, 4–5 mm long, limb 3.5–8 mm across. Nutlets 3.5–4 mm long, marginally with two (rarely three) rows of glochidia on each side of nutlet.

HAB. Screes, sands and grassy slopes; alt. 1000–2000 m; fl. not recorded.
DISTRIB. Reported from Iraq without exact location (Riedl 1967), but no herbarium seen from the country.

Afghanistan, Armenia, Azerbaijan, Georgia, India, Iran, Pakistan, Russia, Syria, Tajikistan, Turkey, Turkmenistan, Ukraine, Uzbekistan.

2. **Lappula microcarpa** (*Ledeb.*) *Gürke* in Engler & Prantl, Nat. Pflanzenfam. 4, 3a: 107 (1897); Popov, Fl. USRR 19: 462 (1953); Rawi in Dep. Agr. Iraq Tech. Bull. 14: 137 (1964); Riedl in Fl. Iranica [K. H. Rechinger] 48: 74 (1967); Y. Nasir in Fl. Pakistan [Ali & Y.Nasir] 191: 143 (1989).

Echinospermum microcarpum Ledeb., Fl. Altaic. 1: 202 (1829).
Echinospermum stylosum Kar. & Kir., Bull. Soc. Imp. Nat. Mosc. 14: 715 (1841).
Echinospermum oligacanthum Ledeb., Fl. Ross. (Ledebour) 3: 161 (1847).
Cynoglossospermum microcarpum (Ledeb.) Kuntze, Revis. Gen. Pl. 2: 437 (1891).

Biennial, up to 50 cm tall. Stems stout, erect, with obliquely oriented branches. Leaves rosulate and cauline, lanceolate or linear, flat or folded on the ventral side, up to 50 mm long and 3 mm wide, villous. Inflorescence elongate, lax. Pedicels 1–3 mm long, ± appressed to the stem. Calyx 1–2 mm long, clasping, its segments ovate or triangular. Corolla blue, 2.5–3 mm long, limb 3–4 mm across, blue. Style 0.4–0.6 mm long. Nutlets

Fig. 106. **Lappula microcarpa**. 1, habit, portion × 1 ½; 2, corolla, opened × 10; 3, fruit × 7. Reproduced with permission from Fl. Pakistan 191: f. 39, A–C (1989). Drawn by S. Hameed. © National Herbarium, Pakistan Agriculture Research Council & University of Karachi, Pakistan.

2.5–3 mm long, oblong or ovate, glochidia located in one row, with a stalk ~0.5 mm. Fig. 106, 1–3.

HAB. Screes and grassy slopes; alt. 1000–2000 m; fl. not recorded.
DISTRIB. **MRO**: Mt. Handren, *N*ábělek 589.

The species was described from East Kazakhstan, and the populations growing in the West Asia deviate from those in Central Asia.

Afghanistan, W. China, Iran, Kazakhstan, Kyrgyzstan, N. India, Pakistan, Russia (Altai), Tajikistan, Turkmenistan, Uzbekistan.

3. **Lappula spinocarpos** (*Forssk.*) *Asch. & Kuntze*, Acta Hort. Petrop. 10: 215 (1887); Blakelock in Kew Bull. 4(4): 524 91849); Rawi in Dep. Agr. Iraq Tech. Bull. 14: 137 (1964); Y. Nasir in Fl. Pakistan [Ali & Y. Nasir] 191: 139 (1989); Abdulridha, Taha & Widad, Ecology & Fl. Basrah: 237 (2016); Haloob & Al-Kaisi (eds), Illtustr. Fl. Lowland Iraq 1: 67 (2016).

Anchusa spinocarpos Forssk., Fl. Aegypt.-Arab.: 41 (1775).
Echinospermum vahlianum Lehm., Pl. Asperif. Nucif. 1: 132 (1818).
Sclerocaryopsis spinocarpos (Forssk.) Brand in Engler, Pflanzenr. 4, 252: 98 (1931).
Echinospermum ceratophorum Popov, Descr. Pl. Nov. Turkestan. [Korovin, Kultiasov & Popov]: 67 (1916).
Lappula ceratophora (Popov) Popov, Fl. USSR 19: 417 (1953).
Lappula spinocarpos (Forssk.) Ascherson & Kuntze subsp. *ceratophora* (Popov) Y.J.Nasir, Fl. Pakistan 191: 139 (1989).

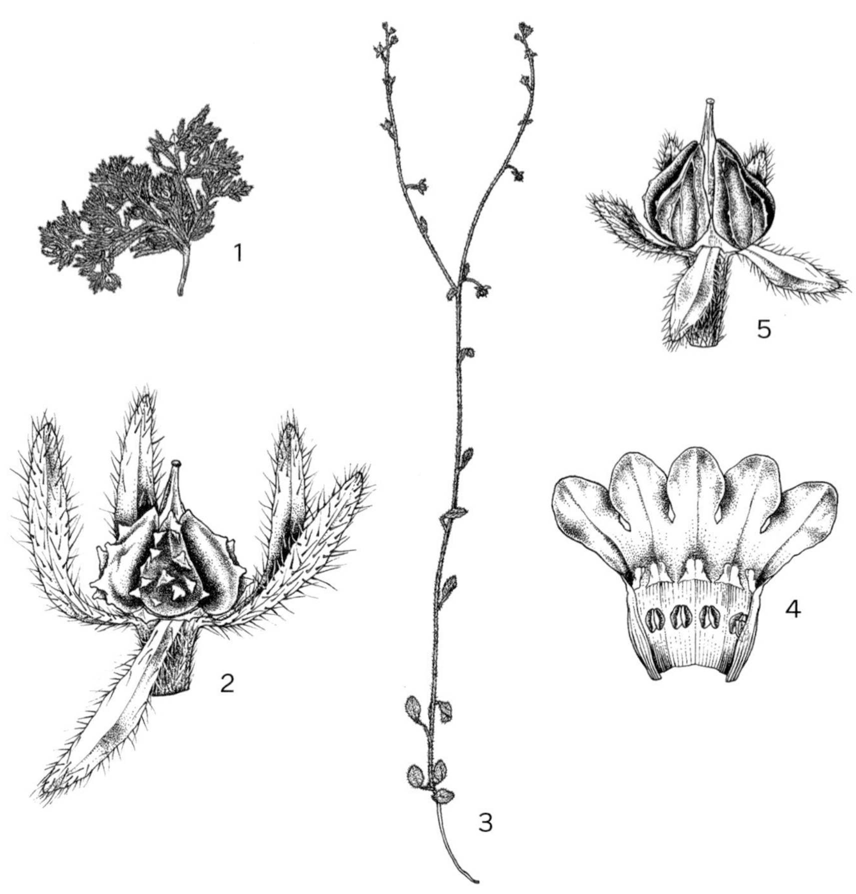

Fig. 107. **Lappula spinocarpos**. 1, habit × ½; 2, fruit with calyx × 5. **Pseudolappula sinaica**. 3, habit, portion of shoot × ½; 4, corolla, opened × 15; 5, fruit × 6. Reproduced with permission from Fl. Pakistan 191: f. 37, A–E (1989). Drawn by S. Hameed. © National Herbarium, Pakistan Agriculture Research Council & University of Karachi, Pakistan.

Annual 5–10(–20) cm tall, with spreading often decumbent branches. Leaves rosulate and cauline, linear or lanceolate, often folded on the ventral side, up to 35 mm long and 3 mm wide, villous. Inflorescence dense. Pedicels short (up to 2 mm), thick. Calyx 3–4 mm long, at fruiting accrescent, up to 7 mm, enclosing the fruit. Corolla 3–4 mm long and 3–4 mm across, blue. Nutlets ± 4 mm long, smooth or tuberculate, without glochidia. Fig. 107, 1–2.

HAB. In desert and gravel plains; alt. 100–1000 m; fl. & fr. Mar.-May.

DISTRIB. **DSD**: Zubair, *Graham* 404 (BM!); nr Jaliba, *Eig & Zohary* 29920 (HUJ!); Shabicha, *Rawi & Gillett* 6292!; 5 km W. of Kerbela, *Rawi & Gillett* 6391!; 23 km S.E. by S. of Zubair, *Guest, Rawi & Rechinger* 16862!; nr Jarishan, *Guest, Rawi & Rechinger* 17147!; nr Jabal Sanam, *Guest, Rawi & Rechinger* 17017!; 10 km S.E. of Jiraibiyat, *Guest, Rawi & Rechinger* 17163!; nr Safai al-Maghif, *Guest, Rawi & Rechinger* 17256!; nr Shabicha, *Guest, Rawi & Rechinger* 19278!; 60 km W. of Shabicha, *Chakravarty & al.* 30050!; 3 km W. of Shabicha, *Botany Staff* 42034!; nr Samah, *Alizzi* 34386!; 20 km S. of Nukhaila, *Alizzi & Omar* 35521!; nr Ansab, *Kasim & al.* 38834!; 20 km from Busaiya, *Karim & al.* 48010!; 55 km E. of Zubair, *Thamer* 47474!; 30 km from Salman to Samawa, *Al-Kaisi, Hamad & Hamid* 48303!; 5 km E. of Jumaima, *Al-Kaisi, Hamad & Hamid* 48196!; Ansab station, *Al-Kaisi, Hamad & Hamid* 48337! **DWD**: between Muhammadi & Ramadi, *Eig & Zohary* s.n. (HUJ!); 260 km N.W. of Ramadi (Ramadi-Rutba highway), *Rawi* 20939!; 12 km W. of Ukhaidhir, *Rawi* 30841!; 70 km W. of Nukhaib, *Rawi* 31027!; Wadi Tibul, 100 km N. of Nukhaib, *Rawi* 31082!; 19 km N.E. of Rutba, *Rawi* 31189!; 65 km N. of Rutba, *Rawi & Khatib* 32268!; 95 km N.E. of Rutba, *Rawi & Khatib* 32328!; 130 km W. of Ramadi, *Rawi & Khatib* 32157!; E. of Rutba, *Khatib & Alizzi* 31595!; Wadi Hauran, 320 m, *Khatib*

& *Alizzi* 31604!; 50 km S.E. of Rutba, *Almokhtar* 33364!; 60 km N.W. of Ramadi, *Alizzi* 35150!; 30 km W. of Kubaisa, *Omar et al.* 44984!; Wadi Hauran, *Omar et al.* 44638!; 205 km from Ramadi to Rutba, *Al-Khayat & Hamad* 51529! **DLJ**: 36 km N.E. of Haditha, *Khatib & Alizzi* 31840!; 70 km N. of Rawah, *Khatib & Alizzi* 31913!; Umm Al-Maten (64 km N.W. the road between Baiji-Haditha), *Khatib & Alizzi* 31902!; nr Baiji, *Rawi & Hamada* 33575! **LEA**: 77 km N.E. Kut al-Imara, Garmashaya village, *Rawi* 18058!; Khider Al-Mai, *Khatib & Alizzi* 32708!; Khider Al-Mai, *Alizzi & Omar* 35756! **LCA**: Hinaidi, *Lazar* 1155!; Baghdad, *Lazar* 3885!; 15 km W. of Falluja, *Chakravarty & Rawi* 29832!; 35 km N.W. of Falluja, *Rawi* 30178!; 15 km W.W.N. of Falluja, *Rawi* 30129!; 15 km W. of Falluja, *Chakravarty & Rawi* 30272! **FPF**: 16 km S. of Badra, *anonymous* 18203!; Badra, *Gillett* 6603!; 30 km S.E. Badra, *Rechinger* 9160 (W!). **FPF/FKI**: Jabal Hamrin, *Bornmüller* 1626 (W!). **FKI**: Kani Domlan hills, 1300 ft, *Guest* 4343!; 26 km S. of Kirkuk, *Barkley* 4492 (W!).

Widely distributed in the Irano-Turanian region, E. Mediterranean and North Africa.

We consider *L. spinocarpos* in a broad sense including *L. ceratophora* (Popov) Popov; the latter species has two well-expressed horny outgrowths at the base of nutlets. This character is present in many plants across Central Asia and Iran.

26. **PSEUDOLAPPULA** Khoshsokhan & Kaz. Osaloo,
Ann. Bot. Fenn. 55: 302 (2018)

A.P. Sukhorukov

Morphologically, very close to *Lappula*, but leaves broader (up to 15 mm), and attachment scar of the nutlets linear (strip-like).

After separation of a new genus *Pseudolappula* based on molecular phylogeny and some morphological characters (Khoshsokhan-Mozaffar & al. 2018), it consisted of a single species, *P. sinaica*. At present, the genus comprises two species, *Pseudolappula sinaica* (A.DC) Khoshsokhan, Sherafati & Kaz.Osaloo and *P. occultata* (Popov) Q.R. Liu & D.H. Liu. The latter was transferred to *Pseudolappula* based on extended analyses including molecular, morphological, palynological and carpological evidence (Liu & al. 2021). Another member of *Lappula*, *L. mogoltavica* Popov ex Czukav., seems to be closely related with both above mentioned species (Ovchinnikova & al. 2017, all species as *Lappula*).

Pseudolappula, derivation from ψευδο, *pseudo*, false, and *Lappa*, false burdock, due to its ambiguity with *Lappula*.

Monotypic; distributed from the Sinai desert through Iraq, Iran, Caucasus, Armenia to Central Asia.

Ovchinnikova S.V., Korolyuk A.Yu., Kupriyanov A.N., Khrustaleva I.A., Lashchinskyi N.N., Ebel A.L. (2017), "New localities of the rare and endemic species of the family Boraginaceae in Kazakhstan Republic", Rastitel'nyi Mir Aziatskoy Rossii 3(27) : 51–63 (In Russ.).
Khoshsokhan-Mozaffar M., Sherafati M. & Kazempour-Osaloo S. (2018). Molecular phylogeny of the tribe Rochelieae (Boraginaceae, Cynoglossoideae) with special reference to *Lappula*. Ann. Bot. Fenn. 55 : 293–308.
Liu D.-H., Xu X.-M., He Y., Liu Q.-R. (2021). A new combination in *Pseudolappula* (Boraginaceae, Rochelieae) based on morphological, molecular and palynological ecidence", PhytoKeys 187(4): 77–92.

1. **Pseudolappula sinaica** (*A.DC.*) *Khoshsokhan, Sherafati & Kaz.Osaloo*, Ann. Bot. Fenn. 55(4–6): 302 (2018).

Echinospermum sinaicum A.DC., Prodr. 10: 141 (1846).
Lappula sinaica (A.DC.) Asch. & Schweinf., Mém. Inst. Égypt. 2 : 111 (1887); Y. Nasir in Fl. Pakistan [Ali & Y. Nasir] 191: 141 (1989).
L. sinaica (A.DC). Grossh.,Věstn. Tiflissk. Bot. Sada 1926–1927. n.s., pt. 3–4 : 21 (1927), isonym.
Cynoglossospermum sinaicum (A.DC.) Kuntze, Revis. Gen. Pl. 2 : 437 (1891).
Echinospermum kotschyi Boiss., Diagn., ser. 1, 7: 29 (1846).
E. divaricatum Bunge, Beitr. Fl. Russl. : 234 (1852).
Cynoglossospermum divaricatum (Bunge) Kuntze, Revis. Gen. Pl. 2 : 437 (1891).
Lappula divaricata (Bunge) B.Fedtsch., Rastit. Turkest.: 663 (1915).

Annual herbs with a taproot, usually branched and spreading from the base, up to 50 cm tall, ± densely covered with simple persistent hairs having a prominent conical podium. Rosulate leaves present or not, 20 – 40 × 5–11 mm, shortly petiolate, oblanceolate or oblong, usually withered at fruiting. Stem leaves distant, almost sessile, shorter than rosulate leaves. Inflorescence spike-like (scorpioid), with distant flowers.. Flowers with pedicel 3–5 mm long, slightly elongated (up to 8 mm long) and recurved at fruiting. Calyx c. 2 mm long, clasping. Corolla campanulate with a tube and 5 spreading lobes; limb 2–3 mm across, blue, with small appendages near the throat. Stamens 5, attached in tube, included. Style 1, gynobasic, terminating with a capitate stigma, exceeding the nutlets. Fruit breaking into 4 stout parts called nutlets (mericarps), marginally rimmed (not winged), oblong or narrowly oblong, adaxial keel expressed through all length of the nutlets. Nutlets tuberculate, inconspicuously glochidiate (seen at highr magnification), their attachment scar linear (strip-like). Seed with a basal embryo; endosperm scanty. Fig. 107, 3–5.

HAB. Mountains, amongst stones; alt. ±1000 m; fl. Jun.
DISTRIB. In the mountainous areas in northern Iraq. **MAM**: Matina, *Rawi* 8730! **MSU**: Pira Magrun, *Gillett* 7763!

Sinai, Jordan, Palestine, Arabian Peninsula, Turkey, Caucasus, Iran, Afghanistan to Pakistan and C. Asia.

27. **ROCHELIA** Rchb.

Flora 7: 234 (1824)
Weigend & al. in Kubitzki, Fam. Gen. Vasc. Pl. 14: 90 (2016)

C.C. Townsend
Revised by Shahina A. Ghazanfar

Annual herbs, pubescent to stiffly hairy. Stem simple or branched. Leaves spatulate to obovate, entire, obtuse, sessile. Flowers in simple or branched racemes. Bracts linear. Calyx (in ours) 5-partite, lobes minute in flower, accrescent to 3.5 mm in fruit, falcately incurved from middle. Corolla blue (rarely white), funnel-shaped, with a short cylindrical tube; faucal scales 5, small. Stamens 5, included; filaments very short, inserted below middle of tube. Style gynobasic, included; stigma capitate. Nutlets 2–4 (1–2 by reduction), transversely ovoid-oblong to pear-shaped with a triangular-ovate disc, attached obliquely near base, sometimes nutlets united with gynophore and upper nutlet wing-cup-shaped, others triangular and wingless.

About 20 species in tropical and temperate Asia, Europe and Africa; 2 species in Iraq.

Khoshsokhan et al. (2018) in their recent analysis of the phylogeny of the tribe Rochelieae have shown two clades exist that correspond to the two subtribes Eritrichiinae and Heterocaryinae (with *Eritrichium, Hacklelia, Lappula* and *Rochelia* in the former and *Heterocaryum, Suchtelenia* and *Pseudoheterocaryum* in the latter). They showed that whereas *Rochelia* and *Hacklelia* were monophyletic, the other three genera in the Eritrichiinae were not. Their analysis showed that *Lappula sessiliflora* and *L. drobovii* are closely allied to *Rochelia* – which they include under *Rochelia*.

Chacón, J., Luebert, F., Hilger, H.H., Ovchinnikova, S., Selvi, F., Cecchi, L., Guilliams, C.M., Hasenstab-Lehman, K., Sutorý, K., Simpson, M.G. & Weigend, M. (2016). The borage family (Boraginaceae s. str.): A revised infrafamilial classification based on new phylogenetic evidence, with emphasis on the placement of some enigmatic genera. Taxon, 65(3): 523–546.
Khoshsokhan-Mozaffar, M., Sherafati, M. & Kazempour-Osaloo, S. (2018). Molecular phylogeny of the tribe Rochelieae (Boraginaceae, Cynoglossoideae) with special reference to *Lappula*. Annales Botanici Fennici 55(4–6): 293–308.

Calyx lobes lanceolate, incurved around the middle in fruit; nutlets
visible . . . . 1. *R. disperma*
Calyx lobes cordate-auriculate, erect to somewhat curved in fruit,
nutlets not visible . . . . 2. *R. cardiosepala*

1. **Rochelia disperma** (*L.f.*) *K.Koch* in Linnaea 22: 649 (1849); Popov in Fl. URSS 19: 551 (1953); Rawi in Dep. Agr. Iraq Tech. Bull. 14: 140 (1964); Y. Nasir in Fl. Pakistan [Ali & Y.

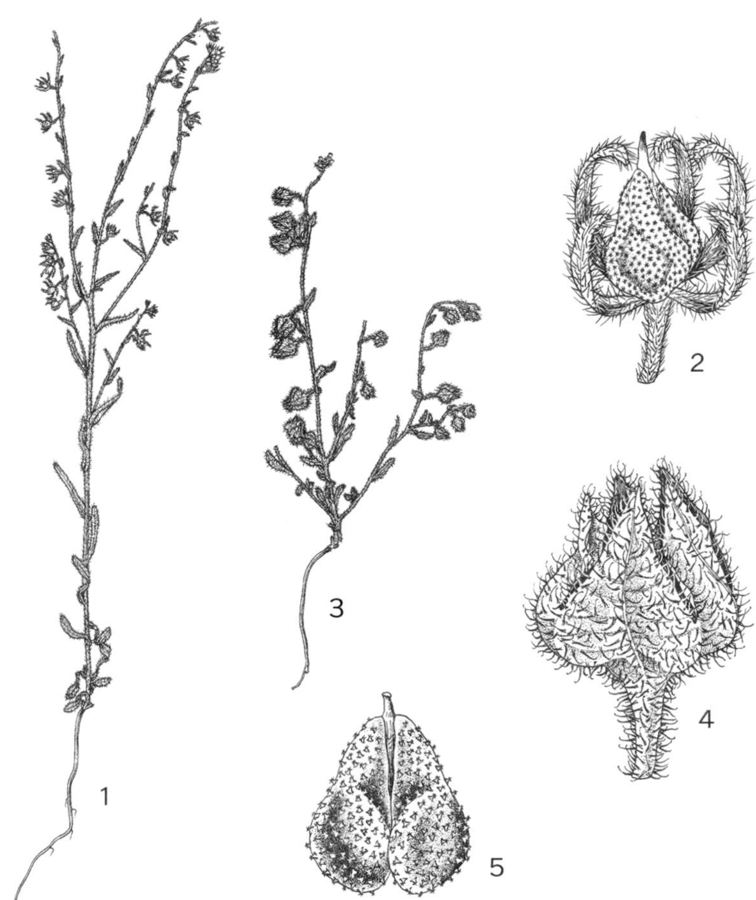

Fig. 108. **Rochelia disperma**. 1, habit × ½; 2, fruit × 7. **Rochelia cardiosepala**. 3, habit × ½; 4, fruiting calyx × 6; 5, fruit × 10. Reproduced with permission from Fl. Pakistan 191: f. 41, C–G (1989). Drawn by S. Hameed. © National Herbarium, Pakistan Agriculture Research Council & University of Karachi, Pakistan.

Nasir] 191: 153 (1989); Riedl in Fl. Iranica [K. H. Rechinger] 48: 94 (1967); Khatamaz, Fl. Iran 39: 299 (2002); Edmondson in Fl. Turkey [P. H. Davis] 6: 262 (1978).

Lithospermum dispermum L.f., Sp. Pl. ed. 2, 1: 191 (1762).
L. retortum Pall., Reise Russ. Reich. 3(2): App. 718 (1776).
Rochelia stellulata Rchb., Iconogr. Bot. Pl. Crit. 1: 13, t. 123 (1824).
R. disperma (L.) Wettst. in Denkschr. Akad. Wiss. Wien l. Ahth. 2, 31 (1885).
R. retorta (Pall.) Lipsky in Trudy Imp. S.-Peterburgsk. Bot. Sada 26: 455 (1910).

Straggly annual, much-branched from base. Basal leaves spatulate to obovate, 10–30 x 3–5 mm, obtuse, petiolate; cauline leaves linear-oblong to narrowly linear, sessile; bracts linear, 3–10 mm. Pedicels 3–5 mm, deflexed, with squarrose hairs. Calyx lobes minute in flower, lanceolate, accrescent to 5–8mm in fruit, incurved from middle, clothed with hooked setules. Corolla sky-blue, 2 mm; limb c. 1.2 mm in diameter. Nutlets 2.5 mm, pear-shaped. Fig. 108, 1–2.

HAB. Rocky mountain, serpentine screes, red stony soil, mountain slope near stream, dry grassy bank in irrigated orchard, by spring, coppiced Quercus forest on limestone; alt. 610-2200 m; fl. Apr.-Jun.
DISTRIB. Widespread in the northern and eastern sectors of the forest zone of Iraq, and extending above. **MRO**: S. slope of Jabal Karoukh, *Kass & Nuri* 27505!; Gali Dergala, 35 km N.W. by N. of Rania, *Rawi, Nuri*

& Kass 28913!; Salahaddin, *Chapman* 11980!; Shaqlawa, *Haines* W. 771!; Ain Shaikh Barakat, N.W. of Haji Omran, *Alkas* 181570!; Haji Omran, *Chapman* 12284!; unloc., "Mesopotamia", *Aucher* 2360! unloc., 'Mesopot., Kurdistan & Mosul', *Kotschy*. **MSU**: Penjwin, *Rawi* 12171!; Penjwin, Malakawa pass, *Rechinger* 12320! **MJS**: Jabal Sinjar, *Kaisi & Hamad* 49166!; Kursi, *Gillett* 10963! S.N.!

N.W. Africa, E. Romania to Central Asia and Pakistan, Oman.

2. **Rochelia cardiosepala** *Bunge* in Mém. Acad. Imp. Sci. St.-Pétersbourg Divers Savans vii. 420 (1851); Boissier, Fl. Orient. 4: 246 (1875); Popov in Fl. URSS 19: 563 (1953); Rawi in Dep. Agr. Iraq Tech. Bull. 14: 140 (1964); Y. Nasir in Fl. Pakistan [Ali & Y. Nasir] 191: 154 (1989); Riedl in Fl. Iranica [K. H. Rechinger] 48: 90 (1967); Khatamaz, Fl. Iran 39: 299 (2002); Edmondson in Fl.Turkey [P. H. Davis] 6: 263 (1978).

Annual herb, with simple or branched stems. Basal leaves spatulate, 8–20 x 2–4 mm, obtuse, shortly petiolate; cauline leaves linear, sessile; bracts linear, 3–10 mm. Pedicels 1–2 mm, elongating and becoming deflexed in fruit. Calyx lobes minute in flower, cordate-auriculate, accrescent in fruit,, clothed with hooked setules, keel prominent. Corolla blue, 2 mm; limb c. 1 mm in diameter. Nutlets 3–4 mm. Fig. 108, 3–5.

HAB. Mountain; alt. 1280–2100 m; fl. Apr.-Jun.
DISTRIB. Forest zone of N. Iraq; not common. **MRO**: Arbil, Haji Omran, *Chapman* 12297. **MSU**: Penjwin, *Rawi* 22546!; *Rechinger* 10456.

Turkey, C. Asia, Caucasus, Iran, Afghanistan, Pakistan, W. Himalaya.

Rochelia persica Boiss. is recorded in Flora Iranica from Mt Avoroman above Tawilla (2100 m, *Rechinger* 10377) and Mosul, Sirsang (1000 m, *Haines* s.n.). *R. persica* is very similar to *R. disperma* with the main distinction of the calyx lobes curved in *R. disperma* and more erect in *R. persica*. I have not seen any of the two specimens cited for Iraq.

28. **CYNOGLOSSUM** L.

Sp. Pl. 1: 134 (1753)
Weigend & al. in Kubitzki, Fam. Gen. Vasc. Pl. Gen. Vasc. Plants 14: 94 (2016)

C.C. Townsend
Revised by Shahina A. Ghazanfar

Perennial and biennial (in Iraq) herbs. Leaves basal and cauline, the basal often with long petioles. Inflorescence with cymes terminal and lateral, ebracteate, elongated in fruit. Flowers pedicellate. Calyx about 2/3-lobed, accrescent in fruit. Corolla rotate, tube shorter than calyx. Faucal scales oblong or subquadrate, ± closing throat. Stamens included, inserted at or above middle of tube; anthers ovoid or narrowly elliptic. Nutlets ovoid to orbicular, dorsal surface (disc) convex or depressed, glochidiate, attached to a narrowly conical gynobase.

About 80–100 species, subcosmopolitan; a single species in Iraq.

Selvi, F. & Jarvis, C.E. (2011). Typification of the name *Cynoglossum creticum* Mill. (Boraginaceae). Taxon 60 (5): 1477.

1. **Cynoglossum creticum** *Mill.*, Gard. Dict. ed. 8, n. 3 (1768); Rawi in Dep. Agr. Iraq Tech. Bull. 14: 136 (1964); Riedl in Fl. Iranica [K. H. Rechinger] 48: 143 (1967); Khatamaz, Fl. Iran 39: 360 (2002).

C. pictum Sol. in Aiton, Hort. Kew. ed. 1, 1: 179 (1789).

Biennial, stems erect, stout, 20–75 cm. Basal leaves lanceolate-elliptic, 9–12 × 1.5–3.5 cm, subacute, tapering into a short petiole, pubescent and tomentose, somewhat greyish; cauline leaves sessile, smaller; apex subacute, base amplexicaul. Cymes ebracteate except at lowermost branches, tightly curled in bud, greatly elongated in fruit. Calyx pubescent, 5-lobed to 2/3, 6–8mm in flower, to 12 mm in fruit. Corolla whitish or pink, becoming bluish and developing reticulate deep blue venation, shortly campanulate to funnel-shaped.

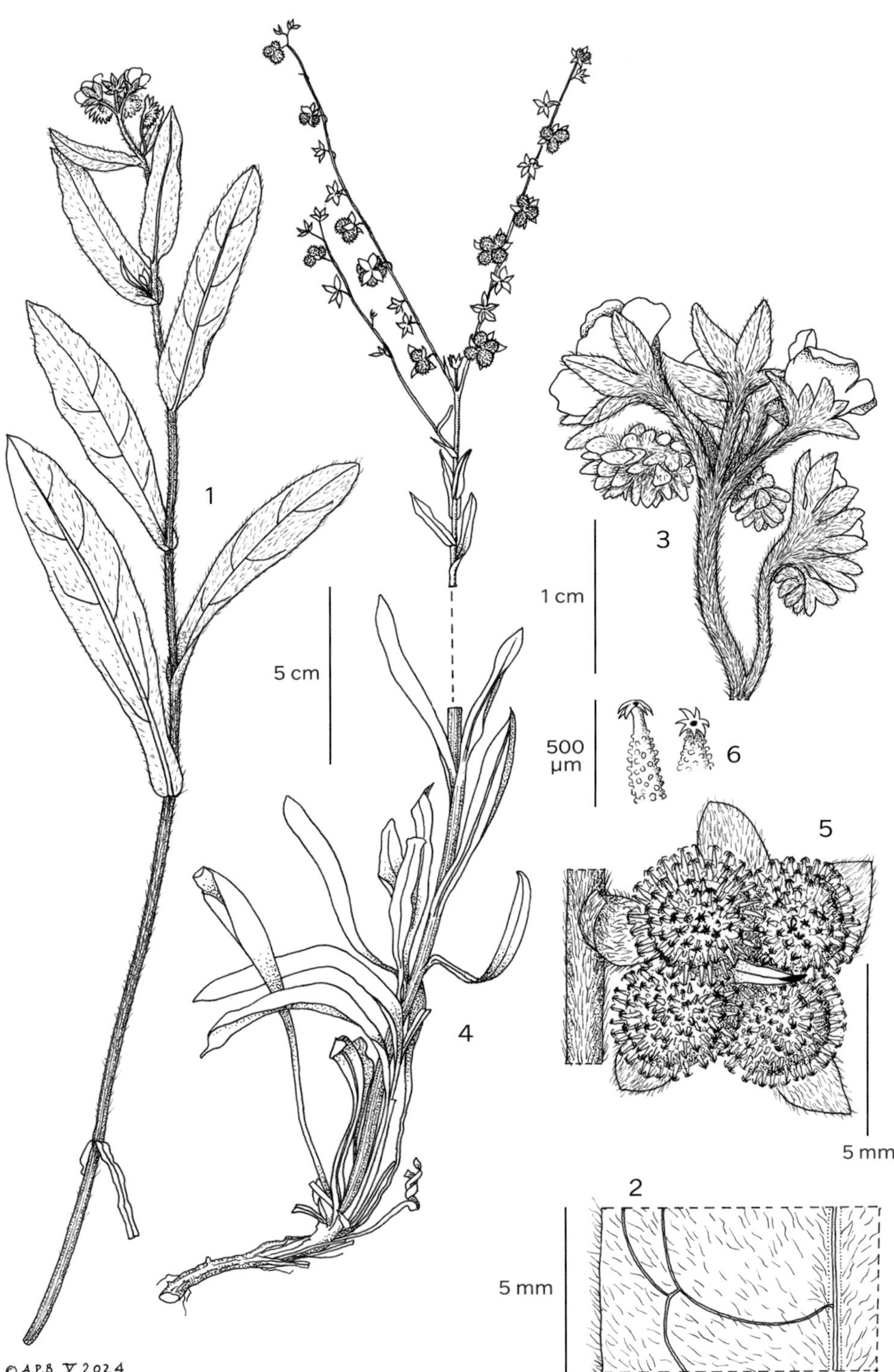

Fig. 109. **Cynoglossum creticum**. 1, flowering stem; 2, adaxial surface of leaf; 3, inflorescences; 4, fruiting stem (the upper and lower parts of separate plants of similar size); 5, mature nutlets seen from above; 6, side and three quarters enlargement of sculpted spines on nutlet. Drawn by © A.P. Brown 2024.

Anthers included. Nutlets 6 mm, ovoid to depressed-globose, densely and evenly glochidiate with tubercles between glochids on dorsal surface. Fig. 109, 1–6.

HAB. Irrigated orchards, hillside, by stream, alt. 750–850 m; fl. Apr.-Jun.
DISTRIB. Mainly in the foothills of the forest zone of N.E. Iraq. **MAM**: Aqra, *Rawi* 11400!; Dohuk, *Lazar* 3347!; 8 km S.W. of Dohuk, Aleka valley, *Barkley* 8285!; Daimka 35 km N.E. of Zakho, *Kaisi, Hamid & Hamad* 45367! **MRO**: Shaqlawa, *Omar, Sahira, Karim & Hamid* 38293! *Guest* 3012! *Karim* 37520!; Salah ad-Din, *Karim, Hamid & Jasim* 40871!; Armota, *Omar, Karim, Hamza & Hamid* 37245! **MSU**: Sarchinar, *Gillett* 7689!; Sarchinar, *Omar & Karim* 37952! **LEA**: Ba'qubah, *Graham* s.n.! **LCA**: Jadriyah, *Haines* W. 1628!; Daura, *Haines* W. 1628a!

NASHKĀ (Kurd.-Dohuk), *Lazar* 3347.

S. Europe, Turkey, Cyprus, Syria, Caucasia, Iran, C. Asia.

29. **PARACARYUM** (DC.) Boiss.

Diagn. Pl. Orient. ser. 1, 11: 128 (1849)
Weigend & al. in Kubitzki, Fam. Gen. Vasc. Pl. 14: 96 (2016)
Omphalodes Moench sect. *Paracaryum* DC., Prodr. 10: 159 (1846); *Mattiastrum* Brand, Feddes Repert. 14: 150 (1915)

Shahina A. Ghazanfar

Annual, biennial and perennial herbs. Inflorescence ebracteate or nearly so. Calyx divided almost to base, accrescent or not. Corolla subcylindrical or salver-shaped, bluish, purplish or brownish, with 5 faucal scales. Anthers included or slightly exserted. Nutlets conspicuously winged, the central portion smooth, aculeolate or glochidiate; wing inflexed, membranous, margin entire or glochidiate.

About 10–15 species from Egypt to C. Asia and W. Himalaya; 6 species in Iraq.

Hilger, H.H., Balzar, M, Frey, W. & Podlech, D. (1985). Heteromerikarpie und Fruchtpolymorphismus bei Microparacaryum, gen. nov., (Boraginaceae). Pl. Syst. Evol. 148: 291–312
Sardar, A. Sh. (2018). A new record of *Paracaryum shepardii* (Post & Beauv.) (Boraginaceae) in Iraq. Diyala J. Pure Sciences 14(2) 175.

1. Plant densely woolly-tomentose, indumentum arachnoid on younger
   branches and leaves; nutlets glochidiate-echinulate; wings denticulate
   with numerous glochids . . . . 6. *P. shepardii*
   Plants hispid, not woolly-tomentose; nutlets smooth or glochidiate;
   wings without glochids. 2
2. Nutlets with disc smooth, margins incurved, rugulose, entire or
   irregularly denticulate . . . . 5. *P. rugulosum*
   Nutlets with disc echinulate or glochidiate 3
3. Basal leaves narrowly lanceolate, margins mostly undulate; nutlet disc
   minutely tuberculate; wing inflexed, dentate . . . . .2. *P. undulatum*
   Basal leaves lanceolate-spatulate, linear-lanceolate to oblong to obovate;
   margins not undulate; nutlet disc smooth, weakly glochidiate, densely
   echinulate with elongate glochids 4
4. Nutlets disc with short scattered glochids; wing not incurved (2–3.5 mm
   broad) . . . . 4. *P. cristatum*
   Nutlets with central area smooth, weakly glochidiate along keel or
   densely echinulate with elongate glochids; marginal wing inflexed, dentate 5
5. Nutlets with central area smooth, weakly glochidiate along keel;
   marginal wing inflexed, dentate . . . . 1. *P. sintenisii*
5. Nutlets with disc densely echinulate with elongate glochids; wing
   strongly incurved, margin denticulate, teeth in 2 rows . . . . 3. *P. hirsutum*

1. **Paracaryum sintenisii** *Hausskn.* ex *Bornm.* in Beih. Bot. Centralbl. Abt. 2, 20(2): 191 (1906).

Cynoglossum sintensii (Bornm.) Greuter & Burdet in Willdenowia 11(1): 36 (1981).
Paracaryum modestum Bornm. in Beih. Bot. Centralbl. 33: 177 (1915), non Boiss.

Densely hispid perennial herb, 20–60 cm, branched. Basal leaves lanceolate-spatulate, petiolate, 5–12 x 0.5–1.5 cm; cauline leaves smaller, sessile. Calyx 3.5–4 mm, deeply divided; lobes lanceolate. Corolla cylindrical-campanulate, 5–6 mm, limb bright blue-violet, equalling tube. Faucal scales oblong, emarginate. Anthers included. Style short. Nutlets ovoid, 7–8 mm, central area smooth or weakly glochidiate along keel, marginal wing inflexed, dentate.

HAB. Rocky limestone ridge, open *Quercus* forest, coppiced *Quercus* forest on limestone; alt. 500–1400 m; fl. Apr.-May.

DISTRIB. Widespread in the northern and eastern sectors of the forest zone of Iraq. MAM: Aqra, *Rawi* 11432!; Bekma, *Gillett* 8263!; Bakarma, *Rawi* 9419-E! 8519C!; Zakho pass, *Guest* 2261!; Zawita gorge, *Guest* 2199! MRO: Salah ad-Din, *Chapman* 11965!; Shaqlawa, *Poore* 358!; mt Sakri Sakran, *Bornmuller* 1616! MJS: Jabal Sinjar, *Gillett & Chapman* 11151!

Turkey (E. Anatolia), W. & S. Iran.

2. **Paracaryum undulatum** *Boiss.*, Diagn. ser. 1, 11: 129 (1849).

Omphalodes stricta C. Koch, Linnaea 19: 302 (1843).
Paracaryum strictum (non Boiss.) Brand in Engler, Pflanzenreich 4, 252: 50 (1921).

Perennial; stems 15–50 cm, patent-hispid, hairs tubercle-based. Basal leaves narrowly lanceolate, margins mostly undulate. Inflorescence corymbose; bracteoles absent. Calyx 3–4 mm. Corolla blue, 4.5–6 mm; faucal scales emarginate. Central area of nutlet minutely tuberculate; wing inflexed, dentate.

HAB. Sandy patches; alt. 1000–1900 m; fl. Apr.-May.

DISTRIB. Rare and very localized in the moyntains of N. Iraq. MAM: Jabal Khantur, N. of Zakho, *Rechinger* 12079, 12090; Aqra, *Rawi* 11432! MRO: Handren Dagh, *Nabélek* 572. DSD: Ruhaba to Zubeida's well, *Agnew, Hadač, Haines & Waleed* s.n. 13.4.1961!

Turkey, Caucasia, N. Iran.

3. **Paracaryum hirsutum** (*DC.*) *Boiss.*, Diagn. ser. 1, 11: 130 (1849).

Omphalodes hirsuta DC., Prodr. [A.P. de Candolle] 10: 159 (1846).
Paracaryum strictum Boiss., Fl. Orient. [Boissier] 4: 256 (1879) pro parte.

Densely canescent-strigose biennial or perennial; stems to 30 cm, erect. Basal leaves linear-lanceolate to oblong, petiolate; cauline ovate, sessile or nearly so. Inflorescence subcorymbose or elongate, branches ± divaricate, white lanate towards tips of cymes. Calyx 4–4.5 mm. Corolla violet-blue, 7–8 mm, tube and limb subequal; scales crescent-shaped, 2.5 mm, apices shallowly emarginate. Style 2–3 mm. Nutlets 7–8 × 4–5 mm, disc densely echinulate with elongate glochids; wing strongly incurved, margin denticulate, teeth in 2 rows.

HAB: Pebbly, stony slopes, stony clay hillside; alt. 850–1000 m; fl. Apr.-May

DISTRIB Occasional in N. Iraq. MAM: Aqra, anon. 6067!; Dihok, *Raoul* (Hb. Haines) W. 1442! MRO: Hab as-Sultan Dagh, *Rawi, Nuri & Kass* 28165! 28213! 11432!

Syria, Turkey, W. Iran.

4. **Paracaryum cristatum** (*Schreber*) *Boiss.*, Diagn. ser. 1, 11: 131 (1849).

Cynoglossum cristatum Schreber in Nova Acta Acad. Leop.-Carol. 3; 476 (1767).
Omphalodes cristata (Schreber) Schrank in Denkschr.Akad. Wiss. München 3: 221 (1812) non DC. (1846).
Rindera cristata (Schreber) G.Don, Gen. Syst. 4: 310 (1838).
Mattiastrum cristatum (Schreber) Brand in Feddes Rep. 14: 154 (1915).

subsp. **carduchorum** *R.Mill* in Notes R.B.G. Edinburgh 35: 306 (1977).

Whitish hirsute perennial or biennial. Stems 12–45 cm. Basal leaves ± narrowly obovate, blade 25–85 × 3–15 mm; cauline obovate to ovate. Inflorescence terminal, bifurcate. Pedicels equalling calyx; fruiting pedicels (4–)6–13 mm. Corolla tube shorter than limb, purplish blue or purple, 3–6 mm. Scales trapeziform, apex incurved. Nutlets suborbicular, 8–12 × 6.5–10 mm, central portion (disc) with scattered short glochids; wing 2–3.5 mm broad.

HAB: On hillside, on stony soil; alt. 1500–1900 m; fl. Jun.
DISTRIB. Rare and scattered in N. Iraq. **MRO**: mt. Hawara Blinda N.E. of Haji Umran, *Rawi, Nuri & Kass* 27801! **FAR**: Khalana, *Rawi* 13714!

subsp. CRISTATUM is endemic to Turkey; subsp. CARDUCHORUM is known from S.E. Anatolia and W. Iran. The type subspecies differs in having shorter fruiting pedicels 1.5–4 mm, and a densely long-glochidiate disc.

5. **Paracaryum rugulosum** (*DC.*) *Boiss.*, Diagn. ser. 1, 11: 129 (1849); Feinbrun-Dothan, Fl. Palaest. 3: 64 (1978); Y. Nasir in Fl. Pakistan [Ali & Y. Nasir] 191:167 (1989).

Omphalodes myosotoides (non Schrank) Fresen. in Mus. Senkenberg 1: 170 (1834); *Omphalodes rugulosa* DC., Prodr. 10: 159 (1846); *P. rubriflorum* Stocks in Hooker's J. Bot. (Kew Gard. Misc.) 4: 175 (1852).

Perennial; stems 10–40 cm, erect, ± branched, densely canescent. Basal leaves linear-lanceolate to spatulate, 50–120 × 2–10 mm, apex obtuse, lamina tapering to petiole; cauline leaves linear, sessile. Inflorescence branched, much accrescent in fruit. Calyx 4–5 mm, densely pubescent; lobes linear-oblong. Corolla purplish-blue, 4–5 mm, cylindrical-campanulate, limb and tube subequal, lobes to 1.5 mm. Faucal scales trapeziform, apex emarginate. Anthers included. Style 2–3 mm, included. Nutlets 5–6 mm, trigonal-ovoid, disc smooth, margins incurved, rugulose, entire or irregularly denticulate. Fig. 110, 1–2.

HAB. Sandy and rocky soil, sandy washes in silty depression, rocky limestone ridge; alt. 350–390 m; fl. Mar.
DISTRIB. Locally common in desert region of Iraq and rare elsewhere. **MRO**: Shaqlawa, *Poore* 358! **DSD**: Sbeicha (Shabicha), *Haines* 14.4.1961 s.n.!; 60 km W. of Shabicha, *Chakravarty, Rawi, Khatib & Tikriti* 30092!; 40 km W.N.W. of Shabicha, *Guest, Rawi & Rechinger* 19296!; 13 km W. by N. of Shabicha, *Guest, Rawi & Rechinger* 19264!; Chibritiya, 25 km S. of Ma'niya, *Botany Staff* 42021!

Turkey (E. Anatolia), Sinai, Palestine, Iran, Pakistan.

6. **Paracaryum shepardii** *Post & Beauv.* in Dinsmore, Pl. Postianae & Dinsmorianae 1:8 (1932); Mill in Fl. Turkey [P. H. Davis] 6: 294 (1978); Sardar in Diyala Journ. Pure Sciences 14(2) 175 (2018).

Perennial herb, 35–45 cm; stems densely woolly-tomentose, arachnoid on younger branches and leaves. Basal leaves greenish yellow, elliptic or narrowly elliptic, 22–32 × 3–5 mm, apex obtuse with small mucro, base truncate, margins entire; petiolate; cauline leaves narrowly elliptic or linear with an attenuate base or subamplexicaul. Inflorescence ebracteate; pedicellate to 5 mm, equalling calyx. Calyx 5-lobed ± to base, lobes narrowly lanceolate, 2–5 mm. Corolla pale purple, tube 1.3–3 mm; limb 1–2 mm, lobes with undulate margins. Faucal scales emarginated. Stamens 5; anthers oblong, included. Nutlets ovoid, 5–11 mm, disc rhombic, glochidiate-echinulate; wing flat, margins incurved, denticulate with numerous glochids.

HAB. On mountains, on rocky soil; alt. ± 1700 m; fl. Jul.-Aug.
DISTRIB. Not common in N. Iraq; only a few individuals seen. **MRO**: Rowanduz district, Hasarost mountain, N.E. of Erbil, *Sardar & Khdir* 7481 (ESUH – Herbarium, College of Education, University of Salahaddin, Iraq).

Turkey.

UNIDENTIFIED SPECIMENS

A specimen from **DSD**: Ruhaba, *Karim, Nuri, Hamid & Kadhim* 39955 cannot be identified to species as it lacks mature fruits. Its desert habitat suggests that it could be *P. rugulosum*.

A specimen named **Paracaryum densum** (Rech.f. & Riedl) D.Heller collected from **FPF**: Koma Sank police station nr Mandali, close to the Iranian frontier, *Rawi* 20650 is heavily grazed and atypical. The status of this species in Iraq requires confirmation. A second specimen cited in *Flora Iranica* under this species was collected from the same area: 10 km E. of Mandali, *Rechinger* 12800.

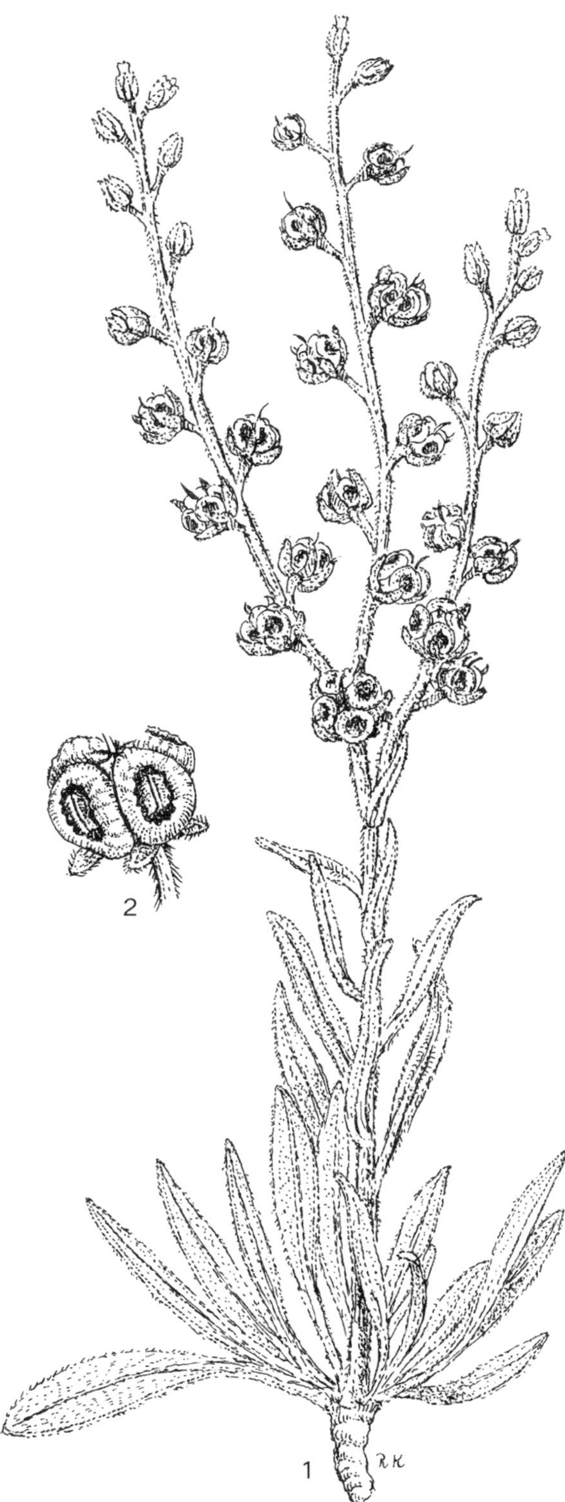

Fig. 110. **Paracaryum rugulosum** 1, habit × 1; 2, fruit × 3. Reproduced with permission from Naomi Feinbrun-Dothan, Flora Palaestina, 3: Plates f. 104 (1977). Drawn by Ruth Koppel. © The Israel Academy of Sciences and Humanities.

30. **RINDERA** Pall.

Reise Russ. Reich. 1: 486 (1771)
Weigend & al. in Kubitzki, Fam. Gen. Vasc. Pl. 14: 96 (2016)
Mattia Schultes, *Cyphomattia* Boiss., *Bilegnum* Brand.

Shahina A. Ghazanfar

Erect perennial herbs with simple stems, ± softly hairy, (rarely glabrescent). Leaves ovate to linear, basal leaves long petiolate. Inflorescence of scorpioid cymes in corymbs. Flowers long pedicellate, reflexed in fruit. Calyx 5-lobed. Corolla cylindrical, longer than the calyx; limb shorter or longer than tube. Faucal scales, distinct, reduced to folds or entirely absent. Filaments included or exserted; anthers oblong. Style long, exceeding calyx, usually exserted from corolla. Nutlets coherent to style, flat, with a broad membranous outer wing and rarely also a narrower ± incurving inner wing.

About 20–25 species, N.W. Africa, S.E. & E. Europe to W. & Central Asia; a single species in Iraq.

Bigazzi, M., Nardi, E. & Selvi, F. (2006) Palynological Contribution to the Systematics of *Rindera* and the Allied Genera *Paracaryum* and *Solenanthus* (Boraginaceae-Cynoglosseae). Willdenowia 36: 37–46.
Weigend, M., Luebert, F., Selvi, F., Brokamp, G. & Hilger, H.H. (2013) Multiple origins for Hounds tongues (*Cynoglossum* L.) and Navel seeds (*Omphalodes* Mill.) the phylogeny of the borage family (Boraginaceae s.str.). Molecular Phylog. Evol. 68: 604–618.

1. **Rindera lanata** (*Lam.*) *Bunge* in Mém. Acad. Imp. Sci. St.-Pétersbourg Divers Savans 7: 415 (1851); Rawi in Dep. Agr. Iraq Tech. Bull. 14: 140 (1964); Riedl in Fl. Iranica [K. H. Rechinger] 48: 127 (1967); Mill in Fl. Turkey [P. H. Davis] 6: 301 (1979); Khatamaz, Fl. Iran 39: 428 (2002).

Cynoglossum lanatum Lam., Encycl. 2: 238 (1786).
Mattia lanata (Lam.) Schultes, Obs. Bot. 31 (1809).
M. umbellata C.Koch in Linnaea 17: 302 (1843), non Schultes (1809).
M. brachyantha Boiss., Diagn. Pl. Orient. ser. 1, 11: 127 (1849).
Rindera pubescens C.Koch in Linnaea 23: 648 (1849).
Rindera punctata Bunge in Mém. Acad. Imp. Sci. St.-Pétersbourg Divers Savans 7: 415 (1851).
R. lanata var. *punctata* (Bunge) Kusn. in Trav. Mus. Bot. Acad. Imp. Sci. St.-Pétersbourg. 7:45 (1910).

Erect perennial herb, 20–50 cm. Stem simple or branched in lower part, sparsely hairy to glabrous. Basal leaves long-petiolate, linear to lanceolate, oblong, elliptic or ovate, 20–90 × 4–20 mm, attenuate at base, softly tomentose or pilose, ± tuberculate; cauline leaves narrower, amplexicaul. Cymes numerous forming a large terminal corymb. Pedicels and internodes greatly elongating in fruit. Calyx 3.5–8mm, lobes ± ovate, densely lanate. Corolla pink, maturing to deep blue, campanulate-cylindrical, 7–12 mm, tube 1/2 – 1/4 length of limb. Faucal appendages quadrate, apex emarginate. Stamens included; filaments equal. Style 7–16 mm, ± exserted. Nutlets orbicular, 15–25 × 15–23 mm, smooth, winged with margin ± undulate, lacking glochids. Fig. 111, 1–8.

HAB. *Quercus* forest, on limestone, clay soil between cracks of rocks, stony mountainside, scree slopes above the treeline; alt. 1100–1900 m; fl. May-Jun.
DISTR. Upper forest and montane zones of Iraq. **MAM**: Matina, *Rawi* 8724! **MRO**: mountain N.E. of Ser-i Hasan Beg, *Guest* 2909!; S. slope of Jabal Karoukh, *Kass & Nuri* 27494!; Mt Avroman, *Gillett* 11824! **MSU**: Penjwin, *Rawi* 12232!; Pira Magrun, *Rawi* 12095! 12105! *Poore* 650!; N. Sulaymaniyah, *Thesiger* 436! **MJS**: Jabal Sinjar, *Kaisi & Hamad* 49200! *Kaisi & Wedad* 52166! *Kaisi, Khayat & Karim* 50936! *Omar, Khayat & Kaisi* 52429!

Turkey, Transcaucasus, N.W. & W. Iran. A variable species with basal leaves ranging from narrow to very broad.

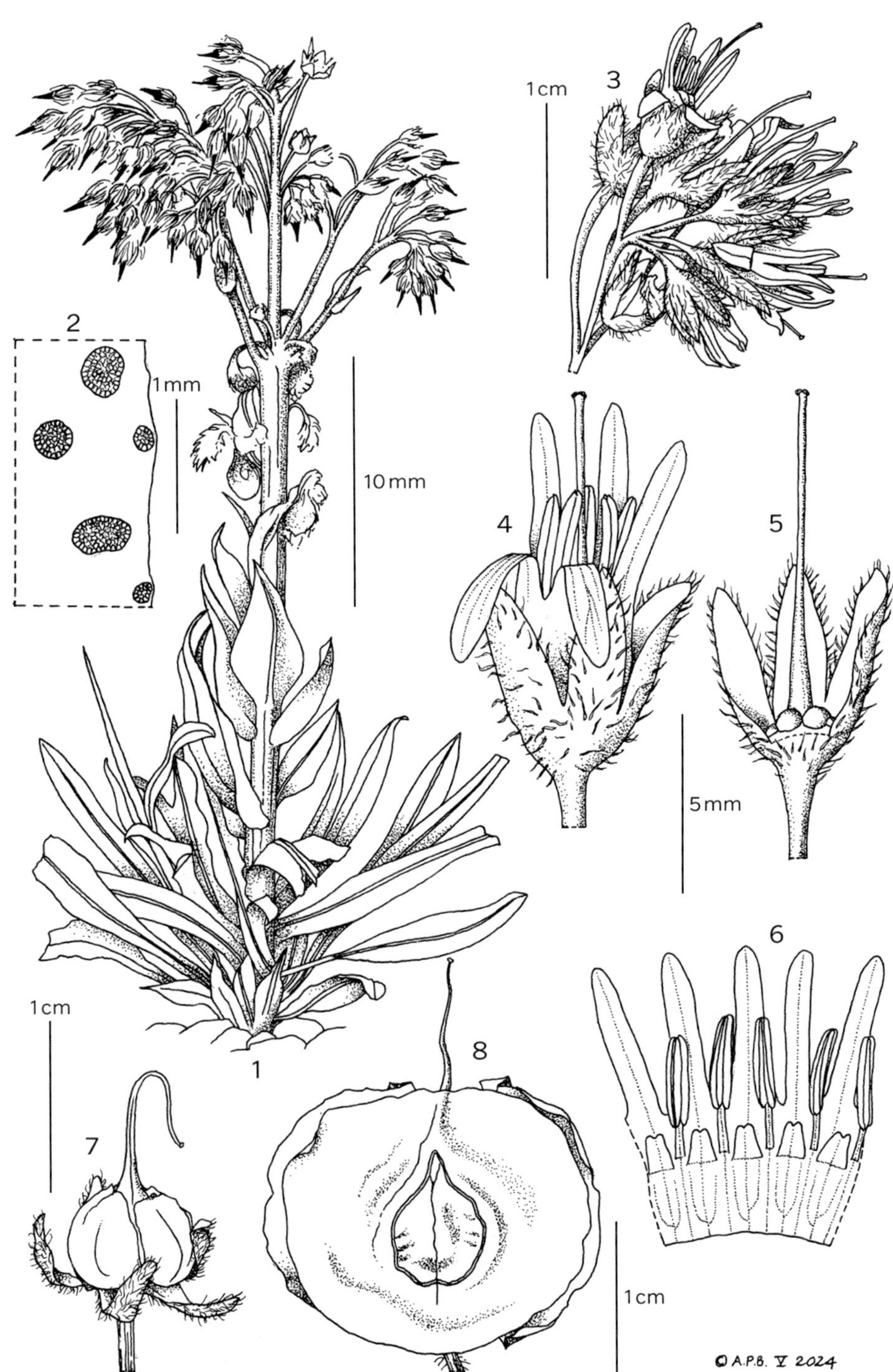

Fig. 111. **Rindera lanata.** 1, habit (from photo – NB scale approximate [based on pressed material]); 2, base of hairs on leaf abaxial surface; 3, part of inflorescence; 4, flower with two corolla lobes folded down to show anthers; 5, flower (dissected) to show ovaries, style and stigma; 6, corolla with stamens in situ; 7, flower with young fruits; 8, mature fruits. Drawn by © A.P. Brown 2024.

31. **SOLENANTHUS** Ledeb.

Ic. Fl. Ross. 8, t.26 (1829)
Weigend & al. in Kubitzki, Fam. Gen. Vasc. Pl. 14: 97 (2016)

C.C. Townsend
Revised by Shahina A. Ghazanfar

Biennial or perennial herbs with a stout rootstock. Stems erect, adpressed-pubescent. Leaves basal and cauline, basal leaves with long petiole. Inflorescence paniculate, made of circinnate cymes. Flowers protandrous. Calyx divided ± to base, lobes oblong-elliptic, enlarging in fruit. Corolla cylindrical or funnel-shaped. Faucal appendages subquadrate to triangular, inserted near middle of tube. Stamens exserted; anthers orbicular-elliptic. Style greatly exceeding calyx, usually exserted; stigma capitate. Nutlets ovoid or subglobose, clothed with spinulose glochids.

About 10 species, S. Europe to Asia; 2 species in Iraq

Basal leaves cordate at base, ± broadly ovate . . . . 1.*S. circinnatus*
Basal leaves tapering into petiole, ± narrowly elliptic to ovate . . . . .2. *S. stamineus*

1. **Solenanthus circinnatus** *Ledeb.*, Fl. Altaic. [Ledebour] 1: 194 (1829); Boissier, Fl. Orient. 4: 270 (1875); Popov in Fl. URSS 19: 642 (1953); Y. Nasir in Fl. Pakistan [Ali & Y. Nasir] 191: 185 (1989); Riedl in Fl. Iranica [K. H. Rechinger] 48: 131 (1967); Khatamaz, Fl. Iran 39: 364 (2002).

Echinospermum circinnatum (Ledeb.) D.Dietr. ex DC., Prodr. [A.P. de Candolle] 10: 164 (1846).
Solenanthus petiolaris DC., Prodr. [A.P. de Candolle] 10: 164 (1846); Rawi in Dep. Agr. Iraq Tech. Bull. 14: 140 (1964).
S. amplifolius Boiss., Diagn. Pl. Orient. ser. 1, 11: 126 (1849).
S. rumicifolius Boiss., Pl. Orient. Nov. Dec. 2: 9 (1875).

Perennial pubescent herb 30–90 cm. Stems unbranched, pubescent. Basal leaves long-petiolate, elliptic- to lanceolate-oblong,10–6 × 5–10 cm, blades truncate or cordate at base; cauline leaves sessile, amplexicaul. Inflorescence much-branched, cymes circinate at flowering. Pedicels 2–3 mm. Calyx 3–5 mm, lobes divided to base, villous. Corolla pale pink to yellowish-red, fading blue, cylindrical to funnel-shaped, c. 5 mm, with short lobes. Filaments unequal, exserted. Nutlets 6–7 mm, ovoid, with marginal glochids much longer than those of dorsal surface. Fig. 112, 1–3.

HAB. Bank on deep earth near stream, shaded old cemetery, in shade of *Quercus* thickets on limestone, beneath shady cliff in rich damp humus, in scree and boulder beds recently uncovered from snow, rocky stony soil; alt. 1400–2500 m; fl. Apr.-Jul.
DISTRIB. Occasional in the upper forest and alpine zones of the mountains of N.E. Iraq. **MAM**: Amadiya, *Guest* 1236! **MRO**: Sefin Dagh, Shaqlawa, *O. Polunin* 5019!; Handren Dagh, *Nabélek* 597; Haji Umran, *C. Boswell* (Hb. Haines) 28–30 Apr. 1957 s.n.!; Shaqlawa, *Haines* W. 770!; S. of Haji Umran, *Poore* 640!; Summit of Jabal Karokh, *Kass & Nuri* 27411! **MSU**: Mt Hawraman above Daimar, *Gillett* 11867!; Mt Hawraman above Tawila, *Rechinger* 10373; Pira Magrun, *Gillett* 7830! *Haussknecht* s.n.; Penjwin, *Rawi* 8805!

KARDO (Kurd.-Pira Magrun), *Gillett* 7830.

Caucasia, Iran, Afghanistan, Pakistan, C. Asia, Siberia.

2. **Solenanthus stamineus** (*Desf.*) *Wettst.* in Denkschr. Akad. Wiss. Wien, Math.-Nat. Kl. 50(2): 88 (1885); Popov in Fl. URSS 19: 652 (1953); Rawi in Dep. Agr. Iraq Tech. Bull. 14: 140 (1964); Y. Nasir in Fl. Pakistan [Ali & Y. Nasir] 191: 186 (1989); Riedl in Fl. Iranica [K. H. Rechinger] 48: 133 (1967); Khatamaz, Fl. Iran 39: 368 (2002).

Cynoglossum stamineum Desf. in Ann. Mus. Natl. Hist. Nat. (Paris) 10: 431 (1807).
Solenanthus tournefortii DC., Prodr. [A.P. de Candolle] 10: 164 (1846).
S. strictissimus Brand in Feddes Repert. 13: 546 (1915).

Erect, softly tomentose perennial. Stems 15–40 cm, densely leafy. Leaves ± densely tomentose to sericeous, narrowly elliptic to ovate, base attenuate into petiole, blade 10–20 × 2.5–4.5 cm, subacute; cauline leaves sessile, elliptic to oblong. Cymes forming a narrowly

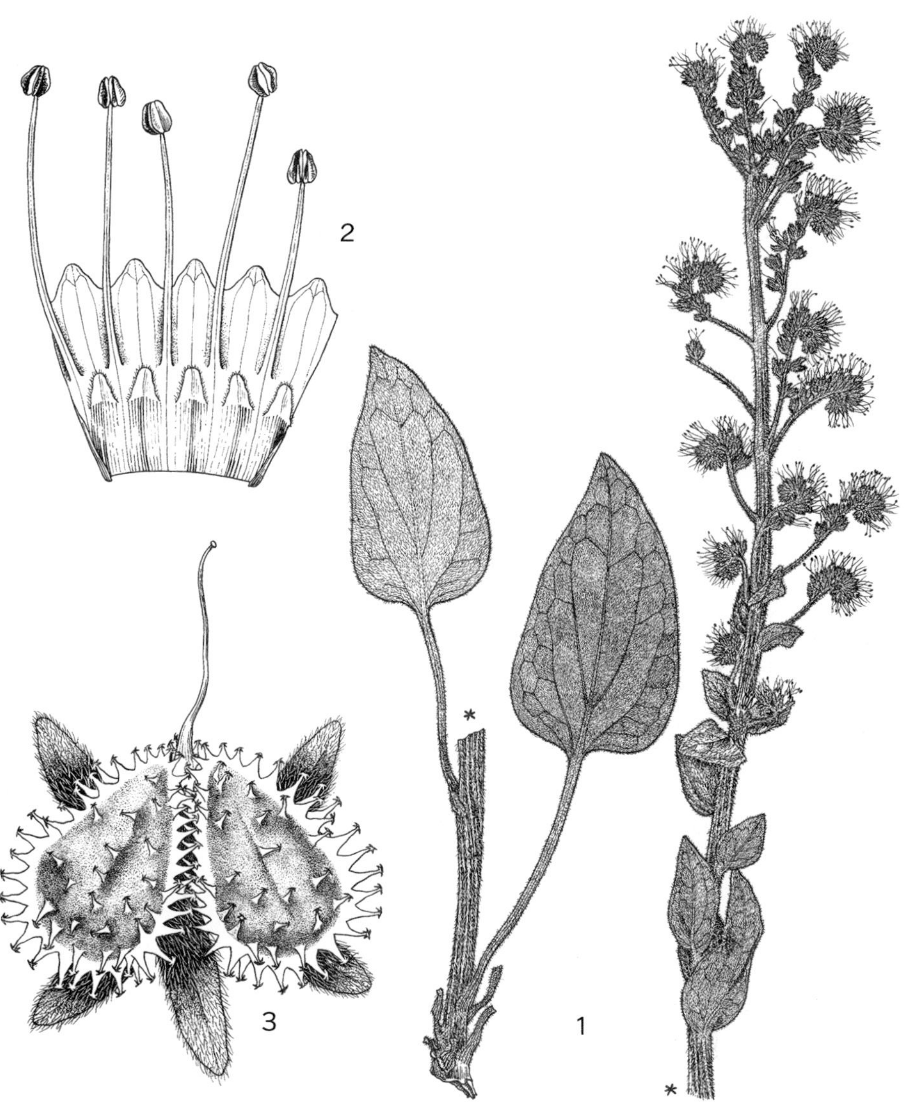

Fig. 112. **Solenanthus circinnatus**. 1, habit, portion of shoot × ½; 2, corolla, opened × 6; 3, fruit × 5. Reproduced with permission from Fl. Pakistan 191: f. 42, D–F (1989). Drawn by S. Hameed. © National Herbarium, Pakistan Agriculture Research Council & University of Karachi, Pakistan.

ovoid panicle. Calyx lobes ovate to lanceolate. Corolla reddish-purple, fading to bluish-violet, 4.5–6.5 mm, limb with short triangular obtuse lobes; scales small, triangular. Filaments slightly unequal; style exserted. Nutlets ovoid, 6.5–10 × 6–11 mm, sparsely glochidiate on dorsal surface, densely glochidiate on margin, glochids similar in length to those on dorsal surface.

HAB. Open rocky scree at head of ravine; alt. ±2500 m; fl. May.
DISTRIB. Alpine zones of the mountains of N.E. Iraq. **MAM**: Sarsang, *Haines* W. 1046! **MSU**: Avroman (Hawraman) & Shahu, *Haussknecht* s.n.!

Greece, Turkey, Lebanon, Syria, Caucasia, Iran, Afghanistan, Pakistan, W. Himalaya, C. Asia,.

SPECIES EXPECTED TO OCCUR IN IRAQ

Echiochilon kotschyi is so far unrecorded from Iraq, but as it occurs on Khark island (its type locality) its occurrence in southernmost Iraq would not be unexpected.

126. **SELAGINACEAE** Choisy

Mem. Soc. Phys. Genèv 2(2), (1822), nom. cons.

Molecular data have indicated that genera formerly included in the family Selaginaceae (together with those included in former families Buddlejaceae and Myoporaceae) are now included in the Scrophulariaceae.

None of the genera formerly included in the Selaginaceae are found in Iraq, most all being restricted to southern Africa. There may be some species cultivated in Iraq, but these are not listed in Husain & Kasim (1975), "Cultivated plants of Iraq and their Importance".

Christenhusz, M.J., Fay, M.F. and Chase, M.W. (2017). *Plants of the world: an illustrated encyclopaedia of vascular plants.* University of Chicago Press.

127. **GLOBULARIACEAE** DC.

in Lam. & DC., Fl. Fr. 3:427 (1805); Wagenitz in Kubitzki (ser. ed.), Fam. Gen. Vasc. Pl. 7: (2004)

J. R. Edmondson

Suffrutescent evergreen herbs and shrublets. Leaves alternate, simple, exstipulate. Flowers in compact globose heads surrounded by involucral bracts. Flowers 5-merous, ± zygomorphic. Calyx 5-fid, ± deeply divided. Corolla blue, bilabiate, upper lip divided into two narrow teeth or lobes, lower longer, ± deeply 3-lobed. Stamens 4, epipetalous, exserted. Ovary superior, 1-locular, with 1 pendent ovule; style exserted, stigma 2-lobed. Fruits dispersed while attached to the calyx.

According to APG IV (2016) the family is now subsumed into the Plantaginaceae, within which it forms the subfamily Globularioideae. As the *Flora of Iraq* follows Hutchinson's system of family delimitation, and the Plantaginaceae has already been described, Globulariaceae is treated separately here. Its two genera, *Globularia* and *Poskea*, are somewhat similar in overall appearance but *Poskea* is confined to Africa and Socotra while *Globularia* has a circum-mediterranean distribution extending into Caucasia.

1. **GLOBULARIA** L.

Sp. Pl. 1: 95 (1753)

Description as for family.

About 28 species found in Macaronesia, Europe, N. Africa, Arabian Peninsula to Iran; 2 species in Iraq.

Globularia, from Lat. *globulus*, a small ball, and *-aria*, a possessive suffix ('belonging to').

Schwarz, O. (1938). Die Gattung *Globularia*. Bot. Jahrb. 69: 318–373.

Capitula numerous, disposed along an arcuate inflorescence; leaves all cauline, not forming a basal rosette . . . . 1. *G. sintenisii*
Capitulum solitary, borne on a scape with cauline leaves much smaller than basal . . . . 2. *G. trichosantha*

1. **Globularia sintenisii** *Hausskn. & Wettst.* ex *Wettst.*, Bull. Herb. Boissier 3: 274, t. 7 fig. 1–5 (1895); Rawi in Dep. Agr. Iraq Tech. Bull. 14: 148 (1964); Edmondson in Fl. Turkey [P. H. Davis] 7: 29 (1982).

A virgate shrublet, basal leaves inconspicuous or lacking. Cauline leaves numerous, lanceolate-elliptic to linear-lanceolate, obtuse to acuminate, subsessile. Capitula numerous, 2.3–3.5 mm in diameter, distributed along upper half of stem. Calyx divided to c. 1/2. Corolla blue-purple, with deeply divided lower lip, segments lanceolate-oblong.

HAB. On limestone rocks, open *Pinus* forest on dolomite, rocky slopes, shady ravine; alt. 915–1520 m; fl. Jul.-Aug.
DISTRIB. Occasional and local in the forest zone of western N. Iraq. MAM. Chiya-e Gara, *Kotschy* 569 A!; Sarsang, *Rechinger* 11695, *Haines* W. 516!; Zawita, *Guest* 3734! 4926! 4934! Zawita, *Rechinger* 33256; 21 km E. of Dohuk towards Sarsang, *Rechinger* 111542! 11572!; Dohuk, *Dabbagh & Hamad* 46140!

Turkey (S.E. Anatolia); a near endemic only found close to the frontier within Turkey.

2. Globularia trichosantha *Fisch. & C.A.Mey.*, Index Seminum [St. Petersburg (Petropolitanus)] 5: 36 (1839); Rawi in Dep. Agr. Iraq Tech. Bull. 14: 148 (1964); Edmondson in Fl. Turkey [P. H. Davis] 7: 29 (1982).

G. *pallida* C.Koch in Linnaea 22: 654 (1849).
G. *vulgaris* Lam. subsp. *trichosantha* (Fisch. & Mey.) Wettst., Bull. Herb. Boissier 3: 286 (1895).

Perennial with branched woody stock forming wide compact cushions. Stems to 25 cm, usually borne singly, occasionally in pairs. Basal leaves forming a rosette, to 20 × 9 mm, narrowly spatulate, tip notched to form 3 small lobes; cauline leaves smaller, linear-lanceolate to elliptic, sessile, acute. Capitula solitary, terminal, 10–20 mm in diameter; involucral bracts lanceolate. Calyx divided to 2/3, lobes narrowly lanceolate, ± erect. Corolla violet to purple or dark blue, lower lip divided into 3 linear segments, upper lip divided into 2 linear segments. Fig. 113, 1–5.

HAB. Mountainside on clay soil, exposed limestone ridge, open rocky ground in *Quercus* and *Pinus* forest; alt. 900–1370 m; fl. Apr.-May; fr. Jul.
DISTRIB. Occasional and local in the forest zone of western N. Iraq. MAM. Zawita nr Sharanish, *Rechinger* 12005!; above Sawara tuka, *Rechinger* 11578, *Guest* 2211! C. *Boswell* (Hb. Haines) W. 1107! *Haines* W. 1109!; Baidi to Kani Masi, Karim, *Dabbagh & Hamad* 44842!; Sarsang, *Al-Kaisi et al.* 43446! *Barkley* 8132! O. *Polunin* 5076! *Haines* W. 1109!; nr Rubal Waterfall 3 km N.E. of Amadiya, *Barkley* 8203!; Atrush, *Guest* 4392!;Ispindar Saddli, Swaratuka, *Chapman* 26332!; *Al-Kaisi et al.* 43446! (BAG).

Balkans, Turkey, Crimea, Caucasia, N.W. Iran.

128. **LAMIACEAE**

Martinov in Tekhno-Bot. Slovar 355 (1820) nom. cons.
Labiatae Juss. in Gen. Pl. [Jussieu] 110 (1789) (nom. alter.: Lamiaceae) (1789) nom. cons.
Harley, Atkins, Budantsev, Cantino, Conn, Grayer, et al. in Kubitzki, ser. ed., J.W. Kadereit
& V. Bittrich, vol. eds, Fam. Gen. Vasc. Pl. 7: 167 (2004).

Tuncay Derminci, Ferhat Celap, Ali Haloob, S.T. Al-Kaisi, Shahina A. Ghazanfar, J. Edmondson & C.C. Townsend

Annual, biennial and perennial herbs, shrubs, sometimes trees (not in Iraq), usually aromatic. Stems and branches often 4-angled in cross-section. Indumentum simple and stellate, often glandular. Leaves opposite or sometimes whorled, simple, without stipules; lamina entire or lobed, margins serrate, crenate or bullate, sometimes inrolled. Inflorescence thyrsoid, terminal or axillary, in whorls, paniculate, racemose, spikate or condensed into heads; cymes 1- many-flowered. Bracts present. Flowers bisexual, ± actinomorphic, zygomorphic, bilabiate. Calyx segments fused at base, unequal, usually persistent. Corolla (4–)5 (or more), bilabiate, usually with 2 upper and 3 lower lobes or with 5 lower petals forming one lip. Stamens 4, didynamous, alternating with petals and fused to the corolla tube; filaments declinate or straight; anthers basifixed or dorsifixed, often versatile and connivent, opening lengthwise; connective often prolonged or modified into a lever-like structure (*Salvia*). Nectar disc entire or lobed. Ovary superior; carpels 2(–5). Style single, gynobasic, terminal or subterminal; bifid at apex; stigma punctate. Fruit schizocarpic with 4 nutlets, 1(–2)-seeded (in Iraq plants). Seeds sometimes mucilaginous.

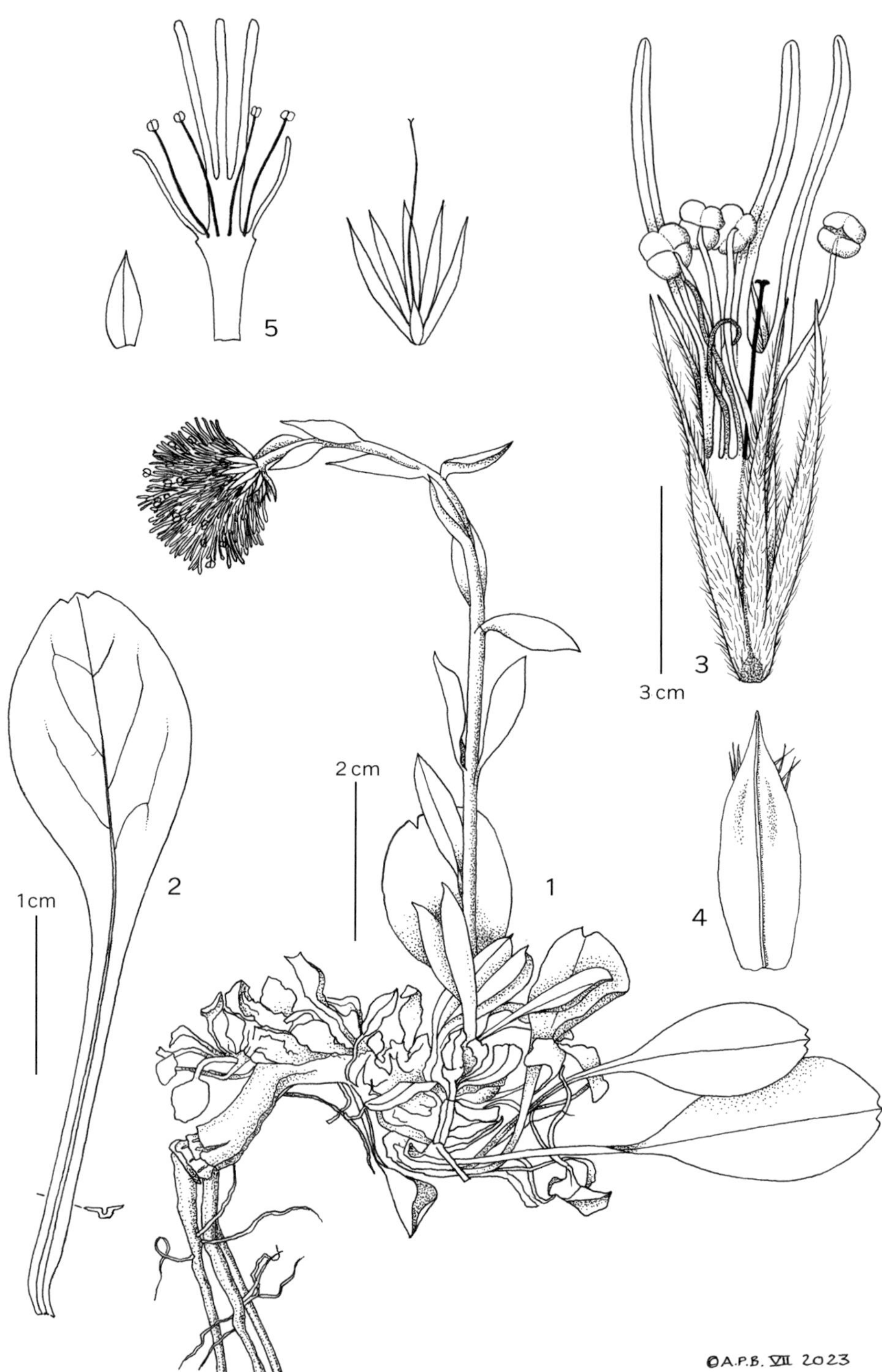

Fig. 113. **Globularia trichosantha**. 1, habit; 2, leaf rosette; 3, flower; 4, bract subtending flower; 5, diagram of flower structure. 1–5 from *Karim et al.* 44842. Drawn by © A.P. Brown 2023.

A large family of 241 genera and more than 6,800 species, cosmopolitan; 32 genera and 168 species in Iraq.

Several members of the Lamiaceae are of wide economic importance used as culinary herbs (basil, oregano, thyme, sage), medicinal herbs (mint, peppermint, thyme), in cosmetic industry in soaps and creams (lavender), and as landscape ornamentals (*Salvia* spp.).

Bendiksby, M., Thorbek, L., Scheen, A.C., Lindqvist, C. and Ryding, O. (2011). An updated phylogeny and classification of Lamiaceae subfamily Lamioideae. Taxon 60(2): 471–484.

Harley, R.M., Atkins, S., Budantsev, A.L., Cantino, P.D., Conn, B.J., Grayer, R., et al. (2004). In: The families and genera of vascular plants (K. Kubitzki, ser. ed.). Lamiaceae (J.W. Kadereit & V. Bittrich, vol. eds), vol. 7: 167–275.

Hedge, I. C. (1990). Labiatae. In: S.I. Ali & Y. J. Nasir (Eds.), Flora of Pakistan. 192: 1–310.

Li, B., Cantino, P.D., Olmstead, R.G., Bramley, G.L., Xiang, C.L., Ma, Z.H., Tan, Y.H. and Zhang, D.X. (2016). A large-scale chloroplast phylogeny of the Lamiaceae sheds new light on its subfamilial classification. Scientific reports, 6(1): 1–18.

Mirzaei, M., Mirzaei, H., Sahebkar, A., Bagherian, A., Masoud Khoi, M.J., Reza Mirzaei, H., Salehi, R., Reza Jaafari, M. and Kazemi Oskuee, R. (2015). Phylogenetic analysis of selected menthol-Producing species belonging to the Lamiaceae family. Nucleosides, Nucleotides and Nucleic Acids. 34(9): 650–657.

Moon, H.K., Smets, E. and Huysmans, S. (2010). Phylogeny of tribe Mentheae (Lamiaceae): The story of molecules and micromorphological characters. Taxon 59(4): 1065–1076.

Oyebanji, O.O., Chukwuma, E.C., Bolarinwa, K.A., Adejobi, O.I., Adeyemi, S.B. and Ayoola, A.O.(2020). Re-evaluation of the phylogenetic relationships and species delimitation of two closely related families (Lamiaceae and Verbenaceae) using two DNA barcode markers. Journal of biosciences 45(1): 1–15.

Xi-wen & Hedge, I. C. (1994). Lamiaceae. In: Z.Y. Wu & P.H. Raven (Eds.), Flora of China. 17: 50–299.

Zhao, F., Chen, Y.P., Salmaki, Y., Drew, B.T., Wilson, T.C., Scheen, A.C., Celep, F., Bräuchler, C., Bendiksby,M., Wang, Q. & Min, D.Z. (2021). An updated tribal classification of Lamiaceae based on plastome phylogenomics. BMC biology 19(1): 1–27.

Harley et al. (2004) recognized 7 Subfamilies, 4 Tribes and 9 Subtribes. Following Harley & al. Iraq Lamiaceae is represented in 4 Subfamilies. Recently Zhao et al. (2021) updated the tribal classification of the Lamiaceae based on plastome phylogenomics. They recognized 12 subfamilies (all monophyletic) and 22 Tribes within Lamiaceae. According to their study Iraq Lamiaceae is represented in 4 Subfamilies and 9 Tribes.

We have followed Harley & al. (2004) for the classification of Lamiaceae in Iraq.

1. Style gynobasic. 2
   Style terminal or subterminal 3
2. Plants usually strongly aromatic; rosmarinic acid present; pollen usually
   hexacolpolate, 3-celled . . . . .Subfamily Nepetoideae
   Plants not aromatic; rosmarnic acid absent; pollen usually tricolpolate,
   2-celled . . . . .Subfamily Lamioideae
3. Corolla zygomorphic, 2-lipped . . . . Subfamily Scutellarioideae
   Corolla ± actinomorphic, 1-lipped . . . . Subfamily Ajugoideae

Subfamily Ajugoideae
   Antherthecae confluent at anthesis .Tribe Teucrieae (*Teucrium*)
   Antherthecae separated at anthesis . . . . .Tribe Ajugeae (*Ajuga*)

Subfamily Scutellarioideae
   Inflorescence thyrsoid with cymes 1-flowered forming a raceme-like
   inflorescence . . . . *Scutellaria*

Subfamily Lamioideae
1. Calyx lobes abruptly narrowed at apex and expanded at corolla
   margins. . . . . *Phlomis, Phlomoides*
   Calyx lobes not abruptly narrowed. 2
2. Leaves lobed; venation palmate; stamens short, included in corolla tube . . . *Lagochilus*
   Leaves not lobed; stamens included or shortly exserted. 3
3. Calyx campanulate or weakly 2-lipped; calyx lobes often spiny; calyx
   throat often hairy; stamens exserted . . . . *Sideritis, Stachys, Thuspeinata, Lamium*
   Calyx widely campanulate; stamens included or shortly
   exserted. . . . . *Ballota, Marrubium, Molucella, Psedodictamnus*

Subfamily Nepetoideae
    Corolla not distinctly 2-lipped; stamens declinate (bent or curved
        downwards or aside), lying along the upper lip of corolla; anthers
        synthecous . . . . Tribe Ocimeae (*Ocimum*)
    Corolla distinctly 2-lipped; stamens divergent, not declinate. . . . .
        Tribe Mentheae (*Clinopodium, Cyclotrichium, Hymenocrater, Lallemantia,*
        Lycopus, Melissa, Mentha, Micromeria, Nepeta, Origanum, Pentapleura,
        Prunella, Salvia, Satureja, Thymbra, Thymus, Ziziphora)

Key to the genera prepared by A. Haloob and S.T. Al-Kaisi
   1. Corolla 1-lipped, upper lip absent or reduced, or if 2-lipped, corolla persistent 2
      Corolla distinctly 2-lipped 3
   2. Lower lip of corolla 3-lobed; corolla tube with hairy ring within, corolla
      persistent . . . . 1. **Ajuga**
      Lower lip of corolla 5-lobed; corolla tube without hairy ring within,
      corolla deciduous. . . . . 2. **Teucrium**
   3. Fertile stamens 2, posterior pair vestigial or absent 4
      Fertile stamens 4 6
   4. Corolla 5-lobed; staminal connective elongate, longer than filament,
      articulated at junction of filament forming a lever-like structure. . . . . 15. **Salvia**
      Corolla 4-lobed; staminal connective not elongate, not articulating. 5
   5. Inflorescence with verticillasters crowded, subcapitate, capitate or
      spicate; corolla > 7 mm . . . . 26. **Ziziphora**
      Inflorescence with verticillasters globose, distant; corolla < 7 mm . . . . 16. **Lycopus**
   6. Plant with axillary spines; bracteoles spinescent . . . . 10. **Lagochilus**
      Plant without axillary spines 7
   7. Calyx bilabiate, lips entire; upper lip of calyx folded about the middle
      along its length making a erect structure (the scutellum) . . . . 3. **Scutellaria**
      Calyx bilabiate or subbilabiate to regular with 5 teeth; upper lip of calyx
      not forming a scutellum 8
   8. Lower lip of corolla 3-lobed with a often concave and crenulate large
      middle and small lateral lobes, as compared to the middle lobe;
      posterior stamen pair longer than anterior pair; antherthecae
      divergent (180°) . . . . 28. **Nepeta**
      Lateral lobes of lower lip of corolla not small compared to the middle
      lobe; posterior stamen pair shorter than anterior. 9
   9. Stamens included in the corolla tube 10
      Stamens exserted from the corolla tube 14
  10. Verticillasters 2-flowered; calyx pendent in fruit. . . . . 4. **Thuspeinanta**
      Verticillasters with more than 3 flowers; calyx not pendent in fruit. 11
  11. Calyx 5-toothed 12
      Calyx 5–15(–30)-toothed. 13
  12. Calyx strongly ribbed and grooved, 5-toothed (without additional
      intervening teeth in Iraq), each tooth ending with an erect mucro;
      bracteoles present; corolla pink . . . . 12. **Ballota**
      Calyx not strongly ribbed nor grooved; bracteoles absent; corolla yellow. . . 9. **Sideritis**
  13. Calyx 10-veined, 6–20-toothed (16 in *P. aucheri*), teeth equal or
      variously unequal, more or less mucronate or gradually narrowed,
      rarely entire, ± spiny; indumentum of simple and glandular hairs (*P.
      aucheri*) . . . . 13. **Pseudodictamnous**
      Calyx 5–10-veined, 5–15(–30)-toothed, straight or spreading-reflexed,
      not mucronate; indumentum of branched and simple hairs . . . . .11. **Marrubium**
  14. Corolla resupinate. 15
      Corolla not resupinate 16
  15. Leaves entire to weakly dentate; calyx not membranous, < than 10 mm;
      teeth subulate. . . . . 25. **Cyclotrichium**
      Leaves dentate, calyx membranous, > than 15 mm, teeth lanceolate 29. **Hymenocrater**
  16. Upper lip of corolla laterally compressed; hooded in Phlomis or Phlomoides 17
      Upper lip of corolla not hooded 20
  17. Plant with stellate and dendroid hairs . . . . 6. **Phlomis**

Plant with simple hairs 18
18. Corolla shorter than the calyx . . . . 14. **Moluccella**
    Corolla longer than calyx 19
19. Corolla tube abruptly dilated; lateral lobes of the lower corolla lip
      reduced; seeds glabrous. . . . . 5. **Lamium**
    Corolla tube not abruptly dilated; lateral lobes of the lower corolla lip
      not reduced; seeds with tufts of hairs at apex . . . . 7. **Phlomoides**
20. Leaves dentate, crenate or serrate 21
    Leaves entire 26
21. Calyx up to 1.5 mm long; corolla 4-lobed; stamens exserated . . . . 17. **Mentha**
    Calyx longer than 2 mm; corolla 5-lobed; stamens usually included. 22
22. Bracteole aristate; upper lip of calyx 3-toothed, the median broader
      and longer than the lateral . . . . 30. **Lallemantia**
    Bracteole if present not aristate, upper lip of calyx not 3-toothed. 23
23. Calyx 5–10-viened 24
    Calyx 13–15-viened 25
24. Calyx teeth equal or subequal, upper lip erect, 2-lips opened at fruiting . . . . 8. **Stachys**
    Calyx teeth unequal, upper lip flattened, shorter than lower lip, 2-lips
      close at fruiting. . . . . 27. **Prunella**
25. Calyx bilabiate, tubular-campanulate, 13-veined, upper lip flattened,
      shortly 3-toothed, lower lip 2-toothed . . . . 31. **Melissa**
    Calyx usually cylindrical to tubular-campanulate, sometimes gibbous,
      usually 13(–11–15)-nerved, lobes often upcurved, posterior 3 often
      partially fused to form a lip, anterior 2 usually free above tube . . . . 24. **Clinopodium**
26. Calyx conspicuously bilabiate. 27
    Calyx teeth equal or subequal. 30
27. Calyx upper lip ovate-orbicular, deflexed in fruit; corolla upper lip
      4-lobed, lower lip entire or crenate . . . . 32. **Ocimum**
    Calyx upper lip subulate, triangular and lanceolate, not deflexed in
      fruit; corolla upper lip 2-lobed 28
28. Calyx dorsally with 2 lateral ciliate keels . . . . 18. **Thymbra**
    Calyx dorsally without 2 lateral ciliate keels. 29
29. Leaves dotted with yellowish or reddish sessile glands; style exserted . . . . . 19. **Thymus**
    Leaves without sessile glands; style included . . . . 24. **Clinopodium**
30. Calyx 5-angled, 5 main nerves along crest of narrow wings . . . . 21. **Pentapleura**
    Calyx not 5-angled. 31
31. Leaves glandular-punctate; sessile glands on calyx 32
    Leaves and calyx without sessile glands 33
32. Verticillasters 2- several-flowered, aggregated in ± dense spike-like
      inflorescences (spicules) which are often arranged in a panicle or
      false corymb; bracts usually imbricate, as long as calyx and covering it 20. **Origanum**
    Inflorescence of 2-many-flowered axillary vertillasters or in lax cymes;
      bracts not imbricate, not covering the calyx . . . . 22. **Satureja**
33. Calyx > than 5 mm, 5–10-veined, ... 8. **Stachys**
    Calyx < 5 mm, 13–15-veined . . . . 23. **Micromeria**

1. **AJUGA** L.

Sp. Pl. ed. 2: 561 (1753)
Harley & al. in Kubitzki (ser. ed.), Fam. Gen. Vasc. Pl. 7: 201 (2004)
Bugula Mill., Gard. Dict. Abr. ed. 4: s.p. (1754); *Chamaepitys* Hill, Brit. Herb.: 371 (1756);
Moscharia Forssk., Fl. Aegypt.-Arab.: 158 (1775); *Phleboanthe* Tausch, Flora 11: 322 (1828);
Abiga St.-Lag., Ann. Soc. Bot. Lyon 7: 85 (1880); *Rosenbachia* Regel, Trudy Imp. S.-
Peterburgsk. Bot. Sada 9: 613 (1886); *Bulga* Kuntze, Revis. Gen. Pl. 2: 512 (1891)

Tuncay Dirmenci & Ferhat Celep

Perennial, biennial and annual herbs, rarely subshrubs. Stems ascending to erect,
sometimes stoloniferous. Basal leaves present or absent. Cauline leaves entire, toothed to
tripartite. Bracts similar to cauline leaves or differentiated from them, green, reddish or

blue. Inflorescence of dense, verticillastrate cymes, sometimes reduced to solitary flowers or merging to form a terminal, spicate or subcapitate thyrse. Verticillasters 2–6-flowered. Calyx campanulate or obovoid, actinomorphic or weakly zygomorphic (posterior lob reduced), not accrescent, 5-lobed, 10-veined. Corolla 2-lipped or 1-lipped and then only anterior lip present, posterior lip short or obsolete, usually 2-lobed or retuse, anterior lip 3-lobed with 2 lateral lobes and a larger emarginate middle lobe, purple to blue, rarely yellow or white, usually annulate tube. Stamens 4, didynamous, exserted or included in corolla tube. Style not gynobasic, inserted well above base of ovary lobes. Fruit usually schizocarp of nutlets; nutlets oblate to obovoid, reticulately rigged to foveolate, usuallay glabrous.

The genus consists of about 60 species, mainly temperate Europe and Asia: 6 species (12 taxa) in Iraq.

Davis, P.H. (1980). Materials for a Flora of Turkey XXXII. Notes Roy. Bot. Gard. Edinburgh 38(1): 23–32.

1. Corolla purplish blue . . . . .6. *A. orientalis*
   Corolla purple, pink, white or yellow. 2
2. Perennial herbs; leaves entire, narrowly lanceolate to elliptic-oblong;
   corolla yellow . . . . .5. *A. oblongata*
   Perennial herbs to suffruticose, leaves shortly 3-lobed to deeply
   3-partite; corolla pink to purple, white or yellow. 3
3. Corolla white; tube much longer than lower lip; nutlets . . . . 4. A. *vestita*
   Corolla yellow, purple or white, sometimes with purple dots and striate;
   tube much shorter than lower lip; nutlets transversely rugose. 4
4. Middle lobe of floral leaves linear to linear-oblong, at least 4-many
   times longer than wide 5
   Middle lobe of floral leaves oblong to broadly obovate-cuneate, and less
   than 4 × longer than wide 6
5. Suffruticose, woody at base; stems long hirsute; nutlets transversely
   rugose. . . . . 1. *A. zakhoensis*
   Annual or perennial herbs; stems glabrous to pilose; nutlets transversely
   rugulose, often foveolate towards apex . . . . .3. A. *chamaepitys*
6. Middle lobes of leaves to 10 mm broad; corolla pink to purple. . . . . . .2. A. *austroiranica*
   Middle lobes of leaves to 3 mm broad; corolla yellow or white, with
   purple dots or purple striate . . . . .3. A. *chamaepitys*

1. **Ajuga zakhoensis** *Rech.f.*, Fl. Iranica [K. H. Rechinger] 150:12 (1982). Type: Iraq, Mosul, ad confines Turciae prov. Hakkari, in ditione pagi Sharanish, in montibus calc. A Zakho septentrionem versus, Jabal Khantur, in saxosis, 4–9.vii.1957, *Rechinger* 10747 (W–1978-0017279! holo. [E-00296989! G-00342365! K-000975104! S-G-288-photo! iso.]).

Suffruticose, caespitose, stems 10–20 cm tall, many stemmed from the base, prostrate to ascending, hirsute-villous, with sparse sessile glands and glandular papillae above. Leaves 12–15(–25) long, cuneate at base, 3-lobed to 2/3; lobes linear to linear-oblong, 6–12 × 1–2 mm, densely hirsute-villous on both surfaces, with sometimes sparsely sessile glands and glandular papillae, dull green. Floral leaves similar to stem leaves in shape and size, mostly longer than verticillasters. Verticillasters remote, uppermost approximate, 2-flowered. Calyx 5–6 mm long, campanulate-infundibular, hirsute-villous, with sparse sessile glands and glandular papillae, green; teeth unequal, one short, lanceolate to subulate, (1–)2.5–3 mm long, longer ones as long as tube or longer; glabrous in throat. Corolla 12–15(–20) mm long, tube long, white, yellowish-white, with purple stripes along nerves, straight, hirsute-pubescent with glandular papillae, annulate or hairy inside; lower lip 3–3.5 mm long, 3-lobed, hairy. Style unequally bilobed. Stamens 4; filaments glabrous to hairy; anthers hairy at base. Nutlets c. 3 × 1 mm, narrowly oblong, transversely rugose.

HAB. Calcareous rocks; alt. 900–1400 m; fl. & fr. Jul.-Aug.
DISTRIB. Known from calcareous habitats around Zakho district in northern Iraq. **MAM**: Mosul, Ad confines Turciae prov. Hakari, in ditione pagi Sharanish, in montibus calc. N. Zakho: Jabal Khantur, in saxosis, *Rechinger* 10747! (type); In rupestribus calc. montis inter Marsis et montem Zawita N. Zakho Sharanish, *Rechinger* 10896! (E, G, W); in jugo inter Dagh al Raziem et Sharanish, *Rechinger* 12121 (BAG-photo); Zakho, Galli Zawita, N.E. of Zakho, nr. Turkish border, in valley, *Rawi* 23650! (BAG-photo).

Ajuga intermedia Boiss. & Orph., Fl. Orient. [Boissier] 4: 802 (1879).
   A. chamaepitys var. *chia* (Schreb.) Batt. in J.A.Battandier & L.C.Trabut, Fl. Algérie, Dicot.: 716 (1890).
   Chamaepitys chia (Schreb.) Holub, Folia Geobot. Phytotax. 9: 269 (1974).

Stems green, with long stiff patent hairs on all sides. Fig. 114, 1–3.

HAB. Stony slopes, mixed grass and shrubland, vineyards, fallow fields, waste and gravelly ground, mountain side rocky clay soil near *Juglans* orchard; alt. 1000–2000 m; fl. & fr. Apr.-Jul.
DISTRIB. Various habitats in northeastern Iraq. **MAM:** Galli Zawita N.E. of Zakho near Turkish border, *Al-Rawi* 23544 (BAG). **MRO:** Warta village, 30 km N.W. & N. of Rania, saxatile, *Rawi, Nuri & Kass* 28866! (BAG-photo); Shaqlawah, *Haines* 656! (E); Shaqlawah, *Haines* 565 (BAG-photo); between Choman and Haji Umran, *Al-Kaisi & A. Haloob* 60659 (BAG)

Greece, Turkey Crimea, Palestine, N.W. & W. Iran, N. Iraq.

1b. var. **ciliata** *Briq.*, in Ann. Cons. Jard. Bot. Geneve 17:401 (1913–1914).

Ajuga comata Stapf. İn Denkchr. Acad. Wien 1, 50 (1885)
A. chamaepitys (L.) Schreber subsp. *ciliata* (Briq.) Smejkal in Preslia 33: 392 (1961).

Stems glabrous on two opposite sides, retrorse-pubescent or puberulous on the other two sides which sometimes also bear long scattered patent hairs.

HAB. Scrub, pastures, screes, steppes, fallow fields, clay banks; alt. 1000–2240 m; fl. & fr. May-Aug.
DISTRIB. Known from only one location in northern Iraq. **MRO:** Arbil: Kuh Safin prope Shaqlawa, *Bornmüller* 1719 (W).

E. Europe, Caucasia, Turkey, Iran.

2. subsp. **laevigata** (*Boiss.*) *P.H.Davis* in Notes Roy. Bot. Gard. Edinburgh 38: 27 (1980).

Ajuga laevigata Boiss., Fl. Orient. 4: 804 (1879).
Chamaepitys laevigata (Boiss.) Holub, Folia Geobot. Phytotax. 9: 269 (1974).

Stems and leaves glabrous (leaves sometimes puberulent); calyx glabrous.

HAB. In shady places, dry open hillsides, stony places; alt. 1000–1200 m; fl. & fr. Apr.-Jun.
DISTRIB. It grows in shady places, dry open hillsides, stony places in nothern Iraq. **MAM:** Yosufka village, N.W. of Sirsang, *Al-Dabbagh & Hamdi* 45772! (BAG-photo); Sikreen, nr Sirsang, dry open grazer hillside in small clumps, *Haines* 1579! (E). **MSU:** Tainal, Suleymaniya to Kirkuk, *Omar & Karim* 37896! (BAG-photo).

Turkey, Syria, Lebanon.

3. subsp. **mardinensis** *P.H.Davis*, Notes Roy. Bot. Gard. Edinburgh 38: 31 (1980).

Cauline and lower floral leaves 5–12 mm long, cuneate-oblong, 3-fid to 1/4–1/2; plant canescent with short slender dense subadpressed hairs; corolla 14–20(–24) mm, yellow.

HAB. Limestone rocks and calcareous steppe; alt. 800–1200 m; fl. & fr. Apr.-May.
DISTRIB. On limestone rocks on mountains of northern Iraq. **MAM:** Zawita gorge, *Guest* 2201!; **MJS:** Jabel Sinjar N. of the town, in limestone rocks, *Gillett* 11115! (BAG-photo); Singarae (Sinjar), *Haussknecht* 828!; Jabal Sinjar, S.E. slope, rocky limestone, dry soil with *Quercus* shrubs, *Widad & Al-Khayat* 53354! (BAG-photo).

Turkey.

4. subsp. **rechingeri** (*Bilik*) *P.H.Davis* in Notes Roy. Bot. Gard. Edinburgh 38: 31 (1980).

Ajuga rechingeri Bilik, Ark. Bot. a.s., 5: 347 (1960).
Chamaepitys rechingeri (Bilik) Holub in Folia Geobot. Phytotax. 9: 269 (1974).

Densely imbricate cauline and floral leaves; inflorescence columnar; corollas 12–14 mm and yellow.

HAB. Rocky steppe, limestone gullies; alt. 500–1000 m; fl. & fr. May-Jul.
DISTRIB. On limestone rocks on mountains of northern Iraq and desert areas of west Iraq. **MAM:** Mosul: Jabel Bekhair, *Rawi* 8460! (BAG-photo); Bekhair mt., on mountain slope, *Rawi* 23072! (BAG-photo); Dohuk, Dohuk gorge, *Chapman* 26265! (BAG-photo). **MRO:** Rawanduz, hillside, *Omar et al.* 38365! (BAG-photo). **DWD:** 160–190 km W.N. of Ramadi to Rutba, by road side, *Rawi* 31350! (BAG-photo).

Typical specimens were collected between Ramadi and Rutba, and other specimens from northern Iraq. It is distinguished from other subspecies by having a columnar inflorescence and shorter corollas.

Turkey.

5. subsp. **cuneatifolia** (*Stapf*) *P.H.Davis* in Notes Roy. Bot. Gard. Edinburgh 38: 30 (1980).

Ajuga cuneatifolia Stapf. Denkschr. Kaiserl. Akad. Wiss., Wien Math.-Naturwiss. Kl. 50:102 (1885).

Stems patent-pilose, with many subsessile glands visible; floral leaves pilose, greenish, conspicuously glandular, broadly cuneate to obtriangular, divided into broader lobes.

HAB. Rocky limestone slopes, mountain slopes, hillsides, dry soils and rocky crevices; alt. 700–2000 m; fl. & fr. Jun.-Aug.
DISTRIB. On rocky mountain slopes of northern Iraq. **MAM**: Mosul, Alho, *Rawi* 8669-a! (BAG-photo); Mosul, Ad confines Turciae prov. Hakkari, in ditione pagi Shranish, in montibus calc., a Zakho seprentrionem versus, in cacumine Zawita, *Rechinger* 12004 (G, W); Zawita, N.E. of Zakho, *Rawi* 23591! (BAG-photo); Sarsang, Gara Dagh, *Chapman* 26392! (BAG-photo). **MRO**: South slope of Karoukh mountain, *Kass* & *Nuri* 27499!

Turkey.

6. subsp. **tridactylites** (*Ging. ex Benth.*) *P.H.Davis* in Notes Roy. Bot. Gard. Edinburgh 38: 29 (1980).

Ajuga tridactylites Ging. ex Benth., Labiat. Gen. Spec.: 699 (1835).
Chamaepitys tridaclylites (Ging. ex Benth.) Holub, Folia Geobot. Phytotax. 9: 269 (1974).

Upper cauline and lower floral leaves broadly cuneate to obtriangular, divided to 1/4–1/2 into divergent ovate to broadly oblong very obtuse lobes; corolla 12–16(–20) mm, yellow or creamy, with reddish spots, usually drying pink or purplish.

HAB. Hillside, crevices of rocks, stony slopes, mountainside, red stony soils; alt. 800–2000; fl. & fr. Mar.-Jun.
DISTRIB. Grows in various habitats in northeastern Iraq. **MAM**: Amadiya, *Botany staff* 58252 (BAG). **MSU**: Pir Omar Gudrum S.W. slopes, *Gillet* 7750! (BAG-photo); Gia Kaverk, mountainside, *Rawi* 8669! (BAG-photo). **MRO**: Ranya, on hillside, *Rawi et al.* 28613! (BAG-photo); Between Sei-Waka and Dargala villages, E. Karoukh, slope of stony mountains, *Al Kas, Nuri* & *Sarhank* 27603! (BAG-photo); Southern slope of Kauroukh mountain, red stony soil, *Kass* & *Nuri* 27499! (BAG-photo); Shaqlawah, in crevices rocks on N. face muntain slope, *Haines* 660!. **MJS**: In lapidopsis montium Dschebel Sindschar supra oppidium Sindschar, *Handel-Mazzetti* 1487 (W); Ad rupes in faucibus El Magharad montium Dschebel supra oppidum Sindchar, calcareo, *Handel-Mazzetti* 1386 (W); 120 km N.W. of Rutba, sandy gravelly soil, *Hamad* 49947! (BAG-photo).

Gillet 11565 (Shaglawa, 950 m) and *Chapman* 26101 (Zawita gorge, at entry, 700 m) at Kew herbarium were identified as intermediate between subsp. *mardinensis* and subsp. *mesogitana* by Davis (1980). These specimens are similar to subsp. *mesogitana* due to their lanate indumentum. However, they differ from subsp. *mesogitana* by their 25–45 cm long stems, lower stem leaves without lobes or slightly indented at the tip, with lax and long inflorescence. According to the label notes on *Gillet* 11565, this situation is observed in shaded and humid places, while it is stated that it has a more compact inflorescence in dry places. However, the same characteristics are observed in *Chapman* 26101 specimen growing in rocky habitat. It is possible that these specimens collected from two different locations may belong to new subspecies of *Ajuga chamaepitys* or a new species of *Ajuga*.

4. **Ajuga vestita** *Boiss.*, Diagn. Pl. Orient., ser. 1, 7: 62 (1846); Dinsmore in Post, Fl. Syria, Palest. & Sinai ed. 2, 2: 401 (1933); Rawi in Dep. Agr. Tech. Bull 14: 149 (1964); Rechinger, Fl. Iranica [K. H. Rechinger] 150: 14 (1982); Davis in Fl. Turkey [P. H. Davis] 7: 52 (1982).

Perennial herb with stout woody rootstock. Stems procumbent or ascending, 4–15 cm tall, simple, densely lanate, densely foliate. Leaves densely and shortly lanate with glandular papillae on both surfaces. Basal leaves (when present) obovate, to 1.5 × 1 cm, toothed above. Cauline and lower floral leaves broadly cuneate-obovate, 1–2 × 0.7–1.5 cm, broadly and obtusely 3-lobed to 1/6–1/3. Inflorescence dense spike, 2–10 cm long, verticillasters 2-flowered. Calyx 5–7 mm long, teeth lanceolate, 3–4 mm long, to 2 × tube, lanate with glandular papillae; hairy with glandular papillae to ½ inside of teeth, otherwise glabrous in throat. Corolla white, 12–15 mm long, tube much longer than calyx, sparsely to densely lanate with glandular papillae on tube and outside of lips, glabrous inside, but annulate; lower lip much shorter than corolla tube, 2–5 mm long, with a small orbicular shortly emarginate middle lobe. Style bilobed, glabrous to sparsely hairy, included in corolla. Stamens 4, included in corolla; filament glabrous to sparsely hairy. Nutlets 2.5–2.7 × 0.7–0.9 mm, oblong, foveolate.

HAB. On calcareous slopes and crevices; alt. 900–1200; fl. & fr. May-Jun.
DISTRIB. On calcareous slopes and crevices in northern Iraq. **MAM**: Mosul: Ad confines Turciae prov. Hakkari, inter Mosul et Zakho, 103 km N. Mosul versus Zakho, in saxosis calc., *Rechinger* 10641! (G, W). **FUJ**: Mosul, in rupestribus inter Jezireh et Mosul, *Kotschy* 320! (W).

S.E. Turkey and N. Iraq.

Easily distinguished from other species by its densely lanate stems, white corolla, with the lower lip much shorter than corolla tube. Among the specimens given under *Ajuga vestita* in Flora of Iranica (Rechinger, 1982), only *Kotschy* 320 has typical characteristic of *A. vestita*. Some specimens identified as *A. vestita* are now placed under *A. chamaepitys* subsp. *cuneatifolia* (*Rechinger* 12004!). We have not seen *Thesiger* 692 and *Low* 188 to correctly identify if they are *A. vestita*.

5. **Ajuga oblongata** *M.Bieb.*, Fl. Taur.-Caucas., Suppl.: 388 (1819); Pis'youkova in Fl. U.S.S.R [Shishkin & Yuzepchuk] 20:19 (1976).

Ajuga salicifolia Steven, Mém. Soc. Imp. Naturalistes Moscou 3: 265 (1812).
A. arenosa Steven ex Ledeb. Fl. Ross. 2: 449 (1844).
Bulga oblongata (M.Bieb.) Kuntze, Revis. Gen. Pl. 2: 512 (1891).

Perennial herb. Stems 10–30(–50) cm tall, robust, ascending to erect, few to many branched and densely foliate, patent-villous with glandular papillae,. Leaves 20–40 × 5–15 mm, narrowly lanceolate to elliptic, subsessile, cuneate at base, obtuse at apex, entire, densely villous with glandular papillae; floral leaves usually longer than flowers (except the uppermost). Inflorescence long, dense, spike-like or verticillasters sometimes distant. Calyx 5–7 mm, rounded-ovoid or campanulate, long pilose with glandular papillae; teeth 2–4 mm long, one short, narrowly lanceolate to oblong, acute to obtuse at apex; hairy with glandular papillae on inside of teeth. Corolla 20–25 mm, yellow, sparsely to densely pilose outside with short glandular hairs, distinctly annulate; upper lip slightly 2-lobed; lower lip large, densely villous with glandular papillae, lateral lobes lanceolate, obtuse at apex; median lobe reniform, obtuse at apex. Style unequally bilobed, sparingly pilose, exserted. Stamens 4, exserted; filaments and anthers hairy. Nutlets 2.7–3 × 1–1.2 mm, oblong, reticulate-rugose, brown to blackish-brown.

HAB. Fallow lands; alt. 40–50 m; fl. & fr. Apr.-Sep.
DISTRIB. Eastern and central alluvial plains of Iraq. **LEA**: Nahrwan, fallow land, *Gerosurveys* 147!; ibid., *VCR* 5/1031!; Aziziya, *Guest* 216! (BM, BAG-photo). **LCA:** Baghdad, prope urban Baghdad, Kasr Naqib, in alveis et fossas, *Handel-Mazzetti* 938 (W!); ibid., *Nábělek* 1629 (SAV-photo); Baghdad, *Kotschy* 1575! (BM); Baghdad, *Olivier* s.n.! Rustamiya, *Guest* 165!; Dialah, *Sutherland* 13!

A. oblongata is distributed in eastern Caucasus (Azerbaijan) and in the central and eastern alluvial plains of Iraq. There is no record of its distribution in Iran and Turkey. It is likely to occur in Iran and near Tigris (Dicle) riverside in the Southeast of Turkey.

E. Caucasus.

6. **Ajuga orientalis** *L.*, Sp. Pl. 2: 561 (1753); Dinsmore in Post, Fl. Syria, Palest. & Sinai ed. 2, 2: 400 (1933); *Rawi* in Dep. Agr. Tech. Bull 14: 149 (1964); Feinbrun-Dothan, Fl. Paleast. 3: 99 (1978); Rechinger, [K. H. Rechinger] Fl. Iranica 150: 21 (1982); Davis in Fl. Turkey [P. H. Davis] 7: 43 (1982); Jamzad in Flora Iran [Assadi & al., in Persian] 76: 41 (2012); Taifour & El-Oqlah, Pl. Jordan Annot. Checkl.: 101 (2017).

Bugula orientalis (L.) Mill. Gard. Dict. Ed. 8:n. °5 (1768).
Bulga orientalis (L.) Kuntze, Revis. Gen. Pl. 2: 512 (1891).
B. obliqua Moench. Methodus: 381 (1794).

Fig. 115. **Ajuga orientalis**. 1, habit with basal petiolate leaves × 1; 2, cauline leaves (sessile) × 1; 3, inflorescence × 1; 4, flower × 3. Reproduced with permission from Feinbrun-Dothan, Fl. Palaestina 3: Plates, f. 157 (1977). Drawn by Ruth Koppel. © The Israel Academy of Sciences and Humanities.

Rhizomatous perennial. Stems 5–20(–30) cm tall, stout, erect to ascending, simple, sparsely to densely lanate-villous, sometimes ± glabrescent towards the base, glandular above. Basal leaves persistent in flowering, to 10 × 5 cm, petiolate to 10 cm long, oblong-elliptic to ovate, attenuate at base, dentate or ± distinctly crenate to obtusely sinuate, glabrescent above; cauline leaves few, 3–8 × 1.5–3 cm, obovate-oblong to oblong, crenate-dentate. Bracts sessile, ± broadly elliptic to obovate or ovate-orbicular, coarsely 3–5-dentate, sparsely to ± densely white-lanate, glandular, exceeding flowers, often tinged with blue or violet. Verticillasters crowded, 4–12-flowered, lower remote, upper congested. Flowers subsessile. Calyx 6–9 mm, campanulate, lanate-villous with glandular hairs, sometimes glabrous at base; teeth unequal, 3–5 mm, as long as or to 2 × tube, narrowly lanceolate, acute at apex, purplish. Corolla 10–16 mm, blue; tube whitish, glabrous, short hairy like annulus; upper lip 2-lobed, lobes ovate; lower lip resupinate, median lobe slightly emarginate; lips lanate outside. Style unequally bilobed, included in tube. Stamens 4; filament short, glabrous, included in corolla. Nutlets 3–3.2 × 1.5–1.8 mm, oblong-obovoid, reticulate-rugose. Fig. 115, 1–3.

HAB. Forests, scrubland, cultivated fields, fallow fields, rocky slopes, pastures; alt. 0–3000 m; fl. & fr. Apr.-Jul.
DISTRIB. In the Amadiya district of northern Iraq. **MAM**: Amedia [Amadiya], in cultivated fields, *Guest* 1229!

Ukraine, Albania, Greece, Italy, Sicily, Cyprus, East Aegean Is., Turkey, Transcaucasus, Iran, Lebanon, Syria, Palestine, Jordan.

2. **TEUCRIUM** L.

Sp. Pl. 2: 562 (1753); Gen. Pl. ed. 5: 247 (1754); Harley & al. in Kubitzki (ser. ed.), Fam. Gen. Vasc. Pl. 7: 201 (2004)

Maysoon Byati
Revised by Shahina A. Ghazanfar

Perennial herbs, subshrubs (rarely annual herbs), often aromatic. Stem simple or branched from the base or above, erect, ascending to decumbent. Leaves opposite, sessile or petiolate, entire, margins dentate to lobed or pinnatisect, revolute or flat. Inflorescence terminal or axillary, of cymes, spicate thyrse, panicles, cymes often reduced to single flowers in leaf axils or in bracteate racemes or heads. Bracts and bracteoles simple or divided. Calyx regular or 2-lipped, with 5 equal or unequal teeth, often ±gibbous at base. Corolla variable in colour, 1-lipped with anterior lip present, the lip 5–lobed, anterior lobe much larger than others, usually concave. Stamens 4, didynamous, exserted; filaments arched or straight, thecae divaricate. Ovary shallow 4-lobed. Style sub-terminal, stigma bifid, lobes equal or unequal. Nutlets 4, obovoid or spherical, reticulately ridged to alveolate, pubescent to glabrous.

A genus of about 250 species almost cosmopolitan, and with centre of diversity in the Mediterranean region; 10 species in Iraq.

Species of *Teucrium* (*T. polium*, *T. chamaedrys*) are commonly called Germander (Eng.)

Greek *Teukrion* (sing.). *Teucrium* is named after Teucer, the first king of Troy (1400–1000 BCE), who was not only a ruler but a miraculous healer as well. Dioscorides named a medicinal herb after Teucer, which was later adopted by Linnaeus in 1753. Several species of *Teucrium* are used in traditional medicine.

Aksoy, A., Özcan, T., Girişken, H., Celik, J. & Dirmenci, T. (2020). A phylogenetic analysis and biogeographical distribution of *Teucrium* Sect. Teucrium (Lamiaceae) and taxonomic notes for a new species from southwest Turkey. Turkish Journal of Botany. 44(3): 322–337.
Stanković, M. ed. (2020). Teucrium species: Biology and Applications. Springer International Publishing.
Jarić, S., Mitrović, M. & Pavlović, P. (2020). Ethnobotanical features of *Teucrium* species. In Teucrium Species: Biology and Applications (pp. 111–142). Springer, Cham.
Navarro, T. (2020). Systematics and biogeography of the genus *Teucrium* (Lamiaceae). In Teucrium species: biology and applications (pp. 1–38). Springer, Cham.
Salmaki, Y., Kattari, S., Heubl, G. & Bräuchler, C. (2016). Phylogeny of non-monophyletic *Teucrium*

(Lamiaceae: Ajugoideae): Implications for character evolution and taxonomy. Taxon, 65(4): 805–822.

1. Stamens included or sub-included, not clearly exserted. 2
   Stamens clearly exserted 3
2. Corolla slightly longer than calyx. . . . . .6. *T. parviflorum*
   Corolla 2–4 x longer than calyx . . . . .9. *T. macrum*
3. Leaves pinnatisect or lobed at apex 4
   Leaves not pinnatisect. 6
4. Leaves lobed at apex, with 3–5(–8) lobes; nutlets with dense woolly
   hairs . . . . .7. *T. oliverianum*
   Leaves pinnatisect; nutlets hairy or glabrous. 5
5. Corolla lateral lobes unequal; style hairy; nutlets with dense short hairs . .8. *T. rigidum*
   Corolla lateral lobes clearly unequal; style glabrous; nutlets glabrous. . . . .5. *T. orientale*
6. Dwarf plant, 5–10 cm; leaves entire or subdentate, petiolate;
   inflorescence spike like; calyx bilabiate . . . . .2. *T. chasmophyticm*
   Plants not dwarfed, 15–90 cm; inflorescence of terminal heads, or
   conical, verticillasters or axillary; calyx not bilabiate. 7
7. Flowers sessile or sub-sessile, in capitate or conical heads; calyx not
   gibbous; hairs branched . . . . 10. *T. capitatum*
   Flowers clearly pedicellate, in terminal or axillary verticillasters; calyx
   gibbous; hairs simple 8
8. Plant with woody base; inflorescence terminal, flowers 4–8 in each
   bracteole; calyx purple . . . . 1. *T. chamaedrys*
   Plant not woody at base; inflorescence axillary; calyx green 9
9. Stems slender, flexible; leaves sessile; indumentum woolly; pedicels
   slender, shorter or equal to calyx . . . . 3. *T. scordium*
9. Stems thick, rigid; leaves petiolate; indumentum velutinous; pedicels
   thick, longer than calyx . . . . .4. *T. melissoides*

1. **Teucrium chamaedrys** *L.*, Sp. Pl. 2: 565 (1753) subsp. **sinuatum** (*Celak.*) *Rech.f.* in Bot. Archiv. 42: 378 (1941). Type: North Iraq, Gara M. Kurdistaniae, 2 viii 1841, *Kotschy* 351 (W, holo.). Rawi in Dep. Agric. Iraq Tech. Bull. 14: 158 (1964); Rechinger in Fl. Iranica [K. H. Rechinger] 150: 33 (1982); Ekim in Fl.Turkey [P. H. Davis] 7:65 (1982).

T. sinuatum Celak. in Bot. Centr.14: 218 (1883); Boissier, Fl. Orient. Suppl. 363 (1888).
T. divaricatum sensu Rech.f. in Ann. Naturh. Mus.Wien.49: 278 (1939).

Perennial herb, 15–45 cm, with a short woody base; indumentum simple, glandular and eglandular, dense; hairs short, retrorse, crispate, white (hairs long and short, not retrorse in subsp. *chamaedrys*). Stems decumbent to ascending, green or brown to greyish at the base, purplish above. Leaves petiolate, ovate, oblong to obovate 7–30 × 5–18 mm, base attenuate, apex rounded to acute, margins entire to crenate-dentate to lobulate; petiole 3–14 mm. Inflorescence of dense or lax racemes; peduncle 40–70 mm. Flowers purple to pink-purple in verticillasters of 4–8 flowers. Lower bracts similar to cauline leaves, 5–2 × 4–12 mm, upper cymbiform, entire near apex or toothed, upper surface purple or green, lower surface usually green, indumentum like cilia on upper surface or on margins, less on lower surface; bracteole 4–7 × 2–5 mm, similar to upper bracts; pedicel 7–11 mm, with dense or few hairs. Calyx 5–9 mm, tubular to campanulate, teeth triangular, apex acute to acuminate, equal 2.5–3.5 mm; tube 2.5–5.5mm, gibbous-subgibbous at base, hairs outside and inside, reddish-purple or green tinged purple (bracts longer than calyx in subsp. *chamaedrys*). Corolla 15–18 mm, tube 5–6 mm, lip 10–12 mm; lateral lobes unequal, lower longer than upper, hairy on the outside, purple or violet with pink. Stamens exserted; filaments with purple hairs, hairs simple, glandular and eglandular; anthers purple. Ovary glabrous. Nutlet obovoid, glabrous, red-blackish, reticulate.

HAB. Limestone, rocky slopes, on cliffs, in open oak forest; *Pinus brutia* forest on limestone; alt. 800–2000 m; fl. May-Sep.
DISTRIB. N. Iraq. **MAM**: Sharanish, 4/7/1957, *Rechinger* 12128 (E); Jabal Khantur, 1200 m, 9/7/1957, *Rechinger*10757! (E); Suwara Tuka, 1000–1500m, 10–12/7/1957, *Rechinger*11565! (E); Zawita, 950–1600 m, *Rechinger* 10910! (E); Rocky slopes under oak, Sersang, 1050 m, 13/7/1957, *Haines* 478!; Zawita, 14/5/1988, Al–*Mayah, Amran* & *Lafta* 36! (BSUH); Ser Amadia, *Guest* 4982 (BAG); Zawita gorge, *Gillett*

& Al-Rawi 8357 (BAG); N.E. Zakho, 25 km S. of Sharanish, *Al-Rawi* 23204 (BAG). **MSU**: Qara Dagh, *Chapman* 12305! (BAG). **FAR**: Salahuddin, Erbil, 1000 m, May 1933, *Regel* 111!

Common germander has been used medicinally (as an infusion of leaves) for abdominal complaints, as a diaphoretic, diuretic, stimulant, tonic, astringent, antiseptic and as vermifuge.

Turkey, Iran.

2. **Teucrium chasmophyticum** *Rech.f.*, Pl. Syst. Evol. 134: 287 (1980). Type: Iraq, Jabal Khantur prope Sharanish, N. Zakho, in fissuris rupium calc. 1200 m, *Rechinger* 12083! (W, holo.; BAG, iso.). Rechinger, Fl. Iranica [K. H. Rechinger]150: 32 (1982).

Perennial herb, 5–10 cm, with short woody base, densely glandular and eglandular hairy. Stem ascending, slender, basally branched, green, weak. Leaves petiolate, ovate, 5–15 × 5–8 mm, ovate, base rounded, apex obtuse, margins entire or entire to dentate in upper half, not revolute, densely pubescent; petiole 4–5 mm. Inflorescence in dense terminal spike-like racemes; flowers arising two opposite bracts; peduncle 10–30 mm. Lower bracts similar to cauline leaves, upper almost sessile, ± cymbiform, 5–11 × 2–4 mm, densely hairy; bracteoles similar to upper bracts, 2–5 × 1.5–2 mm. Calyx 3.5– 5 mm, teeth unequal, upper three teeth triangular, lower two lanceolate, 2.5–3 mm; tube 1–2 mm, gibbous at base, hairs dense outside, green. Corolla 7.5–9 mm, white-purple with glandular and eglandular hairs on the outside, tube not swollen at the base 3.5–4 mm, lip 4–5 mm, lower lobes longer than upper lobes, acuminate, hairy. Stamens exserted, filaments with eglandular hairs. Ovary glabrous. Nutlet obovoid, glabrous, light brown-yellow, reticulate.

HAB. Summit of mountain; alt. 1200–2200 m; fl. Jun.
DISTRIB. Rare, known from two locations in N. Iraq. **MAM**: Jabal Kira N. Zakho, near Sharanish, 19/7/1988, 2000–2200 m. *Z. K. Amran* 43! (BUH); Jabal Khantur prope Sharanish, N. Zakho, *Rechinger* 12083! (type).

Endemic.

3. **Teucrium scordium** *L.*, Sp. Pl. 2: 565 (1753); Meikle, Fl. Cyprus 2: 1335 (1985); Hedge in Fl. Pakistan [Ali & Y. Nasir] 192: 22 (1990); Rechinger, Fl. Iranica [K. H. Rechinger] 150: 33 (1982); Ekim in Fl. Turkey [P. H. Davis] 7: 61 (1982); Chunxing Ke in Illus. Fl. China 17: 133 (1998); Tohme, Illus. Fl. Lebanon: 380 (2008); Navaro, Fl. Iberica 12: 159 (2010).

Perennial herb, 20–90 cm; stems decumbent to ascending, branched at base, flexible, greenish-silvery, pubescence woolly, glandular and eglandular. Leaves sessile, oblong or ovate to oblong, 15–52 × 10–18 mm, base cordate or cuneate, apex rounded or acute, margins flat, not revolute, crenate to dentate or lobed, densely hairy. Inflorescence of axillary verticillasters with 2–6 flowers in axils of each bract; peduncle 20–50 mm; pedicels 3–4.5 mm, with woolly hairs. Bracts overtopping the flowers; lower bracts similar to cauline leaves, 5–2 × 3–15 cm; bracteoles similar to upper bracts, 5–7 × 4–7mm,. Calyx 3–5 mm, teeth equal, triangular, 1–1.5 mm, densely hairy outside and inside; tube 2–3.5 mm, gibbous, green. Corolla 8–8.5 mm, white-purple with glandular and eglandular hairs, tube 4–4.5 mm, not swollen at base, lip 4 mm, lower lobes longer than upper lopes, acuminate, hairy. Stamens slightly exserted; filaments hairy, yellowish-white. Ovary glabrous; style glabrous. Nutlet glabrous, reticulate, blackish- red. Fig. 116, 1–5.

> Leaves of main stem cuneate at the base, not amplexicaul . . . . a. subsp. *scordium*
> Leaves of main stem cordate at the base, usually amplexicaul; dentate
>   or crenate, apex rounded . . . . b. subsp. *scordioides*

a. subsp. **scordium**

Teucrium palustre Lam., Fl. Fr. 2:411 (1779); *T. scordium* L. subsp. *palustre* (Lam.) Gams in Hegi, III, Fl. Mittel-Eur. 5(4): 2531 (1927).

HAB. Hills on moist ground, near the springs in mountain; alt. 900–1350 m; fl. Aug.-Sep.
DISTRIB. N. Iarq. **MAM**: Shikhan valley, *Rawi* 24889 (BAG). **MRO**: Sheqlawa, 900 m, 25/9/1981, *Al-Khayat & Omar* 54982! (BAG). **MSU**: 5 km N.S. Penjwin, by roadside, 1350 m, 21/8/1980, *Al-Musawi & Al- Bermani* 39448! (BUH).

Europe to China.

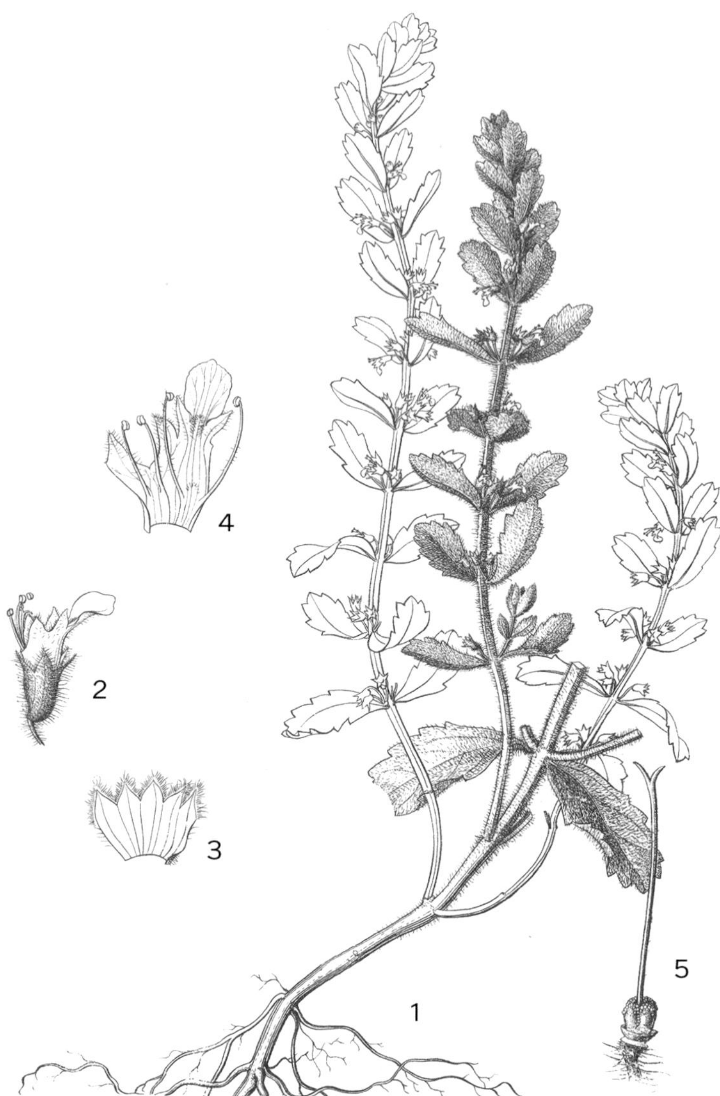

Fig. 116. **Teucrium scordium**. 1, habit × 1; 2, flower × 2; 3, calyx (opened) × 2; 4, corolla, opened with stamens × 2; 5, gynoecium, detail. Reproduced from Flora of China, Plates, 17: f. 133. 1998, with permission from Missouri Botanical Garden Press, St. Louis, and Science Press, Beijing. Drawn by Zeng Xiaolian.

b. subsp. **scordioides** (*Schreber*) *Maire & Petitmengin* in Bull. Soc. Sci. Nancy ser. 3, 9: 411 (1908); Taifour & El-Oqlah, Pl. Jordan Annot. Checklist: 107 (2017).

> *Teucrium scordioides* Schreb., Fl. Vert. Unlab. 37 (1773); *Scordium scorodioides* (Schreb.) Four., Ann. Soc. Linn. Lyon, n.s., 17:138 (1869); *Teucrium scordium* L. var. *scodioides* (Schreb.) Arcangeli, Comp. Fl. Ital. 559 (1882); *Teucrium lanuginosum* Hoffmanns. & Link, Fl. Portug. 1: 84 (1809); *Teucrium amplexicaule* Wallr., Linnaea 14:594 (1840); *Teucrium scordioides* var. *lanuginosum* (Hoffmanns & Link) Nyman, Consp. Fl. Eur.: 565 (1881).

HAB. Moist ground, near the springs in mountain; alt. 900–1350 m; fl. Jul.-Oct.
DISTRIB. N. Iraq. **MAM**: Sersang, Gara M., stream side, 1050 m, 13/7/1961, *Agnew* 676! (E); Sersang, damp slopes, in irrigated forest, 16/7/1955, *Haines* 472! (E); Sersang, oak forest shade, on large isolated boulders, 1200 m, 6/7/1955, *Haines* 580! (E). **MRO**: Shaqlawa, damp shady place, irrigated garden, 900 m, 6/8/1984, *Gillett* 124121! **MSU**: 10 Km N. of Said Sadiq, Sulaimaniya, 24/10/1960,

Agnew, Hadad & Haines 18429 (BUH). **FNI**: nr. Tel-Kaif, clay soil near the water, 19/9/1967, *Alizzi & S. Omar* 35270! (BAG).

Water Germander (English), BODKAH (Kurd.).

Palestine, Jordan, Iran, Turkey.

4. **Teucrium melissoides** *Boiss & Hausskn.* ex *Boiss.*, Fl. Orient. 4: 813 (1879); Zohary in Dep. Agr. Iraq Bull. 31 (1950); Rawi in Dep. Agr. Iraq Tech. Bull. 14: 158 (1964); Rechinger, Fl. Iranica [K. H. Rechinger] 150: 35 (1982).

Perennial herb, 20–50 cm; stems branched, erect to decumbent, rigid, green, pubescence of hairs short, glandular and eglandular hairs. Leaves petiolate, elliptic to ovate, 35–5 × 20–25 mm, base attenuate, apex rounded, margins flat, not revolute, dentate, pubescence short, velutinous. Inflorescence axillary-verticillate, each two opposite bracts bearing 2–4 flowers arranged on the main axis of the stem, or on peduncles. Lower bract 10–45 × 5–25 mm, similar to cauline leaves or entire; bracteoles 4–13 × 3–6 mm; pedicel 8–13 mm. Calyx 6–7 mm, teeth equal, triangular, 1.5–2 mm, with dense, short, hairs, outside and inside; tube 4.5–5 mm, gibbous, green. Corolla 8–15 mm, white to yellowish, glandular and eglandular hairy outside, tube 1–4 mm, not swollen at the base, lip 5–11 mm, lower lobes slightly longer than upper lobes, acuminate, hairy. Stamens exserted; filaments white-yellowish, hairs simple. Ovary glabrous; style glabrous. Nutlet glabrous blackish-red, smooth, glossy.

HAB. Wet hills, near springs, gravelly mountains, moist cracks on mountain slopes, clay soil; alt. 620–2200 m; fl. Jun.-Sep.
DISTRIB. N. Iraq. **MAM**: Sharanish, on wet shady rocks, *Rechinger* 14648! (E); Aqra, side of water, *Chapman* 26169! (BUH); Zawita, Turkish border, *Rechinger* 18351! (E). Zinta gorge, *Chapman* 26103 (BAG). **MRO**: Qandil mt., in Pushtashan, confines Persian border, *Rechinger* 11199! (E); Infra Rawandiz, 700 m, *Rechinger* 11258! (E); Gali Ali Beg, damp rocks in gorge, *Haines* 2161! (BAG); Rawndiz, rocky mt. slopes, *Al-Dabagh & Hamad* 46189! (BAG); Jindian, damp place near spring, *Alizzi* 15880! (BAG), Harrer village, *Sahira* 37542 (BAG); Haibat Sultan, *S. Omar, Sahira, F. Karim & H. Hamid* s.n. (BAG).

Turkey, Iran.

5. **Teucrium orientale** *L.*, Sp. Pl. 2: 562 (1753); Handel-Mazzetti in Ann. Nat. Mus. Wien 27: 409 (1913); Rechinger, Fl. Iranica [K. H. Rechinger] 150: 39 (1982); Ekim in Fl. Turkey [P. H. Davis] 7:58 (1982); Tohme, Illustr. Fl. Lebanon 379 (2007).

Herbaceous perennial with a woody stock, 30–50 cm; stems erect, basally branched, rigid, green with purple tinge, glandular and eglandular hairy, ± glabrous in upper half. Leaves sessile, 20–5 × 25–30 mm, pinnatisect with 5–12 segments, base attenuate, apex acuminate, margins revolute. Inflorescence simple or paniculate racemes, peduncle 30–70 mm bearing many flowers. Lower bracts similar to cauline leaves, 10–25 × 10–15 mm; bracteole 4–5 × 2–3mm, similar to upper bracts. Pedicels 7–10 mm, thick, glabrous or hairy. Calyx campanulate, 4–6 mm, teeth triangular or triangular-lanceolate, 2–2.5 mm, tube 2.5–3.5 mm, not gibbous, glabrous or hairy outside, brownish. Corolla purple, 11–22 mm, usually 3x calyx, tube 3–4.5 mm, laterally swollen, lip 8–17.5 mm, glabrous or hairy outside, two lower lateral lobes wide, apex acuminate, two upper lobes narrow, apex rounded. Stamens exserted; filaments with glandular and eglandular hairs, purple. Ovary glabrous; style glabrous. Nutlet pitted deeply, polygonal-oblong, glabrous.

DISTRIB. of species: Iran, Iraq, Lebanon-Syria, North Caucasus, Palestine, Transcaucasus, Turkey.

    Whole the plant covered with short dense hairs; calyx teeth triangular . .a. subsp. *tylori*
    Lower half of the plant hairy, upper half glabrous; calyx teeth
        triangular-lanceolate . . . . b. subsp. *glabrescence*

a. subsp. **tylori** (*Boiss*) *Rech.f.*, Fl. Iranica [K. H. Rechinger] 150: 41, as ‹taylori› (1982).

Teucrium taylorii Boiss., Diagn. Pl. Orient. 7: 61 (1846).

HAB. Mountains; alt. 1200 m; fl. Jun.
DISTRIB. N. Iraq. **MSU**: Avroman mt., Tawella, 1200 m, *Rechinger* 10229! (BAG).

Transcaucasus, Iran.

b. subsp. **glabrescence** (*Hausskn.* ex *Bornm.*) Rech.f., Fl. Iranica [K. H. Rechinger] 150: 40 (1982).

   Teucrium orientale var. glabrescens Hausskn. ex Bornm., Verh. K. K. Zool.-Bot. Ges. Wien 48: 624 (1898).

HAB. Mountains; alt. ± 1200 m; fl. Jun.
DISTRIB. N. Iraq. MSU: Qara Dagh, 29/6/1964, *Buthaina Maki* s.n.! (BUH); Hauraman, N. slope, N. Tawella, *Wainert* & Al-*Musawi* 27964! (BUH); Qaradagh, *Agnew* 18432! (BUH); Kopi Qaradagh, *Wheeler-Haines* 578! (E).

Turkey, Iran.

6. **Teucrium parviflorum** *Schreber*, Pl. Vert. Unilab. 31, t. 1 (1773); Boissier, Fl. Orient. 4: 810 (1879); Handel-Mazzetti in Ann. Nat. Mus. Wien 27: 409 (1913); Blakelock in Kew Bull. 4: 551 (1950); Rawi in Dep. Agr. Iraq Tech. Bull. 14: 158 (1964); Feinbrun-Dothan, Fl. Palaest. 3: 810 (1978); Ekim in Fl. Turkey [P. H. Davis] 7: 61 (1982); Rechinger, Fl. Iranica [K. H. Rechinger] 150: 37 (1982); Tohme, Illustr. Fl. Lebanon 379 (2007).

Perennial herb, 30–60 cm; stem erect, branched in upper half, rigid, green. Indumentum of simple and dichotomous, dense, appressed, glandular and eglandular hairs. Leaves sessile, 40–85 × 40–60 mm pinnatisect with 10–28 segments, base attenuate, apex acuminate, margins revolute to ± revolute, densely pubescent. Inflorescence simple or paniculate raceme, 20–50 mm; peduncle 100–150 mm. Bract 10–45 × 10–30 mm, similar to cauline leaves but with 1–3 segments at the top of inflorescence; bracteole similar to upper bracts, 8–13 × 4–6 mm. Pedicel 7–18 mm, hairy. Calyx campanulate, 4–7.5 mm, teeth equal, lanceolate, 2–5 mm, tube 2–2.5 mm, not gibbous, hairs short, appressed, purple. Corolla 5–9 mm, tube 2–2.5 mm, laterally swollen, lip 3–6.5 mm, hairy dorsally, lateral upper and lower lobes are similar, with rounded apex, not exserted from the calyx, terminal lobe slightly longer than calyx, purplish-blue. Stamens partially exserted, filaments hairy, pinkish-purple, anthers purple. Ovary with short hairs and sessile glands; style glabrous. Nutlets with short sparse hairs; yellowish-grey pitted deeply. Fig. 117, 1–2.

HAB. Limestone mountains, rocky mountain, degraded *Quercus aegilops* forest, hillside, in field, on clay soil; alt. 600–1650; fl. May-Aug.
DISTRIB. N. Iraq. MJS: Jabal Sinjar N. Slope, *Al-Kaisi, Al-Khayat & F. Karim* 50914 (BAG). MAM: Zakho, *Rawi* 8623 (BAG); Zakho and Sharanish, *Rechinger* 10729! (E); Dohok, *Al-Rawi* 8839! (BAG); S. Omar, *Al-Kaisi & Wedad* 49688 (BAG); Sersang, *Al-Khayat* 1469! (BUH); Kani Masi, 7/7/1976, *Omar & Al-Dabbagh* 45663! (BAG). MRO: Safin mt., above Shaklawa, *Bornmüller* 1714 (E). MSU: Dokan, *Rechinger* 10999! (E); Tainan bet Darbandi Basian and Tasluga, *Gillett & Rawi* 11628 (BAG); Chamchamal, *Al-Rawi* 21676! (BAG). FNI: Mahad (nr. Shaikhan), *Salim Effendi* 2608 (BAG); from Mosul to Aqra, *Al-Kaisi* 49701 (BAG). FKI: Kirkuk, *Rechinger* 10034 (E).

SHIN SHIN (Arb.) & GIA HUSHTER (Kurd. *Rawi* 8623); used as fodder.

Lebanon, Palestine, Iran, Turkey.

7. **Teucrium oliverianum** *Ging. ex Benth.*, Lab. Gen. Sp.: 668 (1835). Type: Inter Aleppo et Baghdad, *Oliver* (G, holo.). Boissier, Fl. Orient [Boissier] 4: 810 (1879); Zohary in Agr. Iraq Bull., 31(1950); Rawi in Agr.Iraq Tech.Bull.14: 158 (1964); Rechinger, Fl. Lowland Iraq: 518 (1964); Rechinger, Fl. Iranica [K. H. Rechinger] 150: 39 (1982); Chaudhary & Hedge Fl. Kingdom Saudi. Arabia 2(2): 16 (2001).

Perennial herb, 20–40 cm; stems erect branching from the woody base and in an inflorescence region, rigid, green-bluish, with simple velvety hairs or hairs dichotomous, dense, adpressed, glandular and eglandular. Leaves sessile, ovate, (20–)30–50(–80) mm, base attenuate, apex 3–5(–8) lobed, sometimes lobes deeply fragmented, margins flat or revolute, appressed pubescent. Inflorescence, simple or paniculate racemes, flowers in pairs; peduncle 70–120 mm. Bracts 5–2 × 2–15mm, similar to cauline leaves or linear and cymbiform, blue; bracteoles 5–8 × 3–6 mm, similar to upper bracts. Pedicel 8–20 mm, hairy. Calyx campanulate, 5–8 mm, teeth equal, 3–4 mm, lanceolate, densely pubescent outside, tube 2–4 mm, green, not gibbous. Corolla 11–20 mm, blue-purple, tube 2.5–3 mm, laterally

Fig. 117. **Teucrium parviflorum**. 1, habit; 2, flower × 5. Reproduced with permission from Feinbrun-Dothan, Fl. Palaestina 3: Plates, f. 163 (1977). Drawn by Esther Huber. © The Israel Academy of Sciences and Humanities.

swollen, lip 5.5–17 mm, hairy outside, lateral lobes equal with acuminate apex. Stamens exserted, purple. Ovary woolly; style glabrous. Nutlet woolly, grey, pitted.

HAB. Dry arid localities, gravel, limestone, gypsum and sandy soils, clay soil; alt. 100–400; fl. Mar.-Jun. & Oct.-Jan.
DISTRIB. Foothills, and western and southern desert. **FNI:** Bardarash, 4 km from Eski-kelek, *Al-Kayat, Jawhar & Abdulla* 58005 (BAG). **FPF:** Jabal Hamrin, 4/5/1957, *Rechinger* 9570! (E); Diyala, towards Iran, in the hills ca. 20 km N. Badra 3/6/1957, *Rechinger* 12721! (E); 20 km S. of Badra between Badra & Shahaby, *A. Haloob & Saif* 58663 (BAG); Jabal Hamrin on Baquba-Khanaqin road, 150 m, gypsum-sandy 11/4/1947, *Gillett & Al-Rawi* 33458! (BUH); Jabal Hamrin S. of Saadiya, *Al-Kaisi* 50481 (BAG); Jabal al-Muwaila, *Guest, Rawi, Rechinger* 17586 (BAG); 19 km S. of Zurbatiya, *Al-Kaisi & Yahya* 45271 (BAG). **DLJ:** 85 km N.N.W. of Faluja, *Al-Rawi* 20301 (BAG). **DWD:** Faluja desert, sandy soil, 16/4/1955, *Wheeler-Haines* 151! (E). **DSD:** 15 km before Al-Rahalia, 3/4/1988, *Amran* 24! (BSUH); 70 km S. W. Nukhaib, 263m, 15/4/1961, *Al-Rawi* 31096! (BUH); 20 km S.E. of Salman, *Guest, Rawi, Rechinger* 14891 (BAG). **LCA:** Diwaniya, sandy soil at the end of the channel, 310 m, 26/4/1952, *Rechinger* 9431! (E).

ZEFRA (زفرة), OMKRAIN (ام كرين); NAWAR (نوار); SHAKARY (شكاري); SHAJRET AL MUR (المر شجرة); SAMARAA (سامراء).

Arabian Peninsula to Iran.

8. **Teucrium rigidum** *Benth,* Prodr. [A. P. de Candolle] 12: 578 (1848). Type: Iraq, Kirkuk, Darband-i-Bazian, *Haussknecht* 822. Zohary in Agr. Iraq Bull., 31 (1950); Blakelock in Kew Bull. 4: 551 (1950); Rawi in Dep. Agr. Iraq Tech. Bull. 14: 158 (1964); Rechinger, Fl. Iranica [K. H. Rechinger] 150: 43 (1982).

Perennial herb, 20–50 cm; stem erect, branched, rigid, thick, green-bluish with simple, rigid, tomentellous, glandular and eglandular. Leaves sessile, 15–5 × 20–35 mm, base attenuate, apex acuminate-acute, margins strongly revolute, pinnatisect with 5–10 segments, Inflorescence simple or paniculate racemes; peduncle 70–120 mm, branched, bearing many flowers. Lower bracts 3–35 × 2–15 mm, similar to cauline leaves; upper bracts 3–5 × 1–2 mm, with 2–3 segments. Pedicel 8–14 mm, pubescent. Calyx conical-campanulate, 6–7 mm, teeth equal, lanceolate, 4–4.5 mm, pubescent outside, tube 2–2.5 mm, green, not gibbous. Corolla 11.5–15 mm, purplish to blue, tube laterally swollen, 3–3.5 mm, lip 8.5–11.5 mm, hairy outside, lateral lobes unequal, apex acuminate. Stamens exserted; filaments hairy, purple; anthers purple. Ovary hairy; style hairy. Nutlet with dense short hairs, pitted-reticulate.

HAB. Moist mountain slopes, dry foothills, sandy-rocky soils; alt. 300–650; fl. May-Jul.
DISTRIB. Mountains, Upper Jazira and the western desert to N. Iraq; endemic. **MSU:** Tasluja, mountain slopes, *Al-Rawi & Kass,* 28942! (BAG); Sulaimania, Rustam farm, *Guest* 813!. **FNI:** Few km W. of Mindan bridge, E. Chapman 26167 (BAG). **FKI:** hills of Bakhtiari conglomerst above Zab river, Guest 4025 (BAG); 2 km N. of Kirkuk, 320 m, stony soil, *Al-Rawi,* 21551! (BAG); 10 km before Kirkuk on the road from Erbil to Kirkuk, *Al-Maya, Amran & Lafta* 41 (BSUH); 8 km E. Kirkuk, *Gillet & Al-Rawi* 11599! (BAG). **FAR:** Makhmour, 750 m, *Gillett* 11834! (K). **FUJ:** 23 km N.W. Hatra, *Bahrucha* 8669! (K). **DWD:** 160–190 km N.W. Ramadi-Rutba, roadside, *Al-Rawi* (BAG, K); Hawran, cultivated patches, *Hunting,* 152! (K).

Endemic.

9. **Teucrium macrum** *Boiss. & Hausskn.* ex *Boiss.,* Fl. Orient. [Boissier] 4: 810 (1879); Rechinger, Fl. Iranica1 [K. H. Rechinger] 50: 38 (1982).

Perennial herb, 30–70 cm, glabrous in upper half, hairs simple, glandular and eglandular. Stem erect, rigid. Leaves sessile, 30–5 × 20–50 mm, base attenuate, apex rounded to acuminate, margin slightly revolute, pinnatisect, 7–15 segments, indumentum dense hairs in upper half of the blade, glabrous in lower half. Inflorescence a simple raceme; peduncle 40–120 mm, bearing many flowers which are concentrated at the fourth quarter in the top, aggregated. Upper bracts similar to cauline leaves10–25 × 6–15 mm; bracteole similar to upper bracts 2–10 × 2 mm. Pedicel 4–10 mm, thin, glabrous. Calyx campanulate, 2.5–3.5mm, teeth 1–1.5mm, equal, triangular, glabrous outside and inside, tube 1.5–2mm, bronze, not gibbous. Corolla 10–15 mm, purple, tube 3–3.5mm, with lateral swollen, lip 7–11.5 mm exserted from the calyx, glabrous at dorsal side, lobes not equal, two lower lobes wide, apex rounded, two upper lobes narrow, apex acuminate. Stamen included or shortly exserted; filaments purple, with glandular and eglandular hairs; anthers purple. Ovary hairy; style glabrous. Nutlet not deeply pitted, short sparsely hairy.

HAB. Mountain side, *Quercus aegilops* forest formation on limestone; alt. 1000–1500; fl. May-Aug.
DISTRIB. Very rare in the mountain region in Iraq. Dohuk, Rawi 8763 (BAG). MSU: Jabal Avraman near Darimar, Gillett 11776 (BAG); Tawella *Rawi* 22297! & 21955! (BAG).

Iran.

10. **Teucrium capitatum** *L.*, Sp. Pl. 2: 566 (1753); Navaro in Fl. Iberica 12: 93 (2010); Bergmeier et. al., Willdenowia 41: 184 (2011).

> *Polium capitatum* (L.) Mill., Gard. Dict. ed. 8: 5 (1768).
> *Chamaedrys capitatum* (L.) Raf., Fl. Tellur. 3: 85 (1837).
> *Teucrium polium* subsp. *capitatum* (L.) Arcang., Comp. Fl. Ital. 559 (1882).

Subshrub, 15–50 cm; stems decumbent to ascending or erect, basally much branched, rigid or flexible, whitish-green or yellowish, covered with dendroid and simple glandular hairs. Leaves sessile, oblong-obovate, 8–30(–40) × 3–12 mm, some in clusters, base attenuate, apex acute to obtuse, margins revolute from the base or the upper half, or flat, crenate or lobed, indumentum dense, branched. Inflorescence of terminal capitate or conical heads; peduncle 5–50 mm, usually branched. Bracts 3–20 × 2–7 mm, longer than flowers, similar to cauline leaves, sometimes with two or three lobes, densely hairy; bracteoles spatulate, 3–8 × 1.5–2.5 mm, with dense hairs. Flowers with short pedicel or sessile. Calyx 3.5–5.5 mm, tubular to campanulate, with tomentose-woolly hairs, teeth triangular, unequal, 0.5–1 mm; tube 3–4.5 mm, not gibbous, throat glabrous, silvery-green. Corolla 7–8 mm, yellowish-white, sometimes with yellow inside the lip, tube swollen slightly above the base, lip 4–4.5mm, the two lower lobes larger and longer than the two upper lobes, dendroid hairy outside. Stamens exserted; filaments with white dendroid hairs. Ovary glabrous; style glabrous. Nutlet reticulate-rugose, light to dark brown.

HAB. Wet hillsides, agricultural land, dry and semi dry desert, on gypsum, limestones; alt. 100–2000 m; fl. Apr.-Aug.
DISTRIB. Mountains, central alluvial plains and desert areas; widespread. MAM: Sharanish, *Rechinger* 12124! (E); Asi, Jabal Bekheir, H. White & *Al-Rawi* 8498 (BAG); Zakho, Turkish-border, *Rechinger* 10667! (E); Sersang, *Rechinger* 11915! (BUH); Sersang, *Winert & Al-Musawi* 28766! (BUH); 2 km N. Atrush, *Al-Shehbaz* 2971! (BUH); Aqra, *Al–Mayah, Amran & Lafta* 38! (BUH); near Zakho, *Al-Rawi* 18326! (E); Pushtashan, *Rechinger* 11070! (E); Jabal Khantur, *Rechinger* 12048! (E); Ser Amadiya (Gali Mazarka), *Al-Dabbagh, Al-Kaisi & H. Hamid* 45963 (BAG). MSU: Derbandi Khan, *Winert & Al-Musawi* 32600! (BUH); Pira Magrun, *Faris,* 5697! (BUH); Kopi Qaradagh, *Agnew,* 18427! (SUH); 8 km S.W. Dokan, *Ali Al-Askari* 3243! (BUH). MRO: Shaqlawa, *Andora,* 1454! (BAG); Rawanduz, *Umer* 38366! (BUH); Piran to Sherwan, Mazin, *Bayati & Haloob* s.n. (BUH); Berrog mt. On road to Qandil, *Al-Rawi & Serhang* 23943 (BAG); Khanzad, *Al-Kaisi & Wedad* 49582 (BAG).MJS: Sinjar, *Al–Mayah, Amran & Lafta* 16! (BUH); Jabal Sinjar N. slope, *Al-Kaisi,* Al-Khayat & F. Karim 50915 (BAG). FNI: Mosul road, 85 km S.E. Mousal near Sherqat city, *Al–Mayah, Amran & Lafta* 3! (BUH). FAR: 15–20 km, before Arbil on Mosul-Arbil Road, *Al–Mayah, Amran & Lafta* 20! (BUH). FKI: Tuz Khurmato, *Rechinger* 10616! (E). FPF: 3 km, S. Mandli, *Barkley & Agnew*18396! (E); 10 km E. Mandli, *Rechinger* 9656! (E); Jabal Hamrin, c.70 km N. of Amara, 100 m, *Guest, Al-Rawi & Rechinger* 17598! (E). DGA:10 km, N. Samara on Baghdad-Tikrit road, *Al–Mayah, Amran & Lafta* 1 (BUH). DWD: 5 km before Ain Al-Tamur in Karbala, *Amran* 23! (BUH); 31 km W. of Ramadi, *Rawi* 20834 (BAG). DSD: 6 km E. of As-Salman, Gillett & *Rawi* 6230 (BAG); Al-Shawiyah depression 35 km E. of Salman, A. Haloob, Saif & Adel 60403 (BAG). LCA: Al-Sudur, *Al-Musawi* 18346! (BUH); *Al-Kaisi* & Khalid 46122 (BAG).

JADDA, YADDA (Basra), MISK-AL-JIN. Used against inflammation of intestines and haemorrhoids in traditional medicine.

Algeria, and from Sinai, Jordan, Saudi Arabia and Palestine to Iran, Afghanistan; Turkey, C. & W. Europe.

SPECIES DOUBTFULLY RECORDED AND /OR NOT PRESENT IN IRAQ

T. multicaule Montb. & Auch. ex Benth., in Ann. Sci. Nat. Ser. 2, 6: 54 (1836).

Handel-Mazzetti (Ann. Nat. Mus. Wien 27: 409 (1913), noted that the species grows in Iraq and Zohary (Agr. Iraq Bull., 31: 124, 1950) mentioned that *T. multicaule* is found in Dohuk which Al-*Rawi* (Dep. Agr. Iraq Tech. Bull. 14: 158, 1964) included in his checklist. It was not included by Rechinger in Fl. Iranica. Zedan in his M.Sc. thesis (1988, University of Basra, unpublished), recorded *T. multicaule* for Iraq, basing his inclusion on a photo of one specimen sent to him from Kew, collected from Mosul, but he could not find the plant from there. In the present account, only one duplicate type specimens is at Kew (Mesopot., Kurdistan and Mosul, *Kotschy* 1841) a photo of which was sent to Zedan. On examination, I saw on one of the two sheets, Montbret wrote in his handwriting in 1836, that the species is found in Halab and Asia Minor, (Syria and Turkey), but did not write Iraq or Mosul. Bentham [Ann. Sci. Nat. Ser. 2, 6: 54, 1836] wrote Aleppum (meaning Halab) and Asia Minor Oriental. Boissier (Fl. Orient. 4: 807, 1879) did not mention Iraq but mentioned Syria and Turkey. I (Byati) did not find any specimens in the herbaria in Iraq, E, K and G, and it has not been collected in Iraq (*Kotschy* 1841 is possibly an error in recording). *T. multicaule* is found in Turkey and N. Syria and therefore not included in the this account for Fl Iraq.

T. divaricatum Sieb., Avis 4:5 (1821).

Blakelock in Kew Bull. (4: p. 550, 1950) recorded *T. divaricatum* from Zawita Badi near Dohuk, and from Dohuk, but all his material is now referred to *T. chamaedrys*. Rechinger also wrote in his handwriting on some specimens to correct the name from *T. divaricatum* to *T. chamaedrys* subsp. *divaricatum*. T. Ekim (botanist), mentioned that *T. divarycatum* sensu Rech.f. (Ann. Nat. Mus. Wien. 49: 278 (1939) non Sieber (1821)) is synonymous to *T. chamaedrys* subsp. *sinuatum* [Fl. Turkey 7: 65 (1982)], and this revision confirms that *T. divaricatum* is not found in Iraq, but is included under *T. chamaedrys*.

T. pruinosum Boiss., Fl. Orient. [Boissier] 4: 808 (1879).

T. pruinosum has been included for Iraq by Handel-Mazzetti, (Ann. Nat. Mus.Wien 27: 409 (1913); Zohary (Agr. Iraq Bull., 31: 124 (1950); Al-Rawi (Dep. Agr. Iraq Tech. Bull. 14: 158 (1964), Rechinger in Fl. Lowland Iraq, (518, 1964) and Ekim (Fl. Turkey 7: 61, 1982).

I have examined specimens in the herbaria in Iraq, K, E, and G but have not found *T. pruinosum* from Iraq. Khalaph in his study in 1980 on Jabal Sinjar did not find it and neither did Zedan in his research in 1988 (Sinjar and Jazeera were mentioned by Handel-Mazettii). I have not included it in this in the present account as either it may have been misreported by Handel-Mazzetti or is now extinct.

T. procerum Boiss et Blanche in Boiss., Fl. Orient [Boissier] 4: 809 (1879).

Zohary (Agr. Iraq Bull. 31: 124 (1950) recorded this species in Sulaimaniya which was given by Al-Rawi (Dep. Agr. Iraq Tech. Bull. 158, 1964) but without any specimens found in the herbaria of Iraq, K, E, or G. It was not included by either Rechinger in Fl. Iranica or Zedan (1988).

3. **SCUTELLARIA** L.

Sp. Pl. 2: 598. 1753; Gen Pl. ed. 5: 260 (1754); Paton, Kew Bull. 45: 399 (1990); Harley & al. in Kubitzki (ser. ed.), Fam. Gen. Vasc. Pl. 7: 241 (2004)

Ali Haloob

Perennial herbs or shrubs, suffruticose at base. Leaves petiolate, cordate or ovate or lanceolate, crenate or pinnatifid at margins. Inflorescence a raceme or spike; flowers solitary in axils of small flat or cucullate bract, secund or not. Calyx bilabiate, lips entire, closed in fruit; upper lip deciduous, with a transverse, rounded, concave, scale-like dorsal appendage (scutellum); lower lip persistent. Corolla bilabiate, tube long-exserted, ascending, curved outwards and gradually widening to throat, narrow slightly raised annulus hairs at the base of the tube; upper lip erect, galeate; lower lip 3-lobed, middle lobe broad, flat to recurved, lateral lobes more closely joined to upper lip than lower. Stamens 4, didynamous, included, arcuate under upper lip, anterior pair longer than posterior pair, thecae ciliate, anterior pair 1-celled with an aborted theca, posterior pair 2-celled with converging thecae. Style subequally bifid. Nutlets ovoid or ellipsoid, smooth, ridged or papillate, with adpressed stellate hairs completely covering the nutlet surface or not.

An almost cosmopolitan genus of about 360 species, poorly represented in tropical lowlands; 9 species in Iraq.

Paton, A. (1990). A global taxonomic investigation of *Scutellaria* (Labiatae). Kew Bull. 45(3): 399–450.

Iraq species are placed in the following two subgenera:

SUBGENUS SCUTELLARIA (Neveski ex Juz.) Juz. emend, Paton, Notes Roy. Bot. Gard. Edinburgh 46: 345 (1990); Paton, Kew Bull. 45: 425 (1990).
Inflorescence one-sided; pedicels round or slightly flattened; flowers opposite or not, sub-tended by leaves or leaf-like bracts. *S. albida, S. megalaspis, S. porphyrantha*

SUBGENUS APELTHANTHUS (Nevski ex Juz.) Juz. emend. Paton, Notes Roy. Bot. Gard. Edinburgh 46:346 (1990); Paton, Kew Bull. 45: 440 (1990).
Inflorescence 4-sided; pedicels strongly flattened and parallel to stems; flowers opposite and decussate, subtended by cucullate bracts that clasp the calyx. *S. bornmuelleri, S. orientalis, S. platystegia, S. pinnatifida, S. tomentosa, S. multicaulis.*

1. Inflorescence one-sided, flowers secund in the axils of small flat bracts
   (subgenus Scutellaria) 2
   Inflorescence 4-sided, flowers opposite and decussate subtended by
   cucullate bracts (subgenus Apelthanthus) 4
2. Stems with dense hispid-hirsute eglandular and glandular hairs,
   fruiting calyx 11–13 mm, corolla pinkish-purple . . . . 2. *S. megalaspis*
   Stems glabrous-glabrescent or pilose, pubescent eglandular and
   glandular, fruiting calyx less than 10 mm, corolla whitish to yellowish-
   white or purple 3
3. Inflorescence less than 15 cm; calyx scutellum green; corolla white or
   yellowish-white . . . . 1. *S. albida*
   Inflorescence more than 20 cm; calyx scutellum purple; corolla purple
   3. *S. porphyrantha*
4. Inflorescence lax 5
   Inflorescence compact 6
5. Stems dense white tomentose with spreading minute glandular hairs;
   inflorescence less than 10 cm long; corolla yellow with purplish-tinged 8. *S. tomentosa*
   Stems with sparsely puberulent eglandular and minute glandular hairs;
   inflorescence more than 11 cm long; corolla purple . . . . .9. *S. multicaulis*
6. Bracts herbaceous, margins incised-crenate; scutellum in fruit more
   than 4.2 mm high. . . . . .7. *S. pinnatifida*
   Bracts membranous or scarious, entire or only the lowest pair with
   crenate margins; scutellum in fruiting less than 4 mm high. 7
7. Stems less than 20 cm; leaves oblong-lanceolate or elliptic or ovate;
   inflorescence less than 3 cm long . . . . 4. *S. orientalis*
   Stems more than 30 cm; leaves triangular-ovate; inflorescence more
   than 3 cm long 8
8. Stems erect-ascending; bracts yellow, oblong to ovate-lanceolate, 12–16
   × 5–8 mm, margins entire . . . . .5. *S. bornmuelleri*
   Stems procumbent-ascending; bract yellow or purple or yellow tinged
   purple, broadly ovate, 10–15 × 7–14 mm, margins entire, only the
   lowest pair with crenate margins. . . . . 6. *S. platystegia*

SUBGENUS SCUTELLARIA (Neveski ex Juz.) Juz. emend Paton

1. **Scutellaria albida** *L.*, Mant. Pl. 2: 248 (1771); Boissier, Fl. Orient. [Boissier] 4: 689 (1879); Zohary in Dep. Agr. Iraq Bull. 31: 125 (1950); Rawi in Dep. Agr. Iraq Tech. Bull. 14: 156 (1964); Ball in Fl. Europ. 3: 135 (1972); Edmondson in Fl. Turkey [P. H. Davis] 7: 82 (1982); Paton in Kew Bull. 45: 430 (1990).

Perennial herb. Stems erect-ascending, 45–85 cm tall, branched, purplish below, glabrescent with sparsely white short eglandular hair on angles or white pubescent eglandular with minute glandular hairs dense on angles or hirsute-villous eglandular and glandular hairs dense on angles. Leaves ovate-triangular or broadly cordate-cordate, 3.5–

6–9.5 × 2.5–4–6 cm, base cordate, truncate-subtruncate, apex acute or obtuse-rounded, margins crenate-serrate, glabrescent to puberulent eglandular hairy; petiole 15–40 mm. Inflorescence a raceme, 8–15 cm, dense white eglandular pubescent-villous with minute glandular hairs. Bracts broadly ovate-elliptic to lanceolate-elliptic, 5–9 × 3–6 mm, margins entire, petiole 2–4.5 mm. Pedicel 3–4 mm. Calyx green, 3–5.5 mm in flower, becoming 4–7 mm with 4–5.5 (–8) mm high scutellum in fruit, white eglandular pubescent with spreading minute glandular hairs outside. Corolla yellowish-white, 14–17(–21) mm, densely puberulent, eglandular outside; tube 8–12(–16) mm. Nutlets ovoid-ellipsoid, ± 1.5 mm.

Subsp. *albida* is found from S.E. Europe to Iran and does not occur in Iraq. It is characterized by its stems with short appressed eglandular hairs; leaves broadly cordate, 2–5 × 1–4 cm, base truncate or subcordate; inflorescence with sparse long patent and very profuse long stipitate glandular hairs; bracts broadly elliptic or ovate-lanceolate; calyx 3–4 mm in flower, becoming 7–8 mm in fruit; corolla 15 mm.

Three subspecies are found in Iraq:

1. Stems with eglandular hairs . . . . a. subsp. *condensata*
   Stems with eglandular and glandular hairs 2
2. Stems densely pilose eglandular and glandular; scutellum more than 6
     mm high in fruit. . . . . b. subsp. *subsimilis*
   Stems white pubescent eglandular and with minute glandular hairs
     dense on angles; scutellum less than 5.5 mm high in fruit. . . . . . . c. subsp. *pycnotricha*

a. subsp. **condensata** (*Rech.f.*) *J.R.Edm.*, Notes Roy. Bot. Gard. Edinburgh 38: 52 (1980); Edmondson in Fl. Turkey [P. H. Davis] 7: 84 (1982).

> *Scutellaria condensata* Rech.f., Bot. Arch. 43: 17 (1941); Blakelock in Kew Bull. 4: 548 (1950); Rawi in Dep. Agr. Iraq Tech. Bull. 14: 156 (1964); Rechinger in Fl. Iranica 150: 50 (1982).

Stems glabrescent with sparsely white short eglandular hair on angles. Leaves broadly cordate, 4.5–6.5 × 4–6 cm, base cordate, apex obtuse to rounded. Inflorescence with white eglandular pubescent with minute glandular hairs; bracts broadly ovate-elliptic. Calyx 3–4.5 mm in flower, becoming 4–5.5 mm with 4.5–5.5 mm high scutellum in fruit.

HAB. Coppiced oak forest, on mountains, in *Astragalus* zone; alt. 850–2380 m; fl. May-Jul.
DISTR. Occasional in the lower forest zone of Iraq. **MAM**: Zawita district(Baphoira N. Zawita), V. C. Robertson 14513 (BAG)!; Shiwarana Pass between Sharanish and Marsis, in Mosul Liwa, *Rechinger*, 0018218! (BUH); Sarsang N. slope of Qara Dagh, Al-*Rawi* 26372 (BAG); Sarsank, 16/6/1970, *S. Omar* 37702 (BAG); Haisi village 10 km S. of Kani Masi, *Al-Dabbagh* & K. Hamad 45545 (BAG)!; Gara mountains, *Al-Dabbagh* & *M. Jasim* 46975! (BAG); **MSU**: Penjwin, *Al-Hashimi*, 0018224! (BUH).

Iraq, E. Turkey & W. Iran.

b. subsp. **pycnotricha** (*Rech.f.*) *A.Haloob*, **stat. & comb. nov**.

> *Scutellaria pycnotricha* Rech.f., Bot. Arch. 43: 19 (1941); Rawi in Dep. Agr. Iraq Tech. Bull. 14: 156 (1964);
> *S. condensata* subsp. *pycnotricha* (Rech.f.) Rech.f., Die Kulturpflanze Beih. 3: 53 (1962); Rechinger, Fl. Iranica 150: 50 (1982).

Stem with white pubescent eglandular and minute glandular hairs dense on stem angles. Leaves ovate-triangular, 6–9.5 × 4–6 cm, base truncate to subtruncate, apex acute to obtuse,. Inflorescence hairs dense than on the stems below; bracts ovate-lanceolate. Calyx 3–5 mm in flower, becoming 4–6 mm long with 4–5.5 mm high scutellum in fruit.

HAB. Roadside, clay soil under oak trees, on wet rocks on mountains; alt. 1200–1370 m; fl. May-Jul.
DISTR. Rare, found in the north sector of the lower forest zone of Iraq. **MAM**: Spindar, *Rawi* 9244 (BAG); 85 km N.W. of Mosul, *Al-Rawi* 23139! (BAG); **MRO**: Serin Mt. on road to Qandil, *Al-Rawi &* *Serhang* 24022! (BAG); Sakran Mt., *A. Haloob* & *M. AL-Bayati* 500360! (BUH); **MSU**: Azmer near police station slope north of Sulaimaniah Chwarta road, *Weinert* & *Al-Musawi*, 0027510! (BUH); Hills around Penjwin in Sulaimaniya Liwa, *Rechinger* 0018220! (BUH).

Iraq & W. Iran.

c. subsp. subsimilis (*Rech.f.*) *A.Haloob*, **stat. & comb. nov**.

Scutellaria velenovskyi subsp. *subsimilis* Rech.f., Bot. Arch. 43: 11 (1941); Rechinger in Fl. Iranica 150: 49 (1982).

Stems pilose eglandular hairs and long glandular hairy. Leaves broadly ovate-cordate, 3.5–6 × 2.5–4 cm, base cordate to subcordate, apex acute to obtuse. Inflorescence hairs dense than on the stems below; bract ovate-lanceolate. Calyx 4–5.5 mm in flower, becoming 5–7 mm with 6–8 mm high scutellum in fruit.

HAB. Clay soil under *Quercus* trees, on roadside under *Quercus* trees on clay soils; alt. 1400–1900 m; fl. Jun.
DISTR. Very rare in the south sector of the lower forest zone of Iraq. **MSU**: Mt. Avroman, *Haussknecht* 788 (W); Hawroman mountain (Hawra Birza), *Al-Rawi, Hosham & Nuri* 29349! (BAG); Azmir, *F. Karim* 39329 (BAG)!; Awi Sart in Tawila in Avroman, *Al-Bayati & A. Haloob* 500359! (BUH).

Iraq & N.W. Iran.

2. **Scutellaria megalaspis** *Rech. f.*, Bot. Arch. 43: 36 (1941); Rawi in Dep. Agr. Iraq Tech. Bull. 14: 156 (1964); Edmondson in Fl. Turkey [P. H. Davis] 7: 86 (1982); Rechinger, Fl. Iranica [K. H. Rechinger] 150: 51 (1982).

Perennial herb. Stems erect to ascending, to 65 cm, branches, purplish below, densely glandular with villous to hirsute eglandular hairs. Leaves long petiolate to 2.5–3.5 cm; lamina broadly ovate or cordate, 5.5–8 × 4.5–6 cm, base cordate to subcordate, apex obtuse to rounded, margins crenate, pubescent to villous eglandular and glandular hairs. Inflorescence a lax raceme, 6–10 cm long; bracts petiolate, lanceolate or elliptic, 10–22 × 5–8 mm, with entire margins. Pedicel 4–5 mm. Calyx green with purplish-tinged scutellum, 4–6 mm long in flower, becoming 5.5–7 mm with 7–10 mm high scutellum in fruit, densely pilose-hirsute eglandular and glandular hairs outside. Corolla pale purple with some white and dark purple markings on middle lobe of lower lip, 16–21 mm, densely pubescent eglandular and glandular hairs outside; tube 10–15 mm. Nutlets broadly ovoid-ellipsoid, ± 2 mm. Fig. 118, 1–4.

HAB. Open *Quercus* forest, pine forest in shade, between rocks under *Quercus* trees, rocky mountain between crack of rocks; alt. 600–1500 m; fl. May-Jul.
DISTR. Occasional in the forest zone of Iraq. **MAM**: Zawita, *Robertson,* 14470! (BAG); Hill above Swara tuka, open *Quercus* forest, *Rechinger* 0018242! (BUH); Khantur Mt., N.E. of Zakho, *Rawi* 23386 (BAG); Bigdaoda village 6 km S. of Kani Masi, *S. Omar & Al-Dabbagh* 45682 (BAG)!; Mergagia village 8 km S. Kani Masi, *Al-Dabbagh & K. Hamid* 45530 (BAG); Bamerny 20 km N.W. Sarsang, *Al-Kaisi & K. Hamad* 45941A (BAG)!; Dohoka village 12 km N.E. Sarsang, *Al-Dabbagh & K. Hamad* 46149 (BAG); **MJS**: Jabal Sinjar E. slope, *S. Omar, Al-Kaisi & Al-Khayat* 52553! (BAG); Jabal Sinjar, E. slope, *S. Omar, Al-Kaisi & Al-Khayat* 52553! (BAG). **MRO**: Khan Mami Sherin N.E., slope of Potin mt., N. of Shirwan Mazin, *Agnew, Hadac, Haines & Kadir* 0018251! (BUH); Sefin Dagh above Shaqlawa, *Al-Rawi* 9085! (BAG).

S.E. Turkey

3. **Scutellaria porphyrantha** *Rech.f.*, Kulturpflanze, Beih. 3: 53 (1962); Rechinger, Fl. Iranica [K. H. Rechinger] 150: 51 (1982).

S. tournefortii Bentham subsp. *porphyrantha* (*Rech.f.*) Edmondson in Fl. Turkey [P. H. Davis] 7: 81 (1982).

Perennial herb. Stems ascending, 30–50 cm, branches purplish below, glabrous with eglandular pubescent hairs on angles. Leaves ovate-triangular, 2–4.5 × 1.5–2.7 cm, base truncate to subcordate, apex acute to obtuse, margins crenate-serrate, glabrescent to puberulent eglandular hairy; petiole 5–10 mm long. Inflorescence a lax raceme, 22–40 cm long, densely eglandular puberulent with minute glandular hairs. Bracts ovate, 5–9 × 3–6 mm, margins entire, with short petiole to 0.5–1.5 mm. Pedicel 3.5–4.5 mm. Calyx green with purplish scutellum, 4.5–5.2 mm in flower, becoming 5.5–6 mm with 4–4.5 mm high scutellum in fruit, fine white eglandular pubescent with spreading minute glandular hairs outside. Corolla dark purple with white markings on middle lobe of lower lip, 19–24

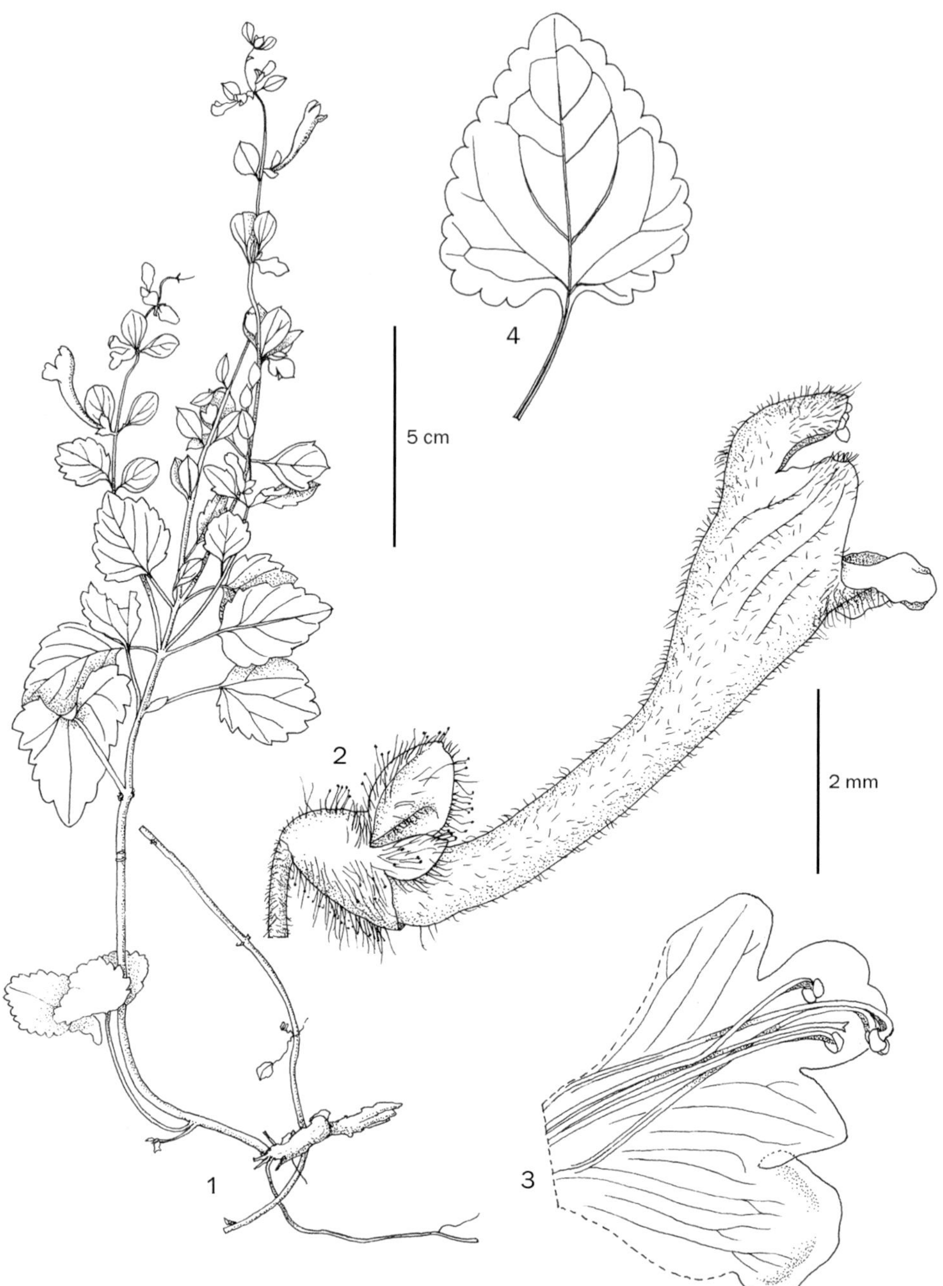

Fig. 118. **Scutellaria megalaspis**. 1, habit; 2, abaxial surface of leaf from a robust specimen; 3, flower (side view); 4, corolla opened. 1, 3, 4 from *Wheeler Haines* W550; 2 from *Rawi* 8703. Drawn by © A.P. Brown, Apr. 2024.

mm, densely puberulent eglandular outside; tube 12.5–16 mm. Nutlets broadly ellipsoid, 1–1.5 mm.

HAB. Slope of mountain, *Quercus* zone on rocky mountain; alt. 1200–1400 m; fl. Jun.-Jul.
DISTR. Rare, found in north sector of the lower forest zone of Iraq. **MAM**: Mosul (Kurdistan) under the district Sharanish, in mountain Jabal Khantur, *Rechinger* 12105; 23 km N.E. of Zakho, 25 km S. of

Sharanish, *Rawi* 23201 (BAG)!; Rakaba Christian Sharanish, *Al-Musawi* & *Al-Sawah* 039469! (BUH); Khantur Mt. N.E. of Zakho, *Al-Rawi*, 23371 (BAG)!; Kani Masi road, Azira village, *Al-Musawi* 43530! (BUH); Kani Masi, *Al-Dabbagh* & *K. Hamad* 45576 (BAG)!; Baidaho village 15 km W. Kani Masi, *S. Omar* 45752 (BAG); MRO: Khan Mam Sherin, N. of Shirwan Mazin, Erbil, *Agnew, Hadac, Haines* & *Kadir* 18250! (BUH); Gali Ali Beg, Erbil Liwa, *F. Kadir* 1205! (BUH).

Turkey.

SUBGENUS APELTHANTHUS (Nevski ex Juz.) Juz. emend. Paton

4. **Scutellaria orientalis** *L.*, Sp. Pl. 598 (1753); Boissier, Fl. Orient. 4: 682 (1879); Zohary in Dep. Agr. Iraq Bull., 31: 125 (1950); Shishkin & Yuzepchuk in Fl. USSR 20: 106 (1976); Rawi in Dep. Agr. Iraq Tech. Bull. 14: 156 (1964); Ball in Fl. Europ. 3: 135 (1972); Edmondson in Fl. Turkey [P. H. Davis] 7: 89 (1982); Paton, Kew Bull. 45: 443 (1990).

Suffruticose herb. Stems prostrate or procumbent to ascending, 4.5–25 cm, branched at base or sometimes above, yellow or purple, sparsely pubescent eglandular, hirsute on angles, minutely glandular. Leaves oblong to lanceolate-elliptic, ovate or broadly ovate to triangular-ovate, 8–15(–22) × 4–6.5(–16) mm, base cuneate to subtruncate, apex acute or subobtuse, margins crenate or incised-crenate, divisions less than half-way to midrib, white tomentose; petiole 2–10 mm. Inflorescence a compact ovoid spike, 1–3 cm long; bract membranous, sessile, pale green or yellow or reddish-purple tinged or whitish-grey, broadly ovate or ovate-lanceolate, margins entire, or only the lowest pair with crenate margins, apex acute to acuminate, sparsely covered with fine short curved eglandular and minute glandular hairs or densely tomentose. Pedicel 1.5–2 mm. Calyx yellow or green reddish-tinged or whitish-grey with purple scutellum, 1.2–2.5 mm in flower with 1.5–2 to 4 mm high scutellum, densely minute glandular hairy or with short spreading eglandular hairs or tomentose, not membranous. Corolla yellow or yellow tinged reddish or purplish, 20–30 mm; tube 15.5–23.5 mm, densely minute glandular hairy outside or with short eglandular hairs. Nutlets ovoid-ellipsoid, ±1.5 mm.

Represented in Iraq by one subspecies:

subsp. **iraqensis** *A.Haloob* **subsp. nov.**

Type: Iraq. Galli Zawita, N.E. of Zakho, nr. Turkish border, 8/7/1957, *Al-Rawi*, 23569 (BAG, holo.)

Affinis subsp. *alpina* (Boiss.) O.Schwarz sed caulibus flavis et purpurascens; pilis sparse pubescens cum hirsutum ad angulos, pilis brevibus glanduliferis sparse. Bracteis flavis vel flavescenti-purpurascens, late ovate, 12–17 × 8–22 mm differt.

The new subsp. is related to subsp. *alpina* (Boiss.) O.Schwarz, but differs in its yellow and purplish stems; hairs sparsely pubescent and sparsely short glandular. Bracts yellow or yellowish-purple, broadly ovate, 12–17 × 8–22 mm.

Stems 4.5–19 cm long, sparsely pubescent with eglandular hairs, hirsute on angles, minutely glandular hairy. Leaves oblong to lanceolate-elliptic, 8–15 × 4–6.5 mm, margins incised-crenate, divisions less than half-way to midrib, tomentose; petiole 6–10 mm. Bracts yellow or with a reddish-purple tinge, broadly ovate, 12–17 × 8–22 mm, margins entire, only the lowest pair crenate, sparsely covered with fine short curved eglandular and minute glandular hairs. Calyx 2–2.5 mm, with 1.5–2 mm high scutellum in flower, densely minute glandular hairy. Corolla 20–25 mm, densely minute glandular hairy outside. Fig. 119, 1–7.

HAB. On summit of mountain; alt. 1880–2060 m; fl. Jul.
DISTR. Very rare, only found once in the forest zone of Iraq. MAM: Galli Zawita, N.E. of Zakho, nr. Turkish border, *Al-Rawi*, 23569! (type).

Endemic.

5. **Scutellaria bornmuelleri** *Hausskn.* ex *Bornm.*, Notizbl. Bot. Gart. Berlin-Dahlem 7: 33 (1917); Rawi in Dep. Agr. Iraq Tech. Bull. 14: 156 (1964); Rechinger, [K. H. Rechinger] Fl. Iranica 150: 57 (1982).

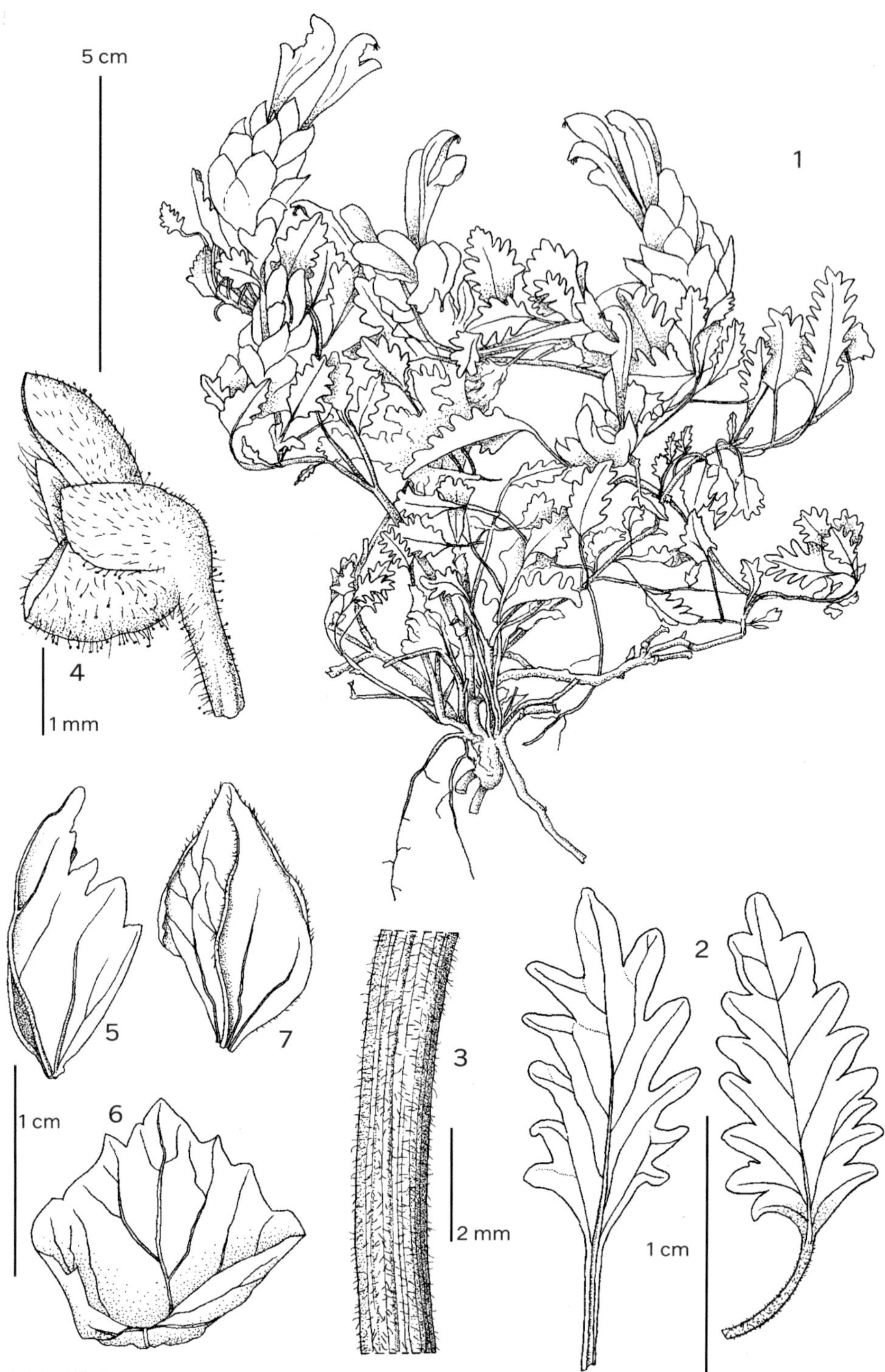

Fig. 119. **Scutellaria orientalis** subsp. **iraqensis**. 1, habit; 2, leaves (adaxial surfaces); 3, detail of stem indumentum; calyx, side view; 4, calyx, side view; 5, lowest bract, side view; 6, lower bract, adaxial view; 7, upper bract, adaxial view. 1–7 from *Rawi* 23569, drawn from photograph of herbarium sheet. Drawn by © A.P. Brown, July 2024.

Scutellaria orientalis subsp. *bornmuelleri* (Hausskn. ex Bornm.) J.R.Edm., Notes Roy. Bot. Gard. Edinburgh 38: 53 (1980).

Suffruticose herb. Stems erect to ascending, 30–42 cm tall, branched in lower part and below the inflorescence, purple to yellowish or green above, eglandular velutinous and papillose-puberulent with minute glandular hairs. Leaves triangular-ovate, 13–24 × 8–18 mm, base cuneate, apex acute, margins shallowly crenate-dentate, white tomentose; petiole 2–6 mm long. Inflorescence an oblong spike, 5–12 cm long; bracts membranous, sessile, yellow, oblong to ovate-lanceolate, 12–16 × 5–8 mm, apex acuminate, concave and curved upwards, margins entire, sparsely covered with eglandular and glandular hairs. Pedicel 1–1.5 mm. Calyx yellow, 2–2.8 mm in flower, becoming 3.5–4 mm with 3.5–3.8 mm high scutellum in fruit, covered with densely white pilose eglandular hairs and short spreading hairs. Corolla yellow with reddish-purple or purple upper lip and lateral lobes of lower lip, 26–32 mm; tube 20–25 mm, densely villous eglandular with minute glandular hairs outside. Nutlets oblong to ellipsoid, 1.9 mm.

HAB. Along rivulets between *Quercus* trees, open *Quercus* forest, on rocks, forest on mountain slope; alt. 1100–1200 m; fl. Jul.
DISTR. Rare, found on only one mountain area in the east sector of forest zone of Iraq. **MRO**: Riwandous in mountain Handarin, Feb. 1895, *Bornmüller* 1691!; N.E. of Rania, lower slopes of Qandil Mts., *Al-Rawi & I. Serhang* 18250! (BAG); Qandil Mts. Pushtanshan, Erbil Liwa, *Rechinger* 18248! (BUH); bet. Pushtashan & Surde, *Serhang & Rawi* 26530 (BAG); Jabal Sakri Sakran, *M. Al-Bayati & A. Haloob* 58708 (BAG).

S.E. Turkey to W. Iran.

6. **Scutellaria platystegia** *Juz.*, Bot. Zhurn. S.S.S.R. 24: 431 (1939); Shishkin & Yuzepchuk in Fl. USSR 20: 108 (1976); Rechinger, Fl. Iranica [K. H. Rechinger] 150: 73 (1982).

Suffruticose herb. Stems procumbent to ascending,12–40 cm long, branched in the lower half, pale yellow below, purple above, densely white puberulent curved and minute glandular hair near inflorescence axil. Leaves triangular-ovate, 8–15 × 5–10 mm, base truncate, apex acute or obtuse, margins incised crenate, upper surface reddish-grey, lower pale green-grey, eglandular white tomentose; petiole 1–7 mm. Inflorescence compact, ovoid or oblong spike, 3.5–8 cm long; bracts membranous, sessile, yellow or purple or yellow tinged purple, broadly ovate, 10–15 × 7–14 mm, apex acuminate, almost curved upwards, margins entire, only the lowest pair with crenate margins, densely curved puberulent with sparse minute glandular hairs. Pedicel 1.5–2 mm. Calyx yellow tinged purplish, 2–2.5 mm in flower, becoming 3–3.5 mm with 2.5–3.2 or 3.5–4 mm high scutellum in fruit, densely minute glandular hairy. Corolla yellow, 27–30 mm; tube 20–24.5 mm, densely minute glandular hairy or with long eglandular hairs and short-stipitate glands outside. Nutlets ellipsoid-ovoid, ± 1.7 mm long.

subsp. *platystegia* is not found in Iraq, but is represented by a single subsp:

subsp. **mazinica** A.Haloob & Al-Bayati **subsp. nov.**

Type: Iraq, Zeta N. of Shirwan Mazin, Erbil liwa, 19/6/1961, *Agnew, Hadac, Haines & Kadir*, 18240! (BUH, holo.).

Affinis subsp. *platystegia* sed caulibus 35–40 cm longus. Calyx cum scutellum 3.5–4 mm altum in fructus, pilis dense brevibus glanduliferis. Corolla cum dense pilis brevibus glanduliferis extra differt.

The new subsp. is similar to subsp. *platystegia*, but differs in the stems 35–40 cm long, calyx with the scutellum 3.5–4 mm high in fruit, densely short glandular hairy. The corolla differs from the type subsp. in its dense short glandular hairs on the outside.

HAB. Dry hillsides; alt. 1060–1370 m; fl. Jun.
DISTR. Very rare, found only once in the forest zone of Iraq. **MAM**: Zeta N, of Shirwan Mazin, Erbil liwa, 19/6/1961, *Agnew, Hadac, Haines & Kadir* 18240 (BUH) (type).

Endemic.

7. **Scutellaria pinnatifida** *A.Ham.*, Esq. Monogr. Scutellaria: 16 (1832); Boissier, Fl. Orient. 4: 683 (1879); Blakelock in Kew Bull., 4: 549 (1950); Zohary in Dep. Agr. Iraq Bull., 31: 125 (1950); Rawi in Dep. Agr. Iraq Tech. Bull. 14: 156 (1964); Rechinger, Fl. Iranica [K. H. Rechinger] 150: 75 (1982).

Suffruticose herb. Stems prostrate-decumbent or ascending or arcuate-erect, 8–24 cm, branched in the lower part, white-grey or green or yellow greenish, glabrescent, sparsely pubescent or white-grey hirsute-tomentose or hirsute with minute glandular hairs. Leaves ovate-triangular to ovate, oblong or lanceolate, 5–28 × 4–15 mm, base truncate, apex acute to acuminate, margins crenate to incised-crenate, indumentum similar to stem; petiole 5–15 mm long. Inflorescence a dense spike, 2–7 cm; bracts herbaceous, sessile, green or green with purple tinge or purple, broadly ovate, 15–27 × 8–22 mm, apex acute-acuminate, margin incised-pinnatifid. Pedicel 2–2.5 mm. Calyx greyish with green or green tinged reddish scutellum, 2–3 mm in flower, becoming 3.5–4.2 mm with 4.2–5 mm high scutellum in fruit, densely pubescent-pilose, with minute glandular hairs. Corolla yellow with yellow or red middle lobe of lower lip, 25–38 mm; tube 21.5–28 mm, densely minute glandular hairs outside. Nutlets ellipsoid, ± 1.8 mm.

Four subspecies are recognized for Iraq:

1. Stems white-grey or purplish. 2
   Stems green or yellow greenish 3
2. Leaves incised-crenate, divisions deeply to half-way to midrib; corolla
   20–25 mm . . . . a. subsp. *pinnatifida*
   Leaves crenate, divisions less than half-way to midrib; corolla more than
   29 mm long . . . . b. subsp. *alpina*
3. Stems green, glabrescent with sparsely pubescent; corolla 25–30 mm
   long. . . . . .c. subsp. *viridis*
   Stems yellow greenish, hirsute with minute glandular hairs. Corolla
   30–32 mm long. . . . . .d. subsp. *pichleri*

a. subsp. **pinnatifida**

Stems ascending, white-grey hirsute to tomentose with glandular hairs. Leaves incised-crenate, divisions deeply to half-way to midrib. Corolla 20–25 mm.

HAB. Not known; alt. 1200–3300 m; fl. June-July
DISTR. Very rare in Iraq, found in the south east sector of the forest zone and in alpine zone. **MRO**: Arl Gird Dagh, *Guest and Ludlow-Hewitt* 2889 (BAG)!; **MSU**: Pir Omar Gudrun, *Rawi* 12110 (BAG); Kirkuk, Jarmo prope Chamchamal, *Haines* 367 (cited in Fl Iranica).

Iran.

b. subsp. **alpina** (Bornm.) Rech.f., Fl. Iranica [K. H. Rechinger] 150: 78 (1982).

Scutellaria pinnatifida var. *alpina* Bornm., Bull. Herb. Boissier, sér. 2, 8: 116 (1908).

Stems prostrate-decumbent or ascending, white-grey hirsute to tomentose with glandular hairs. Leaves incised crenate, divisions less than half-way to midrib. Corolla 30.5–38 mm long.

HAB. Rocky slopes, between rocks, by stream in grove of *Juglans*, limestone mt with coppiced *Quercus aegilops* forest; alt. 1300–3000 m; fl. Jun.
DISTR. Occasional in the eastern sector of the forest zone of Iraq. **MSU**: Jebel Avroman, spur N. of Biyara, Gillett 11821 (BAG); Dara Tri (on the road between Halabja & Tawaila), *Al-Rawi* 22018! (BAG); Kamarspa (on road between Halabja & Tawela), *Al-Rawi* 22216! (BAG). **MRO**: Qandil mts., above Goam-e- Kiromosoran-lake, Erbil Liwa, *Rechinger* 0018231! (BUH).

Iran.

c. subsp. **viridis** (Bornm.) Rech.f., Fl. Iranica [K. H. Rechinger] 150: 78 (1982).

Scutellaria pinnatifida var. *viridis* Bornm., Bull. Herb. Boissier, sér. 2, 8: 115 (1908).

Stems green, glabrescent or sparsely pubescent. Inflorescence axial with hirsute-puberulent and glandular hairs. Corolla 25–30 mm long.

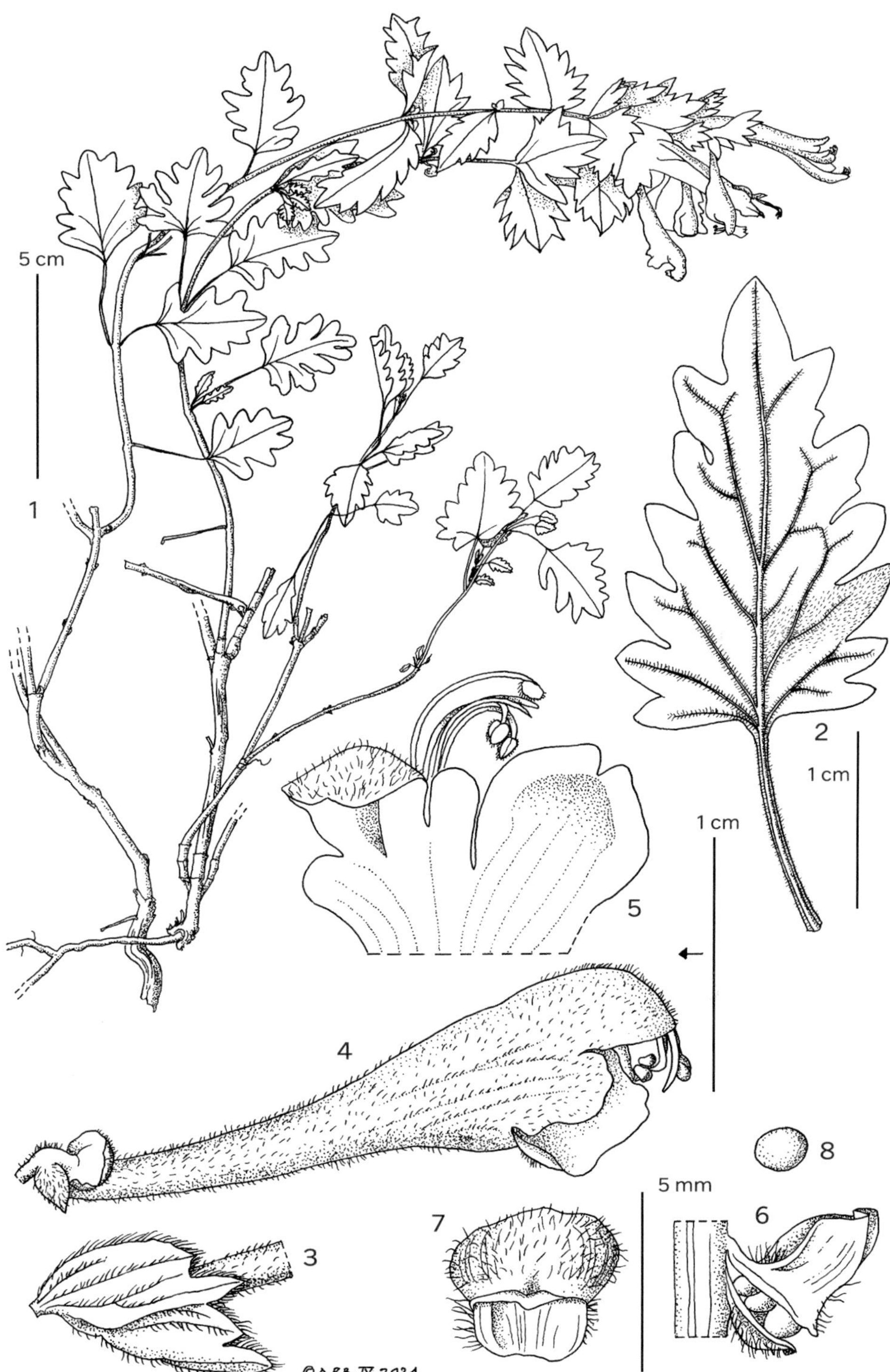

Fig. 120. **Scutellaria pinnatifida** subsp. **pichleri**. 1, habit; 2, abaxial surface of mid-stem leaf; 3, floral bract; 4, flower, side view (bract removed); 5, dissected flower to show corolla lobes; 6, lateral view of dehiscing fruit with nutlets; 7, front view of dehisced (and empty) fruit; 8, seed. 1–5 from *Rechinger* 10369; 6–8 from *Rechinger* 43097. Drawn by © A.P. Brown, Apr. 2024.

Hab. Not recorded; alt. 2100–2700 m; fl. not recorded.

Distr. Very rare found in the forest zone and steppe regions of Iraq. MSU: "M. Avroman et M. Shahu", *Haussknecht* 786 p.p., (cited in Fl. Iranica). FPF: "in subalpinis persiae kurdistanicae prope Mandali", *Noe* 1185 (cited in Fl Iranica).

Turkey & Iran.

d. subsp. **pichleri** (Stapf) Rech.f., Fl. Iranica [K. H. Rechinger] 150: 77 (1982).

Scutellaria pichleri Stapf, Denkschr. Kaiserl. Akad. Wiss., Wien. Math.-Naturwiss. Kl. 50: 47 (1885).
S. orientalis subsp. *pichleri* (Stapf) J.R.Edm., Notes Roy. Bot. Gard. Edinburgh 38: 54 (1980).

Stems ascending or arcuate-erect, yellow greenish, hirsute with minute glandular hairs. Corolla 30–32 mm. Fig. 120, 1–8.

Hab. Cultivated foothill near stream; alt. 1200–2350 m; fl. Jun.

Distr. Occasional in the south east sector of the forest zone of Iraq. MSU: Penjwin, *Al-Rawi* 12254 & 22612! (BAG); Hills around Penjwin in Sulaimaniya Liwa, *Rechinger* 0018232! (BUH); Avroman mountains, Dara Tariq, above Tawilla, Sulaimaniya Liwa, *Rechinger* 0018230! (BUH).

Turkey & W. Iran.

8. **Scutellaria tomentosa** *Bertol.*, Misc. Bot. 2: 11 (1843); *Rawi* in Dep. Agr. Iraq Tech. Bull. 14: 156 (1964); Rechinger, Fl. Iranica [K. H. Rechinger] 150: 72–73 (1982); Edmondson in Fl. Turkey [P. H. Davis] 7: 99 (1982).

Scutellaria amana Bornm., Notizbl. Bot. Gart. Berlin-Dahlem 7: 30 (1917).
Scutellaria cretacea Boiss. & Hausskn., Fl. Orient. [Boissier] 4: 685 (1879).
Scutellaria fruticosa Pers., Syn. Pl. 2: 136 (1806), nom. illeg; Boissier, Fl. Orient 4: 684 (1879); Handel-
    Mazzetti in Ann. Naturh. Mus. Wien 27: 411 (1913); Rawi in Dep. Agr. Iraq Tech. Bull. 14: 156
    (1964).
Scutellaria orientalis subsp. *cretacea* (Boiss. & Hausskn.) J.R.Edm., Notes Roy. Bot. Gard. Edinburgh
    38: 54 (1980).

Suffruticose herb. Stems erect to ascending, 9–25 cm, branched at base and sometimes above, whitish-grey, densely tomentose with or without few minute glandular hairs. Leaves ovate-lanceolate or elliptic, 5–22 × 2–10 mm, base truncate, apex acute to subobtuse, margins crenate to incised-crenate, whitish-grey, densely tomentose; petiole 2–10 mm. Inflorescence lax raceme, 4–10 cm; bract sessile, equalling or shortly exceeding calyx at anthesis, white to grey, ovate-lanceolate, 4–6 × 3.5–4.5 mm, acute to acuminate, margins entire, tomentose with minute glandular hairs. Pedicel 1–1.5 mm. Calyx whitish-grey with purplish-tinged scutellum, 1.5–2.5 mm in flower, becoming 3.5–4.5 mm with 2.5–3 mm high scutellum in fruit, tomentose. Corolla 24–30 mm, yellow, lips purple or yellow, with eglandular and densely minute glandular hairs outside; tube 15–21 mm. Nutlets ovoid-ellipsoid, ± 1.5 mm. Fig. 121, 1–2.

Hab. Limestone slopes with remnants of *Quercus aegilops* forest, limestone rocky, rocky mountain, on sandy soil & rocky soil; alt. 600–800 m; fl. Apr.-May.

Distr. Rare, found only in one mountain in the west sector of the forest zone of Iraq. MJS: Kursi, Jabal Sinjar, *Gillett* 10946! (BAG); Jabal Sinjar, *Al-Kaisi & K.Hamad* 49153! (BAG); Jabal Sinjar, *S. Omar* 56806! (BAG); Between Kersi & Mellak, Sinjar Mt., *Al-Shehbaz, Mayah & Sharifi* 31950! (BUH); 1 km N.W. of Sinjar, *Al-Shehbaz, Hilli & Weinert* 39428! (BUH); 4 km N.E. Sinjar, *Al-Khakani*, 36772! (BUH).

Lebanon, Palestine, Jordan, Syria, Turkey & W. Iran.

9. **Scutellaria multicaulis** *Boiss.*, Diagn. Pl. Orient. 7: 61 (1846); Boissier, Fl. Orient. 4: 685 (1879); Rawi in Dep. Agr. Iraq Tech. Bull. 14: 156 (1964); Rechinger, Fl. Iranica [K. H. Rechinger] 150: 58–60 (1982); Hedge & Paton in Fl. Pakistan [Ali & Qaiser] 192: 39 (1990).

Suffruticose herb. Stems erect to ascending, 22–38 cm, much branched in the lower part, sparsely eglandular puberulent with minute glandular hairs. Leaves ovate-lanceolate, 10–14 × 7–10 mm, base rounded-subrounded, apex acute, margins crenate, densely white pilose with minute glandular hairs; petiole 1–4 mm. Inflorescence lax, 11–25 cm, eglandular puberulous-pilose with minute glandular hairs; bracts sessile, purplish, ovate-lanceolate,

Fig. 121. **Scutellaria tomentosa**. 1, habit × 1; 2, fruiting calyx × 4. Reproduced with permission from Feinbrun-Dothan, Fl. Palaestina 3: Plates, f. 172 (1977). Drawn by Esther Huber. © The Israel Academy of Sciences and Humanities.

Fig. 122. **Scutellaria multicaulis**. 1, habit × ½; 2, flower × 2½; 3, indumentum of stem, detail. Reproduced with permission from Jamzad, Flora of Iran 76: f. 29 (2012). © Ministry of Jihad-e-Agriculture.

5–8 × 3.5–5.5 mm, acute, margins entire, hairs similar to leaves. Pedicel 1–1.5 mm. Calyx purple, 1.2–2 mm in flower, becoming 2.5–3 mm with 1.5–2 mm high scutellum in fruit, densely eglandular pilose and minute glandular hairs outside. Corolla dark purplish or purple, 15 mm long, with spreading curved white eglandular and minute glandular hairs outside; tube ± 10 mm. Nutlets broadly ellipsoid, 1–1.5 mm. Fig. 122, 1–3.

HAB. Dry stony ground, near spring; alt. 140–2200 m; fl. Aug.
DISTR. Very rare, found twice in the lower forest zone of Iraq. **MRO**: Sairo, Erbil liwa, *Gillett* 9663! (BAG); mountains S.E. of Serra and Ser Kurawa, *Gillett* 9693! (BAG).

Iran, Afghanistan, Pakistan.

4. **THUSPEINANTA** T.Durand.

Ind. Gen. Phan. 703 (1888); Harley & al. in Kubitzki (ser. ed.), Fam. Gen. Vasc. Pl. 7: 217 (2004)

Ali Haloob

Annual herbs. Leaves petiolate, entire or absolutely dentate. Verticillasters 2-flowered, axillary in axils of bracts. Calyx campanulate or campanulate-tubular, slightly bilabiate, 5-toothed, 10-veined, inflated and pendant in fruit; teeth triangular, the upper one somewhat shorter. Corolla bilabiate, tube exserted, upper lip erect, shortly 2-lobed, lower lip 3-lobed, medial lobe broad, entire. Stamens 4, didynamous, included in corolla tube, the anterior pair longer than the posterior pair. Style unequally bifid. Nutlets smooth, oblong.

Two species in S.W. Asia; a single species in Iraq.

Salmaki, Y., Heubl, G. and Weigend, M. (2019). Towards a new classification of tribe Stachydeae (Lamiaceae): naming clades using molecular evidence. Bot. J. Linn. Soc. 190(4): 345–358.

1.   **Thuspeinanta persica** (*Boiss.*) *Briq.* in H.G.A.Engler & K.A.E.Prantl, Nat. Pflanzenfam. 4(3a): 229 (1896); Zohary in Dep. Agr. Iraq Bull., 31: 126 (1950); Rechinger, Fl. of Lowland Iraq: 509 (1964); Feinbrun-Dothan, Fl. Palaest. 3: 111 (1978); Rechinger, Fl. Iranica, [K. H. Rechinger] 150: 86 (1982).

Tapeinanthus persicus Boiss. ex Hohen., Exsicc. (Pl. Pers. Bor.) 1843: 18 (1846); Boissier, Fl. Orient. 4: 679 (1879); Rawi in Dep. Agr. Iraq Tech. Bull. 14: 157 (1964)

Annual herb; stem erect, 11–15 cm, simple or branched from base, glandular hairy and viscid. Leaves oblong or lanceolate, 15–18 × 5–10 mm, base attenuate or cuneate, apex obtuse, margins absolutely denticulate, glandular hairy; petiole 2–10 mm. Verticillasters 2-flowered; bracts sessile or with 1 mm long petiole, lanceolate, margins entire or denticulate, densely glandular. Flowers with short pedicel ± 2 mm. Calyx campanulate-tubular, 5–6 mm, teeth short triangular, apex acute, dense glandular hairs outside. Corolla white, 10–12 mm long, tube glabrous outside, lips with glandular hair outside. Nutlets oblong, 2–3 mm. Fig. 123, 1.

HAB. Sandy soil, on road side; alt. 50–510 m; fl. May-Jun.
DISTR. Rare, found twice in the desert region of Iraq. **DGA**: 50 km E. of Samarra, Asila village, *Al-Rawi*, 20444! (BAG). **DWD**: 45 km E. Rutbah, *Rechinger* 9951 (cited in Fl. of Lowland Iraq); 40 km S. of Rutba, *Rawi* 21248 (BAG).

Afghanistan, Turkmenistan, Iran, Syria, Palestine.

5. **LAMIUM** L.

Sp. Pl. 2: 579 (1753); Harley & al. in Kubitzki (ser. ed.), Fam. Gen. Vasc. Pl. 7: 220 (2004)
Orvala L. (1753). *Lamiastrum* Heist. ex Fabr. (1759). *Galeobdolon* Adans., Fam. (1763), nom. superfl. *Polichia* Schrank (1781), non *Pollichia* Aiton. *Wiedemannia* Fisch. & C.A.Mey. (1838)

Ferhat Celep

Annual or rhizomatous or stoloniferous perennial herbs. Leaves ovate to reniform, crenate, dentate to deeply lobed, at least the lower petiolate. Inflorescence thyrsoid. Verticillasters dense or ± remote, 2–12-flowered, in axils of floral leaves. Floral leaves amplexicaul bracts or leaf like. Bracteoles present, equalling or usually shorter than calyx tube. Calyx tubular or campanulate with oblique mouth, 5-lobed, almost equal or subequal, rarely with 3 longer and 2 shorter lobed. Corolla strongly 2-lipped, 4–5-lobed, purple, pinkish purple, pink, yellowish or white; tube with or without a ring of hairs (annulus) inside, tube usually abruptly widened above; upper lip hooded, entire to bifid; lower lip 3-lobed, lateral lobes usually reduced, sometimes appendiculate, median lobe much larger, emarginate, contracted into a substipitate base, crenulate or entire, often with conspicuous coloured blotch or markings. Stamens 4, lower pair longer than upper, not or very shortly exserted from corolla; anthers hairy or glabrous, thecae weakly distinct, bearded or glabrous. Style

Fig. 123. **Thuspeinanta persica**. 1, habit × 1; 2, flower × 5. Reproduced with permission from Jamzad, Flora of Iran 76: f. 38 (2012). Drawn by R. Habibi. © Ministry of Jihad-e-Agriculture.

bifid; stigma lobes subequal or slightly unequal. Nutlets triquetrous, acutely angled, usually ± truncate at apex, smooth, warty or vermiculate.

Lamium L., the type genus of subfamily Lamioideae and the Lamiaceae, comprises 16–38 species or more, depending on the circumscription of the genus (Harley et al., 2004; Bendiksby et al., 2011). *Lamium* species are widely distributed throughout temperate Eurasia, where its center of diversity lies in the Irano-Turanian and Mediterranean regions (Mennema, 1989); 6 species in Iraq.

The latest infrageneric delimitation was done by Mennema (1989), who recognised three subgenera; *Orvala* (L.) Briq., *Galeobdolon* (Adans.) Asch. and *Lamium*. *Lamium* subgenus *Lamium* is in turn divided into three sections: sect. *Amplexicaule* Mennema, sect. *Lamium* and sect. *Lamiotypus* Dumort. However, the infrageneric classification proposed by Mennema (1989) contrasts with his intuitive phylogenetic tree, as well as morphological characters and molecular investigations (Harley et al. 2003; Bendiksby et al., 2011). Therefore, a new infrageneric delimitation is required based on more comprehensive sampling and field observations.

Bendiksby, M., Thorbek, L., Scheen, A.C., Lindqvist, C. and Ryding, O. (2011). An updated phylogeny and classification of Lamiaceae subfamily Lamioideae. *Taxon* 60(2): 471–484.
Harley, R.M., Paton, A.J. and Ryding, O. (2003). New synonymy and taxonomic changes in the Labiatae. *Kew Bulletin*, 485–489.
Mennema, J. (1989). *A taxonomic revision of Lamium (Lamiaceae)* (Vol. 11). Brill Archive.
Ryding, O. (2003). Reconsideration of Wiedemannia and notes on the circumscription of Lamium (Lamiaceae). *Botanische Jahrbücher für Systematik, Pflanzengeschichte und Pflanzengeographie*, pp.325–335.

1. Annual, floral leaves (bracts) amplexicaul. 2
   Perennial, floral leaves (bracts) not amplexicaul 4
2. Calyx teeth clearly× 1.5–4 longer than calyx tube, teeth very narrowly
   triangular, filiform at apex. . . . . .1. *L. macrodon*
   Calyx teeth as long as tube length or slightly longer, teeth triangular. 3
3. Floral leaves ovate-oblong to oblong, leaf lamina oblong to ovate-
   oblong. . . . . 2. *L. aleppicum*
   Floral leaves suborbicular to broadly triangular, leaf lamina ovate-
   reniform to suborbicular. . . . . 3. *L. amplexicaule*
4. Corolla pinkish, purplish or whitish with prominent purple striate
   veins on the corolla tube and usually a large purple blotch on lower
   lip, corolla tube straight at the base (rarely curved) and upper lip of
   corolla shortly or deeply bifid . . . . 4. *L. garganicum*
   Corolla completely white or cream without purple veins on the corolla
   tube, lower lip of corolla sometimes yellowish or cream, corolla
   tube abruptly dilated or ventrally saccate, upper lip of corolla flat or
   slightly entire or slightly emarginate 5
   Stems erect or erect-ascending, not forming mats, (15–)25–90 cm;
   leaves green, glabrous to pilose; usually distributed between 1200–
   2700(–3000) m in Iraq. . . . . 5. *L. album*
   Stem decumbent or ascending, mat forming, 5–25(–30) cm; leaves
   greyish-green, densely and soft hirsute hairy; usually distributed
   between 2700–3600 m in Iraq. . . . . 6. *L. tomentosum*

1. **Lamium macrodon** *Boiss. & A.Huet* in Boiss., Diagn. ser. 2(4):45 (1859); Mill, Fl. Turkey [P. H. Davis] 7: 139 (1982); Mennema, Fl. Iranica [K. H. Rechinger] 150: 331 (1982); Mennema, Leiden Bot. Ser. 11: 1–196 (1989).

L. orthodon Boiss. & A.Huet., nom. nud.
L. ordubadicum Grossh. in Beih. Bot. Centr. 44(2): 236 (1927).
L. plebeium Boiss., Diagn. Pl. Orient. ser. 2, 4: 45 (1859).

Annual. Stem ascending to erect, branched, particularly at the base, 5–25 cm, pubescent to subglabrous, glabrous. Leaves ovate-reniform, suborbicular, 5–25 × (3–)7–25 mm, cordate to obtuse, crenate, doubly crenate or lobed, sparsely to moderate appressed-pubescent; petiole 10–30 mm. Floral leaves (bracts) suborbicular, amplexicaul, lower (4–)7–27(–35) ×

(5–)8–31(–35) mm. Verticillasters 2–5, 6–16-flowered. Calyx 4–12 mm, tubular, moderately to densely pubescent to hirsute, sometimes with short glandular hairs; teeth 1.5–4 × as long as tube, teeth very narrowly triangular, filiform at apex. Corolla purplish, with or without purple spots on lower lip, 12–25 mm; tube 10–18 mm, straight, without annulus; upper lip 2–6 mm, arched, apex rounded, apex emarginate or not, margin usually entire to rarely slightly undulate, outside puberulent to velutinous; Nutlets 2–2.5(–3.0) × 1–1.5 mm, smooth, with raised white reticulate marbling on dark brown background.

HAB. Limestone slopes, mountains slopes, open steppe, cultivated land, roadsides; alt. 1750–3000 m; fl. Apr.-Jun.
DISTRIB. N. Iraq, *Astragalus*-thorn cushion zone. **MRO:** Rost Mountain (Halgurd Mountain), N. of Rost (Rust), 23 km E.N.E. of Rowanduz, *Thesinger* 954! (W); Haji Umran (Haji Omeran), *Astragalus*-thorn cushion zone, *Chapman & Chapman* 12276! (K); Haji Umran (Haji Omeran), at Iranian border, *Mooney* 4269! (K).

Lamium macrodon is similar to *L. amplexicaule* and *L. aleppicum*, but it easily separated from them by its calyx teeth 1.5–4 × as long as tube, teeth very narrowly triangular, filiform at apex.

Turkey, Caucasia, Iran, Lebanon.

2. **Lamium aleppicum** *Boiss. & Hausskn.*, Fl. Orient. [Boissier] 4: 761 (1879); Mill, Fl. Turkey [P. H. Davis] 7: 140 (1982); Mennema, Leiden Bot. Ser. 11: 1–196 (1989).

> *L. amplexicaule* L. var. *aleppicum* (Boiss. & Hausskn.) Bornm. in Beih. Bot. Cehtr. 22(2): 133 (1907); Mennema, Fl. Iranica [K. H. Rechinger] 150: 331 (1982).

Annual. Stems 3–25 cm, puberulent. Leaves with petiole ± equalling lamina; lamina oblong to ovate-oblong, 6–20 (–35) × 4–20 mm, obtuse, crenulate, crenate to lobed, truncate at base, puberulent to pubescent. Floral leaves (bracts) ovate-oblong to oblong, 4–21 × 3–23 mm, amplexicaul, crenate-serrate or crenate. Verticillasters crowded, 2(–5), 10–12–flowered. Calyx 4–7 mm; tube 2–4 mm, teeth slightly shorter or subequal to tube, villous, whitish or silvery. Corolla purplish-pink, 12–25 mm; tube straight, 9–20 mm, without annulus; upper lip oblong-ovate, entire, 2–5 mm. Nutlets brown, 2.2–3.2 × 1.1–1.5 mm, with white confluent raised tubercles.

HAB. Rocky limestone slopes, degraded *Quercus* forest; alt. 1100–1800 m; fl. Mar.-Apr.
DISTRIB. N. Iraq. **MRO**: In montibus prope, near Rayat, 21 Apr. 1936, 1800 m, *Low* 221 (BM). **MSU**: Suleimaniya, Gweija Dagh, forestry enclosure, limestone mt., *Quercus* forest entirely destroyed, *Gillett & Rawi* 10614! (K). **FAR**: Arbil, Kuh Safin prope Shaqlawa, *Bornm*üller 1711! (JE).

Lamium aleppicum was originally described as a distinct species (Boissier, 1879), but it was reduced to a variety under *L. amplexicaule* by Bornmüller (1907). Mennema (1989) followed Bornmüller's treatment in his monograph. Mennema (1989) mentioned that var. *aleppicum* differs from the other varieties in having narrower leaves. Bendiksiby et al. (2011) reported that *L. amplexicaule* is a non-monophyletic species and *L. amplexicaule* var. *aleppicum* does not group together with other *L. amplexicaule* varieties. Therefore, they suggested resurrection of *L. aleppicum*.

S. Anatolia, Syrian Desert, Iran.

3. **Lamium amplexicaule** *L.*, Sp. Pl. 2: 579 (1753); Bentham, Lab. Gen. Spec. 511 (1834); Bentham, Prodr. [A. P. de Candolle] 12: 508 (1848); Mill, Fl. Turkey [P. H. Davis] 7: 138–139 (1982); Mennema, Fl. Iranica [K. H. Rechinger] 150: 328 (1982); Mennema, Leiden Bot. Ser. 11: 1–196 (1989).

> *Galeobdolon amplexicaule* (L.) Moench, Methodus Plantas Horti Botanici et Agri Marburgensis 394 (1794).
> *Pollichia amplexicaulis* Willd., Prod. Fl. Berol. 198 (1787).
> *Lamiella amplexicaulis* (L.) E.Fourn., Ann. Soc. Linn. Lyon sér. 2, 17: 134 (1869).
> *Lamium mesogaeon* Heldr. ex Boiss., Fl. Orient. 4(2): 760 (1879).
> *L. adoxifolium* Hand.-Mazz. in Ann. Naturh. Mus. Wien 27:413 (1913) (syn. of *L. amplexicaule* L. var. *incisum* Boiss., Fl. Orient. [Boissier] 4:761 (1879).
> *L. rumelicum* Velen., Fl. Bulg. 649 (1891).
> *L. amplexicaule* L. var. *nemetzii* Aznav. in Bull. Soc. Bot. Fr. 147 (1899).

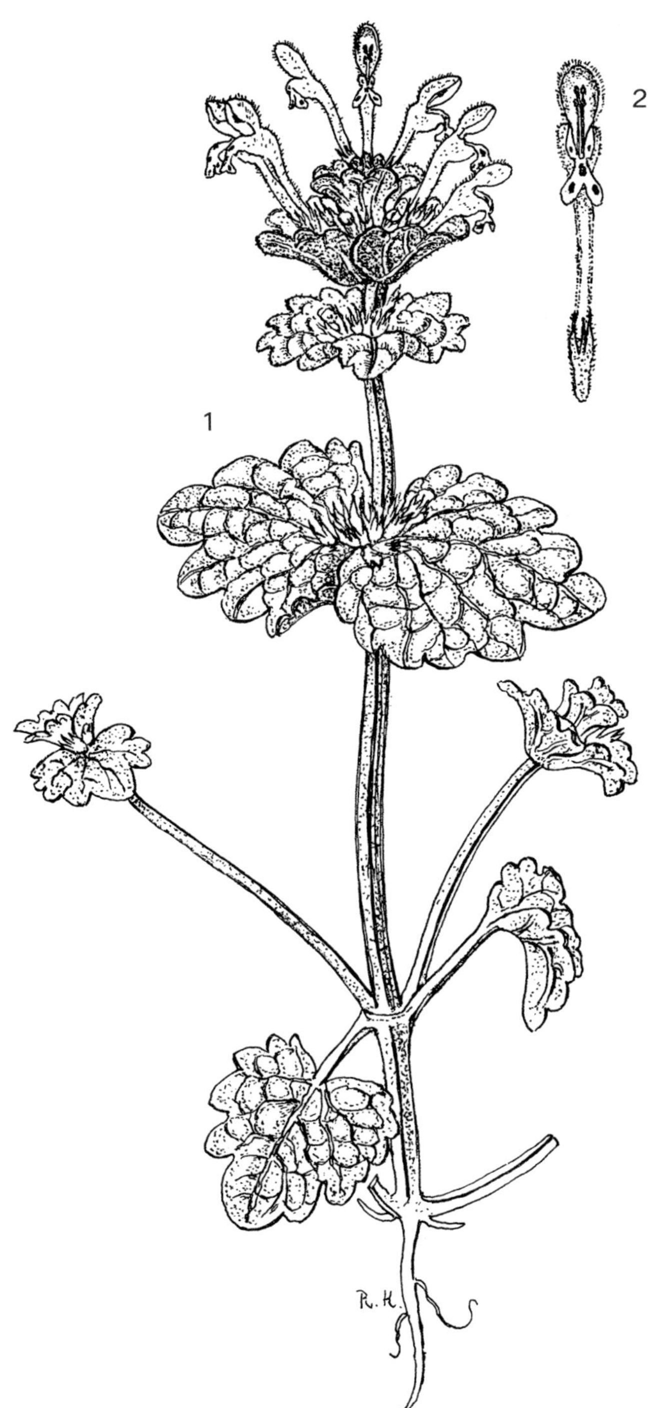

Fig. 124. **Lamium amplexicaule**. 1, habit × 1; 2, flower × 2. Reproduced with permission from Feinbrun-Dothan, Fl. Palaestina 3: Plates, f. 199 (1977). Drawn by Ruth Koppel. © The Israel Academy of Sciences and Humanities.

Lamiopsis amplexicaulis (L.) Opiz, Seznam 56 (1852).
Lamium mauritanicum Gand. Ex Batt., Bull. Soc. Bot. France 56: Ixx (1910).
L. lassithiense Coustur. & Gand., Fl. Cret.: 79 (1916).
L. stepposum Kossko ex Klokov, Fl. URST 9: 644 (1960).
[The following two names are synonyms of *L. amplexicaule* var. *orientale* (Pacz.) Mennema, Leiden
    Bot. Ser. 11: 57. 1989; *L. orientale* (Pacz.) Litv., Maevsh. Fl. ed. 7: 617. 1940 (nom. illegit.); *L.
    palmatum* Sm. in Rees, Cyclop. 20: no. 18 (1811)].

Annual. Stem erect or decumbent, branched, particularly at the base, 3–40 cm, glabrous, subglabrous to pubescent. Leaves ovate-reniform to suborbicular, (2–)6–27(–30) × (2–)7–25(–30) mm, cordate to obtuse, crenate, doubly crenate or palmately lobed, sometimes very deeply incised, sparsely to moderate appressed-pubescent. Floral leaves (bracts) suborbicular, amplexicaul, lower (4–)7–27(–35) × (5–)8–31(–35) mm. Verticillasters 2–5(–7), 4–12-flowered. Calyx 4–8 mm, tubular, moderately to densely pubescent to hirsute, sometimes with short glandular hairs; teeth 1.5–4(–4.5) mm, teeth subequal, as long as the tube, narrowly triangular. Corolla purplish (rarely whitish), with purple spots on lower lip, 12–28 mm; tube whitish, 10–25 mm, straight, without annulus; upper lip 2–6 mm, arched, apex rounded, margin usually entire to rarely slightly undulate, outside puberulent to velutinous; Nutlets (1.5–)1.9–2.5(–3.0) × (0.7–) 0.9–1.5 mm, smooth, dark brown, with or without whitish spots. Sometimes cleistogamous. Cleistogamous flowers ca. 5 mm long. Fig. 124, 1–2.

HAB. Limestone slopes, mountains, under shelter of rocks, hillsides, open steppe, cultivated land, roadsides, waste places; alt. 200–1600 m; fl. Mar.-Jun.
DISTRIB. Mountains, upper plains and foothills and desert plateau. **MAM**: Mosul, Dohuk, 17 Mar. 1956, *Amin* 84! (K); nr. Mosul, on waste ground, *Low* 75! (BM, K); Amadiya, *Rustav* 1224! (K). **MRO**: Salah ad-Din (Salahaddin forestry), *Chapman* 11969! (K); Shaqlawa, *Chapman* 8378 (BAG); Berd Agha Gin, 18 km N.W. of Rania, *Rawi et al.* 28687! (K); in faucibus prope Rawanduz George, *Guest* 596! (G, K). **MSU**: Jarmo, Mound wadi, *Helbaek* 496! (K); Sulaimaniya, Malakawa, *Rawi* 22203! (K); Kajan Mountain, near Penjwin, serpentine rocks, *Rawi* 22697! (K, var. *bornmuelleri*); near Penjwin, in bare mountain, 1600 m, 19/20 June 1957, *Rechinger* 12272! (W, var. *bornmuelleri*); Jarmo, 19 Mar. 1955, *Helbaek* s.n. (K!, var. *incisum*); **MJS**: Khormal, *Rawi* 8862 (BAG). Jabal Sinjar, *Al-Kaisi & K. Hamad* 54082 (BAG). **FUJ**: Jabal Maqlub, nr. Mosul, *Shahwani* 25192! (K). **FNI**: Nineveh, *Anders* 1028! (W, var. *incisum*). **FAR**: Arbil: Kuh Safin (Sefin Dag) prope Shaqlawa, *Polunin* 5027! (K). **DLJ**: Rawa, *Gillett & Rawi* 6979! (K); midway bet Rawa & Haditha, *S. Omar, Al-Kaisi, K. Hamad & H. Hamid* 44627 (BGA). **DWD**: Kalah Island in the Euphrates River at Ana, Ramadi, *Barkley et al.* 671! (K); Al-Bert village bet Nahiya & Qaim, *S. Omar, Al-Kaisi, K. Hamad & H. Hamid* 44398 (BGA). **LCA**: Rustam, *Guest* 367 (BAG); Baghdad, 20 Feb. 1931, *Guest* 1083! (K).

Lamium amplexicaule is non-monophyletic and a very variable species (i.e. size of plant, the indumentum, the length of the calyx teeth and flower colour). Therefore, a lot of infraspecific taxa have been described in the species. Due to high morphological diversity and different taxonomic approaches, the species has very long list of synonyms. Forms with deeply palmatifid leaves have been named *L. adoxifolium* Hand.-Mazz. and *Lamium amplexicaule* var. *incisum* Boiss.. *Lamium palmatum* Sm. var. *nemetzii* Aznav. is a dwarf variant with villous calyces but seems unworthy of formal taxonomic recognition (Mill 1982). Cleistogamous specimens often occur and have been referred to as var. *clandestinum* Reichb. (Mill 1982). In Iraq, a few specimens have been reported as *L. amplexicaule* var. *incisum* by Mennema (1982, 1989). In addition, Mennema (1982, 1989) identified some specimens as *L. amplexicaule* var. *bornmuelleri* from Iraq due to their dwarf habit (≤ 5 cm stem length), very small leaves (5 mm long), only cleistogamous flowers and crowded verticillasters at the top of the stem. The species needs comprehensive phylogenetic, morphologic and taxonomic study; I have used a broad species concept in this account.

Throughout Temperate Eurasia.

4. **Lamium garganicum** *L.*, Sp. Pl. 2 ed. 2: 808 (1763); Bentham, Lab. Gen. Spec. 509 (1834); Bentham, Prodr. [A. P. De Camdolle] 12: 504 (1848); Nyman, Syll. Fl. Eur. 94 (1855); Briq., Lab. Alp. Mar. 2: 291 (1893); Hayek, Prodr. Fl. Penins. Balc. 2: 274 (1929); Mennema, Fl. Iranica [K. H. Rechinger] 150: 326 (1982); Mennema, Leiden Bot. Ser. 11: 1–196 (1989).

Perennial, often sprawling over rocks. Stems 5–45 cm, glabrous to hirsute, pilose, villous or bristly hairy. Leaves broadly ovate to reniform, lamina 3–52 × 3–50 mm, crenate or serrate, villous to ± glabrous. Verticillasters 1–5, (2–)4– 8(–12)-flowered. Floral leaves 0.5–4 cm long. Bracteoles 3–9(–11) mm. Calyx (6–)8–18(–20) mm, tube glabrous to pubescent or villous, 5–10(–11) mm; teeth 2–9 mm. Corolla purplish-pink (rarely white), with purple-striate tube and usually a large purple blotch on lower lip, 22–40(–45) mm; tube straight, without annulus, 14–29 mm; upper lip 6–16 mm, deeply or shortly bifid, emarginate, retuse or ± entire, the lobes themselves sometimes with notched to bifid apices; lower lip 6–20 mm, lateral lobes without or occasionally with a short subulate appendage. Nutlets 2.5–3.5 × 1.3–1.6 mm, blackish-green or brown, usually with a few paler spots.

subsp. **striatum** (*Sm.*) *Hayek*, Repert. Spec. Nov. Regni Veg. Beih. 30(2): 275 (1929). Mill, Fl. Turkey [P. H. Davis] 7: 132 (1982); Mennema, Fl. Iranica [K. H. Rechinger] 150: 326–327; Mennema, Leiden Bot. Ser. 11: 1–196 (1989).

> *Lamium striatum* Sm., Fl. Graec. Prodr. 1(2): 405 (1809).
> *L. garganicum* var. *glabratum* Griseb., Spic. Fl. rumel. 2: 133 (1844).
> *L. bithynicum* Benth., Prodr. (A. P. De Candolle) 12: 505 (1848).
> *L. pictum* Boiss. & Heldr., Diagn. Pl. Orient. ser. 2, 4: 44 (1859).
> *L. striatum* var. *nepetifolium* Boiss. Fl. Orient. 4: 757 (1879)
> *L. striatum* subsp. *nivale* (Boiss. & Heldr.) Nyman, Consp. Fl. Eur. 575 (1881)
> *L. striatum* subsp. *pictum* (Boiss. & Heldr.) Nyman, Consp. Fl. Eur. 575 (1881)
> *L. garganicum* subsp. *glabratum* (Griseb.) Briq., Lab. Alp. Mar.: 295 (1893).
> *L. garganicum* var. *gracile* Briq., Lab. Alp. Mar.: 298 (1893).
> *L. garganicum* L. subsp. *gracile* (Briq.) Greuter & Burdet, Willdenowia 14(2): 300 (1985).
> *L. garganicum* L. subsp. *pictum* (Boiss. & Heldr.) P.W.Ball, Bot. J. Linn. Soc. 65(4): 350 (1972).

Stem usually shorter than 30 cm, glabrate to hirsute or bristly hairy. Leaves usually reniform, cordate or rhomboid, glabrate to pubescent, crenate. Calyx 7–18 mm long. Corolla (25–)30–35 mm, upper lip usually bifid.

HAB. Limestone mountain, mountain sides, scree slopes above tree line, open *Quercus* forest, on rocks, river side alt. 1000–3000 m; fl. Apr.-Aug.

DISTRIB. N. Iraq. **MAM**: N.E. of Zakho, Khantur (Shaxi Xantur), Mountain side, *Rawi* 8595! (glabrous stem) (K), ibid, *Rawi* 8699! (hairy stem) (K); Mosul, Gali Zawita, N.E. of Zakho, near Turkish border, *Rawi* 23545. Ser Amadiya (Amediye), N. of Amadiye, the ridge running East-West, (without flowers), *Rustav* 1536! (K). Prope Sharanish, N.E. of Zakho, *Rechinger* 10905!, 10885!, 11991!. **MRO**: Erbil Liwa: Safin (Sefin) Dagh above Shaqlawa, *Gillett* 8094! (K); Al Gird Dagh (Algurd, Halgurd Mountain), *Guest* 2784! (K); Arbil (Erbil), Algurd (Halgurd) Dagh, *Rawi* 13733! (K); Arbil: Rost Mountain (Halgurd Mountain), N. of Rost (Rust), 23 km E.N.E. of Rowanduz, *Thesiger* 949! (BM); S.W. of Rawanduz, in montibus Karoukh (Korek Mountain), rocky mountain slope, *Serhang, Kass & Nuri* 27351! (K); E. side of Karoukh (Korek) Mountain, Between Sir Waka and Dangala villages, *Kass & Nuri* 27610! (K); Mt. Sakri Sakran prope, near Rawandiz, probably northern end of the Qandil range, alpin region, *Bornmüller* 1709; N.E. of Arbil (Erbil), Supra Shaqlawa, *Haines* W-661! (K); Chiya-i Mandali, Qandil range, *Guest* 2715! (K); Qandil Range, *Rawi & Serhang* 24478! (K); Pushtashan, village in Qandil range, 15 km N.E. of Ranya (Rania), on river edge, *Rawi & Serhang* 24238! (K). **MSU**: Sulaimaniya Liwa: Jebel Avroman (Hawraman) above Darimar, *Gillett* 11854! (K); Sulaimaniya, Jabal Birah Magrun (Pira Magrun), *Poore* 596! (K).

Lamium garganicum is a very variable species. Morphological variations can be seen in the same species and population. To date, many infraspecific taxa have been described under *L. garganicum* (Mennema 1989). Due to high variation in indumentum, flower and leaf morphology, infraspecific classification is very difficult in the species without field and population observations. Mennema (1989) may therefore treated many species and infraspecific taxa as synonyms in *L. garganicum* in his monograph. My field and herbarium observations pointed out that some infraspecific taxa should be resurrected, i.e. *L. garganicum* subsp. *rectum* and subsp. *lasioclades*.

Most of the Iraqi *L. garganicum* species were identified as *L. garganicum* subsp. *striatum* or *L. garganicum* subsp. *pictum* (Mennema 1982). However, Mennema (1989) treated the latter subspecies as a synoym of the former. Though, I agree most of the Iraqi *L. garganicum* species should be placed in *L. garganicum* subsp. *striatum*, some specimens might be evaluated under *L. garganicum* subsp. *rectum* (stem villous, corolla 20–32 mm) or *L. garganicum* subsp. *lasioclades* (stem villous, corolla 32–45 mm) according to Mill (1982). More material and field observations are needed for making precise identifications particularly at subspecies level.

Turkey, Iran, Syria, Lebanon, Israel, Palestine, Greece, Albania, Italy, Balkans, Caucasia.

5. **Lamium album** *L.*, Sp. Pl. 579 (1753); Bentham, Lab. Gen. Spec. 514 (1834); Nyman, Syll. Fl. Eur. 94 (1855); Briq., Lab. Alp. Mar. 2: 312 (1893); Hayek, Prodr. Fl. Penins. Balc. 2: 273 (1929); Mill., Fl. Turkey [P. H. Davis] 7: 143 (1982); Mennema, Flora Iranica [K. H. Rechinger] 150: 322 (1980); Mennema, Leiden Bot. Ser. 11: 1–196 (1989).

Glabrous to pilose stoloniferous perennial. Stems 15–90 cm or more, erect or ascending, strongly quadrangular, often purplish. Leaves with 10–45 mm petioles, lamina broadly ovate, 10–60 × 10–50 mm, obtuse to acute, cordate to truncate, coarsely crenate to dentate or doubly dentate, pubescent-pilose or ± glabrous, sparingly glandular. Bracts lanceolate to ovate, lower usually acute, upper acuminate. Verticillasters 2–8, 8–10-flowered, distant. Bracteoles 1–7 mm. Calyx 9–15 mm; teeth 4–8.5 mm, subequal to or slightly longer than tube, ciliate-setulose, eglandular or with a few or more glands near apex. Corolla white, 20–30 mm; tube 6–20 mm, straight or abruptly dilated with annulus; upper lip 6–12 mm, entire; lower lip 4.5–10 mm. Nutlets 2.8–3.3 × (1.4–)1.6 mm, dark brown or greyish brown.

> Stem and leaves glabrous to sparsely pubescent-pilose; calyx teeth
>   usually as long as or slightly longer than the tube . . . . a. subsp. *album*
> Stem and leaves densely pilose to villous; calyx teeth usually
>   considerably longer than the tube . . . . b. subsp. *crinitum*

a. subsp. **album**

L. album var. *roseum* Dumort., Fl. Belg.: 45 (1827).

L. album var. *parietariifolium* (Benth.) Nyman, Consp. Fl. Eur.: 574 (1881).

L. album subsp. *turkestanicum* (Kuprian.) Kamelin & A. L. Budantsev, Novosti Sist. Vyssh. Rast. 27: 138 (1990).

L. album L. subsp. *hyrcanicum* (A.P.Khokhr.) Menitsky, Bot. Zhurn. (Moscow & Leningrad) 77(6): 69 (1992).

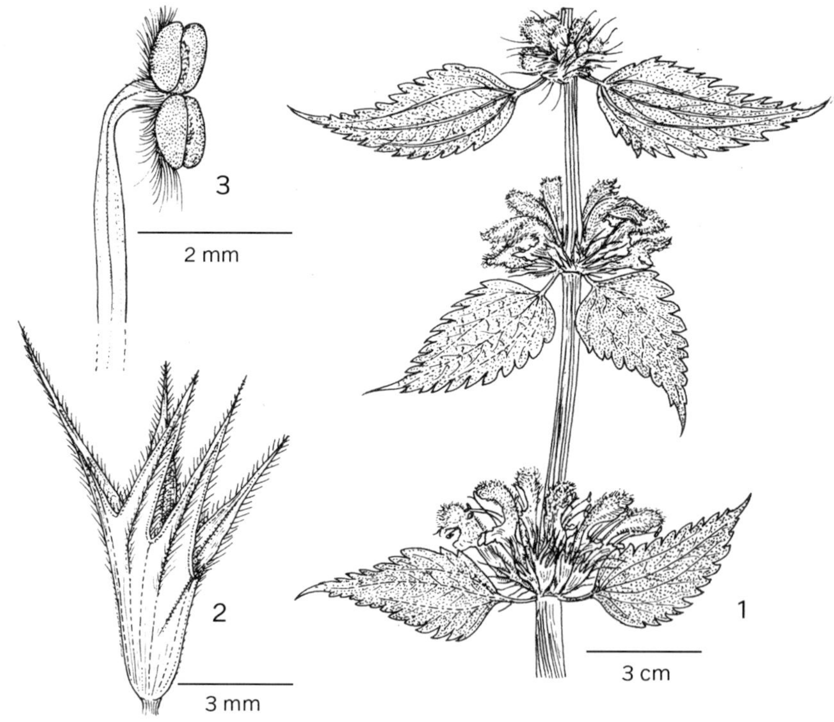

Fig. 125. **Lamium album**. 1, flowering stem; 2, calyx; 3, stamen. Reproduced with permission from Flora of Pakistan 192: f. 20 B–D (1990). Drawn by M. Rafiq. © National Herbarium, Pakistan Agriculture Research Council & University of Karachi, Pakistan.

L. album L. subsp. *sempervirens* (A.P.Khokhr.) Menitsky, Bot. Zhurn. (Moscow & Leningrad) 77(6): 69 (1992).

L. album L. subsp. *transcaucasicum* (A.P.Khokhr.) Menitsky, Bot. Zhurn. (Moscow & Leningrad) 77(6): 69 (1992).

Stem, 30–90 cm long (usually longer than subsp. *crinitum*); stem and leaves glabrous to sparsely pubescent-pilose; calyx teeth usually as long as or slightly longer than the tube. A rather uniform taxon, with some variability in shape and incisions of leaves. Fig. 125, 1–3.

HAB. Shaded places under woodland, forest, roadsides and rocky mountain slopes; alt. ± 1400 m; fl. Apr.

DISTRIB. Rare; known only from one gathering in N. Iraq. **MSU**: Penjwin, *Rawi* 8838! (BM).

Temperate Eurasia, rare in the Mediterranean area, Turkey, Iran, Russia to China, (introduced, naturalized) in Iceland, N. America and New Zealand.

b. subsp. **crinitum** (Montbret & Aucher ex Bentham) Mennema, Fl. Iranica [K. H. Rechinger] 150: 323 (1982); Mill., Fl. Turkey [P. H. Davis] 7: 143 (1982); Mennema, Leiden Bot. Ser. 11: 1–196 (1989).

Lamium petiolatum Royle ex Benth., Bot. Misc. 3: 381 (1833).
L. crinitum Montbret & Aucher ex Benth., Ann. Sci. Nat., Bot II, 6: 48 (1836).
L. robertsonii Boiss., Diagn. Pl. Orient. ser. 1, 7: 54 (1846).
L. persicum Boiss. & Buhse, Nouv. Mém. Soc. Imp. Naturalistes Moscou 12: 180 (1860).
L. setidens Freyn, Oesterr. Bot. Zeitschr. 41: 58 (1891).
L. oreades Azn. in Magyar Bot. Lapok. 17:22 (1919).
L. leucolophum Hausskn. ex R.R.Mill, Notes Roy. Bot. Gard. Edinburgh 38(1): 37 (1980).

Stem, 15–35(–48) cm long (usually shorter than subsp. *album*); stem and leaves densely pilose to villous; calyx teeth usually considerably longer than the tube.

HAB. Rocky slopes, field margins, high subalpine shrubland, cultivated hillsides, *Quercus* woodland; alt. 1280–3000 m; fl. Mar.-Jun.

DISTRIB. N. Iraq. **MRO**: Gali Warta, c. 30 km N.W. of Rania, among *Quercus*-forest, *Rawi et al.* 28735! (K); Arbil, Algurd Dagh, in alpine region, *Bornmüller* 1708! (JE); ibid., in valley above Nowanda, *Rechinger* 11466! (W); Arbil, Qandil Mountains, on E. slope above Pushtashan, in *Astragalus* zone, *Rechinger* 11103! (W); ibid. near rivulet, *Rechinger* 11102! (W); Rowanduz, Sakri Sakran Mountains, *Bornmüller* 1710! (B, JE); Qandil Mountains, on S. slope, *Astragalus* zone, *Rawi & Serhang* 24068! (K); Serin Mountain, on road to Qandil, *Astragalus* zone, *Rawi & Serhang* 24027! (K); Kudu (Kodo), near Haji Umran, mountain side, 9194! (K); E. Karokhi, between Kunar and Kani region, *Alkas et al.* 27637! (K). **MSU**: Near Penjwin, *Rawi* 22577! (K); ibid., *Rawi* 22618! (K); ibid., clay schiefer, in Quercetis, *Rechinger* 10517! (W).

N. Iran, C. & E. Turkey, Azerbaijan, Turkmenistan, Afghanistan, Pakistan, N. India?, Nepal?

6. **Lamium tomentosum** *Willd.*, Sp. Pl. 3: 90 (1800); Mill, Fl. Turkey [P. H. Davis] 7: 145 (1982); Mennema, Flora Iranica [R. H. Rechinger] 150: 325 (1980); Mennema, Leiden Bot. Ser. 11: 1–196 (1989); Jamzad in Fl. Iran [Assadi & al., in Persian] 76: 371 (2012).

Lamium vestitum Benth., Prodr. [A. P. De Candolle] 12: 507 (1848).
L. filicaule Boiss., Diagn. Pl. Orient. ser. 1, 12: 86 (1853).
L. alpestre Trautv., Trudy Imp. S.-Peterburgsk. Bot. Sada ii.: 481 (1873).
L. hakkariense Nábelek, Spisy Prir. Fak. Masarykovy Univ. 70: 67 (1926).
L. sulfureum Hausskn. & Sint. ex R.R.Mill, Notes Roy. Bot. Gard. Edinburgh 38(1): 36 (1980).

Low mat-forming perennial with creeping rootstock. Stems 5–25 cm, slender, tufted, glabrous below, ± purplish, usually villous above with a few, white glands. Leaves greyish or greenish, cordate-ovate to reniform, 3–30 × 6–23 mm, ± rugose, crenate (rarely dentate), villous, tomentose or adpressed hirsute to pubescent, with sessile glands, rarely subglabrous. Bracts similar to leaves. Verticillasters 1–3(–4), 4–12-flowered, crowded. Bracteoles 2–7 mm. Calyx 8–13 mm, greyish-green, villous to tomentose; tube 4–7 mm, teeth 4–8 mm, with patent cilia and white sessile glands. Corolla whitish or yellowish, 18–30 mm; upper lip sometimes pinkish, tube± straight and ventrally saccate, 12–16 mm, with annulus; upper lip retuse, denticulate to subentire, 7–10(–12) mm; lower lip 4–10 mm, lateral lobes with 0.5–2 mm appendage. Anthers pale brownish. Fig. 126, 1–3.

HAB. Rocky volcanic and limestone slopes, shady alpine rocks, crevices in metamorphic rocks, screes; alt. 1800–3000 m; fl. Jul.-Sep.

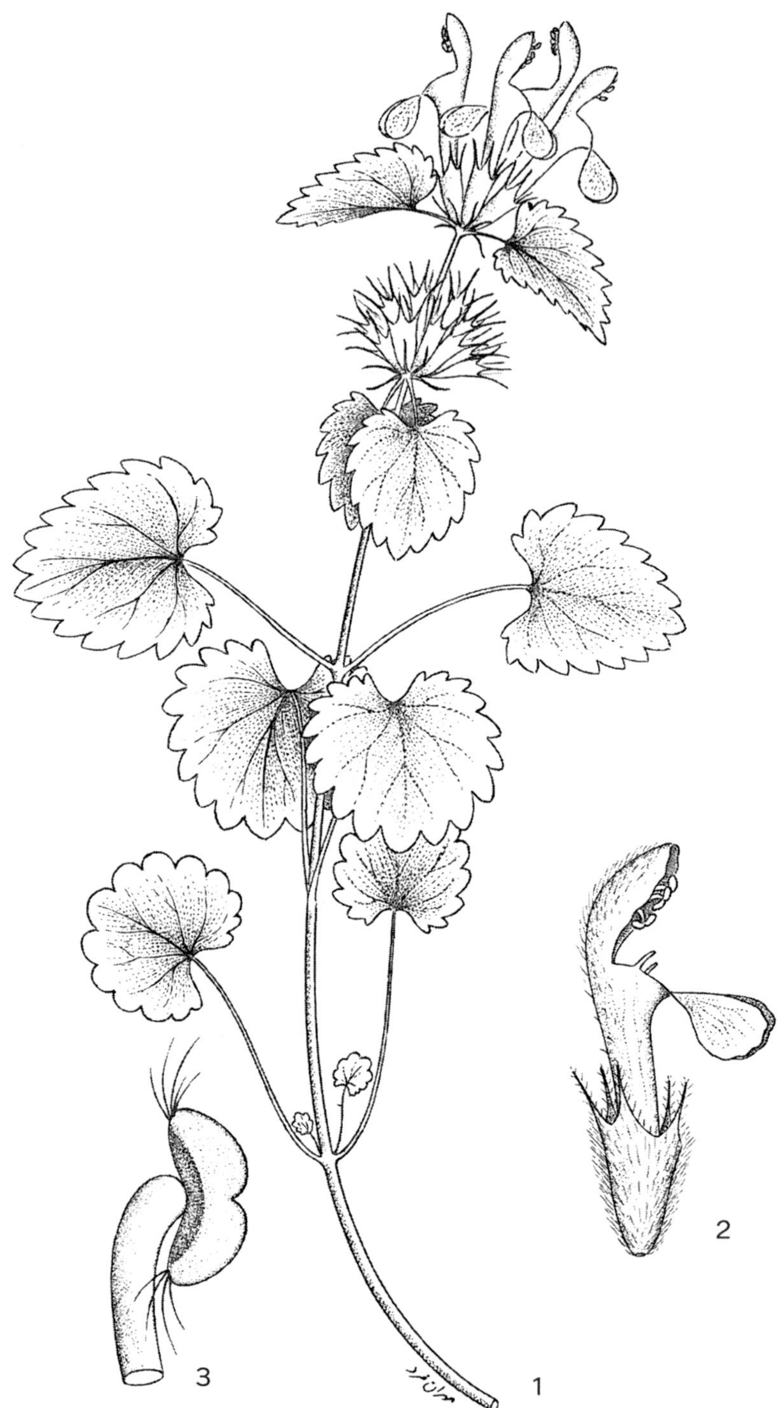

Fig. 126. **Lamium tomentosum**. 1, habit × 1; 2, flower × 3; 3, stamen detail. Reproduced with permission from Jamzad, Flora of Iran 76: f. 110 (2012). Drawn by Mehran. © Ministry of Jihad-e-Agriculture.

Distrib. N. Iraq. **MRO**: N.E. of Qandil, on mountain side, *Rawi & Serhang* 24412! (K); Qandil Mountain, E. of Qala Diza, *Thesiger* 1166 (W); Top of Qandil Mountain between top of Perrish & Bardanas (Iraqi-Persian border), *Rawi & Serhang* 24594! (K); Halgurd Dagh, among rocks, *Guest* 2885! (K); ibid., *Gillet* 12342! (K); N. Halgurd Mountain, Berme Lake, *Rawi & Serhang* 24782! (K); Ser Kurawa, high peak, on the Turkish frontier, *Gillet* 9733! (K).

Lamium tomentosum is a morphologically variable species, particularly in indumentum and leaf morphology. In the Flora of Turkey (Mill 1982), four varities of *L. tomentosum* are given and reported that var. *tomentosum*, var. *hakkariense*, and var. *alpestre* could be found in N. Iraq. However, I have not seen very clear morphological differences among Iraqi specimens during my herbarium studies. *Lamium tomentosum* is related to *L. album* subsp. *crinitum* and it could be easily be misidentified as *L. album* subsp. *crinitum*. *Lamium tomentosum* is usually distributed at high altitude among rocks and screes. It has low mat-forming habit and usually cordate-ovate to reniform leaves. It has smaller leaves and calyx teeth than *L. album* subsp. *crinitum*.

Turkey, Iran, Caucasia.

6. **PHLOMIS** L.

Sp. Pl.: 584 (1753); Harley & al. in Kubitzki (ser. ed.), Fam. Gen. Vasc. Pl. 7: 221 (2004) *Anemitis* Raf., Fl. Tellur. 3: 87 (1837); *Beloakon* Raf., Fl. Tellur. 3: 87 (1837); *Hersilia* Raf., Fl. Tellur. 3: 88 (1837); *Blephiloma* Raf., Fl. Tellur. 3: 95 (1837)

Tuncay Dirmenci & Ferhat Celep

Perennial herbs or small shrubs, pilose or tomentose, with or without glands, with stellate to dendroid hairs. Leaves simple, entire or crenate-dentate, at least lowers petiolate. Verticillasters few to many-flowered, crowded or distant in axil of leaves. Bracteoles present or sometimes absent, few or numerous, subulate to ovate, mostly long. Calyx tubular or narrowly campanulate, 5-toothed, 5–10-veined, lobes equal or unequal to 3/2, sometimes obsolete or triangular-acuminate. Corolla strongly 2-lipped, 4-lobed to 1/3, purple, pink or yellow, tube not exceeding calyx, annulate; upper lip laterally compressed, galeate, emarginate; lower lip patent, 3-lobed. Stamens 4, not or only shortly exserted. Stigma 2-branches, unequal or sometimes equal. Nutlets trigonous, rounded or truncate at apex, glabrous or hairy at apex.

Genus of about 100 taxa and 22 hybrids, distributed from the Mediterranean Region to Central Asia and West Himalaya; 8 species and a single hybrid in Iraq.

Dadandı, M.Y. (2002). The Revision of the genus Phlomis L. (Lamiaceae) of Turkey. Gazi University Institute of Science and Technology (PhD. Thesis, in Turkish).

Jamzad, Z. (2012). Lamiaceae in Assadi M., Mozaffarian V., Maassoumi A.A. (eds). Flora of Iran, 76: 253–300.

Mathiesen, C., Scheen, A.C. & Lindqvist, C. (2011). Phylogeny and biogeography of the lamioid genus Phlomis (Lamiaceae). Kew Bull. 66: 1–17.

Rechinger, K.H. (1940). Kritische Revision von Phlomis Sect. Gymnophlomis Benth. Oesterr. Bot. Z. 89: 257–298.

1. Corolla pink or purple 2
   Corolla yellow. 4
2. Basal leaves clearly cordate at base; bracteoles 20–30 mm; corolla 40–50
   mm long . . . . 6. *P. rigida*
   Basal leaves subcordate to sinuate, rounded to attenuate; bracteoles
   shorter than 20 mm; corolla shorter than 30 mm 3
3. Stems divaricate branched; verticillasters 2–6 flowered; calyx adpressed
   stellate-tomentose; teeth subulate. . . . . 8. *P. herba-venti* subsp. *pungens*
   Stems simple or rarely branched; verticillaster many flowered (more
   than 6); long and short stellate hairy; teeth triangular . . . . 7. *P. anisodanta*
4. Calyx teeth as long as tube or longer, densely plumose . . . . 5. *P. bruguieri*
   Calyx teeth shorter than tube, not plumose 5
5. Cauline and basal leaves clearly cordate at base, 5–14(–20) × 2–8(–10) cm 4. *P. kurdica*
   Cauline leaves subcordate, cuneate, rounded or truncate at base, 2–8 × 1–3 cm. 6

6. Basal leaves length up to twice as long as wide . . . . 2. *P. olivieri*
   Basal leaves length 2.5–4 times as long as wide 7
7. Basal leaves length 2.5–3 times as long as wide . . . . 3. *P. polioxantha*
   Basal leaves length 3–4 times as long as wide. .1. *P. lanceolata*

1. **Phlomis lanceolata** *Boiss & Hohen*, Boiss., Diagn. Pl. Orient., ser. 1, 5: 36 (1844); Rechinger, Fl. Iranica [K. H. Rechinger] 150: 297 (1982); Davis, Fl. Turkey [P. H. Davis] 7: 121 (1982); Jamzad in Flora Iran [Assadi & al., in Persian] 76: 258 (2012).

Perennial herb. Stems 20–45 cm tall, woody at base, stellate hairy with sessile glands. Sterile leaves present, lamina 2–8 × 1–2.5 cm, petiole to 5 cm long, ovate to oblong, cordate to truncate at base, rounded to obtuse at apex. Cauline leaves 3–8 × 1–3 cm, petiole to 5 cm, ovate-oblong to lanceolate and linear-lanceolate, adpressed short stellate hairs, ± canescent or yellowish stellate-tomentose beneath, becoming greenish above, cuneate and tapering towards base, obtuse to subacute at apex, crenulate to crenulate-serrate at margins. Floral leaves linear-lanceolate, shortly petiolate or sessile, lowers 1–2(–3) × verticillasters, uppers longer or shorter. Verticillasters 2–5, remote or upper ± congested, 4–10-flowered. Bracteoles setaceous to subulate, 3–13 mm, stellate-tomentose, as long as or clearly shorter than calyx tube. Calyx 11–17 mm, adpressed or spreading stellate-tomentose, hairy inside of the tube up to ½; teeth subequal, 3–7 mm long, lanceolate, acute to acuminate. Corolla yellow, 22–30 mm, adpressed or spreading stellate-tomentose; tube included in the calyx, hairy at outside and inside, long white hairy near base, annulate, annulus near the base; upper lip laterally compressed, emarginate, densely yellow barbate at margins of inside, otherwise subglabrous to hairy; lower lip 3-lobed; middle lob larger than laterals, emarginate, bearded with minutely glandular. Stamens 4, included in upper lip; filaments hairy. Style included or shortly protruding, glabrous to hairy, unequally bilobed. Nutlets glabrous, mature nutlet unseen.

HAB. Steppes, dry limestone rocks; alt. 1000–2600 m; fl. & fr. Jul.-Aug.
DISTRIB. Common in steppes and stony slopes of mountains in northern and northeast Iraq. **MAM**: Mossul, in ditione Sharanish, in montibus a Zakho septentrionem versus, in declivibus saxosis (Tonschiefer) montis Zawita, *Rechinger* 10919! (E, G, W); Mosul, in ditione pagi Sharanish, in montibus calc. a Zakho septentrionem versus, Jabal Khantur, *Rechinger* 10860! (E, G, W); **MSU**: 20 km Suleymaniya & Arbet, *Al-Kan* 18565!; Suleymaniyah, in ditione Penjwin, *Rechinger* 10490-A! B! (G); Penjwin, *Rechinger* 10490-B (W!); Suleymaniya, montes Avroman ad confines Persiae, in ditione pagi Tawilla, in saxosis, *Rechinger* 10300-C (G, W); Avroman et Schahu, *Haussknecht* 817 (W); Pira Magrun, *Thesiger* 1116. **MRO**: 4 km N.W. of Shaglawah, Sefin Dagh, *Kass & Nuri* 27193!; Mons Helgurd ad confines Persiae, in vale supra pagum Nowanda, *Rechinger* 11408 (E, G, W); Algurd Dagh in valle supra Nowanda, *Rechinger* 11408; N.E. of Algird Dagh, *Gillett* 9531!; Qandil, Pushtashan, *Rechinger* 11087 (E, G, W); Montes Qandil, supra Pushtashan, *Rechinger* 11105 (E, G, W); Newrobar valley-Qandil range, *Rawi & Serhang* 24161!; Serin mountain, on road to Qandil, *Rawi & Serhang* 23999!; Rowanduz, Chia-i-Mandali, *Guest* 2662!; Warshanka-Magar Range, *Rawi & Serhang* 24336! **FUJ**: Mossul, 1841, *Kotschy* 383, 385!, 460!

S.E. Turkey, N. Iraq to W. Iran

2. **Phlomis olivieri** *Benth.*, Labiat. Gen. Spec.: 624 (1834). Type (syntypes): Inter Baghdad et Kermanshah [P00738137, P00738139], et inter Kermanshah et Hamadan (P00738138), *Olivier & Bruiguière* (P-photo!). Rechinger, Fl. Iranica [K. H. Rechinger] 150: 508 (1982); Jamzad in Flora Iran [Assadi & al., in Persian] 76: 260 (2012).

Palomis armeniaca Willd. var. *olivieri* Benth. Prodr. [A. P. de Candolle] 12: 538 (1848).

Perennial herb. Stems 25–40 cm tall, woody at base, many stemmed, ascending, often branched above, pale brown or ivory, stellate-tomentose, or glabrescent below. Sterile leaves present, 3–8 × 1–3 cm (incl. petiole), oblong, subcordate to rounded at base, obtuse at apex, crenate, lanate-floccose, whitish or yellowish. Cauline leaves ovate to oblong, 4–7 × 1–3 cm (incl. petiole), subcordate to attenuate-cuneate at base, crenate to crenulate, obtuse at apex, sometimes discolorous, stellate-tomentose. Bracteoles linear-subulate, 5–9 mm, shorter than calyx tube. Calyx 13–17 mm, tubular, contracted, stellate-tomentose outside, hairy inside of the tube up to ½, stellate-tomentose; teeth unequal, 3–6 mm, lanceolate-acuminate. Corolla yellow, 24–30 mm, stellate-tomentose; tube included in the calyx, hairy at outside and inside, long white hairy near base, annulate, annulus near the base; upper lip 11–15 mm, laterally compressed, densely barbate at margins of inside, otherwise subglabrous to hairy; lower lip 3-lobed; middle lobe larger than lateral, emarginate bearded with minutely glandular. Stamens 4, included in upper lip; filaments hairy. Style included or shortly protruding, glabrous to hairy, unequally bilobed. Nutlets unknown.

HAB. Steppes, rocky slopes; alt. 1400–1600 m; fl. & fr. Jun.-Jul.
DISTRIB. Distributed in the mountains of northern and northwestern Iraq in mixed shrubland and rocky areas; expected to occur at lower elevations as well. MSU: Sulaymaniya, montes vroman ad confines Persiae, in ditione pagi Tawilla, in saxosis, *Rechinger* 10300-B (W); Suleymaniya, in ditione Penjwin, *Rechinger* 10490-C (W); FKI: Tauq bridge, *Guest* 15537!

W. Iran.

3. **Phlomis polioxantha** *Rech.f.*, Kulturpflanze, Beih. 3: 61 (1962). Type: Iraq, Diyala, ad confines Persiae, in collibus conglomeraticis ab oppido Mandali ca. 10 km orientem versus, ad ripam sinistram fluvii, 02.vi.1957, *Rechinger* 9649! (W, holo.; E, G, K, iso.) [W–19790011253! holo., E-00319644! G-00435149! K-000928229! iso.]. Rechinger, Fl. Iranica 150: 300 (1982); Jamzad in Flora Iran [Assadi & al., in Persian] 76: 270 (2012).

Perennial with a thick woody base. Stems 20–30 cm tall, often simple, rarely a few branched, densely short stellate-tomentose. Sterile leaves present, lamina 3–4.5 × 1.5–2.5 cm, petiole to 3 cm, oblong-ovate, yellowish-white, (indumentum the same on all leaves) adpressed stellate-tomentose, canescent or whitish, sometimes yellowish, truncate to subcordate at base, rounded at apex, crenulate at margins. Lower cauline leaves oblong-ovate, lamina to 6 × 2.5 cm, thickish, truncate or sinuate or wide oblique cuneate at base, rounded at apex, slightly or indistinctly crenate at margins. Cauline leaves a few, scarcely or shortly petiolate. Verticillasters 2–4, all remote ca. 6-flowers. Bracteoles 2–6 mm, clearly shorter than tube, subulate. Calyx 15–19 mm. tubular, plicate-sulcate, densely short stellate-tomentose, often canescent, hairy inside of the tube up to ½; teeth unequal, 2–4(–6) mm, triangular to lanceolate. Corolla 20–30 mm, yellow, stellate-tomentose outside of the lips and tube; tube included in the calyx, long white hairy near base, annulate, annulus near the base; upper lip laterally compressed, densely barbate at margins of inside, otherwise subglabrous to hairy; upper lip 10–12 mm, laterally compressed, lower lip 3-lobed; middle lob larger than laterals, emarginate subglabrous to hairy. Stamens 4, usually included in upper lip; filaments hairy. Style usually included in upper lip, subglabrous to hairy, unequally bilobed. Nutlet unknown.

HAB. Mixed shrubland on conglomerate; alt. 800–950 m; fl. & fr. May-Jun.
DISTRIB. Northern Iraq. MSU: Diyala, ad confines Persiae, in collibus conglomeraticis ab oppido Mandali ca. 10 km orientem versus, ad ripam sinistram fluvii, *Rechinger* 9649! (E, G, W). MRO: Arbil: N. of Shaqlawa, Anders 1443 (W). FKI: Kirkuk: Jarmo, *Helbaek* 1706 (W); Jarmo, *Haines* 227!

W. & S.W. Iran.

4. **Phlomis kurdica** *Rech.f.*, Oesterr. Bot. Z. 89: 274 (1940); Rawi in Dep. Agr. Tech. Bull 14: 154 (1964); Feinbrun-Dothan, Fl. Paleast. 3: 119 (1978); Rechinger, Fl. Iranica [K. H. Rechinger] 150: 299 (1982); Davis, Fl. Turkey P. H. Davis 7: 124 (1982); Jamzad in Flora Iran [Assadi & al., in Persian] 76: 268 (2012); Taifour & El-Oqlah, Pl. Jordan Annot. Checkl.: 104 (2017).

Perennial herb. Stems 20–60 cm tall, woody at base, ascending, simple or branched above, canescent or sometimes yellowish, floccose-woolly, with stellate hairs, eglandular or minutely glandular papillae. Leaves 5–14(–20) × 2–8(–10) cm, broadly ovate to ±oblong, petiole 1–6(–11) cm, ± densely (often whitish) stellate-lanate-tomentose, especially beneath, cordate at base, obtuse at apex, crenate to nearly entire. Floral leaves cordate-triangular to lanceolate, cordate or rounded to attenuate at base, petiolate to 2.5 cm long, longer or shorter than verticillasters. Verticillasters 3–8, lowers remote, uppers approximate, 8–10-flowered. Bracteoles few, linear-setaceous to linear, 1.5–5(–12) mm, mostly shorter than pedicel or slightly longer, densely tomentose. Calyx (9–)13–18(–22) mm, densely stellate-tomentose with long spreading hairs and glandular, sparsely hairy inside of the tube up to ½; teeth subequal, subulate, 3–7(–12) mm. Corolla yellow, 25–35 mm, adpressed or spreading stellate-tomentose; tube included in the calyx, short hairy and long white hairy, annulate, annulus is usually located in the half of the tube; upper lip laterally compressed, emarginate, densely barbate at margins of inside, otherwise subglabrous to hairy; lower lip 3-lobed; middle lob larger than laterals, emarginate, bearded with minutely glandular.

Fig. 127. **Phlomis kurdica**. 1, habit × 1; 2, leaf × 1. Reproduced with permission from Feinbrun-Dothan,
Fl. Palaestina 3: Plates, f. 193 (1977). Drawn by Esther Huber. © The Israel Academy of Sciences
and Humanities.

Stamens 4, included in upper lip; filaments hairy. Style included or shortly protruding, glabrous to hairy, unequally bilobed. Nutlets glabrous, mature nutlet unseen. Fig 127, 1–2.

HAB. Mixed shrubland, fallow fields on clay soils; alt. 600–1700 m; fl. & fr. May-Aug.
DISTRIB. Northern Iraq. MAM: Ad confines Truciae prov. Hakkari, inter Mossul et Zakho, in collibus siccis a Mosul 86 km septentrionem versus, *Rechinger* 10629! (W); 86 km N. of Mossul, towards Zakho, dry hill slopes, *Rechinger* 10629! (E, G); inter Dohuk et Amadiya, in Quercetis, infra Sirsank, *Rechinger* 11606 (E, G, W); Ser Amadiya (Gali Mazurka), rocky clay mountain, *Al-Dabbagh, Al-Kaisi & Hamid* 46001!; Zawita, *Robertson* 1065!; Maya mountain 25 km W. Kani-Masi, *Hamid & Fadhil* 45503!; Asi, Jabel Bekhair, *Rawi* 8549-A!; MSU: Suleymaniya: inter Kirkuk et Suleymaniya, in saxosis calc. angustiarum Darband-i Bazian prope Chamchamal, *Rechinger* 10597!; Mela kowa (on Suleymaniya-Penjwin highway), *Rhus-Quercus* forest on hillside, *Rawi* 22459!; 20–25 km E. of Qaranjar (on Kirkuk-Suleymaniya higway), cultivated land on hillside, *Rawi* 21645!; 4 km E. of Qaranjar (on Kirkuk-Suleymaniya higway), *Rawi* 21632!; Qara Anjir, 27 km a Kirkuk orientenm versus, in collibus siccis, *Rechinger* 10015 (E, G, W); Montes Avroman ad confines Persiae, in ditione pagi Tawilla, *Rechinger* 10300-A!; Azmir, on mountain, *Karim* 39353! MRO: Montes Qandil ad confines Persiae, inter Shahilan et Pushtashan, *Rechinger* 11009!; Khulaifan, *Omar, Al-Kaisi & Wedad* 49577! MJS: Jebel Golat between Ain Tellawi and Balad Sinjar, *Field & Lazar* 507!; 50 km W. of Mossul, clay soil, *Al-Dabagh* 43591! Sinjar, Jabal Sinjar, on the top of mountain with *Quercus* tree, *Sharief & Hamad* 50271! FUJ: Mossul, 1841, *Kotschy* 460!; N. of Mosul, *Guest* 3619! FNI: Hinni, Ain Sifni, N. of Mosul, *Guest* 4039!; Mossul, Ain Sifni, N. of Mossul, *Guest* 4037! FAR: Arbil, *Guest* 2997! FKI: Kirkuk, in collibus siccis 37 km a Kirkuk orientem versus, *Rechinger* 12504 (E, G, W).

Phlomis kurdica is easily distinguished from its close relatives by its cordate leaves.

Lebanon, Palestine, Syria, Turkey, Iran.

5. **Phlomis bruguieri** *Desf.*, Mém. Mus. Hist. Nat. 11: 9, figs. 1, 2 (1824). Syntypes: Iran/ Iraq: Hamadan, *Olivier* (FI); [trois journees de Bagdad a] Kermanchah, *d'Olivier & Bruguiére* (G-DC? P-photo!). Rawi in Dep. Agr. Tech. Bull 14: 154 (1964); Rechinger, Fl. Iranica [K. H. Rechinger] 150: 303 (1982); Davis, Fl. Turkey [P. H. Davis] 7: 118 (1982); Jamzad in Flora Iran [Assadi & al., in Persian] 76: 276 (2012).

Perennial herb. Stems 25–80 cm tall, few to many stemmed from the woody rootstock, ascending to erect, unbranched or branched above. Leaves 4–15 × 1–3.5 cm, oblong, lanceolate or linear-lanceolate, eglandular or minutely glandular papillae, densely canescent-tomentose, greyish-green, tapering towards base and acute apex, entire or nearly so, or crenate, lowers petiole to 2 cm long, upper leaves subsessile to sessile. Floral leaves shortly petiolate or sessile, oblong to linear-lanceolate, longer than verticillasters. Verticillasters 2–6, usually ± congested, (2–)6–10-flowered. Bracteoles numerous, subulate, (5–)10–25 mm, densely white-plumose, shorter than calyx, usually longer than tube. Calyx 25–35 mm, densely white-tomentose with long spreading hairs, hairy inside of the tube up to ½–2/3; teeth (10–)15–25 mm, subulate, of unequal length, 1.5–2 × tube. Corolla yellow, 20–30 mm, stellate-tomentose; tube included in calyx, annulate and glandular at base of the filaments in ½ of it, otherwise glabrous inside; upper lip laterally compressed densely barbate at margins of inside, otherwise subglabrous to hairy; lower lip 3-lobed; middle lob larger than laterals, emarginate. Stamens 4, included in upper lip; filaments hairy. Style included or shortly protruding, glabrous to hairy, unequally bilobed. Nutlets glabrous, mature nutlet unseen.

HAB. Mixed shrubland, fallow fields; alt. 350–1200 m; fl. & fr. May-June.
DISTRIB. Northern Iraq to the central alluvial plain districts. MAM: Atrush valley, N. of Mossul, *Guest* 3618!; Mossul, ad confines Turciae province Hakkari, in ditione pagi Sharanish in montibus calc. a Zakho septentrionem versus, Jabal Khantur, *Rechinger* 12097 (E, G, W); Mossul, in ditione oppidi Zakho, ca. 8 km a Zakho merid. Versus, *Rechinger* 12160 (G, W); inter Mossul et Zakho, Dohuk, in collibus siccis 52 km a Mossul septentrionem versus, *Rechinger* 10619 (E, G, W); 20 km N. of Mossul to Dohuk, *Al-Kaisi, Al-Khayat & Karim* 51126!; Dohuk, *Chapman* 26259!; Near Dinarta, dry foothill, *Chapman* 26108!; Zawitah valley, *Guest* 3686! MSU: Suleymaniya, inter Suleymaniya et Dokan, inter Surdash et Dokan, *Rechinger* 10124 (W); Suleymaniyah, *Haines* 1079! (E); Balkha, side of montain rocky land, *Rawi* et al. 29532!; 10 km, N. of Kalar to Derbendi-Khan, clay soil, *Omar, Al-Kaisi & Wedad* 49350!; Jarmo, *Helbaek* 1979! MRO: Khuhzad, rocky clay soil hillside, *Oamar, Al-Kaisi & Wedad* 49601!; Rowandiz gorge, *Field & Lazar* 846!; Haibat Sultan, mountain side, *Omar* et al. 38225! MJS: Karya Bat Khan near Balad Sinjar, *Field & Lazar* 610!; Jabal Sinjar, W. slope, gypsious valley, *Al-Kaisi & K.Hamad* 53735!; Sinjar, Jabal sinjar, rocky clay mountain, *Al-Kaisi* 49739! FUJ: Mosul, 1841, *Kotschy* 321!, 388!, 395!; 49 km W. of Mossul, Kirrana village, *A.Rawi* 22937!; Tal-Afar, weeds in wheat field, *Jabar* 50157!; Near Tel-Afar, *Regel* 32 (G); 4 km E. of Tal-Afar, clay soil with stones, *Al-Dabagh* 43592!; Jebel Golat, between Ain Tellawi and Balad

Sinjar, *Field & Lazar* 508!; Dubardan mountain, 40 km Mossul to Agura, *Al-Dabbagh & M.Jasim* 46817! **FAR**: Arbil, ditionis Erbil in campis ad Ankova, *Bornmuller* 1655!, 1655-B! (G); Arbil, *Guest* 2996!; 10 km from Sanjaq to Arbil, clay soil hillside, *Omar, Al-Kaisi & Wedad* 49519! **FKI**: Near Altun Köprü, on the Zab river, *Guest* 4030 (G); 10–15 km N. of Kirkuk, *Rawi* 21561!; 7 km N.E. of Kirkuk, sand stony hillside, *Rawi* et al. 27915!; Kirkuk, Kifri, ad versuras, *Rechinger* 9967! (W); Kirkuk, in collibus arenaceis, *Rechinger* 9976 (W). **FPF**: 35 km N.E. of Mandali, *Al-Kaisi & Al-Khayat* 50802! **LCA**: Abu Ghraib, 19 km N. of Guayarah, loam of Rolling hills, *Barkley & Haddad* s.n.!;

Phlomis bruguieri is easily distinguished from other species by its long calyx and bracteoles. Among its characteristic features are the congested of verticillasters, too. In some specimens all verticillasters are nearly distinctinct and their bracteoles are often shorter than the tube. Such examples are although they are close to *Ph. bruquieri*, the possibility of being hybrid should not be excluded. It is also one of the parents of the hybrids *Ph.* × *pabotii, Ph.* × *pratervisa* and *Ph.* × *ekimii.*

W. & S.W. Iran, Syria.

6. **Phlomis rigida** *Labill.*, Icon. Pl. Syr. 3: 15 (1809); Rawi in Dep. Agr. Tech. Bull 14: 154 (1964); Dinsmore in Post, Fl. Syria, Palest. & Sinai ed. 2, 2: 398 (1933); Feinbrun-Dothan, Fl. Paleast. 3: 119 (1978); Rechinger, Fl. Iranica [K. H. Rechinger] 150: 304 (1982); Davis, Fl. Turkey [P. H. Davis] 7: 110 (1982); Jamzad in Flora Iran [Assadi & al., in Persian] 76: 281 (2012).

Anemitis rigida (Labill.) Raf., Fl. Tellur. 3: 87 (1837).
Phlomis shepardii Post, Fl. Syria: 658 (1896).

Herb. Stems 30–80 cm tall, unbranched, very rarely branched above, hispid to lanate, hairs patent, swollen and yellowish at base, eglandular below, minutely glandular above. Sterile leaves present, lamina 10–20 × 3–8 cm, petiole to 14 cm, ovate, ovate-oblong to obovate, wrinkled, densely stellate-tomentose, yellowish or pale green above, green or whitish at beneath, ± cordate, rounded to truncate or shortly attenuate and asymmetric at base, obtuse to rounded at apex, crenate. Cauline leaves pale green, 7–16.5 × 3–7 cm, petiole to 10 cm, broadly ovate to ovate-lanceolate, densely stellate-tomentose, obtuse, shortly cuneate, rounded to cordate or truncate at base, obtuse to acute at apex, crenulate to crenate at margins. Lower floral leaves ovate to lanceolate, petiolate, much longer than verticillasters, to 4 × verticillasters, uppers gradually smaller. Verticillasters (2–)3–5, remote, 4–6(–8) cm broad, many-flowered. Bracteoles numerous, subulate, 20–30 mm, densely long-hispid, curved upwards, clearly longer than calyx tube, longer or slightly shorter than calyx. Calyx 15–26 mm, adpressed stellate-tomentose and long hispid; throat hairy; teeth ovate-truncate at base, then subulate, 8–10 mm. Corolla pink, 40–50 mm long, stellate-tomentose to hispid-villous; tube included in the calyx, long white hairy near base, annulate, annulus near the base; upper lip laterally compressed, emarginate densely barbate at margins of inside; lower lip 3-lobed; middle lob larger than laterals, emarginate, ± glabrous. Stamens 4, included in upper lip; filaments hairy. Style included or shortly protruding, glabrous to hairy, unequally bilobed. Nutlets pubescent at apex, mature nutlet unseen.

HAB. Mixed shrubland, *Quercus* scrub, stony slopes; alt. 900–2400 m; fl. & fr. May-July.
DISTRIB. Common on stony slopes and steppes of northern Iraq. **MAM**: Ser Amadiya, Gulli Mazurka, *Guest* 4995!; Khantur mountain, N.E. of Zakho, clay depression under *Quercus, Rawi* 23384!; Mosul, in ditione paigi Sharanish, in montibus calc. a Zakho septentrionem versus, Jabal Khantur, *Rechinger* 10820 (W); Bigdoda village, 6 km S. of Kani Masi, *Omar & Al-Dabagh* 45681!; Kani Masi, *Al-Dabagh & Hamid* 45579!; Gali Mazurka, 2 km N.W. of Amadiya, *Al-Dabbagh & Jasim* 46926!; Sarsang to Amadiya, *Karim, Hamid & Jasim* 40953!; Inter Dohuk et Amadiya, *Rechinger* 11607 (W); Amadiya-Sirsang, *Anders* 2308 (W); Ispindari, Saddle, Suramatuka, *Chapman* 26330! **MSU**: Suleymaniya, *Haussknecht* s.n. (W); 1 km S. of Darbandi Bassian (on Suleymania highway), *Rawi* 21699!; Suleymaniya, Jebel Avroman near Darimar, *Gillett* 11829!; Kopi Qaradagh, *Haines* 1160!; Between Tasluja and Darband, *Al-Katip & Tikriti* 29738!; Kirkuk, A Chamchamal orientem versus declivibus siccis. *Rechinger* 10060 (E, W). **MRO**: Rowanduz, Jebel Baradost near Diana Rowanduz, *Field & Lazar* 891!; Rawanduz, *Regel* 57 (G); Rawanduz, Handren mount, *Regel* 87 (G); Chia-i-Mandali, *Guest* 2661!; Zanita mountain between Rania and Shaglawa, *Karim, Hamid & Jasim* 40793!; Arl-Gird Dagh, *Guest & Ludlow-Hewitt* 2844!; 4 km N.W. of Shaqlawah, Sefin Dagh, *Kass & Nuri* 27190! **FUJ**: Mossul, 1841, *Kotschy* 87! (W); Arbil, in montis Kuh-Sefin, reg. Infer. Supra pagum Schaklava (ditionis Arbil), *Bornmuller* 1560! (G, W); Arbil, in saxosis cal. Jugi inter Dokan et Mirza Rustam, *Rechinger* 11936! (G, W).

Phlomis rigida is distinguished from other related species by its long and pink corolla and long stiff hairs.

Iran, Lebanon, Syria, Turkey.

7. **Phlomis anisodonta** *Boiss.*, Diagn. Pl. Orient. Ser. 1, 5: 37 (1844).

Herb, woody at base, few to many stemmed from the base, stems 15–40 cm tall, ascending to erect, stellate-tomentose, canescent. Sterile leaves present, oblong, 4–12 × 2–4 cm, petiole to 12 cm long, lamina rounded to subcordate at base, or attenuate, ± asymmetric, rounded to obtuse at apex, crenate at margins, white-canescent at beneath, green above. Cauline leaves ovate-lanceolate, 4–8 × 2–4 cm, petiole to 4 cm long, lamina subcordate to attenuate at base, rounded to subacute at apex, crenate to crenulate, usually discolorous, white-canescent at beneath, green above. Floral leaves similar to cauline leaves but small in size, usually longer than verticillasters, lower ones to 4 × verticillasters. Verticillasters 2–4, remote, many flowered. Bracteoles subulate, 12–20 mm, long and short stellate-hairy, incurved, longer or shorter than calyx but usually longer than calyx tube. Calyx 12–17 mm, infundibular-tubular, long and short stellate hairy; teeth unequal, 4–6 mm long, triangular, mucro 1–3 mm, glabrous at apex, subglabrous to hairy in throat. Corolla 18–25 mm, pink, stellate-tomentose, otherwise glabrous outside and inside of the tube; tube included in the calyx, annulate and glandular at base of the filaments; upper lip falcate, tube expand to 8 mm at mouth, emarginate; upper lip laterally compressed, densely barbate at margins of inside, otherwise subglabrous; lower lip 3-lobed; middle lobe larger than laterals, emarginate ± glabrous. Stamens 4, usually included in upper lip; filaments hairy. Style usually included in upper lip, glabrous, unequally bilobed. Nutlet unknown.

HAB. Calcareous; alt. 800–1000 m: fl. & fr. Jun.-Jul.
DISTRIB. Northern Iraq. **MSU**: in montibus calcareis Avroman et Schahu, *Haussknecht* s.n.! (W); Pira Magrun *Thesiger* 111. **FUJ**: Mossul, 1841, *Kotschy* 797 (W).

Iran.

8. **Phlomis herba-venti** *L.* subsp. **pungens** (*Willd.*) *Maire* ex *DeFilipps*, Bot. J. Linn. Soc. 64: 233 (1971); Rawi in Dep. Agr. Tech. Bull 14: 154 (1964); Dinsmore in Post, Fl. Syria, Palest. & Sinai ed. 2, 2: 398 (1933); Feinbrun-Dothan, Fl. Paleast. 3: 121 (1978); Rechinger, Fl. Iranica [K. H. Rechinger] 150: 307 (1982); Davis, Fl. Turkey [P. H. Davis] 7: 108 (1982); Jamzad in Flora Iran [Assadi & al., in Persian] 76: 284 (2012); Taifour & El-Oqlah, Pl. Jordan Annot. Checkl.: 103 (2017).

Phlomis pungens Willd., Sp. Pl. 3: 121 (1800).
Phlomitis pungens (Willd.) Rchb. ex T.Nees, Gen. Fl. Germ. 2(7): 43 (1843).
Phlomis pungens var. *hispida* K.Koch, Linnaea 21: 700 (1849).
Ph. herba-venti var. *pungens* (Willd.) Schmalh., Fl. Sredn. Jushn. Rossii 2: 343 (1897).
Ph. seticalycina Nábelek, Spisy Přír. Fak. Masarykovy Univ. 70: 69 (1926).
Ph. pseudopuogens Knorring in V.L.Komarov, Fl. URSS 21: 647 (1954).

Herb. Stems 30–70 cm tall, branched, rigid, stellate-tomentose, eglandular, rarely glabrescent. Cauline leaves lanceolate to ovate-lanceolate, 5–15 × 1–6 cm, petiole to 10 cm, lamina subcordate to cuneate at base, obtuse, acute to acuminate at apex, denticulate or crenulate-serrate at margins, rarely entire, stellate-tomentose, green above, greyish beneath, discolorous. Floral leaves lanceolate or oblong, clearly longer than verticillasters, to 4 × verticillasters, sessile to shortly petiolate, acute to acuminate. Verticillasters 3–7, 2–6-flowered. Bracteoles subulate, 7–17 mm, stellate-tomentose, shorter or slightly longer than calyx, adpressed stellate-tomentose, with or without undivided hispid hairs. Calyx 12–18 mm, adpressed stellate-tomentose, with or without undivided hispid hairs, hairy in throat; teeth unequal, subulate, (2–)4–7 mm. Corolla purple or pink, 22–27 mm, stellate-tomentose on tube and lips; tube included in calyx, hairy inside, annulate at base of the filaments; upper lip galeate and laterally compressed, emarginate, densely barbate at margins of inside, otherwise glabrous; lower lip 3-lobed; middle lob larger than laterals, emarginate, bearded with glandular papillae. Stamens 4, included in upper lip; filaments hairy. Style included in upper lip, sparsely hairy, unequally bilobed. Nutlets glabrous, mature nutlets unseen.

HAB. Mixed shrubland, fallow fields, dry stony slopes; 1500–2400 m; fl. & fr. Jun.-Aug.

DISTRIB. Northern Iraq. **MAM**: Zakho, *Rawi* 8622!; Between Aqra and Mosul, *Rawi* 11517!; Dori village, Kani Masi, *Omar & Al-Kaisi,* 45429 ! & 45408 (BAG); Kani Masi, *Al-Dabbagh & Hamid* 45581!; Maya mountain, 25 km West of Kani Masi, *Hamid & Fadhil* 45514!; Gara mountain, *Al-Dabbagh & Jasim* 46989! **MSU**: Penjwin, 1957, *Rechinger* 10493 (W!); 11 km North of Penjwin, *Rawi* 29904!; Suleymaniya, in ditione Penjwin, *Rechinger* 10399 (E, W); 8 km North of Kani Spi, Sulaimaniya-Penjwin highway, *Rawi* 22401! **MJS** Jabal Sinjar, *Sharief & Hamid* 50286!; Jabal Sinjar N. Slope, *Al-Kaisi,* Al-Khayat & F. Karim 50913 (BAG). **FUJ**: Mossul, 1841, *Kotschy* 49!

S.E. Europe to Iran, N.W. Africa.

9. **Phlomis × praetervisa** *Rech.f.,* Oesterr. Bot. Z. 89: 296 (1940). Type: [Iraq] In ditione urbis Mossul, in agris ad austro-occid., 250 m, 30.05.1910, Handel-Mazzetti 1300 (W, holo.; W, WU iso.) (W–19130016624! holo., W19130016623! WU-0044127! iso.).

Herbs. Stems loosely stellate at base, densely stellate-tomentose at uppers. Leaves ovate to lanceolate, to 8 × 2 cm, densely-stellate-tomentose, shortly petiolate to subsessile. Floral leaves longer than verticillasters, attenuate at base. Bracteoles c. 10 mm. Calyx c. 25 mm. Corolla c. 30 mm.

HAB. Mixed shrubalnd, fallow fields, clay soils; alt. 500–1200 m; fl. & fr. May-Jun.
DISTRIB. Northern Iraq, on clay soils in areas where its parental distribution overlap. **MAM**: Mosul, in ditione pagi Sharanish in montibus calc. a Zakho septentrionem versus, Jabal Khantur, in saxosis, *Rechinger* 12098 (G). **MSU**: Pir Omar Gudrum, *Haussknecht* s.n. (W); 5 km West of Chemchemal, clay-gravel soil on hillside, *Rawi* 21670! **MRO**: Arbil, in montis Kuh-Sefin reg. infer. Supra Schaklava (ditionis Erbil), *Bornmüller* 1657! & 1658! (G, W); ditionis Erbil ad Ankova, *Bornmüller* 1693! (G). **FUJ**: In ditione urbis Mossul, in agris ad austro-occid., *Handel-Mazzetti* 1300 (W, WU). **FKI**: Kirkuk, in collibus siccis 37 km a Kirkuk orientem versus, *Rechinger* 10028! (E, G, W); Diyala, Jabal Hamrin, *Sutherland* 8. p.p.; Mesopotamia, *Sutherland* 437!

In the protologue Rechinger indicated that this hybrid differed from *Phlomis bruguieri* by its pyramidal branching, yellow pubescence, shorter, wider and cordate leaves, shorter bracteoles and calyx teeth. Also, it differs from *Ph. kurdica* in that its leaves are partially narrower, and bracteoles and calyx teeth longer. Some of the specimens given as *Ph. × praetervisa* (such as *Bornmüller* 1657) are similar to *Ph. kurdica* in general appearance, while some are similar to *Ph. bruguieri* (such as *Rechinger* 10028).

7. **PHLOMOIDES** Moench,

Methodus, 403 (1794); Kamelin, R.V. & Makhmedov, A.M. The system of the genus Phlomoides (Lamiaceae). Bot. Zhurn 75: 241–250 (1990)
Notochaete Benth. in N.Wallich, Pl. Asiat. Rar. 1: 63 (1830); *Clueria* Raf., Fl. Tellur. 3: 87 (1837); *Eremostachys* Bunge in C. F. von Ledebour, Fl. Altaic. 2: 414 (1830); *Lamiophlomis* Kudô, Mem. Fac. Sci. Taihoku Imp. Univ. 2: 210 (1929); *Pseuderemostachys* Popov, Novye Mem. Moskovsk. Obshch. Isp. Prir. 19: 148 (1940 publ. 1941); *Paraeremostachys* Adylov, Kamelin & Makhm., Novosti Sist. Vyssh. Rast. 23: 112 (1986); *Phlomidopsis* Link, Handbuch 1: 479 (1829), nom. superfl.; *Phlomitis* Rchb. ex T.Nees, Gen. Fl. Germ. 2(7): 43 (1843), nom. superfl.; *Orlowia* Gueldenst. ex Georgi, Beschr. Russ. Reich. 3(5): 1089 (1800); *Trambis* Raf., Fl. Tellur. 3: 87 (1837).

Tuncay Dirmenci & Ferhat Celep

Perennial herbs, usually with woody rhizomes and/or tuberous rootstock. Stems erect, simple or branched, 20–120 cm tall. Basal leaves oblong, oblong-orbicular to cordate, 7–15(–25) × 3–15(–20) cm, undivided or laciniate to pinnatisect with toothed margins; lobes 4–7 per side. Cauline leaves sessile or petiolate, oblong, ovate-orbicular or oblong-lanceolate, 2–15 × 1–10 cm, undivided or divided, lobes 1–5 per side, incised-dentate, dentate, crenate or double crenate. Floral leaves sessile, elliptic, oblong-lanceolate or ovate-elliptic, 1.5–8 × 0.5–4 cm, incised dentate, dentate or entire. Inflorescence thyrsoid to racemose with 2–20 flowers arranged in opposite axillary cymes, forming verticillasters with bracteoles. Bracteoles linear-lanceolate or subulate, 5–30 × 1.5–3 mm. Calyx tubular-campanulate or broadly funnel-shaped, 12–40 mm; calyx lobes equal to subequal, sometimes broad at base and abruptly narrowed to short spinose at apex, or rarely hooked. Corolla 10–40 mm, pale yellow, white, yellowish-white or pinkish, red to dark purple, strongly 2-lipped;

the posterior lip hooded (often deeply concave and dome-shaped) and bearded, 10–18 mm; lower lip flabelliform-trilobate or triangular-cordate, white, yellowish, orange-yellow or buff-orange, pink, red to purple, the middle lobe obcordate, ovate-cordate, ovate-elliptic, oblong, the lateral lobes oblong, ovate-orbicular, orbicular-elliptic, ovate-elliptic or obcordate, corolla tube cylindrical and sometimes hairy at the throat. Stamens 4; upper filaments longitudinal, fimbriate; those of lower filaments squamose, with tooth-like ring of hairs at base. Stigma with equal/unequal lobes. Nutlets truncate or sub-truncate ad mostly bearded at apex.

About 170 species found in eastern Central Europe to Korea and Himalaya; 2 species in Iraq.

Adylov, T.A. & Makhmedov, A.M. (1987). *Phlomoides* Moench. Pp. 82–107 in: Adylov, T.A. (ed.), Conspectus florae Asiae Mediae, vol. 9. Tashkent: Izdatel'stvo Akademii Nauk SSSR.

Bendiksby M., Thorbek, T., Scheen, A., Lindqvist, C. & Ryding, O. (2011). An updated phylogeny and classification of Lamiaceae subfamily Lamioideae. Taxon 60(2): 471–484.

Salmaki, Y., Zarre, S., Ryding, O., Lindqvist, C., Scheunert, A., Bräuchler, C., Heubl, G. (2012). Phlogeny of the tribe Phlomideae (Lamioideae:Lamiaceae) with special focus on *Eremostachys* and *Phlomoides*: New insights from nucleae and chloroplast sequences. Taxon 61(1): 161–179.

Salmaki, Y., Zarre, S. & Heubl, G. (2012). The Genus *Phlomoides* Moench. (Lamiaceae; Lamioideae; Phlomideae) in Iran: An Updated Synopsis. Iran. J. Bot. 18(2): 207–219.

Phlomoides (Lamioideae, Lamiaceae) as currently circumscribed contains about 170 species depending on the narrow or broad concept followed by the taxonomists and is one of the largest genera of tribe Phlomideae. The distribution area of the genus extends from central Europe to Korea and the Himalayas. The major centres of diversity of *Phlomoides* are Central Asia, the Iranian highlands (with a diversity hotspot in Iran and Afghanistan), China, and Mediterranean Europe with less importance in terms of species number.

> Basal leaves pinnatifid to pinnatisect; fruiting calyx not accrescent . . . . . 1. *P. laciniata*
> Basal leaves simple, undivided rounded; fruiting calyx accrescent,
>   35–45 mm in diameter in fruit . . . . 2. *P. molucelloides*

1. **Phlomoides laciniata** (*L.*) *Kamelin & Makhm.*, Bot. Zhurn. (Moscow & Leningrad) 75: 249 (1990); Dinsmore in Post, Fl. Syria, Palest. & Sinai ed. 2, 2: 399 (1933); Rawi in Dep. Agr. Tech. Bull 14: 150 (1964); Feinbrun-Dothan, Fl. Paleast. 3: 117 (1978); Rechinger, Fl. Iranica [K. H. Rechinger] 150: 261 (1982); Davis, Fl. Turkey [P. H. Davis] 7: 102 (1982).

Phlomis laciniata L., Sp. Pl.: 585 (1753).
Eremostachys laciniata (L.) Bunge in C.E. von Ledebour, Fl. Altaic. 2: 416 (1830); Jamzad in Flora Iran [Assadi & al., in Persian] 76: 303 (2012).
Clueria laciniata (L.) Raf., Fl. Tellur. 3: 87 (1837).
Eremostachys nerimani Stapf, Denkschr. Kaiserl. Akad. Wiss., Wien. Math.-Naturwiss. Kl. 50: 50 (1885).
E. cilicica Gand., Bull. Soc. Bot. France 65: 65 (1918).

Stems erect, simple (rarely branched at inflorescence), robust, 30–100 cm tall, to 1 cm diameter, subglabrous, villous to densely white-lanate, long eglandular hairy and sparse glandular hairs. Basal leaves with lamina oblong-elliptic in outline, to 30–50 × 10–24 cm, petiolate to 8–(–20) cm long, pinnatisect, segments 4–5 per side, oblong, lanceolate, alternate, ± sparsely pubescent to villous with minutely glandular and sessile glands; cauline leaves few (1–2 pairs), lower similar to basal, upper sessile, pinnatifid to pinnatisect. Inflorescence 15–42 cm long at main stems. Verticillasters 5–10, many-flowered, remote below, approximate above, densely white-lanate, glandular. Bracteoles 8–16 mm, lanceolate to linear-subulate, lanate-villous and glandular. Calyx 14–21 mm, tubular-campanulate, ± densely lanate-villous, glandular papillae, not accrescent in fruit, hairy and glandular papillate inside of tube; teeth triangular, spine 1.5–2 mm long. Corolla 20–30(–35) mm, white or yellowish, tube annulate; upper lip shorter than lower lip; lower lip 3-lobed, lateral lobs usually oblong or ovate-orbicular, middle lobe of lower lip ovate-cordate to obcordate and often tinged orange or speckled with red; hood pubescent with glandular papillae outside, whitish-bearded within. Stamens 4, included in corolla, posteriors stamens longer than the anterior; filaments hairy, the lower part of the filament of the posterior stamens thickened. Style included in corolla, unequally bilobed; ovary hairy. Nutlets oblong-obovate to trigonous, 8–10 × 4–6 mm, densely barbate. Fig. 128, 1–3.

HAB. Mixed shrubland, *Quercus* scrub, heavy clay loam of valley in hills, roadsides, meadows, field margins; alt. 700–2200 m; fl. & fr. Apr.-Aug.

DISTRIB. On diverse habitats of the mountains of northern and northeastern Iraq. **MAM**: Mosul, ad confines Turciae province Hakkari, inter Dohuk et Amadiya, supra Sirsang, *Rechinger* 11657 (W); Mosul, In quercetis jugi 8 km south of Zakho, *Rechinger* 12168! **MRO**: Berd Aghagin village, *Rawi et al.* 28698!; Kani Watman, 20 km S.W. of Dukan, *Al-Kaisi* & *Hamad* 49339!; Southern slope of Karoukh mountain, *Kass* & *Nuri* 27531!; Jebel Baradost near Diana Rowandız, *Field* & *Lazar* 872!; Erbil, in montis Kuh-Sefin, prope Shaqlawa, *Bornmuller* 1676! (G, W); Shaqlawa, *Haines* 392! (E); Gali Dargala, 35 km N.W. by N. of Rania, *Rawi*, Nuri & Kass 28910 (BAG). **MSU**: Suleymaniya, montes Avroman ad confines Persiae, in ditione pagi Tawilla, *Rechinger* 12387 (G, W); Gweija, *Rawi* 8966B (BAG); Avroman mountain, N. of Halabja, on Persian border, *Rawi* 23094!; In planitie Sulaymaniyah, *Thesiger* 350-a; Chemchemal, 5 km from Chemchemal to Kirkuk, *Omar* & *Karim* 37882!; Qaranjir, between Kirkuk and Chemchemal, *Gillett* & *Rawi* 7541!. **FKI**: Kirkuk, 33 km E. of Kirkuk, *Barkley* & *Barkley* 5810! (W); Kirkuk: Inter Kirkuk et Arbil, *Thesiger* 468–453; Kirkuk, Altın Köprü, *Guest* 1984!

Phlomoides laciniata is a very variable species with a wide distribution. The delimitation of species in the *P. laciniata* complex, based on solely on morphological characters, is problematical. Various botanists have tried to recognize taxa among its forms such as *P. azerbaijanica* in N.W. Iran and *P. laevigata* in W. Iran and Iraq. According to Rechinger (1982), some of these forms may represent independent taxa, but here we prefer to include all these forms found in Iraq in the polymorphic *P. laciniata.*

NEMENLOKA (Kurdish, Gweija, *Rawi* 8966B (BAG).

From East Mediterranean to Afghanistan.

2. **Phlomoides molucelloides** *(Bunge) Salmaki,* Taxon 61: 176 (2012); Dinsmore in Post, Fl. Syria, Palest. & Sinai ed. 2, 2: 399 (1933); Rawi in Dep. Agr. Tech. Bull 14: 150 (1964); Feinbrun-Dothan, Fl. Paleast. 3: 117 (1978); Rechinger, Fl. Iranica [K. H. Rechinger] 150: 292 (1982); Davis, Fl. Turkey [P. H. Davis] 7: 102 (1982); Jamzad in Flora Iran [Assadi & al., in Persian] 76: 333 (2012).

> *Eremostachys molucelloides* Bunge in C.E. von Ledebour, Fl. Altaic. 2: 415 (1830); Taifour & El-Oqlah, Pl. Jordan Annot. Checkl.: 102 (2017).
> *E. macrophylla* Montbret & Aucher ex Benth., Ann. Sci. Nat., Bot., sér. 2, 6: 54 (1836).
> *E. macrochila* Jaub. & Spach, Ill. Pl. Orient. 5: 13 (1855).
> *E. pyramidalis* Jaub. & Spach, Ill. Pl. Orient. 5: t. 462 (1855).
> *E. molucelloides* var. *intermedia* Regel, Trudy Imp. S.-Peterburgsk. Bot. Sada 9: 570 (1886).
> *E. molucelloides* var. *macrophylla* (Montbret & Aucher ex Benth.) Regel, Trudy Imp. S.-Peterburgsk. Bot. Sada 9: 570 (1886).
> *Moluccella lanata* Post, Pl. Post. 2: 19 (1891).
> *Eremostachys molucelloides* subsp. *macrophylla* (Montbret & Aucher ex Benth.) Takht. in A.L.Takhtajan & A.A.Fedorov, Fl. Erevana: 256 (1972).

Root napiform. Stems erect or ascending, usually branched in upper 1/2, robust, 20–50 cm tall, sparsely to densely lanate-villous, short glandular hairy, rarely sparsely hairy with dense glandular hairs. Basal leaves orbicular to broadly elliptic, petiolate to 14 cm long, lamina 5–18 × 4–17 cm, serrate-crenate, truncate to subcordate at base, acute to rounded at apex, veins impressed on upper surface, lower surface thickly lanate with glandular papillae, rarely subglabrous; cauline leaves similar to basal leaves in shape and pubescence, 3–8 × 3–8 cm, villous-lanate, rarely subglabrous, sessile. Verticillasters 4–6-flowered, approximate, densely white-lanate, with glandular papillae. Bracteoles 2, 9–22 mm, subulate to linear-filiform, long hairs with short glandular hairs. Calyx 30–40 mm, broadly campanulate-rotate, tube narrow, densely white-lanate, ±lax villous, glandular hairy, ±rarely subglabrous; limb 25–40 mm in diameter at anthesis, accrescent to 35–45 mm in diameter in fruit, reticulate-veined, membranous; limb teeth triangular, spinulose to 3–4 mm, throat sparsely long hairy or not. Corolla 18–30 mm, shorter than calyx, pale lemon-yellow with orange-yellow to buff-orange lower lip; hood pubescent to villous outside, long bearded within, annulate above the middle of tube; lower lip with obcordate middle lobe and oblong lateral lobes. Nutlets oblong, 6–8 × 2–3 mm, striate, blackish, densely barbate.

HAB. Mixed shrubland, limestone hills, shale and igneous slopes, sandy depresion; alt. 600–1800 m; fl. & fr. Apr.-Jul.

DISTRIB. Usually distributed in North and Northeast Iraq, with some records from W. Iraq. **MAM**: Mosul, ad confines Turciae province Hakkari, in ditione pagi Sharanish in montimus calc. a Zakho

Fig. 128. **Phlomoides molucelloides.** 1, habit, basal part × ½; 2, inflorescence × ½; 3, flower with calyx × ½. Reproduced with permission from Jamzad, Flora of Iran 76: f. 88 (2012). Drawn by Farh. © Ministry of Jihad-e-Agriculture.

septentrionem versus, Jabal Khantur, *Rechinger* 10830 (W); Khantur mountain, N.E. of Zakho, *Rawi* 23423!; Mosul: In decl. australibus M. Zawita, substr. Tonschiefer, *Rechinger* 12013! **MSU**: Penjwin, *Rawi* 12186!; Suleymaniya, Gweija Dagh, near Suleymaniya, *Gillett & Rawi* 11671!; Suleymaniya, Basuga, *E.Chapman* 1880!; Suleymaniya, montes Avroman ad confines Persiae, in ditione pagi Tawilla, *Rechinger* 10238 (W); Gweija mountain [Al-Ḥawīja], *Omar & Karim* 37944!; Gweija, 1200 m, *A.Rawi* 8849-A!; Arbil: W. Qala Diza, *Thesiger* 1141-a. **MRO**: Hawara Blinda mountain, N.E. of Haji Omran, *Rawi et al.* 27789! **FAR**: Mollakhort mountains, 1700 m, *Rawi et al.* 29579! **DWD**: 50 km N. of Rutba, Tall-Al Nisr, *Rawi & Nuri* 27120!; Tal Al-Nisr, 48 km W. of Rutba, *Al-Kaisi & Hamad* 46590!; Al-Math 20 km E. of Rutba, *S. Omar*, H. Abass, *Al-Kaisi* & K. Hamad 54668A (BAG); 50 km N. of Rutba, Nubian sandstone, Gara depression, *Barkley* 5355! (W).

West and Central Asia to Mongolia.

8. **STACHYS** L.

Sp. Pl. ed. 2, 580 (1753); Gen. Pl. ed. 5, 253 (1754)
Harley & al. in Kubitzki (ser. ed.), Fam. Gen. Vasc. Pl. 7: 223 (2004)

F. Patzak
Revised by Shahina A. Ghazanfar

Perennial subshrubs or annual herbs, erect or ascending-erect or ± pendulous from clefts of rocks, often aromatic. Leaves simple, usually petiolate, and with toothed margins. Verticillasters few- or many-flowered, distant or the upper approximated, or confluent in long dense spikes; flowers pedicellate. Bracts and bracteoles present or absent. Calyx funnel-shaped to campanulate, 5- or 10-veined, with equal or oblique 5-toothed throat (in ours), rarely somewhat 2-lipped; tube throat often barbate Corolla 2-lipped, tube cylindrical, included or rarely somewhat exserted from calyx, with a transverse or oblique hairy ring, or rarely naked within; upper lip erect, entire or retuse, or rarely bifid, usually somewhat arched, lower lip spreading, 3-lobed, middle lobe the largest. Stamens didynamous, not exserted, lower longer, with 2 divergent cells. Style bifid. Nutlets obovoid, ± trigonous, compressed, rounded at apex.

About 300 species, widespread except for Australia and New Zealand, centered mainly in the Mediterranean region of Europe and Turkey; 20 species in Iraq.

Several species of *Stachys* are cultivated as ornamentals; *S. affinis* is cultivated for its edible tubers.

Akçiçek, E., Dirmenci, T. and Dündar, E. (2012). Taxonomical notes on Stachys sect. Eriostomum (Lamiaceae) in Turkey. Turkish Journal of Botany, 36(3): 217–234.
Bhattacharjee, (1980). Notes Royal Bot. Gard. Edinburgh 38: 65 (1980).
Al-Musawi, A. H. E. (1990). Key to the genus *Stachys* L. (Labiatae) in Iraq with three new records. Bull. Iraq Nat. His. Mus. 8 (3): 53–59.
Aun, Z.A. and Haloob, A., (2020). A New Species Of Genus Stachys .(Lamiaceae) in Iraq. Plant Archives, 20(2): 1131–1134.
Güner, Ö. & Ferrer-Galego, P.P. (2021). Nomenclatural and taxonomic notes on some *Stachys* taxa (Lamiaceae). Turkish Journal of Botany 45: 69–82. https://doi-org.ezp.lib.cam.ac.uk/10.3906/bot-2010-12
Salmaki, Y., Zarre, S., Govaerts, R., & Bräuchler, C. (2012). A taxonomic revision of the genus Stachys (Lamiaceae: Lamioideae) in Iran. Botanical Journal of the Linnean Society 170(4): 573–617.

1. Annual herbs 2
   Suffrutescent or perennial herbs 4
2. Corolla yellowish to whitish-yellow, 14–15 mm; calyx teeth lanceolate,
   somewhat spiny, rigid, upcurved, shorter than tube . . . . 16. *S. annua*
   Corolla pink or purple, 10–12 mm. 3
3. Calyx teeth ± equalling tube; corolla pink, c. 10 mm, tube exserted;
   nutlets minutely tuberculate-punctate . . . . 17. *S. melampyroides*
   Calyx teeth c. ⅓-½ as long as tube; corolla purple, c. 12 mm, tube
   included; nutlets nearly smooth . . . . 18. *S. burgsdorffioides*
4. Verticillasters usually more than 10-flowered; bracts more than half as
   long as to ± equalling calyx 5
   Verticillasters 2–10-flowered; bracts usually very short or absent 9
5. Corolla 12–16 mm 6
   Corolla 6–8 mm 8
6. Upper part of stem, bracts and calyx glandular-hairy, clothed with
   long, spreading, englandular and short glandular hairs; leaves
   membranous . . . . 21. *S. zakhoensis*
   Stems, bracts and calyx eglandular, densely white-lanate; leaves not membranous . . . 7
7. Calyx 10–12 mm, teeth c. ½ as long as tube, recurved in fruit. . . . . 1. *S. cretica*
   Calyx 6–7 mm, teeth more than ½ as long as tube, erect in fruit. . . . . . . .2. *S. spectabilis*
8. Verticillasters all distant, in long racemes; calyx clothed with spreading
   hairs, teeth triangular, 3–4 × shorter than tube, recurved in fruit . . . . . . .3. *S. setifera*
   Verticillasters crowded into a long spike, often 20–40 cm; calyx densely
   tomentellous-canescent, teeth lanceolate-subulate, scarcely shorter
   than tube, erect in fruit . . . . 4. *S. longespicata*

9. Pannose, white-felted or white-tomentose or white-canescent, clothed
    with appressed, short and long spreading hairs. 10
    Hairy or glabrescent 13
10. White-canescent, clothed with appressed short and long spreading
    hairs; calyx 15–18 mm, teeth plumose, lanceolate-subulate from a
    narrowly lanceolate base, 5–6 × longer than the short campanulate
    tube. . . . . 15. *S. lavandulaefolia*
    Pannose, densely white-felted or white-tomentose all over. 12
12. Calyx 12–15 mm, densely white-felted, pannose, much inflated in fruit,
    teeth very short, ovate, subacute; corolla 15–18 mm. . . . . 13. *S. inflata*
    Calyx 8–10 mm, densely short-tomentellous, teeth ½ as long as tube,
    subspinescent, divergent; corolla 10–12 mm . . . . 14. *S. kotschyi*
13. Erect perennials 30–50 cm tall, virgate, rather rigid. 14
    Perennials with stems usually pendulous from clefts of rocks, fragile 15
14. Corolla purple, 14–18 mm; calyx teeth triangular-lanceolate,
    acuminate, tapering into a long, weak, spiny mucro, c. ½ as long as
    tube or somewhat shorter; whorls 6–8(–10)-flowered . . . . 12. *S. iberica*
    Corolla white or yellowish, 8–22 mm; calyx teeth triangular to lanceolate 15
15. Calyx 6–8 mm, ± pubescent; teeth lanceolate, keeled, subpungent,
    erect, ½ as long as tube . . . . 10. *S. kurdica*
    Calyx with stalked and sessile glands; calyx teeth triangular 16
16. Corolla white, 11–13 mm, tube markedly exceeding the calyx . . . .19. *S. bhattacharjeeeae*
    Corolla yellow, 20–22 mm, tube as long as or shorter than calyx. . . . . . .20. *S. multicaulis*
17. Corolla white, lower lip purple-spotted 18
    Corolla yellow. 19
18. Felted-lanate, less glandular; calyx teeth c. ½ as long as tube; tube of
    corolla c. 17 mm. . . . . .5. *S. lanigera*
    Densely glandular, and clothed with spreading hispid hairs; calyx teeth
    about as long as tube; tube of corolla c. 14 mm. . . . . .6. *S. fragillima*
19. Calyx 10–12 mm in flower, 11–14 mm in fruit, teeth about as long as
    tube or somewhat shorter . . . . .11. *S. megalodonta*
    Calyx 6–10 mm, teeth ½-⅓ as long as tube. 20
20. Calyx 8–10 mm, glandular-viscid, clothed with sessile glands and short
    spreading glandular white hairs; short grey-pubescent herb, clothed
    with short spreading rough hairs and glandular-viscid above. . . . . . . 9. *S. benthamiana*
    Calyx 6–8 mm. 21
21. Bracts subulate, very short; corolla 12(–13) mm. . . . . 10. *S. kurdica*
    Bracts oblong- or linear-lanceolate, about as long as calyx tube; corolla 15–20 mm. . .22
22. Calyx 5-veined; corolla c. 20 mm, tube c. 12 mm; leaves crenate; petiole
    of cauline leaves 5–10 mm. . . . . .7. *S. mardenensis*
    Calyx 10-veined; corolla c. 15 mm, tube c. 10 mm; leaves serrate; petiole
    of cauline leaves to 5 mm . . . . 8. *S. graveolens*

    1. **Stachys cretica** *L.*, Sp. Pl. 2: 581 (1753) subsp. **garana** (*Boiss.*) *Rech.f.*, Ann. Naturh. Mus. Hofmus. Wien 48: 176 (1937). Type: Iraq, in lapidosis montis Gara Kurdistaniae, *Th. Kotschy* 413 (G-BOIS!, holo; G, K!, W, iso. K!, G, W!); Blakelock in Kew Bull. 1949: 549 (1950); Zohary in Dep. Agr. Iraq Bull. 31: 127 (1950); Rawi in Dep. Agr. Iraq Tech. Bull. 14: 157 (1964); Rechinger, Fl. Iranica 150: 363 (1982); Bhatacharjee in Fl. Turkey [P. H. Davis] 7: (1982); Akçiçek & al., Turk. J. Bot. 36(3): 227 (2011); Salmaki & al. in Bot. J. Linn. Soc. 170(4): 591 (1200); Taifour & El-Oqlah, Pl. Jordan Annot. Checklist 106 (2017).

    S. garana Boiss., Diagn. Pl. Orient. ser. 1, 12: 76 (1853).
    S. cretica L. var. *garana* (Boiss.) Boiss., Fl. Orient. [Boissier] 4: 719 (1879); Dinsmore in Post, Fl. Syria,
      Palest. & Sinai ed. 2, 2: 378 (1933).
    S. germanica L. subsp. *italica* (Miller) Briq. var. *garana* (Boiss.) Briq., Lab. Alp. Marit. 223 (1893).

    Perennial, up to 1 m or more tall. Stems erect, usually simple, stout, strictly short-branched towards apex, sparsely and very shortly appressed-lanate, canescent or pale green. Basal and lower cauline leaves densely and long white-lanate-tomentose, especially below, greenish and glabrescent above, slightly wrinkled; petiole 4–7 cm; lamina oblong, 6–8 × 2–4 cm, base cordate, apex tapering, obtuse, margins crenate-dentate; intermediate cauline leaves gradually shorter-stalked, lamina gradually smaller, oblong-lanceolate, acute or subacute; uppermost and floral leaves on c. 5 mm petioles, lamina oblong-lanceolate, 30–40 × 15 mm, subacute, crenulate-denticulate. Verticillasters many-flowered, lower ones at least distant, upper confluent. Bracts 5–7 mm, oblong, spiny-mucronate, ½ as long as calyx, lanate. Calyx 10–12 mm, densely lanate, tube erect, teeth lanceolate, spiny-tipped, c. ½ as long as tube, recurved in fruit. Corolla 14–16 mm, pink or flesh-coloured, tube included, upper lip retuse, hairy. Nutlets c. 1.5 mm, oblong, trigonous, blackish-brown, minutely punctulate, glabrous.

HAB. Hills and stony places, near top of mountain slopes, *Quercus* forest, rocky ground, sides of stream; alt. 1000–1500 m; fl. & fr. Jun.-Jul.

DISTRIB. Occasional in the lower forest zone of Iraq: **MAM**: Qurnago valley N. of Pushtashan, *Rawi* 26637!, 26684!; Jabal Khantur N.E. of Zakho, *Rawi* 23407!, *Rechinger* 10838!; Zawita, *Guest* 3732!, 4583!; Kani Masi, *Al-Kaisi* et al. 43892 (BAG); Ser Amadiya (Kani Sinji), *Al-Dabbagh, Al-Kaisi & H. Hamid* 46040 (BAG); Swara Tuka, *Al-Kaisi,* Al-Khayat & F. Karim 51000 (BAG); Sarsang, *Al-Kaisi,* Al-Khayat & F. Karim 51028 (BAG). **MRO**: Rayat to Al Gird (Helgord) Dagh, *Gillett* 9477! **MSU**: Penjwin, *Rechinger* 10549!; Qara Dagh, *Kotschy* 413!; Sarsang on N. slopes of Qara Dagh, *Chapman* 26375!

Turkey (E. Anatolia), W. Iran. Distribution of species: S.E. Europe from Italy to Crimea, Turkey (E. & S.E. Anatolia), Syria.

2. **Stachys spectabilis** *Choisy* ex DC., Mém. Soc. Phys. Genève 1(2): 457 1823); Boissier, Fl. Orient. 4: 723 (1879); Dinsmore in Post, Fl. Syria, Palest. & Sinai ed. 2, 2: 379 (1933); Zohary in Dep. Agr. Iraq Bull. 31: 127 (1950); Rawi in Dep. Agr. Iraq Tech. Bull. 14: 157 (1964); Feinbrun-Dothan, Fl. Palaest. 3: 129, pl. 209 (1978); Rechinger, Fl. Iranica [K. H. Rechinger] 150: 364 (1982); Salmaki & al. in Bot. J. Linn. Soc. 170(4): 609 (2012); Taifour & El-Oqlah, Pl. Jordan Annot. Checklist 106 (2017).

Perennial, 50–100 cm tall. Stems simple or branched from base, canescent-tomentellous. Basal and lower cauline leaves withered at flowering time; middle cauline leaves with 1–1.5 cm long petioles; lamina ovate-cordate to ovate-lanceolate, 10–12 × 4–5 cm, base cordate, apex acute, margins crenulate-dentate, scallops often mucronate, not wrinkled, green to canescent above, pannose below; upper and floral leaves gradually smaller, the latter 2(–5) × 0.8(–2) cm, oblong-lanceolate, acute, margins denticulate or entire, subsessile. Verticillasters 10–12-flowered, distant or the upper ones confluent. Bracts 4–5 mm, elliptic-rhomboid, cuspidate, shorter than calyx. Calyx 6–7 mm, densely lanate, tube erect, teeth lanceolate, spiny-tipped, more than ½ as long as tube, erect in fruit. Corolla purple or flesh-coloured or variegated with whitish-yellow, 12–14 mm; upper lip entire, hairy. Nutlets 1–1.2 mm, oblong, trigonous, brownish-black, minutely punctulate, glabrous. Fig. 129, 1–3.

HAB. Damp thickets and near streams, damp bank of irrigation canal, mountain slopes, in valley; alt. 1000–1800 m; fl. & fr. Aug.

DISTRIB. Occasional in lower forest zone of Iraq (N.E. sector). **MAM:** Kani Masi, *Al-Kaisi et al.* 43892 (BAG) **MRO:** mt. Helgord above Nawanda, *Rechinger* 11474!; bet. Rayat and Ari Gird-Dagh, Gillett 9477 (BAG); Shakh Wa Shikhan vally, *Rawi & Serhang* 24869 (BAG); Qenna Qaw vally N. of Pushtashan, *Rawi & Serhang* 26637 (BAG). **MSU:** Qara Dagh, *Kotschy* 328!; Sarsang on N. slopes of Qara Dagh, *Haines* W. 497!,W. 1574!

N. Syria, Turkey, Azerbaijan, Armenia, Georgia, Iran.

3. **Stachys setifera** *C.A.Mey.*, Verz. Pfl. Casp. Meer. (C.A. von Meyer) 94. (1831); Boissier, Fl. Orient. 4: 724 (1879); Rawi in Dep. Agr. Iraq Tech. Bull. 14: 157 (1964); Rechinger, Fl. Iranica [K. H. Rechinger] 150: 365 (1982); Bhatacharjee in Fl. Turkey [P. H. Davis] 7: 229 (1982); Salmaki & al. in Bot. J. Linn. Soc. 170(4): 607 (2012).

S. daenensis Gand. in Bull. Soc. Bot. France 65: 68 (1918).
S. sintenisii Gand. in Bull. Soc. Bot. France 65: 68 (1918).

Perennial, 15–25 cm tall, pale green or canescent, softly hairy. Stems erect or erect-ascending, somewhat flexuous, branched from base. Basal and lowermost cauline leaves withered at flowering time, lower and middle cauline leaves subsessile, lamina oblong, 1.5–2 × 0.5–1 cm, base cuneate or rounded, apex acuminate or acute and spiny; upper and floral leaves gradually smaller, sessile, overtopping whorls. Verticillasters (8–)9–10-flowered, all distant, lower ones 2–3 cm, upper somewhat less, arranged in long racemes. Bracts c. 4(–5) mm, lanceolate, aristate. Calyx 5–6(–7) mm, somewhat canescent, softly hairy, clothed with

Fig. 129. **Stachys spectabilis**. 1, habit × ½; 2, flower × 2; 3, calyx × 2. Reproduced with permission from Jamzad, Flora of Iran 76: f. 44 (2012). © Ministry of Jihad-e-Agriculture.

spreading hairs; tube erect, accrescent in fruit, teeth triangular, subulate, spiny, 3–4 × shorter than tube, ± recurved in fruit. Corolla 6–8 m, pinkish or crimson, glabrescent, upper galea subretuse. Nutlets c. 1.2 mm, obovate, trigonous, compressed, minutely punctulate, glabrous.

HAB. Thickets, damp banks, stream sides; alt. 1800–2500 m; fl. & fr. Aug.-Sep.
DISTRIB. Occasional and rather local in lower forest zone of Iraq (N.E. sector). MRO: Algird Dagh, Guest 2783 (BAG); Gunda Zhor valley E. of Helgord Dagh, *Gillett* 9485!, 12320! MSU: Sarsang on N. slopes of Qara Dagh, *Haines* W.1223!

Agnew 800 from Garagu, nr. Sarsang, identified as *S. iraqensis* by Bhattacharjee as sp. nov. [https://data.rbge.org.uk/herb/E00319631] is referrable to *S. setifera*. More material is needed to fully access its taxonomic status.

Turkey, Transcaucasia, N. & C. Iran, Afghanistan, Pamirs and Tian-Shan.

4. **Stachys longespicata** *Boiss. et Kotschy* in Boissier, Fl. Orient. 4: 725 (1875); Dinsmore in Post, Fl. Syria, Palest. & Sinai ed. 2, 2: 379 (1933); Zohary in Dep. Agr. Iraq Bull. 31: 127 (1950); Rawi in Dep. Agr. Iraq Tech. Bull. 14: 157 (1964); Feinbrun-Dothan, Fl. Palaest. 3: 130, pl. 211 (1978); Bhatacharjee in Fl. Turkey [P. H. Davis] 7: 222 (1982); Taifour & El-Oqlah, Pl. Jordan Annot. Checklist: 106 (2017).

Perennial, 60–100 cm, densely soft-tomentellous, canescent. Stems erect, somewhat flexuous, much branched from base; stems and branches quadrangular, densely soft-felted, clothed with long retrorse hairs. Basal and lowermost cauline leaves withered at flowering time, middle cauline leaves few with petioles 0–5–15 mm; lamina oblong to oblong-lanceolate, 15–20 × 5–10 mm, base truncate or rounded, apex acute, margin minutely crenate-dentate, softly tomentellous on both sides, upper and floral leaves gradually smaller, linear to linear-lanceolate, shortly petiolate to sessile. Verticillasters (8–)10-flowered, except for a few lower ones crowded into a dense spike often 10–40 cm; few lower ones often rather distant, their internodes 2–5 cm, upper ones approximated, confluent. Bracts 4–5 mm, subulate, acute. Calyx 4–5 mm, tube erect, teeth lanceolate at base, subulate, tapering into a long spiny mucro, scarcely shorter than tube, erect in fruit. Corolla pink, 6–8 mm, limb hairy, upper lip retuse. Nutlets 1.2–1.5 mm, oblong, trigonous, compressed, minutely punctulate, glabrous.

HAB. Thickets, near streams; alt. 300–800 m; fl. & fr. not recorded.
DISTRIB. Rare and needs confirmation. Recorded only by Zohary for Iraq, without locality.

Turkey, Syria, Lebanon.

5. **Stachys lanigera** (*Bornm.*) *Rech.f.* in Bot. Jahrb. Syst. 71(4): 528 (1941); Blakelock in Kew Bull. 1949: 550 (1950); Rawi in Dep. Agr. Iraq Tech. Bull. 14: 157 (1964); Rechinger, Fl. Iranica [K. H. Rechinger] 150: 373 (1982); Salmaki & al. in Bot. J. Linn. Soc. 170(4): 598 (2012).

S. fragillima Bornm. var. *lanigera* Bornm. in Beih. Bot. Centralbl. 33, 2: 189 (1915).

Perennial, 10–15 cm, densely white felted-lanate, sparsely glandular. Stems numerous, herbaceous, simple or rarely somewhat branched, pendulous from clefts of rocks, fragile. Basal and lowermost cauline leaves withered at flowering time; lower and intermediate cauline leaves with petioles 3–5 mm; lamina ovate-reniform to ovate, 15–25 × 10–25 mm, obtusely and coarsely crenate-dentate, base cordate, apex obtuse; floral leaves similar in size, the uppermost triangular-ovate, subacute, about as long as flowers. Verticillasters 4–6-flowered, distant; internodes 2–5 cm, uppermost somewhat approximated. Bracts narrowly linear, obtusely and coarsely crenate-dentate, longer than the c. 2 mm pedicels. Calyx 8–10 mm, clothed with sessile glands and short spreading white hairs, teeth broadly lanceolate, acute, not spiny, c. ½ as long as tube. Corolla white, lower lip purple-spotted; tube exserted, c. 17 mm, twice as long as calyx. Nutlets not seen.

HAB. Rocks, in crevices, shady cliffs, near waterfalls, stream sides; alt. 400–1000 m; fl. & fr. Jul.
DISTRIB. Occasional and local in the lower forest zone of N.E. Iraq. **MRO**: Rowanduz gorge, *Guest* 2973!, 13608!; Bakhad falls nr Rowanduz, *Guest* 15911!; Helgord Dagh, Gali Ali Beg, *Rawi* 24893!, 26801!; Helgord Dagh, near last bridge on way to Rowanduz, *Kass & Nuri* 27214!; Serkupkan village, 70 km N. by W. of Rania, *Rawi, Nuri & Kass* 28512!; Shaqlawa, Sahira 37550 (BAG); Bikhal, *S. Omar* 37635 (BAG).

Salamki & al. (l.c.) note that this species is an endemic, very rare in Iran and is known only from the type gathering (IRAN, Kermanshah: Kerind, in monte Noa Kuh). It is found in N.E. Iraq and is different from the endemic *S. fragillima*.

S.W. Iran.

6. **Stachys fragillima** *Bornm.* in Bull. Herb. Boissier 7: 118 (1899); Type: Kurdistania, in montis Kuh-Sefin, 16 May 1893, Iter Persico-turcicum, no. 1666. (https://data.rbge.org.uk/herb/E00319627) Rawi in Dep. Agr. Iraq Tech. Bull. 14: 157 (1964); Rechinger, Fl. Iranica [K. H. Rechinger] 150: 373 (1982).

Perennial, 20–50 cm tall, dull grey-green, minutely but densely glandular and clothed with spreading hispid hairs. Stems numerous, herbaceous, simple, erect or erect-ascendent, somewhat branched from base, pendulous from clefts of rocks, fragile. Basal and lowermost stem leaves withered at flowering time, lower cauline leaves with 2–3 mm petioles; lamina ovate-orbicular or ovate, 15–20 × 10–15 mm, base cordate, apex subacute, obtusely and coarsely crenate-dentate, about as long as flowers. Verticillasters 4–6-flowered, distant, internodes 3–5 cm, upper ones somewhat approximated. Bracts narrowly linear, crenate-serrate, much longer than the rather short 1–2 mm pedicels. Calyx 5–6 mm, glandular-viscid, campanulate, teeth broadly lanceolate, acute, about as long as tube. Corolla white, lower lip purple-spotted, tube exserted, c. 14(–15) mm, nearly twice as long as calyx. Nutlets c. 1.5 mm, obovate, trigonous, compressed, 3–angled, dark brown, minutely punctulate, glabrous.

HAB. Rocky cliffs, in crevices, in gorges; alt. 600–1500 m; fl. & fr. May-Jul.
DISTRIB. Rare, found in the lower forest zone of N.E. Iraq. **MRO**: Sefin Dagh above Shaqlawa, *Bornm*üller 1666!, *Gillett* 8077!, *Rawi* 9079! **FNI**: Dohuk, *Rechinger* 11493!

Endemic.

The hybrid *S. fragillima* × *graveolens* Blakelock, Kew Bull. 1949: 549 (1950) was recorded from a single locality: **MAM**: Zawita, 1050 m, *Guest* 4731! This differs from the typical *S. fragillima* in its oblong-lanceolate leaves, smaller leaves subtending verticillasters, and smaller corollas (10–14 mm). The longer and narrower calyx teeth distinguish it from *S. graveolens*.

7. Stachys mardinensis (*Post*) *R.Mill*, Notes Roy. Bot. Gard. Edinburgh 38: 57 (1980); Bhatacharjee in Fl. Turkey [P. H. Davis] 7: 232 (1982).

Nepeta mardinensis Post, Bull. Herb. Boissier 7:159 (1899) (!); *Stachys glechomifolia* Nábělek, Spisy Přír. Fak. Masarykovy Univ. 70: 61, t. 7 (1926); *S. nephrophylla* Rech.f., Pl. Syst. Evol. 134: 291 (1980), Type: Iraq, Erbil, *Rechinger* 11250 (G).

Suffrutescent, with woody rootstock, 20–30 cm tall. Stems numerous, herbaceous, slender, erect to erect-ascendant, ± pendulous from clefts of rocks, fragile, densely lanate, clothed with spreading glandular and eglandular hairs to 1 mm. Sterile stems few, slender, scarcely 5 cm long. Basal and lowermost cauline leaves withered at flowering time, lower and intermediate cauline leaves long-petiolate, broadly ovate to semi-orbicular, base broadly cordate, apex rounded or subretuse, margins coarsely and deeply crenate, densely lanate on both sides; intermediate cauline leaves on 5–10 mm petioles, 2(–2.5) × 2.5–3 cm, upper gradually smaller, uppermost and floral leaves sessile, minute, as long as calyx or slightly longer, acute or subacute. Leaves of sterile stems minute, long-petiolate, orbicular, 5–10 × 5–10 mm, base cordate, obsoletely crenate or subentire. Verticillasters 2–4-flowered, lower distant, upper 3–4 approximated or confluent. Bracts narrowly oblong-lanceolate, 6–7 mm, about as long as calyx tube, short-acuminate, acute, glandular-hairy. Flowers on 3 mm pedicels. Calyx 6–8 mm, tube c. 5 mm, prominently 5-nerved, clothed with spreading eglandular and glandular hairs, teeth triangular-lanceolate, acute, upper 3 scarcely half as long as tube, c. 2 mm, lower 2 deeply bifid. Corolla c. 20 mm, yellowish, tube exserted, twice as long as calyx, c. 12 mm, upper lip long-hairy outside. Nutlets c. 2.5 mm, obovate, trigonous, compressed, blackish-brown, minutely punctulate, glabrous.

HAB. Limestone rocks, in crevices, in gorges; alt. 700–800 m; fl. & fr. Aug.
DISTRIB. Rare and local, found only in the N.E. sector of the lower forest zone of Iraq. **MRO**: Rowanduz, *Haley* 42! *Rechinger* 11250! (type of *S. nephrophylla*).

Turkey (S.E. Anatolia).

As noted in Fl Turkey (l.c.) *S. mardinensis* is closely related to *S. fragillima* Bornm. but differs by its usually ± subsessile, ovate cauline leaves with crenate-dentate margins, usually shorter yellow corolla and nutlets usually without prominent marginal wings and deeply notched apex. Besides these, the species shows transitions between pendent and erect habit, while *S. fragillima* is always pendent.

8. Stachys graveolens *Nábělek*, Spisy Přír. Fak. Masarykovy Univ. 70: 62 (1950); Blakelock in Kew Bull. 1949: 549 (1950); Zohary in Dep. Agr. Iraq Bull. 31: 127 (1950); Rawi in Dep. Agr. Iraq Tech. Bull. 14: 157 (1964); Rechinger, Fl. Iranica [K. H. Rechinger] 150: 375 (1982).

Suffrutescent, with woody rootstock, 20–40 cm. Stems numerous, herbaceous, slender, erect or erect-ascendent to ± pendulous from clefts of rocks, fragile. Basal and lowermost cauline leaves withered at flowering time, lower and intermediate cauline leaves with short petioles to 5 mm; lamina ovate to ovate-orbicular, 2–3.5 × 2–3 mm, base cordate, apex acute, margins coarsely serrate, softly hairy on both sides, greenish, prominently veined below, floral leaves truncate, uppermost cuneate at base, c. 6 × 4 mm, as long as calyx. Verticillasters many-flowered, distant, upper approximated or confluent. Bracts linear-lanceolate, plumose-hirsute and glandular, 5–6 mm, about as long as calyx tube. Pedicels very short. Calyx 6–8 mm, tube c. 5 mm, prominently 10-veined, glandular-viscid and clothed with spreading hairs outside, teeth subequal, triangular-lanceolate, acute, scarcely ½ as long as tube, c. 2 mm. Corolla yellowish, c. 15 mm, tube exserted, c. 10 mm, upper lip with purple markings. Nutlets c. 2 mm, obovate, trigonous, compressed, blackish-brown, minutely punctulate, glabrous.

HAB. Rocks, in crevices, in gorges; alt. 400–1000 m; fl. & fr. Jun.-Jul.
DISTRIB. Rare and local in the lower forest zone of N. Iraq. **MAM**: Dohuk, *Guest* 1587 (BAG); Dohuk gorge, *Guest* 3709!; Mar Ya'qub village near Simel, *Nábělek* 1507!

Endemic.

9. **Stachys benthamiana** *Boiss.*, Fl. Orient. [Boissier] 4: 734 (1875); Zohary in Dep. Agr. Iraq Bull. 31: 128 (1950);

> *Rawi* in Dep. Agr. Iraq Tech. Bull. 14: 156 (1964); Rechinger, Fl. Iranica 150: 370 (1982); Salmaki & al. in Bot. J. Linn. Soc. 170(4): 589 (2012).

Suffrutescent, with rather woody rootstock, 10–30 cm, many-stemmed, pale green, short grey-pubescent, clothed with short spreading rough hairs and glandular-viscid above. Stems slender, virgate, ± flexuous, ± pendulous from clefts of rocks, fragile. Basal and lowermost cauline leaves withered at flowering time, lower and intermediate cauline leaves with 2–5 mm long petioles; lamina ovate-oblong, 10–25 × 8–15 mm, base cordate or truncate, apex rounded or subacute, margins deeply and coarsely crenate-serrate; floral leaves overtopping Verticillasters 4–6-flowered, lower distant, internodes 1.5–3 cm, upper approximated. Bracts subulate, very short. Calyx 8–10 mm, campanulate, glandular-viscid, clothed with sessile glands and short-spreading glandular white hairs, teeth lanceolate, acute, about half as long as tube, c. 3 mm. Corolla c. 15–18 mm, yellowish, viscid-hirsute, tube exserted, c. 1.5 × as long as calyx. Nutlets c. 1.5 mm, oblong, trigonous, compressed, blackish-brown, minutely punctulate, glabrous. Fig. 130, 1–2.

HAB. Rocks, in crevices, in gorges; alt. 1000–1600 m; fl. & fr. Jun.-Jul.
DISTRIB. Rare and local, found only in the N.E. sector of the forest zone of Iraq. **MRO**: Jabal Baradost, nr Diana, *Field & Lazar* 883!, 908!; between Erbil and Ruwandaz, *Nábělek* 1573 (SAV). **MSU**: Qara Dagh, *Buthaina Makki* 449 p.p.

S. & W. Iran.

10. **Stachys kurdica** *Boiss. & Hohen.*, Diagn. Pl. Orient. ser. 1, 5: 31 (1844). Type: Iraq, Kurdistan, in rupestribus totius m. Gara Kurdistaniae, versus cacumen, 1841, *Th. Kotschy* 390 (G-BOIS, holo.; G, LE, M, W!, iso.); Blakelock in Kew Bull. 1949: 550 (1950); Rawi in Dep. Agr. Iraq Tech. Bull. 14: 157 (1964); Rechinger in Fl. Iranica [K. H. Rechinger] 150: 371 (1982); Salmaki & al. in Bot. J. Linn. Soc. 170(4): 596 (2012).

> *S. ballotiformis* Vatke, Bot. Zeitung (Berlin) 33: 448 (1875). Type: Iraq, Iter Syriaco-Armeniacum, Kurdistan, Pir Omar Gudrun, 4000 ped., 6.6.1867, *Haussknecht* 806 (W, holo.; JE, iso.); Blakelock in Kew Bull. 1949: 549 (1950); Rawi in Dep. Agr. Iraq Tech. Bull. 14: 156 (1964); Rechinger in Fl. Iranica 150: 370 (1982).
> *S. subnuda* Montbret. & Aucher ex Benth., Ann. Sci. Nat., Bot. Ser. 2, 6: 50 (1836) var. *kurdica* (Boiss. & Hohen.) Boissier, Fl. Orient. [Boissier] 4: 734 (1875); Zohary in Dep. Agr. Iraq Bull. 31: 128 (1950); Rawi in Dep. Agr. Iraq Tech. Bull. 14: 157 (1964) (as *subnuda* Montbret. & Auch. ex Benth.
> *S. kurdica* Boiss. et Hohen. var. *glaber* Blakelock in Kew Bull. 1949: 550 (1950).

Perennial, 20–60 cm tall, many-stemmed, pale green, minutely and sparsely pubescent, rarely glabrescent. Stems numerous, erect to erect-ascendent to pendulous, virgate, strictly much-branched from base. Basal and lowermost cauline leaves withered at flowering time; lower cauline leaves with 5–15 mm long petioles; lamina oblong or ovate-oblong, 15–20(–

Fig. 130. **Stachys benthamiana**. 1, habit × ½; 2, flower × 2. Reproduced with permission from Jamzad, Flora of Iran 76, f. 2012. © Ministry of Jihad-e-Agriculture.

30) × 8–10(–15) mm, base cuneate, apex obtuse, remotely crenate-dentate; upper gradually smaller, oblong to oblong-lanceolate, acute, entire obscurely crenate-dentate, floral ones shorter than calyx. Verticillasters 3–6-flowered, mainly distant, lower internodes 4–6(–7) cm, upper shorter, uppermost verticillasters approximated, confluent. Bracts subulate or absent. Calyx 6–8 mm, campanulate, erect, ± pubescent, teeth lanceolate, keeled, erect, about ½ as long as tube. Corolla yellowish, 10–13 mm, pubescent, tube shortly exserted. Nutlets c. 1.2 mm, trigonous, compressed, brownish-black, minutely punctulate or nearly smooth, glabrous.

HAB. Dry mountain slopes and hillsides, on rocky ground, in crevices, in ravines; alt. 600–3200 m; fl. & fr. Jun.-Aug.

DISTRIB. Common in the lower forest zone of Iraq. **MAM**: Jabal Khantur N.E. of Zakho, *Rawi* 23308!; Sulaf, *Rustam Hb.* 3774!; Sarsang, Dohuk to Amadiya, *Haines* W. 1048!, *Rechinger* 11653!; Aqra, *Rawi* 11486!; Jabal Khantur, N. of Zakho, *Rechinger* 10746!; above Marsis, N. of Zakho, *Rechinger* 10893!; above Sarsang, Dohuk to Amadiya, *Rechinger* 11674!; above Sawara Tuka, Dohuk to Amadiya, *Rechinger* 11961!; Amadiya, *Guest* 3774! **MRO**: Qurnago valley E. of Pushtashan, *Rawi & Serhang* 26642!; Gara Dagh, *Kotschy* 390!; above Pushtashan, mt. Qandil, *Rechinger* 11097! 11096!; above Gom-i Kirmosoran, mt. Qandil, *Rechinger* 11784!; Kermasur lake, *Rawi & Serhang* 24140!; mt. Berrog (Perrish?), *Rawi* 23936!; between Saran village and Kilkil, Jabal Karoukh, *Rawi & Serhang* 27327!; S. slope of Jabal Karoukh, *Kass & Nuri* 27497!; Sala Khal, *Rawi & Serhang* 24694!; Walza nr Chiya-i Mandali, *Guest* 2703!; Chia-i Mandali, *Guest* 2766 (BAG); Jabal Sakri Sakran, *Bornmüller* 1669!; Rayat, *Guest* 13064!; Jabal Baradost, nr Diana, *Field & Lazar* 883!; Khalifan, Rowanduz gorge, *Emberger & Guest* 15465!; Rowanduz gorge, *Guest* 2974!; Jindian nr Rowanduz, *Guest* 13007!; below Rowanduz, *Rechinger* 11251!; Rowanduz gorge, *Guest* 2974!; Serderian traject., Arbil to Rowanduz, *Nábĕlek* 1573! **MSU**: Pir Omar Gudrun, *Hausskn.* 806!; Avroman (Hauraman) and Shahu mts., *Haussknecht* s.n.!; mt. Zalim, *Rawi & Nuri* 29386!; Tairan, between Darband-i Basian and Tasluja, *Gillett & Rawi* 11643!; Tawila, mt. Hawraman, *Gillett* 11789!; Zewiya, Pira Magrun, *Rawi* 12066!, *Gillett* 7778!; Amoret, nr Qara Dagh, *Haines* W. 1077!; Qara Dagh, *Chapman* 26418!, *Rawi* 9259! **FAR**: Ser Kurawa, *Gillett* 9786!; Haji Umran, *Chapman* 11925! **FKI**: Darband-i Bazian nr Chemchemal, *Rechinger* 10605!; 10606!;

Turkey (S.E. Anatolia), S.W. Iran.

11. **Stachys megalodonta** *Hausskn. & Bornm.* ex *P.H.Davis*, Notes Roy. Bot. Gard. Edinburgh 21: 46 (1956); Rechinger, Fl. Iranica [K. H. Rechinger] 150: 369 (1982).

Perennial, somewhat woody at base, 15–40 cm, many-stemmed. Stems simple or branched from base, ± slender, virgate, somewhat flexuous sometimes, erect or erect-ascendent to ± pendulous from clefts of rocks, fragile, glandular-viscid and clothed with hispid eglandular hairs, reflexed or somewhat spreading; branches ascending to pendulous. Basal and lowermost cauline leaves withered at flowering time; lower cauline leaves rather small, gradually increasing in size upwards, intermediate cauline leaves with 5–15 mm long petioles; lamina ovate or ovate-oblong, 4–5 × 3–4 cm, base cordate, apex obtuse to subacute, coarsely crenate-dentate, appressed-hairy; upper cauline leaves gradually smaller, ovate-oblong to oblong, shortly petiolate to subsessile; floral leaves ovate-lanceolate, acute or acuminate, crenate or crenate-dentate, as long as calyx or slightly longer. Verticillasters 6–8-flowered, few, lower ones rather distant, internodes 4–10 cm, upper approximated, confluent. Bracts linear-lanceolate to filiform, c. 2 mm. Flowers short pedicellate. Calyx 10–12 mm in flower, 11–14 mm in fruit, glandular papillose, clothed with spreading eglandular hairs, tube narrowly funnel-shaped, 5–7 mm, teeth lanceolate-subulate, acute, about as long as tube or somewhat shorter, 4–7 mm. Corolla yellow with purple markings at throat, 12–14 mm, tube scarcely exserted. Nutlets c. 1.5 mm, oblong, trigonous, compressed, brownish-black, minutely punctulate, glabrous.

HAB. Rocks, in crevices, in gorges; alt. 600–1500 m; fl. & fr. Jun.-Aug.

DISTRIB. Occasional in N. & N.E. sector of lower forest zone of Iraq. Khantur mountain N.E. of Zakho, *Rawi*, Tikreti & Nuri 29011 (BAG). **MRO**: between mt. Karoukh and Dargala, *Nuri & Kass* 27764!; Sefin Dagh above Shaqlawa, *Bornmüller* 1672!, 1674!, *Gillett* 8079!, *Rawi* 9082! ?Kaina nr Shar, *Rawi* 8602!; Rawandoz, *S. Omar*, Sahira, F. Karim & *H. Hamid* 38364 (BAG); 27 km from Koi-Sanjak to Erbil, Wadi Shawrash, *S. Omar, Al-Kaisi* & Wedad 49526 (BAG). **MRO/FAR**: Salahaddin forestry, *Gillett* 11277!

Endemic.

12. **Stachys iberica** *M.Bieb.*, Fl. Taur.-Caucas. 2: 51 (1808) subsp. **georgica** *Rech.f.*, Repert. Spec. Nov. Regni Veg. 53: 84 (1944); Rechinger, Fl. Iranica [K. H. Rechinger] 150: 376 (1982); Salmaki & al., Bot. J. Linn. Soc. 170(4): 592 (2012).

Perennial, 30–50 cm tall, green, hirsute. Stems ascendent or erect-ascendent, virgate, simple or branched from base, clothed with long spreading or floccose hairs. Basal and lowermost cauline leaves withered at flowering time; lower and intermediate cauline leaves with short petioles 4–8 mm; lamina narrowly oblong, 3–4 × 1–1.5 cm, gradually tapering towards base, clothed with few rather long hairs; uppermost and floral leaves linear-lanceolate, 2–4 × 4–6 mm, base rather broad, apex acute or subacute, muticous. Verticillasters 6–10-flowered, lower ones distant, internodes 2–2.5 cm, upper verticillasters approximated, confluent. Bracts ± absent. Calyx 6–8 mm, hirsute, tube campanulate, teeth ± acuminate from a triangular-lanceolate base, tapering into a long weak ± spiny mucro c. ½ as long as tube or somewhat longer. Corolla purple, 14–18 mm, tube 8–10 mm, exserted, hirsute. Nutlets c. 1.2 mm, trigonous, compressed, brownish-black, nearly smooth, subglabrous.

HAB. Mountain slopes, in *Quercus* forests; alt. 1200–2000(–2600) m; fl. & fr. Jun.-Aug.
DISTRIB. Occasional in the lower forest zone of Iraq. **MAM**: gorges above Masis, *Rechinger* 10880!; mt. Helgord, gorges above Nowanda, *Rechinger* 11393!: Penjwin, *Rechinger* 10477!

Turkey, Azerbaijan, Armenia, Georgia, W. Iran.

13. **Stachys inflata** *Benth.*, Lab. Gen. Spec. 562 (1834); Boissier, Fl. Orient. 4: 739 (1879); Hand.-Mazz. in Ann. Natur. Mus. Wien 27: 414 (1913); Zohary in Dep. Agr. Iraq Bull. 31: 128 (1950); Rawi in Dep. Agr. Iraq Tech. Bull. 14: 157 (1964); Rechinger, Fl. Iranica [K. H. Rechinger] 150: 391 (1982); Bhatacharjee in Fl. Turkey [P. H. Davis] 7: 259 (1982); Salmaki & al. in Bot. J. Linn. Soc. 170(4): 593 (2012).

Suffrutescent, with very stout woody rootstock, many-stemmed, stems 15–25 cm tall, densely appressed somewhat greyish white-lanate to white-tomentose. Stem and branches erect or erect-ascendant, somewhat flexuous, simple or sparsely branched at base. Basal and lowermost cauline leaves minute, subsessile, narrowly oblong to narrowly ovate, 12–20 × 4–7 mm, apex obtuse, rarely subacute; intermediate cauline leaves gradually larger, oblong-elliptic, 20–30 × 10–12 mm, obtuse to subacute, upper oblong-linear to linear; floral leaves much longer than calyx, leaves all densely appressed somewhat greyish white-tomentose, prominently veined below, margins slightly revolute. Verticillasters 6–8-flowered, lower distant, upper much approximated, confluent. Bracts linear-filiform, 5–6 mm. Calyx 12–15 mm, densely appressed canescent white-tomentose to pannose, tube cylindrical, teeth very short, ovate, subacute, ¼–⅕ as long as tube; fruiting calyx much inflated. Corolla 15–18 mm, purplish, silky outside, tube included. Nutlets c. 1.5 mm, oblong, trigonous, compressed, brownish-black, minutely punctulate, glabrous.

HAB. Mountain slopes, limestone rocks, denuded *Quercus* forest, gypsum valley; alt. 600–1000 m; fl. & fr. Jun.-Aug.
DISTRIB. Local in lower forest zone of Jabal Sinjar in N.W. Iraq. **MJS**: above Kursi, *Gillett* 10900!; *S. Omar*, Al-*Khayat & Al-Kaisi*, 52496 (BAG); Jabal Sinjar, *Haussknecht* s.n.!; Jabal Sinjar W. Slope, *Al-Kaisi & K. Hamad* 53738 (BAG); Wadi Shilu and above Bara, *Handel-Mazzetti* 1560!

N.E. Turkey, Armenia, Azerbaijan, Georgia, Iran.

14. **Stachys kotschyi** *Boiss. & Hohen.*, Diagn. Pl. Orient. [Boissier] ser. 1, 5: 32 (1844). Type: Iraq, in rupestribus Mts Gara Kurdistan, versus Cacumen, *Kotschy* 367 (G-BOISS, holo; G, M, W!, iso.); Boissier, Fl. Orient. 4: 741 (1879); Nábělek, Spisy Přír. Fak. Masarykovy Univ. 70: 66 (1926); Blakelock in Kew Bull. 1949: 549 (1950); Zohary in Dep. Agr. Iraq Bull. 31: 128 (1950); Rawi in Dep. Agr. Iraq Tech. Bull. 14: 157 (1964); Rechinger, Fl. Iranica [K. H. Rechinger] 150: 394 (1982); Salmaki & al. in Bot. J. Linn. Soc. 170(4): 594 (2012).

S. haussknechtii Vatke, Bot. Zeti. 33: 46 (1875).
S. babylonica Hamodie & Wilcock, Al-Mustansiriyah J. Sci 29(4): (2018).

Suffrutescent, with stout woody rootstock, stems numerous, 15–30 cm tall. Stems and branches ascendant, ± flexuous, much-branched at base, densely appressed greyish-white lanate-tomentose. Basal and lowermost cauline leaves withered at flowering time; intermediate cauline leaves with 3–6 mm long petioles; lamina ovate to ovate-oblong, 1.5–3 × 1.2–2.2 cm, base rounded or truncate, rarely cuneate, apex rounded or obtuse, entire, densely greyish-white lanate-tomentose and prominently veined below, greenish-canescent silky-tomentellous with impressed veins above; cauline leaves usually increasing upwards, to

Fig. 131. **Stachys kotschyi**. 1, habit; 2, flower; 3, calyx (opened); 4, corolla. Illustration from Bot. Lin. 170: Fig. 6, p. 595 (2012). © 2012 The Linnean Society of London, Botanical Journal Linnean Society.

4 × 2.5 cm; floral leaves somewhat smaller, others rather minute, oblong, obtuse or rounded, rarely subacute, ± equaling fruiting calyx, 1–1.2(–1.5) cm. Verticillasters 4–6-flowered, in short, lax, 4–6 cm racemes, lower verticillasters somewhat distant, upper confluent. Bracts filiform, very short or absent. Pedicels 1–2 mm. Calyx 8–10 mm, densely short-tomentose, tube campanulate, teeth lanceolate, acute, subspinescent, divergent, c. ½ as long as tube,

3–4 mm. Corolla 10–12 mm, pink, silky, tube included. Nutlets c. 1.5 mm, oblong, trigonous, compressed, dark brown, minutely punctulate, glabrous. Fig. 131, 1–4.

HAB. Rocky ground, limestone cliffs, *Quercus* forest; alt. 800–1800 m; fl. & fr. Jun.-Aug.
DISTRIB. Occasional in lower forest zone of Iraq. **MAM**: Matina, Zakho to Amadiya, *Rawi* 8723!; Sulaf gorge, *Guest* 3772!; Gara Dagh, *Kotschy* 367!; Mazurka village, nr Amadiya, *Nábělek* 1532!; Sharifa, nr Amadiya, *Polunin* 5130!; 3 km W. of Amadiya, *Rechinger* 11628!; Zawita gorge, Chapman 26104 (BAG); Yosifka village, 18 k N.W. of Sarsang, *Al-Dabbagh & H. Hamid* 45785 (BAG); Ser Amadiya kani Siji, *Al-Dabbagh, Al-Kaisi & K.* Hamad 53789 (BAG); **MSU**: Pira Magrun, *Rawi* 12125!; *Haussknecht* s.n.! Amoret, Qara Dagh area, *Haines* 1129!; Qara Dagh, *Poore* 587! **FNI**: Aqra, *Rawi* 11460!

Rawi 12125 was invalidly published as a new species, *S. babylonica* by Hamoudie & Wilcock based on a single collection. I have included it under *S. kotschyi*, as the material falls within the variation of the vegetative characters of *S. kotschyi*. With more material it may prove to be a distinct species.

W. Iran.

15. **Stachys lavandulifolia** *Vahl*, Symb. Bot. (Vahl) 1: 42 (1790); Boissier, Fl. Orient. 4: 743 (1879) Blakelock in Kew Bull. 1949: 550 (1950); Zohary in Dep. Agr. Iraq Bull. 31: 128 (1950); Rawi in Dep. Agr. Iraq Tech. Bull. 14: 157 (1964); Rechinger, Fl. Iranica [K. H. Rechinger] 150: 386 (1982); Bhatacharjee in Fl. Turkey [P. H. Davis] 7: 239 (1982).

Suffrutescent, with rather stout woody rootstock, stems numerous, 15–25 cm tall. Stems much-branched at base, with numerous short stems and branches, somewhat greenish to white-canescent, ± densely long-villous, clothed with short silky and often villous hairs. Short stems and branches ascendant, simple, some sterile, very short, close-clustered, leafy, some floriferous. Leaves conspicuously parallel-grooved; basal and lowermost cauline leaves withered at flowering time; lower and intermediate cauline leaves gradually narrowed to 2–6 mm long petioles; lamina oblong to linear-lanceolate, 30–40 × 8–12 mm, apex obtuse or subacute; upper and floral leaves gradually smaller, sessile, ovate-elliptic, acute, deflexed, somewhat longer than calyx tube. Verticillasters 4–6-flowered, approximated in rather oblong, 3–6 cm racemes, rarely lower verticillasters somewhat distant. Bracts absent or nearly so. Flowers subsessile. Calyx 15–18 mm, densely long white-villous, clothed with ± spreading, to 4(–5) mm long hairs; tube very short, campanulate, teeth plumose, lanceolate-subulate from a narrow lanceolate base, spreading, c. ⅝ times longer than tube, 12–15 mm. Corolla 12–15 mm, purple, hairy, shorter than calyx. Nutlets c. 1.5 mm, oblong, trigonous, compressed, brownish-black, minutely punctulate, glabrous.

HAB. Rocky ground, mountain slopes, stream sides, in *Quercus* and *Rhus* forest; alt. 1200–3000(–3200) m; fl. & fr. Jun.-Jul.
DISTRIB. Common in the forest and thorn-cushion zones of Iraq. **MAM**: Zawita pass N. of Zakho, *Rechinger* 10941!; Sarsang, Dohuk to Amadiya, *Haines* W.1355!; Qara Dagh, *Kotschy* s.n.! **MRO**: Qandil mts., *Rawi & Sarhang* 24075!; Sewok village, Jabal Karoukh, *Kass, Nuri & Serhang* 27584!; between Saran and Kilkil, Jabal Karoukh, *Kass & Nuri* 27343!; Chia-i Mandali, *Guest* 2713!; Dargala, c. 335 km N.W. by N. of Rania, *Rawi, Nuri & Kass* 28891!; Helgord Dagh, *Guest* 2872!; above Shaqlawa, Sefin Dagh, *Gillett* 8108!, *Bornmüller* 1665!; N. of Helgord range nr Bermasand lake, *Rawi & Serhang* 24786!; Helgord range, E. side, *Rawi & Serhang* 24750! **MSU**: Gweija Dagh, *Gillett & Rawi* 11678!, *Mooney* 4332!; Malakawa, *Rawi* 22457!, *Rechinger* 10431!; Jabal Kajan nr Penjwin, *Rawi* 22674!; Azmir Dagh, N. of Sulaimaniya, *Poore* 336!; Tawila, Hauraman Dagh, *Rechinger* 10311!, *Haussknecht* s.n.!

Turkey, Transcaucasia, Iran, Turkmenistan.

16. **Stachys annua** (*L.*) *L.*, Sp. Pl. ed. 2, 813 (1763); Smith, Fl. Graec. Prodr. 1(2): 410 (1809); Boissier, Fl. Orient. 4: 745 (1879); Bhattacharjee in Notes Roy. Bot. Gard. Edinburgh 33: 288 (1974); Rechinger, Fl. Iranica [K. H. Rechinger] 150: 378 (1982); Bhatacharjee in Fl. Turkey [P. H. Davis] 7: 248 (1982). Salmaki & al. in Bot. J. Linn. Soc. 170(4): 586 (2012).

Betonica annua L., Sp. Pl. 573 (1753).
Stachys betonica Crantz, Stirp. Austr. ed. 2(4): 264 (1769).
S. pubescens Ten., Fl. Nap. 1, Prodr. 34 (1811); Blakelock in Kew Bull. 1949: 550 (1950); Rawi in Dep. Agr. Iraq Tech. Bull. 14: 157 (1964); Rechinger, Fl. Iranica 150: 377 (1982).

Annual, 15–80 cm tall, rather glabrous below, minutely pubescent above. Stems erect or erect-ascendant, simple or branching from base; stems and branches slender, virgate, glabrous below, minutely pubescent above, ± glandular. Basal and lowermost cauline leaves withered at flowering time; lower and intermediate cauline leaves with 5–10 mm petioles; lamina oblong to oblong-spatulate to oblong-ovate, 20–40 × 15–20 mm; upper cauline leaves gradually smaller, oblong-lanceolate to oblong-linear, major ones up to 30–40 × 5–6 mm, subacute, crenulate or coarsely crenate-dentate, as long as or slightly longer than verticillasters; leaves sparsely pubescent to glabrescent, upper minutely ciliate or sometimes long-ciliate at margins. Verticillasters c. 6-flowered, lower distant, in spikes or spike-like racemes, upper approximated to confluent. Bracts setaceous, 1–2 mm or absent. Flowers subsessile. Calyx 6–12 mm, erect, tube campanulate, prominently 10-veined; teeth lanceolate, 2.5–3 mm, tapering, somewhat spiny, rigid, upcurved, shorter than tube, densely villous to sometimes glabrescent, ± glandular hairy. Corolla 14–15 mm, yellowish or whitish yellow, somewhat hairy, tube exserted. Nutlets c. 1.5 mm, oblong, trigonous, compressed, brownish-black, minutely punctulate to smooth, glabrous.

HAB. Mountain slopes, in crevices and gorges, mountain caves, by streams, in cornfields on muddy soil; alt. 900–2200 m; fl. & fr. Jun.-Aug.
DISTRIB. Occasional in the forest and lower forest zone of N. Iraq. MAM: Sharanish, c. 25 km N.E. of Zakho, *Rawi* 23236! MRO: Pushtashan, Qandil mts., *Rechinger* 11036!; Dargala village, Jabal Karoukh, *Kass & Nuri* 27688, 27716! !; between Kujar and Kani GRawi region E. of Jabal Karoukh, *Kass & Nuri* 27632!; Helgord Dagh, *Guest* 2817!; Walza, Chia-i Mandali, *Guest* 2739! Gali Warta, *Rawi, Nuri & Kass* 28808! FAR/MRO Merga Dreija nr Haji Umran, *Rawi* 9137! MSU: Penjwin valley, *Rawi* 12154!

A widespread species with several variable characters especially pubescence of leaves. Bhatacharjee (l.c. 1982) recognizes annual plants with glabrescent, glandular stems as subsp. *annua*, under which most of our material falls.

Europe to Siberia and Iran.

17. **Stachys melampyroides** *Hand.-Mazz.* in Ann. Nat. Hofmus. Wien 27: 415 (1913). Type: Iraq, Kurdistan, Riwandous, in mt. Sakri-Sarkhan, 1700 m, *Bornmüller* 1663 (B! lecto. designated by Salmaki & al. in Bot.J. Linn. Soc. 170(4): 600 (2012); K, W!, WU!, isolect.); Nábělek, Spisy Přír. Fak. Masarykovy Univ. 70: 66 (1926); Zohary in Dep. Agr. Iraq Bull. 31: 128 (1950); Rawi in Dep. Agr. Iraq Tech. Bull. 14: 157 (1964); Rechinger, Fl. Iranica [K. H. Rechinger] 150: 381 (1982); Bhatacharjee in Fl. Turkey [P. H. Davis] 7: 256 (1982); Salmaki & al. in Bot. J. Linn. Soc. 170(4): 596 (2012).

Annual, slender, 15–20 cm tall. Stem ± simple, erect or erect-ascendant, somewhat branched at base, rarely several stems from base. Stem and branches slender, somewhat flexuous, rubescent, minutely and sparsely glandular-pubescent, sometimes glabrescent, branches usually arcuate-ascendant. Basal and lowermost cauline leaves usually withered at flowering time; lower and intermediate cauline leaves sessile, linear to narrowly linear-spatulate, obtuse or subacute, to 15(–20) × 1–2 mm, subentire or remotely crenulate, glabrous; upper cauline leaves gradually smaller, acute or subacute, floral ones ovate, acute, short crisp glandular-pubescent and long-ciliate at margin. Verticillasters 4–6-flowered, lower distant, upper approximated to confluent. Bracts absent. Flowers on very short, thick (c. 1 mm) pedicels, upper subsessile. Calyx c. 8 mm, glandular-pubescent, erect, tube campanulate, conspicuously 10-veined, teeth lanceolate-subulate, c. 4 mm, somewhat spiny, rigid, awn-like and recurved in fruit, about as long as tube. Corolla pink, c. 10 mm, tube exserted. Nutlets scarcely 1 mm, oblong, trigonous, compressed, brownish-black, minutely tuberculate-punctate, glabrous. Fig. 132, 1–4.

HAB. Calcareous rocks, mountain slopes, *Quercus* forest; alt. 700–1800 m; fl. & fr. May-Jul.
DISTRIB. Common in the N. & N.E. sectors of the lower forest zone of Iraq. MAM: between Dagh-al-Radziem and Sharanish, *Rechinger* 12117!; MAM, Bakirman nr Zakho, *Rawi* 8512!; mt. Bekhair, *Rawi* 23028!, *Field & Lazar* 774!; c. 8 km S. of Zakho, *Rechinger* 10698!; c. 85 km N.W. by N. from Mosul to Zakho, *Rawi* 23135! MRO: Saran village, Jabal Karoukh, *Nuri & Kass* 27273!; Dergala village, Jabal Karoukh, *Kass & Nuri* 27711!; N. of Haibat Sultan Dagh, *Rawi, Nuri & Kass* 28334!; Jabal Baradost nr Diana, *Field & Lazar* 918!; Rowanduz gorge, *Field & Lazar* 854!; Jabal Sakri Sakran, *Bornmüller* 1663!; Zab Ala, *Haussknecht* s.n.! FNI: Aqra, *Rawi* 11479!; Dohuk, *Haines* W. 1228!, W. 1433!;

Turkey (S.E. Anatolia), Iran.

18. **Stachys burgsdorffioides** (*Benth.*) *Boiss.*, Diagn. Pl. Orient. [Boissier] ser. 1, 12: 85 (1853); Nábělek, Spisy Přír. Fak. Masarykovy Univ. 70: 66 (1926); Blakelock in Kew Bull.

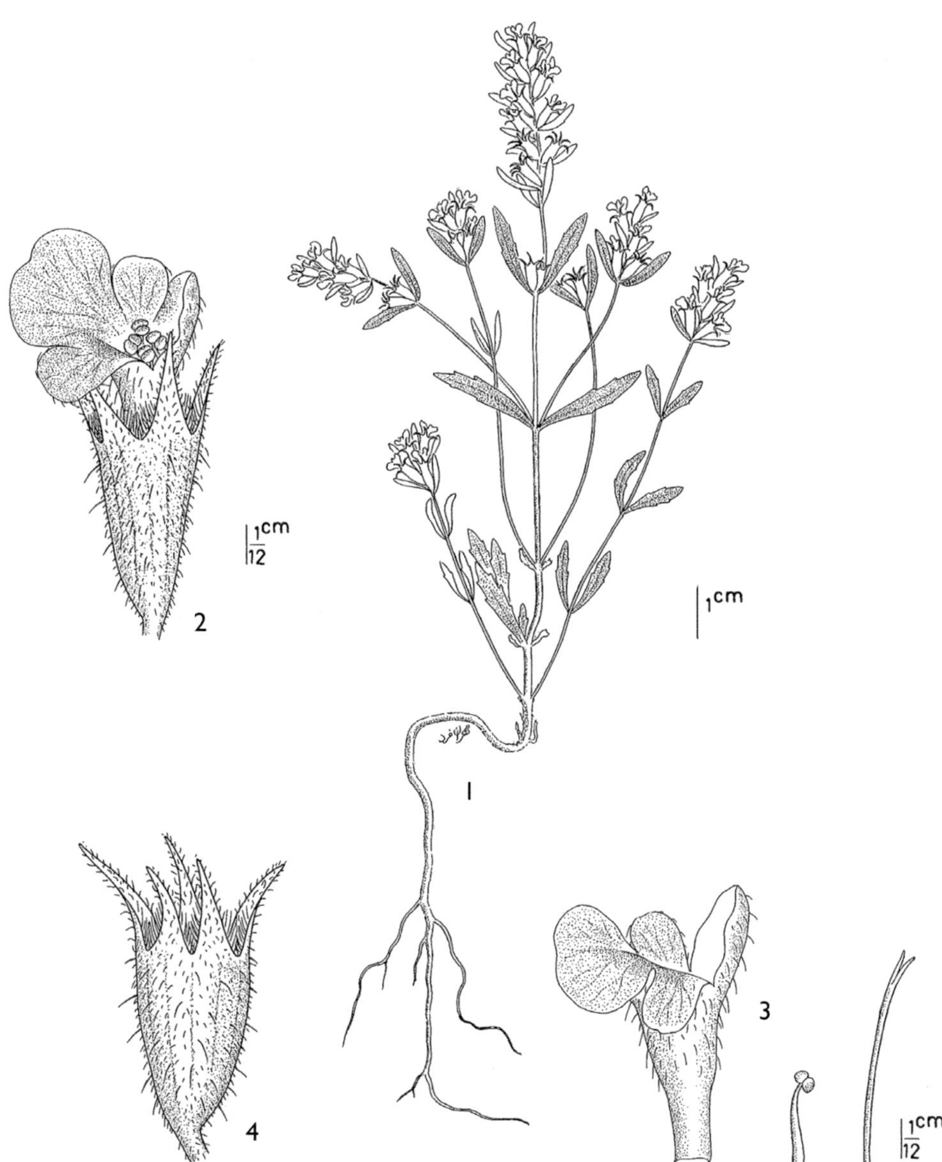

Fig. 132. **Stachys melampyroides**. 1, habit; 2, flower; 3, calyx opened; 4, corolla. Illustration from Bot. Lin. 170: Fig. 7, p. 601 (2012). © 2012 The Linnean Society of London, Botanical Journal Linnean Society.

1949: 549 (1950); Zohary in Dep. Agr. Iraq Bull. 31: 128 (1950); Rawi in Dep. Agr. Iraq Tech. Bull. 14: 157 (1964); Rechinger, Fl. Iranica [K. H. Rechinger] 150: 380 (1982); Bhatacharjee in Fl. Turkey [P. H. Davis] 7: 256 (1982).

S. burgsdorffioides (Benth.) Boiss. subsp. *ladanoides* Hand.-Mazz in Ann. Naturh. Mus. Wien 27: 417 (1913).

Annual, slender, 10–50 cm tall. Stems erect or erect-ascending, ± simple or branching from base; stems and branches slender, ± virgate, ± rigid, often rubescent, minutely and sparsely glandular-pubescent. Basal and lowermost cauline leaves withered at flowering time; lower and intermediate cauline leaves short-stalked or subsessile, oblong-spatulate to oblong-linear, 20–35 × 3–4 mm, obtuse or subacute, subentire or obtusely coarsely denticulate;

upper cauline leaves gradually smaller, oblong-lanceolate to oblong-linear, 15–25 × 2–3 mm, obtuse or subacute, apiculate; floral leaves linear-lanceolate, overtopping flowers, leaves all ± glandular-pubescent and long-ciliate at margin. Verticillasters c. 4-flowered, distant, in long spikes or spike-like racemes, uppermost approximated and confluent. Bracts absent or nearly so. Flowers subsessile or sessile. Calyx 8–10 mm, glandular-hispid, erect, tube funnel-shaped, prominently 10-veined, teeth lanceolate-subulate, 2.5–3.5(–4) mm, somewhat spiny, rigid, awn-like and erect-spreading in fruit, ⅓-½ as long as tube. Corolla purple, 12–14 mm, tube included. Nutlets 1.0–1.2 mm, oblong, trigonous, compressed, brownish-black, ± smooth, shining.

HAB. Mountain slopes, in ravines, hillside, clay rocky mountain side; alt. 320–1200 m; fl. & fr. Jun.-Jul. DISTRIB. Rare and localised. **MAM**: Ain Nuni village between Amadiya and Ashuti (Turkey), *Nábělek* 1512! **MJS:** Jabal Sinjar, *Al-Kaisi, Al-Khayat & F. Karim* 50882 (BAG); Jabal Sinjar S.W. slope, *Al-Kaisi, Al-Khayat & F. Karim* 50985 (BAG); Jabal Sinjar N. slope, *Al-Kaisi, Al-Khayat & F. Karim* 50919 (BAG). **MAM/FNI**: Mahad nr Shekhan and Punkah Dim Hok (?), *Salim Effendi* 2611!; Mar Ya'qub village nr Simel, *Nábělek* 1504!; **FNI**: 30 km from Mosul to Aqra, *Al-Kaisi* 49723 (BAG).

Turkey (S.E. Anatolia), Syria.

19. **Stachys bhattacharjeeae** *Ö.Güner*, Feddes Repert. 133(4): 258 (2022). Type: Iraq, Kaiwa rush, near Rania, rocky mountain, 670 m, 27.07.1957, *Rawi* (K, holo.).

Perennial; stems 75 cm, erect, branched, glabrous, sparsely glandular hairy at below. Cauline leaves sessile, ovate to ovate-lanceolate, upper ovate, 40–50 × 16–28 mm, gradually becoming smaller above, apex acute, base subcordate, margins dentate to crenate, faintly serrate to entire throughout above. Floral leaves similar to cauline leaves but smaller. Verticillasters 4–6-flowered, the lower remote, the upper gradually approximate. Bracteoles few, setaceous, 0.5–1 mm. Pedicels 0.5 mm long. Calyx tubular to sub-campanulate, 5–6 mm, with dense stalked and sessile glands, teeth equal. Corolla white, 11–13 mm, tube markedly exceeding the calyx; limb bilabiate, upper lip 2 mm, the lower lip 3-lobed, middle lobe larger than 2 lateral lobes, 4 mm long. Style not exceeding the upper lip, apex equally bifid into subulate stigmas. Stamens 4, within corolla tube. Nutlets oblong-obovoid, 1.5–2 × 0.5 mm, slightly winged near base, reticulate on surface, brown.

HAB. Foot of rocky mountain of NE Iraq, known from a single location; alt. ± 600m; fl.& fr. May-Jun. DISTRIB. N.E. Iraq. **MRO**: Kaiwa rush, near Rania, *Rawi & Serhang* 23774!

Endemic.

20. **Stachys multicaulis** Benth., Prodr. [A. P. de Candolle] 12: 486 (1848); Salmaki et al. (2012); Aun & Haloob, l.c. 1132 (2020).

Perennial, cushion-shaped with erect to ascending stems, 24–30 cm, stems woody below, branched, sparsely pubescent. Cauline leaves elliptic, 17–23 × 2–5 mm, sessile, apex acute, base attenuate, margins entire, sparsely covered with appressed and sessile glandular hairs. Basal leaves elliptic, 6–11 × 2–4 ; caluline leaves oblong. Verticillasters 2-flowered, remote. Pedicels c. 1 mm. Calyx campanulate-tubular, 10–11 mm; calyx tube c. 6 mm, sparsely covered with short hairs, calyx teeth c. 5 mm, triangular, erect, acute, spinescent, covered with sessile glandular hairs on margins. Corolla yellow, 20–22 mm, corolla-tube shorter or as long as calyx tube, upper lip 10–11 mm, lower lip 9–11 mm. [Nutlets not seen].

HAB: On clay and limestone; alt. ± 770 m; fl. Jun., Jul. DISTRIB. Foothills of mountains, localized and seen at one location. **FPF**: E. of Baghdad, nr. Gliejall, below Derbendikhan lake, coll. 2015, *Aun* s.n (BAG).

Known from a single collection. The flowers in Iraq material are larger than those in the Iranian material.

Iran (12-18 mm as recorded by Salmaki et al. l.c.).

21. **Stachys zakhoensis** (Meikle & Patzak) Ghaz.

Type: Iraq, Zawita pass N. of Zakho, Rechinger 10936 (W, holo.).

S. zakhoensis Meikle & Patzak in sched.

Perennial 80–100 cm tall. Stems erect, quadrangular, branched from base, densely white-lanate, clothed with spreading, long hairs and sessile papillae and glands. Basal leaves often withered at flowering time, lowermost cauline leaves long-petiolate; petiole 5–10 cm; lamina ovate-oblong, 8–12 × 6–8 mm, membranous, crenate-dentate, rounded or cordate at base, conspicuously net-veined, ± densely white felted on both sides; intermediate cauline leaves gradually reducing in petiole length, lamina gradually smaller, ovate-oblong to oblong, narrowed or rounded at base, acuminate and subacute at apex; uppermost and floral leaves sessile or with short petioles, oblong-lanceolate to lanceolate, 3–4 × 12–18 mm, base somewhat caudate, apex acute; lower floral leaves with serrate-crenate margins, the others with entire margins. Verticillasters 10–12-flowered, the lower ones distantly placed with internodes 4–5 cm, gradually shorting, uppermost Verticillasters approximated. Bracts filiform, 10–15 mm; bracts and calyx white-lanate, clothed with long spreading eglandular hairs and short glandular hairs. Calyx 10–12 mm, tube campanulate, teeth lanceolate-ovate, acuminate, short-spiny at tip, c. ½ as long as tube. Corolla reddish-brown, 15–16 mm, tube somewhat exserted, upper lip entire. Nutlets c. 1.5 mm, obovoid, trigonous, compressed, smooth, glabrous.

S. zakhoensis differs from *S. alpina* L., its closest ally within subsect. Germanicae Boiss., by its smaller corolla and purely membranous leaves as well as its rather elongate narrowly filiform bracts, 1–1.5 cm. From *S. pinetorum* Boiss. et Bal., another closely allied species from the Cilician Taurus and southern Turkey, it differs by its dense felty-tomentose stems, and leaves white-silky hairy below.

S. pinetorum and/or *S. alpina* are not recorded from Iraq.

HAB. *Quercus* forests, rocky slopes; alt. 400–3500 m; fl. & fr. Jun.-Jul.
DISTRIB. Occasional in the forest zone of the N. and N.E. sectors of Iraq. **MAM**: Zawita pass N. of Zakho, *Rechinger* 10936! (type). **MAM**: Jabal Khantur N. of Zakho, *Rechinger* 10813! **MRO**: E. of Qala Diza, mt. Qandil, *Thesiger* 1164.

The description included here was made on an earlier draft (1980s) of *Stachys* possibly by Patzak who worked at the Natural History Museum in Vienna.

Endemic.

SPECIES DOUBTFULLY RECORDED

Stachys brantii *Benth.*, Prodr. [A. P. de Candolle] 12: 463 (1848).

Imperfectly known from an inadequate specimen of uncertain locality. Perennial, woody at base, sericeous-tomentellous, clothed with soft and appressed long eglandular and short glandular hairs. Stem simple, rather thick, erect-ascendent to pendulous from clefts of rocks, bearing a dense, ovate head surrounded by leaves. Leaves short-petiolate, lamina ovate, base cordate, apex subacute, ± serrate-dentate, teeth rather deep and subacute; floral leaves acuminate, overtopping calyx. Bracts linear, as long as calyx. Calyx tomentellous, teeth lanceolate-subulate, ± recurved, 4 × shorter than tube. Corolla sulphur-yellow, hairy. Nutlets unknown.

Recorded only once from "Kurdistan", *Brant*. Much of the material collected by Captain James W. Brant, who was British Consul in Erzurum (Turkey), is believed to have been collected from E. Anatolia; the record is therefore doubtful for Iraq.

<h1 align="center">9. SIDERITIS L.</h1>

Sp. Pl. 2: 574 (1753); Harley & al. in Kubitzki (ser. ed.), Fam. Gen. Vasc. Pl. 7: 224 (2004)

Ali Haloob

Perennial or annual herbs or small shrubs, usually aromatic. Leaves petiolate, entire or obsoletely crenate. Verticillasters 4–6-flowered, distant or ± crowded, extending in upper half or almost the whole length of stem. Bracts leaf-like or different; bracteoles absent. Calyx tubular-campanulate, 5–10-veined, equal to subequally 5-toothed, zygomorphic to almost regular. Corolla bilabiate, yellow; tube included within calyx; upper lip flat, straight, entire or 2-lobed, lower lip 3-lobed, middle lobe the largest. Stamens 4, didynamous, included in corolla tube; anterior pair longer than posterior pair; the anterior anthers often reduce or deformed. Style included in the tube of corolla, apex with 2-unequal lobes. Nutlets glabrous, ovoid or oblong ovoid, apex rounded.

About 140 species from Macaronesia, Mediterranean region to Russia, Tibet and W. China; 2 species in Iraq.

Dülgeroğlu, C. (2017). A preliminary intra phylogeny of the genus *Sideritis* by morphology. Int. J. Agric. Environ. Res., 3: 3901–3909.

Pérez de Paz, P.L. & Negrin Sosa, L. (1992). Revisión Taxonómica de Sideritis L. Subgénero Marrubiastrum-(Moench) Mend.-Heuer (Endemismo Macaronésico).

Small shrubs, perennial; calyx teeth acute to acuminate; corolla longer
   than calyx . . . . .1. *S. libanotica*
Annual herbs; calyx teeth mucronate; corolla shorter than calyx . . . . 2. *S. montana*

1. Sideritis libanotica *Labill.*, Ic. Pl. Syr. 4: 13 (1812); Boissier, Fl. Orient. 4: 718 (1879); Handel-Mazzetti in Ann. Naturh. Mus. Wien 27: 411 (1913); Dinsmore in Post, Fl. Syria, Palest. & Sinai ed. 2, 2: 643 (1933); Rawi in Dep. Agr. Iraq Tech. Bull. 14: 156 (1964); Ball in Fl. Europ. 3: 162 (1972); Huber-Morath in Fl. Turkey [P. H. Davis] 7: 196 (1982).

Small shrubs. Stem simple or branched above, erect, 40–110 cm tall, white tomentose with minute glandular hairs. Leaves lanceolate or linear, 15(–80)–110 × 4(–15)–18 mm, apex acute to mucronate, margins entire or serrulate to obsolete crenulate, abaxial surface with conspicuous reticulate veins, tomentose hairs with minute glandular hairs, lower two pairs leaves petiolate to 15–30 mm, middle petiolate to 1–5 mm. Bracts sessile, lanceolate-ovate or ovate-reniform, apex mucronate to mucronate-acuminate. Verticillaster 1–7 cm distant, 6-flowered. Calyx campanulate, 10-veined, 6–10 mm, teeth 2–4 mm acute to acuminate, throat hairy, tomentose with glandular hairs. Corolla yellow, 9–14 mm, throat hairy, tube tomentose outside with a ring of hairs inside, upper lip straight, erect, 2-lobed, 4–5 mm, with a line of hair on top; lower lip 3-lobed, 4 mm, the middle one largest. Nutlet glabrous, 3–3.5 mm, ovoid-oblong, brownish.

Represented in Iraq by 3 subspecies:

1. Bracts lanceolate-ovate, 10–12 × 6–10 mm. . . . . .c. subsp. *microchlamys*
   Bracts broadly ovate-reniform 9–16 × 8–16 mm 2
2. Bracts apex with mucro 1–3 mm long . . . . .b. subsp. *kurdica*
   Bracts apex mucronate to acuminate, acumen 4–6 mm long . . . . . . a. subsp. *libanotica*

a. subsp. **libanotica**

Sideritis ambigua Fenzl in P.de Tch., As. Min., Bot. 2: 163 (1860).
Sideritis bourgaei Boiss., Fl. Orient. [Boissier] 4: 712 (1879).
Sideritis libanotica var. *incana* Boiss., Fl. Orient. [Boissier] 4: 712 (1879).
Sideritis microstegia Boiss. & Hausskn., Fl. Orient [Boissier] 4: 712 (1879).

Leaves 20–80 × 6–15 mm. Bracts broadly ovate-reniform, 9–16 × 8–16 mm, apex mucronate to acuminate, acumen 4–6 mm. Calyx 8–10 mm, teeth 3–4 mm.

HAB. Mountain side, on mountain slope, rocky clay mountain under oak trees, between cracks on mountain, in orchard of clay soil; alt. 800–1300 m; fl. Jun.-Jul.
DISTRIB. Occasional in the lower forest zone of N. of Iraq. MAM: Tirwanish 10 km E. Kani Masi, *Al-Kaisi, H. Hamid & K. Hamad* 45730! (BAG); Bamerny 20 km N.W. Sarsang, *Al-Kaisi & K. Hamad* 45920! (BAG); Spindar near Garadagh, *Rawi* 9248! (BAG); Sandur, *Al-Kaisi & Khalid* 46107! (BAG).

E. Mediterranean, S. Turkey.

b. subsp. **microchlamys** (*Hand.-Mazz.*) *Hub.-Mor.*, Bauhinia 6: 290 (1978); Huber-Morath in Fl. Turkey [P. H. Davis] 7: 197 (1982).

Sideritis libanotica var. *microchlamys* Hand.-Mazz. in Ann. Naturh. Mus. Wien 27: 411 (1913).
S.microchlamys (Hand.-Mazz.) Bornm., Magyar Bot. Lapok. 31: 140 (1932).

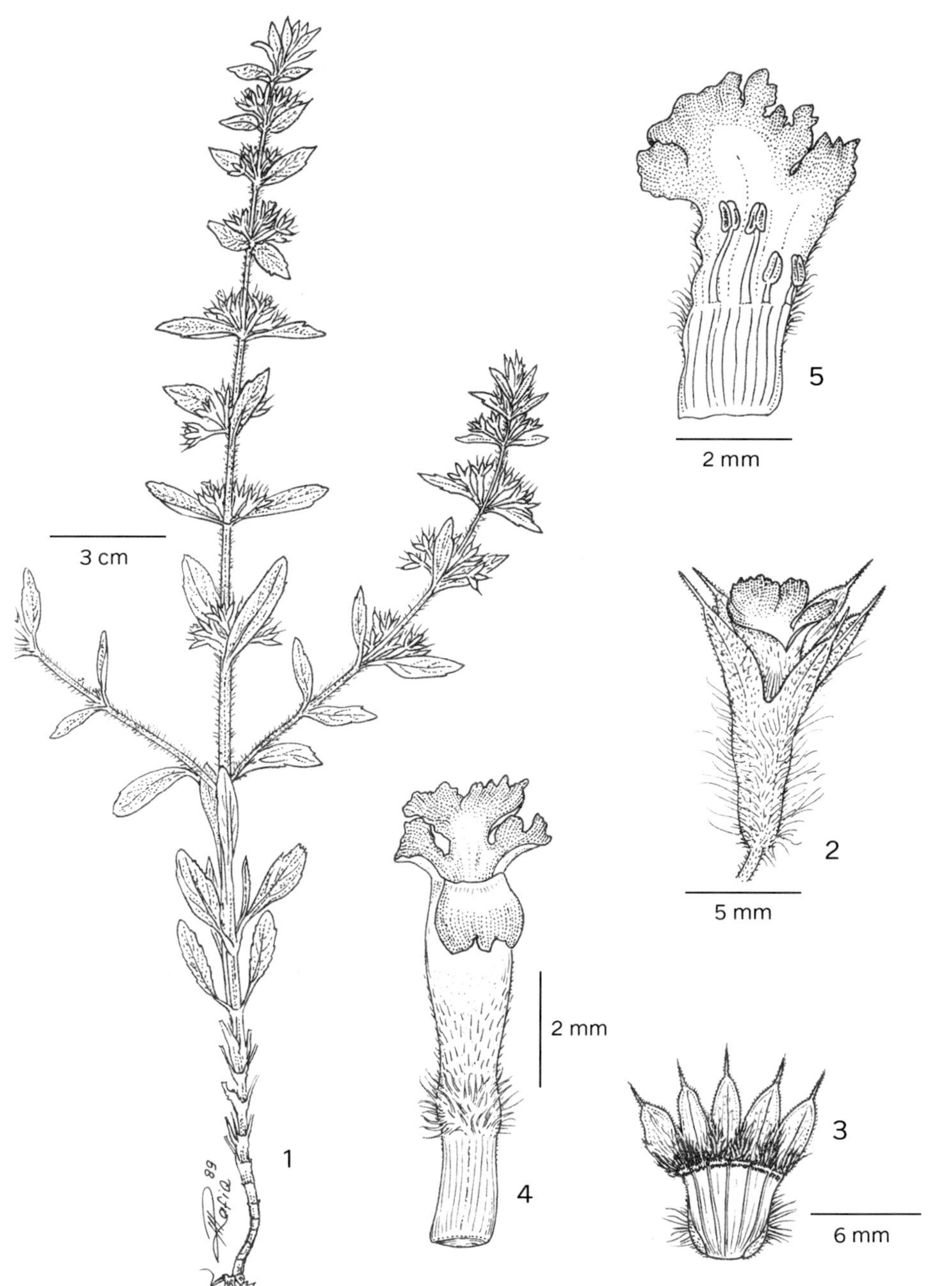

Fig. 133. **Sideritis montana**. 1, habit; 2, flower; 3, calyx, opened; 4, corolla; 5, corolla opened with stamens. Reproduced with permission from Flora of Pakistan 192: f. 7, A–D1 (1990). Drawn by M. Rafiq. © National Herbarium, Pakistan Agriculture Research Council & University of Karachi, Pakistan.

Leaves 15–70 × 4–15 mm. Bracts lanceolate-ovate, 10–12 × 6–10 mm, apex mucronate to acuminate, acumen 2–4 mm. Calyx 6–8 mm, teeth 2–3 mm.

HAB. Sand stone in valley, in mountain cave, on mountain slope; alt. 700–1800 m; fl. Jul. fr. Sep.
DISTRIB. Rare in the lower forest zone of N. sector of Iraq. MAM: Khantur mountain N.E. of Zakho,

Rawi, Tikriti & Nuri 29021! (BAG); Sharanish 25 N.E. of Zakho, *Rawi* 23211! (BAG); Bikhair mountain N. Zakho, *Rawi* 23095 (BAG).

S. Turkey, N.W. Iran.

c. subsp. **kurdica** (*Bornm.*) *Hub.-Mor.*, Bauhinia 6: 290 (1978); Huber-Morath in Fl. Turkey [P. H. Davis] 7: 197 (1982).

Sideritis kurdica Bornm., Magyar Bot. Lapok 31: 138 (1932); Zoh. in Dep. Agr. Iraq Bull., 31: 125 (1950); Blakelock in Kew Bull., 4: 549 (1950); Rechinger, Fl. Iranica 150: 105 (1982).

Leaves 20–110 × 4–18 mm. Bracts broadly ovate to reniform, 9–12 × 10–12 mm, apex with mucro 1–3 mm long. Calyx ±8 mm; teeth 2.5–3.5 mm.

HAB. Hillside and mountain slope between oak trees, rocky mountain slope; alt. 1000–1250 m; fl. Jul. fr. Sep.
DISTRIB. Frequently in the lower forest zone of N. sector of Iraq. **MAM**: Daimka village 35 km N.E. Zakho, *S.Omar & Al-Dabbagh* 45391! (BAG); Baidhaho village 15 km W. Kani Masi, *S. Omar* 45754! (BAG); Zawitah, *Guest* 4582! (BAG); Sarsang *Al-Dabbagh* 46164! (BAG); Sulaf, *S. Omar* 37707! (BAG); Shaikh Adi, *Guest* 3674A! (BAG).

S. Turkey, N.W. Iran.

2. **Sideritis montana** *L.* Sp. Pl. 2: 575 (1753); Boissier, Fl. Orient. 4: 706 (1879); Dinsmore in Post, Fl. Syria, Palest. & Sinai ed. 2, 2: 642 (1933); Zohary in Dep. Agr. Iraq Bull. 31: 125 (1950); Rawi in Dep. Agr. Iraq Tech. Bull. 14: 156 (1964); Ball in Fl. Europ. 3: 143 (1972); Huber-Morath in Fl. Turkey [P. H. Davis] 7: 183 (1982); Rechinger, Fl. Iranica [K. H. Rechinger] 150: 106 (1982); Hedge in Fl. Pakistan [Ali & Y. Nasir] 192: 57 (1990).

Annual herbs. Stem simple or branched below, erect, 15–30 cm long, villous with minute glandular hairs. Leaves spatulate-lanceolate or elliptic, 15–19 × 3–5 mm, apex acute, margins entire-obsolete serrulate, pilose with minute glandular hairs; petiole 2–4 mm. Bracts sessile to 3 mm, ovate-oblong, apex mucronate. Verticillasters ± crowded, extending almost the whole length of stem, 4–6-flowered. Pedicel ± 1 mm. Calyx campanulate, ± 9 mm, teeth ± equal, ovate-oblong, 4–4.5 mm, calyx throat hairy, puberulent and pilose with minute glandular hairs. Corolla yellow, brownish-black on tip of limbs, shorter than calyx, 6.5–5 mm; tube with a ring of hairs inside; upper lip straight, 0.7–1.5 mm, apex entire; lower lip 3-lobed, the middle lobe largest to 2 mm long. Nutlet glabrous, ovoid, 2 mm, brown. Fig. 133, 1–5.

HAB. Dry slopes (igneous or metamorphic rock), loamy mountain, roadside, rocky clay soil near *Juglans* orchard; alt. ±1600 m; fl. May-Aug.
DISTRIB. Occasional in the N. and N.E. sector of lower forest zone of Iraq. **MAM**: Bamerni-Tashish high way, *Al-Musawi & Addai* 44216! (BUH). **MRO**: 5–30 km S.E. Meirose, N.E. Shirwan Mazin, *Al-Musawi & Addai* 43712! (BUH); Between Lolan & Ari Gird Dagh, *Gillett* 12463! (BAG); bet Karookh mt and Dargala village, *Rawi, Nuri &Kass* 27716 (BAG); betweeb Choman & Haji Omran, *A. Haloob & Al-Kaisi* 60655 (BAG).

Southern Europe, N.W. Africa, Jordan, Syria, Caucasus, Turkey, Iran, Afghanistan, Pakistan, China, Russia,.

10. **LAGOCHILUS** Bunge ex Benth.

Labiat. Gen. Spec. 640 (1834)
Harley & al. in Kubitzki (ser. ed.), Fam. Gen. Vasc. Pl. 7: 225 (2004)
Yermoloffia Bél., Voy. Indes Or.: t. s.n. (1846); *Chlainanthus* Briq., H.G.A.Engler &
K.A.E.Prantl, Nat. Pflanzenfam. 4(3a): 257 (1896); *Lagochilopsis* Knorring, Novosti Sist.
Vyssh. Rast. 1966: 197 (1966).

Ferhat Celep & Tuncay Dirmenci

Shrubs, subshrubs or perennial herbs with woody rootstock, with or without spines in leaf-axils, glabrous or sparsely hirsute with simple hairs. Leaves entire or mostly palmatifid with spinose mucronate or rounded lobes. Inflorescence thyrsoid or racemoid, with up to

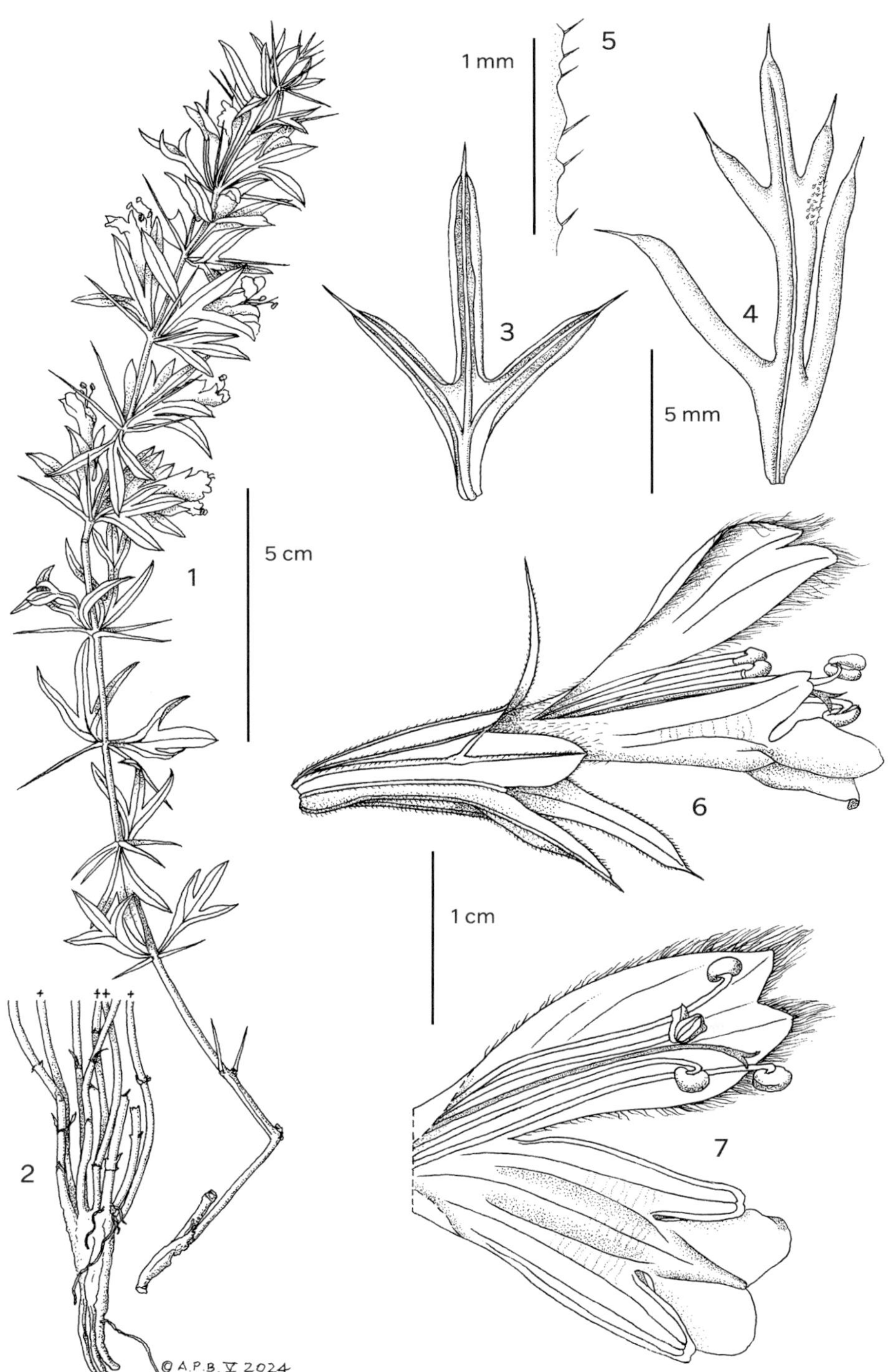

Fig. 134. **Lagochilus aucheri**. 1, habit flowering stem; 2, base of plants showing stems and root; 3, trilobed leaf from upper part of stem, abaxial surface; 4, five-lobed leaf from mid part of stem, adaxial surface; 5, profile of hairs on adaxial surface; 6, flower, side view; 7, corolla flattened. Drawn by © A.P. Brown, May 2024.

5-flowered cymes; bracteoles spinescent. Calyx slightly zygomorphic, (4–)5-lobed, lobes subequal or unequal (3/2), posterior often longer, lobes spreading or erect, oblong to broadly ovate, often spinose-mucronate, equalling or longer than tube. Corolla strongly 2-lipped, 5-lobed (2/3), white, yellowish-white, rarely pink, posterior lip long, straight, shallowly hooded, densely villous outside, 2-lobed, lobes sometimes 2-toothed, anterior lip with lateral lobes acute or emarginate, median lobe obcordate; tube densely pilose-annulate inside. Stamens 4, not or shortly exserted from corolla tube; anthers glabrous or hairy, thecae distinct. Stigma lobes equal. Nutlets trigonous, obovoid, apex truncate or rounded, glabrous, pilose or glandular.

About 40 species distributed from eastern Iraq, Iran to Mongolia, Russia (S. Siberia), N.W. China and N. Pakistan; a single species in Iraq.

Jamzad, J. (1988). The genus *Lagochilus* (Labiatae) in Iran. Iran. J. Bot. 4(1): 91–103.

1. **Lagochilus aucheri** *Boiss.*, Boiss., Diagn. Pl. Orient., ser. 1, 5: 38 (1844); Rechinger, Fl. Iranica [K. H. Rechinger] 150: 338 (1982); Jamzad in Flora Iran [Assadi & al., in Persian] 76: 337 (2012).

Lagochilus aucheri var. *elegans* Jamzad Iran J. Bot. 4: 96 (1988).
L. aucheri subsp. *heterophyllus* Jamzad Iran J. Bot. 4: 97 (1988).
L. aucheri var. *tomentosus* Jamzad Iran J. Bot. 4: 97 (1988).
L. lasiocalyx (Stapf) Jamzad Iran. J. Bot. 4: 97 (1988).

Suffruticose, caespitose perennial. Stems numerous, 6–30 (–40) cm tall, whitish, quadrangular, ascending to erect, simple or sparingly branched, glabrous below, puberulous with sparse sessile glands above. Lower leaves subpetiolate, 1.5–4 × 0.8–1.5 cm, 1–3-pinnatisect, scabrid or ± glabrous, lobes linear to linear-lanceolate, 1.2–2 mm wide, rounded or acute at apex, margins narrowly revolute; leaf axils 2-spined, spines 10–20 mm, thin but strong and sharp, patent. Calyx 20–25 (–30) mm, glabrous to pilose with sessile glands and glandular papillae, sometimes scabridulous; tube 5–8 mm, teeth 13–18 × 2–4 mm, linear-lanceolate, acute with a terminal spine 1–2(–3) mm long, margins minutely antrorse scabrid hairy, 10-veined, the veins to the teeth clearly thick, throat glabrous or sparsely hairy. Verticillasters 2–4(–6)-flowered, each whorl with 4-spines (modified bracts), often congested above, sometimes distinct. Corolla 2.5–3.5 cm, pinkish white; upper lip emarginate; lower lip 3-fid; upper lip densely villous outside and within, shallowly hooded, lower lip sparsely hairy inside and outside with minutely glands; tube retrorsely hairy in the calyx; annulate. Stamens 4, two stamens included in corolla and two stamens shortly exserted or protruding; filaments and anthers hairy, sometimes glandular. Style exserted or included in corolla, unequally bilobed, glabrous or hairy. Fig. 134, 1–7.

HAB. Limestone mountain, *Quercus* forest; alt. 1400–1700 m: fl. & fr. Jun.-Jul.
DISTRIB. Rare and known so far from two locations in north and north east Iraq. **MAM**: Molla Khort Mountain, *Rawi, Alizzi, Nuri* 29570! **MSU**: Gweija Dagh (Goizha), *Quercus* forest, nr. Suleimaniya (rare), *Gilette & Rawi* 11707!

The species is widely distributed in Iran, and both Iraq specimens are from locations close to the Iranian border.

(The Iraq specimens at K were identified as *L. kotschyanus* (synonym of *L. hispidus*); *L. aucheri* differs from *L. kotschyanus* in its narrow (usually 2 mm wide) laciniate calyx. The type, *Aucher-Eloy* 1779 from Iran is incorrectly given in Flora Iranica as "*Aucher-Eloy* 1979").

W. Iran.

11. **MARRUBIUM** L.

Sp. Pl. 2: 583 (1753); Harley & al. in Kubitzki (ser. ed.), Fam. Gen. Vasc. Pl. 7: 226 (2004)

Ali Haloob & S.T. Al-Kaisi

Perennial herbs (in Iraq); stem base woody, erect-ascending, dense stellate pilose or white lanate, hairs branched and simple. Leaves petiolate, crenate, densely stellate pilose or lanate. Verticillasters axillary, many-flowered, remote. Bracts almost like leaves but smaller. Bracteoles subulate. Calyx 5–10-veined, tubular or obconical, teeth 5–15(–30-lobed, subulate-lanceolate, usually unequal, straight or spreading-reflexed. Corolla bilabiate, yellowish-white or purple or crimson, tube included in calyx; upper lip 2–lobed, bifid, erect; lower 3–lobed, spreading, middle lobe broadest. Stamens 4, didynamous, anterior pair longer, included in corolla tube; anther thecae divergent. Stigma 2–lobed, equal, obtuse, included in corolla tube. Nutlets triquetrous-ovoid or oblong.

The genus includes 51 species, native to Europe, N. Africa to central Asia; 6 species in Iraq.

Siadati, S., Salmaki, Y., Mehrvarz, S.S., Heubl, G. & Weigend, M. (2018). Untangling the generic boundaries in tribe Marrubieae (Lamiaceae: Lamioideae) using nuclear and plastid DNA sequences. Taxon 67(4): 770–783.

1. Calyx teeth 5; corolla purple or reddish-purple 2
   Calyx teeth more than 10; corolla yellowish-white 4
2. Bracteole deep purple; calyx 7–9 mm, tube purple, teeth 2–4 mm,
   apex hairy; corolla equal or shorter than calyx, upper lip 1.5–2 mm,
   middle lobe of lower lip 1.5–2 mm broad . . . . 1. *M. eriocephalum*
   Bracteole yellowish green; calyx ≤ 7 mm, tube yellowish-green, teeth
   shorter, apex glabrous; corolla longer than calyx, upper lip more
   than 2 mm long, middle lobe of lower lip less than 2 mm broad. 3
3. Plant with eglandular hairs; corolla purple, 8–10 mm, upper lip 3–4
   mm, middle lobe of lower lip 4–5 mm broad; seed 1.5–2 mm long,
   glabrous with stellate hairs at base . . . . 2. *M. astracanicum*
   Plant with eglandular and glandular hairs; corolla reddish-purple, 6–8
   mm long, upper lip 2 mm middle lobe of lower lip 2.5–3 mm broad;
   seed 2–2.5 mm, glabrous . . . . 3. *M. cordatum*
4. Calyx with dense stellate hairs and sessile glands, the upper 1/3 of
   teeth glabrous; corolla upper lip 1.5–2 mm, middle lobe of lower lip
   ± 2 mm wide . . . . 4. *M. vulgare*
   Calyx with dense stellate hairs, less than 1/4 of the teeth glabrous;
   corolla upper lip 2–2.5 mm, middle lobe of lower lip 1.2–1.5 mm wide 5
5. Calyx 7–8 mm, teeth 10–15, 2–3 mm long, base united for 1 mm in
   fruit; corolla shorter than calyx. . . . . 5. *M. crassidens*
   Calyx 4–6 mm, teeth 10–30, 1–2 mm, base united for 2 mm in fruit;
   corolla longer than calyx. . . . . 6. *M. cuneatum*

1. **Marrubium eriocephalum** *Seybold*, Stuttgarter Beitr. Naturk., A 310: 25 (1978). Type: Iraq, Perrish mt., 3340 m, 27-08–1957, *Al-Rawi & Serhang* 24522! (K, holo. [K000249641]; BAG, iso.). Seybold in Fl. Iranica [K. H. Rechinger] 180: 104 (2015).

Perennial. Stems ascending-erect, 20–40 cm, branched from base, tomentose-lanate. Leaves petiolate, ovate-orbicular, 10–20 × 10–25 mm, base cuneate-truncate, apex obtuse, margins irregularly crenate, densely stellate (appearing lanate); petiole 5–20 mm. Verticillasters 2–4, many-flowered, 15–25 mm diameter, 15–45 mm distance between verticillasters. Bracts ovate-orbicular, 8–18 × 6–17 mm, apex obtuse, base cuneate, irregularly crenate, hairy as the leaves; petiole 3–6 mm; bracteole subulate, 6–8 mm, deep purple, densely yellowish grey pilose. Pedicel ± 1 mm long. Calyx 7–9 mm, purple, densely pilose with white sessile glands, teeth 5, straight or slightly spreading, subulate, 2–4 mm, apex straight, hairy. Corolla purple, 7–8 mm, equal or shorter than calyx, upper lip 1.5–2 mm, erect, bifid, middle lobe of lower lip 1.5–2 mm wide, lips glabrous inside, pubescent outside.

HAB. Side of mountain slope, serpentine rocks on mountain slope, black and brown stones, on rocky cliff; alt. 2600–3340 m; fl. Aug.-Sep.
DISTRIB. Rare; collected from the alpine region in the Qandil range N.E. of Iraq. **MRO**: Qandil range, *Al-Rawi & Serhang* 24449! (BAG); Perrish Mt., *Al-Rawi & Serhang* 24522! (type, BAG); top of Qandil range bet. top of Perrish & Bardanas (Iraqi-Persian border), *Al-Rawi & Serhang* 24589! (BAG); Qandil range, *Al-Rawi & Serhang* 26723! (BAG).

Endemic.

2. **Marrubium astracanicum** *Jacq.*, Icon. Pl. Rar. 1: 11, t. 109 (1787); Rawi in Dep. Agr. Iraq Tech. Bull. 14: 151 (1964); Cullen in Fl. Turkey [P. H. Davis] 7: 176 (1982); Seybold in Fl. Iranica [K. H. Rechinger 180: 102 (2015).

M. kotschyi Boiss. et Hohen., Diagn. Pl. Orient. ser. 1, 5 : 33 (1844); *M. kotschyi* Boiss. & Hohen. var. *brachyodon* Boiss., Fl. Orient. [Boissier] 4: 696 (1879).

Perennial. Stems ascending-erect, 25–55 cm, branched from base, with dense white eglandular stellate pilose hairs. Basal leaves crowded, petiolate, obovate, 5–15 × 5–10 mm, almost covered with dense white eglandular stellate pilose hairs, cauline leaves obovate-orbicular, crenate, apex obtuse, base cuneate, stellate pilose eglandular hairs; petiole 10–30 mm long. Verticillasters up to 5, many flowers, 15–20 mm diameter, 10–60 mm distance between verticillasters. Bracts obovate or oblong-lanceolate, 10–35 × 5–20 mm, apex obtuse, base cuneate, crenate, hairy as the leaves; petiole up to 8 mm long. Bracteole subulate, 5–8 mm, yellowish green, with sparsely whitish eglandular pilose hairs, apex stout, glabrous. Pedicel ± 1 mm long. Calyx 5–7 mm long, yellowish green, sparsely eglandular stellate pilose hairs with sessile glands, teeth 5, straight, lanceolate, 1–2 mm long, apex straight, spiny, yellowish purple, glabrous. Corolla purple, 8–10 mm long, longer than calyx, upper lip 3–4 mm long, erect, deeply bifid, middle lobe of lower lip 4–5 mm broad. Nutlets brownish, oblong, 1.5–2 mm long, glabrous with stellate hairs at base.

HAB. Stony mountain side, serpentine rocks, muddy soil; alt. 1500–2400 m. fl. & fr. Jun.-Jul.
DISTRIB. Occasional in the mountain region of Iraq. **MRO**: Cha-i-Mandali, *Guest* 2723! (BAG); Kods near Haji Omran, *Rawi* 9182! (BAG); between slope of Karoukh mt. & Dargala village, *Kass & Nuri* 27737! (BAG). **MSU**: Gara Dagh, *Rawi* 9261! (BAG); Zewiya in Pir omar Gudrun, *Rawi* 12036! (BAG); Kajan mountain nr. Penjwin, *Rawi* 22697! (BAG).

RAIHANA (Kurd., *Rawi* 12036).

Iran, Transcaucasus, Turkey.

3. **Marrubium cordatum** *Nábelek*, Spisy Přír. Fak. Masarykovy Univ. 70: 58, f. 12 (1926); Cullen in Fl. Turkey [P. H. Davis] 7: 177 (1982); Seybold in Fl. Iranica [K. H. Rechinger 180: 103 (2015).

Perennial. Stems ascending, 20–25 cm, branching from base, densely white pilose with minute glandular hairs. Basal leaves crowded, petiolate, variable in shape from reniform to obovate, 4–9 × 6–8 mm, almost covered with dense white eglandular pilose hairs; cauline leaves reniform or orbicular, apex obtuse, base cordate or cuneate, margins irregularly crenate, stellate pilose; petiole 10–30 mm. Verticillasters 1–3, 10–20-flowered, 13–20 mm in diameter; 10–30 mm distance between verticillasters. Bracts rounded-obovate to oblong-lanceolate, 10–16 × 5–15 mm, apex obtuse, base cuneate, margins crenate, hairy as leaves; petiole up to 5 mm long. Bracteoles subulate, 5–8 mm, yellowish green, covered with dense whitish long pilose hairs, apex stout, glabrous. Calyx 5–6 mm, tube yellowish green, whitish long pilose with minute glandular hairs, teeth 5, subulate-lanceolate, straight to slightly curved, ± 2 mm, purple; apex spiny, yellowish purple, glabrous. Corolla crimson, longer than calyx, 6–8 mm, tube exserted from calyx tube, upper lip 2 mm long, erect to curving upward, deeply bifid, middle lobe of lower lip 2.5–3 mm wide. Nutlets dark brown, oblong, 2–2.5 mm long, glabrous.

HAB. On rocks, among metamorphic rocks, in *Astragalus* thorn cushion zone, on mountain slope, on road side; alt. 2500–3600 m; fl. Jul.-Sep.
DISTRIB. Rare; almost in alpine range of N.E. of Iraq. **MRO**: Arl gird Dagh, *Guest* 3054! (BAG); Ser Kurawa, *Gillett* 9753! (BAG, K); Helgurd range, *Al-Rawi & Serhang* 24910! (BAG); road bet. Bardanas and Qandil ranges, *Al-Rawi & Serhang* 24623! (BAG).

Iran, Turkey.

4. **Marrubium vulgare** *L.*, Sp. Pl.: 583 (1753); Rawi in Dep. Agr. Iraq Tech. Bull. 14: 151 (1964); Rawi & Chakrawarty in Dep. Agr. Tech. Bull. 15: 63 (1964); Cullen in Fl. Turkey [P. H. Davis] 7: 168 (1982); Seybold in Fl. Iranica [K. H. Rechinger] 180: 90 (2015); Taifour & El-Oqlah, Pl. Jordan Annot. Checklist: 102 (2017).

Fig. 135. **Marrubium vulgare**. 1, habit × ½; 2, flower × 5; 3, calyx (opened) × 5; 4, corolla opened, with stamens × 5; 5, gynoecium, detail; 6, nutlet (adaxial view), detail. Reproduced from Flora of China, Plates, 17: f. 180 (1998), with permission from Missouri Botanical Garden Press, St. Louis, and Science Press, Beijing.Drawn by Zeng Xiaolian.

Perennial. Stems erect-ascending, 30–85 cm, branched almost from lower 1/3 of stem, lanate, whitish. Leaves aromatic, broadly ovate-orbicular, 20–40 × 15–35 mm, base truncate-subtruncate, apex obtuse, margins irregularly crenate, stellate pilose with sessile glands, upper surface greenish, lower surface whitish grey; petiole 5–30 mm long. Verticillasters 4–8, many-flowered, 10–15 mm in diameter, 15–65 mm distance between verticillasters.

Bracts similar to leaves. Bracteoles subulate, 4–6 mm, dense whitish pilose hairs but the upper 1/3 glabrous, apex hooked, stout, yellow. Calyx 5–7 mm, green, densely stellate with sessile glands, teeth 10, unequal, 1–2 mm long, subulate, spreading-reflexed, base united for 0.2–0.3 mm, apex hooked, stout, yellow, the upper ± 1/3 glabrous. Corolla creamy, 6–7 mm, equal or longer to calyx, tube included or slightly exerted from calyx tube, upper lip 1.5–2 mm, erect, bifid, middle lobe of lower lip ± 2 mm broad. Fig. 135, 1–6.

HAB. On clay-gypsum soils, clay soil on cliffs of valley; alt. 150–180 m; fl. Apr.-May.
DISTRIB. Rare in the north area of the western desert district in Iraq. **DWD**: Ana, *Al-Khayat, S. Omar & Adil* 55457! (BAG); Ana, *S. Omar, H. Abass & Al-Kaisi* 56387! (BAG); 1.5 km S. of Al-Baghdadi, *A. Haloob* 60688! (BAG).

Rawi & Chakrawarty list the names HASHISHAT AL-KALIB and QUTAINAH (Kurd.) for this and note that it is chiefly used as a domestic remedy as a tonic, against coughs and colds, as a carminative and stimulant.

Europe, N. Africa to C. Asia.

5. **Marrubium crassidens** *Boiss.*, Diagn. Pl. Orient. 5: 35 (1844); Rawi in Dep. Agr. Iraq Tech. Bull. 14: 151 (1964); Seybold in Fl. Iranica [K. H. Rechinger 180: 93 (2015).

Perennial. Stems erect to ascending, 35–100 cm, branched, lanate, whitish. Leaves broadly oblong-oblong or ovate, 10–40 × 7–20 mm, base truncate or cuneate, apex acute or obtuse, margins crenate, densely stellate pilose, upper surface light green, lower surface whitish grey; petiole 7–40 mm. Verticillasters 4–9, 10–25 mm in diameter many-flowered, verticillasters with 15–100 mm distance between each. Bracts somewhat similar to leaves, shortly petiolate. Bracteoles subulate, 3–5 mm, densely whitish stellate pilose, apex glabrous, yellow. Calyx subconical, 7–8 mm, yellowish green, densely stellate hairy, teeth 10–15, unequal, subulate-sublanceolate, 2–3 mm, base united for 1 mm, spreading or reflexed, apex glabrous, hooked or curved, stout, yellow; calyx slightly enlarged in fruiting. Corolla yellowish-white, 5–6 mm, shorter than calyx, tube exserted from the calyx tube, upper lip 2–2.5 mm, erect, bifid, middle lobe of lower lip 1.2–1.5 mm wide. Nutlets triquetrous-ovoid, 2.5–2.9 × 1.2–1.5 mm, with rough surface.

HAB. Rocky slope, limestone; *Quercus* forest formation overgrazed near village, mountain sides, on cliff of cave, *Quercus* zone on mountain slope, sandstone in valley, dry stony slopes, hillside, under pine trees, rocky mountain near water; alt. 700–1360 m; fl. Mar.-Jul.; fr.: May-Aug.
DISTRIB. Frequent in the mountain region of N. Iraq. **MAM**: Zawitah gorge, *Guest* 3716! (BAG); Ser Amadia, *Guest* 4986! (BAG); Aqra, *Rawi* 11334! (BAG); Bekheir, *Rawi* 8633! (BAG); Khantur Mt. N.E. of Zakho, *Rawi* 23292! (BAG); Bikhair mountain near Zakho, *Rawi* 23094! (BAG); Zinta gorge, *E. Chapman* 26118! (BAG); Zawita, *S. Omar, Al-Dabbagh & Al-Kaisi* 45335! (BAG); Daiwka village 35 km N.E. Zakho, *S. Omar & Al-Dabbgh* 45381! (BAG); Dori village 5 km E. Kani Masi, *S. Omar & Al-Kaisi* 45436! (BAG); Maya mountain 25 km W. Kani Masi, *H. Hamid & Fadhil* 45512! (BAG). **MRO**: Shaqlawa, *Gillett* 8080! (BAG); Jindian by Rawandiz, *Guest* 13006! (BAG); Pushtashan, 15 km N.E. of Rania lower slope of Qandil range, *Rawi & Serhang* 23838! (BAG); Gali Warta, 30 km N.W. by N. of Rania, *Rawi, Nuri & Kass* 28830! (BAG); Rayat, *S. Omar, Sahira, F. Karim & H. Hamid* 38448! (BAG); Khulaifan, *S. Omar, Al-Kaisi & Wedad* 45436! (BAG); Wadi Malekan (Erbil), *Botany staff* 59621! (BAG). **MSU**: Azmir, *Fawzi Karim* 39359! (BAG).

REHANIPALA (Kurd. *Rawi* 8633)

Iran.

6. **Marrubium cuneatum** [*Soland*] Nat. Hist. Aleppo, ed. 2 [A.Russell] 2: 255 (1794); Rawi in Dep. Agr. Iraq Tech. Bull. 14: 151 (1964); Cullen in Fl. Turkey [P. H. Davis] 7: 170 (1982); Seybold in Fl. Iranica [K. H. Rechinger 180: 95 (2015); Taifour & El-Oqlah, Pl. Jordan Annot. Checklist: 102 (2017).

M. radiatum Delile in Benth., Lab. Gen. Sp. 591 (1834); *M. polyodon* Boiss., Prodr. [A. P. de Candolle] 12: 453 (1848) ; *M. cuneatum* Russell var. *spinulosum* Boiss., Fl. Orient. [Boissier] 4: 704 (1879).

Perennial. Stems erect-ascending, 20–50 cm, branched, lanate, whitish. Leaves broadly oblong-oblong or oblanceolate-obovate or ovate, 10–30 × 7–20 mm, base truncate or cuneate, apex obtuse, margins crenate, densely stellate pilose, upper surface light green, lower surface whitish grey; petiole 5–25 mm. Verticillasters 4–12, 10–15 mm in diameter, many-flowered, 10–45 mm distance between verticillasters. Bracts smaller than leaves, with shorter petiolate. Bracteoles subulate, 1.5–2.5 mm, dense whitish stellate pilose, apex yellow, tip glabrous. Calyx subconical, 4–6 mm, yellowish-green, densely stellate hairy, teeth 10–30, unequal, 1–2 mm long, subulate-lanceolate, base united for 0.5–1 mm, straight to curved; teeth expanded in fruiting to ± 2 mm, apex glabrous, straight or curved, stout, yellowish-red. Corolla yellowish-white, 5–7.5 mm, longer than calyx, tube exserted from the calyx tube, upper lip 2–2.5 mm, erect, bifid, middle lobe of lower lip 1.2–1.5 mm wide. Nutlets triquetrous-ovoid, 2.5–3 × 1.2–1.5 mm, with rough surface.

HAB. Cultivated fields, irrigated cereal fields near river, dry shrubland, moist *Quercus* forest, open oak forest on northern mountain slope, loamy soil, *Daphne* & *Astragalus* region, *Rhus-Quercus* forest on hillside, clay soil on hillside, on road side, in depressions, clay soil on plain; alt. 65–1830 m; fl. Mar.-Jun.; fr. Jun.-Aug.

DISTRIB. Common in the mountain region and occasional in the upper plains, foothills and desert regions of Iraq. **MSU**: nr. Chamchamal, *F. A. Rogers* 237! (BAG); Sarchinar nr. Suleimaniya, *K. Feddo* 5352! (BAG); Qaradagh-Jaaferan, *Gillett* 7900! (BAG); Gweija Dagh nr. Suleimaniya, *Rawi* 8880! (BAG); Jebel Avroman, spur N. of Biyara, *Gillett* 11764! (BAG); Penjwin, *Guest* 12935! (BAG); Avroman mountain N. of Halabja, *Rawi* 22082! (BAG); Mela Kowa, *Rawi* 22435! (BAG); Mella Khort mountain, *Rawi, Hosham & Nuri* 29365! (BAG); Zalem mountain, *Rawi, Hosham & Nuri* 29404! (BAG); Hawara Barza mountain, *Rawi, Hosham & Nuri* 29505! (BAG); Jasina, *S. Omar* 42722! (BAG); Abu-Abidah near Halabcheh, *S. Omar, F. Karim, R. Hamza & H. Hamid* 37465! (BAG); Bakrajo, *S. Omar & F. Karim* 37969! (BAG); Talan nr. Dokan, *F. Karim* 39290! (BAG); Azmir, *F. Karim* 39358! (BAG); 10 km from Sulimaniya to Choarta, *M. Noori & K. Hamid* 41253! (BAG); Zewiya, Pira Magrun, *Botany staff* 58937! (BAG); Mirgaban Eastern slope in Pira Magrun, *Botany staff* 58974! (BAG). **MRO**: Khanzad, *S. Omar, Al-Kaisi & Wedad* 49598! (BAG). **MJS**: Kursi, Jebel Sinjar, *Gillett* 10852! (BAG); Sinjar, *Al-Kaisi & K. Hamad* 49112! (BAG); Sinjar, *Widad & Al-Khayat* 53271! (BAG). **FUJ**: Jazira N. of Sinjar, *Guest* 13341! (BAG); 14 km S. of Sinjar, *Chakravarty, Rawi, Khatib & Alizzi* 33110! (BAG); Hamam Alil, *Anders* 35963! (BAG); near Ba'aj, *S. Omar & H. Hamid* 36543! (BAG); bet Tal-Afer & Sinjar, *F. Karim* 37486! (BAG); Tal Afar, *Al-Kaisi & K. Hamad* 49227! (BAG). **FAR**: Erbil, *Gillett* 8002 (BAG); 7 km from Altun Kopri to Erbil, *Botany staff* 43227! (BAG). **FKI**: Tuz, *Geust* 1407! (BAG); Gara Tepe, *Geust* 1485! (BAG); c. 10 km N. of Qara Ghan, *Gillett & Rawi* 7370! (BAG); 4 km S. of Taktak bridge, Kirkuk-Koi, *Rawi, Nuri & Kass* 28064! (BAG). **DLJ**: 6 km above Rawa on Euphrates, *Gillett & Rawi* 7007! (BAG). **DGA**: 4 km E. Samarra, *Rawi* 20354! (BAG). **DWD**: 40 km N. of Rutba, *Al-Kaisi & K. Hamad* 46585! (BAG). **DSD**: Al-Takhaded, 60 km S. of Salman, *Botany staff* 42209! (BAG).

TALISHKA (Kurd. *Rawi* 8880); HAMHAMKA (Kurd. *Guest* 13341); BETTINESA (Kurd. *Widad & Al-Khayat* 53271).

Turkey, Iran, Syria and Jordan.

12. **BALLOTA** L.

Sp. Pl.: 582 (1753); Gen. Pl. ed. 5, 253 (1754)
Siadati, Salmaki, Saeidi Mehrvarz, Heubl & Weigend in Taxon 67(4): 778 (2018)

C.C. Townsend
Revised by Shahina A. Ghazanfar

Perennial herbs, branched from base. Stems erect to ascending; indumentum simple or glandular. Leaves simple, petiolate, irregularly coarsely crenate-dentate. Flowers in axillary many-flowered whorls; bracteoles herbaceous. Calyx tubular-campanulate, 5-lobed (without additional intervening teeth in Iraq); lobes ± equal, acuminate to mucronate. Corolla strongly 2-lipped; upper lip hooded or almost flattened, entire; lower spreading, 3-lobed, with middle lobe retuse. Stamens 4, not or only shortly exerted from corolla. Style bifid. Nutlets rounded, apically subtruncate, glabrous.

About 30 species in Europe, North Africa, West and Central Asia; introduced into other temperate regions of the world; a single species in Iraq.

Harley, R.M. et al. (2004). Labiatae, in The Families and Genera of Vascular Plants (series ed. K. Kubitzki) 6: 167–275. Springer-Verlag, Berlin, Heidelberg, New York.
Siadati, S., Salmaki, Y., Saeidi Mehrvarz, S., Heubl, G. & Weigend, M. (2018). Untangling the generic

boundaries in tribe Marrubieae (Lamiaceae: Lamioideae) using nuclear and plastid DNA sequences. Taxon 67(4): 778 .

1. **Ballota nigra** *L.*, Sp. Pl. 2: 582 (1753); Boissier, Fl. Orient. 4: 775 (1879); Blakelock in Kew Bull. 1949: 539 (1950); Rawi in Dep. Agr. Iraq Tech. Bull. 14: 149 (1964).

subsp. **kurdica** *Davis* in Notes Roy. Bot. Gard. Edinburgh 21: 62 (1962); Patzak in Ann. Naturh. Mus. Wien 62: 68 (1958); Rechinger, Fl. Iranica 150: 352 (1982).

Perennial, branched from base, to 1.2 m tall. Stems and branches rather strict, erect, obtusely quadrangular, lower internodes 10–12 cm, upper gradually shorter, glabrescent or glabrous below, pubescent above. Lower leaves with petioles c. 2 cm, 6–7 × 4–5 cm, triangular-lanceolate, base cuneate, apex acute, irregularly coarsely crenate-dentate, silky-tomentellous on both surfaces, especially below, sometimes glabrescent. Flowers subsessile, in axillary, many-flowered whorls, lower distant, upper approximated. Bracts linear-subulate, ½–2.3 as long as calyx. Calyx 7–8(–9) mm, strongly ribbed and grooved, pubescent; teeth 5, straight, subequal, triangular-lanceolate, (1.5–)2–2.5(–2.8) mm, ⅓–¼ as long as tube, with a 0.5–0.8 mm erect mucro. Corolla pink, c. 14 mm, tube exserted. Nutlets nor seen. Fig. 136, 1–8.

HAB. Shady places, in crevices, stream sides, *Astragalus* zone; alt. 1500–2800 m; fl. & fr. Aug.
DISTRIB. Occasional in upper forest zone of Iraq: **MAM**: mt. Gara, *Kotschy* 354! **MRO**: Ser-i Hasan Beg, *Guest* 2913!; mt. Qandil near Pushtashan, *Rechinger* 11747 (W); Baski Hawaran mt., Al-*Rawi* 23974 (BAG); Serin mt. On road to Qandil, AL-*Rawi* 24023 (BAG).

N. Syria, Turkey (S.E. Anatolia), W. Iran. Distribution of species: Europe, N. Africa, S.W. Asia.

13. **PSEUDODICTAMNUS** Fabr.

Enum. [Fabr.] 56 (1759); Siadati, Salmaki, Saeidi Mehrvarz, Heubl & Weigend in Taxon
67(4): 770–783 (2018)
Zapateria Pau, Not. Bot. Fl. Españ. 1: 17 (1887)

Shahina A. Ghazanfar

Perennial herbs, suffruticose (woody at base) or small shrubs; indumentum of branched and stellate and/or simple hairs. Leaves mostly petiolate, with dentate or crenate margins. Inflorescence thyrsoid or racemoid; verticillasters with few to many cymes; bracteoles membranous or herbaceous. Calyx campanulate or hypocrateriform, tube 10-veined, limb undulate, 6–10(–20)-toothed or crenate, lobes equal or variously unequal, more or less mucronate or gradually narrowed, rarely entire, ± spiny. Corolla purple to white, strongly 2-lipped (one lobe forming upper lip, 2 or 3 lobes forming the lower lip); upper lip long hooded or almost flattened, entire, 2-fid or dentate; corolla-tube shorter than or equalling the calyx, with a ring of hairs inside. Stamens not or only shortly exerted. Nutlets rounded or subtruncate at apex, glabrous.

The resurrected genus *Pseudodictamnus* is morphologically characterized by a 6–20-toothed or crenate calyx limb which is usually expanded, and stellate or branched multinodal hairs on calyx and leaves.

Thirteen species from Mediterranean region to Iran and the Arabian Peninsula, Chad, S. Africa; a single species in Iraq.

Siadati, S., Salmaki, Y., Saeidi Mehrvarz, S., Heubl, G. & Weigend, M. (2018). Untangling the generic boundaries in tribe Marrubieae (Lamiaceae: Lamioideae) using nuclear and plastid DNA sequences. Taxon 67(4): 770–783.

1. **Pseudodictamnus aucheri** (*Boiss.*) *Salmaki & Siadati*, Taxon 67(4): 779 (2018).

Ballota aucheri Boiss., Diagn. Pl. Orient. ser. 1, 5: 39 (1844); Boissier, Fl. Orient. 4: 773 (1879); Handel-Mazzetti in Ann. Naturh. Mus. Wien 27: 414 (1913); Zohary in Dep. Agr. Iraq Bull. 31: 127 (1950); Patzak in Ann. Naturh. Mus. Wien 63: 63 (1959); Rawi in Dep. Agr. Iraq Tech. Bull. 14: 149 (1964); Rechinger, Fl. Iranica 150: 353 (1982).

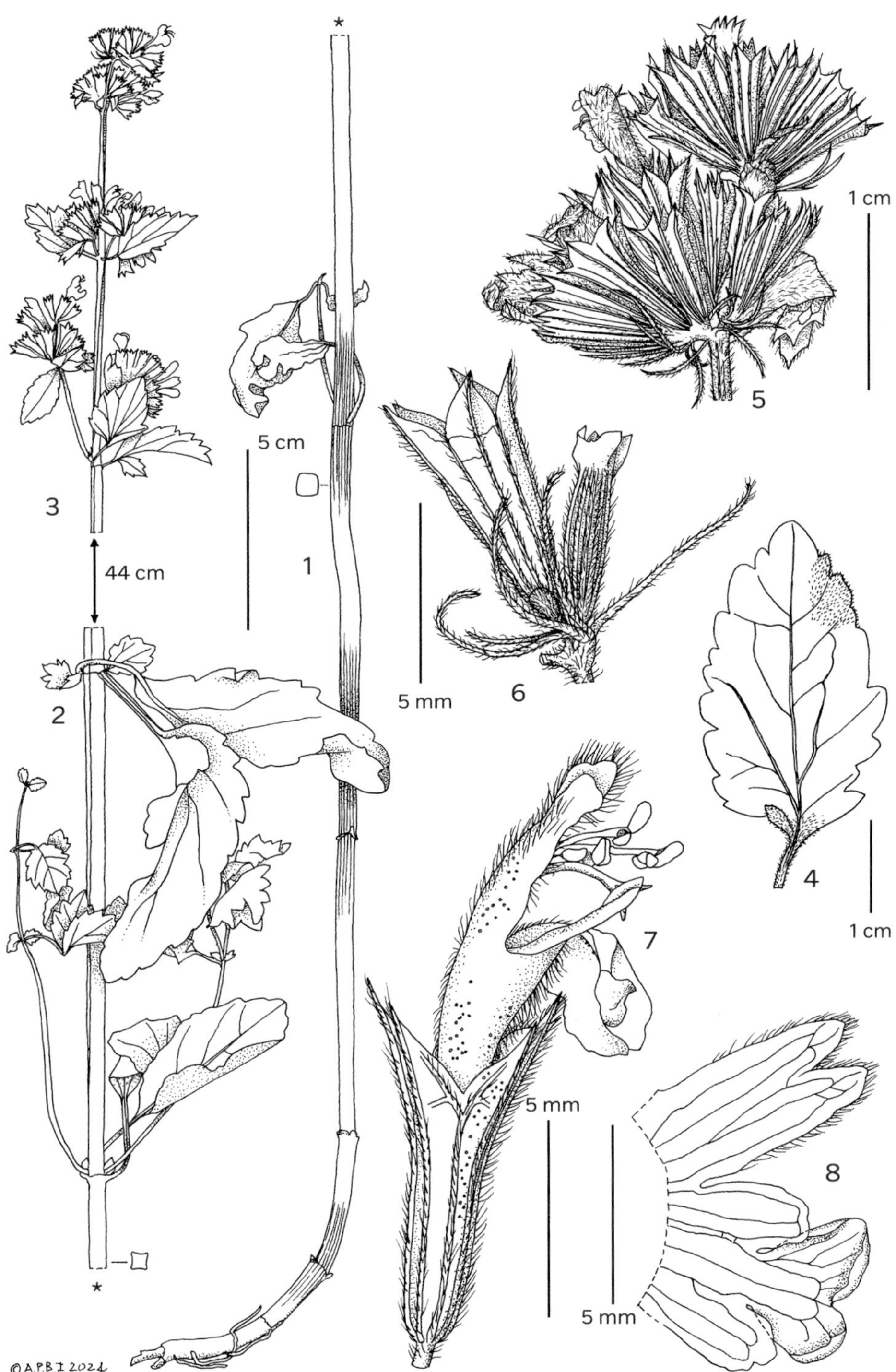

Fig. 136. **Ballota nigra** subsp. **kurdica**. 1–3, habit; 4, leaf, adaxial surface; 5, terminal and subterminal inflorescences; 6, bracts with calyx; 7, flower, side view; 8, upper and lower lobes of corolla, flattened. 1–8 from *Guest* 2953. Drawn by © A.P. Brown 2024.

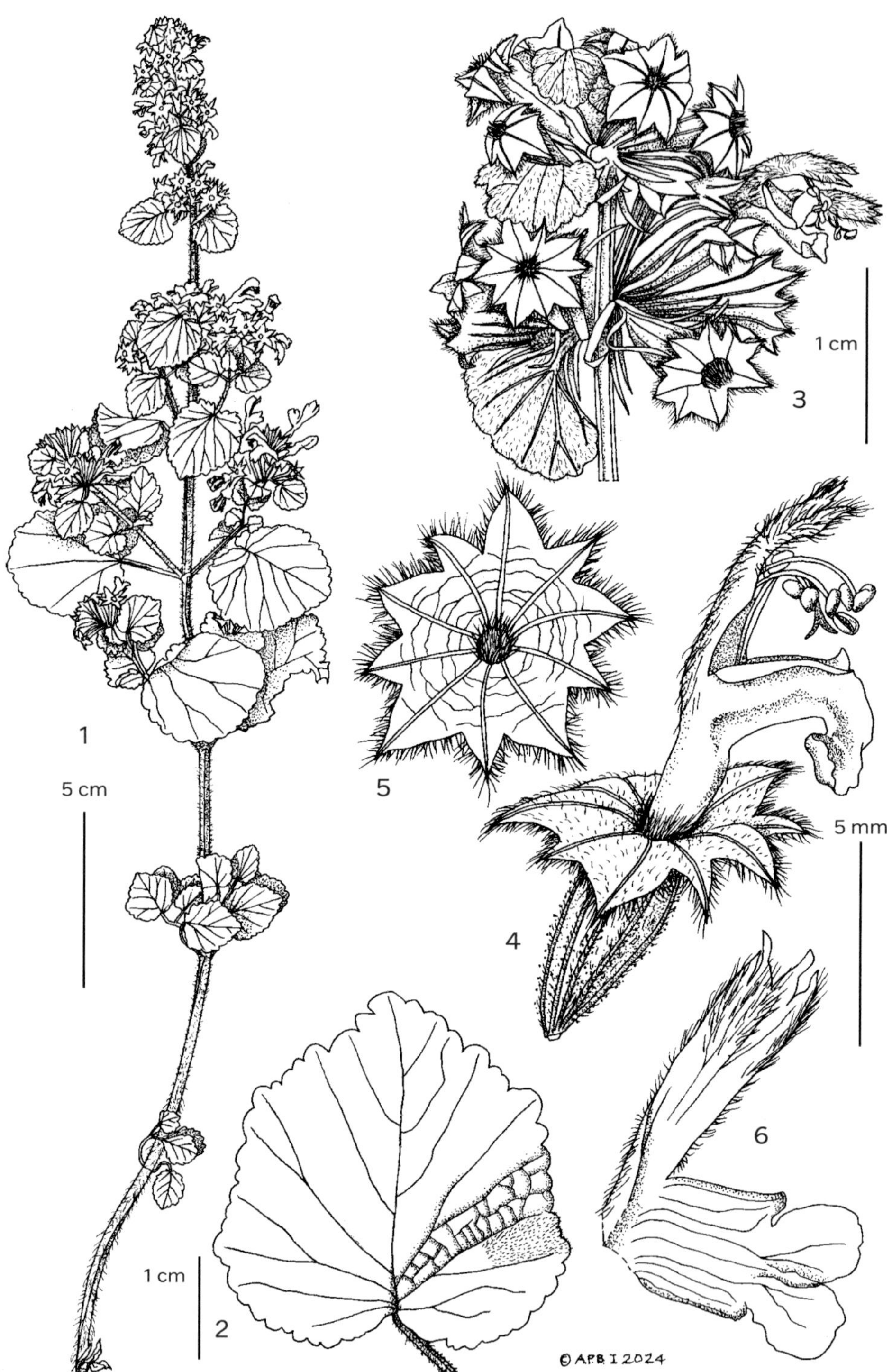

Fig. 137. **Pseudodictamnous aucheri**. 1, habit; 2, leaf, adaxial surface; 3, terminal and subterminal inflorescences; 4, flower, side view; 5, calyx tube, face view; 6, upper and lower lobes of corolla, flattened. 1–6 from *Wheeler Haines* W2081. Drawn by © A.P. Brown 2024.

Fig. 138. **Molucella laevis**. 1, habit × 1. Reproduced with permission from Feinbrun-Dothan, Fl. Palaestina 3: Plates, f. 202 (1977). Drawn by Esther Huber. © The Israel Academy of Sciences and Humanities.

Suffrutescent, branched from base, many-stemmed, 40–50 cm tall, ± densely canescent-tomentose, clothed with glandular or eglandular hairs. Stems and branches erect or erect-ascending, obtusely quadrangular, lower internodes 2.5–4.5 cm, upper gradually shorter, hispid with ± spreading hairs. Lower leaves with petioles to 2.5 cm, to 3.5 × 4 cm, broadly ovate-rounded or reniform-orbicular, base rounded or cuneate, ± irregularly obscurely crenate-dentate, densely canescent-tomentose. Flowers subsessile, in many-flowered axillary whorls, lower distant, upper approximated. Bracts linear-subulate, ⅓-½ as long as calyx. Calyx 9–11 mm, strongly ribbed and grooved, constricted above, limb about ½ as long as tube, densely canescent-hirsute, 5–7 mm in diameter (when flattened); teeth 16, spreading or somewhat recurved, or unequal length. Corolla pink, 12 mm, tube included. Fig. 137, 1–6.

HAB. Rocks, in crevices, in gorges, mountain side, in *Rhus coriaria* & *Juglans* orchards, on clay, and rocky clay soil; alt. 700–2400 m; fl. & fr. Jun.-Aug.

DISTRIB. Occasional in lower forest zone of Iraq. **MRO:** Sakran mountain 15 km S.E. Choman, *Al-Dabbagh & K. Hamad* 46255 (BAG); Malikan valley S. Of Rawandoz, Riyadh, *Haydar, Shiymaa & Adil* 58757 & 58758 (BAG); Sakran mountain, *A. Haloob & Al-Kaisi* 60684 (BAG). **MRO/FAR:** Arbil to Rowanduz, *Bornmüller* 1683, 1751!; Mt Zernakew (Qurnaqo) N. of Rania, *Thesiger* 1196! **MSU:** Sulaimaniya, *Zohary* s.n.; Sahvan, *Haley* 214. **MJS:** El Magharad gorge, *Handel-Mazzetti* 1389!; Shakhrakhan Pira Magroon, *Al-Hashimi* 43730 (BAG); Pira Magron Dagh, *A. Haloob, Ikhlas & R. Hamushkan* 60218 (BAG).

S. & S.W. Iran.

14. **MOLUCCELLA** L.

Sp. Pl. 2: 587 (1753); Gen. Pl. ed. 5: 1211 (1754); Harley & al. in Kubitzki (ser. ed.), Fam. Gen. Vasc. Pl. 7: 227 (2004); Bendiksby & al. Taxon 60(2): 471 (2011)

Ali Haloob

Perennial shrublets, or annual or short-lived herbs. Stems and leaves glabrescent or hairy with short hairs. Leaves petiolate, crenate or incised. Inflorescence of many-flowered verticillasters. Bracts similar to cauline leaves; bracteoles subulate to spiny. Calyx zygomorphic, mostly 2-lipped, tube strongly or slightly expanded; limb membranous in fruit, greatly enlarged, with small mucronate projections at the edges, reticulately veined. Corolla bilabiate, tube oblique, included in the calyx, upper lip erect, hooded; lower lip 3-lobed, medial lobe broad, notched. Stamens 4, didynamous, anterior (outer) pair longer than posterior (inner) pair, ascending under upper lip; anthers 2-celled divaricate. Style bifid. Ovary glabrous. Nutlets acutely triquetrous, truncate at apex.

A genus of eight species in the circumscription of the genus as given by Bendiksby & al. (2011), found from S. Europe to C. Asia, Pakistan and Kashmir; introduced into parts of N. & S. America and other warm regions of the world; a single species in Iraq.

Bendiksby, M., Thorbek, L., Scheen, A.C., Lindqvist, C. and Ryding, O. (2011). An updated phylogeny and classification of Lamiaceae subfamily Lamioideae. Taxon 60(2): 471–484.

1. **Molucella laevis** *L.*, Sp. Pl. 2: 587 (1753); Boissier, Fl. Orient. 4: 768 (1879); Handel-Mazzetti in Ann. Naturh. Mus. Wien 27: 414 (1913); Zohary in Dep. Agr. Iraq Bull. 31: 127 (1950); Blakelock in Kew Bull. 4: 542 (1950); Rawi in Dep. Agr. Iraq Tech. Bull. 14: 152 (1964); Rechinger, Fl. of Lowland Iraq: 523 (1964); Ball in Fl. Europ. 3: 149 (1972); Feinbrun-Dothan, Fl. Paleast. 3: 124 (1978); Mill in Fl. Turkey [P. H. Davis] 7: 155 (1982); Rechinger, Fl. Iranica 150: 344 (1982); Hedge in Fl. Pakistan [Ali & Y. Nasir] 192: 175 (1990).

Annual, 20–60 cm, erect or erect-ascending, branched below, glabrous or glabrescent (with spreading minute hair on angles only). Leaves ovate-orbicular, 20–45 × 20–50 mm, base rounded, truncate or subcordate, apex obtuse or subacute, margins crenate or incised, lower surface with hairs on veins; petiole 15–30 mm. Bracts similar to leaves; bracteoles subulate, spiny, 6–10 mm, much shorter than calyx. Verticillasters 6–8-flowered, arranged in long spikes, the lower distant, the upper congested; flowers pedicel ± 1.5 mm. Calyx 25–30 mm, tube obliquely campanulate, glabrous, limb membranous in fruit, greatly enlarged and almost enveloping the corolla, 25–35 mm across, reticulately veined, pentagonal. Corolla white, shorter than the calyx, ± 15 mm, tube glabrous, upper lip entire, densely short eglandular hairy outside. Nutlets ± 2 mm, brown. Fig. 138, 1.

HAB. Rocky mountain slopes, sandy clay slopes, roadsides, cultivated fields and hillsides; alt. 350–1700 m; fl.: May-Jun.

DISTRIB. Found in the lower forest zone and N.W. Sector of the upper plains of Iraq; not common. **MAM**: Gali mazurka (2 km N.W. Amadya), *Al- Dabbagh & Jasim* 46919!; Dohuk, *Yussaf Lazar* 3346! **MRO**: Mirza Rustam et Rania, *Rechinger* 11230! **MSU**: Bakrajo, *Ali Askari* 1054; Qara dagh, *Haines*, 1505! **MJS**: Jabal Sinjar, *Al-Kaisi* 49745!; 30 km east of Karsi, *Bharucha & Hikmat Abbas-Al-Ani* 8809!; Jabal Singar S.W. slope, *Al-Kaisi, Al-Khayat & F. Karim* 50970! **FUJ**: Kanispan (60 km from Mosul to Aqura), *Al-Dabbagh & M. Jasim* 46825! **FNI**: 16 km west of Mosul, *Rawi* 22926!

Cultivated ornamental, commonly called Bells-of-Ireland; BAIT LAHAM (common Arabic name).

Cyprus, Lebanon, Palestine, Syria, Turkey, Iran, Afghanistan, Turkmenistan, Caucasus, Russia.

15. **SALVIA** L.

Sp. Pl. 1: 23 (1753) & Gen. Pl. ed. 5: 15 (1754); Harley in Kubitzki, Fam. Gen. Vasc. Pl. 7: 235 (2004)

Schraderia Medicus, Philos. Bot. 2: 40 (1791); *Polakia* Stapf, Denkschr. Akad. Wiss. Wien Math.-Nat. Kl. 50: 43 (1885); *Arischra*da Pobed., Nov. Syst. Pl. Vasc. Leningrad 9: 247 (1972)

Ferhat Celep

Herbaceous, suffruticose, chamaephytes or shrubby perennials, biennials or rarely annual, aromatic plants. Leaves undivided, lobed or pinnatisect. Inflorescence of variously arranged cymes or often in false whorls (verticillasters). Verticillasters (1–) 2–8 (–40)-flowered distant or approximate. Bracts always present, bracteoles sometimes present. Calyx tubular, campanulate or infundibular, bilabiate, more or less distinctly 2–labiate, upper lip short 3–dentate, lower lip 2–dentate. Fruiting calyces slightly or clearly expanded and then sometimes membranous. Corolla white, yellow, pink, purple, lilac or violet, bilabiate. Upper lip straight to falcate (hood), lower lip (labellum) 3–lobed with a broad concave middle lobe and two small lateral lobes; tube straight or curved, or ventricose, annulate (hairy ring) or not, squamulose or not. Stamens 2; three stamen types are distinguished: stamens articulating at junction of filament and connective, a short filament and a short or slightly elongated connective bearing upper and lower fertile theca (stamen type A); stamens articulating at junction of filament and connective, upper theca fertile but lower theca sterile with variously shaped sterile tissue (dolabriform) (stamen type B); stamens not articulating with filament and connective, and lower theca sterile with subulate shape (stamen type C); staminodes (posterior pair of stamens) always present, small. Style 2 lobed. Nutlets glabrous or hairy, ovoid, oblong, spherical, obovate and ovate, trigonous or rounded, often producing mucilage on wetting (myxocarpy) (Hedge 1972; 1982a; 1982b). Fig. 139, 1–3.

A large genus with 1023 species, cosmopolitan; introduced to N. Europe and Canada; 32 species in Iraq.

Salvia, Lat. (sage) based on *salvus*, healthy (referring to the medicinal use of the plant).

Rose, J.P., Kriebel, R., Kahan, L., DiNicola, A., González-Gallegos, J.G., Celep, F., Lemmon, E.M., Lemmon, A.R., Sytsma, K.J. & Drew, B.T. (2021). Sage insights into the phylogeny of *Salvia*: dealing with sources of discordance within and across genomes. Frontiers in Plant Science 12, https://doi.org/10.3389/fpls.2021.767478.

Walker, J.B., Sytsma, K.J., Treutlein, J. & Wink, M. (2004). *Salvia* (Lamiaceae) is not monophyletic: implications for the systematics, radiation, and ecological specializations of Salvia and tribe Mentheae. American Journal of Botany 91(7): 1115–1125.

Zhao, F., Chen, Y.P., Salmaki, Y., Drew, B.T., Wilson, T.C., Scheen, A.C., Celep, F., Bräuchler, C., Bendiksby, M., Wang, Q. and Min, D.Z. (2021). An updated tribal classification of Lamiaceae based on plastome phylogenomics. BMC biology 19: 1–27.

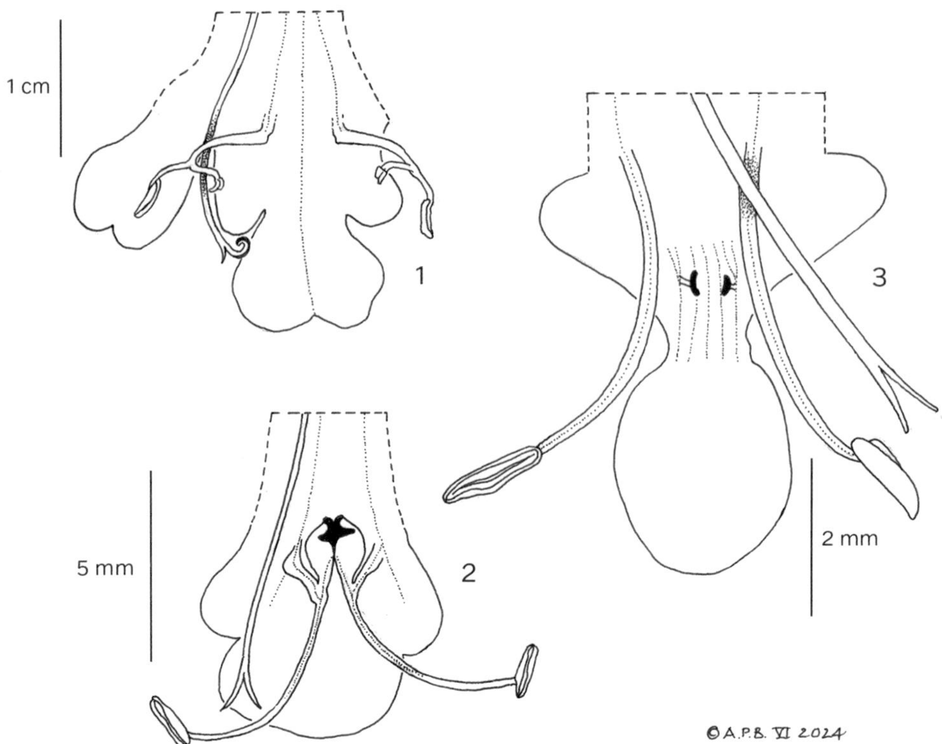

Fig. 139. Stamen types in *Salvia*. 1, stamen type A; 2, stamen type B; 3, stamen type C. Drawn by © A.P. Brown 2024.

1. Annual herb, stems often topped by a coloured coma of sterile bracts;
      pedicels deflexed in fruit. . . . . 10. *S. viridis*
    Perennial herbs or shrubs. 2
2. Fruiting calyces membranous-reticulate, infundibular with widely
      spreading rounded lobes. 3
    Fruiting calyces thick-textured, tubular to infundibular with erect or
      spreading acute lobes 4
3. Stem branched at base; leaf indumentum of dendroid to dendroid-
      stellate hairs; inflorescence unbranched; lower theca fertile . . . . 9. *S. multicaulis*
    Stem unbranched at base; leaf indumentum simple; inflorescence
      widely paniculate, lower theca sterile . . . . 8. *S. compressa*
4. Staminal connective not articulating with filament; lower theca subulate
      sterile (Stamen type C); verticillasters 20–40-flowered; upper lip of
      corolla clearly narrowed at base 5
    Staminal connective articulating with filament; lower theca dolabriform
      (sterile) or producing pollen grains (fertile); verticillasters
      1–8-flowered; upper lip of corolla falcate or straight 6
5. Leaves linear-oblong; calyx teeth not mucronate . . . . 32. *S. russellii*
    Leaves oblong to ovate; calyx teeth mucronate. . . . . 31. *S. verticillata*
6. Lower theca of the staminal connective producing pollen grains
      (fertile, Stamen type A), connective tissue shorter than or equal to
      filaments; upper lip of corolla straight, corolla tube within often annulate 7
    Lower theca of the staminal connective sterile (dolabriform plate
      of sterile tissue, Stamen type B), connectives clearly longer than
      filaments; upper lip of corolla somewhat falcate to strongly falcate;
      corolla tube usually not annulate 13

7. Much branched suffruticose herb; corolla up to 8 mm; of arid to
      desertic habitats . . . . 1. *S. aegyptiaca*
      Perennial herbs; corolla more than 20 mm; not of arid to desertic habitats 8
8. Leaves simple 9
      Leaves pinnate 10
9. Bracts clearly longer than calyces, bracts membranous; verticillasters
      2-flowered; leaves ovate to elliptic . . . . 2. *S. macrochlamys*
      Bracts shorter than calyces, bracts thick textured; verticillasters
      1–2-flowered; leaves cordate . . . . 3. *S. kurdica*
10. Corolla yellow; terminal leaf segment linear-oblong or lanceolate . . . . 7. *S. suffruticosa*
      Corolla pink; terminal leaf segment broadly oblong to ovate 11
11. Bracts clearly shorter than calyces; calyx prominently urceolate . . . . 4. *S. pinnata*
      Bracts clearly longer than calyces; calyx tubular-infundibular 12
12. Stems with very long spreading eglandular hairs; fruiting calyces
      (15–)20–25 mm . . . . 6. *S. trichoclada*
      Stems with or without long spreading eglandular hairs; fruiting calyx
      ca. 16 mm . . . . 5. *S. bracteata*
13. Corolla tube not squamulose and not invaginate 14
      Corolla tube squamulose and invaginate 23
14. Corolla 6–15 mm; bracts always shorter than calyx 15
      Corolla 16–35 mm; bracts only slightly shorter, equal or longer than the calyx 17
15. Corolla white; pedicel 3–8 mm in flower; calyx upper lip not concave
      and bisulcate in fruit . . . . 11. *S. syriaca*
      Corolla violet or lilac, rarely white; pedicel 1–3 mm in flower; calyx
      upper lip concave and bisulcate in fruit 16
16. Leaves pinnatisect with narrow linear ultimate segments . . . . 30. *S. lanigera*
      Leaves simple to pinnatifid with broad lobes . . . . 29. *S. virgata*
17. Leaves linear to linear lanceolate; bracts herbaceous with a thick white
      indumentum . . . . 12. *S. montbretii*
      Leaves oblong to ovate; bracts not herbaceous with a thick white indumentum 18
18. Stem below densely hirsute; leaves oblong to oblong ovate, lyrate to
      pinnatifid, flowers always lilac to lilac-white . . . . 18. *S. palaestina*
      Stem below pilose to villous; leaves elliptic, ovate or orbicular, not lyrate
      or pinnatifid; flowers usually white, rarely lilac 19
19. Calyx campanulate to tubular-infundibular 20
      Calyx narrowly or broadly tubular 21
20. Stem below appressed pubescent, eglandular; leaves widely ovate; bracts
      22–35 × 10–22 mm . . . . 17. *S. sclareopsis*
      Stem below glandular and eglandular hairy; leaves oblong to ovate;
      bracts 12–20 × 12–16 mm . . . . 16. *S. persepolitana*
21. Calyx narrowly tubular, 4–5 mm wide; calyx teeth spinulose in fruit;
      leaves elliptic to ovate . . . . 14. *S. macrosiphon*
      Calyx widely tubular, 6–7 mm wide; calyx teeth rigidly spinulose in fruit;
      leaves widely elliptic or ovate or suborbicular 22
22. Bracts clearly involucrant the calyces, widely ovate to suborbicular . . . . . 15. *S. reuterana*
      Bracts not involucrant the calyces, widely ovate . . . . 13. *S. spinosa*
23. Leaves pinnatifid with spreading linear segments; corolla yellow . . . . 20. *S. ceratophylla*
      Leaves not pinnatifid; corolla white or lilac 24
24. Corolla c. 30 mm long, lower lip dark violet; calyx truncate at apex . . . . . . 28. *S. indica*
      Corolla usually 12–30 mm long, lower lip white, cream or light lilac,
      calyx not truncate at apex 25
25. Bracts clearly longer than the calyces 26
      Bracts shorter than the calyces 28
26. Bracts lilac or white, to 35 mm long; corolla lilac or pink 20–30 mm . . . . . 19. *S. sclarea*
      Bracts green, 10–20 mm; corolla white or rarely pale lilac, 12–22 mm 27
27. Calyx obtriangular; corolla 12–16 mm . . . . 27. *S. poculata*
      Calyx tubular-campanulate; corolla 15–22 mm . . . . 26. *S. urmiensis*
28. Corolla 12–16 mm, somewhat falcate . . . . 25. *S. frigida*
      Corolla 17–30 mm, clearly falcate 29

29. Leaves linear or linear oblong, 5–30 mm wide . . . . .22. *S. atropatana*
    Leaves oblong to ovate 30
30. Calyx shortly papillose or glandular-pilose; inflorescence ± slender;
      leaves pubescent to densely pannose, subentire to erose . . . . 21. *S. candidissima*
    Calyx with long spreading glandular hairs; inflorescence sturdy; leaves
      lanate to pannose, irregularly lobed or dentate to subentire 31
31. Calyx campanulate-infundibular, 9–13 mm in flower, 13–18 mm in fruit;
      stems to 30 cm . . . . .24. *S. xanthocheila*
    Calyx campanulate, 9–12 mm in flower, 11–14 mm in fruit; stems to 100
      cm . . . . .23. *S. microstegia*

1. **Salvia aegyptiaca** *L.*, Sp. Pl. 23 (1753); Dinsmore in Post, Fl. Syria, Palest. & Sinai ed 2, 2: 360 (1933); Rawi in Dep. Agr. Iraq Tech. Bull. 14: 154 (1964); Hedge, Rev. *Salvia* 33 (1): 37–40 (1974); Feinbrun-Dothan, Fl. Paleast. 3: 143 (1978); Hedge in Flora Iranica [K. H. Rechinger] 150: 432 (1982); Jamzad in Flora Iran [Assadi & al., in Persian] 76: 819 (2012).

S. pumila Benth., Lab. Gen. Sp. 726 (1835).
S. aegyptiaca var. *pumila* sensu Hook., Fl. Ind. 4: 656 (1885).
S. aegyptiaca var. g*landulosissima* Bornm. & Kneucker Allgem. Bot. Zeitschr. 22: 5 (1916).
S. gabrieli Rech. f. Bot. Jahrb. 71: 538 (1941).

Much branched suffuticose herb, 10–20(–40) cm. Stem above and below with short or longer retrorse eglandular hairs. Leaves narrow linear-elliptic, rarely ovate-oblong, up to 55 × 8 mm, crenate to serrate, sessile or narrowed into an indistinct petiole; above and below with very short eglandular hairs and at leaf base eith long spreading eglandular hairs. Verticils up to 8, 2–6-flowered, up to 3 cm apart below and less above. Floral leaves ovate-lanceolate, up to 4.5 × 2 mm, bracts present. Pedicels up to 5 mm. Calyx 13-veined, with a prominent indumentum of glandular capitate hairs and eglandular hairs; upper lip of three closely connivent teeth up to 0.3 mm, concave in fruit; lower lip with two c. 3 mm acuminate-subulate teeth. Corolla pale lilac or lavender up to 8 mm; upper lip fairly broad, ± straight or somewhat reflexed; lower lip longer than upper; median lobe clearly bifid; tube with a thin annulus. Stamens Type A. Staminal cennectives c. 2 mm; filaments c. 2.5 mm; lower theca fertile. Nutlets black, trigonous, 2 × 1 mm, mucilaginous on wetting. Fig. 140. 1–3.

HAB. Arid to desertic habitats, in gravel and sandy gravel soils; alt. 50–100 m; fl. & fr. Jan.-Apr.
DISTRIB. S.E. Iraq. **LEA:** W. of Basra, *Botany Staff* 42409 (K). **DSD:** Kuwaibda wells, near Zubair, *Eig & Zohary* 5052 (K); Shu'aiba, near Basra, *Rogers* 09 (K).

Canary Islands, N. Africa, Arabian Peninsula, Iran, Pakistan, Afghanistan,

2. **Salvia macrochlamys** *Boiss. & Kotschy* ex *Boiss.* Fl. Orient. [Boissier] 4(2): 595 (1879); Hedge in Fl. Iranica [K. H. Rechinger] 150: 415 (1982); Hedge in Fl. Turkey [P .H. Davis] 7: 415 (1982); Jamzad in Flora Iran [Assadi & al., in Persian] 76: 824 (2012).

Perennial herb. Stems many, 30–50 cm tall, procumbent, branched above, leafy, densely glandular-villous. Leaves simple, ovate to elliptic, 4–7 × 2–4 cm, cordate or rounded, crenulate to serrulate, glandular-pilose, and with few sessile glands; petiole 2–4 cm. Inflorescence shortly paniculate or not; verticillasters 2-flowered, approximate. Bracts ± enclosing flowers, ovate to ovate-oblong, 30–40 × 15–20 mm, membranous, prominently veined, greenish white. Calyx tubular-ovate, c. 20 mm, glandular villous; upper lip tridentate, acuminate. Corolla pink or white, lower lip flecked or all violet, c. 35 mm; tube straight below, widening above, annulate; upper lip ± straight, truncate. Stamens type A. Nutlets rounded trigonous, broadly ovate, c. 5 × 4 mm.

HAB. Rocky mountain slopes, open slopes in *Quercus* scrub; alt. 1200–1500 (–2300) m; fl. & fr. Jul.
DISTRIB. Mountains slopes N. W. Iraq. **MAM:** Mosul, Jabal Khantur, *Rechinger* 10752 (K); 25 km N.E. of Zakho, Shivara village (Sharanish), *Rawi* 23455 (K).

A very handsome and distinctive species on account of the very large bracts, 2-flowered verticillasters and thickly glandular-villous stems. Without any close allies.

N.W. Iran, E. Turkey.

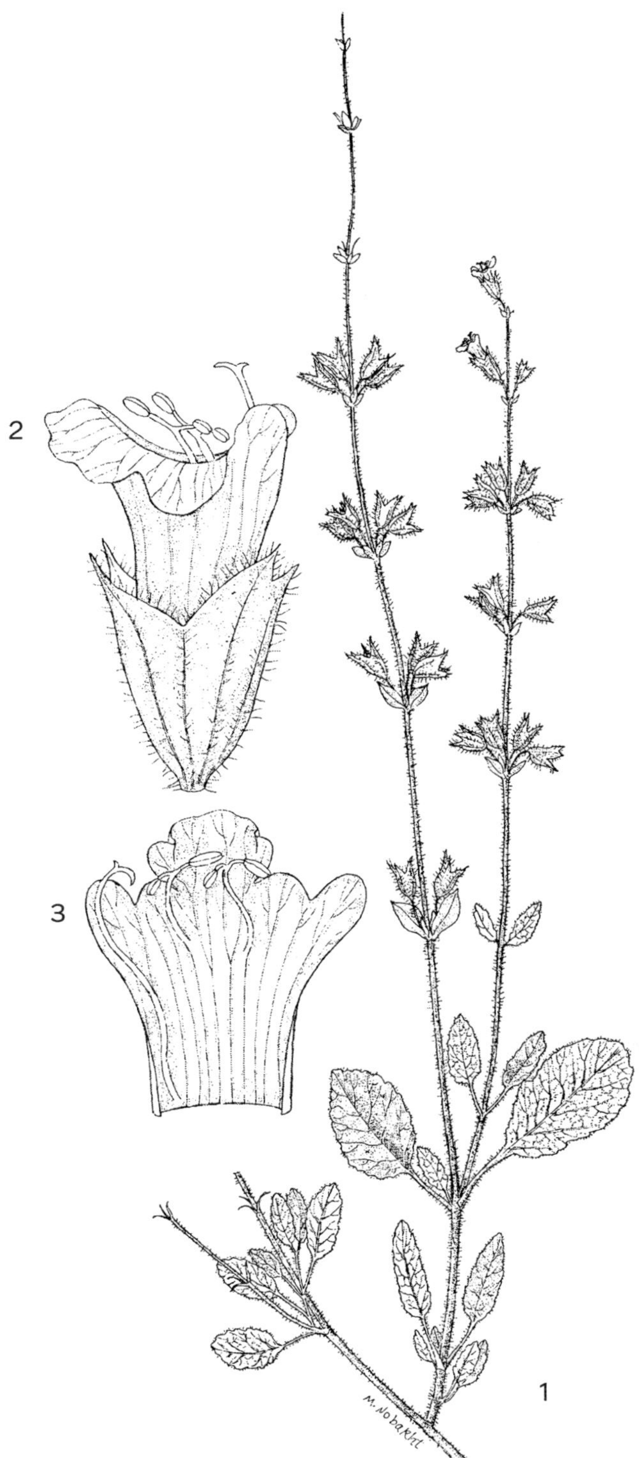

Fig. 140. **Salvia aegyptiaca**. 1, habit × 1; 2, flower × 5; 3, corolla, opened × 5. Reproduced with permission from Jamzad, Flora of Iran 76: f. 253 (2012). © Ministry of Jihad-e-Agriculture.

3. **Salvia kurdica** *Boiss. & Hohen.* ex *Benth.*, DC., Prodr. [A.P. de Candolle] 12: 268 (1848). Type: Iraq ad rupes prope Amadia, 31 July1841, *Kotschy* 347 (G!, holo.; BM! K! HAL! iso, photo); Rawi in Dep. Agr. Iraq Tech. Bull. 14: 155 (1964); Hedge in Fl. Iranica [K. H. Rechinger] 150: 416 (1982); Hedge in Fl. Turkey [P. H. Davis] 7: 415 (1982).

Perennial suffruticose herb. Stems to 50 cm, brittle ± densely pilose to villous with capitate glandular hairs. Leaves simple, broadly ovate to suborbicular, 4–7 × 3–5 cm, cordate, glandular-pilose, crenulate, rugulose; petiole 2–4 cm. Inflorescence paniculate, few-flowered, leafy; verticillasters 1–2-flowered, pedunculate from upper leaf axils, distant. Pedicels 1–5 mm. Calyx tubular-infundibular, 15–20 mm, widening in fruit, glandular-villous, upper lip shortly tridentate, longer than lower. Corolla bright pink, c. 30 mm; tube straight below, widening above, incompletely annulate; upper lip ± straight. Stamens type A. Nutlets ± spherical, c. 5 × 4.5 mm.

HAB. Rocky cliffs, clay soil; alt. 700–1530 m; fl. & fr. May-Jul.
DISTRIB. Lower hills and mountains of N.W. Iraq. **MAM**: Mosul: Ad confines Turciae prov. Hakkari, Amadiya, 900 m, *Rechinger* 11637 (K); ibid. *Agnew* 666; Zawita George, saxatile rocks about road, 8 June 1953, 700 m, *E. Chapman* 26102 (K); Jabal Khair, 32 km W. of Aqra, 1530 m, *Thesiger* 613; Arbil: Infra Rawanduz, 700 m, *Rechinger* 11243; E. of Akre (Aqra), Bujal, Sipa waterfall, in urban area, roadside, clay soil, 8 July 2017, 705 m, *Hamshkan, Riyadh, Dalaf & Shyimaa*, National Herbarium of Iraq 59182! **FKI/ FPF**: Diyala, Jabal Hamrin, *Sutherland* 2.

A near endemic, with no clear allies; the species is distinct in its strongly cordate leaves, few flowered pedunculate verticillasters and short axillary inflorescences.

E. Turkey.

4. **Salvia pinnata** *L.*, Sp. Pl. 27 (1753). Ic: Rouy, III. PI. Eur. Rar. 11: t. 269 (1899); Dinsmore in Post, Fl. Syria, Palest. & Sinai ed 2, 2: 351 (1933); Rawi in Dep. Agr. Iraq Tech. Bull. 14: 154 (1964); Hedge, Rev. *Salvia* 33 (1): 37–40 (1974); Feinbrun-Dothan, Fl. Paleast. 3: 143 (1978); Hedge in Fl. Turkey [P. H. Davis] 7: 417 (1982).

Salvia orientalis Mill. Gard. Dict., ed. 8. n. 8 (1768).
Ormiastis pinnata Raf. Fl. Tellur. 3: 93 (1837).
Salvia ertekinii Yıldırımlı, Ot Sistematik Botanik dergisi 15: 1–8 (2008).

Perennial herb. Stems 18–60 cm, simple or sparingly branched, all leafy, procumbent to erect, quadrangular, viscid, densely glandular villous. Leaves irregularly pinnatisect, with an ovate-oblong terminal segment to 4–8 × 2–5 cm and 2–5 pairs of sessile or petiolate lateral segments irregularly arranged, somewhat membranous, crenate-serrate, glandular-pilose; petiole 4–12 cm. Inflorescence raceme. Verticillasters 4–8, 4–6(–9)-flowered, distant. Bracts c. 6 mm, oblong-ovate, soon deciduous. Pedicels 5–15(–20) mm, erecto-patent. Calyx urceolate, often dark brownish purplish, 8–15 mm, scarcely expanding in fruit, densely glandular-villous; upper lip truncate, obsoletely tridentate. Corolla purplish-pink, 20–30 mm; tube gradually widening towards throat, longitudinally pilose within; upper lip ± straight, shorter than lower. Stamens type A. Fig. 141, 1–3.

HAB. Fallow fields, dry meadows; alt. 700–1530 m; fl. & fr. Mar.-May.
DISTRIB. N.W. Iraq. **MAM**: Zawita, June 1955, *Unknown collector* 134 RC 41Z (K).

Turkey, Cyprus, Palestine, Lebanon, Syria, Israel, Bulgaria (Greuter et al., Med–Checklist 3, 1986).

5. **Salvia bracteata** Banks & Sol. in Russell, Aleppo 2(2): 242 (1794); Dinsmore in Post, Fl. Syria, Palest. & Sinai ed 2, 2: 352 (1933); Rawi in Dep. Agr. Iraq Tech. Bull. 14: 154 (1964); Feinbrun-Dothan, Fl. Paleast. 3: 136 (1978); Hedge in Fl. Iranica [K. H. Rechinger] 150: 417 (1982); Hedge in Fl. Turkey [P. H. Davis] 7: 418 (1982); Jamzad in Flora Iran [Assadi & al., in Persian] 76: 824 (2012); Taifour & El-Oqlah, Pl. Jordan Annot. Checklist: 104 (2017).

Perennial herb, more or less suffruticose at the base. Stems several, 20–80 cm, often purplish, ascending or erect, densely glandular pilose–villous, sometimes with long eglandular-villous hairs. Leaves pinnatisect, with an ovate to oblong terminal segment 2.5– 7(–11) × 1.5–3.5(–6) cm and 1–2(–5) pairs of smaller lateral segments, ± densely eglandular-

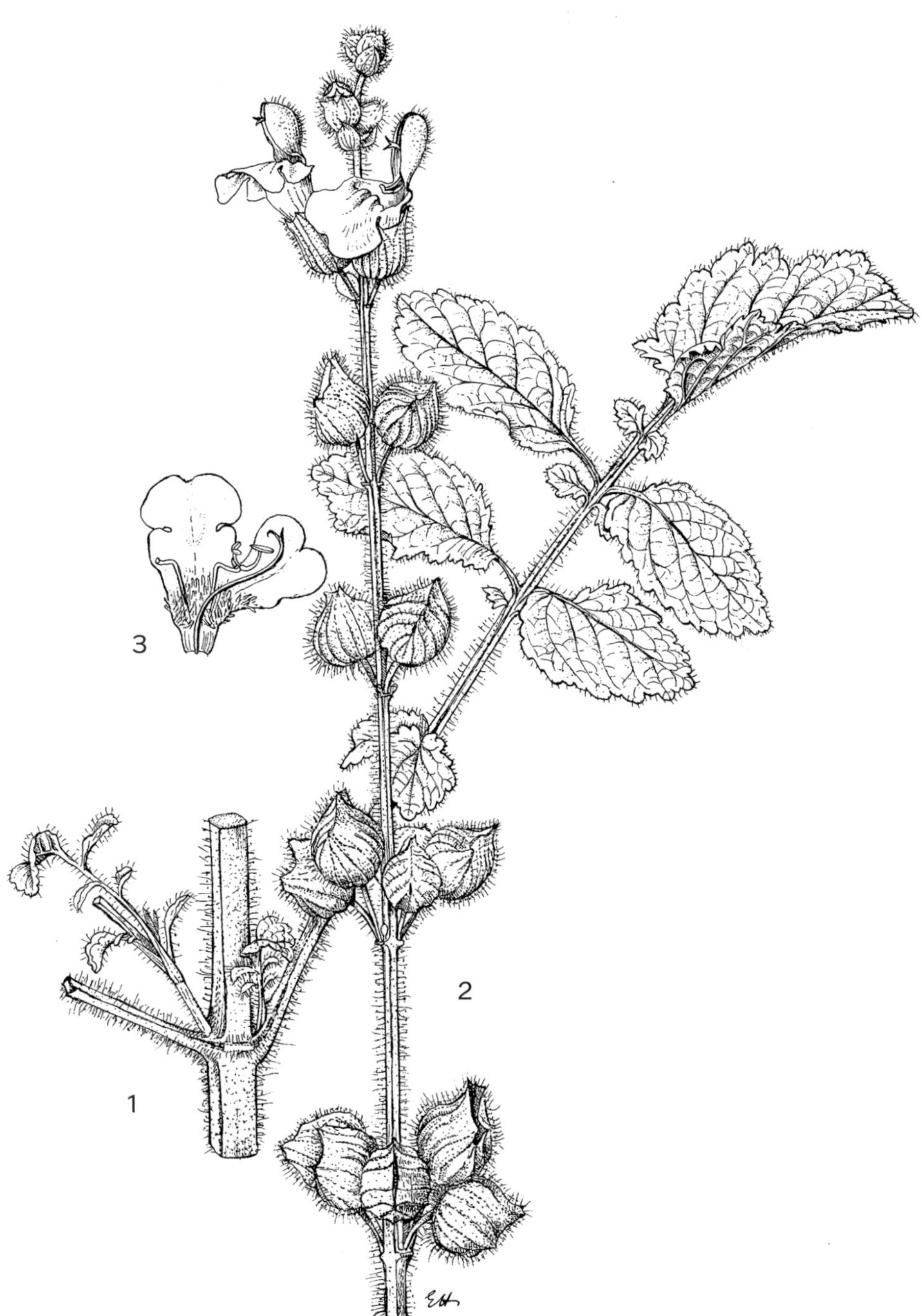

Fig. 141. **Salvia pinnata**. 1, habit with leaves × 1; 2, flowering stem × 1; 3, flower, opened showing stamens and style × 1. Reproduced with permission from Feinbrun-Dothan, Fl. Palaestina 3: Plates, f. 220 (1977). Drawn by Esther Huber. © The Israel Academy of Sciences and Humanities.

Fig. 142. **Salvia bracteata**. 1, habit with leaves × 1; 2, flowering stem × 1; 3, calyx × 1. Reproduced with permission from Feinbrun-Dothan, Fl. Palaestina 3: Plates, f. 221 (1977). Drawn by Esther Huber. © The Israel Academy of Sciences and Humanities.

pilose, serrulate; petiole 1–5 cm, sparsely ciliate. Inflorescence paniculate. Verticillasters 5–10-flowered, distant, ± enclosed by sub-membranous floral leaves, 15–30 × 9–17 mm. Bracts many, purplish; bracteoles present. Pedicels 1–5 mm. Calyx tubular-infundibular, 10–16 mm, scarcely expanding in fruit, glandular-villous. Corolla pink to purplish, 20–35(–42) mm; tube 14–25 mm, irregularly annulate; upper lip straight. Stamens type A. Fig. 142, 1–3.

HAB. Rocky slopes, with *Quercus* woodland, fallow fields, roadsides, clay soils, waste places; alt. 900–2000(–2400) m; fl. & fr. Jun.
DISTRIB. N.W. mountains. **MAM**: Zawita, Scrub area, June 1955, *V.C. Robertson* 143.RC. 22 (K). **MRO**: Kuh Safin supra Shaqlawa, N.E. of Arbil, 1100 m, *Bornmüller* 1721! **MSU**: M. Avroman prope Tawilla, 1500 m, *Rechinger* 10242!; Avroman Mountain, 24 June 1960, *Rawi et al.* 29514 (K); Kamasrpa between Halabja and Tawela, *Rawi* 22177 (K); Penjwin, *Rawi* 12260 (K).

N. & N.E. Iraq, Turkey, Iran, Syria, Palestina, Israel, Jordan and Lebanon.

6. **Salvia trichoclada** *Bentham* in DC., Prodr. [A.P. de Candolle] 12:267 (1848); Rawi in Dep. Agr. Iraq Tech. Bull. 14: 155 (1964); Hedge in Fl. Iranica [K. H. Rechinger] 150: 418 (1982); Hedge in Fl. Turkey [P. H. Davis] 7: 419 (1982).

S. trichoclada is very similar to *S. bracteata* but stems have very long spreading eglandular hairs and glandular hairs; floral leaves smaller (to 12–20 × 6–11 mm); bracts smaller; verticillasters with fewer (2–8) flowers; pedicels longer, to 3–6 mm; fruiting calyces larger with diverging lips, 15–20(–25) mm; corollas longer, 20–40 mm.

HAB. Rocky limestone slopes, hillsides, open forest *Quercus* scrubs, shades of *Quercus* in deep soil, grassy places, fallow fields, rocky clay soils; alt. 900–1700) m; fl. & fr. Apr.-Jun.
DISTRIB. N.W. mountains. **MAM**: Jabal Bakhair prope Zakho, *Field & Lazar* 788!, ibid., *Rawi* 8452! (K); Sarsang, *Haines* 533! (K); Sarsang, slopes of Qara Dagh, *Chapman* 26381! (K); Zakho pass, *Guest* 2248! (K); nr. Zakho, *Karim, Al Dabbagh & Hamad* 44905! (K); Sharanish, N. Zakho prope Marsis, *Rechinger* 10883! (W); Mosul, 8 km S. Zakho, *Rechinger* 10700! (W); Dohuk, *Guest* 2153! (K); N.N.E. of Mosul, Shaik Adi village prope Ayn Sifni, *Field & Lazar* 701! (K); Khantur Mountain, *Rawi* 23312! (K); Jabal Khantur, *Rechinger* 10785!; ibid, *Rawi* 8614! (K); Amadiya, *Anders* 2274 (K); Zawita, 21 km E. Dohuk to Amadiya, *Rechinger* 11545 (W); Swara Tuka, mountain pass, c. 25 km W.S.W. of Amadiya, rocky clay soil, 30 May 1977, *Al Dabbagh & Jasim* 46842! (K); 13 km from Amadiya toward Sarsang, between *Quercus* trees, *Sharief & Hamad* 50331! (K); Sandur village, near Dohuk, on the road to Zawita, *Omar et al.* 49686! (K). **MRO**: Arbil, Kuh Safin prope Shaqlawa, *Bornmüller* 1721! (K); ibid, *Haines* 652! (K); Chinarock (Chinaruk) village, c. 6 km N.E. by E. of Koi Sanjak, N. of Haibat Sultan dagh, *Rawi et al.* 28291! (K); near Dargala village, 4–5 km S.S.W. of Razinok, *Rawi et al.* 28879! (K); Gali Warta, N.W. by N. of Rania, *Rawi et al.* 28728! (K); Galli ali Beg, near the last bridge to Rowanduz, *Kass, Nuri & Serhang* 27205! (K). Mt. Sakri Sakran prope Rawanduz, *Bornmüller* 1736! (K); Haji Omran, *Rechinger* 11317 (W). **MSU**: S. Kanian mountain left to Dokan, in the valley between rocks, *Botany Staff* 43128! (K); Qara Dagh, *Rawi* 9288! (K); Pir Omar Gudrun (Jabal Birah Makrun) above Qarachatan village, *Gillett* 7728! (K); ibid., Zawiya, Pir Omar Gudrun, *Rawi* 11534! (K); S.W. Sulaymaniya, *Stutz* 1396! (K); between Suleymaniyah and Dokan, prope Surdash, *Rechinger* 10103 (W). **MJS**: Dair Asi, *Rawi* 8504! (K).

A problematical species which intergrades with *S. bracteata*. It may be restricted to more natural habitats than *S. bracteata*. The distribution areas of *S. bracteata* and *S. trichoclada* are similar, but *S. bracteata* extends into most of Anatolia (the Asian part of Turkey). *S. bracteata* and *S. trichoclada* are closely related species, and several intermediate forms are present. Intermediate forms based on the character of very long eglandular spreading hairs on the stem of *S. trichoclada* are recorded in S.E. & E. Anatolia and N. Iraq. With further studies *S. trichoclada* could be evaluated under *S. bracteata* as a subspecies.

N.W. & W. Iran, S.E. Turkey.

7. **Salvia suffruticosa** *Montbret & Aucher*, in Ann. Sci. Nat. ser. 2, 6: 39 (1836); Hedge in Fl. Iranica [K. H. Rechinger] 150: 419 (1982); Hedge in Fl. Turkey [P. H. Davis] 7: 426 (1982) Jamzad in Flora Iran [Assadi & al., in Persian] 76: 831 (2012).

S. alexandri Pobed. in Fl. URSS 21:653 (1954).
S. vanensis Boiss. & Noë, Diagn. Pl. Orient. Ser. 2, 4: 17 (1859).

Perennial herb, somewhat suffruticose at base. Stems many, ascending, 30–60 cm, branched above, often yellowish-green, usually glabrous, occasionally pilose. Leaves pinnatisect, ovate in outline, terminal segment lanceolate, cuneate at base, acute, sharply dentate, 2.5–5.5 × 0.5–2 cm, with 2–3(–4) pairs of smaller lateral segments, pilose on veins with some 2–3(–4) pairs of smaller lateral segments, pilose on veins with some sessile glands, serrulate to serrate; petiole 0.6–3 cm with ciliate hairs. Inflorescence axis with short stipitate glands. Verticillasters 2–8(–10)-flowered, distant. Bracts ovate-acuminate, 8–17 × 4–8 mm; bracteoles present. Pedicels 3–8 mm. Calyx campanulate, 8–12 mm, to c. 15 mm in fruit and broadening, subglabrous to pilose-villous with dark subsessile glands; upper lip tridentate. Corolla sulphur yellow or purplish, 22–25(–30) mm; tube c. 15 mm, annulate; upper lip ± straight. Stamens type A.

HAB. Mixed shrubland, fallow fields, among shrubs; alt. 900–1400 m; fl. & fr.
DISTRIB. MSU: Halabja, *Rawi* 8920! (K). MJS: 36 km East of Karsi (Kursi), on North slope of Jabal Sinjar, *Barkley & Brahim* 8003! (K).

The species is characterized by large yellow flowers, more or less glabrous stems and pinnatisect leaves. Some specimens with characters intermediate between *S. suffruticosa* and *S. bracteate* have been recorded from East Anatolia and Syria; these may be found in N. Iraq.

N.W. & W. Iran, Turkey, Armenia, Syria.

8. **Salvia compressa** Vent., Descr. Pl. Nouv. 59 (1801). Lectoype: Iraq: Hort. Cels, *Ventenat s.n.* (G, photo!). Rawi in Dep. Agr. Iraq Tech. Bull. 14: 155 (1964); Hedge in Fl. Iranica [K. H. Rechinger] 150: 426 (1982); Jamzad in Flora Iran [Assadi & al., in Persian] 76: 878 (2012).

S. polyclonos Boiss., Diagn. Pl. Or. Nov. Ser. 1, 7: 47 (1846).
Stiefia compressa (Vent.) Soják, Čas. Nár. Muz. Praze, Rada Přír. 152(1): 22 (1983).

Perennial herb with a thick rootstock. Stems usually single, 12–45 cm, eglandular-pilose and arachnoid hairy. Leaves simple, mostly basal, broadly ovate, ovate, elliptic, 3–14 × 2–10 cm, sparsely eglandular-pilose, cordate or rounded at base, margin erose or irregularly crenulate; petiole 0–6 cm. Inflorescence widely paniculate. Verticillasters 4–10-flowered, usually distant. Bracts broadly ovate and ovate-oblong, greenish or purplish, 12 × 14 mm; bracteoles present. Pedicels 2–4 mm long. Calyx more or less infundibular, purplish or greenish, 10–12 mm in flower and looks thick texture, to c. 20 mm diameter in fruit, expanding and membranous, eglandular-pilose and villous with some sessile glands, purplish, rarely yellowish-green; lobes short and obtuse. Corolla pinkish-purple to dark purplish, c. 15; tube slightly ventricose, 12–14 mm, inside glabrous; upper lip ± straight. Stamens type A. Nutlet ovoid, 2.5 × 2 mm, light brown, venose. Fig. 143, 1–6.

HAB. Clay and sandy gravelly hill, gypseous hills, stony hills, mixed shrubland with *Artemisia* and *Ranunculus,* on hillside with *Achillea, Ranunculus* and *Poa,* in rich loam and conglomerate rocks on hillside; alt. 60–1000 m; fl. & fr. Apr.
DISTRIB. MRO: 31 km N. of Kirkuk to Koysinjaq, *Rawi, Noori & Al-Kaisi* 27950! (K). MSU: Kirkuk, Prope Darbandikhan, *Haines* 1927! (K); valley near the Silwan River, 7 km south of Derbendikhan, Sulaimaniya Liwa, *Barkley & Haddad* 5145 (K); Mt. Avroman prope Sosakan, *Rechinger* 10196 (W). FKI: Near Baba Gurgur (Kurkar), 20 km W. of Kirkuk, *Barkley & Brahim* 1332! (K); Kifri, *Gillett & Rawi* 7379! (K); between Tuz Khurmatu and Tauq, *Gillett & Rawi* 7456! (K). FKI/FPF: Jabal Hamrin, *Surol?* 1433! (K); Jabal Hamrin, *Haines s.n.* (K). FPF: Diyala, 10 km S. Khanaqin, *Rechinger* 9070! (W); S.E. of Baghdad, Bagsaya, *Rawi & Haddad* 25563! (K); Badra, *Gillett* 6639! (K); Mandali, *Kaisi & Khayat* 50787! (K); 10 km E. Mandali, *Rechinger* 12763 (W); 5 km from Saadiya to Khanaqin, *Kaisi & Hamid* 46636! (K); Naft Khina (Khanaqin), *Rust* 1853! (K). DGA: Sudur (near Mansur, Delli Abbas), *Omar, Fawzi, Al-Kaisi & Noori* 44322! (K). LEA: Kut al Imara, 16–30 km S.E. Badrah, *Rechinger* 9170, 13934 (W).

The name *Salvia compressa* was validated in descriptions in 1801 based on a specimen cultivated in Cels's garden originating from a Bruguière and Olivier collection from Iraq (between Mosul and Baghdad). The only specimen in Ventenat's herbarium from Cels's garden is designated as the lectotype with a duplicate in C. Four specimens collected between Mosul and Baghdad by Bruguière and Olivier have been located in G–BOISS [G00330751] and P [P02888639, P02888640, P02888641]. They are considered as elements of original material by Callmander et al. (2017). Hedge (1982: 426) in his revision for Fl. Iranica did not specify the herbarium where the type specimen has been

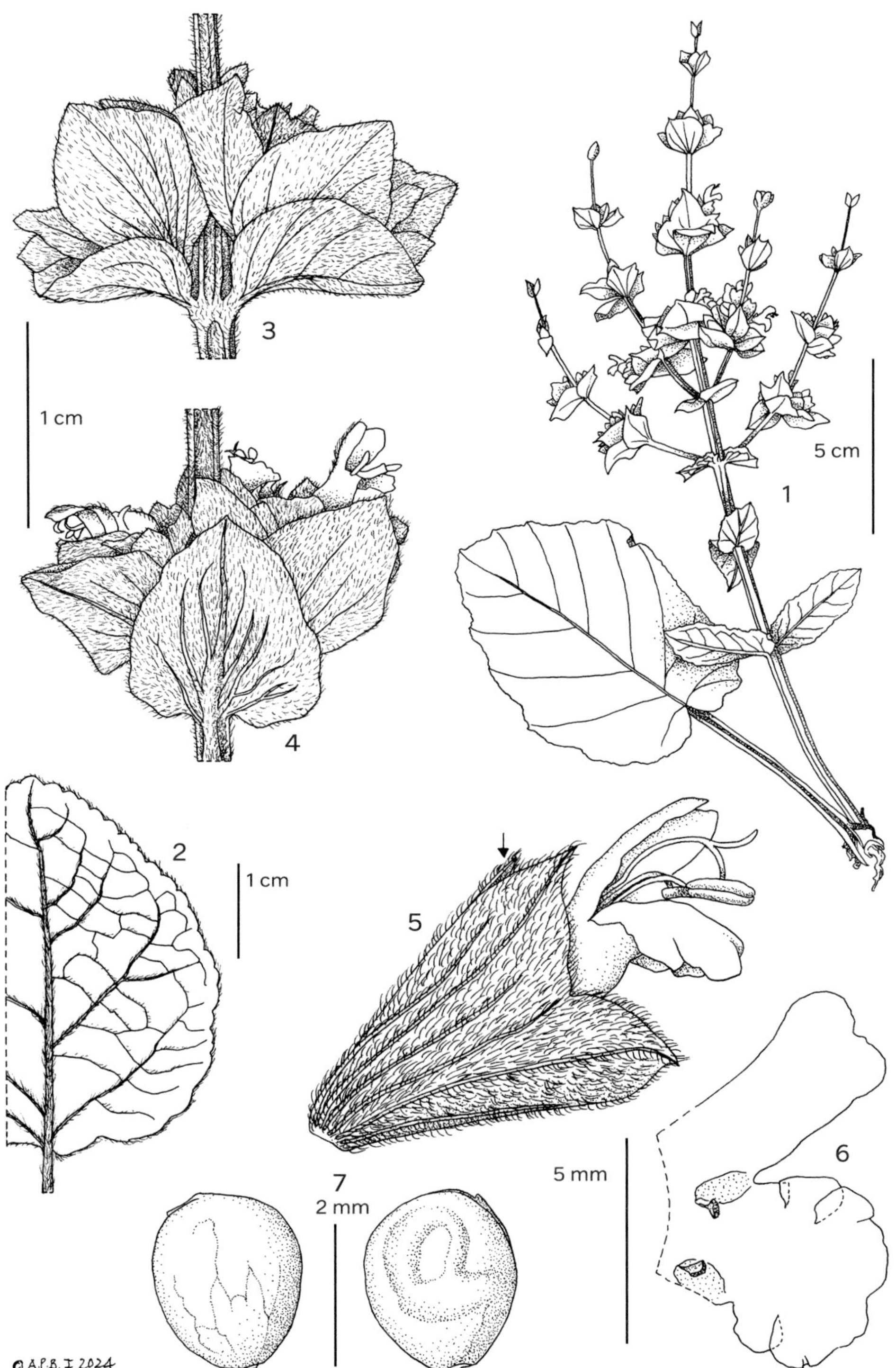

Fig. 143. **Salvia compressa**. 1, habit; 2, basal leaf; 3,4, inflorescences, detal; 5, flower lateral view; 6, upper and lowe lobes of corolla flattened. 1 from *Barkley et al.* 1332; 2 from *Omar et al.* 37870; 3–6 from *Wheeler Haines* W1927; 7 from *Rawi et al.* 25563. Drawn by © A.P. Brown 2024.

deposited; this cannot be considered as an implicit lectotypification (Callmander et al. 2017).

Iran, Afghanistan.

9. **Salvia multicaulis** *Vahl,* Enum. Pl. 1: 225 (1805); Dinsmore in Post, Fl. Syria, Palest. & Sinai ed 2, 2: 353 (1933); Feinbrun-Dothan, Fl. Paleast. 3: 136 (1978); Hedge in Fl. Iranica [K. H. Rechinger] 150: 424 (1982); Hedge in Fl. Turkey [P. H. Davis] 7: 432 (1982); Jamzad in Flora Iran [Assadi & al., in Persian] 76: 839 (2012); Taifour & El-Oqlah, Pl. Jordan Annot. Checklist: 105 (2017).

S. acetabulosa Vahl, op. Cit.227 (1805) non L. (1767) excl.
S. szovitsiana Bunge, Lab. Pers. 43 (1873).
Schraderia 'acetabulosa' (Vahl) Pobed. in Fl. URSS 21:369 (1954).
Arischrada multicaulis (Vahl) Pobed. in Novit. Syst. Pl. Vasc. (Leningrad) 9:247 (1972).
Stiefia multicaulis (Vahl) Soják in Cas. Nár. Mus., Odd. Prîr. 152: 22 (1983). Ic: Fl. URSS 21: t.18 f. 1 (1954), as *Schraderia acetabulosa.*

Perennial herb, mat-forming, with a woody rootstock. Stems ascending to erect, 12–55 cm, usually glandular-pilose to villous, especially above, rarely glabrous, occasionally with dendroid hairs. Leaves simple, rarely with 1–2 pairs of small basal lobes, broadly ovate-elliptic to suborbicular, 2–4.5 (–7) × 1–3.5 cm, rugose, crenulate, with a dense indumentum of adpressed dendroid to dendroid-stellate hairs; petiole 1.5–6 cm. Verticillasters 4–10-flowered, usually distant. Bracts broadly ovate, 15–10 mm; bracteoles present. Pedicels 2–4 mm, erecto-patent. Calyx campanulate, 12–17 mm, to 22 mm in fruit and broadening, entire, sparsely to densely glandular-pilose or villous, purplish, rarely yellowish-green; upper lip indistinctly 3-lobed. Corolla purplish, rarely white, 16–22 mm; tube straight, 12–14 mm, annulate; upper lip straight. Stamens type A. Fig, 144, 1–3.

HAB. Rocky limestone and igneous slopes, shale and sandy slopes, clay soil, scree dry slopes, fallow fields, in *Quercus* scrub, in *Artemisia* zone, mountain plain with *Quercus* and *Crataegus*; 300–1700 m; fl. & fr. Mar.-Jun.
DISTRIB. N.W. mountains and lower hills of northern Iraq. **MAM**: Dohuk, *Guest* 13251! (K); Amadiya, clay with organic matter in rocky mountain, *Botany Staff* 43415! (K); Sarsang, *Botany Staff* 43444! (K); Khantur Mountain, N.E. of Zakho, *Rawi* 23293! (K); Zawita Gorge, *Guest* 2205! (K). **MRO**: Near Dergala (Dargala) village, 5 km S.S.W. of Razanuk (Razinok), *Rawi et al.* 28881! (K); N.E. of Arbil, Shaqlawa, *Haines* W659! (K); Arbil, 19 km S.W. of Salahaddin, *F.A. Barkley & E.D. Barkley* 5633! (K); Kawriesh, E. Side of Karoukh Mountain, *Kass & Nuri* 27665! (K); Khanzad, hill pass, c. 20 km E. of Arbil, Rowanduz, *Guest* 2008! (K). **MSU**: S.W. of Dukan, Shewasür (Waddy Al-Ahmet, N.E. of Kani Hanjir), *Omar et al.* 37118! (K); Bakrajo, *Omar & Kasim* 38005! (K); nr. Chemchemal, *Rogers* 0236! (K); 5 km S.W. of Chemchemal, *Gillett & Rawi* 7581! (K); Sulaimania, Azmir, *Omar & Kasim* 38043! (K); 3 km S. of Sulaimaniye, *Barkley & Haddad* 7480! (K); Jarmo, *Helbaek* 320! (K). **MJS**: Jabal Sinjar, *Kaisi & Hamad* 49134! (K); ibid, Kursi, Jabal Sinjar, *Gillett* 11007! (K). **FUJ**: 30 km N. of Mosul, *Al-Dabbagh* 46745! (K). **FNI**: Mosul, Jabal Maqlub, *Polunin et al.* 85! (K); 33 km N.E. of Mosul, *Tawfig, Dabbagh & Lafta* 44120! (K). **FAR**: Arbil, South of Bastora Wadi, near Khanzad, *Barkley & Abbas Al-Ani* 7174! (K). **FPF**: Miqdadiyah, Jazira, *Omar* 42807! (K); North of Khanaqin, near Chia Surtek (Hanekin), left bank of the Diyala, *Poore* 386! (K).

Iran, Turkey, Syria, Palestina, Sinai, Jordan.

10. **Salvia viridis** *L.,* Sp. Pl. 1: 24 (1753); Hedge in Fl. Iranica [K. H. Rechinger] 150: 434 (1982); Hedge in Fl. Turkey [P. H. Davis] 7: 434 (1982); Jamzad in Flora Iran [Assadi & al., in Persian] 76: 849 (2012); Taifour & El-Oqlah, Pl. Jordan Annot. Checklist: 105 (2017).

S. horminum L., Sp. Pl. 24(1753)! Ic: Sibth. & Sm., Fl. Graeca 1: t. 19 & t. 20 (1806), as *S. horminum;* Hegi, 111. Fl. Mittel–Eur. 5(4): f. 2488, 2489 (1927).
Sclarea viridis (L.) Soják Cas. Nár. Muz. Praze, rada Přir. 152(1): 22 (1983).

Annual. Stems simple or branched, (7–) 25 (–45) cm, glandular or eglandular pilose, topped by a usually purplish or pink coma (rarely white) or coma absent. Leaves simple, oblong-ovate, 2–7 × 1–3 cm, finely crenulate, obtuse, rugose; petiole 2–5 cm. Verticillasters 4–6-flowered, distant or approximating. Bracts broadly ovate, c. 10×10 mm; bracteoles present. Verticillasters 4–6-flowered. Pedicels 2–3 mm, erect, flattened. Calyx tubular, 7–12 mm, to c. 13 mm in fruit and strongly deflexed, eglandular-pilose or glandular; upper lip ± truncate. Corolla pinkish-purple to white, 12–15 mm; tube straight, glabrous within; upper lip broad, ± falcate. Stamens type B.

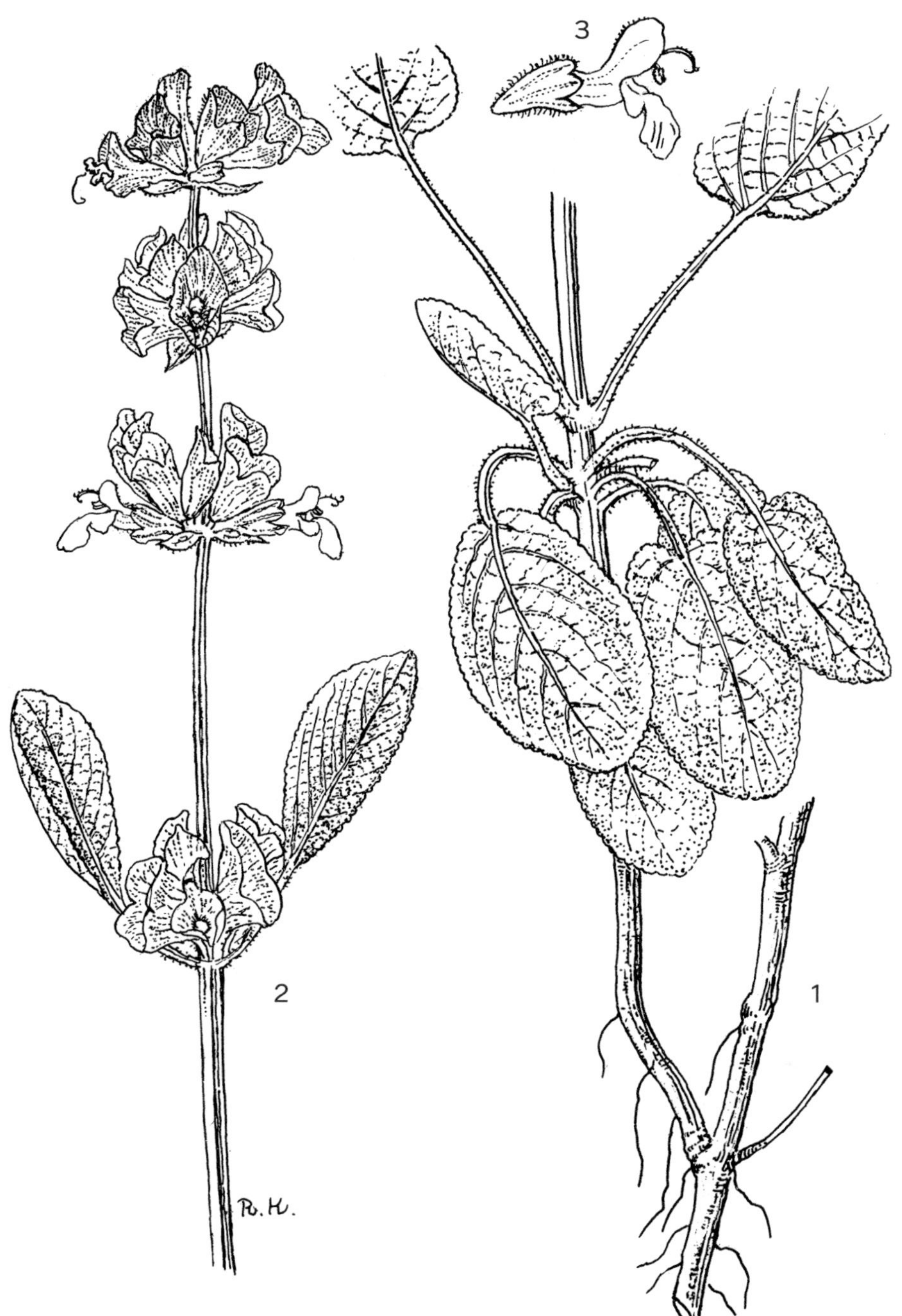

Fig. 144. **Salvia multicaulis**. 1, habit with leaves × 1; 2, flowering stem × 1; 3, flower × 2. Reproduced with permission from Feinbrun-Dothan, Fl. Palaestina 3: Plates, f. 222 (1977). Drawn by Ruth Koppel. © The Israel Academy of Sciences and Humanities.

HAB. Gravelly and clay soil, slopes; alt. 400–500 m; fl. & fr. Apr.
DISTRIB. N. Iraq. **MSU**: Kirkuk, Darbandikhan, *Haines* 2158! (K); 1 km south of Darbandikan, *Barkley & Ihsan Ali Shehbaz* 7718! (K).

N. Iraq, most of Mediterranean area, Iran, Turkey, Crimea, Caucasia, Turcomania.

11. **Salvia syriaca** *L.*, Syst. Nat., ed. 10: 854 (1759); Rawi in Dep. Agr. Iraq Tech. Bull. 14: 155 (1964); Dinsmore in Post, Fl. Syria, Palest. & Sinai ed 2, 2: 354 (1933); Feinbrun-Dothan, Fl. Paleast. 3: 137 (1978); Hedge in Fl. Iranica [K. H. Rechinger] 150: 435 (1982); Hedge in Fl. Turkey [P. H. Davis] 7: 434 (1982); Jamzad in Fl. Iran [Assadi & al., in Persian] 76: 849 (2012); Taifour & El-Oqlah, Pl. Jordan Annot. Checklist: 105 (2017).

Sclarea syriaca Mill. Gard. Dict., ed. 8. n. 5. (1768).
Salvia parviflora Vahl. Enum. Pl. 1: 268 (1804).
S. varia Vahl. Enum. Pl. 1: 273 (1804).

Perennial herb; rhizomatous. Stems 15–35 (–80) cm, yellowish-green, erect, simple or branched, eglandular-pubescent to villous below, denser above (and rarely glandular). Leaves simple, obtuse or acute, ovate, cordate, 5–16 × 3–10 cm, rugose, slightly erose to serrulate, shortly eglandular or glandular-pubescent below with sessile glands, rugose; petiole 3–6 cm. Verticillasters 2–6-flowered, distant. Bracts ovate, c. 5 × 5 mm. Pedicels 3–4 mm, erecto-patent. Calyx tubular, 5–10 mm, 10–12-veined, densely eglandular or glandular-pubescent, upper lip straight, tridentate. Calyx scarcely expanding in fruit. Corolla white to cream, 8–12 mm; upper lip slightly falcate, tube straight, glabrous within; upper lip ± straight to slightly falcate. Stamens type B. Fig. 145, 1–4.

HAB. Mixed shrubland, marly banks, fallow and cultivated fields, in *Quercus* forest; alt. 270–1180 m; fl. & fr. Apr.
DISTRIB. Mountains and foothills of N. Iraq. **MAM**: Between Zakho and Dohuk, *Guest* 2231! (K); Dohuk, *Guest* 2162! (K). **MRO**: Arbil, Kuh Safin prope Shaqlawa, *Bornmüller* 1723 (K); Salahaddin, *Gillett* 8179! (K). **MSU**: Kirkuk, Darband-i-Bazian, prope Chamchamal, *Rechinger* 10588 (W); ibid, *Gilllett & Rawi* 7623! (K); Halabja, *Rawi* 8921! (K). **FUJ**: Tal ash Shaur, between Tal Afar and Balad Sinjar, *Field & Lazar* 594 (K); Mahad prope Shaikhan, *Salim Effendi* 2610 (K); Tal Afar, *Ahmad & Jabar* 50156! (K). **FNI**: Tal Kaif village, *Al-Kaisi & Hamad* 49272! (K); 4 km to Tal Kaiff, *Botany Staff* 43395! (K). **FAR**: Arbil, *Bornmüller* 1724 (K); ibid, *Radhi* 3850! (K).

Turkey, Iran, Caucasus, Armenia, Palestine, Jordan.

12. **Salvia montbretii** *Benth.*, Ann. Sci. Nat. ser. 2, 6:42 (1836); Hedge in Fl. Iranica [K. H. Rechinger] 150: 436 (1982); Hedge in Fl. Turkey [P. H. Davis] 7: 436 (1982).

S. montbretii sensu Boiss., Fl. Orient. 4:611 (1879) p.p.

Perennial herb forming tufts to 60 cm in diameter. Stems erect, sturdy, 25–50(–60) cm, usually unbranched, eglandular, arachnoid to lanate below, glandular pilose above. Leaves simple, linear to linear-oblong, mostly basal, greenish above, white lanate below, margins subentire or suberose, 4–13 × 1–2 cm, including the indistinct petiole. Verticillasters 4–8-flowered, clearly distant. Bracts broadly ovate, 20–30 × 25–35 mm, acuminate, lower surface lanate. Pedicels 2–4 mm, ± erect. Calyx tubular-ovate, c. 15 mm in flower, to c. 18 mm in fruit, lanate and glandular; upper lip tridentate with mucronate teeth. Corolla pinkish purple to purplish-blue, 22–28 mm; tube straight, slightly ventricose above; upper lip falcate. Stamen type B.

HAB. Hillside, calcareous slopes, *Quercus* scrub, road side, fallow fields; alt. 700–1150 m; fl. & fr. Apr.-May.
DISTRIB. N. Iraq. **MAM**: Mosul, Sirsang, *Anders* 2333!, *Haines* 1047! (K); Sarsang to Sawara Tuka, Karim, *Hamid & Jasim* 41059! (K). **FUJ**: Mosul, *Kotschy* 1841! (K).

S.E. Anatolia, Syria.

13. **Salvia spinosa** *L.*, Mant. Pl. Altera 511 (1771); Dinsmore in Post, Fl. Syria, Palest. & Sinai ed 2, 2: 355 (1933); Rawi in Dep. Agr. Iraq Tech. Bull. 14: 155 (1964); Feinbrun-Dothan, Fl. Paleast. 3: 137 (1978); Hedge in Fl. Iranica [K. H. Rechinger] 150: 444 (1982); Hedge in Fl. Turkey [P. H. Davis] 7: 436 (1982); Jamzad in Flora Iran [Assadi & al., in Persian] 76: 875 (2012); Taifour & El-Oqlah, Pl. Jordan Annot. Checklist: 105 (2017).

Salvia aegyptiaca L., Mant. 26 (1767) non L. (1753).
Horminum suaveolens Moench, Methodus 377 (1794).
Salvia praecox Vahl, Enum. Pl. i. 274 (1804).

Fig. 145. **Salvia syriaca**. 1, habit × 1; 2, flowering stem × 1; 3, flower × 3; 4, calyx × 3. Reproduced with permission from Feinbrun-Dothan, Fl. Palaestina 3: Plates, f. 223 (1977). Drawn by Esther Huber. © The Israel Academy of Sciences and Humanities.

Fig. 146. **Salvia spinosa**. 1, habit, leaf × 1; 2, flowering stems × 1. Reproduced with permission from Feinbrun-Dothan, Fl. Palaestina 3: Plates, f. 224 (1977). Drawn by Esther Huber. © The Israel Academy of Sciences and Humanities.

Sclarea spinosa Raf., Fl. Tellur. 3: 94 (1837).
Salvia doryphora Stapf, Denkschr. Kaiserl. Akad. Wiss., Wien. Math.-Naturwiss. Kl. 1. 41 (1885).
S. distincta Grossh. Věstn. Tiflissk. Bot. Sada 35: 41 (1914).

Perennial herb. Stems single, erect, quadrangular, 30–50 cm, branched above, glandular-pilose to villous below, densely glandular above. Leaves simple, ± thin-textured, broadly ovate to ovate-oblong, 8–12 × 4.5–8 cm, ± tomentose, cordate to rounded, subentire to erose; petiole 3–11 mm. Inflorescence paniculate; verticillasters 2–6-flowered, distant. Bracts broadly ovate, acuminate, 20 × 16 mm. Pedicels c. 3 mm, erecto-patent. Calyx tubular, c. 20 mm, greenish yellow, slightly expanding and hardening in fruit to c. 22 mm, glandular-hirsute; upper lip equally tridentate, spiny in fruit. Corolla white, c. 24 mm; tube straight, c. 17 mm, slightly widening above, glabrous within; upper lip falcate. Stamens type B. Nutlets ± spherical, rounded trigonous, c. 3 × 2.7 mm. Fig. 146, 1–2.

HAB. Disturbed mixed shrubland, fallow fields, roadsides, sandy, silty and clay soil, associated with *Capparis*, among *Tamarix* forest; alt. 10–650 m; fl. & fr. Mar.-May.
DISTRIB. Widely distributed in the foothills, plains and desert areas of Iraq. **MSU**: Sulaymaniya, *Thesiger* 348! (K). **FUJ**: Between Mosul and Tal Afar, on road side, *Karim* 37473! (K); Mosul, *Lazar* 3342! (K). **FAR**: Inter Kirkuk et Arbil, *Thesiger* 451 (K). **FKI**: Kirkuk, 10–18 km E. Kirkuk, *Rechinger* 10004 (W); inter Kirkuk et Altin Köprü, *Bornmüller* 1732! (K); Kirkuk, *Bornmüller* 1731! (K). between Kirkuk and Baba Gurgur, *Guest* 1353! (K); 25 km south of Tuz Khurmatu on the way to Khalis, *Al-Ani* 9812! (K); Tuz Khurmatu, *Roger* 0218! (K). **FKI/FPF**: Jabal Hamrin, near Injanai, *Guest* 1434 (K). **DLJ**: Tikrit, 20 km W. of Tikrit, *Alizzi & Husain* 33775! (K); 60 km N.W. of Ramadi, *Alizzi* 35142! (K). **DWD**: 37 km S.E. of T1 (oil pumping station, c. 100 km W. by N. of Haditha) to K3 (c. 8 km S. of Haditha), *Chakravarty, Rawi, Khatib & Alizzi* 31803! (K); 36 km N.E. of Haditha, rocky land, *Chakravarty, Rawi, Khatib & Alizzi* 31838! (K); 87 km from Al Bussayyah to Al Salman, *Chakravarty, Rawi, Khatib & Kirty* 29995! (K); 5 km S. of Rutba, *Chakravarty, Rawi, Khatib & Alizzi* 32884! (K); Rutba, *Field & Lazar* 131! (K); Ramadi-Rutkah Road (20 km W. of Ramadi), *Omar* 34305! (K). **DSD**: Az Zubair, *Rawi* 25967! (K). **LEA**: Nahrwan, *VCR* 1031! (K); between Baghdad and Aziziyah, *Alizzi* 34339! (K); Basra, southern desert, 40 km S.W. of Basra, Haswa, *Rechinger* 8833! (K); Kut al Imara, Badrah, *Rechinger* 9198 (W); Amarah, Shattat-Tib, 70 km N. Amarah, *Rechinger* 14182 (W). nr. Baghdad, Rustam Farm, 15 April 1933, *Lazar* 213! (K). **LCA**: 22 km W. of Falluja, *Chakravarty & Rawi* 30243! (K).

S.E. Turkey, Iran, Transcaucasica, Afghanistan; Egypt, Palestine, Jordan, Lebanon, Syria.

14. **Salvia macrosiphon** *Boiss.* Diagn. Pl. Orient. Nov. Ser. 1, 5: 11 (1844); Hedge in Fl. Iranica [K. H. Rechinger] 150: 441 (1982); Hedge in Fl. Turkey [P. H. Davis] 7: 437 (1982); Jamzad in Flora Iran [Assadi & al., in Persian] 76: 854 (2012).

Salvia kotschyi Boiss., Diagn. Pl. Orient. Nov. 46 (1846).
S. macrosiphon var. *cabulica* Benth. in Prodr. [A.P. Candolle] 12: 282 (1848).
S. macrosiphon var. *kotschyi* (Boiss) Boiss. Fl. Orient. 4: 615 (1879).
S. cuspidatissima Pau, Trab. Mus. Nac. Cienc. Nat. Ser. Bot. Madrid 14: 33 (1918).
S. albifrons Nab., Publ. Fac. Sci. Univ. Masaryk (Brno) 70: 49, tab. 5 (1926).
S. macrosiphon var. *glandulosissima* Bornm., Bot. Jahrb. 62: 238 (1934).
S. macrosiphon var. *brachycalycina* Bornm., Bot. Jahrb. 62 : 238 (1934).
S. nachiczevanica Pobed., Fl. URSS. 21: 657 (1954).

Perennial herbs with a woody rootstock at base. Stems erect, 15–45 (–90) cm, branched above, below scarcely eglandular pilose to tomentose, densely glandular pilose above. Leaves simple, ± thin-textured, elliptic to ovate-oblong, 4–16(–22) × 1.5–9.5(–11) cm, eglandular tomentose, margins irregular serrate to erose. Petiole 2–10 cm. Inflorescence panicle, densely glandular-pilose. Verticillasters 2–6-flowered, internodes 2–2.5 cm. Bracts broadly ovate, 10–27 × 7–20 mm, acuminate, eglandular pilose. Calyx tubular, 12–17 mm, up to 22 mm in fruit, 4–5 mm in diameter, scarcely expanding in fruit, glandular hirsute with some sessile glands, teeth spiny. Corolla white, 18–35 mm, tube 12–22 mm, upper lip falcate, not squamose. Stamens 2, staminal connectives clearly longer than filaments. Style glabrous 20–35 mm, long exerted from corolla lips and divided in two parts at apex. Nutlets 2.8 × 2.3 mm. Fig. 147, 1–3.

HAB. Clayey soils, fallow fields; alt. m; fl. & fr. Apr.-May.
DISTRIB. Mountains and lower hills of N. Iraq. **MAM**: Amedi, *Stapf* s.n.! (K). **MSU**: Kirkuk, Jarmo, *Helbaek* 1170! (K). **FNI**: Mosul, Nimrud (Nimrod), Helbaek 978! (K); *Thesiger* 781! (K).

Iran, S.E. Turkey, Afghanistan, Pakistan, Transcaucasus.

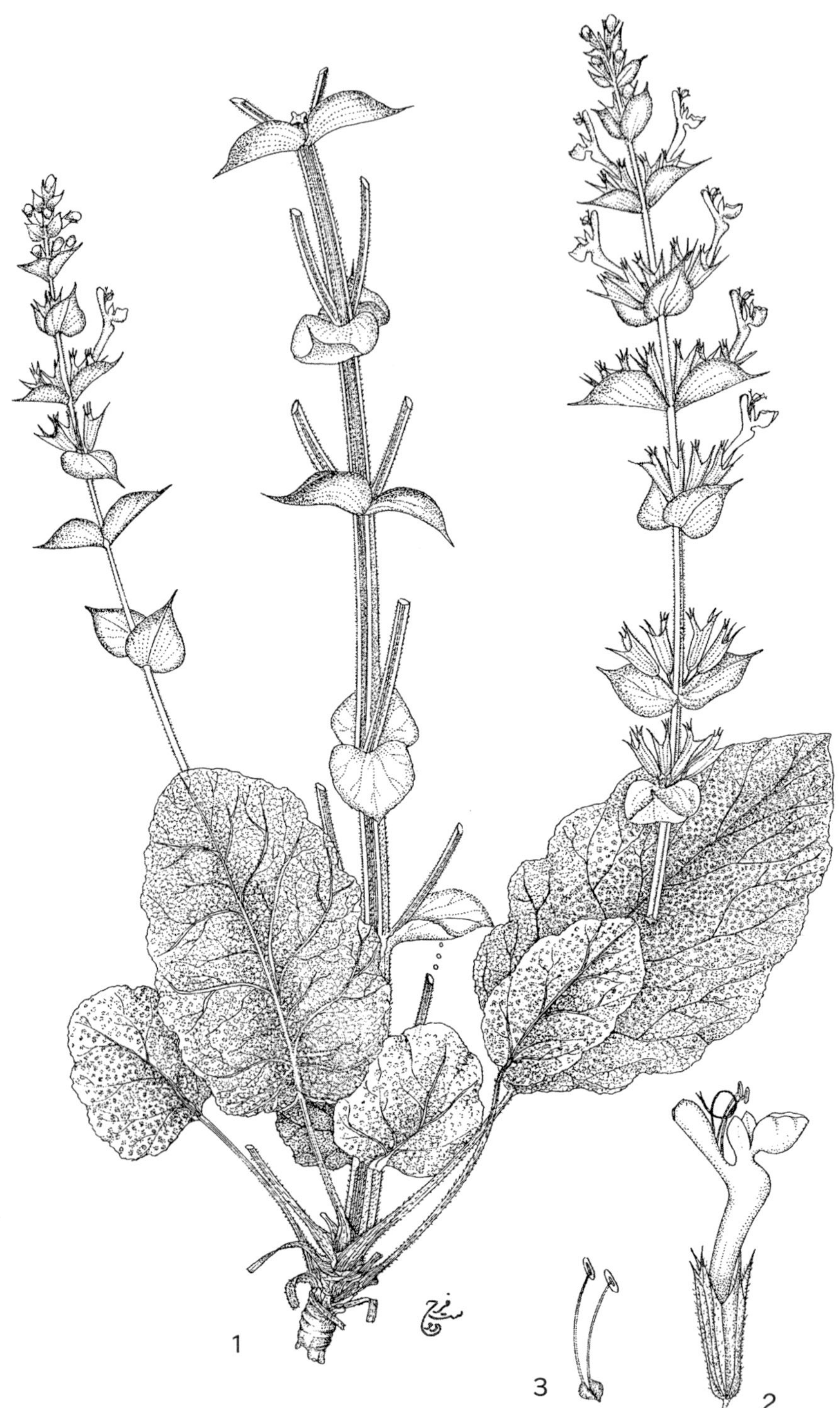

Fig. 147. **Salvia macrosiphon**. 1, habit × 1; 2, flower × 1; 3, stamens detail. Reproduced with permission from Jamzad, Flora of Iran 76: f. 264 (2012). © Ministry of Jihad-e-Agriculture.

15. **Salvia reuterana** *Boiss.*, Diagn. Pl. Orient. Nov. Ser. 1, 5: 10 (1844); Hedge in Fl. Iranica [K. H. Rechinger] 150: 445 (1982); Jamzad in Flora Iran [Assadi & al., in Persian] 76: 861 (2012).

Perennial herb. Stems single, erect, quadrangular, 20–100 cm, dense eglandular-villous below, densely glandular capitate above. Leaves simple, ovate-oblong or suborbicular, 5–13 (–24) × 4–11 (–14) cm, below dense appressed eglandular-pilose and some sessile glandular-pilose, base cordate to rounded, subentire to coarsely serrate; petiole 3–14 cm.

Fig. 148. **Salvia reuterana**. 1, habit × ½; 2, flower × 2; 3, calyx × 2; 4, stamens, detail. Reproduced with permission from Jamzad, Flora of Iran 76: f. 266 (2012). © Ministry of Jihad-e-Agriculture.

Inflorescence widely paniculate; verticillasters 1–6-flowered, distant. Bracts broadly ovate, acuminate, 15–24 × 14–22 mm, 27 × 40 mm at fruting, membranous, yellowish-green or purplish. Pedicels 2–3 mm, erecto-patent. Calyx tubular, 18–20 mm at flowering, slightly expanding and hardening in fruit to 25–30 mm, densely capitate glandular with some eglandular villous and sessile glandular hairy; upper lip tridentate, spiny in fruit. Corolla white or yellowish-white, 24–30 mm; tube straight, 15–20 mm, slightly widening above, glabrous within; upper lip falcate. Stamens type B. Nutlets ± spherical c. 3 × 2.8 mm. Fig. 148, 1–4.

HAB. Clay soils, in plains; alt. 10–650 m; fl. & fr. Jun.
DISTRIB. Mountain and lower hills of N. Iraq. **MAM**: N. Iraq, Bamrni (probably Bamerni, Bamerne, N.E. Sirte, close to Turkish border), *Omar* 37748! (K). **LEA**: Kut al- Imara, 30 km S.E. Badrah, *Rechinger* 9169 (W).

Iran, probably in E. & S.E. Turkey.

16. **Salvia persepolitana** *Boiss.*, Diagn. Pl. Orient. Ser. 1, 12: 60 (1853); Hedge in Fl. Iranica [K. H. Rechinger] 150: 446 (1982); Jamzad in Flora Iran [Assadi & al., in Persian] 76: 864 (2012).

Salvia arabica Al-Musawi et Weinert, Feddes Repert. 100: 451 (1989).

Perennial herb, very aromatic. Stem single, erect, 20–50 cm long, long white glandular hairs and short glandular hairs with eglandular hairs. Leaves simple, oblong to ovate, 4–14 × 2–10 cm, densely villous with stalked glandular and sessile glandular hairs, base rounded to subcordate, margins erose dentate to serrate; petiole 4–11 cm, upper leaves sessile. Inflorescence widely paniculate; verticillasters 1–6-flowered. Bracts broadly ovate, shortly acuminate, 12–20 × 12–16 mm, membranous, yellowish-green. Pedicels 2–3 mm. Calyx campanulate to tubular-campanulate, 10–14 mm at flowering, slightly expanding in fruit to 18 mm, long and short capitate glandular hairs with sessile glandular hairs and eglandular hairs, ± white-scabrid; upper lip tridentate, small teeth closed or distinct small mucronate. Corolla white or pale yellow (or pale lilac), 16–22 mm; tube ± erect, c. 10 mm, not squamulose, ± glabrous; upper lip ± falcate. Stamen type B. Nutlets almost spherical 2.4 × 2 mm

HAB. Low hills, on sand and gravel, conglomerate hills with salty loam and clay soils; alt. 200–250 m; fl. & fr. Apr.-May.
DISTRIB. Foothills of N. Iraq. **FPF**: Wasit, Prov. Zurbatiya, near Iranian border, *Al Bermani* 39190! (BUH); Zurbatiya region, 16 km S. Zurbatiya, *Al Kaisi & Yahya* 45264! (BAG); 8–12 km S.E. Zurbatiya, *Al Musawi* 42111! (BUH); nr. Zurbatiya, *Al Musawi & Addai* 44721! (BUH) & 57955! (BUH).

17. **Salvia sclareopsis** *Bornm. ex Hedge*, Hedge in Fl. Iranica [K. H. Rechinger] 150: 447 (1982); Jamzad in Flora Iran [Assadi & al., in Persian] 76: 870 (2012).

S. reuteriana Boiss. var. *depilata* Bornm., Beih. Bot. Centrbl. 33, 2: 185 (1915).

Perennial herb. Stem single, 15–30(–40) cm long, below appressed short eglandular-pubescent with sessile glandular hairs, above glandular-pilose and short eglandular hairy. Leaves simple, ovate to broadly ovate, 4–10 × 3–8 cm, below softly pubescent and ± pannose with sessile glandular hairs, rugose, base rounded to subcordate, margins small crenate-serrate; petiole 1–10 cm. Inflorescence shortly branched; verticillasters 2–4-flowered. Bracts broadly ovate, 22–35 × 10–22 mm, membranous, whitish, yellowish green and pinkish-purple. Pedicels 2–3 mm. Calyx tubular-infundibular, 14–15 mm at flowering, slightly expanding in fruit, with short capitate glandular hairs and eglandular hairs on nerves; upper lip subequally tridentate with small spinolose teeth. Corolla white (?), 16–28 mm, tube 11–16 mm, gradually broadening, not squamulose. Stamen type B. Fig. 149, 1–4.

HAB. Gravelly wadis, on sandy and silty soils; alt. 120–450 m; fl. & fr. Apr.-May.
DISTRIB. Mountains and lower hills in N. Iraq. **MRO**: Qara village, on road to Kai Sanjak, *Rawi, Nuri & Al Kass* 28021! (K). **MSU**: S.W. of Dukan, Shewasŭr (Wadi Al Ahmet, N.E. of Kani Hanjir), *Omar et al.* 37120! (K). **FPF**: Nr Badra, probably between Badra and Mandali, *Al-Kaisi & Khayat* 50741 (BAG, photo!, det. Ali Haloob, unusual specimen).

Fig. 149. **Salvia sclareopsis**. 1, habit × 1; 2, flower × 2; 3, calyx × 2; 4, stamens, detail. Reproduced with permission from Jamzad, Flora of Iran 76: f. 270 (2012). © Ministry of Jihad-e-Agriculture. Fl. Iran.

18. **Salvia palaestina** *Benth.*, Lab. Gen. Spec. 718 (1835); Dinsmore in Post, Fl. Syria, Palest. & Sinai ed 2, 2: 355 (1933); Rawi in Dep. Agr. Iraq Tech. Bull. 14: 155 (1964); Feinbrun-Dothan, Fl. Paleast. 3: 137 (1978); Hedge in Fl. Iranica [K. H. Rechinger] 150: 448 (1982); Hedge in Fl. Turkey [P. H. Davis] 7: 437 (1982); Taifour & El-Oqlah, Pl. Jordan Annot. Checklist: 105 (2017).

> *Salvia sinaica* Delile ex Benth. Labiat. Gen. Spec. 718 (1835).
> *S. lorentii* Hochst. in Lorent, Wanderungen 333 (1845).
> *S. sieberi* C. Presl. Abh. Königl. Böhm. Ges. Wiss. ser. 5, 3: 530 (1845); Bot. Bemerk. (C. Presl): 100 (1846).
> *S. rassami* Boiss. Fl. Orient. 4(2): 615 (1879).
> *S. alliaria* Parsa, Kew Bull. 3(2): 224 (1948).

Perennial herb. Stems 18–70 cm, hirsute with long flattened eglandular hairs below and densely glandular pilose (sometimes with long flattened eglandular hairs) above. Leaves 4–15(–20) × 1–7(–9) cm, oblong to ovate, margins erose, tomentose; petiole 2–15 cm. Inflorescence 10–50 cm. Verticillasters (2–)3–6-flowered. Bracts 13–26 × 10–20 mm, often tinged pink or purple. Pedicels 2–6 mm. Calyx almost tubular, often pink or purple (rarely green), 12–16 mm in flower and 17–25 mm in fruit; papillose-glandular with some longer hairs; upper lip equally tridentate and spinulose; teeth 2–4 mm. Corolla 20–37 mm, pinkish or whitish-pink; tube 10–20 mm long and not squamulose. Style is 25–45 mm. Stamen type B. Fig. 150, 1–2.

HAB. Limestone and igneous rocky slopes, cliffs, in *Quercus* scrub, clay and loamy soil, pasture, road side, fallow fields; alt. 300–1500 m; fl. & fr. Apr.-Jun.

DISTRIB. Mountains, upper plains and foothills of N. Iraq. **MAM**: In faucibus Bekma, *Helbaek* 761 (K); Dohuk, *Rechinger* 11980 (W); inter Zakho et Sharanish, *Rechinger* 10730 (W). **MRO**: Arbil: Kuh Safin supra Shaqlawa, 1100 m, *Bornmüller* 1726 (K); 45 km S.E. of Arbil, *Barkley & Haddad* 5774 (K); inter Arbil et Salahaddin, *Haines s.n.*; ibid, K*arim* 37539! (K); Haibat Sultan Mountain, Koi Sanjak, *Omar* et al. 37185 ! (K) ; ibid., *Rawi et al.* 28128! (K). **MSU**: Sulaimaniye, Sosakan prope Tawila, *Rechinger* 10194 (W); ca. 10 km W. of Tawela, on road between Halabja and Tawela, *Rawi* 22163! (K); 29 km a Sulaymaniyah versus Kirkuk, *Rechinger* 10061 (W); prope Sulaymania, *Regel* 91 (K); inter Sulaymania et Dokan, *Rechinger* 10091 (W); Kirkuk, Darband i Bazian prope Chamchamal, *Rechinger* 12199 (W); Jarmo, *Helbaek* 1172! (K); prope Darbandikhan, *Barkley et al.* 51101! (K); ibid., *Haines* 1928! (K); Durbendikan, Tuni Baba Amara hills, 10 km south of Durbendikan, *Barkley & Haddad* 7441! (K). **FUJ**: 20 km S. Mosul, *Anders* 1236 (K); between Mosul and Tal Afar, *Karim* 37470! (K); 19 km N. of Qaiyara, Mosul, *Barkley & Haddad* 5490! (K). **FNI**: Mosul, inter Mosul et Zakho, *Rechinger* 10650 (W); 8 km S. Zakho, *Rechinger* 10699 (W). **FAR**: 5 km N.W. of Arbil, Ankawa, *Bornmüller* 1727 (K). **FKI**: Kirkuk province, *Rustam* 3941! (K); 9 km a Kirkuk versus Altin Köprü, *Erdtman & Goedemans in Rechinger* 15500 (W). **FPF**: Diyala, Jabal Hamrin, *Bornmüller* 1725 (K); ibid *Rechinger* 9563! (K).

Iran, S. & S.E. Turkey, Jordan, Palestine, Saudi Arabia, Egypt.

19. **Salvia sclarea** *L.*, Sp. Pl. 27 (1753); Dinsmore in Post, Fl. Syria, Palest. & Sinai ed 2, 2: 356 (1933); Feinbrun-Dothan, Fl. Paleast. 3: 139 (1978); Hedge in Fl. Iranica [K. H. Rechinger] 150: 450 (1982); Hedge in Fl. Turkey [P. H. Davis] 7: 438 (1982). Ic: Reichb., Ic. Fl. Germ. 18: t. 1249 (1856); Huxley & Taylor, Fls. Greece t. 239 (1977); Jamzad in Flora Iran [Assadi & al., in Persian] 76: 887 (2012); Taifour & El-Oqlah, Pl. Jordan Annot. Checklist: 105 (2017).

> *Salvia haematodes* Scop., Fl. Carniol., ed. 2. 1: 29 (1771), nom. illeg.
> *Sclarea vulgaris* Mill., Gard. Dict., ed. 8. n. 1. (1768).
> *Salvia foetida* Lam., Tabl. Encycl. i. 69 (1791).
> *S. coarctata* Vahl, Enum. Pl. i. 253 (1804).
> *S. simsiana* Schult., Maut. i. 210 (1822).
> *Aethiopis sclarea* (L.) Fourr., Ann. Soc. Linn. Lyon sér. 2, 17: 134 (1869).
> *Salvia calostachya* Gand., Fl. Lyon. 171 (1875).
> *S. turkestanica* Noter, Rev. Hort. 77: 502 (1905).
> *S. lucana* Cavara & Grande, Bull.Orto Bot. regia Univ. Napoli 3: 436 (1913).
> *S. pamirica* Gand., Bull. Soc. Bot. France 60: 26. (1913).
> *S. altilabrosa* Pan, Trab. Mus. Nac. Ci. Nat., Ser. Bot. No. 14, 33 (1918).

Biennial or short-lived perennial aromatic herb with erect rather coarse quadrangular stems to 120 cm, much branched above, pubescent to hirsute below, glandular above. Leaves simple, broadly ovate to ovate-oblong, 5–20 × 4–15 cm, cordate at base, pubescent,

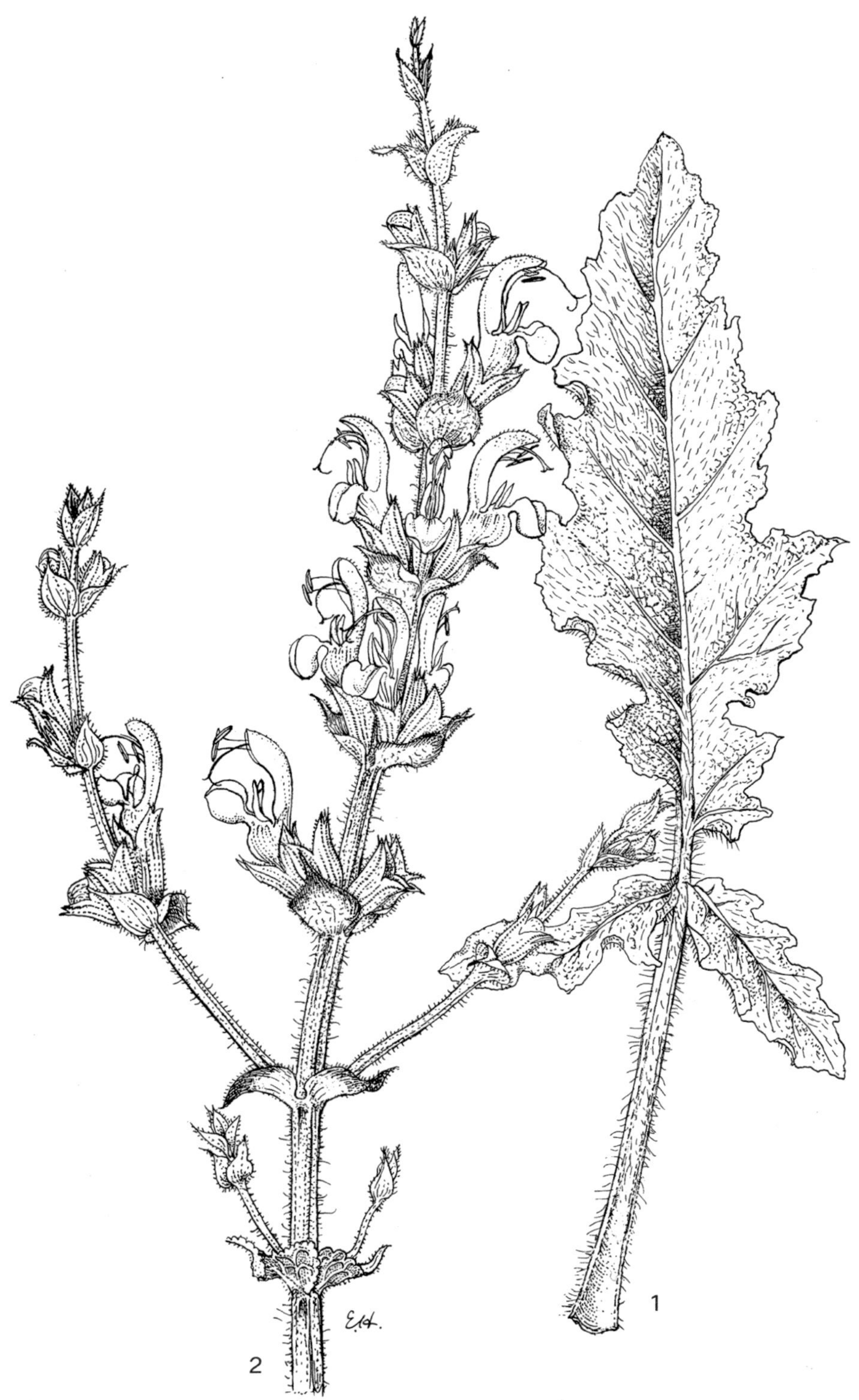

Fig. 150. **Salvia palaestina**. 1, habit, leaf × 1; 2, flowering stem × 1. Reproduced with permission from Feinbrun-Dothan, Fl. Palaestina 3: Plates, f. 225 (1977). Drawn by Esther Huber. © The Israel Academy of Sciences and Humanities.

Fig. 151. **Salvia sclarea**. 1, habit, leaf × 1; 2, flowering stem × 1. Reproduced with permission from Feinbrun-Dothan, Fl. Palaestina 3: Plates, f. 228 (1977). Drawn by Esther Huber. © The Israel Academy of Sciences and Humanities.

irregularly crenate-erose; petiole 3–9 mm. Inflorescence paniculate, many-flowered; Verticillasters 2–6-flowered, distant. Bracts ± obscuring flowers, pink to mauve, membranous, ovate, acuminate, 10–38 × 6–30 mm. Pedicels 2–3 mm, erecto-patent. Calyx ovate-campanulate, c. 10 mm, to 15 mm in fruit, scabrid, glandular-punctuate and sparsely white-pubescent with sessile glands; upper lip tridentate, mucronate. Corolla with a lilac, purplish-pink or pale blue upper lip and cream lower lip, 20–30 mm; upper lip strongly falcate, tube abruptly ventricose, squamulose. Stamens type B. Fig. 151, 1–2.

HAB. Rocky igneous slopes, woodland, shale banks, fields, roadsides; alt. 800–2000 m; fl. & fr. Apr.-Jul. DISTRIB. Mountains and upper plains of N. Iraq. **MAM**: 36 km N.E. of Zakho, Mergaqus (Marsis ?) village, in valley, near Turkish border, 8 July 1957, *Rawi* 23532! (K). **MRO**: Between the slope of Karaoukh Mountain and Dargala village, *Kass & Nuri* 27692! (K); Al Gird Dagh, near Nawanda village, *Guest* 2852! (K). **MSU**: Biyara, Jebel Avroman, *Gillett* 11744! (K). **FUJ**: Mosul, *Yussof Lazar* 3342! (K).

Distinct in its habit and the large coloured bracts, often exceeding the flowers. Previously used type specimen LINN 42/45 corresponds to *S. argentea*.

Most of Europe, S.W. & C. Asia, Transcaucasica, Morocco, Algeria, Ukraine, Crimea.

20. **Salvia ceratophylla** *L.*, Sp. Pl. 27 (1753); Dinsmore in Post, Fl. Syria, Palest. & Sinai ed 2, 2: 356 (1933); Rawi in Dep. Agr. Iraq Tech. Bull. 14: 154 (1964); Feinbrun-Dothan, Fl. Paleast. 3: 139 (1978); Hedge in Fl. Iranica [K. H. Rechinger] 150: 452 (1982); Hedge in Fl. Turkey [P. H. Davis] 7: 440 (1982); Jamzad in Flora Iran [Assadi & al., in Persian] 76: 891 (2012); Taifour & El-Oqlah, Pl. Jordan Annot. Checklist: 105 (2017).

Sclarea ceratophylla Mill., Gard. Dict., ed. 8. n. 8. (1768).
Salvia semilanata Czerniak., Repert. Spec. Nov. Regni Veg. 27: 278 (1930).

Perennial or biennial herb, with a stout woody root. Stems sturdy, 20–65 cm, glandular-villous, densely above. Leaves mostly basal, deeply pinnatifid, "oblong" in outline, with spreading linear segments, the segments alternate, narrow, linear, obtuse, coarsely toothed to suboblate, the terminal segments usually several times as long as the lateral segments, white-lanate when young, 7–25 × 4–8 cm, rachis winged, rugose; petiole almost absent. Inflorescence pyramidal and spreading paniculate, yellowish-green. Verticillasters 2–6-flowered, distant. Bracts ovate, cuspidate, c. 15 × 12 mm. Pedicels erecto-patent, 2–4 mm. Calyx ovate-campanulate, 10–12 mm, to c. 16 mm in fruit with diverging lips, glandular-villous; upper lip tridentate with cuspidate teeth. Corolla cream or yellow, rarely with pinkish upper lip, 15–30 mm; tube abruptly ventricose, squamulose; upper lip falcate. Stamens type B.

HAB. Limestone, igneous and gypsium slopes, corn and fallow fields, waste ground; alt. 300–1000 m; fl. & fr. Mar.-Jul. DISTRIB. Upper plaina nd lower hills of N. Iraq. **MJS**: Karsi (Kursi), Jebel Sinjar, *Gillett* 10927! (K). **FUJ**: Mosul, *Kotschy* 1841! (K); Muwasul Tiatan Mukzuk Nuwar (vicinity of Jabal Qilat?), *Field & Lazar* 469! (K). **FNI**: Mindan (Mandan) district (Mosul–Aqra road), *Bradbranee?* (62) 9026! (K). **FKI**: Kirkuk: Baba Gurgur, 20 km W. of Kirkuk, *Barkley & Brahim* 1414! (K).

Iran, Turkey, Syria, Lebanon, Armenia, Afghanistan, Israel-Jordan, Transcaucasica.

21. **Salvia candidissima** *Vahl*, Enum. Pl. 1:278 (1804); Rawi in Dep. Agr. Iraq Tech. Bull. 14: 154 (1964); Hedge in Fl. Iranica [K. H. Rechinger] 150: 461 (1982); Hedge in Fl. Turkey [P. H. Davis] 7: 447 (1982).

Salvia albida Jacq., Observ. Bot. 1 : 10 (1764).
S. albida Spreng. Index Seminum [Halle] 53: (1807).
S. argentea Hort. ex Benth. Labiat. Gen. Spec. 223 (1833), nom. illeg.
S. armena K.Koch., Linnaea 21 : 654 (1849).
S. candidissima (Vahl) Soják, Čas. Nár. Muz. Praze, Rada Přír. 152(1): 21 (1983).
S. candidissima Vahl var. *cordifolia* Náb. in Publ. Fac. Sci. Univ. Masaryk Brno 70: 52, tab. 4 f. 3 (1926).
    Ic: Jacq., Eclog. Pl. Rar. t. 16 (1811), as *S. odorala* Willd.
S. crassifolia Sm. Fl. Graec. Prodr. 1(1): 17 (1806).
S. odorata Willd. Enum. Pl. [Willdenow] 1: 43 (1809).
S. pycnophylla Greuter & Burdet, Willldenowia 14(2): 301 (1985) nom. nov.
S. semilunata Hausskn. nomen.

Perennial herb. Stems 30–60(–90) cm, erect, branched above, pilose to lanate below with few sessile glands, densely pilose to glandular-papillose above. Leaves simple, oblong to broadly ovate, 2.5–10(–14) × 1–9 cm, pubescent to densely pannose, subentire to erose, cordate to rounded; petiole 3–11 cm. Inflorescence paniculate, often yellowish-green; Verticillasters 2–6-flowered, distant. Bracts ovate-acuminate, 4–10 × 3–6 mm. Pedicels 2–4 mm. Calyx tubular-campanulate, 12–15 mm, to 18 mm and scarcely widening in fruit, densely pilose to glandular-papillose; upper lip with 3 closely connivent mucronate teeth. Corolla totally white or white with yellow lower lip, 20–32 mm; tube c. 12 mm, ventricose, squamulose; upper lip strongly falcate. Stamens type B.

HAB. Slopes, road sides, rocky limestone, mixed shrubland; alt. 1000–1200 m; fl. & fr. May.
DISTRIB. Mountains of N.W. Iraq. **MAM**: Mosul, Amadiyah, *Anders* 2268! (K). **MRO**: Mawelean Mt., c. 10 km E. of Rowanduz, *Rawi, Nuri & Kass* 28922! (K); Arbil, Mt. Qandil, Pushtashan, *Rechinger* 11052 (W); inter Erbil et Rowanduz, *Bornmüller* 1729! (K).

A polymorphic species with variations in indumentum and leaf shape.

Turkey.

22. **Salvia atropatana** *Bunge*, Mém. Acad. Imp. Sci. St.-Petersbourg, Ser. 7. 21(1): 47 (1873); Rawi in Dep. Agr. Iraq Tech. Bull. 14: 154 (1964); Hedge in Fl. Iranica [K. H. Rechinger] 150: 456 (1982); Hedge in Fl. Turkey [P. H. Davis] 7: 444 (1982); Jamzad in Flora Iran [Assadi & al., in Persian] 76: 895 (2012).

Salvia argentea L. var. *angustifolia* Benth. in DC., Prodr. 12: 285 (1848).
S. *bachtiarica* Bunge, Mém. Acad. Imp. Sci. St.-Pétersbourg, Sér. 7. 21(1): 47 (1873).
S. *montbretii* Benth. var. *virescens* Freyn in Bull. Herb. Boiss. ser. 2, 1:278 (1901)!
S. *atropatana* var. *glandulosa* Bornm., Beih. Bot. Centrbl. 22, 2: 122 (1907).
S. *kopetdaghensis* Kudr., Mat. Izuch. Shalf. Sr. Az.: 26 (1937).
S. *linczevskii* Kudr., Mat. Izuch. Shalf. Sr. Az.: 22 (1937).
S. *kourossia* Parsa, Kew Bull. 3(2): 224 (1948).
S. *lurorum* Rech.f., Oesterr. Bot. Z. 99: 57 (1952).
S. *linguifolia* Hedge & Hub.-Mor. in Notes R.B.G. Edinburgh 22:181 (1957)!

Perennial herb with a woody rootstock. Stems several, 25–70 cm, erect, usually densely glandular-pilose (or arachnoid-egandular with sessile glands). Leaves mostly basal, linear to linear-oblong, 10–15(–20) × 0.5–3(–5) cm, arachnoid-tomentose with sessile glands to ± lanate, entire (on linear leaves), serrate to suberose, cuneate; petiole 3–9 cm. Inflorescence paniculate, narrowly obpyramidal. Verticillasters 4–6-flowered, approximating above. Bracts ovate acuminate, 8–15 × 7–10 mm. Pedicels 2–4 mm. Calyx more or less campanulate, 8–10 mm, densely glandular, to 11–13 mm in fruit; upper lip indistinctly tridentate. Corolla white, or with a yellow lip, 20–25 mm; tube 7–8 mm ventricose, squamulose; upper lip strongly falcate. Stamens type B. Nutlets ovoid, c. 3 × 2.5 mm.

HAB. Rocky slopes in mixed shrubland, sloping meadows; alt. 1300–2100 m; fl. & fr. Jun.
DISTRIB. Mountains of N. Iraq. **MAM**: Mosul: Jabal Khantur prope Sharanish, *Rechinger* 12062 (W). **MRO**: Mt. Sakri Sakran prope Rawandiz, 2100 m, *Bornmüller* 1893/1733! (K). **MSU**: Mt. Avroman prope Tawilla, *Rechinger* 12421 (W); Mt. Avroman et M. Shahu, *Haussknecht* 768; spur of Jebel Avroman, North of Biyara (Biara) village, *Gillett* 1184! (K); Penjwin, *Rechinger* 10530 (W). Cultivated hillroad near stream, *Rawi* 22634! (K).

E. Turkey, Iran & Turcomania.

23. **Salvia microstegia** *Boiss. & Bal.* in Boissier, Diagn. Pl. Nov. ser. 2(4):17 (1859); Hedge in Fl. Iranica [K. H. Rechinger] 150: 458 (1982); Hedge in Fl. Turkey [P. H. Davis] 7: 442 (1982); Jamzad in Flora Iran [Assadi & al., in Persian] 76: 900 (2012).

Salvia verbascifolia Bieb. var. *cana* Boiss., Fl. Orient. 4: 619 (1879).
S. *chnoodes* Stapf in Denkschr. Akad. Wiss. Wien, Math.-Nat. Kl. 50: 98 (1885). Ic: J. Roy. Hort. Soc. 85(10): t. 140 (1960).

Perennial herb with a woody rootstock. Stems few or several, erect, 20–120 cm, densely pilose-villous glandular, often eglandular lanate below. Leaves mostly basal, variable in size and shape, ovate to oblong, (3.5–) 7–17 (–20) × 3–8(–14) cm, white or grey lanate, obtusely lobed, irregularly serrate or erose, subcordate; petiole 2–16(–20) cm, tomentose.

Inflorescence usually a widely spreading panicle, often yellowish-green. Verticillasters 2–6-flowered, usually distant. Bracts variable, broadly ovate, 9–17 × 8–14 mm. Pedicels 2–5 mm. Calyx± campanulate, 9–13 mm, to 11–14 mm in fruit, densely glandular with capitate glandular hairs and sessile glands; upper lip shortly tridentate, median tooth much shorter; lips ± equal. Corolla white, lower lip fading yellow, (17–)20–30 mm, tube 5–15 mm, ventricose, squamulose; upper lip compressed, strongly falcate. Stamens type B.

HAB. Rocky limestone, serpentine, metamorphic and igneous slopes, screes; alt. 2300–3000 m; fl. & fr. May-Sep.
DISTRIB. Mountains of N. Iraq. **MRO**: N.W. Haji Umran, 45 km E. by N. Rowanduz, *Al Kas* 18575! (K); Ser Kurawa, *Gillett* 9760! (K); Qandil Range, Germasur Lake, N. side of mountain slope, *Rawi & Serhang* 26716! (K); N. of Algurd Dagh (Helgard, Halgurd), near Berma Sard (Bermasand) Lake, *Rawi & Serhang* 24796! (K); ibid., *Rawi & Serhang* 24767! (K).

The species is closely related to *S. argentea*, *S. xanthocheila* and the Caucasian *S. verbascifolia*. The differences between most of them are often slight. *Salvia argentea* is distributed in mainly Mediterranean or submediterranean habitats. *Salvia verbascifolia* differs from *S. microstegia* by its much thinner indumentum on all parts.

Turkey, Lebanon, Latakia (Syria), probably in Iran & Armenia.

24. **Salvia xanthochelia** *Boiss. ex Benth.*, Prodr. [A.P. de Candolle] 12: 284 (1848); Hedge in Fl. Iranica [K. H. Rechinger] 150: 458 (1982); Hedge in Fl. Turkey [P. H. Davis] 7: 443 (1982); Jamzad in Flora Iran [Assadi & al., in Persian] 76: 906 (2012).

Perennial herb with a thick woody rootstock. Stems 15–30 cm, sturdy, erect densely glandular-pilose. Leaves mostly basal, broadly ovate to ovate-oblong, 2.5–8.5(–11) × 1.8–6 cm, lanate to pannose, irregularly lobed or dentate to subentire, ± cordate; petiole 1–6.5 cm. Inflorescence a short sturdy panicle. Verticillasters 4–6-flowered, crowded. Bracts broadly ovate, acuminate, 13–18 × 10–14 mm. Pedicels 2–4 mm. Calyx campanulate-infundibular, 9–13 mm, to 13–18 mm in fruit with widely diverging lips, densely glandular; upper lip shortly tridentate. Corolla white, 22–26 mm; tube c. 9 mm, ventricose, squamulose; upper lip clearly falcate. Stamens type B. Nutlets ovoid, c. 4 × 2.5 mm.

HAB. Rocky igneous slopes, in *Astragalus* zone and alpine habitat; alt. 2700–2900 m; fl. & fr. Jul.
DISTRIB. N. Iraq. **MRO**: Kandil (Qandil) range, N.E. of Rania, *Rawi & Serhang* 18266! (K).

Most closely related to *S. microstegia* and essentially a dwarf higher alpine version of it, but differing by its larger bracts and fruiting calyces, and densely glandular hairy stem at base.

E. Turkey, Iran (not in Caucasia), Talish region; probably found in Armenia.

25. **Salvia frigida** *Boiss.*, Diagn. Pl. Nov. Ser. 1(5): 10 (1844); Hedge in Fl. Iranica [K. H. Rechinger] 150: 459 (1982); Hedge in Fl. Turkey [P. H. Davis] 7: 444 (1982); Jamzad in Flora Iran [Assadi & al., in Persian] 76: 928 (2012).

Salvia spinulosa Montbret & Aucher ex Benth., Prodr [A.P. Candolle] 12: 283 (1848).
S. oreades Schott & Kotschy ex Boiss., Fl. Orient. [Boissier]. 4(2): 621 (1879).
S. frigida Boiss. var. *oblongifolia* Boiss., Fl. Orient. 4: 621 (1879)!
S. frigida Boiss. var. *albiflora* Bornm. in Beih. Bot. Centrablat. 24(2): 487 (1909).
S. gilliatii Turrill, Bull. Misc. Inform, Kew 1930: 456 (1930).

Perennial herb with a thick woody rootstock. Stems solitary or several, erect, 8–40 (–50) cm, below pilose to villous with sessile glands, densely glandular above with capitate glands. Leaves mostly basal, variable, ovate to narrowly oblong, 2–15 × 1.2–5 cm, crenulate to erose, rugulose, arachnoid to ± lanate with many sessile glands; petiole 1.5–8 cm. Inflorescence little branched to widely paniculate. Verticillasters 2–6-flowered, usually distant. Bracts ovate to orbicular, 8–12 × 7–11 mm. Pedicels 2–3 mm. Calyx campanulate to infundibular, 7–10 mm, to 10–12 mm in fruit and widening, densely capitate-glandular; teeth usually prominently spinulose; upper lip tridentate, median tooth much shorter, truncate or not. Corolla white, lower lip yellow, 12–16 mm; tube c. 8 mm, ventricose, squamulose; upper lip narrow, somewhat falcate, scarcely bifid. Stamens type B.

HAB. Limestone slopes and crevices, meadows; alt. 1700 m; fl. & fr. May.

DISTRIB. Mountains and upper plains of N. Iraq. **MAM**: Khantur, mountainside, *Rawi* 8782! (K). **FUJ**: Mosul, 1841, *Kotschy* 231! (K, leaves narrowly oblong).

The species varies widely in its indumentums and leaf shapes. *Kotschy* 231 (K) from Mosul has narrowly oblong leaves and seems to be intermediate between *S. frigida* and *S. atropatana*. *Rawi* 8782 is similar to *S. microstegia* but it has smaller corolla, therefore it was identified as *S. frigida* in this Flora. *Salvia frigida* differs from *S. microstegia*, *S. xanthochelia* and *S. atropatana* by its smaller corolla (12–16 mm). It is also similar to Turkish endemics *S. yosgadensis*, *S. modesta* and *S. ekimiana*.

N.W. Iran, Turkey.

26. **Salvia urmiensis** *Bunge*, Mém. Acad. Imp. Sci. St.-Petersbourg, Sér. 7, 21: 48 (1873); Hedge in Fl. Iranica [K. H. Rechinger] 150: 460 (1982); Jamzad in Flora Iran [Assadi & al., in Persian] 76: 922 (2012).

Salvia brachysiphon Stapf, Denkschr. Akad. Wiss. Wien Math.–Nat. Kl. 50: 41 (1885).
S. ali-askaryi S.A.Ahmad, Harvard Papers in Botany 21(2): 227 (2016), Type: Sulaimani Province, Azmar Mountain, near Harwta forest station, eroded places, sandy soil, 1238 m, 15 May 2015, *S. A. Ahmad, A. Hama & S. Babarasul*, 14-818 (KBFH, holo.).

Perennial herb. Stems solitary, erect, 15–40(–50) cm, below eglandular pilose and coarsely villous (with sessile glands?), above densely capitate-glandular and eglandular-villous hairy. Leaves simple, 8–25 × 4–15 cm, ovate to ovate-oblong, somewhat erose to irregular dentate, indumentum variable, subglabrous to pilose to villous; petiole 2–11 cm. Inflorescence paniculate. Verticillasters 2–6(–15)-flowered, lower ones distant. Bracts ovate and semi–orbicular, 10–20 mm long. Pedicels 2–3 mm. Calyx tubular-campanulate to infundibular, 8–10 mm, to 10–12 mm in fruit and widening, capitate-glandular and densely eglandular hairy; upper lip tridentate, teeth subequal and spinulose. Corolla white and upper lip pinkish (often), lower lip yellowish cream, 15–22 mm; tube 7–8 mm, ventricose, squamulose; upper lip falcate, glandular capitate hairy. Stamens type B.

HAB. Eroded places, mixed shrubland, *Daphne* & *Astragalus* zone, fields; alt. 900–1800 m; fl. & fr. Apr.-Jun.
DISTRIB. N. Iraq. **MSU**: Avroman (Hawraman) mountain, N. of Halabja (on Persia border), *Rawi* 22068! (K). Kamarspa, on the road between Halabja and Tawela, *Rawi* 22176! (K). Sosa Daratea (Dara Tri), between Halabja and Tawela, Barley field, (unusual specimen between *S. agg. frigida* and *S. urmiensis*), *Botany Staff* 43016! (K); Sulaimani Province, Azmar Mountain, near Harwta forest station, eroded places, sandy soil, *S. A. Ahmad, A. Hama & S. Babarasul* 14-818 (KBFH, type of *S. ali-askaryi*).

Salvia urmiensis has clear affinities with *S. poculata* and *S. frigida*. *Rawi* 22176 seems to an intermediate specimen between *S. agg. frigida* and *S. urmiensis*. Recently published new species, *S. ali-askaryi*, from Sulaimani Province seems not to be different from *S. urmiensis* due to its fully overlapping morphological description and distribution range. Therefore, I have reduced *S. ali-askaryi* as a synonym of *S. urmiensis*.

W. Iran (possibly also present in adjacent areas of Turkey, Iraq and Iran borders).

27. **Salvia poculata** *Nábělek*, Spisy Přír. Fak. Masarykovy Univ. 70: 50 (1926). Type: Iraq, Serizor, Handrian Dar (Handren Dagh) supra Rowanduz dit. Erbil, in humosis, c. 1700 m, 23 May 1910, *Nábělek* 1570 (BRA!, lecto.). Rawi in Dep. Agr. Iraq Tech. Bull. 14: 155 (1964); Hedge in Fl. Iranica [K. H. Rechinger] 150: 461 (1982); Hedge in Fl. Turkey [P. H. Davis] 7: 446 (1982); Jamzad in Flora Iran [Assadi & al., in Persian] 76: 926 (2012).

Salvia brevidens Hedge & Hub.-Mor. in Notes R.B.G. Edinburgh. 22:183 (1957).

Perennial herb. Stems erect, 25–50 cm, branched above, glandular-pilose below, densely glandular-pilose to villous above. Leaves simple, ovate to oblong, 6–15 × 3–6.5 cm, arachnoid floccose (± lanate on young growth), crenate to erose; petiole 2.5–8.5 cm. Inflorescence paniculate; verticillasters 2– 8-flowered, clearly distant. Bracts broadly ovate to ±semicircular, abruptly acuminate, 17–22 × 20–25 mm. Pedicels 1–3.5 mm, erecto-patent. Calyx obtriangular, 8–10 mm, sometimes tinged purplish, glandular-villous, to c. 14 mm in fruit and broadening; upper lip tridentate, median tooth clearly shorter than laterals. Corolla white or pale purple, 12–16 mm; tube ± straight, c. 6 mm, squamulose, ventricose above; upper lip scarcely falcate. Stamens type B. Nutets rounded trigonous, ± spherical, c. 4.5 × 4 mm.

HAB. Rocky limestone and igneous slopes, sloping meadows; alt. 1700–2300 m; fl. & fr. May-Jul.
DISTRIB. N. Iraq. **MAM**: Zawita prope Sharanish, N. Zakho, *Rechinger* 10942! (K); Jabal Khantur, *Rechinger* 10865 (W). **MRO**: Arbil, Handren (Chiyi-i-handren, Handre Dagh) supra Rowanduz, in humosis, *Nábelek* 1570 (JE); Algurd Dagh (M. Helgurd) supra Rowandaz, *Rechinger* 11371 (W). **MSU**: Sulamaniya, M. Avroman prope Tawilla (Tawela), *Rechinger* 10366 (W).

N.W. Iran, E. Turkey.

28. **Salvia indica** *L.*, Sp. Pl. 26 (1753); Rawi in Dep. Agr. Iraq Tech. Bull. 14: 155 (1964); Dinsmore in Post, Fl. Syria, Palest. & Sinai ed 2, 2: 357 (1933); Feinbrun-Dothan, Fl. Paleast. 3: 140 (1978); Hedge in Fl. Iranica [K. H. Rechinger] 150: 466 (1982); Hedge in Fl. Turkey [P. H. Davis] 7: 451 (1982); Jamzad in Flora Iran [Assadi & al., in Persian] 76: 928 (2012); Taifour & El-Oqlah, Pl. Jordan Annot. Checklist: 105 (2017). Ic: Bot. Mag. t. 395 (1798). Neotype: Herb. Tournefort No. 1079 (P).

> *Sclarea indica* Mill. Gard. Dict., ed. 8. N. 9 (1768).
> *Salvia elongata* Salisb. Prodr. Stirp. Chap. Allerton 74 (1796).
> *Hematodes indica* (L.) Raf., Fl. Telluer. 3: 93 (1837).
> *Larnastyra indica* Raf., Fl. Tellur. 3: 92 (1837).
> *Salvia brachycalyx* Boiss., Fl. Or. 4:625 (1879).

Perennial herbs. Stems 60–150 cm, erect, branched above, sparsely eglandular-pilose below, densely or sparsely glandular- or eglandular-pilose above. Leaves simple, 12–32 × 9–24 cm, broadly ovate, truncate, veins reticulate, margins erose-dentate, glaucous or eglandular-pilose especially on veins, cordate; cauline leaves ± sessile. Petiole 2–12 cm. Inflorescence paniculate. Verticillasters 4–8-flowered, clearly distant. Bracts 8–10 × 6–8 mm, ovate, acuminate. Pedicels 2–5(–7) mm, erect. Calyx campanulate, 10–12 × 7–9 mm, truncate at apex, densely glandular-villous and sparsely eglandular, slightly broadening and up to 15 mm in fruit; upper lip shortly tridentate. Calyx teeth mucronate, connivent. Corolla upper lip lilac, lower lip dark purplish-pink, spotted with purple, 25–32 mm, strongly compressed laterally; tube 9–13 mm, whitish, short, squamulose within, abruptly ventricose above. Stamens 2. Upper thecae clearly longer than filaments; filaments 3–4 mm, fertile anther 4–5 mm, upper thecae 17–21 mm. Style glabrous, 38–45 mm, exerted from corolla lips and divided in two parts at apex. Fig. 152, 1–3.

HAB. Rocky limestone slopes, clay mountain side, clay soil, stony muddy soil and open shrubby places, at high altitudes with *Astragalus* and *Amygdalus*; alt. 500–2000 m; fl. & fr. Apr.-Jun.
DISTRIB. Mountains and upper plains of N. Iraq. **MAM**: Amadiya, *Raphael & Hewitt* 1501! (K); Amadiya, clay with organic matter in rocky mountain, *Botany Staff* 43414! (K); Sharifa near Amadiya, limestone ledges and crevices, *Polunin* 5119! (K); Zakho pass, *Guest* 2249! (K), ibid near Zakho, hill side, *Karim, Al–Dabbagh & Hamad* 44808! (K); Bakirma–Bakurman, Mosul province, *Rawi* 8539! (K). **MRO**: Shaqlawa, *Hainess* W651! (K); Chiya-i Marmarut, near Rowanduz, *Squadr & Cuckney* 3826! (K); Rowanduz george, *Guest* 2024! (K); 25 km N.W. of Rania, Zewa village, mountain slope, *Rawi et al.* 28932! (K); Karokh (Kurek, Karaukh) Mountain, stony muddy soil, *Al-Kaisi, Sarhank, Nori* 27289! (K); Salah ad Din (Salahaddin), ridge of calcereous limestone between Erbil and Shaqlawa, grass, shrubs and trees, *Helbaek* 739! (K). **MSU**: Jebel Avroman above Darimar, limestone above tree level with *Astragalus* and *Amygdalus* sp., *Gillett* 11849! (K); North of Qara Dagh ridge near Bani Khailan (Bani Khelan), barley field, *Poore* 659! (K); 4 km north of Durbendikan dam, *Barkley & Haddad* 7466! (K); Sulaimaniya, Dokan, Kanizar *Omar et al.* 37346! (K); Sekanian mountain, S. of Dokan, rocky clay mountain side, *Botany Staff* 43072! (K); 7 km W. of Tawela (Balcha, Balkha village), cultivated valley, *Rawi* 22373! (K). **MJS**: Sinjar, clay soil hillside in wheat field, *Al Kaisi & Hamad* 49091 (K). For additional records see Flora Iranica 150: 466–467.

Morphological characteristics such as leaf size and corolla characteristics are important for the dentification of the species. A very attractive species with great potential as an ornamental.

S. & S.E. Turkey, W. Iran, Syria, Lebanon, Palestine, Jordan.

29. **Salvia virgata** *Jacq.*, Hort. Vindob. 1:14, 1.37 (1770); Hedge in Fl. Iranica [K. H. Rechinger] 150: 469 (1982); Hedge in Fl. Turkey [P. H. Davis] 7: 454 (1982); Jamzad in

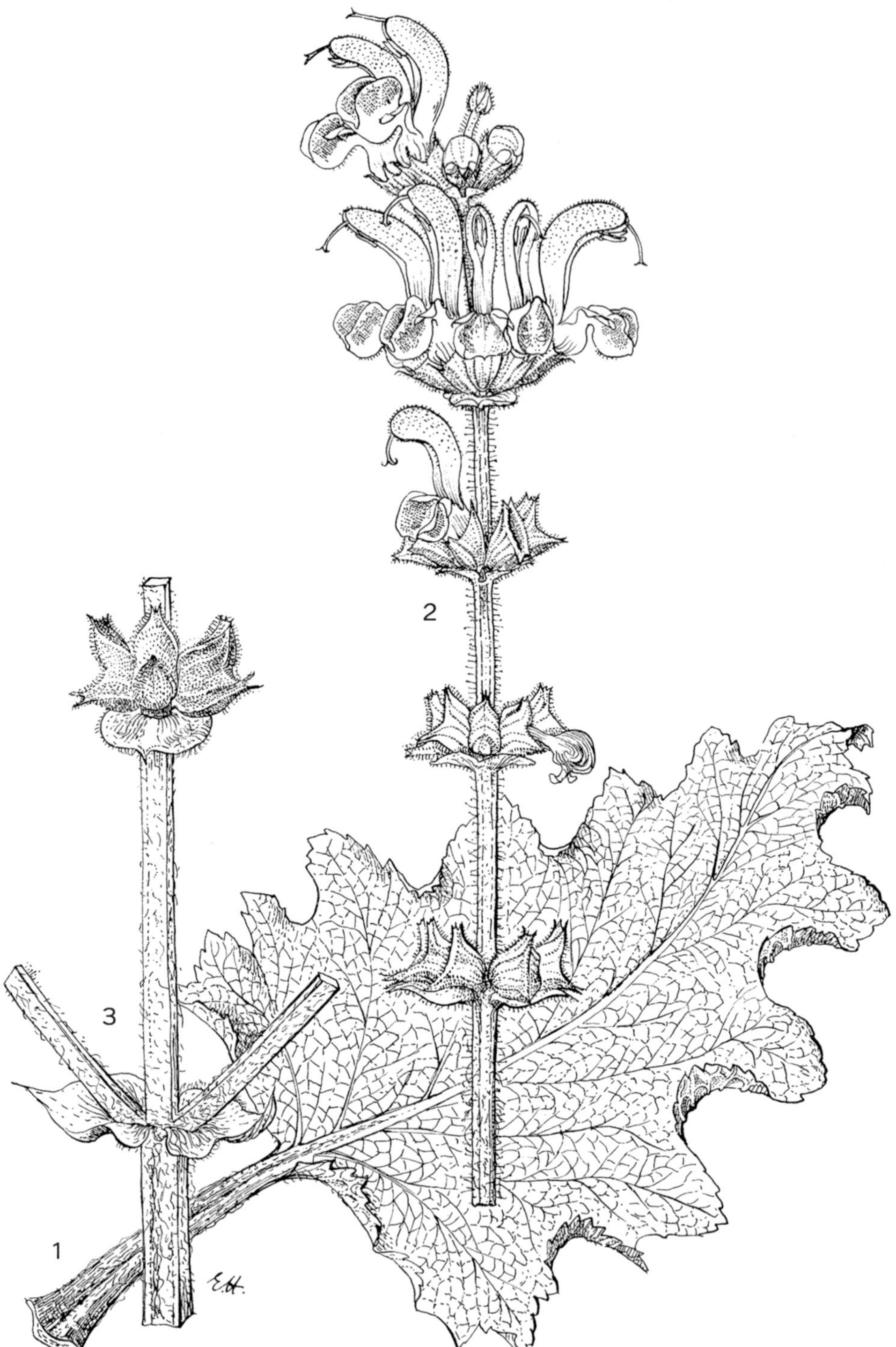

Fig. 152. **Salvia indica**. 1, habit, leaf × 1; 2, flowering stem × 1; 3, stem with fruiting calyx × 1. Reproduced with permission from Feinbrun-Dothan, Fl. Palaestina 3: Plates, f. 230 (1977). Drawn by Esther Huber. © The Israel Academy of Sciences and Humanities.

Flora Iran [Assadi & al., in Persian] 76: 934 (2012); Taifour & El-Oqlah, Pl. Jordan Annot. Checklist: 105 (2017). Ic: Sibth. & Sm., Fl. Graeca 1: t. 22 (1806), as S. *sibthorpii.*

Salvia sibthorpii Sm., Fl. Graec. Prodr. 1(1): 15. 1806; Fl. Graeca 1:17, t. 22. (1806).
S. campestris M.Bieb., Fl. Taur.-Caucas. 1: 20 (1808).
S. praecox Loisel., Not. Fl. France 6 (1810).
S. caduca Vahl ex Hornem., Hort. Bot. Hafn. i. 30 (1813).
S. mollis J.Jacq., Ecl. Pl. Rar.: 56 (1813).
S. gigantea Desf., Tabl. Ècole Bot., ed. 2: 68 (1815).
S. affinis Spreng. ex Steud., Nomencl. Bot. [Steudel] 724 (1821).
S. hypanica Andrz., Bess. Enum. Pl. Volh. 3 (1821).
S. rubra Spreng., Syst. Veg., ed. 16 [Sprengel] 4(2, Cur. Post.): 17 (1827).
S. caucasica Schrank, Syll. Pl. Nov. ii. 58 (1828).
S. garganica Ten., Index Seminum (NAP, Neapolitano) 1829: 17 (1829).
S. amplexicaulis Hort. ex Benth., Labiat. Gen. Spec. 236 (1833).
S. barrelieri Hort. ex Benth., Labiat. Gen. Spec. 235 (1833).
S. bauhinia Hort. ex Benth., Labiat. Gen. Spec. 235 (1833).
S. grandidentata Ten., Index Seminum (NAP, Neapolitano) 1833: 15 (1833).
S. quercifolia Hort. ex Benth., Labiat. Gen. Spec. 235 (1833).
Euriples rugosa Raf., Fl. Tellur. 3: 94 (1837).
Salvia oblonga K.Koch, Linnaea 19: 24 (1846).
S. nudicaulis K.Koch, Linnaea 19: 24 (1846), nom. illegit.
S. utilis Braun ex Engl., Abh. Königl. Akad. Wiss. Berlin 1891: 367 (1892).
S. similata Hausskn. in Mitth. Thüring. Bot.Verins. n.s., xi: 36 (1897).
S. virgata Jacq. var. *densiflora* Nab. in Publ. Fac. Sci. Univ. Masaryk Brno 70:52, t. 15, f. 4 (1926).
S. virgata Jacq. var. *canovelutina* Rech.f., in Ann. Naturh. Mus. Wien 51:420 (1941).
S. extersa Klokov, Fl. URSR 9: 654 (1960).
Sclarea sibthorpii (Sm.) Soják, Čas. Nár. Muz. Praze, Rada Přír. 152 (1): 22 (1983).
S. virgata (Jacq.) Soják, Čas. Nár. Muz. Praze, Rada Přír. 152 (1): 22 (1983).

Perennial. Stems erect, (10–)30–130 cm, stem solitary, erect, leafy, much branched above or not, indumentum variable, pilose to tomentose, glandular or eglandular. Leaves simple, distributed over stem or rarely restricted to basal rosettes, ovate-oblong, elliptical oblong to broadly ovate, 5–30 × 2–15 cm, eglandular-pilose with numerous sessile glands, obtuse or rounded at apex, subcordate at base, rugulose, erose, crenate, serrate to subentire; petiole 1–15 cm. Inflorescence a widely branched panicle with long ± slender secondary branches; verticillasters 2–6-flowered, distant, rarely condensed. Bracts ovate-acuminate, 3–9 × 3.5–7 mm. Pedicels 1–3 mm. Calyx ± tubular campanulate, 6–10 mm, to 10–12 mm in fruit with a strongly recurved bisulcate upper lip, glandular- or eglandular-pilose or villous. Corolla purplish-blue to purplish-pink, rarely white, 10–18 mm; tube 6–12 mm, ventricose, not squamulose; upper lip falcate. Stamens type B. (Plants with male sterile flowers are frequent).

HAB. Shrubland, mountain slopes, roadsides, woodland, meadows, fallow fields, in orchard, near water; alt. 1400–2000 m; fl. & fr. Mar.?
DISTRIB. Mountains of N.W. Iraq. **MAM**: Mosul, Amadiya, prope Sarsang, *Haines* 1231! (lK) ; ibid, *Agnew & Haines* 799 ! (K). **MRO**: Haji Omran, *Dabbagh & Hamad* 46268! (K); ibid, *Al-Kaisi & Hamad* 43553! (K); Magar Range, Haji Omran, south side of mountain slope, *coll. ignot.*, 1957! (K); Arbil, M. Baradost, *Thesiger* 1221! (BM).

Turkey, N. Iran, Greece, Albania, Bulgaria, Crimea, Ukraine, Cyprus (?), Balkans, Italy, Transcaucasica, Afghanistan, C. Asia.

30. **Salvia lanigera** *Poir.* in Lamark, Encyclop. Méthod. Suppl. 5: 49 (1817); Dinsmore in Post, Fl. Syria, Palest. & Sinai ed 2, 2: 359 (1933); Täckholm, Students Fl. Egypt 146 (1956); Quezel & Santa, Nouv. Fl. Algér. 2: 795 (1963); Rawi in Dep. Agr. Iraq Tech. Bull. 14: 155 (1964); Feinbrun-Dothan, Fl. Paleast. 3: 142 (1978); Hedge in Fl. Iranica [K. H. Rechinger] 150: 472 (1982); Jamzad in Flora Iran [Assadi & al., in Persian] 76: 941 (2012); Taifour & El-Oqlah, Pl. Jordan Annot. Checklist: 105 (2017).

Ic.: Bouloumoy, Fl. Liban et Syrie t. 322 (1930)– as S. *controversa* Ten.

S. lanigera Desf., Tabl. l'école Mus. Hist. Nat. Ed. 3: 95, 394 (1829).
S. controversa auctt. non Ten., Syll. Fl. Neap. (1831).

Perennial with a woody rootstock. Stems erect, simple or branched, below with numerous long or short white spreading eglandular hairs, above similar with some short capitate glands. Leaves deeply pinnatisect, oblong in outline, with irregular lobed linear bullate segments, up to 9 × 3.5 cm; above and below with eglandular hairs and oil globes. Verticils up to 12, c. 5-flowered, distinct. Floral leaves broadly ovate, acuminate, c. 4.5 × 4 mm; bracts absent. Pedicels erect, spreading, c. 3 mm. Calyx tubular-campanulate, expanding slightly in fruit to c. 8 mm, with a dense indumentum of long and short eglandular hairs, short capitate glandular hairs; teeth of upper lip subequal, c. 0.3 mm; teeth of lower lip c. 3.5 mm. Corolla deep purplish-blue to purple, variable in lenght, up to 17 mm; upper lip somewhat falcate; lower lip shorter than upper; tube c. 10 mm, glabrous within. Stamens type B. Nutlets round to trigonous, c. 1.5 × 2.5 mm.

HAB. Stony and sandy desert, sandy plain, fields, gravelly, clay and loamy soils; alt. 85–650 m; fl. & fr. Feb.-Apr.

DISTRIB. Plains and desert regions. **FKI**: Ghurfah Plain (Tuz Khurma) Plain, Injana, *Guest, Eig & Zohary* 5080! (K). Kirkuk, desertis, *Bornmueller* 1741! (K); 7 km from Altun Kopri (Altun Kupri) to Erbil (?), *Botany Staff* 43259! (K). **FUJ/DLJ**: 4 km West of Wadi Thirthir (Tharthar) on pipeline, *Gillette & Rawi* 7126! (K); Rawah to Sinjar, *Chakravarty, Rawi, Khatib & Alizzi* 31995! (K). **FPF**: between Tursaq and Mandali, *Al-Kaisi, Thamer & Salah* 51421! (K); Bigdaoda village, 6 km S. of Kani Masi, in plain under *Quercus* trees, *Omar & Dabbagh* 45651! (K). **DLJ**: near Tikrit, *Osman & Hamid* 36290! (K); 15 km N.E. of Rummana to Rawa, *Khayat & Hamad* 51627! (K); Samadan 95 km N.E. of Haditha, *Al-Khayat & Hamad* 51855! (K). **DWD**: 28 km S. of Nakhaib (Nukhayb), *Botany Staff* 41979! (K); Rutba, Tall-Al-Niser, 48 km N.E. of Rutba, *Rawi* 31220! (K); nr Rutba, roadside, *Alizzi & Husain* 24116! (K); 5 km S.E. of Rutba, *Jenan Al-mokhtan* 33331! (K); 28 km West of Hoklaniya, Ramadi Liwa, *Fred A. Barkley* 4251! (K); 26 km W. of Ramadi, *Chaksavarty, Rawi, Khatil, Alizzi* 32827! (K); 15 km W. of Ramadi on road to Hit, subdesert with scattered *Astragalus spinosus* on gypsaceous hills, *Gillett & Rawi* 6771! (K); 62 km Husaiba to Akashat, *Al-Khayat & Hamda* 51570! (K). **DSD**: Al Shabakah (Shabicha), stony, sandy slopes, subdesert, 25 February 1947, 250 m, *Gillett & Rawi* 6249 (K). 5 km from Al-Salman to Samah, rocky sandy hills, *Al-Kaisi et al.* 48133! (K). **LCA**: Falluja, Road side, *Fauzi & Noori* 39643! (K).

N. Africa, Sinai, Cyprus, Palestine, Jordan, Syria, Lebanon?, Arabia.

31. Salvia verticillata *L.*, Sp. Pl. 1: 26. 1753; Rawi in Dep. Agr. Iraq Tech. Bull. 14: 155 (1964); Hedge in Fl. Iranica [K. H. Rechinger] 150: 473 (1982); Hedge in Fl. Turkey [P. H. Davis] 7: 458 (1982); Jamzad in Flora Iran [Assadi & al., in Persian] 76: 944 (2012).

Perennial herb; stems several, simple or branched, erect or ascending, 15–75 cm, many, branched above or not, pilose to villous below with sessile glands. Leaves simple, oblong to ovate, 2–14 × 2–10 cm, or lyrate with one or two pairs of unequal basal lobes, pilose to villous with many sessile glands, subentire to serrate, rounded to cordate, petiole 1–9 cm. Verticillasters (8–)15–40-flowered, clearly distant. Bracts ovate-acuminate, c. 7 × 3 mm, deciduous. Pedicels 2–10 mm, some ± deflexed. Calyx tubular, 5–7 mm to c. 8 mm in fruit with a bisulcate upper lip, violet-blue, pilose to villous with sessile glands, teeth mucronate. Corolla violet-blue, lilac, rarely white; c. 12 mm; tube straight, c. 8 mm with a V- shaped annulus; upper lip straight, narrowed at base. Stamens type C. Nutlets c. 2.2 × 1.3 mm.

Two subspecies are recognized:

<pre>
Leaves broadly ovate, clearly cordate; stem, leaf and calyx indumentum
    of villous soft hairs; plants to 75 cm, much branched . . . . a. subsp. *verticillata*
Leaves oblong, elliptic, or oblong-ovate, rounded or subcordate;
    indumentum of short more or less scabridulous hairs; plant 15–50
    cm, little branched. . . . . a. subsp. *amasiaca*
</pre>

subsp. **verticillata** *L.*; Hedge in Fl. Iranica [K. H. Rechinger] 150: 473 (1982); Hedge in Fl. Turkey [P. H. Davis] 7: 458 (1982).

Horminum verticillatum Mill., Gard. Dict., ed. 8. n. 3 (1768).
Covola verticillata Medik., Philos. Bot. (Mdeikus) 2: 67 (1791).
Salvia mollis Donn, Hort. Cantabrig., ed. 3. 7 (1804).
S. lampsanifolia Vahl ex Steud., Nomencl. Bot. [Steudel] 727 (1821).
S. peloponnesiaca Boiss. & Heldr., Diagn. Pl. Orient. Ser. 1, 7: 47 (1846).
Hemisphace verticillata Opiz, Seznam 50 (1852).
Salvia regeliana Trautv., Index seminum [St. Petersburg] 93 (1866).
Sphacopsis verticillata Briq., Lab. Alp. Marit. i. 184 (1891).
Salvia uberrima Recf.f., Bot. Jahrb. Syst. 71(4): 543 (1941).

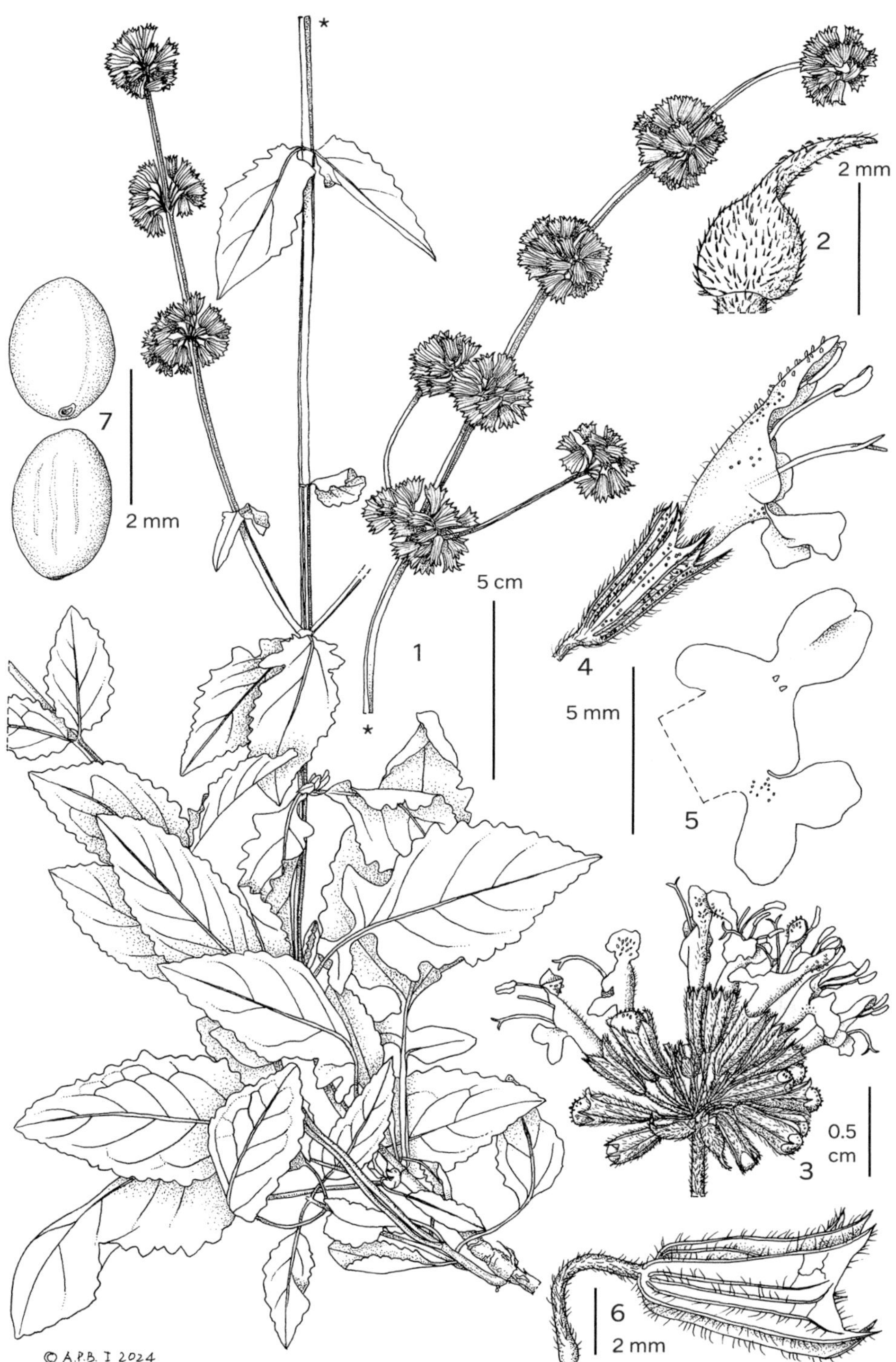

Fig. 153. **Salvia verticillata** subsp**. amasiaca**. 1, habit, plant in fruit; 2, bract subtending inflorescence, outer face; 3, inflorescence; 4, flower, side view; 5, upper and lower lobes of corolla, flattened; 6, calyx at fruiting stage; 7, nutlets, both faces. 1, 6, 7 from *Guest et al.* 2929; 2–5 from VCR 5/1065 June 1955 (K004726387). Drawn by © A.P. Brown, Feb. 2024.

HAB. Woodlands, meadows, roadsides, damp shady place, mesophytic habitats, igneous or metamorphic rocks, by stream at upper limit of forest; alt. 1400–2300 m; fl. & fr. Aug.

DISTRIB. Mountains of N. Iraq. **MAM**: Mosul, in montibus N. Zakho, Basingera prope Sharanish, *Rechinger* 11527! (K); Mosul, montibus Qara, *Kotschy* 355! (K); Zawita, *Rechinger* 10925 (W). **MRO**: Mountains, S.E. of Serva (Zirva), 50–55 km N. of Rowanduz, on Turkish frontier, *Gillett* 9686! (K); Arbil, Algurd Dagh (Helgurd) supra Nowanda, *Rechinger* 11332 (W); Gunde Sher, Darband near Algurd Dagh, damp shady with *Juglans* in *Quercus libani* and *Q. infectoria* forest, *Gillett* 12390! (K).

Europe, N. Iran, Turkey, Caucasia, Russia, Ukraine, Crimea; naturalized in N. Europe and N. America.

b. subsp. **amasiaca** (*Freyn & Bornm.*) *Bornm.*, Bull. Herb. Boiss. ser. 2, 8:110 (1908); Hedge in Fl. Iranica [K. H. Rechinger] 150: 474 (1982); Hedge in Fl. Turkey [P. H. Davis] 7: 459 (1982).

Fig. 153, 1–7.

S. amasiaca Freyn & Bornm. in Oesterr. Bot. Zeitschr. 41: 58 (1891).
S. paalii Pénzes in Borbásia 5–6 (1–3): 14 (1946).

HAB. In a wide variety of habitats, mountain side, medows, mixed shrubland, woodland, in mountains crevices, stream banks; alt. 110–2150 m; fl. & fr. May-Jul.

DISTRIB. Mountains of N. Iraq. **MAM**: Ser Amadiya (Gali Mazurka), gorge of Mazurka, *Al-Dabbagh, Al-Kaisi & Hamid* 45964! (K); Zawita, bank of stream, *VCR* 353 (K!). **MRO**: Arbil, Rawandiz (Rowanduz), M. Sakri Sakran, *Bornmueller* 1742 (K!); Algird Dagh (Halgurd), near Rust, *Guest & Hewitt* 2929! (K); Kodo (Kede?) near Haji Omran, *Rawi* 9174! (K). **FPF**: Badra, in orchard, *Al-Kaizi & Yahya* 45299! (K).

32. **Salvia russellii** *Benth.*, Labiat. Gen. Spec. 312 (1833); Rawi in Dep. Agr. Iraq Tech. Bull. 14: 155 (1964); Hedge in Fl. Iranica [K. H. Rechinger] 150: 474 (1982); Hedge in Fl. Turkey [P. H. Davis] 7: 460 (1982); Jamzad in Flora Iran [Assadi & al., in Persian] 76: 948 (2012).

Salvia sinapifolia K.Koch, Linnaea 21: 659 (1849).
S. bornmuelleri Hausskn., Mitt. Bot. Ver. Jena ix. 21 (1891).
S. margasurica Al-Musawi & Al-Hussaini, Journal of Global Pharma Technology 10(9): 291–292 (2017), nom illegit. Type: Margasur-Shanidar, *Al-Musawi & Al- Hussaini* 46393 (BUH).

Perennial herb. Stems 20–60 cm, erect, usually simple, many, arising from a woody rootstock, eglandular-pubescent. Leaves simple, linear-oblong, sometimes sublyrate or deeply lobed at base, 5–8 × 1–1.5 cm, rugulose, cuneate, eglandular-pilose especially on veins and with numerous sessile glands; petiole 2–8(–10) cm. Verticillasters 20–30-flowered, clearly distant. Bracts c. 6 × 2 mm, ovate-acuminate. Pedicels 2–6 mm, erecto-patent. Calyx tubular, 5–7 mm, purplish-blue, eglandular spreading pilose and with sessile glands, ovate in fruit with a concave-bisulcate upper lip, teeth not mucronate. Corolla pinkish-blue, 9–11 mm, tube straight with annulus; upper lip straight, compressed, narrowed at base. Stamens type C.

HAB. Rocky slopes, grassy meadows amongst *Quercus*, fallow and cultivated fields, banks of streams in clearing; alt. 1000–1600 m; fl. & fr. May-Jul.

DISTRIB. Mountains and upper plains of N. Iraq. **MAM**: 20 km from Zawita toward Shekan, *Sharief & Hamad* 50358! (K); Atrush-Rabetki, *Chapman* 9336! (K); Zawita, *RC.* 43Z! (K); Sirsang (Sarsang), *Haines* 1819! (K); Sulaf, c. 1 km N.W. of Amadiya, *Omar* 37710! (K); between Sulaf and Amadiya, *Kaisi, Al-Khayat & Kerim* 51078! (K); Mosul, Amadiya, *Anders* 2277! (K); Mosul, Amadiya, *Al-Kaisi & Hamad* 53796! (K). **MAM/FNI**: Road between Aqra and Mosul, *Rawi* 11511! (K). **MRO**: Soran village, N. of Kani Kawan, Water Spring, Karookh, *Nuri & Kass* 27283! (K); Jebel Baradost (Bradost), near Diana Rowanduz, *Field & Lazar* 930! (K); Arbil, Kuh-e Sefin supra Shaqlawa, *Bornmueller* 1743! (K); Margasur-Shanidar, *Al-Musawi & Al- Hussaini* 46393 (BUH, type of *S. margasurica*. **MSU**: Sulamaniya, Penjwin, *Rechinger* 12252 (W). **FUJ**: Mosul, 1841, *Kotschy* 102! (K).

A clear relative of the much commoner and widespread *S. verticillata* but more compact in habit with narrowly oblong leaves. The recently published new species, *S. margasurica*, from Margasur Province seems not to be different from *S. russellii* due to its fully overlapping morphological description and distribution range. Therefore, I have placed *S. margasurica* as a synonym of *S. russellii*.

Iran, Turkey, Syria.

CULTIVATED SPECIES

Salvia farnacea Benth.
LCA: Abu Gharib, cultivated, *Sakira & Tenam* 270! (K).

Salvia splendens Sellow ex Nees.
LEA: Baghdad, cultivated, 16 October 1966, *J. Mokhtar* 34757!; *H. Ahmad* 10108!; *Hamad* 9839A! (K).

THE FOLLOWING SPECIES MAY ALSO BE FOUND IN IRAQ.

Salvia officinalis L. (in cultivation); *Salvia aethiopis* L.; *Salvia pachystachys* Trautv.; *Salvia limbata* C.A.Mey.;
Salvia staminea Montbret & Aucher ex Benth.; *Salvia nemorosa* L.

16. **LYCOPUS** L.

Sp. Pl. 1: 21 (1753) & Gen. Pl. ed. 5: 12 (1754)
Henderson, Amer. Midl. Nat. 68: 95–138 (1962); Harley & al. in Kubitzki (ser. ed.), Fam. Gen. Vasc. Pl. 7: 237 (2004)

Ali Haloob
Revised by Ferhat Celep & Tuncay Dirmenci

Perennial non-aromatic herbs with creeping rhizome. Leaves coarsely dentate to pinnatisect; floral leaves similar to stem leaves. Verticillasters sessile and globose, many-flowered, small, remote. Calyx campanulate, ± regular, glabrous inside; teeth 4 or 5, equal or 1 larger. Corolla campanulate, 2-lipped, throat intricately villous; upper lip entire or emarginate; lower lip 3-lobed, middle lobe larger than lateral lobes. Anterior 2 stamens fertile, slightly exserted from corolla, straight, posterior 2 rudimentary or filiform, apex clavate or capitate; filaments glabrous; anther cells 2, parallel, becoming divergent. Style exserted, apex 2–cleft; lobes flattened, acute, equal or posterior smaller. Nutlets trigonous, truncate at apex.

20 species in the temperate Northern Hemisphere to Northwest Africa to East & Southeast Australia; a single species in Iraq.

Henderson, N.C. (1962). A Taxonomic Revision of the Genus *Lycopus* (Labiatae). Amer. Midl. Nat. 68: 95–138.

1. **Lycopus europaeus** *L.*, Sp. Pl. 21 (1753); Dinsmore in Post, Fl. Syria, Palest. & Sinai ed 2, 2: 331 (1933); Rawi in Dep. Agr. Tech. Bull 14: 151 (1964); Rechinger, Fl. Lowland Iraq: 530 (1964); Feinbrun-Dothan, Fl. Paleast. 3: 156 (1978); Rechinger, Fl. Iranica 150: 554 (1982); Mill in Fl. Turkey [P. H. Davis] 7: 394 (1982); Xiwen & Hedge in Fl. China 17: 240 (1994); Jamzad in Flora Iran [Assadi & al., in Persian] 76: 633 (2012).

Lycopus alboroseus Gilib. Fl. Lit. Inch. i. 71. (1782). nom. inval.
L. albus Mazziari, Ionios Anthologia 1(2): 446. (1834); nom. inval.
L. aquaticus Moench, Methodus (Moench) 370 (1794).
L. decrescens C.Koch in Linnae 21: 646 (1848).
L. europaeus subsp. *europaeus* var. *glabrescens* Schmidely, Bull. So Bot. Geneve 3: 129 (1884).
L. europaeus subsp. *europaeus* var. *pubescens* Benth. Prodr. [A. P. de Candolle] 12: 17 (1848).
L. europaeus subsp. *europaeus* subsp. *mollis* (Kerner) Rothm. ex Skalicky, Acta Mus. Nat. Pra 24B: 203 (1968).
L. europaeus subsp. *europaeus* subsp. *menthifolius* (Mabille) Skalicky, op. cit. 206 (1968 Ic: Jav. & Csap., Ic. Fl. Hung. t. 441 f. 2992 (1932); Polunin, Fls. Europe 116 no. 1166 (1969).
L. mollis A.Kern., Oesterr. Bot. Z. 16: 371 (1866).
L. niger Gueldenst., Reis. Russland (Gueldenst.) ii. 65. (1791).
L. palustris Burm.f., Fl. Ind. (N. L. Burman) Prodr. Fl. Cap.: 1. (1768).
L. riparius Salisb., Prodr. Stirp. Chap. Allerton 72. 1796.
L. solanifolius Lojac., Fl. Sicul. (Lojacono) 2(2): 194. (1907).
L. souliei Sennen, Bol. Soc. Aragonesa Ci. Nat. xv. 250 (1916)
L. vulgaris Pers., Syn. Pl. [Persoon] 1: 24. (1805)

Perennial. Rhizomes transverse, producing long stolons enlarged at apex, with scale-like leaves. Stems 15–90(–120) cm tall, sparingly puberulent to shortly pilose, subglabrous,

Fig. 154. **Lycopus europaeus**. 1, habit × ½; 2, flower × 10; 3, calyx × 10. Reproduced with permission from Jamzad, Flora of Iran 76: f. 198 (2012). © Ministry of Jihad-e-Agriculture.

unbranched or apically branched. Leaves 20–90(–150) × 10–40(–60) mm, ovate-lanceolate, oblong-elliptic to lanceolate-elliptic, coarsely dentate or crenate, less commonly pinnatisect at base or ± throughout their length, lobes 1–7(–35) mm deep, subglabrous or sparsely pubescent, with sessile glands. Verticillasters globose, 8–10 mm in diameter; floral leaves subsessile; outer bracteoles to 5 mm long, inner ones ca. 3 mm long, linear-subulate, spinescent. Calyx 3–4.5 mm long, puberulent, ± conspicuously 10–15-veined; teeth 4 or

5, 1.5–2.5 mm long, erect, linear-subulate, spinescent at apex, throat naked. Corolla white with small dark purple spots, 3–4 mm long, puberulent, tube c. 2.5 mm long, intricately white villous inside; limb obscurely 2-lipped; upper lip circular, emarginate, lobes subequal. Anterior stamens exserted, posterior 2 lacking or reduced to staminodes. Nutlets 4-sided, 1.5–1.8 × 0.9–1.2 mm, adaxially slightly swollen, sparsely to densely glandular at middle, base slightly attenuate, apex rounded; areolae basal, white. Fig. 154, 1–3.

HAB. Wet places by streams, pools, lakes and marshes, field margins, streamside, grasslands rarely drier banks, often in shade; common; alt. 75–1070 m; fl. & fr. Jan.-Oct.

DISTRIB. In mountains and lower plains of north, east and southeast Iraq. MAM: 25 km from Zakho to Kani Masi, *Botany staff* 43805 (BAG). MRO: Alana valley, 12 km S.W. Rowanduz, Riyadh, Hydar, Noor, Shayma & Adil 58713 (BAG); Gali Ali Beg, Riyadh, Hydar, Noor, Shayma & Adil 58792 (BAG). MSU: Ahmad Kulwan, village near Persian frontier, c. 6 km S.S.W. of Penjwin, coll. ignot., 5328! (W, BAG); Diyala, *Sutherland* 445 (W); 5 km N. of Saiyid Sadiq, *Al-Khayat, Al-Kaisi & Thamer* 46309! (BAG); 15 km S. of Chuwarta (Tagran), *Al-Kaisi & Al-Khayat* 51143 (BAG); MAM/FNI: Mindan Bridge on Klaazir river between Mosul & Aqra, *Alizzi & S. Omar* 35293! (BAG); FNI: Sheik Hums Zimma near Mosul, *Mashtouf* 5198! (BAG); LCA: Hurriya, near Baghdad, *Haines* 804! FPF:10 km E. of Al-Sudur, *Botany staff* 47188! (BAG); 3 km N. of Sa'adiya, 1/9/1979, *Al-Kaisi* 50497 (BAG); LSM: Hor al- Shain, S. of Majarr, Alkabbeer, *Thamer* 46654!; Amarah marshes, 13/12/1931, *Guest* 1626 (BAG).

Variable in leaf morphology and indumentum. Therefore, numerous infraspecific taxa have been recognised under the species (see webpages: IPNI and POWO [www.plantsoftheworldonline.com). The degree of leaf dissection in *L. europaeus* is dependent on the amount of moisture, specimens from dry habitats having small dentate leaves, while those from wetter places have larger leaves. Indumentum of the species is also variable in different habitats. Based on different indumentum types, several infraspecific taxa were published. Since the subordinate taxa all intergrade and the characters on which they are based are partly dependent on habitat, therefore the species is treated here in a broad sense.

BARCHIBAH برجيبة or TARMAHI ترماهي (Arab., Amarah marshes, *Guest* 1626).

Most of Europe, China, Japan, Kazakhstan, Kyrgyzstan, Russia, Tajikistan, Turkmenistan, Uzbekistan; S.W. Asia (Turkey, Iran, W. Syria), Europe; introduced into North America.

17. **MENTHA** L.

Sp. Pl. 2: 576 (1753)
Harley & al. in Kubitzki (ser. ed.), Fam. Gen. Vasc. Pl. 7: 237 (2004)
Pulegium Mill., Gard. Dict. Abr. ed. 4: s.p. (1754); *Preslia* Opiz, Naturalientausch 8: 86 (1824); *Audibertia* Benth., Edwards's Bot. Reg. 15: t. 1282 (1829); *Menthella* Pérard, Bull. Soc. Bot. France 17: 205 (1870). *Minthe* St.-Lag., Ann. Soc. Bot. Lyon 7: 130 (1880)

Tuncay Dirmenci & Ferhat Celep

Shrubs, subshrubs, or perennial (rarely annual) herbs, of damp places, strongly aromatic, usually gynodioecious. Rhizomes creeping. Stems ascending to erect, simple or branched. Leaves opposite, petiolate to sessile, simple, dentate to crenate or serrate, cordate to attenuate at base, with characteristically scented epidermal glands. Bracteoles lanceolate to linear-subulate, filiform. Flowers few to many in axillary cymes, pedunculate or not, often either in dense verticillasters along stem, in axils of leaf-like bracts, or congested in terminal spicate thyrses in axils of reduced bracts; bracteoles inconspicuous. Calyx actinomorphic or sub-bilabiate, 5–(rarely 4)-lobed, lobes subequal, triangular to subulate or acicular, or slightly 2-lipped; anterior lobes often longer; tube cylindrical to infundibuliform, 10–15-veined, with 5 subequal or unequal teeth, throat hairy or glabrous. Corolla ± actinomorphic, weakly 2-lipped, with four subequal lobes, upper lobe wider, usually emarginate; tube lilac, pink or white, ± cylindrical or rarely gibbous on anterior side, usually included within calyx tube. Stamens 4, subequal, divergent or ascending under upper lip, exserted from corolla (except in female flowers and most hybrids, where stamens are reduced or absent), anthers ellipsoidal, thecae parallel, distinct. Stigma lobes subequal, shortly spreading; disc ± symmetrical. Nutlets ovoid, smooth, foveolate, reticulate or rugulose, glabrous to sub scabrous, rarely pilose at apex, brown.

The genus *Mentha* is very complex due to hybridization, gynodioecism, different ploidy levels, vegetative propagation and morphological plasticity. With 23 species (more than 40 taxa) and 15 hybrids, it is a genus with a wide distribution, mainly in Europe and Asia, as well as the Cape area (1 species) and New Zealand (1 species). The species prefer moist habitats. The number of chromosomes numbers varies between 18–132. It is an economically valuable genus with several species used as tea and as a spice; 2 species and one cultivated hybrid in Iraq.

1. Inflorescence interrupted and slender, a few verticillasters usually
      approximate above. . . . . 1. *M. spicata*
      Inflorescence congested or with a few verticillasters interrupted only
      below and rather thick or rarely slender 2
2. All Leaves petiolate; usually subglabrous, dark green plants (cultivated
      in Iraq) . . . . *M.* ×*piperita*
      Leaves sessile or with very short petiolate, green- or grey- or sometimes
      white-tomentose beneath, with simple curved hairs with a felted
      appearance when dry (natural in Iraq) . . . . 2. *M. longifolia*

1. **Mentha spicata** *L.*, Sp. Pl.: 576 (1753); Rechinger, Fl. Iranica 150: 569 (1980); Davis, Fl. Turkey 7: 391 (1982); Jamzad in Flora Iran [Assadi & al., in Persian] 76: 774 (2012).

subsp. **spicata**

Mentha crispa L., Sp. Pl.: 576 (1753).
M. spicata var. *viridis* L., Sp. Pl.: 576 (1753).
M. viridis (L.) L., Sp. Pl. ed. 2: 804 (1763).
M. romana Garsault, Fig. Pl. Méd.: t. 378 (1764).
M. glabra Mill., Gard. Dict. ed. 8: n.° 2 (1768).
M. tenuis Michx., Fl. Bor.-Amer. 2: 2 (1803).
M. laevigata Willd., Enum. Pl.: 609 (1809)
M. ocymiodora Opiz, Naturalientausch 4: 21 (1823).
M. rosanii Ten., Fl. Neapol. Prodr. App. 5: 18 (1826).
M. spicata subsp. *glabrata* (Lej. & Courtois) Lebeau, Nouv. Fl. Belg., Grand-Duché Luxemb., Nord
    France: 759 (1973).
M. spicata var. *oblongifolia* (Wimm. & Grab.) Lebeau, Bull. Soc. Échange Pl. Vasc. Eur. Occid. Bassin
    Médit. 17: 67 (1979).
M. spicata var. *undulata* (Willd.) Lebeau, Bull. Soc. Échange Pl. Vasc. Eur. Occid. Bassin Médit. 17:
    68 (1979).

Rhizomatous perennial herbs. Stems 15–100 cm, green, ascending to erect, usually branched, glabrous to coarsely grey-villous with sparse to dense sessile glands and glandular papillae. Lower and middle stem leaves shortly petiolate, upper subsessile to sessile; lamina ovate-oblong to oblong-lanceolate, 1.5–8 × 0.5–3 cm, rounded to cordate at base, obtuse or acute at apex, margins serrate, flat or undulate, glabrous to coarsely grey-villous (*Gillet* 9802), hairs simple with sessile glands and glandular papillose. (Leaves of cultivated plants of near cultivations are subglabrous and green). Verticillasters many-flowered, forming a terminal spike 1.5–14 × 0.5–1 cm, usually interrupted or at least below, approximate above; cymes shortly pedunculate; pedicel 1–2 mm. Calyx campanulate, 1.5–2.5 mm, glandular papillose with sessile glands, slightly constricted in fruit; teeth subequal to equal, triangular-lanceolate, acuminate; glabrous at throat. Corolla white or pink, 2.5–3.2 mm, glabrous with a few sessile glands. Stamens 4, included in corolla. Style unequally bilobed, long exserted. Nutlets not seen.

HAB. By irrigation channels in shady forest and cultivated areas; alt. ± 1000 m; fl. & fr. May-Sep.
DISTRIB. Mostly collected around Baghdad, in cultivated fields and orchards; only one record from the wild. **MRO**: Between Zerwa and Chamaluya, *Gillet* 9802! **LCA**: Baghdad, cult., *Tenar* 487!; Baghdad liwa, drug garden near Agar Ghuf (cult.), *Barkley* & *Brahim* 2524!; Abu Ghraib (cult.), *Makhtar* 35840! (K!). **LSM**: Amara exp. Station.

Europe to China.

2. **Mentha longifolia** (*L.*) *L.*, Amoen. Acad., Linnaeus ed. 4: 485 (1759); Dinsmore in Post, Fl. Syria, Palest. & Sinai ed. 2, 2: 330 (1933); Rawi in Dep. Agr. Tech. Bull 14: 152 (1964); Feinbrun-Dothan, Fl. Paleast. 3: 157 (1978); Rechinger, Fl. Iranica 150: 557 (1982);

Fig. 155. **Mentha longifolia**. 1, habit × 1; 2, flower × 3. Reproduced with permission from Feinbrun-Dothan, Fl. Palaestina 3: Plates, f. 261 (1977). Drawn by Esther Huber. © The Israel Academy of Sciences and Humanities.

Davis, Fl. Turkey 7: 388 (1982); Jamzad in Flora Iran [Assadi & al., in Persian] 76: 774 (2012); Taifour & El-Oqlah, Pl. Jordan Annot. Checkl.: 102 (2017).

Mentha spicata var. *longifolia* L., Sp. Pl.: 576 (1753).
M. spicata subsp. *longifolia* (L.) Tacik, Fl. Polska 11: 216 (1967).

Perennial herbs, rhizomes mainly hypogeal with scale-like leaves. Stems 30–80 cm tall, erect, a few to many branched or rarely simple, variously simple hairy, hirsute-puberulent to greyish tomentose, densely villous to pilose, with sessile glands and glandular papillose, especially in the upper part whitish-villous. Leaves very variable in shape and size, ovate-oblong to oblong-lanceolate, or lanceolate, broadest at or above middle, sessile or petiolate, 3–8 × 1–3 cm, lamina smooth or very weakly rugose, green- to grey-tomentose above, green- to white-tomentose beneath or villous with densely sessile glands and glandular papillose, (discolorous in subsp. *longifolia*), rounded to subcordate at base, usually acute, or obtuse, acuminate at apex, margin sharply serrate with many irregular and often spreading teeth. Floral leaves bractaeformis, linear-subulate. Verticillasters many, congested, rarely the lowest 1–2 verticillasters distant, forming a terminal often much branched spike (1.5–)3–9 × 0.7–1.5 cm. Flowers pedicels to 2 mm long, tomentose. Calyx 1.5–3 mm long, narrowly campanulate, tomentose to villous with sessile glands and glandular papillose; teeth subequal, 1–1.2 mm long, triangular-acuminate to linear-subulate, equalling to or shorter than tube, ciliate; glabrous at throat. Corolla 2.5–5 mm long, lilac, pink to purple or white, puberulous outside, glabrous within; lobes ± equal; upper lobe oblong-ovate, emarginate or crenate, the others narrow, oblong, flat. Stamens 4, filaments long exserted from corolla, sterile if included in corolla. Style unequally bilobed, long exserted from corolla. Nutlets c. 0.5–0.8 × 0.5– 1 mm long, ovoid, reticulate, pilose at apex. Fig. 155, 1–3.

HAB. Streamside and marshy fields, rivers, alt. 300–2600 m; fl. & fr. Jun.-Sep.
DISTRIB. Macaronesia, Temperate Eurasia to S. Africa

Mentha longifolia is a species with a very wide distribution and wide morphological variations. It is divided into many sub-taxa. Here, however, all taxa previously recorded from Iraq (Rechinger 1982, pp. 557–565) were broadly evaluated under two subspecies, subsp. *noeana* (included var. *petiolata* (Boiss.) Rech.f.) and subsp. *thyphoides* (included var. *asiatica* (Boiss.) Rech.f., var. *amphilema* (Briq.) ex Rech.f., var. *chlorodictya* Rech.f.)

Leaves subsessile to sessile. . . . . a. subsp. **typhoides**
Leaves distinctly petiolate . . . . b. subsp. **noeana**

a. subsp. **typhoides** (*Briq.*) *Harley*, Notes Roy. Bot. Gard. Edinburgh 38: 38 (1980).

Mentha calliantha Stapf, Denkschr. Kaiserl. Akad. Wiss., Wien. Math.-Naturwiss. Kl. 50: 36 (1885).
M. sylvestris subsp. *typhoides* Briq., Bull. Trav. Soc. Bot. Genève 5: 90 (1889).
M. cyprica Heinr.Braun, Verh. K. K. Zool.-Bot. Ges. Wien 39: 217 (1889).
M. sylvestris subsp. *calliantha* (Stapf) Briq. Bull. Trav. Soc. Bot. Genève 5: 86 (1889).
M. longifolia subsp. *cyprica* (Heinr.Braun) Harley, R.D.Meikle, Fl. Cyprus 2: 1897 (1985)
M. longifolia subsp. *calliantha* (Stapf) Briq., H.G.A.Engler & K.A.E.Prantl, Nat. Pflanzenfam. 4(3a): 321 (1897).

HAB. Near water on metamorphic rocks, on river edges, under shade of *Juglans* sp. by stream, in orchards under *Populus* trees, rocky mountain slopes near streams; alt. 100–2600 m; fl. & fr. May-Sep.
DISTRIB. It is a widespread taxon that grows near water and moist areas in Northern and Northeastern Iraq and Eastern and Central alluvial plain district of Iraq. **MAM**: Bet Qayara and Mosul, *Alizzi & Omar* 35261!; Zakho, *Botany Staff* 43783!; Sersang, *Haines* 560!; Sharanish village, 25 km N.E. of Zakho, *Nuri* et al. 29032!; 25 km from Zakho to Kani Masi, *Botany Staff* 43832!; **MSU**: Kirkuk, a Chamchamal orientem versus in secus rivulos, *Rechinger* 12495 (W); Qara Dagh mountains, Suleymaniya liwa, *Makki* 495 (W); Halabja, *Noori & Hamad* 41204! Between Erbil and Shaqlawa, *Rawi* et al. 19681! **MRO**: Kodo near Haji Omran, *Rawi* 9231!; Erbil, Haji Omran, in declivibus siccis, *Rechinger* 11301 (G, W); Haji Omran, *Al-Dabbagh & K.Hamad* 46288!; Haji Omran, on Persian border, *Rawi* 24939!; Sula Khal, black-brownish rocks, *Rawi & Serhang* 24686!; Pushtashan, 15 km N.E. of Rania, lower slope of Qandil range, *Rawi & Serhang* 24199!; Hassan Beg, *Guest* 3024!; Sakran mountain, 15 km S.E. of Ghuman, *Al-Dabbagh & K.Hamad* 46246! Newrobar valley-Qandil range, *Rawi & Serhang* 24162!; Montes Qandil ad confines Persiae, Pustashhan, *Rechinger* 11206 (G, W); Qandil, in declivibus orientem supra Pushtashan, secus rivulos, *Rechinger* 11107!; Pushtashan, 15 km N.E. of Rania, *Rawi* 23919!; Serin mountain, on road to Qandil, *Rawi & Serhang* 24019! Erbil liwa, Koi Sanjak, *Gillett* 5223!; Gali Ali Beg, *Gillett* 9437!; Erbil, mons Helgurd ad confines Persiae, in vale supra pagum Nowanda, *Rechinger*

11388 (W); Valley E. of Algird Dagh, *Gillett* 9506! **FUJ**: Abu-Tina, 2 km E. of Hadhr, *Omar & Aluzzi* 35253! **DLJ**: Kirkuk liwa, 40 km N.E. of Beije Fatha, near lake, *Brahim & Mohammes* 3651! **FNI**: Near Tel-Kaif, Alizzi & S.Omar 35265! Rich bottomland of wadi Bakok, 35 km N. of Mosul, *Barkley & Ani* 8985 (W); **FKI**: Hills and Lake 10 km N. of Kirkuk, *Brahim* 6148! **FPF**: Al Saadiya, 18 mm, along irrigation canal, *Omar & Thamer* 47125! 10 km E. of Al-Sudur, *Botany Staff* 47186! & 47156! **LEA**: 20 km N.E. of Baquba, *Botany Staff* 47146! **LCA**: Baghdad, Pig Island, *Haines* 1823!; Al-Sudur, gravelly sandy soil, *Botany Staff* 47156!

b. subsp. **noeana** (*Briq.*) *Briq.*, in H.G.A.Engler & K.A.E.Prantl, Nat. Pflanzenfam. 4(3a): 322 (1897).

M. royleana Benth subsp. *noeana* Briq. Bull. Trav. Soc. Bot. Genève 5: 80 (1889).
M. sylvestris L. var. *petiolata* Boiss., Fl. Orient. 4:543 (1879).

HAB. Near water, rocky mountain slopes near streams; alt. 200–1500 m; fl. & fr. Apr.-Aug.
DISTRIB. It is a widespread taxon that grows near water and moist areas in Northern and Northeastern Iraq and Eastern and Central alluvial plain district of Iraq. **MAM**: Ad rivulos et scaturigines montain Gara, *Kotschy* 395! **MSU**: Hauraman (Avroman) mountain, *Rawizchak* et al. 19817!; Rawanduz, *Guest* 13616!; Sipa water fall in Urban area, R. Hamishkan, Riyadh, Najwan & Shyimaa 59197 (BAG). **MRO**: Suwara-Gukkah, *Guest* 1579!; 20 km N.E. of Kirkuk to Kai Sanjang, *Rawi* 26453!; Lower part of Rewanduz gorge, *Gillett* 9438!; Rowanduz, *Guest* 456!; Pushtashan, N.E. of Qandil range, *Rawi & Serhang* 26564; Mons Qandil, Pushtashan, *Rechinger* 11043 (G, W); Bala, 20 km N.E. of Rewanduz, *Guest* 15855!; Kanirush, *Serhang & Rawi* 26670!; Alana valley, Rawawish village S. of Khalifan, Riyadh, Hydar, Noor, Shayma & Adil 58750 & 58755 (BAG). **FNI**: N. of Mosul, *Anders* 1579! **LEA**: Diyala, Baquba, *Rechinger* 8361 (W); Baquba, *Haines* 2! **LCA**: Al-Sudur, *Omar* et al 36975! (K). **LBA**: Abul-Khasib, Thamer 47400 (BAG).

3. **Mentha × piperita** *L.* Sp. Pl.: 576 (1753); Ball & Harley in Fl. Europ. [Tutin et al.] 3: 185 (1972); Borisova in Fl. U.S.S.R [Shishkin & Yuzepchuk] 20: 445 (1976). Davis, Fl. Turkey 7: 387 (1982); Jamzad in Flora Iran [Assadi & al., in Persian] 76: 772 (2012).

Selected synonyms:

Mentha × nigricans Mill., Gard. Dict. ed. 8: n.° 12 (1768).
M. × glabrata Vahl, Symb. Bot. 3: 75 (1794).
M. × piperita var. *officinalis* Sole, Menth. Brit.: 15 (1798).
M. × hircina Hull, Brit. Fl. 1: 127 (1799).
M. × balsamea Willd., Enum. Pl.: 608 (1809).
M. × piperita var. *officinalis* W.D.J.Koch, J.C.Röhling, Deutschl. Fl., ed. 3, 4: 250 (1833).
M. × piperoides Malinv., Bull. Soc. Bot. France 54: 653 (1907).

Perennial herbs, rhizomes horizontal. Stems 30–100 cm tall, erect or ascending, branched, green or reddish, usually subglabrous or rarely hairy to grey-tomentose. Leaves with petiole 6–8 mm long, 3–8 × 1.5–4 cm, ovate-lanceolate to lanceolate, rounded to subcordate at base, usually serrate at margins, acuminate at apex, glabrous or with short setiform hairs on the veins beneath, dark green, densely punctate-glandular beneath. Floral leaves resembling the cauline, smaller. Inflorescences at ends of stems and branches, capitate-spicate, short and broad, interrupted at base, 30–80 × 12–18 mm; bracts narrow, setaceous, ciliate, glabrous at base, the lower longer than verticillasters, the upper shorter. Calyx 3–4 mm long, tubular, glabrous, violet-tinged, punctate-glandular, the teeth erect, not connivent in fruit, ciliate, one-third the length of the tube. Corolla pink or lilac; glabrous, the tube whitish, about as long as calyx. Stamens shorter than corolla. Style exserted. Nutlets obovoid, ca. 0.75 × 0.5 mm, dark brown, glandular at apex.

HAB. Cultivated and occasionally naturalized; fl. & fr. Jul.-Sept.
DISTRIB. **LCA**: Abu Ghraib, Department of Agriculture farm, *Barkley & Abbas* 3876-B! (W); Abu Ghraib, *Alizzi* 39177! **LSM**: Amara, *Omar* 37761!

Mentha piperita, a hybrid between *Mentha aquatica* and *M. spicata*, has been cultivated since a long time in many countries where it has also become naturalized. It is commonly known as peppermint; the leaves are popularly used in an infusion as peppermint tea; the leaves are also used as a spice.

18. **THYMBRA** L.

Sp. Pl. 2: 569 (1753); Gen. Pl. 5: (1754); Harley & al. in Kubitzki (ser. ed.), Fam. Gen. Vasc. Pl. 7: 238 (2004)

Ali Haloob

Small perennial shrubs. Leaves sessile, lanceolate-linear, entire, with sessile red glands. Inflorescence spicate, verticillasters dense, oblong. Bracts similar to leaves; bracteoles lanceolate, with ciliate margins. Flowers many. Pedicel less than 1 mm. Calyx tubular, 13-veined, flattened dorsally with conspicuous 2 lateral ciliolate veins; throat hairy; limb teeth ciliate; upper lip with 3 short triangular teeth; lower lip with 2 upcurved lanceolate teeth. Corolla tube exserted, straight, upper lip erect, flattish, apex emarginated, lower lip 3-lobed. Stamens 4, didynamous, the anterior 2 filaments longer than the 2 posterior, ascending under upper lip, included; anther 2-celled, parallel. Style apex bifid, subequal. Ovary glabrous. Nutlets ovoid.

About 4 species in the Mediterranean region; 2 species in Iraq.

Corolla pink or pinkish-purple, 14–6 mm, tube hairy inside, glabrous
  outside; bracteoles long ciliate . . . . .1. *T. spicata*
Corolla white, 10–12 mm, tube glabrous inside, hairy outside;
  bracteoles short ciliate . . . . 2. *T. sintenisii*

1. **Thymbra spicata** *L.*, Sp. Pl. 2: 569 (1753); Boissier, Fl. Orient. 4: 1190 (1879); Handel-Mazzetti in Ann. Nat. Mus. Wien 27: 420 (1913); Nábělek, Spisy Přír. Fak. Masarykovy Univ. 35: 43 (1923); Guest in Dep. Agr. Iraq Bull. 27: 100 (1933); Zohary in Dep. Agr. Iraq Bull. 31: 131 (1950); Blakelock in Kew Bull., 4: 552 (1950); Rawi in Dep. Agr. Iraq Tech. Bull. 14: 158 (1964); Heywood & DeFilipps in Fl. Europ. 3: 170 (1972); Feinbrun-Dothan, Fl. Palaest. 3: 152 (1978); Davis, Fl. Turkey [P. H. Davis] 7: 383 (1982); Rechinger, Fl. Iranica [K. H. Rechinger] 150: 523 (1982).

Suffrutescent, woody at base, 15–40 cm. Stems erect or ascending, branched below, rigid, internodes ± 10 mm, white puberulent along two opposite sides. Leaves sessile, linear or lanceolate, 12–20 × 2–4 mm, obtuse or acute apex, margins entire, glabrous. Inflorescence 15–50 mm, verticillasters 5–10-flowered, except a few upper, lowermost internodes 5–8 mm, the upper ones congested. Bracts similar to leaves, long ciliate; bracteoles sessile, linear or lanceolate, 7–10 × 1–3 mm, longer than calyx, acute-acuminate apex, densely long ciliate, purplish. Calyx 5–7 mm, with long cilia on 2 lateral dorsal veins and limb. Corolla pink or pinkish-purple, 14–16 mm, tube exserted, twice as long as calyx, glabrous outside but with minute eglandular hairs inside; lower lip with 3 equal to subequal lobes, entire. Nutlets ± 2 mm, brown. Fig. 156, 1–2.

HAB. Rocky clay mountain slopes, rocky gypsum mountain, mud banks, near spring in pine forest, oak forest, open limestone slopes, clay hillsides, roadsides, rocky clay soils; alt. 320–1400 m; fl. Apr.-Jul.
DISTRIB. Common in the N. & N.E. forest zone and occasional in N.W. sector of the upper plains of Iraq.

MAM:, Jabal Bekhair near Zakho, *Rawi* 23054!; Zakho, *Rechinger* 10639!; between Sulaf & Sarsang, *S. Omar* 37729!; Zawita, *Al-Dabbagh & Al-Kaisi* 45324!; S. of Atrush, *Chapman* 9327!; Ain Sifni, *Salim* 2569! MRO: Khalana, *S. Omar, Al-Kaisi & Wedad* 49462!; Saladdin, *F. Karim, H. Hanide & M. Sasim* 40874!; Koi Sanjak, *S. Omar* 37600!; Haibat Sultan, *S. Omar, Sahira, F. Karim & H. Hammd* 38233! MSO: Dokan, *Salah & K. Hamid* 52663! FNI: Jarmo, *Wheelar-Haines* 233! (W); 30 km from Mosul to Aqra, *Al- Kaisi* 49703! FAR: Bekhme Gorge, *Gillett* 8225!

PUNK DAIMI بونك دايمي (Kurd. in Guest in Dep. Agr. Iraq Bull. 27: p. 100, 1933), JATA جاتي or ZAITAR زعتر (*Salim* 2569); a good fodder plant.

Greece, Cyprus, Lebanon, Palestine, Syria, Turkey, Iran.

2. **Thymbra sintenisii** *Bornm. & Aznavour*, Repert. Spec. Nov. Regni Veg. 10: 471 (1912); Handel-Mazzetti in Ann. Nat. Mus. Wien 27: 420 (1913); Nábělek, Spisy Přír. Fak. Masarykovy Univ. 35: 43 (1923); Zohary in Dep. Agr. Iraq Bull. 31: 131 (1950); Rawi in Dep. Agr. Iraq Tech. Bull. 14: 158 (1964); Davis, Fl. Turkey [P. H. Davis] 7: 384 (1982); Rechinger, Fl. Iranica [K. H. Rechinger] 150: 524 (1982).

Fig. 156. **Thymbra spicata**. 1, habit × 1; 2, flower × 5. Reproduced with permission from Jamzad, Flora of Iran 76: f. 236 (2012). Drawn by M. Nobakht. © Ministry of Jihad-e-Agriculture.

Suffrutescent, 10–25 cm; stems erect or ascending, branched at base, rigid, internodes 5–8 mm, with white puberulent. Leaves sessile, linear or lanceolate, 10–18 × 2–3 mm, apex acute, entire. Inflorescence dense, 10–40 mm, verticillasters 5–10-flowered, lowermost internodes 5 mm long, the upper congested. Bracts similar to leaves, short-ciliate; bracteoles linear or lanceolate, 8–10 × 1–2 mm, apex acute-acuminate, short-ciliate, green, longer than calyx. Calyx 4–6 mm, short-ciliolate on 2 lateral dorsal veins, 2 conspicuous ventral veins and limb. Corolla white, 9–11 mm, tube exserted, glabrous inside, hairy outside. Nutlets less than 2 mm. brown.

HAB. Rocky mountain sides, mountain slope, rocky clefts, valley, oak forest, on clay soils and clay hillside; alt. 800–1800 m; fl. Jun.-Sep.

DISTRIB. Locally common in the N. of forest zone of Iraq (only in MAM district). **MAM**: Khantur mountain N.E. of Zakho, *Rawi, Tikriti & Nuri* 28994!; Kani Masi, *K. Hamad & Fadhil* 45644!; Gara Dagh S. of Amadia, *Rawi* 9292!; Sarsank, *Haines* 536! (W); Zawita, 21 km from Dohuk, *Rechinger* 11544!

S. E. Turkey.

19. **THYMUS** L.

Sp. Pl. 2: 590 (1753); Morales, in Stahl-Biskup & Sáez F (eds), The genus *Thymus*, 1–44 (2002); Harley & al. in Kubitzki (ser. ed.), Fam. Gen. Vasc. Pl. 7: 238 (2004)
Mastichina Mill., Gard. Dict. Abr. ed. 4: s.p. (1754); *Serpyllum* Mill., Gard. Dict. Abr. ed. 4: s.p. (1754); *Cephalotos* Adans., Fam. Pl. 2: 189 (1763).

Tuncay Dirmenci & Ferhat Celep

Perennial, subshrubs or shrubs, woody at base, but often herbaceous above, aromatic. Stems erect to prostrate, sometimes caespitose, usually ± quadrangular, hairy all around, on two opposite sides or only on angles; stems, bracts, calyces and especially leaves with sessile colourless to bright red glands (oil dots). Leaves small, simple, entire or sometimes toothed, revolute, or flat, sessile or petiolate, glabrous or hairy, often ciliate towards base of lamina, very variable in indumentum. Inflorescence of verticillasters forming a terminal, condensed, often spicate or interrupted thyrse, 2-many-flowered. Bracts leaf like or not, lanceolate to broadly ovate, sometimes coloured; bracteoles inconspicuous. Calyx 2-lipped, 5-lobed to 2/3, sometimes nearly actinomorphic, ± campanulate or cylindrical, 10-nerved, lobes triangular or setaceous, rarely posterior lip entire, anterior lip upcurved or spreading, throat bearded. Corolla 2-lipped, or rarely actinomorphic, 4-lobed to 1/3, white, cream, pink, purple or violet, posterior lip ± rounded, emarginate, straight, anterior rectangular to suborbicular, rounded, spreading, tube ± cylindrical, sometimes very long; stamens 4, sometimes reduced or not (gynodioecious), inserted in upper half of the tube, exserted or not; thecae parallel, distinct. Stigma lobes ± equal. Nutlets ovoid, smooth.

About 220 species and 90 hybrids from Europe to Japan, mostly in the Mediterranean basin with a few in Macaronesia, Greenland and Ethiopia; 7 species in Iraq.

Thymus is a taxonomically complex genus. In the identification of species, leaf characters (shape, length/width ratio, flat or revolute, and veins), indumentum characters (whether hairy all round, on adaxial and abaxial sides or along the four angles alone), floral characters (calyx size, corolla size and colour, etc.) play an important role, but sometimes the vast majority of these characters are useless among closely related species. In addition to morphological variations, chemotypes are also present. In addition, hybridization easily occurs between closely related species or even species in different sections. This makes it more difficult to determine the boundaries of species.

Jalas, J. (1980). Turkish taxa of *Thymus* (Labiatae) described as new or revised. Ann. Bot. Fennici 17: 315–324.
Morales, R. (2002). The history, botany and taxonomy of the genus *Thymus*. In: Stahl-Biskup E. and Sáez F (eds). The genus *Thymus*. Pp.: 1–44, London & New York, Taylor & Franchis.
Rechinger, K.H. (1954). Die Gattung *Thymus* in Persien und angrenzenden Gebieten. Phyton, 5: 280–303.
Velenovsky, J. (1906). Vorstudien zu einer Monographie der Gattung *Thymus* L. Beih. Bot. Centr. 19 (B2): 271–287.

1. Corolla white and almost whole plant covered with only dense red oil dots. 2
    Corolla whitish-pink to purple, if white, whole plant not covered with
        only dense red oil dots. 3
2. Leaves ± elliptic, usually widest in the middle, 4–12 × 1.5–4 mm; calyx
        3.5–5 mm; corolla 4–5.5 mm long (adapted to desert areas) . . . . .5. *T. musilii*
    Leaves narrowly lanceolate to elliptic or linear-subulate, 8–22 × 1.2–3(–
        4) mm; calyx 4.8–7 mm long; corolla 6.5–9 mm long . . . . 6. *T. syriacus*
3. Leaves elliptic, ovate to linear-lanceolate, longer than 3 × as long as wide 4
    Leaves broadly ovate, obovate or orbicular, usually less than 3 × as long as wide 5
4. Flowering stems 6–30 cm, subglabrous to pilose or villous; leaves linear-
        lanceolate to ovate-lanceolate, 9.5–30 × 1.7–8 mm . . .7. *T. daenensis* subsp. *lancifolius*
    Flowering stems 2–11 cm, puberulent to hirsute, hairs patent to
        retrorse; leaves ovate to lanceolate-elliptic, 5.5–12 × 1.8–3 mm . . . . . . .4. *T. pubescens*
5. Leaves of flowering stems clearly growing upwards; calyx tube usually
        long and densely white-pilose . . . . .2. *T. erioclayx*
    Leaves all similar; calyx tube subglabrous to pilose. 6
6. Upper stem leaves ovate to broadly ovate . . . . 1. *T. kotschyanus*
    Upper stem leaves usually orbicular. . . . . 3. *T. carmanicus*

1. **Thymus kotschyanus** *Boiss.* & *Hohen.*, Diagn. Pl. Orient. Ser. 1, 5: 16 (1844). Type: Iraq in regione media montis Gara Kurdistaniae in rupestribus, 25.vii.1841, *Kotschy* 327 (W! holo., G! H! P-photo! W! iso.); Dinsmore in Post, Fl. Syria, Palest. & Sinai ed. 2, 2: 336 (1933); Rawi in Dep. Agr. Tech. Bull 14: 159 (1964); Rechinger, Fl. Iranica [K. H. Rechinger] 150: 540 (1982); Jalas in Fl. Turkey [P. H. Davis] 7: 368 (1982); Jamzad in Flora Iran [Assadi & al., in Persian] 76: 728 (2012).

> *Thymus arthoclados* Stapf, in Denkschr. Akad. Wien I: 36 (1885).
> *T. eriophorus* Ronniger, in Fl. Kavkaza [A.A. Grossheim] 3: 336 (1932).
> *T. kotschyanus* var. *elbursensis* Rech.f., Phyton (Horn) 5: 294 (1954).
> *T. kotschyanus* var. *behboudianus* Rech.f., Phyton (Horn) 5: 295 (1954).
> *T. kotschyanus* var. *pseuderiophorus* Rech.f., Phyton (Horn) 5: 292 (1954).
> *T. kotschyanus* var. *eriophorus* (Ronniger) Jalas, Ann. Bot. Fenn. 17: 321 (1980).
> *T. kotschyanus* var. *intercedens* Heinr.Braun, Verh. Zool.-Bot. Ges. Wien 39: 219–220 (1889).

Perennial, woody at base, herbaceous above, many stemmed from the base and lacking prostrate basal branches; flowering stems 3–15 cm tall, with 4–6 pairs of leaves, variously hairy all round, hairs ± reflexed. Leaves 9–13 × 4–9 mm, including 1–2 mm petiole, ovate to broadly ovate, truncate to rounded at base, obtuse to acute at apex, flat, glabrous to puberulent or shortly pilose on both sides, sessile gland numerous, pale or red, ciliate at base; venation prominent beneath, with 2–3 pairs of lateral veins, the lowest marginal vein and often also the other two joining apically to form a marginal thickening. Inflorescence usually a dense head, 1–2.5 × 1–1.5 cm, with subtending bracts. Bracts similar to leaves. Bracteoles 0.5–2.5(–3) mm, linear-setaceous, ± as long as or shorter than pedicels. Pedicels 2–4 mm. Calyx 4–5.5(–6) mm, tube subcylindric to ± campanulate, usually shorter than lips, subglabrous to densely pilose with numerous sessile glands and minutely glandular papillae; upper lip wider than tube, usually equal to or somewhat longer than lower teeth, densely hairy in throat; upper teeth 0.8–2 mm, triangular to lanceolate, acuminate to aristate, ciliate or not; lower teeth linear-subulate, 2–4 mm, ciliate. Corolla white or pale pinkish to purplish, 5–7.5 mm, included or slightly exserted from calyx. Stamen 4, sometimes reduced, included or slightly protruding. Style exserted from corolla, unequally bilobed. Nutlets 1–1.2 × 0.9–1 mm, rounded, dark brown.

HAB. Stony mountain slopes; alt. 600–3300 m; fl. & fr. May-Aug.
DISTRIB. Mountains and lower slopes in N. and N.E. Iraq. **MAM**: Dori village, 5 km E. of Kani-Masi, rocky mountain near water, *Omar* & *Al-Kaisi* 45443!; Ibid, *Omar* & *Al-Kaisi* 45398!; Bamerny, 20 km N.W. of Sarsang, *Al-Kaisi* & *K.Hamad* 45934!; Sarsang, *Haines* 461! (E); **MSU**: Sulaimaniya: dry slope, *Regel* s.n. (G); **MRO**: Sie-waka village (near S. of foot of Karoukh), *Al-Kaisi, Nuri* & *Sarhan* 27591!; Gali Ali Beg, mountain side, *Karim, Hamid* & *Jasim* 40911!; Gali Ali Beg, *Omar, Al-Kaisi* & *Wedad* 49557!; Rawanduz gorge, in rocks, *Guest* 607!; Rawanduz gorge, rocky clift, *Guest* 2989!; Rowanduz gorge, c. *Luckuer* 3828!; Rowanduz gorge, *Gillett* 8316!; Ibid, *Gillett* 8317!; Chia-i-Mandali, *Guest* 2789!; Ibid, *Guest* 2676!; Kermasur Lake-Qandil range, *Rawi* & *Serhang* 24129!; Arl-Gird Dagh, *Gillett* 12373!; Helgurd ad confines Persiae, *Rechinger* 11413 (G, W); Bekhal, *Omar et al* 38406! Erbil, in montis Kuf-Sefin (ditionis

Erbil), regione superiore, *Bornmuller* 1700!, Hawara Blinda mountain, N.E. of Haji Omran, *Rawi, Nuri & Kass* 27792!; Baski Hawaram mountain, *Rawi & Serhang* 23983! **FUJ**: Mossull, Kotschy 282! (type of *Th. kotschyanus* var. *intercedens*).

Thymus kotshcyanus is widely distributed and has wide morphological variations, such as in leaf shape, size and venation, and calyx and calyx teeth characteristics. It can be broadly regarded as a species complex. Some current species (*T. migricus, T. eriocalyx, T. fetsdchenkoii* var. *handelii* etc.) are also been evaluated as variations of *T. kotschyanus*.

Syria, Lebanon, C., E. and S. Turkey to N.W. Iran.

2. **Thymus eriocalyx** (*Ronniger*) *Jalas* in Fl. Iranica [K. H. Rechinger] 150: 542 (1982); Jalas in Fl. Turkey (suppl. 1) [P. H. Davis] 10: 209 (1988); Jamzad in Flora Iran [Assadi & al., in Persian] 76: 740 (2012).

Thymus kotschyanus var. *eriocalyx* Ronniger, Spisy Prír. Fak. Masarykovy Univ. 70: 42 (1926).

Perennial ± subshrubs, woody at base, herbaceous above, many stemmed from base. Flowering stems 3–20 cm tall, ascending to erect, slender, retrorsely puberulent to pilose. Cauline leaves 3–5 pairs, 6–15 × 4–8 mm, incl. 1–2.5 mm petiole, becoming larger towards the capitate inflorescence, truncate to rounded at base, obtuse to acute at apex, pale green to subglaucous, usually sparsely pilose to glabrescent above, oil dots pale yellow, or rarely reddish; lateral veins 2–3 pairs, thickened or evanescent towards margin. Inflorescence globose-capitate, 1–2.5 × 1–1.5 mm, verticillasters short pedunculate. Bracteoles 1.5–2.5 mm long, as long as or longer than pedicels, linear-setaceous, long-ciliate. Pedicels 1–2 mm long. Calyx (3.5–)4.5–6 mm, greenish, usually with long patent pilose with sessile glands, tube subcylindrical to campanulate, straight, densely hairy in throat; upper teeth usually 0.9–2 mm long, triangular to triangular-lanceolate, long-ciliate; lower teeth ± equalling upper lip, or slightly shorter, lower teeth linear-subulate, 3–4 mm, long ciliate, as long as or longer than tube. Corolla pale pink, 6–7 mm long. Stamens 4, if the stamens are well developed, the two stamens protrude distinctly from corolla. Style usually exserted from corolla, bilobed. Nutlet globose to subglobose, 1.2 × 1 mm, slightly tuberculate. Fig. 157, 1–9.

HAB. Rocky calcareous slopes, steppes; alt. 1300–3200 m; fl. & fr. Jul.-Aug.
DISTRIB. On calcareous slopes on mountains and lower hills of N. and N.E. Iraq. **MAM**: Mosul, in ditione pagi Sharanish, in montibus calc. a Zakho septentrionem versus, in faucibus Pushtashan, *Rechinger* 11216-A! (E, G, W), 11216-B (G); Mosul, Ad confines Turciae province Hakkari, in ditione pagi Sharanish, in montibus cal. a Zakho septentrionem versus, in saxosis cacuminis Zawiata, *Rechinger* 10970-A, 10870 (W); Ibid *Rechinger* (W); Amadiya, Garadagh South of Amadiye, mountain side, *Rawi* 9282! **MRO**: Montes Qandil, in declivibus orientes supra Pushtashan, *Rechinger* 11085! (G, W); Qandil, in saxosis calc. supra lacum Goame Kirmosoran, *Rechinger* 11146-A! (E, G); Arbil: M. Qandil, in saxosis calc. supra lacum Goam-e Kirmosoran, Rechinger 11136, 11762, Inter Arbil et Rawandiz, trajectus Serderrian, Nabalek 1572!; Mons Helgurd ad confines Persiae, in vale supra pagum Nowanda, *Rechinger* 11360-b (E, G), 11369-b; Helgurd, *Rawi* 13813!; M. Helgurd, *Haley* 114, *Thesiger* 1012, 1017

Thymus eriocalyx, T. fallax, T. kotschyanus, T. migricus and *T. pubescens* are a group of species that are closely similar to each other. In particular, for *T. kotschyanus* and *T. eriocalyx*, leaf, colour of oil dots, and leaf and stem pubescence are not distinctive characters being highly variable in both. This makes the identification of two species difficult. The most important distinguishing characters that separate *T. eriocalyx*'s from *T. kotschyanus* are its inflorescence and calyx with long and dense white pilose and slender flowering stems.

GATRA (Kurd., Helgurd, *Rawi* 13813).

S. E. Turkey, W. Iran.

3. **Thymus carmanicus** *Jalas*, Fl. Iranica [K. H. Rechinger] 150: 547 (1982); Jamzad in Flora Iran [Assadi & al., in Persian] 76: 746 (2012).

Subshrubs, caespitose, with woody rootstock. Stems 2–12 cm tall, ascending to erect, subglabrous to pilose-villous or retrorsely puberulent. Flowering stems leaves orbicular to ovate-elliptic, rounded to attenuate at base, rounded to subacute at apex, sometimes apiculate, axillary leaves present, 4–12 × 2–6 mm (incl. 1 mm petiole), puberulent to pilose with sessile pale or red glands; veins 2–3 pairs, evanescent. Bracts similar to leaves, verticils

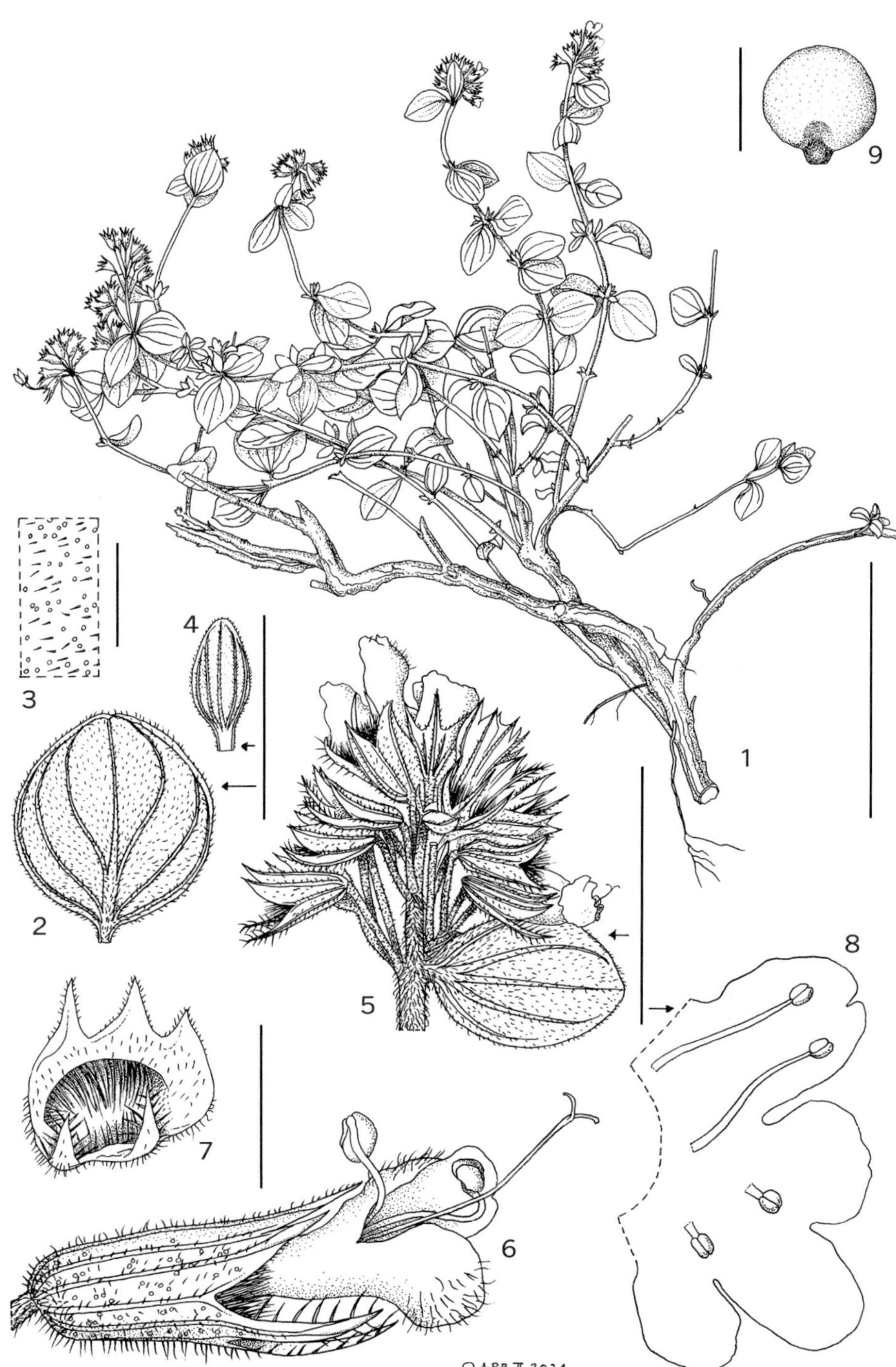

Fig. 157. **Thymus eriocalyx**. 1, habit; 2, abaxial leaf; 3, surface of leaf showing hairs and resin glands; 4, abaxial surface of lowest bract within inflorescence; 5, inflorescence; 6, flower, side view; 7, face view of fruiting calyx; 8, corolla opened showing dorsal and ventral lobes; 9, nutlet. 1–9 from *Omar et al.* 38502. Drawn by © A.P. Brown, Feb. 2024.

pedunculate; bracteoles 1–2 mm, linear-setaceous, usually shorter than pedicels. Pedicels 1.5–4 mm. Inflorescence capitate, 1–2 × 0.6–1.5 cm. Calyx 4–5.5 mm, tube cylindric, green or partly purple, subglabrous to pilose with sessile glands, densely hairy in throat; upper teeth 0.5–1.5 mm long, triangular to lanceolate, acute-acuminate; lower teeth 2–3 mm long, usually longer than uppers, linear-subulate; ciliate or scabrid. Corolla pinkish or purplish, 5–7 mm, puberulous with sessile glands on tube and outside of lips. Style unequally bilobed, exserted from corolla. Stamens exserted from corolla. Nutlets not seen.

HAB. Alpine, stony places; alt. 2600–3800 m; fl. & fr. Jul.-Aug.
DISTRIB. In the Suleymaniyah disctrict of N. Iraq. **MRO**: Top of Qandil range between top of Parrish & Bardanas (Iraq-Persian border), black and brown stones, *Rawi & Serhang* 24565!, 24595!; Qandil range, *Rawi & Serhang* 26736! N.E., of Qandil, *Rawi & Serhang* 24369! 24419! 24445! North of Helgurd dagh, *Rawi & Serhang* 24791!; Helgurd range, *Rawi & Serhang* 24735!; Helgurd range, *Rawi & Serhang* 24899!; Algurd Dagh (M. Helgurd), *Rechinger* 11413 (W).

Iran.

4. **Thymus pubescens** *Boiss.* & *Kotschy* ex *Celak.*, Flora 66: 152 (1883); Jalas in Fl. Iranica [K. H. Rechinger] 150: 546 (1982); Jalas in Fl. Turkey [P. H. Davis] 7: 375 (1982); Jamzad in Flora Iran [Assadi & al., in Persian] 76: 746 (2012).

> *Thymus fallax* var. *pubescens* (Boiss. & Kotschy ex Celak.) Rech.f., Phyton 5: 287 (1954).
> *T. balansae* Boiss. & Kotschy ex Čelak. var. *pubescens* (Boiss. & Kotschy ex Čelak.) Bornm. in Beih. Bot. Centr. 33(2):184 (1915).
> *T. xylorrhizus* Boiss. & Kotschy in Fl. Orient. [Boissier] 4: 556 (1879), not validly publ.

Subshrub, many stemmed from woody base, procumbent to suberect; flowering stems 2–11 cm tall, puberulent to hirsute, hairs patent to retrorse with sparse sessile glands, shorter than stem diameter. Flowering stem leaves ovate to lanceolate-elliptic, 5.5–12 × 1.8–3 mm, ± rounded to cuneate at base, obtuse to acute at apex, ± fleshy, glabrous, glabrescent to densely puberulent or villous, oil dots pale or rarely red, lateral veins 2 pairs, prominent. Inflorescence capitate, 8–20 × 8–15 mm. Bracts similar to leaves, usually green, longer or shorter than verticillasters; bracteoles 0.5–2 mm, linear-setaceous, as long as pedicel or shorter. Calyx 3.5–5.5 mm, green to purplish, tube cylindric to campanulate, sparsely to densely hirsute-villous with pale or red sessile glands, densely hairy in throat; upper teeth 0.6–1.5 mm long, triangular to lanceolate, acute-acuminate; lower teeth 2–3 mm long, slightly or clearly longer than upper, sometimes shorter, linear-subulate; all teeth usually ciliate or scabrid. Corolla pinkish or purplish, 5–7 mm, puberulous with sessile glands on tube and outside of lips. Style unequally bilobed, slightly or clearly exserted from corolla. Stamens slightly or clearly exserted from corolla. Nutlets not seen.

HAB. Lower hills, rocky slopes; alt. ± 2000 m; fl. & fr. Jun.-Aug.
DISTRIB. Known from a single gathering near the Turkish-Iraq border in N. Iraq. **MAM**: Mosul, ad confines Turciae prov. Hakkari, la ditione pagi Sharanish, in montibus calc. a Zakho septentrionem versus, in cacuminis Zawita, *Rechinger* 10970-B (W).

E. & S. E. Turkey, Iran.

5. **Thymus musilii** *Velen.*, Sitzungsber. Königl. Böhm. Ges. Wiss., Math.-Naturwiss. Cl. 11: 5 (1911 publ. 1912).

Subshrub with a woody taproot, many stemmed from base; flowering stems woody, 4–12 cm tall, ascending to erect, curved pubescent to densely hirsute with red sessile glands. Leaves elliptic-oblong, 4–12 × 1.5–4 mm, widest in middle, usually with axillary fascicles of small fleshy leaves, attenuate at base, acute at apex, leaf margins flat or slightly revolute towards apex, shortly petiolate, subglabrous to scabrid-puberulent with densely red sessile glands on both surfaces, ciliate at base; veins 2–3 pairs, prominent or evanescent. Inflorescence ± capitate or lower ones distant, 5–15 × 4–12 mm. Bract similar to leaves, upper ones broader, elliptic to ovate; bracteoles 0.5–1 mm, setaceous. Flowers shortly pedicellate to subsessile. Calyx 3.5–5 mm, subcylindric to campanulate, short pubescent with densely red sessile glands, densely hairy in throat; upper lip wider than tube; upper teeth 0.5–1.5 mm, triangular, ciliate; lower teeth 2–3 mm, linear-subulate, as long as or longer than upper teeth, long ciliate. Corolla 4–5.5 mm, white, included in or slightly exserted from calyx.

Style and stamens slightly exserted or included in corolla, sometimes style and two stamens long exserted. Nutlet 0.8–0.9 × 1–1.1 mm, oblong to obovoid, brown.

HAB. On sand dunes, on sandy-gravelly soils, valley beds in the desert; alt. 150–600 m; fl. & fr. Mar.-May. DISTRIB. C, W. and S.W. deserts of Iraq. **DWD**: Ud as Saba (2 km S.W. of H2), *Chakravarty* 31448!; 160–190 km W.N. of Ramadi to Rutba, *Rawi* 31343!; Rutba range manigment station, *Omar, Al-Kaisi & Hamid* 43945!; Rutba, *Field & Lazar* 132!; 10 km S.W. of Rutba, *Rawi* 21077!; 12 km S.E. of Hawran wadi, *Rawi & Alizzi* 31652! Ajrumiya, *Botany Staff* 41872!; Afaiyif, *Alizzi & Husain* 34057! **DSD**: 15 km W. of Shabicha, *Hazim* 30677!; 6 km S.E. of Shabicha, *Guest, Rawi & Rechinger* 19230!; 7 km W.N. of Shabicha, *Guest, Rawi & Rechinger* 19260-A!; 6 km S.E. Shabicha, in alveo arenoso calc., 26.04.1957, *Rechinger* 9436 (E); 25 km W. of Ma'niya, sandy gravelly soil, *Hazim* 32473! **LCA:** Abu Ghraib, 62 km W. of H1, in course gravel of wadi, *Fred* (531) & *Barkley* (5379)!

Lebanon, Palestine, Jordan, Saudi Arabia.

6. **Thymus syriacus** *Boiss.*, Diagn. Pl. Orient. 12: 47–48 (1853). Syntypes: [Syria] Hab. in siccis Caelesyriae prope Balbeck, circa Aleppum (Bekaa et Alep), Mai–July 1846, *Boissier* (G-BOIS! K!); [Iraq] In Mesopotamia, Mossul, 1841, *Kotschy* 428 (G-BOIS! K! W!). Rawi in Dep. Agr. Tech. Bull 14: 159 (1964); Dinsmore in Post, Fl. Syria, Palest. & Sinai ed. 2, 2: 336 (1933); Rechinger, Fl. Iranica [K. H. Rechinger] 150: 535 (1982); Duman in Fl. Turkey [Güner et al. (eds)] 11 (suppl 2): 209 (2000); Taifour & El-Oqlah, Pl. Jordan Annot. Checkl.: 103 (2017).

Shrubs, flowering stems ascending to erect, 6–30 cm tall, sparsely (retrorsely) pubescent to hirsute, with red sessile glands above, sometimes with glandular papillae above. Leaves narrowly lanceolate to elliptic or linear-subulate, usually fasciculate, 8–22 × 1.2–3(–4) mm, usually longer than internode, rigid or not, sessile, ciliate at base, glabrous to sparsely pubescent with densely red sessile glands, flat, acute at base, acute-acuminate to subulate at apex; 3-pairs lateral veins evanescent. Inflorescence subcylindrical-capitate, 1–3 × 1–2 cm, lower verticillasters pedunculate. Pedicels to 2 mm long. Bracts wider than leaves, sometimes lowermost similar, 7–10 × 2.5–4(–6) mm, ovate to ovate-lanceolate, with 5–7-veins, ±glabrous or pubescent to shortly pilose, with densely red glands, acute to acuminate; bracteoles 1–2.5 mm, setaceous. Calyx 4.8–7 mm, tube cylindrical, sometimes curved, densely pubescent with red sessile glands, glandular papillae present or not; upper lip broadly ovate to oblong, 2–3 mm wide, with three very narrow teeth or edentate, if teeth present, 0.6–1.1 mm, middle teeth longer than laterals, ciliate; lower lip bidentate, longer than uppers, 3.5–4.5 mm long, ciliate; densely hairy in throat. Corolla 6.5–9 mm, white, included in or exserted from calyx. Style ±included to long exserted from corolla, ± equally biloded. Stamens 4, two stamens subincluded to exserted from corolla, other two stamens included to subincluded in corolla. Nutlets oblong to obovoid, c. 1.2 × 0.9 mm, brown.

Two varieties are recognized:

Upper lip of calyx tri-dentate . . . . a. var. **syriacus**
Upper lip of calyx edentate. . . . . .b. var. **edentatus**

a. var. **syriacus**

Origanum syriacum (Boiss.) Kuntze, Revis. Gen. Pl. 2: 528 (1891), nom. illeg.
Thymus densus Celak., Flora 66: 151 (1883).
T. fallax var. *densus* (Celak.) Rech.f., Phyton (Horn) 5: 288 (1954).
Thymbra neurophylla Rech.f., Pl. Syst. Evol. 133: 107 (1979).
Thymus neurophyllus (Rech.f.) R.Morales, Anales Jard. Bot. Madrid 45(2): 562 (1989), syn. nov.

HAB. Clay hillside, rocky slopes, steppes; alt. 400–1000 m; fl. & fr. May-Jun. DISTRIB. Northern and northeastern Iraq. **MAM**: Zakho: Ad medium fluvium Chabur in steppis lapidopsis inter Abed et Gharra, *Handel-Mazzetti* 1713 (W); In Mesopotamia, Mossul, 1841, *Kotschy* 428! (G-BOIS, W); Mosul, *Rawi* 9180-A!; Suwaratuka, *Rawi* 9233 (BAG). **MRO**: Khanzad, *Omar, Kaisi & Wedad* 49623! Arbil, *Radhi* 3846!. **MSU**: Chamchamal, *Haines* 1511! (E); Jarro, nr Chamchamal, *Haines* 229-A! (E); Jarmo, *Halbaek* 1724! 1892!, 1906!; Suleymaniya Liwa, *Regel* 4 (G); Qaranjir, bentween Kirkuk and Chamchamal, *Katib & Tikriti* 29732!; Between Kirkuk and Suleymaniya, *Rawi* 21483!; 3 km from Kirkuk to Suleymaniyah, *Karim, Hamid & Jasim* 40476!; Kirkuk, Gara Anjir, 27 km a Kirkuk orientem versus in collibus siccis, *Rechinger* 10017 (type of *Thymus neurophyllus*) (E, W). Tasluja, *Omar & Karim* 47076! Kirkuk, S. Kirkuk, Uvarov s.n; Haibat Sultan mountain, *S. Omar, F. Karim R. Hamza & H. Hamid* 37197 (BAG). **FUJ**: Sinjar, Jabel Khatchra, near Balad Sinjar, *Field & Lazar* 629!; Ad rupes in faucibus El Magharad montium Dschebel Sindschar supra oppidum Sindschar, *Handel-Mazzetti* 1382 (W);

Dubardan mountain, 40 km from Mosul to Agura, *Dabbagh & Jasim* 46805! **FNI**: Mosul, 47 km from Mosul to Aqra, *Al-Kaisi* 49705! **FKI**: Bawanur, *Poore* 551! Altın Kupri, *Guest* 4031!

Jordan, Lebanon, Syria, S. Turkey.

b. var. **edentatus** *Jalas*, in Fl. Iranica [K. H. Rechinger] 150: 536 (1982). Type [Iraq] Arbil, in deserto prope Arbil, 05.1867, *Haussknecht* s.n.! (W, JE-photo!).

HAB. Rocky slopes, steppes; alt. not recorded; fl. & fr. May.
DISTRIB. Known only from type specimen location. **FAR**. Arbil, *Haussknecht* s.n. (type).

Var. *edentatus* is only known from the type locality. The characteristic feature of var. *edentatus* is that the upper calyx lip is edentate. In var. *syriacus*, the upper calyx lip is prominently tridentate and the middle tooth is significantly longer than the two lateral teeth, and all of them are ciliate.

Endemic.

7. **Thymus daenensis** *Celak.*, Flora 66: 150 (1883) subsp. **lancifolius** (*Celak.*) *Jalas*, Fl. Iranica [K. H. Rechinger] 150: 537 (1982).

Thymus lancifolius Celak., Flora 66: 149 (1883).

Subshrubs, woody at base; gynodioecious. Flowering stems 6–30 cm tall, ascending, few to many stemmed from base, subglabrous to pilose or villous, with red sessile glands. Leaves sessile, linear-lancelate to ovate-lanceolate, 9.5–30 × 1.7–8 mm, ± patent, longer or shorter than internode, ± attenuate at base and acute at apex, entire, glabrous to pilose, red sessile glands numerous; lateral veins 2-pairs, evanescent. Inflorescence usually capitate, rarely elongate, 0.8–2 × 0.8–2 cm. Verticillasters short pedunculate, to 5 mm long; flowers pedicellate to 3 mm. Bracts ovate to lanceolate, lowest similar to ± leaves, 3–10 × 1.5–5 mm, subglabrous to pilose with densely red glands; bracteoles 1–2 mm, setaceous. Calyx cylindric to campanulate, 4–5.5 mm, puberulent to pilose with red sessile glands on tube, mouth hairy; upper lip wider than tube; upper teeth triangular, 0.5–0.8 mm; lower teeth linear-subulate, 2.5–3 mm, longer than upper teeth, ciliate. Corolla 4.5–6(–7) mm long, pinkish-purple, included in calyx or slightly exserted. Style 2-lobed, exserted from corolla. Stamens 4, fertile or staminode, included in corolla if staminode, long exserted if not. Nutlets oblong to obovoid, 1–1.2 × c. 0.8, pale-brown.

HAB. Rocky slopes; alt. 2100–2400 m; fl. & fr. May-Jun.
DISTRIB. **MAM**: Mosul: Chiya-e Gara, *Kotschy* 560; Mesopotomia, *Aucher* s.n.! **MSU**: Suleymaniya, *Haussknecht* s.n. (JE).

Iran.

20. **ORIGANUM** L.

Sp. Pl. 2: 590 (1753); Ietswaart, Taxon. Rev. Gen. *Origanum* (1980); Harley & al. in Kubitzki (ser. ed.), Fam. Gen. Vasc. Pl. 7: 239 (2004)
Dictamnus Mill., Gard. Dict. Abr. ed. 4: s.n. (1754), nom. illeg.; *Majorana* Mill., Gard. Dict. Abr. ed. 4. [829] (1754); *Marum* Mill., Gard. Dict. Abr. ed. 4. (1754); *Amaracus* Hill, Brit. Herb.: 381 (1756); *Dictamnus* Zinn, Cat. Pl. Hort. Gott.: 316 (1757); cf. Sprague in Kew Bull. 1934, 218, nom. illeg.; *Hofmannia* Heist. ex Fabr., Enum. [Fabr.]. 61 (1759); *Beltokon* Raf., Fl. Tellur. 3: 86 (1837); *Onites* Raf., Fl. Tellur. 3: 86 (1837).
Zatarendia Raf., Fl. Tellur. 3: 86 (1837); *Schizocalyx* Scheele, Flora 26: 575 (1843); × *Origanomajorana* Domin, Preslia 13–15: 197 (1935), in adnot.; × *Majoranamaracus* Rech.f., Repert. Spec. Nov. Regni Veg. 45: 95 (1938)

Tuncay Dirmenci & Ferhat Celep

Suffruticose or herbaceous perennials, hairy or glabrous (often glaucous). Stems several, ascending or erect, usually branched. Leaves subsessile or ± petiolate, elliptic, ovate, cordate or suborbicular, entire or ± toothed, apex obtuse or acuminate. Verticillasters 2- several-flowered, aggregated in ± dense spike-like inflorescences (spicules) which are often arranged in a panicle or false corymb. Bracts always distinct from leaves in shape and size, usually imbricate, 1/2–3 × as long as calyx, either membranous and partly purple or yellowish-green, or resembling leaves in texture and colour. Flowers hermaphrodite or gynodioecious. Calyx variable, ± actinomorphic and 5-toothed, or zygomorphic and 1–2-lipped, 13- or 10-veined; throat usually with a ring of hairs. Corolla purple, pink or white, ± equally 2-lipped, tube sometimes saccate or flattened; upper lip emarginate or shortly bilobed; lower lip 3-lobed. Stamens 4, lower pair longer, exserted from or ± included in corolla, ascending under upper lip, straight or divergent; filaments ± unequal; thecae divergent. Nutlets small, ovoid, brown.

(Including *Amaracus* Gledit., *Majorana* Mill. Gynodioecism is common in sections *Chilocalyx, Majorana, Origanum* and *Prolaticorolla*).

The genus *Origanum* is one of the most commercially important genera within the family Lamiaceae. It is found exclusively in the Northern Hemisphere of the Old World, with 90% of taxa found in the Mediterranean region (42 species (49 taxa) and 22 hybrids); 2 species in Iraq.

Celep, F., Özcan, T., Jeffrey, P.R., Türker, Y. & Dirmenci, T. (2021). Phylogeny, Biogeography and Character Evolution of *Origanum* L. (Lamiaceae), in Dirmenci, T. (ed.), "Oregano" The genus *Origanum* (Lamiaceae): Taxonomy, Cultivation, Chemistry and Uses. Pp.: 49–104. Nova Science Publishers. New York.

Dirmenci, T., Özcan, T., Açar, M., Arabacı, T., Yazıcı, T. & Martin, E. (2019). A rearranged homoploid hybrid species of *Origanum* (*Lamiaceae*): *O.* × *munzurense* Kit Tan & Sorger. Botany Letters 166:153–162.

Dirmenci, T., Özcan, T., Arabacı, T., Çelenk, S. & Martin, E. (2020). An important hybrid zone: evidence for two natural homoploid hybrids among three *Origanum* species – Ann Bot Fenn 57(1–3): 143–157.

Dirmenci, T., Arabacı, T., Özcan, T., Drew B., Yazıcı, T., Celep, F. & Martin, E. (2021). Taxonomy of *Origanum* L. (Lamiaceae), in Dirmenci, T. (ed.) "Oregano" The genus *Origanum* (Lamiaceae): Taxonomy, Cultivation, Chemistry and Uses. Pp.: 1–48. Nova Science Publishers. New York.

Ecevit-Genç, G., Özcan, T., Arabacı, T. & Dirmenci, T. (2020). Mericarp morphology of the genus *Origanum* (Lamiaceae) and its taxonomic significance in Turkey. Phytotaxa 460(1):1–21.

Ietswaart, J.H. (1980). A taxonomic revision of the genus *Origanum* (Labiatae). Leiden University Press (Leiden Botanical Series No.4).

Spicules ± globose, sometimes cylindric, 10–45 × 10–30 mm, pendulous;
    bracts 7–22 × 6–20 mm; calyx 5–7.5 mm, 2-lipped . . . . .1. *O. acutidens*
Spicules ovoid to cylindrical, 4–30 × 3–4 mm, erect-ascending; bracts
    3–6 × 1–2 mm; calyx 2–4 mm, regular . . . . .2. *O. vulgare* subsp. *gracile*

1. **Origanum vulgare** *L.* subsp. **gracile** (*K.Koch*) *Ietsw.*, Leiden Bot. Ser. 4: 111 (1980); Rawi in Dep. Agr. Tech. Bull 14: 153 (1964); Rechinger, Fl. Iranica [K. H. Rechinger] 150: 530 (1980); Davis, Fl. Turkey [P. H. Davis] 7: 310–312 (1982); Delghandi, Pl. Herb. Min. Ir. Agr., 20: 77 (1991); Jamzad in Flora Iran [Assadi & al., in Persian] 76: 710 (2012).

Origanum gracile K.Koch, Linnaea 21(6): 661 (1849).
O. tyttanthum Gontsch., Delect. Sem. Inst. Bot. Tadshik. Akad. Sc. URSS 12 (1934).
O. kopetdaghense Boriss., Bot. Mater. Gerb. Bot. Inst. Komarova Akad. Nauk S.S.S.R. 16: 280 (1954).
O. glaucum Rech.f. & Edelb., Biol. Skr. 8(1): 76 (1955).
O.vulgare var. *glaucum* (Rech.f. & Edelb.) Hedge & Lamond, Notes Roy. Bot. Gard. Edinburgh 28: 124 (1968).

Perennial herb, to 90 cm tall, usually glabrescent, sometimes somewhat pilose, usually glaucous. Branches slender, up to 10 pairs per stem. Leaves up to 25 pairs per stem, ovate, obtuse to acute at apex, 3–40 × 5–22 mm, petioles up to 15 mm, glabrescent or slightly pilose, usually glaucous, densely sessile glands on both surfaces, margins entire or remotely serrulate. Inflorescence thyrsoid. Spikes usually cylindrical, 4–30 × 3–4 mm. Bracts 2–26 pairs per spike, ovate-obovate or oval, 3–6 × 1–2 mm, herbaceous or ± membranous, glabrous or pilose, green or slightly purple, usually glaucous. Flowers 2 per verticillaster, bisexual or female. Calyx regular, 2–4 mm, glabrescent to puberulent, with sessile glands; teeth 0.8–1.2 mm; throat hairy. Corolla 5–8 mm, usually white, exserted from calyx, puberulent to pilose with sessile glands. Stamens 4, filaments 2.5–3.5 mm, two stamens clearly or shortly protruding and two stamens (sub-)included in the corolla. Style unequally bilobed, exserted from corolla. Nutlets oblong-ovoid, 0.8–1 × 0.3–0.5 mm, brown, glabrous, slightly tuberculate. Fig. 158, 1–4.

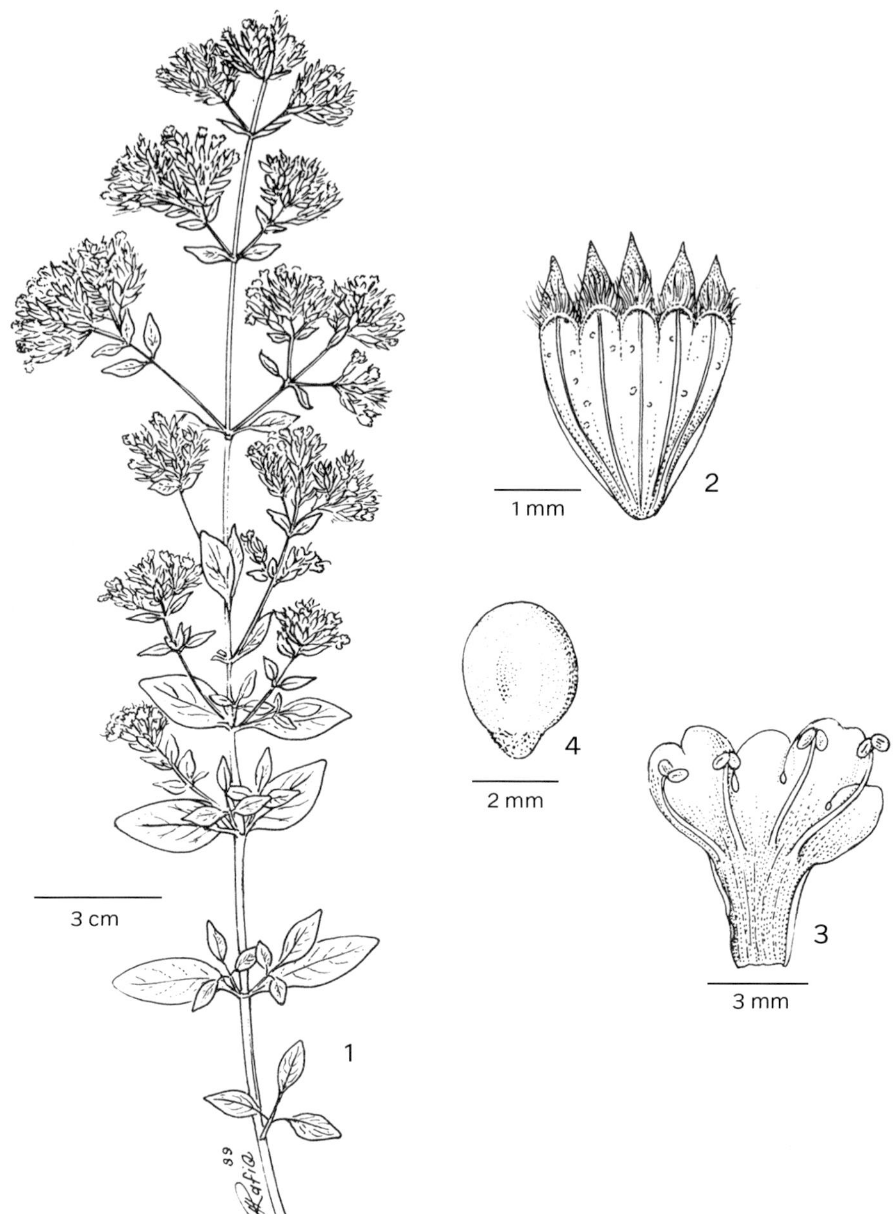

Fig. 158. **Origanum vulgare**. 1, habit; 2, calyx (opened); 3, corolla, opened; 4, nutlet. Reproduced with permission from Flora of Pakistan 192: f. 30, A–C1 (1990). Drawn by M. Rafiq. © National Herbarium, Pakistan Agriculture Research Council & University of Karachi, Pakistan.

HAB. Various habitats, dry hills, rocky slopes, near stream banks, gravel terraces, mountain side clay soil in *Rhus coriaria* & *Juglans* orchards; alt. 600–2100 m; fl. & fr. Jul.-Sep.

DISTRIB. Northern Iraq, bordering Turkey and Iran. **MAM**: Zawita, c. 15 km E.N.E. of Dohuk, *Guest* 4805! (BAG-photo); 25 km N.E. of Zakho, *Rawi* 23507! (BAG-photo); Dairalok village, 16 km E. of Amadiya, *Al-Dabbagh* & *K.Hamid* 45855! (BAG-photo); 10 km S. of Sarsang, *Al-Kaisi* & *K.Hamad* 45828! (BAG-photo); Sarsang, *Al-Dabbagh* 46180!; ibid., *Al-Dabbagh* 46180 (BAG-photo). **MSU**: In montibus calcareis Avroman et Schahu, *Haussknecht* 961 (G, W). **MRO**: Rowanduz Gorge, *Guest* 2979! (BAG-photo); Querna, Qaw Valley, N. of Pushtashan, *Rawi* & *Serhang* 26610 (BAG-photo); Ibid., *Rawi* &

Serhang 26618! (BAG-photo); *Rawi & Serhang* 26632 (BAG-photo); Pushtashan, N.E. of Rania, lower slope of Qandil range, *Rawi & Serhang* 26583! (BAG-photo); Ibid, *Rawi & Serhang* 26483! (BAG-photo); Pushtashan, 15 km N.E. of Rania, *Rawi & Serhang* 23828! (BAG-photo); Ibid., *Rawi & Serhang* 23858 (BAG-photo); Ibid., *Rawi & Serhang* 24210! (BAG-photo); Ibid., *Rawi & Serhang* 24243! (BAG-photo); Sakran Mountain, near Rowanduz, *Al-Dabbagh & Hamad* 46213! (BAG-photo); Sakran mountain, A. Haloob & *Al-Kaisi* 60685 (BAG); Sula Khal, N.W. of Haji Omran, *Rawi* 24690! (BAG-photo); N. of Haji Omran, Persian border, *Rawi* 24959! (BAG-photo); Ibid, *A.Rawi* 24965! (BAG-photo); Chia-i Mandali, *E.R.Guest* 2744! (BAG-photo); Arbil: M. Qandil in decl. orientalibus supra Pushtashan in astragaletis, *Rechinger* 11071; Berrog mountain, on to Qandil road, *Rawi & Serhang* (BAG-photo).

Origanum vulgare is cosmopolitan distributed from Azores to Taiwan and throughout much of Eurasia. It is divided into six subspecies worldwide; only subsp. *gracile* is found in Iraq. Subsp. *gracile* is close to subsp. *hirtum*, but it differs from it by its slender habit, glabrescent and glaucous stems and leaves.

E. & S. E. Turkey, Iran to central U.S.S.R., Afghanistan.

2. **Origanum acutidens** (*Hand.-Mazz.*) *Ietsw.*, Leiden Bot. Ser. 4: 62 (1980); Leiden Bot. Ser. 4: 62 (1980); Rechinger, Fl. Iranica [K. H. Rechinger] 150: 530 (1980); Davis, Fl. Turkey [P. H. Davis] 7: 304 (1982).

Amaracus haussknechtii var. *acutidens* Hand.-Mazz., Ann. K. K. Naturhist. Hofmus. 27: 420 (1913). Ic: Ietswaart, op. cit. 69, t. 11; 70, t. 12 (1980).

Subshrubs, 20–50 cm tall, ascending to erect a few to many stemmed, ± glabrous, branched or unbranched above, 2–15 cm long, sessile glandular. Leaves subsessile, ovate, 5–30 × 4–24 mm, obtuse, glaucous, coriaceous, with sessile glands on both surfaces. Spicules ± globose, sometimes cylindric, 10–45 × 10–30 mm, pendulous. Bracts suborbicular, obovate or elliptic, 7–22 × 6–20 mm, 3–12 on per spicules, acute or obtuse, yellowish-green, rarely tinged purple. Verticillasters 2–12-flowered, ± sessile. Calyx 5–7.5 mm long; upper lip with broadly triangular-ovate or triangular teeth to c. ½, 0.5–1 mm; lower lip as long as upper lip or slightly longer, triangular-lanceolate, acute-acuminate teeth, 1.5–2.5 mm long. Corolla white or tinged pink, 10–16 mm, not saccate. Stamens 4, filaments glabrous, upper filaments 1–3 mm, included in upper lip, lower 6–10 mm, exceeding to upper lip. Style unequally bilobed, included in upper lip or exceeding. Nutlets oblong-ovate, 1–1.5 × 0.7–1.1 mm, brown.

HAB. Calcareous and non-calcareous rocks, slopes and screes; alt. 1000–3000 m; fl. & fr. Jul.-Aug.
DISTRIB. Known from Iraq only from a single location in east-northeast of Amadiya. **MAM**: Mosul/ Amadiya: Inter Hora et Baibu in Raikan prope, near Wazi, Turkish frontier, *Thesiger* 1241 (BM).

Origanum acutidens is mainly distributed in east and southeast Turkey. It is only known from North Iraq from a locality close to the Turkish border.

E. & S.E. Turkey.

Origanum majorana *L.*, Sp. Pl.: 590 (1753) is cultivated in farms and gardens.

DISTRIB. **DWD**: AL-Habbaniyah, in garden, *Kalf et al.* 916! **LEA**: Baghdad, Besrengesh, *Rawi* 9090B! (BAG-photo); **LCA**: Zafariniya, c. 15 km S.W. by S. of Baghdad, site of Government Horticultural Research Station, cultivated, *Rawi & Shahwane* 25323! (BAG-photo).

21. **PENTAPLEURA** Hand.-Mazz.

Oesterr. Bot. Z. 63: 225 (1913); Harley & al. in Kubitzki (ser. ed.), Fam. Gen. Vasc. Pl. 7: 239 (2004)

Tuncay Dirmenci & Ferhat Celep

Small twiggy fastigiate shrub, 20–50 cm tall, aromatic. Stems hirsute with eglandular hairs and very shortly stalked glands. Leaves broadly ovate, entire. Inflorescence of 2–8-flowered axillary cymes, shortly pedunculate or sessile, forming dense verticillasters on upper part of stem, in axils of leaf-like bracts, often somewhat congested, bracteolate. Calyx actinomorphic, long-cylindrical, 5-lobed, lobes subulate, pentagonal, with 5 main veins along crest of narrow wings, and 5 weak veins in lower half of membranous intercostal furrows; throat hairy within. Corolla ± actinomorphic to weakly 2-lipped, 5-lobed to 2/3, pink or pinkish purple, 1/3 longer than calyx, tube straight and very narrow; annulus absent. Stamens 4, included in corolla; filaments straight, anterior slightly longer, anthers ellipsoidal, thecae free but nearly parallel. Stigma lobes subequal. Nutlets oblong, glabrous.

Monotypic; distributed in the border region of Northern Iraq and Turkey.

According to Brauchler et al. (2010), *Pentapleura* is related to *Zataria* more that it is to *Origanum* and *Cyclotricium* based on ITS and *trn*L-F squences. It differs from *Zataria* with longer calyx and five winged calyx veins. It is additionally far from *Zataria* in terms of geographical distribution. While *Zataria* is distributed in Iran, Afghanistan, Pakistan and Oman, *Pentapleura* is distributed only in the border region of Turkey and Northern Iraq. Monotypic.

Brauchler, C., Meimberg, H. & Heubl, G. (2010). Molecular phylogeny of Menthinae (Lamiaceae, Nepetoideae, Mentheae) taxonomy, biogeography and conflicts. Mol. Phylogenet. Evol. 55: 501–523.
Handel-Mazzetti, H. (1913). *Pentapleura*. Ann. K. K. Naturhistorischen Hofmuseums 27: 421–422.
Satıl, F., Kaya, A. & Dirmenci, T. (2018). Investigations on a monoptypical genus: *Pentapleura* Hand.-Mazz. (Lamiaceae). Bangladesh J. Bot. 47(1):1–8.

1. **Pentapleura subulifera** *Hand.-Mazz.* in Ost. Bot. Zeitschr. 63(6): 225–226 (1913). Lectotype: [Iraq] Ubi legi prope conventum Mar Yakub supra vicum Simel ad septentriones ab urbe Mossul, substrato calcareo, ca. 400–800 m, 24.viii.1910, *Handel-Mazzetti* 3092 ! (W, holo.; JE-photo; W, WU, iso). Rechinger, Fl. Iranica [K. H. Rechinger] 150: 553 (1982); Davis, Fl. Turkey [P. H. Davis] 7: 313 (1982).

Small shrub. Stems 10–50 cm tall, hirsute with eglandular hairs and very shortly stalked glands. Leaves 10–22 × 6–16 mm, diminishing from base to inflorescence, broadly ovate to ovate-elliptic, shortly petiolate to subsessile, tomentose to pilose on both surfaces, densely very shortly stalked glandular hairy with sessile glands, entire, rounded to cuneate at base, obtuse at apex. Inflorescence simple or rarely branched. Verticillasters many, remote below, approximate at above in axils of floral leaves, shortly pedunculate to sessile, 2–8-flowered. Bracteoles 2, 3–6 mm, lanceolate-subulate. Calyx 7–10 mm, regular, narrowly tubular, straight, pentagonal, with 5 main veins along crest of narrow wings, and 5 weak veins in lower half of membranous inter-costal furrows, densely stalked glandular hairy with sessile glands, throat hairy; teeth 2–2.5 mm, aristate tip. Corolla 8–13 mm long, pink or purplish-pink, ± 1/3 longer than calyx, tube straight and very narrow, eglandular hairy outside and inside, shortly stalked glandular hairy outside, obscurely bilabiate; annulus absent; upper lip bilobed; lower lip trifid. Stamens 4, included in corolla; filaments glabrous; thecae free but nearly parallel. Stylar branches subulate, subequal, glabrous, included in or longer than corolla. Nutlets c. 1.5 × 0.7 mm, oblong, brown, glabrous. Fig. 159, 1–5.

HAB. Dry limestone slopes; alt. 400–1000 m; fl. & fr. Jul.-Sep.
DISTRIB. Slopes of limestone cliffs in Northern Iraq. **MAM**: 20–25 km N.E. of Zakho, *Rawi* 23240! (BAG-photo); Ad confines Turciae province Hakkari, in ditione pagi Sharanish, in montibus calca. A Zakho, septentrionem versus, Jabal Khantur, *Rechinger* 12085 (E, W).

22. **SATUREJA** L.

Sp. Pl. 2: 568 (1753); Harley & al. in Kubitzki (ser. ed.), Fam. Gen. Vasc. Pl. 7: 241 (2004) *Thymbra* Mill., Gard. Dict. Abr. ed. 4: s.p. (1754), nom. illeg.; *Saturiastrum* Fourr., Ann. Soc. Linn. Lyon, n.s., 17: 133 (1869); *Euhesperida* Brullo & Furnari, Webbia 34: 433 (1979); *Argantoniella* G.López & R.Morales, Anales Jard. Bot. Madrid 61: 25 (2004)

Tuncay Dirmenci, Ferhat Celep & Taner Özcan

Low shrubs, suffruticose herbs, rarely annual, strongly aromatic. Leaves usually cuneate below, entire, subsessile, usually conduplicate at least when young, glandular punctate, young leaves in axillary fascicles. Inflorescence of 2–many-flowered axillary verticillasters or in ±lax cymes. Bracts usually leaf-like, bracteoles inconspicuous. Calyx actinomorphic with ±equal teeth to sub-bilabiate with 3 upper teeth distinctly shorter than lower 2; posterior lobes straight to upcurved, anterior lobes longer; tube 10–13-nerved, tubular to campanulate, calyx-throat bearded. Corolla 2-lipped, 4-lobed, white, lilac, pink or pale-purple; posterior

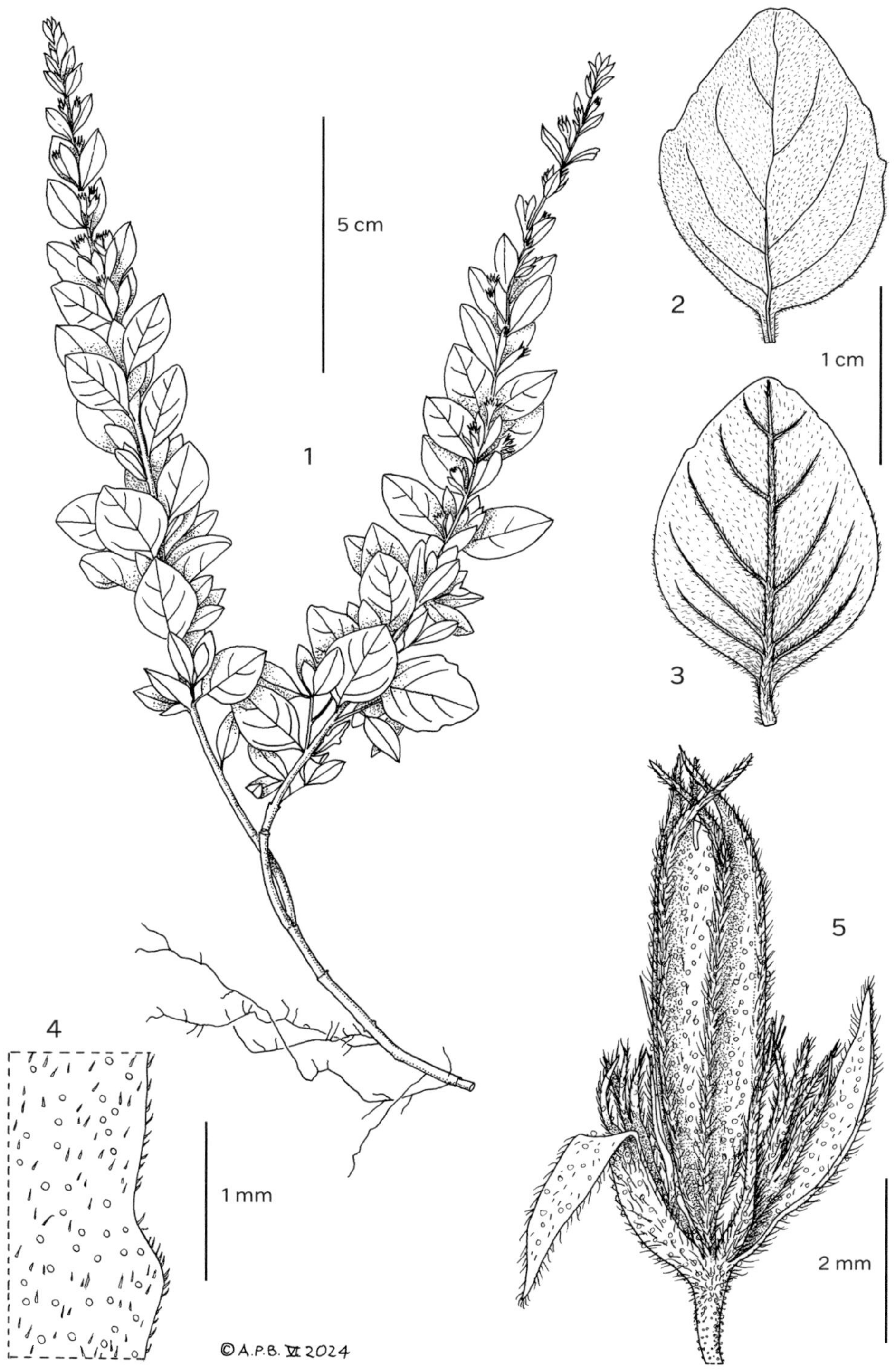

Fig. 159. **Pentapleura subulifera**. 1, habit; 2, 3, lower leaf, adaxial view; 4, indumentum of leaf margin; 5, young inflorescences. 1–5 from *Rawi* 23240. Drawn by © A.P. Brown, Feb. 2024.

lobes concave, ±rounded, emarginate; anterior lip with median lob oblong, spreading, tube ±cylindrical, straight. Stamens 4, scarcely didynamous or anterior pair longer; filaments included in or exserted, curved; thecae ±divaricate or parallel, ellipsoidal. Style branches subequal, subulate. Nutlets ovoid, sometimes elongate or apiculate, smooth or muricate, sometimes hairy.

About 43 species and 4 hybrids in the Mediterranean region, Europe, W. Asia, Caucasia, W. Himalaya and Xinjiang; 4 species in Iraq.

Dorosenko, A. (1985). Taxonomic Studies on the Satureja complex (Labiatae) Ph.D. dissertation. Edinburgh University.
Dirmenci, T., Yıldız, B. & Öztekin, M. (2019). A new Record For The Flora Of Turkey: *Satureja metastasiantha* Rech.f. (Lamiaceae). Bagbahce Bilim Dergisi, 6: 54.

1. Annual herbs . . . . 1. *S. hortensis*
   Suffruticose perennials 2
2. Calyx 4–6 mm; corolla pinkish-purple to purple, 15–27 mm . . . . 2. *S. macrosiphonia*
   Calyx 3–3.5 mm; corolla white, 4–8 mm 3
3. Inflorescence usually lax; cymes usually clearly pedunculate; peduncle
   to 8 mm; pedicel 1–3 mm; calyx 3–3.5 mm; corolla 4–8 mm . . . . . 3. *S. metastasiantha*
   Inflorescence tightly arranged; cymes and flowers sessile; calyx 1.5–2.5
   mm; corolla 3–4 mm . . . . 4. *S. bachtiarica*

1. **Satureja hortensis** L. Sp. Pl. 2: 568 (1753); Dinsmore in Post, Fl. Syria, Palest. & Sinai ed. 2, 2: 338 (1933); Davis in Fl. Turkey [P. H. Davis] 7: 322 (1982).

Clinopodium hortense (L.) Kuntze, Revis. Gen. Pl. 2: 515 (1891).
Satureja officinarum Crantz, Inst. Rei Herb. 1: 526 (1766).
S. viminea Burm.f., Fl. Indica: 126 (1768), nom. illeg.
S. brachiata Stokes, Bot. Mat. Med. 3: 300 (1812).
S. pachyphylla K.Koch, Linnaea 17: 295 (1844).
S. laxiflora K.Koch, Linnaea 21: 668 (1849).
S. hortensis var. *distans* K.Koch, Linnaea 21: 667 (1849).
S. filicaulis Schott ex Boiss., Fl. Orient. [Boissier] 4: 562 (1879).
S. litwinowii Schmalh. ex Lipsky, Fl. Caucasi: 108 (1899).
S. altaica Boriss., Bot. Mater. Gerb. Bot. Inst. Komarova Akad. Nauk S.S.S.R. 15: 326 (1953).
S. zuvandica D.A.Kapan., Soobshch. Akad. Nauk Gruzinsk. S.S.R. 119: 375 (1985).
S. laxiflora subsp. *zuvandica* (D.A.Kapan.) D.A.Kapan., Fl. Gruzii, ed. 2, 11: 188 (1987).

Annual. Stem single, widely branched, (5–)10–50 cm tall, curved-pubescent or puberulent with sessile glands above. Leaves shortly petiolate, linear to linear-oblanceolate, 8–30 × 1–4.5 mm, attenuate at base, obtuse at apex acute to ± obtuse, margins entire, sparsely to densely scabrid pubescent on both surfaces with densely red sessile glands. Inflorescence lax, elongate. Verticillasters remote or approximate, 2–20-flowered, congested to lax; if lax, cymes pedicellate to 5 mm long. Floral leaves 1–2 × calyces. Bracteoles linear-setaceous, 1–1.5 mm. Calyx sub-bilabiate to 1/2–3/4, 3–4.5 mm long, teeth lanceolate to subulate, apiculate, 2 lower teeth longer than upper 3, infundibular-campanulate, short pilose; lower teeth as long as or to 2(–3) × longer than tube, 2–3 mm, lanceolate-subulate; glabrous in throat. Corolla pinkish, purplish or white, 6–10 mm; tube included in calyx or exserted to 2 × calyx, pubescent; upper lip shortly 2-lobed; lower lip 3-lobed, bearded. Style included or slightly exserted from corolla, unequally bilobed. Stamens 4, two stamens sub-included and other two included in corolla, sometimes all included; anterior stamens longer than posterior; filaments glabrous; anthers purplish. Nutlets 1–1.2 × 0.5–0.8 mm, rotund-ovoid to oblong, slightly tuberculate, black, glandular or not.

HAB. Rocky or eroded slopes, screes, gravelly places, coastal dunes, fallow fields, roadsides; alt. s.l.–2000 m; fl. & fr. Jun.-Sep.
DISTRIB. Known only from Sulaymaniyah district. **MRO**: In montibus calcareis Avroman et Shahu, *Haussknecht* s.n.; **LCA**: Baghdad, Zafariniya, horticulture, *Rawi & Shahwani* 25327! (BAG-photo) (cultivated).

We've only seen a cultivated specimen (*Rawi & Shahwani* 25327) and have not seen *Haussknecht*'s specimen given in Flora Iranica (Rechinger, 1982; p. 503). *S. hortensis*, which is naturally distributed in S.E. Turkey is also likely to be widely present in Northern Iraq.

Satureja hortensis shows a wide variation, particularly in its inflorescence. The verticillasters are 2-many flowered, in lax or congested inflorescences. In lax inflorescences, the cymes are clearly pedunculate and usually 2(–4)-flowered. Specimens with these characters were named *Satureja laxiflora* K.Koch. Here we include this taxon in *Satureja hortensis*.

Cosmopolitan; S.E. Europe to S.W. Siberia

2. **Satureja macrosiphonia** *Bornm.*, Repert. Spec. Nov. Regni Veg. 6: 114 (1908). Type: Iraq, Assyria, in siccis ad Mar Jako ditionis urbis Mosul, ix.1879, *Barré De Lancy* s.n. (JE-photo! holo.). Rawi in Dep. Agr. Tech. Bull 14: 156 (1964); Rechinger, Fl. Iranica [K. H. Rechinger] 150: 501 (1982); Jamzad in Flora Iran [Assadi & al., in Persian] 76: 693 (2012).

Suffrutescent, many stemmed from base. Stems 30–50 cm tall, virgate, minutely setulose to scabrid-canescent or retrorsely puberulent, with minutely glandular papillae and sessile glands, paniculately lax branched above. Leaves greyish-green, petiolate, oblong-linear, oblong-elliptic or obovate, 5–25 × 1–10 mm, attenuate at base, obtuse to acute at apex, margins entire, sparsely to densely scabrid-puberulent on both surfaces with dense sessile glands. Inflorescence lax, 4–15 cm, sometimes paniculately branched. Verticillasters 2–4-flowered, cymes clearly pedunculate to 1 cm. Bracteoles 0.8–1.2 mm, setaceous. Pedicel as long as or shorter than calyx, to 4 mm long, scabridulous with minutely glandular papillae. Calyx (3–)4–6.5 mm, tubular-campanulate, scabrous with pale sessile glands and glandular papillae, purplish or green, obscurely bilabiate; teeth unequal, slightly recurved; upper lip 3-dentate, teeth 1–1.5 mm, triangular-lanceolate; lower lip 2-dentate, teeth 1.5–2.5(–3) mm, linear-subulate, longer than upper, glabrous to sparsely hairy in throat. Corolla purple, (7–)13–22 mm, tube long, narrow and slender, glabrous to puberulent; lower lip bearded inside. Style and stamens included or barely exserted from corolla. Nutlets oblong, 1.25–1.6 × 0.8–1 mm, blackish-brown. Fig. 160, 1–6.

HAB. Rocky slopes, calcareous; alt. 500–800 m; fl & fr. Aug.-Oct.
DISTRIB. Northern mountains. **MAM**: Mosul liwa, mountainside of Gali Dohuk, *Barkley & Ischaq* 9031 (W); Bikhair mt., near Zakho, 90 km N.W. of Mosul, *Rawi* 23023! (BAG-photo) (without flowers); Zawitah, N. of Dohuk, *Guest* 1565!; Dohuk, Mosul liwa, *Jackson* 3528! (BAG-photo); Dohuk, rocky slope, *Haines* 1222-B! (with *S. metastasiantha* on the same sheet) & 1222 (E). **MRO**: Rawanduz, Bakhal water fall, *Al-Dabbagh & Hamid* 46187! (BAG-photo); ibid., *Guest* 15925!; Rawanduz, *Guest* 1551!; Rawanduz gorge, *Guest* 1591! (BAG-photo); Rawanduz gorge, upper part, *Gillett* 9448! (BAG-photo); Rowanduz, *Rechinger* 11233 (E, W); Galli Ali Beg, *Rawi* 24253! (BAG-photo).

E. & S.E. Turkey.

3. **Satureja metastasiantha** *Rech.f.*, Fl. Iranica 150: 497 (1982). Type: Iraq, Arbil, M. Qandil, in saxosis calc. supra lacum Goam-e Kirmosoran, 3200 m, 28. vii.- 1. viii. 1957, *Rechinger* 11142! (W, holo.; G, E. iso.). Dirmenci et al. in Bagbahce 6: 55 (2019).

Subshrub, many stemmed from base. Stems 20–30 cm tall, ascending to erect, simple to branched above, short papillose to retrorsely puberulent with sparsely sessile glands, scabridulous. Leaves petiolate, herbaceous to coriaceous, concolorous, greyish-green, narrowly lanceolate to oblanceolate, 5–25 × 1–5 mm, attenuate at base, acute at apex, margins entire scarcely to densely short scabrid-puberulent with sessile glands, sometimes glandular punctate. Verticillasters lax, 2–4-flowered, cymes often pedunculate to 8 mm long, longer or shorter than floral leaves; pedicel 1–3 mm, as long as or shorter than calyx. Bracteoles minute, 1–2 mm, linear-setaceous, as long as or shorter than pedicel. Calyx 2.5–3.5 mm, tubular, veins slightly visible, white setulose-puberulous with dense sessile glands, slightly bilabiate; teeth unequal; upper 3-teeth triangular-lanceolate, c. 1 mm long, slightly recurved; lower 2-teeth lanceolate, c. 1.5 mm long; glabrous in throat. Corolla white, 4–8 mm, tube included in or exserted from calyx. Stamens 4; two stamens slightly exserted from corolla, the other two included in corolla. Style included or exserted from corolla. Nutlet oblong to trigonous, 1.5–2 × 1–1.2, brown to black.

HAB. Rocky slopes, calcareous, rocky clay mountain side; alt. 1000–3200 m; fl. & fr. Aug.-Sep.
DISTRIB. Mountains of northern Iraq. **MAM**: Dohuk, rocky slope, *Haines* 1222-A! (with *S. macrosiphonia* (1222-B) on the same sheet); in cacumine mt. Gara, *Kotschy* 409! (paratype) (G, W); Mosul, ad confines

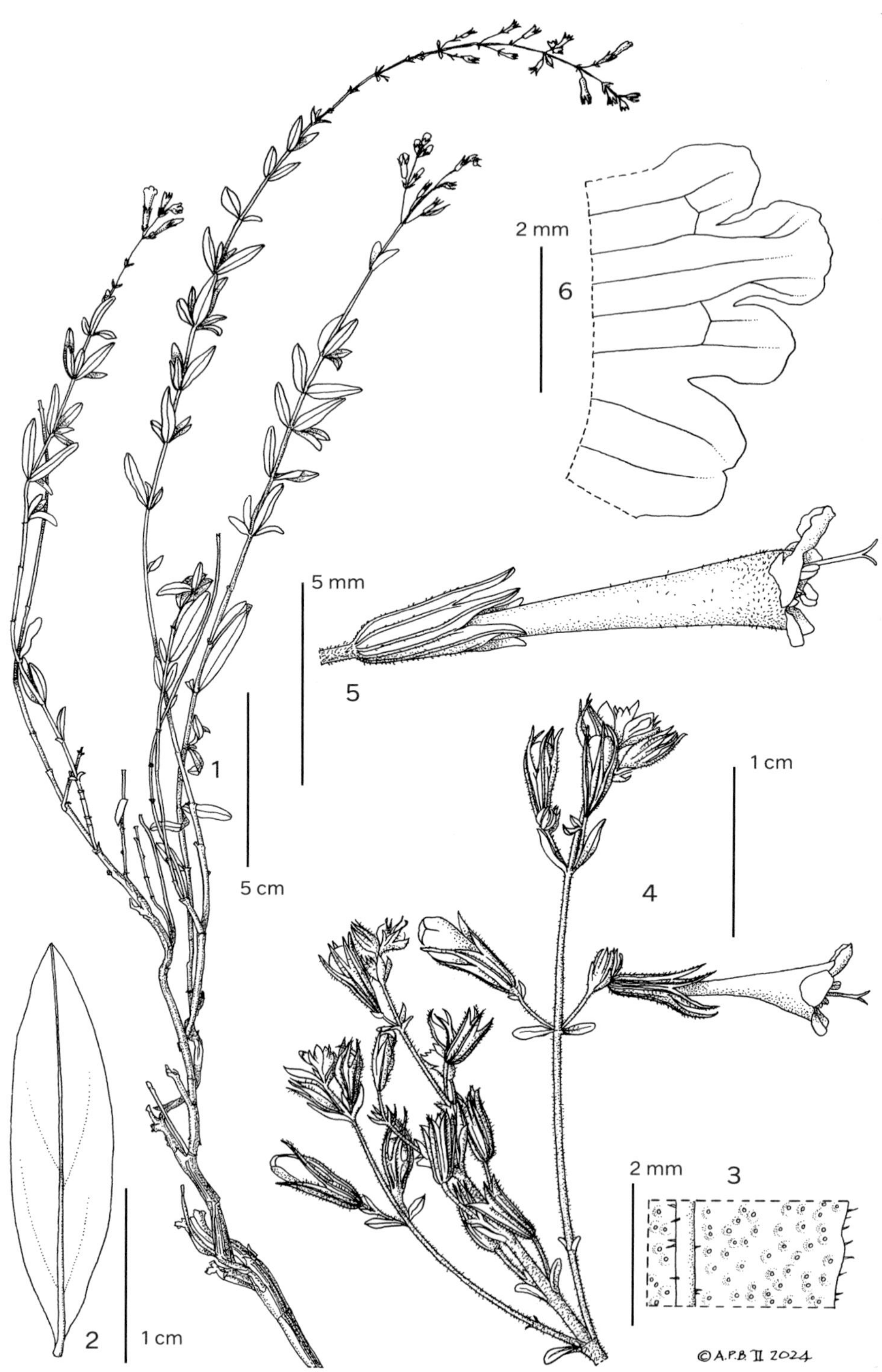

Fig. 160. **Saturegia macrosiphonia**. 1, habit; 2, abaxial leaf; 3, surface of leaf showing resin glands; 4, inflorescence; 5, flower, side view; 6, corolla opened showing dorsal and ventral lobes. 1–4 from *Guest* 1551; 5, 6 from *Gillett* 9448. Drawn by © A.P. Brown, Feb. 2024.

Turciae province Hakkari, in ditione pagi Sharanish, in montibus calc. a Zakho septentrionem versus, Jabal Khautur, in saxosis, *Rechinger* 10773 (W) (syntype); Sirsang, *Haines* 1225! (E) (syntype). **MRO**: Erbil, valley to east of Algird Dagh, *Gillett* 9517! (syntype); Querna valley, N. of Pushtashan, *Serhang &* *Rawi* 26606! (BAG-photo); S.W. of Akoyan S. of Rawanduz, *Al-Kaisi, K. Qader & A. Haloob* 60615 (BAG). **MSU**: Pira Magrun, *Haines* s.n. (E).

Turkey, Iran.

4. **Satureja bachtiarica** *Bunge* Labiat. Persic.: 37 (1873); Rechinger, Fl. Iranica [K. H. Rechinger] 150: 503 (1982); Jamzad in Fl. Iran [Assadi & al., in Persian] 76: 684 (2012).

Suffruticose, 10–15 cm tall, canescent-puberulous to hirsute with sessile glands, glandular papillate above. Flowering stems erect, thin, simple or branched. Leaves oblong-spatulate to oblong-linear, longitudinally plicate, 5–13 × 1.5–4(–5) mm, recurved, attenuate at base, obtuse at apex, margins entire, canescent, white hirsute-puberulent with sessile glands, lower leaves usually punctate. Verticillasters usually sessile, 4–8-flowered, remote, upper ones approximate. Calyx 1.5–2.5 mm, sub bilabiate, campanulate, hirsute with sessile glands and glandular papillae, main veins visible and interval veins ± obsolete; slightly hairy between the teeth in throat; teeth unequal; upper 3-teeth triangular-lanceolate, 0.7–1 mm; lower 2-teeth lanceolate to linear-lanceolate, 1–1.5 mm. Corolla 3–4 mm, white, (sub) included in the calyx, hirsute; lower lip bearded. Stamens 4; 2-stamens included in corolla. Style bilobed, included or slightly exserted. Nutlets unknown.

HAB. Rocky slopes, calcareous; alt. 1000–3000 m; fl. & fr. Aug.-Sep.
DISTRIB. Known from one location from close by West Iran border of Iraq. **MSU**: Suleymaniya; Hauraman mountain, rocky mountain, *Rawi, Nuri & Alizzi* 19772! (BAG-photo); Suleymaniya: Hauraman (Avroman) Montain, *Rawi et al.* 19759! (BAG-photo); ibid., *Rawi et al.* 19789! (BAG-photo);

Rawi, Nuri & Alizzi 19772 was identified as *Satureja cuneifolia* in Flora Iranica (Rechinger, 1982). However, it differs from the typical *S. cuneifolia* by its smaller verticillasters, ± sessile cymes, smaller calyx and corolla. Also, it differs from *S. metastasiantha* by its verticillasters 4–6-flowered and ±*sessile,* tightly arranged cymes.

Iran.

SPECIES DOUBTFULLY RECORDED

Satureja cuneifolia Ten., Fl. Napol. 1(Prodr.): XXXIII (1812).

Suffruticose perennial. Many stemmed from base, unbranched or branched above, erect or ascending, 12–50 cm tall, retrorsely puberulent to short hirsute-pubescent. Leaves herbaceous-coriaceous, cuneate-oblanceolate, 6–16 × 2–5 mm, attenuate at base, rounded at apex, shortly mucronate, without visible lateral veins below, scabrid-pubescent, dense sessile glands on both surfaces, greyish. Inflorescence usually long, 2–20 cm long. Verticillasters 2–6-flowered, lower cymes short pedunculate, uppers subsessile, upper verticillasters sometimes congested. Floral leaves 1–2 × calyces, bracteoles minute, lanceolate. Calyx 3–5 mm, sub-bilabiate nearly to middle, campanulate-infundibular, glabrous to hirsute pubescent; teeth 1/3–1/2 × calyx, lanceolate-subulate. Corolla white (rarely pale purple-pink), 6–8 mm long. Nutlets 1–1.4 mm long, obovate-oblong, obtuse to rounded.

The existence of *Satureja cuneifolia* in N. Iraq is doubtful. Specimen collected by *Rawi, Nuri & Alizzi* 19772 identified as *S. cuneifolia* in Flora Iranica is referrable to *S. bachtiarica. Satureja cuneifolia* is distributed from S. and E. Spain to the Central Taurus Mountains (S. Turkey) and not found further in the southeast neighbouring Iraq. In addition, the another specimen *Rechinger* 11390 (K, W) cited under *S. cuneifolia* differs from *S. bachtiarica* in that its calyx (4–4.5 mm) and corolla size (5–6 mm). In our opinion this specimen closer to *S. metastasiantha* rather than *S. cuneifolia.* However, these specimens differ from *S. metastasiantha* by the number of flowers in the verticillasters (2–8), the verticillasters tightly arranged and lower calyx teeth subulate. *Rechinger* 11390 (K) does not have the typical characters of *S. cuneifolia* nor of *S. metastasiantha.* More examples need to be examined to clarify the status of these related taxa.

23. **MICROMERIA** Benth.

Edward's Bot. Reg. 15:t.1282 (1829) nom. cons.
Harley & al. in Kubitzki (ser. ed.), Fam. Gen. Vasc. Pl. 7: 241 (2004); Brauchler & al.,
Willd. 38: 363 (2008)
Sabbatia Moench, Methodus [Moench] 386 (1794); *Zygis* Desv. ex Ham., Prodr. Pl. Ind.
Occid. [Hamilton] 15(46) (1825); *Piperella* (C.Presl ex Rchb.) Spach, Hist. Nat. Vég.
[Spach] 9: 164 (1838); *Apozia* Willd. ex Steud., Nomencl. Bot., [Steudel] ed. 2, 1: 114
(1840); *Cuspidocarpus* Spenn. Gen. Fl. Germ. [T.Nees] Gamop., 2: 18 (1843); *Micronema*
Schott, in Oesterr. Bot. Wochenbl. 7: 95 (1857); *Tendana* Rchb.f., in Oesterr. Bot.
Wochenbl. 7: 160 (1857).

Tuncay Dirmenci

Perennial herbs, subshrubs or shrubs, rarely annual herbs (*M. cymuligera*), 2–130 cm tall, often aromatic. Leaves 2–30 × 1–11 mm, entire or toothed, linear to ovate, flat or revolute, margin thickened, petioles usually distinct but short. Inflorescence in axillary verticillasters, thyrsoid or sometimes raceme-like, lax or dense or spike-like; bracts usually similar to leaves but mostly gradually smaller upwards in the inflorescence; cymes 1-many-flowered; peduncles to 12 mm long; bracteoles present, small, rather narrow. Calyx 5-lobed, scarcely accrescent, almost actinomorphic to distinctly 2-lipped, lobes ± equal or (3/2), anterior often longer, sometimes curved, spreading, tube 13–15-veined with prominent veins, throat bearded or not. Corolla 2-lipped, purple, pink, violet or white, sometimes shorter and cleistogamous, rarely female; posterior lip emarginate; anterior lip almost flat, 3-lobed with the mid-lobe broader. Stamens 4, didynamous, ascending under the posterior corolla lip, included or exserted from tube, filaments straight or connivent, thecae divaricate. Style lobes subequal, subulate. Nutlets brown or brownish, matt, glabrous or rarely hairy at apex, longer than broad, apex acuminate, acute, subacute or rounded to obtuse, with a ± distinct areole at the scar, producing mucilage when wet.

Micromeria s.l., is a taxonomically complex genus. In recent changes to the circumscription of the genus *Micromeria*, Sections *Xenopoma* and *Hesperothymus* pro parte were transferred to the genus *Killickia* Bräuchler, Heubl & Doroszenko by Bräuchler et al. (2008a), species in Section *Pseudomelissa* were transferred to *Clinopodium* by Bräuchler et al. (2005) and Ryding (2006). After these transfers, only the species included in Section *Micromeria* have remained within the genus *Micromeria* s.str.

The genus *Micromeria* Benth. s.str. is currently circumscribed as monophyletic and is represented by nearly 70 species, distributed from the Macaronesian-Mediterranean region to southern Africa, India, and China; a single species in Iraq, one other recorded doubtfully, is included in the key.

Arabacı, T., Dirmenci, T. & Celep, F. (2010). Morphological character analysis in Turkish *Micromeria* Benth. (Lamiaceae) species with a numerical taxonomic study. Turk. J. Bot. 34: 379–389.
Bräuchler, C., Meimberg, H. & Heubl, G. (2006). New names in Old World *Clinopodium* – the transfer of the species of *Micromeria* sect. *Pseudomelissa* to *Clinopodium*. Taxon 55: 977–981.
Brauchler, C., Ryding, O. & Heubl, G. (2008). The genus *Micromeria* (Lamiaceae), a synoptical update. Willd. 38: 363–410.
Martin, E., Çetin, Ö., Dirmenci, T. & Ay, H. (2011). Karyological studies of *Clinopodium* L. (Sect. *Pseudomelissa*) and *Micromeria* Benth. s.str. (Lamiaceae) from Turkey. Caryologia 64(1): 398–404.
Morales, R. (1993). Sinopsis y distribución del género *Micromeria* Bentham. Bot. Complut. 18: 157–168.
Ryding, O. (2006). Revision of the *Clinopodium abyssinicum* group (Lamiaceae). Bot. J. Linn. Soc. 150: 391–408.

Leaves ovate to ovate-lanceolate, 5–11 × 3–7 mm; verticillasters densely globose-hemispherical and congested above, many-flowered; calyx 3.5–4 mm . . . . 1. *M. myrtifolia*
Leaves lanceolate, 5–7 × 2–3 mm; verticillasters 3–5-flowered; calyx 2–3.5 mm . . . . 2. *M. persica*

1. **Micromeria myrtifolia** *Boiss. & Hohen.* Boiss., Diagn. Pl. Orient., ser. 1, 5: 19 (1844)).
Type: Iraq, in rupestribus umbrosis ad aquaeductus pr. pagum Gara Kurdistaniae, 24.vii.1844, *Kotschy* 305! (G, holo., BM, G, GOET, K, [K000193718, K000193719], P–photo! W. TCD,

Fig. 161. **Micromeria myrtifolia**. 1, habit × 1; 2, flower × 6; 3, inflorescence with calyx and bracts. Reproduced with permission from Feinbrun-Dothan, Fl. Palaestina 3: Plates, f. 246 (1977). Drawn by Esther Huber. © The Israel Academy of Sciences and Humanities.

WU, iso.); Dinsmore in Post, Fl. Syria, Palest. & Sinai ed. 2, 2: 340 (1933); Rawi in Dep. Agr. Tech. Bull 14: 152 (1964); Feinbrun-Dothan, Fl. Paleast. 3: 149 (1978); Rechinger, Fl. Iranica [K. H. Rechinger] 150: 508 (1982); Davis, Fl. Turkey [P. H. Davis] 7: 341 (1982); Jamzad in Flora Iran [Assadi & al., in Persian] 76: 701 (2012); Taifour & El-Oqlah, Pl. Jordan Annot. Checkl.: 103 (2017).

> *Micromeria juliana* var. *myrtifolia* (Boiss. & Hohen.) Boiss., Fl. Orient. [Boissier] 4: 570 (1879).
> *Micromeria lycia* Stapf, Denkschr. Kaiserl. Akad. Wiss., Wien. Math.-Naturwiss. Kl. 50: 94 (1885).
> *M. obtusiflora* Gand., Fl. Cret.: 80, 1419 (1916).
> *M. minoa* Coustur. & Gand., Bull. Soc. Bot. France 63: 12 (1916 publ. 1917).
> *Satureja myrtifolia* (Boiss. & Hohen.) Greuter & Burdet, Willdenowia 14: 305 (1984 publ. 1985).

Suffruticose perennial. Stems 20–40 cm tall, erect-ascending, recurved-pubescent. Leaves ovate to ovate-lanceolate, 5–11 × 3–7 mm, petiole c. 1 mm long, rounded to cuneate at base, acute to acuminate at apex, margin entire, flat to narrowly revolute towards apex, pubescent to shortly hirsute. Inflorescence long, (6–)10–20 cm long, verticillasters remote below, approximate towards apex, internodes to 2–3 × verticillasters. Verticillasters densely globose-hemispherical and typically many-flowered, ½–1 × floral leaves; cymules and flowers subsessile. Bracteoles linear-subulate, 3–4 mm long, c. as long as or slightly shorter than calyx tube, sometimes longer. Calyx cylindrical, 3–4 mm long, veins distinct, shortly hispid-puberulous, teeth subequal or upper 3 half as long as lower 2, 1/5–1/4 as long as calyx, narrowly lanceolate, divergent; bearded in throat. Corolla 3–5 mm long, pink, included in calyx or slightly exceeding. Nutlets oblong, 0.7–1 × 0.3–0.4 mm. Fig. 161, 1–3.

HAB. Rocky slopes and crevices (often limestone); alt. 500–1450 m; fl. & fr. Apr.-Jul.
DISTRIB. North and Northeast Iraq. **MAM**: Bikhair mountain, near Zakho, *Rawi* 22986! (BAG-photo); Jabal Bekhair, *Rawi* 8463 (BAG-photo); Jabal Bekhair prope Asi, *Rawi* 8489-b (BAG-photo), 8496 & 8505 (BAG-photo); Amadiya, *Al-Dabbagh, Al-Kaisi & Hamid* 46020! (BAG-photo); Amadiya, Kani Sinji, *Al-Dabbagh, Al-Kaisi & Hamid* 46036! (BAG-photo); Inter Dohuk et Amadiya, 3 km ab Amadiya occidentem versus, *Rechinger* 11945 (E); Zawita, *Omar et al.* 45325! (BAG-photo); ibid, *S.Omar et al.* 49646! (BAG-photo); inter Sirsang et Amadiyah, *Anders* 2222; Dohuk gorge, *Chapman* 26264 (BAG); 5 km from Zawita to Sersang, *Kaisi, Khayat & Karim* 51107 (BAG-photo) & 51108 (BAG-photo); Yosifka village, 10 km N.W. Sarsang, *Al-Dabbagh & K.Hamid* 45797! (BAG); 10 km S. of Sarsang, *Al-Kaisi & Hamid* 45826! (BAG); Sulaf, *Omar* 37714! (BAG-photo); Armotah, *Omar* 37247!; Bawagi, *Omar et al.* 37269! (BAG); Zinta gorge, *Chapman* 26098 (BAG-photo) & 26107 (BAG-photo); Dadi, *Omar, Al-Kaisi & Wedad* 49644 (BAG-photo); Mosul: In montibus calc. N. Zakho septentrionem versus, infra pagum Sharanish, *Rechinger* 15825, 11983 (W); 25 km N.E. of Zakho, *Rawi* 23226! (BAG); N.E. of Zakho, *Rawi* 23416 (BAG); Mosul, ad confines Turciae prov. Hakkari, inter mosul et Zakho, 103 km a Mosul septentrionem versus, *Rechinger* 10654 (E); in faucibus Shaikh Adi, *Guest* 4409! (BAG-photo); in faucibus Bekma, *Gillet* 8231! (BAG-photo); Aqrah, *Rawi* 11325! (BAG) & 11456! (BAG); Mosul, Jabal Pika Sar prope Aqrah, *Field & Lazar* 840;. **MRO**: Bel Hayat, Rowanduz, *Omar et al.* 38509! (BAG-photo); Arbil: In faucibus Barsarini prope Rawandiz, *Guest* 15884 (BAG-photo); Querna Qaw valley, N. of Pushtashan, *Rawi & I.Serhang* 26639!; Pushtashan, 15 km N.E. of Rania, *Rawi* 23898! (BAG-photo); Kaiwa rush, near Rania, *Rawi & Serhang* 23763! (BAG-photo); Galli Ali Beg, *Rawi* 26796! (BAG); Galli Ali Beg region, *Omar et al.* 38330! (BAG); Berrog mountain, on road to Qandil, *Rawi & Serhang* 23941 (BAG-photo). **MSU**: Suleymaniya, inter Kirkuk et Suleymaniya, In saxosis calc. angustiarum Darband-i Bazian prope Chamchamal, *Rechinger* 10591 (E); Suleymaniya, montes Avroman ad confines Persiae, in ditione pagi Tawilla, Sosakan, *Rechinger* 10201 (E); c. 10 km W. of Tawela (on road between Halabja & Tawela), *Rawi* 22168! (BAG-photo); Darbandi Basian, on the road between Kirkuk & Suleymaniya, *Rawi* 22847! (BAG); Azmir, *Karim* 39326 (BAG-photo); Pira Magrun, *Haussknecht s.n.*; In montibus cal careis Avroman et Shahu, *Haussknecht s.n.* **FNI**: Jabal Maglup, near Mosul, *Shahwani* 25187! (BAG-photo); Hinnis prope Ayn Sifni, *Guest* 3614!; Atrush, *Guest* 3627! (BAG-photo);

Micromeria myrtifolia, of the eastern Mediterranean phytochoria, is widely distributed from Greece to West Iran. The abundant, tightly arranged flowers in verticillasters are the most characteristic features of this species. Occasionally, stems with lax verticillasters (less flowering) are found together with this characteristic feature. It differs from *M. persica* by many-flowered verticillasters and approximate verticillasters above.

East Mediterranean to Iran (Greece, Turkey, Iran, Cyprus, Syria, Lebanon, Jordan, Saudi Arabia, Sinai).

DOUBTFULLY RECORDED SPECIES

Micromeria persica *Boiss.*, Diagn. Pl. Orient. Ser. 1, 7: 48 (1846); Rawi in Dep. Agr. Tech. Bull 14: 152 (1964); Rechinger in Fl. Iranica, [K. H. Rechinger] 150:510 (1982); Jamzad in Flora Iran [Assadi & al., in Persian] 76: 702 (2012).

Satureja persica (Boiss.) Briq. in H.G.A.Engler & K.A.E.Prantl, Nat. Pflanzenfam. 4(3a): 299 (1896). Icon.: Tab. 402.

Suffruticose perennial, many stemmed from base, 20–40 cm tall. Stems erect to ascending, greyish-velutinous, sometimes glabrescent. Leaves erect or adpressed to stem, subsessile to sessile, 5–7 × 2–3 mm, lanceolate, attenuate at base, acute at apex, margin revolute, veins distinct at margin. Verticillasters 4–10 long, 3–5-flowered, remote, shortly pedunculate. Bracteoles subulate, as long as pedicel or longer. Pedicel 0.5–1.5 mm long, Calyx 2.5–3.5 mm long, tubular, veins slightly visible, puberulous or ± glabrescent; teeth subequal, lanceolate, ¼ of calyx tube. Corolla c. 4 mm long, purple; tube included in calyx or slightly exceeding.

HAB. Rocky slopes and crevices (often limestone); alt. 500–1450 m; fl. & fr. Apr. to Jul.
DISTRIB. W. Iran.

Micromeria persica differs from *M. myrtifolia* by its slender stems and 3–5-flowered verticillasters. *M. persica* was reported to be distributed in south Turkey, northern Iraq (Mosul) and Iran. In detailed studies for Turkey, some variations of inflorescence in *Micromeria myrtifolia* do not represent the typical inflorescence characteristics of that species, and verticillasters such as those in *M. persica* have few flowers and slender stems. These specimens were identified as *M. persica.* Rechinger's specimens (15825 and 11983) cited in Flora Iranica (Rechinger, 1982) were examined at W. and I identified these variations of *M. myrtifolia* rather than *M. persica. M. persica* has a narrow distribution; the type locality and general distribution area is Shiraz and its surroundings and it is unlikely that *M. persica* is found in Iraq or Turkey.

24. **CLINOPODIUM** L.

Sp. Pl. 2: 597 (1753); Gen. Pl. ed. 5: 256 (1754)
Harley & al. in Kubitzki (ser. ed.), Fam. Gen. Vasc. Pl. 7: 241 (2004)
Acinos Mill., Gard. Dict. Abr. ed. 4: s.p. (1754); *Calamintha* Mill., Gard. Dict. Abr. ed. 4: s.p. (1754); *Gardoquia* Ruiz & Pav., Fl. Peruv. Prodr.: 86 (1794); *Rizoa* Cav., Anales Ci. Nat. 3: 132 (1801); *Xenopoma* Willd., Mag. Neuesten Entdeck. Gesammten Naturk. Ges. Naturf. Freunde Berlin 5: 399 (1811); *Diodeilis* Raf., Fl. Tellur. 3: 82 (1837); *Nostelis* Raf., Sylva Tellur.: 76 (1838); *Rafinesquia* Raf., New Fl. 3: 51 (1838); *Bancroftia* R.K.Porter, Gaceta Venez. 370: 3 (25 Feb. 1838); *Faucibarba* Dulac, Fl. Hautes-Pyrénées: 402 (1867); *Oreosphacus* Phil. in F.Leybold, Escurs. Pampas: 45 (1873); *Ceratominthe* Briq., Bull. Herb. Boissier 4: 875 (1896); *Antonina* Vved., Bot. Mater. Gerb. Inst. Bot. Akad. Nauk Uzbeksk. S.S.R. 16: 16 (1961); × *Calapodium* Holub, Folia Geobot. Phytotax. 11: 82 (1976); *Drymosiphon* Melnikov, Novosti Sist. Vyssh. Rast. 46: 178 (2015)

Tuncay Dirmenci, Ali Haloob & Ferhat Celep

Annual and perennial herbs and shrubs, aromatic, pubescent to hirsute-pilose. Leaves petiolate to subsessile, simple, ovate to lanceolate to orbicular, margins variously toothed or entire, sometimes revolute. Inflorescence of axillary cymes, these sometimes reduced to solitary flowers or merging to form a terminal panicle to a spike, verticillaster, or a subcapitate thyrse. Calyx actinomorphic to 2-lipped, not accrescent, usually cylindrical to tubular-campanulate, straight to sigmoid, sometimes gibbous, usually 13–(11–15)-nerved, 5-lobed, lobes often upcurved, posterior 3 often partially fused to form a lip, anterior 2 usually free above tube, throat hairy or not. Corolla white to blue, lavender, red, or orange (rarely yellow), 2-lipped, upper 2-lobed, lower 3-lobed, lower lip usually notched, tube straight or curved. Stamens usually 4 (rarely the posterior pair reduced to staminodes) didynamous, included or exserted, filaments usually glabrous, thecae parallel to divaricate, separate at dehiscence; stigma-lobes equal or unequal; disc usually symmetrical. Nutlets ellipsoid to ovoid or subglobose, often somewhat trigonal, smooth or minutely sculptured, glabrous or puberulent.

Clinopodium consist of approximately 165 species worldwide; general distribution of the genus is temperate and tropical New world, temperate Eurasia, Africa, tropical Asia and Indomalaysia; 5 species in Iraq.

The boundaries of the genus *Clinopodium* have expanded with the recent studies, especially with the transfer of *Acinos, Calamintha, Micromeria* Sect. *Pseudomelissa* and *Satureja* species of the New World to *Clinopodium* (Bräuchler, 2006; Harley & Granda, 2000; WCVP, 2022).

Arabacı, T., Dirmenci, T. & Celep, F. (2010). Morphological character analysis in Turkish *Micromeria* Benth. (Lamiaceae) species with a numerical taxonomic study. Turk. J. Bot. 34: 379–389.

Bräuchler C., Meimberg H., Abele T. & Heubl G. (2005). Polyphyly of the genus *Micromeria* (Lamiaceae) — evidence from cpDNA sequence data. Taxon 54 (3): 639–650.

Bräuchler, C., Meimberg, H. & Heubl, G. (2006). New names in Old World *Clinopodium.* – the transfer of the species of *Micromeria* sect. *Pseudomelissa* to *Clinopodium.* Taxon 55: 977–981.

Bräuchler, C. (2018). And now for something completely different—new names in *Clinopodium* with comments on some types. Phytotaxa 356(1): 71–80).

Cantino, PD. & Wagstaff, SJ. (1998). A re-examination of North American *Satureja* s.l. (Lamiaceae) in light of molecular evidence. Brittonia 50: 63–70.

Dirmenci, T., Haloob, A., Celep, F. & Ghazanfar, S.A. (2024). Revision of *Clinopodium* (Lamiaceae) in Iraq with a new species, *Clinopodium dokanicum,* and an identification key. Kew Bull. 79(1): 131–141.

Harley, RM. & Granda, A. (2000). List of species of tropical American *Clinopodium* (Labiatae), with new combinations. Kew Bulletin 55: 917–927

Martin E., Çetin, E., Dirmenci, T. & Ay, H. (2011). Karyological studies of *Clinopodium* L. (Sect. *Pseudomelissa*) and *Micromeria* Benth. s.str. (Lamiaceae) from Turkey. Caryologia 64(1): 398–404.

1. Annual; calyx gibbous . . . . 2. *C. graveolens*
    Perennial herbs or suffruticose; calyx not gibbous, straight or curved 2
2. Leaves lanceolate-oblong to ovate, 20–55 × 8–25 mm; verticillasters
    globose; calyx 10–12 mm long; corolla 10–15 mm long . . . . 1 *C. vulgare*
    Leaves usually broadly ovate to suborbicular; 3–20(–25) × 3–15 mm,
    verticillasters arranged in lax thyrsoid or raceme-like inflorescence;
    calyx 1.6–5 mm long; corolla 3–10 mm long 3
3. Calyx bilabiate, 13-veined; corolla 8–10 mm . . . . 3. *C. brevifolium*
    Calyx with 5 ± equal teeth (± actinomorphic), 15-veined; corolla 3–8 mm 4
4. Stems slender, flexuous, 5–30 cm tall; inflorescence thyrsoid, lax; calyx
    3–5 mm; corolla 5–8 mm . . . . .4. *C. mole*
    Stems sturdy, 40–70 cm tall; inflorescence verticillate; calyx 1.6–2.2 mm;
    corolla 3–4 mm . . . . .5. *C. dokanicum*

1. **Clinopodium vulgare** *L.*, Sp. Pl. 2: 587 (1753) subsp. **orientale** Bothmer, Bot. Not. 120: 206 (1967); Rawi in Dep. Agr. Tech. Bull 14: 150 (1964); Feinbrun-Dothan, Fl. Paleast. 3: 151 (1978); Rechinger, Fl. Iranica [K. H. Rechinger] 150: 519 (1982); Davis, Fl. Turkey [P. H. Davis] 7: 329 (1982); Jamzad in Flora Iran [Assadi & al., in Persian] 76: 667 (2012).

Satureja vulgaris subsp. *orientalis* (Bothmer) Greuter & Burdet, Willdenowia 14: 306 (1984 publ. 1985).
Clinopodium orientale (Bothmer) Melnikov, Novosti Sist. Vyssh. Rast. 44: 204 (2013).

Perennial herbs. Stem erect to ascending, 20–70 cm tall, retrorsely puberulent to densely pilose with sometimes glandular papillae. Leaves lanceolate to ovate or oblong-ovate, 2–55 × 8–25 mm, petiole to 10 mm long, obtuse or obtuse-truncate at base, shallowly or conspicuously serrate or crenate-dentate at margins, obtuse to acute at apex, thin veined on lower surface, sparsely to densely pilose on both surfaces with sometimes glandular papillae and sessile glands; floral leaves exceeding the cymes. Verticillasters dense, remote or sometimes approximate above, many flowered, globose; cymes usually shortly pedunculate, sometimes lowermost long pedunculate. Bracteoles linear, 8–14 mm long, shorter than calyx, subulate, plumose with densely scabrous. Calyx bilabiate, 10–12 mm long, tube curved, 13-nerved, sparsely to densely hirsute-pilose, with glandular-papillose and sessile glands, sometimes stalked glandular-hairy; lower teeth. 3–5.5 mm long, incurved, subulate; upper teeth 2–4 mm long, recurved; throat narrowed, hairy between the teeth. Corolla 10–15 mm long, pink to purple, included in calyx or exceeds, upper lip and middle lobe of lower lip emarginate, hirsute with glandular-papillose on tube, pilose-villous outside of the lips, bearded on lower lip. Stamens and style included in corolla. Nutlets subglobose, 1–1.2 × 1–1.1 mm, smooth, brown. Fig. 162, 1–2.

HAB. Rocky mountain slope, river or shady banks, scrub, dry limestone slope, grass and fields, shades in *Quercus* forest near stream, mountain side; alt. 800–1600 m; fl. & fr. Jun.-Aug.

DISTRIB. Near stream banks and shady areas in Amadiya, Rowanduz and Sulaymaniyah districts. **MAM**: Daimka village, 35 km N.E. of Zakho, *Al-Kaisi, H.Hamid & K.Hamad* 45361! (BAG); Bighaoda village, 6 km S. of Kani Masi, *Omar & Dabbagh* 45680! (BAG); Sulaf (c. 1 km N.W. of Amadiya), *Omar* 37711! (BAG); Sersang, rocky slopes, *Haines* 1350!: Dohoka village, 12 km N.E. of Sarsang, *Al-Dabbagh & K.Hamid* 46128 (BAG) & 46133 (BAG-photo); **MSU**: 11 km N. of Penjwin, *Rawi* 22914! (BAG); Penjwin, *Al-Hashimi* 43731 (BAG); Suleymaniya, montes Avroman ad confines Persiae, in ditione pagi Tawilla, *Rechinger* 12377! (W); Biara valley, *Al-Shehbaz, Al-Khaikani & Ibrahem* 37647 (BUH); **MRO**: Algird Dagh, near Nawanda, *Guest* 2778! (BAG); Pishtashan, N.E. of Rania, lower slopes of Qandil range, *Rawi & Serhang* 26554! (BAG) & 23855 (BAG); Ari, *Gillett* 9702! (BAG)

S. France to W. Asia

2. **Clinopodium graveolens** (*M.Bieb.*) *Kuntze*, Revis. Gen. Pl. 2: 515 (1891); Feinbrun-Dothan, Fl. Paleast. 3: 152 (1978); Rechinger, Fl. Iranica [K. H. Rechinger] 150: 521 (1982); Davis in Fl. Turkey [P. H. Davis] 7: 334 (1982); Jamzad in Flora Iran [Assadi & al., in Persian] 76: 638 (2012); Taifour & El-Oqlah, Pl. Jordan Annot. Checkl.: 101 (2017).

[as *Acinos graveolens* or *A. rotundifolius*]

subsp. **graveolens**

Thymus graveolens M.Bieb., Fl. Taur.-Caucas. 2: 60 (1808).
Th. canus Steven ex M.Bieb., Fl. Taur.-Caucas. 2: 60 (1808).
Th. exiguus Sm. in J.Sibthorp & J.E.Smith, Fl. Graec. Prodr. 1: 421 (1809).
Th. patavinus var. *graveolens* (M.Bieb.) Trevir., Index Seminum (WROCL, Wratislaviensi) 1818: 7 (1818). *Acinos exiguus* G.Don ex Steud., Nomencl. Bot. 1: 17 (1821).
A. graveolens (M.Bieb.) Link, Enum. Hort. Berol. Alt. 2: 117 (1822).
A. canus (Steven ex M.Bieb.) Rchb., Fl. Germ. Excurs.: 327 (1831).
Satureja graveolens (M.Bieb.) Caruel in F.Parlatore, Fl. Ital. 6: 143 (1884).
Melissa graveolens (M.Bieb.) Benth., Labiat. Gen. Spec.: 390 (1834).
M. graveolens (M.Bieb.) Benth., Labiat. Gen. Spec.: 390 (1834).
Calamintha cana (Steven ex M.Bieb.) Heynh., Alph. Aufz. Gew.: 104 (1846).
C. graveolens (M.Bieb.) Benth. in A.P.de Candolle, Prodr. 12: 231 (1848).
Clinopodium canum (Steven ex M.Bieb.) Kuntze, Revis. Gen. Pl. 2: 515 (1891).
Satureja exigua (Sm.) Grande, Boll. Soc. Bot. Ital. 1912: 178 (1912.
Calamintha exigua (Sm.) Holmboe, Stud. Veg. Cyprus: 161 (1914).
Acinos fominii Des.-Shost., Symp. Mem. Fomin: 39 (1938).
Satureja crassinervis H.Lindb., Årsbok-Vuosik. Soc. Sci. Fenn. 20B(7): 6 (1942).
Acinos exiguus (Sm.) Meikle, Fl. Cyprus 2: 1281 (1985).
Ziziphora fominii (Des.-Shost.) Melnikov, Bot. Zhurn. (Moscow & Leningrad) 101: 89 (2016).
Z. graveolens (M.Bieb.) Melnikov, Bot. Zhurn. (Moscow & Leningrad) 101: 89 (2016).

Annual. Stems 5–20 cm tall, erect to decumbent, simple or branched below, green or purplish above, stems often crispate and hirsute-villous or retrorsely puberulent, glandular or not. Leaves lanceolate-ovate or obovate to orbicular, 5–20 × 3–15 mm, petiole to 15 mm long, puberulous to crispate or densely hirsute-villous with glandular papillae, ±truncate to cuneate at base, upper part shallowly or conspicuously serrate, apex mucronate. Verticillasters 1–10, 2–12-flowered, lower distant, uppers approximate or spike-like, sometimes subcapitate. Bracteoles lanceolate-subulate. Flowers pedicellate, 2–5 mm long, pedicels flattened. Calyx tubular, bilabiate, 8–10 mm long, tube slightly sigmoid, gibbous towards base, constricted near middle, sparsely to densely long-hirsute with densely stalked glandular hairy; lower teeth 2–3.5 mm long, as long as or shorter than upper teeth, lanceolate-subulate, incurved, ciliate; upper teeth 1–2.5 mm long, triangular-acuminate to lanceolate-subulate, spreading to recurved, ciliate; densely long white hairy in throat. Corolla 7–10 mm long, pale blue to purple or pink, slightly exceeding the calyx, upper lip shortly obtuse; lower lip lobe subequal. Stamens and style included in corolla. Nutlets oblong, 1.8–2.1 × 0.8–1 mm, pale brown, slightly tuberculate. Fig. 163, 1–6.

HAB. Limestone slopes, rocky mountain, mountain slope, fallow fields; alt. 600–1500 m; fl. & fr. May-Jun.
DISTRIB. Rocky slopes, steppes and follow fields in northern and northeastern Iraq. **MJS**: Jabal Sinjar above Kursi, *Gillett* 11048! (BAG); Jabal Sinjar, *Omar, Al-Kayat & Al-Kaisi* 52424 (BAG); Jabal Sinjar, *Al-Khayat, Al-Kaisi & F. Karim* 51195 (BAG); 12 km N.W. Jabal Sinjar, *Al-Khakani* 0036735 (BUH); Jabal

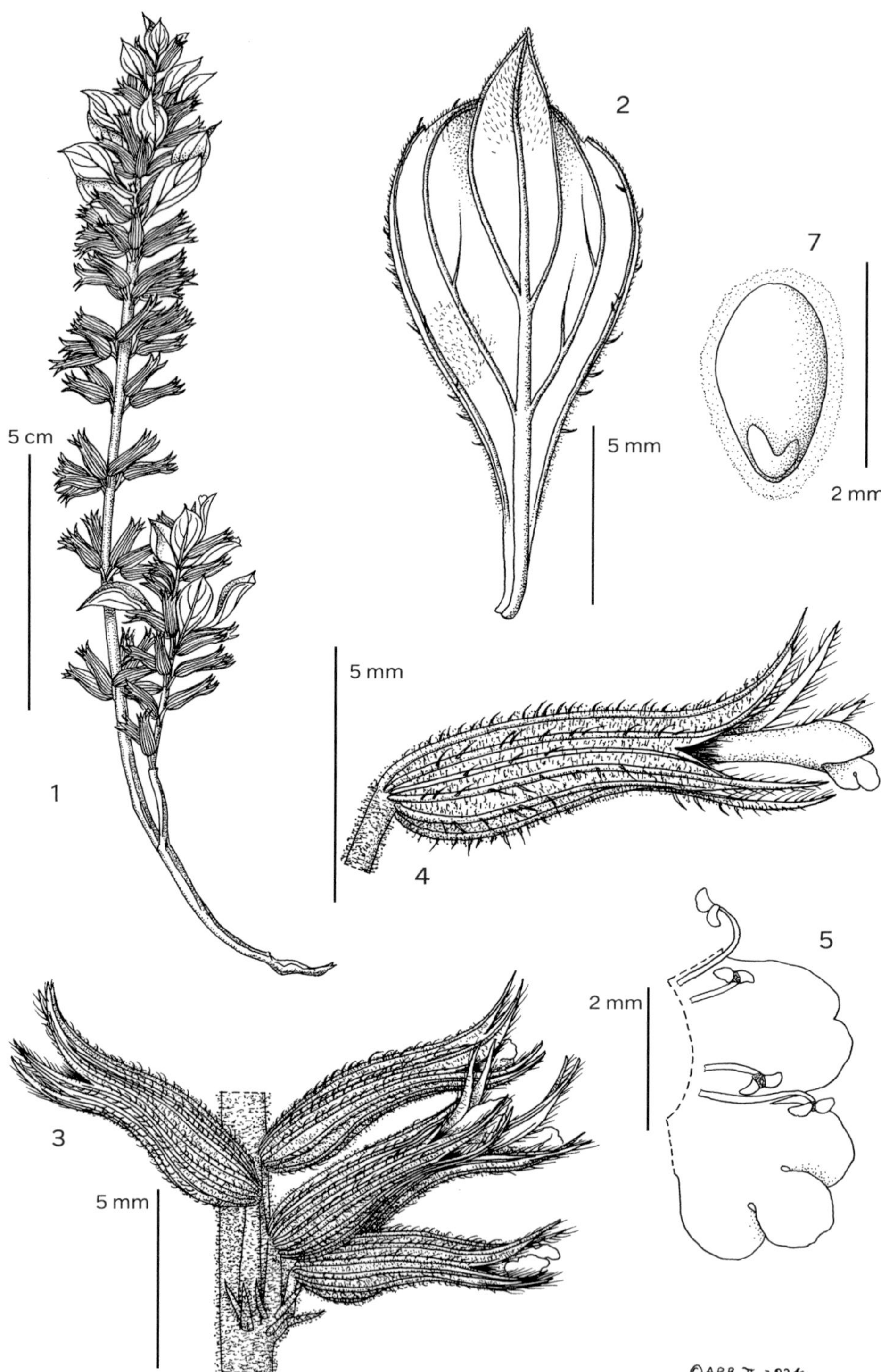

Fig. 162. **Clinopodium graveolens**. 1, habit; 2, abaxial leaf surface; 3, Inflorescence (half verticillaster); 4, flower side view; 5, corolla opened showing dorsal and ventral lobes; 6, nutlet, hydrated showing mucilaginous coat. 1–6 from *Wheeler Haines* W1563. Drawn by © A.P. Brown, Feb. 2024.

Fig. 163. **Clinopodium vulgare** L. subsp. **orientale**. 1, habit × 1; 2, calyx with bracts × 1. Reproduced with permission from Feinbrun-Dothan, Fl. Palaestina 3: Plates, f. 250 (1977). Drawn by Esther Huber. © The Israel Academy of Sciences and Humanities.

Sinjar above Ain Kursi, *Al-Khakani* 0036738 (BUH); Sinjar between Malak & Gora Mishy, *Al-Shehbaz, Mayah & Sharifi* 0030019 (BUH); **MAM**: Mosul: Jabal Khantur N. Zakho prope Sharanish, *Rechinger* 12075 (W); **MSU**: Qara Dagh, Sulaimaniya, *Haines* 1563!

E. Mediterranean to C. Asia.

Previously included in the genus *Acinos*. It differs from other previously *Calamintha* and Sect. *Pseudomelissa* species with the calyx being bilabiate, the calyx tube slightly sigmoid, gibbous towards base and constricted near middle.

3. **Clinopodium brevifolium** (*Wahlenb.*) *Bräuchler* & *Hjertson*, Phytotaxa 356: 74 (2018); Dinsmore in Post, Fl. Syria, Palest. & Sinai ed. 2, 2: 343 (1933); Feinbrun-Dothan, Fl. Paleast. 3: 150 (1978); Rechinger, Fl. Iranica [K. H. Rechinger] 150: 510 (1982); Davis in Fl. Turkey [P. H. Davis] 7: 328 (1982); Taifour & El-Oqlah, Pl. Jordan Annot. Checkl.: 101 (2017) [as *Calamintha incana* or *Clinopodium insulare*]

Thymus brevifolius Wahlenb, J.Berggren, Resor Eur. Österländ. 2(Bih.): 61 (1826)
Micromeria insularis Candargy, Bull. Soc. Bot. France 44: 149 (1897).
Thymus incanus Sm. in J.Sibthorp & J.E.Smith, Fl. Graec. Prodr. 1: 421 (1809).
Melissa incana (Sm.) Benth., Labiat. Gen. Spec.: 386 (1834).
Clinopodium incanum (Sm.) Kuntze. Revis. Gen. Pi. 2:515 (1891).
Calamintha incana (Sm.) Boiss., Fl. Orient. 4: 578 (1879.
Satureja incana (Sm.) Briq. in H.G.A.Engler & K.A.E.Prantl, Nat. Pflanzenfam. 4(3a): 301 (1896).
S. insularis Greuter & Burdet, Willdenowia 14: 304 (1984 publ. 1985).
Clinopodium insulare (Candargy) Govaerts, World Checkl. Seed Pl. 3(1): 17 (1999).

Suffruticose, few to many stemmed from base. Stems ca. 30 cm tall, ascending to decumbent, longitudinally articulate, often glandular and very densely grey-velutinous-tomentose. Leaves ovate-orbicular, 6–13 × 6–12 mm, lower leaves petiolate to 5 mm long, upper leaves subsessile, rounded to truncate-subcordate at base, rounded to obtuse at apex, entire or scarcely notched, canescent-tomentose, glandular papillose. Verticillasters 6–15, scarcely approximate, 4–6-flowered, to 5 mm pedicellate on very short or obsolescent peduncles. Bracteoles linear-lanceolate, 0.5–1.5 mm long, velutinous. Calyx bilabiate, 3–5 mm, tube straight, glandular and densely velutinous-puberulous, 13-veined, hairs of throat exserted; upper teeth 0.3–0.8 mm, triangular; lower teeth 1–2 mm, as long as or longer than uppers, lanceolate and long-ciliate. Corolla 8–10 mm, mauve, short hairy with sessile glands. Stamen and style included in corolla. Nutlets ±globose, ca. 1 × 1 mm, dark brown, glandular papillose.

HAB. Rocky calcareous places; alt. 600 m; fl. & fr. Jun.-Aug.
DISTRIB. Known only from a single location from northern Iraq. **MAM**: Mosul, Dohuk, in fissuris rupium calc., *Rechinger* 11489! (W).

East Mediterranean to N. Iraq.

4. **Clinopodium molle** (*Benth.*) *Kuntze*, Revis. Gen. Pl. 2: 515 (1891 Rawi in Dep. Agr. Tech. Bull 14: 152 (1964); Rechinger, Fl. Iranica [K. H. Rechinger] 150: 506 (1982); [as *Micromeria flacca* or *M. mollis*]. Type: [Iraq] Mossul, 1841, *Kotschy* 552a! (K, holo. (K000193473); G, W, iso. (G00769725, W0027466).

Micromeria mollis Benth., Prodr. [A.P.de Candolle] 12: 226 (1848).
Satureja flacca Nábelek, Spisy Prír. Fak. Masarykovy Univ. 70: 44 (1926).*Micromeria flacca* (Nábelek) Hedge, Notes Roy. Bot. Gard. Edinburgh 25: 51 (1965); Davis in Fl. Turkey [P. H. Davis] 7: 339 (1982).

Caespitose-suffruticose, saxatile perennial with stout woody stock, many stemmed. Stems slender, flexuous, fragile, 5–30 cm tall, simple or branched, pubescent to villous with sessile glands with or without glandular papillae. Leaves ovate to ovate-orbicular, 5–25 × 5–15 mm, petiole to 10 mm, truncate to rounded or broadly cuneate at base, subentire to 3-toothed on per side, obtuse at apex, flat, tomentellous to pilose on both surfaces, sparsely to densely sessile glands on beneath, sometimes glandular-papillose on both surfaces. Inflorescence thyrsoid or raceme, lax, verticillasters 2–7-flowered, peduncles 4–17 mm, longer or shorter than floral leaves. Bracteoles filiform, 1–2.5 mm. Pedicels as long as calyx or longer, sometimes shorter. Calyx ±actinomorphic, 3–4.5 mm, tubular-obconical to campanulate, hirtellous, densely glandular papillose, hairy between teeth, 13-veined, veins slightly or clearly visible, teeth triangular, 0.5–2 mm long, ½–1/3(–1/4) as long as calyx, acute to shortly acuminate, to triangular-subulate, erect or recurved, ciliate or not. Corolla lilac, 5–8 mm, tube exserted from calyx, hirtellous; upper lip retuse; lower lip 3-lobed, median lobe larger than lateral 2 lobes, obovate, bearded, lateral 2 lobes rounded. Stamens 4, included in corolla. Style sub-included in corolla. Nutlets 1–1.1 × 0.4–0.5 mm, oblong, subacute-mucronate, glandular-hairy.

HAB. Crevices of limestone rocks, cliffs, *Quercus* open forest, on rocks in valley; alt. 600–1800 m; fl. & fr. Jun.-Sept.

DISTRIB. Rocky crevices in northern and northeastern Iraq mountains. **MAM**: Sarsang, Gara mount, summit cliffs, *Agnew* 691 (E); Sarsang, cracks in shady rock face in deep wooded gorge, *Haines* 459! (E); Amadiya, Summit Sarsang mountain, rock, *Agnew* 694 (E); Zawita, *Guest* 4621!; Dohuk and Zawita gorges, cliffs, *Agnew* 734 (E); Zawitah, *Guest* 4730! **MSU**: Sulaimaniya, Kupi Qara Dagh, rocky cliff, near summit ridge, *Haines* 2093! (E) & 1076! (E); **MRO**: Rowanduz gorge, *Guest* 8315 & 9446! & 13092! (BAG); Rowanduz gorge limestone rocks, *Quercus* open forest, *Gillett* 8313! (BAG); Gali Ali Beg (Rowanduz Gorge), on rocks in valley, *Rawi* 26794! (BAG); In flssuris rupium calc. faucium infra Rawandiz, *Guest* 2972! (BAG); Gali Ali Beg, *Hadac* & *al.* 6244; Erbil, in fissuris rupium calc. faucim infra Rowanduz, *Rechinger* 11263 (E, W); Dalow Kaioter, road to Qandil, *Rawi* & *Serhang* 26477!

S. and S.E. Turkey.

5. **Clinopodium dokanicum** *Dirmenci* in Kew Bull. 79(1): 3 (2024). Type: (Iraq), Sulaimaniya, Dokar, cliffs bordering gorge, Zab river below dam, 21 xi 1961, *Agnew* & *Haines* 6409! (E, holo. [E00952170]; BAG, iso. (accession no. 46491).

Perennial. Stems 40–70 cm tall, widely much branched, lower branches to 30 cm long, densely soft pilose to villous with sessile glands and glandular papillose, internodes clearly longer than leaves. Leaves broadly ovate to ovate-orbicular, 8–18 × 4–16 mm, lower leaves petiolate to 3 mm long, upper subsessile, densely tomentose to villous with sessile glands and glandular papillose, greyish-green, truncate to attenuate at base, ±entire or obscurely crenate at margins, teeth 1–3 pairs per side, obtuse to rounded at apex. Verticillasters 3–13, clearly remote, uppers 2–3 verticillasters approximate, not columnar, many-flowered, cymes shortly pedunculate. Bracteoles linear to setaceous, 1–1.5 mm long, densely hairy with sessile glands, ciliate, shorter than calyx, longer than pedicel, green or partly purplish. Calyx tubular-subcampanulate to campanulate, cup shaped in fruiting time, 1.6–2.2 mm, densely villous with sessile glands and glandular-papillose, veins visible, green or partly purplish; teeth 0.4–0.6 mm, broadly triangular to acuminate; densely long white hairy in throat, hairs clearly visible. Corolla white, sometimes purple dots present on lower and upper lips, 3–4 mm, exserted from calyx, hairy with sessile glands. Stamens 4, included or subincluded in corolla. Style usually exserted from corolla. Nutlets oblong-trigonous, c. 1 × 0.6 mm, mucronate, glandular and slightly tuberculate.

HAB. Cliffs bordering gorge; alt. not recorded; fl. Nov.

DISTRIB. Known only from the type gathering. **MSU**: Sulaimaniya, Dokar, *Agnew* & *Haines* 6409! (type). *RAWI* & *Serhang* 26477 (Dolaw, Road to Qandil) is closely related to *C. dokarensis*, but its verticillasters are more lax than in *C. dokanicum*.

25. **CYCLOTRICHIUM** (Boiss.) Manden. & Scheng.

Not. Syst. Leningrad 15: 336 (1953); Harley & al. in Kubitzki (ser. ed.), Fam. Gen. Vasc. Pl. 7: 242 (2004)

Ali Haloob

Suffrutescent perennial herbs. Stem simple or branch above. Leaves petiolate, ovate or suborbicular or lanceolate, margins entire or serrulate. Verticillasters 2-many flowered, Bract sessile or short petiolate; Bracteoles linear-lanceolate. Calyx tubular, 13-veined, bilabiate or sub-bilabiate, straight or slightly curved, with a dense white pilose ring in throat; limb 5-toothed, teeth subulate-lanceolate, upper 3 shorter than the lower 2 teeth. Corolla bilabiate, upper lip 2-lobed, emarginate, lower lip with 3-subequal lobes, tube resupinate (twisted at base) which make upper lip lower (abaxial), and lower lip upper (adaxial); tube exserted, with ring of hair in throat. Stamens 4, exserted from corolla; anther 2-celled.

Style exserted, longer than stamens, branches unequal, subulate. Nutlets glabrous, oblong-ovoid.

A genus of eight species found in Turkey, Iran, Iraq and Syria; 4 species in Iraq.

1. Calyx 8.5–10 mm, upper teeth 2.5–3 mm; corolla 13–18 mm . . . . 1. *C. longiflorum*
   Calyx less than 8 mm, upper teeth less than 2.5 mm; corolla less than 12 mm. 2
2. Stems with dense minute glandular hairs; leaves glabrescent; lower
   teeth of calyx 3–3.5 mm. . . . . .2. *C. stamineum*
   Stems with spreading eglandular hairs, hirsute or densely pilose-villous
   and with dense glandular minute hairs; leaves hirsute or densely
   pilose-villous; lower teeth of calyx less than 2.5 mm 3
3. Stems densely white pilose-villous; leaves ovate-suborbicular, 10–13 mm,
   with dense white eglandular pilose-villous hairs; calyx with spreading
   villous eglandular and densely minute glandular hairs. . . . . .3. *C. straussii*
   Stems patent-hirsute; leaves ovate-lanceolate, 12–18 mm, patent hirsute,
   eglandular; calyx with dense minute glandular hairs . . . . .4. *C. leucotrichium*

1. **Cyclotrichium longiflorum** *Leblebici,* Bitki 1: 406 (1974); Rechinger, Fl. Iranica [K. H. Rechinger] 150: 514 (1982).

Suffrutescent herb. Stems erect, 20–40 cm long, simple, with dense white eglandular pilose and minute glandular hairs. Leaves ovate, 10–20 × 8–15 mm, base rounded-subcordate or truncate, apex obtuse to acute, margins entire, densely eglandular pilose with sessile yellow glands on abaxial side; petiole ± 1 mm. Verticillaster 6–24-flowered. Bracts ovate, 5–12 × 3–7 mm, acute; bracteolate lanceolate-linear, 5–8 mm, densely white pilose with sessile yellow glands. Calyx 8.5–10 mm, upper teeth 2.5–3 mm, lower teeth 3.5–4.5 mm, densely eglandular hirsute-pilose with minute glandular hairs. Corolla 13–18 mm, pale purple. Nutlets oblong, ± 1.5 mm.

HAB. Limestone gorge, rocky mountain, on summit, open pine forest on slopes, in *Quercus* forest on rocky mountain slope; alt. 1360–2060 m.; fl. Jun.-Sep.
DISTRIB. Rare in the lower forest & thorn-cushion zone of northern Iraq. **MAM**: Kantur mountain N.E. of Zakho, *Rawi* 23343! (BAG); Marsis near Sharanish, *Rechinger* 10891! (BUH); above Basingera, near Sharanish, *Rechinger* 11572! (BUH); near Mt. Zawita, *Rechinger* 11992! [K000509384]; Zawita N.E. of Zakho, *Rawi* 23616! (BAG). **MJS:** Jabal Sinjar E. slope, *S. Omar, Al-Kaisi & Al-Khayat* 52554! (BAG).

Turkey, Iran.

2. **Cyclotrichium stamineum** (*Boiss. & Hohen.*) *Manden. & Scheng.*, Bot. Mater. Gerb. Bot. Inst. Komarova Akad. Nauk S.S.S.R. 15: 337 (1953); Rechinger, Fl. Iranica [K. H. Rechinger] 150: 513 (1982).

Suffrutescent herbs. Stem erect to ascending, 20–40 cm, simple or branched above, with dense minute glandular hairs. Leaves ovate-lanceolate, 14–25 × 8–12 mm, base rounded or obtuse, apex acute to obtuse, margins entire or a few serrulate, adaxially glabrescent, abaxially with spreading sessile yellow glands; petiole 1–3 mm. Verticillaster 2–18- flowered. Bract ovate, 7–12 × 4–9 mm, apex acuminate; bracteole lanceolate-linear, 4–6 mm, puberulent-pilose with dense minute glandular hairs and sessile glands. Calyx 6–8 mm, upper teeth 2–2.5 mm, lower teeth 3–3.5 mm, hirsute with dense minute glandular hairs. Corolla 10–12 mm, purple. Nutlets oblong, ± 1.3 mm.

HAB. Limestone rocks, mountain slopes, valleys; alt. 1000–1500 m.; fl. Jun.-Jul.
DISTRIB. Occasional in in the lower forest zone of northern Iraq. **MAM**: Bamerny 20 km N.W. Sarsang, *Al-Kaisi & K. Hamad* 45931; Gorge, 3 km from Lamadiya, *K.H.Rechinger* 11618! (BAG); Sulaf, *S. Omar* 37717! (BAG); 10 km S. of Sarsang, *Al-Kaisi & K.Hamad* 45832! (BAG); Sarsang-slopes of Gara dagh, *E. Ehap* 26409! (BAG); Gara Dagh, *Rawi* 9293! (BAG); Dohoka village 12 km N.E. Sarsang, *Al-Dabbagh & K. Hamad* 46135! (BAG); Dori village 5 km E. Kani Masi, *S. Omar & Al-Kaisi* 45401! (BAG); Zawita, *C. Cuear* 4622! (BAG); Zawita, 21 km on the road from Dohuk to Sarsang, *K. H. Rechinger* 11541! (BUH).

E. Turkey.

3. **Cyclotrichium straussii** (*Bornm.*) *Rech.f.*, Fl. Iranica [K. H. Rechinger] 150: 516 (1982).

Calamintha straussii Bornm., Beih. Bot. Centralbl. 22(2): 119 (1907).

Suffrutescent herbs. Stem erect to ascending, 9–23 cm, simple, densely white pilose-villous and with dense glandular minute hairs. Leaves ovate-orbicular, 10–13 × 7–9 mm, apex acute, base obtuse-rounded, margins ± serrulate, with dense white eglandular pilose-villous and sessile glands on abaxial side; petiole 1–3 mm. Verticillaster 2-many-flowered. Bract ovate, 4–11 × 2–9 mm, apex acute; bracteole lanceolate-linear, 3–5 mm, eglandular white villous and with sessile glands. Calyx 5.5–6 mm, upper teeth 1–1.2 mm, lower teeth 2 mm, with spreading eglandular villous and dense minute glandular hairs. Corolla c. 10 mm, pale pink (in herbarium specimen).

HAB. Clay soil between rocks on mountain slope; alt. 800–900 m; fl. May.
DISTRIB. Very rare, found only once in the lower forest zone in northern Iraq. **MRO**: Chewa Rash N.E. of Rania, *Rawi, Nuri & Kass* 28480! (BAG).

W. Iran

4. **Cyclotrichium leucotrichum** (*Stapf ex Rech.f.*) *Leblebici*, Bitki 1: 405 (1974); Edmondson in Fl. Turkey [P. H. Davis] 7: 348 (1982); Rechinger, Fl. Iranica [K. H. Rechinger] 150: 515 (1982).

Suffrutescent herbs. Stem erect to ascending, 15–35 cm, simple or branch above, with spreading eglandular hirsute and dense glandular minute hairs. Leaves ovate-lanceolate, 12–18 × 7–9 mm, base rounded or obtuse, apex acute, margins entire or ± serrulate, eglandular patent hirsute with sessile glands on abaxial side; petiole 1–4 mm. Verticillaster 2–8-flowered. Bracts ovate, 5–7 × 3 mm, apex acute; bracteoles lanceolate-linear, 2–3.5 mm, eglandular puberulent with dense minute glandular hairs and sessile glands. Calyx ± 6 mm, upper teeth 1–1.5 mm, lower teeth 2–2.5 mm, with dense minute glandular hairs. Corolla 10–12 mm, pinkish-pale purple. Nutlets oblong, ± 1.5 mm. Fig. 164, 1–2.

HAB. Limestone rocks, rocky mountain slope, shady cliffs, rocks near water fall, open *Quercus* forest; alt. 500–880 m; fl. Jul.-Sep.
DISTRIB. Occasional in N.E. Sector in the lower forest zone. **MRO**: Rowanduz Gorge, *Gillett* 9444! (BAG); Rowanduz Gorge, *G. Throham* 15909! (BAG); Bakhal water fall nr. Rowanduz, *Al-Dabbagh & K. Hamad* 46181! (BAG); Bakhal, *S. Omar, Sahira, F. Karim & H. Hamid* 38408! (BAG); Gali Ali Beg, *Rawi* 26795! (BAG)

JANTARA جنتارة (Kurd.-Rowanduz Gorge, *G. Throham* 15909).

S.E. Turkey, W. Iran

26. **ZIZIPHORA** L.

Sp. Pl. 1: 21 (1753)
Harley & al. in Kubitzki (ser. ed.), Fam. Gen. Vasc. Pl. 7: 242–243 (2004)
Zwingeria Heist. ex Fabr., Enum.: 59 (1759); *Faldermannia* Trautv., Bull. Cl. Phys.-Math. Acad. Imp. Sci. Saint-Pétersbourg, sér. 2, 6: 185 (1840)

Ali Haloob
Revised by Tuncay Dirmenci & Ferhat Celep

Annual or perennial herbs or subshrubs, strongly aromatic. Stems ascending to erect or rarely prostrate. Leaves ±entire, subsessile to shortly petiolate, ovate, oblong to lanceolate, elliptic or rarely orbicular, veins distinct at beneath. Verticillasters distant or crowded, subcapitate, capitate or spicate, 2–12-flowered, cymes sessile or pedunculate in axils of leaf-like bracts; flowers shortly pedicellate, pedicels sometimes flattened. Bracts patent or imbricate, conspicuous, similar to cauline leaves or not; bracteoles inconspicuous. Calyx ±actinomorphic, 5-lobed (±3/2), weakly bilabiate, lobes triangular, posterior lobes shorter than straight or curved anterior lobes, calyx-tube cylindrical, with converging teeth, sometimes curved, gibbous below, 13-nerved, accrescent, throat bearded, closed. Corolla 2-lipped, ±4-lobed (1/3), white, rose-red or purple, posterior lip rounded, sometimes emarginate, straight, anterior lobes ±rounded, corolla tube cylindrical, short, sometimes scarcely exserted from calyx, exannulate. Often gynodioecious; fertile stamens 2 (posterior pair staminode or absent), anterior pair ascending under posterior corolla-lip, exserted from tube, anthers often cohering at margins, with fertile theca ellipsoidal, divaricate, lower theca sterile and forming a small appendage. Stigma-lobes unequal; disc symmetrical.

Fig. 164. **Cyclotrichium leucotrichum**. 1, habit × ½; 2, flower × 2. Reproduced with permission from Jamzad, Flora of Iran 76: f. 237 (2012). © Ministry of Jihad-e-Agriculture.

Nutlets 1.2–2 × 0.5–0.9 mm, oblong, rarely obovoid-oblong and trigonous, obtuse or obtuse-rounded at apex. smooth or granulate, areole clearly bilobed and white coloured.

17 species (29 taxa) in S. Europe, N.W. Africa and Asia up to the Himalayas and Altai mountains; adapted to open areas and generally xeric habitats; 4 species in Iraq.

Species in *Ziziphora* are strongly aromatic and are used as a source of herbal teas, medicinals, flavouring, spice and as essential oils.

Hedge, I.C. (1961). Some remarks on the perennial species of *Ziziphora* with particular reference to those in Turkey. Notes R.B.G. Edinb. 23: 209–221.
Kaya, A. & Dirmenci, T. (2012). Nutlet morphology of Turkish *Ziziphora* L. (Lamiaceae). Pl. Biosystems 146(3): 560–563.
Kaya, A., Satil, F., Dirmenci, T. & Selvi, S. (2013). Trichome micromorphology in Turkish species of *Ziziphora* (Lamiaceae). Nordic J. Bot. 31: 270–277.
Rechinger, K.H. (1951). Die ausdauernden *Ziziphora*-Aiten des Iranischen Hochlandes und seiner Nachbargebiete. Phyton (Horn) 3: 161–172.
Selvi, S. (2011). Morphological and anatomical studies on *Ziziphora* L. Species in Turkey. Balikesir University Institute of Science (Ph.D. Thesis-in Turkish, unplulished).
Selvi, S., Satil, F., Martin, E., Celenk, S. & Dirmenci, T. (2015). Some evidence for infrageneric classification in *Ziziphora* L. (Lamiaceae: Mentheae). Pl. Biosystems 149(2): 415–423.

1. Perennial herbs or suffruticose; inflorescence always capitate . . . . . . .1. *Z. clinopodioides*
   Annual; inflorescence spicate, subcapitate or capitate 2
2. Inflorescence capitate, partially enveloped by broad ovate bracts, bracts
   longer or shorter than terminal heads, abruptly acuminate. . . . . .2. *Z. capitata*
   Inflorescence spicate or subcapitate; bracts narrow, not enveloping
   heads, generally longer than verticillasters, abruptly or gradually attenuate 3
3. Leaves linear to linear-lanceolate; inflorescence an oblong terminal
   spike; calyx 6–10 mm; corolla 8–16 mm, c. 2 × calyx. . . . . .3. *Z. taurica*
   Leaves linear; inflorescence ± narrowly oblong, spicate; calyx 5–8 mm;
   corolla 8–11 mm, not more than 1.5 × calyx . . . . .4. *Z. tenuior*

1. **Ziziphora clinopodioides** *Lam.*, Tabl. Encycl. 1: 63 (1791); Dinsmore in Post, Fl. Syria, Palest. & Sinai ed. 2, 2: 346 (1933); Rawi in Dep. Agr. Tech. Bull 14: 159 (1964); Rechinger, Fl. Iranica [K. H. Rechinger] 150: 481 (1982); Davis, Fl. Turkey 7: 396 (1982); Jamzad in Flora Iran [Assadi & al., in Persian] 76: 637 (2012).

Suffruticose, often mat-forming perennial. Stems prostrate to erect, usually much branched from base, 5–50 cm tall, glabrous to hirsute, sometimes pilose above. Leaves very variable in size and shape, 4–15 × 2–6 mm, linear-lanceolate to ovate-oblong, rarely suborbicular to orbicular, petiolate to 2.5 mm long, attenuate, cuneate or rounded at base, acute to acuminate at apex, entire or slightly serrulate, glabrous to ± pilose or velutinous with sessile glands. Inflorescence a dense terminal head, verticillasters 8–17-flowered. Bracts usually broader than leaves, 3–13 × 1.8–8.5 mm, ovate to ovate-lanceolate or ovate-elliptic or spatulate, sessile, entire, densely pilose or velutinose at margins, acute to acuminate at apex. Bracteoles linear-lanceolate or subulate, 1–2 mm. Calyx (3–)4–6(–8) mm, weakly bilabiate, green or partly purple, 13-veined, glabrous to densely hirsute-villous, with sessile glands, hairy in throat; teeth triangular-subulate, 1–1.5 mm. Corolla 7–12 mm, purplish pink (rarely white), tube included in calyx (in females) or shortly exserted (in hermaphrodites); corolla bilabiate, upper lip emarginate; lower lip 3-lobed, median lobe truncate, lateral lobes rounded. Fertile stamens 2, as long as upper lip or slightly longer; filaments 3–5 mm, glabrous, appendage absent. Style unequally bifid, 5–11 mm, included in corolla or longer. Nutlets 1.2–1.8 × 0.5–0.8 mm, obovoid-oblong, slightly papillate, brown.

General distribution of species: Turkey (except for western Turkey) to W. Himalaya.

Ziziphora clinopodioides has a wide distribution from Turkey (except for western Turkey) to Afghanistan and shows a wide variation of morphological characters; 4 subspecies are recognized for Iraq:

1. Plant suffruticose; stems to 50 cm tall, erect; leaves narrowly lanceolate b. subsp. *rigida*
   Plants herbaceous to woody at base; stems to 15 cm tall, procumbent to
   arcuate-ascending; leaves ovate-lanceolate to orbicular 2

2. Stems, leaves and calyx purplish; flowering stems retrorsely pubescent .
  c. subsp. *elbursensis*
  Stems, leaves and calyx green; flowering stems spreading hirsute. 3
3. Calyx 5–6 mm long; stamens exserted from corolla throat . . . . .a. subsp. *kurdica*
  Calyx 4–4.5 mm long; stamens included in corolla throat . . . . . . .d. subsp. *brevistamina*

a. subsp. **kurdica** (*Rech.f.*) *Rech.f.*, Fl. Iranica 150: 487 (1982). Type [Iraq] Ad rupes in mediis regionibus m. Gara Kurdist. 25.07.1841, *Kotschy* 326! (W, holo.; G, K, iso.).

Ziziphora kurdica Rech.f., Phyton (Horn) 3: 167 (1951).

Flowering stems to 20 cm tall, decumbent to erect, branched from base, pubescent to spreading hirsute-villous. Median leaves elliptic-lanceolate to ovate-lanceolate, or ovate, attenuate-cuneate at base, rarely ± rounded, acuminate at apex, light green, veins not prominent beneath, spreading hirsute. Calyx 5–6 mm long, sparsely to densely hirsute-villous, whitish-green.

HAB. Rocky slopes and screes, mixed shrubland on mountain; alt. 1500–3400 m; fl & fr. Jul.-Sep.
DISTRIB. Rare in the upper forest and lower thorn-cushion zones, and in the lower forest zone of Iraq. MAM: Zawita Mts., on the Turkish border, between the pass & the Western peak, in Mosul Liwa, *Rechinger* 18504! (BUH); Har Rost, *Haley* 123! (BAG); Mosul, in ditione pagi Sharanish, in montibus calc. a Zakho septentrionem versus, Zawita, *Rechinger* 10980-A! (G, W), 10980-B! (G, W). MRO: Chia-i-Mandali, *Guest & Ludlow-Hewitt* 2679! (BAG); Riwandous, in montes Sakri-Sakran, *Bornmuller* 1704 (W); Newrobar valley-Qandil range, 1/8/1957, *Al-Rawi & Serhang*, 24145 (BAG); Erbil, montes Qandil ad confines Persiae, supra Pushtashan, *Rechinger* 11086! (W). MSU: Suleymaniyah. Pira Magrun, W. of Suleymaniyah, *Thesiger* 1110; Ibid, *Poore* 593!; In montis calcaresi Avroman et Schahu, *Haussknecht* s.n.! (W); In montibus calcareis Avroman et Schahu, *Haussknecht* s.n.!; Avroman mountain, *Rawi et al.* 19768!

E. Turkey, N.W. Iran.

b. subsp. **rigida** (*Boiss.*) *Rech.f.*, Fl. Iranica 150: 483 (1982).

Ziziphora clinopodioides Lam. var. *rigida* Boiss., Fl. Orient. [Boissier] 4: 586 (1879).
Z. rigida (Boiss.) Heinr.Braun, Verh. K. K. Zool.-Bot. Ges. Wien 39: 239 (1889).
Z. fasciculata K.Koch ex Boiss., Fl. Orient. [Boissier] 4: 586 (1879).
Z. fasciculata K.Koch ex Rech.f., Phyton (Horn) 3: 168 (1951).

Flowering stems erect, simple or branched, thin and long to 50 cm tall, sparsely short pubescent to villous or glabrescent. Leaves narrowly lanceolate, subglabrous to pubescent, flat, attenuate at base, acute to acuminate at apex. Flowers short pedicellate, verticillasters forming a densely subglobose capitate head. Calyx (3–)4–5 mm, whitish green, pubescent to villous, sometimes glabrescent.

HAB. Open dry rocky slopes, stony hillsides, calcareous, *Rhus-Quercus* forest on hillside; alt. 1300–2300 m; fl. & fr. Jul.-Aug.
DISTRIB. Region of Sulaymaniyah province close to the Iranian border in Iraq. MSU: Penjwin, *Guest,* 12960! (BAG), 12961! (BAG); Mola Kowa (on Sulaimaniya-Penjwin highway), *Rawi* 22484! (BAG); Penjwin, open dry rocky slopes, *Haines* 2040! (E); Pass, 7 km before reaching Penjwin, in Sulaimaniya Liwa, *Rechinger* 18506! (BUH); 1.5 km N.W. Basneh village, open area with clay soil, *Haloob, Ikhlas & Hamshkan* 60240 (BAG); in ditione pagi Penjwin, in glareosis jugi Malakawa, substr. Serpentine, *Rechinger* 10413! (G, W); Pengwin, *Haines,* 18508! (BUH); Penjwin, *Alhashimi* 18503 (BUH); in montibus calcareis Avroman et Schahu, *Haussknecht* s.n. (W).

Afghanistan, Iran, Transcaucasus, Turkmenistan.

c. subsp. **elbursensis** (*Rech.f.*) *Rech.f.*, Fl. Iranica 150: 487 (1982).

Ziziphora elbursensis Rech.f., Phyton (Horn) 3: 169 (1951).

Flowering stems to 10 cm tall, arcuate-ascending, green to purplish, pubescent. Leaves broadly ovate to elliptic, or orbicular, rounded at base, often acuminate-apiculate at apex, sparsely papillose-puberulent to glabrescent, veins prominent beneath. Calyx 5–6 mm, often purplish above half of the tube, sparsely short hirtellous and glandular punctate.

HAB. High alpine steppes, calcareous: alt. 3000–3800 m; fl. & fr. Aug.
DISTRIB. Occasional in north and central sector of the alpine and thorn-cushion zones of Iraq. MAM: Galli Zawita, N.E. of Zakho nr. Turkish border, *Rawi* 23621 (BAG). MRO: Arl Gird Dagh, *Guest* 3058 (BAG); Qandil range, *Rawi & Serhang* 24461 (BAG); N.E. of Qandil, *Rawi & Serhang* 24419A (BAG);

Qandil range, *Rawi & Serhang* 24505 (BAG); Kandil range (N.E. of Rania), *Rawi & Serhang* 18284 (BAG); Qandil range, in rocky cliff, *Serhang & Rawi* 26735 & 26736 (BAG); Newrabor valley-Qandil range, *Rawi & Serhang* 24145 (BAG); Helgurd Peak, Erbil Liwa, *Rechinger* 18518 (BUH); Erbil, Mons Helgurd ad confinel Persiae, in declivibus occiedentalis summi montis, *Rechinger* 11424 (E, G).

E. Turkey, W. and N. Iran.

Subsp. *elbursensis* is adapted to the high mountain steppes. It generally grows above 3000 m height. It is distinguished from other subspecies by its low and reddish stems, purple calyx, broadly ovate to orbicular and dark green leaves.

d. subsp. **brevistamina** *Haloob* in Phytotaxa 592:282 (2023). Type: Iraq. S. Pir Omar Gudrun, 6/6/1948, *A. Rawi,* 12031 (BAG, holo.); Pir Omar Gudrun, 7/6/1948, *Rawi* 12087 (BAG, syn.).

Stem ascending or decumbent, 15–20 cm tall, branched from base and in upper half, dense white eglandular pubescent-villous hairs. Leaves ovate-suborbicular or lanceolate, 4–10 × 4–7 mm, apex acute or obtuse, base rounded or cuneate, dense white eglandular pubescent hairs and sessile yellow glands; petiole ±1 mm. Bracts ovate-lanceolate, 3–7 × 1–4 mm, petiolate. Pedicel c. 1 mm. Calyx 4–4.5 mm, green-whitish, densely pubescent-hirsute with sessile yellow glands outside. Corolla c. 7 mm, white-yellowish with purple dots on lower lip (in dry material), with sessile yellow glands on outside of lips. Stamens not exserted from corolla throat; filaments 0.3–0.5 mm long, not coherent.

HAB. Mountain side; alt. 1800–2000 m; fl. & fr. Jun.-Jul.
DISTRIB. Very rare found only twice in one location in southern sector of thorn-cushion zone of Iraq.
MSU: S. Pir Omar Gudrun, *Rawi* 12031! (type, BAG); Pir Omar Gudrun, *Rawi* 12087 (type, BAG).

GIAPAPULA or NA'NA' (Kurd., *Rawi* 12031 & 12087).

Endemic.

2. **Ziziphora capitata** *L.*, Sp. Pl. 1:21 (1753); Rawi in Dep. Agr. Tech. Bull. 14: 159 (1964); Dinsmore in Post, Fl. Syria, Palest. & Sinai ed. 2, 2: 346 (1933); Feinbrun-Dothan, Fl. Palaest. 3: 146 (1978); Edmondson in Fl. Turkey [P. H. Davis] 7: 397 (1982); Rechinger, Fl. Iranica [K. H. Rechinger] 150: 489 (1982); Jamzad in Flora Iran [Assadi & al., in Persian] 76: 645 (2012); Taifour & El-Oqlah, Pl. Jordan Annot. Checkl.: 107 (2017).

Ziziphora clinopodioides Lam. var. *integrifolia* K.Koch, Linnaea 17: 293 (1844).
Z. alpina Mill., Gard. Dict. ed. 8.n.4 (1768).
Z. compacta Friv., Flora 18(1): 336 (1835).
Z. glomerata Friv. ex Ledeb., Fl. Ross. 3(1,9): 370 (1849).
Z. capitellata Juz. in V.L.Komarov, Fl. URSS 21: 671 (1954).

Annual herb. Stems 5–25 cm tall, erect, simple to much branched from below, hispid, retrorsely puberulent. Leaves 6–30 × 1.5–14 mm, linear-lanceolate to elliptic below, ovate-acuminate above, uppermost broadly ovate, shortly petiolate, glabrous to hispidulous with glandular papillae on the upper surface, scabrid at veins on the lower surface, attenuate at base, acute to acuminate at apex, margins entire. Inflorescence a globose terminal head, 12–20 × 10–16 mm, partly enclosed by broadly ovate to orbicular bracts. Bracts 6–25 × 4–20 mm, shorter or longer than heads, broadly ovate to orbicular or ovate-lanceolate, shortly hispid and glandular-papillose on both surfaces or pilose on upper surface, acuminate at apex, margins ciliate, prominently veined. Bracteoles 1–2 mm, subulate. Calyx 6–11 mm, cylindric, prominently 13-veined, pilose to hispid with sessile glands and glandular papillae, hairy in throat, sub-bilabiate; teeth 1.5–2.5 mm long, triangular-lanceolate, margins scabrid to pilose. Corolla 8–13 mm, violet, purple or purplish-pink (rarely pink or white), exserted from the calyx, hispidulous to pilose with sessile glands and glandular-papillose. Fertile stamens 2, included in or slightly exserted from corolla, filaments 0.5–2 mm, glabrous; anthers not appendiculate. Style 5–11 mm, glabrous, included in or slightly exserted from corolla. Nutlet 1.6–2 mm × 0.6–0.8 mm, oblong-ovoid, brown to pale brown, slightly papillate.

Distrib. of genus. Mediterranean basin to Europe, Caucasia, and Iran.

Two subsp. are recognized for Iraq:

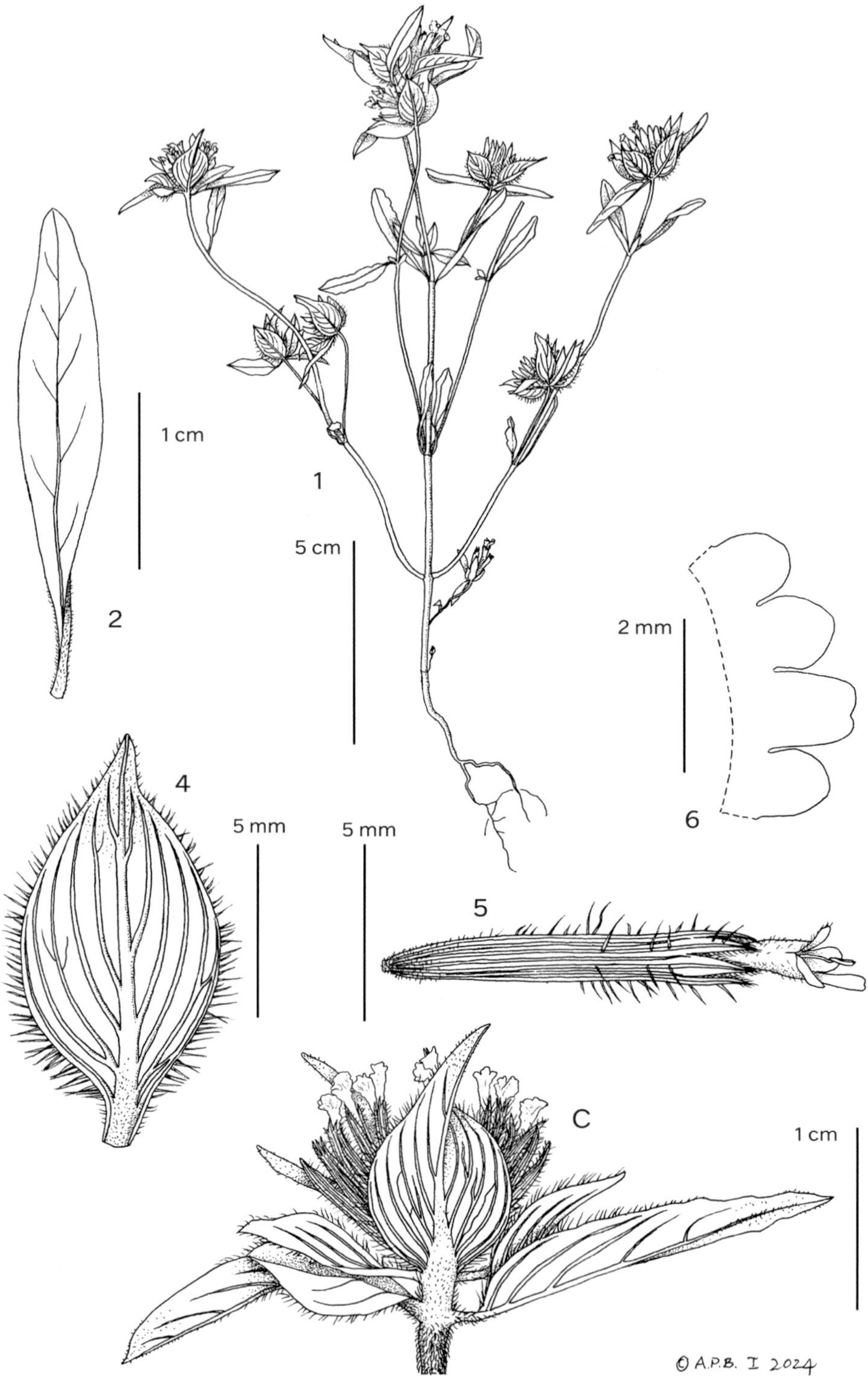

Fig. 165. **Ziziphora capitata** subsp. **capitata**. 1, habit; 2, leaf adaxial surface; 3, inflorescence; 4, abaxial surface of basal bract of inflorescence; 5, flower, side view; 6, dorsal and ventral lobes of corolla. 1–6 from *Guest* 13248. Drawn by © A.P. Brown, Feb. 2024.

Floral leaves clearly longer to ±2 × capitula, lanceolate to oblong, long
   acuminate at apex . . . . a. subsp. *capitata*
Floral leaves as long as or shorter than capitula, suborbicular to broadly
   ovate, short acuminate. . . . . b. subsp. *orientalis*

a. subsp. **capitata**

Leaves lanceolate-elliptic, 6–25 × 1.5–7 mm. Bracts ovate-lanceolate, 7–20 × 4–12 mm, clearly longer than heads, apex acuminate-attenuate. Fig. 165, 1–6.

HAB. Roadside, loam soil, fields, rocky hillside, dark brownish hillside, clay plain, in wheat field; alt. 600–2200 m; fl. Mar.-May.

DISTRIB. Occasional in the forest zone in Iraq, rare in N. sector of moist steppe zone. **MAM**: Bakerma, *Rawi* 8519-B (BAG); Zakho, *Rawi* 8621 (BAG); S.W. of Aqra, *Anders* 36144 (BAG); Sarsang to Amadiya, 23/5/1973, *F. Karim, Hamid & Jasim* 40987 (BAG); between Zakho & Turkish border, *Karim, Al-Dabbagh & Hamad* 44783 (BAG); S.W. Aqrah, *Anders* 1275 (W); Shaikh Adi N.W. Ayn Sifni, *Thesiger* 688; Hinnis 10 km N. Ayn Sifni, *Thesiger* 641. **MRO**: 2 km N. of Chmarock, N. of Haibat Sultan Dagh, *Rawi, Nuri & kass* 28847 (BAG); 6 Km S.W. of Rania, *Rawi, Nuri & Kass* 28435 (BAG); Shaqlawah, *Haines* 665 (E). **MSU**: Suleymaniya, inter Suleymanita et Dokan, A bifurcationae viae Suleymaniya-Dokan ca. 4 km bor.-occid. Versus, in collibus siccis, *Rechinger* 12479 (W); Halabcha, *Rawi* 8874 (BAG). **FUJ**: Mosul, Dorud Sefiddasht, *Riazi* 9765 (W). **FNI/ MAM**: Mosul, *Lazar* 3323 (BAG); Mosul-Zakho, *Guest* 13248 (BAG); 20 km S. of Zakho, *Al-Kaisi et al* 43318 (BAG).

ZOFAH زوفة (Mosul, *Y. Lazar*, 3323).

Balkans, Russia, Caucasia, Cyprus, Syria, Lebanon, Turkey, Iran.

b. subsp. **orientalis** *Rech.f.*, Ark. Bot., a.s., 2,1: 535 (1951).

Leaves ovate-lanceolate to broadly elliptic, 12–30 × 5–14 mm. Bracts broadly ovate to orbicular, 9–25 × 7–20 mm, apex acute-acuminate, as long as or shorter than flowering heads. Fig. 166, 1–3.

HAB. On mountains limestones slopes, near top of mountain slope with *Quercus*, dry open places, cultivated valley steppes, clay soil on hillside, stony hillside, clay loam soil; alt. 600–2400; fl. & fr. Mar.-Jul.

DISTRIB. Very common in the forest and lower thorn-cushion zones of Iraq, and rare in moist steppe zone. **MAM**: Mosul, In jugo inter Dagh al Radzjiem et Sharanish, in saxosis calc., *Rechinger* 12111 (W); Mosul, in ditione pagi Sharanish, in montibus calc. a Zakho septentrionem versus, Jabal Khantur, *Rechinger* 12076 (W); Mosul, inter Dohuk et Amadiya, supra Sawara Tuka, *Rechinger* 11970 (W); Mosul, 5 km a Zakho, *Rechinger* 12124 (W); Zakho, *Guest* 2277 (BAG); Dohuk, *Lazar* 3324 (BAG); Betas near Zakho, *Rawi* 8469 (BAG); Zawita, *Chapman* 9331 (BAG); between Oqur between Swaratuka & Screen, Al-Kas 18667 (BAG); Sarsing (near water spring), *Al-Kas* 18687 (BAG); Zawita mountain between Rania & Shaqlawa, *Karim, Hamid & Jasim* 40801 (BAG); Aqra, *Rawi* 11416 (BAG); Bikhair Mt. nr. Zakho (90 km N.W. of Mosul), *Rawi* 23073 (BAG); Khantur Mt. N.E. of Zakho, *Rawi* 23409 (BAG); Zakho to Sharanish, *Karim, Hamid & Jasim* 41100 (BAG); 30 km N. Mosul, *Al-Dabbagh* 46763 (BAG). **MRO**: M. Algurd Dagh E. Rost, *Thesiger* 1003; Montes Qandil ad confines Persiae, Pushtashan, *Rechinger* 11221 (W!). Salahaddin, Abu Ghraib, *Barkley* 5683 (W); Sefin Dagh above Shaqlawa, *Gillett* 8083 (BAG); Pushtashan N.E. of Rania, *Rawi* 23881 (BAG); Between Sei-Waka & Darsala villages E. Karoukh, *Al-Kas, Nuri & Sarhank* (BAG); Haibat Sultan Dagh mountain N. of Koi Sanjaq, *Rawi, Nuri & Kass* 28186 (BAG); Gali Warta c. 30 km N.W. by N. of Rania, *Rawi, Nuri & Kass* 28844 (BAG); Salah al-Din, *Rawi* 36210 (BAG); Kuh Sefin prope Shaqlawa, *Bornmüller* 1754; 3 km S.W. Salahad-Din, *Rechinger* 15565; In valle Rawandiz, *Thesiger* 843. **MSU**: 6 km E. of Qaranjir, Kirkuk-Chemchemal road, *Gillett & Rawi* 7563 (BAG); Guaiga, *Rawi* 8879 B (BAG); nr. Pir Omar Gudrun, *Rawi* 11537 (BAG); 56 km N.W. of Sulaimaniya, *Rawi* 21809 (BAG); Tawela, *Rawi* 21870 (BAG); Dara Tri (on the road between Halabja & Tawaila), *Rawi* 22003 (BAG); c. 10 km W. of Tawela (on road between Halabja & Tawela), *Rawi* 22130 (BAG); Kamarapa (on road between Halabja & Tawela), *Rawi* 22272 (BAG); 7 km W. of Tawela (Balcha village), *Rawi* 22385 (BAG); Mela Kowa (on Sulaimaniya-Penjwin highway), *Rawi* 22475 (BAG); Penjwin, *Rawi* 22606 (BAG); Suleymaniya: In ditione pagi Penjwin, Malakawa, *Rechinger* 12293 (W); Suleymaniya, montes Avroman ad confines Persiae, in ditione pagi Tawilla, *Rechinger* 12442! (E); Abu-Abidah near Halabaja, *Omar, Karim, Hamza & Hamid* 37456 (BAG); Bakrajo, *Omar & Karim* 37999 (BAG). **MJS**: Jabal Sinjar, *Al-Kaisi & K. Hamad* 49181 (BAG); Jabal Sinjar, *Omar, Al-Khayat & Al-Kaisi* 52427 (BAG); Top of Sinjar, *Omar, Al-Khayat & Al-Kaisi* 52521 (BAG); Jabal Sinjar S.E. hills, road to Karsi, *Widad & Al-Khayat* 53558 (BAG); Sinjar, *Anders* 746 (W). Sinjar, jabal Sinjar, *Anders* 2055 (W). **FAR**: 4 km S. of Taktak bridge on road to Kirkuk, *Rawi, Nuri & Kass* 28074 (BAG). **FNI**: nr. Ain Sifni, *Salim Effendi* 2589 (BAG).

KIA'AI BI ZHINH (Kurd., *Salim Effendi* 2589), GUL ZANIANA (Kurd., *Rawi* 8879); KLAOB SHILA (Kurd. *Rawi* 11537)

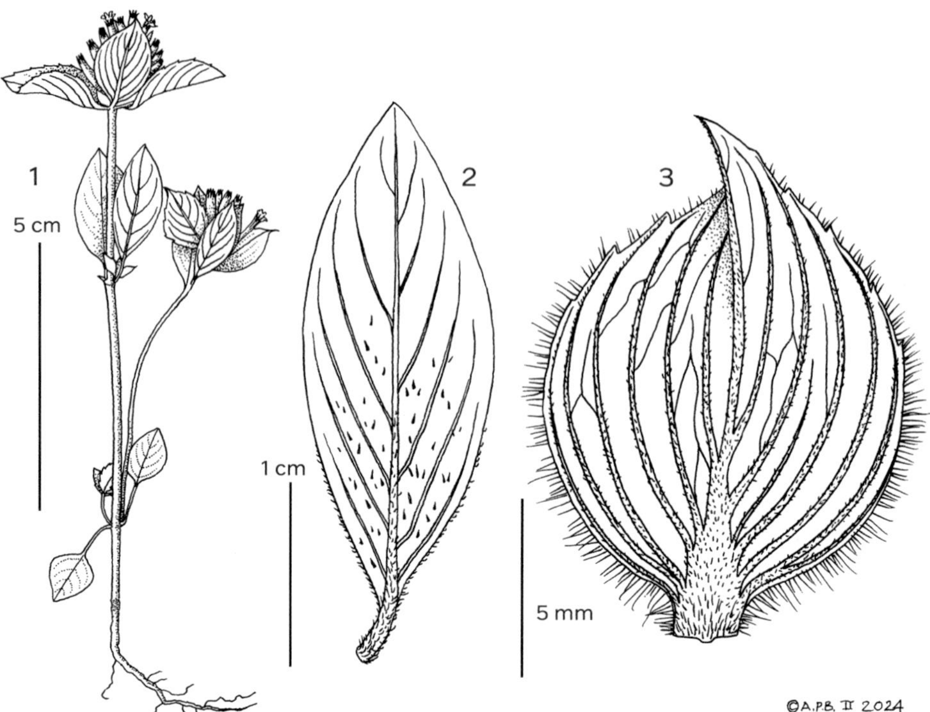

Fig. 166. **Ziziphora capitata** subsp. **orientalis**. 1, habit; 2, leaf below inflorescence, adaxial surface; 3, abaxial surface of basal bract of inflorescence. 1–3 from *Barkley* 8005. Drawn by © A.P. Brown, Feb. 2024.

Iran, Lebanon, Syria, Turkey, Turkmenistan.

3. **Ziziphora taurica** *M.Bieb.*, Fl. Taur.-Caucas. 1: 414 (1808); Dinsmore in Post, Fl. Syria, Palest. & Sinai ed. 2, 2: 347 (1933); Edmondson in Fl. Turkey [P. H. Davis] 7: 399 (1982).

subsp. **taurica**

Dracocephalum cuniloides Pall., Reise Statth. Russ. Reich. 2: 221 (1773).
D. odoratissimum Poir. in J.B.A.M.de Lamarck, Encycl., Suppl. 2: 521 (1812).
Ziziphora lanceolata Boiss., Fl. Orient. [Boissier] 4(2): 588 (1879).
Z. alpina Freyn, Bull. Herb. Boissier 4: 134 (1896), nom. illeg.
Z. abd-el-asisii Hand.-Mazz., Ann. Nat. Hofmus. Wien. 27: 418 (1913). Type: Prope vici ruina Gharra in medio pedis septentrionalis montium Dschebel Abd el Asis, *Handel-Mazzetti* 1764 W, holo; WU, iso.).

Annual herbs. Stems 5–10 cm tall, simple or branched, retrorsely puberulent to shortly hispid. Leaves linear to linear-lanceolate, ovate-lanceolate or elliptic, 7–20 × 1.3–8 mm, puberulent to pilose with sessile glands, sessile to subsessile, attenuate at base, acute to acuminate at apex, entire, prominently nerved. Inflorescence a dense ± elongated oblong terminal spike, 1–5 × c. 1 cm. Bracts usually similar to leaves, 5–25 × 3–7 mm, as long as or longer than verticillasters, linear to linear-lanceolate, puberulous, densely glandular-papillose with sessile glands, scabrid on veins on lower surface. Bracteoles 0.8–2 mm, linear-subulate. Calyx (6–)7–10 mm, tubular, hispid to pilose with glandular papillae and sessile glands, slightly oblique at mouth, scarcely curved at fruiting time; teeth triangular-lanceolate to lanceolate, 1–1.5 mm long, scabrid to hispid at margins. Corolla (8–)10–14(–16) mm, c. 2 × calyx, reddish purple or pinkish, tube included in calyx, limp expanded to 2 × calyx, puberulous to pilose with sessile glands or glandular-papillose. Fertile stamens 2; filaments 0.5–2.5 mm, glabrous; anthers appendiculate, included in corolla. Style 8–13 mm long, slightly exserted from or included in corolla. Nutlets 1.5–1.8 × 0.6–0.8 mm, oblong, brown, slightly papillate.

HAB. Steppe, rocky slopes, waste places; alt. 200–900 m; fl. & fr. Apr.-Jun.

DISTRIB. Very rare in the desert region of Iraq. **DLJ**: 15 km E. of Rawa, rocky clay valley, *Omar, Al-Kaisi, K Hamad & H. Hamid*, 45132!; 5 km S.W. of Ana, *Omar, Al-Kaisi, K. Hamad & H. Hamid* 44998! **DWD**: Prope vici ruina Gharra in medio pedis septentrionalis montium Dschebel Abd el Asis, *Handel-Mazzetti* 1764 (type of *Z. abdel-asisii*) (W, WU).

4. **Ziziphora tenuior** *L.* Sp. Pl. 1: 21 (1753); Rawi in Dep. Agr. Tech. Bull. 14: 159 (1964); Dinsmore in Post, Fl. Syria, Palest. & Sinai ed. 2, 2: 347 (1933); Feinbrun-Dothan, Fl. Palaest. 3: 146 (1978); Edmondson in Fl. Turkey [P. H. Davis] 7: 398 (1982); Rechinger, Fl. Iranica [K. H. Rechinger] 150: 490 (1982); Jamzad in Flora Iran [Assadi & al., in Persian] 76: 651 (2012).

> *Faldermannia tenuior* (L.) Ter-Chatsch., Zametki Sist. Geogr. Rast. 17: 73 (1953).
> *Ziziphora spicata* Lag. & Rodr., Anales Ci. Nat. 4: 259 (1801).
> *Z. serpyllacea* Ten., Syll. Pl. Fl. Neapol.: 16 (1831), nom. illeg.
> *Z. acutifolia* Montbret & Aucher ex Benth., Ann. Sci. Nat., Bot., sér. 2, 6: 43 (1836).
> *Faldermannia parviflora* Trautv., Bull. Cl. Phys.-Math. Acad. Imp. Sci. Saint-Pétersbourg, sér. 2, 7: 22 (1840).
> *Ziziphora parviflora* (Trautv.) Des.-Shost., Bot. Mater. Gerb. Bot. Inst. Komarova Akad. Nauk S.S.S.R. 8: 156 (1940).

Annual herbs. Stems 5–20(–30 cm) tall, simple to much-branched, shortly hispid, retrorsely pubescent. Leaves 7–25 × 2–6 mm, linear to linear-lanceolate; lower leaves petiolate, to 7

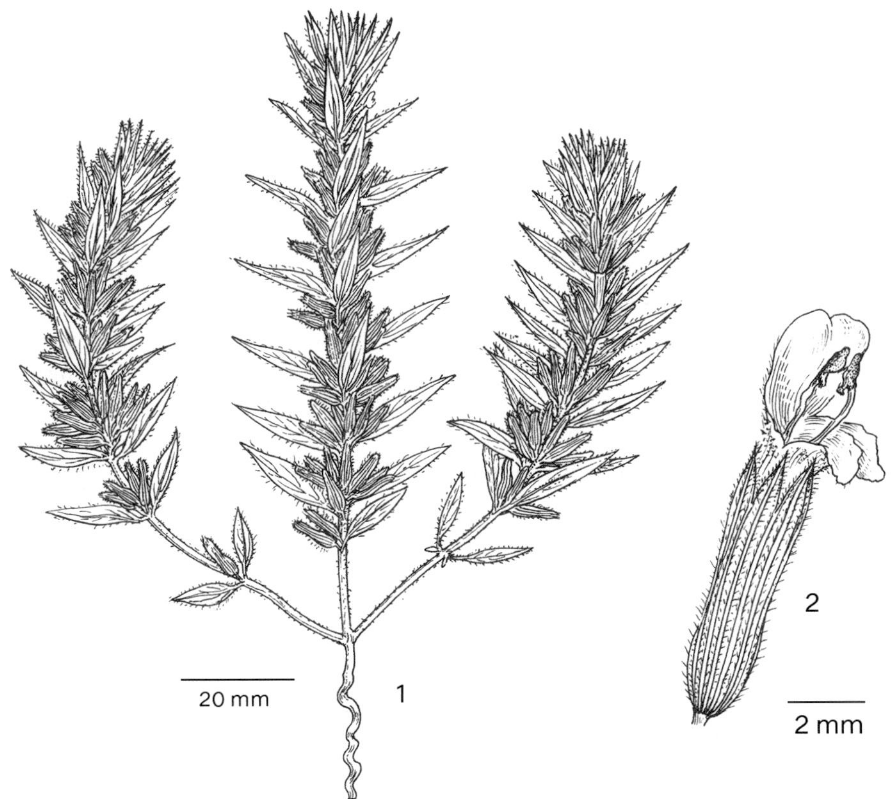

Fig. 167. **Ziziphora tenuior**. 1, habit; 2, flower. Reproduced with permission from Flora of Pakistan 192: f. 28, A–B, 1990. Drawn by M. Rafiq. © National Herbarium, Pakistan Agriculture Research Council & University of Karachi, Pakistan.

mm long, broadly lanceolate (upper leaves narrow lanceolate), attenuate at base; acute to acuminate at apex, hispid, ciliate at margins, prominently nerved. Inflorescence a dense ± elongate narrowly oblong terminal spike, 2–16 cm long, rarely less remote, pedicels short. Bracts equalling or longer than leaves, linear to linear-lanceolate, scabrid to hispid with sessile glands and glandular papillate on lower surfaces, ciliate. Calyx 5–8 mm, narrowly cylindric, ± hirsute; teeth short, ovate-triangular, 1–1.5 mm long. Corolla 8–11 mm, pink or lilac, tube included in calyx or slightly exserted, with limb not very broadly expanded, extending at most to 1.5 × calyx. Fertile stamens 2; anthers appendiculate or rarely not. Style slightly exserted from or included in corolla. Nutlets 1.4–1.6 × 0.5-0.6 mm, oblong, pale brown, slightly papillate. Fig. 167, 1–2.

HAB. Steppe, rocky slopes, gypseous hills, fallow fields, on hillside, sandy and gravel banks; alt. 100–680 m; fl. & fr. Mar.-Jun.

DISTRIB. Occasional in the north and west sectors of the forest region, very common in north, central and west sectors of mixed shrubland and north west sector of the desert regions of Iraq. **MAM**: Dohuk, *Mekki Beg Sidqi* 3279 (BAG); Serdeian nr. Aqra, *Memanyan* 10787 (BAG); Mahad (nr. Shaikhan), *Salim Efendi* 2609! (BAG). **MSU**: Jarmo, Halbaek 552, 713; **MRO**: Diyala, Mandali, Noe 710; **MJS**: Rawah-Sinjar, 115 km S.W. of Sinjar, sandy soil, *Rawi et al.* 31988; Sinjar, *Rawi* 5737 (BAG). **FUJ**: Mosul, in ditione urbis Mossul, in agris ad austro-occid., *Hand.-Mazz.* 1297 (W); Mosul, Kalaat Schergat (assur veterum) ad Tigridem infra urbem Mossul, quatuor horas in cirsuitu substrato calcareo; ca., *Maresch* 96 (W); Jabal Makhul, *Guest* 13190 (BAG); Jazira (north of Sinjar), *Guest* 13319 (BAG; Jazira (South of Sinjar: near Ba'aj), *Guest* 13393 (BAG); 5 km, N. of Shaabani to Sinjar, *Al-Kaisi & Hamad* 48971 (BAG); 50 km S.W. of Hadar, *Chakravarty, Rawi, Khatib & Alizzi* 33058 (BAG); 5 km S. of Hadar, *Alizzi & Husain* 33862 (BAG); 100 km to Mosul from Beiji, *Omar, Al-Khayat & Al-Kaisi* 52364 (BAG). **FAR**: Erbil, *Al-Radhi* 3845!; Erbil, *Gillett* 7998 (BAG); **FNI**: Baskaya, *Rawi & Sh. Haddad* 25599 (BAG). **FKI**: Kirkuk, 9 km a Kirkuk versus Altun Köprü, inter segetes, *Erdtman* & *Goedemans* 15530 (W); Kifri, *Gillett & Rawi* 7421 (BAG); Jebel Hamrin on Baghdad-Kirkuk road, *Gillett & Rawi* 10243 (BAG); 10–15 km N. of Kirkuk, *Rawi* 21562 (BAG); 7 km N.E. of Kirkuk, *Rawi, Al-Kass & Nuri* 27903 (BAG); Tuz Khurmato, *Shermatov* 47574 (BAG); Rawah-Sinjar, 115 km S.W. of Sinjar, *Chakravarty, Rawi, Khatib & Alizzi* 31988! (BAG). **DLJ**: Tekret, *Omar, Al-Khayat & Al-Kaisi* 52354 (BAG); 5 km N. of Manayif, *Al-Kaisi & K. Hamad* 49007 (BAG); Um Al-maten 74 km N.W. the road between Beiji and Haditha, *Chakravarty, Rawi, Khatib & Alizzi* 32044! (BAG). **DWD**: Ramadi, ad Euphrate medium inter Abukemal et Ramadi, in desertis, prope Nahije, ca., *Hand.-Mazz.* 730 (W); Rawa, *Rawi* 5738 (BAG); 4 km N.W. of Rawa, *Gillett & Rawi* 7077 (BAG); Ras Chpap between H2 and T1, *Chakravarty, Rawi, Khatib & Alizzi* 31711 (BAG); 12 km N. T1 (road to Husabah), *Chakravarty, Rawi, Khatib & Alizzi* 31725! (BAG); 12 km W. of Rawa, *S. Omar, Al-Kaisi, K Hamad & H. Hamid* 44489 (BAG); 24 km N. of Rawa to Shaabani, *Al-Khayat & K. Hamad* 51802 (BAG); Ana–Wadi Al-Qaser, *Omar, Al-Kaisi, Hamad & Hamid* 44375-A (BAG); 10 km W. of Ana, 2/4/1976, *Omar, Al-Kaisi, Hamad & Hamid* 45028 (BAG); 13 km of K3, *Chakravarty, Rawi, Khatib & Alizzi* 32953 (BAG); Between Ana and Al-Qaium, *Hamid* 39176 (BAG); Abu-Amad, 110 km N.E. of Rawa, *Al-Khayat & Hamad* 51836 (BAG); 100 km N.E. of Haditha, *Al-Khayat & Hamad* 51839 (BAG).

Ziziphora tenuior and *Z. taurica* both are widely distributed species, with wide variations in their inflorescence, leaf shape and calyx and corolla sizes. Although inflorescence and the calyx/corolla ratio are used as distinguishing characters, these characters are not always consistent. Further studies on these two species are needed.

KHUZAMAH or KHUZEIMAH خزامة(*Gillett and Rawi* 7077), NA'NA' (*Guest* 13393), ZATTARA (Kurd. *Gillett* 7998).

Sinai, Palestine, Syria, Arabian Peninsula to Turkey, C. Asia, S. Russia, W. Siberia, Xinjiang.

27. **PRUNELLA** L.

Sp. Pl. 2: 600 (1753); Harley & al. in Kubitzki (ser. ed.), Fam. Gen. Vasc. Pl. 7: 249 (2004)

Ali Haloob

Perennial herbs. Leaves petiolate, simple, entire, lobed or pinnatifid-pinnatisect. Inflorescence of dense terminal spikes; verticillasters 6-flowered. Bracts sessile, different from cauline leaves, broadly ovate-orbicular; bracteoles reduced or absent. Calyx bilabiate, campanulate, 10-veined; upper lip flattened, 3-toothed, shorter than lower lip; lower lip 2-toothed, lanceolate, throat glabrous, ± closed in fruit. Corolla tube exserted slightly from calyx, throat with a ring of hairs; upper lip 2-lobed, erect, concave-galeate; lower lip 3-lobed, middle lobed the largest, denticulate. Stamens 4, curved under upper lip, included, anterior pair longer than posterior pair; filament 2-branched, one branch attached with anther, the other forming an appendage; anther 2-celled, divaricate. Style apex equally branched. Nutlets glabrous, apex rounded.

About seven species in Europe, N.W. Africa and Asia; 2 species in Iraq.

Plant glabrescent to eglanduler pilose; margins of leaves entire to obsoletely crenate, corolla 10–14 mm . . . . 1. *P. vulgaris*
Plant densely white eglanduler hirsute; margins of leaves lobed to pinnatifid; corolla 15–20 mm . . . . 2. *P. orientalis*

1. **Prunella vulgaris** *L.*, Sp. Pl. 2: 600 (1753); Blakelock in Kew Bull., 4: 546 (1950); Rawi in Dep. Agr. Iraq Tech. Bull. 14: 154 (1964); Ball in Fl. Europ. 3: 162 (1972); Edmondson in Fl. Turkey [P. H. Davis] 7: 295 (1982); Rechinger, Fl. Iranica [K. H. Rechinger] 150: 494 (1982); Meikle, Fl. Cyprus 2: 1326–1327 (1985); Hedge in Fl. Pakistan [Ali & Y. Nasir] 192: 133 (1990).

Brunella vulgaris Moench, Meth. 414 (1794); Boissier, Fl. Orient. 4: 691 (1879); Handel-Mazzetti, Ann. Naturh. Mus. Wien 27: 412 (1913); Nábělek in Publ. Fac. Sci. Univ. Masarky 35: 85 (1923); Zohary in Dep. Agr. Iraq Bull. 31: 125 (1950).

Perennial herb. Rhizome, simple or branched, erect or ascending-decumbent, 10–50 cm, glabrescent to eglandular pilose. Leaves petiolate, lanceolate or ovate-oblong, 20–70 × 10–30 mm, apex acute to obtuse, base cuneate or rounded or subcordate, margins entire to obsolete crenate, glabrescent to eglandular pilose; petiole 5–35 mm. Bracts white with green veins and margins, sessile, broad ovate-orbicular, ciliate, cuspidate. Inflorescence a terminal spike; verticillasters 6-flowered. Pedicel flattened, 1–2 mm. Calyx purple or green, 7–9 mm, lateral upper teeth less than 0.5 mm long, sinus between the lateral and middle tooth rounded; lower lip with 2-lanceolate teeth, 2–4 mm long. Corolla purplish-pink to purplish-blue, 10–14 mm, tube glabrous outside, throat hairy; upper lip 2-lobed concave-galeate, with a line of hairs on top; lower lip 3-lobed, the middle one largest, denticulate. Nutlet glabrous, 1.5–2 mm, ovoid-ellipsoid, brownish. Fig. 168, 1–4.

HAB. Valley in *Salix* & *Franxinus* forest, wet meadow by stream, near water spring, in orchard near water, clay soil under apple trees, shady rocky forest slopes, slope of mountain, rocky limestone, metamorphic rocks, *Quercus libani* forest by spring, under shade of *Juglans* by stream, on reddish muddy and clay soils; alt. 700–2250 m; fl. Jun.-Aug.

DISTRIB. Frequently in the lower forest zone of Iraq. **MAM**: Sulaf N. of Amadiya, *E. Weinert & A. Mousawi* 0028141! (BUH); Maya village 20 km W. Kani Masi, *H. Hamid & Fadhil* 45486! (BAG); Kani Masi, *Al-Kaisi* et al. 43899 (BAG); Bamarli near Sersang, *E. Weinert & A. Mousawi* 0029862! (BUH); Mazna between Diana and Mirgasour, *Agnew, Hadac, Haines & Kadir* 6279! (BUH); Daimka village 35 km N.E. Zakho, *Al-Kaisi, H. Hamid & K. Hamad* 45353! (BAG). **MRO**: Haji Omran, *Z. Chalabi* 3431! (BUH); *Al-Kaisi & K. Hamad* 43559 (BA); Merga dreija near Haji Omran, *Rawi* 9119! (BAG); east of Arl Gird Dagh, *Gillett* 9495! (BAG); Kani Khanjer Khan N.E. of Helgord, *Rawi & Serhang* 24708! (BAG); Gallalah, *S. Omar* 37650 (BAG); Sakran Mt 15 km S.E. Chuman, *Al-Dabbagh & K. Hamad* 46233! (BAG); Jindian, *Rawi* 9231 A! (BAG); Shaglawa, *S. Omer, Sahira, F. Karim & H. Hamid* 38299! (BAG); Pushtashan, 15 km N.E. of Rania, *Rawi & Serhang* 23856! (BAG). **MSU** : Penjwin, *Al-Hashimi* 0018013! (BUH); Biara, *I. Al-Shehbaz, M. Al-Khaikani & K. Ibrahem* 37637! (BUH); Zalam near Khormal, *Rawi, Husham & Nuri* 29498 (BAG)

GIARIGYAWA كيارىجياوا (Kurd., *Rawi* 9119).

Throughout most Europe, N. Africa and Asia.

2. **Prunella orientalis** *Bornm.* in Verh. Zool.-Bot. Ges. Wie'n 48:620 (1898); Edmondson in Fl. of Turkey [P. H. Davis] 7: 296 (1982).

Perennial herbs. Rhizome, simple or branched, erect or ascending-decumbent, 10–30 cm, densely white eglanduler hirsute. Leaves petiolate, lyrate-lanceolate, 30–60 × 10–25 mm, apex acute to obtuse, base cuneate-truncate, margins lobed to pinnatifid, densely white eglanduler hirsute; petiole 5–55 mm. Bracts white with conspicuous green veins, sessile, broadly ovate-orbicular, ciliate, cuspidate. Inflorescence a terminal spike, verticillasters 6-flowered. Pedicel flattened ±1 mm. Calyx green, 9–10 mm, lateral upper teeth 1 mm, sinus between the lateral and middle tooth notched; lower lip with 2 lanceolate teeth, 4–5 mm. Corolla pale purple (yellowish in dry specimens), 15–20 mm, tube glabrous outside

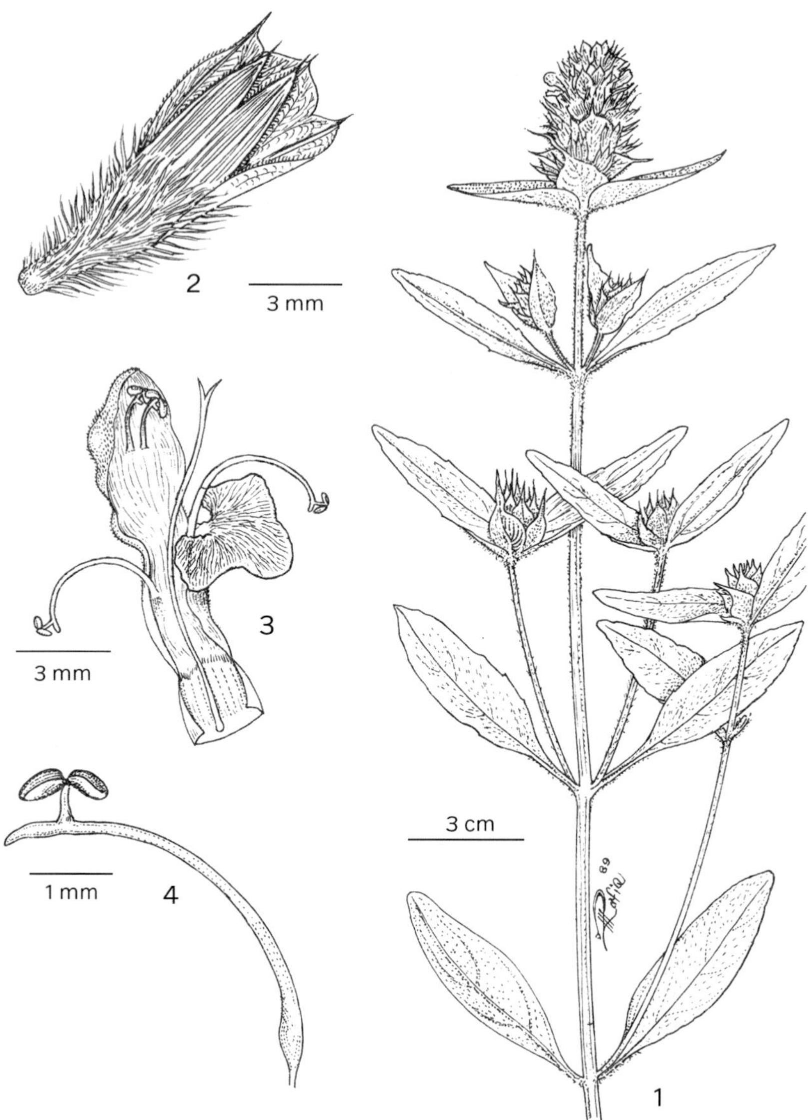

Fig. 168. **Prunella vulgaris**. 1, habit; 2, calyx; 3, corolla, opened, with stamens; 4, stamen. Reproduced with permission from Flora of Pakistan 192: f. 16, A–D, 1990. Drawn by M. Rafiq. © National Herbarium, Pakistan Agriculture Research Council & University of Karachi, Pakistan.

with hairy throat; upper lip 2-lobed, concave-galeate, with line of pilose hair on top; lower lip 3-lobed, the middle one largest, denticulate.

HAB. Clay rocky hillsides, rocky mountain slopes, in orchard, near water; alt. 950–1200 m; fl. Jun.-Aug. DISTR. Occasional in the N. sector of lower forest zone of Iraq. **MAM**: Kani Masi, *Al-Dabbagh & K. Hamad* 45565! (BAG); Deheek Cany Masy, *Al-Berwary* 0039528! (BUH); Sersang, *R. Wheeler Haines* 0018026! (BUH); Sarsang village, *Chapman* 26431! (BAG); 10 km S. of Sarsang, *Al-Kaisi, Al-Khayat & F. Karim* 51018! (BAG); Sarsang, *Al-Dabbagh* 46171 (BAG).

Lebanon, Syria, Turkey.

28. **NEPETA** L.

Sp. Pl. 2: 570 (1753); Harley & al. in Kubitzki (ser. ed.), Fam. Gen. Vasc. Pl. 7: 250 (2004) *Cataria* Adans., Fam. Pl. 2: 192 (1763); *Saccilabium* Rottb., Acta Lit. Univ. Hafn. 1: 294 (1778); *Saussuria* Moench, Methodus: 388 (1794); *Marmoritis* Benth., Bot. Misc. 3: 377 (1833); *Oxynepeta* Bunge, Mém. Acad. Imp. Sci. Saint Pétersbourg, Sér. 7, 21(1): 58 (1878); *Schizonepeta* (Benth.) Briq. in Engler & Prantl, Nat. Pflanzenfam. 4(3a): 235 (1896); *Phyllophyton* Kudô, Mem. Fac. Sci. Taihoku Imp. Univ. 2: 225 (1929); *Pseudolophanthus* Levin, Trudy Bot. Inst. Akad. Nauk S.S.S.R., Ser. 1, Fl. Sist. Vyssh. Rast. 5: 294 (1941)

Tuncay Dirmenci, Ali Haloob, Ferhat Celap & Shahina A. Ghazanfar

Perennial or annual herbs, aromatic; stems erect, ascending or procumbent; eglandular or glandular; hermaphrodite, gynodioecious or dioecious; leaves simple, entire or crenate to serrate, lowers ± petiolate, upper usually sessile; inflorescence thyrsoid, cymes lax to congested, pedunculate or sessile, axillary, distant or crowded into spike-like or ovoid heads; bracts leaf-like or reduced, sometimes longer than calyx; bracteoles equal to longer than or shorter than calyces; calyx tubular to campanulate, regularly 15-veined, actinomorphic to strongly 2-lipped, 5-lobed, upper lip tridentate, lower lip bidentate or rarely with lips entire, lobes subequal to unequal, throat straight to strongly oblique; corolla 2-lipped, 5-lobed (2/3), blue, violet, pink, yellow or white, posterior lip straight or curved, anterior lip with median lobe concave or ± flat, undulate or entire, tube included in or exserted from calyx, gradually or ± abruptly dilating upwards; stamens 4, rarely 2 with anterior pair absent, posterior (upper) pair longer than anterior (lower); anthers with widely divergent thecae, usually shorter than corolla; style shortly 2-lobed, subequal; nutlets ellipsoid to obovoid, rounded to acuminate, smooth to tuberculate, oblong to ±spherical, trigonous, glabrous or hairy at apex, mucilaginous or not.

A large genus of about 320 species distributed in Temperate Eurasia, tropical Africa, Arabian Peninsula to S.E. China. 19 taxa in Iraq, of which four are endemic.

Kaya, A. & Dirmenci, T. (2008). Nutlet Surface Micromorphology of the Genus *Nepeta* L. (Lamiaceae) in Turkey. Turkish Journal of Botany 32: 103–112.

Serpooshan F., Jamzad Z., Nejadsattari, T. & Mehregan, I. (2018). Molecular phylogenetics of *Hymenocrater* and allies (Lamiaceae):new insights from nrITS, plastis trnL intron and trnL-F intergenic spacer DNA sequences. Nord. J. Bot. 36(1–2):1–13.

Jamzad, Z., Chase, M., Ingrouille, M., Simmonds M., and Jalili, A. (2003). Phylogenetic relationships among *Nepeta* L. (Lamiaceae) species and related genera based on sequence data from internal transcribed spacers (ITS) of nuclear ribosomal DNA. Taxon 52: 21–32.

1. Annual **2**
   Perennial 4
2. Verticillasters congested into terminal elongated spikes . . . . 15. *N. iraqo-iranica*
   Verticillasters distant or upper 2–3 approximate. 3
3. Bracteoles 4–10 mm, lanceolate to subulate; calyx 6–9.5 mm, with a
      deep cleft to 1/2, teeth oblong to triangular-lanceolate, acuminate,
      mouth oblique; corolla 11–14 mm . . . . 13. *N. humilis*
   Bracteoles 4–7 mm, lanceolate to linear-lanceolate, obtuse to acute at
      apex; cleft indistinct, teeth lanceolate, acute, mouth straight; corolla
      7–10 mm long. . . . . 14. *N. petraea*
4. Inflorescence thyrsoid with wide-spreading branches arising from leaf
      axils; plants polygamous; calyx campanulate and straight 5
   Inflorescence of distant, approximating or conferted verticillasters,
      rarely paniculate, if paniculately branched or thyrsoid, flowers all
      hermaphrodite or all male-sterile; calyx straight or scarcely to clearly oblique. 6
5. Plants pale green, or yellowish-green when dry; lower leaves crenate;
      bracteoles broadly elliptic to lanceolate; corolla white or cream; teeth
      as long as tube . . . . 1. *N. congesta*
   Plants tinged purplish-blue at least at inflorescence; lower leaves entire;
      bracteoles narrowly elliptic to linear-subulate; calyx teeth shorter
      than or equalling tube; corolla lilac to purple. . . . . 2. *N. stricta*
6. Flowers white with pink or purple dots on lower lip. 7
   Flowers deep reddish-purple, lilac to violet. 9

7. Bracteoles 6–12 mm, ± equal to or longer than calyces, rigid and
　　incurved; bracteoles and calyx teeth with wide white-membranous
　　margins; calyx 6–10 mm, teeth unequal, 3–4 mm long. . . . . .3. *N. italica*
　　Bracteoles 2–5 mm, shorter than calyces, not rigid and incurved;
　　bracteoles and calyx teeth with narrow white membranous margins;
　　calyx 4–7 mm; teeth 2–3 mm 8
8. Upper cauline leaves sessile; calyx tube straight with erect teeth of
　　equal length . . . . .5. *N. nuda*
　　All leaves petiolate; calyx tube ± curved; teeth spreading, ± unequal . . . . . 4. *N. cataria*
9. Whole plant stellate-tomentose or plurifurcate hairy 10
　　Plants variously hairy and glandular, but not stellate-tomentose or
　　plurifurcate hairy. 11
10. Whole plant white-stellate tomentose; lower cauline leaves 1.5–3 ×
　　0.8–2 cm, oblong to ovate-lanceolate, cuneate to narrowly rounded
　　at base, crenate to obtuse-dentate at margins, acute at apex; usually
　　longitudinally plicate. . . . . 8. *N. menthoides*
　　Whole plant sparsely to densely short adpressed pluri-furcate hairy;
　　leaves 3–5.5 × 1.5–3 cm, ovate to lanceolate; subcordate to truncate
　　or rarely ± cordate at base, crenate-dentate to dentate at margins;
　　always flat . . . . 9. *N. elymaitica*
11. Whole plant eglandular, with sessile glands or with glandular papillae. 12
　　Whole plant or at least axis of inflorescence with clearly viscid, stalked
　　glandular hairs. . . . . 6. *N. autraniana*
12. Stems densely pilose to villous, leaves broadly ovate-orbicular; calyx
　　straight and expanded at mouth, teeth broadly triangular. . . . . .10. *N. iraqense*
　　Stems glabrescent to finely-pilose to villous; leaves oblong-elliptic to
　　lanceolate or ovate to suborbicular; calyx usually curved, slightly to
　　clearly oblique at mouth, teeth oblong to triangular 13
13. Calyx slightly to clearly bilabiate with a deep cleft on lower lip 14
　　Calyx regular to slightly bilabiate without a deep cleft on lower lip 17
14. Inflorescence verticillate, verticillasters remote or approximate at above,
　　or conferted into terminal heads 15
　　Inflorescence lax with most cymes pedunculate 16
15. Inflorescence conferted into ± sessile verticillasters and/or terminal
　　heads; bracteoles elliptic to linear-lanceolate; 4–6 mm. . . . . .19. *N. lamiifolia*
　　All verticillasters remote, lowers clearly pedunculate 1–1.5 cm;
　　bracteoles 2–4 mm, linear to subulate . . . . 18. *N. iodantha*
16. Inflorescence lax, cymes all pedunculate, loosely flowered; calyx 8–12
　　mm; corolla 20–26(–30) mm. . . . . 16. *N. macrosiphon*
　　Inflorescence broad, loose panicle, sometimes repeatedly dichotomous
　　or bostryciform; the axes of first order to 3 cm long or more, calyx
　　7–9 mm; corolla 13–17 mm. . . . . 17. *N. trautvetteri*
17. Inflorescence an oblong 'spike'; calyx 13–20 mm; teeth 4–8 mm,
　　usually aristate; corolla 20–25 mm, dark reddish purple, tube narrow,
　　straight . . . . .7. *N. trachonitica*
　　Verticillasters remote or at least upper verticillasters condensed into a
　　spike; calyx less than 13 mm 18
18. Leaves spreading; all verticillasters usually remote or the upper
　　approximate. . . . . .12. *N. haussknechtii*
　　Leaves ± adpressed to stem; verticillasters condensed into a spike,
　　sometimes the lower remote . . . . .11. *N. betonicifolia*

　　1. **Nepeta congesta** *Fisch. & C.A.Mey.* subsp. **cryptantha** (*Boiss. & Hausskn.*) *Dirmenci & Yildiz* in A.Güner & al. (eds.), Türk. Bitkileri List.: 565 (2012); Dinsmore in Post, Fl. Syria, Palest. & Sinai ed. 2, 2: 366 (1933); Feinbrun-Dothan, Fl. Paleast. 3: 116 (1978); Rechinger, Fl. Iranica [K. H. Rechinger] 150: 189 (1982); Davis, Fl. Turkey [P. H. Davis] 7: 285–286 (1982); Jamzad in Flora Iran [Assadi & al., in Persian] 76: 544 (2012); Taifour & El-Oqlah, Pl. Jordan Annot. Checkl.: 103 (2017) [as *Nepeta cryptantha*, *N. congesta* var. *cryptantha* and *N. involucrata*]

　　N. cryptantha Boiss. & Hausskn., Fl. Orient. [Boissier] 4: 669 (1879).

Fig. 169. **Nepeta congesta**. 1, habit × 1; 2, flower × 4. Reproduced with permission from Feinbrun-Dothan, Fl. Palaestina 3: Plates, f. 188. (1977). Drawn by Esther Huber. © The Israel Academy of Sciences and Humanities.

Glechoma cryptantha (Boiss. & Hausskn.) Kuntze, Revis. Gen. Pl. 2: 518 (1891).
Nepeta congesta var. *cryptantha* (Boiss. & Hausskn.) Hedge & Lamond, Notes Roy. Bot. Gard. Edinburgh 38: 43 (1980)
Oxynepeta involucrata Bunge, Mém. Acad. Imp. Sci. Saint Pétersbourg, Sér. 7, 21(1): 59 (1878).
Nepeta involucrata (Bunge) Bornm., Magyar Bot. Lapok 11: 9 (1912).
N. erivanensis Grossh., Beih. Bot. Centralbl. 44(2): 234 (1927).

Perennial. Stems single, green to yellowish-green, erect, 20–40(–50) cm tall, branching paniculate, with a ± dense indumentum of spreading villous multicellular hairs, shorter hairs sometimes present, eglandular or with sessile glands and glandular papillose. Leaves ovate-oblong to elliptic, 2–4 × 1–2 cm, lower leaves petiolate to 1.5 cm long, upper subsessile, crenate, uppermost rarely entire, truncate or cuneate at base, acute to obtuse at apex, ± puberulous or villous with sessile glands at beneath, subglabrous above. Inflorescence a widely paniculate thyrse, pale green; flowers clustered; lower cymes pedunculate, upper ± sessile. Bracteoles 5–12 mm, elliptic to lanceolate, shorter or longer than calyx, acuminate, pilose with sessile glands. Calyx 5–10 mm, ± campanulate, tube 3–4.5 mm, veins very prominent, villous, with sessile glands and ± glandular papillose; teeth ± equalling to tube, narrowly triangular, straight, rarely recurving. Corolla 6–8(–9) mm white or yellowish-white, shorter or longer than calyx teeth; hairy in throat. Nutlets 2–2.2 × 1.8–2 mm, broadly oblong, rounded-trigonous, usually finely tuberculate above. Fig. 169, 1–2.

HAB. Sandy soils, roadsides, clay soils, fallow or wheat fields; alt. 300–2100? m; fl. & fr. Apr.-Jun.
DISTRIB. N. and N.W. of Iraq. **MAM**: Hammam Ali, *Anders* 2480! (W); **FUJ**: Mosul, *Omar & Hamid* 36462!; Mesopotamia *Aucher-Eloy* 1792! (syntype of *N. cryptantha*); Midway Tal-Afar & Sinjar, by roadside, *Omar & Hamid* 36474!; 40 km E. of Sinjar, *Rawi* 33138!

Nepeta congesta is divided into two subspecific taxa, subsp. *congesta* and subsp. *cryptantha*. Subsp. *congesta* is glabrous to puberulous; it is endemic to Turkey (distributed in central Anatolia). Subsp. *cryptantha* usually has long spreading multicellular hairs and a wider distribution than subsp. *congesta*.

E. and S.E. Turkey, Syria, Palestine, Lebanon, Iran and Transcaucasus.

2. **Nepeta stricta** (*Banks & Sol.*) *Hedge & Lamond*, Notes Roy. Bot. Gard. Edinburgh 38: 45 (1980); Davis, Fl. Turkey [P. H. Davis] 7: 286 (1982); Feinbrun-Dothan, Fl. Paleast. 3: 116 (1978) [as *N. calycina*].

Satureja stricta Banks & Sol., A.Russell, Nat. Hist. Aleppo, ed. 2, 2: 255 (1794).
Glechoma calycina (Fenzl) Kuntze, Revis. Gen. Pl. 2: 518 (1891).
Nepeta calycina Fenzl, Flora 26: 400 (1843).

Perennial. Stems several, 15–30 cm tall, divaricately branched, short puberulous to hirsute, with sessile glands or glands absent. Leaves ovate to oblong-elliptic, 1.2–3 × 0.5–1 cm, petiole short above, entire, truncate or cuneate at base, acute at apex, hirsute-puberulous, usually with sessile glands below, glandular papillose; floral leaves narrowly lanceolate, acute at apex, sessile. Inflorescence a widely branched thyrse, often bluish tinged, flowers clustered or lax. Bracteoles 4–6 mm, shorter than or equalling to calyx, lanceolate-subulate. Calyx ±campanulate, 6–8 mm, sparsely to densely puberulous to tomentose, hirsute; teeth 1–2 mm, narrowly to broadly triangular, straight; hairy in throat. Corolla pale lilac to blue, 5–8 mm, included in calyx. Stamens and style included in calyx. Nutlets not seen. Fig. 170, 1–3.

HAB. There is no habitat and altitude information on the single collection from Mosul, but probably found in fallow or cultivated fields; alt.; 300?; fl. & fr. Apr.-Jun.
DISTRIB. Known from one location in northern Iraq. **FUJ?**: Mosul, 1841, *Kotschy* 568.145!

Iran, Lebanon-Syria, Palestine, Turkey.

3. **Nepeta italica** *L.*, Sp. Pl.: 571 (1753); Dinsmore in Post, Fl. Syria, Palest. & Sinai ed. 2, 2: 363 (1933); Rawi in Dep. Agr. Tech. Bull 14: 152 (1964); Rechinger, Fl. Iranica [K. H. Rechinger] 150: 175 (1982); Hedge in Fl. Turkey [P. H. Davis] 7: 267 (1982); Taifour & El-Oqlah, Pl. Jordan Annot. Checkl.: 103 (2017).

Nepeta orientalis Mill., Gard. Dict. ed. 8: n.º 10 (1768).
N. canescens J.F.Gmel., Syst. Nat. ed. 13[bis]: 900 (1792).
Cataria canescens Moench, Methodus: 388 (1794).

Fig. 170. **Nepeta stricta**. 1, habit, lower part of stem × 1; 2, habit, upper portion of stem × 1; 3, flower × 3. Reproduced with permission from Feinbrun-Dothan, Fl. Palaestina 3: Plates, f. 187 (1977). Drawn by Esther Huber. © The Israel Academy of Sciences and Humanities.

Nepeta incana Willd., Sp. Pl. 3: 52 (1800), nom. illeg.

N. macrostachya Jan ex Benth., Labiat. Gen. Spec.: 736 (1835).

Glechoma italica (L.) Kuntze, Revis. Gen. Pl. 2: 518 (1891).

G. canescens (J.F.Gmel.) Kuntze, Revis. Gen. Pl. 2: 518 (1891).

Nepeta teucrioides Lam., Encycl. 1: 711 (1785).

N. marrubioides Willd., Enum. Pl.: 603 (1809).

N. italica var. *incanescens* Fisch., Index Seminum (LE, Petropolitanus) 7: 53 (1841).

N. leucostegia Boiss. & Heldr., Diagn. Pl. Orient. 12: 62 (1853); Dinsmore in Post, Fl. Syria, Palest. & Sinai ed. 2, 2: 363 (1933); Rechinger, Fl. Iranica150: 175 (1982); Jamzad in Flora Iran [Assadi & al., in Persian] 76: 469 (2012); *Glechoma leucostegia* (Boiss. & Heldr.) Kuntze, Revis. Gen. Pl. 2: 518 (1891).

Perennial, few to many stemmed from base, stems ascending to erect, flowering stems usually sparsely branched, 20–90 cm tall, forming tufts, tomentose or adpressed pilose with sessile glands, densely glandular with papillose or ± eglandular with pubescent to sublanate hairs and sessile glands above. Leaves ovate to ovate-oblong, 1.5–4 × 1–3 cm, clearly petiolate to 2 cm long; pubescent to canescent with numerous sessile glands and glandular paplillose, cordate at base, crenate, acute to obtuse at apex; lower floral leaves longer than verticillasters. Verticillasters distant below, approximate above, many-flowered, scarcely pedunculate. Bracteoles linear-lanceolate to linear-elliptic, 6–12 × 1–2 mm, usually as long as calyx, often ± rigid and incurved, mucronate, usually with a ± broad white (sometimes purple) membranous margin; short hirsute with glandular. Calyx 6–10 mm, ± tubular, slightly curved, scarcely oblique at mouth, tube green, shortly hirsute, glandular-papillose; teeth unequal, 3–4 mm, linear-lanceolate, acuminate, with white or purplish membranous margin. Corolla white, purple spotted on lower lip, 11–13 mm; tube narrow, wider above, curved, exserted from calyx, pubescent to pilose, glandular-papillose; middle lobe of lower lip concave and irregularly crenate, bearded. Stamens 4, sometimes staminode, posterior ones sometimes sligthly protruding; filaments glabrous. Style unequally bilobed, slightly or clearly exserted. Nutlets oblong, 1.7–2 × 0.8–1.2 mm, tuberculate.

HAB. In *Quercus* forests, rocky limestone and volcanic slopes, dried stream beds, shady banks, roadsides; alt. 500–1900 m; fl. & fr.; May-Jul.

DISTRIB. Northern Iraq; widespread. **MAM**. Khantur mountain, N.E. of Zakho, near top of mountain slope, with *Quercus, Rawi* 8591! (K, BAG) & 23413! (K, BAG) & 23466! (K, BAG); Mosul, ad confines Turciae prov. Hakkari, in ditione pagi Sharanish, in montibus calc. a Zakho septentrionem versus, Jabal Khantur, *Rechinger* 10840! (K, G, W); 25 km N.E. of Zakho, *Rawi* 23453! (K); Dohuk gorge, stony slopes, *Chapman* 26280! (K, BAG); Mosul, ad confines Turciae prov. Hakkari, Dohuk, in fissuris rupium calc., *Rechinger* 11490 (G, W); Dohuk, *Rawi* 8650! (K, BAG-photo); Aary, *Rawi* 8485! (K, BAG-photo); Gara, *Kotschy* 408! (K, W); Garadagh, *Rawi* 9260! (K, BAG-photo); Sarsang, slopes of Garadagh, *Chapman* 26440! (K); Agra, *Rawi* 11473! (K, BAG). **MSU**: Suleymaniya, Zewiya valley, pir omar Gudrum, *Rawi* 12071! (K, BAG); Kopi Karadagh, *Haines* 1067! (K); In montibus calcareis Avroman et Shahu, *Haussknecht* s.n.! (BM, K). **MRO**: Kanurish, Qerna valley, N. of Pushtashan, *Serhang & Rawi* 2666! (K, BAG). **MJS**: Sinjar, Jebel Sinjar, north of the town, *Gillett* 11141! (K, BAG); Kursi, Jebel Sinjar, *Gillett* 10875! (K, BAG); Karsi, N.W. slope, *Omar, Khyat, Kaisi* 52460! (BAG). **FUJ**: Mosul, *Kotschy* 400! (K).

Italy, Greece, Syria, Lebanon, Turkey, Iran.

4. **Nepeta cataria** *L.*, Sp. Pl.: 570 (1753); Dinsmore in Post, Fl. Syria, Palest. & Sinai ed. 2, 2: 363 (1933); Rechinger, Fl. Iranica [K. H. Rechinger] 150: 141 (1982); Hedge in Fl. Turkey [P. H. Davis] 7: 267 (1982); Jamzad in Flora Iran [Assadi & al., in Persian] 76: 493 (2012): Taifour & El-Oqlah, Pl. Jordan Annot. Checkl.: 103 (2017).

Nepeta vulgaris Lam., Fl. Franç. 2: 398 (1779).

Cataria vulgaris Gaterau, Descr. Pl. Montauban: 105 (1789).

Glechoma cataria (L.) Kuntze, Revis. Gen. Pl. 2: 518 (1891).

Nepeta minor Mill., Gard. Dict. ed. 8: n.º 2 (1768).

Cataria tomentosa Gilib., Fl. Lit. Inch. 1: 78 (1782).

Nepeta ruderalis Boiss., Fl. Orient. 4: 643 (1879).

Perennial herbs. Stems erect, 30–100 cm tall, robust, branched above, retrorsely eglandular pilose with short hairs and sessile glands, glandular papillose above. Leaves ovate to triangular-ovate, 2–7 × 1–4 cm, clearly petiolate to 4 cm long, finely adpressed pilose with many sessile glands and glandular papillose, often greyish beneath, cordate or truncate at base, serrate at margins, acute at apex. Lower floral leaves clearly petiolate. Inflorescence widely paniculate, cymes pedunculate; verticillasters ± distant below, condensed above,

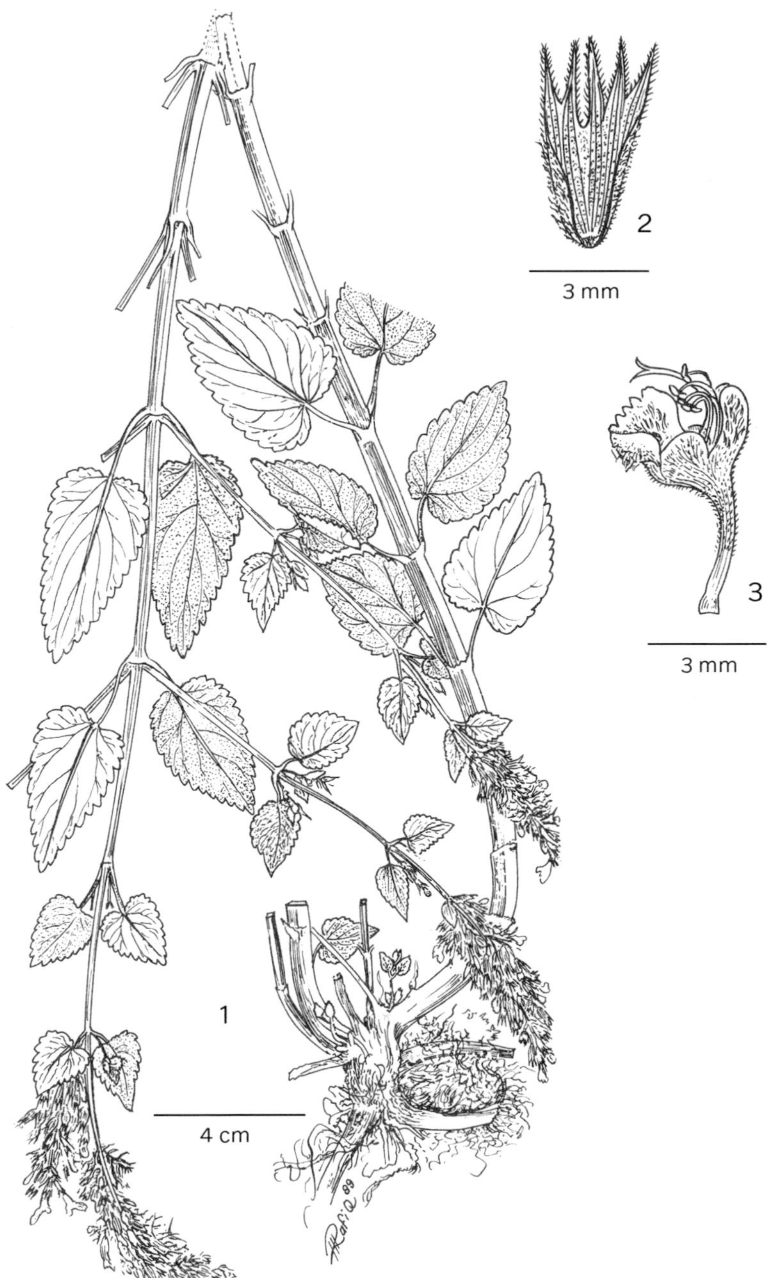

Fig. 171. **Nepeta cataria**. 1, habit; 2, calyx; 3, corolla. Reproduced with permission from Flora of Pakistan 192: f. 12, D–F (1990). Drawn by M. Rafiq. © National Herbarium, Pakistan Agriculture Research Council & University of Karachi, Pakistan.

many-flowered, flowers pedicellate. Bracteoles 2–4(–5) mm, linear, clearly shorter than calyces. Calyx tubular, 5–7 mm ± curved, scarcely oblique or not at mouth, ± densely pilose-pubescent and with sessile glands; teeth linear-lanceolate, 2–3 mm; hairy in throat. Corolla white with blue-purple spots, 6–8(–10) mm, tube scarcely exceeding calyx or included in calyx, pilose with short hairs and sessile glands; middle lobe of lower lip concave and crenate, bearded. Stamens 4, included in upper lip of corolla, sometimes sterile. Style unequally bilobed, included in corolla. Nutlets broadly ellipsoid to oblong, 1.3–1.5 × c.1 mm, dull, obsoletely to clearly tuberculate at apex, areole straight. Fig. 171, 1–3.

HAB. Humid habitats, fallow fields, near the stream banks, moist areas; alt. 600–1500 m; fl. & fr. Jun.-Aug.;

DISTRIB. Northern and northeastern Iraq. **MAM**: Kani-Masi, near water spring, *Botany Staff* 43883! (K); Sersang, *Haines* 1358-A! (BUH, K). **MSU**: Suleymaniya, M. Avroman et M. Shahu, *Haussknecht* 778; Alyasafra, W. Penjwin, *A. Haloob, Ikhlas & Hamishkan* 60186 (BAG). **FKI/DGA**: between Khalis and Karkuk, gravely soil, *Rawi et.al.* 19705! (BAG, K).

Europe to Southwest and Central Asia, Himalayas to Korea.

5. **Nepeta nuda** L. subsp. **albiflora** (Boiss.) Gams, in G.Hegi, Ill. Fl. Mitt.-Eur. 5(4): 2372 (1927); Dinsmore in Post, Fl. Syria, Palest. & Sinai ed. 2, 2: 365 (1933); Rawi in Dep. Agr. Tech. Bull 14: 153 (1964); Hedge in Fl. Turkey [P. H. Davis] 7: 272 (1982); Jamzad in Flora Iran [Assadi & al., in Persian] 76: 537 (2012). – Lectotype: designated here by Dirmenci, "Plantae in prov. Musch ad radices australes Bimgoel montis ad Gumgum in districtu Warto lectae, in valle Merga Sauk, 6000', 21.08.1859, *Kotschy* 368 (G-BOISS [G00787649]; isolectotypes. (JE00018411, K)

Nepeta nuda var. *albiflora* Boiss., Fl. Orient. [Boissier] 4: 663 (1879)/
N. marrubioides Boiss. & Heldr. ex Benth., Prodr. [A.P. de Candolle] 12: 387 (1848), nom. illeg.
Glechoma marrubioides Kuntze, Revis. Gen. Pl. 2: 518 (1891).
Nepeta bithynica Bornm., Mitth. Thüring. Bot. Vereins, n.f., 20: 41 (1905).

Perennial, few to many stemmed from base, erect, 50–90 cm tall, simple or a few branched above, subglabrous to adpressed short eglandular pilose with sessile glands, sometimes glandular papillose above. Leaves ovate to ovate-oblong to elliptic-lanceolate, 1.5–7 × 1–3(–3.5) cm, sessile, adpressed pubescent with numerous sessile glands, crenate to dentate at margins, cordate or truncate at base; obtuse to acute at apex, veins prominent. Floral leaves sessile, lower similar to leaves and usually longer than verticillasters, upper short and lanceolate to elliptic-lanceolate. Inflorescence paniculate of numerous many-flowered verticillasters, rarely thyrsoid, verticillasters distinct below, approximate above, short pedunculate. Bracteoles linear-subulate with narrow white membranous margins, 2–3 mm, shorter than calyx tube or teeth, subequal, Calyx tubular, 4–5(–7) mm, yellowish-green, tube and mouth straight, adpressed eglandular-pilose with sessile glands, sometimes glandular papillose; teeth 2–3 mm erect, linear-lanceolate to linear-oblong, blunt or acute, soft-tipped, separated by rounded sinuses, often with membranous margins. Corolla white or yellowish-white, (6–)8–10(–15) long, included or slightly exserted from calyx, tube slightly curved, pilose; middle lobe of lower lip concave and crenate, bearded. Stamens 4, included in upper lip of corolla, sometimes sterile. Style unequally bilobed, sligthly or clearly exceeding upper lip of corolla. Nutlets ovoid-oblong, 1.5–2.2 × 1–1.2 mm, tuberculate, prominently so at apex with crowded projecting tubercles, brown, areole bilobed.

HAB. Woodlands, meadows, streamsides, mountain steppes, field sides; alt. 1300–1650 m; fl. & fr. Jun.-Aug.

DISTRIB. **MAM**: Mosul, ad confines Turciae prov. Hakkari, in ditione pagi Sharanish, in montibus calc. a Zakho septentrionem versus, in jugo Shiwarana inter Sharanish, *Rechinger* 10877 (G, W); 25 km N.E. of Zakho, Shivara Hina village, *Rawi* 23453! (K). **MSU**: Suleymaniya, Avroman Dagh, spur north of Biyara, *Gillett* 11818! (K); Kopi Karadagh, *Haines*! 1159 (K). **FUJ**: Mosul, *Kotschy* 170! (K).

Balkan Peninsula, Greece, Lebanon-Syria, Turkey.

6. **Nepeta autraniana** Bornm., Bull. Herb. Boissier 7: 248 (1899). Type: Iraq, Riwandous (ad fines Pers.) in monte Handren (Handarin), 1300 m, 06.06.1893, *Bornmüller* 1684-a (holo. B; iso K, [B100277268-photo!; K001222554 iso.]); Rechinger, Fl. Iranica [K. H. Rechinger] 150: 178 (1982).

Perennial. Stems 60 cm tall or longer, erect, simple, paniculately branched above, quadrangular, sparsely to densely viscid glandular-hairy. Leaves oblong, ovate-oblong to lanceolate towards to inflorescence, 5–10 × 3–4 cm, green, sessile, auriculate-amplexicaule, crenate to coarsely crenate-dentate at margins, obtuse at apex, glandular papillose-hirtellous on both sides, densely sessile glandular at beneath. Verticillasters many-flowered, cymes pedunculate, laxly spicate, rarely interrupted. Bracteoles linear-lanceolate, 5–7 mm, often as long as calyx tube or longer, partly purple membranous–Margined, densely glandular papillae, ciliate. Pedicel short. Calyx 7–9 mm, straight, mouth oblique, densely glandular hairy and papillate with sessile glands; teeth lanceolate-subulate, 3.5–4.5 mm, subequal, hyaline-membranous, purplish. Corolla blue, 11–15 mm, tube exserted from calyx, abruptly enlarged in throat, sparsely hairy and glandular papillose on tube, long hairy outside of the lips; middle lobe of lower lip concave and crenate-undulate, bearded. Stamens 4; anthers black, posterior ones shortly exserted. Style often protruding. Nutlet triangular-ovate, tuberculate.

HAB. Not specified in the protologue and sheet label, but possibly moist areas; alt. ± 1300 m; fl. & fr. Jun.
DISTRIB. Endemic, known only from type location in Rawanduz, N. Iraq. MAM: Riwandus (ad fines Pers.) in monte Handren (Handarin), 1300 m, 06.06.1893, *Bornmüller* 1684-a (type, B, K)

7. **Nepeta trachonitica** *Post*, J. Linn. Soc., Bot. 24: 439 (1888); Dinsmore in Post, Fl. Syria, Palest. & Sinai ed. 2, 2: 364 (1933); Rawi in Dep. Agr. Tech. Bull 14: 153 (1964); Rechinger, Fl. Iranica [K. H. Rechinger] 150: 162 (1982); Hedge in Fl. Turkey [P. H. Davis] 7: 279 (1982); Jamzad in Flora Iran [Assadi & al., in Persian] 76: 523 (2012).

Nepeta purpurea Nábelek, Spisy Přír. Fak. Masarykovy Univ. 70: 55 (1926).

Perennial herbs. Stems 40–85 cm tall, single or a few stemmed from the same root, usually unbranched, erect, sturdy, quadrangular, glabrous or very rarely finely pilose, eglandular. Leaves ovate-triangular to oblong-lanceolate, 3–7 × 2–4 cm, green, petiole to 25 mm long, median and upper subsessile to sessile, cordate to truncate at base, coarsely crenate-serrate, acute at apex, glabrous to finely eglandular-pilose. Inflorescence an (ovate-)oblong spike, 3–8 × 2–3.5 cm; verticillasters many-flowered, conferted, rarely the lowest one distant. Bracteoles linear-filiform to narrowly lanceolate, 5–8(–10) mm, clearly shorter than calyces. Calyx 13–20 mm, tubular, tube and mouth straight, pilose; teeth variable in length, 4–8 mm, usually aristate; sparsely hairy between teeth. Corolla 20–25 mm, dark reddish-purple, tube narrow, straight, clearly exserted from calyx, or sub-included, pilose and sparsely sessile glands; upper lip 2-lobed, 3.5–4 mm; lower lip 3-lobed, 4–5 mm; median lobe concave, crenate, bearded. Stamens 4, sometimes posteriors slightly protruding. Style included in corolla or slightly protruding. Nutlets ± oblong, trigonous, 2.5–3 × 1.5–2 mm, distinctly tuberculate.

HAB. Rocky slopes, in *Quercus* scrub, alt.: 1200–2000 m.; fl. & fr. May-Jul.
DISTRIB. Northern Iraq. MAM: Mosul, ad confines Turciae province Hakkari, in ditione pagi Sharanish in montibus calc. a Zakho septentrionem versus, *Rechinger* 10786! (K, W) & 10943 (BUH, W); Zawita, *Guest* 4464! (K); Khantur mountain, N.E.. of Zakho, *Quercus* zone, on mountain slope, *Rawi* 23361! (K). MRO: Hawara Blinda mountain, *Kass* & *Nuri* 27816! (BAG, K)

S.E. Turkey, Syria, Lebanon.

8. **Nepeta menthoides** *Boiss. & Buhse*, Nouv. Mém. Soc. Imp. Naturalistes Moscou 12: 174 (1860); Rechinger, Fl. Iranica [K. H. Rechinger] 150: 143 (1982); Jamzad in Flora Iran [Assadi & al., in Persian] 76: 484 (2012).

Perennial, scarcely hardened, all plant white-stellate tomentose. Many stemmed from base, 10–40 cm tall, stout, arcuate-ascending, short articulating, densely foliate, simple to branched. Lower cauline leaves oblong to ovate-lanceolate, 15–30 × 8–20 mm, petiolate to 10 mm, cuneate to narrowly rounded at base, crenate to obtuse-dentate at margins, acute at apex, usually longitudinally plicate, white-stellate tomentose with sessile glands and glandular papillose; upper cauline leaves smaller, lanceolate, sessile, usually longer than verticillasters. Bracteoles linear-lanceolate, 5–7 mm, as long as or shorter than calyx tube. Lower verticillasters remote, the uppers approximate or spicate. Calyx somewhat incurved, 6.5–8 mm, mouth oblique, densely stellate-tomentose with sessile glands and glandular papillose; teeth triangular-lanceolate, 1.5–2.5 mm; hairy in throat. Corolla purplish-blue, 13–18 mm long, puberulous, tube included in calyx or exserted, abruptly expanded in throat, to 5 mm wide; upper lip 2 mm, emarginate; median lobe of lower lip c. 2.5 mm wide, concave, undulate, bearded. Stamens 4, included in corolla. Style unequally bilobed, exserted from corolla. Nutlet ovate-trigonous, glabrous.

HAB. On metamorphic and serpentine rocks; alt. 2600–3800 m; fl. & fr. Aug.-Sep.

DISTRIB. High mountain belt of northern Iraq. **MRO**: Arbil, mons Helgurd ad confines Persiae, in valle supra pagum Nowanda, *Rechinger* 11470-a! (G, K, W); mons Helgurd ad confines Persiae, in declivibus occidentalibus summi montis, schist, metamorph. et serpentin, *Rechinger* 11886 (W).

W. Iran.

9. **Nepeta elymaitica** *Bornm.*, Russk. Bot. Zhurn. 1911: 6 (1911); Rechinger, Fl. Iranica [K. H. Rechinger] 150: 177 (1982); Jamzad in Flora Iran [Assadi & al., in Persian] 76: 484 (2012).

> *Nepeta ludlow-hewittii* Blakelock, Kew Bull. 4: 543 (1949 publ. 1950). syn. nov. Type: Lecto. (designated by Dirmenci), Iraq, Erbil: Algird Dagh, among rocks, 3000–3300 m, 22.07.1932, *Guest & Ludlow-Hewitt* 2871 lecto. K [K000910840]; isolecto. K [K000910837, K000910838, K000910839].

Perennial, a few to many stemmed from woody rootstock. Stems erect or ascending, 20–50 cm tall, robust, sometimes a few branched above, leafy along all length, all plant sparsely to densely short adpressed pluri-furcate hairy with sparsely sessile glands, densely glandular papillose above. Leaves ovate to lanceolate, herbaceous, 3–5.5 × 1.5–3 cm, green to slightly greyish, concolor, veins pinnately reticulate; lower and middle cauline leaves petiolate, 10–15 mm, with brown membranous scales at base, to 1 cm, subcordate to truncate or rarely ± cordate at base, crenate-dentate to dentate at margins, obtuse to acute at apex; floral leaves sessile, oblong-lanceolate. Verticillasters 3–7, many flowered, lower verticillasters remote, uppers approximate or spicate. Bracteoles linear to lanceolate-subulate, 5–7 mm. Cymes shortly pedunculate, lower peduncles to 1 cm. Flowers pedicellate to 3 mm. Calyx tubular, 6–9 mm, mouth oblique and expanded, sometimes purplish at mouth and teeth; upper teeth lanceolate, 1–2.5 mm; lower teeth triangular, 1–3 mm; hairy with glandular papillose in throat. Corolla violet-blue, 15–20 mm, tube curved, distinctly exserted from the calyx, widened in the upper part, to 6 mm width, outside hairy; upper lip short, 2-lobed, emarginate; lower lip 3-lobed, median lobe widely rounded, bearded and papillose. Four stamens and style included in corolla, Nutlet oblong-trigonous, 2.5–2.8 × 1–1.2 mm, brown, densely glandular-papillose.

HAB. Calcareous rocks, metamorphic and igneous and serpentine rocks; alt. 2900–3500 m; fl. & fr. Jul.-Aug.

DISTRIB. High mountain belt of Qandil and Algird mountains in northern/northeastern Iraq. **MRO**: Erbil, Ser Kurawa, *Gillett* 9756! (BAG, K); Erbil, Algird Dagh, among rocks, *Guest & Ludlow-Hewitt* 2871! (type of *N. ludlow-hewittii*) (K); Erbil, N.E. slopes of Arlgird Dagh, *Gillet* 9567! (BAG, K); Algurd Dagh (Helgurd), *Haley* 137! (BM); N. of Helgord range, E. of Berma sand lake, *Serhang & Rawi* 24775-A! (K); Qandil range, *Serhang & Rawi* 24067! (K); montes Qandil ad confines Persiae, insaxosis calc. supra lacum Goam-e Kirmosoran, *Rechinger* 11123! (G, W); Perrish mountain, cerpentine rocks, *Rawi* 24549! (K).

Western Iran.

10. **Nepeta iraqensis** *Dirmenci*, Turk. J. Bot. 47: 178, f. 4 (2023). Type: Iraq, Qandil range, *Astragalus* zone, on S. side of mountain slope, 3880 m, 30.07.1957, *Rawi & Serhang* 24071! (K, holo. [K001222565]).

Perennial herb. Stems erect, c. 30 cm tall, single or a few branched from below, leafy along all length, densely villous with sparsely sessile glands, and densely glandular papillose above. Leaves herbaceous, broadly ovate to orbicular, 1.4–3.2 × 1.6–3.2 cm, all cauline leaves short to clearly petiolate, petioles 2–10 mm and densely villous; leaves cordate at base, crenate at margins, rounded at apex, villous with sessile glands and glandular papillose, denser at beneath; lower floral leaves similar to cauline leaves and longer than verticillasters. Verticillasters many-flowered, lower verticillasters remote and cymes pedunculate, uppers approximate. Bracteoles linear, 2–3 mm, shorter than calyx, densely villous and glandular papillae, partly purple. Flowers shortly pedicellate. Calyx tubular-campanulate, 5–6 mm, mouth expanded, purplish at mouth and teeth, densely long villous with sessile glands and glandular papillose; teeth ± equal, c. 2 mm, triangular-lanceolate to ovate-acuminate, 1–2.5 mm, densely hairy, glandular papillose and ciliate; hairy in throat. Corolla purplish? densely long hairy outside of lips. Nutlets unknown.

HAB. On mountains in the *Astragalus* zone, on slopes amongst small rocks; alt. ± 3880 m; fl. & fr. Jul.-Sep. DISTRIB. Known only from type material. **MAM**: Qandil range, *Astragalus* zone, *Rawi & Serhang* 24071! (type).

Endemic. There are no species in Iraq, Turkey and Iran that *Nepeta iraqensis* is close to. It grows at the highest elevation amongst the *Nepeta* species in Iraq. In its calyx characters it comes closest to *N. elymaitica* and *N. menthoides*, but different from these two species in its lack of pluri-furcate hairs.

11. **Nepeta betonicifolia** *C.A.Mey.* in Verz. Pfl. Casp. Meer.: 92 (1831)

subsp. **strictifolia** (*Pojark.*) *Menitsky*, Fl. Armen. 8: 47 (1987); *Rawi* in Dep. Agr. Tech. Bull 14: 152 (1964); Rechinger, Fl. Iranica [K. H. Rechinger] 150: 177 (1982); Hedge in Fl. Turkey [P. H. Davis] 7: 277 (1982); Jamzad in Flora Iran [Assadi & al., in Persian] 76: 522 (2012); Taifour & El-Oqlah, Pl. Jordan Annot. Checkl.: 103 (2017).

> *Nepeta strictifolia* Pojark., Bot. Mater. Gerb. Bot. Inst. Komarova Akad. Nauk S.S.S.R. 15: 303 (1953).
> *N. speciosa* Boiss. & Noë, Diagn. Pl. Orient., ser. 2, 4: 34 (1859).
> *Glechoma speciosa* (Boiss. & Noë) Kuntze, Revis. Gen. Pl. 2: 519 (1891).
> *Nepeta buhsei* Pojark., Bot. Mater. Gerb. Bot. Inst. Komarova Akad. Nauk S.S.S.R. 15: 299 (1953).

Perennial herb. Stems few to many, erect, unbranched, 25–55 cm tall, finely pilose with very short eglandular hairs, glandular papillose above. Leaves erect, ± adpressed to stem, especially upper, oblong to ovate-lanceolate, 1.5–3.5 × 0.5–2.2 cm, bright green, truncate or subcordate at base, serrate at margins, obtuse to acute at apex, sessile or subsessile, with a sparse indumentum of very short hairs on both surfaces, and sparsely to densely sessile glands at beneath. Inflorescence a ± condensed oblong 'spike', usually one or two lowermost verticillaster remote. Bracteoles linear to subulate, 4–10 mm, shorter than or as long as calyx, violet. Pedicel to 1 mm. Calyx narrowly tubular, 8–10 mm, purple, shortly eglandular-pilose, tomentellous, sometimes long pilose with glandular papillae and sessile glands, straight or almost so, mouth slightly oblique; teeth subequal, narrowly triangular-acuminate, 1.5–3.5 mm; hairy in throat. Corolla purplish or purplish-blue, 13–17 mm, tube narrowly expanded above, ± straight or curved, puberulous on the outside; median lobe of lower lip concave, crenate, bearded. Stamens 4, included in corolla. Style exserted from corolla. Nutlets broadly oblong, trigonous, 2–3 × 1–1.2 mm, tuberculate.

HAB. Limestone slopes, *Quercus* forests, alpine meadows, fallow fields; alt. 1100–3100 m; fl. & fr. May-Aug.
DISTRIB. Northern Iraq. **MAM**: Mossul, Matina, *Rawi* 8849-B! (BAG, K). **MSU**: In montibus calcareis Avroman et Schahu, *Haussknecht* s.n. (K). **MRO**: Arbil: mt. Rayat, *Thesiger* 1054! (BM); Gali Warta, N.W. & N. of Rania, *Rawi, Nuri & Kass* 28796! (K).

E. Turkey, Caucasus, Iran.

12. **Nepeta haussknectii** *Bornm.*, Bull. Herb. Boissier 7: 249 (1899). Type: Iraq, Riwandous (ad fines Persicos) in alpibus Sakri-Sakran, 2200 m, 23.06.1893, *Bornmüller* 1686! (G, lecto. designated by Dirmenci [G00424356]; isolecto. B, E, G, HBG, JE, K, LE, W. [B100277249, B100277250, E00319651, G00424355, G00424357, HBG518441-photo! JE00018457-photo! JE 00018458-photo! K000910891, LE00016672, W1895-0001705]); Rawi in Dep. Agr. Tech. Bull 14: 152 (1964); Feinbrun-Dothan, Fl. Paleast. 3: 149 (1978); Rechinger, Fl. Iranica [K. H. Rechinger] 150: 184 (1982); Jamzad in Flora Iran [Assadi & al., in Persian] 76: 505 (2012).

> *Nepeta racemosa* subsp. *haussknectii* (Bornm.) A.L. Budantzev, Bot. Zhurn. (Moscow & Leningrad) 77(1): 121 (1992). *syn. nov.*

Perennial herb. Stems 25–55 cm tall, ascending to erect, few to many stemmed from the base, branched or not, finely pilose to villous with eglandular hairs or white lanate with sessile glands, glandular-papillose above. Leaves spreading to ± erecto-patent, ovate to ovate-elliptic, or oblong to oblong-lanceolate, 15–40 × 5–25 mm, truncate to ± cordate at base, crenate to serrate, obtuse at apex, green to greyish-green, finely to densely pilose and with few to numerous sessile glands, sometimes glandular-papillose; lower leaves shortly petiolate, upper sessile; lower floral leaves longer than verticillasters. Verticillasters ± few to many flowered, distant, or approximate above. Bracteoles linear-lanceolate, 2–4 mm, clearly shorter than calyx tube. Calyx tubular, 6–9 mm, at least at mouth and teeth dark purplish, densely pilose-villous to tomentose with sessile glands and glandular-papillose, or lanate; slightly curved and oblique; teeth subequal, triangular, acuminate, 1.5–2.5 mm; hairy in throat. Corolla purplish to purplish-blue, 10–18 mm; tube slender, slightly exserted from calyx, abruptly widening above; lower lip median lobe concave, crenate, bearded. Stamens 4, well-developed or sterile. Style slightly to clearly exserted from corolla. Nutlets oblong, trigonous, 2–2.2 × 8–1.2 mm, tuberculate.

HAB. Rocky igneous, limestone and shady slopes, banks, meadows, near streams; alt. 1300–2600 m; fl. & fr. May-Aug.

DISTRIB. Northern Iraq, Sulimaniyah District. **MRO**: Riwandous (ad fines Persicos) in regione alpina montis Helgurd, *Bornmüller* 1685! (B, JE, syn., [B100277251, JE 00018459-photo!]; Erbil, Montes Qandil ad confines Persiae, in decl. orient. supra Pushtashan, *Rechinger* 11076 (G, W); Qandil range, *Serhang & Rawi* 26724! (BAG, K); Erbil, mons Helgurd ad confines Persiae, in valle supra pagum Nowanda, *Rechinger* 11354! (G, K, W); N.E. slopes of Algird Dagh, *Gillet* 9530! (BAG, K); Helgurd range, *Rawi* 24916 (BAG); Chia-i-Mandali, *Guest* 2740! (BAG, K); Kawriech, E. side of Kanoukla mountain, *Kass & Nuri* 27675! (BAG, K); Gali Warta, c. 30 km N.W. of Rania, *Quercus* forest, *Nuri, Kass & Rawi* 28746! (K).

Nepeta racemosa, N. haussknechtii, N. crassifolia Boiss. & Buhse, *N. stenantha* Kotschy & Boiss. ex Boiss., *N. transcaucasica* Grossh., with several intermediates form a species complex where several taxa have been described. It would be beneficial to review the species in this group distributed from the Caucasus to northern Iraq.

Transcaucasus, E. Turkey, Iran.

13. **Nepeta humilis** Benth., Prodr. [A.P. de Candolle] 12: 382 (1848); Rawi in Dep. Agr. Tech. Bull 14: 152 (1964); Rechinger, Fl. Iranica [K. H. Rechinger] 150: 207 (1982); Duman in Fl. Turkey [Guner et. al] 11 (suppl. 2): 207 (2000); Jamzad in Flora Iran [Assadi & al., in Persian] 76: 556 (2012).

Annual. Stem erect, branched or unbranched, (5–)10–45 cm tall, glandular-papillose and puberulous. Leaves rotundate to triangular, 10–25 × 10–25 mm, lower leaves petiolate, cordate to truncate at base, crenate-dentate at margins, rarely subentire, obtuse at apex, minutely puberulent and glandular-papillose, sometimes glabrous above; petiole to 1.5 cm. Inflorescence paniculate; verticillasters many-flowered, lower remote and cymes long-pedunculate, upper approximate. Bracteoles lanceolate to subulate, 4–10 mm, equal or slightly shorter than calyx, densely ciliate or canescent, with densely glandular-papillae. Calyx tubular, 6–9.5 mm, tube slightly curved, with a deep cleft to 1/2, mouth oblique, densely glandular-papilose with simple hairs; teeth straight, 1–3 mm, oblong to triangular-lanceolate, acuminate, shorter than tube, sometimes long ciliate; sparsely hairy in throat. Corolla violet, 11–14 mm, short-puberulous, tube ± straight, exserted beyond calyx teeth; middle lobe constricted from base, transversely dilated, crenate-incised, bearded. Stamens 4, upper ones slightly protruding or included in upper lip. Style included in corolla. Nutlets oblong-trigonous, 1.8–2 × 1–1.2 mm, black, smooth. Fig. 172, 1–7.

HAB. Wet places, slopes, eroded slopes; alt. 800–3200 m.; fl. & fr. Jun.-Aug.

DISTRIB. Widely distributed northern and northeastern Iraq. **MAM**: Mossul, Gara Dagh, *Rawi* 9272! (BAG, K); Mosul, ad confines Turciae prov. Hakkari, in ditione pagi Sharanish, in montibus cal. A Zakho septentrionem versus, in saxosis cacuminis Zawiata, *Rechinger* 10964! (G, K); Galli Zawita, N.E. of Zakho, near Turkish border, *Rawi* 23574! (BAG, K); Mosul, ad confines Truciae prov. Hakkari, inter Dohuk et Amadiya, in quercetis saxosis supra Sirsank, calc., *Rechinger* 11694! (G). **MSU**: Sulaymaniyah, montes Avroman ad confines Persiae, in ditione pagi Tawilla, in saxosis calc. *Rechinger* 10338-a (G); Suleymaniyah, monte Avroman, ad confines Persiae, in ditione pagi Tawilla, *Rechinger* 12414 (G); in montibus calcareis Avroman et Schahu, *Haussknecht* s.n.! (BM, K, W); Suleymaniya, Jebal Avroman, spur north of Biyara, on screes, *Gillett* 11795! (BAG, K); Hawraman mountain, *Rawi* 29525! (K); Avroman mountain, N. of Halabje (on Persian border), *Rawi* 22051! (BAG, K); Sulaymaniyah, in ditione pagi Panjwin, in glareosis serpenticis jugi Malakawa, *Rechinger* 10423! (G); Suleymaniyah, Qara Dagh, *Makki* 449! (W); Suleymaniya, Gulashina, Pir Omar Gudrum, *Rawi* 12113! (BAG, K); Pir Omar Gudrun, *Haussknecht* s.n.! ; 12 km east of Chemchemal, *Gillett & Rawi* 11617! (BAG, K); Jarmo, *Halbaek* 1291! (K) & 1825! (K); Mela Kowa, Suleymaniya-Penjwin highway, *Rawi* 24447! (BAG, K); Dara, on the road between Halabja and Tawaila, *Rawi* 21991! (BAG, K). **MRO**: Rawandiz, *Bornmuller* 1687! (K); Hawawra Blinda mountain (N.E. of Haji Omran), *Kass & Nuri* 27839! (BAG, K); Riwandous, *Bornmuller* 1688! (K); Riwandous (ad fines Persiae), *Bornmuller* 1689! (type of *N. kurdica*, K, W); Rowandiz gorge, *Field & Lazar* 851! (K); Erbil, mons Helgurd ad consfines Persia, in valle supra pagum Nowanda, ca. *Rechinger* 11333! (G); Erbil, Algird Dagh, *Guest* 2921! (K); Between Rawanduz and Agoyan village, *Kass & Nuri* 27283! (BAG, K); Erbil, montes Qandil ad confines Persiae, supra Pushtashan, *Rechinger* 11227 (G); Erbil: montes Qandil ad confines Persiae, Pushtashan, *Rechinger* 11737 (G); Qandil range, Baisar village, *Rawi & Serhang* 24166! (K); Erbil, montes Qandil ad confines Persiae, supra lacum Goam-e Kirmosoran, calc., *Rechinger* 11138! (G); Lower slope of Qandil range, *Rawi & Serhang* 24696! (K); Pustashan, 15 km N.E. of

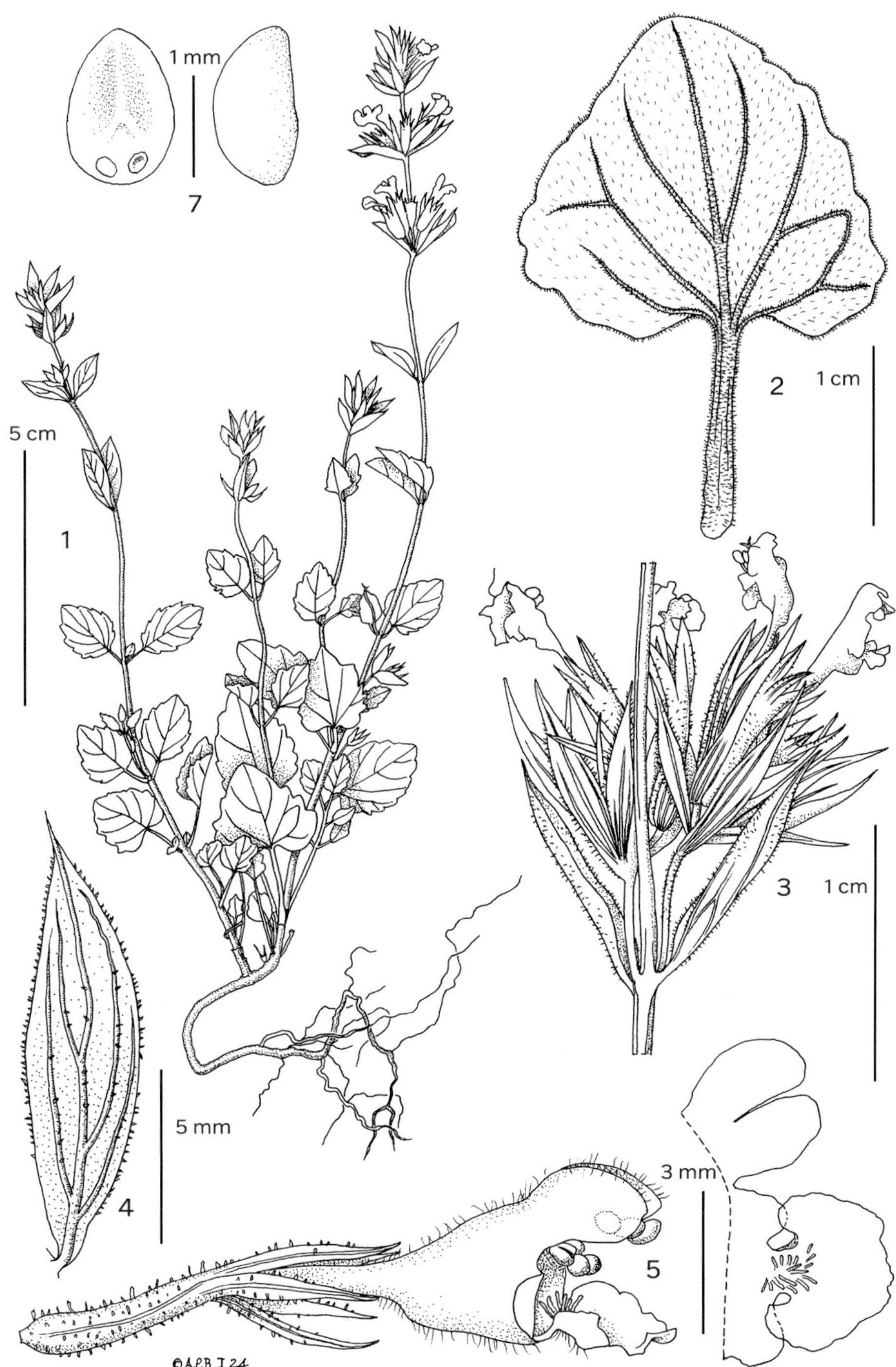

Fig. 172. **Nepeta humilis.** 1, habit; 2, near basal leaf, abaxial surface; 3, inflorescence, mid stem; 4, abaxial surface of bract of inflorescence; 5, flower, side view; 6, upper and lower lobes of corolla (flattened); 7, mature nutlets inner face (left), side view (right). 1–2 from *Helbaek* 1291; 3-6 from *Garden* s.n.; 7 from *Field et al.* 851. Drawn by © A.P. Brown, Jan. 2024.

Rania, lower slope of Qandil range, *Rawi & Serhang* 24193! (BAG, K); Serin mountain, on road to Qandil, *Rawi* 23987! (BAG); in steppa ad colles aridos prope pagum Der Harir inter Erbil (Arbela) et Rewanduz, *Nábĕlek* 1624 (SAV-photo). **FKI**: Kirkuk, conglomerate rock in narrow gorge in Tuni Baba Amera hills 10 km south of Durbendiken, *Barkley & Haddad* 7440! (K) & 7443! (K, W); Near Pewaz, *Poore* 599! (K).

S.E. of Turkey, Iran.

14. **Nepeta petraea** *Benth.*, Prodr. [A.P. de Candolle] 12: 394 (1848); Rechinger, Fl. Iranica [K. H. Rechinger] 150: 193 (1982); Jamzad in Flora Iran [Assadi & al., in Persian] 76: 525 (2012).

Glechoma petraea (Benth.) Kuntze, Revis. Gen. Pl. 2: 518 (1891).

Annual. Stems 15–40–(60) cm tall, erect, branched from base to inflorescence, laxly crisp villous, densely glandular-papillose. All leaves white-lanate with glandular papillae; lower leaves ovate, 10–25 × 10–20 mm, petiolate to 1 cm, truncate to cuneate at base, obtuse to rounded at apex; upper leaves narrowly lanceolate, petiole gradually shorter to sessile, attenuate at base, crenate to obtusely dentate at margin, acute at apex. Cymes finely long pedunculate, capitulate, 3–7-flowered. Pedicel 3–10 mm. Bracteoles lanceolate to linear-lanceolate, 4–7 mm, shorter than calyx, obtuse at apex, white lanate-ciliate. Calyx tubular, 4.5–6 mm, mouth straight, cleft indistinct, sparsely scabridulous to white-lanate, with minutely glandular papillae and sessile glands; teeth erect, subequal, 1.5–2.5 mm, lanceolate, acute, ciliate or not; glabrous in throat, but hairy between the teeth. Corolla white to pinkish-purple, 7–10 mm, pubescent, tube exserted from calyx; lower lips deflexed, median lobe flabellate-dilatate, concave, crenate-incised, bearded. Stamens and style included in corolla. Nutlet ovoid-oblong, 1.5 × 1 mm, dark brown, glabrous, areola narrowly transverse.

HAB. Stony hillsides, open herbaceous vegetation; alt. 1200–2500 m; fl. & fr. Jun.-Jul.
DISTRIB. Mountains of northern and northeastern Iraq. **MSU**: in montibus calcareis Avroman et Schahu, *Haussknecht* 780! (BM, K, W); Tawela, *Rawi* 21940! (K); Suleymaniya, Pira Magrun (Pir Omar Magrun), *Haussknecht* s.n.! (K). **MRO**: Arbil: Rost, *Haley* 187! (BM).

Iran.

15. **Nepeta iraqo-iranica** *Haloob, Bordbar & Qader*, Phytotaxa 550(2): 196 (2022). Type: Iraq, N.E., 38 km from N.E. Sulaimaniya, N. Basneh village, near Rushikani, Shahrbazhar (Chuwarta), 1575 m, 24.05.2021, *Omar* 60380 (BAG, holo. & iso.).

Annual, 30–60 cm tall; stem erect, simple or branched from lower half, with dense puberulent-crispate and sparsely glandular papillose hairs. Leaves bright green, covered with dense pubescent or ± villous and sparsely glandular papillose hairs, densee on lower surface; lower leaves petiolate, petiole 10–35 mm, broadly triangular-ovate, 25–35 × 20–30 mm, cordate-subcordate at base, crenate at margins, rounded at apex, sometimes mucrunulate, middle leaves with petiole 5–10 mm, lamina ovate, 11–18 × 8–15 mm, crenate-dentate at margins, subcordate-subtruncate at base, obtuse at apex; upper leaves sessile, lanceolate, 10–15 × 3–8 mm, cuneate at base, apiculate-mucronate at apex. Verticillasters congested into a terminal elongated spike, 30–60 mm long, 10–15 mm wide; flowers many, pedicel 1–2 mm. Bracts broadly lanceolate, 11–13 × 2–4 mm, sessile, green, purplish in the upper half, apex rostrate, adaxial surface glabrous, abaxial surface with villous-crispate and sparsely glandular papillose hairs. Calyx straight, tubular, 8–9 mm, oblique at throat, tube 5–6 mm, green, glabrous inside, covered with villous-crispate and sparsely glandular papillose hairs outside; upper teeth lanceolate, 2.0–2.5 mm; lower teeth narrowly lanceolate 2.5–3.0 mm, teeth purple, margins ciliate. Corolla pinkish-purple, 13–14 mm, exserted from the calyx, tube 7–9 mm; upper lip 2 × 2 mm; lateral lobes of lower lip 1 mm long, middle lobe 4–5 mm long, margins crenate; corolla villous outside, glabrous inside, internally with a dense tuft of short hairs at throat, upper lip covered with white pilose hairs inside. Stamens 4, posterior pair longer than anterior; filaments glabrous; anthers purple, glabrous. Style shortly 2-branched, branches unequal. Nutlets oblong, ca. 1.7 × 0.8 mm, brown, surface glabrous, areole 0.3–0.4 mm, lobes oblique.

HAB. Colluvial soils of hillslopes; alt. 1575–1790 m; fl. & fr. May-July.
DISTRIB. N.E. Iraq. **MSU**: 38 km from N.E. Sulaimaniya, N. Basneh village, near Rushikani, Shahrbazhar (Chuwarta), 1790 m, *Haloob, et al.* 60236! (BAG).

Nepeta iraqo-iranica is similar to *N. wettsteinii* distributed in Gilan and Azerbaijan Province of Iran. But it differs mainly in plant height, size of the lower leaf blade, inflorescence shape, length of bract, upper teeth shape of the calyx, indumentum of bract and calyx. *Nepeta iraqo-iranica* also similar morphologically to *N. humilis* and *N. petraea*, the two other sympatric annual species, from west of Iran and northeast of Iraq. It can be readily distinguished from them by the difference in the shape of the lower leaf blade, inflorescence shape, shape and length of bract, length of calyx, and corolla.

W. Iran.

16. **Nepeta macrosiphon** *Boiss.*, Diagn. Pl. Orient. 7: 51 (1846); Rechinger, Fl. Iranica [K. H. Rechinger] 150: 153(1982); Hedge, Fl. Turkey [P. H. Davis] 7: 284 (1982); Jamzad in Flora Iran [Assadi & al., in Persian] 76: 553 (2012).

Glechoma macrosiphon (Boiss.) Kuntze, Revis. Gen. Pl. 2: 518 (1891).
Nepeta glandulosa Blakelock, Kew Bull. 4: 542 (1949 publ. 1950).

Perennial herbs, 20–60 cm tall; stems few to many from a woody rootstock, erect or ascending, glabrescent to canescent, glandular-papillose or not. Leaves ovate to triangular, 1–4 × 1–3(–4) cm, cordate at base, coarsely crenate, obtuse at apex, sparsely to densely pilose or canescent, glandular papillose or not; lower leaves clearly petiolate to 20 mm, upper subsessile to sessile. Inflorescence lax, cymes all pedunculate, loosely flowered. Bracteoles elliptic to linear, 2–4 mm, much shorter than calyx. Calyx tubular, 8–12 mm, clearly bilabiate with a deep cleft on lower lip; tube long and usually partially or completely deep purple, somewhat curved, mouth oblique, veins prominent, ± sparsely pilose to scabridulous, glandular or not; teeth 1–2.5 mm; upper teeth oblong-triangular; lower teeth lanceolate-triangular, lower lip deeply cleft; sparsely hairy inside of the tube up to ½. Corolla purplish-blue, 20–26(–30) mm, tube curved or not, narrow, long exserted from calyx teeth, tube subglabrous to puberulous, pilose at lips; upper lip c. 2 mm; lower lip 3–4 mm, median lobe concave, crenate, bearded. Stamens 4, included. Style unequally bilobed, included in corolla. Nutlets oblong, ± trigonous, 2.5–2.8 × 1–1.2 mm, ± smooth.

HAB. Rocky slopes and screes; alt. 2000–3600 m; fl. & fr. Jul.-Sep.
DISTRIB. Rocky slopes and screes of the northern Iraq mountains. **MRO**: Erbil, mons Helgurd ad confines Persiae, in valle supra pagum Nowanda, *Rechinger* 11343! (BUH, G, K, W); Helgord range, E. of Berma sand lake, cerpentine rocks, *Serhang & Rawi* 24774! (BAG-photo); Arbil, Algurd Dagh, *Guest* 3067! (syntype of *N. glandulosa*) (BAG); N.E. slopes of Arl Gird Dagh, *Gillett* 9586! (BAG, K); Ibid., *Gillet* 9588! (BAG, K); Erbil, Algird Dagh, among rocks, *Gillet* 12365! (BAG, K); Ser Kurawa, *Gillet* 9755 (BAG); Ser-i Khazni, *Haley* 253 (BM, BUH). **MSU**: in montibus calcaresi Avroman et Schahu, *Haussknecht* s.n.! (BM, K, W).

E. and S.E. Turkey, W. Iran.

17. **Nepeta lamiifolia** *Willd.*, Enum. Pl.: 602 (1809); Poyarkova in Fl. U.S.S.R [Shishkin & Yuzepchuk] 20: 272 (1976); Hedge, Fl. Turkey [P. H. Davis] 7: 282 (1982).

Glechoma lamiifolia (Willd.) Kuntze, Revis. Gen. Pl. 2: 518 (1891), nom. illeg.
Nepeta iodantha Nábelek, Spisy Prír. Fak. Masarykovy Univ. 70: 53 (1926).

Perennial with ± woody rhizome. Stems ascending, 10–22 cm tall, scabridulous to pilose, with or without glandular papillae or sessile glands. Leaves ovate-triangular to broadly triangular, 1–2 × 1–2.2 cm, cordate to ± truncate at base, obtuse at apex, coarsely crenate-dentate, ± puberulous to pilose with glandular papillae and sessile glands; lower leaves petiolate to 1.3 cm; upper subsessile. Inflorescence verticillate, verticillasters usually congested at apex, sometimes lowermost verticillaster remote, flowers congested. Bracteoles elliptic to linear-lanceolate, 4–6 × 0.3–1 mm, usually shorter than tube. Calyx tubular, 6–7 mm, straight or somewhat curved, expanded above, clearly bilabiate with a deep cleft on lower lip; usually partially or completely deep purple, mouth oblique, pilose with glandular papillose. Corolla pale purplish-blue, 12–16 mm, expanded at mouth; tube somewhat curved, exserted from calyx, tube subglabrous to puberulous or sparsely long hairy; upper lip 2–3 mm; lower lip 5–6 mm, median lobe concave, crenate, bearded. Stamens 4, included in corolla. Style included in corolla or slightly protruding. Nutlets ovate, trigonous, 2–2.6 × 1.2–1.4 mm, ± smooth. Fig. 173, 1–6.

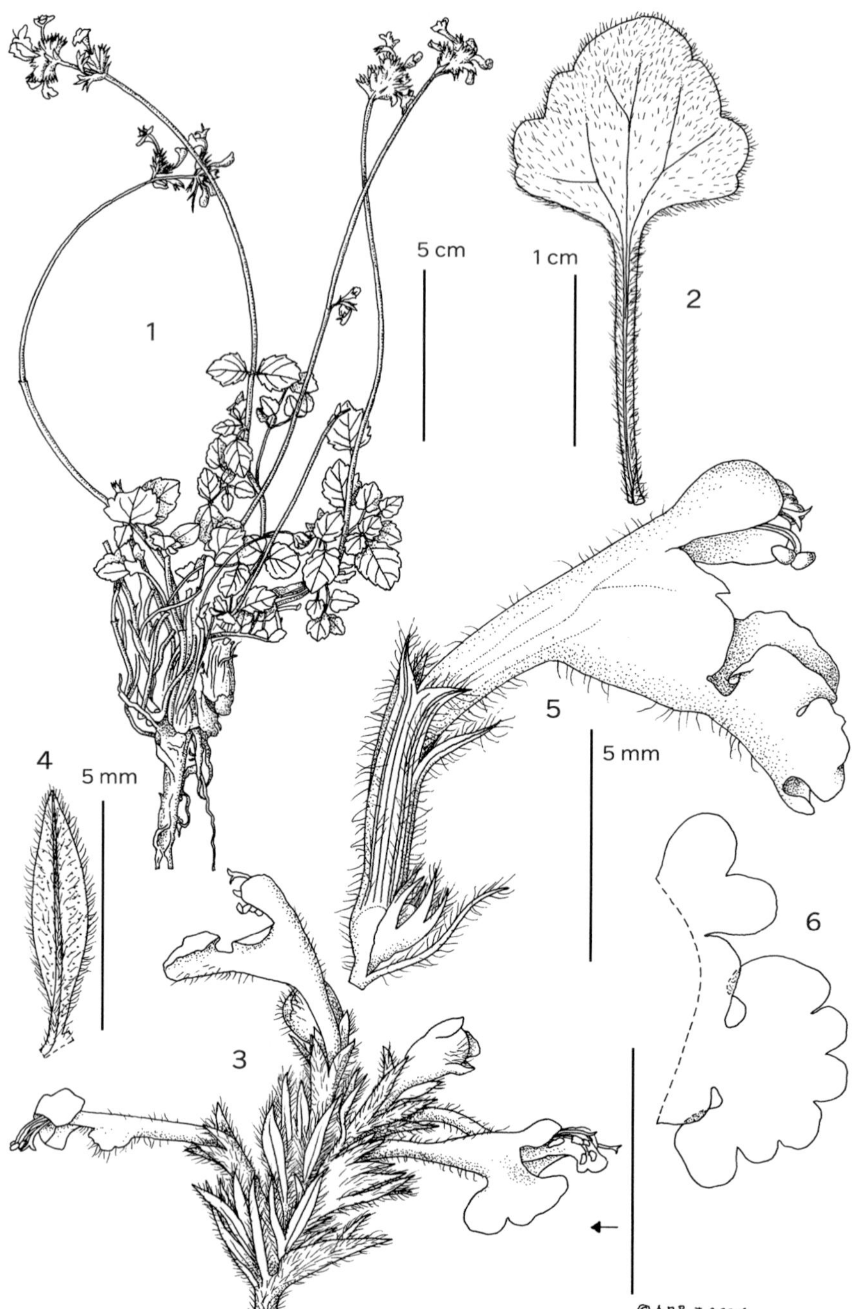

Fig. 173. **Nepeta lamiifolia.** 1, habit; 2, near basal leaf, adaxial surface; 3, inflorescence; 4, bract of inflorescence; 5, flower, side view; 6, upper and lower lobes of corolla (flattened). 1, 4 from *Rawi et al.* 24515; 2–3, 5–6 from *Rawi et al.* 24523. Drawn by © A.P. Brown, Jan. 2024.

HAB. Igneous screes, usually volcanic soils; alt. 2900–3340 m; fl. & fr. Jul.-Aug.

DISTRIB. High mountain range N. Iraq, usually to volcanic soils and igneous scree habitats. **MRO**: Qandil range, N.E. of Rania, brownish-black rocky land, *Serhang & Rawi* 26786! (K); ibid., *Serhang & A.Rawi* 26810! (K); Qandil range, *Rawi* 24515! (BAG, K); Perrish mountain, *Serhang & Rawi* 24523! (K).

Nepeta lamiifolia has not been recorded for Iraq by either Rechinger or Al Rawi. According to Hedge (1982: 282) its distribution in northern Iraq was doubtful. *Nepeta lamiifolia* is Armenian in origin, described by Willdenow (B-W–10742-010!, type in B-Willd.), distributed along from the Southern Caucasus and the high mountain belt in the Northeast Turkey to the mountains of northern Iraq. It is a species with a short capitate inflorescence, often densely pilose. Found on volcanic soils in Iraq. Although it is accepted as closely related to *Nepeta brevifolia* C.A.Mey. by some authors, it differs from it in its capitate inflorescence with a dense indumentum. A large number of specimens consistent with the type specimen of *N. lamiifolia* were collected in the Caucasus, Turkey, and Iraq. It differs from *N. macrosiphon* in Iraq, to which it is also closely related, by its shorter calyx and corolla and its capitate inflorescence.

E. & S.E. Turkey to S. Transcaucasia.

18. **Nepeta trautvetteri** *Boiss. & Buhse*, Nouv. Mém. Soc. Imp. Naturalistes Moscou 12: 175 (1860); Rawi in Dep. Agr. Tech. Bull 14: 153 (1964).

Glechoma trautvetteri (Boiss. & Buhse) Kuntze, Revis. Gen. Pl. 2: 519 (1891).

Perennial, 40–85 cm tall; stems mostly branching from base and leafy in lower part, axillary branches long, with few pairs of cymes, densely covered with coarse white spreading hairs, glandular papillose. Leaves broadly ovate to suborbicular-ovate, 1–3.5 × 1–3 cm, deeply cordate at base, crenate at margins, rounded or obtuse at apex; lower cauline leaves petiolate to 1.5 cm long, upper short petiolate to subsessile, sparsely to densely hirsute with few to many sessile glands; floral leaves small; sessile, oblong-ovate to narrowly elliptic, entire or lower with few teeth. Inflorescence broad, loose panicle, dichotomously branched; cymes remote, the axes of first order to 3 cm long or more, sometimes repeatedly dichomotous or bostryciform. Bracteoles 2–4 mm, elliptic to linear. Calyx tubular, 7–9 mm, purplish, densely covered with whitish or yellowish sessile glands and sparse spreading 3-jointed simple hairs with glandular papillose, frontally incised to the middle; teeth with scarious margins, upper teeth 1–2 mm long, oblong-ovate, mucronate; lower teeth 1.5–2.5 mm long, lanceolate to narrowly lanceolate, acuminate; tube hairy inside. Corolla purplish-blue, 13–17 mm long, tube slender and curved, exserted from calyx, tube and lips short hairy and glandular papillose outside; middle lobe of lower lip concave and crenate, bearded. Stamen and style included in corolla. Nutlets ellipsoid, 1.5–2.1 × 1–1.2 mm, brown, minutely tuberculate.

HAB. Dry stony ground, on rocks, eroded slopes; alt. 1000–2000 m; fl. & fr. Jun.-Aug.
DISTRIB. N. Iraq, known only from around Amadia. **MAM**: Mountains of S.E. of Serva, *Jillet* 9696! (BAG, K); Amadia, *Guest* 3770! (K); Sulaf, *Haines* 2067! (K).

N.E. Turkey, Transacaucasus.

19. **Nepeta iodantha** *Nábělek*, Spisy Prír. Fak. Masarykovy Univ. 70: 53 (1926).

Perennial, many stemmed from base, and usually few branched from lower part, ascending to erect, 30–45(–80) cm tall, base densely foliate, puberulent to hirsute with glandular papillose. Leaves cordate-ovate to suborbicular, 1.5–3 × 1.5–3 cm, petiolate to 2 cm long, cordate at base, crenate at margins, obtuse at apex, tomentose with sessile glands; upper cauline leaves short petiolate to sessile, ovate-cordate to oblong; floral leaves small. Cymes 3–5-flowered, 1 or 2 lower cymes clearly pedunculate, peduncle 1–1.5 cm, upper cymes short pedunculated or sessile, panicles contracted; all verticillasters distant. Bracteoles linear to subulate, 2–4 mm, green or violet, hirsute, ciliate. Calyx 6–8 mm, tube slightly incurved, partly or completely purple, mouth oblique, densely strigose-hirsute with glandular papillae; teeth unequal, 1.5–2 mm, triangular to triangular-lanceolate, acute-acuminate; hairy in throat. Corolla pinkish-purple, 14–20 mm, hirsute with glandular papillose, tube slightly incurved; median lobe of lower lip concave, crenate, bearded. Stamens 4, included in or slightly exserted. Style slightly exserted. Nutlet oblong, 2.5–3 × 1–1.2 mm, brown, smooth.

HAB. Serpentine rocks; alt. 1060–2100; fl. & fr. Aug.-Sept.
DISTRIB. N. Iraq. **MRO**: Pushtashan, 15 km E. of Ranin, lower slope of Qandil range, *Rawi & Serhang* 24203! (BAG, K); Sula Khal, black-brownish rocks, *Serhang & Rawi* 24679! (BAG, K); Kanirush, Qerna valley, N. of Pushtashan, near streams, *Rawi & Serhang* 26648! (BAG, K).

Nepeta iodantha originates from Turkey, first described by Nábělek (1926). It is distinguished from *N. macrosiphon* and *N. trautvetteri*, to which it is closely related to in its more tightly arranged verticillasters and shorter pedunculated cymes. *Nepeta iodantha* var. *parviflora* (*Ludlow-Hewitt* & *Guest* 2726!-2727!, in K and BAG), described by Blakelock (1949), adapted to higher altitudes than *Nepeta iodantha*, related to *Nepeta lamiifolia*. However, it differs from it in that its verticillasters which are remote from each other.

S.E. Turkey.

species doubtfully recorded for Iraq

Nepeta wettsteinii *Heinr.Braun,* Verh. K. K. Zool.-Bot. Ges. Wien 39: 226 (1889); Rechinger, Fl. Iranica [K. H. Rechinger] 150: 212 (1982); Jamzad in Flora Iran [Assadi & al., in Persian] 76: 498 (2012).

The existence of *Nepeta wettsteinii* in Iraq is doubtful. *Nepeta kurdica* which is known from Iraq was placed as a synonym of *Nepeta wettsteinii* by Rechinger (1982: 212). Thus, specimens belonging to *Nepeta kurdica* were included as *Nepeta wettsteinii* there. We studied the type specimens of *Nepeta wettsteiinii, N. humilis* and *N. kurdica,* and concluded that *Nepeta kurdica* is a synonym of *Nepeta humilis* (as treated here). All specimens identified as *Nepeta wettsteinii* in Flora Iranica are referred to as *Nepeta humilis.*

Nepeta teucriifolia Willd., Enum. Pl.: 602 (1809).

Specimens representing *Nepeta teucriifolia* (formerly *N. fissa* in Flora of Turkey, p. 283 and Flora Iranica, p. 152) are characterized by the inflorescence being lax and dichotomously branched. Flowering branches usually end with two flowers in *N. teucriifoia.* However, in some specimens, flowering branches can carry up to 8 flowers horizontally or bostryciform and these features are seen in *Nepeta trautvetteri* (*Buhse* 756-photo!, in P). Specimens with these inflorescence characteristics collected from Northern Iraq are referred to as *N. trautvetteri.* However, it is known that *N. teucriifolia* is also distributed in regions very close to the Iraq-Turkey border, therefore, it is probable that *N.teucriifolia* is also be found in northern Iraq.

29. **HYMENOCRATER** Fisch. & C.A. Mey.

Index Sem. Hort. Bot. Petropol. 2: 39 (1835).
Harley & al. in Kubitzki (ser. ed.), Fam. Gen. Vasc. Pl. 7: 250 (2004)

Tuncay Dirmenci & Ferhat Celep

Low shrubs or perennial herbs. Leaves simple, ovate-oblong or lanceolate, cordate to truncate. Inflorescence thyrsoid, spike-like or interrupted, cymes long to short-pedunculate or sessile; bracts leaf-like, oblong-elliptic to ± ovate, bracteoles inconspicuous. Flowers sometimes resupinate. Calyx tubular, campanulate, straight or curved, with cylindrical tube, annulate inside, 5-lobed, 15-veined, enlarged after anthesis, membranous, often pale green, rose or purplish. Corolla purple, pale purple or blue, tube elongate, gradually widening towards the throat; upper lip erect, bifid; lower lip patent, 3-lobed. Stamens 4, upper pair longer than lower, with divergent thecae. Nutlets ovoid or elliptic, smooth or tuberculate.

About 11 species, Iran, Turkey, Iraq, Pakistan and Afghanistan; a single species in Iraq.

Hymenocrater was transferred into *Nepeta* by Serpooshan et al. (2018) on the basis of morphological and molecular evidence. However, *Hymenocrater* has some clear morphological differences from *Nepeta*, i.e. the membranous, enlarged and coloured calyx lobes; it is treated as a distinct genus in many floras. Here too we treat *Hymenocrater* as a separate genus.

Budantsev, A.L. (1992). A synopsis of the tribe Nepeteae (Lamiaceae). The Genera *Meehania, Glechoma, Drepanocaryum, Marmoritis* and *Hymenocrater*. Bot. Zhurn. 77(12): 118–128.
Serpooshan F., Jamzad Z., Nejadsattari, T. & Mehregan, I. (2018). Molecular phylogenetics of *Hymenocrater* and allies (Lamiaceae): new insights from nrITS, plastis trnL intron and trnL-F intergenic spacer DNA sequences. Nord. J. Bot. 36(1–2):1–13.
WCVP (2021). World Checklist of Vascular Plants, version 2.0. http://wcvp.science.kew.org/

1. **Hymenocrater longiflorus** *Benth.* in DC., Prodr. [A. P. de Candolle] 12:407 (1848); *Rawi* in Dep. Agr. Tech. Bull 14:150 (1964); Rechinger, Fl. Iranica [K. H. Rechinger] 150: 240–241 (1980); Jamzad in Flora Iran [Assadi & al., in Persian] 76: 610 (2012).

> *Hymenocrater haussknechtii* Boiss. & Reut. Fl. Orient. [Boissier] 4: 677 (1879). Type: Iraq, Sulaimaniya, in montibus calcareis Avroman et Schahu, *Haussknecht* s.n.! (E!, JE-photo!, P-photo!, W!).
> *Nepeta avromanica* Jamzad & Serpooshan, Nordic J. Bot. 36(1–2): 10 (2018).

Perennial herb with a woody base. Stems erect, 20–60 cm tall, ascending to erect, the upper part of the inflorescence paniculate, subglabrous to short crispy and densely glandular-papillose hairy. Leaves 2.5–5 × 1.5–4 cm, cordate at base, apex obtuse, margins crenate; petiole up to 20 mm long on lower leaves, on upper leaves diminishing or sessile, indumentum papillose, with densely short stalked and sessile glands. Floral leaves ovate to oblong-lanceolate, sessile, acute. Bracteoles 5–10 mm, shorter than calyx tube, herbaceus, elliptic to lanceolate, acute. Peduncle 5–20 mm. Flowers subsessile to shortly pedicellate. Calyx 18–25 mm, accrescent in fruit, tube 8–11 mm, cylindiric, shortly pilose on nerves, densely glandular papillose; lobes 12–18 mm, 5–10 mm wide, obtuse, acute, or mucronate, reddish-purple, membranous. Corolla blue, 21–28 mm, tube 3.5 mm wide at throat, upper lip 5–6 mm, lower lip 5 mm. Stamens and style clearly exserted from corolla. Nutlets oblong, 2.7–3 × 1.8–2 mm, brown, tuberculate. Fig. 174, 1–3.

HAB. Calcareous rocks; alt. 1190–2000 m; fl. & fr. Jun.-Jul.
DISTRIB. On calcareous soils on mountain slopes in northeast Iraq. MSU: Sulaimaniya, in montibus calcareis Avroman et Schahu, *Haussknecht* s.n.! (type of *H. Hausskenctii*, E! JE-photo! P-photo! W!); Avroman mountain, Tawella, in saxosis calcerous, *Rechinger* 10374! (E, W); Jebel Avroman, above Darimar, *Gillett* 11864!; Avroman mountain, *Rawi et al.* 29520!; Molla Khort mountain, *Rawi et al.* 29541!; 7 km W. of Tawela, Balkha village, cultivated valley, *Rawi* 22366!; Kamarspa (on road bet Halabja & Tawella), *Rawi* 22208 (BAG).

W. Iran.

30. **LALLEMANTIA** Fisch. & C.A.Mey.

Index Seminum [St.Petersburg (Petropolitanus)] vi. 52 (1840); Harley & al. in Kubitzki (ser. ed.), Fam. Gen. Vasc. Pl. 7: 252 (2004)

Ali Haloob

Annual, biennial or perennial herbs. Stems simple or branched. Leaves petiolate. Inflorescence loosely spicate, verticillasters 4–8-flowered; pedicel erect, flattened. Bract petiolate or nearly sessile; bracteoles dentate, aristate. Calyx bilabiate, tubular, 15-veined, straight; upper lip 3-toothed, the median broader and longer than the lateral teeth, lower lip 2-toothed, teeth lanceolate. Corolla bilabiate, purple or white or yellow or yellow with purple limb, tube narrow, straight, included or slightly exserted, upper lip 2-lobed, concave, apex emarginate, with 2-longtudinal folds inside, lower lip 3-lobed, middle lobe broad and reniform. Stamens 4, didynamous, included, ascending under the upper lip, posterior pair longer than anterior pair; anther 2-celled, divaricate. Style apex with 2-unequal lobes. Ovary glabrous. Fruiting calyx closed by limb teeth. Nutlets oblong, smooth, dark brown.

A genus of 5 species found from Sinai to W. Siberia and W. Himalaya; 3 species in Iraq.

1. Bracteole orbicular, > 9-aristate . . . . .1. *L. peltata*
   Bracteole obovate-flabellate, 5–9-aristate. 2
2. Bract serrulate-entire; calyx 13–15 mm . . . . .2. *L. iberica*
   Bract crenate with 2 or 4 teeth (short aristae) near the base; calyx
   6–8 mm. . . . . .3. *L. royleana*

1. **Lallemantia peltata** *Fisch. & C.A.Mey*, Index Seminum [St. Petersburg (Petropolitanus)] vi. 52 (1840); Boissier, Fl. Orient. 4: 674 (1879); Rawi in Dep. Agr. Iraq Tech. Bull. 14: 150 (1964); Edmondson in Fl. Turkey [P. H. Davis] 7: 291 (1982); Rechinger, Fl. Iranica [K. H. Rechinger] 150: 233 (1982).

Annual; stem simple or branched at base, erect, 20–40 cm, puberulent or glabrescent. Leaves ovate-oblong, 30–40 × 10–20 mm, base cuneate, apex obtuse, margins serrulate;

Fig. 174. **Hymenocrater longiflorus**. 1, habit × 1; 2, flower × 1; 3, seed detail. Reproduced with permission from Jamzad, Flora of Iran 76: f. 189 (2012). © Ministry of Jihad-e-Agriculture.

petiole 10–20 mm. Verticillasters 6–8-flowered, upper whorls congested, the lower loose with 30–50 mm apart. Bract lanceolate, 20–60 × 5–15 mm, margins serrulate; bracteoles orbicular, 8–10 × 8–11 mm, light green with conspicuously reticulate veins, puberulent with yellow sessile glands, many (more than 9) aristate; aristae ± 5 mm. Pedicel 2–4 mm, densely puberulent. Calyx 11–13 mm, teeth mucronate, pubescent with yellow sessile glands, the upper tooth ovate, ± 4 mm, lateral oblong, lower teeth lanceolate. Corolla pale purple or white with a pale purple limb, 14–17 mm, tube glabrous outside, puberulent inside; limb with dense minute glandular and eglandular hairs on the outside. Nutlets oblong, 3–3.5 mm, glabrous.

HAB. Limestone mountain, waste ground near spring; alt. 1300–2000 m; fl. Jun.-Jul.
DISTRIB. Rare in N.E. sector in the lower forest zone. MRO: Khan Mami Sherin N. of Shirwan Mazin, *Agnew, Hadac, Huine & Kadir* 6186! (BUH); Sarcal, Helgurd Mt., *Hadac* 2664! (BUH). MSU: Avroman Mt. Dara Tariq above Tawilla, *Rechinger* 10328! (BUH); Kamarspa (on the road between Halabja & Tawela), *Rawi* 22215!

Iran, Turkey, Russia.

2 **Lallemantia iberica** *Fisch. & C.A.Mey,* Index Seminum [St. Petersburg (Petropolitanus)] vi. 52 (1840); Boissier, Fl. Orient. 4: 674 (1879); Zohary in Dep. Agr. Iraq Bull. 31: 126 (1950); Blakelock in Kew Bull., 4: 540 (1950); Rawi in Dep. Agr. Iraq Tech. Bull. 14: 150 (1964); Rechinger, Fl. Lowland Iraq: 521 (1964); Feinbrun-Dothan, Fl. Palaest. 3: 117 (1978); Edmondson in Fl. Turkey [P. H. Davis]. 7: 292 (1982); Rechinger, Fl. Iranica [K. H. Rechnger] 150: 235 (1982).

Annual; stem simple or branched at base, erect, 15–45 cm puberulent or glabrescent. Leaves ovate-oblong, 10–40 × 10–20 mm, base rounded or cuneate, apex obtuse, margins crenate to serrulate; petiole 5–10 mm. Verticillaster 6–8-flowered, upper whorls congested, the lower loose, 20–50 mm apart. Bract linear-lanceolate, 20–45 × 3–10 mm, margins entire to serrulate; bracteoles obovate-flabellate, 4–7 × 2–4 mm, light green with conspicuously reticulate veins, puberulent with yellow sessile glands, 5–9-aristate; aristae 6–8 mm. Pedicel 2–4 mm. Calyx 13–15 mm, teeth mucronate, pubescent with yellow sessile glands, the upper tooth ovate, ± 5 mm, lateral oblong, lower teeth lanceolate. Corolla pale purplish-blue or white or white with a pale purple limb, 15–18 mm. long, tube glabrescent or glabrous; limb with dense minute glandular and eglandular hairs on the outside. Nutlets oblong, ± 4 mm. long, glabrous. Fig. 175, 1–4.

HAB. Limestone rocky mountain, mountain and hill slopes, *Quercus* forest, wheat and cultivated field, under oak trees, on clay and sandy clayey soil, open dry wasteland, near roadside; alt. 270–1700 m; fl. Mar.-Jul.
DISTRIB. Occasional in the lower forest zone and N.W. sector of the upper plains of Iraq. MJS: Albaiyder, 78 km S. of Sinjar, *Khatib & Alizzi* 32109; Sinjar, *Al-Kaisi & K. Hamad* 49076!; Jabal Sinjar 2 km E. of Kursi, *Al-Khakani* 0036739! (BUH); MAM: Matine, *Rawi* 8715! MRO: near Silca village, between Meirose and Shirwan Mazin, *Al-Musawi & Addai* 44732! (BUH); Kawricsh E. side of Karoukh mountain, *Kass & Nuri* 27652!; Gali Warta c. 30 km. N.W. of Rania, *Nuri & Kass* 28724!; Shaqlawah, *R. Wheeler-Haines* 654! MSU: 30 km from Kirkuk to Sulaimaniya, *F. Karim, H. Hamid & M. Jasim* 40521!; near Chamchemal, *F.A. Rogers* 244!; Kopi Qaradagh, *Agnew* 4903! (BUH). FNI: Namrood, *W. Hashimi* 1071! (BUH); Tal Kaif, *Al-Kaisi & K. Hamad* 49271!; Mosul, *Guest* 1343!

SIMSIM BARRI (Arb. in Chakravarty, Plant Wealth of Iraq 1: 318, 1976).

Lebanon, Palestine, Syria, Turkey, Iran, Russia.

3. **Lallemantia royleana** *Benth.,* Prodr. [A. P. de Candolle] 12: 404 (1848); Boissier, Fl. Orient. 4: 674 (1879); Handel-Mazzetti in Ann. Naturh. Mus. Wien 27: 412 (1913); Zohary in Dep. Agr. Iraq Bull. 31: 126 (1950); Rawi in Dep. Agr. Iraq Tech. Bull. 14: 150 (1964); Rechinger, Fl. Lowland Iraq: 520 (1964); Rechinger, Fl. Iranica [K. H. Rechinger] 150: 236 (1982); Hedge in Fl. Pakistan [Ali & Y. Nasir] 192: 130 (1990).

Annual; stem simple or branched at basal, erect, 10–30 cm, pubescent. Leaves ovate, 5–13 × 5–14 mm, base truncate to subtruncate, apex obtuse, margins crenate; petiole 14–18 mm. Verticillaster 4–6-flowered, upper whorls congested, the lower loose, 15–35 mm apart. Bract ovate-oblong, 8–20 × 5–15 mm, margins crenate with 2 or 4 teeth (short aristae) near the base; bracteoles obovate to oblong, 5–10 × 3–5 mm, puberulent with yellow sessile glands, 5–7-aristate; aristae ± 3 mm. Pedicel 2–3 mm, densely puberulent. Calyx 6–8 mm, teeth acute, densely pubescent with yellow sessile glands, the upper tooth ovate, 1.5–2.5 mm, lateral oblong, lower teeth lanceolate-linear. Corolla blue, 8–10 mm, glabrous outside, pubescent inside; limb with pubescent and glandular hairs outside. Nutlets oblong, ± 2–3 mm, glabrous.

Fig. 175. **Lallemantia iberica**. 1, habit; 2, flower × 2; 3, calyx × 2; 4, bract × 2. Reproduced with permission from Feinbrun-Dothan, Fl. Palaestina 3: Plates, f. 189 (1977). Drawn by Esther Huber. © The Israel Academy of Sciences and Humanities.

HAB. Rocky and stony mountains and hillsides, limestone hills, dry rocky valleys, in wheat fields, on sandy clayey soils, near Euphrates river; alt. 85–680 m; fl. Mar.-May.
DISTRIB. Common in desert region. **DLJ**: 5 km E. of Rawa in Jazira, *Omar, Al-Kaisi, K. Hamad & H. Hamid* 45131!; 50 km from Haditha to Baji, *Noori, Omar, Hamid, Kh. Muhamad & Jasim* 41479! (BAG); Rawa – Sinjar 115 S.W. of Sinjar, *Khatib & Alizzi* 31976! **DWD**: 10 km S.W. of Rutba, *Rawi* 21061; Rutba, *Hadac* 4369! (BUH); 5 km S.W. of Ana, *Omar, Al-Kaisi, K. Hamad & H. Hamid* 44999! **DSD**: Basrah, *Wheeler-Haines* 1434!; 30 km from Salman to Samawa, *Al-Kaisi, K. Hamad & H. Hamid* 48290!; c. 40 km N.W. of Shabicha, *Guest, Rawi & Rechinger* 19312!

Syria, Iran, C. Asia, Pakistan, W. China, S. Russia.

31. **MELISSA** L.

Sp. Pl. 2: 592 (1753); Harley & al. in Kubitzki (ser. ed.), Fam. Gen. Vasc. Pl. 7: 253 (2004)

Ali Haloob

Perennial herbs, aromatic. Stems quadrangular. Leaves petiolate, ovate, with crenate or serrate margins. Verticillasters axillary. Bract petiolate, similar to cauline leaves; bracteolate minute. Calyx bilabiate, tubular campanulate, 13-veined, upper lip flattened, shortly 3-toothed, lower lip 2-toothed, throat with or without ring of hairs. Corolla bilabiate, tube exserted or slightly exserted from calyx, curved, upper lip 2-lobed, apex emarginate, straight or slightly hooded, lower lip 3-lobed. Stamens 4, anterior pair longer than posterior pair, included or the anterior pair shortly exserted from corolla, arcuate under upper lip; anther 2-celled (bilocular), divergent. Style branches subequal. Nutlets basally attenuate and smooth.

Genus with four species, distributed from southern Europe, N.W. Africa to W. Malesia; a single species in Iraq.

1. **Melissa officinalis** *L.*, Sp. Pl. 2: 592 (1753); Boissier, Fl. Orient. 4: 584 (1879); Handel-Mazzetti in Ann. Naturh. Mus. Wien 27: 420 (1913); Nábělek, Publ. Fac. Sci. Univ. Masarky 35: 46 (1923); Blakelock in Kew Bull., 4: 541 (1950); Rawi in Dep. Agr. Iraq Tech. Bull. 14: 154 (1964); Rechinger, Fl. Lowland Iraq: 529 (1964); Ball in Fl. Europ. 3: 162 (1972); Feinbrun-Dothan, Fl. Paleast. 3: 146 (1978); Rechinger, Fl. Iranica [K. H. Rechinger] 150: 494 (1982); Mill in Fl. Turkey [P. H. Davis] 7: 262 (1982); Meikle, Fl. Cyprus 2: 1284 (1985); Hedge, Fl. Pakistan [Ali & Y. Nasir]192: 229 (1990).

Perennial. Stems 25–100 cm, branched, erect or ascending, densely glandular-puberulent with spreading eglandular villous or hirsute hairs or glabrescent. Leaves cordate, ovate to broadly ovate to broadly elliptic, 13–80 × 7–50 mm, base cuneate or truncate-subcordate, apex acute, margins crenate; petiole 10–50 mm, subglabrous to pubescent. Verticillaster 4–12-flowered. Bract similar to cauline leaves, margins crenate-serrate, hairs more dense than cauline leaves; bracteolate oblong-subelliptic, 3–7 × 1–3 mm, entire with petiole ± 1 mm. Calyx 7–9 mm, upper lip flattened, with 3-mucronate teeth, middle tooth broadly triangular or obsolete, lower lip 2-toothed, lanceolate-aristate, 3–4 mm, puberulent to villous or minutely glandular. Corolla white or pale yellow or pinkish, 12–14 mm, tube exserted from calyx, curved, glabrous outside and with rigid eglandular hairs inside; upper lip 2-lobed, apex emarginate, straight or slightly hooded, lower lip 3-lobed, middle lobe entire and broader than the 2-lateral lobes. Stamens with anterior pair shortly exserted from corolla, arcuate under upper lip; anther 2-celled, divergent. Style branches subequal. Nutlets ± 3 mm, dark-brown, basally attenuate, smooth. Fig. 176, 1–3.

> Stems densely glandular-puberulent mixed with patent eglandular
> hairs; middle tooth of upper lip of calyx distinct broadly
> triangular . . . . a. subsp. *officinalis*
> Stems glabrescent to eglandular villous or hirsute; middle tooth of
> upper lip of calyx obsolete . . . . .b. subsp. *inodora*

a. subsp. **officinalis**

HAB. In shaded *Quercus* forest, valleys, under shade of *Juglans* near stream, on road side, in orchard between trees, in between rocks near stream, bank of stream, on clay soil; alt. 1000–1500 m; fl. Jun.-Aug.

Fig. 176. **Melissa officinalis**. 1, habit × 1; 2, calyx × 4; 3, flower × 4. Reproduced with permission from Jamzad, Flora of Iran 76: f. 244 (2012). © Ministry of Jihad-e-Agriculture.

DISTR. Occasional in the lower forest zone of Iraq.

MAM: Amadia, *J.F. Tilereet* 16397! (BAG); Sevara Gaurie Valley, N.E. of Zakho, between Sharanish & Zawita, *Rawi* 23668! (BAG). MRO: Ari near Ser Kurawa, *Gillett* 9705! (BAG); Shaglawah, *Guest* 3010! (BAG); Pushtashan, N.E. of Rania, *Serhang & Rawi* 26503! (BAG); Bekhal, *S.Omar, Sahira, F.Kurim & H.Hamid* 38399! (BAG); 11 km N. of Shaglawa, *Thamer* 47875A! (BAG); near Jabal Seffen, *Sahira* 37578! (BAG). MSU: Pakrajo, *Ali Askari* 1041! (BUH); Biara Valley, *Al-Shehbaz, M. Al-Khaikani & K. Ibrahem* 37654! (BUH); Penjwin, *W. Al-Hashimi* 17842! (BUH).

S. Europe, Turkey, N. Iran, Caucasia, Pakistan.

b. subsp. **inodora** *Bornm.*, Beih. Bot. Centralbl. 31(2): 250 (1914); Rechinger, Fl. Iranica [K. H. Rechinger] 150: 495 (1982); Mill in Fl. Turkey [P. H. Davis] 7: 263 (1982).

HAB. On road sides, in orchard, palm groves; alt. ± 110 m.; fl. May-Jul.
DISTR. Occasional in W. sector of the upper plains of Iraq. **PFP**: Khanaqin, *H.P. Parauibue* 85! (BAG); Baquba, *S.Omar, J.Mokhtar & S. Abdilly* 36999! (BAG); Badra, *Al-Kaisi & Yahya* 45299! (BAG); Baquba, *K. H. Rechinger* 9143! (BUH).

Lebanon, N. Syria, S. & E. Turkey.

32. **OCIMUM** L.

Sp. Pl.: 597 (1753); Gen. Pl., ed. 5: 259 (1754)
Paton in Kubitzki (ser. ed.) Fam. Gen. Vasc. Pl. 7: 260 (2004)
Becium Lindl. in Edwards' Bot. Reg. 28:42 (1842); *Erythrochlamys* Gürke in Bot. Jahrb. Syst. 19: 222 (1894); *Hesperaspis* Briq. in Bull. Herb. Boiss. sér. 2, 3: 975 (1903); *Nautochilus* Bremek. in Ann. Transvaal Mus. 15: 253 (1933).

Ferhat Celep & Tuncay Dirmenci

Annual or perennial herbs, or small shrubs, usually aromatic. Stems erect, usually branching above, often woody at base. Leaves opposite or occasionally ternate, petiolate or rarely subsessile. Verticillasters in bracteate terminal racemes. Bracteoles absent. Calyx funnel shaped to tubular, bilabiate, 5-lobed to 1/4, lobes unequal, deflexed in fruit; often densely hairy at throat; upper lip larger than lower, ovate-orbicular, entire, membranous, with decurrent margins; lower lip 4-toothed, teeth ovate, acute. Corolla whitish or pinkish, bilabiate, upper lip 4-lobed, lower lip entire or crenate. Stamens 4, declinate above lower lip, exserted from corolla; upper pair with appendiculate filaments. Style exserted, shortly bifid at apex. Nutlets ovoid or subglobose, mucilaginous.

About 100–150 species in temperate regions worldwide, especially in Africa and South America, in drier areas, grassland and open woodland. Some species are commonly grown for their aromatic fragrance, or as culinary or medicinal herbs; single cultivated species from Iraq.

1. **Ocimum basilicum** *L.*, Sp. Pl. 597 (1753); Rawi in Dep. Agr. Tech. Bull 14: 153 (1964); Rechinger, Fl. Iranica [K. H. Rechinger] 150: 575 (1982); Davis, Fl. Turkey [P. H. Davis] 7: 462 (1982); Jamzad in Flora Iran [Assadi & al., in Persian] 76: 955 (2012); Paton in Fl. Trop. E. Africa, Lamiaceae 137–147 (2009).

Ocimum thyrsiflorum L., Mant. Pl. alt.: 84 (1767).
O. album L., Mant Pl. Alt.: 84 (1767).
O. medium Mill., Gard. Dict., ed. 8, no. 3 (1768).
O. hispidum Lam., Encycl. Méth. Bot. 1 (2): 384 (1785).

Aromatic, annual or short lived perennial herb. Stems rounded-quadrangular, erect or ascending, branching above, 15–60 cm tall, shortly pubescent to subglabrous, sometimes densely pilose above. Leaves narrowly ovate, elliptic to lanceolate, 15–70 × 5–25 mm, usually remotely and shallowly dentate, subglabrous or puberulous, glandular-punctate with densely sessile glands on both surfaces, glaucous (sometimes heavily tinged purple), long-petiolate, 2–40 mm long. Inflorescence lax, verticils up to 12 mm apart; bracts deciduous or not, narrowly ovate to elliptical, 3–8 ×1–3 mm, acute to cuspidate at apex, cuneate at base, often coloured purple. Verticillasters 6-flowered. Calyx ± downward-pointing, 3–5 mm in flower, 6–8 mm in fruit, posterior lip ± glabrous, tube and anterior lip pubescent or pilose, sparsely gland-dotted, interior with a dense ring of hairs from base to throat with eglandular hairs and glandular papillae. Corolla white, creamy yellow or pink, 7–8–(–10) mm, tube straight, funnel shaped, scarcely exceeding calyx. Stamens 6–8.5 mm, exceeding corolla by 2–3 mm. Nutlet black, ovoid, longer than broad, 2–2.5 mm long, ± smooth to minutely tuberculate. Fig. 177, 1–3.

HAB. Cultivated in gardens, orchards and in fields.
DISTRIB. Native to tropical Asia, Africa and America, cultivated in Europe and S.W. Asia. **MRO:** Shaqlawa, 37 km N.E. of Arbil, in orchard, *Omar et al.* 38291!; Magron range, Haj Omran, *Rawi & Serhang* 24303

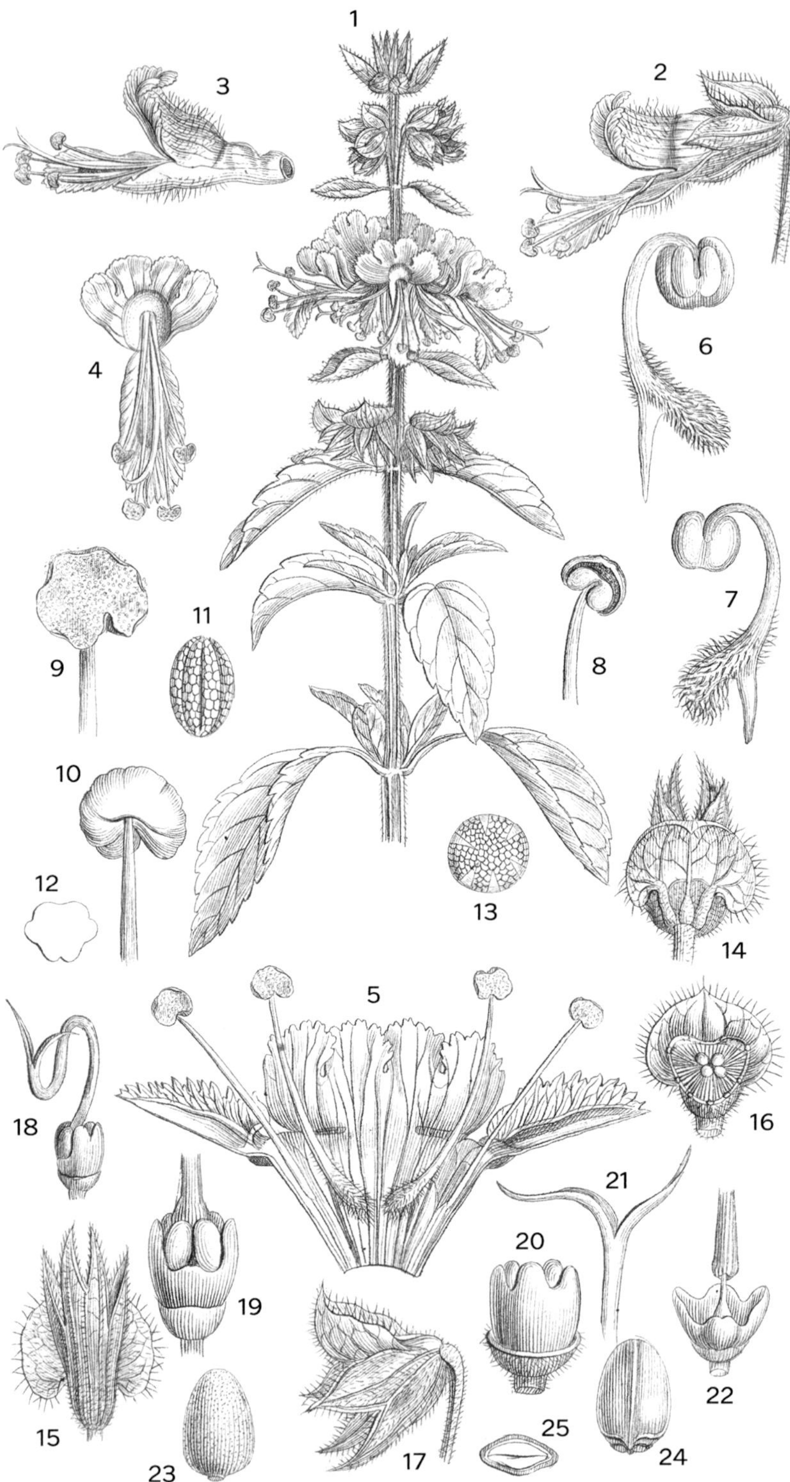

Fig. 177. **Ocimum basilicum**. 1, habit; 2–4, flower, different views; 5, flower opened; 6–8, stamen detail; 9–10, anther detail; 11–13, pollen; 14–17, calyx different views & detail; 18–22, ovary with style detail; 23–25, seed detail. Reproduced from Genera Plantarum Florae Germanicae Iconibus et Descriptionibus Illustrata by Nees von Esenbeck, 1843, vol 4, t.356. © Kew Archives.

(BAG). **fuj**: Tal afar, 60 km W. of Mosul, on the hill, *Rahman* 38149!. **lca:** Baghdad (Abu Ghraib), Salah Askar 14532 (BAG); Nasiriyah, Fawzi & *H. Hamid* 37789 (BAG). **lba:** Bassrah, E. Guest 295 (BAG).

The species is grown for its essential oils. The leaves are used in salads and sauces.

129. **BIEBERSTEINIACEAE** Endlicher

Enchiridion botanicum: 618 (1841); Muellner in Kubitzki (ser. ed.), Fam. Gen. Vasc. Pl. 10: 72 (2011)

Shahina A. Ghazanfar

Perennial herbs, woody at base, with a characteristic odour and viscous glandular pubescence; rhizomes sometimes tuberous. Leaves alternate, pinnate or pinnatisect, stipules adnate to petiole. Inflorescence rhizomes paniculate or spicate; pedicels with prophylls. Flowers 5-merous, hermaphrodite, regular, white or yellow. Sepals and petals imbricate or sometimes contorted. Nectar glands 5, alternating with petals. Stamens 10, filaments shortly connate. Carpels 5, ovary deeply 5-lobed, on a short gynophore; ovule 1 per locule; styles filiform connected above into a capitate stigma. Fruit schizocarpic, dehiscing into 5, 1-seeded mericarps; mericarps rugose. Seeds rugulose.

Five species from Greece to the Caucasus, western Himalayas, Central Asia and southwestern China; a single species in Iraq.

Biebersteinia, named for Friedrich Marschall von Bieberstein, 1768–1826, German-born soldier, botanist and explorer; author of *Flora Taurico-Caucasica*.

Biebersteina has been placed in the Geraniaceae by several authors (see Bakker et al. 1998 for summary), but recognised as a separate order and family by Takhtajan (1997) based on pollen characters, one ovule per locule and in having a gynophore. Based on molecular evidence it is placed as a distinct family Biebersteiniaceae in the Order Sapindales by APG IV 2016 classification. Since Geraniaceae is also treated in this volume, we are keeping in line with the APG IV classification and treating *Biebersteina* in its own family.

Muellner, A. N., Vassiliades, D. D., & Renner, S. S. (2007). Placing Biebersteiniaceae, a herbaceous clade of Sapindales, in a temporal and geographic context. Pl. Syst. Evol. 266 (3–4), 233–252.
Bakker, F.T., Vassiliades, D.D., Morton, C. & Savolainen, V. (1998). Phylogenetic relationships of Biebersteinia Stephan (Geraniaceae) inferred from rbc L and atp B sequence comparisons. Bot. J. Linn. Soc. 127(2): 149–158.

1. **BIEBERSTEINIA** Stephan

Zap. Obshch. Isp. Prir. Imp. Moskovsk. Univ. **1: 126 (1806)**

Characters same as those of the family.

A monogeneric family with 5 species in southern Caucasus and Central Asia to the borders of China; a single species in Iraq.

1. **Biebersteinia multifida** *DC.*, Prodr. [A.P. de Candolle] 1: 708 (1824); Boissier, Fl. Orient. 1: 899 (1867); Blakelock in Kew Bull. 1948: 408 (1949); Zohary in Dep. Agr. Iraq Bull. 31: 97 (1950); Rawi in Dep. Agr. Iraq Tech. Bull. 14: 58 (1964).

Perennial herb with a tuberous rootstock. Stem solitary, 20–35 cm, or 2–3 together, unbranched, clothed with broad, brown, scarious stipules at base, covered with black-tipped stalked glands, and sparsely lanate hairy. Leaves lanceolate, acute in outline, narrowed at base, petiole obsolete, with 2 conspicuous brown scarious stipules; lamina 8–25 × 6–10 cm, irregularly pinnate, pinnae bi-pinnatisect, sparsely lanate. Inflorescence a raceme of 1–4-flowered cymes, rather diffuse; pedicels as long as or shorter than flowers, with 2 bracteoles. Sepals ovate, obtuse, herbaceous, 9–12 mm, lanate and with stalked black-tipped glands. Petals golden-yellow with black nerves, scarcely longer than sepals, obovate, fimbriate at tip into 5–9oblong, single-veined lobes or entire. Stamens ⅓ as long as sepals, connate at base for ¼ their length. Achenes 4–5 × 3 mm, keeled within, triangular in cross

Fig. 178. **Biebersteinia multifida**. 1, habit; 2, detail of leaf; 3, young flower and floral bracts; 4, indument detail; 5, mature flower with front sepal removed. 1–4 from *Emberger et al.* 15493; 5 from *Rawi et al.*, 28720. Drawn by © A.P. Brown, 2023.

section, hemispherical in longitudinal section, with obscurely reticulate faces and back. Fig. 176, 1–5.

HAB. Dry rocky soil, in clearings in high altitude forest, or more commonly above the tree line; alt. (1200–) 1800–2400 m; fl. Apr.-Jun.

DISTRIB. An uncommon plant of mountainous areas of N. Iraq. **MRO**: Haji Umran, *Gillett* 11185; Rayat, *Low* 209; Handren Dagh E. of Rowanduz, *Bornmüller* 1007; *Thesiger* 1075; Gali Dargala, c. 35 km W. by N. of Rania, *Rawi, Nuri & Kass* 28915; Mermarut mt., *Guest* 2086! (BAG); Haji Umran, *Chapman* 12290; Karoukh Mountain (between Saran village & Kilkil), *Al- Kass, Nuri & Sarhang* 27334! (BAG); Gali Warta, c. 30km N.W. by N. of Rania, *Rawi, Nuri & Kass* 28720! (BAG). **MSU**: Sulaimaniya, Gweija Dagh, *Rawi* 8878!; *Ludlow-Hewitt* 1937!; Pira Magrun, *Rawi* 12091; *Haussknecht*, 1897, s.n.!; Tawila, mt. Avroman, *Rechinger* 10354!; **MJS**: Karsi, *Al-Kaisi & K. Hamad* 54103! (BAG).

HARSAIA, BONKHOSHA.

A medicinal plant whose root has been used for treating abdominal conditions.

Lebanon, Syria, Armenia, N. & W. Iran, Afghanistan, Turkmenia to the Pamirs.

INDEX TO VERNACULAR NAMES

Vernacular names are cited in the text along with their source, and in some cases are translated into English. Arabic and Kurdish names are capitalised in this list; English names are given in lower case letters. In this volume, the Arabic version of the script is employed on the authority of the author of the account; most information on local names is derived from herbarium labels which lack the Arabic version.

AHWAR, 105
ALGA, 74
ARAIFIYĀN, 87
ARGED, 87
AUSAJ, 15
AWSAJ, 15
BADINJAN, 7
BAIT LAHAM, 381
BATATA, 8
BARCHIBAH, 417
BERDUSHKA, 213
BETTINESA, 375
BINJ, 22
BONKHOSHA, 491
BUHNSHINK, 250
CHABAND, 19
CHAHAL, 204, 206
CHAHĀL, 206
CHAHALL, 206
CHAMAL, 232
CHAUHAR, 22
DUTALAN, 273
ENAB-ADH-DHIB, 1
ENAB AL-THA'LAB, 7
GARNA, 105
GIA HUSHTER, 312
GIAMARANA, 22
GIAPAPULA, 455
GIA RASHA, 63
GIA RIGYAWA, 461
GUL ZANIANA, 475
GŪRISK, 242
HAB AL-A'LAM, 7
HABĀT, 231
HALUK, 136
HALUK-RIHI, 132
HAMĀT, 231
HAMHAMKA, 375
HAMHAN, 237
HAMEDH, 180
HANDAKUKI, 97
HARSAIA, 491
HASHISHAT AL-KALIB, 374
HASSAR, 206
HAUWA ZATTA, 115
HELIMA, 206
HEMHEM
INI'MAH, 246
JADDA, 315
JADHA, 275

JANTARA, 451
JATA, 422
JEDHA, 275
KARAM MARTAB, 180
KASHISHAT, 77
KAVALPIR, 65
KHASIAN AL-BAGHL, 213
KHATONAN, 246
KHUZAIMAH, 197, 206
KHUZAMAH, 206, 460
KHUZEIMAH, 460
KIA'AI BI ZHINH, 257
KIA BAR RUZH HOR, 247
KLAOB SHILA, 457
KULA ZWANA, 237
KUZRIK, 236
LISĀN AL-THŪR, 237
LISĀN AT THŌR, 135
MASI JARRAK, 65
MĪSHMAIZHŬK, 224
MISK-AL-JIN, 315
MISMISOH, 216
NA'NA', 455, 460
NASHKĀ, 286
NAWAR, 314
NEMENLOKA, 348
NORA ZALA, 5
OMKRAIN, 314
PUNK DAIMI, 422
QOLA ZWAN, 236
QURNA, 108
QURQAH, 135
QUTAINAH, 374
RAIHANA, 372
REHANIPALA, 374
RIM RĀM, 253
SAMARAA, 314
SARIM, 12
SATIH, 246
SAYKARĀN, 10
SHAJRET AL MUR, 314
SHAKARY, 314
SHEHAIMA, 105
SHIN SHIN, 312
SHIRRABAND, 22
SIMRAH, 65
SIMSIM BARRI, 483
SULAJA, 74
TALISHKA, 375
TAMATAM, 7

INDEX TO FAMILIES, GENERA AND SPECIES

Scientific names of accepted species and infraspecific taxa are in **bold**, synonyms and misapplied names are in *italics* and subgeneric and sectional names are in SMALL CAPITALS. Where a name appears on more than one page, the principal page reference is given in **bold**. Names of species not recorded from Iraq, but mentioned in the text, are given in plain text.

Acanthaceae Juss., 122
Acanthus L., 123
  dioscoridis L., 123
    var. *boissieri* Bornm., 125
    var. *boissieri* Freyn, 125
    var. *grandiflorus* Bornm., 125
    var. *straussii* Hausskn. ex Bornm., 125
  grandiflorus Bornm., 125
Abiga St.-Lag., 299
Acinos Mill., 443
  canus (Steven ex M.Bieb.) Rchb., 445
  exiguus (Sm.) Meikle, 445
  fominii Des.-Shost., 445
  graveolens (M.Bieb.) Link, 445
Aethiopis sclarea (L.) Fourr., 402
Ajuga L., 299
  arenosa Steven ex Ledeb., 305
  austroiranica Rech.f., 301
  chamaepitys (L.) Schreb., 301
    subsp. *ciliata* (Briq.) Smejkal, 303
    var. *chia* (Schreb.) Batt., 302
    subsp. **chia** (Schreb.) Arcang., 302
      subsp. **laevigata** (Boiss.) P. H. Davis, 303
      subsp. **mardinensis** P. H. Davis, 303
      subsp. **rechingeri** (Bilik) P.H.Davis, 303
      subsp. **cuneatifolia** (Stapf) P. H. Davis, 304
      subsp. **tridactylites** (Ging. ex Benth.) P. H. Davis, 304
      var. **ciliata** Briq., 303
  chia, 302
  comata Stapf, 303
  cuneatifolia Stapf., 304
  intermedia Boiss. & Orph., 303
  laevigata Boiss., 303
  oblongata M.Bieb., 305
  rechingeri Bilik, 303
  salicifolia Steven, 305
  tridactylites Ging. ex Benth., 304
  vestita *Boiss.*, 304
  zakhoensis Rech.f., 300
Ajuga orientalis L., 305
Albraunia Speta, 87
  fugax (Boiss. & Noë) Speta, 89
  psilosperma Speta, 89
Alkanna Tausch, 194
  amplexicaulis DC., 197
  brachysolen Boiss., 197

bracteosa Boiss., 197
frigida Boiss., 199
heterophylla Vatke, 197
hirsutissima (Bertol.) A.DC., 195
kotschyana DC., 197
orientalis (L.) Boiss., 195
trichophila Hub.-Mor.
  subsp. **trichophila**, 199
  subsp. mardinensis Hub.-Mor., 199
Alkekengi Mill., 26
  officinarum Moench, 26
Amaracus Hill, 430
Amaracus haussknechtii var. *acutidens* Hand.-Mazz., 433
Anchusa L., 234
  aegyptiaca (L.) A.DC., 237
  arvensis (L.) M.Bieb.
    subsp. **orientalis** (L.) Nordh., 238
  aucheri A.DC., 238
  azurea Mill., 234
    var. *kurdica* (Gusul.) Chamberlain
  echioides (L.) M.Bieb., 260
  flava Forssk., 237
  hispida Forssk., 239
  hispidissima, 204
  italica Retz., 236
    var. *kurdica* Gusuleac, 236
    var. *macrocarpa* (Boiss. & Hohen.) Gusuleac, 236
    subsp. *macrocarpa* (Boiss. & Hohen.) Chamberlain, 236
  macrocarpa Boiss. & Hohen., 234
  neglecta A.DC. in DC., 240
  orientalis (L.) Reichb., 238
  ovata Lehm., 238
  palmyrensis Post, 260
  paniculata Aiton, 236
  picta M.Bieb., 259
  spinocarpos Forssk., 279
  strigosa Banks & Soland., 236
  strigosa Labill., 237
    var. *decalvata* Bég. & A.Vacc., 237
  ventricosa Sm., 260
Anemitis rigida (Labill.) Raf., 344
Antirrhinum L.
  aegyptiacum L., 91
  albifrons Sm., 100
  calycinum Soland., 104
  ceratotheca Nábělek, 89
  chalepense L., 96

oxyloba (Reut.) Soják, 133
ramosa (L.) Pomel, 131
umqasrensis Al-Mayah & Al-Asadi, 132
Phlomidopsis Link, 346
Phlomis L., 339
   anisodonta Boiss., 345
   armeniaca Willd. var. *olivieri* Benth., 340
   bruguieri Desf., 343
   herba-venti L. subsp. **pungens** (Willd.)
     Maire ex DeFilipps, 345
     var. *pungens* (Willd.) Schmalh., 345
   kurdica Rech.f., 341
   laciniata L., 347
   lanceolata Boiss & Hohen. ex Boiss.,
     340
   olivieri Benth., 340
   polioxantha Rech.f., 341
   × **praetervisa** Rech.f., 346
   pseudopungens Knorring, 345
   pungens Willd., 345
   pungens (Willd.) Rchb. ex T.Nees
     var. *hispida* K.Koch, 345
   pungens Willd., 345
   rigida Labill., 344
   seticalycina Nábělek, 345
   shepardii Post, 344
Pseuderemostachys Popov, 346
Phlomitis Rchb. ex T.Nees, 346
Phlomoides Moench, 346
   laciniata (L.) Kamelin & Makhm., 347
Phlomoides molucelloides (Bunge)
     Salmaki, 348
Phyllocara aucheri (DC.) Gusuleac, 238
Phyllophyton Kudô, 463
Physalis L., 8
   alkekengi L., 26
   angulata L., 8
   divaricata D.Don, 8
   franchetii Mast., 26
   halicacabum Crantz, 8
   peruviana L., 29
   somnifera L., 10
Pinguicula L., 155
Podonosma Boiss., 199
Podonosma orientalis (L.) Feinbrun, 200
   sindjarensis (Riedl) L.Cecchi & Hilger
   syriaca (Labill.) Boiss., 200
Polakia Stapf, 381
Polium capitatum (L.) Mill., 315
Pollichia amplexicaulis Willd., 332
Preslia Opiz, 417
Prunella L., 460
   orientalis Bornm., 461
   vulgaris L., 461
Pseuderemostachys Popov, 346
Pseudodictamnus Fabr., 376
   aucheri (Boiss.) Salmaki & Siadati, 376
Pseudoheterocaryum Kaz.Osaloo &
     Saadati, 274

szovitsianum (Fisch. & C.A.Mey.) Kaz.
     Osaloo & Saadati, 275
subsessile (Vatke) Kaz.Osaloo &
     Saadati, 275
rigidum (A.DC.) Kaz.Osaloo & Saadati,
     277
Pseudolappula Khoshsokhan & Kaz.
     Osaloo, 281
   sinaica (A.DC.) Khoshsokhan, Sherafati
     & Kaz.Osaloo, 281
Pseudolophanthus Levin, 463

Rafinesquia Raf., 443
Rhabdotosperma Hartl, 55
Rhinanthus elephas L., 153
   trifidus Vahl, 153
Rhynchocorys Griseb., 149
   elephas (L.) Griseb., 153
     subsp. **carduchorum** R.B.Burb &
      I.Richardson, 153
   kurdica Nábělek, 151
   odontophylla R.B.Burb. & I.Richardson,
     151
Rindera Pall., 290
   cristata (Schreber) G.Don, 287
   lanata (Lam.) Bunge, 290
     var. *punctata* (Bunge) Kusn., 290
   pubescens K.Koch, 290
   punctata Bunge, 290
Rizoa Cav., 443
Rochelia Rchb., 282
subtribe ERITRICHIINAE, 282
subtribe HETEROCARYINAE, 282
   cardiosepala Bunge, 284
   disperma (L.f.) K.Koch, 282
   disperma (L.) Wettst., 283
   persica Boiss., 284
   retorta (Pall.) Lipsky, 283
   stellulata Rchb., 283
Rosenbachia Regel, 299

Sabbatia Moench, 440
Saccilabium Rottb., 463
Salvia L., 381
   acetabulosa Vahl, 392
   aegyptiaca L., 384
   aegyptiaca L. non L. (1753), 394
     aegyptiaca var. *glandulosissima* Bornm.
      & Kneucker, 384
     aegyptiaca var. *pumila* sensu Hook.,
      384
   aethiopis L., 415
   affinis Spreng. ex Steud., 411
   albida Jacq., 405
   albida Spreng., 405
   albifrons Nábělek, 397
   alexandri Pobed., 389
   ali-askaryi S.A.Ahmad, 408
   alliaria Parsa, 402

sclareopsis Bornm. ex Hedge, 400
semilanata Czerniak., 405
semilunata Hausskn., 405
sibthorpii Sm., 411
sieberi C. Presl, 402
similata Hausskn., 411
simsiana Schult., 402
sinaica Delile ex Benth., 402
sinapifolia K.Koch, 414
spinosa L., 394
spinulosa Montbret & Aucher ex Benth., 407
splendens Sellow ex Nees., 415
staminea Montbret & Aucher ex Benth., 415
suffruticosa Montbret & Aucher, 389
syriaca L., 394
szovitsiana Bunge, 392
trichoclada Bentham, 389
turkestanica Noter, 402
uberrima Rech.f., 412
urmiensis Bunge, 408
utilis Braun ex Engler, 411
    subsp. **amasiaca** (Freyn & Bornm.) Bornm., 414
    subsp. **verticillata** L., 412
vanensis Boiss. et Noë, 389
varia Vahl, 394
verbascifolia Bieb., 406
var. *cana* Boiss., 406
verticillata L., 412
virgata *Jacq.*, 409
virgata Jacq. var. *canovelutina* Rech.f., 411
virgata Jacq. var. *densiflora* Nábělek, 411
viridis L., 392
xanthochelia Boiss. ex Benth., 407
Satureja L., 434
altaica Boriss., 436
bachtiarica Bunge, 439
brachiata Stokes, 436
crassinervis H.Lindb., 445
cuneifolia Ten., 439
exigua (Sm.) Grande, 445
filicaulis Schott ex Boiss., 436
flacca Nábělek, 448
graveolens (M.Bieb.) Caruel, 445
hortensis L., 436
    var. *distans* K.Koch, 436
laxiflora K. Koch, 436
    subsp. *zuvandica* (D.A.Kapan.) D.A.Kapan., 436
litwinowii Schmalh. ex Lipsky, 436
macrosiphonia Bornm., 437
metastasiantha Rech.f., 437
myrtifolia (Boiss. & Hohen.) Greuter & Burdet, 442
officinarum Crantz, 436
pachyphylla K.Koch, 436

persica (Boiss.) Briq., 443
stricta Banks & Sol., 466
viminea Burm.f., 436
    vulgaris subsp. *orientalis* (Bothmer) Greuter & Burdet, 444
zuvandica D.A.Kapan., 436
Saturiastrum Fourr., 434
Saussuria Moench, 463
Schizocalyx Scheele, 430
Schizonepeta (Benth.) Briq., 463
Schraderia Medicus
    'acetabulosa' (Vahl) Pobed., 392
Sclarea spinosa Raf., 397
Sclarea ceratophylla Mill., 405
    indica Mill., 409
    sibthorpii (Sm.) Soják, 411
    syriaca Mill., 394
    virgata (Jacq.) Soják, 411
    viridis (L.) Soják, 392
    vulgaris Mill., 402
Sclerocaryopsis spinocarpos (Forssk.) Brand, 279
Scordium scordioides (Schreb.) Four., 310
Scrophularia L., 72
alata Gilib., 79
    var. *cordata* Gilib., 79
amadiyana Ghaz. & Haloob, 85
amatiana Sawah, 85
amplexicaulis Benth., 87
atroglandulosa Grau, 82
azerbaijanica Grau, 75
boissierana Jaub. & Spach, 81
catariifolia Boiss. & Heldr., 82
crenophila Boiss., 84
decipiens Boiss. & Kotschy, 75
deserti Delile, 73
    var. *foliata* Eig, 73
ericiflora Rech.f., 85
gileadense Post, 75
gracilis Blakelock, 85
guestii Eig, 80
haematantha Boissier & Heldr., 84
hajariana Parsa, 87
hispidula Boiss. & Balansa, 75
hypericifolia Wydler, 87
juncea C.Richter ex Stapf, 75
kollakii S.A.Ahmad, 80
kurdica Eig, 81
    subsp. **glabra** Grau, 81
libanotica Boiss., 77
    subsp. **libanotica** Grau, 77
    var. *libanotica* Lall & Mill, 79
macrophylla Boiss., 79
marginata Boiss., 73
nervosa Benth., 81
orientalis Ehrenb. ex Boiss., 75
orientalis Boiss. non L., 81
pegaea Hand.-Mazz., 84
persica Benth., p.p., 75

NEW TAXA, COMBINATIONS AND SYNONYMS IN THIS VOLUME

Linaria iraqensis *Al-Kaisi & A.Haloob* **sp. nov.** Type: Iraq, Serva, 9/8/1947, *Gillett* 9697 (BAG, holo.).

Stachys zakhoensis (*Meikle & Patzak*) *Ghaz.* **sp. nov.** Type: Iraq, Zawita pass N. of Zakho, *Rechinger* 10936 (W, holo.).
S. zakhoensis Meikle & Patzak in sched.

Convolvulus leptocladus *Boiss.*, Diagn. Pl. Orient. ser. 1, 7: 25 (1846) subsp. **glabrosepalus** *Kandemir* **subsp. nov.** Type: Iraq, Qarati, Khanaqin, *Rawi* 5720 (K, holo.).

Linaria genistifolia (*L.*) *Mill* subsp. **qandilensis** *A.Haloob & Al-Kaisi* **subsp. nov.** Type: Iraq, NE of Qandil, 25/8/1957, *Al-Rawi and Serhang* 24367 (BAG, holo.).

Scutellaria orientalis *L.*, Sp. Pl. 598 (1753) subsp. **iraqensis** *A.Haloob* **subsp. nov.** Type: Iraq. Galli Zawita, N.E. of Zakho, nr. Turkish border, 8/7/1957, *Al-Rawi*, 23569 (BAG, holo.).

Scutellaria platystegia *Juz.*, Bot. Zhurn. S.S.S.R. 24: 431 (1939) subsp. **mazinica** *A.Haloob & Al-Bayati* **subsp. nov.** Type: Iraq, Zeta N of Shirwan Mazin, Erbil liwa, 19/6/1961, *Agnew, Hadač, Haines & Kadir*, 18240 (BUH, holo.).

Erodium moschatum (*L.*) *L'Her.* ex *Aiton* subsp. **touchyanum** (*Delile ex Godr.*) *Ghaz.*, **stat. & comb. nov.**
Erodium touchyanum Delile ex Godr., Mem. Acad. Montp. (Sect. Medic.) i. 423 (1853).

Scutellaria albida *L.* subsp. **pycnotricha** (*Rech.f.*) *A.Haloob*, **stat. & comb. nov.**
Scutellaria pycnotricha Rech.f., Bot. Arch. 43: 19 (1941);

Scutellaria albida *L.* subsp. **subsimilis** (*Rech.f.*) *A.Haloob*, **stat. & comb. nov.**
Scutellaria velenovskyi subsp. *subsimilis* Rech.f., Bot. Arch. 43: 11 (1941).

Nepeta elymaitica *Bornm.*, Russk. Bot. Zhurn. 1911: 6 (1911)
Nepeta ludlow-hewittii Blakelock, Kew Bull. 4: 543 (1949 publ. 1950). **syn. nov.** Type: Lecto. (designated by Dirmenci), Iraq, Erbil: Algird Dagh, among rocks, 3000–3300 m, 22.07.1932, *Guest & Ludlow- Hewitt* 2871 lecto. K [K000910840]; isolecto. K [K000910837, K000910838, K000910839].